Function Keys

- F1 Displays online help
- F2 Toggles text & graphics windows
- F3 Toggles object snap
- F4 Toggles tablet mode
- F5 Cycles through isoplanes
- F6 Cycles coordinate display
- F7 Toggles grid display
- F8 Toggles ortho mode
- F9 Toggles snap mode
- F10 Toggles polar tracking
- F11 Toggles object snap tracking
- F12 Toggles dynamic input

- Ctrl F4 Closes drawing
- Ctrl F6 Switches to next drawing

- Alt F4 Closes AutoCAD
- Alt F8 Runs VBA
- Alt F11 Opens VBA IDE

Mouse Buttons

(***Sketch*** *command buttons*)

① Pick objects (*raises, lowers pen*)

② Displays cursor menu (*draws line*)

③ Selects (*records sketch*)

Ctrl ② Displays osnap cursor menu

Acad.exe Startup Switches

/b	Runs *.scr* script file
/c	Specifies path to alt *.cfg* hardware
/layout	Displays named layout
/ld	Loads ARX or DBX app
/nologo	Suppresses AutoCAD logo
/nossm	Suppresses Sheet Set Manager
/p	Loads user-defined profile
/r	Restores default pointer
/s	Sets paths to support folders
/set	Loads *.dst* sheetset file
/t	Loads *.dwt* template file
/v	Displays named view
/w	Shows workspace at startup

Co

- Ctrl
- Ctrl
- Ctrl B Toggles snap mode
- Ctrl C Copies to the Clipboard
- Ctrl Shift C Copies with base point
- Ctrl D Cycles coordinate display
- Ctrl E Cycles through isoplanes
- Ctrl F Toggles object snap mode
- Ctrl G Toggles grid display
- Ctrl H Toggles pick style
- Ctrl K Displays Hyperlinks dialog box
- Ctrl L Toggles ortho mode
- Ctrl N Opens new drawing
- Ctrl O Opens drawing files
- Ctrl P Displays Plot dialog box
- Ctrl Q Toggles log file
- Ctrl R Cycles through viewports.
- Ctrl S Saves drawing
- Ctrl Shift S Displays Save As dialog box
- Ctrl T Toggles tablet mode
- Ctrl U Toggles polar tracking
- Ctrl V Pastes from the Clipboard
- Ctrl Shift V Pastes with insertion point
- Ctrl X Cuts to the Clipboard
- Ctrl Y Redoes last undo (MRedo)
- Ctrl Z Undoes last command
- Ctrl 0 Toggles cleanscreen mode
- Ctrl 1 Toggles Properties window
- Ctrl 2 Toggles DesignCenter window
- Ctrl 3 Toggles Tool Palettes window
- Ctrl 4 Toggles SheetSet Manager
- Ctrl 5 Toggles Info Palette window
- Ctrl 6 Toggles dbConnect Manager
- Ctrl 7 Toggles Markup Set Manager
- Ctrl 8 Toggles QuickCalc window
- Ctrl 9 Toggles the command line

Other Keys

- Esc Cancels commands and grips
- Delete Deletes selected objects
- Enter Executes and repeats commands

THE Illustrated AutoCAD® 2007 Quick Reference

Ralph Grabowski

Autodesk

THOMSON DELMAR LEARNING™

Australia • Canada • Mexico • Singapore • Spain • United Kingdom • United States

Autodesk

The Illustrated AutoCAD® 2007 Quick Reference

Ralph Grabowski

Vice President, Technology and Trades SBU:
Dave Garza

Senior Acquisitions Editor:
Jim Gish

Marketing Director:
Deborah S. Yarnell

Channel Manager:
Guy Baskaran

Marketing Coordinator:
SueLaine Frongello

Production Director:
Patty Stephan

Senior Content Project Manager:
Stacy Masucci

Editorial Assistant:
Niamh Matthews

Book Design and Typesetting:
Ralph Grabowski

Cover Images:
Getty Images

Printed in the United States of America
1 2 3 4 5 XX 06 05 04

For more information contact Thomson Delmar Learning Executive Woods, 5 Maxwell Drive, PO Box 8007, Clifton Park, NY 12065-8007 Or find us on the World Wide Web at www.delmarlearning.com

Library of Congress Cataloging-in-Publication Data:

ISBN: 1-4180-4892-5

NOTICE TO THE READER

About This Book

The Illustrated AutoCAD 2007 Quick Reference presents concise facts about all commands found in AutoCAD 2007 and earlier. The clear format of this reference book illustrates commands with over 500 figures. Each command starts on its own page, and includes one or more of the following: command line options, dialog box and palette options, toolbar icons, shortcut menus, related commands, tand related system variables. Plus:

- Variations of commands, such as the **View** and **-View** commands.
- Dozens of AutoCAD 2007 commands and system variables not documented by Autodesk.
- Icons, such as ☑ and ⊙, that indicate default settings of check boxes, radio buttons, and other contols in dialog boxes and palettes.
- "Quick Start" mini-tutorials that help you get started quickly with selected commands.
- Over 100 definitions of acronyms and hard-to-understand terms.
- Nearly 1,000 context-sensitive tips.
- Express Tool command names and descriptions, listed in Appendix A.
- Obsolete commands that no longer work in AutoCAD, listed in Appendix B.
- All system variables including those not listed by the **SetVar** command, listed in Appendix C.

The name of each command shown in mixed upper and lower case to help you understand the name, which is often condensed. For example, the **VpClip** command is short for "ViewPort CLIP." Each command includes all alternative methods of command input:

- Alternate command spelling, such as **Donut** and **Doughnut**.
- ' (the apostrophe prefix) indicating transparent commands, such as **'Blipmode**.
- All aliases, such as **L** for the **Line** command.
- Pull-down menu picks, such as **Draw** ↳ **Construction Line** for the **XLine** command.
- Control-key combinations, such as **CTRL+E** for the **Isoplane** toggle.
- Function keys, such as **F1** for the **Help** command.
- Alt-key combinations, such as **ALT+TE** for the **Spell** command.
- Table menu coordinates, such as **M2** for the **Hide** command.

The version or release number indicates when the command first appeared in AutoCAD, such as **Ver. 1.0**, **Rel. 9**, or **2007** — useful when working with older versions of AutoCAD.

A special thank you to Stephen Dunning for editing the text, and to Bill Fane for his technical editing of this book. *Soli Deo Gloria!*

Ralph Grabowski

Abbotsford, British Columbia, Canada

May 1, 2006

Table of Contents

A

Indicates commandd new to AutoCAD 2007

B

C

D

Dimensioning

E

F

G

H

I

J

L

M

N

O

P

T

Z

3

Appendices

'About

Rel.12 Displays the version number and other information about AutoCAD.

Command	Alias	Ctrl+	F-key	Alt+	Menu Bar	Tablet
about	...	...	...	HA	Help ↳About	...

Command: about

Displays dialog box:

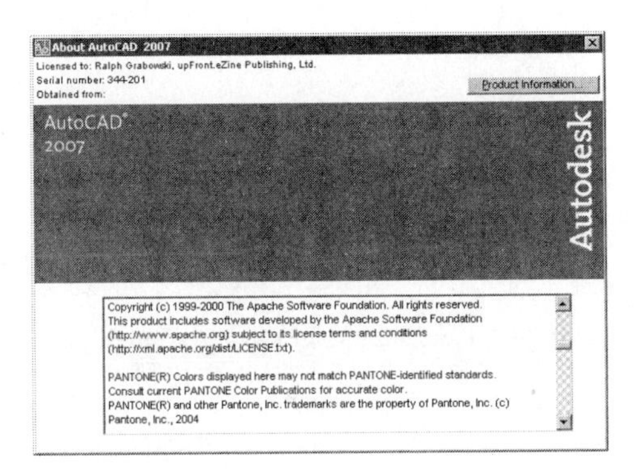

ABOUT AUTOCAD DIALOG BOX

☒ dismisses dialog box; alternatively, press ESC.

Product Information displays the Product Information dialog box:

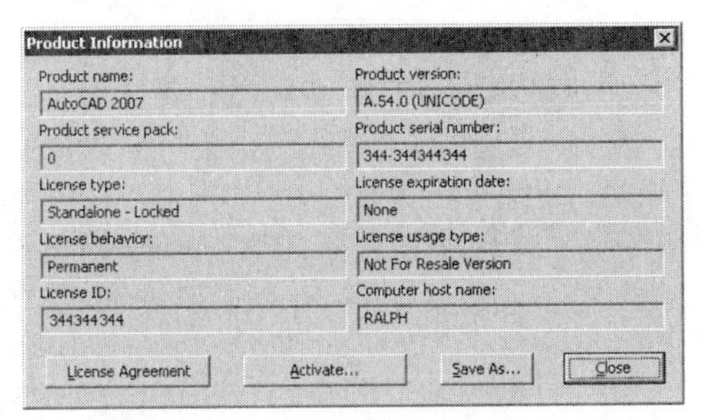

License Agreement opens the computer's default word processor, and then displays the Autodesk Software License Agreement document (*license.rtf*).

Activate runs the Product Activation wizard.

Save As displays the Save As dialog box, which records the information displayed above.

Close returns to AutoCAD.

RELATED COMMANDS

Properties displays information about selected objects.

Status displays information about the drawing and environment.

RELATED SYSTEM VARIABLES

_PkSer displays the AutoCAD software serial number.

_Server displays the network authorization code.

AcisIn

Rel. 13 Imports *.sat* files into drawings to create 3D solids, 2D regions, and bodies.

Command	Alias	Ctrl+	F-key	Alt+	Menu Bar	Tablet
acisin	...	...	...	IA	Insert ↳ACIS File	...

Command: acisin

Displays Select ACIS File dialog box. Select a .sat file, and then click **Open**.

DIALOG BOX OPTIONS

Cancel dismisses the dialog box.

Open opens the selected *.sat* file.

RELATED COMMANDS

AcisOut exports solid objects — 3D solids, 2D regions, and bodies — to *.sat* files for import into ACIS-aware CAD software.

AmeConvert converts AME v2.0 and v2.1 solid models and regions into solids.

RELATED FILE

***.sat** is the ASCII format of ACIS model files; short for "save as text."

TIPS

- See the **Open** command for options related to the Select ACIS File dialog box.
- When system variable **FileDia** is turned off (set to 0), this command prompts:

  Enter SAT file name: *(Enter file name.)*

  Enter the tilde (~) to force the display of the dialog box.
- "ACIS" comes from the first names of the original developers, "Andy, Charles, and Ian's System." In Greek mythology, Acis was the lover of the goddess Galatea; when Acis was killed by the jealous Cyclops, Galatea turned the blood of Acis into a river.
- ACIS is the name of solids modeling technology from the Spatial Technologies division of Dassault Systemes, and is used by numerous 3D CAD packages. As of AutoCAD 2004, Autodesk uses its own ACIS-derived solids modeler, called ShapeManager.

AcisOut

Rel.13 Exports 3D solids, 2D regions, and bodies in SAT format.

Command	Alias	Ctrl+	F-key	Alt+	Menu Bar	Tablet
acisout	...	...	...	**FEE**	**File**	...
				↳ACIS	**↳Export**	
					↳ACIS	

Command: acisout
Select objects: *(Select one or more objects.)*
Select objects: *(Press Enter to end object selection.)*

Displays Create ACIS File dialog box. Enter a name for the .sat file, and then click ***Save.***

DIALOG BOX OPTIONS

Cancel dismisses the dialog box.

Save saves the selected object(s) in a *.sat* file.

RELATED COMMANDS

AcisIn imports *.sat* files, and creates 3D solids, 2D regions, and bodies.

StlOut exports solid models in STL format for use by stereolithography devices.

3dsOut exports solid models as 3D faces for import into 3D Studio software.

RELATED SYSTEM VARIABLES

AcisOutVer specifies the version number of ACIS.

UseAcis toggles the use of ACIS (undocumented).

RELATED FILE

***.sat** is the ASCII format of ACIS model files.

TIPS

- **AcisOut** exports 3D solids, 2D regions, and bodies only.
- When system variable **FileDia** is turned off (set to 0), this command prompts:

  Select objects: *(Select one or more objects.)*
  Select objects: *(Press* ENTER.*)*
  Enter SAT file name <Drawing1.sat>: *(Enter a file name.)*
- Other ACIS-based software can read *.sat* files.

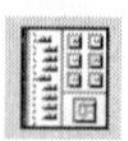

'AdCenter, 'AdcClose

2000 Opens and closes the Design Center palette (short for Autocad Design CENTER).

Command	Aliases	Ctrl+	F-key	Alt+	Menu Bar	Tablet
adcenter	**adc**	**2**	**...**	**TPG**	**Tools**	**X12**
	content				**↳Palettes**	
	dc, dcenter				**↳DesignCenter**	
adcclose	**...**	**2**	**...**	**TPG**	**Tools**	**X12**
					↳Palettes	
					↳DesignCenter	

Command: adcenter

Displays DesignCenter palette:

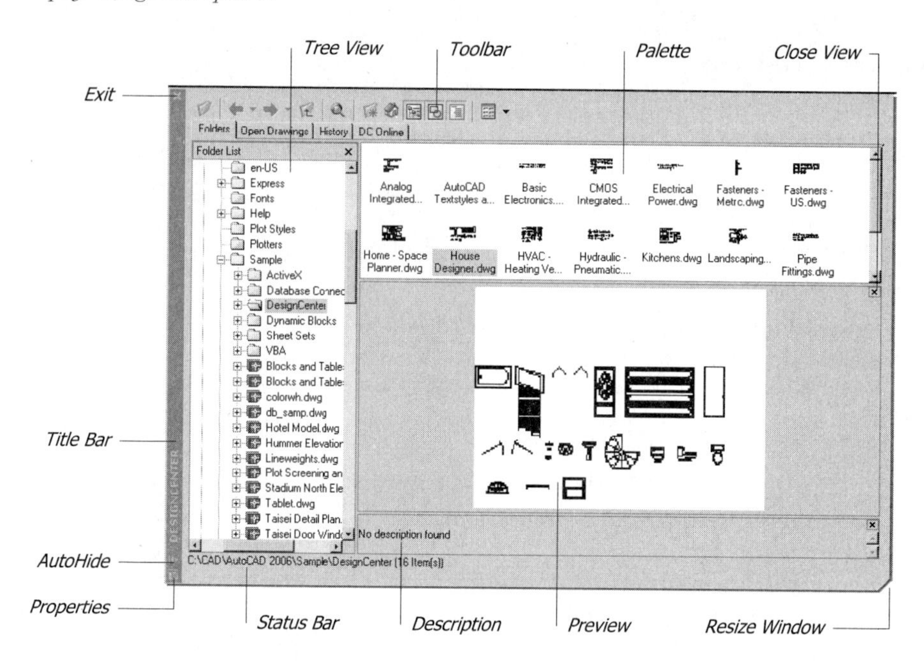

Command: adcclose

Closes the DesignCenter palette.

TABS

Folders displays the content of the local computer, as well as networked computers.

Open Drawings displays the content of drawings currently open in AutoCAD.

History displays the drawings previously opened.

DC Online displays the content available from Autodesk via the Internet.

Palette toolbar

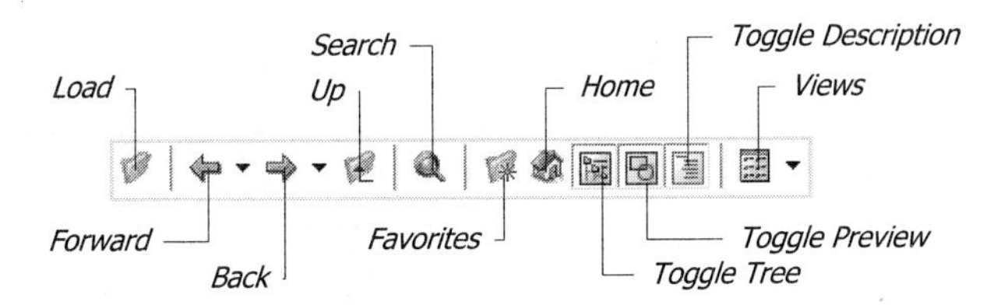

Load displays the Open dialog box to open the following vector and raster file types: *.dwg*, *.dws*, *.dwf*, *.dxf*, *.bil*, *.bmp*, *.cal*, *.cg4*, *.dib*, *.flc*, *.fli*, *.gif*, *.gp4*, *.ig4*, *.igs*, *.jpg*, *.jpeg*, *.mil*, *.pat*, *.pcx*, *.png*, *.rlc*, *.rle*, *.rst*, *.tga*, and *.tif*.

Back returns to the previous view.

Forward goes to the next view.

Up moves up one folder level.

Search opens the Search dialog box to search for AutoCAD files (on the computer) and the following objects (in drawings): blocks, dimstyles, drawings, hatch patterns and *.pat* files, layers, layouts, linetypes, text styles, table styles, and xrefs.

Favorites displays the files in the *\documents and settings\<username>\favorites\autodesk* folder.

Home displays the files in the *\autocad 2007\sample\designcenter* folder.

Tree View Toggle hides and displays the Folders and Open Drawings tree views.

Preview toggles the display of the preview image of *.dwg* and raster files.

Description toggles the display of the description area.

View changes the display format of the palette area.

RELATED COMMAND

AdcNavigate specifies the initial path for DesignCenter.

RELATED SYSTEM VARIABLE

AdsState reports whether DesignCenter is open or not.

TIPS

- Use Design Center to keep track of drawings and parts of drawings, such as block libraries.
- You can drag blocks, table styles, and other drawing parts from the Design Center palette into the drawing, and blocks onto Tool palette.
- Design Center allows you to share text and dimension styles between drawings.
- The Design Center palette can switch between floating and docked modes by right-clicking and selecting the option from the menu.
- **Ctrl+2** opens and closes the Design Center palette each time you press the keys.
- This command cannot be used transparently (within another command) until it has been used at least once non-transparently.

AdcNavigate

2000 Specifies the initial path for Design Center to access content.

Command	Alias	Ctrl+	F-key	Alt+	Menu Bar	Tablet
adcnavigate	...	...	...	...	...	...

Command: adcnavigate

Opens DesignCenter, if not already open.

Enter pathname <>: *(Enter a path, and then press Enter.)*

COMMAND LINE OPTION

Enter pathname specifies the path, such as *c:\program files\autocad2007\sample.*

RELATED COMMAND

AdCenter opens the Design Center palette.

TIPS

- AutoCAD uses the path specified by **AdcNavigate** to locate content displayed by DesignCenter's Desktop option.
- You can enter the path to a file, folder, or network location:

Example of folder path:	`c:\program files\autocad 2007\sample`
Example of file path:	`c:\design center\welding.dwg`
Example of network path:	`\\downstairs\c\project`

AecToAcad

2000 Converts Architectural Desktop custom objects to AutoCAD objects; short for "Architecture Engineering Construction TO AutoCAD."

Command	Alias	Alt+	Menu Bar	Tablet
aectoacad	-exporttoautocad	...	...	...

Command: aectoacad
File format: 2007
Bind xrefs: Yes
Bind type: Insert
Filename prefix:
Filename suffix:
Export options [Format/Bind/bind Type/Maintain/Prefix/Suffix/?] <Enter for filename>: *(Enter an option.)*
Export drawing name <d:\path\filename.dwg>: *(Press Enter.)*

COMMAND OPTIONS

Format specifies the DWG version.

Bind determines whether xrefs are bound.

bind Type specifies the type of xref binding.

Maintain specifies how to deal with blocks in ADT custom objects.

Prefix specifies a prefix for file names.

Suffix specifies a suffix for file names.

? lists the setting for each option.

Format options
Enter file format [r14/2000/2004/2007] <2007>:

r14 saves drawing in Release 14 DWG format.

2000 saves drawing in 2000/2000i/2001 DWG format.

2004 saves drawing in 2004/2005/2006 DWG format.

2007saves drawing in 2007 DWG format.

Bind option
Bind xrefs [Yes/No] <Yes>:

Yes binds externally-referenced files to the drawing.

No stores links to xref files.

bind Type option
Bind type [Bind/Insert] <Insert>:

Bind makes xrefs part of the drawing.

Insert inserts xrefs into the drawing as if they were blocks.

Maintain option

Maintain resolved properties [Yes/No] <Yes>:

Yes explodes blocks in AEC custom objects.

No keeps blocks as unexploded xrefs.

Prefix and Suffix options

Filename prefix <>:

Filename suffix <>:

Prefix and **suffix** add text in front of and behind the current drawing's file name.

RELATED SYSTEM VARIABLE

ProxyNotice controls the display of custom objects in drawings:

0 — Custom objects are not displayed.

1 — Custom objects are displayed as proxy objects.

2 — Custom objects are displayed by a bounding box.

TIPS

- AutoCAD and related software have the ability to create *custom objects* which have special properties not otherwise found in AutoCAD. These objects are controlled by programming code written by in ARX. Autodesk products, such as Architectural Desktop and Mechanical Desktop, use custom objects. When drawings made with these programs are opened in AutoCAD, custom objects are displayed as *proxy objects*, which cannot be edited by AutoCAD. (The
- This command explodes custom objects created in ADT so that they can be edited by AutoCAD, and then saved to file by another name in order to preserve the original drawing.
- Exploded custom objects lose their intelligence.
- The following commands are also available for dealing with ADT custom objects:

  • **-AecDwgUnits** sets the units for inserting ADT objects into AutoCAD drawings.

  • **AecDisplayManagerConfigsSelection** changes the display configuration.

  • **AecFileOpenMessage** toggles display of dialog box warning of opening drawings from older releases of ADT.

  • **AecFileSaveMessage** toggles display of warning message when saving to previous versions.

  • **AecObjectCopyMessage** toggles display of warning mesage when saving to newer versions.

  • **AecObjRelDump**, **AecObjrelDumpEnhancedReferences**, and **AecObjRelShowEnhancedReferences** perform AEC custom object database dumps to the text screen.

  • **AecObjRelShow** displays data about which ADT objects are related to each other.

  • **AecObjRelUpdate** updates ADT objects.

  • **AecPostDxfinFix** fixes problems with imported ADT drawings that cannot be repaired by AutoCAD's Audit command.

  • **AecSetXrefConfig** allows the same xref to be displayed several times.

  • **AecVersion** reports the version number of the ADT object enabler.

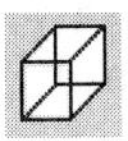 Ai_Box/Cone/Dish/Dome/Mesh/Pyramid/Sphere/Torus/Wedge

Draws 3D boxes as mesh objects (undocumented commands); they do not draw surfaces.

Command	Alias	Ctrl+	F-key	Alt+	Menu Bar	Tablet
ai_box	...	...	...	...	...	...

Command: ai_box
Specify corner point of box: *(Pick a point.)*
Specify length of box: *(Pick a point.)*
Specify width of box or [Cube]: *(Pick a point, or type* **C***.)*
Specify height of box: *(Pick a point.)*
Specify rotation angle of box about the Z axis or [Reference]: *(Specify the rotation angle, or type* **R***.)*

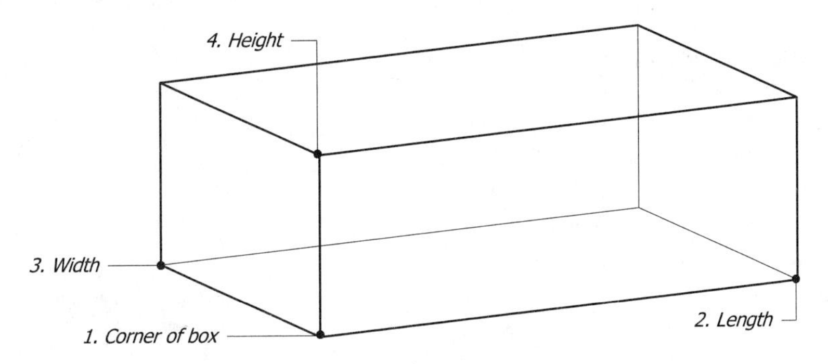

COMMAND LINE OPTIONS

Corner of box specifies the initial corner of the box.

Length specifies the length of one side of the box along the x-axis.

Cube creates a cube based on **Length**.

Width specifies the width of the box, along the y-axis.

Height specifies the height of the box, along the z-axis.

Rotation angle specifies the angle the box rotates about the z-axis.

Reference prompts you to pick two points that represent the new angle.

 Ai_Cone command

Command: ai_cone
Specify center point for base of cone: *(Pick a point.)*
Specify radius for base of cone or [Diameter]: *(Pick a point, or type* **D***.)*
Specify radius for top of cone or [Diameter] <0>: *(Specify the radius, or type* **D***; enter* **0** *for a pointy top.)*

Specify height of cone: *(Pick a point.)*
Enter number of segments for surface of cone <16>: *(Enter a number between 3 and 255, or press* **ENTER***.)*

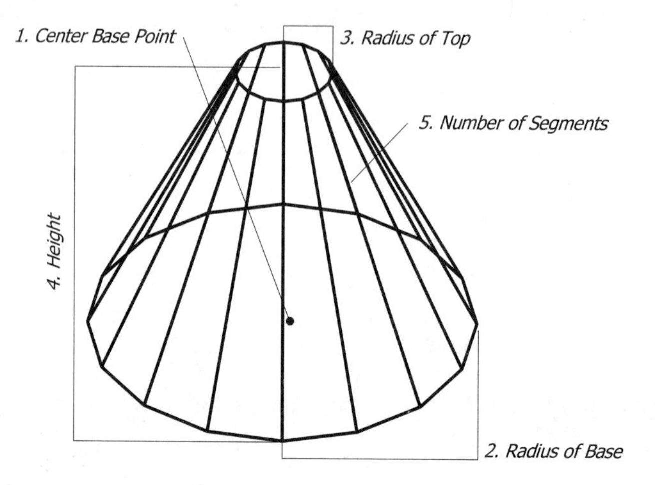

COMMAND LINE OPTIONS

Base center point specifies the center point of the base of the cone.

Diameter of base specifies the diameter of the base.

Radius of base specifies the radius of the base.

Diameter of top specifies the diameter of the top of the cone.

Radius of top specifies the radius of the top of the cone; 0 = cone with a point.

Height specifies the height of the cone.

Number of segments specifies the number of "lines" defining curved surfaces; default = 16.

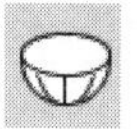 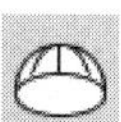 Ai_Dish and Ai_Dome commands

Command: ai_dish
Specify center point of dish: *(Pick a point.)*
Specify radius of dish or [Diameter]: *(Pick a point, or type* **D***.)*
Enter number of longitudinal segments for surface of dish <16>: *(Enter a value between 3 and 255, or press* **ENTER***.)*
Enter number of latitudinal segments for surface of dish <8>: *(Enter a value between 3 and 255, or press* **ENTER***.)*

COMMAND LINE OPTIONS

Center of dish specifies the center of the dish's base.

Diameter specifies the diameter of the dish.

Radius specifies the radius of the dish.

Number of longitudinal segments specifies the number of "lines" that define the curved surface in the vertical direction; default = 16.

Number of latitudinal segments specifies the number of "lines" that define the curved surface in the horizontal direction; default = 8.

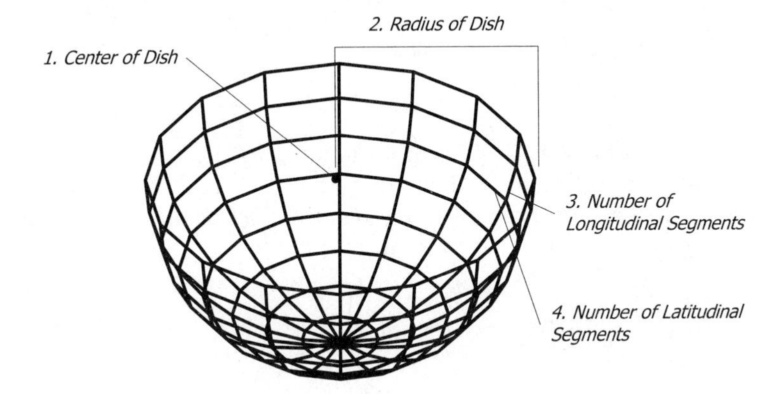

Ai_Mesh command

Command: ai_mesh
Specify first corner point of mesh: *(Pick point 1.)*
Specify second corner point of mesh: *(Pick point 2.)*
Specify third corner point of mesh: *(Pick point 3.)*
Specify fourth corner point of mesh: *(Pick point 4.)*
Enter mesh size in the M direction: *(Specify a number.)*
Enter mesh size in the N direction: *(Specify a number.)*

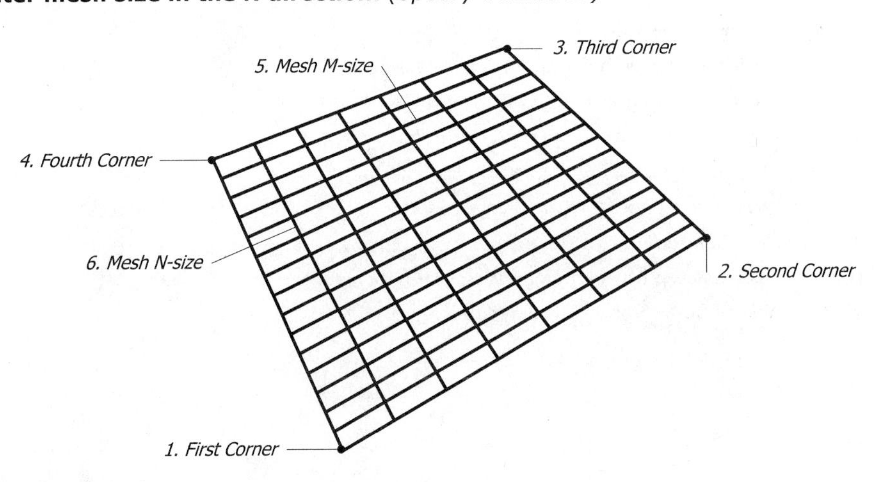

COMMAND LINE OPTIONS

First corner specifies the mesh's first corner.

Second corner specifies the mesh's second corner.

Third corner specifies the mesh's third corner.

Fourth corner specifies the mesh's last corner.

Mesh size M direction specifies the number of horizontal "lines" that define the mesh's surface.

Mesh size N direction specifies the number of vertical "lines" that define the mesh's surface.

Command: ai_pyramid
Specify first corner point for base of pyramid: *(Pick point 1.)*
Specify second corner point for base of pyramid: *(Pick point 2.)*
Specify third corner point for base of pyramid: *(Pick point 3.)*
Specify fourth corner point for base of pyramid or [Tetrahedron]: *(Pick point 4, or type* **T***.)*

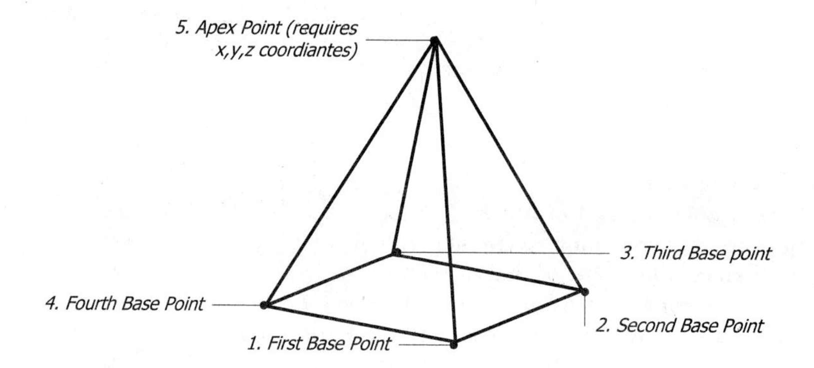

COMMAND LINE OPTIONS

First base point specifies the pyramid's first base point.
Second base point specifies pyramid's second base point.
Third base point specifies the pyramid's third base point.
Fourth base point specifies the pyramid's last base point.
Tetrahedron draws a pyramid with triangular sides.
Ridge specifies a ridge-top for the pyramid; see figure below.
Top specifies a flat-top for the pyramid; see figure below.
Apex point specifies a point for the pyramid's top

Ai_Sphere command

Command: ai_sphere
Specify center point of sphere: *(Pick point 1.)*
Specify radius of sphere or [Diameter]: *(Pick point 2, enter a radius, or type* **D***.)*
Enter number of longitudinal segments for surface of sphere <16>: *(Enter a value.)*
Enter number of latitudinal segments for surface of sphere <16>: *(Enter a value.)*

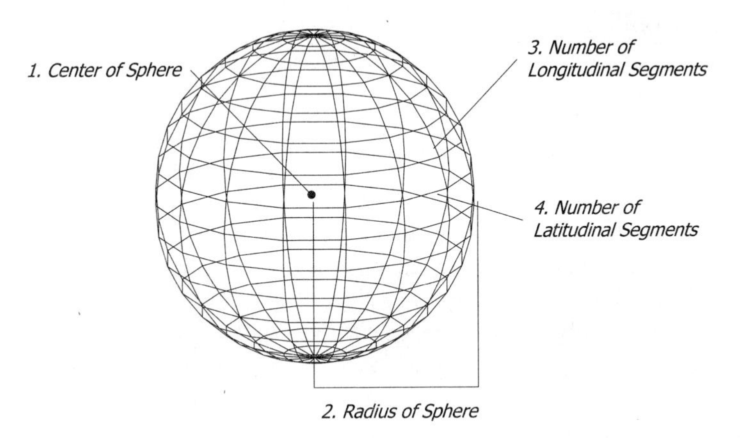

COMMAND LINE OPTIONS

Center of sphere specifies the center point of the sphere.

Diameter specifies the diameter of the sphere.

Radius specifies the radius of the sphere.

Number of longitudinal segments specifies the number of "lines" that define the curved surface in the vertical direction; default = 16.

Number of latitudinal segments specifies the number of "lines" that define the curved surface in the horizontal direction; default = 16.

 Ai_Torus command

Command: ai_torus
Specify center point of torus: *(Pick point 1.)*
Specify radius of torus or [Diameter]: *(Pick point 2, specify the radius, or type* **D***.)*
Specify radius of tube or [Diameter]: *(Pick point 3, specify the radius, or type* **D***.)*
Enter number of segments around tube circumference <16>: *(Press* ENTER*.)*
Enter number of segments around torus circumference <16>: *(Press* ENTER*.)*

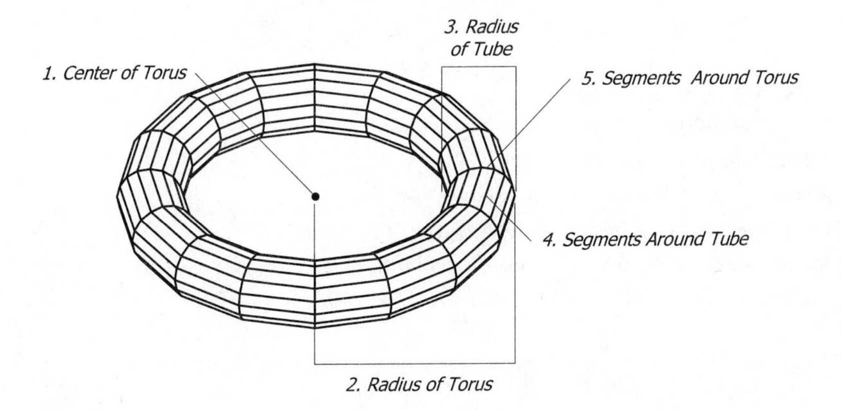

COMMAND LINE OPTIONS

Center of torus specifies the center point of the torus.

Diameter of torus specifies the diameter of the torus as measured across the center of the tube.

Radius of torus specifies the radius of the torus, as measured from the center of the torus to the center of the tube.

Diameter of tube specifies the diameter of the tube cross-section.

Radius of tube specifies the radius of the tube cross-section.

Segments around tube circumference <16> specifies the number of faces defining the curved surface; default = 16, range = 3 to 255.

Segments around torus circumference <16> specifies the number of faces defining the curved surface; default = 16, range = 3 to 255.

Ai_Wedge Command

Command: ai_wedge
Specify corner point of wedge: *(Pick point 1.)*
Specify length of wedge: *(Pick point 2.)*
Specify width of wedge: *(Pick point 3.)*
Specify height of wedge: *(Pick point 4.)*
Specify rotation angle of wedge about the Z axis: *(Enter a rotation angle, or type* **0** *for no rotation.)*

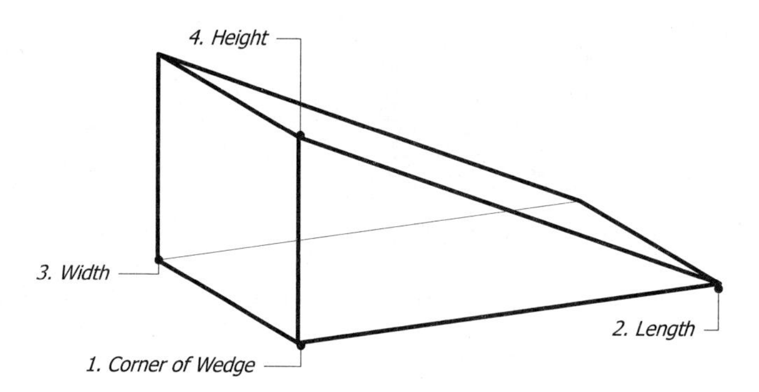

COMMAND LINE OPTIONS

Corner of wedge specifies the corner of the wedge's base.

Length specifies the length of the wedge's base.

Width specifies the width of the wedge's base.

Height specifies the height of the wedge.

Rotation angle specifies the angle the box rotates about the z-axis.

*The following items apply to the **Ai_** surface model commands:*

RELATED SYSTEM VARIABLES

SurfU specifies the surface mesh density in the m-direction.

SurfV specifies the surface mesh density in the n-direction.

TIPS

- As of AutoCAD 2007, Autodesk changed the name of these objects from "surface" to "mesh" objects.
- Types of 3D objects drawn by the **Ai_** series of commands:

Command	Objects Created
Ai_Box	Rectangular boxes, and cubes.
Ai_Dish	Bottom half of a sphere.
Ai_Dome	Top half of a sphere.
Ai_Cone	Pointy cones, and truncated cones.
AI_Mesh	Non-planar polyface meshes.
Ai_Pyramid	Pyramids, truncated pyramids, tetrahedrons, truncated tetrahedrons, and roof shapes.
Ai_Sphere	Spheres.
Ai_Torus	Tori (donuts).
Ai_Wedge	Wedges.

- This command creates 3D mesh (hollow) objects; to draw solid models, use commands such as Box and Sphere.
- When specifying the **Width** and **Height** in Ai_Box and Ai_Wedge, move the cursor back to the **Corner of box** point; otherwise the box may have a different size than you expect.
- You *cannot* perform Boolean operations (such as intersect, subtract, and union) on 3D mesh objects. These mesh objects are made of a single polymesh; Explode converts each side to an independent 3D face object.
- If no z coordinate is specified, then the base of mesh objects is drawn at the current setting of the Elevation system variable; the centers of spheres and tori are at the current elevation, unless you specify the z coordinate.
- Mesh *m* and *n* sizes (SurfU and SurfV) are limited to values between 2 and 256. The meaning of these varies with the type of mesh model created.
- You can use the Hide, VisualStyle, and Render commands on these mesh model shapes.
- To draw a 2D pyramid, enter no z-coordinate for the **Ridge**, **Top**, and **Apex point** options; use the **.xy** filter to specify the z-coordinate for these options.
- The point you pick for the **Corner of wedge** option determines the taller end of the wedge.
- The Ai_Torus command draws donut shapes.; the tube diameter cannot exceed the torus radius.
- Use the **.xy** filter first to specify the x,y coordinate, followed by the z coordinate, as follows:

  First corner:.xy
  of *(Pick a point.)*
  need Z: *(Specify z.)*

- The Ai_Dish command draws the bottom half of a sphere; Ai_Dome draws the top half.

'Ai_CircTan

Draws circles tangent to three points (an undocumented command).

Command	Alias	Ctrl+	F-key	Alt+	Menu Bar	Tablet
ai_circtan	...	...	...	...	...	...

Command: ai_circtan
Enter Tangent spec: *(Pick an object.)*
Enter second Tangent spec: *(Pick an object.)*
Enter third Tangent spec: *(Pick an object.)*

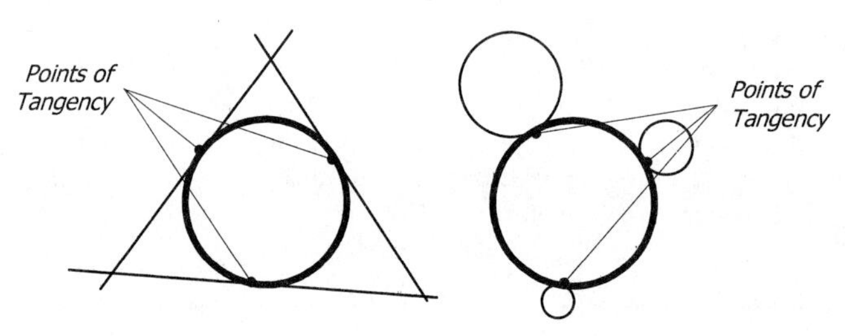

***Left**: Circle drawn tangent to three lines.*
***Right**: Circle drawn tangent to three circles.*

COMMAND LINE OPTION

Enter Tangent spec picks the objects to which the circle will be made tangent.

TIPS

- This command is meant for use in toolbar and menu macros.
- If the circle cannot be drawn between three tangents, AutoCAD complains, "Circle does not exist."
- This command is an alternative to the Circle command's **3P** option with the TANgent object snap.

Ai_Custom_Safe/Product_Support _ Safe/Training_Safe/Access_SubCtr

Accesses Autodesk support Web pages (undocumented commands).

Commands	Alias	Ctrl+	F-key	Alt+	Menu Bar	Tablet
ai_custom_safe	...	...	...	**HRD**	**Help** ↳**Additional Resources** ↳**Online Development Center**	...
ai_product_support_safe				**HRK**	**Help** ↳**Online Resources** ↳**Support Knowledge Base**	
ai_training_safe				**HRT**	**Help** ↳**Online Resources** ↳**Online Training Resources**	
access_subctr						

Command: ai_custom_safe

_.browser Enter Web location (URL) <http://www.autodesk.com> http://www.www.autodesk.com/developautocad

Opens Web browser, and accesses Autodesk's Web site.

Command: access_subctr

Enter subscription resource request ID (1-4): *(Enter a digit.)*

Opens Web browser, and accesses Autodesk's developer center Web site.

COMMAND LINE OPTIONS

None.

TIPS

- The ai_-commands access the following Web pages:

Command	Web Page Accessed
ai_custom_safe	AutoCAD Developer Center
ai_product_support_safe	AutoCAD Support
ai_training_safe	AutoCAD Training

- The options of the Access_SubCtr command access the following Autodesk Subscription Web pages:

Option	Web Page Accessed
1	e-Learning Catalog
2	Create Support Request
3	View Support Requests
4	Edit Subscription Center Profile

- These commands are meant for use in menu and toolbar macros.

AiDimFlipArrow

Reverses the direction of selected arrowheads (undocumented command).

Command	Alias	Ctrl+	F-key	Alt+	Menu Bar	Tablet
aidimfliparrow	...	...	...	...	...	...

Command: aidimfliparrow
Select objects: *(Select one or more arrowheads.)*
Select objects: *(Press Enter .)*

Top: *Dimension and arrowheads as drawn by AutoCAD's* ***DimLinear*** *command.*
Above: *Both arrowheads flipped by the* ***AiDimFlipArrow*** *command.*

COMMAND LINE OPTIONS

Select objects selects one or more arrowheads.

TIPS

- In some cases, AutoCAD places arrowheads differently from your preference. This command allows you retroactively to change the direction arrowheads point.
- Selecting one arrowhead allows you to flip it independently of the companion arrowhead.
- This command has no effect on arrowheads in leaders.

AiDimPrec

Changes the displayed precision of existing dimensions (undocumented command).

Command	Alias	Ctrl+	F-key	Alt+	Menu Bar	Tablet
aidimprec	...	...	...	...	...	...

Command: aidimprec

Enter option [0/1/2/3/4/5/6] <4>: *(Enter a digit.)*

Select objects: *(Select one or more dimensions; non-dimensions are ignored.)*

Select objects: *(Press Enter .)*

*Before (left) and after (right) applying **AiDimPrec = 1** to a decimal dimension.*

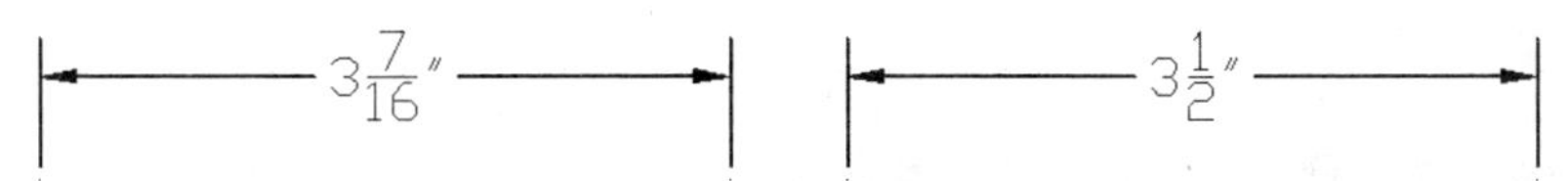

*Before (left) and after (right) applying **AiDimPrec = 1** to a fractional dimension.*

COMMAND LINE OPTIONS

Enter option specifies the precision (number of decimal places, or fractional equivalent); enter a number between 0 and 6.

Select objects selects one or more dimensions.

TIPS

- This command allows you retroactively to change the displayed precision of selected dimensions.
- Alternatively, you can select a dimension, right click, and then select **Precision** from the shortcut menu.
- Zero to six decimal places can be specified; fractional units are rounded to the nearest fraction:

AiDim	Prec Architectural Units
0	Rounded to the nearest unit.
1	1/2"
2	1/4"
3	1/8"
4	1/16"
5	1/32"
6	1/64"

- *Caution!* Because **AiDimPrec** rounds off dimensions, it can display false values. The dimension line below measures 3.4375", but setting **AiDimPrec** to 0 rounds it down to 3".

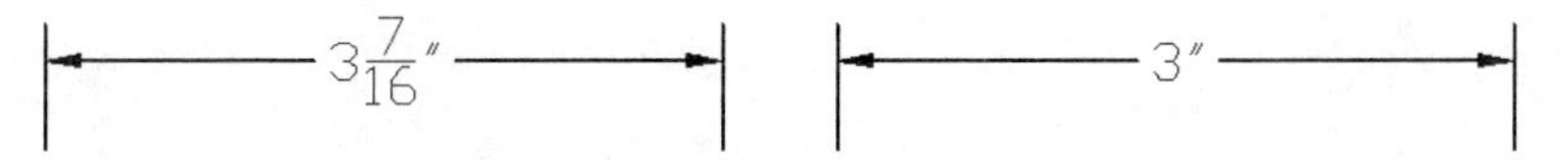

*Applying **AiDimPrec = 0** to a 3 7/16" dimension.*

AiDimStyle

Saves and applies preset dimension styles (undocumented command).

Command	Alias	Ctrl+	F-key	Alt+	Menu Bar	Tablet
aidimstyle	...	...	...	...	...	...

Command: aidimstyle
Enter option [1/2/3/4/5/6/Other/Save] <1>: *(Specify option, then press Enter .)*
Select objects: *(Select one or more dimensions.)*
Select objects: *(Press Enter .)*

COMMAND LINE OPTIONS

Enter option specifies a predefined dimension style, numbered 1 through 6.

Other applies a named dimension style to selected dimension(s).

Save saves the style of the selected dimension(s).

Select objects selects one or more dimensions.

Other option
Enter option [1/2/3/4/5/6/Other/Save] <1>: o

Displays dialog box after selecting objects:

Save option
Enter option [1/2/3/4/5/6/Other/Save] <1>: s

Displays dialog box after selecting exactly one dimension:

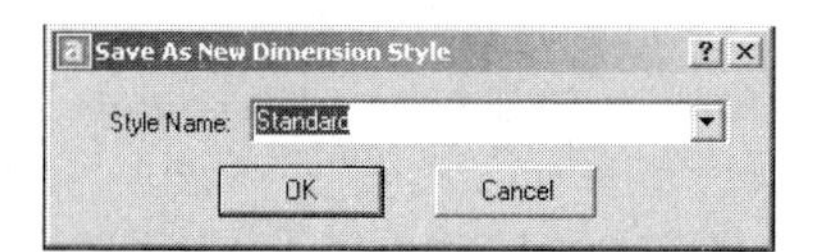

OK saves style to the selected dimension style name.

TIPS

- This command quickly applies and saves dimensions styles. It is used by the right-click shortcut menu: left-click a dimension to select it, right click, and then select **Dim Style** from the shortcut menu.
- *Caution!* The **Save** option overwrites existing dimstyles; it does not create new style names. It does warn you, and asks for permission to make the change.

Ai_Dim_TextAbove/Center/Home

Moves dimension text relative to dimension lines (undocumented commands).

Command	Alias	Ctrl+	F-key	Alt+	Menu Bar	Tablet
ai_dim_textabove	...	...	...	...	...	...
ai_dim_textcenter						
ai_dim_texthome						

Command: ai_dim_textabove
Select objects: *(Select one or more dimensions.)*
Select objects: *(Press Enter.)*

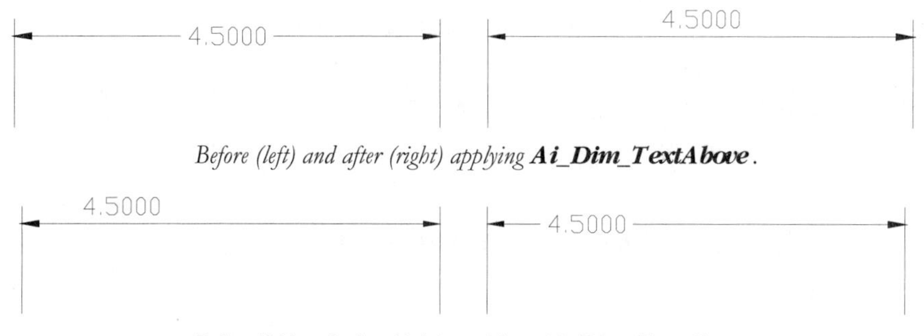

Before (left) and after (right) applying ***Ai_Dim_TextAbove***.

Before (left) and after (right) applying ***Ai_Dim_TextCenter***.

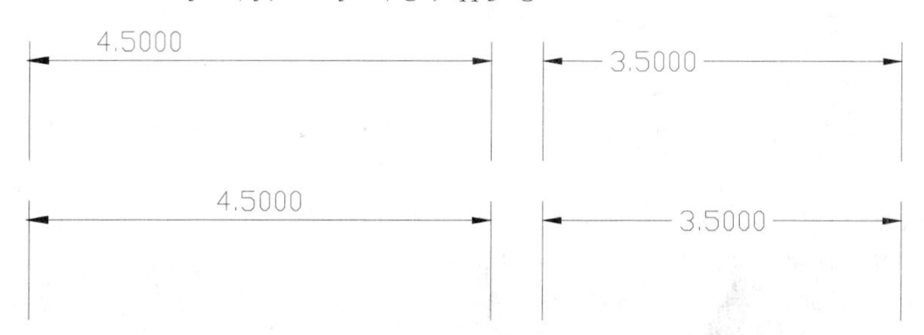

Before (top) and after (bottom) applying ***Ai_Dim_TextHome***.

COMMAND LINE OPTION

Select objects selects one or more dimensions.

TIPS

- Select a dimension, right-click, and then select **Dim Text Position** from the menu:
  Ai_Dim_TextAbove makes dimensions compliant with JIS dimensioning.
  Ai_Dim_TextCenter centers text vertically on the dimension line, but not horizontally.
  Ai_Dim_TextHome centers text horizontally on the dimension line, but not vertically.
- Use the **DimTEdit** command to align text to the left, center, or right on horizontal dimensions.

AiDimTextMove

Moves dimension text (undocumented command).

Command	Alias	Ctrl+	F-key	Alt+	Menu Bar	Tablet
aidimtextmove	...	...	...	...	...	...

Command: aidimtextmove
Enter option [0/1/2] <2>: *(Enter an option, and then press Enter .)*
Select objects: *(Select one dimension.)*
Select objects: *(Press Enter .)*

*Before (left) and after (right) applying **AiDimTextMove = 0** to dimension text.*

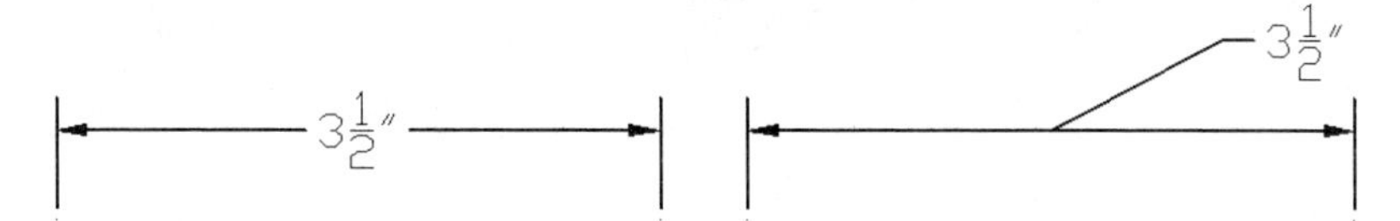

*Before (left) and after (right) applying **AiDimTextMove = 1** to dimension text.*

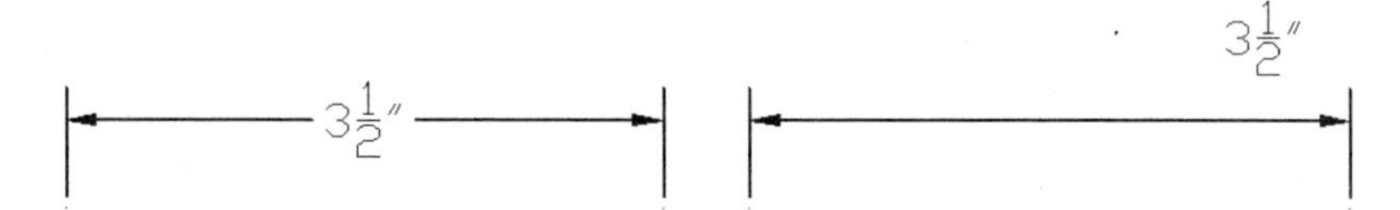

*Before (left) and after (right) applying **AiDimTextMove = 2** to dimension text.*

COMMAND LINE OPTIONS

Enter option specifies the nature of text movement:

0 — Text moves with the dimension line.

1 — Adds a leader to the moved text.

2 — Text moves independently of dimension line and leader (default).

Select objects selects one or more dimensions.

TIPS

- This command allows you retroactively to change the position of dimension text. It is used by the right-click shortcut menu: select a dimension, right-click, and then select **Dim Text Position** from the shortcut menu.
- Although the command allows you to select more than one dimension, it operates on the first-selected dimension only.

Ai_Fms

Switches to layout mode, and then to floating model space (short for Floating Model Space; an undocumented command).

Command	Alias	Ctrl+	F-key	Alt+	Menu Bar	Tablet
ai_fms	...	...	...	...	...	...

Command: ai_fms

Switches to the last active layout, then to the first floating model viewport.

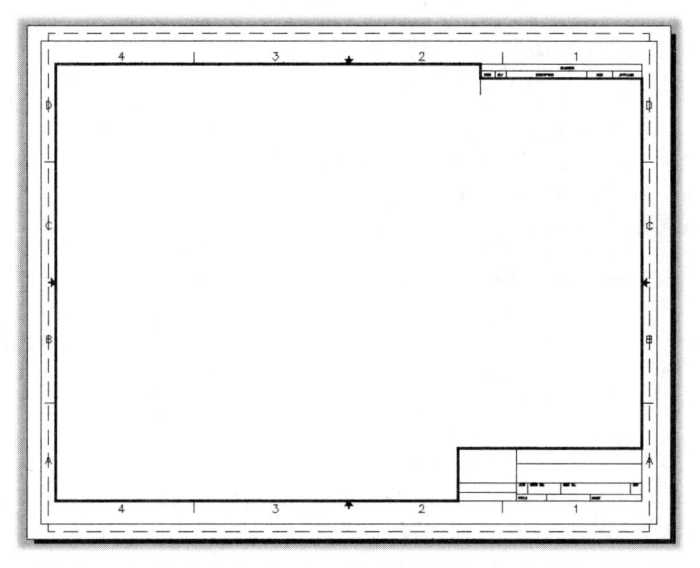

The heavy border indicates the currently-active floating viewport in model space.

COMMAND LINE OPTIONS

None.

TIPS

- This command combines two commands: Tilemode **0** followed by MSpace.
- This command is meant for use with menu and toolbar macros.

Ai_Molc

Changes the current layer to the one on which the selected object is located (short for Make Object Layer Current; undocumented command).

Command	Alias	Ctrl+	F-key	Alt+	Menu Bar	Tablet
ai_molc	...	...	...	...	...	...

Command: ai_molc
Select object whose layer will become current: *(Pick an object.)*

COMMAND LINE OPTION

Select object selects a single object.

RELATED COMMANDS

Layer displays the Layer Properties Manager dialog box.

LayerP reverts to the previous layer.

MatchProp matches the properties of one object to other objects.

RELATED SYSTEM VARIABLE

CLayer holds the name of the current layer.

TIPS

- This command is activated by the **Make Object's Layer Current** button on the Layers toolbar.
- Click the button, and then select an object: the object's becomes the current layer.
- Merely selecting an object displays its layer on the Object Properties toolbar without changing the current layer.

Ai_SelAll

Selects all objects in drawings in the current space, including those off-screen (undocumented command).

Command	Alias	Ctrl+	F-key	Alt+	Menu Bar	Tablet
'ai_selall	**...**	**a**	**...**	**...**	**...**	**...**

Command: ai_selall
Selecting objects...done.

COMMAND LINE OPTIONS

None.

TIPS

- This command is meant for use in menu macros and toolbars.
- Use the CTRL+A shortcut to select all objects in the drawing, other than those on frozen layers.
- The opposed command to selecting all is **(ai_deselect)**, an AutoLISP routine.

Align

Rel.12 Moves, transforms, and rotates objects in three dimensions.

Command	Alias	Ctrl+	F-key	Alt+	Menu Bar	Tablet
align	**al**	...	...	**M3L**	**Modify** ↳**3D Operation** ↳**Align**	**X14**

Command: align
Select objects: *(Select one or more objects to be moved.)*
Select objects: *(Press Enter .)*
Specify first source point: *(Pick a point.)*
Specify first destination point: *(Pick a point.)*
Specify second source point: *(Pick a point.)*
Specify second destination point: *(Pick a point, or press Enter .)*
Specify third source point or <continue>: *(Pick a point, or press Enter .)*
Specify third destination point: *(Pick a point.)*

COMMAND LINE OPTIONS

First point moves object in 2D or 3D when one source and destination point is picked.

Second point moves, rotates, and scales object in 2D or 3D when only two source and destination points are picked.

Third point moves objects in 3D when three source and destination points are picked.

Continue option
Scale objects based on alignment points? [Yes/No] <N>: *(Type* **Y** *or* **N**.*)*

Specifies that the distance between the first and second source points is to be scaled to the distance between the first and second destination points.

RELATED COMMANDS

Mirror3d mirrors objects in three dimensions.

Rotate3d rotates objects in three dimensions.

TIPS

- Enter the first pair of points to define the move vector (distance and direction):

  Specify first source point: *(Pick a point.)*
  Specify first destination point: *(Pick a point.)*
  Specify second source point: *(Press* ENTER.*)*

- Enter two pairs of points to define a 2D (or 3D) transformation, scaling, and rotation:

Points	Alignment Defined
First	Base point for alignment.
Second	Rotation angle.
Third	Planes aligned by source and destination points.

- The third pair defines the 3D transformation.

AmeConvert

Rel.13 Converts PADL solid models and regions created by AME v2.0 and v2.1 (AutoCAD Releases 11 and 12) to ShapeManager solid models.

Command	Alias	Ctrl+	F-key	Alt+	Menu Bar	Tablet
ameconvert	...	...	...	...	...	...

Command: ameconvert
Select objects: *(Select one or more objects.)*
Processing Boolean operations.

COMMAND LINE OPTION

Select objects selects AME objects to convert; ignores non-AME objects, such as the ACIS solids produced by AutoCAD Release 13 through 2002, and ShapeManager solids from AutoCAD 2004-2007.

RELATED COMMAND

AcisIn imports ACIS models from a *.sat* file.

TIPS

- After the conversion process, the AME model remains in the drawing in the same location as the solid model. Erase, if necessary.
- AME holes may become blind holes (holes that are solid cylinders) in the solid model.
- AME fillets and chamfers may be placed higher or lower in the solid model.
- Once Release 12 PADL drawings are converted to AutoCAD 2007 solid models, they cannot be converted back to PADL format.
- This command ignores objects that are neither AME solids nor regions.
- Old AME models are stored in AutoCAD as anonymous block references.

DEFINITIONS

ACIS — solids modeling technology used by AutoCAD in Release 13 through 2002.

AME — short for "Advanced Modeling Extension," the solids modeling module used by AutoCAD in Releases 10 through 12.

PADL — short for "Parts and Description Language," the solids modeling technology used by AutoCAD Releases 10 through 12.

ShapeManager — the solids modeling technology used by AutoCAD 2004 through 2007.

AniPath

2007 Specifies animation paths along which cameras move; can be saved as movie files; short for "ANImation PATH."

Command	Alias	Ctrl+	F-key	Alt+	Menu Bar	Tablet
anipath	...	...	...	VM	View ⇘Motion Path Animation	...

Command: anipath

Displays the Motion Path Animation dialog box:

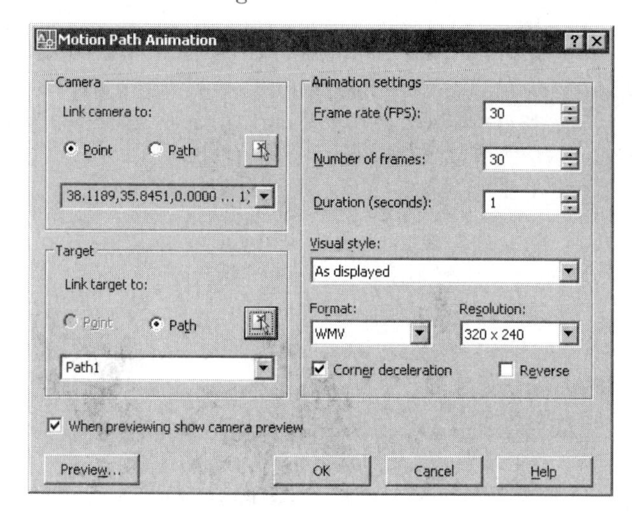

DIALOG BOX OPTIONS

Camera

⊙ **Point** places the camera in the drawing.

○ **Path** guides the camera along an object in the drawing: a line, arc, elliptical arc, circle, polyline, spline, or 3D polyline.

Pick picks the camera point or selects the path in the drawing.

List lists previously selected points and paths.

Target

○ **Point** points the camera at a target in the drawing.

⊙ **Path** points the camera along a path defined by an object (line, circle, etc.) in the drawing.

Pick picks the target point or selects the path in the drawing.

List lists previously selected target points and paths.

Animation Settings

Frame Rate (FPS) specifies the speed of the animation; ranges from 1 to 60 frames per second.

Number of Frames specifies the total number of frames to be captured for the animation.

Duration (seconds) specifies duration of the animation in seconds; linked to Number of Frames.

Visual Style selects the preset visual style or rendering quality for the animation: as displayed, rendered, 3D hidden, 3D wireframe, conceptual, realistic, draft, low, medium, high, presentation.

Format specifies the animation file format: AVI, MOV, MPG, or WMV.

Resolution provides a list of resolutions, ranging from 160x120 to 1024x768.

☑ **Corner Deceleration** reduces the camera's speed around curves.

☐ **Reverse** reverses the animation direction.

☑ **When Previewing Show Camera Preview** displays the Animation Preview dialog box.

Preview previews the animation.

Animation Preview dialog box

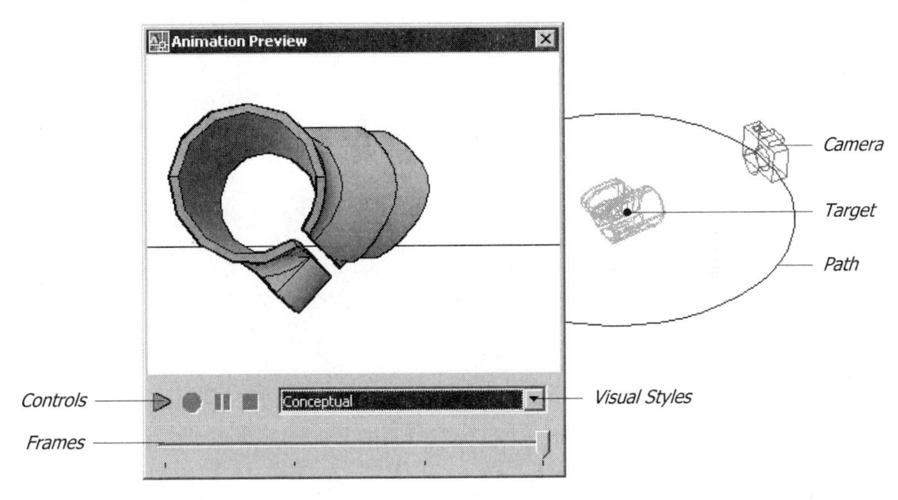

Animation Preview controls

▶ **Play** plays the animation once.

● **Record** records the animation beginning with the current frame (non-operational).

❚❚ **Pause** pauses the animation at the current frame; click Play to continue.

■ **Save** displays the Save As dialog box; saves the animation in AVI, MOV, MPG, or WMV format.

Visual Styles selects the visual style for the animation.

Slider moves through the animation by frames.

RELATED COMMAND

VisualStyles creates and edits custom visual styles.

TIPS

- Cameras are created automatically when the motion path is specified.
- Deleting objects that define motion paths also deletes the path.
- Targets must be linked to paths when cameras are linked to points..
- Changing the frame rate determines the duration, and vice versa.
- MOV format requires QuickTime player; WMV format requires Media Player 9.

'Aperture

V. 1.3 Sets the size (in pixels) of the object snap target height, or box cursor.

Command	Alias	Ctrl+	F-key	Alt+	Menu Bar	Tablet
aperture	...	...	...	...	...	...

Command: aperture
Object snap target height (1-50 pixels) <10>: *(Enter a value.)*

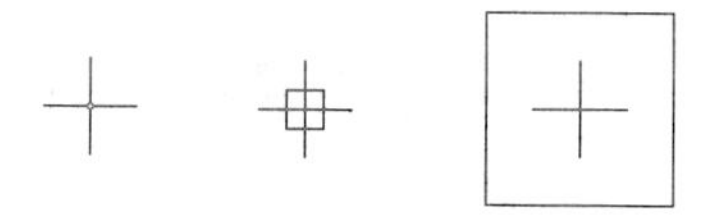

Aperture size = 1 (left), 10 (center), and 50 pixels (right).

COMMAND LINE OPTION

Height specifies the height of the object snap cursor's target.

RELATED COMMAND

Options allows you to set the aperture size interactively (**Drafting** tab).

RELATED SYSTEM VARIABLES

ApBox toggles the display of the aperture box cursor (*undocumented*).

Aperture contains the current target height, in pixels:

Aperture	Meaning
1	Minimum size.
10	Default size, in pixels.
50	Maximum size.

TIPS

- Besides the aperture cursor, AutoCAD has two other similar-looking cursors: the *osnap* cursor appears only during object snap selection; the *pick* cursor appears anytime AutoCAD expects object selection. The size of both cursors can be changed; to change the size of the pick cursor, use the Pickbox command.
- By default, the box cursor does not appear. Nevertheless, it determines how close you have to be to an object for AutoCAD to "snap" to it.
- To display the box cursor, use the Options command, select the Drafting tab, and then enable the **Display AutoSnap Aperture box** option. Alternatively, use the undocumented ApBox system variable.
- Use the Options command to change the size of the aperture visually: select the Drafting tab, and then move the Aperture Size slider.

AppLoad

Rel.12 Creates a list of LISP, VBA, ObjectARx, and other applications to load into AutoCAD (short for APPlication LOADer).

Command	Alias	Ctrl+	F-key	Alt+	Menu Bar	Tablet
appload	**ap**	...	...	**TL**	**Tools** ↳**Load Applications**	**V10**

Command: appload

Displays dialog box:

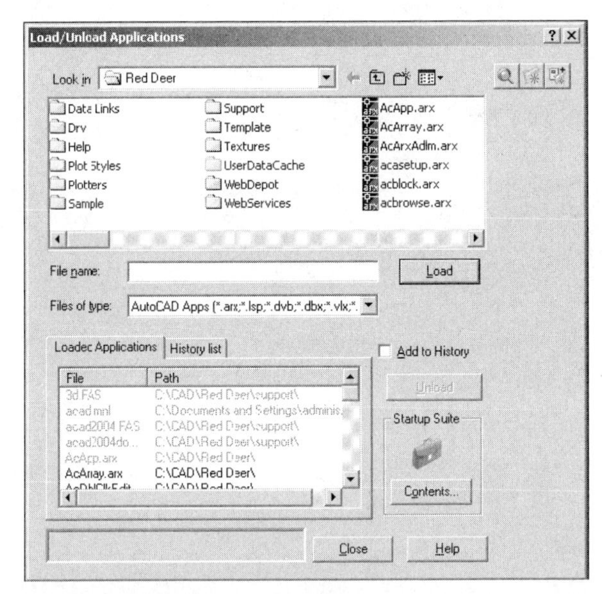

LOAD/UNLOAD APPLICATIONS DIALOG BOX

Look in lists the names of drives and folders available to this computer.

File name specifies the name of the file to load.

Files of type displays a list of file types:

Filetype	Meaning
ARX	objectARX.
DVB	Visual Basic for Applications (VBA).
DBX	objectDBX.
FAS	FASt load autolisp.
LSP	autoLiSP.
VLX	Visual Lisp eXecutable.

Load loads all or selected files into AutoCAD.

Loaded Applications displays the names of applications already loaded into AutoCAD.

History List displays the names of applications previously saved to this list.

☐ **Add to History** adds the file to the History List tab.

Unload unloads all or selected files out of AutoCAD.

Close exits the dialog box.

Startup Suite

Contents displays dialog box:

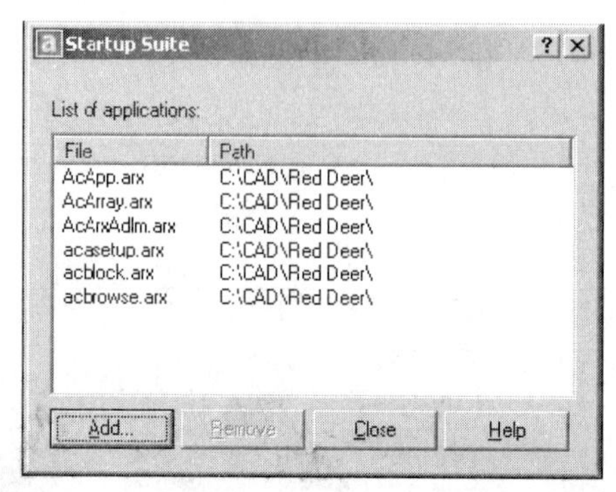

List of applications lists the file names and paths of applications to be loaded automatically each time AutoCAD starts. AutoLISP related files (*.lsp*, *.fas*, and *.vlx*) are loaded whenever a drawing is loaded or a new drawing created. All others are loaded when AutoCAD starts.

Add displays the Add File to Startup Suite dialog box; allows you to select one or more application files.

Remove removes the application from the list.

Close returns to the Load/Unload Applications dialog box.

RELATED COMMANDS

Arx lists ObjectARX programs currently loaded in AutoCAD.

VbaLoad loads VBA applications.

RELATED AUTOLISP FUNCTIONS

(load) loads an AutoLISP program.

(autoload) predefines commands to load related AutoLISP programs.

STARTUP SWITCH

/ld loads an ARX or DBX application when AutoCAD starts up.

TIPS

- Use **AppLoad** when AutoCAD does not automatically load a command; alternatively, you can drag files from Windows Explorer into the **Loaded Applications** list.
- ObjectARX, VBA, and DBX applications are loaded immediately; FAS, LSP, and VLX files are loaded after this dialog box closes.
- This command is limited to loading 50 applications at a time.
- Load the *asdkTPCtest.arx* application for undocumented Tablet PC support.
- The *acad2007doc.lsp* file establishes autoloader and other utility functions, and is loaded automatically each time a drawing is opened; the *acad2007.lsp* file is loaded only once per AutoCAD session. Use the AcadLspAsDoc system variable to control whether these files are loaded with AutoCAD.

Arc

V. 1.0 Draws 2D arcs of less than 360 degrees, by eleven methods.

Command	Alias	Ctrl+	F-key	Alt+	Menu Bar	Tablet
arc	a	...	...	DA	Draw ↳Arc	R10

Command: arc
Specify start point of arc or [CEnter]: *(Pick a point, or enter the* **CE** *option.)*
Specify second point of arc or [CEnter/ENd]: *(Pick a point, or enter an option.)*
Specify end point of arc: *(Pick a point.)*

COMMAND LINE OPTIONS

SSE (start, second, end) arc options

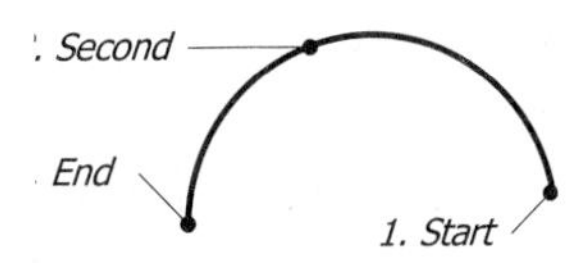

Start point indicates the start point of a three-point arc.
Second point indicates a second point anywhere along the arc.
Endpoint indicates the end point of the arc.

SCE (start, center, end), SCA (start, center, angle), and SCL (start, center, length) options

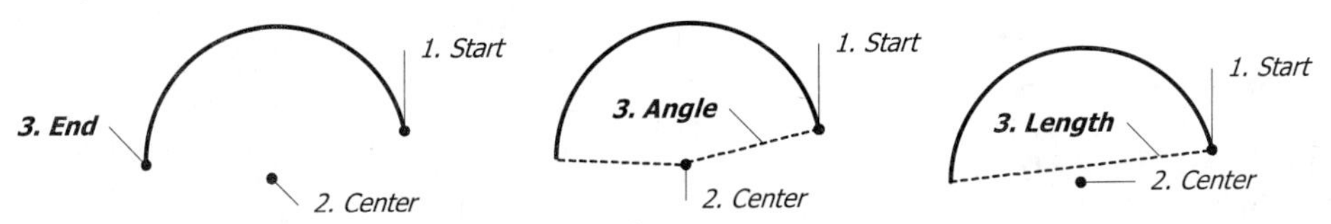

Start point indicates the start point of a two-point arc.
Center indicates the arc's center point.
Angle indicates the arc's included angle.
Length of chord indicates the length of the arc's chord.
Endpoint indicates the arc's end point.

SEA (start, end, angle), SED (start, end, direction), SER (start, end, radius), and SEC (start, end, center) options

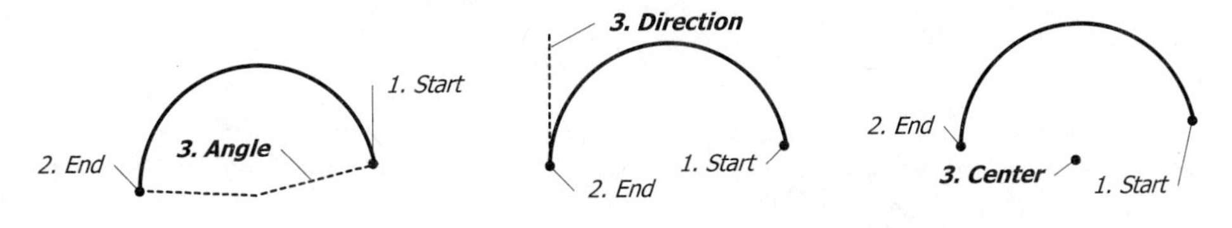

Start point indicates the start point of a two-point arc.

End indicates the arc's end point.

Center point indicates the arc's center point.

Angle indicates the arc's included angle.

Direction indicates the tangent direction from the arc's start point.

Radius indicates the arc's radius.

CSE (center, start, end), CSA (center, start, angle), and CSL (center, start, length) options

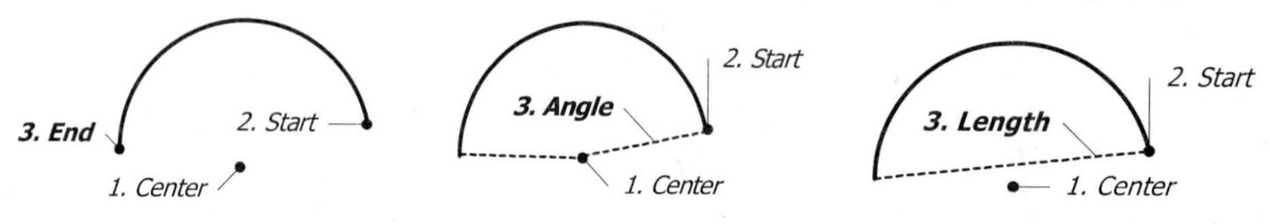

Center indicates the center point of a two-point arc.

Start point indicates the arc's start point.

Endpoint indicates the arc's end point.

Angle indicates the arc's included angle.

Length of chord indicates the length of the arc's chord.

Continued Arc option

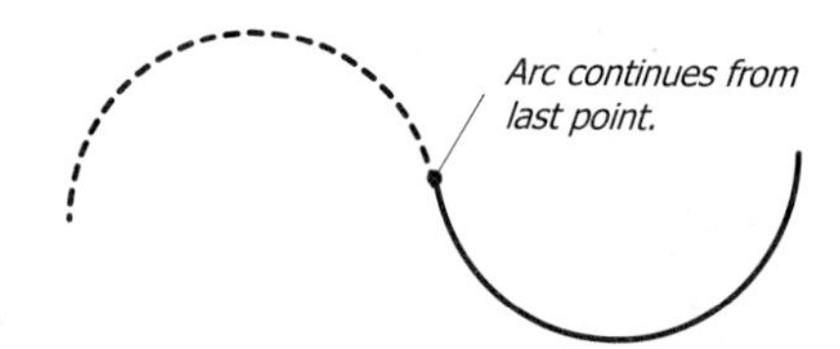

Enter continues the arc tangent from the end point of last-drawn line or arc.

RELATED COMMANDS

Circle draws an "arc" of 360 degrees.

DimArc dimensions the length of the arc.

DimCenter places center marks at the arc's center.

Ellipse draws elliptical arcs.

Polyline draws connected polyline arcs.

ViewRes controls the roundness of arcs.

RELATED SYSTEM VARIABLE

LastAngle saves the included angle of the last-drawn arc (read-only).

WhipArc controls the smoothness in display of circles and arcs.

TIPS

- Arcs are drawn counter clockwise.
- In most cases, it is easier to draw a circle, and then use the Trim command to convert the circle into an arc.
- It may be easier to use the Fillet command to create arcs tangent to lines.
- To start an arc precisely tangent to the end point of the last line or arc, press Enter at the 'Specify start point of arc or [CEnter]:' prompt.
- You can drag the arc only during the last-entered option.
- Specifying an x,y,z-coordinate as the starting point of the arc draws the arc at the z-elevation.
- The components of AutoCAD arcs:

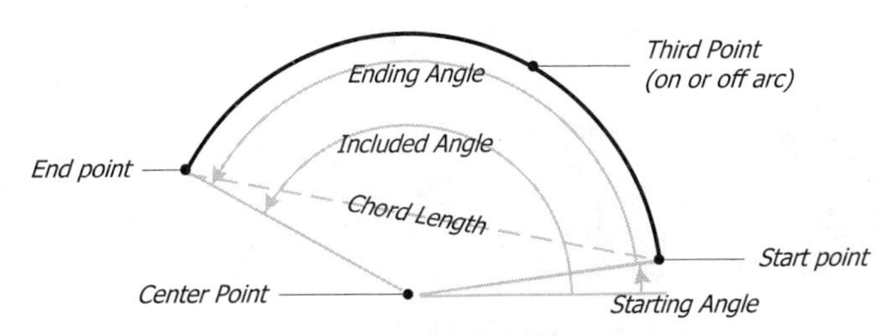

- AutoCAD 2006 adds three more stretch handles (shown as triangles) to arcs:

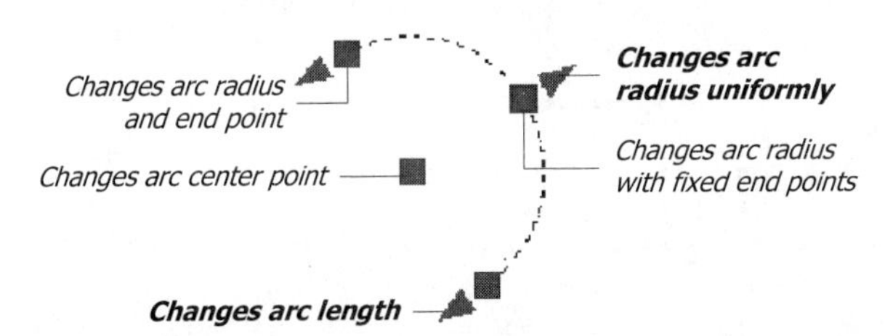

- When the chord length is positive, the minor arc is drawn counterclockwise from the start point; when negative, the major arc is drawn counterclockwise.

Archive

2005 Packages all files related to the current sheet set.

Command	Alias	Ctrl+	F-key	Alt+	Menu Bar	Tablet
archive	...	...	...	...	...	...
-archive						

Command: archive

*When no sheet sets are open, AutoCAD complains, "No Sheet Set is Open," and terminates the command. (To open a sheet set, use the **OpenSheetset** command.)*

When at least one sheet set is open, AutoCAD displays the following dialog box:

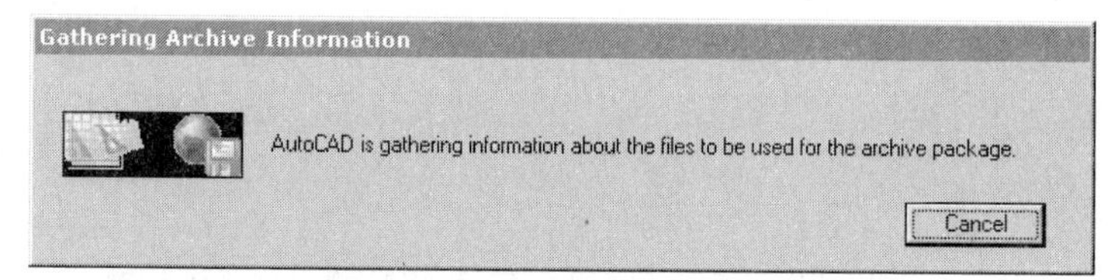

After a moment, AutoCAD displays the next dialog box:

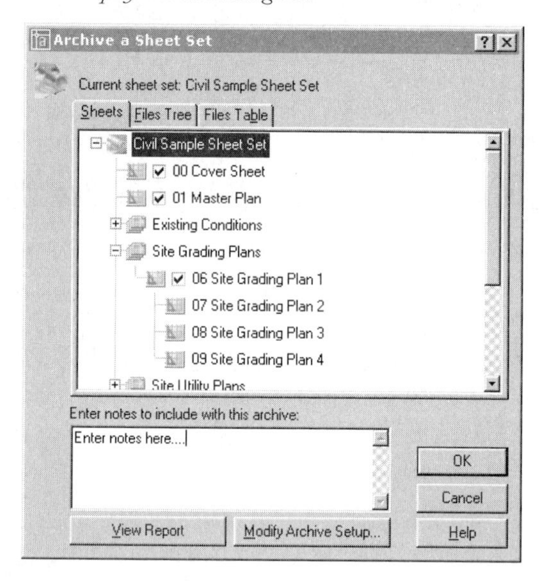

ARCHIVE A SHEET SET DIALOG BOX

Sheets tab displays sheets included with the archive.

Files Tree tab displays names of drawing and support files, grouped by category.

Files Table tab displays file names in alphabetical order.

Enter notes to include with this archive provides space for entering notes.

View Report displays the View Archive Report dialog box.

Modify Archive Setup displays the Modify Archive Setup dialog box.

Files Tree tab

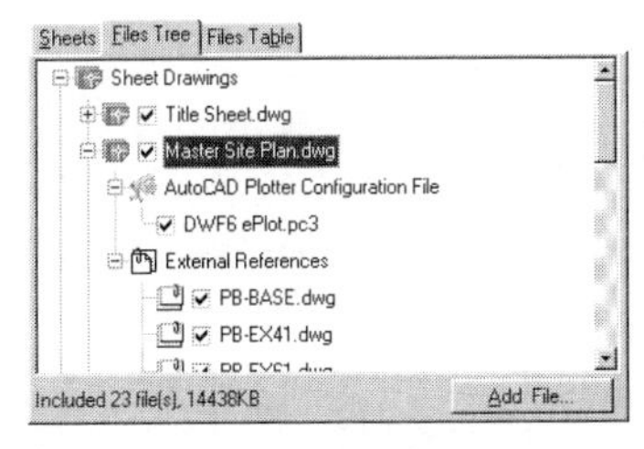

☑ File is included in archive.

☐ File excluded from archive.

Add File displays the Add File to Archive dialog box, which adds files to the archive set.

Files Table tab

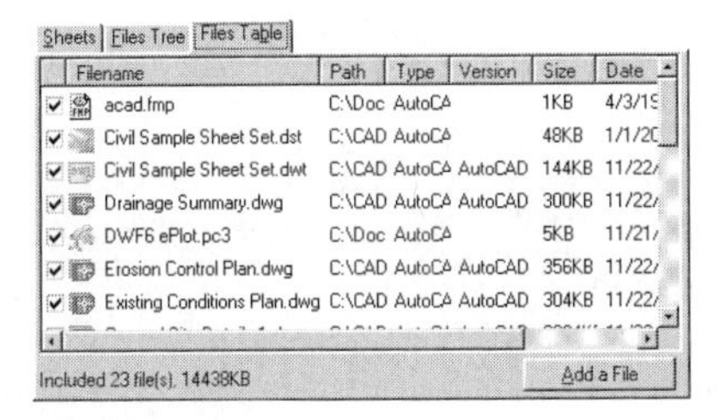

VIEW REPORT DIALOG BOX

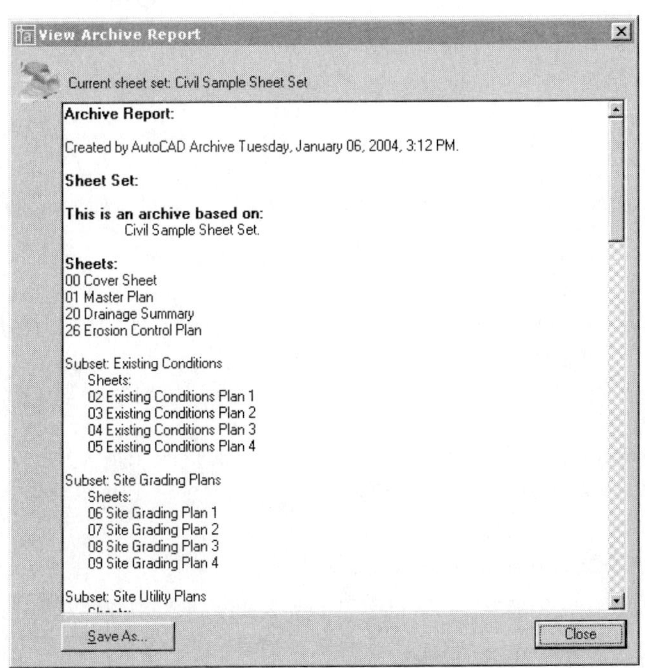

Save As saves the report in *.txt* format (plain ASCII text).

Close closes the dialog box, and then returns to the previous dialog box.

MODIFY ARCHIVE SETUP DIALOG BOX

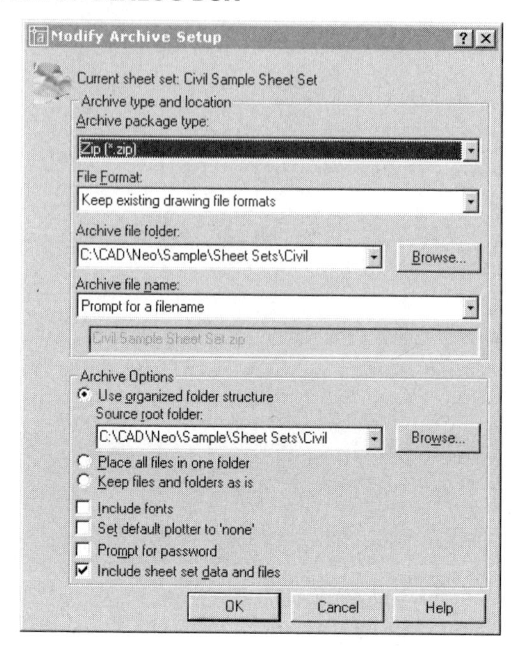

Archive Type and Location

Folder (set of files) archives uncompressed files in new and existing folders. This is the best option when archiving to CD or FTP.

Self-Extracting Executable (*.exe) archives files in a compressed, self-extracting file. Uncompress by double-clicking the file.

Zip (*.zip) (*default*) archives files as a compressed ZIP file. Uncompress the file using PkZip or WinZip. This is the best option when sending by email.

File Format

Keep existing drawing file formats archives files in their native format.

AutoCAD 2007/LT 2007 Drawing Format archives files in AutoCAD 2007 format.

AutoCAD 2004/LT 2004 Drawing Format archives files in AutoCAD 2004/5/6 formats.

AutoCAD 2000/LT 2000 Drawing Format archives files in AutoCAD 2000/i/1 formats. *Warning!* Some objects and properties not found in earlier AutoCAD versions may be changed or lost.

Archive File Folder specifies the location in which to archive the files. When no location is specified, the archive is created in the same folder as the *.dst* file.

Archive File Name

Prompt for a File Name prompts the user for the file name.

Overwrite if Necessary uses the specified file name, and overwrites the existing file of the same name.

Increment File Name if Necessary uses the specified file name, and appends a digit to avoid overwriting the existing file of the same name.

⊙ **Use Organized Folder Structure** preserves the folder structure in the archive, but makes the changes listed below; allows you to specify the name of the folder tree. The option is unavailable when saving archives to the Internet. Autodesk reminds you of the following:

- Relative paths remain unchanged.
- Absolute paths inside the folder tree are converted to relative paths; absolute paths outside the folder tree are converted to "No Path," and are moved to the folder tree.
- A *Fonts* folder is created when font files are included in the archive.
- A *PlotCFG* folder is created when plotter configuration files are included.
- A *SheetSets* folder is created for sheet set support files; the *.dst* sheet set data file is placed in the root folder.

○ **Place All Files in One Folder** locates all files in a single folder.

○ **Keep Files and Folders As Is** preserves the folder structure in the archive. The option is unavailable when saving archives to the Internet.

☐ **Include Fonts** includes all *.ttf* and *.shx* font files in the archive.

☐ **Set Default Plotter to 'None'** resets the plotter to None for all drawings in the archive.

☐ **Prompt for Password** displays a dialog box for specifying a password for the archive; not available when the Folder archive type is selected.

☑ **Include Sheet Set Data (DST) File** includes the *.dst* sheet set data file with the archive.

-ARCHIVE Command

Command: -archive

Sheet Set name or [?] <Sheet Set>: *(Enter a name or* **?***.)*

Enter an option [Create archive package/Report only] <Create>: *(Type* **c** *or* **r***.)*

Gathering files ...

Sheet Set name specifies the name of the sheet set to archive.

? lists the names of sheet sets open in the current drawing.

Create archive package creates the archive in ZIP format.

Report only displays the Save Report File As dialog box for saving the archive report as a text file; also prompts you for a Transmittal Note.

RELATED COMMANDS

eTransmit creates an archive of only the current drawing and its support files.

SheetSet creates and controls sheet sets.

TIPS

- This command works only when at least one sheet set is open in the current drawing.
- Archive files can become large when they hold many drawing files.
- The -Archive is useful for creating a script or macro to repeatedly archive the same group of drawings and other files.

Area

V. 1.0 Calculates the area and perimeter of objects.

Command	Alias	Ctrl+	F-key	Alt+	Menu Bar	Tablet
area	**aa**	...	...	**TQA**	**Tools** **↳Inquiry** **↳Area**	**T7**

Command: area

Specify first corner point or [Object/Add/Subtract]: *(Pick a point, or enter option.)*

Specify next corner point or press ENTER for total: *(Pick a point, or press Enter .)*

Sample response:

Area = 1.8398, Perimeter = 6.5245

COMMAND LINE OPTIONS

First corner point specifies the first point to begin measurement.

Object selects the object to be measured.

Add switches to add-area mode.

Subtract switches to subtract-area mode.

ENTER indicates the end of the area outline.

RELATED COMMANDS

Dist returns the distance between two points.

Id lists the x,y,z coordinates of a selected point.

List reports on the area and other properties of objects.

MassProp returns surface area, and so on, of solid models.

RELATED SYSTEM VARIABLES

Area contains the most recently-calculated area.

Perimeter contains the most recently-calculated perimeter.

TIPS

- Before subtracting, you must use the Add option. Picking points or using the Object option first prevents you from subtracting.
- At least three points must be picked to calculate an area; AutoCAD "closes the polygon" with a straight line before measuring the area.
- The **Object** option returns the following information:

Object	Measurement Returned
Circle, ellipse	Area and circumference.
Planar closed spline	Area and circumference.
Closed polyline, polygon	Area and perimeter.
Open objects	Area and length.
Region	Net area of all objects in region.
2D solid	Area.

- Areas of wide polylines are measured along center lines; closed polylines must have only one closed area.

Array

V. 1.3 Creates 2D linear, rectangular, and polar arrays of objects.

Commands	Aliases	Ctrl+	F-key	Alt+	Menu Bar	Tablet
array	**ar**	...	...	**MA**	**Modify** ↳**Array**	**V18**
-array	**-ar**					

Command: array

Displays dialog box:

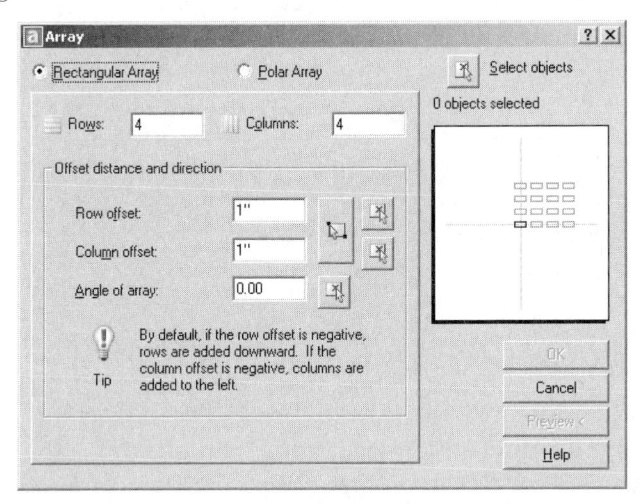

ARRAY DIALOG BOX

⊙ **Rectangular Array** displays the options for creating a rectangular array.

○ **Polar Array** displays the options for creating a circular array.

Select objects dismisses the dialog box temporarily, so that you can select the objects in the drawing:

Select objects: *(Select one or more objects.)*

Select objects: *(Press* **ENTER***.)*

Press Enter, or right-click to return to the dialog box.

Preview dismisses the dialog box temporarily, so that you can see what the array will look like.

Rectangular Array

Rows specifies the number of rows; minimum=1, maximum=32767.

Columns specifies the number of columns; minimum=1, maximum=32767.

Row offset specifies the distance between the center lines of the rows; use negative numbers to draw rows in the negative x-direction (to the left).

Column offset specifies the distance between the center lines of the columns; use negative numbers to draw rows in the negative y-direction (downward).

Angle of array specifies the angle of the array, which "tilts" the x and y axes of the array.

Polar Array

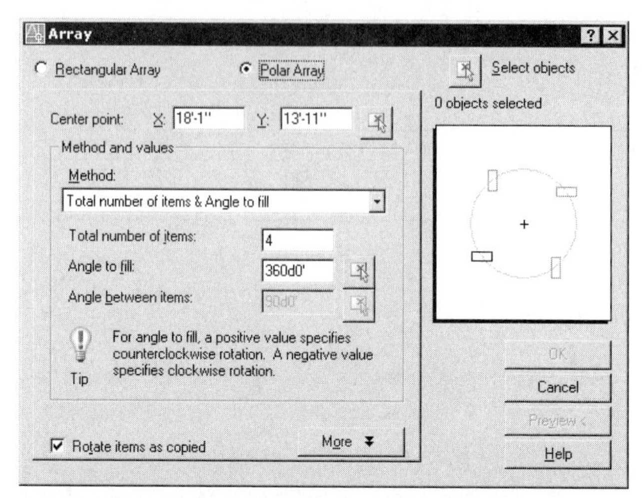

Center point specifies the center of the polar array.

Method specifies the method by which the array is constructed:

- Total number of items and angle to fill.
- Total number of items and angle between items.
- Angle to fill and angle between items.

Total number of items specifies the number of objects in the array; minimum=2.

Angle to fill specifies the angle of "arc" to construct the array; min=1 deg; max=360 deg.

Angle between items specifies the angle between each object in the array.

Rotate items as copied

☑ Objects are rotated so that they face the center of the array.

☐ Objects are not rotated.

More displays additional options for constructing a polar array.

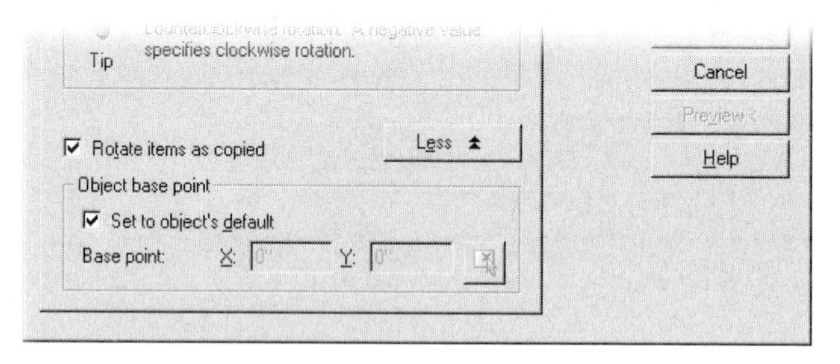

Object Base Point

Set to object's default:

☑ The base point of the object is used, as listed below.

☐ The base point is specified by the user; if objects are not rotated, select a base point to avoid unexpected results.

Object	Default base point
Arc, circle, ellipse	Center point.
Polygon, rectangle	First vertex.
Line, polyline, 3D polyline, ray, spline	Start point.
Donut	Start point.
Block, mtext, text	Insertion point.
Xline	Midpoint.
Region	Grip point.

Base point specifies the x and y coordinates of the new base point.

-ARRAY Command

Command: -array

Select objects: *(Select one or more objects.)*

Select objects: *(Press Enter .)*

Enter the type of array [Rectangular/Polar] <R>: *(Type* **R** *or* **P**.*)*

Rectangular options

Enter the number of rows (---) <1>: *(Enter a value, or press Enter .)*

Enter the number of columns (|||) <1>: *(Enter a value, or press Enter .)*

Enter the distance between rows or specify unit cell (---): *(Enter a value.)*

Specify the distance between columns (|||): *(Enter a value.)*

Polar options

Specify center point of array or [Base]: *(Select one or more objects, or enter* **B**.*)*

Enter the number of items in the array: *(Enter a value.)*

Specify the angle to fill (+=ccw, -=cw) <360>: *(Enter a value, or press Enter .)*

Rotate arrayed objects? [Yes/No] <Y>: *(Enter* **Y** *or* **N**.*)*

COMMAND LINE OPTIONS

R creates a rectangular array from the selected object.

P creates a polar array from the selected object.

B specifies base point of objects, and the center point of the array.

Center point specifies the center point of the array.

Rows specifies the number of horizontal rows.

Columns specifies the number of vertical columns.

Unit cell specifies the vertical and horizontal spacing between objects.

RELATED COMMANDS

3dArray creates a rectangular or polar array in 3D space.

Copy creates one or more copies of the selected object.

MInsert creates a rectangular block array of blocks.

RELATED SYSTEM VARIABLE

SnapAng determines the default angle of a rectangular array.

TIPS

- To create a rectangular array at an angle, use the **Rotation** option of the **Snap** command.
- Rectangular arrays are drawn upward in the positive x-direction, and to the right in the positive y-direction; to draw the array in the opposite directions, specify negative row and column distances.
- Polar arrays are drawn counterclockwise; to draw the array clockwise, specify a negative angle.

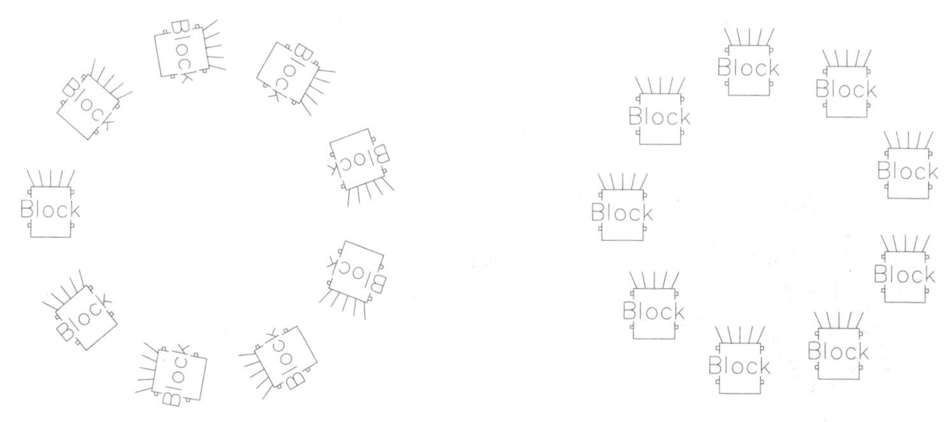

Nine-item polar arrays — rotated (left) and unrotated (right).

- For linear arrays, enter 1 for the number of rows or columns, or use **Divide** or **Measure**.

Arx

Rel.13 Displays information regarding currently loaded ObjectARX programs.

Command	Alias	Ctrl+	F-key	Alt+	Menu Bar	Tablet
arx	...	...	...	...	...	...

Command: arx

Enter an option [?/Load/Unload/Commands/Options]: *(Enter an option.)*

COMMAND LINE OPTIONS

? lists the names of currently loaded ObjectARX programs.

Load loads the ObjectARX program into AutoCAD.

Unload unloads the ObjectARX program out of memory.

Commands lists the names of commands associated with each ObjectARX program.

Options options

CLasses lists the class hierarchy for ObjectARX objects.

Groups lists the names of objects entered into the "system registry."

Services lists the names of services entered in the ObjectARX "service dictionary."

RELATED COMMAND

AppLoad loads LISP, VBA, ObjectDBX, and ObjectARX programs via a dialog box.

RELATED AUTOLISP FUNCTIONS

(arx) lists currently loaded ObjectARX programs.

(arxload) loads an ObjectARX application.

(autoarxload) predefines commands that load the ObjectARX program.

(arxunload) unloads an ObjectARX application.

RELATED FILE

***.arx** are objectARX program files.

TIPS

- Use the **Load** option to load external commands that do not seem to work.
- Use the **Unload** option of the **Arx** command to remove ObjectARX programs from AutoCAD to free up memory.

Removed Commands

The following ASE (AutoCAD SQL Extension) commands were removed from AutoCAD 2000: **AseAdmin**, **AseExport**, **AseLinks**, **AseRows**, **AseSelect**, and **AseSqlEd**. They were replaced by **DbConnect**.

'Assist, 'AssistClose

2000i Opens and closes Info palette, providing real-time assistance.

Command	Alias	Ctrl+	F-key	Alt+	Menu Bar	Tablet
assist	...	5	...	HI	**Help** ↳**Info Palette**	...
assistclose	...	5	...	...	**Tools** ↳**Info Palette**	...

Command: assist

Displays palette that updates as you enter commands (unless Lock is turned on):

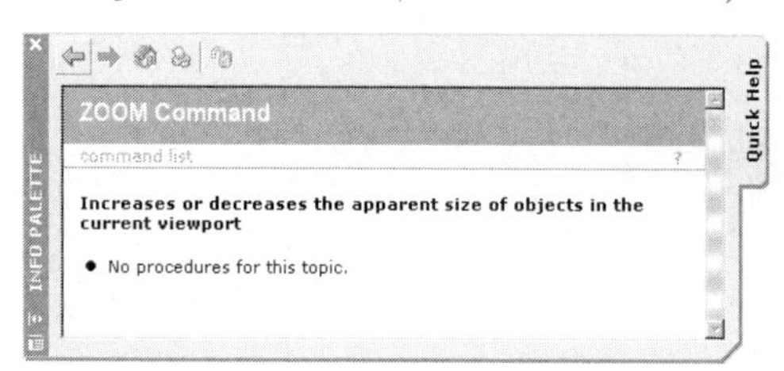

Command: assistclose

Closes palette.

Info Palette Toolbar

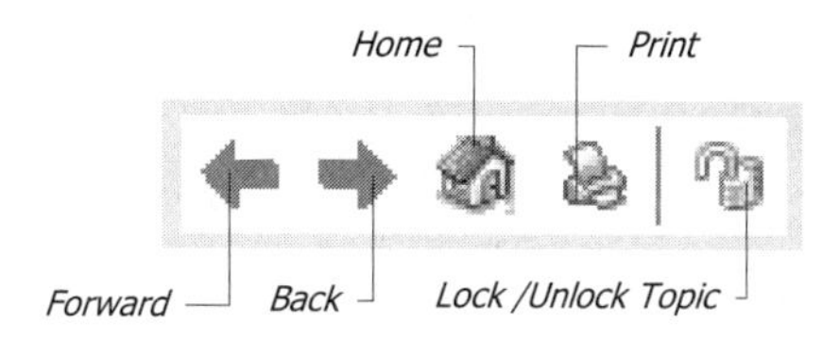

SHORTCUT MENU OPTIONS

Forward displays the next topic, if the **Back** option has been used.

Back displays the previous topic.

Home displays the "home" page.

Print prints the topic.

Lock prevents contents from updating when a different command is started.

RELATED COMMAND

Help displays online help in a window.

RELATED SYSTEM VARIABLE

AssistState reports whether the Info Palette is active.

TIPS

- This menu provides real-time help for commands, dialog boxes, and system variables.
- Turn off for better performance during scripts, AutoLISP routines, and VBA macros.
- This command cannot be used transparently (within another command) until it has been used at least once non-transparently.

AttachURL

Rel.14 Attaches hyperlinks to objects and areas.

Command	Alias	Ctrl+	F-key	Alt+	Menu Bar	Tablet
attachurl	...	...	...	...	...	...

Command: attachurl
Enter hyperlink insert option [Area/Object] <Object>: *(Type* **A** *or* **O***.)*
Select objects: *(Select one or more objects.)*
Select objects: *(Press Enter .)*
Enter hyperlink <current drawing>: *(Enter an address.)*

COMMAND LINE OPTIONS

Area creates rectangular hyperlinks by specifying two corners of a rectangle.

Object attaches hyperlinks after you select one or more objects.

Enter hyperlink requires you to enter a valid hyperlink.

Area Options

First corner picks the first corner of the rectangle.

Other corner picks the second corner of the rectangle.

RELATED COMMANDS

Hyperlink displays a dialog box for adding a hyperlink to an object.

SelectUrl selects all objects with attached hyperlinks.

TIPS

- The hyperlinks placed in the drawing can link to *any* other file: another AutoCAD drawing, an office document, or a file located on the Internet.
- Autodesk recommends that you use the following URL (uniform resource locator) formats:

File Location	Example URL
Web Site	**http:**//*servername*/*pathname*/*filename*.**dwg**
FTP Site	**ftp:**//*servername*/*pathname*/*filename*.**dwg**
Local File	**file:**///*drive*:/*pathname*/*filename*.**dwg**
or	**file:**////*localPC*/*pathname*/*filename*.**dwg**
Network File	**file:**//*localhost*/*drive*:/*pathname*/*filename*.**dwg**

- The URL (hyperlink) is stored as follows:

Attachment	URL
One object	Stored as xdata (extended entity data).
Multiple objects	Stored as xdata in each object.
Area	Stored as xdata in a rectangular object on layer URLLAYER.

- The **Area** option creates a layer named URLLAYER with the default color of red, and places a rectangle object on this layer; do not delete the layer.

AttDef

V. 2.0 Defines attribute modes and prompts (short for ATTribute DEFinition).

Commands	Aliases	Ctrl+	F-key	Alt+	Menu Bar	Tablet
attdef	**att** **ddattdef**	...	...	**DKD**	**Draw** ↳**Block** ↳**Define Attributes**	...
-attdef	**-att**					

Command: attdef

Displays dialog box:

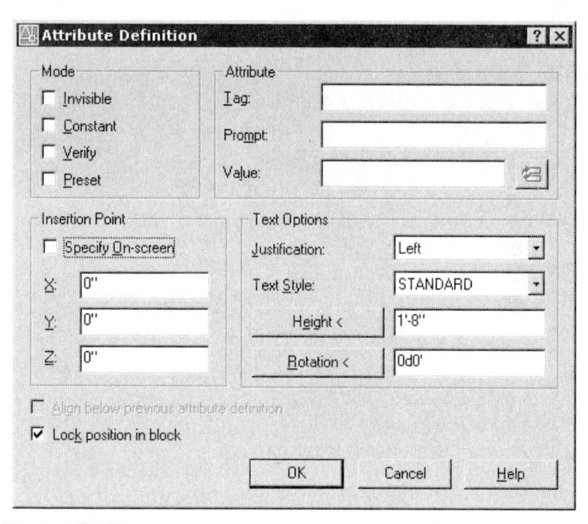

After you click **OK**, *AutoCAD prompts:*

Specify start point: *(Pick a point to locate the attribute text.)*

ATTRIBUTE DEFINITION DIALOG BOX

Mode

☐ **Invisible** makes the attribute text invisible.

☐ **Constant** uses constant values for the attributes.

☐ **Verify** verifies the text after input.

☐ **Preset** presets the variable attribute text.

Attribute

Tag identifies the attribute.

Prompt prompts the user for input.

Value sets the default value for the attribute.

Insert Field displays the Field dialog box; select a field, and then click **OK**. See the Field command.

Insertion Point

Pick point picks the insertion point with cursor.

X specifies the x coordinate insertion point.

Y specifies the y coordinate insertion point.

Z specifies the z coordinate insertion point.

Text

Justification sets the text justification.

Text style selects a text style.

Height specifies the height.

Rotation sets the rotation angle.

☐ **Align below previous attribute definition** places the text automatically below the previous attribute.

☑ **Lock location in block** prevents attributes from being repositioned; this is required to include attributes in dynamic blocks in the selection set of actions.

-ATTDEF Command

Command: -attdef

Current attribute modes: Invisible=N Constant=N Verify=N Preset=N

Enter an option to change [Invisible/Constant/Verify/Preset/Lock position] <done>: *(Enter an option.)*

Enter attribute tag: *(Enter text, and then press Enter .)*

Enter attribute prompt: *(Enter text, and then press Enter .)*

Enter default attribute value: *(Enter text, and then press Enter .)*

Specify start point of text or [Justify/Style]: *(Pick a point, or enter an option.)*

Specify height <0.200>: *(Enter a value.)*

Specify rotation angle of text <0>: *(Enter a value.)*

COMMAND LINE OPTIONS

Attribute mode selects the mode(s) for the attribute:

- **I** toggles visibility of attribute text in drawing (short for Invisible).
- **C** toggles fixed or variable value of attribute (short for Constant).
- **V** toggles confirmation prompt during input (short for Verify).
- **P** toggles automatic insertion of default values (short for Preset).

Start point indicates the start point of the attribute text.

Justify selects the justification mode for the attribute text.

Style selects the text style for the attribute text.

Height specifies the height of the attribute text; not displayed if the style specifies a height other than 0.

Rotation angle specifies the angle of the attribute text.

RELATED COMMANDS

AttDisp controls the visibility of attributes.

EAttEdit edits the values of attributes.

EAttExt extracts attributes to disk.

AttRedef redefines an attribute or block.

Block creates blocks with attributes.

Insert inserts blocks; if block has attributes, prompts for their values.

RELATED SYSTEM VARIABLE

AFlags holds the default value of modes in bit form:

0 — No attribute mode selected.

1 — Invisible.

2 — Constant.

4 — Verify.

8 — Preset.

TIPS

- Constant attributes cannot be edited.
- Attribute tags cannot be null (have no value); attribute values may be null.
- You can enter any characters for the attribute tag, except a space or an exclamation mark. All characters are converted to uppercase.
- When you press ENTER at 'Attribute Prompt,' AutoCAD uses the attribute *tag* as the prompt.
- When you press ENTER at the 'Starting point:' prompt, **AttDef** automatically places the next attribute below the previous one.

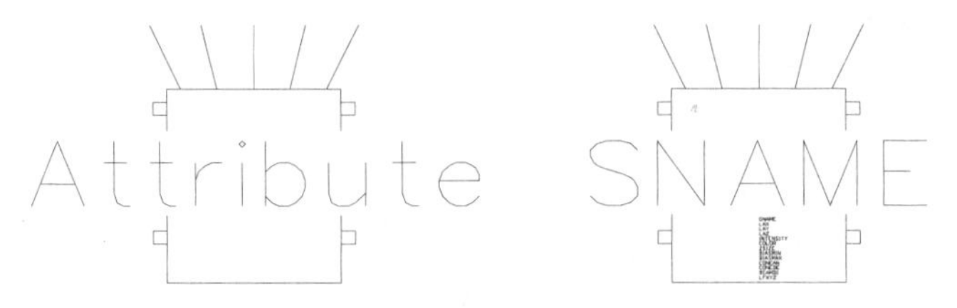

*Left: Block with attribute **value** ("attribute").*
*Right: Attribute **tags** ("sname" and so on).*

- 'Attribute Prompt' and 'Default Attribute Value' are not displayed when constant mode is turned on. Instead, AutoCAD prompts 'Attribute Value.'
- AutoCAD does not prevent you from using the same tag over and over again. During attribute extraction, you may have difficulty distinquishing between tags.

'AttDisp

V. 2.0 Controls the display of all attributes in the drawing (short for ATTribute DISPlay).

Command	Alias	Ctrl+	F-key	Alt+	Menu Bar	Tablet
attdisp	...	...	...	VLA	View ↳Display ↳Attribute Display	L1

Command: attdisp

Enter attribute visibility setting [Normal/ON/OFF] <Normal>: *(Enter an option.)*

Regenerating drawing.

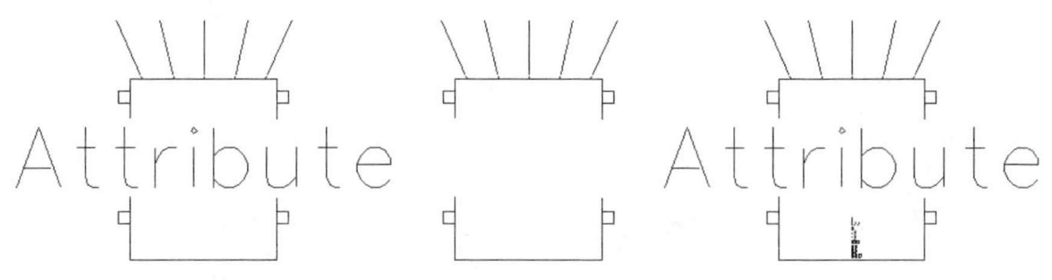

Attribute display ***Normal*** *(left),* ***Off*** *(center), and* ***On*** *(right).*

COMMAND LINE OPTIONS

Normal displays attributes according to **AttDef** setting.

ON displays all attributes, regardless of **AttDef** setting.

OFF displays no attributes, regardless of **AttDef** setting.

RELATED COMMAND

AttDef defines new attributes, including their default visibility.

RELATED SYSTEM VARIABLE

AttMode holds the current setting of **AttDisp**:

0 — Off: no attributes are displayed.

1 — Normal: invisible attributes are not displayed.

2 — On: all attributes are displayed.

TIPS

- When RegenAuto is off, use Regen after AttDisp to change to attribute display.
- When you define invisible attributes, use AttDisp to view them.
- Use AttDisp to turn off the display of attributes, which increases display speed and reduces drawing clutter.

AttEdit

V. 2.0 Edits attributes in drawings (short for ATTribute EDIT).

Commands	Aliases	Ctrl+	F-key	Alt+	Menu Bar	Tablet
attedit	**ate**	...	...	**MOAS**	**Modify**	**Y20**
	ddatte				**↳Object**	
					↳Attribute	
					↳Single	
-attedit	**-ate**			**MOAG**	**Modify**	
	atte				**↳Object**	
					↳Attribute	
					↳Global	

Command: attedit

Select block reference: *(Pick a block.)*

Displays dialog box:

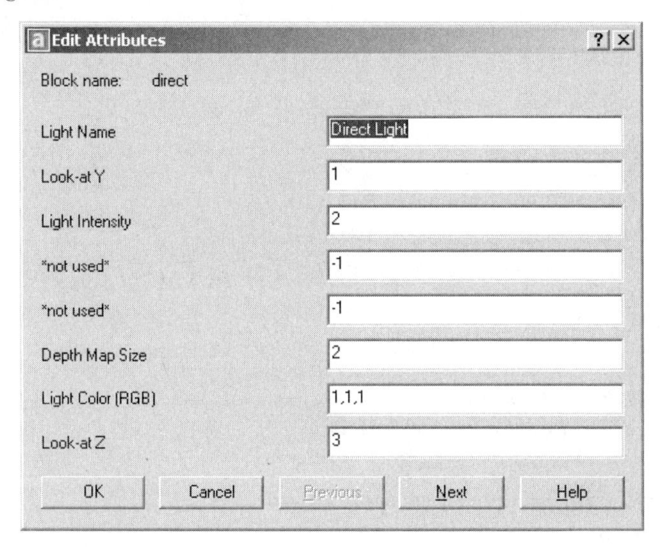

EDIT ATTRIBUTES DIALOG BOX

Block Name names the selected block.

Attribute-specific prompts allow you to change attribute values.

OK accepts the changes and closes the dialog box.

Cancel discards the changes and closes the dialog box.

Previous displays the previous list of attributes, if any.

Next displays the next list of attributes, if any.

-ATTEDIT Command

Command: -attedit

One-at-time Attribute Editing options

Edit attributes one at a time? [Yes/No] <Y>: *(Enter **Y**.)*

Enter block name specification <*>: *(Press Enter to edit all.)*

Enter attribute tag specification <*>: *(Press Enter to edit all.)*

Enter attribute value specification <*>: *(Press Enter to edit all.)*

Select Attributes: *(Select one or more attributes.)*

Select Attributes: *(Press Enter .)*

Enter an option [Value/Position/Height/Angle/Style/Layer/Color/Next] <N>: *(Enter an option, and then press Enter .)*

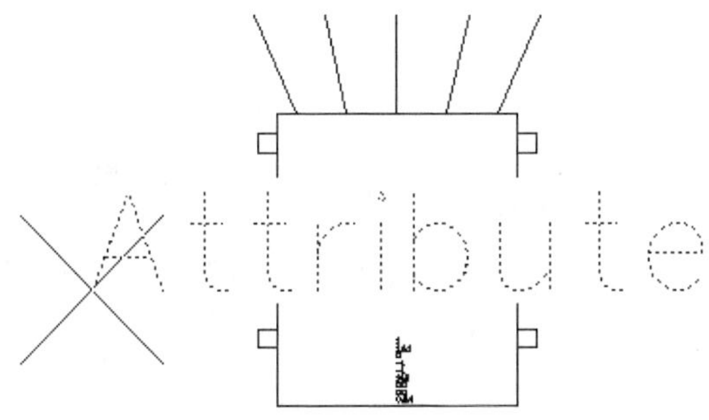

*During single attribute editing, **AttEdit** marks the current attribute with an 'X.'*

Global Attribute Editing options

Edit attributes one at a time? [Yes/No] <Y>: *(Enter **N**.)*

Performing global editing of attribute values.

Edit only attributes visible on screen? [Yes/No] <Y>: *(Press Enter .)*

Enter block name specification <*>: *(Press Enter .)*

Enter attribute tag specification <*>: *(Press Enter .)*

Enter attribute value specification <*>: *(Press Enter .)*

Select Attributes: *(Select one or more attributes.)*

Select Attributes: *(Press **ENTER**.)*

Enter string to change: *(Enter existing string, and then press Enter .)*

Enter new string: *(Enter new string, and then press Enter.)*

COMMAND LINE OPTIONS

Value changes or replaces the value of the attribute.

Position moves the text insertion point of the attribute.

Height changes the attribute text height.

Angle changes the attribute text angle.

Style changes the text style of the attribute text.

Layer moves the attribute to a different layer.

Color changes the color of the attribute text.

Next edits the next attribute.

RELATED SYSTEM VARIABLE

AttDia toggles use of **AttEdit** during the **Insert** command.

RELATED COMMANDS

AttDef defines an attribute's original value and parameter.

AttDisp toggles an attribute's visibility.

AttRedef redefines attributes and blocks.

EAttEdit edits all aspects of attributes.

Explode reduces an attribute to its tag.

TIPS

- Constant attributes cannot be edited with **AttEdit**.
- The **DdEdit** command also displays this dialog box.
- You can only edit attributes parallel to the current UCS.
- Unlike other text input to AutoCAD, attribute values are case-sensitive.
- To edit null attribute values, use **-AttEdit**'s global edit option, and enter \ (backslash) at the 'Enter attribute value specification' prompt.
- The wildcard characters **?** and * are interpreted literally at the 'Enter string to change' and 'Enter new string' prompts.
- To edit the different parts of an attribute, use the following commands:

Command	Edit Attribute
Attedit	Selects non-constant attribute *values* in one block.
-AttEdit	Selects attribute *values* and *properties* (such as position, height, and style) in one block or in all attributes.

- When selecting attributes for global editing, you may pick the attributes, or use the following selection modes: **Window**, **Last**, **Crossing**, **BOX**, **Fence**, **WPolygon**, and **CPolygon**.
- You may use wildcards in the block name, tag, and value specifications:

  # matches any single numeric character.
  @ matches any single alphabetic character.
  . matches any single non-alphabetic character.
  * matches any string.
  ? matches any single character.
  ~ matches anything but the following pattern.
  [ ] matches any single character enclosed.
  [~] matches any single character not enclosed.
  [-] matches any single character in the enclosed range.
  ' treats the next character as a non-wild-card character.

- **AttEdit** does not trim leading and trailing spaces from attribute values. Be sure to avoid entering them to reduce unexpected results.

AttExt

V. 2.0 Extracts attribute data from drawings to files on disk (short for ATTribute EXTract).

Commands	Alias	Ctrl+	F-key	Alt+	Menu Bar	Tablet
attext	ddattext	...	...	...	...	...
-attext						

Command: attext

Displays dialog box:

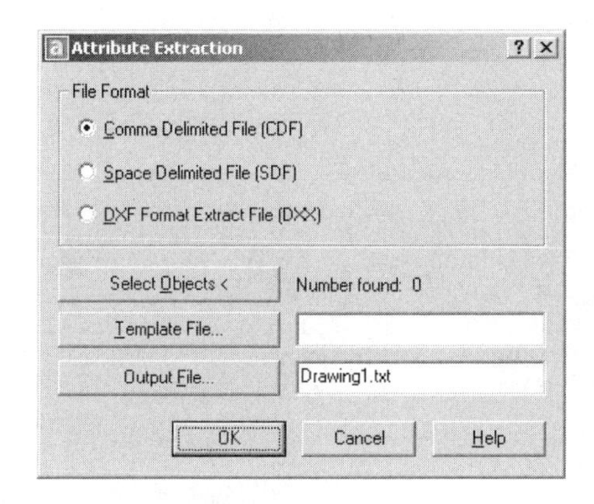

ATTRIBUTE EXTRACTION DIALOG BOX

File Format

⊙**Comma Delimited File (CDF)** creates a CDF text file, where commas separate fields.

○**Space Delimited File (SDF)** creates an SDF text file, where spaces separate fields.

○**DXF Format Extract File (DXX)** creates an ASCII DXF-format file.

Select Objects returns to the graphics screen to select attributes for export.

Template File specifies the name of the TXT template file for CDF and SDF files.

Output File specifies the name of the attribute output file, *.txt* for CDF and SDF formats, or *.dxx* for DXF format.

-ATTEXT Command

Command: -attext

Enter extraction type or enable object selection [Cdf/Sdf/Dxf/Objects] <C>: *(Enter an option.)*

*Displays the **Select Template File** dialog box; select the template file.*

*Displays the **Create Extract File** dialog box.*

COMMAND LINE OPTIONS

Cdf outputs attributes in comma-delimited format.

Sdf outputs attributes in space-delimited format.

Dxf outputs attributes in DXF format.

Objects selects objects from which to extract attributes.

RELATED COMMANDS

AttDef defines attributes.

EAttExt provides a smoother interface for extracting attributes.

RELATED FILES

*** *.txt*** required extension for template file; extension for CDF and SDF files.

*** *.dxx*** extension for DXF extraction files.

TIPS

- It is easier to use the EAttExt command than this command for attribute extraction.
- **CDF** is short for "Comma Delimited File"; it has one record for each block reference; a comma separates each field; single quotation marks delimit text strings.
- **SDF** is short for "Space Delimited File"; it has one record for each block reference; fields have fixed width padded with spaces; string delimiters are not used.
- **DXF** is short for "Drawing Interchange File"; it contains only block reference, attribute, and end-of-sequence DXF objects; no template file is required.
- CDF files use the following conventions:

  Specified field widths are the maximum width.
  Positive number fields have a leading blank.
  Character fields are enclosed in ' ' (single quotation marks).
  Trailing blanks are deleted.
  Null strings are '' (two single quotation marks).
  Use spaces; do not use tabs.
  Use the C:DELIM and C:QUOTE records to change the field and string delimiters to another character.

- To output the attributes to the printer, specify:

Logical Filename	Meaning
CON	Displays on text screen.
PRN *or* **LPT1**	Prints to parallel port 1.
LPT2 *or* **LPT3**	Prints to parallel ports 2 or 3.

- Before you can specify the SDF or CDF option, you must create a template file.

AttRedef

Rel.13 Redefines blocks and attributes (short for ATTribute REDEFinition).

Command	Alias	Ctrl+	F-key	Alt+	Menu Bar	Tablet
attredef	at	...	...	...	...	...

Command: attredef
Name of Block you wish to redefine: *(Enter name of block.)*
Select objects for new Block...
Select objects: *(Select one or more objects.)*
Select objects: *(Press* ENTER.*)*
Insertion base point of new block: *(Pick a point.)*

COMMAND LINE OPTIONS

Name of Block you wish to redefine specifies the name of the block to be redefined.

Select objects selects objects for the new block.

Insertion base point of new block picks the new insertion point.

RELATED COMMANDS

AttDef defines an attribute's original value and parameter.

AttDisp toggles an attribute's visibility.

EAttEdit edits the attribute's values.

Explode reduces an attribute to its tag.

TIPS

- Existing attributes retain their values.
- Existing attributes not included in the new block are erased.
- New attributes added to an existing block take on default values.

AttSync

2002 Updates blocks with new attribute definitions (short for ATTribute SYNChronization).

Command	Alias	Ctrl+	F-key	Alt+	Menu Bar	Tablet
attsync	...	...	...	...	...	...

Command: attsync
Enter an option [?/Name/Select] <Select>: *(Specify an option.)*
Select a block: *(Pick a block.)*
ATTSYNC block name? [Yes/No] <Yes>: *(Enter* **Y** *or* **N**.*)*

COMMAND LINE OPTIONS

? lists the names of all blocks in the drawing.

Name enters the name of the block.

Select selects a single block with the cursor.

RELATED COMMANDS

AttDef defines an attribute.

BattMan edits the attributes in a block definition.

EAttEdit edits the attributes in block references.

TIPS

- This command is used together with other attributed-related commands, in the following order:
    1. **AttDef** and **Block** define attributes, and attach them to blocks.
    2. **Insert** inserts the block, and gives values to the attributes.
    3. **EAttEdit** changes the attributes in selected blocks, adding, deleting, or modifying the attribute definitions.
    4. **BAttMan** changes the attributes in the orignal blocks.
    5. **AttSync** updates the attributes to the new definition. (**BAttMan** also performs this task.)
- This command does not operate if the drawing lacks blocks with attributes. AutoCAD complains, "This drawing contains no attributed blocks."

Audit

Rel.11 Examines drawing files for structural errors.

Command	Alias	Ctrl+	F-key	Alt+	Menu Bar	Tablet
audit	...	...	...	**FUA**	**File** **↳Drawing Utilities** **↳Audit**	...

Command: audit

Fix any errors detected? [Yes/No] <N>: *(Type* **Y** *or* **N**.*)*

Sample output

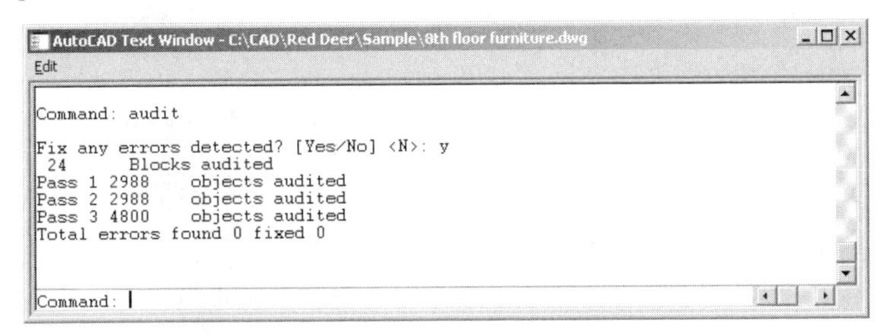

COMMAND LINE OPTIONS

N reports errors found in drawing files, but does not fix errors.

Y reports and fixes errors.

RELATED COMMANDS

DrawingRecover displays a palette listing recovered drawings.

Save saves recovered drawings to disk.

Recover recovers damaged drawing files.

RELATED SYSTEM VARIABLE

AuditCtl creates *.adt* audit log files when set to 1.

RELATED FILE

***.adt** is the audit log file, which records the auditing process.

TIPS

- The **Audit** command is a diagnostic tool for validating and repairing the contents of *.dwg* files.
- Objects with errors are placed in the **Previous** selection set. Use an editing command, such as **Copy**, to view the objects.
- If **Audit** cannot fix a drawing file, try the **Recover** command.

Removed Commands

Background is replaced by the Background option of the **View** command, as of AutoCAD 2007.

'Base

V. 1.0 Changes the insertion point of the drawing, located by default at (0,0,0).

Command	Alias	Ctrl+	F-key	Alt+	Menu Bar	Tablet
base	...	...	...	DKB	Draw ↳Block ↳Base	...

Command: base
Enter base point <0.0,0.0,0.0>: *(Pick a point.)*

COMMAND LINE OPTION

Enter base point specifies the x, y, z coordinates of the new insertion point.

RELATED COMMANDS

Block specifies the insertion point of new blocks.

Insert inserts another drawing into the current drawing.

Xref references other drawings.

RELATED SYSTEM VARIABLE

InsBase contains the current setting of the drawing's insertion point.

TIPS

- Use this command to shift the insertion point of the current drawing.
- This command does not affect the current drawing. Instead, the relocated base point comes into effect when you insert or xref the drawing into another drawing.

BAttMan

2002 Edits all aspects of attributes in blocks; works with one block at a time (short for Block ATTribute MANager).

Command	Alias	Ctrl+	F-key	Alt+	Menu Bar	Tablet
battman	...	...	...	**MOAB**	**Modify ↳Object ↳Attribute ↳Block Attribute Manager**	...

Command: battman

When the drawing contains no blocks with attributes, displays error message, "This drawing contains no attributed blocks."

When drawing contains at least one block with attributes, displays dialog box:

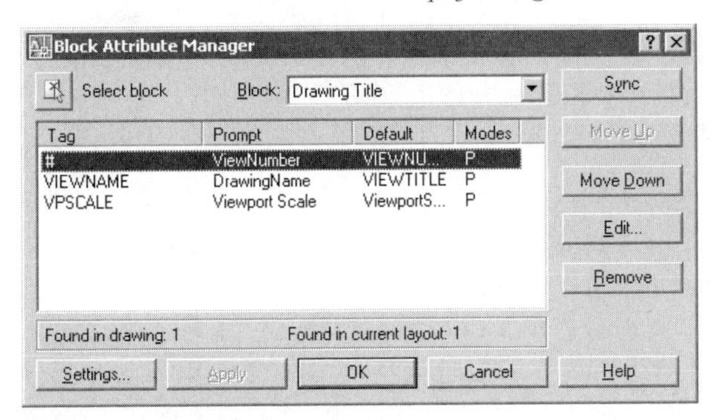

BLOCK ATTRIBUTE MANAGER DIALOG

Select block hides the dialog box, and then prompts, 'Select a block:'.

Block lists the names of blocks in the drawing, and displays the name of the selected block.

Sync changes the attributes in block insertions to match the changes made here.

Move Up moves the attribute tag up the list; constant attributes cannot be moved.

Move Down moves the attribute tag down the list.

Edit displays the Edit Attribute dialog box; see the **EAttEdit** command.

Remove removes the attribute tag and related data from the block; it does not operate when the block contains a single attribute.

Settings displays the Settings dialog box.

Apply applies the changes to the block definition.

SETTINGS DIALOG BOX

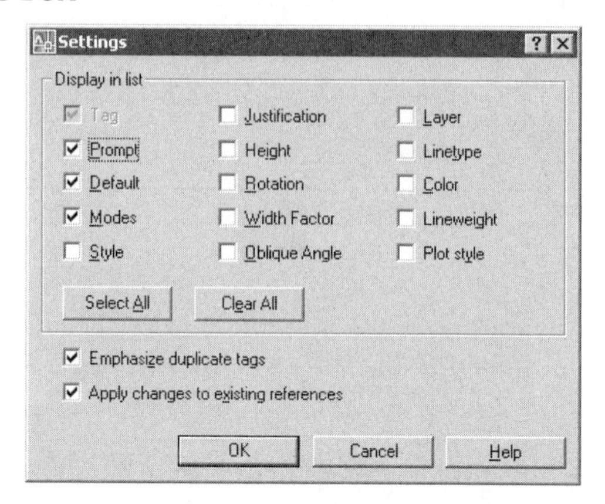

Display In List

☑ **Tag** toggles (turns on and off) the display of the column of tags.

☑ **Prompt** toggles display of the column of attribute prompts.

☑ **Default** toggles the display of the attribute's default value.

☑ **Modes** toggles the display of the attribute's modes: invisible, constant, verify, and/or preset.

☐ **Style** toggles the display of the attribute's text style name.

☐ **Justification** toggles the display of the attribute's text justification.

☐ **Height** toggles the display of the attribute's text height.

☐ **Rotation** toggles the display of the attribute's text rotation angle.

☐ **Width Factor** toggles the display of the attribute's text width factor.

☐ **Oblique Angle** toggles the display of the attribute's text obliquing angle (slant).

☐ **Layer** toggles the display of the attribute's layer.

☐ **Linetype** toggles the display of the attribute's linetype.

☐ **Color** toggles the display of the attribute's color.

☐ **Lineweight** toggles the display of the attribute's lineweight.

☐ **Plot style** toggles the display of the attribute's plot style name (available only when plot styles are turned on).

Select All selects all display options.

Clear All clears all display options, except tag name.

Emphasize duplicate tags:

☑ Highlights duplicate attribute tags in red.

☐ Does not highlight duplicate tags.

Apply changes to existing references:

☑ applies changes to all block instances that reference this definition in the drawing.

☐ applies the new attribute definitions only to newly-inserted blocks.

RELATED COMMANDS

AttDef defines attributes.

Block binds attributes to a symbol.

Insert inserts a block, and then allows you to specify the attribute data.

TIPS

- Use this command to edit and remove attribute definitions, as well as to change the order in which attributes appear.
- The **Sync** option does not change the values you assigned to attributes.
- When an attribute has a mode of Constant, it cannot be moved up or down the list.
- Turning on all the options displays a lot of data. To see all the data columns, you can stretch the dialog box.

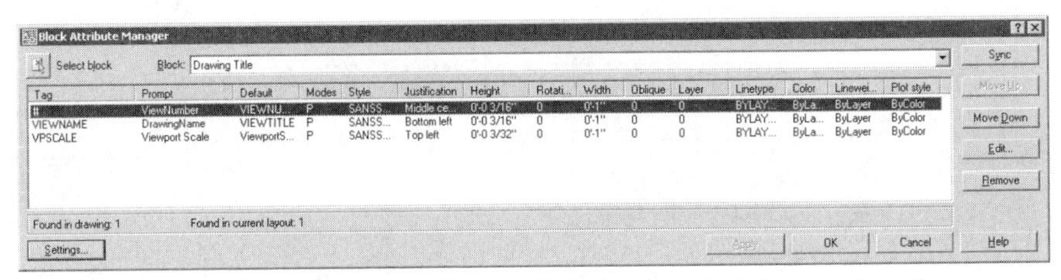

With the cursor, grab the edge of the dialog box to make it larger and smaller.

- An attribute definition cannot be changed to Constant via the Edit Attribute dialog box.
- The **Remove** option does not work when the block contains a single attribute.

2006

BEdit

Edits blocks and external references; assigns actions to blocks to create dynamic blocks (short for Block EDITor).

Command	Alias	Ctrl+	F-key	Alt+	Menu Bar	Tablet
bedit	**be**	...	...	**TB**	**Tools** ↳**Block Editor**	...
-bedit						

Command: bedit

Displays dialog box:

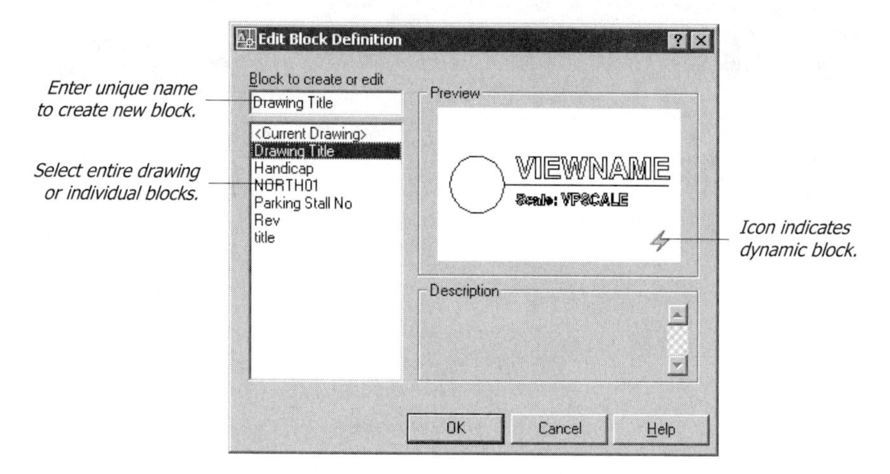

EDIT BLOCK DEFINITION DIALOG BOX

Block to create or edit

- To create a new block, enter a name.
- To edit an existing block, select its name from the list.
- To edit the drawing as a block, select "<Current Drawing>."

OK causes AutoCAD to display the Block Editor.

Cancel returns to the drawing editor.

-BEDIT Command

Command: -bedit

Enter block name or [?]: *(Enter the name of a block or type* **?**.*)*

Enter block name specifies the block to edit.

? lists the names of blocks in the drawing, if any.

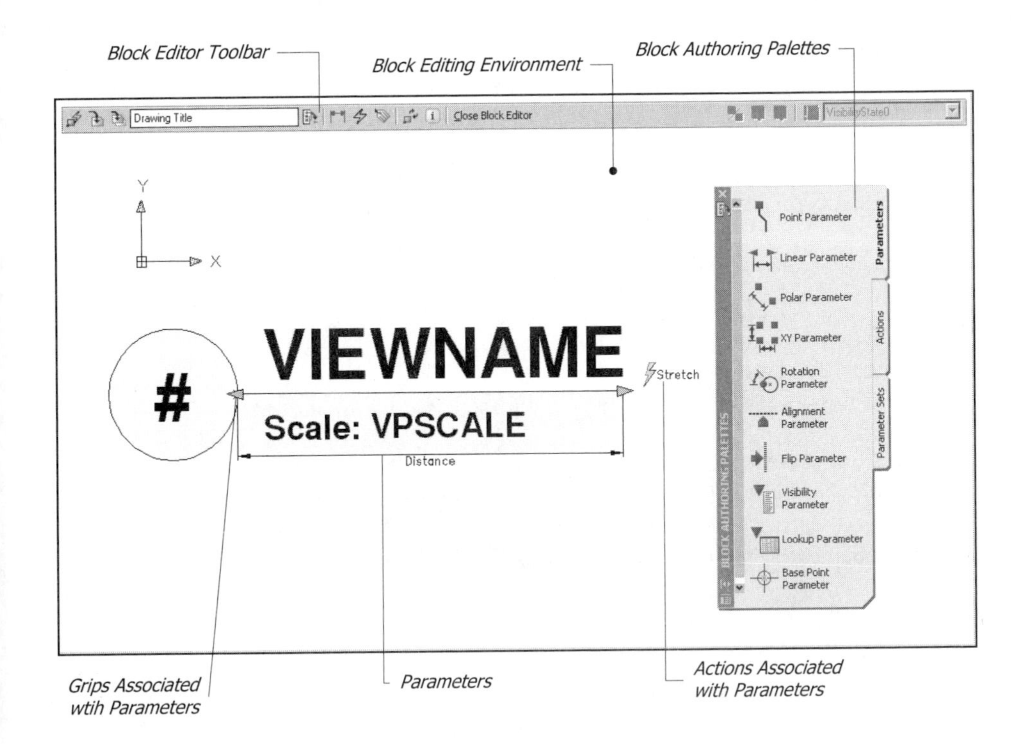
Block Editor Toolbar
Block Editing Environment
Block Authoring Palettes
Drawing Title
Close Block Editor
VisibilityState0
Point Parameter
Linear Parameter
Polar Parameter
XY Parameter
Rotation Parameter
Alignment Parameter
Flip Parameter
Visibility Parameter
Lookup Parameter
Base Point Parameter
Parameters
Actions
Parameter Sets
BLOCK AUTHORING PALETTES
VIEWNAME
#
Scale: VPSCALE
Stretch
Distance
Grips Associated wtih Parameters
Parameters
Actions Associated with Parameters

Block Editor Toolbar

Left end of toolbar:

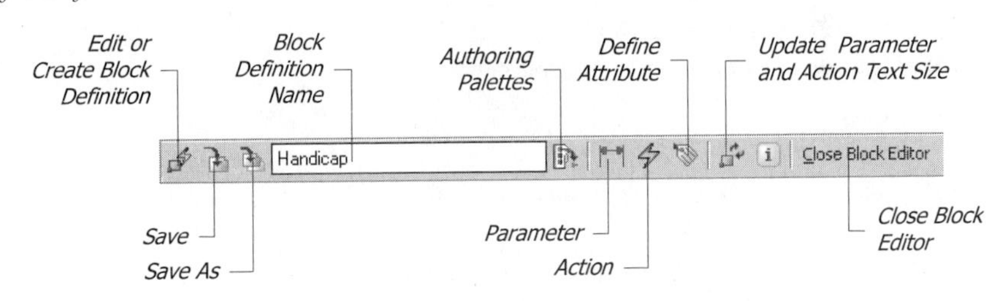
Edit or Create Block Definition
Block Definition Name
Authoring Palettes
Define Attribute
Update Parameter and Action Text Size
Handicap
Close Block Editor
Save
Save As
Parameter
Action
Close Block Editor

Right end of toolbar:

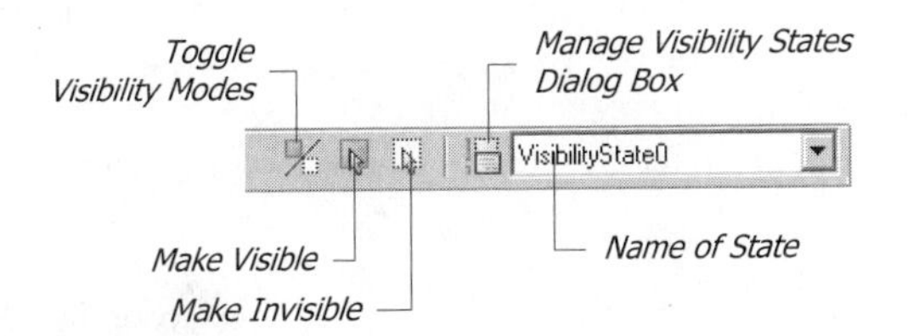
Toggle Visibility Modes
Manage Visibility States Dialog Box
VisibilityState0
Make Visible
Make Invisible
Name of State

Block Editor: **BAction**

Adds actions to dynamic block definitions (alias: **ac**).

Command: baction
Select parameter: *(Select one parameter with which to associate an action.)*
Select action type [Array/Move/Scale/Stretch/Polar Stretch]: *(Enter an option.)*

Notes: the actions available depend on the parameter selected.

Block Editor: **BActionSet**

Respecifies selection set of objects associated with actions.

Select action: *(Select an action from the dynamic block definition.)*
Specify selection set for action object [New/Modify] <New>: *(Enter an option.)*

Modify modifies the existing selection set by adding and subtracting objects.

New creates a new selection set.

Block Authoring Palette: **BActionTool**

Defines changes to geometry of dynamic block references by changing custom properties.

Select action type [Array/Lookup/Flip/Move/Rotate/Scale/sTretch/Polar stretch]: *(Enter an option.)*
Select parameter: *(Select a parameter to associate with the action.)*
Specify selection set for action
Select objects: *(Select one or more objects.)*

COMMAND OPTIONS

Array adds array actions to dynamic block definitions (selected objects array).

Flip adds flip actions to dynamic block definitions (objects flip about a mirror line).

Lookup adds lookup actions to the dynamic block definitions (lookup actions display the Property Lookup Table dialog box).

Move adds move actions to dynamic block definitions (objects move in linear or polar directions).

Rotate adds rotate actions to dynamic block definitions (prompts "Specify action location or [Base type]").

Scale adds resize actions to dynamic blocks (objects resize in linear, polar, or x,y directions).

 **Stretch** adds stretch actions to dynamic blocks (objects stretch in point, linear, polar, or x,y directions).

Polar Stretch adds stretch actions to dynamic blocks (stretches by distance and angle).

Several actions use these options:

Base Type specifies the type of base point to use for the action:

- **Dependent** rotates dynamic block about the associated parameter's base point.
- **Independent** rotates dynamic block about a specified base point.

Multiplier changes the associated parameter value by the specified factor; prompts "Enter distance multiplier <1.0>."

Offset increases and decreases angles by the specified number of degrees; prompts "Enter angle offset <0>."

XY specifies distances in the x-, y-, and xy-directions from the parameter base point; prompts "Enter XY distance type [X/Y/XY]."

Block Editor: **BAssociate**

Associates orphaned actions with parameters.

Select action object: *(Select one action not associated with parameters.)*
Select parameter to associate with action: *(Select the parameter to associate.)*

Block Editor: **BAttOrder**

Controls the order in which attributes are listed.

Displays dialog box:

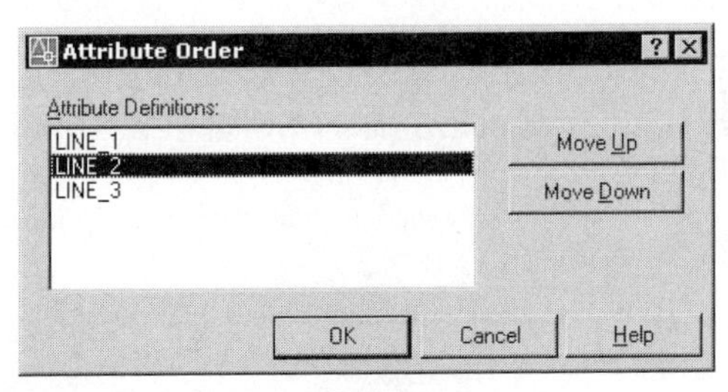

Move Up moves the selected attribute up the list.

Move Down moves the selected attribute down the list.

OK exits the dialog box, and returns to the Block Editor.

Cancel cancels changes, and returns to the Block Editor.

Block Editor: **BAuthorPalette** and **BAuthorPaletteClose**

Toggles the display of the Block Authoring Palettes window.

Block Editor: **BClose**

Closes the Block Editor environment, and returns to the drawing editor (alias: **bc**)

Block Editor: **BCycleOrder**

Changes the cycle order of grips in dynamic block references.

Displays dialog box:

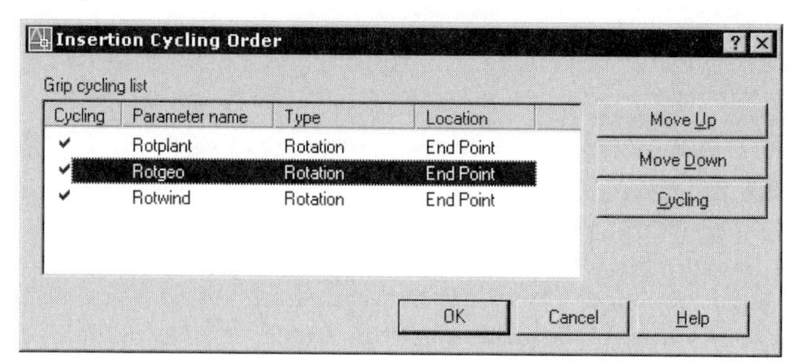

Move Up moves the selected parameter up the list.

Move Down moves the selected parameter down the list.

Cycling toggles cycling of the selected parameter.

Block Editor: **BGripSet**

Creates, resets, and deletes grips associated with parameters.

Select parameter: *(Select one parameter in the dynamic block definition.)*
Enter number of grip objects for parameter or reset position [0/1/2/4/ Reposition]: *(Enter number of grips to display, or type* **R**.*)*

0 - 4 specifies the number of grips to display; number varies, depending on the parameter. Enter **0** to hide the display of grips.

Reposition repositions grips to their default positions (a.k.a. reset).

Grip	Name	Grip Manipulation	Associated Parameters
	Alignment	Aligns with objects within 2D planes.	Alignment
	Flip	Flips (mirrors).	Flip
	Linear	Moves along defined directions or axes.	Linear
	Lookup	Displays lists of items.	Visibility, Lookup
	Rotation	Rotates about axes.	Rotation
	Standard	Moves in any direction within 2D planes.	Base, Point, Polar, XY

Creates lookup tables for dynamic block definitions.

If dynamic block definition contains a lookup action and at least one lookup parameter, displays dialog box:

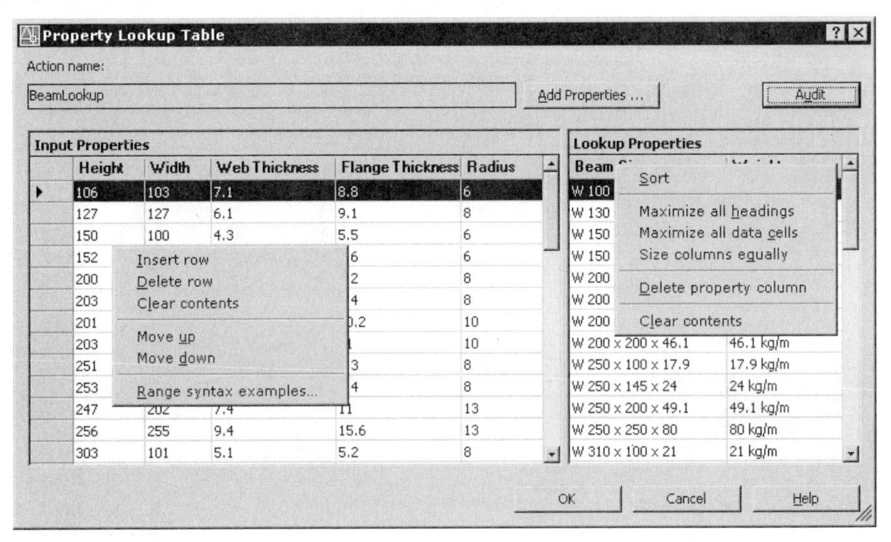

Add Properties displays the Add Parameter Properties dialog box for adding parameter properties to this table.

Audit checks data to ensure that rows are unique.

Right-click data or rows:

Insert Row adds a row above the selected row.

Delete Row erases the row and its data.

Clear Contents erases the data in the selected cell or row.

Move Up moves the row up.

Move Down moves the row down.

Range Syntax Examples displays help on writing ranges.

Right-click headings (columns):

Sort sorts data in the selected column.

Maximize All Headings makes each column wide enough to show the heading's wording.

Maximize All Data Cells makes each column wide enough to show the data's wording.

Size Columns Equally makes all columns the same width, fitted to the dialog box width.

Delete Property Column erases the selected column.

Clear Contents erases the data in the selected cell or row.

Block Editor: **BParameter**

Adds parameters with grips to dynamic block definitions (alias: **param**).

Enter parameter type [Alignment/Base/pOint/Linear/Polar/XY/Rotation/Flip/Visibility/looKup] <last>: *(Enter an option.)*

Select a parameter, and then associate an action with the parameter:

Parameter	Options	Action(s)	Comments
Alignment	Perpendicular Base point	...	Automatically rotates and aligns with other objects.
Base	Location	...	Defines base points for dynamic blocks.
Flip	Reflection Line	Flip	Flips objects about their reflection lines.
Linear	Start Point End Point Chain Midpoint List Increment	Array Move Scale Stretch	Constrains grip movement to preset angles.
Lookup	Action Name Properties Audit Input Lookup	Lookup	Defines custom properties that evaluate values from lists and tables you define.
Point	Location Chain	Move Stretch	Defines x,y locations in drawings.
Polar	Base Point Endpoint Chain List Increment	Array Move Polar Stretch Scale Stretch	Constrains grip movement to distances and angles.
Rotation	Base Point Radius Base Angle Default Angle Chain List Increment	Rotate	Defines angles.
Visibility	Display	...	Toggles the visibility states of blocks.
XY	Base Point Endpoint Chain List Increment	Array Move Scale Stretch	Shows x,y distances from the base points of blocks.

Block Editor: **BSave** and **BSaveAs**

Saves block definitions.

BSave saves the changes to the block definition. Changes are also saved when exiting the Block Editor (alias: **bs**)

BSaveAs saves the block definition by another name; displays dialog box.

Blocks are saved in the drawing, not to disk.

Block Editor: **BvHide** and **BvShow**

Changes the visibility state of objects in dynamic block definitions.

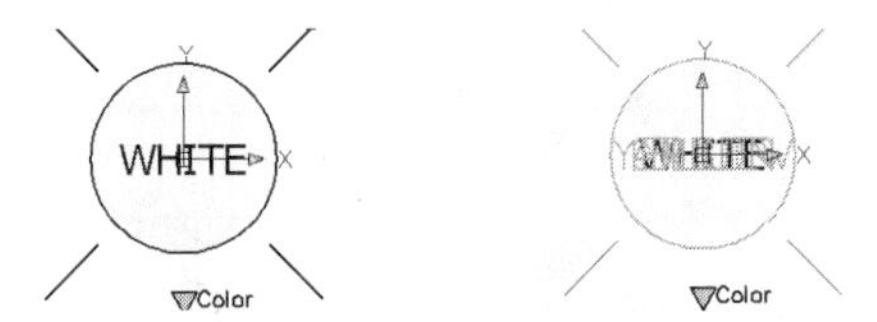

***Left:** Visibility state set to **Current** (White).*

***Right:** Visibility state set to **All** (White, Green, Red, Blue, Yellow, and Amber).*

Select objects to hide/show: *(Select objects to hide or show.)*

Hide/Show for current state or all visibility states [Current/All]: *(Enter an option.)*

Current sets the visibility of selected objects to the current visibility state.

All sets the visibility of selected objects to all visibility states, as defined by the block.

Block Editor: **BvState**

Creates, sets, renames, and deletes visibility states in dynamic blocks (alias: **vs**)

Displays dialog box:

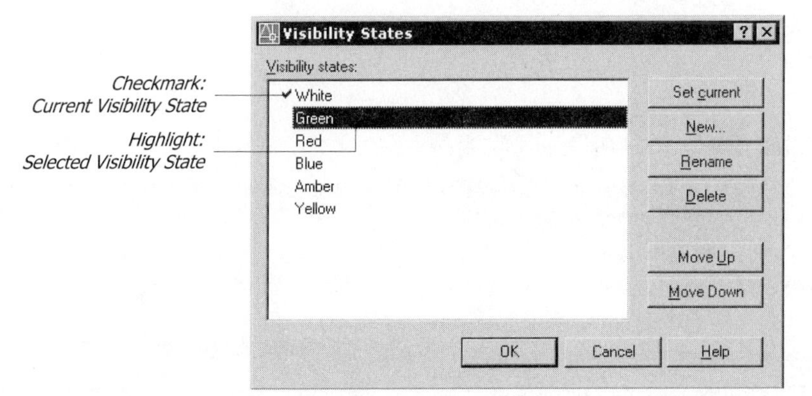

Set Current sets the selected visibility state as current in the Block Editor; does not affect the default visibility state of blocks in drawings.

Rename changes the name of the selected visibility state.

Delete erases the selected visibility state.

Move Up moves the selected visibility state up the list.

Move Down moves the selected visibility state down the list.

New displays the New Visibility State dialog box.

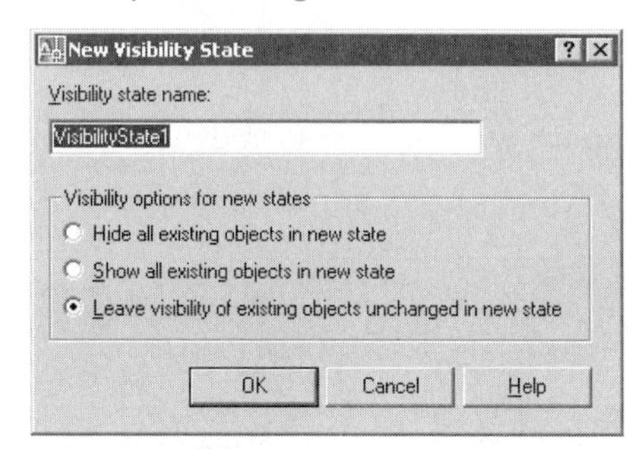

Visibility State Name specifies the name of the new visibility state.

○ **Hide All Existing Objects in New State** hides all objects in this visibility state.

○ **Show All Existing Objects in New State** shows all objects in this visibility state.

⊙ **Leave Visibility of Existing Objects Unchanged in New State** retains the visibility state of objects as the current visibility state.

RELATED COMMANDS

AttDef defines attributes for blocks.

Block makes blocks.

Insert inserts blocks into drawings.

RefEdit edits blocks and xrefs in-place.

WBlock writes blocks to individual files on disk.

RELATED SYSTEM VARIABLES

BlockEditLock toggles editing of dynamic blocks in the Block Editor.

BlockEditor reports whether the Block Editor is active.

The following system variables affect the Block Editor only:

BActionColor specifies the text color of actions.

BDependencyHighlight determines whether dependent objects are highlighted when parameters, actions, and grips are selected.

BGripObjColor specifies the color of grips.

BGripObjSize specifies the size of custom grips relative to the window size.

BParameterColor specifies the color of parameters.

BParameterFont specifies the font for parameters and actions.

BParameterSize specifies the size of parameter text and features relative to the window size.

BVMode toggles the display of hidden objects between invisible or dimmed.

TIPS

- Selecting a block before entering the **BEdit** or **-BEdit** command causes AutoCAD immediately to enter the Block Editor environment. Alternatively, you can double-click a dynamic block to open it in the Block Editor.
- Entering the name of a nonexistent block creates a new block for the Block Editor.
- Most of AutoCAD's drawing and editing commands are available in the Block Editor.
- Values in lookup tables are limited to 256 characters per cell, including distances and angles for points, linear, polar, XY, and rotation parameters; text string parameters; properties for flip; visibility parameter values; and architectural and mechanical units, such as 23'1/2". Invalid values are automatically reset to the last valid value.
- Commas are delimiters between values. Brackets [ ] specify inclusive ranges; parentheses ( ) specify exclusive ranges, as follows:

[8,31]	Any values between 8 and 31 inclusive.
(9,15)	Any values between 9 and 15, excluding 9 and 15.
[,25]	Any values up to and including 25.
(56,)	Any values larger than 56.

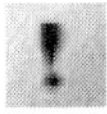 When this icon appears, it means that no action is associated with the parameter.

DEFINITIONS

Actions — determine how the geometry of dynamic blocks change when their custom properties are modified; actions are associated with parameters.

Parameters — define custom properties for dynamic block references; actions must be associated with parameters.

Orphaned Actions — arise when parameters are removed from block definitions, orphaning their associated actions.

Renamed Command

The **BHatch** and **-BHatch** commands are now found under the **Hatch** and **-Hatch** commands.

'Blipmode

V. 2.1 Turns the display of pick-point markers, known as "blips," on and off.

Command	Alias	Ctrl+	F-key	Alt+	Menu Bar	Tablet
blipmode	...	...	...	...	...	...

Command: blipmode
Enter mode [ON/OFF] <OFF>: on

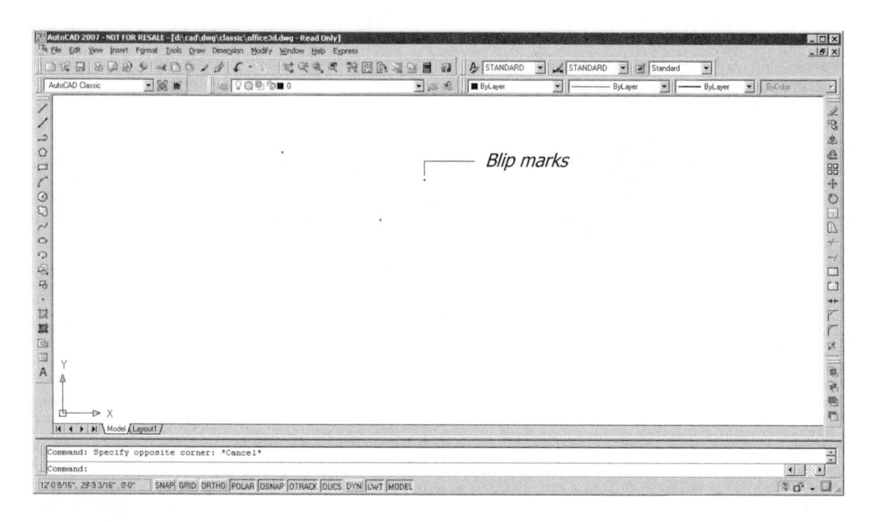

COMMAND LINE OPTIONS

ON turns on the display of pick-point markers.

OFF turns off the display of pick-point markers.

RELATED COMMANDS

Options allows blipmode toggling via a dialog box.

Redraw cleans blips off the screen.

RELATED SYSTEM VARIABLE

Blipmode contains the current blipmode setting.

TIPS

- You cannot change the size of the blip mark.
- Blip marks are erased by any command that redraws the view, such as Redraw, Regen, Zoom, and Vports.

Block

V. 1.0 Defines a group of objects as a single named object; creates symbols.

Commands	Aliases	Ctrl+	F-key	Alt+	Menu Bar	Tablet
block	**b**	...	...	**DKM**	**Draw**	**N9**
	bmake				↳**Block**	
	bmod				↳**Make**	
	acadblockdialog					
-block	**-b**					

Command: block

Displays dialog box:

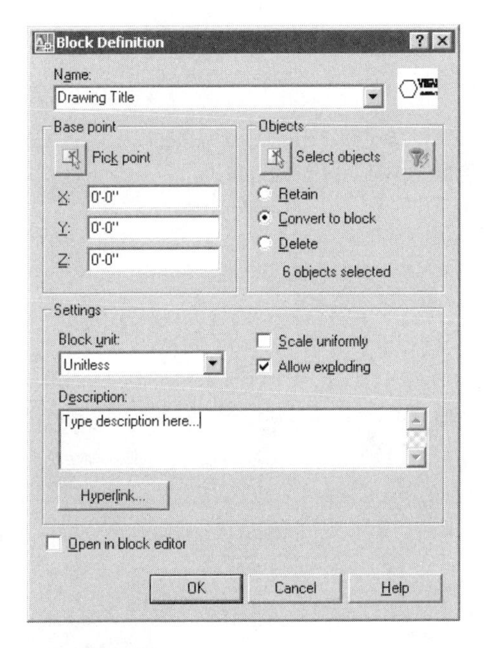

BLOCK DEFINITION DIALOG BOX

Name names the block (*maximum = 255 characters, including spaces, numbers, and other special characters, except \ | / * ? : < or >).*

Base Point

Pick point dismisses the dialog box; AutoCAD prompts, 'Specify insertion base point:'. Pick a point that specifies the block's insertion point, usually the lower-left corner.

X, **Y**, and **Z** specify the x, y, z coordinates of the insertion point.

Objects

Select objects dismisses the dialog box; AutoCAD prompts, 'Select objects:'. Select one or more objects that make up the block:

○ **Retain** leaves objects in place after the block is created.
⊙ **Convert to Block** erases objects making up the block, and replaces them with the block.
○ **Delete** erases the objects making up the block; the block is stored in drawing.

Quick Select displays the Quick Select dialog box; see the QSelect command.

Preview Icon

⊙ **Do not include an icon** does not create an icon.

○ **Create icon from block geometry** creates previews of block for Design Center.

Insert units selects the units for the block when dragged from the Design Center.

Description describes the block.

Hyperlink displays the Insert Hyperlink dialog box; see the Hyperlink command.

Open in block editor opens the block in the Block Editor.

-BLOCK Command

Command: -block
Enter block name or [?]: *(Enter a name, or type* **?***.)*
Specify insertion base point: *(Pick a point.)*
Select objects: *(Select one or more objects.)*
Select objects: *(Press Enter.)*

COMMAND LINE OPTIONS

Block name allows you to name the block.

? lists the names of blocks stored in the drawing.

Insertion base point specifies the x, y coordinates of the block's insertion point.

Select objects selects the objects and attributes that make up the block.

RELATED COMMANDS

BEdit opens blocks in the Block Editor environment; makes dynamic blocks.

BlockIcon generates icons of blocks defined in earlier releases of AutoCAD.

Explode reduces blocks to their original objects.

Insert adds blocks or other drawings to the current drawing.

Oops returns objects to the screen after creating the block.

Purge removes unused blocks from drawings.

RefEdit edits blocks and xrefs in-place.

WBlock writes blocks as drawings to *.dwg* files on disk.

XRef displays another drawing in the current drawing.

RELATED FILES

All *.dwg* drawing files can be inserted as blocks.

RELATED SYSTEM VARIABLES

InsName default block name.

InsUnits drawing units for blocks dragged from the DesignCenter:

InsUnits	Meaning
0	Unitless.
1	Inches.
2	Feet.
3	Miles.
4	Millimeters.
5	Centimeters.
6	Meters.
7	Kilometers.
8	Microinches.
9	Mils.
10	Yards.

InsUnits	Meaning
11	Angstroms.
12	Nanometers.
13	Microns.
14	Decimeters.
15	Decameters.
16	Hectometers.
17	Gigameters.
18	Astronomical Units.
19	Light Years.
20	Parsecs.

TIPS

- Blocks consist of these parts:

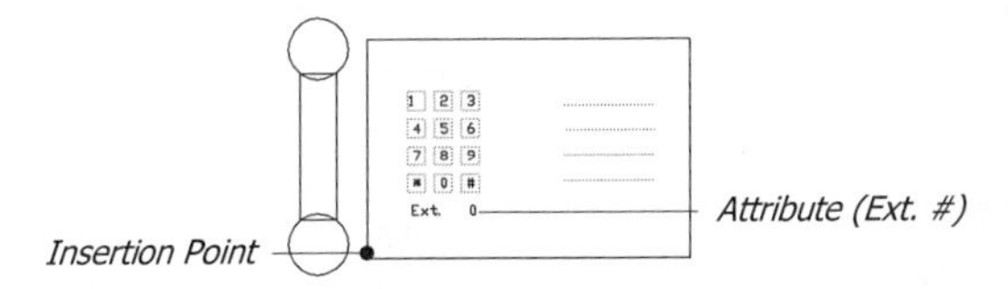

- The names of blocks have up to 255 alphanumeric characters, including $, -, and _.
- Use the **INSertion** object snap to select the insertion point of blocks.
- Objects within a block definition take on the properties specified by their layers, with one exception: if the objects were created on layer 0, then upon insertion they take on the properties specified by the host layer.
- AutoCAD has five types of blocks:

Block	Meaning
User block	Named blocks created by users.
Nested block	Blocks inside other blocks.
Unnamed block	Blocks created by AutoCAD.
Xref	Externally-referenced drawings.
Dependent block	Blocks in externally-referenced drawings.

- AutoCAD sometimes creates unnamed blocks, also called "anonymous blocks":

Name	Meaning
***A*n**	Groups.
***D*n**	Associative dimensions.
***U*n**	Created by AutoLISP or ObjectARx apps.
***X*n**	Hatch patterns.

- AutoCAD automatically purges unreferenced anonymous blocks when drawings are first loaded.

BlockIcon

2000 Creates preview images for all blocks in drawings created with AutoCAD Release 14 or earlier.

Command	Alias	Ctrl+	F-key	Alt+	Menu Bar	Tablet
blockicon	...	...	...	FUU	File ↳Drawing Utilities ↳Update Block Icons	...

Command: blockicon

Enter block names <*>: *(Enter a name, a wildcard pattern, or press Enter for all names.)*

n **blocks updated.**

COMMAND LINE OPTION

Enter block names specifies the blocks for which to create icons; press ENTER to add icons to all blocks in the drawing.

RELATED COMMANDS

AdCenter displays the icons created by this command:

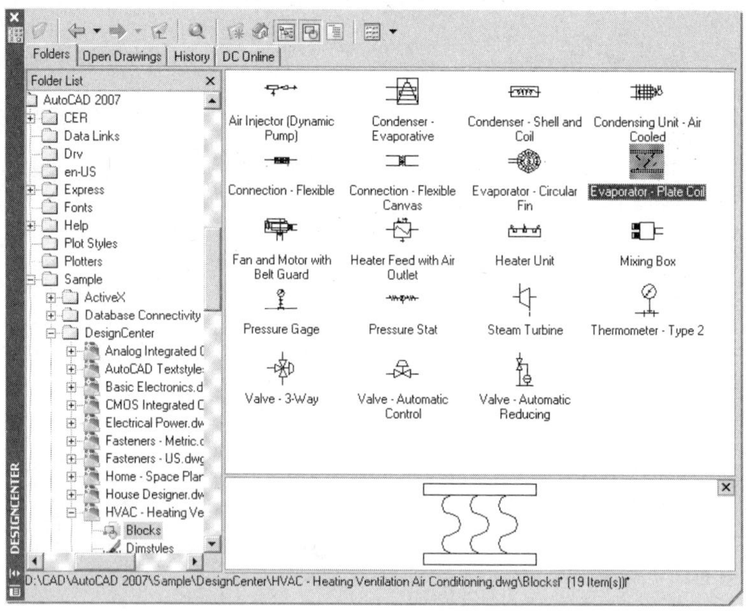

Block creates new blocks and their icons.

Insert inserts blocks; dialog box displays icons generated by this command.

TIP

- AutoCAD creates icons of blocks automatically; this command is necessary only for drawings older than AutoCAD 2000 containing blocks.

BmpOut

Rel.13 Exports the current viewport as a raster image in BMP bitmap format.

Command	Alias	Ctrl+	F-key	Alt+	Menu Bar	Tablet
bmpout	...	...	...	...	...	...

Command: bmpout

Displays Create BMP File dialog box. Enter a file name, and then click ***Save***.

Select objects or <all objects and viewports>: *(Press Enter to select all objects, or select individual objects.)*

DIALOG BOX OPTION

Save saves drawings as BMP format raster files.

RELATED COMMANDS

JpgOut exports objects and viewports in JPEG format.

PngOut exports objects and viewports in PNG format.

TifOut exports objects and viewports in TIFF format.

WmfOut exports selected objects in WMF format.

RELATED WINDOWS COMMANDS

PRT SCR saves screen to the Clipboard.

ALT+PRT SCR saves the topmost window to the Clipboard.

TIPS

- The *.bmp* extension is short for "bitmap," a raster file standard for Windows.
- This command creates uncompressed *.bmp* files.

Boundary

Rel.12 Creates boundaries as polylines or 2D regions.

Command	Aliases	Ctrl+	F-key	Alt+	Menu Bar	Tablet
boundary	**bo**	...	...	**DB**	**Draw**	**Q9**
	bpoly				↳**Boundary**	
-boundary	**-bo**					

Command: boundary

Displays dialog box, redesigned in AutoCAD 2006:

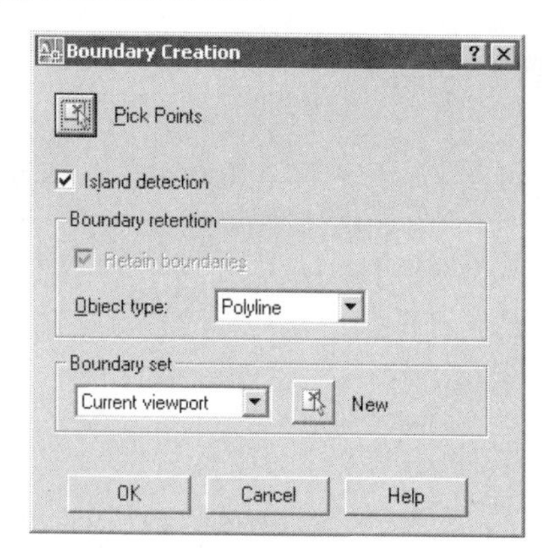

BOUNDARY CREATION DIALOG BOX

Pick Points picks points inside of closed areas.

Island detection:

☑ Island detection is turned on.

☐ Island detection is turned off.

Boundary Retention

Retain Boundaries is grayed out (unavailable), because the point of this command is to create boundaries, not discard them!

Object Type constructs the boundary from:

- **Polylines** forms the boundary from a polyline.
- **Region objects** forms the boundary from a 2D region.

Boundary Set

Boundary Set defines how objects are analyzed for defining boundaries (default = current viewport).

New creates new boundary sets; AutoCAD dismisses the dialog box, and then prompts you to select objects.

-BOUNDARY Command

Command: -boundary

Specify internal point or [Advanced options]: *(Pick a point, or type* **A**.*)*

COMMAND LINE OPTIONS

Specify internal point creates a boundary based on the point you pick.

Advanced options options

Enter an option [Boundary set/Island detection/Object type]: *(Enter an option.)*

Boundary set defines the objects **-Boundary** analyzes when defining a boundary from a specified pick point. It chooses from a new set of objects, or all objects visible in the current viewport.

Island detection:

- **On** uses objects within the outermost boundary as boundary objects.
- **Off** fills objects within the outermost boundary.

Object type specifies polyline or region as the boundary object.

RELATED COMMANDS

Hatch creates boundaries for hatches and fills.

PLine draws polylines.

PEdit edits polylines.

Region creates 2D regions from a collection of objects.

RELATED SYSTEM VARIABLE

HpBound specifies the default object used to create boundary:

0 — Draw as region.

1 — Draw as polyline (default).

TIPS

- Use this command to measure irregular areas:
    1. Apply the **Boundary** command to an irregular area.
    2. Use the **Area** command to find the area and perimeter of the boundary.
- Use **Boundary** together with the **Offset** command to help create *poching*, areas partially covered by hatching.

Box

Rel.13 Draws a 3D box as a solid model.

Command	Alias	Ctrl+	F-key	Alt+	Menu Bar	Tablet
box	...	...	...	DMB	Draw ↳Modeling ↳Box	J7

Command: box
Specify first corner or [Center]: *(Pick point 1, or type the* **C** *option.)*
Specify other corner or [Cube/Length]: *(Pick point 2, or enter an option.)*
Specify height or [2Point] <-1.2757>: *(Pick point 3, or type the* **2** *option.)*

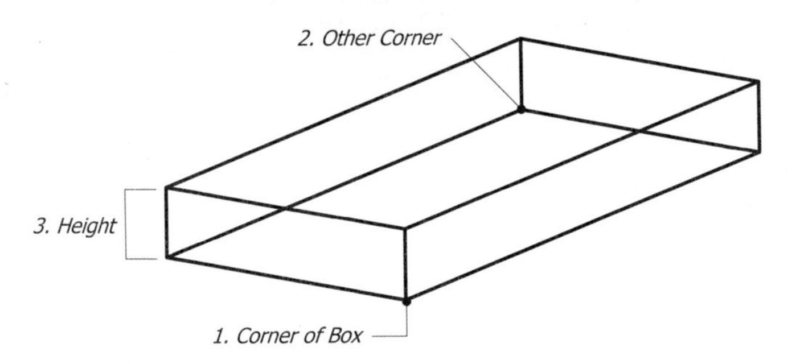

COMMAND LINE OPTIONS

First corner specifies one corner for the base of box.

Center draws the box about a center point.

Other corner specifies the second corner for the base of box.

Cube draws a cube box — all sides have the same length.

Length specifies the length along x axis, width along y axis, and height along z axis.

Height specifies the height of the box.

2point specifies height as the distance between two points *(new to AutoCAD 2007)*.

RELATED COMMANDS

Ai_Box draws 3D surface model boxes.

Cone draws 3D solid cones.

Cylinder draws 3D solid tubes.

Sphere draws 3D solid balls.

Torus draws 3D solid donuts.

Wedge draws 3D solid wedges.

RELATED SYSTEM VARIABLES

DispSilh displays 3D objects as silhouettes after hidden-line removal and shading.

DragVs specifies the visual style during construction of 3D solid primitives *(new to AutoCAD 2007)*.

TIP

- Once the box is placed in the drawing, you can edit it interactively. Select the box, and then notice the grips. Select a grip, which turns red, and then move it. Press ESC to exit direct editing.

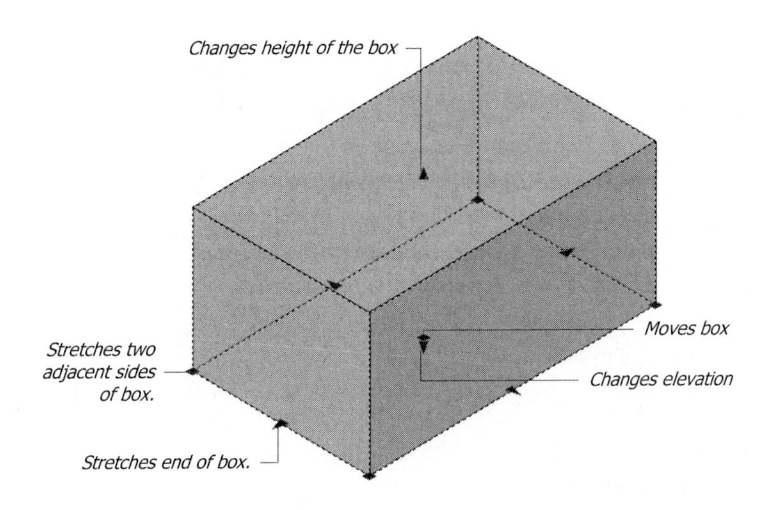

Break

V. 1.4 Removes portions of objects.

Command	Alias	Ctrl+	F-key	Alt+	Menu Bar	Tablet
break	**br**	...	...	**MK**	**Modify** ⇨**Break**	**W17**

Command: break
Select object: *(Select one object — point 1.)*
Specify second break point or [First point]: *(Pick point 2, or type* **F***.)*

1. First Point *2. Second Point*

Breaking a polyline at two points.

COMMAND LINE OPTIONS

Select object selects one object to break; the pick point becomes the first break point, unless the **F** option is used at the next prompt.

@ uses the first break point's coordinates for the second break point.

First Point options
Enter first point: *(Pick a point.)*
Enter second point: *(Pick a point.)*

First point specifies the first break point.

Second point specifies the second break point.

RELATED COMMANDS

Change changes the length of lines.

PEdit removes and relocates vertices of polylines.

Trim shortens the length of open objects.

TIPS

- Use this command to convert circles into arcs; pick the break points clockwise to keep the portion between pick points. Use the Join command to convert arcs back to circles.
- This command can erase a portion of an object (as shown in the figure above) or remove the end of an open object.
- The second point does not need to be on the object; AutoCAD breaks the object at the point nearest to the pick point.
- The Break command works on the following objects: lines, arcs, circles, polylines, ellipses, rays, xlines, and splines, as well as objects made of polylines, such as donuts and polygons.

BRep

2007 Removes construction history from 3D solids; short for "Boundary REPresentation."

Command	Alias	Ctrl+	F-key	Alt+	Menu Bar	Tablet
brep	...	...	...	...	...	...

Command: brep
Select 3D solids: *(Select one or more solid models.)*
Select 3D solids: *(Press Enter to exit command.)*

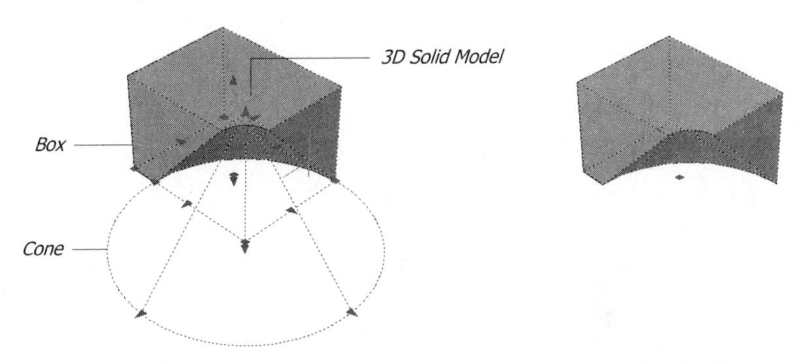

***Left**: Before applying BRep, the model's original primitives can be selected.*
***Right**: After applying BRep, the history of original primitives disappears.*

COMMAND LINE OPTION

Select 3D solids selects one or more objects from which to remove history.

RELATED SYSTEM VARIABLES

ShowHist toggles the preservation of history in solid models.

SolidHist toggles the display of primitive history entities.

Browser, Browser2

Rel.14 Prompts for a Web address, and then launches the Web browser.

Commands	Alias	Ctrl+	F-key	Alt+	Menu Bar	Tablet
browser	...	...	...	...	...	**Y8**
browser2						

Command: browser

Enter Web location (URL) <http://www.autodesk.com>: *(Enter a Web address.)*

Launches browser:

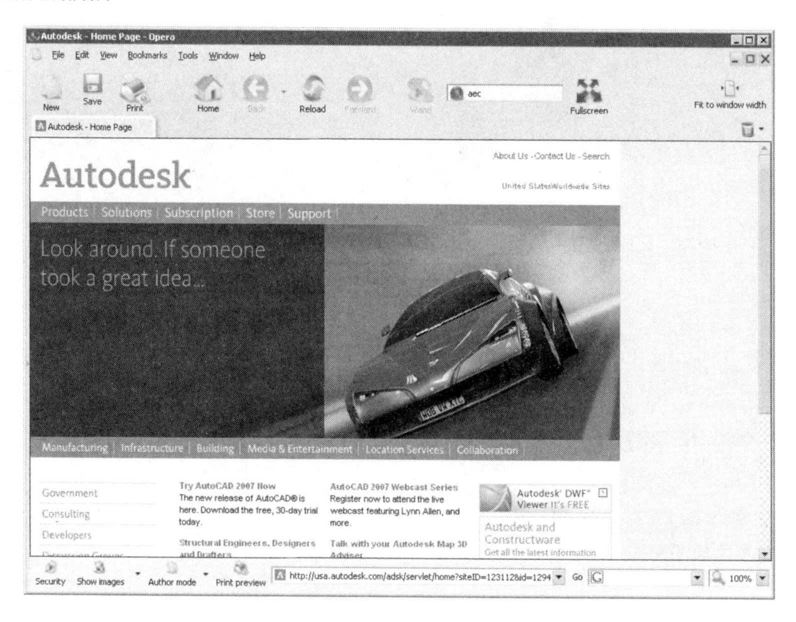

COMMAND LINE OPTION

Enter Web location specifies the URL; see the AttachURL command for information about URLs.

RELATED SYSTEM VARIABLE

InetLocation contains the name of the default URL.

TIPS

- *URL* is short for "uniform (or universal) resource locator," the universal file-naming system used on the Internet; also called a link or hyperlink. An example of a URL is http://www.autodeskpress.com, the Autodesk Press Web site.
- Many file dialog boxes also give you access to the Web browser; see the Open command.
- The undocumented Browser2 command also opens URLs.

'Cal

Rel.12 Calculates algebraic and vector geometry at the command line (*short for CALculator*).

Command	Alias	Ctrl+	F-key	Alt+	Menu Bar	Tablet
cal	...	...	...	...	...	...

Command: cal

>>Expression: *(Enter an expression, and then press* **ENTER.***)*

COMMAND LINE OPTIONS

( )	Grouping of expressions.
[ ]	Vector expressions.
+	Addition.
-	Subtraction.
*	Multiplication.
/	Division.
^	Exponentiation.
&	Vector product of vectors.
sin	Sine.
cos	Cosine.
tang	Tangent.
asin	Arc sine.
acos	Arc cosine.
atan	Arc tangent.
ln	Natural logarithm.
log	Logarithm.
exp	Natural exponent.
exp10	Exponent.
sqr	Square.
sqrt	Square root.
abs	Absolute value.
round	Round off.
trunc	Truncate.
cvunit	Converts units using *acad.unt.*
w2u	WCS to UCS conversion.
u2w	UCS to WCS conversion.
r2d	Radians-to-degrees conversion.
d2r	Degrees-to-radians conversion.
pi	The value PI (3.14159).
xyof	x,y coordinates of a point.
xzof	x,z coordinates of a point.
yzof	y,z coordinates of a point.
xof	x coordinate of a point.
yof	y coordinate of a point.
zof	z coordinate of a point.
rxof	Real x coordinate of a point.
ryof	Real y coordinate of a point.
rzof	Real z coordinate of a point.
cur	x, y, z coordinates of a picked point.
rad	Radius of object.
pld	Point on line, distance from.
plt	Point on line, using parameter *t*.
rot	Rotated point through angle about origin.
ill	Intersection of two lines.
ilp	Intersection of line and plane.
dist	Distance between two points.
dpl	Distance between point and line.
dpp	Distance between point and plane.
ang	Angle between lines.
nor	Unit vector normal.
vec	Vector translation between two points.
vec1	Unit vector direction.
dee	Distance between two endpoints.
ille	Intersection of two lines defined by endpoints.
mee	Midpoint between to endpoints.
nee	Unit vector in the x,y-plane normal to two endpoints
vee	Vector from two endpoints.
vee1	Unit vector from two endpoints.
end	Endpoint object snap.
ins	Insertion point object snap.

int	Intersection object snap.	**qua**	Quadrant object snap.
mid	Midpoint object snap.	**per**	Perpendicular object snap.
cen	Center object snap.	**tan**	Tangent object snap.
nea	Nearest object snap.	**ESC**	Exits Cal mode.
nod	Node object snap.		

RELATED COMMAND

QuickCal displays a graphical user interface for the **Cal** command.

RELATED SYSTEM VARIABLES

UserI1 — **UserI5** store integers.

UserR1 — **UserR5** store real numbers.

TIPS

- To use Cal, type expressions at the >> prompt.

  For example, to find the area of a circle (pi*r^2) with radius of 1.2 units, enter the following:

  Expression >> pi*(1.2^2)
  4.52389

- **Cal** recognizes these prefixes:

  * Scalar product of vectors.

  & Vector product of vectors.

- And the following suffixes:

  r Radian (degrees is the default).

  g Grad.

  ' Feet (unitless distance is the default).

  " Inches.

- Because 'Cal is a transparent command, it can perform calculations in the middle of other commands, and then return the value to that command.

  For example, to set the offset distance to the radius of a circle:

  Command: offset
  Specify offset distance or [Through] <Through>: 'cal
  >> Expression: rad
  >> Select circle, arc or polyline segment for RAD function: *(Pick circle.)* 2.0
  Select object to offset or <exit>: *(And so on.)*

- This command works with real numbers and integers. The smallest and largest integers are -32768 and 32767.

 Camera

2000 Creates 3D perspective views; significantly changed in AutoCAD 2007.

Command	Alias	Ctrl+	F-key	Alt+	Menu Bar	Tablet
camera	**cam**	...	...	**VT**	**View** **↳Create Camera**	...

Command: camera

Current camera settings: Height=0'-0" Lens Length=50.0 mm

Specify camera location: *(Pick a point, or enter coordinates.)*

Specify target location: *(Pick another point, or enter coordinates.)*

Enter an option [?/Name/LOcation/Height/Target/LEns/Clipping/View/ eXit]<eXit>: *(Enter an option, or press Enter to exit the command.)*

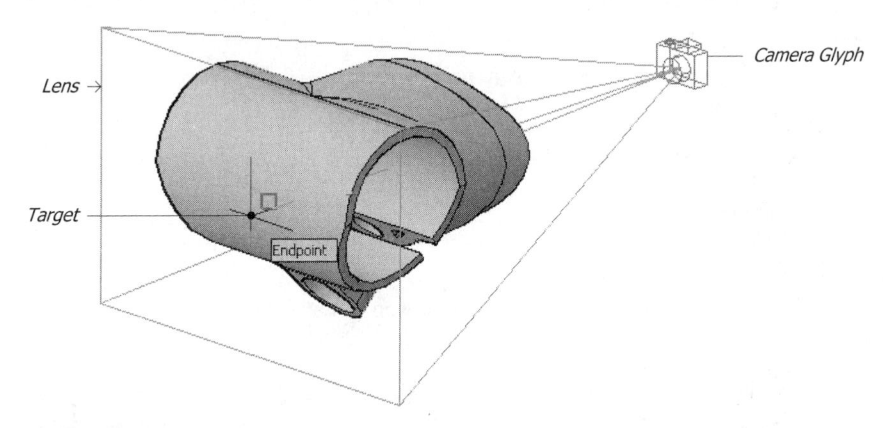

Note: The camera glyph is not displayed, by default; turn it on with the CameraDisplay system variable.

COMMAND LINE OPTIONS

Specify camera location specifies the x, y, z coordinates for the camera ("look from").

Specify target location specifies the x, y, z coordinates for the target ("look at").

? option

Enter camera name(s) to list <*> lists the names of cameras stored in the drawing, which are accessed through the **View** command.

```
Camera Name
-----------------
Camera1
Camera2
```

Name option

Enter name for new camera specifies the name for the current camera-target position, which it saves for later reuse.

LOcation option

Specify camera location locates the camera using x,y,z coordinates.

Height option

Specify camera height elevates the camera above the x,y-plane; saved in the CameraHeight system variable.

Target option

Specify target location locates the target point and determines the camera's direction.

LEns option

Specify lens length in mm specifies the angle of view; 50mm is normal; smaller numbers (such as 35) widen the view but also distort it, while larger numbers (such as 100) narrow the view and foreshorten it.

Clipping options

Enable front clipping plane? toggles the front clipping plane, which is used to cut off part of the view.

Specify front clipping plane offset from target plane specifies the location of the plane.

Enable back clipping plane? toggles the rear clipping plane.

Specify back clipping plane offset from target plane specifies the location of the plane.

View option

Switch to camera view? changes the viewpoint to that of the camera.

RELATED COMMANDS

DView has a **CAmera** option, which interactively sets a new 3D viewpoint.

View lists the names of saved camera views.

VPoint sets a 3D viewpoint through x, y, z coordinates or angles.

3dOrbit sets a new 3D viewpoint interactively.

RELATED SYSTEM VARIABLES

CameraDisplay toggles the display of the camera glyphs.

CameraHeight specifies the height of the camera above the x,y-plane.

TIPS

- Colors of the camera and field of view can be changed in the Options dialog box.
- Saved cameras are listed in the **View** command's dialog box.
- After placing a camera, select its glyph to edit the viewpoint and to display the Camera Preview window:

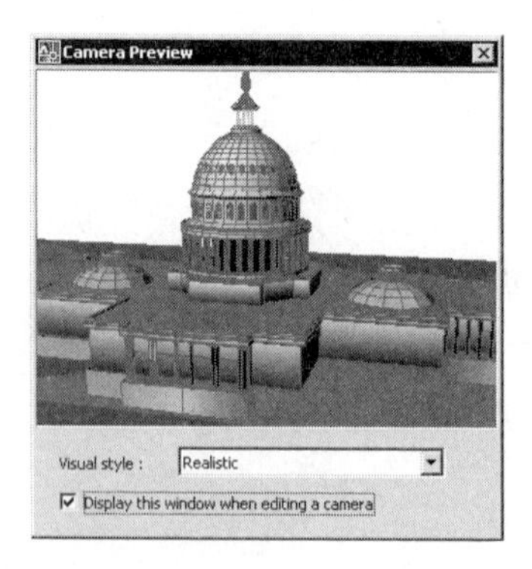

- Grips are used interactively to adjust the camera, target, and field of view. If clipping planes were specified, they can also be adjusted using grips.

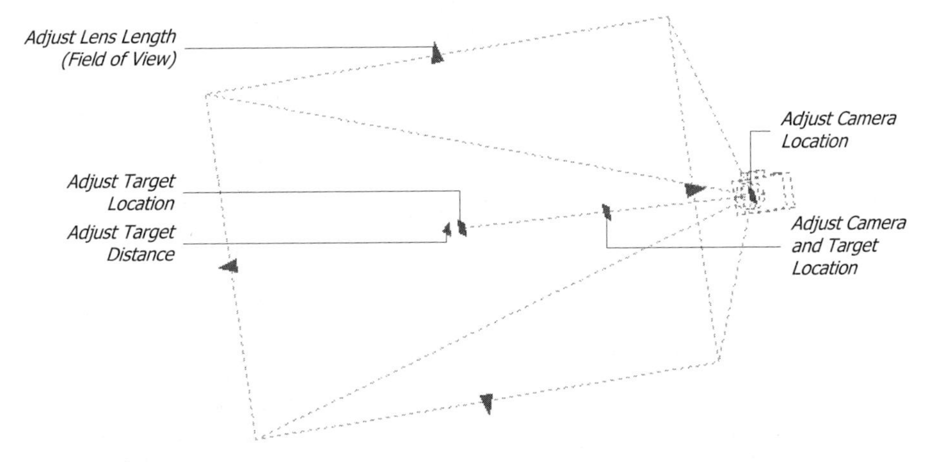

- Right-click a camera glyph, and then select **Properties** from the shortcut menu.

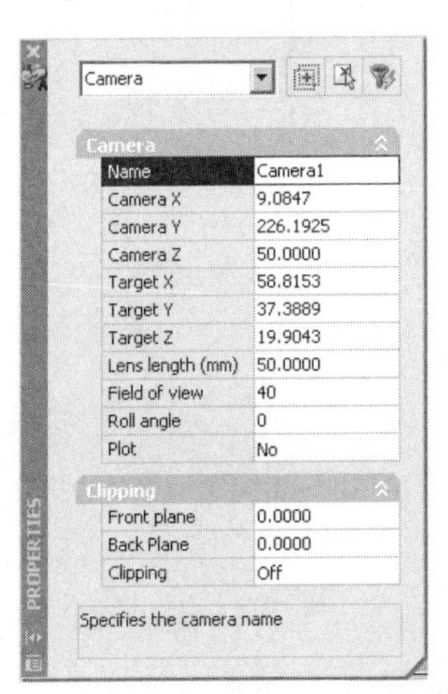

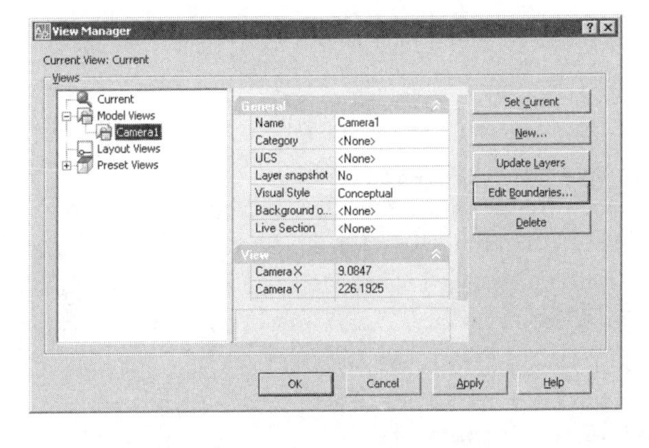

***Left**: Camera properties in the Properties palette.*
***Right**: Camera properties in the View Manager dialog box.*

- The **View** command also lists the names of cameras and their properties, but provides access to many more options.
- The 3D Navigation toolbar contains a droplist with the names of stored camera views. The Camera Adjustment toolbar has nothing to do with this command.
- Entering the **Camera** command automatically turns on the perspective viewpoint.

Chamfer

V. 2.1

Bevels the intersection of two lines, all vertices of 2D polylines, and the faces of 3D solid models.

Command	Alias	Ctrl+	F-key	Alt+	Menu Bar	Tablet
chamfer	cha	...	...	MCC	Modify ↳Chamfer	W18

Command: chamfer

(TRIM mode) Current chamfer Dist1 = 0.0, Dist2 = 0.0

Select first line or [Undo/Polyline/Distance/Angle/Trim/mEthod/Multiple]: *(Pick an object, or enter an option.)*

Select second line: *(Pick an object.)*

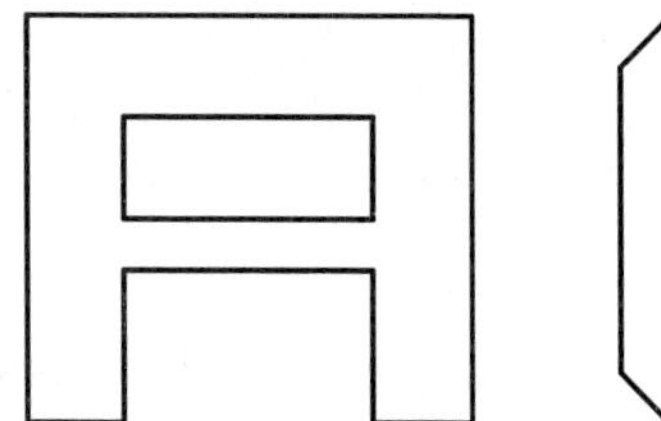

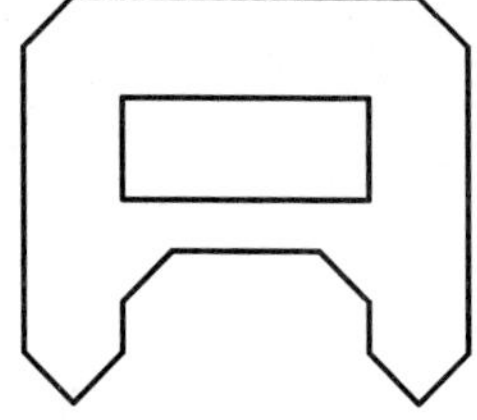

***Left**: Original object drawn with polylines.*
***Right**: Chamfered with the Polyline option.*

COMMAND LINE OPTIONS

Select first line selects the first line, arc, face, or edge.

Select second line selects the second line, arc, face, or edge.

Undo reverses the last chamfer operation within the command.

Multiple allows more than one pair of lines to be chamfered.

Polyline options

Select 2D polyline: *(Pick a polyline.)*

***n* lines were chamfered**

Select 2D polyline chamfers *all* segments of a 2D polyline; if the polyline is not closed with the **Close** option, the first and last segments are not chamfered.

Distance options

Specify first chamfer distance <0.5000>: *(Enter a value.)*

Specify second chamfer distance <0.5000>: *(Enter a value.)*

First distance specifies the chamfering distance along the line picked first.

Second distance specifies the chamfering distance along the line picked second.

Angle options

Specify chamfer length on the first line <1.0000>: *(Enter a distance.)*

Specify chamfer angle from the first line <0>: *(Enter an angle.)*

Chamfer length specifies the chamfering distance along the line picked first.

Chamfer angle specifies the chamfering angle by an angle from the line picked first.

Trim options

Enter Trim mode option [Trim/Notrim] <Trim>: *(Type* **T** *or* **N***.)*

Trim trims or extends lines, edges, and faces after chamfer.

No trim does not trim lines, edges, and faces after chamfer.

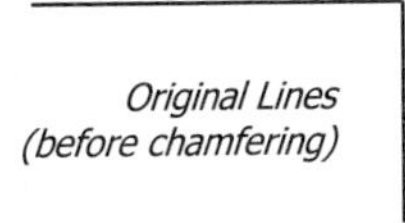

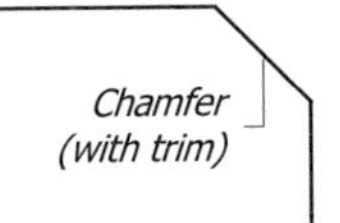

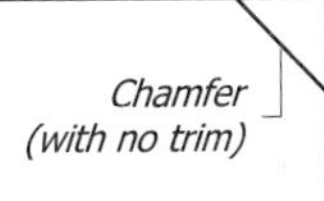

Intersecting lines before and after chamfering.

Method options

Enter trim method [Distance/Angle] <Angle>: *(Type* **D** *or* **A***.)*

Distance determines chamfer by two specified distances.

Angle determines chamfer by the specified angle and distance.

Chamfering 3D Solids — Edge Mode Options

Command: chamfer

(TRIM mode) Current chamfer Dist1 = 0.0, Dist2 = 0.0

Polyline/.../<Select first line>: *(Pick a 3D solid model.)*

Select base surface: *(Pick surface edge on the solid model — see 1, below.)*

Next/<OK>: *(Enter* **N***, or select the next surface, or* **OK** *to end selection.)*

Enter base surface distance <0.0>: *(Enter a value.)*

Enter other surface distance <0.0>: *(Enter a value.)*

Loop/<Select edge>: *(Select an edge — see 2, below.)*

Loop/<Select edge>: *(Press Enter .)*

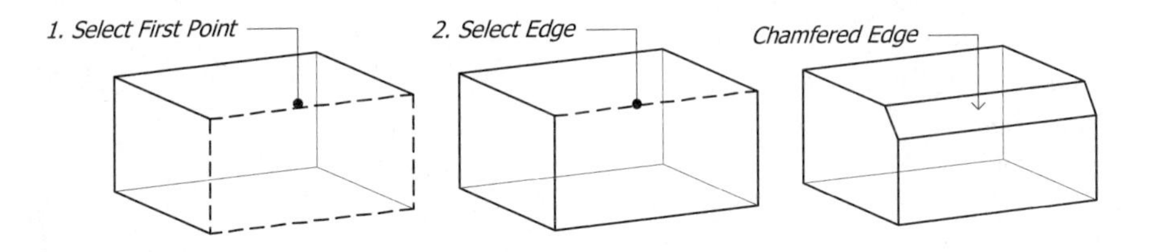

Chamfering 3D Solids — Loop Mode Options

Command: chamfer

(TRIM mode) Current chamfer Dist1 = 0.0, Dist2 = 0.0

Polyline/.../<Select first line>: *(Pick a 3D solid model.)*

Select base surface: *(Pick a surface edge — see 1, following.)*

Next/<OK>: *(Enter* **N***, or select the next surface, or* **OK** *to end selection.)*

Enter base surface distance <0.0>: *(Enter a value.)*

Enter other surface distance <0.0>: *(Enter a value.)*

Loop/<Select edge>: *(Enter* **L.***)*
Edge/<Select edge loop>: *(Pick a surface edge — see 2, below.)*
Edge/<Select edge loop>: *(Press Enter.)*

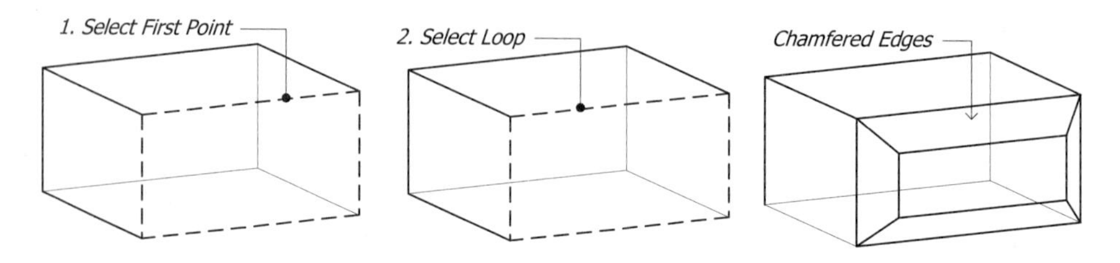

COMMAND LINE OPTIONS

Select first line selects the 3D solid.

Next selects the adjacent face, or press **ENTER** to accept face.

Enter base surface distance specifies the first chamfer distance (default = 0.5).

Enter other surface distance specifies the second chamfer distance (default = 0.5).

Loop selects all edges of the face.

Select edge selects a single edge of the face.

RELATED COMMANDS

Fillet rounds the intersection with a radius.

SolidEdit edits the faces and edges of solids.

RELATED SYSTEM VARIABLES

ChamferA is the first chamfer distance (default = 0.5).

ChamferB is the second chamfer distance (default = 0.5).

ChamferC is the length of chamfer (default = 1).

ChamferD is the chamfer angle (default = 0).

ChamMode toggles chamfer measurement:

0 — Chamfer by two distances (default).

1 — Chamfer by distance and angle.

TrimMode determines whether lines/edges are trimmed after chamfer:

0 — Do not trim selected edges.

1 — Trim selected edges (default).

TIPS

- When TrimMode is set to 1 and lines do not intersect, Chamfer extends or trims the lines to intersect before chamfering.
- When the two objects are not on the same layer, Chamfer places the chamfer line on the current layer; the chamfer line is placed on the current layer.

Change

V. 1.0 Modifies the color, elevation, layer, linetype, linetype scale, lineweight, plot style, material, and thickness of most objects, and additional properties of lines, circles, blocks, text, and attributes.

Command	Alias	Ctrl+	F-key	Alt+	Menu Bar	Tablet
change	-ch	...	...	...	...	...

Command: change
Select objects: *(Pick one or more objects.)*
Select objects: *(Press Enter.)*
Specify change point or [Properties]: *(Pick a point, an object, or type* **P**.*)*
Enter property to change [Color/Elev/LAyer/LType/ltScale/LWeight/ Thickness/Plotstyle/Material]: *(Enter an option.)*

COMMAND LINE OPTIONS

Specify change point selects the object to change:

(pick a line) indicates the new length of lines.

(pick a circle) indicates the new radius of circles.

(pick a block) indicates the new insertion point or rotation angle of blocks.

(pick text) indicates the new location of text.

(pick an attribute) indicates the attribute's new text insertion point, text style, height, rotation angle, text, tag, prompt, or default value.

ENTER changes the insertion point, style, height, rotation angle, and text of text strings.

Properties options

Color changes the color of objects.

Elev changes the elevation of objects.

LAyer moves the object to different layers.

LType changes the linetype of objects.

ltScale changes the scale of linetypes.

LWeight changes the lineweight of objects.

Thickness changes the thickness of objects, except blocks and 3D solids.

Plotstyle changes the plot style of objects (available only when plot styles are turned on).

Material changes the material assigned to objects (*new to AutoCAD 2007*).

RELATED COMMANDS

AttRedef changes blocks and attributes.

ChProp contains the properties portion of the **Change** command.

Color changes the current color setting.

Elev changes the working elevation and thickness.

LtScale changes the linetype scale.

Properties changes most aspects of all objects.

PlotStyle sets the plot style.

Materials assigns materials and textures to objects.

RELATED SYSTEM VARIABLES

CeColor contains the current color setting.

CeLType contains the current linetype setting.

CelWeight contains the current lineweight.

CircleRad contains the current circle radius.

CLayer contains the name of the current layer.

CMaterial contains the name of the default material.

CPlotstyle contains the name of the current plot style.

Elevation contains the current elevation setting.

LtScale contains the current linetype scale.

TextSize contains the current height of text.

TextStyle contains the current text style.

Thickness contains the current thickness setting.

TIPS

- The **Change** command cannot change:

  The size of donuts.

  The radius and length of arcs.

  The length of polylines.

  The justification of text.

- Use this command to change the endpoints of groups of lines to a common vertex:

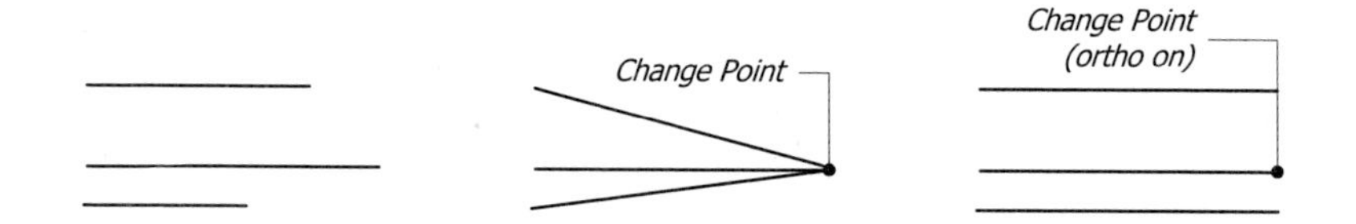

Left: *Original lines.*
Center: *Line endpoints changed with ortho mode turned off.*
Right: *Line endpoints changed wiht ortho mode turned on.*

- Turn on ortho mode to extend or trim a group of lines, without needing a cutting edge (as do the **Extend** and **Trim** commands).
- The **PlotStyle** option is not displayed when plot styles are not turned on.

'CheckStandards

2001 Checks drawings for adherence to standards previously specified by the Standards command.

Command	Alias	Ctrl+	F-key	Alt+	Menu Bar	Tablet
checkstandards	**chk**	...	...	**TSK**	**Tools** ↳**CAD Standards** ↳**Check**	...

Command: checkstandards

*When the **Standards** command has not set up standards for the drawing, AutoCAD displays this error message:*

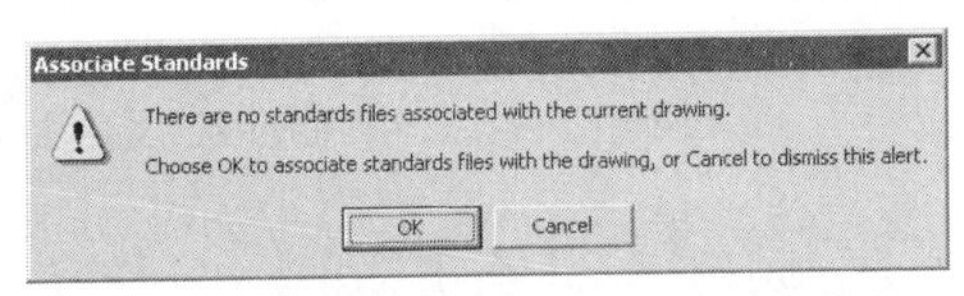

When standards have been set up for the drawing, this command checks the drawing against the CAD standards, and then displays this dialog box:

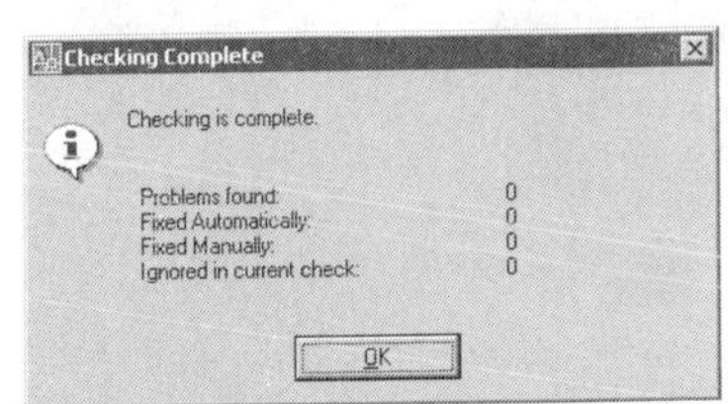

*Click **OK**.*

*To change settings in the **Configure Standards** dialog box, click **Settings**.*

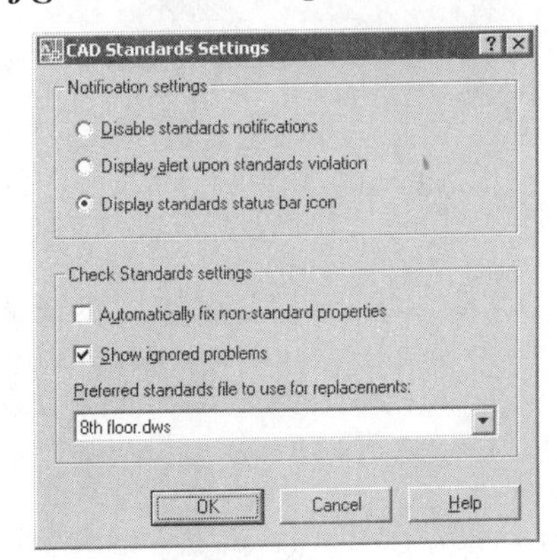

CAD STANDARDS SETTINGS DIALOG BOX OPTIONS

Notification Settings

- ○ Disables standards notifications.
- ○ Displays alert upon standards violation.
- ⦿ Displays standards status bar icon.

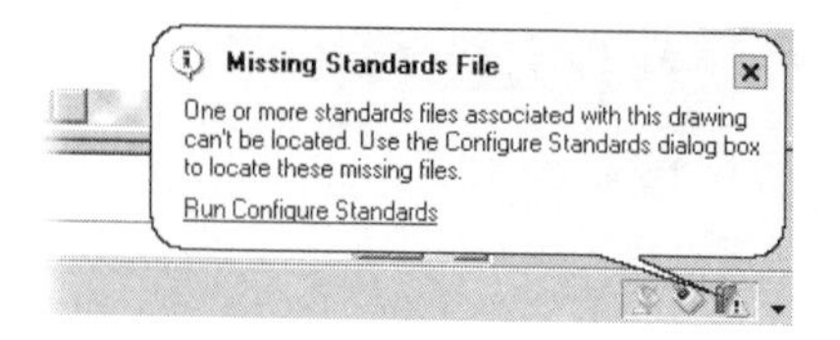

Check Standards Settings

Automatically fix non-standard properties:

☑ Fixes properties not matching the CAD standard automatically.

☐ Steps manually through properties not matching the CAD standard.

Show ignored problems:

☑ Displays problems marked as ignored.

☐ Does not display ignored problems.

Preferred standards file to use for replacements selects the default *.dws* file.

CHECK STANDARDS DIALOG BOX

When an object does not match the standards, AutoCAD displays this dialog box:

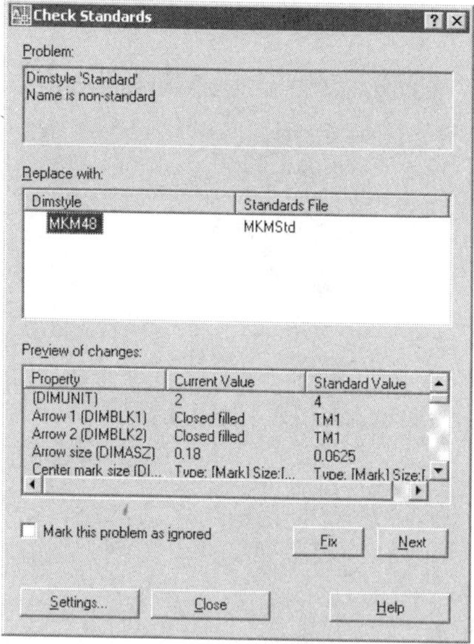

Problem describes properties in drawings that do not match the standard; this dialog box displays one problem at a time.

Replace With lists linetypes, text styles, etc. found in the *.dws* standards file.

Mark this problem as ignored:

☑ Ignores nonstandard properties, and marks them with the user's login name; some errors are always ignored by AutoCAD, such as settings for layer 0 and DefPoints.

☐ Does not ignore nonstandard properties.

Fix replaces the nonstandard property with the selected standard; the color check mark icon means fixes are available.

Next displays the next nonstandard property.

Settings displays the Check Standards Settings dialog box.

Close closes the dialog box; displays this warning dialog box if standards are not fully checked:

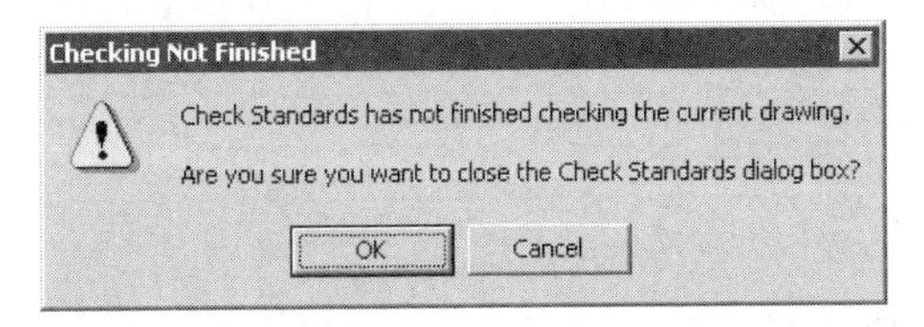

TRAY ICON

When drawings contain drawings standards, AutoCAD displays a book icon on the tray:

Click the icon to start the ***CheckStandards*** *command. Right-clicking the icon displays the shortcut menu:*

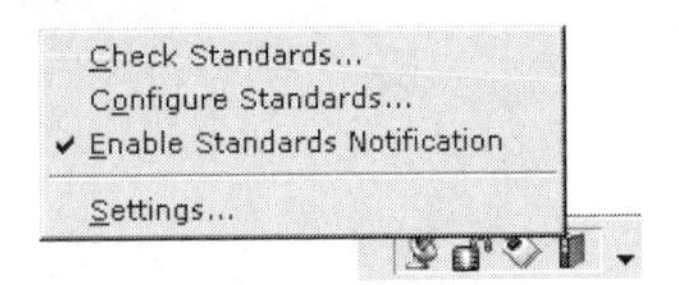

Check Standards runs the **CheckStandards** command.

Configure Standards runs the **Standards** command.

Enable Standards Notification toggles the display of the balloon.

Settings displays the CAD Standards Settings dialog box.

RELATED COMMANDS

Standards selects *.dws* standards files.

LayTrans translates layers between drawings.

RELATED SYSTEM VARIABLE

StandardsViolation determines whether alerts are displayed when a CAD standard is violated in the current drawing.

RELATED PROGRAM

DwgCheckStandards.Exe is the external Batch Standards Checker program that checks one or more drawings at a time. It checks drawings without needing AutoCAD.

RELATED FILES

***.dws** drawing standards file; stored in DWG format.

***.chs** standard check file; stored in XML format.

TIP

- While this dialog box is open, you can use the following shortcut keys:

Shortcut	Meaning
F4	Fix problem.
F5	Next problem.

ChProp

Rel.10 Modifies the color, layer, linetype, linetype scale, lineweight, plot style, and thickness of most objects (short for CHange PROPerties).

Command	Alias	Ctrl+	F-key	Alt+	Menu Bar	Tablet
chprop	...	...	...	...	...	...

Command: chprop
Select objects: *(Select one or more objects.)*
Select objects: *(Press Enter.)*
Enter property to change [Color/LAyer/LType/ltScale/LWeight/Thickness/Plotstyle/Material]: *(Enter an option.)*

COMMAND LINE OPTIONS

Color changes the color of objects.

LAyer moves objects to a different layer.

LType changes the linetype of objects.

ltScale changes the linetype scale.

LWeight changes the lineweight of objects.

Thickness changes the thickness of all objects, except blocks.

PlotStyle changes the plot style assigned to the objects.

Material changes the material assigned to objects (*new to AutoCAD 2007*).

RELATED COMMANDS

Change changes lines, circles, blocks, text and attributes.

Color changes the current color setting.

Elev changes the working elevation and thickness.

LtScale changes the linetype scale.

LWeight sets the lineweight options.

Properties changes most aspects of all objects.

PlotStyle sets the plot style.

RELATED SYSTEM VARIABLES

CeColor specifies the current color setting.

CeLType specifies the current linetype setting.

CelWeight specifies the current lineweight.

CLayer contains the name of the current layer.

CPlotstyle contains the name of the current plot style.

LtScale contains the current linetype scale.

Thickness specifies the current thickness setting.

TIPS

- Use the **Change** command to change the elevation of an object.
- The **Plotstyle** option is not displayed when plot styles are not turned on.

ChSpace

2007 Moves objects between paper and model space, scaling them appropriately; a former Express Tool (short for CHange SPACE).

Command	Alias	Ctrl+	F-key	Alt+	Menu Bar	Tablet
chprop	...	...	...	...	...	...

Command: chprop

*In model tab, AutoCAD responds, "** Command not allowed in Model Tab **."*

In layout tabs, the command continues:

Select objects: *(Select one or more objects.)*

Select objects: *(Press* **ENTER**.*)*

n **object(s) changed from PAPER space to MODEL space.**

Objects were scaled by a factor of 192.000000708072 to maintain visual appearance.

COMMAND LINE OPTION

Select objects selects the objects to be move. *Warning*! Objects are moved, not copied, which means they no longer appear in the other space.

RELATED COMMAND

SpaceTrans converts the height of text between model and paper space.

Circle

V. 1.0 Draws 2D circles by five different methods.

Command	Alias	Ctrl+	F-key	Alt+	Menu Bar	Tablet
circle	c	...	...	DC	Draw ↳Circle	J9

Command: circle

Specify center point for circle or [3P/2P/Ttr (tan tan radius)]: *(Pick a point, or enter an option.)*

Specify radius of circle or [Diameter]: *(Pick a point, or type* **D***.)*

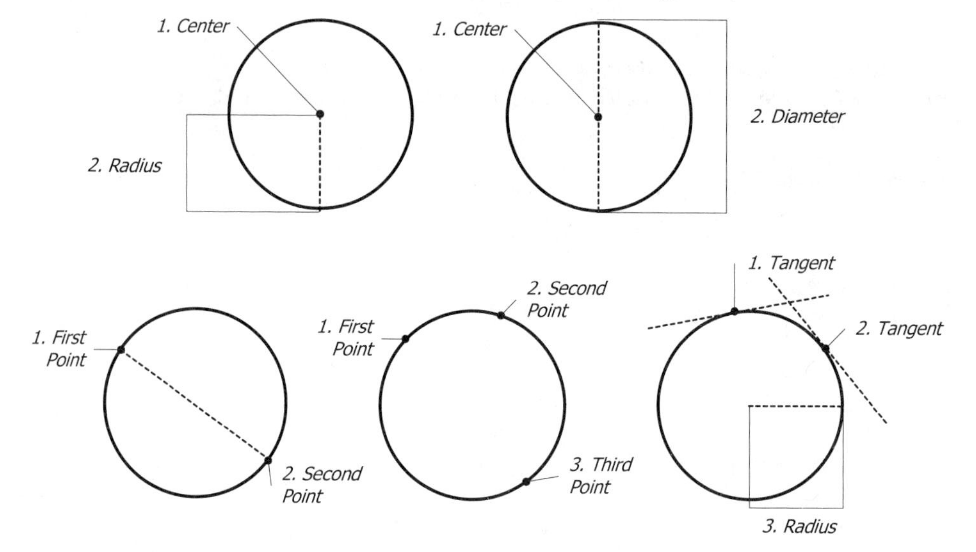

COMMAND LINE OPTIONS

Center and Radius or Diameter Options

Specify center point for circle or [3P/2P/Ttr (tan tan radius)]: *(Pick a point, or enter an option.)*

Specify radius of circle or [Diameter] <0.5>: *(Pick a point, or type* **D***.)*

Specify diameter of circle <0.5>: *(Enter a value.)*

Center point indicates the circle's center point.

Radius indicates the circle's radius.

Diameter indicates the circle's diameter.

3P (three-point) Options

Specify first point on circle: *(Pick a point.)*

Specify second point on circle: *(Pick a point.)*

Specify third point on circle: *(Pick a point.)*

First point indicates the first point on the circle.

Second point indicates the second point on the circle.

Third point indicates the third point on the circle.

2P (two-point) Options

Specify first end point of circle's diameter: *(Pick a point.)*

Specify second end point of circle's diameter: *(Pick a point.)*

First end point indicates the first point on the circle.

Second end point indicates the second point on the circle.

TTR (tangent-tangent-radius) Options

Specify point on object for first tangent of circle: *(Pick a point.)*

Specify point on object for second tangent of circle: *(Pick a point.)*

Specify radius of circle <0.5>: *(Enter a radius.)*

First tangent indicates the first point of tangency.

Second tangent indicates the second point of tangency.

Radius indicates the first point of radius.

Tan Tan Tan *(menu-only option)*:

*From the **Draw** menu, select **Circle | Tan,Tan,Tan**. AutoCAD prompts:*

Command: _circle Specify center point for circle or [3P/2P/Ttr (tan tan radius)]: _3p Specify first point on circle: _tan to *(Pick an object.)*

Specify second point on circle: _tan to *(Pick an object.)*

Specify third point on circle: _tan to *(Pick an object.)*

AutoCAD draws a circle tangent to the three points, if possible.

RELATED COMMANDS

Ai_CircTan draws circles tangent to three objects.

Arc draws arcs.

Donut draws solid-filled circles or donuts.

Ellipse draws elliptical circles and arcs.

Sphere draws 3D solid balls.

RELATED SYSTEM VARIABLE

CircleRad specifies the default circle radius.

TIPS

- The **3P** (three-point) circle defines three points on the circle's circumference.
- When drawing **TTR** (tangent, tangent, radius) circles, AutoCAD draws the circles with tangent points closest to the pick points; note that more than one circle placement is possible.
- Using the **TTR** option automatically turns on the TANgent object snap.
- Giving circles thickness turns them into hollow cylinders.

 CleanScreenOn, CleanScreenOff

2004 Maximizes the drawing area.

Command	Alias	Ctrl+	F-key	Alt+	Menu Bar	Tablet
cleanscreenon	...	**0** *(zero)*	...	**VCC**	**View** ↳**Clean Screen**	...
cleanscreenoff	...	**0**	...	**VCC**	**View** ↳**Clean Screen**	...

Command: cleanscreenon

AutoCAD turns off the title bar, toolbars, and window edges, and maximizes the AutoCAD window to the full size of your computer's screen:

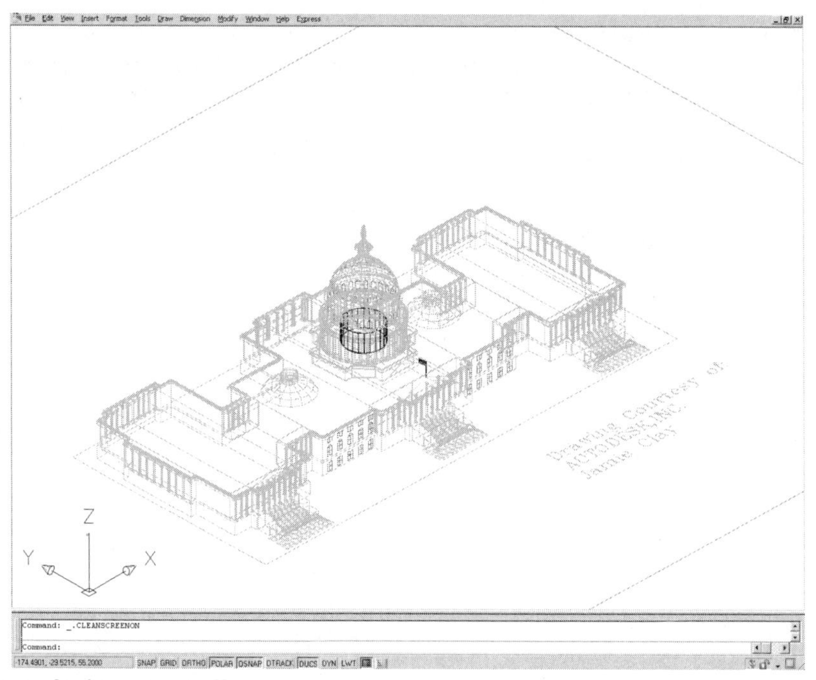

Command: cleanscreenoff

AutoCAD returns to normal.

TIPS

- Alternatively, you can click the Cleanscreen icon in the tray on the status bar (*new to AutoCAD 2007*).
- For an even larger drawing area, turn off the scroll bars and layout tabs through the **Options | Display** dialog box, and then drag the command prompt area into a window.
- To toggle this command, press CTRL+**0** (zero).

Close, CloseAll

2000 Closes the current drawing, or all drawings; does not exit AutoCAD.

Command	Alias	Ctrl+	F-key	Alt+	Menu Bar	Tablet
close	...	**F4**	...	**FC**	**File** ↳**Close**	...
closeall	...	...	...	**WL**	**Window** ↳**Close All**	...

Command: close

When drawings are not saved since last change, displays dialog box:

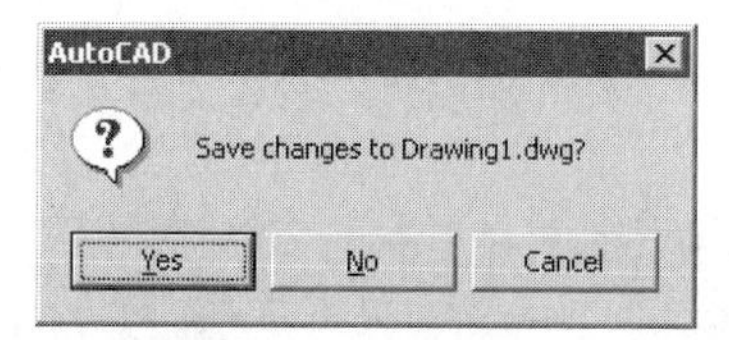

AutoCAD DIALOG BOX

Yes displays the Save Drawing As dialog box, or saves the drawing if it has been previously saved.

No exits the drawing without saving it.

Cancel returns to the drawing.

RELATED COMMANDS

Quit exits AutoCAD.

Open opens additional drawings, each in its own window.

CloseAll closes all drawings, displaying the same warning dialog box as above.

TIP

- As an alternative to this command, you can click the **x** button on the title bar:

'Color

V. 2.5 Sets the new working color.

Commands	Aliases	Ctrl+	F-key	Alt+	Menu Bar	Tablet
color	ddcolor	...	...	OC	Format	U4
	col				↳Color	
	colour					
-color						

Command: color

Displays dialog box:

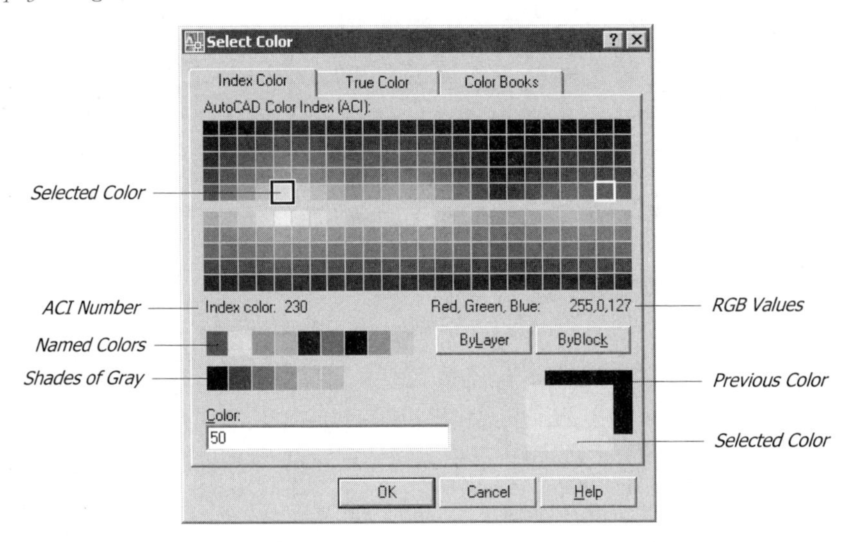

SELECT COLOR DIALOG BOX

Index Color Tab

AutoCAD Color Index (ACI) selects one of AutoCAD's colors (short for "AutoCAD Color Index").

Bylayer sets color to BYLAYER (color 256); color 257 is ByEntity in some situations.

Byblock sets color to BYBLOCK.

Color sets the color by number or name.

True Color Tab

Hue selects the hue (color), ranging from 0 (red) to 360 (violet).

Saturation selects the saturation (intensity of color), ranging from 0 (gray) to 100 (color).

Luminance selects the luminance (brightness of color), ranging from 0 (white) to 100 (black).

Color specifies the color number as hue, saturation, luminance.

Color model selects the type of color model:

- **HSL** is hue, saturation, luminance.

- **RGB** is red, green, blue.

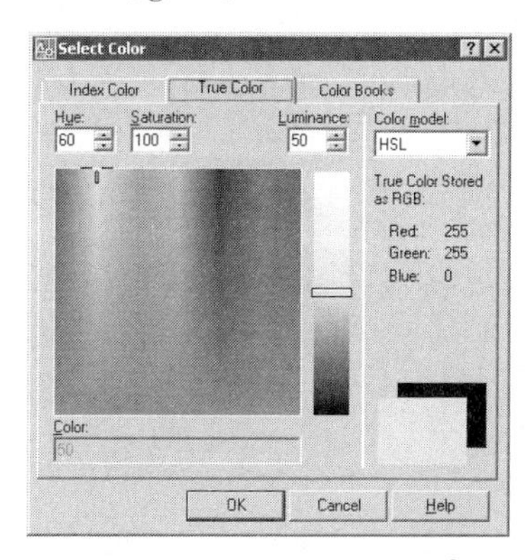

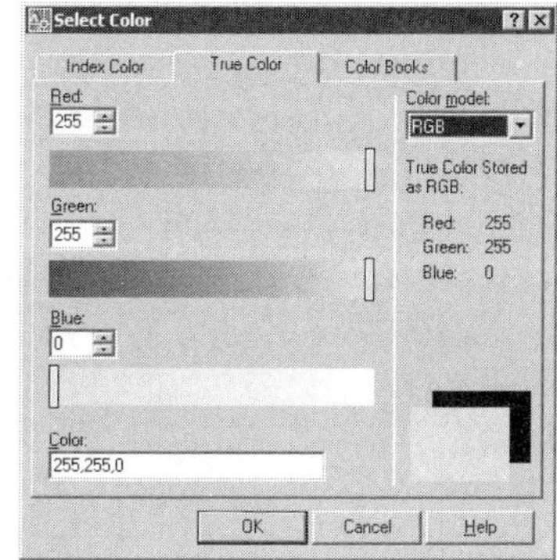

Left: *HSL color picker.*
Right: *RGB color picker.*

Red selects the range of red from 0 to 255.

Green selects the range of green from 0 to 255.

Blue selects the range of blue from 0 to 255.

Color specifies the color number as red, green, blue.

Color Book Tab

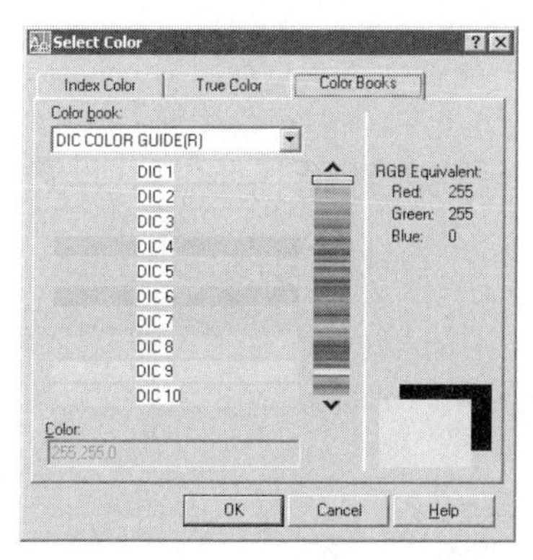

Color book selects a predefined collection of colors.

Color specifies the name of the Pantone, RAL, and DIC colors.

Historical notes: The **RAL** color system was designed in 1927 to standardize colors by limiting the number of color gradations, at first to just 30, but now to over 1,600. RAL (Reichs Ausschuß für Lieferbedingungen – German for "Imperial Committee for Supply Conditions") is administered by the German Institute for Quality Assurance and Labeling <www.ral.de>.

The **Pantone Color System** was designed in 1963 to specify color for graphic arts, textiles, and plastics, based on the assumption that colors are seen differently by different individuals. Designers typically work with Pantone's fan-format book of standardized colors <www.pantone.com>.

The **DIC Color Guide** is the Japanese standard for colors, developed by Dainippon Ink and Chemicals <www.dic.co.jp/eng/index.html>.

-COLOR Command

Command: -color

Enter default object color [Truecolor/COlorbook] <BYLAYER>: *(Enter a color number or name, or enter an option.)*

COMMAND LINE OPTIONS

BYLAYER sets the working color to the color of the current layer.

BYBLOCK sets the working color of inserted blocks.

color number sets the working color using number (1 through 255), name, or abbreviation:

Color Number	Color Name	Abbreviation	Comments
1	Red	R	
2	Yellow	Y	
3	Green	G	
4	Cyan	C	
5	Blue	B	
6	Magenta	M	
7	White	W	Or black.
8 - 249			Additional colors.
250 - 255			Shades of gray.

Truecolor Options

Red, Green, Blue: *(Specify color values separated by commas.)*

Red, Green, Blue specifies color by red, green, and blue in the range from 0 to 255:

Red, Green, Blue: 255,128,0

COlorbook Options

Enter Color Book name: *(Specify a name, such as* **Pantone***.)*
Enter color name: *(Enter a color name, such as* **11-0103TC***.)*

Enter Color Book name specifies the name of a color book.

Enter color name specifies the name of the color.

RELATED COMMANDS

ChProp changes the color via the command line in fewer keystrokes.

Properties changes the color of objects via a dialog box.

RELATED SYSTEM VARIABLES

CeColor is the current object color setting.

TIPS

- 'BYLAYER' means that objects take on the color assigned to that layer. When BYLAYER objects on layer 0 are part of a block definition, they take on the properties of whichever layer the block was inserted on, or the layer to which the block was moved.
- 'BYBLOCK' objects in a block definition take on the color assigned to the block. This may be the color in effect at the time the block was inserted, or the color to which the block was changed.
- When more than one method is used to assign colors to objects in a block, results may be confusing.
- White objects display as black when the background color is white.
- Color "0" cannot be specified; AutoCAD uses it internally as the background color.
- The *colorwh.dwg* drawing has 255- and 16.7 million-color wheels.

CommandLine, CommandLineHide

2006 Toggles the display of the command line window.

Command	Alias	Ctrl+	F-key	Alt+	Menu Bar	Tablet
commandline	**cli**	**9**	**...**	**T**	**Tools** **↳Command Line**	**...**
commandlinehide		**9**				

Command: commandline

Displays the command line window:

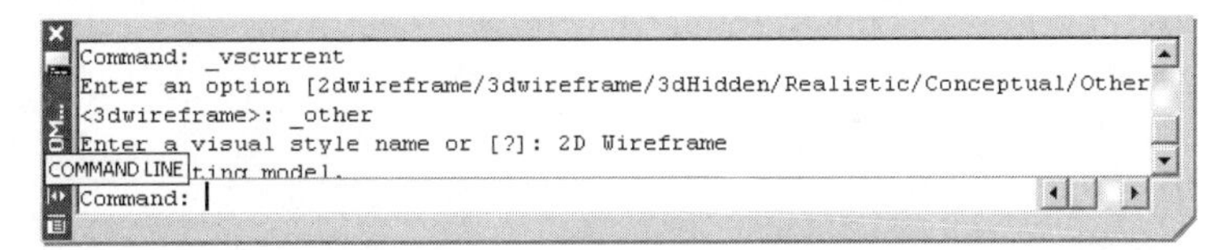

Command: commandlinehide

Hides the command line window.

RELATED SYSTEM VARIABLES

CmdInputHistory specifies the maximum number of previous commands remembered by AutoCAD.

CmdNames reports the name of the current command.

TIPS

- When the command line window is missing, you can type commands and see them in the drawing area when the DYN button is depressed on the status bar, and system variable **DynMode** is set to 1 or 3.
- The command line window normally shows three lines of history; you can stretch the window to show a greater or lesser number of lines.
- The window can be made translucent: right-click the title bar, and then select **Transparency** from the shortcut menu.
- Press **F2** to display the text window, which displays the last 400 lines of command history.

Compile

Rel.12 Compiles *.shp* shape and font files and *.pfb* font files into *.shx* files.

Command	Alias	Ctrl+	F-key	Alt+	Menu Bar	Tablet
compile	...	...	...	...	...	...

Command: compile

Displays Select Shape or Font File dialog box. Select a .shp or .pfb file, and then click ***Open.***

DIALOG BOX OPTIONS

Open opens the *.shp* or *.pfb* file for compiling.

Cancel closes the dialog box without loading the file.

RELATED COMMANDS

Load loads compiled *.shx* shape files into the current drawing.

Style loads *.shx* and *.ttf* font files into the current drawing.

RELATED SYSTEM VARIABLE

ShpName is the current *.shp* file name.

RELATED FILES

.shp are AutoCAD font and shape source files.

.shx are AutoCAD compiled font and shape files.

.pfb are PostScript Type B font files.

TIPS

- As of AutoCAD Release 12, **Style** converts *.shp* font files on-the-fly; it is only necessary to use the **Compile** command to obtain *.shx* font files.
- Prior to Release 12, this command was an option on the number menu system that appeared when AutoCAD started up.
- As of Release 14, AutoCAD no longer directly supports PostScript font files. Instead, use the **Compile** command to convert *.pfb* files to *.shx* format.
- TrueType fonts are not compiled.

Cone

Rel.11 Draws 3D solid cones with circular or elliptical bases.

Command	Alias	Ctrl+	F-key	Alt+	Menu Bar	Tablet
cone	...	...	...	DMO	Draw ↳Modeling ↳Cone	M7

Command: cone

This command has new prompts in AutoCAD 2007.

Specify center point of base or [3P/2P/Ttr/Elliptical]: *(Pick center point 1.)*

Specify base radius or [Diameter] <2">: *(Specify radius 2, or type* **D**.*)*

Specify height or [2Point/Axis endpoint/Top radius] <2">: *(Specify height 3, or type* **A**.*)*

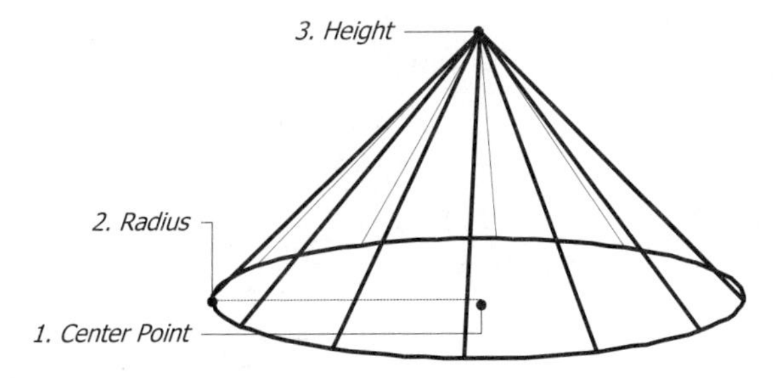

COMMAND LINE OPTIONS

Center point of base specifies the x,y,z coordinates of the center point of the cone's base.

3P picks three points that specify the base's circumference.

2P specifies two points on the base's circumference.

Ttr specifies two points of tangency (with other objects) and the radius.

Elliptical creates cone with an elliptical base.

Base radius specifies the cone's radius.

Diameter specifies the cone's diameter.

Height specifies the cone's height.

2Points picks two points that specify the z orientation of the cone.

Axis endpoint picks the other end of an axis formed from the center point.

Top radius creates a cone with a flat top.

RELATED COMMAND

Ai_Cone draws 3D surface model cones.

RELATED SYSTEM VARIABLES

DragVS specifies the visual style during cone creation.

IsoLines specifies the number of isolines on solid surfaces.

TIPS

- You define the elliptical base in two ways: by the length of the major and minor axes, or by the center point and two radii.
- To draw a cone at an angle, use the **Axis** option.

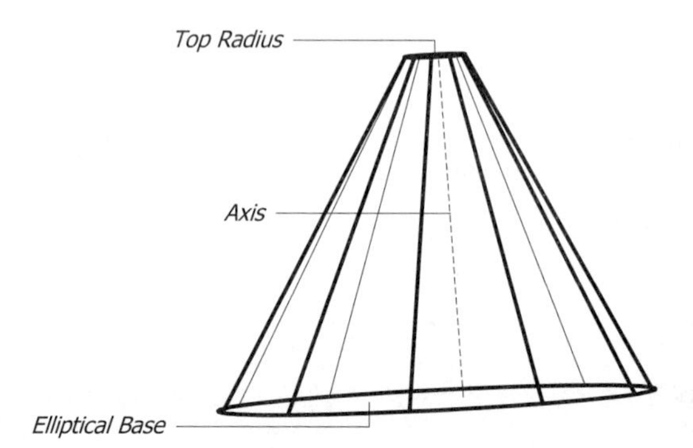

- Silhouette lines are displayed when IsoLines is set to 0 and DispSihl is set to 1. There is no need to hide or shade the cone to see them.
- Once the cone is placed in the drawing, you can edit it interactively. Select the cone, and then select a grip, which turns red, meaning it can be dragged. Press ESC to exit direct editing.

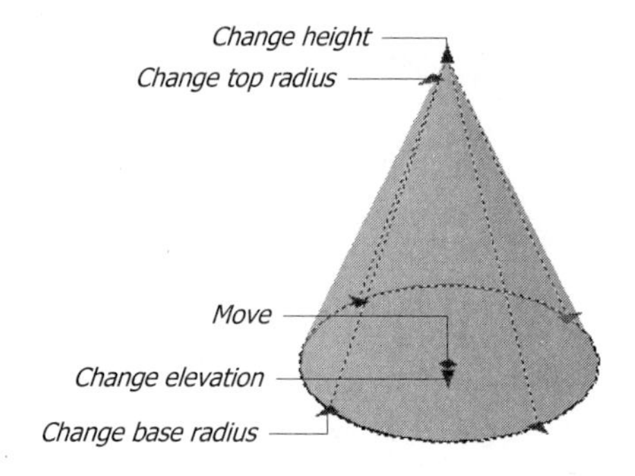

Changed Command

The **Config** command now displays the Options dialog box; see the **Options** command.

Convert

Rel.14 Converts 2D polylines and associative hatches (created in R13 and earlier) to an optimized "lightweight" format.

Command	Alias	Ctrl+	F-key	Alt+	Menu Bar	Tablet
convert	...	...	...	...	...	...

Command: convert
Enter type of objects to convert [Hatch/Polyline/All] <All>: *(Enter an option.)*
Enter object selection preference [Select/All] <All>: *(Type* **S** *or* **A***.)*

COMMAND LINE OPTIONS

Hatch converts associative hatch patterns from anonymous blocks to hatch objects; displays warning dialog box:

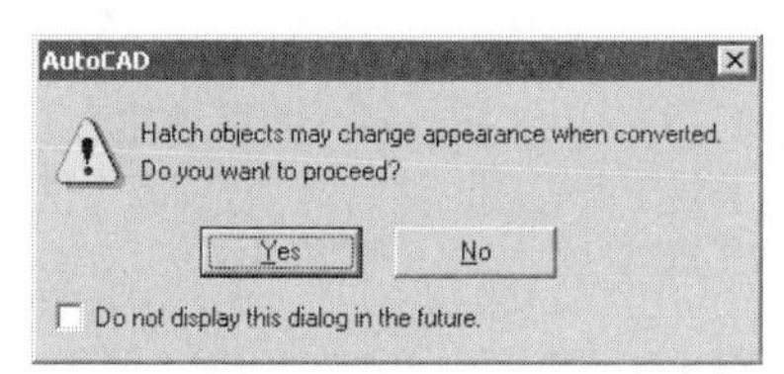

Yes converts hatch patterns to hatch objects.

No does not convert hatch patterns.

Polyline converts 2D polylines to Lwpolyline objects.

All converts all polylines and hatch patterns.

Select selects the hatch patterns and 2D polylines to convert.

RELATED COMMANDS

BHatch creates associative hatch patterns.

PLine draws 2D polylines.

RELATED SYSTEM VARIABLE

PLineType determines whether pre-Release 14 polylines are converted in AutoCAD.

0 — Not converted; **PLine** creates old-format polylines.

1 — Not converted; **PLine** creates lwpolylines.

2 — Converted; **PLine** creates lwpolylines (default).

TIPS

- When a Release 13 or earlier drawing is opened, AutoCAD automatically converts most (not all) 2D polylines to lwpolylines; hatch patterns are not automatically updated.
- Hatch patterns are automatically updated the first time the HatchEdit command is applied, or when their boundaries are changed.
- Polylines are not converted when they contain curve fit segments, splined segments, extended object data in their vertices, or 3D polylines.
- PLineType affects the following commands: Boundary (polylines), Donut, Ellipse (PEllipse = 1), PEdit (when converting lines and arcs), Polygon, and Sketch (SkPoly = 1).

ConvertCTB

2001 Converts a plot style file from CTB color-dependent format to STB named format (short for CONVERT Color TaBle).

Command	Alias	Ctrl+	F-key	Alt+	Menu Bar	Tablet
convertctb	...	...	...	...	...	...

Command: convertctb

*Displays **Select File** dialog box.*

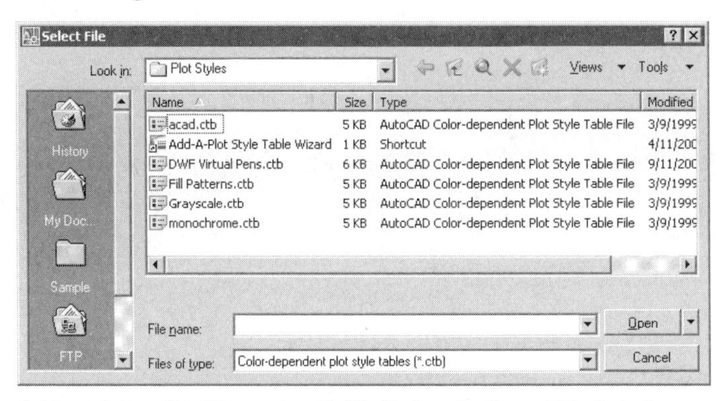

*1. Select a .ctb file, and then click **Open**. AutoCAD displays the Create File dialog box.*

*2. Specify the name of an .stb file, and then click **Save**. When AutoCAD completes the conversion, it displays this dialog box:*

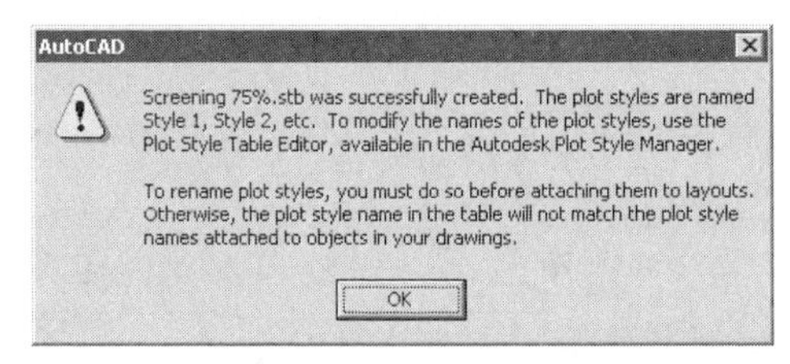

RELATED COMMANDS

ConvertPStyles converts a drawing between color-dependent and named plot styles.

PlotStyle sets the plot style for a drawing.

RELATED FILES

**.ctb* are color-dependent plot style table files.

**.stb* are named plot style tables.

TIPS

- "Color-dependent plot styles" are used by older versions of AutoCAD, where the color of the object controls the pen selection.
- "Named plot styles" is the alternative introduced with AutoCAD 2000, which allows plotter-specific information to be assigned to layers and objects.

ConvertOldLights ConvertOldMaterials

2007 Converts lights and materials from drawings created in AutoCAD 2006 and earlier.

Command	Alias	Ctrl+	F-key	Alt+	Menu Bar	Tablet
convertoldlights	...	...	...	...	...	...
convertoldmaterials						

Command: convertoldlights

Lights defined in drawings created by AutoCAD 2006 and earlier are converted to the new format.

Command: convertoldmaterials

Materials defined in drawings created by AutoCAD 2006 and earlier are converted to the new format.

RELATED COMMANDS

LightList displays the Light List palette, which lists and edits lights in the current drawing.

Materials displays the Materials palette, which modifies materials.

TIP

- Autodesk warns that the conversion may not be correct in all cases. For example, the intensity of lights may need to be corrected and the mapping of materials may need to be adjusted.

ConvertPStyles

2001 Converts a drawing from color-dependent to named plot styles (short for Convert Plot STYLES).

Command	Alias	Ctrl+	F-key	Alt+	Menu Bar	Tablet
convertpstyles	...	...	...	...	...	...

Command: convertpstyles

Displays warning dialog box:

When AutoCAD has completed the conversion, it displays the message, "Drawing converted from Named plot style mode to Color Dependent mode."

AUTOCAD DIALOG BOX

OK proceeds with the conversion.

Cancel prevents the conversion.

RELATED COMMANDS

ConvertCTB converts a plot style file from CTB color-dependent format to STB named format.

PlotStyle sets the plot style for a drawing.

TIPS

- "Color-dependent plot styles" was used by older versions of AutoCAD, where the color of the object controlled the pen selection.
- "Named plot styles" is the alternative introduced with AutoCAD 2000, which allows plotter-specific information to be assigned to layers and objects.

ConvToSolid, ConvToSurface

2007 Converts certain objects to solids, and open objects to surfaces.

Command	Alias	Ctrl+	F-key	Alt+	Menu Bar	Tablet
convtosolid	...	...	...	...	...	...
convtosurface						

Command: convtosolid

Converts these objects to solids: closed, uniform-width polylines with thickness, closed zero-width polylines with thickness, and circles with thickness.

Select objects: *(Select one or more objects.)*
Select objects: *(Press Enter.)*

Command: convtosurface

Converts open objects to surfaces: 2D solids, regions, open zero-width polylines with thickness, lines with thickness, arcs with thickness, and planar 3D faces.

Select objects: *(Select one or more objects.)*
Select objects: *(Press Enter.)*

RELATED COMMANDS

Extrude extrudes 2D objects into 3D solids or surfaces.

Revolve rotates 2D objects into 3D solids or surfaces.

Thicken thickens 2D objects into 3D solids or surfaces.

Sweep rotates 2D objects into 3D solids or surfaces.

Loft converts 2D objects into lofted 3D solids or surfaces.

RELATED SYSTEM VARIABLE

DelObj determines whether to retain or erase objects used to create the surfaces and solids.

TIP

- Use the Explode command to convert 3D solids with curved surfaces into surfaces.

Copy

V. 1.0 Creates one or more copies of objects.

Command	Alias	Ctrl+	F-key	Alt+	Menu Bar	Tablet
copy	co	...	...	MY	Modify	V15
	cp				↳Copy	

Command: copy
Select objects: *(Select one or more objects — point 1.)*
Select objects: *(Press Enter.)*
Specify base point or [Displacement] <Displacement>: *(Pick point 2.)*
Specify second point of displacement or <use first point as displacement>: *(Point 3.)*
Specify second point or [Exit/Undo] <Exit>: *(Press Enter.)*

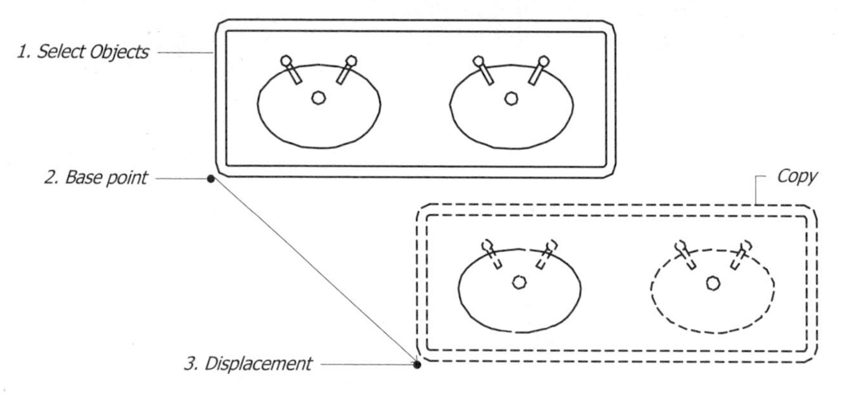

COMMAND LINE OPTIONS

Base point indicates the starting point; does not need to be on the object.

Second point indicates the point to which to copy.

Displacement prompts for two points to use as the displacement.

Undo undoes the last copy.

Exit exits the command.

RELATED COMMANDS

Array draws a rectangular or polar array of objects.

MInsert places an array of blocks.

Offset creates parallel copies of lines, polylines, circles, and arcs.

TIPS

- Turn on **Ortho** to copy objects in a precise horizontal and vertical direction, or **Polar** to achieve precise incremental angles; use **OSnap** to copy objects precisely to a geometric feature.
- Inserting a block multiple times is more efficient than placing multiple copies.
- To mimic the single-copy behavior of AutoCAD 2004 and earlier, add the following AutoLISP code to the *acad2007doc.lsp* file in the *\autocad\support* folder:

```
(command ".undefine" "copy")(princ)(defun c:copy()(command ".copy"))
```

CopyBase

2000 Copies selected objects to the Clipboard with a specified base point (short for COPY with BASEpoint).

Command	Alias	Ctrl+	F-key	Alt+	Menu Bar	Tablet
copybase	...	Shift+C	...	EB	Edit ↳Copy with Base Point	...

Command: copybase
Specify base point: *(Pick a point.)*
Select objects: *(Select one or more objects.)*
Select objects: *(Press Enter.)*

COMMAND LINE OPTIONS

Specify base point specifies the base point.

Select objects selects the objects to copy to the Clipboard.

RELATED COMMANDS

CopyClip copies selected objects to the Clipboard with a base point equal to the lower-left extents of the selected objects.

PasteBlock pastes objects from the Clipboard into drawings as blocks.

PasteClip pastes objects from the Clipboard into drawings.

TIPS

- When PasteBlock pastes objects previously selected with the CopyBase command, AutoCAD prompts you 'Specify insertion point:', and then pastes the objects as a block with a name similar to A$C7E1B27BE.
- When specifying the **All** option at the 'Select objects:' prompt, CopyBase selects only objects visible in the current viewport.
- As of AutoCAD 2004, you can use the CTRL+SHIFT+C shortcut for this command.

CopyClip

Rel.12 Copies selected objects from the drawing to the Clipboard (short for COPY to CLIPboard).

Command	Alias	Ctrl+	F-key	Alt+	Menu Bar	Tablet
copyclip	...	C	...	EC	Edit ↳Copy	T14

Command: copyclip
Select objects: *(Select one or more objects.)*
Select objects: *(Press Enter.)*

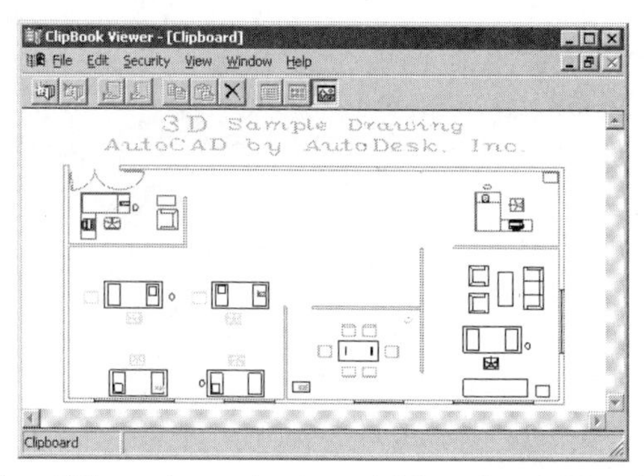

The Clipboard Viewer shows a drawing copied from AutoCAD to the Clipboard.

COMMAND LINE OPTION

Select objects selects the objects to copy to the Clipboard.

RELATED COMMANDS

CopyBase copies objects to the Clipboard with a specified base point.

CopyHist copies Text window text to the Clipboard.

CopyLink copies the current viewport to the Clipboard.

CutClip cuts selected objects to the Clipboard.

PasteBlock pastes objects from the Clipboard into the drawing as a block.

PasteClip pastes objects from the Clipboard into the drawing.

RELATED WINDOWS COMMANDS

PRT SCR copies the entire screen to the Clipboard

ALT+PRT SCR copies the topmost window to the Clipboard.

TIPS

- Contrary to the AutoCAD *Command Reference*, text is *not* copied to the Clipboard in text format; instead, text is copied as AutoCAD objects. (Text copied from the command line is copied as plain text.)
- When the **All** option is specified at the 'Select objects' prompt, CopyClip selects only objects visible in the current viewport.

CopyHist

Rel.13 Copies all of the Text window text to the Clipboard (short for COPY HISTory).

Command	Alias	Ctrl+	F-key	Alt+	Menu Bar	Tablet
copyhist	...	...	...	...	...	...

Command: copyhist

AutoCAD copies all text in the history window to the Clipboard.

COMMAND LINE OPTIONS

None.

RELATED COMMAND

CopyClip copies selected text from the drawing to the Clipboard.

RELATED WINDOWS COMMAND

ALT+PRT SCR copies the Text window to the Clipboard in graphics format.

TIPS

- To copy a selected portion of Text window text to the Clipboard, highlight the text first, then select Copy from the Text window's Edit menu bar.

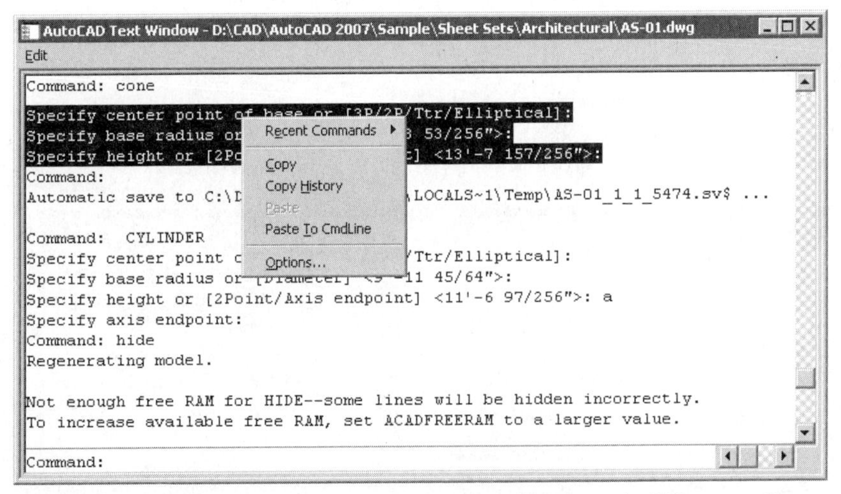

- To paste text to the command line, select Paste to Cmdline from the Edit menu. However, this only works when the Clipboard contains text — not graphics.
- As an alternative, right-click in the **Text** window to bring up the cursor menu.

Quick Start Tutorial

Getting the Right Clipboard Result

You can use the Clipboard to display AutoCAD drawings in other Windows applications, such as word processing, desktop publishing, and paint programs. AutoCAD has two primary commands for copying objects from the drawing to the Clipboard: **CopyLink** and **CopyClip**. The effect of using these commands, however, may surprise you, since each command produces a different result, depending on whether AutoCAD is in model or layout mode.

You can see the content of the Clipboard in the Windows Clipboard Viewer. Enter the following in the Taskbar's Run command: c:\windows\system32\clipbrd.exe.

- **CopyClip** (Edit | Copy) copies selected objects from the drawing, as follows:

  Command: copyclip
  Select objects: all
  Select objects: *(Press Enter.)*

Warning! When you select **All** objects, AutoCAD selects only those objects visible in the current viewport.

- **CopyLink** (Edit | Copy Link) copies everything visible in the current viewport. Use this command to capture a layout view with *all* viewports.

The two commands take on different meanings, as described below.

Model Mode

In *model mode*, AutoCAD copies only objects visible in the current viewport. If the drawing contains more than one viewport, you must select the correct viewport before copying objects to the Clipboard. CopyClip and CopyLink have the same effect:

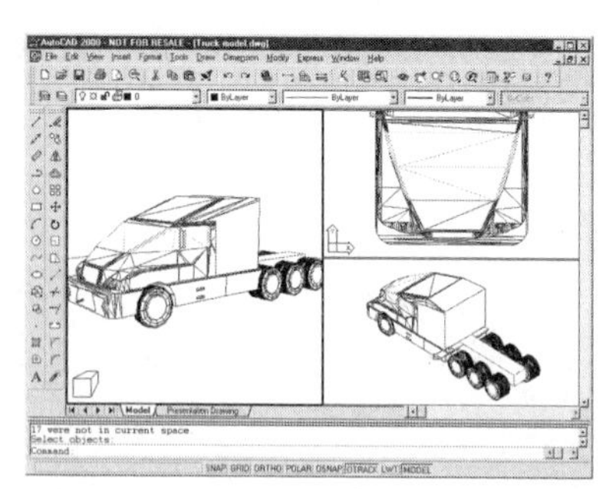

AutoCAD viewports in model mode.

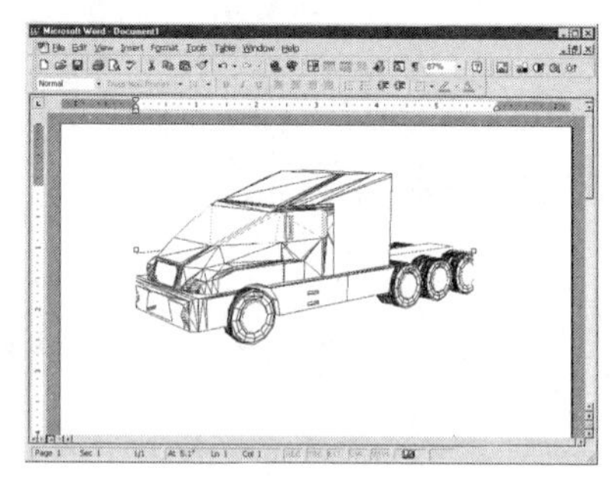

Drawing pasted in Word after either CopyClip or CopyLink.

Layout Mode

In layout mode's PAPER space, CopyClip copies objects drawn only in paper space — *nothing* drawn in model space is copied to the Clipboard, even if it is visible.

CopyLink copies *all* visible objects, whether drawn in paper space or model space. Notice that this command also copies the margin lines and gray background.

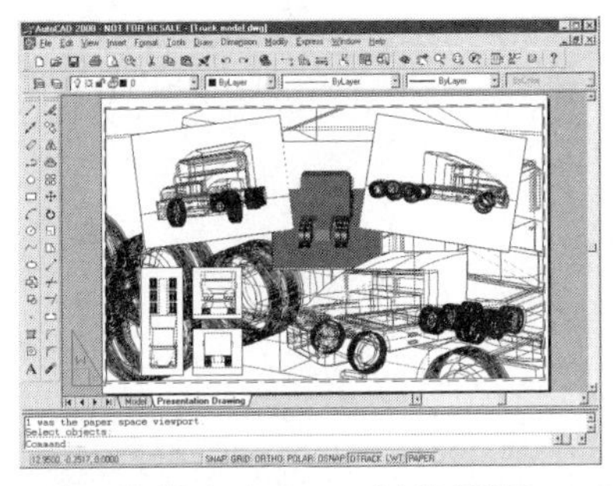

AutoCAD in layout mode's PAPER space.

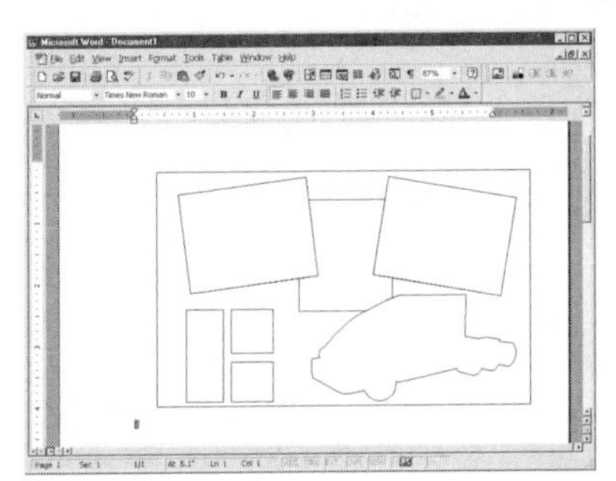

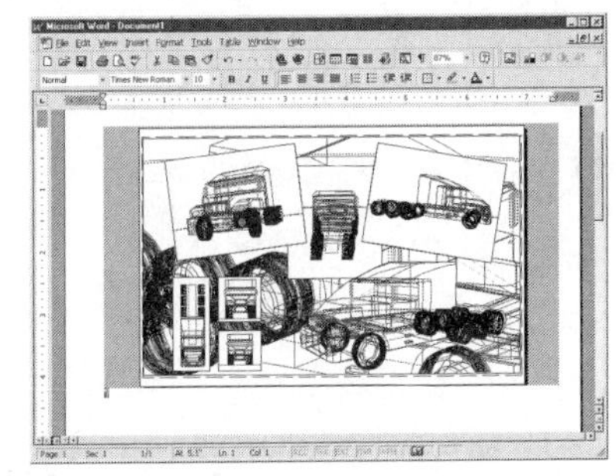

Result in Word after using AutoCAD's ***CopyClip*** *(left) and* ***CopyLink*** *(right).*

In layout mode's MODEL space, AutoCAD only copies objects drawn in model space that are visible in the selected viewport. CopyClip and CopyLink produce the same result when pasted in Word and other Windows applications.

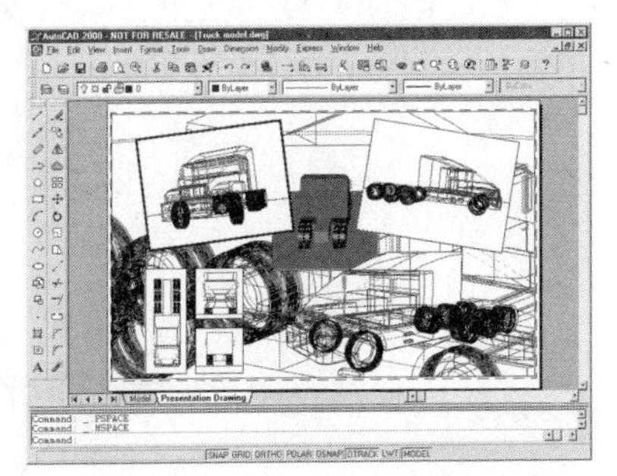

A viewport in layout mode's MODEL space.

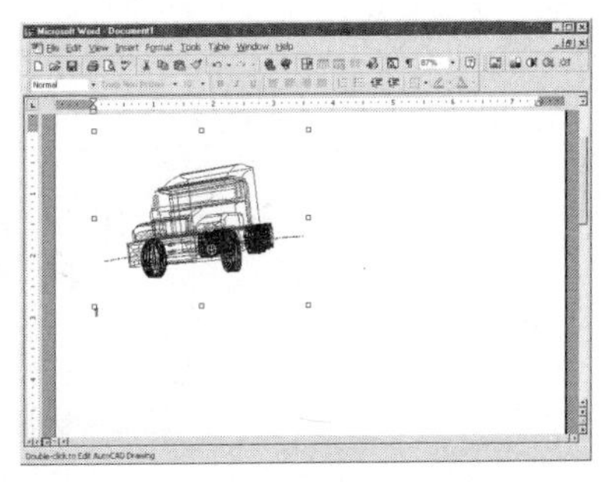

Drawing pasted in Word after either CopyClip or CopyLink.

CopyLink

Rel.13 Copies the current viewport to the Clipboard; optionally allows you to link the drawing to AutoCAD.

Command	Alias	Ctrl+	F-key	Alt+	Menu Bar	Tablet
copylink	...	...		EL	Edit ↳Copy Link	...

Command: copylink

COMMAND LINE OPTIONS

None.

RELATED COMMANDS

CopyClip copies selected objects to the Clipboard.

CopyEmbed copies selected objects to the Clipboard.

CopyHist copies Text window text to the Clipboard.

CutClip cuts selected objects to the Clipboard.

PasteClip pastes objects from the Clipboard into the drawing.

RELATED WINDOWS COMMANDS

PRT SCR copies the entire screen to the Clipboard

ALT+PRT SCR copies the topmost window to the Clipboard.

TIPS

- In the other application, select Paste Special from the Edit menu to paste the AutoCAD image into the document; to link the drawing to AutoCAD, select the **Paste Link** option.
- AutoCAD does not let you link a drawing to itself.
- This command copies everything in the current viewport (if in model space) or the entire drawing (if in paper space).
- CopyEmbed is identical to CopyLink, except that CopyEmbed prompts you to select objects.
- If you use Paste Special | AutoCAD Drawing when pasting the drawing into a document, you can then double-click the drawing, which launches AutoCAD so that you can edit the drawing.

CopyToLayer

2007 Copies selected objects to the layer of another object; formerly an Express Tool.

Command	Alias	Ctrl+	F-key	Alt+	Menu Bar	Tablet
copytolayer	...	...		OAP	**Format** ⇘**Layer Tools** ⇘**Copy Objects to New Layer**	...
-copytolayer						

Command: copytolayer
Select objects to copy: *(Pick one or more objects.)*
Select objects to copy: *(Press Enter.)*

Select object on destination layer or [Name] <Name>: *(Pick an objects, or type* **N***.)*
n **object(s) copied and placed on layer "***layername***".**
Specify base point or [Displacement/eXit] <eXit>: *(Enter an option or press Enter.)*

COMMAND LINE OPTIONS

Select objects to copy selects the objects to be copied.

Select object on destination layer specifies the layer through the selection of another object.

Name displays the Copy to Layer dialog box with names of all layers in drawing.

Specify base point specifies the starting point of the copy.

Displacement specifies the displacement distance, from original to copy.

eXit exits the command.

Name dialog box

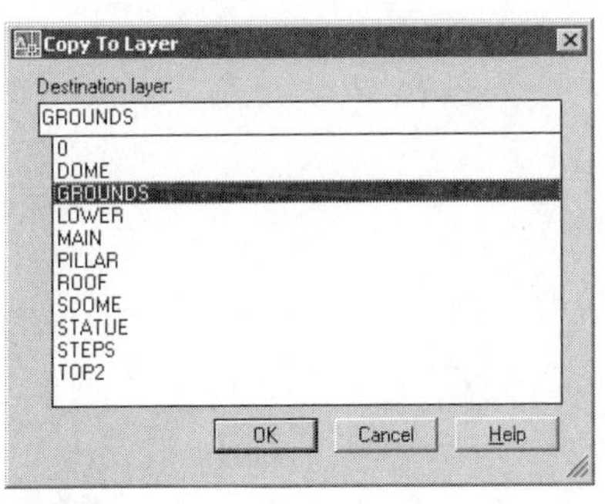

Select a layer name, and then click ***OK***

-COPYTOLAYER COMMAND

Command: -copytolayer

Select objects to copy: *(Pick one or more objects.)*

Select objects to copy: *(Press Enter.)*

Specify the destination layer name or [?/= (select object)] <0>: *(Enter the name of a layer, or enter an option.)*

n **object(s) copied and placed on layer "***layername***".**

Specify base point or [Displacement/eXit] <eXit>: *(Enter an option or press Enter.)*

Specify second point of displacement or <use first point as displacement>:

COMMAND LINE OPTIONS

Select objects to copy selects the objects to be copied.

Select object on destination layer specifies the layer through the selection of another object.

? lists the name os layers in the drawing.

= prompts to select an object whose layer should be used; AutoCAD prompts, "Select an object with the desired layer name."

Specify base point specifies the starting point of the copy.

Displacement specifies the displacement distance, from original to copy.

eXit exits the command.

RELATED COMMANDS

Copy copies objects onto the same layer.

Ai_Molc makes the layer of the selected object current.

ChSpace moves objects from paper to model space.

CUI

2006 Customizes many aspects of the user interface; replaces the Menu, Toolbar, and Customize commands (short for Customize User Interface).

Command	Alias	Ctrl+	F-key	Alt+	Menu Bar	Tablet
cui	**toolbar**	...	...	**TCI**	**Tools**	...
	tbconfig				**↳Customize**	
	to				**↳Interface**	

Command: cui

Displays the ***Customize*** *tab of the dialog box:*

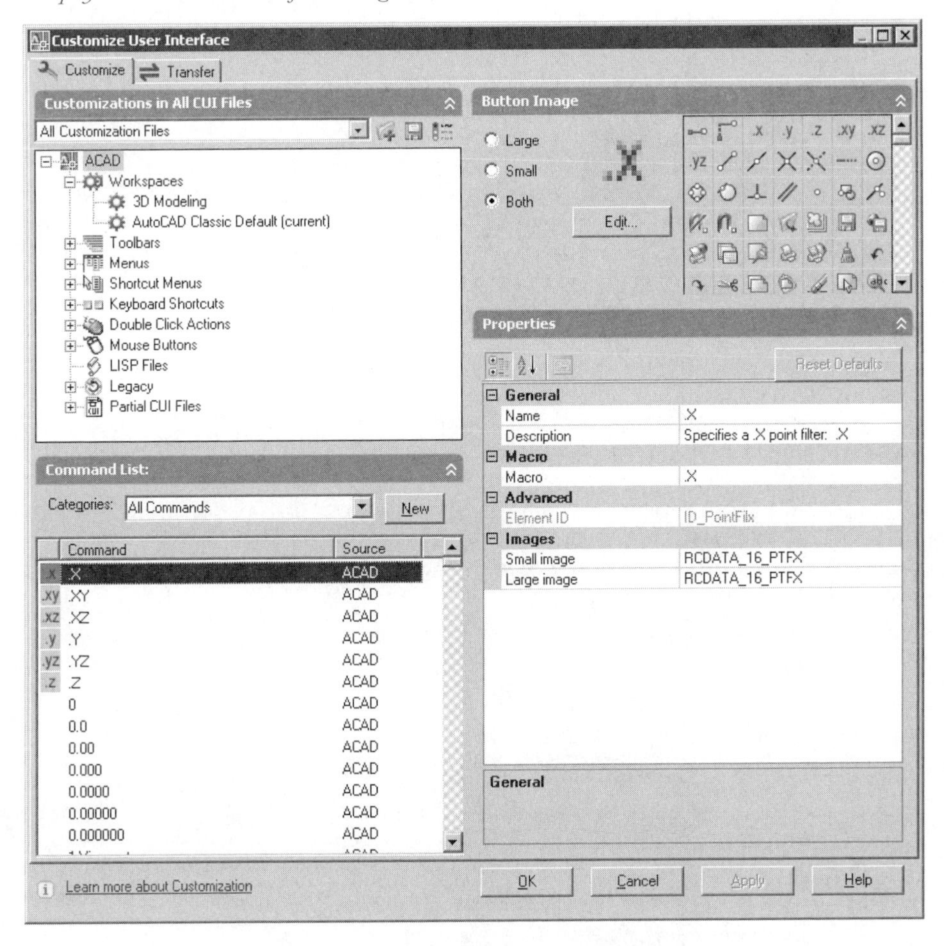

Customizations In pane

A "pane" is an area of a dialog box.

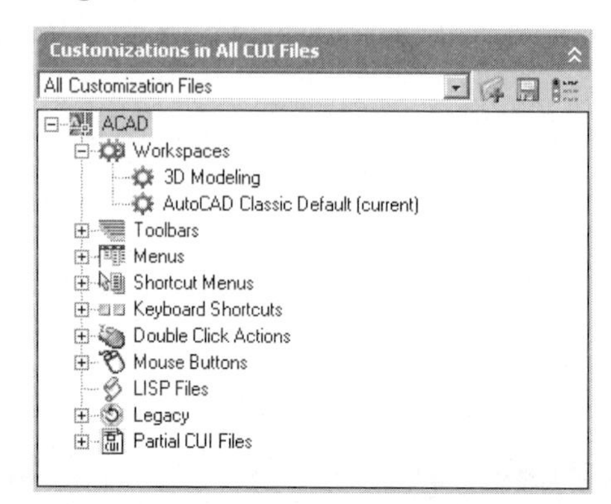

Right-click items to access commands on shortcut menus.

Added to AutoCAD 2007: Double Click Actions.

Command List pane

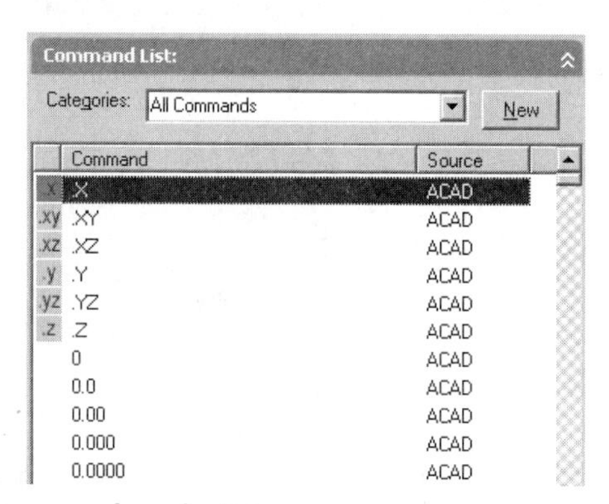

Right-click items to access commands on shortcut menus.

Categories lists categories of commands.

New creates new commands.

Properties pane

Content varies, depending on the item being edited:

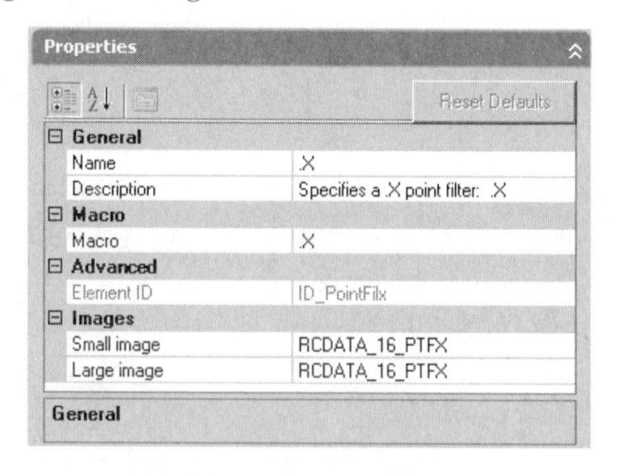

Name specifies the name of the command or macro.

Description specifies the help text displayed on the status bar.

Macro executes the code when the command, menu item, toolbar button, or shortcut keystroke is invoked; click the **...** button to display the Long String Editor dialog box:

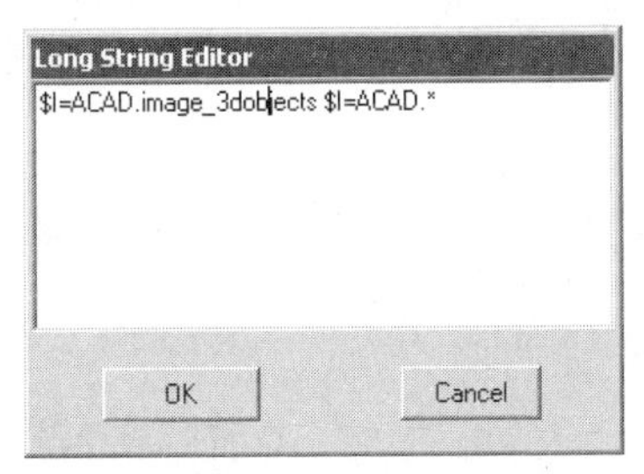

ElementID reports the internal identification assigned to the command by AutoCAD.

Small image shows the 16x16-pixel bitmap image displayed on toolbars and so on; click the **...** button to display the Select Image File dialog box for loading *.bmp* image files.

Large image shows the 24x24-pixel bitmap image displayed on toolbars and so on.

Button Image pane

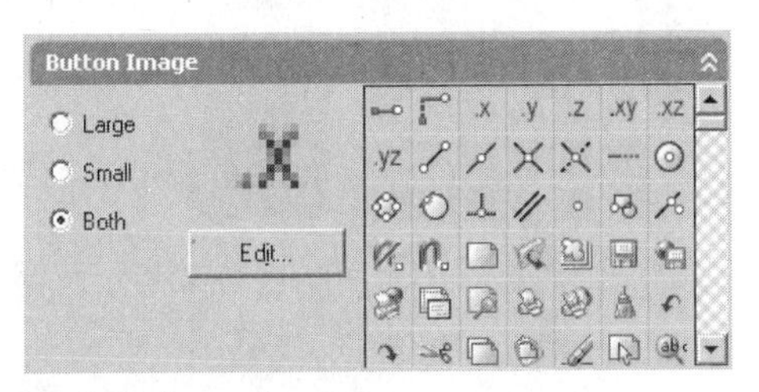

Small works with 16x16-pixel bitmap images.

Large works with 24x24-pixel bitmap images.

Both works with both sizes of images.

Edit displays the Button Editor dialog box.

BUTTON EDITOR DIALOG BOX

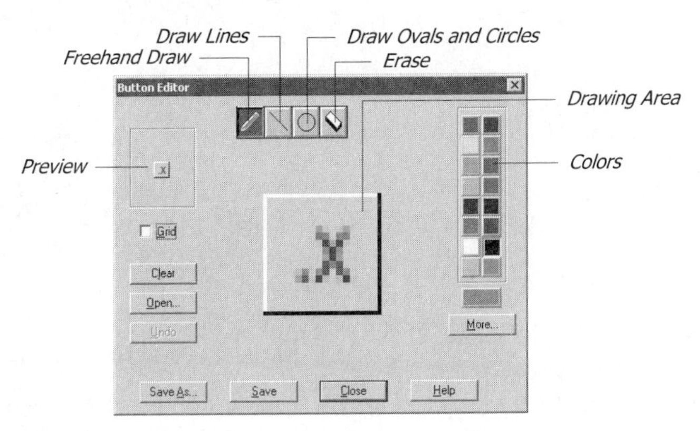

Grid:

☑ Displays a grid in the icon drawing area.

☐ Hides the grid.

Clear erases the icon.

Open opens a *.bmp* (bitmap) file, which can be used as an icon.

Undo undoes the last operation.

Save As saves the icon as a *.bmp* file.

Save saves the changes to the icon in AutoCAD's internal storage.

Close closes the dialog box.

More displays the Select Color dialog box.

Dialog Box Toolbar

Customizations In pane

 Loads partial *.cui* and *.mnu* customization files; displays Open dialog box.

 Saves all current customization files; does not display a dialog box.

 Opens Display Filters dialog box:

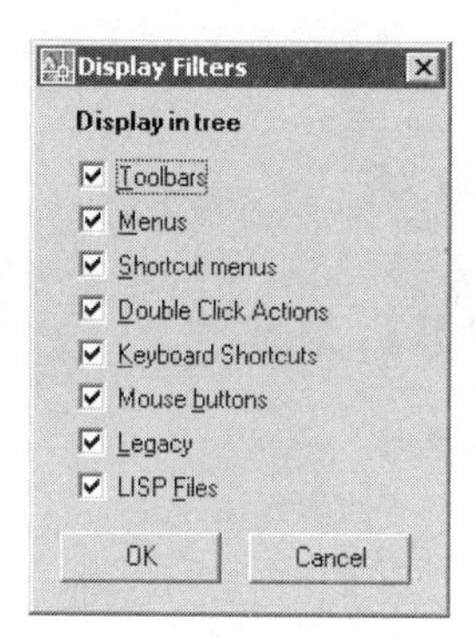

Properties pane

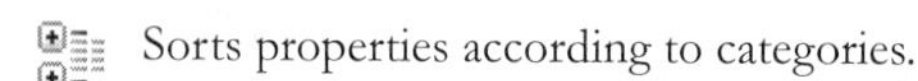
Sorts properties according to categories.

Sorts properties alphabetically.

Toggles the display of the Tips box below the Properties grid.

SHORTCUT MENUS

Customizations In pane

Content of shortcut menus varies, depending on the item selected:

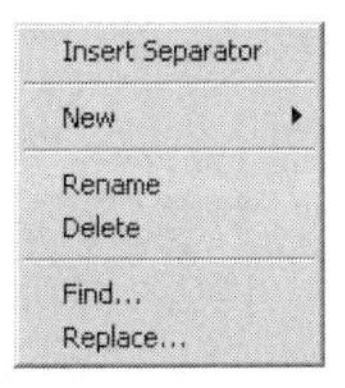

Insert Separator adds separation lines between items.

New adds an item specific to the section; examples include new toolbars and flyouts, new workspaces, and new menus and sub-menus.

Rename changes the name of the item.

Delete erases the item; AutoCAD displays a warning dialog box.

Find displays the Find tab of the Find and Replace dialog box.

Replace displays the Replace tab of the Find and Replace dialog box.

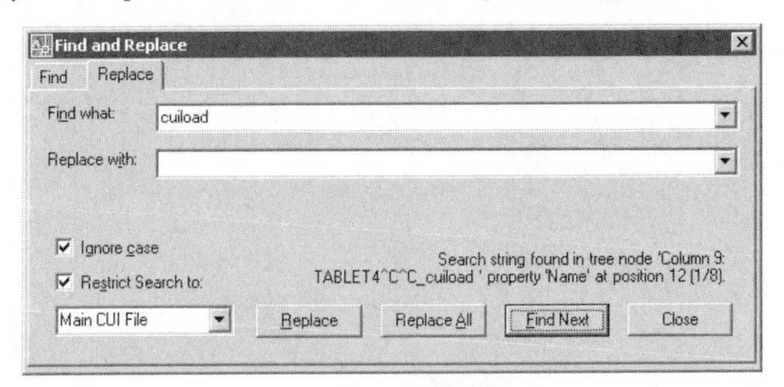

Command List pane

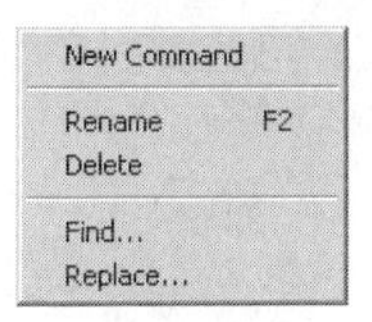

Edit Properties displays the Properties pane.

New Command adds new "commands" (macros, actually) to AutoCAD.

Delete erases the item (available only for user-created commands).

Rename displays the Rename Command dialog box for renaming items.

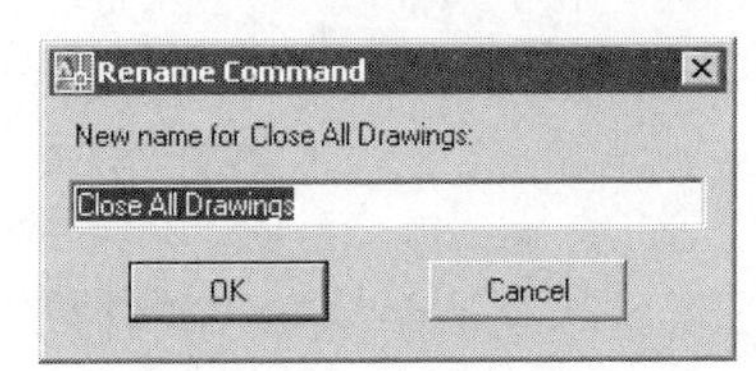

RELATED COMMANDS

CuiExport and **CuiImport** transfer customization settings between the local *acad.cui* and enterprise and partial *cui* files.

CuiLoad and **CuiUnload** loads and unloads *.cui* files; replaces the MenuLoad command.

Customize rearranges tool palettes.

Tablet calibrates, configures, and toggles digitizing tablet menus.

Workspace creates, modifies, and saves workspaces.

RELATED SYSTEM VARIABLES

CuiState reports whether the Customize User Interface dialog box is open or not.

EnterpriseMenu specifies the path and name of the *.cui* file used for the enterprise.

LockUi prevents the windows and toolbars from being moved; hold down the **Ctrl** key to override the lock.

TempOverrides turns on and off temporary override keys.

Tooltips toggles the display of tooltips.

DEFINITIONS

CUI — customization user interface.

Enterprise CUI — customization data shared by all users in an office; typically controlled by the CAD manager.

TIPS

- CUI replaces the Menu, Toolbar, and Customize commands as of AutoCAD 2006.
- The *.cui* file replaces the *.mnu*, *.mns*, and *.mnc* files.
- This command handles the customization of the following:

  Commands.
  Double-click actions (*new to AutoCAD 2007*).
  Image tile menus (rarely used).
  Mouse buttons.
  Pull-down menus.
  Screen menus (rarely used).
  Shortcut keys and temporary override keys.
  Shortcut menus (right-click or context-sensitive menus).
  Status bar help messages.
  Tablet buttons and menus.
  Toolbars and toolbar buttons.
  Workspaces.

- The LockUi system variable prevents the windows and toolbars from being moved; to override the lock, hold down the **Ctrl** key while dragging them. Alternatively click the padlock icon in the tray.

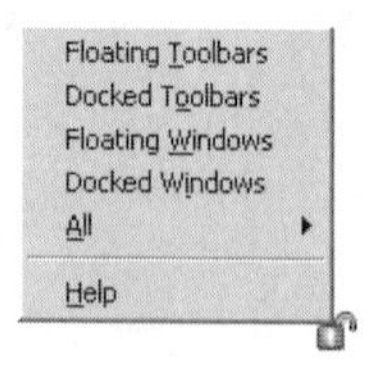

CuiExport, CuiImport

2006 Transfers user interface customization data between *.cui* files.

Command	Alias	Ctrl+	F-key	Alt+	Menu Bar	Tablet
cuiexport	...	...	...	TC	**Tools** ↳**Customize** ↳**Export Customizations**	...
cuiimport	...	...	...	TC	**Tools** ↳**Customize** ↳**Import Customizations**	...

Command: cuiexport *or* cuiimport

*Both commands display the same **Transfer** tab of the dialog box:*

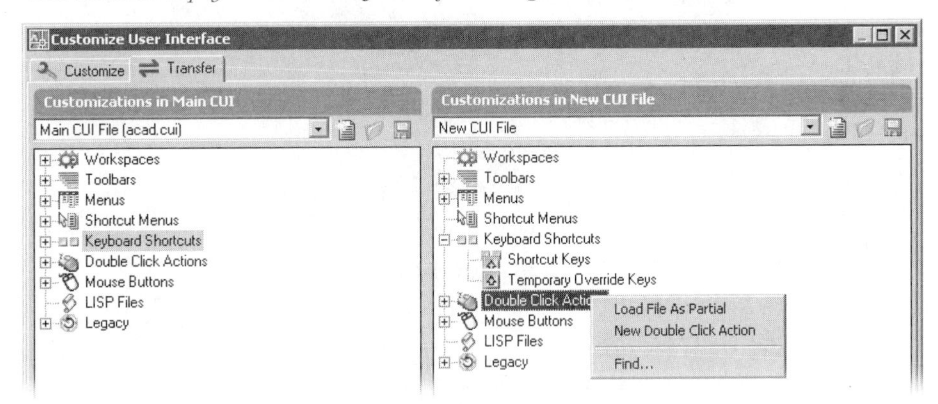

CUSTOMIZE USER INTERFACE DIALOG BOX

To import and export items, follow these steps:

1. Open a *.cui* or *.mnu* file in one panel.
2. Open another *.cui* or *.mnu* file in the second panel.
3. Drag items between the two panels.
4. Click the **Save** (diskette) button.

Dialog Box Toolbar

 Creates new customization files.

 Opens customization and menu files; displays the Open dialog box.

 Saves the current customization file; does not display a dialog box.

RELATED COMMANDS

Cui customizes elements of the user interface.

CuiLoad adds *.cui* and *.mnu* files to the user interface.

CuiLoad, CuiUnload

2006 Loads (and unloads) "partial" *.cui* files.

Command	Alias	F-key	Alt+	Menu Bar	Tablet
cuiload	menuload	...	...	...	...
cuiunload	menuunload				

Command: cuiload *or* cuiunload

Both commands display the same dialog box:

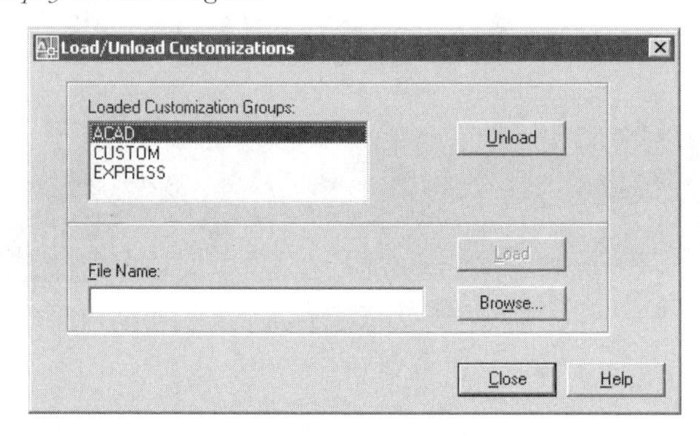

LOAD/UNLOAD CUSTOMIZATIONS DIALOG BOX

Loaded Customization Groups lists the names of loaded menu groups and files.

Unload unloads selected menu group.

Load loads the selected menu group into AutoCAD.

File Name displays the name of the *.cui* file.

Browse displays the Select Customization File dialog box.

RELATED COMMANDS

CUI manipulates the user interface customization.

Tablet configures digitizing tablets for use with overlay menus.

RELATED FILE

.cui is the customization user interface file; stored in XML format.

TIPS

- The CUILoad command allows you to add *partial* user interfaces, without replacing the entire UI structure.
- The *custom.cui* file is meant customizing menus independently of *acad.cui.*
- As of AutoCAD 2006, these two commands replace the MenuLoad and MenuUnload commands of earlier releases of AutoCAD.

Customize

2000i Manages tool palettes and creates palette groups.

Command	Aliases	Ctrl+	F-key	Alt+	Menu Bar	Tablet
customize	...	...	...	TCP	Tools ↳Customize ↳Tool Palettes	T13

Command: customize

Displays dialog box:

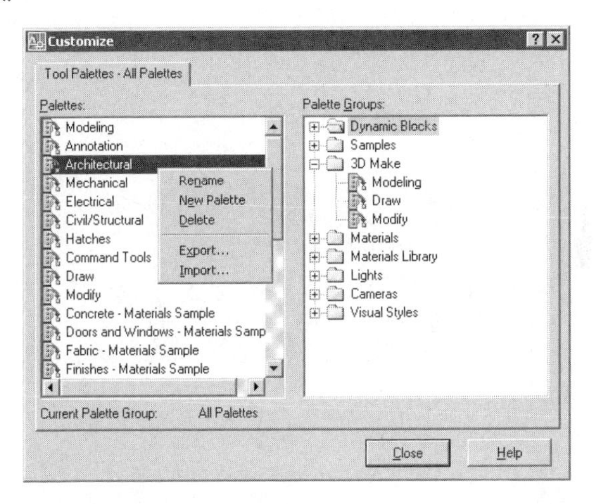

SHORTCUT MENUS

Palettes

Rename renames the selected palette.

New Tool Palette creates a new, blank palette.

Delete removes the selected palette.

Import imports palettes (*.xtp* files).

Export exports the selected palette in an XML-like *.xtp* file.

Palette Groups

Right-click a folder for the shortcut menu:

New Group creates new groups and subgroups.

Rename renames the selected group.

Delete removes the group.

Set Current makes the palette group current.

RELATED COMMAND

ToolPalette toggles the display of the Tool Palette.

TIP

- As of AutoCAD 2006, the CUI command customizes menus, toolbars, and shortcuts.

Rel.12 Cuts the selected objects from the drawing to the Clipboard (short for CUT to CLIPboard).

Command	Alias	Ctrl+	F-key	Alt+	Menu Bar	Tablet
cutclip	...	X	...	ET	Edit ↳Cut	T13

Command: cutclip
Select objects: *(Select one or more objects.)*
Select objects: *(Press Enter.)*

COMMAND LINE OPTION

Select objects selects the objects to cut to the Clipboard.

RELATED COMMANDS

BmpOut exports selected objects in the current view to a *.bmp* file.

CopyClip copies selected objects to the Clipboard.

CopyHist copies the Text window text to the Clipboard.

CopyLink copies the current viewport to the Clipboard.

PasteClip pastes objects from the Clipboard into the drawing.

RELATED WINDOWS COMMANDS

PRT SCR copies the entire screen to the Clipboard

ALT+PRT SCR copies the topmost window to the Clipboard.

TIPS

- When the **All** option is specified at the 'Select objects:' prompt, **CutClip** selects objects visible in the current viewport only.
- In the other application, use the **Edit | Paste** or **Edit | Paste Special** commands to paste the AutoCAD image into the document; the **Paste Special** command lets you specify the pasted format.
- You can use the **Undo** command to return the "cut" objects to the drawing.

Cylinder

Rel.12 Draws a 3D solid cylinder with a circular or elliptical cross-section.

Command	Alias	Ctrl+	F-key	Alt+	Menu Bar	Tablet
cylinder	**cyl**	...	...	**DMC**	**Draw** ↳**Modeling** ↳**Cylinder**	**L7**

Command: cylinder

This command has new prompts in AutoCAD 2007:

Specify center point of base or [3P/2P/Ttr/Elliptical]: *(Pick center point 1, or enter an option.)*

Specify base radius or [Diameter] <3>: *(Specify radius 2, or type* **D**.*)*

Specify height or [2Point/Axis endpoint] <1">: *(Specify height 3, or type* **C**.*)*

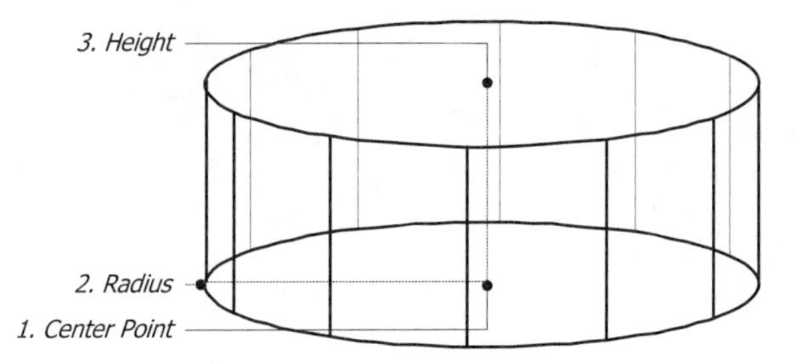

COMMAND OPTIONS

Center point of base specifies the x,y,z coordinates center point of the cylinder's base.

3P picks three points that specify the base's circumference.

2P specifies two points on the base's circumference.

Ttr specifies two points of tangency (with other objects) and the radius.

Elliptical creates elliptical cylinders.

Base radius specifies the cylinder's radius.

Diameter specifies the cylinder's diameter.

Height specifies the cylinder's height.

2Points picks two points that specify the z orientation of the cylinder.

Axis endpoint picks the other end of an axis formed from the center point.

RELATED COMMANDS

Ai_Cylinder creates cylinders made of 3D meshes.

Extrude creates cylinders and other extruded shapes with arbitrary cross-sections and sloped walls.

RELATED SYSTEM VARIABLES

DragVs determines the visual style while creating 3D solid objects.

IsoLines determines the number of isolines on curved solids.

TIP

- The **Axis endpoint** option rotates the cylinder.
- Once the cylinder is placed in the drawing, you can edit it interactively:
  1. Select the cylinder.
  2. Select one of the grips, which turns red.
  3. Drag the grip to change the cylinder's size or position.
  4. When done, press ESC to exit direct editing.

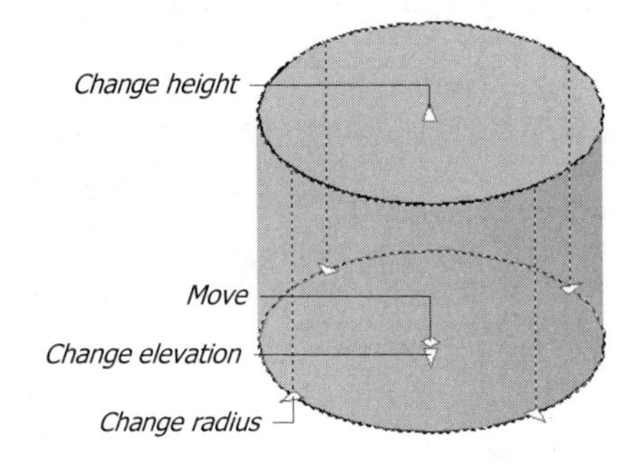

'Dashboard, 'DashboardClose

2007 Opens and closes the Dashboard palette; contains 3D commands.

Command	Alias	Ctrl+	F-key	Alt+	Menu Bar	Tablet
dashboard	...	...	...	**TP** ↳*Enter*	**Tools** ↳**Palettes** ↳**Dashboard**	...
dashboardclose	...	...	...	**TP** ↳*Enter*	**Tools** ↳**Palettes** ↳**Dashboard**	...

Command: dashboard

Displays palette with control panels:

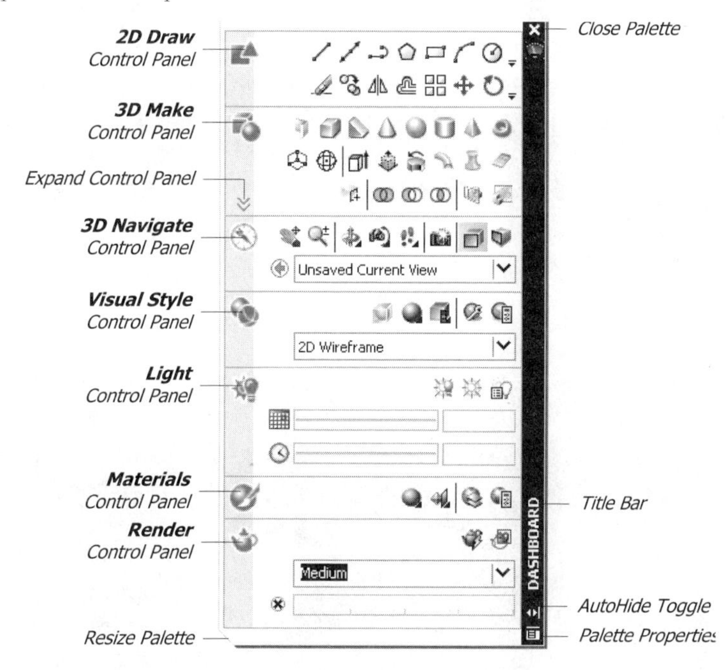

Command: dashboardclose

Closes the palette.

2D Draw control panel

This panel is normally closed; to open, right-click the Dashboard, and then select ***Control panels | 2D Draw control panel****.*

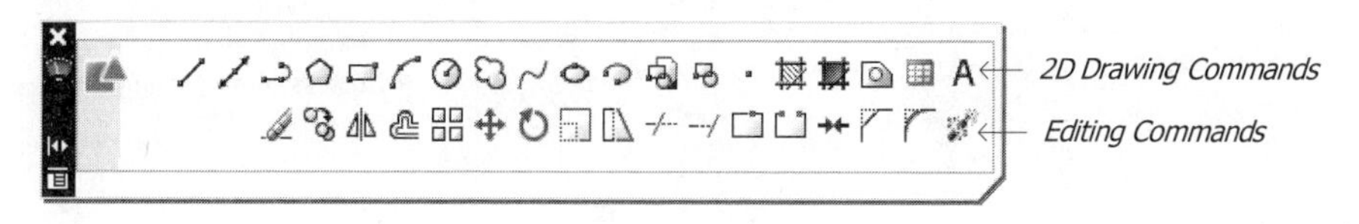

3D Make control panel

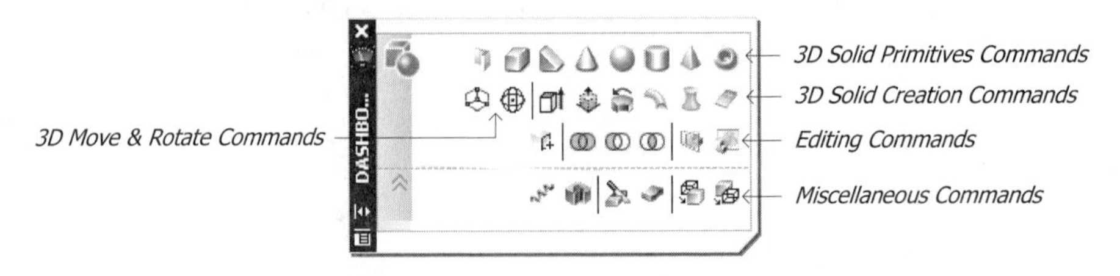

3D Navigate control panel

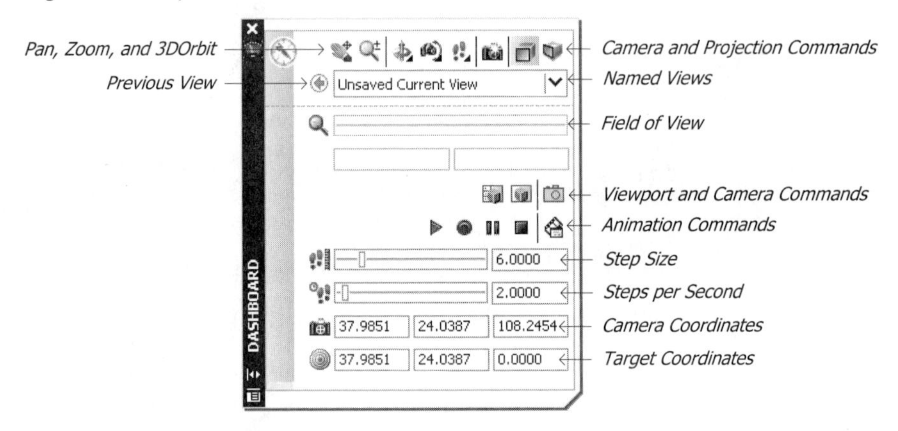

Visual Style control panel

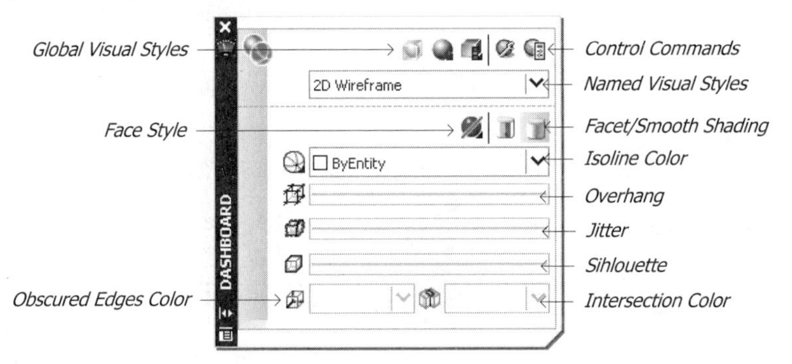

Light control panel

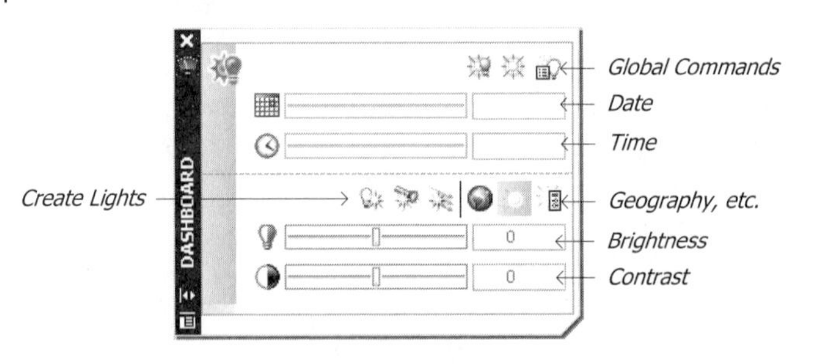

Materials control panel

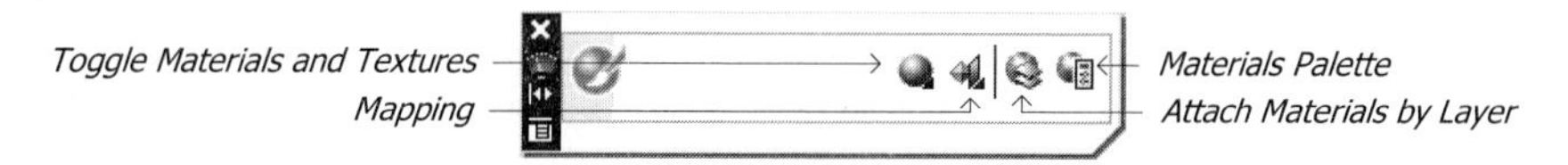

Render control panel

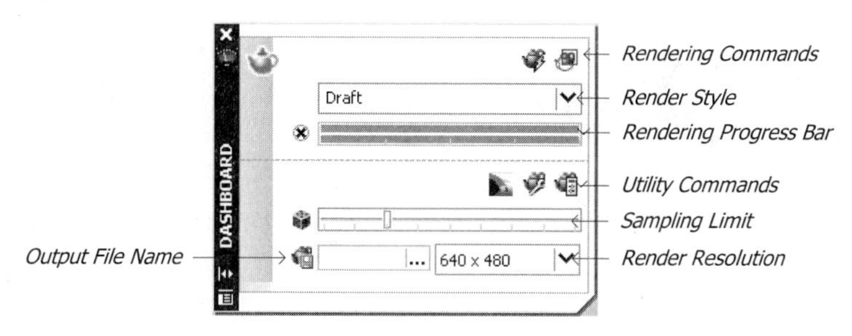

RELATED SYSTEM VARIABLE

DashboardState reports whether the Dashboard palette is open or closed.

TIPS

- The Dashboard lists commands in workflow order: 2D drawing and editing, 3D modeling and editing, navigating the drawing, applying visual styles, inserting lights, adding materials, and finally rendering the model.
- To determine the purpose of a command, pause the cursor over its icon until the tooltop appears.

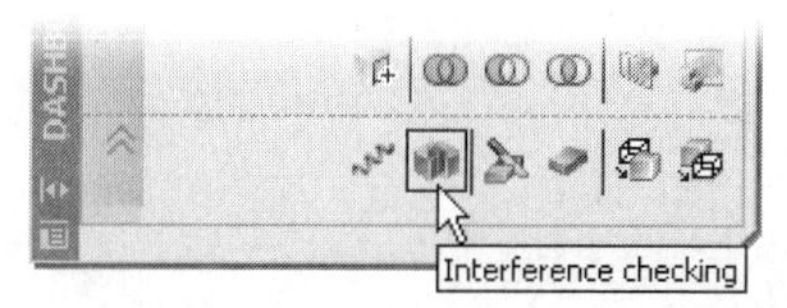

- To turn control panels on and off, right-click the Dashboard, and then select Control Panels from the shortcut menu:

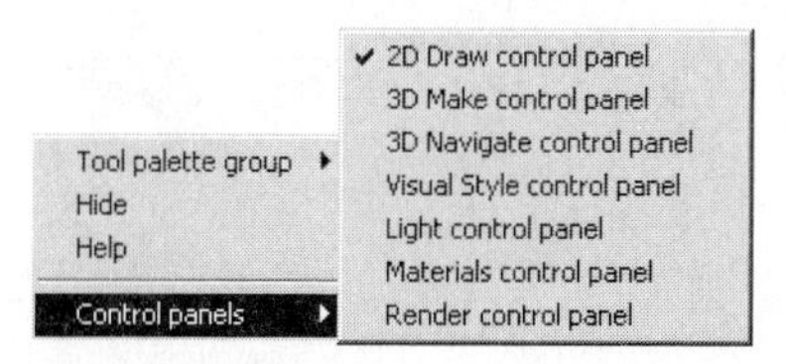

- The Dashboard cannot be customized.
- This command cannot be used transparently (within another command) until it has been used at least once non-transparently.

'DbConnect, DbClose

2000 Opens and closes the dbConnect Manager window to connect objects with rows in external database tables (short for Data Base CONNECTion).

Command	Alias	Ctrl+	F-key	Alt+	Menu Bar	Tablet
dbconnect	**dbc**	**6**	**...**	**TPD**	**Tools ↳Palettes ↳dbConnect**	**W12**
dbcclose	**...**	**6**	**...**	**TPD**	**Tools ↳Palettes ↳dbConnect**	**...**

Command: dbconnect

Displays window. (If a red × appears, the database is disconnected from the drawing.)

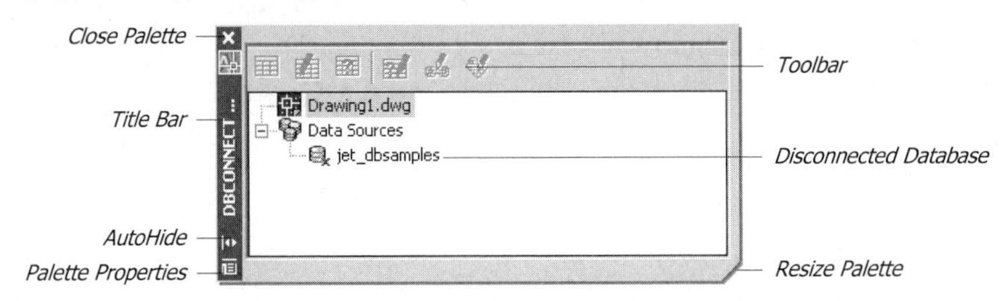

To connect the drawing with the database, right-click the database icon, and then select ***Connect****.*

The icons have the following meaning:

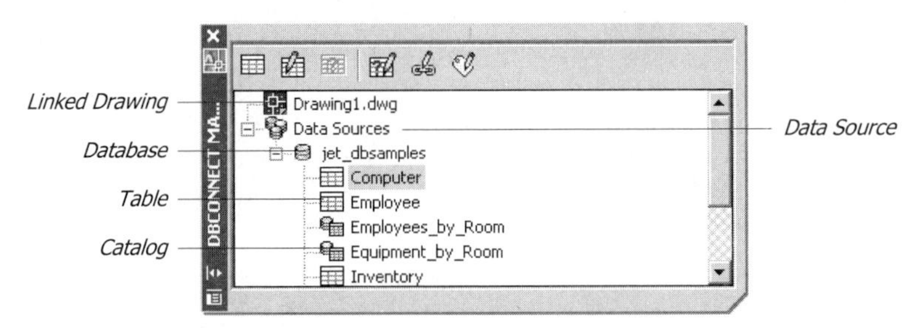

dbConnect Manager Toolbar

View Table opens an external database table in *read-only* mode; select a table, link template, or label template to make this button available.

Edit Table opens an external database table in *edit* mode; select a table, link template, or label template to make this button available.

Execute Query executes a query; select a previously-defined query to make this button available.

New Query displays the New Query dialog box when a table or link template is selected; displays the Query Editor when a query is selected.

New Link Template displays the New Link Template dialog box when a table is selected; displays the Link Template dialog box when a link template is selected; not available for link templates with links already defined in a drawing.

New Label Template displays the New Label Template dialog box when a table or link template is selected; displays the Label Template dialog box when a label template is selected.

VIEW TABLE AND EDIT TABLE WINDOWS

*The **View Table** and **Edit Table** windows are identical, with the exception that **View Table** is read-only; hence all text in columns is grayed-out.*

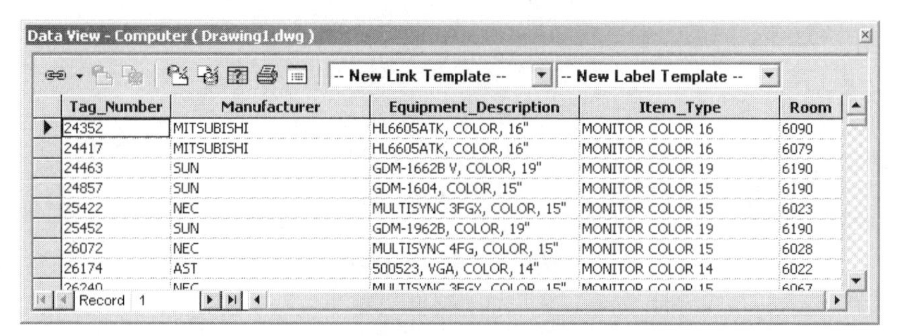

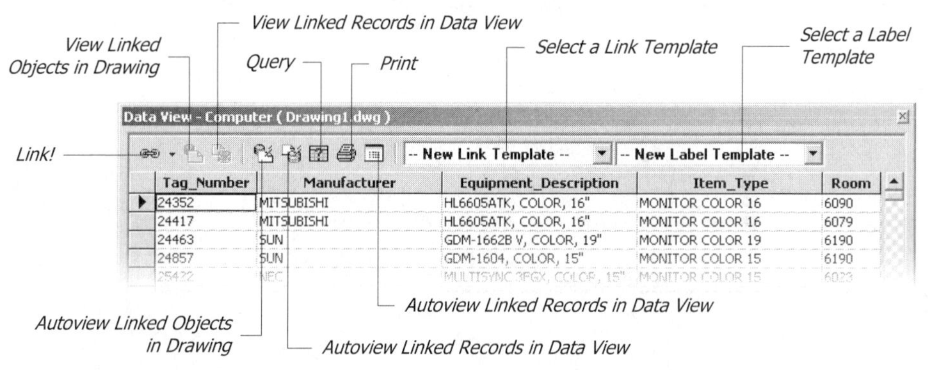

Link! creates a link or a label; click the drop list to select an option:

- **Create Links** turns on link creation mode.
- **Create Attached Labels** turns on the attached label creation mode.
- **Create Freestanding Labels** turns on freestanding label creation mode.

View Link Objects in Drawing highlights objects linked to selected records.

View Linked Records in Data View highlights records that are linked to selected objects in the drawing.

AutoView Linked Objects in Drawing automatically highlights objects that are linked to selected records.

AutoView Linked Records in Data View highlights records linked to selected objects in the drawing.

Query displays the New Query dialog box.

Print Data View prints the data in this window.

Data View and Query Options displays the Data View and Query Options dialog box.

Select a Link Template lists the names of previously-defined link templates.

Select a Label Template lists the names of previously-defined label templates.

New Query dialog box

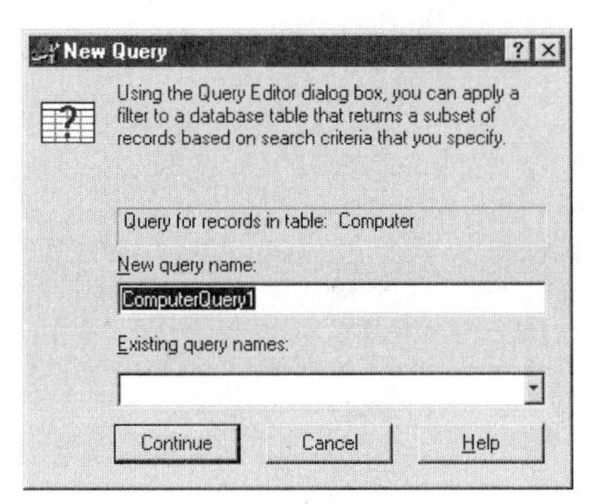

New query name specifies the name of the query.

Existing query names uses an existing query.

Continue displays the Query Editor dialog box.

QUERY EDITOR DIALOG BOX

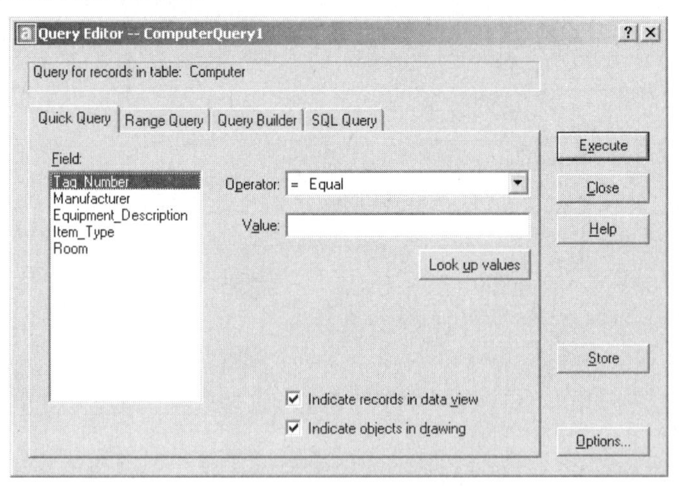

Execute executes the query.

Close closes the dialog box.

Store saves the settings.

Options displays the Data View and Query Options dialog box.

Quick Query Tab

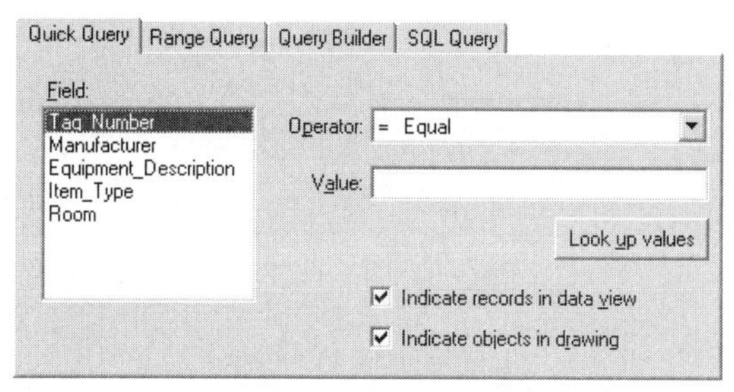

Field selects a field name.

Operator specifies a conditional operator:

Operator	Meaning
=	**Equal** — Exactly match the value (default).
<>	**Not equal** — Does not match the value.
>	**Greater than** — Greater than the value.
<	**Less than** — Less than the value.
>=	**Greater than or equal** — Greater than or equal to the value.
<=	**Less than or equal** — Less than or equal to the value.
Like	Contains the value; use the **%** wild-card character (equivalent to * in DOS).
In	Matches two values separated by a comma.
Is null	Does not have a value; used for locating records that are missing data.
Is not null	Has a value; used for excluding records that are missing data.

Value specifies the value for which to search.

Look up values displays a list of existing values:

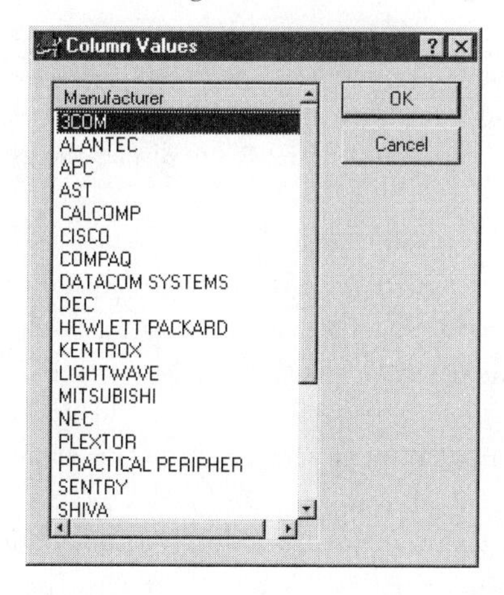

Range Query Tab

Field lists the names of fields in the current table.

From specifies the first value of the range.

Look Up Values displays the Column Values dialog box.

Through specifies the second value of the range.

☑ **Indicate Records in Data View** highlights records that match the search criteria in the Data View window.

☑ **Indicate Objects in Drawing** highlights objects that match the search criteria in drawings.

Query Builder Tab

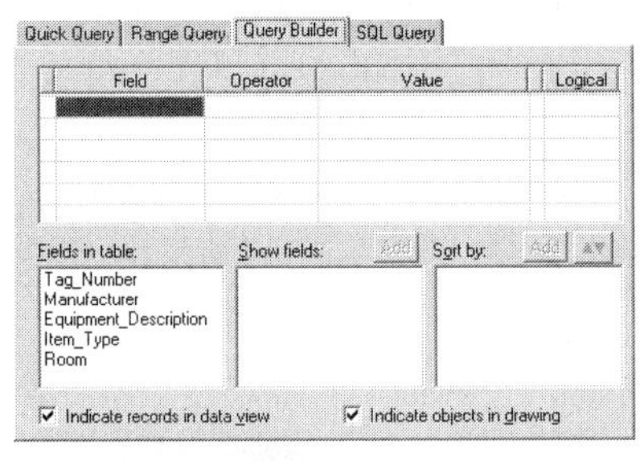

(specifies an opening parenthesis, which groups search criteria with parentheses; up to four sets can be nested. **)** specifies a closing parenthesis.

Field specifies a field name; double-click the cell to display the list of fields in the current table.

Operator specifies a logical operator; double-click to display the list of operators.

Value specifies a value for the query; click **...** to display a list of current values.

Logical specifies an And or Or operator; click once to add And; click again to change to Or.

Fields in table displays the fields in the current table; when no fields are selected, the query displays all fields from the table; double-click a field to add it to the list.

Show fields specifies the fields displayed by the Data View window; drag the field out of the list to remove it.

Add adds a field from the Fields in Table list to the Show Fields list.

Sort By specifies the sort order: the first field is the primary sort; to change the sort order, drag the field to another location in the list; press DELETE to remove a field from the list.

Add adds a field from the Fields in Table list to the Sort By list (default = ascending).

▲▼ reverses sort order.

☑ **Indicate Records in Data View** highlights records that match the search criteria in the Data View window.

☑ **Indicate Objects in Drawing** highlights objects that match the search criteria in drawings.

SQL Query Tab

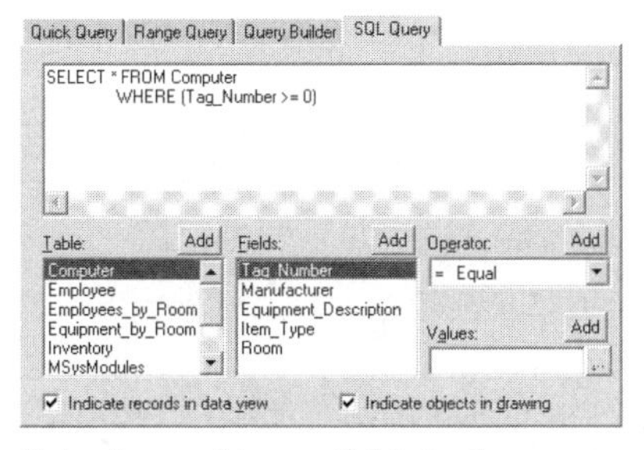

Table lists the names of all database tables available in the current data source.

Add adds the selected table to the SQL text editor.

Fields displays a list of field names in the selected database table.

Add adds the selected field to the SQL text editor.

Operator specifies the logical operator, which is added to the query (default = Equal).

Add adds the selected operator to the SQL text editor.

Values specifies a value for the selected field.

Add adds the value to the SQL text editor.

... lists available values for the field.

☑ **Indicate Records in Data View** highlights records that match the search criteria in the Data View window.

☑ **Indicate Objects in Drawing** highlights objects that match the search criteria in drawings.

DATA VIEW AND QUERY OPTIONS DIALOG BOX

Opened by the Query Editor dialog box:

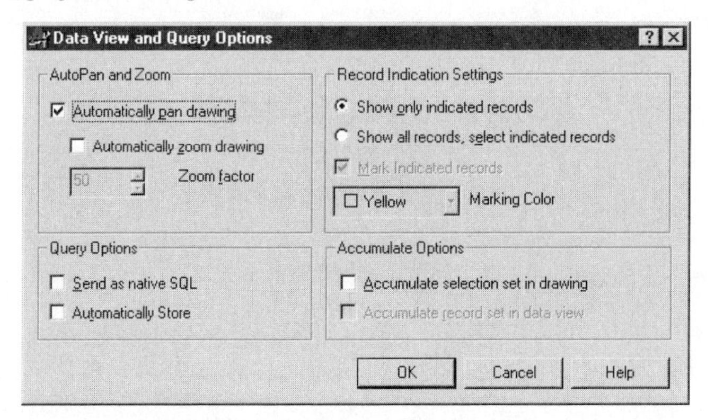

AutoPan and Zoom

☑ **Automatically Pan Drawing** causes AutoCAD to pan the drawing automatically to display associated objects.

☐ **Automatically Zoom Drawing** causes AutoCAD to zoom the drawing automatically to display associated objects.

Zoom Factor specifies the zoom factor as a percentage of the viewport area:

Zoom Factor	Meaning
20	Minimum.
50	Default.
90	Maximum.

Query Options

Send as Native SQL makes queries to database tables in:

☑ The format of the source table.

☐ SQL 92 format.

☐ **Automatically Store** automatically stores queries when they are executed (default = off).

Record Indication Settings:

⊙ **Show Only Indicated Records** displays the records associated with the current AutoCAD selections in the Data View window (default).

○ **Show All Records, Select Indicated Records** displays all records in the current database table.

☑ **Mark Indicated Records** colors linked records to differentiate them from unlinked records.

Marking Color specifies the marking color (default = yellow).

Accumulate Options

Accumulate Selection Set in Drawing

☑ Adds objects to the selection set as data view records are added.

☐ Replaces the selection set each time data view records are selected.

Accumulate Record Set in Data View

☑ Adds records to the selection set as drawing objects are selected.

☐ Replaces the selection set each time drawing objects are selected.

New Link Template dialog box

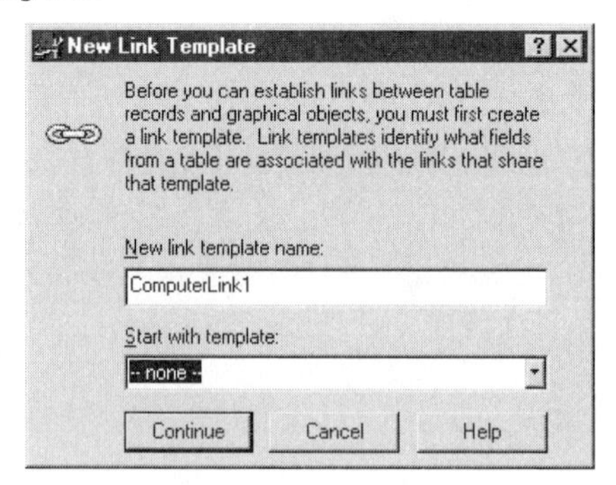

New link template name specifies the name of the link template.

Start with template reuses an existing template.

Continue displays the Link Template dialog box.

LINK TEMPLATE DIALOG BOX

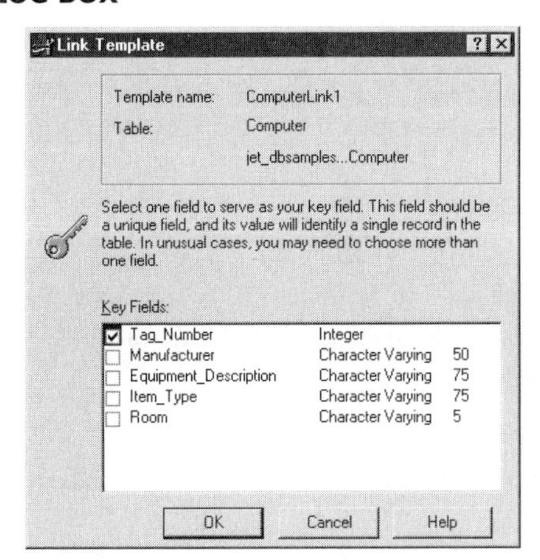

Key Fields selects one field name; you may select more than one field name, but AutoCAD warns you that too many key fields may slow performance.

NEW LABEL TEMPLATE DIALOG BOX

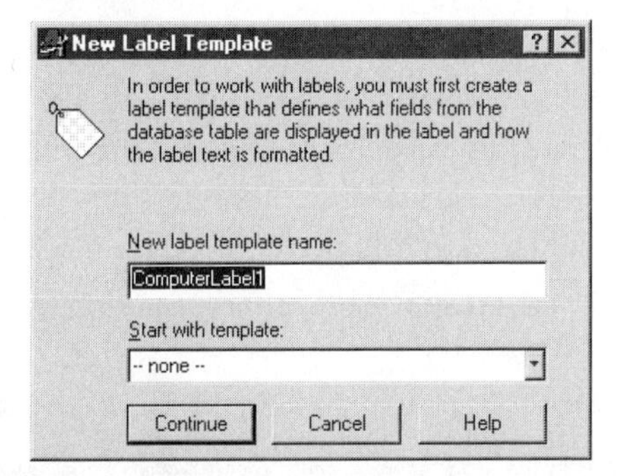

New label template name specifies the name of the label template.

Start with template reuses an existing template.

Continue displays the Label Template dialog box.

LABEL TEMPLATE DIALOG BOX

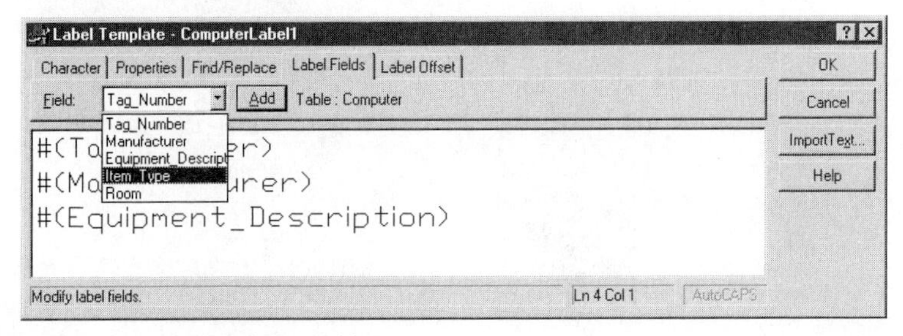

Field specifies the names of the fields.

Add adds the field to the label.

Label Offset displays the Label Offset tab:

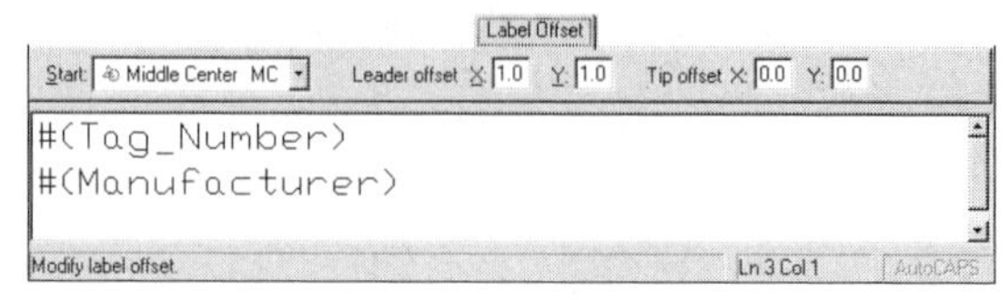

Start specifies the justification of the label starting point.

Leader offset specifies the x,y distance between the label's starting point and the leader line.

Tip offset specifies the offset distance to the leader tip or label text.

*See the **MText** command for more information.*

RELATED COMMANDS

AttDef creates an attribute definition, akin to an internal database.

EAttExt exports attributes.

TIPS

- Query searches are case sensitive: "Computer" is not the same as "computer."
- OLE DB v2.0 must be installed before you use the **dbConnect Manager**.
- Leaders must have a length; to get rid of a leader, use a freestanding label.
- SQL is short for "structured query language," a standard method of querying databases.
- The properties of a link template can only be edited if it contains no links, and if the drawing is fully (not partially) loaded.
- Before you can edit a record with an SQL Server table, you must define a *primary key*.
- This command cannot be used transparently (within another command) until it has been used at least once non-transparently.
- The figure below illustrates a database connected with the *db_samp.dwg* sample drawing:

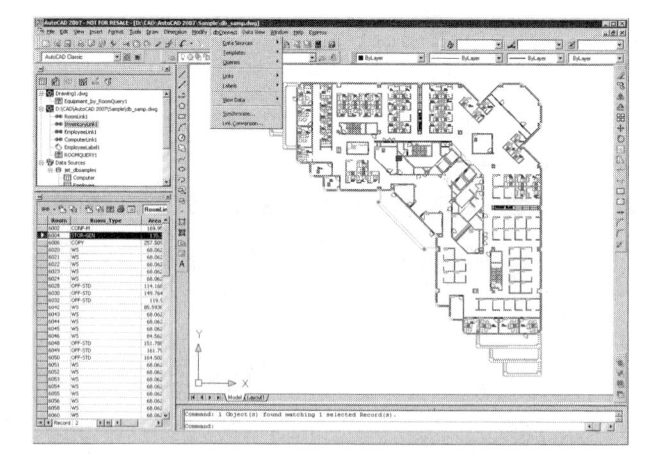

Quick Start Tutorial

Constructing Your First Query

Step 1

From the **Tools** menu, select **Palettes | dbConnect**. AutoCAD adds the **dbConnect** item to the menu and opens the dbConnect Manager palette.

Step 2

In dbConnect Manager, right-click "jet_dbsamples" and then select **Connect** from the menu.

Step 3

From the **dbConnect** menu, select **Queries | New Query on an External Table**.

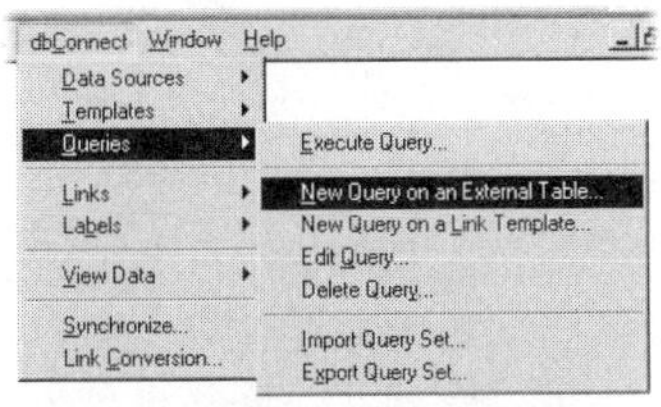

AutoCAD displays the Select Data Object dialog box.

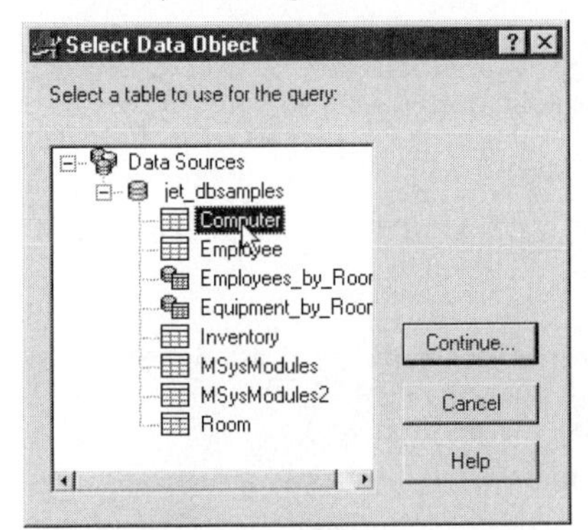

Step 4

Select a table, and then click **Continue**. AutoCAD displays the New Query dialog box.

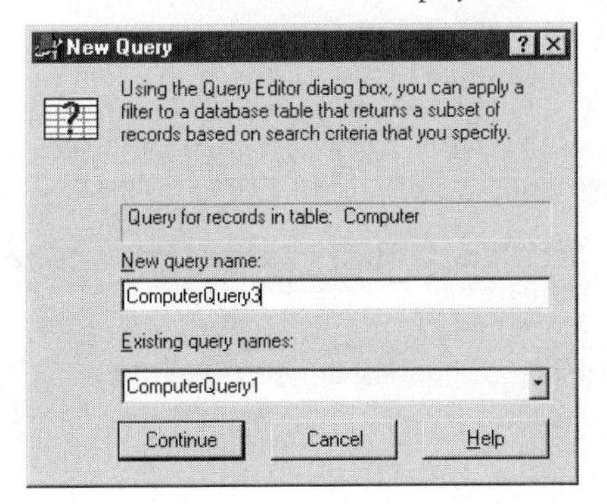

Step 5

Enter a name for the query in **New query name** text field, or select an existing name.

Step 6

Click **Continue**. AutoCAD opens the Query Editor dialog box.

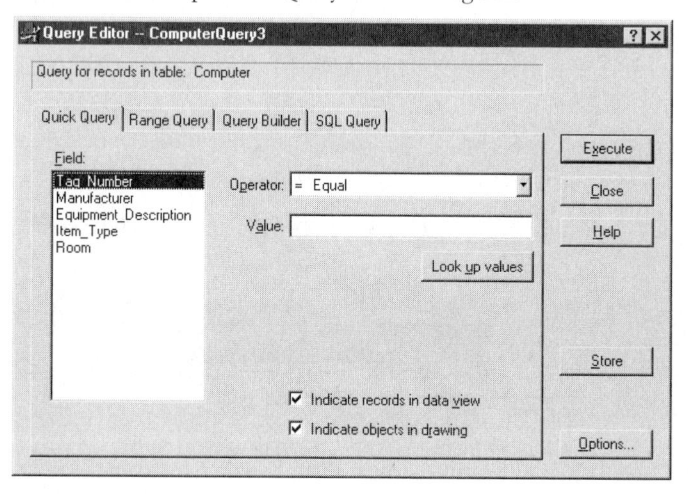

Step 7

Ensure the **Quick Query** tab is showing. Also, make sure that **Indicate records in data view** and **Indicate objects in drawing** are both checked.

Step 8

Select a field name from the **Field** list by highlighting it.

Step 9

Select an operator from the **Operator** list. For example, to match a value, select **= Equal**.

Step 10

Enter a value in the **Value** field, or click **Look up values**, and then select a value from the list of values already in the database, and then click **OK**.

Step 11

Click **Store** to save the query for reuse in the future.

Step 12

Click **Execute** to run the query. AutoCAD closes the dialog box, and then displays the records that match your selection in the **Data View** window.

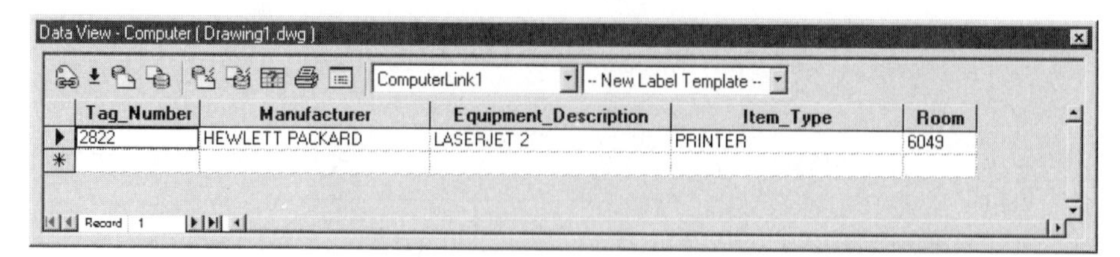

DblClkEdit

2000i Determines whether objects are edited by double-clicking the mouse button (short for DouBLe CLicK EDITing).

Command	Alias	Ctrl+	F-key	Alt+	Menu Bar	Tablet
dblclkedit	...	...	...	...	...	...

Command: dblclkedit
Enter double-click editing mode [ON/OFF] <ON>: *(Enter* **ON** *or* **OFF***.)*

COMMAND LINE OPTIONS

ON enables double-click editing.

OFF disables double-click editing.

RELATED COMMANDS

(Right-click) displays a shortcut menu with editing commands.

Properties displays the Properties palette.

TIPS

- When this command is turned on, double-clicking objects is the equivalent of entering the related editing command:

Object	Command	Dialog Box or Palette
Attribute Definition	**DdEdit**	Edit Attribute
Attribute	**EAttEdit**	Enhanced Attribute Editor
Block	**BEdit**	Block Editor
Dynamic Block	**BEdit**	Block Editor
Hatch	**HatchEdit**	Hatch Edit
Image	**ImageAdjust**	Image Adjust
Mline	**MlEdit**	Multiline Edit Tools
Mtext	**MtEdit**	Multiline Text Editor
Polyline	**PEdit**	*Command-line prompts.*
Section	**LiveSection**	...
Spline	**SplinEdit**	*Command-line prompts.*
Table Cell	**TablEdit**	Multiline Text Editor
Text	**DdEdit**	Edit Text
Viewport	**VpMax**	...
Xref	**RefEdit**	Reference Edit

- For other objects, double-clicking them displays the Properties palette, and highlights them with grips.
- The CUI command's Double-click Edit section allows you to customize the double-click action for any object type (*new to AutoCAD 2007*).
- In the User Preferences tab of the Options command, you can toggle (turn on or off) double-click editing.

DbList

V. 1.0 Lists information on all objects in the drawing (short for Data Base LISTing).

Command	Alias	Ctrl+	F-key	Alt+	Menu Bar	Tablet
dblist	...	...	...	...	...	...

Command: dblist

Sample listing:

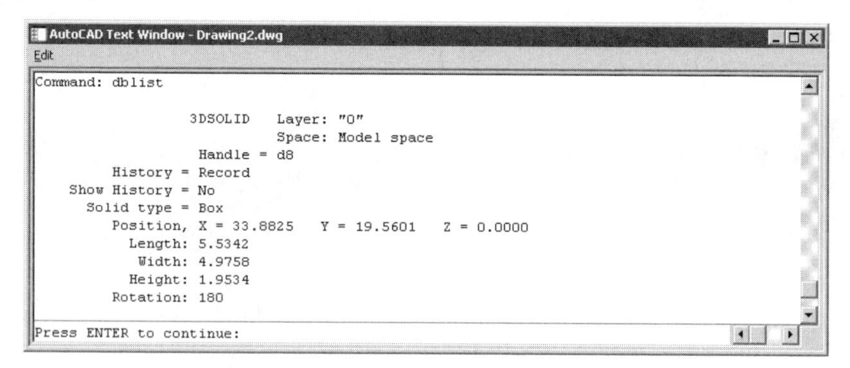

COMMAND LINE OPTIONS

ENTER continues display after pausing.

ESC cancels database listing.

RELATED COMMANDS

Area lists the area and perimeter of objects.

Dist lists the 3D distance and angle between two points.

Id lists the 3D coordinates of a point.

List lists information about selected objects in the drawing.

TIP

- This command is typically used to debug, and has little application for most users.

Renamed Commands

DdAttDef was replaced by **AttDef** in AutoCAD 2000.

DdAttE was replaced by **AttEdit**. in AutoCAD 2000.

DdAttExt was replaced by **AttExt** in AutoCAD 2000.

DdChProp was replaced by **Properties** in AutoCAD 2000.

DdColor was rreplaced by **Color** in AutoCAD 2000.

DdEdit

Rel.11 Edits a single-line text, multiline text blocks, attribute values, and geometric tolerances (short for Dynamic Dialog EDITor).

Command	Alias	Ctrl+	F-key	Alt+	Menu Bar	Tablet
ddedit	**ed**	...	...	**MOTE**	**Modify ↳Object ↳Text ↳Edit**	**Y21**

Command: ddedit

Select an annotation object or [Undo]: *(Select a text object, or type* **U***.)*

Select single-line text placed by the Text command, and the direct editing field is displayed.

Select paragraph text placed by the MText command, and the in-place text editor is displayed.

Select an attribute definition (not part of a block definition), and the Edit Attribute Definition dialog is displayed.

Select a block with attributes, and the Enhanced Attribute Editor dialog box is displayed.

Select geometric tolerances, and the Geometric Tolerance dialog box is displayed.

Select an annotation object or [Undo]: *(Press* ESC *to exit command.)*

COMMAND LINE OPTIONS

Undo undoes editing.

ESC ends the command.

SHORTCUT MENU

Right-clicking the text displays the shortcut menu:

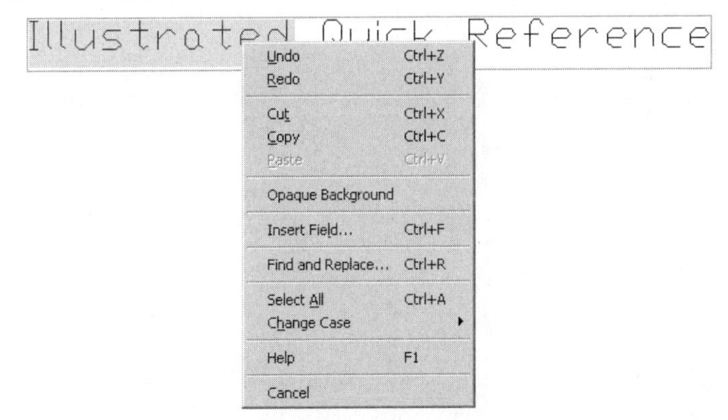

Undo undoes the last action.

Redo redoes the last undo.

Cut copies the selected text to the Clipboard, and then erases it.

Copy copies the selected text to the Clipboard.

Paste pastes text from the Clipboard; unavailable if Clipboard contains no text.

Opaque Background changes the background color to gray.

Insert Field displays the Field dialog box; see the Field command.

Find and Replace displays the Find and Replace dialog box; see the Find command.

Select All selects all text.

Change Case changes the selected text between all UPPERCASE or all lowercase.

Help displays online help for this command.

Cancel ignores editing changes for the current annotation object, and dismisses the dialog box.

RELATED COMMANDS

EAttEdit edits all text attributes connected with a block.

Field places automatically-updatable field text.

Find searches for text in drawings.

MtEdit edits paragraph text.

Properties edits all text *properties*, including the text itself.

Spell checks the spelling of words in drawings.

RELATED SYSTEM VARIABLES

DTextEd toggles between direct in-place editing and the old Edit Text dialog box.

FontAlt specifies the name of the font to be used when the required font file cannot be located by AutoCAD.

FontMap maps fonts.

MirrText determines how the Mirror command affects text.

TextFill toggles the fill of TrueType fonts during plotting and rendering.

TIPS

- **DdEdit** automatically repeats; press ESC to cancel the command.
- Between AutoCAD 2000 and 2004, this command did edit attribute text.
- As an alternative to this command, double-click text to bring up the appropriate editor.
- AutoCAD 2006 changed this command from a dialog box format to direct editing in drawings.
- Text placed with the Field, Leader, and QLeader is mtext; double-click the text to edit.
- Text can also be edited through the Properties window.

Renamed Commands

DdGrips was replaced by the **Selection** tab of the **Options** command in AutoCAD 2000.

DDim was rreplaced by **DimStyle** in AutoCAD 2000.

DdInsert was replaced by **Insert** in AutoCAD 2000.

DdModify was was replaced by **Properties**. in AutoCAD 2000.

'DdPtype

Rel.12 Sets the style and size of points (short for Dynamic Dialog Point TYPE).

Command	Alias	Ctrl+	F-key	Alt+	Menu Bar	Tablet
ddptype	...	...	...	**OP**	**Format** **↳Point Style**	**U1**

Command: ddptype

Displays dialog box:

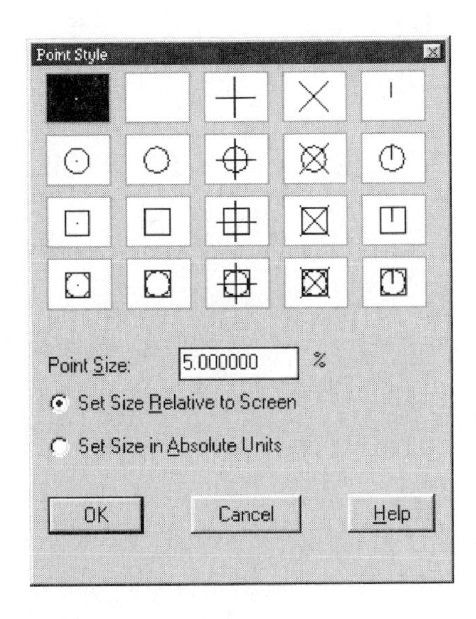

POINT STYLE DIALOG BOX

Point size sets the size in percentage or pixels.

⊙ **Set Size Relative to Screen** sets the size as a percentage of the total viewport height.

○ **Set Size in Absolute Units** sets the size in drawing units.

RELATED COMMANDS

Divide draws points along an object at equally-divided lengths.

Point draws points.

Measure draws points a measured distance along an object.

Regen displays the new point format with a regeneration.

RELATED SYSTEM VARIABLES

PdSize specifies the size of points:

0 — Point is 5% of viewport height (default).

positive — Absolute size in drawing units.

negative — Percentage of the viewport size.

PdMode determines the look of points:

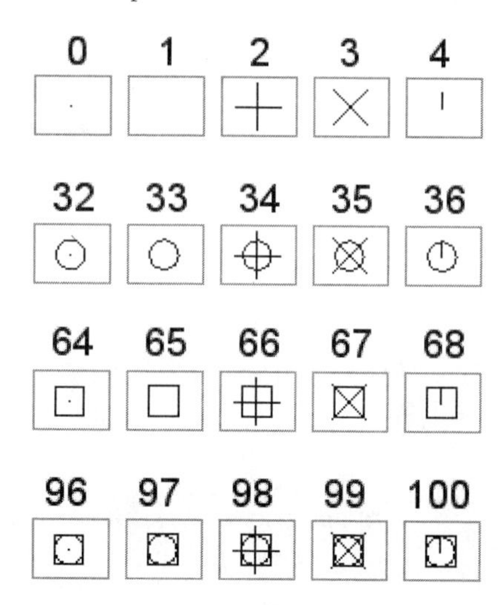

TIPS

- Points often cannot be seen in the drawing. To make them visible, change their mode and size.
- The two system variables listed above affect *all* points in the drawing.

Removed Commands

DdEModes was removed from AutoCAD Release 14; it was replaced by the **Object Properties** toolbar.

DdLModes was removed from AutoCAD 2000; it was replaced by **Layer**.

DdLtype was removed from AutoCAD 2000; it was replaced by **Linetype**.

DdRename was removed from AutoCAD 2000; it was replaced by **Rename**.

DdRModes was removed from AutoCAD 2000; it was replaced by **DSettings**.

DdSelect was removed from AutoCAD 2000; it was replaced by the **Selection** tab of the **Options** command.

DdUcs and **DdUcsP** were removed from AutoCAD 2000; they were replaced by **UcsMan**.

DdUnits was removed from AutoCAD 2000; it was replaced by **Units**.

DdView was removed from AutoCAD 2000; it was replaced by **View**.

DdVPoint

Rel.12 Changes the 3D viewpoint through a dialog box (short for Dynamic Dialog ViewPOINT).

Command	Alias	Ctrl+	F-key	Alt+	Menu Bar	Tablet
ddvpoint	**vp**	...	...	**V3I**	**View** ↳**3D Views** ↳**Viewpoint Presets**	**N5**

Command: ddvpoint

Displays dialog box:

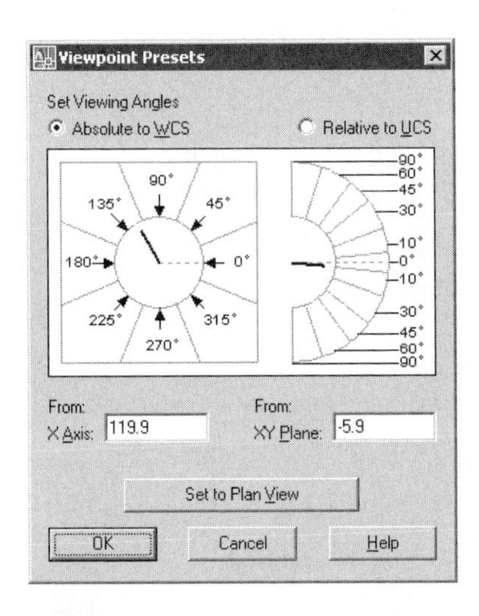

VIEWPOINT PRESETS DIALOG BOX

Set Viewing Angles options

⊙ **Absolute to WCS** sets the view direction relative to the WCS.

○ **Relative to UCS** sets the view direction relative to the current UCS.

From X Axis measures the view angle from the x axis.

From XY Plane measures the view angle from the x,y plane.

Set to Plan View changes the view to plan view in the specified UCS.

RELATED COMMANDS

VPoint adjusts the viewpoint from the command line.

3dOrbit changes the 3D viewpoint interactively.

RELATED SYSTEM VARIABLE

WorldView determines whether viewpoint coordinates are in WCS (1) or UCS (0).

TIPS

- After changing the viewpoint, AutoCAD performs an automatic zoom extents.
- In the image tile shown below, the black arm indicates the new angle.
- In the image tile, a second arm indicates the current angle.
- To select an angle with your mouse:

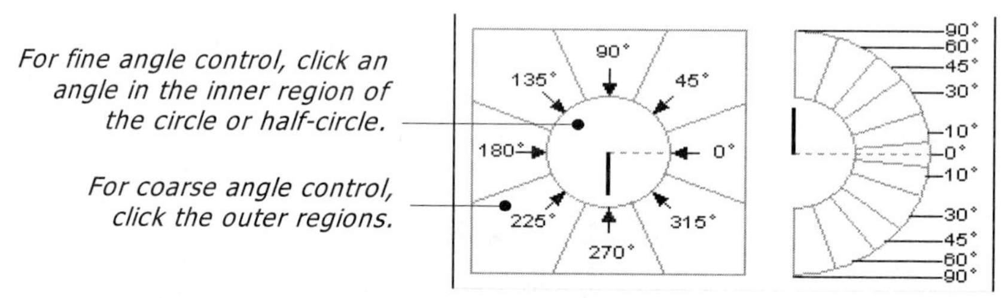

- WCS is short for "world coordinate system."
- UCS is short for "user-defined coordinate system."
- As an alternative, use the 3dOrbit command to set the 3D viewpoint.
- This command is not permitted in paper space (layout mode).

'Delay

v. 1.4 Delays the next command, in milliseconds.

Command	Alias	Ctrl+	F-key	Alt+	Menu Bar	Tablet
delay	...	...	...	...	...	...

Command: delay
Delay time in milliseconds: *(Enter the number of milliseconds.)*

COMMAND LINE OPTION

Delay time specifies the number of milliseconds by which to delay the next command.

RELATED COMMAND

Script initiates scripts.

TIPS

- Use **Delay** to slow down the execution of a script file.
- The maximum delay is 32767, just over 32 seconds.

DetachURL

Rel.14 Removes URLs from objects and areas.

Command	Alias	Ctrl+	F-key	Alt+	Menu Bar	Tablet
detachurl	...	...	...	...	...	...

Command: detachurl
Select objects: *(Select one or more objects.)*
Select objects: *(Press Enter .)*

COMMAND LINE OPTION

Select objects selects the objects from which to remove URL(s).

RELATED COMMANDS

AttachUrl attaches a hyperlink to an object or an area.

Hyperlink attaches and removes hyperlinks via a dialog box.

SelectUrl selects all objects with attached hyperlinks.

TIPS

- When you select a hyperlinked area to detach, AutoCAD reports:

```
1. hyperlink ()
Remove, deleting the Area.
1 hyperlink deleted...
```

- When you select an object with no hyperlink attached, AutoCAD reports nothing.
- A URL (short for "universal resource locator") is the universal file naming convention of the Internet; also called a link or hyperlink.

Dim

V. 1.2 Changes the prompt from 'Command' to 'Dim', allowing access to AutoCAD's original dimensioning mode (short for DIMensions).

Command	Alias	Ctrl+	F-key	Alt+	Menu Bar	Tablet
dim	...	...	...	...	...	...

Command: dim
Dim: *(Enter a dimension command from the list below.)*

COMMAND LINE OPTIONS

Aliases for the dimension commands are shown in UPPERCASE letters, such as "al" for ALigned.

ALigned draws linear dimensions aligned with objects (*first introduced with AutoCAD version 2.0*); replaced by DimAligned.

ANgular draws angular dimensions that measure angles (*ver. 2.0*); replaced by DimAngular.

Baseline continues dimensions from base points (*ver. 1.2*); replaced by QDim and DimBaseline.

CEnter draws '+' center marks on circles' and arcs' centers (*ver. 2.0*); replaced by DimCenter.

COntinue continues dimensions from previous dimensions' extension lines (*ver. 1.2*); replaced by the QDim and DimContinue commands.

Diameter draws diameter dimensions on circles, arcs, and polyarcs (*ver. 2.0*); replaced by the QDim and DimDiameter commands.

Exit returns to 'Command' prompt from 'Dim' prompt (*ver. 1.2*).

HOMetext returns dimension text to its original position (*ver. 2.6*); replaced by the DimEdit command's Home option.

HORizontal draws horizontal dimensions (*ver. 1.2*); replaced by DimLinear.

LEAder draws leaders (*ver. 2.0*); replaced by the Leader and QLeader commands.

Newtext edits text in associative dimensions (*ver. 2.6*); replaced by the DimEdit command's New option.

OBlique changes the angle of extension lines in associative dimensions (*rel. 11*); replaced by the DimEdit command's Oblique option.

ORdinate draws x- and y-ordinate dimensions (*Rel. 11*); replaced by QDim and DimOrdinate.

OVerride overrides current dimension variables (*Rel. 11*); replaced by DimOverride.

RAdius draws radial dimensions on circles, arcs, and polyline arcs (*ver. 2.0*); replaced by the QDim and DimRadius commands.

REDraw redraws the current viewport (same as 'Redraw; *ver. 2.0*).

REStore restores dimensions to the current dimension style (*Rel. 11*); replaced by the -DimStyle command's Restore option.

ROtated draws linear dimensions at any angle (*ver. 2.0*); replaced by DimLinear.

SAve saves the current settings of dimension styles (*Rel. 11*); replaced by the -DimStyle command's Save option.

STAtus lists the current settings of dimension variables (*ver. 2.0*); replaced by the -DimStyle command's Status option.

STYle defines styles for dimensions (*ver. 2.5*); replaced by DimStyle.

TEdit changes the location and orientation of text in associative dimensions (*Rel. 11*); replaced by DimTEdit.

TRotate changes the rotation angle of text in associative dimensions (*Rel. 11*); replaced by the DimTEdit command's Rotate option.

Undo undoes the last dimension action (*ver. 2.0*); replaced by the Undo command.

UPdate updates selected associative dimensions to the current dimvar setting (*ver. 2.6*); replaced by the -DimStyle command's Apply option.

VAriables lists the values of variables associated with dimension styles, *not* dimvars (*Rel. 11*); replaced by the -DimStyle command's Variables option.

VErtical draws vertical linear dimensions (*ver. 1.2*); replaced by DimLinear.

RELATED DIM VARIABLES

Dim*xxx* specifies system variables for dimensions; see the DimStyle command.

DimAso determines whether dimensions are drawn associatively.

DimScale determines the dimension scale.

TIPS

- Most dimensions consist of the basic components illustrated below:

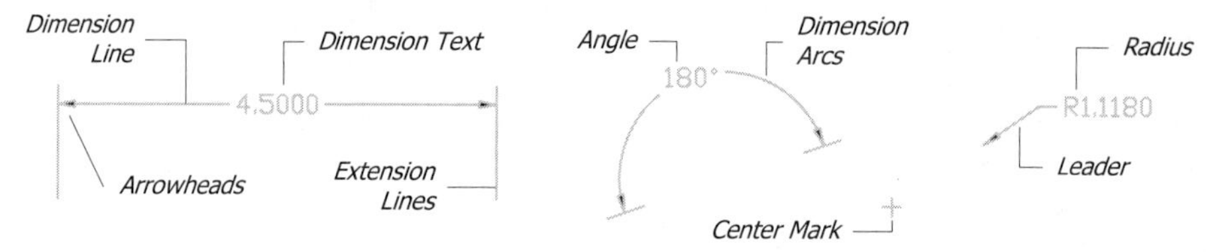

- The 'Dim' prompt dimension commands are included for compatibility with AutoCAD Release 12 and earlier. Only transparent commands and dimension commands work at the 'Dim' prompt. To use other commands, you must exit the 'Dim' prompt with the Exit command or by pressing ESC, which returns you to the 'Command' prompt.

- *Defpoints* (short for "definition points") are used by earlier AutoCADs to locate extension lines. Defpoints appear as small dots on the DefPoints layer. When stretching dimensions, ensure defpoints are included; otherwise the dimensions are not updated automatically.

- The DefPoints layer *never* plots, making it useful for objects and notes that should never plot and for great practical jokes.

- As of AutoCAD 2002, defpoints are not used when DimAssoc = 2 (default); dimensions are instead attached directly to objects.

- Dimension text can have a colored background, which is set globally through the DimStyle command, or overridden locally with the Properties command.

- All the components of an *associative dimension* are treated as a single object; components of a nonassociative dimension are treated as individual objects.

- For additional control over dimensions, the following undocumented commands are described earlier in this book: AiDimFlipArrow, AiDimPrec, AiDimStyle, Ai_Dim_TextAbove, Ai_Dim_TextCenter, Ai_Dim_TextHome, and AiDimTextMove.

Dim1

Displays the 'Dim' prompt for a single dimensioning command, and then returns to the 'Command' prompt (short for DIMension once).

Command	Alias	Ctrl+	F-key	Alt+	Menu Bar	Tablet
dim1	...	...	...	...	...	...

Command: dim1
Dim: *(Enter a dimension command.)*

COMMAND LINE OPTION

Accepts all "original" dimension commands; see the ***Dim*** *command for the complete list.*

RELATED COMMANDS

DimStyle displays the dialog box for setting dimension variables.

Dim switches to AutoCAD's original dimensioning mode and remains there.

RELATED DIM VARIABLES

DimAso determines whether dimensions are drawn associatively.

DimTxt determines the height of text.

DimScale determines the dimension scale.

TIPS

- Use **Dim1** when you require just a single original dimension command.
- This command is rarely used.

DimAligned

V. 2.5 Draws linear dimensions aligned with objects.

Command	Aliases	Ctrl+	F-key	Alt+	Menu Bar	Tablet
dimaligned	dal dimali	...	...	NG	Dimension ↳Aligned	W4

Command: dimaligned

Specify first extension line origin or <select object>: *(Pick a point, or press* **ENTER** *to enable object selection, as shown by the following prompt.)*

Select object to dimension: *(Select an object.)*

Specify dimension line location or [Mtext/Text/Angle]: *(Pick a point, or enter an option.)*

Dimension text = *nnn*

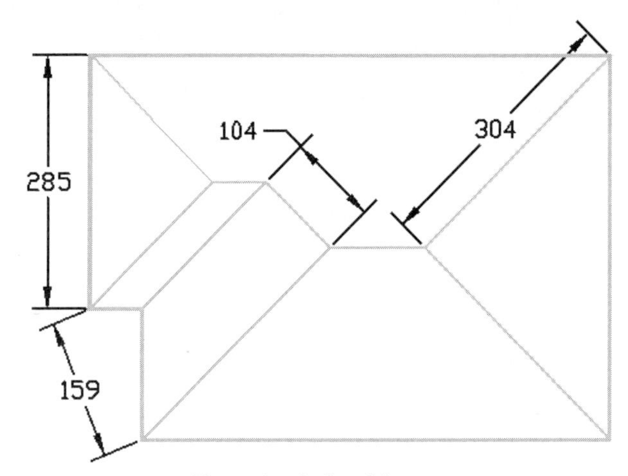

Examples of aligned dimensions.

COMMAND LINE OPTIONS

Specify first extension line origin picks a point for the origin of the first extension line.

Specify second extension line origin picks a point for the origin of the second extension line.

Select object selects an object to dimension, after pressing **ENTER**.

Select object to dimension picks a line, circle, arc, polyline, or explodable object; individual segments of polylines are dimensioned.

Specify dimension line location picks a point from which to locate the dimension line and text.

Mtext changes the wording of the dimension text.

Text changes the position of the dimension text.

Angle changes the angle of the dimension text.

RELATED DIM COMMAND

DimRotated draws rotated dimension lines with perpendicular extension lines.

DimAngular

Rel 13 Draws dimensions that measure angles.

Command	Aliases	Ctrl+	F-key	Alt+	Menu Bar	Tablet
dimangular	dan	...	...	NA	Dimension	X3
	dimang				↳Angular	

Command: dimangular

Select arc, circle, line, or <specify vertex>: *(Select an object, or pick a vertex.)*

Specify second angle endpoint: *(Pick a point.)*

Specify dimension arc line location or [Mtext/Text/Angle]: *(Pick a point, or enter an option.)*

Dimension text = *nnn*

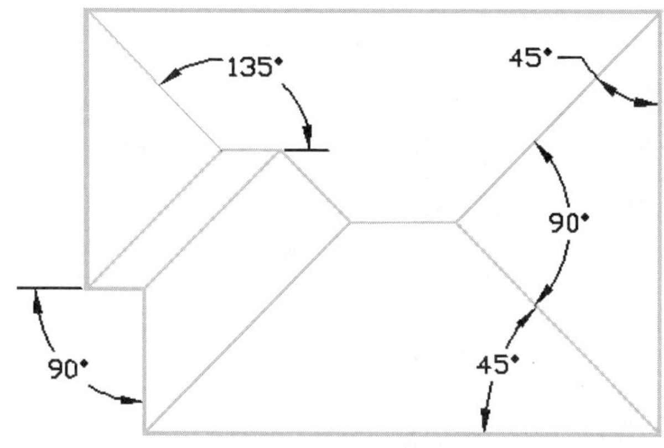

Examples of angular dimensions.

COMMAND LINE OPTIONS

Select arc measures the angle of the arc.

Circle prompts you to pick two points on the circle.

Line prompts you to pick two lines.

Specify vertex prompts you to pick points to make an angle.

Specify dimension arc/line location specifies the location of the angular dimension.

Mtext changes the wording of the dimension text.

Text changes the position of the text.

Angle changes the angle of the dimension text.

RELATED DIM COMMANDS

DimArc dimensions the lengths of arcs.

DimCenter places a center mark at the center of an arc or circle.

DimRadius dimensions the radius of an arc or circle.

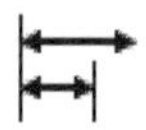

DimBaseline

Rel 13 Draws linear dimensions based on previous starting points.

Command	Aliases	Ctrl+	F-key	Alt+	Menu Bar	Tablet
dimbaseline	**dba**	...	...	**NB**	**Dimension**	...
	dimbase				**↳Baseline**	

Command: dimbaseline

Specify a second extension line origin or [Undo/Select] <Select>: *(Pick a point, or enter an option.)*

Dimension text = *nnn*

Specify a second extension line origin or [Undo/Select] <Select>: *(Pick a point, enter an option, or press* ESC *to exit command.)*

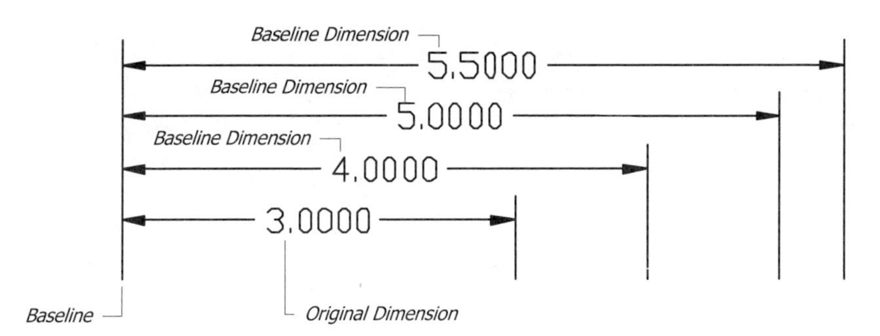

COMMAND LINE OPTIONS

Specify a second extension line origin positions the extension line of the next baseline dimension.

Select prompts you to select the base dimension.

Undo undoes the previous baseline dimension.

ESC exits the command.

RELATED DIM COMMANDS

Continue continues linear dimensioning from the last extension point.

QDim creates continuous or baseline dimensions quickly.

RELATED DIM VARIABLES

DimDli specifies the distance between baseline dimension lines.

DimSe1 suppresses the first extension line.

DimSe2 suppresses the second extension line.

DimArc

2006 Dimensions the length along arcs and polyline arcs.

Command	Alias	Ctrl+	F-key	Alt+	Menu Bar	Tablet
dimarc	dar	...	...	NH	Dimension ↳Arc Length	...

Command: dimarc
Select arc or polyline arc segment: *(Select an arc.)*
Specify arc length dimension location, or [Mtext/Text/Angle/Partial]: *(Pick a point, or enter an option.)*
Dimension text = 15'-9 3/16"

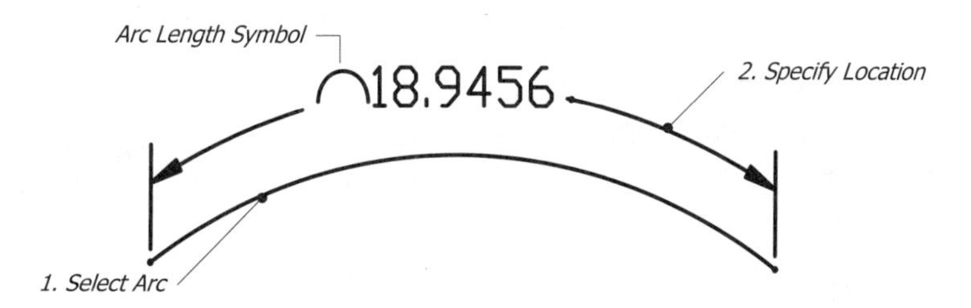

COMMAND LINE OPTIONS

Select arc measures the length of arcs.

Specify dimension arc length location specifies the location of the arc length dimension.

Mtext changes the wording of the dimension text.

Text changes the position of the text.

Angle changes the angle of the dimension text.

Partial prompts you to pick two points for a partial dimension:

Specify first point for arc length dimension: *(Pick a point.)*
Specify second point for arc length dimension: *(Pick another point.)*

RELATED SYSTEM VARIABLE

DimArcSym specifies the location of the dimension arc symbol.

TIPS

- Specify the placement of the arc length symbol in the Symbols and Arrows tab of the Dimension Style Manager. The arc symbol can be displayed above or in front of the dimension text.
- The extension lines can be orthogonal (angle less than 90 degrees) or radial (more than 90 degrees).

DimCenter

Rel 13 Draws center marks and lines on arcs and circles.

Command	Alias	Ctrl+	F-key	Alt+	Menu Bar	Tablet
dimcenter	dce	...	...	NM	Dimension ↳Center Mark	X2

Command: dimcenter
Select arc or circle: *(Select an arc, circle, or explodable object.)*

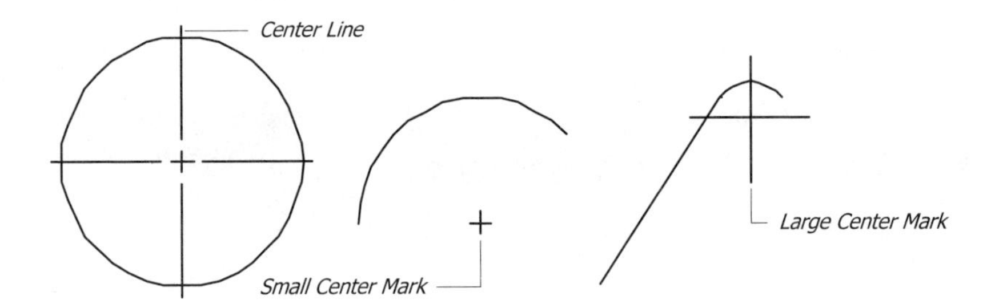

COMMAND LINE OPTION

Select arc or circle places the center mark at the center of the selected arc, circle, or polyarc.

RELATED DIM COMMANDS

DimAngular dimensions arcs and circles.

DimDiameter dimensions arcs and circles by diameter value.

DimRadius dimensions arcs and circles by radius value.

RELATED DIM VARIABLE

DimCen specifies the size and type of the center mark:

negative value — Draws center marks and lines.

0 — Does not draw center marks or center lines.

positive value — Draws center marks.

0.09 — Default value.

TIPS

- Changing the center mark size for a dimension style does not update existing center lines and marks.
- The center mark length defined by DimCen is from the center to one end of the mark; the center line length is the size of gap and extension beyond the circle or arc.

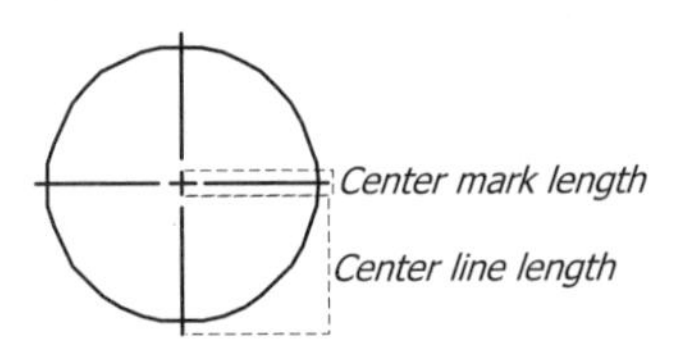

DimContinue

Rel 13 Continues dimension from the second extension line of previous dimensions.

Command	Aliases	Ctrl+	F-key	Alt+	Menu Bar	Tablet
dimcontinue	**dco**	...	...	**NC**	**Dimension**	...
	dimcont				**↳Continue**	

Command: dimcontinue
Specify a second extension line origin or [Undo/Select] <Select>: *(Pick a point, enter an option, or press* ENTER *to select a base dimension.)*
Dimension text = *nnn*
Specify a second extension line origin or [Undo/Select] <Select>: *(Pick a point, enter an option, or press* ESC *to exit command.)*

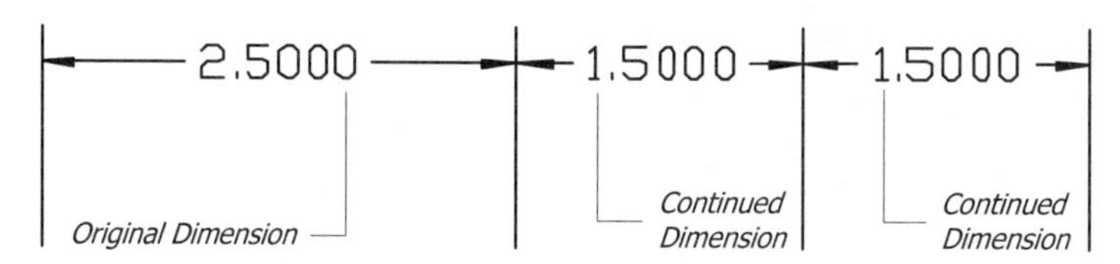

COMMAND LINE OPTIONS

Specify a second extension line origin positions the extension line of the next continued dimension.

Select prompts you to select the originating dimension.

Undo undoes the previous continued dimension.

ESC exits the command.

RELATED DIM COMMANDS

DimBaseline continues dimensioning from the first extension point.

QDim creates continuous or baseline dimensions quickly.

RELATED DIM VARIABLES

DimDli sets the distance between continuous dimension lines.

DimSe1 suppresses the first extension line.

DimSe2 suppresses the second extension line.

DimDiameter

Rel 13 Draws diameter dimensions on arcs, circles, and polyline arcs.

Command	Aliases	Ctrl+	F-key	Alt+	Menu Bar	Tablet
dimdiameter	ddi	...	...	ND	Dimension	X4
	dimdia				↳Diameter	

Command: diameter
Select arc or circle: (Select *an arc, circle, or explodable object.)*
Dimension text = *nnn*
Specify dimension line location or [Mtext/Text/Angle]: *(Pick a point, or enter an option.)*

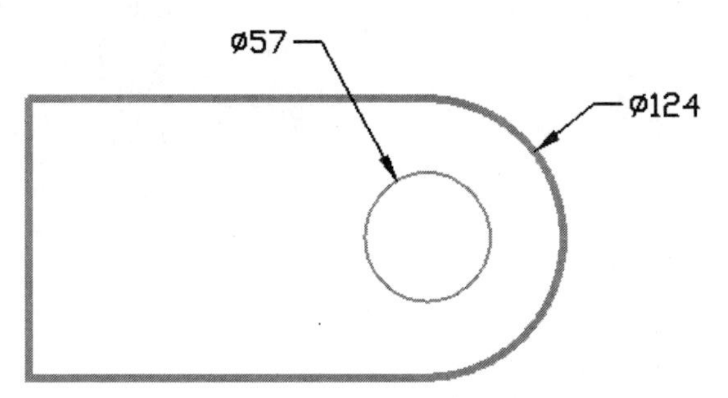

Examples of diameter dimensions.

COMMAND LINE OPTION

Select arc or circle selects an arc, circle, or polyarc.

Specify dimension line location specifies the location of the angular dimension.

Mtext changes the wording of the dimension text.

Text changes the position of the dimension text.

Angle changes the angle of the dimension text.

RELATED DIM COMMANDS

DimCenter marks the center point of arcs and circles.

DimRadius draws the radius dimension of arcs and circles.

TIP

- The diameter symbol is automatically included. But if you need to add it manually, use the **%%d** code or the Unicode **\U+2205**.

DimDisassociate

2002 Converts associative dimensions to non-associative (the command that's most difficult to spell correctly).

Command	Alias	Ctrl+	F-key	Alt+	Menu Bar	Tablet
dimdisassociate	**dda**	...	...	...	...	...

Command: dimdisassociate
Select dimensions to disassociate...
Select objects: *(Select one or more dimensions, or enter* **All** *to select all dimensions.)*
Select objects: *(Press Enter to end object selection.)*
nn **disassociated.**

COMMAND LINE OPTION

Select objects selects dimensions to convert to non-associative type.

RELATED DIM COMMANDS

DimReassociate converts dimensions from non-associative to associative.

DimRegen updates the location of associative dimensions.

RELATED SYSTEM VARIABLE

DimAssoc determines whether newly-created dimensions are associative:

0 — Dimension is created "exploded," so that all parts (such as dimension lines, arrow heads) are individual, ungrouped objects.

1 — Dimension is created as a single object, but is not associative.

2 — Dimension is created as a single object, and is associative (default).

TIPS

- As you select objects, this command ignores non-dimensions, dimensions on locked layers, and those not in the current space (model or paper).
- The command reports filtered and disassociated dimensions.
- The effects of this command can be reversed with the U and the DimReassociate commands.

DimEdit

Rel 13 Applies editing changes to dimension text.

Command	Aliases	Ctrl+	F-key	Alt+	Menu Bar	Tablet
dimedit	**ded**	...	...	**NQ**	**Dimension**	**Y1**
	dimed			↳*Enter*	↳**Oblique**	

Command: dimedit

Enter type of dimension editing [Home/New/Rotate/Oblique] <Home>: *(Enter an option, or press* **ENTER** *for Home.)*

Select objects: *(Select one or more objects.)*

Select objects: *(Press Enter .)*

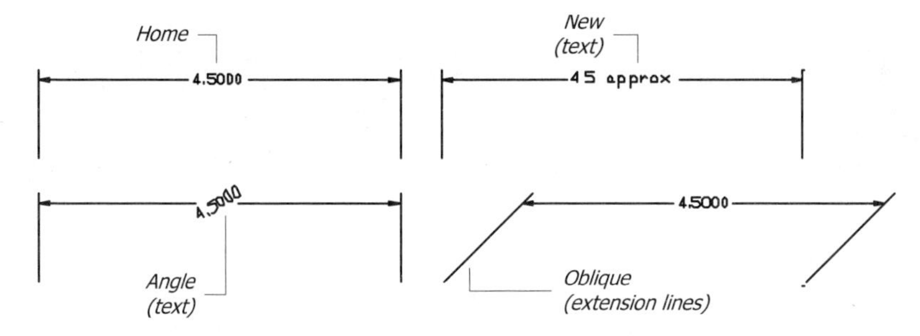

COMMAND LINE OPTIONS

Angle rotates the dimension text.

Home returns the dimension text to its original position.

Oblique rotates the extension lines.

New allows editing of the dimension text.

RELATED DIM COMMANDS

All.

RELATED DIM VARIABLES

Most.

TIPS

- DimEdit operates differently when used from the command line than from the menu bar: Oblique is the default option when selected from the menu bar.
- When you enter dimension text with the DimEdit command's New option, AutoCAD recognizes <> as *metacharacters* representing existing text.
- Use the Oblique option to angle dimension lines by 30 degrees, suitable for isometric drawings; use the Style command to oblique text by the same angle. See the Isometric command.

DimHorizontal

2004 Draws horizontal dimensions (an undocumented command).

Command	Aliases	Ctrl+	F-key	Alt+	Menu Bar	Tablet
dimhorizontal	...	...	...	...	...	...

Command: dimhorizontal

Specify first extension line origin or <select object>: *(Pick a point, or press Enter to select one object.)*

Specify second extension line origin: *(Pick a point.)*

Specify dimension line location or [Mtext/Text/Angle]: *(Pick a point, or select an option.)*

Dimension text = *nnn*

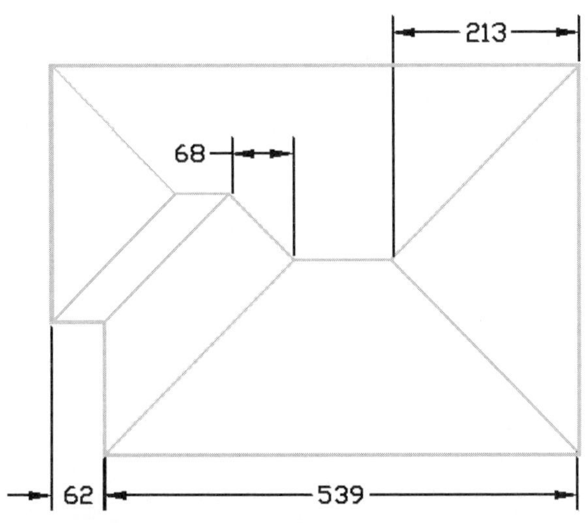

Examples of horizontal dimensions.

COMMAND LINE OPTIONS

Specify first extension line origin specifies the location of the first extension line's origin.

Select object dimensions a line, arc, or circle automatically.

Specify second extension line location specifies the location of the second extension line.

Specify dimension line location specifies the location of the dimension line.

Mtext displays the Text Formatting bar, which allows you to modify the dimension text; see the **MText** command.

Text prompts you to replace the dimension text on the command line: 'Enter dimension text <*nnn*>'.

Angle changes the angle of the dimension text: 'Specify angle of dimension text:'.

RELATED DIM COMMANDS

All.

RELATED DIM VARIABLES

Most.

DimJogged

2006 Places jogged radial dimensions; used for very large radii.

Command	Aliases	Ctrl+	F-key	Alt+	Menu Bar	Tablet
dimjogged	jog djo	...	...	NJ	Dimension ↳Jogged	...

Command: dimjogged
Select arc or circle: *(Pick an arc or a circle.)*
Specify center location override: *(Pick point 2.)*
Dimension text = 28.7205
Specify dimension line location or [Mtext/Text/Angle]: *(Pick point 3.)*
Specify jog location: *(Pick point 4.)*

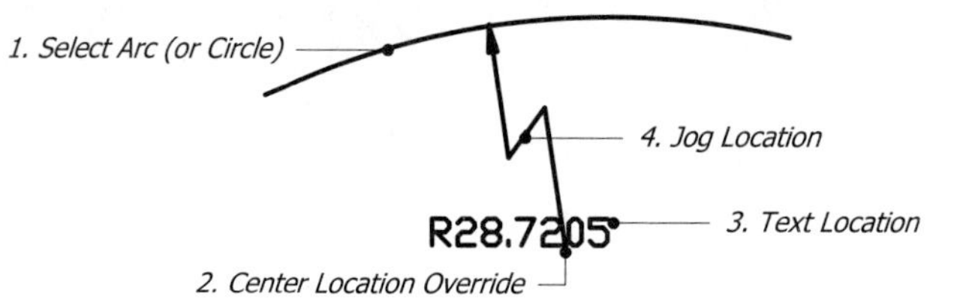

COMMAND LINE OPTIONS

Select arc or circle measures the radius of arcs and circles.

Specify center location override relocates the center from which to draw the dimension.

Specify dimension arc length location specifies the location of the arc length dimension.

Mtext changes the wording of the dimension text.

Text changes the position of the text.

Angle changes the angle of the dimension text.

Specify job location positions the jog.

RELATED SYSTEM VARIABLE

DimJogAngle specifies the jog angle; default = 45 degrees.

TIPS

- The *transverse angle* of the jog can be set in the Dimension Style Manager.
- Use this command when the center of arcs and circles is located outside the drawing area. It is generally used with very shallow arcs.
- Jogged dimensions can be edited using grips.

DimLinear

Rel 13 Draws linear dimensions.

Command	Aliases	Ctrl+	F-key	Alt+	Menu Bar	Tablet
dimlinear	**dli**	...	...	**NL**	**Dimension**	**W5**
	dimlin				**↳Linear**	

Command: dimlinear

Specify first extension line origin or <select object>: *(Pick a point, or press* ENTER *to select an object.)*

Specify second extension line origin: *(Pick a point.)*

Specify dimension line location or [Mtext/Text/Angle/Horizontal/Vertical/ Rotated]: *(Pick a point, or enter an option.)*

Dimension text = *nnn*

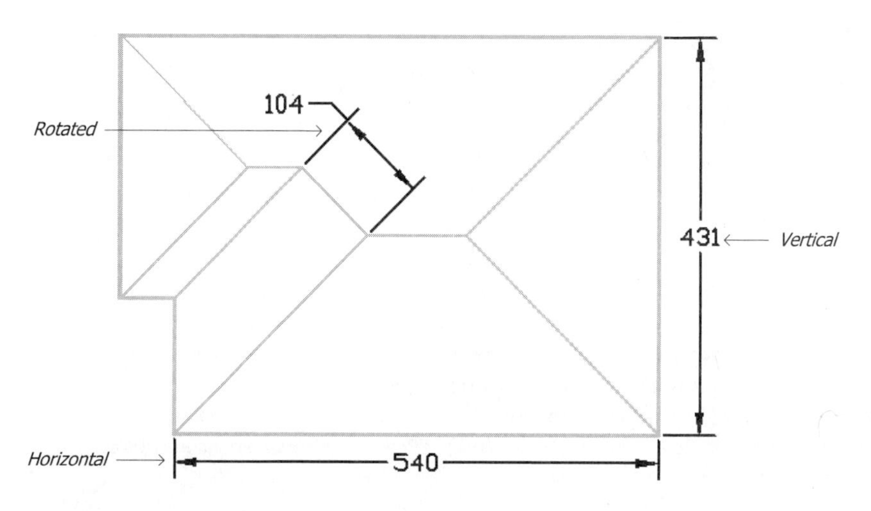

Examples of linear dimensions.

COMMAND LINE OPTIONS

Specify first extension line origin specifies the origin of the first extension line.

Select object dimensions a line, arc, or circle automatically.

Specify second extension line location specifies the location of the second extension line.

Specify dimension line location specifies the location of the dimension line.

Mtext displays the Text Formatting bar, which allows you to modify the dimension text; see the **MText** command.

Text Option

Enter dimension text <*nnn*>: *(Enter dimension text, or press* ENTER *to accept default value.)*

Enter dimension text prompts you to replace the dimension text on the command line.

Angle option
Specify angle of dimension text: *(Enter an angle.)*

Specify angle changes the angle of dimension text.

Horizontal option
Specify dimension line location or [Mtext/Text/Angle]: *(Pick a point or enter an option.)*

Horizontal option forces dimension to be horizontal.

Vertical option
Specify dimension line location or [Mtext/Text/Angle]: *(Pick a point or enter an option.)*

Vertical option forces dimension to be vertical.

Rotated option
Specify angle of dimension line <0>: *(Enter an angle.)*

Rotated option rotates the dimension.

Select Object Options

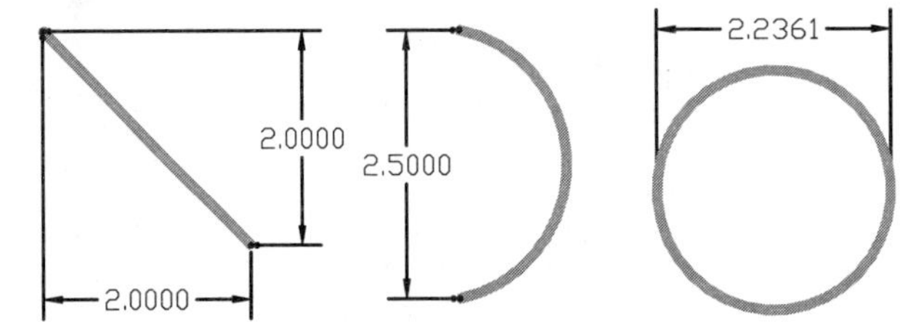

Examples of dimensioned objects: line, arc, and circle.

Specify first extension line origin or <select object>: *(Press* ENTER.*)*
Select object to dimension: *(Select one object.)*
Specify dimension line location or
[Mtext/Text/Angle/Vertical/Rotated]: *(Pick a point, or enter an option.)*
Dimension text = *nnn*

RELATED DIM COMMANDS

DimAligned draws linear dimensions aligned with objects.

QDim dimensions objects quickly.

DimOrdinate

Rel 13 Draws x and y ordinate dimensions.

Command	Aliases	Ctrl+	F-key	Alt+	Menu Bar	Tablet
dimordinate	dor	...	...	NO	Dimension	W3
	dimord				↳Ordinate	

Command: dimordinate

Specify feature location: *(Pick a point.)*

Specify leader endpoint or [Xdatum/Ydatum/Mtext/Text/Angle]: *(Pick a point, or enter an option.)*

Dimension text = *nnn*

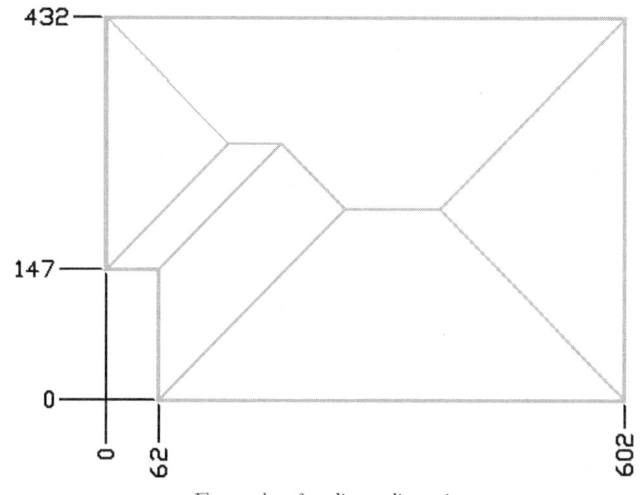

Examples of ordinate dimensions.

COMMAND LINE OPTIONS

Xdatum forces x ordinate dimension.

Ydatum forces y ordinate dimension.

Mtext displays the Text Formatting toolbar for modifying dimension text; see **MText** command.

Text prompts you to replace dimension text on the command line: 'Enter dimension text <*nnn*>'.

Angle changes the angle of dimension text: 'Specify angle of dimension text:'.

RELATED DIM COMMANDS

Leader draws leader dimensions.

Tolerance draws geometric tolerances.

TIPS

- Strictly speaking, "ordinate" is the distance from the x-axis only, but AutoCAD and the rest of the drafting world use it to mean from both the x and y axes; distance from y axis is "abcissa."
- The 0,0 point is determined by the UCS origin. Use the UCS Origin command to relocate 0,0.
- Define the x=0 and y=0 ordinate dimensions, and then use DimBaseline for additional ones.

DimOverride

Rel 13

Overrides the current dimension variables.

Command	Aliases	Ctrl+	F-key	Alt+	Menu Bar	Tablet
dimoverride	dov dimover	...	...	NV	Dimension ↳Override	Y4

Command: dimoverride

Enter dimension variable name to override or [Clear overrides]: *(Enter the name of a dimension variable, or type* **C***.)*

Enter new value for dimension variable <*nnn*>: *(Enter a new value.)*

Select objects: *(Select one or more dimensions.)*

Select objects: *(Press Enter .)*

COMMAND LINE OPTIONS

Dimension variable to override requires you to enter the name of the dimension variable.

Clear removes the override.

New Value specifies the new value of the dimvar.

Select objects selects the dimension objects to change.

RELATED DIM COMMAND

DimStyle creates and modifies dimension styles.

RELATED DIM VARIABLES

All dimension variables.

DimRadius

Rel 13 Draws radial dimensions on circles, arcs, and polyline arcs.

Command	Aliases	Ctrl+	F-key	Alt+	Menu Bar	Tablet
dimradius	**dra**	...	...	**NR**	**Dimension**	**X5**
	dimrad				**↳Radius**	

Command: dimradius
Select arc or circle: *(Select an arc, circle, or explodable object.)*
Dimension text = *nnn*
Specify dimension line location or [Mtext/Text/Angle]: *(Pick a point, or enter an option.)*

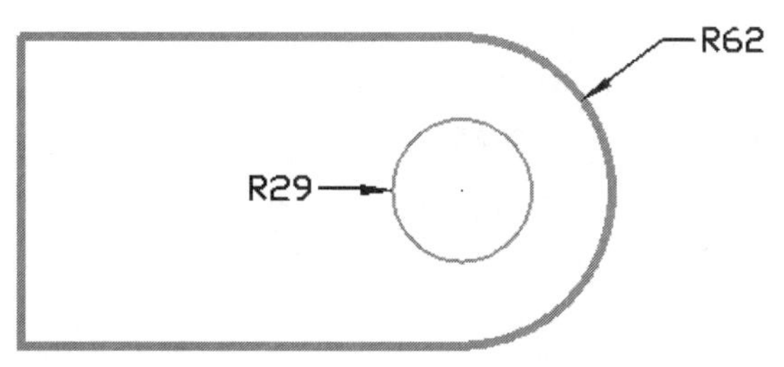

Examples of radial dimensions.

COMMAND LINE OPTIONS

Select arc or circle selects the arc, circle, or polyarc to dimension.

Specify dimension line location specifies the location of the angular dimension.

Mtext displays the Text Formatting bar, which allows you to modify the dimension text; see the **MText** command.

Text prompts you to replace the dimension text at the command line, 'Enter dimension text <*nnn*>'.

Angle changes the angle of dimension text, and prompts, 'Specify angle of dimension text:'.

RELATED DIM COMMANDS

DimCenter draws center marks on arcs and circles.

DimDiameter draws diameter dimensions on arcs and circles.

RELATED DIM VARIABLE

DimCen determines the size of the center mark.

DimReassociate

2002 Associates dimensions with objects.

Command	Alias	Ctrl+	F-key	Alt+	Menu Bar	Tablet
dimreassociate	dre	...	...	NN	Dimension ↳Reassociate Dimensions	X4

Command: dimreassociate

Select dimensions to reassociate...

Select objects: *(Select one or more dimensions, or enter* **All** *to select all dimensions.)*

nn **found.** *nn* **were on a locked layer.** *nn* **were not in current space.**

Select objects: *(Press* **ENTER** *to end object selection.)*

The prompts that follow depend on the type of each dimension selected.

Specify first extension line origin or [Select object]: *(Pick a point, or select an object.)*

Select arc or circle <next>: *(Press* **ENTER** *to end object selection.)*

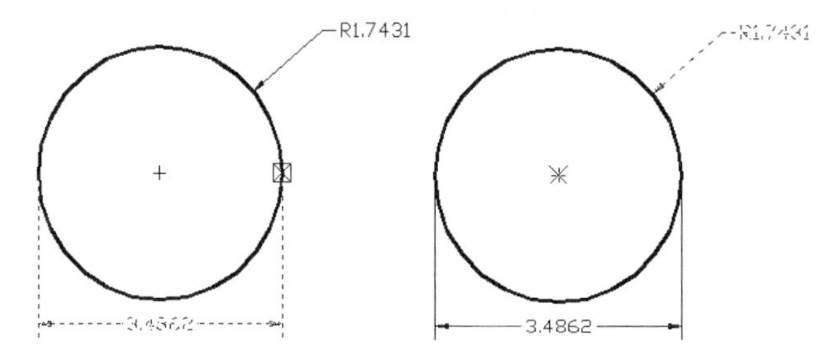

Left: *Boxed X indicates that the line dimension is associated with the circle.*
Right: *X indicates the radius dimension is not associated.*

COMMAND LINE OPTIONS

Select objects selects the dimensions to reassociate.

Specify first extension line origin picks a point with which to associate the dimension.

Select arc or circle picks a circle or arc with which to associate the dimension.

Next goes to the next circle or arc.

RELATED DIM COMMANDS

DimDisassociate converts dimensions from associative to non-associative.

DimRegen updates the locations of associative dimensions.

TIPS

- As you select objects, this command ignores non-dimensions, dimensions on locked layers, and those not in the current space (model or paper).
- AutoCAD displays a boxed **X** to indicate the object with which the dimension is associated; an unboxed **X** indicates an unassociated dimension. The markers disappear when a wheelmouse performs a zoom or pan.

DimRegen

2002 Updates the locations of associative dimensions.

Command	Alias	Ctrl+	F-key	Alt+	Menu Bar	Tablet
dimregen	...	...	...	...	...	...

Command: dimregen

COMMAND LINE OPTIONS

None.

RELATED DIM COMMANDS

DimDisassociate converts dimensions from associative to non-associative.

DimReassociate converts dimensions from non-associative to associative.

TIP

- This command is meant for use after three conditions:

  The drawing has been edited by a version of AutoCAD prior to 2004.

  The drawing contains dimensions associated with an external reference, and the xref has been edited.

  The drawing is in layout mode, model space is active, and a wheelmouse has been used to pan or zoom.

DimRotated

2004 Draws rotated dimensions (an undocumented command).

Command	Alias	Ctrl+	F-key	Alt+	Menu Bar	Tablet
dimrotated	...	...	...	...	...	...

Command: dimrotated
Specify angle of dimension line <0>: *(Enter a rotation angle.)*
Specify first extension line origin or <select object>: *(Pick a point, or press Enter to select one object.)*
Specify second extension line origin: *(Pick a point.)*
Specify dimension line location or [Mtext/Text/Angle]: *(Pick a point, or select an option.)*
Dimension text = *nnn*

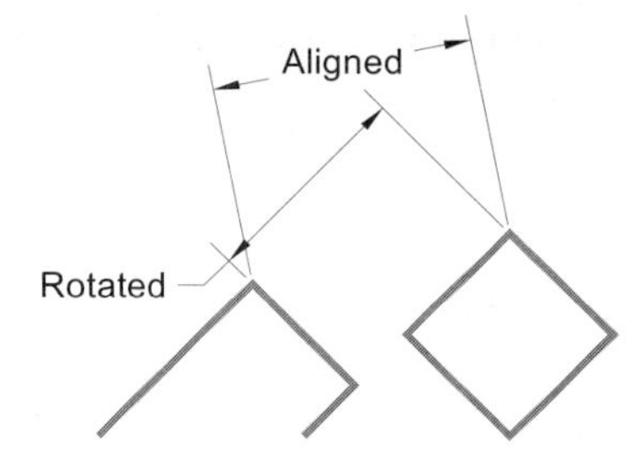

The aligned dimension is aligned with the two corner points.
The rotated dimension is placed at 45 degrees.

COMMAND LINE OPTIONS

Specify angle of dimension line specifies the angle at which the dimension line is rotated.

Specify first extension line origin specifies the origin of the first extension line.

Select object dimensions a line, arc, or circle automatically.

Specify second extension line location specifies the location of the second extension line.

Specify dimension line location specifies the location of the dimension line.

Mtext displays the Text Formatting bar, which allows you to modify the dimension text; see the **MText** command.

Text prompts you to replace the dimension text on the command line: 'Enter dimension text <*nnn*>'.

Angle changes the angle of dimension text: 'Specify angle of dimension text:'.

TIPS

- The DimRotated command draws dimensions at a specified angle, while the DimAligned command draws dimensions that align with the two pick points.
- You can specify the rotation angle by picking two points.

'DimStyle

Rel 13 Creates and edits dimstyles (short for DIMension STYLE).

Commands	Aliases	Ctrl+	F-key	Alt+	Menu Bar	Tablet
dimstyle	**d**	...	...	**NS**	**Dimension**	**Y5**
	dst				**↳Style**	
	dimsty			**OD**	**Format**	
	ddim				**↳Dimension Style**	
-dimstyle	...	...	...	**NU**	**Dimension**	
					↳Update	

Command: dimstyle

Displays dialog box:

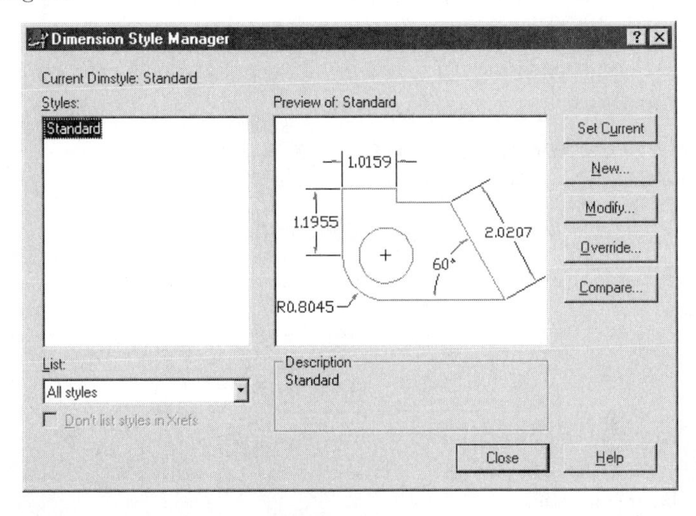

DIMENSION STYLE MANAGER DIALOG BOX

Styles lists the names of dimension styles in the drawing.

List modifies the style names listed under Styles:

- **All styles** lists all dimension style names stored in the current drawing (default).
- **Styles in use** lists only those dimstyles used by dimensions in the drawing.

Don't list styles in Xrefs

☑ Does not list dimension styles found in externally-referenced drawings.

☐ Lists xref dimension styles under Styles (default).

Set Current sets the selected style as the current dimension style.

New creates a new dimension style via Create New Dimension Style dialog box.

Modify modifies an existing dimension style; displays the Modify Dimension Style dialog box.

Override allows temporary changes to a dimension style; displays the Override Dimension Style dialog box.

Compare lists the differences between dimension variables of two styles; displays the Compare Dimension Styles dialog box.

SHORTCUT MENU

Right-click a dimension style name under ***Styles*****:**

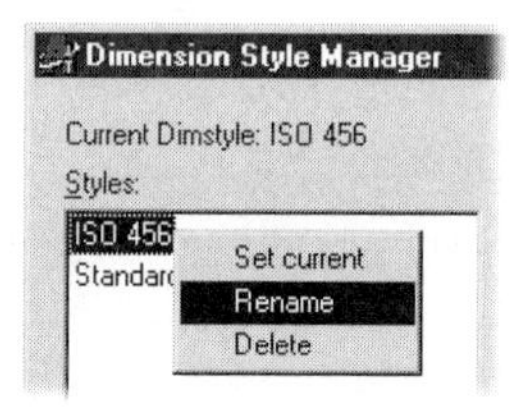

Set Current sets the selected dimension style as the current style.

Rename renames dimension styles.

Delete erases selected dimension styles from the drawing; you cannot erase the Standard style, or styles that are in use.

CREATE NEW DIMENSION STYLE DIALOG BOX

When creating a new dimension style, typically you change an existing style.

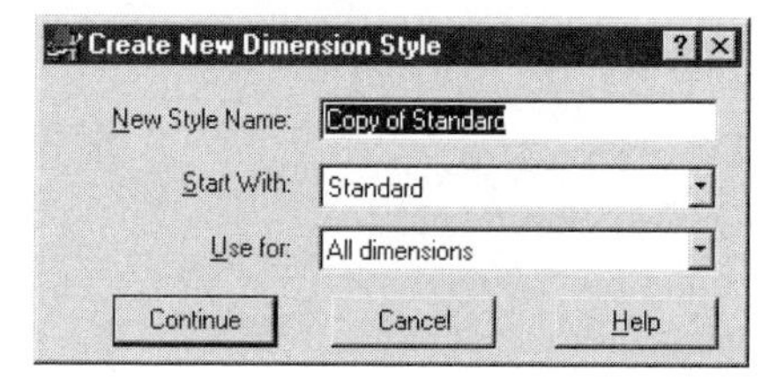

New Style Name specifies the name of the new dimension style.

Start With lists the names of the current dimension style(s), which are used as the template for the new dimension style.

Use for creates a substyle that applies to a specific type of dimension type: linear, angular, radius, diameter, ordinate, leaders and tolerances.

Continue continues to the next dialog box, New Dimension Style.

Cancel dismisses this dialog box, and returns to the Dimension Style Manager dialog box.

NEW DIMENSION STYLE DIALOG BOX

OK records the changes made to dimension properties, and returns to the Dimension Manager dialog box.

Cancel cancels the changes, and returns to the Dimension Manager dialog box.

Lines Tab

Sets the format of dimension lines and extension lines.

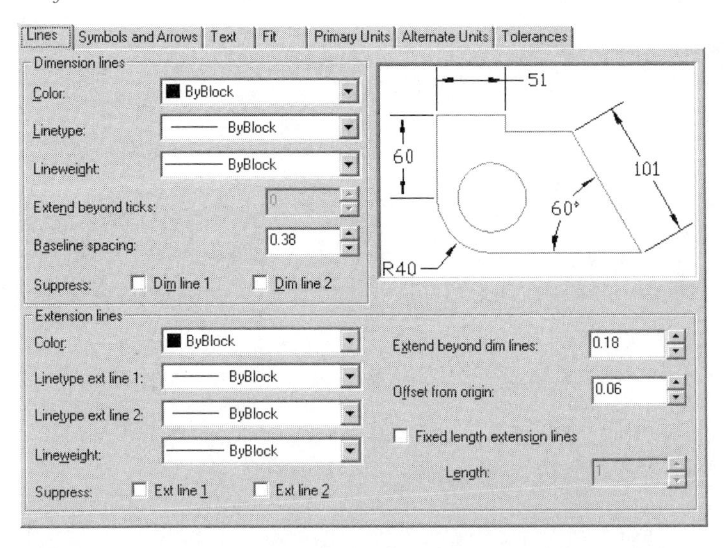

Dimension Lines

Color specifies the color of the dimension line; select Other to display the Select Color dialog box (stored in dimension variable **DimClrD**; default = ByBlock).

Linetype specifies the linetype of the dimension line (**DimLtype**).

Lineweight specifies the lineweight of the dimension line (**DimLwD**; default = ByBlock).

Extend beyond ticks specifies the distance the dimension line extends beyond the extension line; used with oblique, architectural, tick, integral, and no arrowheads (**DimDlE**; default = 0).

Baseline spacing specifies the spacing between the dimension lines of a baseline dimension (**DimDlI**; default = 0.38).

Suppress suppresses dimension lines when outside the extension lines (**DimSD1** and **DimSD2**):

- ☐ **Dim line 1** for portion of dimension line left of text.
- ☐ **Dim line 2** for portion of dimension line right of text.

Extension Lines

Color specifies the color of the extension line; select Other to display the Select Color dialog box (stored in dimension variable **DimClrE** ; default=ByBlock).

Linetype ext 1 specifies the linetype of the first extension line (**DimLtEx1**).

Linetype ext 2 specifies the linetype of the second extension line (**DimLtEx2**).

Lineweight specifies the lineweight of the extension line (**DimLwE**; default=ByBlock).

Suppress suppresses extensions lines (**DimSE1** and **DimSE2**):

- ☐ **Ext line 1** for left extension line.
- ☐ **Ext line 2** for right extension line.

Extend beyond dim lines specifies the distance the extension line extends beyond the dimension line; used with oblique, architectural, tick, integral, and no arrowheads (**DimExe**; default = 0.18).

Offset from origin specifies the distance from the origin point to the start of the extension lines (**DimExO**; default = 0.0625).

☐ **Fixed length extension lines** specifies the length of extension lines (**DimFxlOn**):

- ☐ **Length** is 0.18 by default (**DimFXL**).

Symbols and Arrows Tab

Sets the format of arrowheads, center marks, and arc lengths.

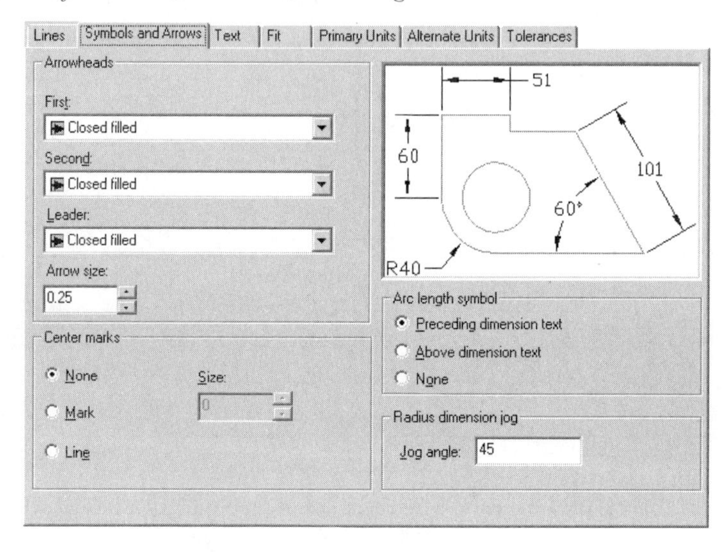

Arrowheads

First specifies the name of the arrowhead to use for the first end of the dimension line (**DimBlk1**; default = closed filled).

*To use a custom arrowhead, select **User Arrow** to display the Select Custom Arrow Block dialog box:*

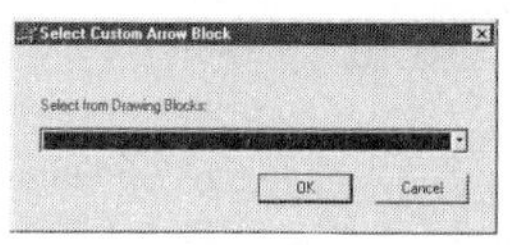

Second specifies the arrowhead for the second dimension line; select **User Arrow** to display the Select Custom Arrow Block dialog box (**DimBlk2**; default = closed filled).

Leader specifies the arrowhead for the leader; select **User Arrow** to display the Select Custom Arrow Block dialog box (**DimLdrBlk**; default = closed filled).

Arrow Size specifies the size of arrowheads (**DimASz**; default = 0.18).

Closed filled
Closed blank
Closed
Dot
Architectural tick
Oblique
Open
Origin indicator
Origin indicator 2
Right angle
Open 30
Dot small
Dot blank
Dot small blank
Box
Box filled
Datum triangle
Datum triangle filled
Integral
None

Center Marks

○ **None** places no center marks or center lines (**DimCen**=0).

⊙ **Mark** places center marks (**DimCen** > 0).

○ **Line** places center marks and center lines (**DimCen** < 0).

Size specifies size of center marks or center lines (**DimCen**; default = 0.09).

Arc Length Symbol

⊙ **Preceeding dimension text** places symbol in front of text (**DimArcSym**).

O **Above dimension text** places symbol above text.

O **None** displays no symbol.

Radius Dimension Jog

Jog Angle specifies the default jog angle for the **DimJogged** command (**DimJogAngle**).

Text Tab

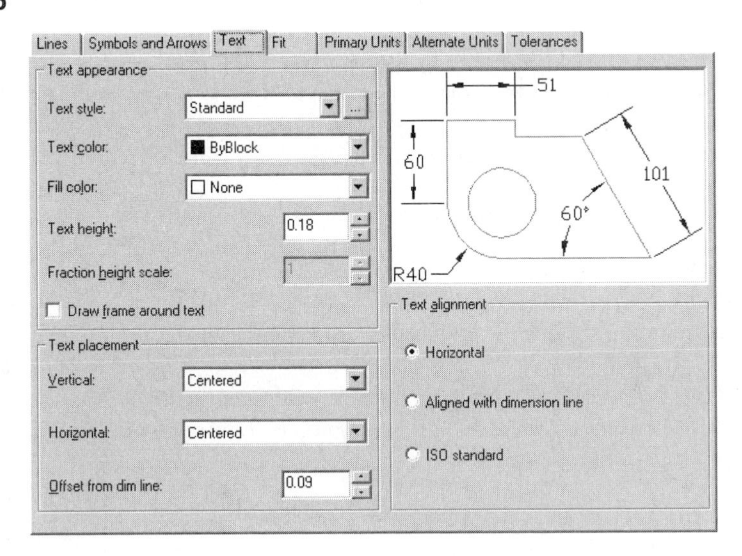

Text Appearance

Text style specifies the text style name for dimension text (**DimTxSty**; default = Standard).

... displays the Text Style dialog box; see the **Style** command.

Text color specifies the color of the dimension line; select **Other** to display the Select Color dialog box (**DimClrT**; default = ByBlock).

Fill color specifies the color of the background behind the text (**DimTFill** and **DimTFillClr**).

Text height specifies the height of the dimension text, when the height defined by the text style is 0 (**DimTxt**; default = 0.18).

Fraction height scale scales fraction text height relative to dimension text; AutoCAD multiples this value by the text height (**DimTFac**; default = 1.0).

☐ **Draw frame around text** draws a rectangle around dimension text; when on, dimension variable **DimGap** is set to a negative value (**DimGap**).

Text Placement

⊙ **Vertical** specifies the vertical justification of dimension text relative to the dimension line:

Centered — Centers dimension text in the dimension line (**DimTad** = 0).

Above — Places text above the dimension line (**DimTad** = 1).

Outside — Places text on the side of the dimension line farthest from the first defining point (**DimTad** = 2).

JIS — Places text in conformity with JIS (**DimTad**= 3).

○ **Horizontal** specifies the horizontal justification of dimension text along the dimension and extension lines (**DimJust**; default = 0):

Centered — Centers dimension text along the dimension line between the extension lines (**DimJust** = 0).

1st Extension Line — Left-justifies the text with the first extension line (**DimJust** = 1).

2nd Extension Line — Right-justifies the text with the second extension line (**DimJust** = 2).

Over 1st Extension Line — Places the text over the first extension line (**DimJust** = 3).

Over 2nd Extension Line — Places the text over the second extension line (**DimJust** = 4).

○ **Offset from dimension line** specifies the text gap, the distance between the dimension text and the dimension line (**DimGap**; default = 0.09).

Text Alignment

Horizontal forces dimension text always to be horizontal (**DimTih** = on; **DimToh** = on).

Aligned with dimension line forces dimension text to be aligned with the dimension line (**DimTih** = off; **DimToh** = off).

ISO Standard forces text to be aligned with the dimension line when inside the extension lines; forces text to be horizontal when outside the extension lines (**DimTih** = off; **DimToh** = on).

Fit Tab

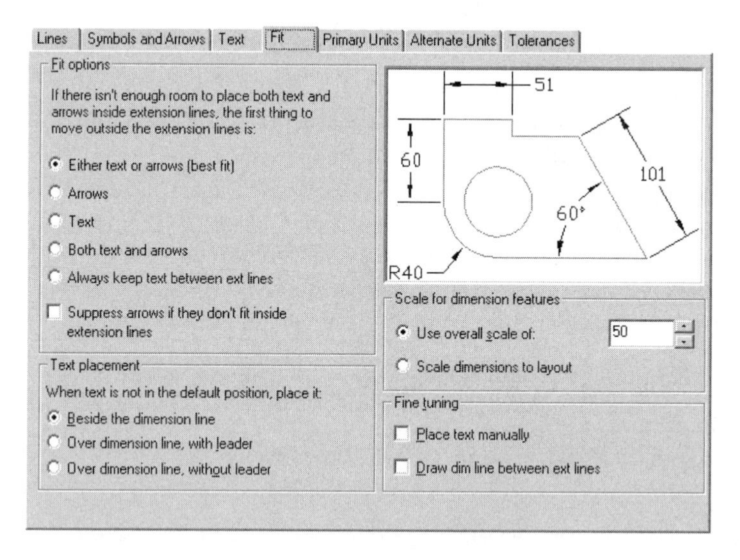

Fit Options

If there isn't enough room to place both text and arrows inside extension lines, the first thing to move outside the extension lines is:

⊙ **Either the text or the arrows, whichever fits best** places dimension text and arrowheads between the extension lines when space is available; when space is not available for both, the text or the arrowheads are placed outside the extension lines, whichever fits best; if there is room for neither, both are placed outside the extension lines (**DimAtFit** = 3; default).

○ **Arrows** places arrowheads between the extension lines when there is not enough room for arrowheads and dimension text (**DimAtFit** = 2).

○ **Text** places text between extension lines when there is not enough room for arrowheads and dimension text (**DimAtFit** = 1).

○ **Both text and arrows** places both outside the extension lines when there is not enough room for dimension text and arrowheads (**DimAtFit** = 0).

○ **Always keep text between ext lines** forces text between the extension lines (**DimTix**; default = off).

☐ **Suppress arrows if they don't fit inside the extension lines** suppresses arrowheads when there is not enough room between the extension lines.

Text Placement:

⊙ **Beside the dimension line** places dimension text beside the dimension line (**DimTMove** =0; default).

○ **Over the dimension line, with a leader** draws a leader when dimension text is moved away from the dimension line (**DimTMove** = 1).

○ **Over the dimension line, without a leader** does not draw a leader when dimension text is moved away from the dimension line (**DimTMove** = 2).

Scale for Dimension Features

⊙ **Use overall scale of** specifies the scale factor for all dimensions in the drawing; affects text and arrowhead sizes, distances, and spacing (**DimScale**; default = 1.0).

○ **Scale dimension to layout (paper space)** determines the scale factor of dimensions in layout mode; based on the scale factor between the current model space viewport and the layout (**DimScale** = 0; default = off).

Fine Tuning

☐ **Place text manually when dimensioning** places text at the position picked at the 'Dimension line location' prompt (**DimUpt**).

☐ **Always draw dim line between ext lines** forces the dimension line between the extension lines (**DimTofl**).

Primary Units Tab

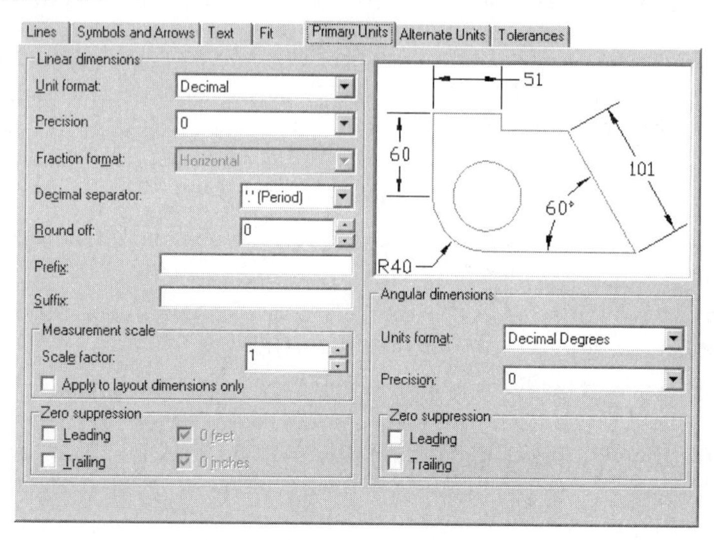

Linear Dimensions

Unit Format specifies the linear units format; does not apply to angular dimensions; **DimLUnit**:

1 — Scientific.
2 — Decimal (default).
3 — Engineering.
4 — Architectural.
5 — Fractional.
6 — Windows desktop setting.

Precision specifies the number of decimal places (or fractional accuracy) for linear dimensions (**DimDec**; default = 4).

Fraction Format specifies the stacking format of fractions; **DimFrac**:

0 — Horizontal stacked: $\frac{1}{2}$ (default).
1 — Diagonal stacked: ½
2 — Not stacked: 1/2.

Decimal Separator specifies the separator for decimal formats (**DimDSep**; default = .).

Round Off specifies the format for rounding dimension values; does not apply to angular dimensions (**DimRnd**; default = 0).

Prefix specifies a prefix for dimension text (**DimPost**; default = nothing); you can use the following control codes to show special characters:

Control Code	Meaning
%%nnn	Character specified by ASCII number *nnn*.
%%o	Turns on and off overscoring.
%%u	Turns on and off underscoring.
%%d	Degrees symbol (°).
%%p	Plus/minus symbol (±).
%%c	Diameter symbol (Ø).
%%%	Percentage sign (%).

Suffix specifies a suffix for dimension text (**DimPost**; default = nothing); you can use the control codes listed above to show special characters.

Measurement Scale

Scale factor specifies a scale factor for linear measurements, except for angular dimensions (**DimLFac**; default = 1.0); for example, use this to change dimension values from imperial to metric.

☐ **Apply to layout dimensions only** specifies that the scale factor is applied only to dimensions created in layout mode or paper space (stored as a negative value in **DimLFac**; default = off).

Zero Suppression options:

☐ **Leading** suppresses leading zeros in all decimal dimensions (**DimZin** = 4).

☐ **Trailing** suppresses trailing zeros in all decimal dimensions (**DimZin** = 8).

☑ **0 Feet** suppresses zero feet of feet-and-inches dimensions (**DimZin** = 0).

☑ **0 Inches** suppresses zero inches of feet-and-inches dimensions (**DimZin** = 2).

0 — Suppresses zero feet and precisely zero inches (*default*).
1 — Includes zero feet and precisely zero inches.
2 — Includes zero feet and suppresses zero inches.
3 — Includes zero inches and suppresses zero feet.
4 — Suppresses leading zeros in decimal dimensions.
8 — Suppresses trailing zeros in decimal dimensions.
12 — Suppresses leading and trailing zeros.

Angular Dimensions options

Units Format specifies the format of angular dimensions; **DimAUnit**:

0 — Decimal degrees (*default*).
1 — Degrees/minutes/seconds.
2 — Grads.
3 — Radians.

Precision specifies the precision of angular dimensions (**DimADec**; default = 0).

Zero Suppression

*Same as for linear dimensions (**DimAZin**; default = 0).*

Alternate Units Tab

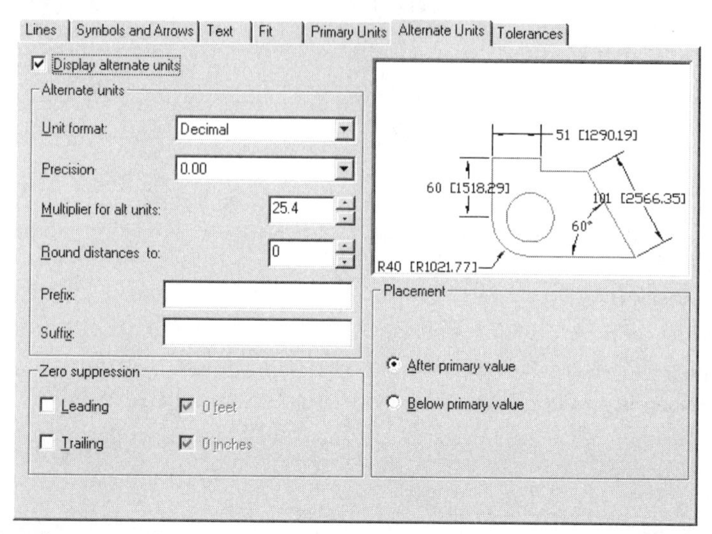

☐ **Display alternate units** adds alternate units to dimension text (**DimAlt**).

Alternate Units

Unit Format specifies the alternate units format (**DimAltU**; default = Decimal).

Precision specifies the number of decimal places or fractional accuracy (**DimAltD**; default=2).

Multiplier for alt units specifies the conversion factor between primary and alternate units (**DimAltF**; default = 25.4).

Round distances to specifies the format for rounding dimension values; does not apply to angular dimensions (**DimAltRnd**; default = 0.0000).

Prefix specifies a prefix for dimension text (**DimAPost**; default = nothing); you can use control codes to show special characters.

Suffix specifies a suffix for dimension text (**DimAPost**; default = nothing).

Zero Suppression

☐ **Leading** suppresses leading zeros in all decimal dimensions (**DimAltZ** = 4).

☐ **Trailing** suppresses trailing zeros in all decimal dimensions (**DimAltZ** = 8).

☑ **0 Feet** suppresses zero feet of feet-and-inches dimensions (**DimAltZ** = 0).

☑ **0 Inches** suppresses zero inches of feet-and-inches dimensions (**DimAltZ** = 2).

Placement:

⊙ **After primary units** places alternate units behind the primary units (**DimAPost**).

○ **Below primary units** places alternate units below the primary units.

Tolerances Tab

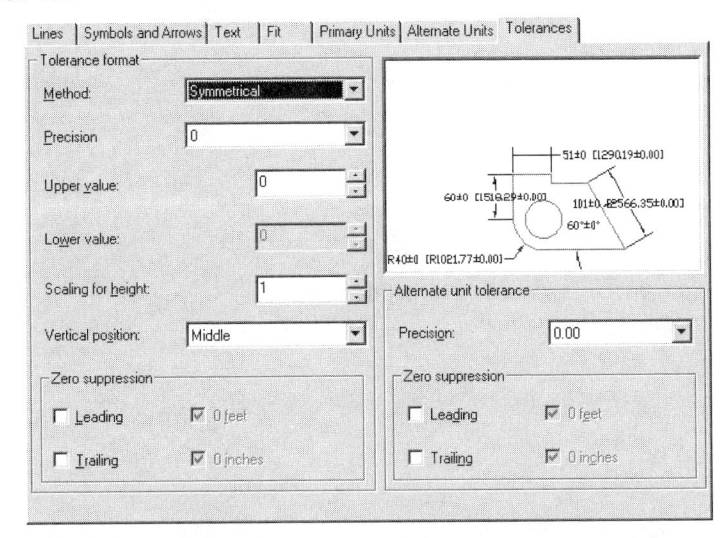

Tolerance Format

Method specifies the tolerance format:

- **None** does not display tolerances (**DimTol** = 0; default).
- **Symmetrical** places ± after the dimension (**DimTol**=0; **DimLim**=0).
- **Deviation** places + and – symbols (**DimTol** = 1; **DimLim** = 1).
- **Limits** places maximum over minimum value (**DimTol**=0; **DimLim**=1):

  Maximum value = dimension value + upper value.

  Minimum value = dimension value - lower value.
- **Basic** boxes the dimension text (**DimGap**=negative value).

Precision specifies the number of decimal places for tolerance values (**DimTDec**; default=4).

Upper value specifies the upper tolerance value (**DimTp**; default = 0).

Lower value specifies the lower tolerance value (**DimTm**; default = 0).

Scaling for height specifies the scale factor for tolerance text height (**DimTFac**; default = 1.0).

Vertical position specifies the vertical text position for symmetrical and deviation tolerances:

Vertical	Meaning
Top	Aligns the tolerance text with the top of the dimension text (**DimTolJ** = 2).
Middle	Aligns the tolerance text with the middle of the dimension text (**DimTolJ** = 1).
Bottom	Aligns the tolerance text with the bottom of the dimension text (**DimTolJ** = 0).

Zero Suppression

☐ **Leading** suppresses leading zeros in all decimal dimensions (**DimTZin** = 4).

☐ **Trailing** suppresses trailing zeros in all decimal dimensions (**DimTZin** = 8).

☑ **0 Feet** suppresses zero feet of feet-and-inches dimensions (**DimTZin** = 0).

☑ **0 Inches** suppresses zero inches of feet-and-inches dimensions (**DimTZin** = 2).

Alternate Unit Tolerance

Precision specifies the precision — the number of decimal places — of tolerance text (**DimAltTd**; default = 2).

Zero Suppression options: the same as for tolerance format; stored in **DimAltTz**.

MODIFY DIMENSION STYLE DIALOG BOX

This dialog box is identical to the New Dimension Style dialog box.

OVERRIDE DIMENSION STYLE DIALOG BOX

This dialog box is identical to the New Dimension Style dialog box.

COMPARE DIMENSION STYLES DIALOG BOX

*The list is blank when AutoCAD finds no differences. When **With** is set to **<none>** or the same style as **Compare**, AutoCAD displays all dimension variables.*

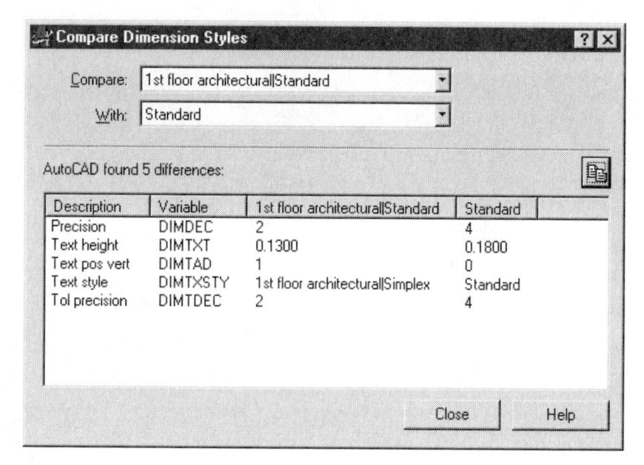

Compare displays the name of one dimension style.

With displays the name of the second dimension style.

Copy to Clipboard copies the style comparison text to the Clipboard, which can be pasted in another Windows application.

Description describes the dimension variable.

Variable names the dimension variable.

Close closes the dialog box.

-DIMSTYLE Command

Command: -dimstyle
Current dimension style: Standard
Enter a dimension style option
[Save/Restore/STatus/Variables/Apply/?] <Restore>: *(Enter an option.)*
Current dimension style: Standard
Enter a dimension style name, [?] or <select dimension>: *(Enter an option.)*
Select dimension: *(Select a dimension in the drawing.)*
Current dimension style: Standard

COMMAND LINE OPTIONS

Save saves current *dimvar* (dimension variable) settings as a named *dimstyle* (dimension style).

Restore retrieves dimvar settings from a named dimstyle.

STatus lists dimvars and current settings.

Variables lists dimvars and their current settings.

Apply updates selected dimension objects with current dimstyle settings.

? lists names of dimstyles stored in drawing.

INPUT OPTIONS

~ ***dimvar*** (*tilde prefix*) lists the differences between current and selected dimstyles.

ENTER lists the dimvar settings for the selected dimension object.

RELATED DIM COMMANDS

DDim changes dimvar settings.

DimScale determines the scale of dimension text.

RELATED DIM VARIABLES

All.

DimStyle contains the name of the current dimstyle.

TIPS

- At the 'Dim' prompt, the Style command sets the text style for the dimension text, and does *not* select a dimension style.
- Dimstyles cannot be stored to disk, except in a drawing.
- Read dimstyles from other drawings with the XBind Dimstyle command, or by dragging dimstyle names from Design Center.

DimTEdit

Rel 13 Changes the location and orientation of text in dimensions dynamically.

Command	Alias	Ctrl+	F-key	Alt+	Menu Bar	Tablet
dimtedit	dimted	...	...	NX	Dimension ↳Align Text	Y2

Command: dimtedit

Select dimension: *(Select a dimension in the drawing.)*

Specify new location for dimension text or [Left/Right/Center/Home/Angle]: *(Pick a point, or enter an option.)*

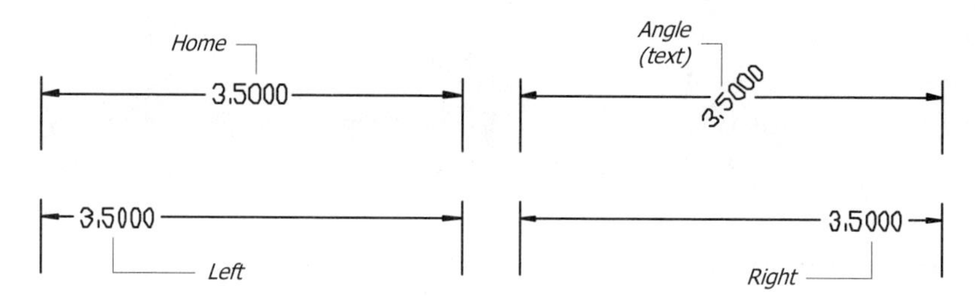

COMMAND LINE OPTIONS

Select dimension selects the dimension to edit.

Angle rotates the dimension text.

Center centers the text on the dimension line.

Home returns the dimension text to the original position.

Left moves the dimension text to the left.

Right moves the dimension text to the right.

RELATED DIM VARIABLES

DimSho specifies whether dimension text is updated dynamically while dragged.

DimTih specifies whether dimension text is drawn horizontally or aligned with the dimension line.

DimToh specifies whether dimension text is forced inside the dimension lines.

TIPS

- This command works only with dimensions created with DimAssoc = 1 or 2; use the DdEdit command to edit text in non-associative dimensions.
- An angle of 0 returns dimension text to its default orientation.

DimVertical

2004 Draw vertical dimensions (an undocumented command).

Command	Alias	Ctrl+	F-key	Alt+	Menu Bar	Tablet
dimvertical	...	...	...	...	...	...

Command: dimvertical

Specify first extension line origin or <select object>: *(Pick a point, or press Enter to select one object.)*

Specify second extension line origin: *(Pick a point.)*

Specify dimension line location or [Mtext/Text/Angle]: *(Pick a point, or select an option.)*

Dimension text = *nnn*

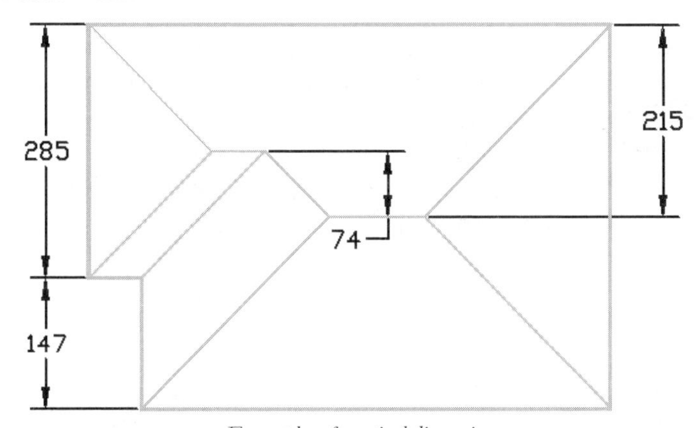

Examples of vertical dimensions.

COMMAND LINE OPTIONS

Specify first extension line origin specifies the origin of the first extension line.

Select object dimensions a line, arc, circle, or explodable object automatically.

Specify second extension line location specifies the location of the second extension line.

Specify dimension line location specifies the location of the dimension line.

Mtext displays the Multiline Text Editor dialog box, which allows you to modify the dimension text; see the **MText** command.

Text prompts you to replace the dimension text on the command line: 'Enter dimension text <*nnn*>'.

Angle changes the angle of dimension text: 'Specify angle of dimension text:'.

RELATED DIM COMMANDS

All.

RELATED DIM VARIABLES

Most.

'Dist

V. 1.0 Lists the 3D distances and angles between two points (short for DISTance).

Command	Alias	Ctrl+	F-key	Alt+	Menu Bar	Tablet
dist	**di**	...	...	**TQD**	**Tools** ↳**Inquiry** ↳**Distance**	**T8**

Command: dist
Specify first point: *(Pick a point.)*
Specify second point: *(Pick another point.)*

Sample result:
Specify first point: Specify second point:
Distance = 19, Angle in XY Plane = 22, Angle from XY Plane = 0
Delta X = 18, Delta Y = 7, Delta Z = 0

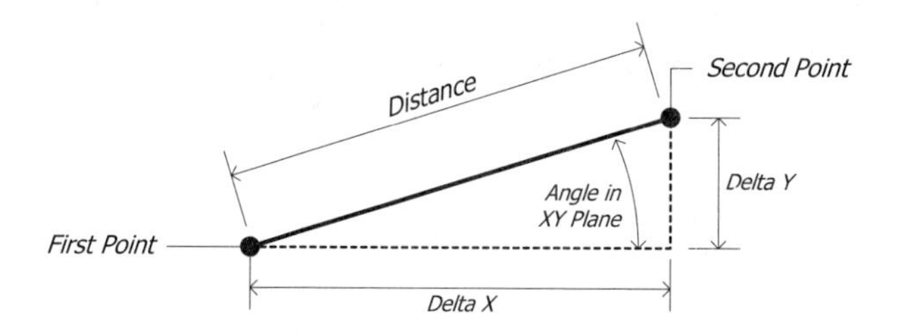

COMMAND LINE OPTIONS

Specify first point determines the start point of distance measurement.

Specify second point determines the endpoint.

RELATED COMMANDS

Area calculates the area and perimeter of objects.

Id lists the 3D coordinates of points.

RELATED SYSTEM VARIABLE

Distance specifies the last calculated distance.

TIPS

- Use object snaps to measure precisely the distance between two geometric features.
- When the z-coordinate is left out, Dist uses the current elevation for z.
- Object snaps give 3D points, and therefore the 3D distance.

Aliased Command

Details about the **DistantLight** command are found under the **Light** command.

Divide

V. 2.5 Places points or blocks at equal distances along objects.

Command	Alias	Ctrl+	F-key	Alt+	Menu Bar	Tablet
divide	**div**	...	...	**DOD**	**Draw** ↳**Point** ↳**Divide**	**V13**

Command: divide
Select object to divide: *(Select one object.)*
Enter the number of segments or [Block]: *(Enter a number, or enter* ***B****.)*

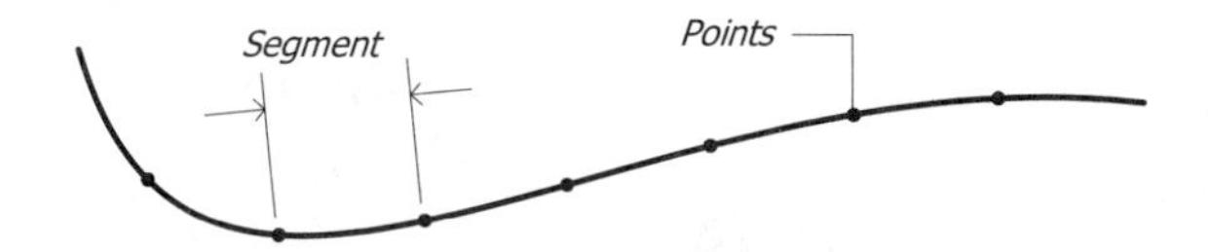

COMMAND LINE OPTIONS

Select object to divide selects a single open or closed object.

Enter the number of segments specifies the number of segments; must be a number between 2 and 32767.

Block Options
Enter name of block to insert: *(Enter the name of a block.)*
Align block with object? [Yes/No] <Y>: *(Enter* ***Y*** *or* ***N****.)*
Enter the number of segments: *(Enter a number.)*

Enter name of block to insert specifies the block to insert. *Caution*! The block must be defined in the current drawing prior to starting this command.

Align block with object aligns the block's x axis with the object.

RELATED COMMANDS

Block creates the block to use with the Divide command.

DdPType controls the size and style of points.

Insert places a single block in the drawing.

MInsert places an array of blocks in the drawing.

Measure places points or blocks at measured distances.

TIPS

- The first dividing point on a closed polyline is its initial vertex; on circles, the first dividing point is in the 0-degree direction from the center.
- The points or blocks are placed in the **Previous** selection set, so that you can select them with the next 'Select Objects' prompt.
- Objects are unchanged by this command.
- Use the DdPType command to make points visible.

Donut

V. 2.5 Draws solid-filled circles as a wide polyline consisting of a pair of arcs.

Commands	Alias	Ctrl+	F-key	Alt+	Menu Bar	Tablet
donut	do doughnut	...	...	DD	Draw ⇘Donut	K9

Command: donut
Specify inside diameter of donut <0.5000>: *(Enter a value.)*
Specify outside diameter of donut <1.0000>: *(Enter a value.)*
Specify center of donut or <exit>: *(Pick a point.)*
Specify center of donut or <exit>: *(Press Enter to exit command.)*

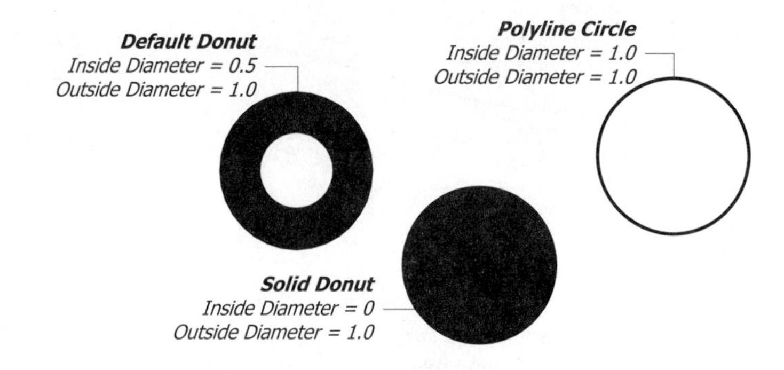

COMMAND LINE OPTIONS

Inside diameter specifies the inner diameter by entering a number or picking two points.

Outside diameter specifies the outer diameter.

Center of donut determines the donut's center point by specifying coordinates, or picking points.

Exit exits the command.

RELATED COMMAND

Circle draws circles.

RELATED SYSTEM VARIABLES

DonutId specifies the default internal diameter.

DonutOd specifies the default outside diameter.

Fill toggles the filling of the donut, as well as wide polylines, hatch, and 2D solids.

TIPS

- This command repeats itself until canceled.
- Donuts are made of two polyline arcs.

'Dragmode

V. 2.0 Controls the display of objects during dragging operations.

Command	Alias	Ctrl+	F-key	Alt+	Menu Bar	Tablet
dragmode	...	...	...	...	...	...

Command: dragmode
Enter new value [ON/OFF/Auto] <Auto>: *(Enter an option.)*

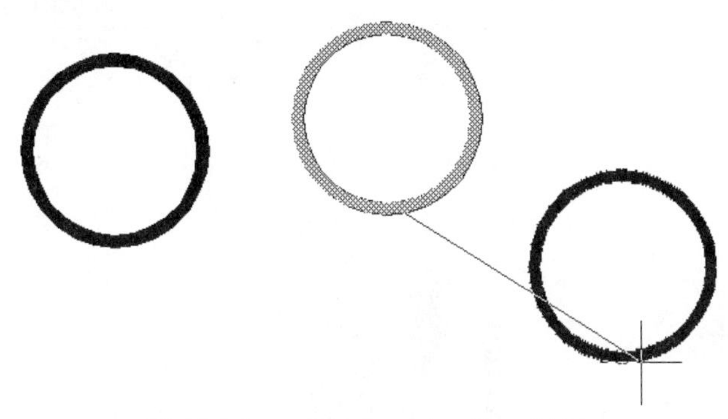

Highlight image (center) and drag image (right).

COMMAND LINE OPTIONS

ON enables dragging display only with the **Drag** option.

OFF turns off all dragging display.

Auto allows AutoCAD to determine when to display drag image.

COMMAND MODIFIER

Drag displays drag images when DragMode = on.

RELATED SYSTEM VARIABLE

DragMode is the current drag setting:

- **0** — No drag image.
- **1** — On if required.
- **2** — Automatic.

TIP

- Turn off **DragMode** and **Highlight** in very large drawings to speed up editing.

2006 DrawingRecovery, DrawingRecoveryHide

Displays and hides the Drawing Recovery Manger palette.

Command	Alias	Ctrl+	F-key	Alt+	Menu Bar	Tablet
drawingrecovery	**drm**	...	...	**FUD**	**File** ↳**Drawing Utilities** ↳**Drawing Recovery Manager**	...
drawingrecoveryhide						

Command: drawingrecovery

Displays palette (also displayed automatically following a software crash):

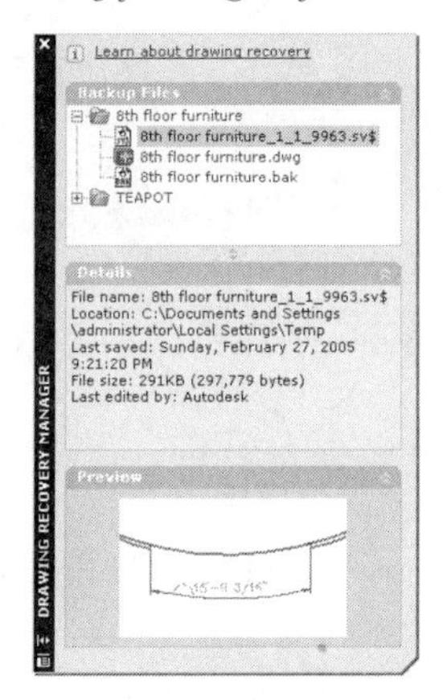

SHORTCUT MENU

Right-click drawing names for shortcut menu:

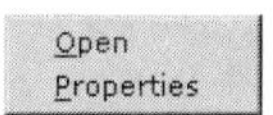

Open opens the drawing in AutoCAD; see the Open command.

Properties displays the Properties dialog box; see the DwgProps command.

Command: drawingrecoveryclose

Closes the palette.

RELATED COMMANDS

Audit determines if drawing files have problems.

Recover recovers drawings at the command line.

RELATED SYSTEM VARIABLES

DrState reports whether the Drawing Recovery palette is open.

RecoveryMode specifies when the Drawing Recovery palette appears, if at all.

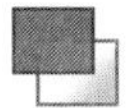 DrawOrder

Rel. 14 Controls the display of overlapping objects.

Command	Alias	Ctrl+	F-key	Alt+	Menu Bar	Tablet
draworder	dr	...	...	TO	Tools ↳Draw Order	T9

Command: draworder
Select objects: *(Select one or more objects.)*
Select objects: *(Press Enter.)*
Enter object ordering option
[Above objects/Under objects/Front/Back] <Back>: *(Enter an option.)*

Left: *Text* ***under*** *solid.*
Right: *Text* ***above*** *solid.*

COMMAND LINE OPTIONS

Select objects selects the objects to be moved.

Above object forces selected objects to appear above the reference object.

Under object forces selected objects to appear below the reference object.

Front forces selected objects to the top of the display order.

Back forces selected objects to the bottom of the display order.

RELATED COMMANDS

BHatch displays hatching in front of or behind other objects.

TextToFront displays text and/or dimensions on top of overlapping objects.

RELATED SYSTEM VARIABLES

DrawOrderCtl controls draw order:

- **0** — Turns off draw order.
- **1** — Specifies that objects displayed with draw order.
- **2** — Specifies that new objects take on draw order of the object selected first.
- **3** — Combines options 1 and 2 (default).

HpDrawOrder specifies display order of hatch patterns and fills; see the BHatch command.

TIPS

- When you pick more than one object for reordering, AutoCAD maintains the relative display order of the selected objects. The order in which you select objects has no effect on display order.
- When DrawOrderCtl is set to 3, editing operations may take longer.
- *Draw order inheritance* means that new objects created from objects with draw order are assigned the display order of the object selected first.

'DSettings

2000 Controls the most common settings for drafting operations (short for Drafting SETTINGS).

Commands	Aliases	Ctrl+	F-key	Alt+	Menu Bar	Tablet
dsettings	ds	...	...	TF	Tools	W10
	se, ddrmodes				↳Drafting Settings	
	os, osnap					
+dsettings						

Command: dsettings

Displays dialog box (substantially changed in AutoCAD 2007):

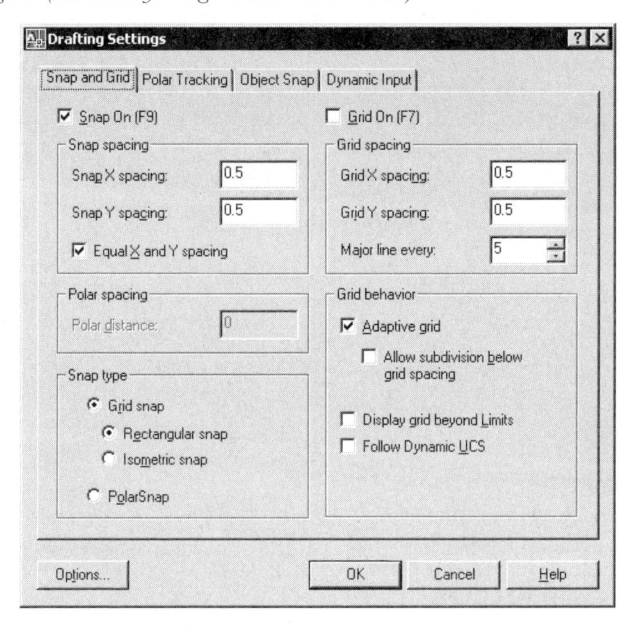

DRAFTING SETTINGS DIALOG BOX

Options displays the Options dialog box; does not work during transparent commands.

Snap and Grid Tab

☐ **Snap On (F9)** turns on and off snap mode. (Stored in system variable **SnapMode**; default = off.)

☐ **Grid On (F7)** turns on and off the grid display (**GridMode**; default = off).

Snap

Snap X spacing specifies the snap spacing in the x direction (**SnapUnit**; default = 0.5).

Snap Y spacing specifies the snap spacing in the y direction (**SnapUnit**; default = 0.5).

☑ **Equal X and Y spacing** sets the y spacing to that of x.

Polar Spacing

Polar distance specifies the snap distance, when Snap type & style is set to Polar snap; when 0, the polar snap distance is set to the value of Snap X spacing (**PolarDist**; default = 13).

Snap Type & Style

⊙ **Grid snap** specifies non-polar snap (**SnapType**; default = on).

⊙ **Rectangular snap** specifies rectangular snap (**SnapStyl**; default = on).

○ **Isometric snap** specifies isometric snap mode (**SnapStyl**; default = off).

○ **Polar snap** specifies polar snap (**SnapType**; default = on).

SnapType	SnapStyl	Meaning
0 (off)	**0** (off)	Rectangular snap.
0 (off)	**1** (on)	Isometric snap.
1 (on)	**0** (off)	Polar snap.

Grid

Grid X spacing specifies the spacing of grid lines in the x direction; when 0, the grid spacing is set to the value of Snap X spacing (**GridUnit**; default = 0.5).

Grid Y spacing specifies the spacing of grid lines in the y direction; when 0, the grid spacing is set to the value of Snap Y spacing (**GridUnit**; default = 0.5).

Major line every specifies the distance between major grid lines (*new to AutoCAD 2007*; **GridMajor**; default = 3).

Grid Behavior

New to AutoCAD 2007:

☑ **Adaptive grid** changes the grid spacing, depending on the zoom level (**GridDisplay** = 2).

☐ **Allow subdivision below grid spacing** draws sub-grid lines (**GridDisplay** = 4).

☐ **Display grid beyond limits** draws the grid beyond the extents set by the Limits command (**GridDisplay** = 0 or 1).

☐ **Follow Dynamic UCS** causes the grid to rotate with the dynamic UCS (**GridDisplay** = 8).

Polar Tracking Tab

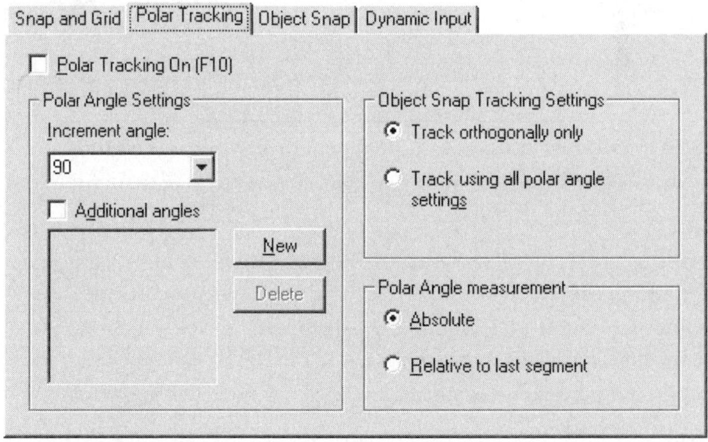

☑ **Polar Tracking On (F10)** turns on and off polar tracking (**AutoSnap**; default = off).

Polar Angle Settings

Increment angle specifies the increment angle displayed by the polar tracking alignment path; select a preset angle — 90, 60, 45, 30, 22.5, 18, 15, 10, or 5 degrees — or enter a value (**PolarAng**).

☑ **Additional angles** allows you to set additional polar tracking angles (**PolarMode**; default=off).

New adds up to ten polar tracking alignment angles (**PolarAddAng**; default = 0;15;23;45).

Delete deletes added angles.

Object Snap Tracking Settings

⊙ **Track orthogonally only** displays orthogonal tracking paths when object snap tracking is on (**PolarMode**).

○ **Track using all polar angle settings** tracks cursor along polar angle tracking path when object snap tracking is turned on (**PolarMode**).

Polar Angle Measurement options

⊙ **Absolute** forces polar tracking angles along the current user coordinate system (**UCS**).

○ **Relative to Last Segment** forces polar tracking angles on the last-created object.

Object Snap Tab

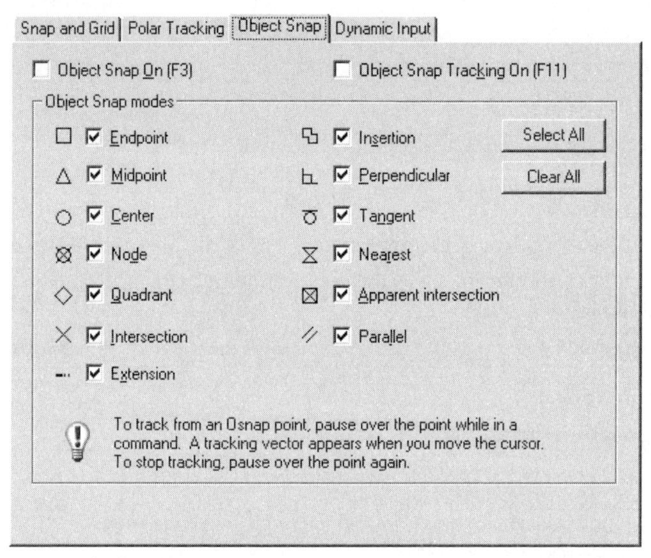

☐ **Object Snap On (F3)** turns on and off running object snaps (**OsMode**).

☐ **Object Snap Tracking On (F11)** toggles object snap tracking (**AutoSnap**).

Object Snap Modes

☑ **ENDpoint** snaps to the nearest endpoint of a line, multiline, polyline segment, ray, arc, and elliptical arc; and to the nearest corner of a trace, solid, and 3D face.

☐ **MIDpoint** snaps to the midpoint of a line, multiline, polyline segment, solid, spline, xline, arc, ellipse, and elliptical arc.

☑ **CENter** snaps to the center of an arc, circle, ellipse, and elliptical arc.

☐ **NODe** snaps to a point.

☐ **QUAdrant** snaps to a quadrant point (90 degrees) of an arc, circle, ellipse, and elliptical arc.

☑ **INTersection** snaps to the intersection of a line, multiline, polyline, ray, spline, xline, arc, circle, ellipse, and elliptical arc; edges of regions; it does not snap to the edges and corners of 3D solids.

☑ **EXTension** displays an extension line from the endpoint of objects; snaps to the point where two objects would intersect if they were infinitely extended; does not work with the edges and corners of 3D solids; automatically turns on intersection mode. (Do not turn on apparent intersection at the same time as extended intersection.)

☐ **INSertion** snaps to the insertion point of text, block, attribute, or shape.

☐ **PERpendicular** snaps to the perpendicular of a line, multiline, polyline, ray, solid, spline, xline, arc, circle, ellipse, and elliptical arc; snaps from a line, arc, circle, polyline, ray, xline, multiline, and 3D solid edge; in this case *deferred perpendicular mode* is automatically turned on.

☐ **TANgent** snaps to the tangent of an arc, circle, ellipse, or elliptical arc; deferred tangent snap mode is automatically turned on when more than one tangent snap is required.

☐ **NEArest** snaps to the nearest point on a line, multiline, point, polyline, spline, xline, arc, circle, ellipse, and elliptical arc.

☐ **APParent intersection** snaps to the apparent intersection of two objects that do not actually intersect but appear to intersect in 3D space; works with a line, multiline, polyline, ray, spline, xline, arc, circle, ellipse, and elliptical arc; does not work with edges and corners of 3D solids.

☐ **PARallel** snaps to a parallel point when AutoCAD prompts for a second point.

Clear All turns off all object snap modes.

Select All turns on all object snap modes.

Dynamic Input Tab

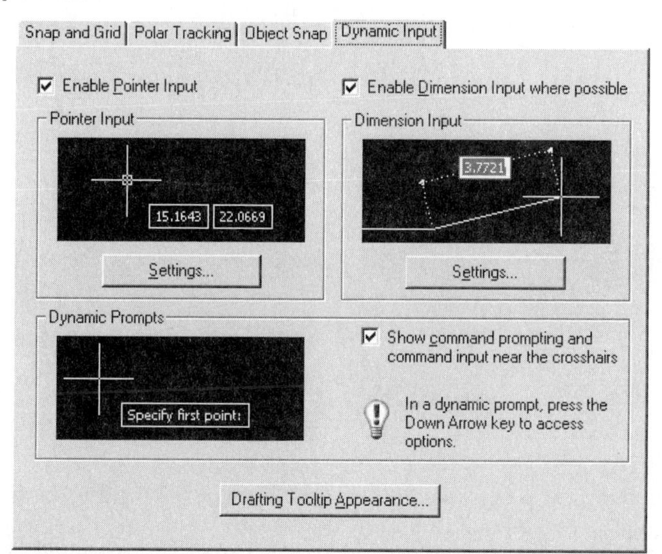

☑ **Enable Pointer Input** toggles coordinate input at the drawing cursor, as illustrated below. Press the **Tab** key to switch between the x and y coordinates of the cursor position.

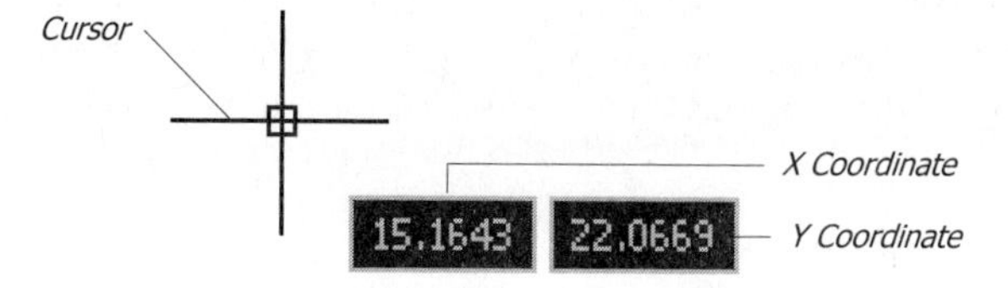

☑ **Enable Dimension Input where possible** toggles the display of lengths and angles when drawing and editing. Press **Tab** to switch between distance and angle.

☑ **Show command prompting near crosshairs** switches the display of commands and prompts between the 'Command:' prompt and the drawing area. Press the **Down Arrow** key to display additional options.

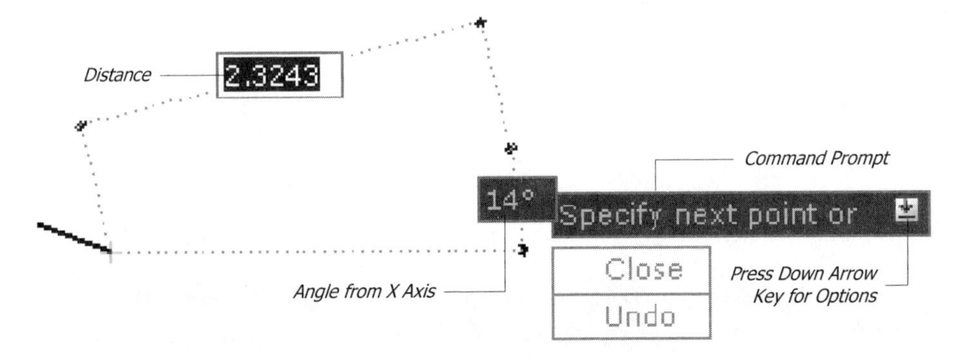

POINT INPUT SETTINGS DIALOG BOX

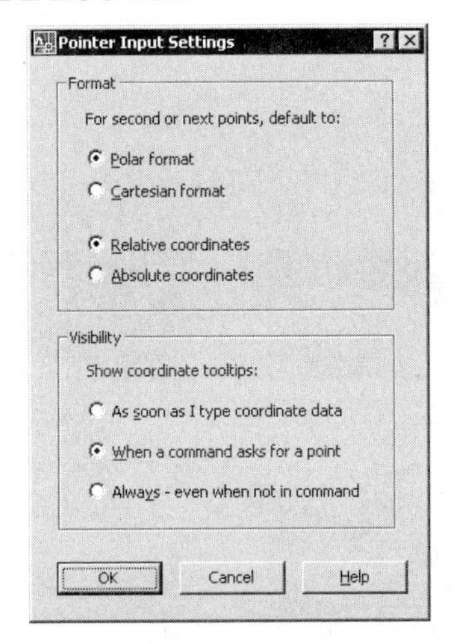

Format

For the second or next points, default to (DynPiFormat and **DynPiCoords)**:

⊙ **Polar Format** shows distance and angle; press **,** to switch to Cartesian format.

○ **Cartesian Format** shows x and y distance; press **<** to change to polar format.

⊙ **Relative Coordinates** shows coordinates relative to the last point; press **#** to switch to absolute format.

○ **Absolute Coordinates** shows coordinates relative to the origin; press **@** to switch to relative format; direct distance entry is disabled when this option is turned on.

Visibility

Show coordinate tooltips (DynPiVis):

○ **As Soon As I Type Coordinate Data** displays tooltips when you enter coordinates.

⊙ **When a Command Asks for a Point** displays tooltips when commands prompt for points.

○ **Always - Even When Not in a Command** always displays tooltips.

DIMENSION INPUT SETTINGS DIALOG BOX

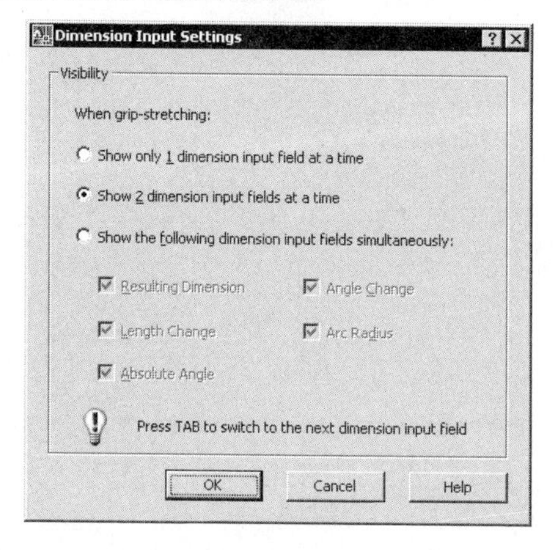

Visibility

When grip stretching (DynDiVis):

○ **Show Only 1 Dimension Input Field at a Time** displays only the length dimension.

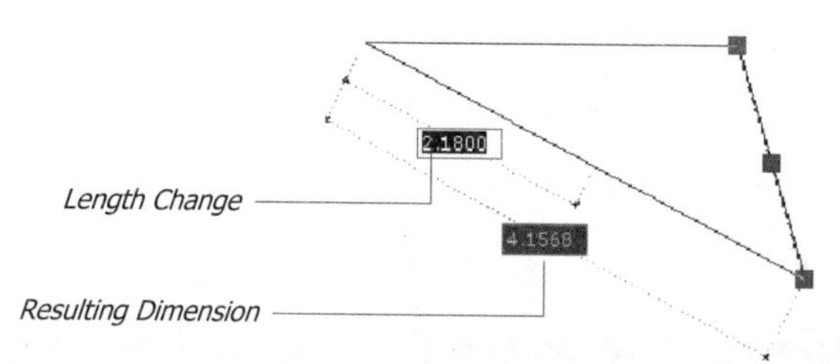

⊙ **Show 2 Dimension Input Fields at a Time** displays the length and resulting dimension.

○ **Show the Following Dimension Input Fields Simultaneously (DynDiGrip):**

☑ **Resulting Dimension** displays the length dimension tooltip.

☑ **Length Change** displays the change in length.

☑ **Absolute Angle** displays the angle dimension tooltip.

☑ **Angle Change** displays the change in the angle.

☑ **Arc Radius** displays the radius of arcs.

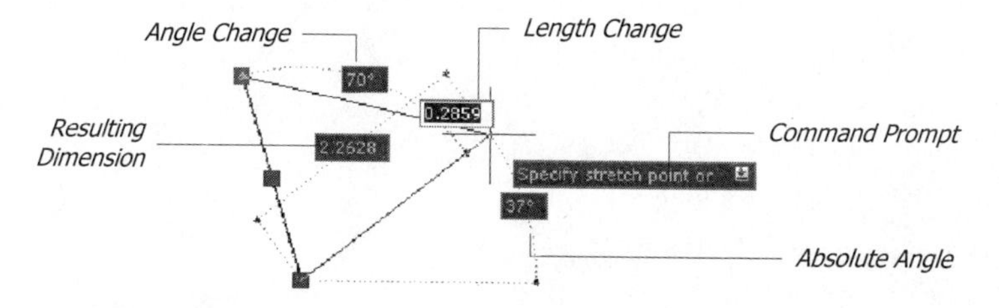

DRAFTING TOOLTIP APPEARANCE DIALOG BOX

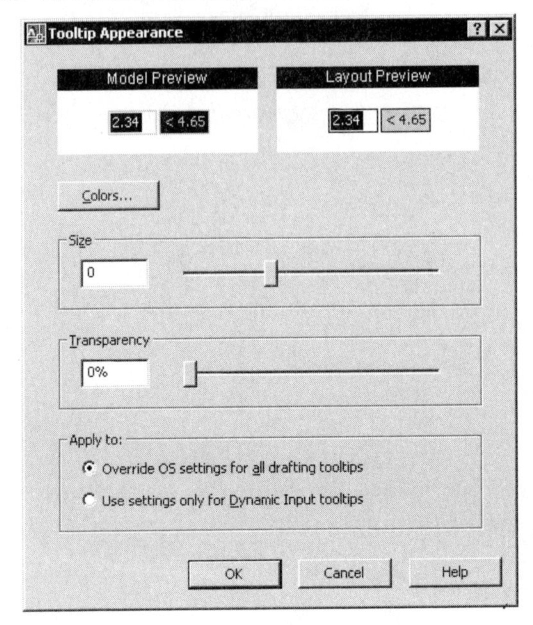

Color

Model Color displays the Select Color dialog box for specifying tooltip colors in model space.

Layout Color displays the same dialog box, but for tooltips in paper space.

Size specifies tooltip size; 0 = default size.

Transparency controls transparency of tooltips; 0% = opaque.

Apply To

⊙ **Override OS Settings for All Drafting Tooltips** applies settings to all tooltips, overriding settings in the operating system.

○ **Use Settings Only for Dynamic Input Tooltips** applies settings to drafting tooltips only.

+DSETTINGS Command

Command: +dsettings

Tab Index <0>: *(Enter a digit.)*

COMMAND LINE OPTION

Tab Index displays the Drafting Settings dialog box with the associated tab:

0 — Displays Snap and Grid tab (*default*).

1 — Displays Polar Tracking tab.

2 — Displays Object Snap tab.

3 — Displays Dynamic Input tab.

RELATED COMMANDS

Grid sets the grid spacing and toggles visibility.

Isoplane selects the working isometric plane.

Ortho toggles orthographic mode.

Snap sets the snap spacing and isometric mode.

RELATED SYSTEM VARIABLES

AutoSnap controls AutoSnap, polar tracking, and object snap tracking.

CmdInputHistoryMax limits the maximum number of commands stored in history.

CrossingAreaColor specifies the color for crossing area selections.

DynDIGrip determines which dynamic dimensions are displayed during grips editing.

DynDIVis determines which dynamic dimensions are displayed during grips editing.

DynMode toggles dynamic input.

DynPICoords switches pointer input between relative and absolute coordinates.

DynPIFormat switches between polar and Cartesian format for coordinates.

DynPIVis toggles pointer input.

DynPrompt toggles prompts in dynamic input.

DynTooltips determines which tooltips are affected by appearance settings.

GridMode indicates the current grid visibility.

GridUnit indicates the current grid spacing (default = 0.0).

InputHistoryMode specifies where each command history is displayed.

MButtonPan toggles the middle mouse button between panning and *.cui* definition.

PolarAddAng specifies user-defined polar angles, separated by semicolons.

PolarAng specifies the increments of the polar angle.

PolarDist specifies the polar snap distance.

PreviewEffect specifies whether selection preview applies dashes and/or thickness to objects.

PreviewFilter excludes xrefs, tables, mtext, hatches, groups, and objects on locked layers.

SnapAng specifies the current snap rotation angle (default = 0).

SnapBase sets the base point of the snap rotation angle (default = 0,0).

OsMode holds the current object snap modes.

PolarMode holds the settings for polar and object snap tracking.

SelectionArea toggles area selection coloring.

SelectionAreaOpacity specifies the transparency of the area selection color.

SelectionPreview determines when selection preview is active.

SnapIsoPair specifies the current isoplane.

SnapMode sets the current snap mode setting.

SnapStyl specifies the snap style setting.

SnapUnit sets the current snap spacing (default = 1,1).

WindowAreaColor specifies the color for windowed area selections.

TIPS

- Use snap to set the cursor movement increment.
- Use the grid as a visual display to help you better gauge distances.
- Use object snaps to draw precisely to geometric features.
- **m2p**, a running object snap, snaps to the midpoint of two picked points.

Relocated Command

Look for the **Dtext** command under **Text**.

'DsViewer

Rel.13 Displays the bird's-eye view palette; provides real-time pan and zoom (short for DiSplay VIEWer).

Command	Alias	Ctrl+	F-key	Alt+	Menu Bar	Tablet
dsviewer	av	...	...	VW	View ↳Aerial View	K2

Command: dsviewer

Displays the Aerial View palette:

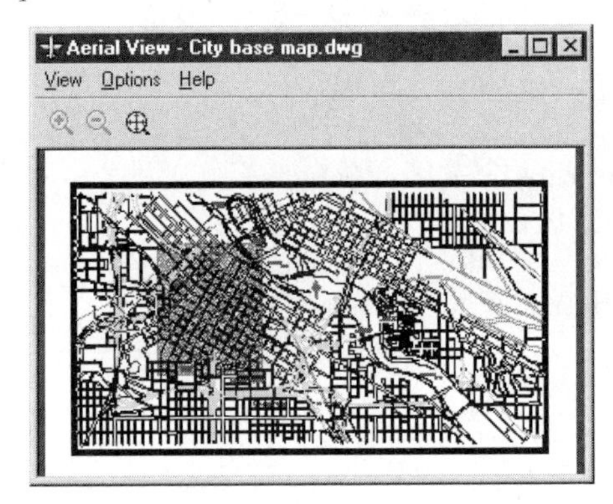

MENU BAR

View Menu

Zoom In increases centered zoom by a factor of 2.

Zoom Out decreases centered zoom by a factor of 2.

Global displays entire drawing in Aerial View palette.

Options Menu

Auto Viewport updates the Aerial View automatically with the current viewport.

Dynamic Update updates the Aerial View automatically with editing changes in the current viewport.

Realtime Zoom updates the drawing in real time as you zoom in the Aerial View palette.

Palette Toolbar

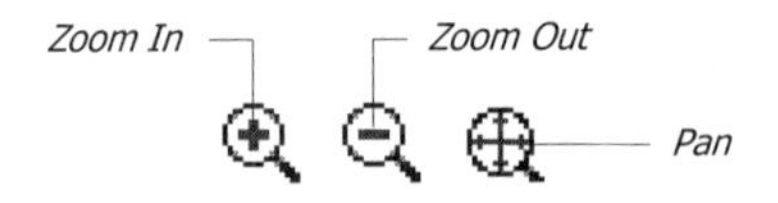

RELATED COMMANDS

Pan moves the drawing view.

View creates and displays named views.

Zoom makes the view larger or smaller.

TIPS

- The purpose of the Aerial View is to let you see the entire drawing at all times, and to zoom and pan without entering the Zoom and Pan commands, or selecting items from the menu.
- The parts of the Aerial View palette:

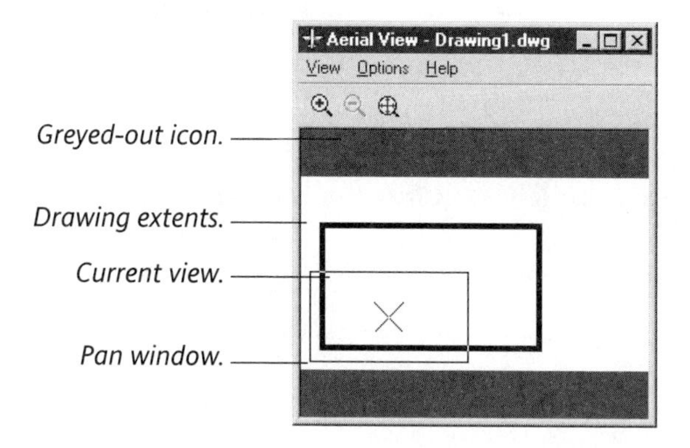

- *Warning!* When in paper space, the Aerial View palette shows only paper space objects.

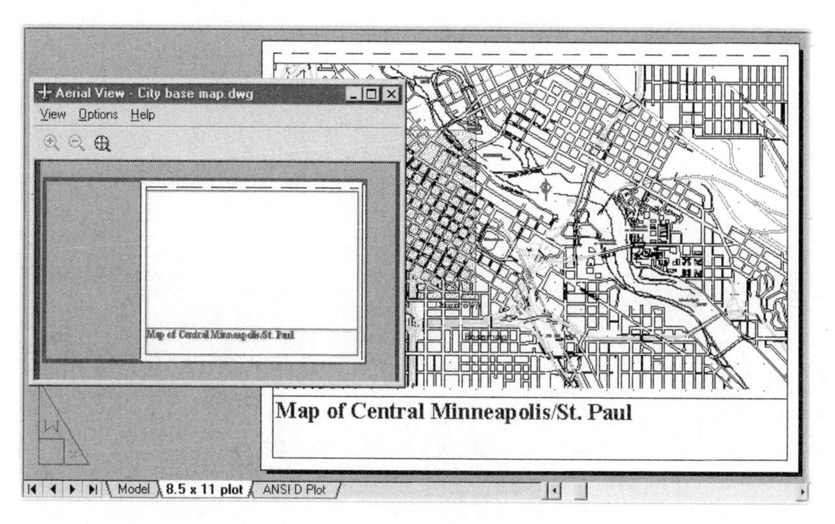

- To switch quickly between Pan (default) and Zoom modes, click on the Aerial View palette.
- Right-click or press **Enter** to lock the Aerial View palette and return cursor to editing window.

Renamed Command

DText was reanmed in AutoCAD 2000 as an alias of the **Text** command.

DView

Rel.10 Zooms and pans 3D drawings dynamically, turns on perspective mode for 3D drawings; superceded by the **3dOrbit** command (short for Dynamic VIEW).

Command	Alias	Ctrl+	F-key	Alt+	Menu Bar	Tablet
dview	dv	...	...	...	...	...

Command: dview

Select objects or <use DVIEWBLOCK>: *(Select objects, or press Enter.)*

Enter option [CAmera/TArget/Distance/POints/PAn/Zoom/TWist/CLip/Hide/Off/Undo]: *(Enter an option.)*

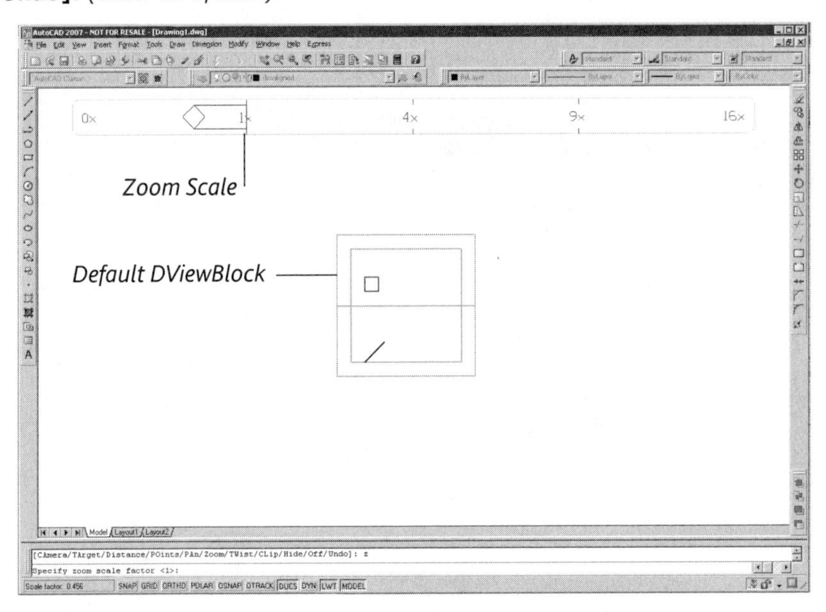

COMMAND LINE OPTIONS

CAmera indicates the camera angle relative to the target.

Toggle switches between input angles (*available after starting CAmera or TArget*).

TArget indicates the target angle relative to the camera.

Distance indicates the camera-to-target distance; turns on perspective mode.

POints indicates both the camera and target points.

PAn pans the view dynamically.

Zoom zooms the view dynamically.

TWist rotates the camera (the view).

CLip options

Enter clipping option [Back/Front/Off] <Off>:

Back clip options

Specify distance from target or [ON/OFF] <0.0>:

ON turns on the back clipping plane.

OFF turns off the back clipping plane.

Distance from target locates the back clipping plane.

Front Clip options

Specify distance from target or [set to Eye(camera)/ON/OFF] <1.0>:

Eye positions the front clipping plane at the camera.

Distance from target locates the front clipping plane.

Off turns off view clipping.

Hide removes hidden lines.

Off turns off the perspective view.

Undo undoes the most recent **DView** action.

eXit exits **DView**.

RELATED COMMANDS

Hide removes hidden lines from non-perspective views.

Pan pans non-perspective views.

VPoint selects non-perspective viewpoints of 3D drawings.

Zoom zooms non-perspective views.

3dOrbit creates 3D views interactively, in parallel or perspective mode.

RELATED SYSTEM VARIABLES

BackZ specifies the back clipping plane offset.

FrontZ specifies the front clipping plane offset.

LensLength specifies the perspective view lens length, in millimeters.

Target specifies the UCS 3D coordinates of target point.

ViewCtr specifies the 2D coordinates of current view center.

ViewDir specifies the WCS 3D coordinates of camera offset from target.

ViewMode specifies the perspective and clipping settings.

ViewSize specifies the height of view.

ViewTwist specifies the rotation angle of current view.

RELATED SYSTEM BLOCK

DViewBlock is the alternate viewing object displayed during DView.

TIPS

- The view direction is from the camera to target.
- Press **Enter** at the 'Select objects' prompt to display the DViewBlock house. You can replace the house block with your own by redefining the DViewBlock block.
- To view a 3D drawing in one-point perspective, use the Zoom option.
- Menus and transparent zoom and pan are not available during DView. Once the view is in perspective mode, you cannot use any command that requires pointing with the cursor.

Renamed Commands

DwfOut was was made part of **Plot** with AutoCAD 2004.

DwfOutD was removed from AutoCAD 2000; it was combined with **DwfOut**.

Two- and 3-point Perspectives

In two-point perspective, the camera and the target are at the same height. Vertical lines remain vertical. In three-point perspective, the camera and target are at different heights.

Two-Point Perspectives

Place Camera and Target

1. Start **DView** and select all objects:

    Command: dview
    Select objects: all
    Select objects: *(Press* **ENTER**.*)*

2. The **POints** option combines the **TArget** and **CAmera** options:

    CAmera/TArget/Distance/POints/.../Undo/<eXit>: po

3. Use the **.xy** filter to pick the target point:

    Enter target point <0.4997, 0.4999, 0.4997>: .xy
    of *(Pick target point.)*

4. Enter a number for your eye height, such as 5'10" or 180cm:

    (need Z): *(Enter height.)*

5. Use the **.xy** filter to pick the camera point:

    Enter camera point <0.4997, 0.4999, 1.4997>: .xy
    of *(Pick camera point.)*

6. Type the same z coordinate for the camera height:

    (need Z): *(Enter same height as in #3, above.)*

Turn On Perspective Mode

1. The **Distance** option turns on perspective mode:

    CAmera/TArget/Distance/POints/.../Undo/<eXit>: d

2. In perspective mode, the UCS icon becomes a perspective icon. Use the slider bar to set the distance while in **Distance** mode:

    New camera/target distance <1.0943>: *(Move slider bar.)*

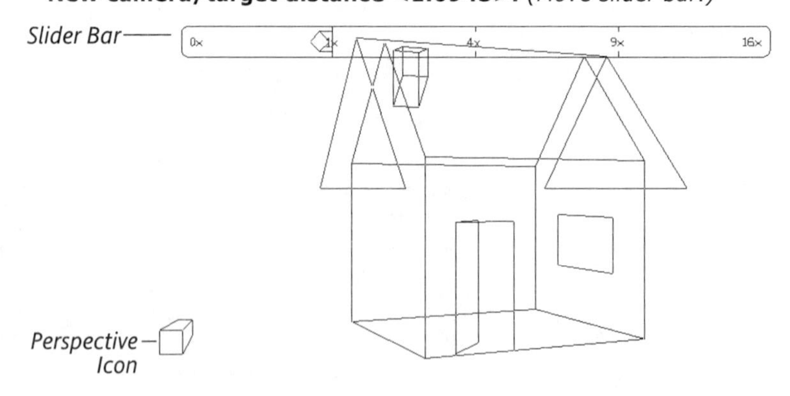

Three-Point Perspective

In three-point perspective, the target and camera heights differ. Most commonly, the camera is higher than the target, so that you look down on the 3D scene.

1. Follow the earlier steps, but change the camera and target heights, as follows:

   Command: dview
   Select objects: all
   1 found Select objects: *(Press* **ENTER**.*)*
   CAmera/TArget/Distance/POints/.../Undo/<eXit>: po

2. For target height, enter the height of an object you are looking at, such as a window or table:

   Enter target point <0.4997, 0.4999, 0.4997>: .xy
   of *(Pick target point.)*
   (need Z): *(Enter a height.)*

3. For the camera height, enter your eye height or a larger number for a bird's-eye view:

   Enter camera point <0.4997, 0.4999, 1.4997>: .xy
   of *(Pick camera point.)*
   (need Z): *(Enter a height greater than in #2, above.)*
   CAmera/TArget/Distance/POints/.../Undo/<eXit>: d
   New camera/target distance <1.0943>: *(Adjust distance.)*

4. Use the **Hide** option to create a hidden-line view:

   CAmera/TArget/Distance/POints/.../Hide/Off/Undo/<eXit>: h

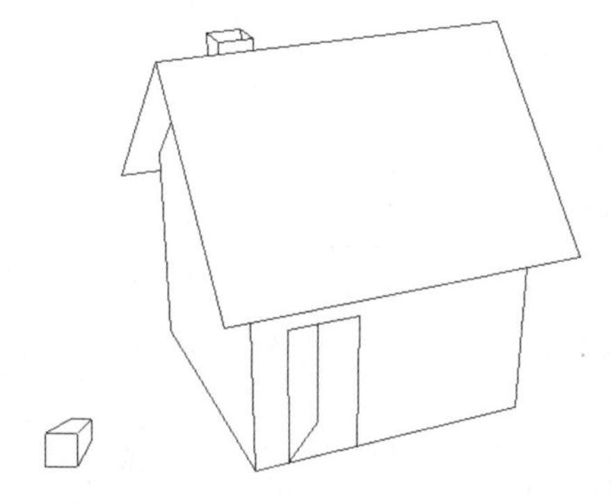

Hidden-line view in three-point perspective mode.

Exiting Dview and Perspective Mode

1. Exit **DView**:

   CAmera/TArget/Distance/POints/.../Undo/<eXit>: *(Press* **ENTER**.*)*

   The view remains in perspective mode. While in perspective mode, the scroll bars do not work, nor does any command requiring points specified in the drawing.

2. To exit perspective mode, use the **Plan** command.

DwfAdjust

2007 Adjusts the contrast, fade, and color of DWF underlays.

Command	Alias	Ctrl+	F-key	Alt+	Menu Bar	Tablet
dwfadjust	...	...	...	...	...	...

Command: dwfadjust
Select DWF underlay: *(Select objects.)*
Select DWF underlay: *(Press Enter.)*
Enter DWF Underlay option [Fade/Contrast/Monochrome] <Fade>: *(Enter option.)*

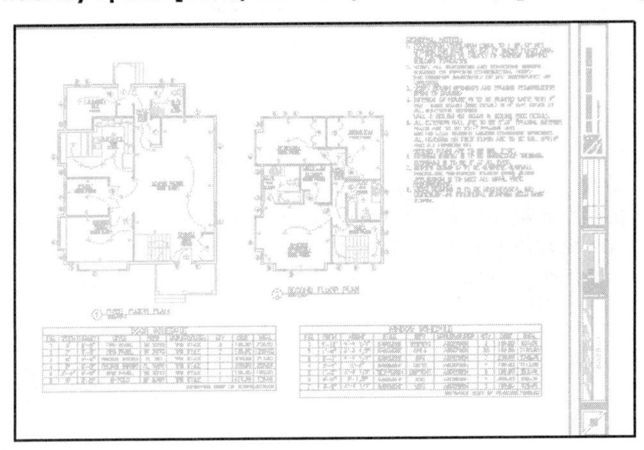

DWF underlay with fade set to maximum (80).

COMMAND LINE OPTIONS

Select DWF underlay selects the underlay to adjust.

Enter fade value specifies the amount of fade:

0 — no fade.

25 — default value.

80 — maximum fade.

Enter contrast value specifies the change in contrast, ranging from 0 to 100.

Monochrome toggles monochrome colorization.

RELATED COMMANDS

DwfAttach attaches *.dwf* files as underlays.

DwfClip clips DWF underlays.

TIPS

- This command has no effect on the frame around the DWF underlay. Use the DwfFrame system variable to toggle its display.
- Use the **Monochrome** option to change all of the underlay's colors to black; use the **Fade** option to change the colors to gray.

DwfAttach

2007 Attaches 2D *.dwf* files as xref-like underlays to the current drawing..

Command	Alias	Ctrl+	F-key	Alt+	Menu Bar	Tablet
dwfattach	...	...	...	**IDD**	**Insert** ↳**DWF Underlay**	...
-dwfattach						

Command: dwfattach

Displays the Select DWF File dialog box.

Select a file, and then click ***Open.***

Displays the Attach DWF Underlay dialog box:

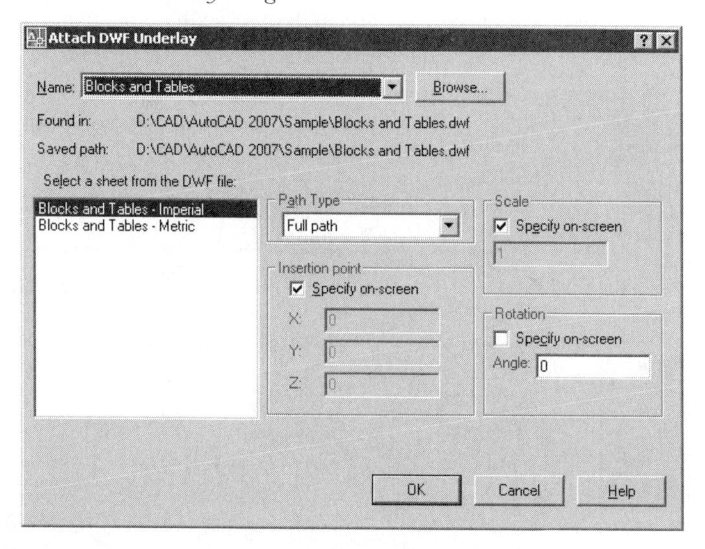

DIALOG BOX OPTIONS

Name specifies the *.dwf* file to attach to the current drawing; click the droplist for other *.dwf* files. More than one *.dwf* file can be attached to drawings. (This droplist is disabled when the *.dwf* file is stored in Autodesk's Vault software.)

Browse displays the Select DWF File dialog box to select another *.dwf* file.

Select a sheet from the DWF file lists the names of sheets in the *.dwf* file; 3D *.dwf* sheets are not listed. Only one sheet can be attached at a time.

Path Type specifies the type of path to the .dwf file:

- **Full Path** saves the full path name.
- **Relative Path** saves the relative path.
- **No Path** saves the file name only.

Insertion Point specifies the *.dwf* underlay's insertion point (lower left corner); default =0,0,0.

Specify On-Screen:

☑ prompts you for the insertion point at the command line after closing the dialog box.

☐ enters the x, y, z coordinates in the dialog box.

Scale specifies the DWF underlay's scale factor; default = 1.0

Specify On-Screen:

☑ prompts you for the scale factor at the command line after closing the dialog box.

☐ enters the scale factor in the dialog box.

Angle specifies the DWF underlay's angle of rotation; default = 0 degrees.

Specify On-Screen:

☑ prompts you for the rotation angle at the command line after closing the dialog box.

☐ enters the angle in the dialog box.

RELATED COMMANDS

DwfAdjust adjusts the contrast, fade, and color of DWF underlays.

DwfClip clips DWF underlays.

Markup places marked-up (redlined) *.dwf* files in the current drawing.

Plot creates single-sheet *.dwf* files of the current drawings.

Publish creates multi-sheet *.dwf* files of multiple drawings.

3dDwfOut creates 3D *.dwf* files.

RELATED SYSTEM VARIABLES

DwfFrame toggles display of the rectangular frame around DWF underlays.

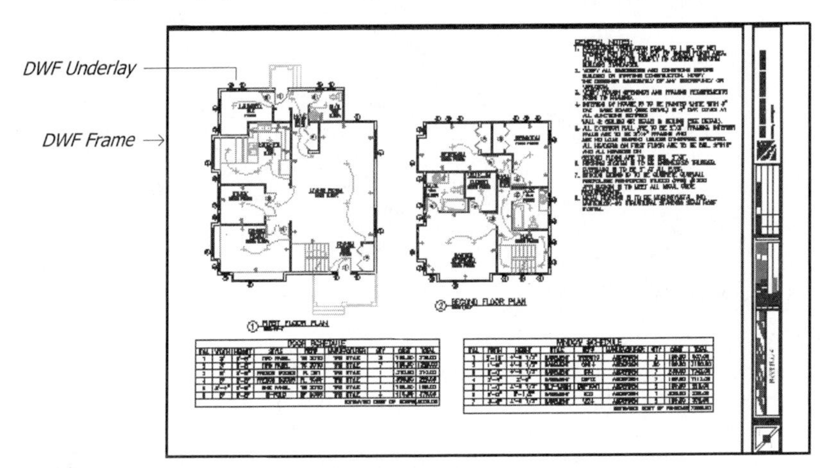

DwfOsnap toggles whether osnap recognizes the frame of DWF underlays.

TIPS

- This command cannot import 3D *.dwf* files.
- Double-click the underlay to display its properties.
- To remove the underlay from the drawing, simply select it, and then press **Del**.

DwfClip

2007 Clips DWF underlays to isolate portions of the drawing.

Command	Alias	Ctrl+	F-key	Alt+	Menu Bar	Tablet
dwfclip	...	...	...	...	...	...

Command: dwfclip

Select DWF to clip: *(Select one underlay.)*

Enter DWF clipping option [ON/OFF/Delete/New boundary] <New boundary>: *(Enter an option.)*

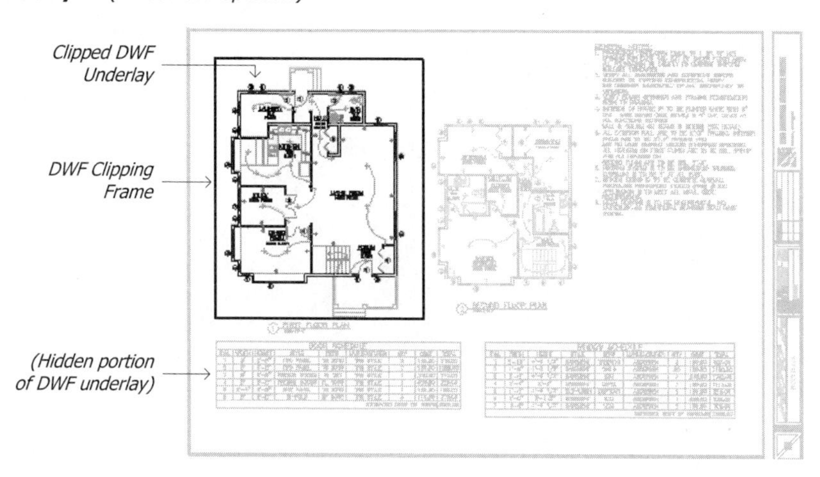

COMMAND LINE OPTIONS

ON turns on a previously-created clipping boundary.

OFF turns off the clipping boundary, so that it is hidden.

Delete removes the clipping boundary, restoring the DWF underlay.

New boundary creates a new clipping boundary:

- **Polygonal** creates a multi-side boundary using prompts similar to the PLine command.
- **Rectangular** creates a rectangular boundary based on two pick points.

RELATED COMMANDS

DwfAdjust adjusts the contrast, fade, and color of DWF underlays.

DwfAttach attaches *.dwf* files as underlays.

RELATED SYSTEM VARIABLE

DwfFrame toggles the display of the rectangular frame around DWF underlays.

TIPS

- After the clipping boundary is in place, you can use grips editing to adjust its size.
- This command can be started by selecting the DWF border, right-clicking, and then selecting **DWF Clip** from the shrotcut menu.

DwgProps

2000 Records and reports information about drawings (short for DraWinG PROPertieS).

Command	Alias	Ctrl+	F-key	Alt+	Menu Bar	Tablet
dwgprops	...	...	...	**FI**	**File ⇨Drawing Properties**	...

Command: dwgprops

Displays tabbed dialog box; see below.

DRAWING PROPERTIES DIALOG BOX

Displays information about the drawing obtained from the operating system:

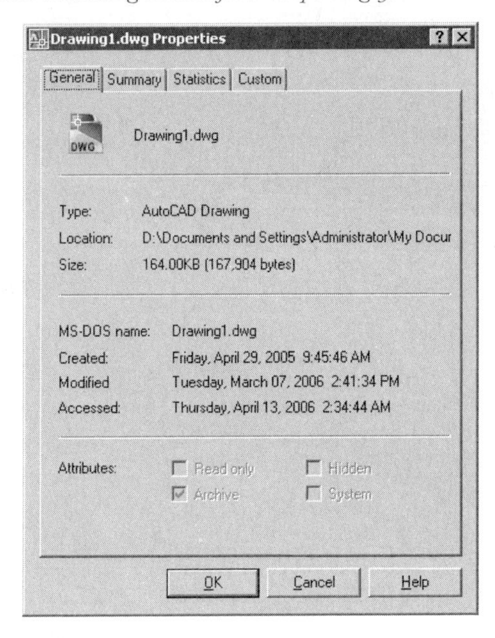

General

File Type indicates the type of file.

Location indicates the location of the file.

Size indicates the size of the file.

MS-DOS Name provides the MS-DOS file name truncated to eight characters with three-letter extension.

Created indicates the date and time the file was first saved.

Modified indicates the date and time the file was last saved.

Accessed indicates the date and time the file was last opened.

Attributes

Read-Only indicates the file cannot be edited or erased.

Archive indicates the file has been changed since it was last backed up.

Hidden indicates the file cannot be seen in file listings.

System indicates the file is a system file; DWG drawing files never have this attribute turned on.

OK records the changes, and exits the dialog box.

Cancel discards the changes, and exits the dialog box.

Summary Tab

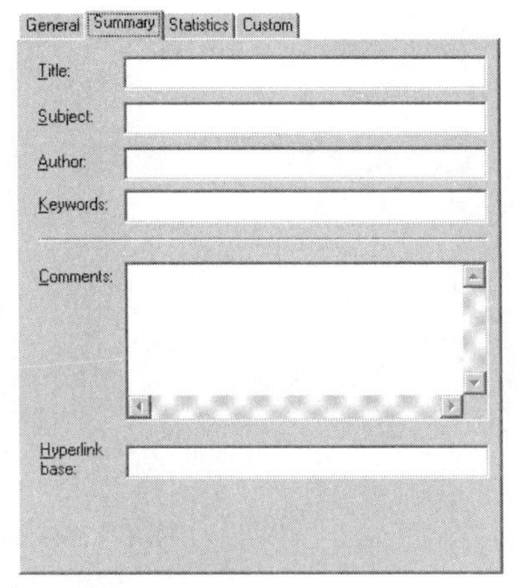

Title specifies a title for the drawing; is usually different from the file name.

Subject specifies a subject for the drawing.

Author specifies the name of the drafter of the drawing.

Keywords specifies keywords used by the operating system's Find or Search commands to locate drawings.

Comments contains comments on the drawing.

Hyperlink Base specifies the base address for relative links in the drawing, such as http://www .upfrontezine.com; may be an operating system path name, such as *c:\autocad 2007*, or a network drive name. (Stored in system variable HyperlinkBase.)

Statistics Tab

Displays information about the drawing obtained from the drawing:

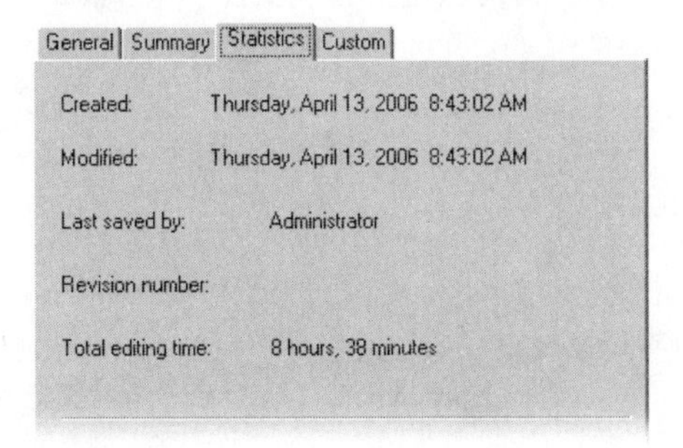

Created indicates the date and time the drawing was first opened (**TdCreate**).

Modified indicates the date and time the drawing was last opened or modified (**TdUpdate**).

Last saved by indicates who last accessed the drawing (**LoginName**).

Revision number indicates the revision number; usually blank.

Total editing time indicates the total amount of time that the drawing has been open (**TdInDwg**).

Custom Tab

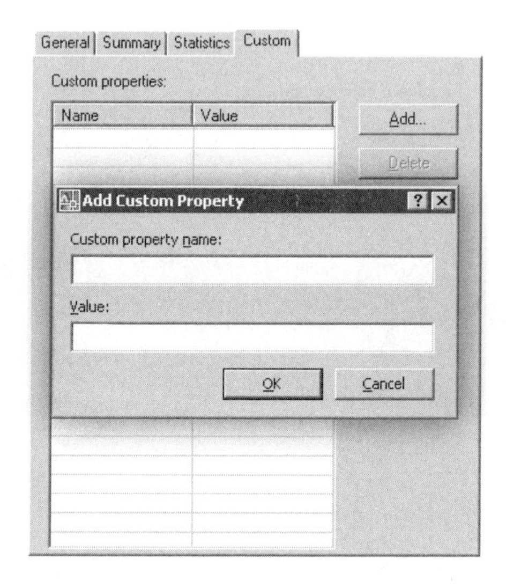

Custom Properties options

Add displays the Add Custom Property dialog box; enter a name and a value. To edit values, click them under **Value**.

Delete removes the custom property.

RELATED COMMANDS

Properties lists information about objects in the drawing.

Status lists information about the drawing.

TIPS

- Use the Custom Properties tab to include project data in drawings.
- Use the **Find** button in the Design Center (or Search button in Windows Explorer) to search for drawings containing values stored in the Custom Properties tab, as well as for text in the drawing.
- You can use Google to search for text in *.dwg* files located on the Internet.

DxbIn

Ver. 2.1 Imports *.dxb* files into drawings (short for Drawing eXchange Binary INput).

Command	Alias	Ctrl+	F-key	Alt+	Menu Bar	Tablet
dxbin	...	...	...	IE	Insert ↳Drawing Exchange Binary	...

Command: dxbin

Displays Select DXB File dialog box: Select a .dxb file, and then click ***Open***.

DIALOG BOX OPTION

Open opens the *.dxb* file, and inserts it in the drawing.

RELATED COMMANDS

DxfIn reads DXF-format files.

Plot writes DXB-format files when configured for an ADI plotter.

TIPS

- To produce *.dxb* files, configure AutoCAD with the ADI plotter driver: after starting the PlotterManager command's Add-a-Plotter wizard, select "AutoCAD DXB File" as the manufacturer.
- This command was created for an early software product named CAD\camera, which converted raster scans into the DXB vector format. CAD\camera had the distinction of being Autodesk's first software release following the success of AutoCAD — and Autodesk's first failure following AutoCAD.

DxfIn, DxfOut

V. 2.0 Opens and exports *.dxf* files into and from drawings (short for Drawing interchange Format INput / OUTput; undocumented commands).

Command	Alias	Ctrl+	F-key	Alt+	Menu Bar	Tablet
dxfin	...	...	...	**FO**	**File**	...
				↳DXF	**↳Open**	
					↳DXF	
dxfout	...	...	...	**FA**	**File**	...
				↳DXF	**↳Save As**	
					↳DXF	

Command: dxfin

Displays Select File dialog box. Select a .dxf file, and then click ***Open.***

DXFOUT COMMAND

Command: dxfout

Displays Save Drawing As dialog box. Enter a file name, and then click ***Save.***

DIALOG BOX OPTIONS

Save saves the drawing as a *.dxf* file.

Files of type creates a *.dxf* file compatible with these versions of AutoCAD:

- AutoCAD 2007.
- AutoCAD 2004 (compatible with 2005 and 2006).
- AutoCAD 2000 (compatible with 2000, 2000i, and 2002, and AutoCAD LT).
- AutoCAD Release 12 and AutoCAD LT Release 2 (compatible with Release 13 and 14).

SAVEAS OPTIONS DIALOG BOX

From the Tool menu, select Options:

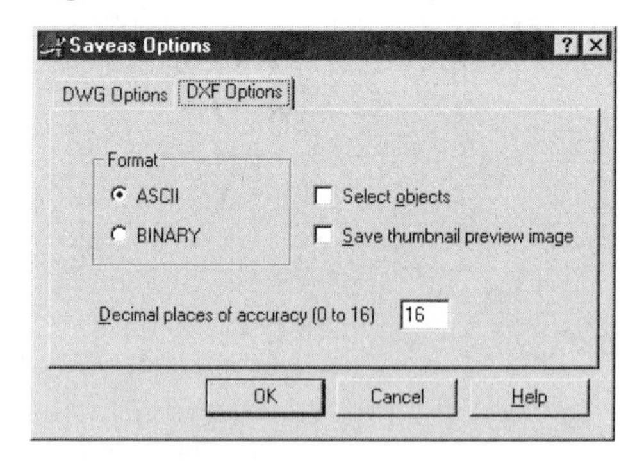

Format

⊙**ASCII** creates files in text format, readable by humans, and imported by most applications.

○**Binary** creates binary files with a smaller file size, but cannot be read by all applications.

Additional

☐ **Select objects** selects objects to export, instead of the entire drawing.

☐ **Save thumbnail preview image** includes a preview image in the *.dxf* file.

Decimal places of accuracy (0 to 16) specifies the decimal places of accuracy.

RELATED COMMANDS

DxbIn reads DXB-format files.

AcisOut saves solid model objects in the drawing as ACIS-compatible SAT format.

SaveAs writes drawings in DWG and DXF formats.

TIPS

- Alternatively, you can use the Open command to open *.dxf* files.
- The *.dxf* file comes in two styles: *complete* and *partial*:

  Complete *.dxf* files contain all data required to reproduce a complete drawing.

  Partial *.dxf* files must be imported into existing drawings.

  To load a complete *.dxf* file, AutoCAD requires the current drawing to be empty. Partial *.dxf* files can be imported or inserted into any drawing, empty or not.
- Prior to AutoCAD 2006, DxfIn required a new drawing. (To create an empty drawing, use New with the **Start from Scratch** option.) As of AutoCAD 2006, *.dxf* files are opened in new drawings automatically.
- If you need to import the complete *.dxf* file into a non-empty drawing, use the Insert command, and insert the *.dxf* file with the **Explode** option turned on.
- Use the ASCII DXF format to exchange drawings with other CAD and graphics programs. Some applications, such as those for CNC (computer numerically controlled) machines, require 4 decimal places.
- Binary *.dxf* files are much smaller and are created much faster than ASCII binary files; few applications, however, read binary *.dxf* files.
- The AutoCAD Release 12 dialect of DXF is the most compatible with other applications.
- *Warning!* When saving a drawing in DXF format for earlier releases, AutoCAD 2007 erases or converts some objects into simpler objects.
- Autodesk documents the DXF format at usa.autodesk.com/adsk/servlet/item?siteID=123112&id=752569

EAttEdit

2002 Edits attribute values and properties in a selected block (short for Enhanced ATTribute EDITor).

Command	Alias	Ctrl+	Key	Alt+	Menu Bar	Tablet
eattedit	...	...	...	**MOAS**	**Modify ⇘Object ⇘Attribute ⇘Single**	...

Command: eattedit

If the drawing contains no blocks with attributes, AutoCAD complains, "This drawing contains no attributed blocks," and the command exits.

When attributes exist, the command continues:

Select a block: *(Select a single block.)*

Displays dialog box:

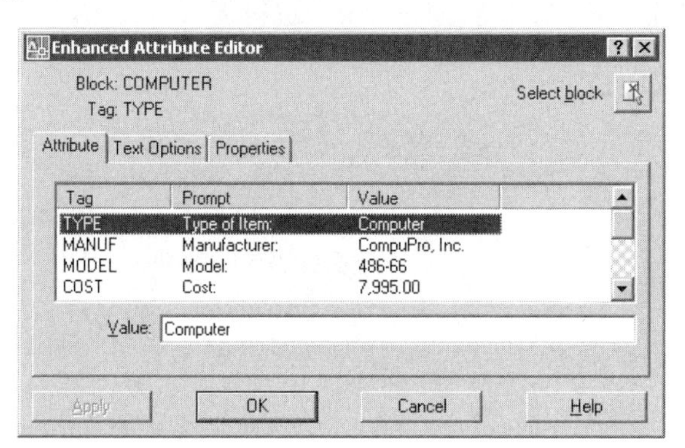

ENHANCED ATTRIBUTE EDITOR DIALOG BOX

Select block selects another block for attribute editing.

Apply applies the changes to the attributes.

Value modifies the value of the selected attribute; neither the tag nor the prompt can be modified by this command.

Text Options tab

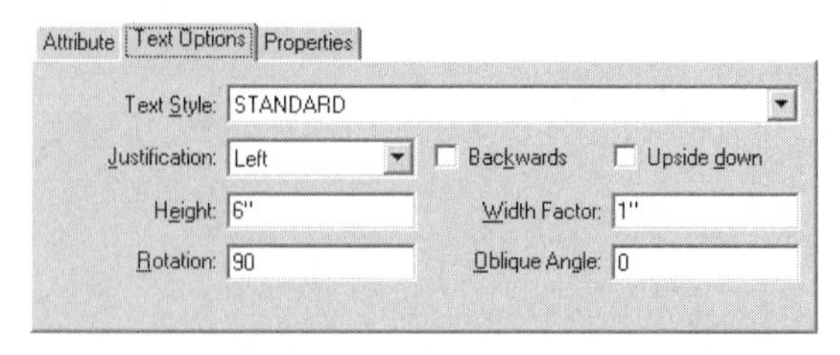

Text Style selects a text style name from the list; text styles are defined by the Style command; default is Standard.

Justification selects a justification mode from the list; default is left justification.

Height specifies the text height; can be changed only when height is set to 0.0 in the text style.

Rotation specifies the rotation angle of the attribute text; default = 0 degrees.

☐ **Backwards** displays the text backwards.

☐ **Upside Down** displays the text upside down.

Width Factor specifies the relative width of characters; default = 1.

Oblique Angle specifies the slant of characters; default = 0 degrees.

Properties tab

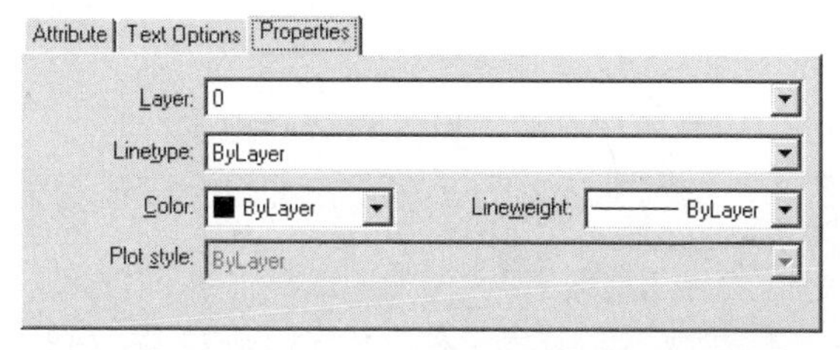

Layer selects a layer name from the list; layers are defined by the **Layer** command.

Linetype selects a linetype name from the list; linetypes are loaded into the drawing with the **Linetype** command.

Color selects a color from the list; to select from the full 255-color spectrum, select **Other**.

Lineweight selects a lineweight from the list; to display lineweights, click **LWT** on the status bar.

Plot style selects a plot style name from the list; available only if plot styles are enabled in the drawing.

RELATED COMMANDS

AttDef creates attribute definitions.

Block attaches attributes to objects.

BAttMan manages attributes.

EAttExt extracts attributes to a file.

TIPS

- This command edits only attribute values and their properties; to edit all aspects of an attribute, use the BAttMan command.
- If you select a block with no attributes, AutoCAD complains, 'The selected block has no editable attributes.' When you select an object that isn't a block, AutoCAD complains, 'Error selecting entity.'
- Blocks on locked layers cannot be edited.

EAttExt

2002 Extracts attribute data to file via a step-by-step procedure (short for Enhanced ATTribute EXTraction).

Command	Alias	Ctrl+	Key	Alt+	Menu Bar	Tablet
eattext	...	...	...	**TX**	**Tools** ↳**Attribute Extraction**	...
-eattedit						

Command: eattext

Displays dialog box.

ATTRIBUTE EXTRACTION DIALOG BOX

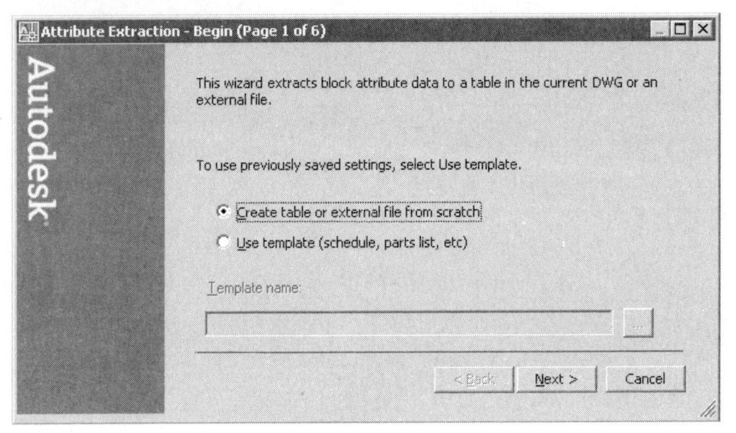

⊙**Create table or external file from scratch** creates them from scratch. (Select this option when a template has not yet been created.)

○**Use template** uses a predefined *.blk* template file.

Template name specifies the name of the *.blk* file; click the **...** button for the Open Template dialog box.

Back goes back to the previous step.

Next proceeds to the next step.

Cancel exits the dialog box.

Select Drawings page

*After clicking **Next**, this dialog box appears:*

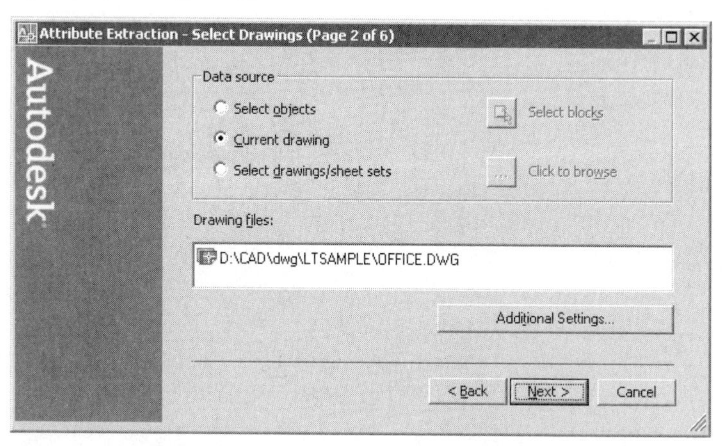

Data Source specifies the location of the blocks containing attributes:

○ **Select objects** selects the specific blocks in the current drawing.

Select Blocks dismisses the dialog box to pick blocks in the drawing.

⊙ **Current drawing** selects all blocks containing attributes in the current drawing.

○ **Select drawings** selects drawings located on your computer or network.

... Click to Browse displays the Select Files dialog box.

Drawing Files lists the names of drawing files from which attributes will be extracted.

Additional Settings displays dialog box:

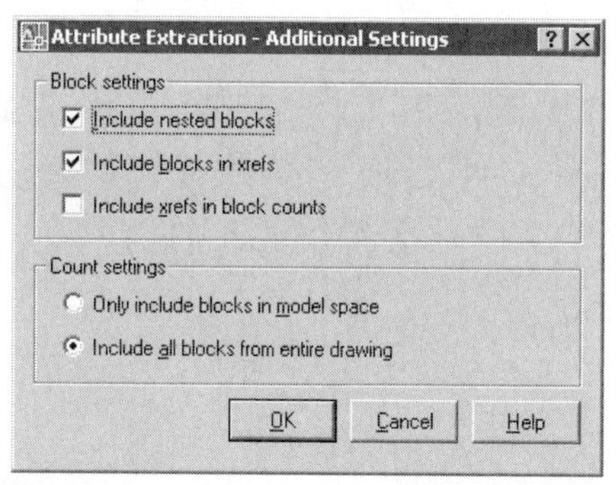

☑ **Include Nested Blocks** includes *nested* blocks (blocks defined in blocks).

☑ **Include Blocks in Xrefs** includes blocks stored in externally-referenced drawing files.

☐ **Include Xrefs in Block Counts** counts xrefs as blocks.

○ **Only Include Blocks in Model Space** excludes blocks inserted in paper space (layouts).

⊙ **Include All Blocks from Entire Drawing** counts blocks in the entire drawing.

Select Attributes page

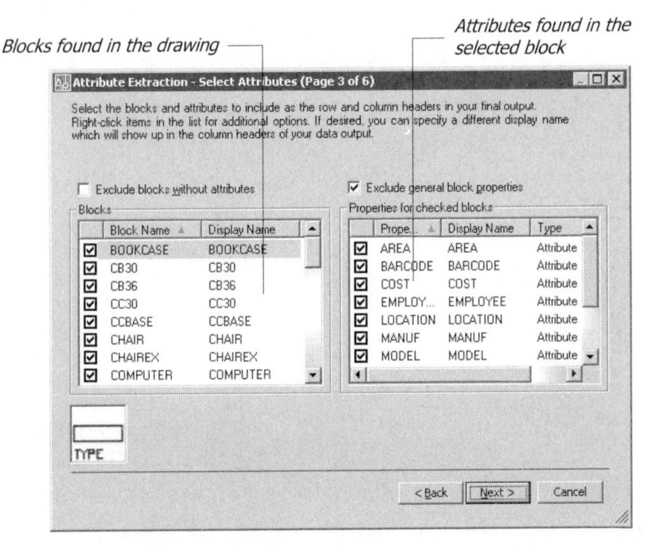

☐ **Exclude blocks without attributes** leaves out blocks that lack user-defined attributes.

☑ **Exclude general block properties** leaves out attributes defined by AutoCAD.

Blocks and **Properties for Checked Blocks**:

☑ *or* ☐ Includes or excludes blocks and attributes by checking or unchecking boxes in left or right columns, respectively.

Note: This dialog box calls attributes "properties." AutoCAD 2006 removes the ability to give aliases to blocks and attributes.

Finalize Extraction page

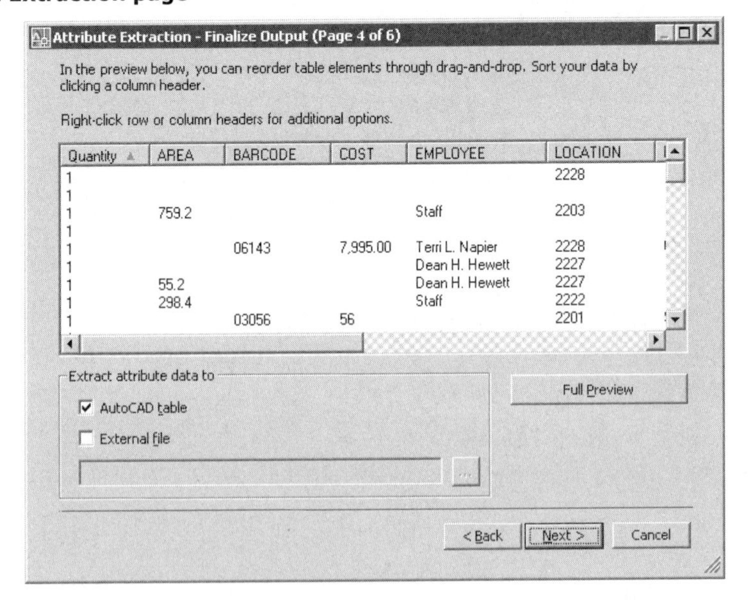

Columns can be resized and re-ordered by dragging. Right-click columns for options.

☑**AutoCAD Table** extracts data to tables in the drawing; displays the Table Style page after clicking **Next**.

☐**External File** specifies the file name to which the information is exported in comma-separated file (CSV) or tab-separated file (TXT) formats.

Full Preview previews the output in a window. (The Alternate View option was removed from AutoCAD 2006.) Press **ESC** to return.

Attribute Extraction - Full Preview

Quantity	AREA	BARCODE	COST	EMPLOYEE	LOCATION	MANUF	MODEL	N.
1					2228			CH
1								CE
1	759.2			Staff	2203			RM
1								VE
1		06143	7,995.00	Terri L. Napier	2228	CompuPro, Inc.	486-66	CC
1				Dean H. Hewett	2227			PH
1	55.2			Dean H. Hewett	2227			RM
1	298.4			Staff	2222			RM
1		03056	56		2201	Sierra Furniture	Guest	CH
1								RA
1		03037	325		2201	Sierra Furniture, Inc.	L1304	DE
1		03025	345		2201	Sierra Furniture	Executive	CH
1								CV
1	582.7			Staff	2223			RM
1		06142	7,995.00	Dean H. Hewett	2227	CompuPro, Inc.	486-66	CC
1		03003	640.00		2201	Colorado Sofa Factory	CU990	SC
1					2227	Rocky Mtn. Desks	S74	DE
1								DI

Table Style Page

*Displayed only when **AutoCAD Table** option is selected in the previous step.*

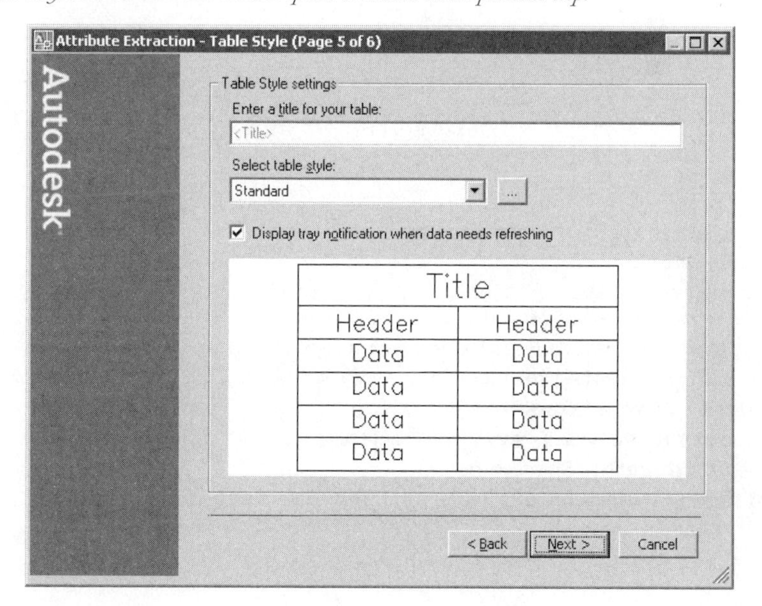

Enter a title for your table specifies the table's title.

Select table style specifies the style of table; click **...** to display the Table Style dialog box; see the **TableStyle** command.

☑ **Display tray notification when data needs refreshing** displays a yellow balloon when attribute data changes, and the table should be updated.

Swedish Page

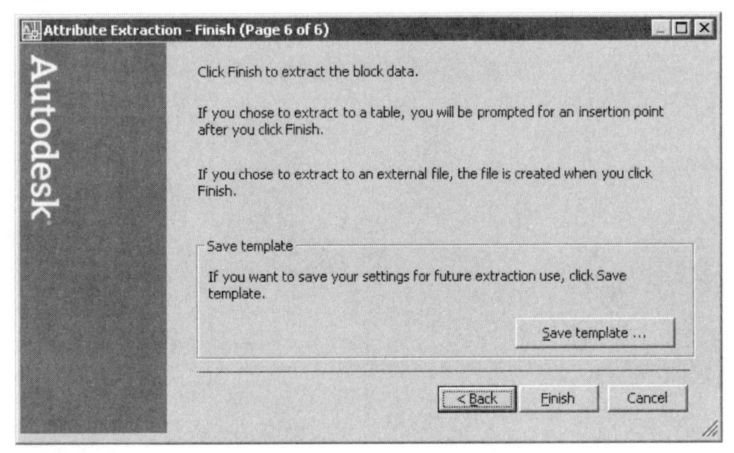

Save template displays the Save As dialog box to save the template in *.blk* format.

Finish completes the attribute data extraction process.

When ***AutoCAD Table*** *option was selected earlier, AutoCAD prompts:*

Specify insertion point: *(Pick a point to locate the table.)*

Quantity	AREA	BARCODE	COST	EMPLOYEE	LOCATION	MANUF	MODEL	Name	PHONE	PURCHDT	TYPE
1					2228			CHAIR			Chair
1								CB30			
1	759.2			Staff	2203			RMTAG	2523		Room tag
1								VENDING			
1		06143	7,995.00	Terri L. Napier	2228	CompuPro, Inc.	486-66	COMPUTER	214	08-20-93	Computer
1				Dean H. Hewett	2227			PHONE	218		Telephone
1	55.2			Dean H. Hewett	2227			RMTAG	4356		Room tag
1	298.4			Staff	2222			RMTAG	4565		Room tag
1		03056	56		2201	Sierra Furniture	Guest	CHAIR		04-10-93	Chair
1								RANGE			

EATTEXT Command

Command: -eattext

Enter the extraction type [Objects/Currentdwg/Selectdwg/Template] <Currentdwg>: *(Enter an option.)*

Extract data from external reference drawing? [Yes/No] <Yes>: *(Type* **Y** *or* **N***.)*

Extract data from nested blocks? [Yes/No] <Yes>: *(Type* **Y** *or* **N***.)*

Do you want to use a template? [Yes/No] <Yes>: *(Type* **Y** *or* **N***.)*

Enter the output filetype [Csv/Txt/Xls/Mdb] <Csv>: *(Enter an option.)*

Enter output filepath: *(Specify the path and filename.)*

External file "c:\filename.ext" was successfully created.

COMMAND LINE OPTIONS

Objects selects blocks in the drawing.

Currentdwg selects all blocks in the drawing.

Selectdwg selects drawings from disk.

Template specifies the name of a *.blk* template file.

Extract data from external reference drawing?

Yes extracts attributes from xrefs in addition to the drawing.

No does not extract attributes from xrefs.

Extract data from nested blocks?

Yes extracts attributes from nested blocks (blocks within blocks), as well as the drawing.

No does not extract attributes from xrefs.

Do you want to use a template?

Yes specifies the name of the *.blk* template file.

No exports the attributes in *.csv*, *.txt*, *.xls*, or *.mdb* format.

Enter the output filetype

Csv exports the attributes as comma-separated values.

Txt exports the attributes as tab-separated values.

Xls exports the attributes in Excel spreadsheet format.

Mdb exports the attributes in Microsoft database format.

Enter output filepath specifies the full path and filename, such as "c:\filename."

RELATED COMMANDS

AttExt exports in DXF format, as well as to comma- and tab-delimited formats; it is an older method of attribute extraction.

AttDef creates attribute definitions.

Block attaches attributes to objects.

BAttMan manages attributes.

Table creates tables.

TIPS

- This command does not export attribute data in DXF format; if you require this format, use the AttExt command.
- The -EAttExt command is undocumented by Autodesk.
- When the attribute data has been exported, you may open the file in another program, such as WordPerfect (word processing), Lotus 1-2-3 (spreadsheet), or Access (database).
- A drawing (and its xrefs) can contain many attributes. For example, the *1st Floor.dwg* sample drawing contains nearly a thousand blocks, which take up nearly 8,000 rows in a spreadsheet.

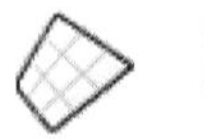

Edge

Rel.12 Toggles the visibility of 3D faces.

Command	Alias	Ctrl+	F-key	Alt+	Menu Bar	Tablet
edge	...	...	...	DMME	Draw ↳Modeling ↳Meshes ↳Edge	...

Command: edge
Specify edge of 3dface to toggle visibility or [Display]: *(Pick an edge, or type* **D***.)*
Enter selection method for display of hidden edges [Select/All] <All>: *(Type* **S** *or* **A***.)*
Select objects: *(Select one or more objects.)*
Specify edge of 3dface to toggle visibility or [Display]: *(Press* ESC *to end the command.)*

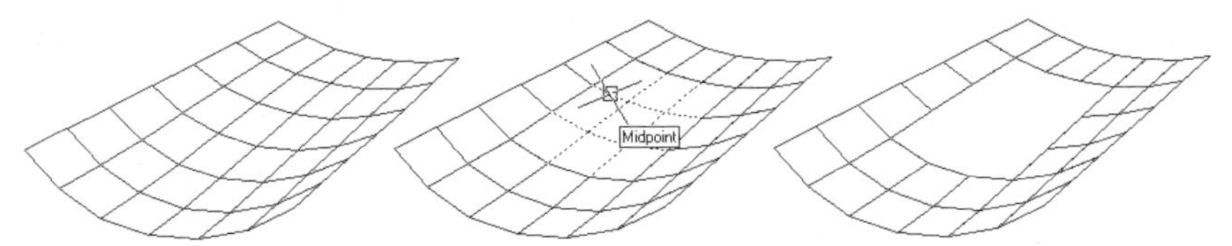

From left to right: *3D faces; edges selected with* ***Edge****; and invisible edges.*

COMMAND LINE OPTIONS

Specify edge selects edge to make invisible.

Display Options

Select highlights invisible edges.

All selects all hidden edges, and regenerates them.

RELATED COMMAND

3dFace creates 3D faces.

RELATED SYSTEM VARIABLE

SplFrame toggles visibility of 3D face edges.

TIPS

- Make edges invisible to improve the appearance of 3D objects.
- **Edge** applies only to objects made of 3D faces; it does not work with polyface meshes or solid models.
- Use the **Explode** command to convert meshed objects into 3D faces.
- Re-execute **Edge** to display an edge that has been made invisible.
- The command repeats until you press **Enter** or ESC at the 'Specify edge of 3dface to toggle visibility or [Display]:' prompt.
- When you hide all edges of a 3D face, it can be difficult to unhide them.

EdgeSurf

Rel.10 Draws 3D meshes as Coons patches between four boundaries (short for EDGE-defined SURFace); does not draw surfaces.

Command	Alias	Ctrl+	F-key	Alt+	Menu Bar	Tablet
edgesurf	...	...	...	**DFD**	**Draw** **↳Modeling** **↳Meshes** **↳Edge Mesh**	**R8**

Command: edgesurf

Current wire frame density: SURFTAB1=6 SURFTAB2=6

Select object 1 for surface edge: *(Pick an object.)*

Select object 2 for surface edge: *(Pick an object.)*

Select object 3 for surface edge: *(Pick an object.)*

Select object 4 for surface edge: *(Pick an object.)*

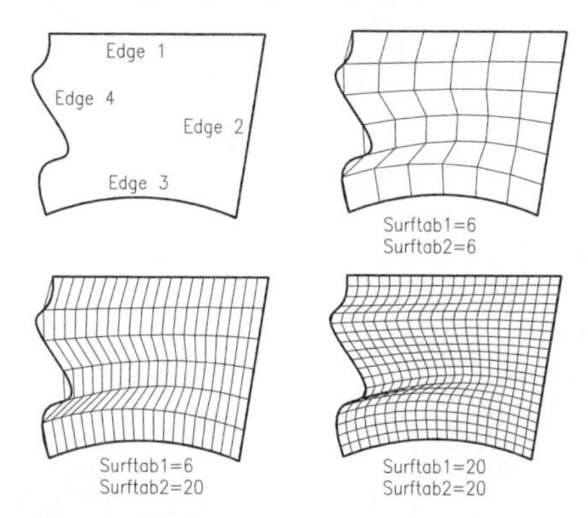

COMMAND LINE OPTION

Select object picks an edge.

RELATED COMMANDS

3dMesh creates 3D meshes by specifying every vertex.

3dFace creates 3D meshes of irregular vertices.

PEdit edits meshes created by EdgeSurf.

TabSurf creates tabulated 3D surfaces.

RuleSurf creates ruled 3D surfaces.

RevSurf creates 3D surfaces of revolution.

RELATED SYSTEM VARIABLES

SurfTab1 stores the current *m*-density of meshing; the maximum mesh density is 32767.

SurfTab2 stores the current *n*-density of meshing; maximum is 32767.

TIP

- The four boundary edges can be made from lines, arcs, and open 2D and 3D polylines; the edges must meet at their end pints.

'Elev

V. 2.1 Sets elevation and thickness for creating extruded 3D objects (short for ELEVation).

Command	Alias	Ctrl+	F-key	Alt+	Menu Bar	Tablet
elev	...	...	...	...	...	...

Command: elev
Specify new default elevation <0.0000>: *(Enter a value for elevation.)*
Specify new default thickness <0.0000>: *(Enter a value for thickness.)*

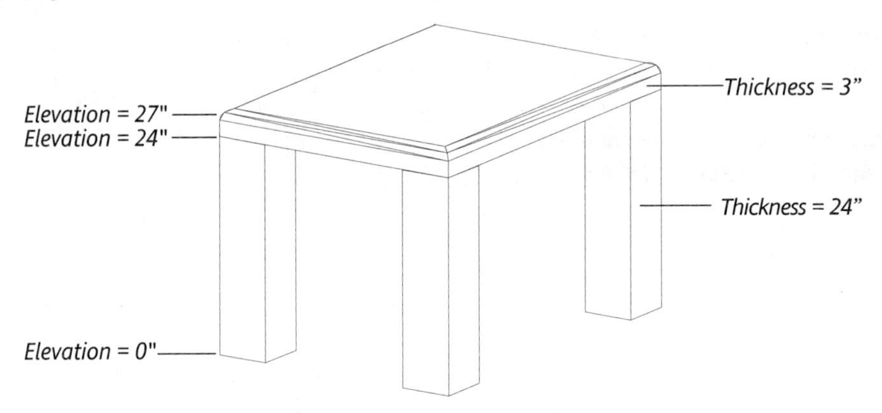

COMMAND LINE OPTIONS

Elevation changes the base elevation from z = 0.

Thickness extrudes new 2D objects in the z-direction.

RELATED COMMANDS

Change changes the thickness and z coordinate of objects.

Move moves objects, even in the z direction.

Properties changes the thickness of objects.

RELATED SYSTEM VARIABLES

Elevation stores the current elevation setting.

Thickness stores the current thickness setting.

TIPS

- The current value of elevation is used whenever the z coordinate is not supplied.
- Thickness is measured up from the current elevation in the positive z-direction.

Ellipse

V. 2.5 Draws ellipses by four different methods, and draws elliptical arcs and isometric circles.

Command	Alias	Ctrl+	F-key	Alt+	Menu Bar	Tablet
ellipse	**el**	...	...	**DE**	**Draw** ⇘**Ellipse**	**M9**

Command: ellipse
Specify axis endpoint of ellipse or [Arc/Center]: *(Pick a point, or enter an option.)*
Specify other endpoint of axis: *(Pick a point.)*
Specify distance to other axis or [Rotation]: *(Pick a point, or type* **R***.)*

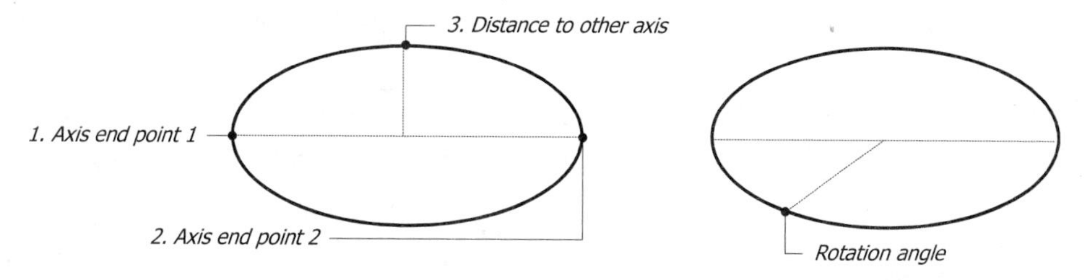

COMMAND LINE OPTIONS

Specify axis endpoint of ellipse indicates the first endpoint of the major axis.

Specify other endpoint of axis indicates the second endpoint of the major axis.

Specify distance to other axis indicates the half-distance of the minor axis.

Elliptical Arcs

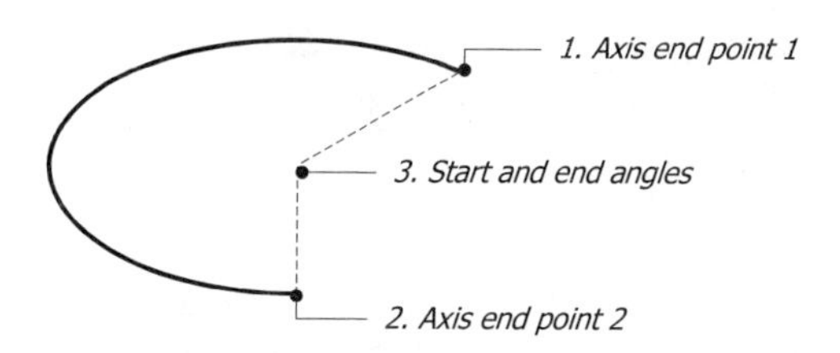

Command: ellipse
Specify axis endpoint of elliptical arc or [Center]: *(Pick a point, or type* **C***.)*
Specify other endpoint of axis: *(Pick a point.)*
Specify distance to other axis or [Rotation]: *(Pick a point, or type* **R***.)*
Specify start angle or [Parameter]: *(Enter an angle, or type* **P***.)*
Specify end angle or [Parameter/Included angle]: *(Enter an angle, or enter an option.)*

COMMAND LINE OPTIONS

Specify start angle indicates the starting angle of the elliptical arc.

Specify end angle indicates the ending angle of the elliptical arc.

Parameter indicates the starting angle of the elliptical arc; draws the arc with this formula:

p(u)=**c**+(**a***cos(u))+(**b***sin(u))

Parameter	Meaning
a	Major axis.
b	Minor axis.
c	Center of ellipse.

Included angle indicates an angle measured relative to the start angle, rather than 0 degrees.

Center Option

Specify center of ellipse: *(Pick a point.)*

Specify center of ellipse indicates the center point of the ellipse.

Rotation Option

Specify rotation around major axis: *(Enter an angle.)*

Specify rotation around major axis indicates a rotation angle around the major axis:

Rotation	Meaning
0 degrees	Minimum rotation: creates round ellipses, like circles.
89.4 degrees	Maximum rotation: creates very thin ellipses.

Isometric Circles

*This option appears only when system variable **SnapStyl** is set to **1** (isometric snap mode). Use **F5** to switch between the three isometric drawing planes.*

Command: ellipse

Specify axis endpoint of ellipse or [Arc/Center/Isocircle]: *(Type **I**.)*

Specify center of isocircle: *(Pick a point.)*

Specify radius of isocircle or [Diameter]: *(Pick a point, or type **D**.)*

COMMAND LINE OPTIONS

Specify center indicates the center point of the isocircle.

Specify radius indicates the radius of the isocircle.

Diameter indicates the diameter of the isocircle.

RELATED COMMANDS

IsoPlane sets the current isometric plane.

PEdit edits ellipses (when drawn with a polyline).

Snap controls the setting of isometric mode.

RELATED SYSTEM VARIABLES

PEllipse determines how the ellipse is drawn:

0 — Draws ellipse with the ellipse object (default).

1 — Draws ellipse as a series of polyline arcs.

SnapIsoPair sets the current isometric plane:

0 — Left (default).

1 — Top.

2 — Right.

SnapStyl specifies regular or isometric drawing mode:

0 — Standard (default).

1 — Isometric.

TIPS

- Previous to AutoCAD Release 13, Ellipse constructed the ellipse as a series of short polyline arcs. The PEllipse system variable controls how ellipses are drawn. When 0, true ellipses are drawn; when 1, a polyline approximation of an ellipse is drawn.
- When PEllipse = 1, the **Arc** option is not available.
- Use ellipses to draw circles in isometric mode. When Snap is set to isometric mode, Ellipse's **Isocircle** option projects a circle into the working isometric drawing plane. Press CTRL+E or F5 to switch between isoplanes.
- The **Isocircle** option only appears in the option prompt when Snap is set to isometric mode.

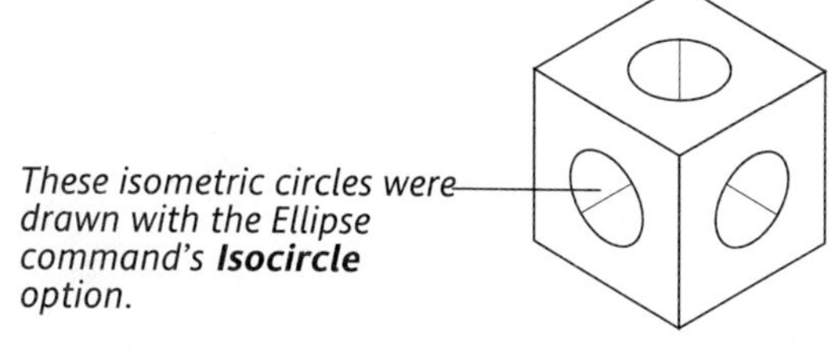

*These isometric circles were drawn with the Ellipse command's **Isocircle** option.*

- See the Isoplane command for a tutorial on creating isometric objects.

Removed Commands

End was removed from AutoCAD Release 14; it was replaced by **Quit**.

EndToday was removed from AutoCAD 2004.

Erase

V. 1.0 Erases objects from drawings.

Command	Alias	Ctrl+	Key	Alt+	Menu Bar	Tablet
erase	e	...	Del	ME	Modify ↳Erase	V14
					Edit ↳Clear	U14

Command: erase
Select objects: *(Select one or more objects.)*
Select objects: *(Press Enter to end object selection.)*

COMMAND LINE OPTION

Select objects selects the objects to erase.

RELATED COMMANDS

Break erases a portion of a line, circle, arc, or polyline.

Fillet cleans up intersections by erasing selected segments.

Oops returns the most-recently erased objects to the drawing.

Trim cuts off the end of a line, arc, and other objects.

Undo returns the erased objects to the drawing.

TIPS

- The **Erase L** command erases the last-drawn item visible in the current viewport.
- **Oops** brings back the most-recently erased objects; use **U** to bring back other erased objects.
- Objects on locked and frozen layers cannot be erased. To erase them, change the layers to unlocked and thawed.
- *Warning!* The **Erase All** command erases all objects in the current space (model or layout), except on locked, frozen, and/or off layers.

eTransmit

2000i Transmits drawings and related files as email messages (short for Electronic TRANSMITtal).

Commands	Alias	Ctrl+	Key	Alt+	Menu Bar	Tablet
etransmit	...	...	...	**FT**	**File** ↳**eTransmit**	...
-etransmit						

Command: etransmit

If the drawing has not been saved since the last change, displays error dialog box:

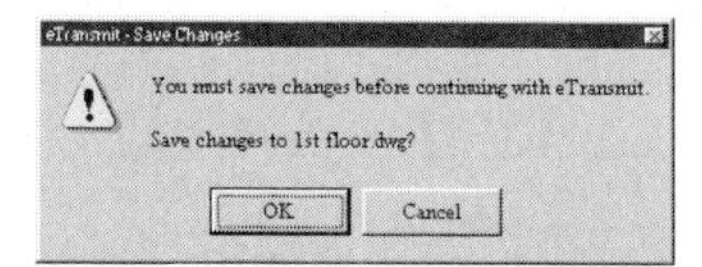

OK saves the drawing, and then displays the Create Transmittal dialog box.

Cancel cancels the **eTransmit** command.

CREATE TRANSMITTAL DIALOG BOX

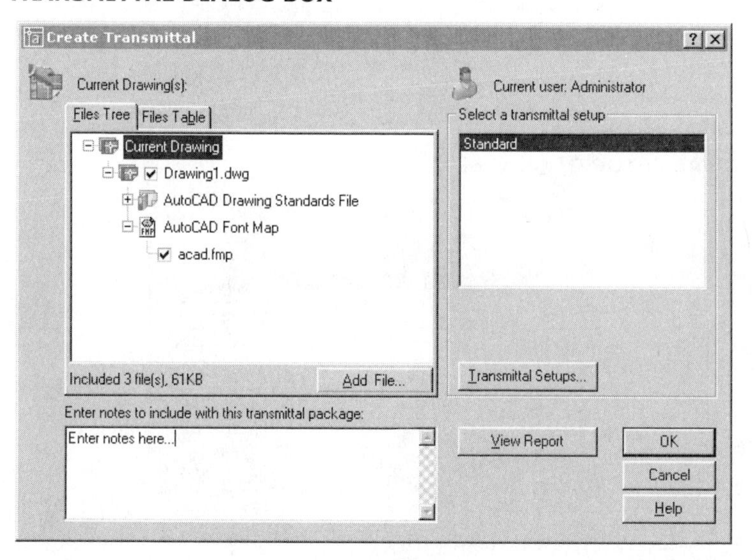

Files Tree Tab

Sheets tab displays sheets included with transmittal.

Files Tree tab displays names of drawing and support files, grouped by category.

Files Table tab displays file names in alphabetical order.

Select a Transmittal Setup lists pre-defined setups.

Transmittal Setups displays the Transmittal Setups dialog box.

Enter notes to include with this transmittal package provides space for entering notes.

View Report displays the View Transmittal Report dialog box.

Files Tree Tab

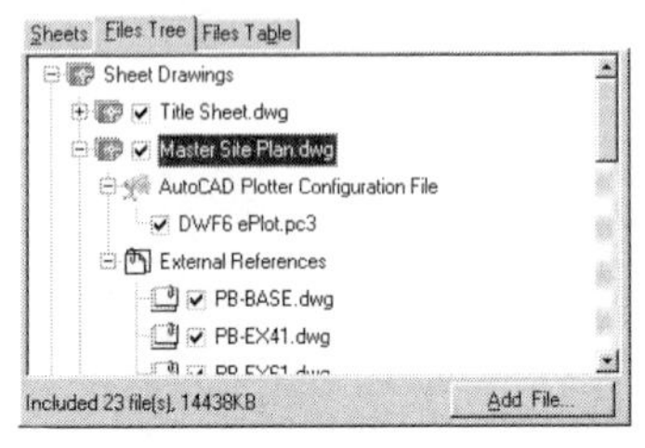

☑ File included in transmittal.

☐ File excluded from transmittal.

Add File displays the Add File to Transmittal dialog box, to add files to the transmittal.

Files Table Tab

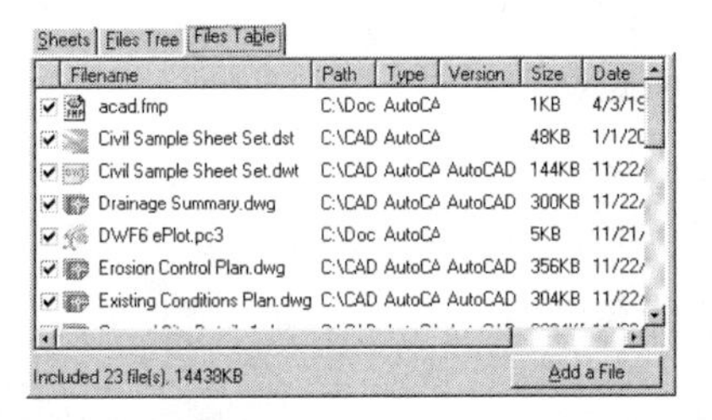

TRANSMITTAL SETUP DIALOG BOX

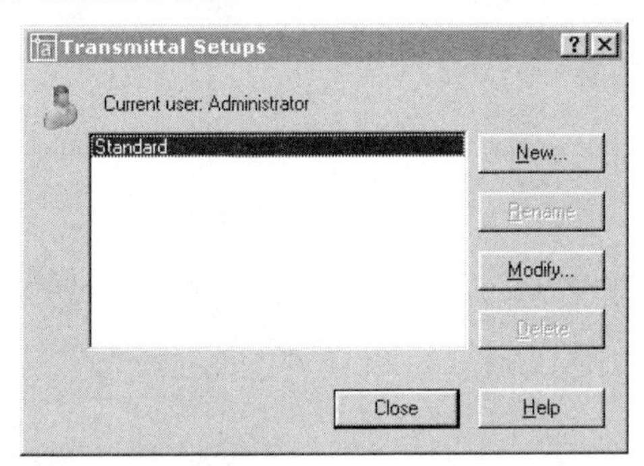

New creates new setups; displays New Transmittal Setup dialog box.

Rename renames setups, except for "Standard."

Modify changes setups; displays Modify Transmittal Setup dialog box.

Delete removes setups, except for "Standard."

NEW TRANSMITTAL SETUP DIALOG BOX

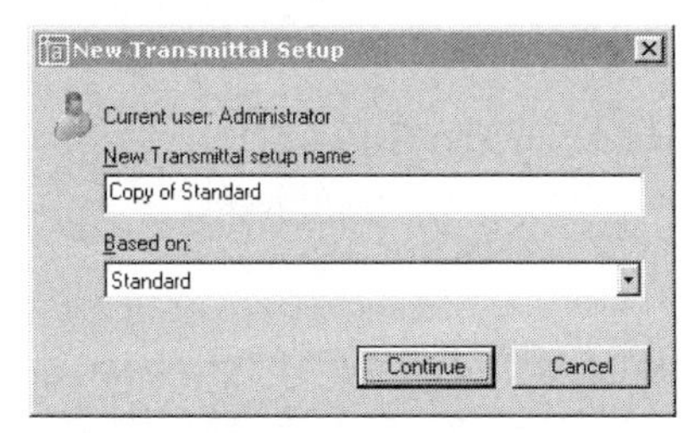

New Transmittal Setup Name specifies the name of the new setup.

Based on selects the setup to copy.

Continue displays the Modify Transmittal Setup dialog box.

MODIFY TRANSMITTAL SETUP DIALOG BOX

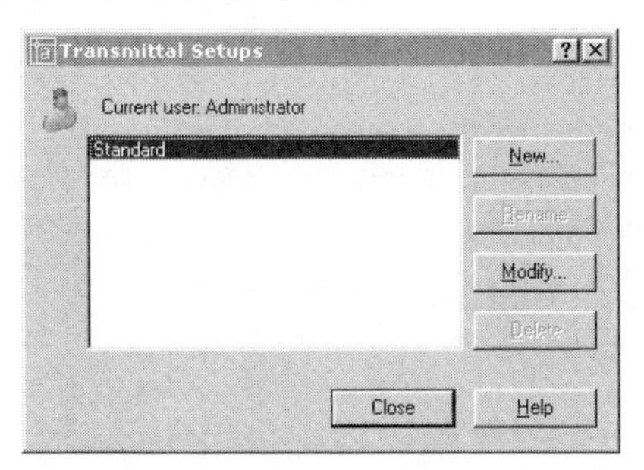

New creates new setups; displays New Transmittal Setup dialog box.

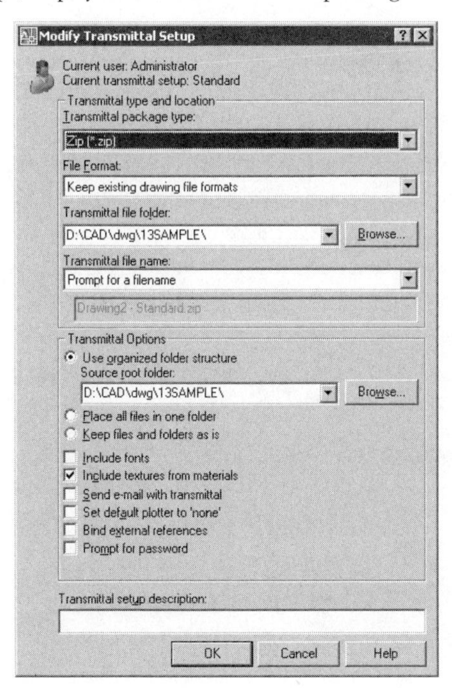

Transmittal Type and Location

Transmittal Package Type:

- **Folder (set of files)** transmits uncompressed files in new and existing folders. Best option for transmitting files on CDs or by FTP.
- **Self-Extracting Executable (*.exe)** transmits files in a compressed, self-extracting file. Uncompress by double-clicking the file.
- **Zip (*.zip)** (*default*) transmits files as a compressed ZIP file. Uncompress the file using the PkZip or WinZip programs. Best option when sending by email.

File Format:

- **Keep existing drawing file formats** transmits files in their version.
- **AutoCAD 207/LT 2007 Drawing Format** transmits files in AutoCAD 2007 format.
- **AutoCAD 2004/LT 2004 Drawing Format** transmits files in AutoCAD 2004/5/6 format.
- **AutoCAD 2000/LT 2000 Drawing Format** transmits files in AutoCAD 2000/i/1 format. *Warning!* Those objects and properties not found in AutoCAD 2000-2006 may be changed or lost.

Transmittal File Folder specifies the location in which to collect the files. When no location is specified, the transmittal is created in the same folder as the *.dst* file.

Transmittal File Name:

- **Prompt for a File Name** prompts the user for the file name.
- **Overwrite if Necessary** uses the specified file name, and overwrites the existing file of the same name.
- **Increment File Name if Necessary** uses the specified file name, and appends a digit to avoid overwriting the existing file of the same name.

Transmittal Options

⊙ **Use Organized Folder Structure** preserves the folder structure in the transmittal, but makes the changes listed below; allows you to specify the name of the folder tree.

○ **Place All Files in One Folder locates** all files in a single folder.

○ **Keep Files and Folders As Is** preserves the folder structure in the transmittal.

☐ **Include Fonts** includes all *.ttf* and *.shx* font files in the transmittal.

☑ **Include textures from materials** includes material files (*new in AutoCAD 2007*).

☐ **Send email with transmittal** opens the computer's default email software.

☐ **Set Default Plotter to 'None'** resets the plotter to None for all drawings in the transmittal.

☐ **Bind External References** merges all xrefs into the drawing.

☐ **Prompt for Password** displays a dialog box for specifying a password; not available when the Folder archive type is selected.

Transmittal setup description describes the setup.

-ETRANSMIT command

Command: -etransmit

Enter an option [Create transmittal package/Report only/CUrrent setup/ CHoose setup/Sheet set] <Report only>: *(Enter an option.)*

COMMAND LINE OPTIONS

Create Transmittal Package creates transmittal packages from the current drawing and all support files; uses settings in the current transmittal setup.

Report Only displays the Save Report File As dialog box; enter a file name, and then click **Save**. Saves the report in plain ASCII text format; does not create the transmittal package.

Current Setup lists the name of the current transmittal setup.

Choose Setup selects the transmittal setup.

Sheet Set specifies the sheet set and transmittal setup to use for the transmittal package; available only when a sheet set is open in the drawing.

RELATED COMMANDS

Archive creates archive sets of drawings, support files, and sheet sets.

Publish creates a *.dwf* file with drawing sheets.

PublishToWeb saves the drawing as an HTML file.

TIPS

- The **Password** button is *not* available when you select the **Folder** option.
- Including a TrueType font (*.ttf*) is touchy, because sending a copy of the font might infringe on its copyright. All *.shx* and *.ttf* files included with AutoCAD may be transmitted. For this reason, the **Include Fonts** option is turned off by default. In addition, not including fonts saves some file space; recall that smaller files take less time to transmit via the Internet.
- "Self-extracting executable" means that the files are compressed into a single file with the *.exe* extension. The email recipient double-clicks the file to extract (uncompress) the files. The benefit is that recipients do not need to have a copy of PkUnzip or WinZip on their computers; the drawbacks are that a virus could hide in the *.exe* file, and some e-mail programs prevent receipt of attachments with *.exe* extensions..
- The benefit to converting the drawing to AutoCAD 2000 format is that clients with older versions of the software can read the file; the drawback is that R2007-specific objects are either erased or modified to a simpler format. Two problems that can occur include: (1) lineweights are no longer displayed, and (2) database links and freestanding labels are converted to Release 14 links and displayable attributes. (Lineweights are restored when the drawing is opened again in AutoCAD 2007.)
- If your computer cannot compress the files (or if your recipient cannot uncompress them), you are probably lacking the software needed to uncompress the *.zip* file. (Do not confuse *.zip* files with Iomega's ZIP disk drive; the two having nothing in common, except the name.) You can obtain the PkUnzip and WinZIP utilities as freeware or shareware.
- Files occasionally become corrupted when sent by email; the transmittal may need to be resent. If you continue to have problems, you may need to change a setting in your email software. Try changing the attachment encoding method from BinHex or Uuencode to MIME.

Explode

V. 2.5 Explodes polylines, blocks, associative dimensions, hatches, multilines, 3D solids, regions, bodies, and meshes into their constituent objects.

Command	Alias	Ctrl+	F-key	Alt+	Menu Bar	Tablet
explode	x	...	...	MX	Modify ↳Explode	Y22

Command: explode
Select objects: *(Select one or more objects.)*
Select objects: *(Press Enter to end object selection.)*

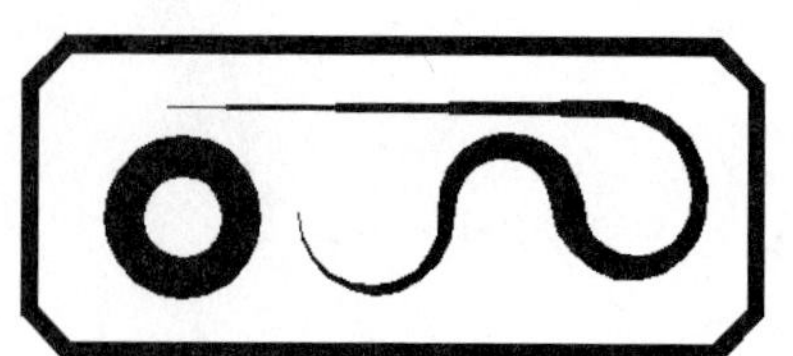

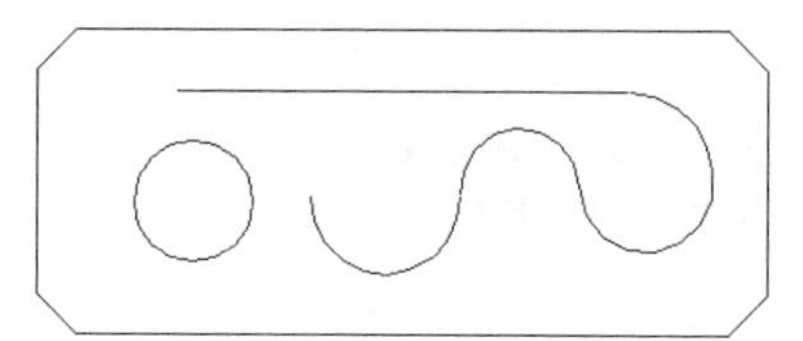

Left: *Polylines.*
Right: *Exploded into lines and arcs.*

COMMAND LINE OPTION

Select objects selects the objects to explode.

RELATED COMMANDS

Block recreates a block after an explode.

PEdit converts a line into a polyline.

Region converts 2D objects into a region.

Undo reverses the effects of this command.

Xplode provides control over the explosion process.

RELATED SYSTEM VARIABLE

ExplMode toggles whether non-uniformly scaled blocks can be exploded.

TIPS

- As of Release 13, AutoCAD can explode blocks inserted with unequal scale factors, mirrored blocks, and blocks created by the MInsert command.
- You cannot explode xrefs and dependent blocks (blocks from xref drawings), or single-line text.
- Parts making up exploded blocks and associative dimensions of BYBLOCK color and linetype are displayed in color White (or Black when the background color is white) and Continuous linetype.
- This command is not a transparent command, but Xplode is.

- The Explode command alters objects, as follows:

Objects	Exploded Into
Arcs in non-uniformly scaled blocks	Elliptical arc.
Associative dimensions	Lines, solids, and text.
Blocks	Constituent parts.
Circles in non-uniformly scaled blocks	Ellipse.
Mtext and field text	Text
Multilines	Lines.
Polygon meshes	3D faces.
Polyface meshes	3D faces, lines, and points.
Tables	Lines.
2D polylines	Lines and arcs; width and tangency information lost.
3D polylines	Lines.
3D solids	Regions and surfaces.
Regions	Lines, arcs, ellipses, and splines.
Bodies	Single bodies, regions, and curves.

- Resulting objects become the previous selection set.
- Flat faces of 3D solids become regions; curved faces become surfaces (*new to AutoCAD 2007*).

Export

Rel.13 Saves drawings in formats other than DWG and DXF.

Command	Alias	Ctrl+	F-key	Alt+	Menu Bar	Tablet
export	**exp**	...	...	**FEE**	**File** ↳**Export**	**W24**

Command: export

Displays dialog box:

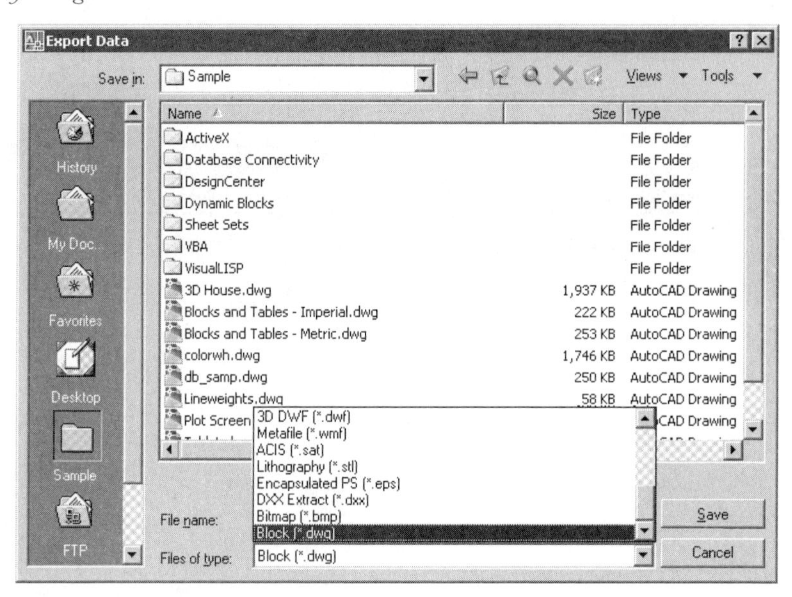

EXPORT DATA DIALOG BOX

Save in selects the folder (subdirectory) and drive into which to export the file.

Back returns to the previous folder (ALT+1).

Up One Level moves up one level in the folder structure (ALT+2).

Search the Web displays a simple Web browser that accesses the Autodesk Web site (ALT+3).

Delete erases the selected file(s) or folder (DEL).

Create New Folder creates a new folder (ALT+5).

Views displays files and folders in a list or with details.

Tools lists several additional commands, including the Options dialog box — available only with encapsulated PostScript; see the PsOut command.

File name specifies the name of the file, or accepts the default.

Save as type selects the file format in which to save the drawing.

Save saves the drawing.

Cancel dismisses the dialog box, and returns to AutoCAD.

RELATED COMMANDS

AttExt exports attribute data in the drawing in CDF, SDF, or DXF formats.

CopyClip exports the drawing to the Clipboard.

CopyHist exports text from the text screen to the Clipboard.

Import imports several vector and raster formats.

LogFileOn saves the command line text as ASCII text in the *acad.log* file.

MassProp exports the mass property data as ASCII text in an *.mpr* file.

MSlide exports the current viewport as an *.sld* slide file.

Plot exports the drawing in many vector and raster formats.

SaveAs saves the drawing in AutoCAD's DWG format.

SaveImg exports the rendering in TIFF, Targa, or BMP formats.

TIPS

- The **Export** command exports the current drawing in the following formats:

Extension	File Type	Related Command
3D DWF	3D Studio file	**3dDwf**
BMP	Device-independent bitmap	**BmpOut**
DWG	AutoCAD drawing file	**WBlock**
DXX	Attribute extract DXF file	**AttExt**
EPS	Encapsulated PostScript file	**PsOut**
SAT	ACIS solid object file	**AcisOut**
STL	Stereolithography	**StlOut**
WMF	Windows metafile	**WmfOut**

- This command acts as a "shell": it launches other AutoCAD commands that perform the actual export function, as noted above.
- When drawings are exported in formats other than *.dwg* or *.dxf*, information is lost, such as layers and attributes.
- AutoCAD 2007 removes 3DS (3D Studio) and WMF formats, but adds 3D DWF.

Removed (and Restored) Command

The **ExpressTools** (also known as "bonus CAD tools" in earlier versions of AutoCAD) were removed from AutoCAD 2002, but returned to AutoCAD 2004.

Extend

V. 2.5

Extends the length of lines, rays, open polylines, arcs, and elliptical arcs to boundary objects.

Command	Alias	Ctrl+	F-key	Alt+	Menu Bar		Tablet
extend	ex	...	...	MD	Modify ⇘Extend	—	W16

Command: extend
Current settings: Projection=UCS Edge=None
Select boundary edges ...
Select objects: *(Select one or more objects.)*
Select objects: *(Press* **ENTER**.*)*
Select object to extend or shift-select to trim or [Project/Edge/Undo]: *(Select an object, or enter an option.)*
Select object to extend or shift-select to trim or [Project/Edge/Undo]: *(Press* **ENTER** *to end the command.)*

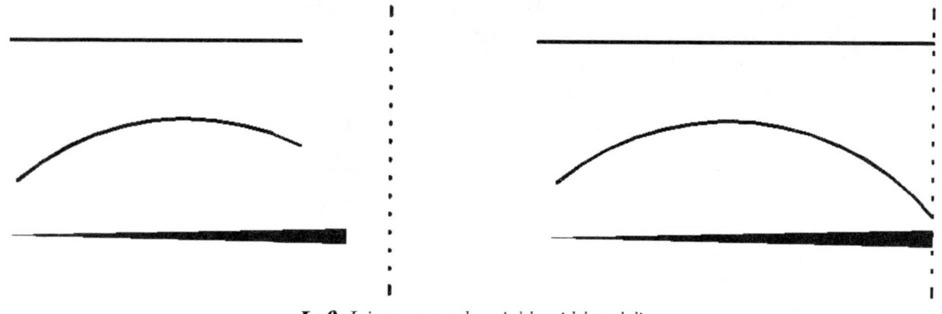

Left: *Line, arc, and variable-width polyline.*
Right: *Extended to dotted line.*

COMMAND LINE OPTIONS

Select objects selects the objects to use for the extension boundary.

Select objects to extend selects the objects to be extended.

Shift-select to trim trims objects when you hold down the SHIFT key.

Undo undoes the most recent extend operation.

Project options

Enter a projection option [None/Ucs/View] <Ucs>: *(Enter an option.)*

None extends objects to boundary (Release 12-compatible).

Ucs extends objects in the x,y-plane of the current UCS.

View extends objects in the current view plane.

Edge options

Enter an implied edge extension mode [Extend/No extend] <No extend>:

(Enter an option.)

Extend extends to implied boundary.

No extend extends only to actual boundary (Release 12-compatible).

RELATED COMMANDS

Change changes the length of lines.

Lengthen changes the length of open objects.

SolidEdit extends the face of a solid object.

Stretch stretches objects wider and narrower.

Trim reduces the length of lines, polylines and arcs.

RELATED SYSTEM VARIABLES

EdgeMode toggles boundary mode for the **Extend** and **Trim** commands:

- **0** — Use actual edges; Release 12 compatible (default).
- **1** — Use implied edge.

ProjMode toggles projection mode for the **Extend** and **Trim** commands:

- **0** — None; Release 12 compatible.
- **1** — Current UCS (default).
- **2** — Current view plane.

TIPS

- The following objects (even when inside blocks) can be used as boundaries:

2D polyline	Line
3D polyline	Ray
Arc	Region
Circle	Spline
Ellipse	Text
Floating viewport	Xline
Hatch	

- When a wide polyline is the edge, **Extend** extends to the polyline's centerline.
- Pick the object a second time to extend it to a second boundary line.
- Circles and other closed objects are valid edges: the object is extended in the direction nearest to the pick point.
- Extending a variable-width polyline widens it proportionately; extending a splined polyline adds a vertex.

ExternalReferences, ExternalReferencesClose

2007 Displays the External References palette for centrally managing all xref'ed drawings, images, and DWF underlays; the longest command name in AutoCAD.

Command	Aliases	Ctrl+	F-key	Alt+	Menu Bar	Tablet
externalreferences	**er**	**...**	**...**	**TPN**	**Modify**	**W16**
	image, im				**↳Palettes**	
	xref, xr				**↳External References**	
externalreferencesclose						

Command: externalreferences

Opens the External References palette:

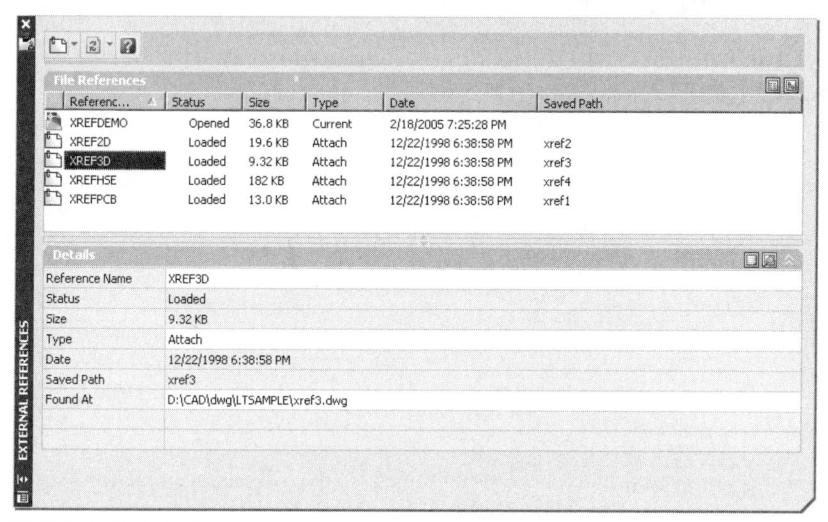

Command: externalreferencesclose

Closes the External References palette.

PALETTE TOOLBARS

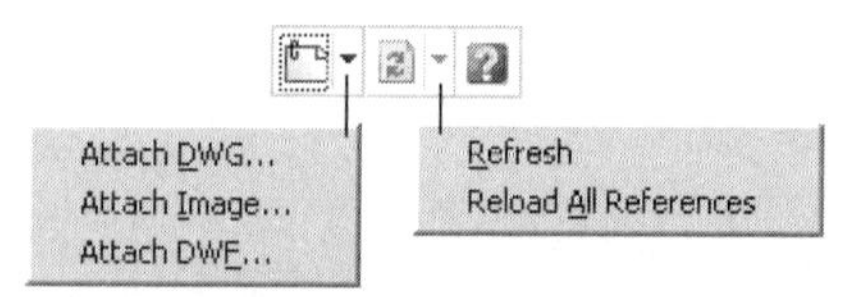

Attach DWG attaches *.dwg* files as external references; displays the Select Reference File dialog box, followed by the External Reference dialog box. See XAttach command.

Attach Image attaches raster files as underlays; displays the Select Image File dialog box, followed by the Image dialog box. See ImageAttach command.

Attach DWF attaches *.dwg* files as external references; displays the Select DWF File dialog box, followed by the Attach DWF Underlay dialog box. See DwfAttach command.

File References

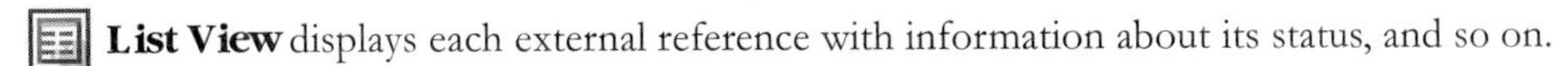

List View displays each external reference with information about its status, and so on.

Tree View shows the connections between external references.

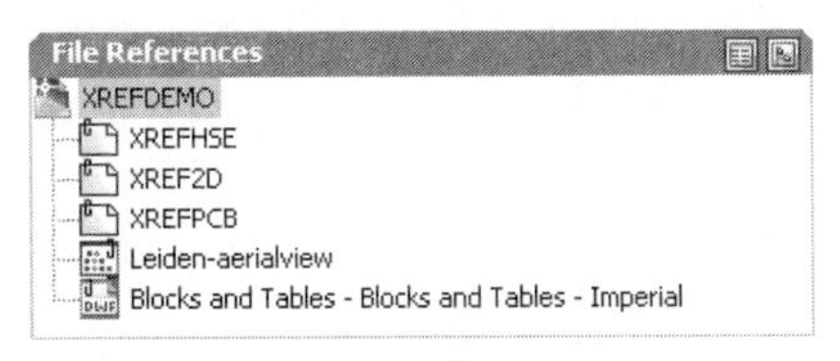

Right-click an externally-referenced file to display this shortcut menu:

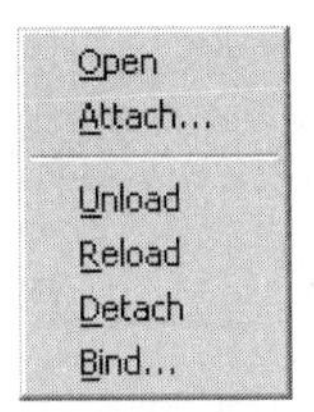

Open opens the selected external reference so that it can be edited:

- *.dwg* files are opened in another AutoCAD window; see the XOpen command.
- Image files are opened in the default external paint program, such as PaintShop Pro.
- *.dwf* files are opened in the external DWF Viewer program.

Attach displays the related dialog box so that changes can be made to the attachment status:

- *.dwg* files display the External Reference dialog box.
- Image files display the Image dialog box.
- *.dwf* files display the Attach DWF Underlay dialog box.

Unload prevents the external reference from being displayed in the drawing, but preserves attachment information; use the **Reload** option to bring it back.

Reload updates the external reference, if it has been unloaded, changed its content or moved to another folder.

Detach removes the external reference from the drawing.

Bind merges the external reference with the drawing (for *.dwg* files only); see the XBind command.

Details

Details lists information about the selected external reference.

Preview shows a preview image of the selected external reference; no preview is available for old versions of *.dwg* files.

RELATED COMMANDS

DwfAttach attaches *.dwf* files.

ImageAttach attaches raster files.

XAttach attaches *.dwg* files.

RELATED SYSTEM VARIABLES

ErState reports whether this palette is open or closed.

TIPS

- To reload all external references at once, right-click a blank area in the File References pane, and then select **Reload All References** from the shortcut menu.
- If you have access to Autodesk's Vault software, the **Attach from Vault** option allows you to attach files from the Vault.
- *.dwg* xrefs can be attached or overlaid; see the XRef command.
- The **Status** column displays brief reports that have the following meaning:

Status	Meaning
Loaded	File is attached to the drawing.
Unloaded	File will be unloaded.
Not Found	File cannot be found on search paths.
Unresolved	File cannot be read.
Orphaned	File is attached to an unresolved file.

Extrude

Rel.11 Creates 3D solids and surfaces by extruding 2D objects.

Command	Alias	Ctrl+	F-key	Alt+	Menu Bar	Tablet
extrude	**ext**	...	...	**DMX**	**Draw** **↳Modeling** **↳Extrude**	**P7**

Command: extrude

This command changed significantly in AutoCAD 2007.

Select objects: *(Select one or more objects.)*

Select objects: *(Press Enter.)*

Specify height of extrusion or [Direction/Path/Taper angle] <1>: *(Enter a height or an option.)*

COMMAND LINE OPTIONS

Select objects selects the 2D objects to extrude, called the "profile."

Specify height of extrusion specifies the extrusion height. As of AutoCAD 2007, the extrusion dynamically changes height as you move the cursor.

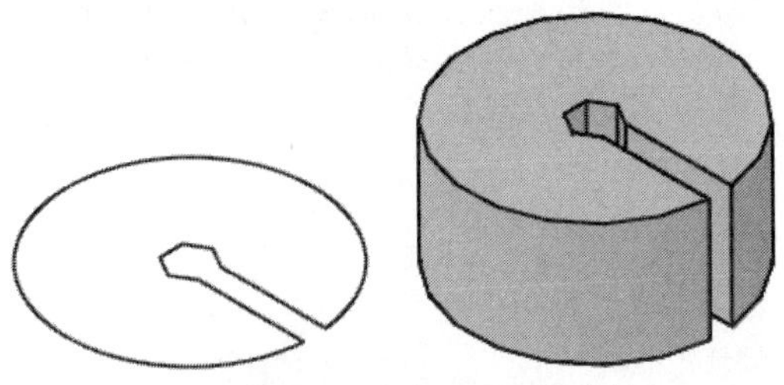

***Left**: Original object (the profile).*
***Right**: Extrusion by height.*

Direction specifies the extrusion height, direction, and angle by picking two points.

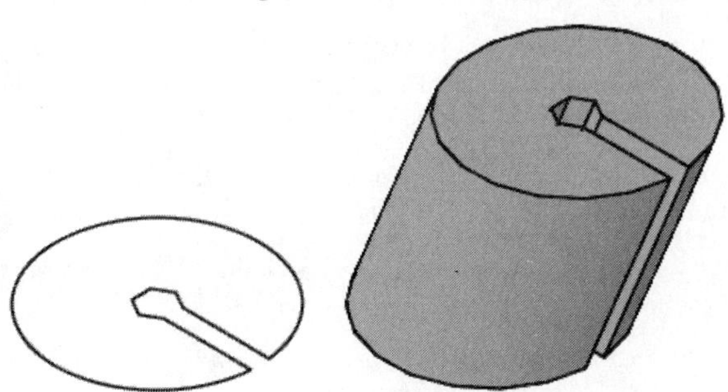

Extrusion by direction (picking two points).

Path extrudes the profile along another object, called the "path."

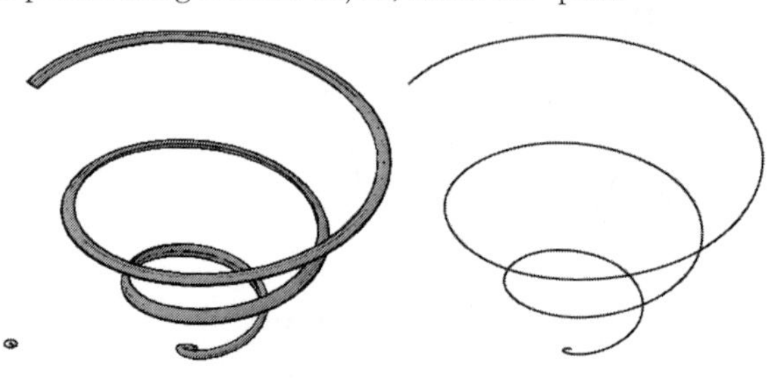

Left: *Original object (the profile).*
Center: *Extrusion by path.*
Right: *The path object (a helix).*

Taper extrudes the profile at an angle; ranges from -90 to +90 degrees.

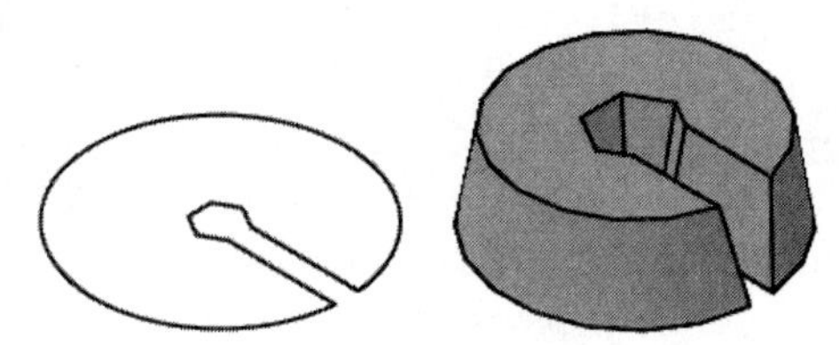

Extrusion by 10-degree taper (specifying an angle for the walls).
Notice that the outside tapers in, while the inside tapers out.

RELATED COMMANDS

Revolve creates a 3D solid by revolving a 2D object.

Elev gives thickness to non-solid objects to extrude them.

Thicken creates 3D solids by thickening surfaces.

TIPS

- Prior to AutoCAD 2007, this command worked only with closed objects, creating 3D solids. It now works with open and closed objects: open ones, such as arcs, create 3D surfaces, while closed ones, such as circles, create 3D solids.
- When extruding an object along a path, AutoCAD relocates the path to the object's centroid.
- This command extrudes the following objects: lines, traces, arcs, elliptical arcs, ellipses, circles, 2D polylines, 2D splines, 2D solids, regions, planar 3D faces, planar surfaces, and planar faces on solids. To select the face of a solid object, hold down the **Ctrl** key when picking.
- This command can use the following objects as extrusion paths: lines, circles, arcs, ellipses, elliptical arcs, 2D and polylines, 2D and 3D splines, helixes, solid edges, and edges of surfaces.
- You *cannot* extrude polylines with less than 3, or more than 500, vertices; similarly, you cannot extrude crossing or self-intersecting polylines.
- Objects within a block cannot be extruded; use the Explode command first.
- The taper angle must be between 0 (default) and +/-90 degrees; positive angles taper in from the base; negative angles taper out.
- This command also does not work if the combination of angle and height makes the object's extrusion walls intersect.

Field

2005 Places automatically updatable field text in drawings.

Command	Alias	Ctrl+	F-key	Alt+	Menu Bar	Tablet
field	...	...	...	**IF**	**Insert** ↳**Field**	...

Command: field

Displays dialog box, the content of which varies depending on the field name selected:

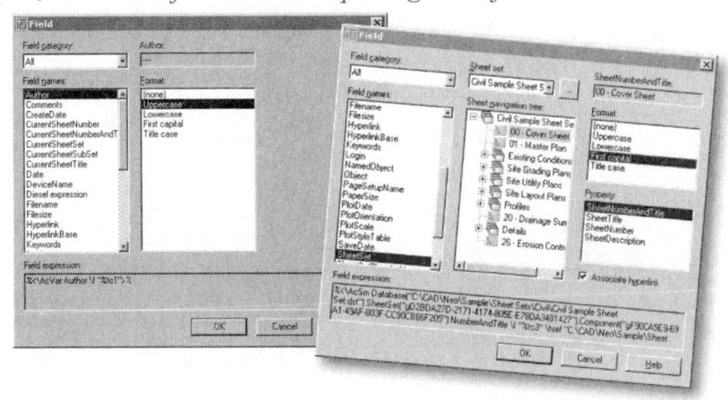

Select a field name and its options.

Click ***OK****. AutoCAD places the field as mtext:*

MTEXT Current text style: "Standard" Text height: 0.2000

Specify start point or [Height/Justify]: *(Pick a point, or enter an option.)*

DRAWING1.DWG

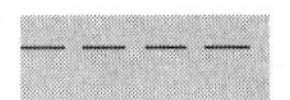

If the field has no value, AutoCAD inserts dashes as placeholders

FIELD DIALOG BOX

Field category lists groups of fields.

Field names lists the names of fields that can be placed in drawings.

Format lists optional formats available.

Field Expression illustrates field codes inserted in the drawing; cannot be edited directly.

Additional options appear, depending on the field name selected.

RELATED COMMANDS

UpdateField forces selected fields to update their values.

Find finds field text in drawings.

MText places text in drawings.

AttDef defines attribute text.

RELATED SYSTEM VARIABLES

FieldDisplay toggles the gray background to fields (default = 1, on).

FieldEval determines when fields are updated (default = 31, all options on).

TIPS

- When **FieldEval**=31 (all on), all fields are supposed to update whenever you use the **Plot**, **Save**, and other *trigger* commands. In practice, this occurs when the field is also a system variable. For example, if the field accesses the Date system variable, it updates correctly; when it is the Date option of the Date & Time category, **UpdateField** must be used to force the update.
- To edit: (1) double-click a field; (2) in the mtext editor, right-click the field text; (3) from the shortcut menu, select **Edit Field**. To convert fields to mtext, select **Convert Field to Text**.
- The gray background helps identify field text in drawings.
- AutoCAD supports the following field names:

Field Category	Field Names	Comments
Date & Time	Create Date	Date drawing created.
	Date	Current date and time.
	Plot Date	Date last plotted.
	Save Date	Date last saved.
Document	Author	Data stored in the Drawing Properties Comments dialog box; see DwgProps command.
	Filename	
	Filesize	
	HyperlinkBase	
	Keywords	
	LastSavedBy	
	Subject	
	Title	
Linked	Hyperlink	Links to other drawings and files.
Objects	Block	Placeholder; can be used in Block Editor only.
	NamedObject	Blocks, layers, linetypes, etc.
	Object	Object selected from drawing.
	Formula	Average, Sum, Count, Cell, and Formula; for use in tables only.
Other	Diesel Expression	Allows use of Diesel macros.
	LISP Variable	All AutoLISP and VLISP variable names.
	System Variable	Access to all system variables.
Plot	DeviceName	Name of the plotter.
	Login	Login name.
	PageSetup	Page setup name.
	PaperSize	Size of paper.
	PlotDate	Same as under Date & Time.
	PlotOrientation	Orientation of the plot.
	PlotScale	Scale of the plot.
	PlotStyleTable	Name of the plot style table.
Sheetset	SheetSet	Information about sheet sets.
	SheetSetPlaceholder	Placeholder for future sheet sets.
	SheetView	CurrentSheetSetSubset
	CurrentSheetNumber	CurrentSheetSet
	CurrentSheetNumberandTitle	CurrentSheetSubset
	CurrentSheetTitle	CurrentSheetCategory
	CurrentSheetIssuePurpose	CurrentSheetRevisionDate
	CurrentSheetRevisionNumber	CurrentSheetCategory
	CurrentSheetCustom	CurrentSheetDescription
	Current SheetSetCustom	CurrentSheetSetDescription
	CurrentSheetProjectMilestone	CurrentSheetSetProjectNumber
	CurrentSheetProjectName	CurrentSheetSetProjectPhase

- The following fields were removed from AutoCAD 2007: CurrentSheetProjectMilestone, CurrentSheetProjectName, CurrentSheetProjectNumber, and CurrentSheetProjectPhase.

FileOpen

Rel.12 Opens drawing files without dialog boxes (an undocumented command).

Command	Alias	Ctrl+	F-key	Alt+	Menu Bar	Tablet
fileopen	...	...	...	...	...	...

Command: fileopen
Enter name of drawing to open <filename.dwg>: *(Enter a file name.)*

COMMAND LINE OPTIONS

Enter name of drawing specifies the name of the *.dwg* file to open.

RELATED COMMANDS

Close closes the current drawing.

Open opens multiple drawing files.

RELATED SYSTEM VARIABLES

DbMod detects whether the drawing was changed since being opened.

SDI allows only one drawing at a time to be opened in AutoCAD .

TIPS

- Use this command in menu and toolbar macros to open a drawing file when you don't want to display a dialog box.
- This command can only be used in SDI (single drawing interface) mode; if you use this command when two or more drawings are open, AutoCAD complains, "The SDI variable cannot be reset unless there is only one drawing open. Cannot run FILEOPEN if SDI mode cannot be established."
- Use QSave before using FileOpen; otherwise, FileOpen displays the following dialog box:

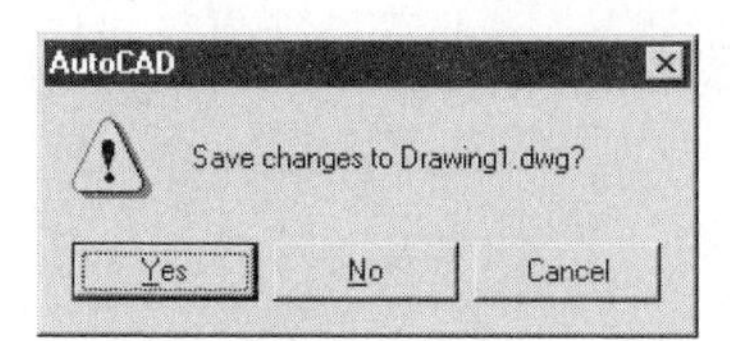

When you click **Yes**, AutoCAD closes the current drawing, and displays the Select Drawing File dialog box.

Removed Command

The **Files** command was removed from Release 14. In its place, use Windows Explorer.

'Fill

V. 1.4 Toggles whether hatches and wide objects — traces, multilines, solids, and polylines — are displayed and plotted with fills or as outlines.

Command	Alias	Ctrl+	F-key	Alt+	Menu Bar	Tablet
fill	...	...	...	...	...	...

Command: fill
Enter mode [ON/OFF] <ON>: *(Type* **ON** *or* **OFF***.)*

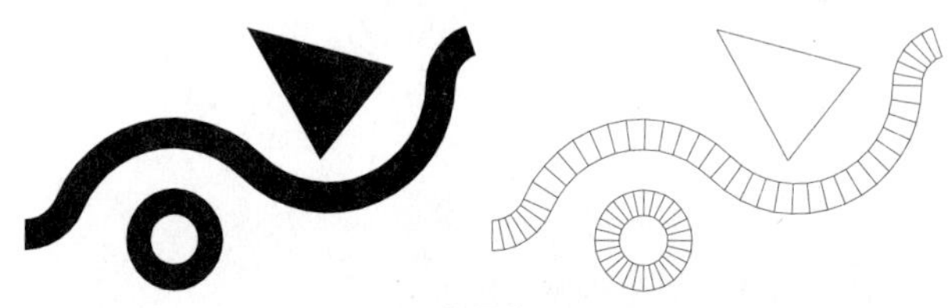

***Left**: Fill on.*
***Right**: Fill turned off with a wide polyline, donut, and 2D solid.*

COMMAND LINE OPTIONS

ON turns on fill after the next regeneration.

OFF turns off fill after the next regeneration.

RELATED SYSTEM VARIABLE

FillMode holds the current setting of fill status:

0 — Fill mode is off.

1 — Fill mode is on (default).

RELATED COMMAND

Regen changes the display to reflect the current fill or no-fill status.

TIPS

- The state of fill (or no fill) does not come into effect until the next regeneration:

  Command: regen
  Regenerating model.

- Traces, solids, and polylines are only filled in plan view, regardless of the setting of Fill. When viewed in 3D, these objects lose their fill.
- Fill affects objects derived from polylines, including donuts, polygons, rectangles, and ellipses, when created with PEllipse = 1.
- Fill does *not* affect TrueType fonts, which have their own system variable — TextFill — which toggles their fill/no-fill status for plots only.
- Fill does not toggle in rendered mode.

Fillet

V. 1.4 Joins intersecting lines, polylines, arcs, circles, and faces of 3D solids with a radius.

Command	Alias	Ctrl+	F-key	Alt+	Menu Bar	Tablet
fillet	**f**	...	...	**MF**	**Modify** **↳Fillet**	**W19**

Command: fillet

Current settings: Mode = TRIM, Radius = 0.0

Select first object or [Polyline/Radius/Trim/mUltiple]: *(Select an object, or enter an option.)*

Select second object: *(Select another object.)*

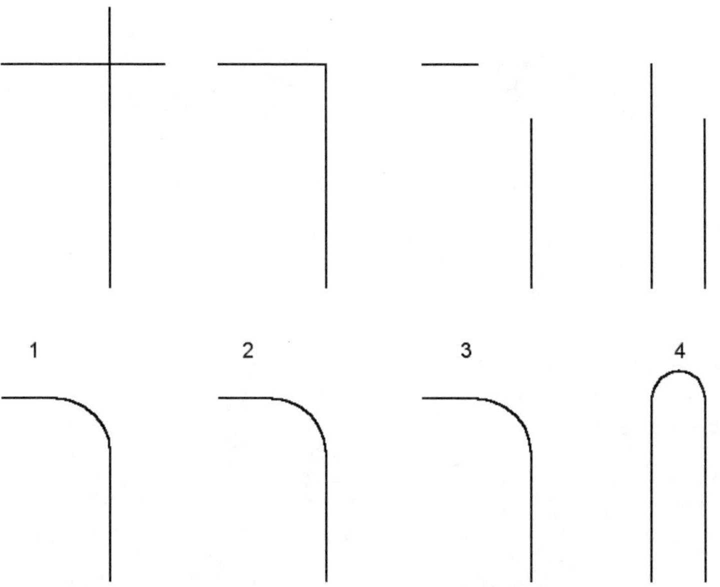

*Lines before (top) and after (bottom) applying the **Fillet** command:*
1. Crossing lines. 2. Touching lines. 3. Non-intersecting lines. 4. Parallel lines.

COMMAND LINE OPTIONS

Select first object selects the first object to be filleted.

Select second object selects the second object to be filleted.

mUltiple prompts you to select additional pairs of objects to fillet.

Polyline option

Select 2D polyline: *(Pick a polyline.)*

Select 2D polyline fillets all vertices of a 2D polyline; 3D polylines cannot be filleted.

Radius option

Specify fillet radius <0.0>: *(Enter a value.)*

Specify fillet radius specifies the filleting radius.

Trim option

Enter Trim mode option [Trim/No trim] <Trim>: *(Type* **T** *or* **N***.)*

Trim trims objects when filleted.

No trim does not trim objects.

Filleting 3D Solids

Select an edge or [Chain/Radius]: *(Pick an edge, or enter an option.)*

Select an edge or [Chain/Radius]: *(Press Enter.)*

n **edge(s) selected for fillet.**

Left: *Box before filleting.*
Right: *After applying* ***Fillet*** *to the 3D solid.*

Edge option

Select an edge or [Chain/Radius]: *(Pick an edge.)*

Select edge selects a single edge.

Chain option

Select an edge or [Chain/Radius]: *(Type* **C***.)*

Select an edge chain or [Edge/Radius]: *(Pick an edge.)*

Select an edge chain selects all tangential edges.

Radius option

Select an edge or [Chain/Radius]: *(Type* **R***.)*

Enter fillet radius <1.0000>: *(Enter a value.)*

Enter fillet radius specifies the fillet radius.

RELATED COMMANDS

Chamfer bevels intersecting lines or polyline vertices.

SolidEdit edits 3D solid models.

RELATED SYSTEM VARIABLES

FilletRad specifies the current filleting radius.

TrimMode toggles whether objects are trimmed.

TIPS

- Pick the end of the object you want filleted; the other end will remain untouched.
- The lines, arcs, or circles need not touch.
- As a faster alternative to the Extend and Trim commands, use the Fillet command with a radius of zero.
- If the lines to be filleted are on two different layers, the fillet is drawn on the current layer.
- The fillet radius must be smaller than the length of the lines. For example, if the lines to be filleted are 1.0m long, the fillet radius must be less than 1.0m.
- Use the **Close** option of the PLine command to ensure a polyline is filleted at all vertices.

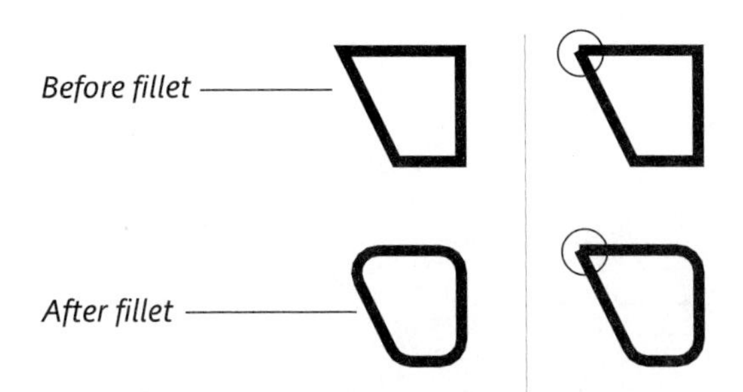

Left: *Polyline closed with Close option...*
Right: *...and without.*

- Filleting polyline segments from different polylines joins them into a single polyline.
- Filleting a line or an arc with a polyline joins it to the polyline.
- Filleting a pair of circles does not trim them.
- As of AutoCAD Release 13, the Fillet command fillets a pair of parallel lines; the radius of the fillet is automatically determined as half the distance between the lines.
- You cannot apply internal and external fillets in the same operation, as illustrated below:

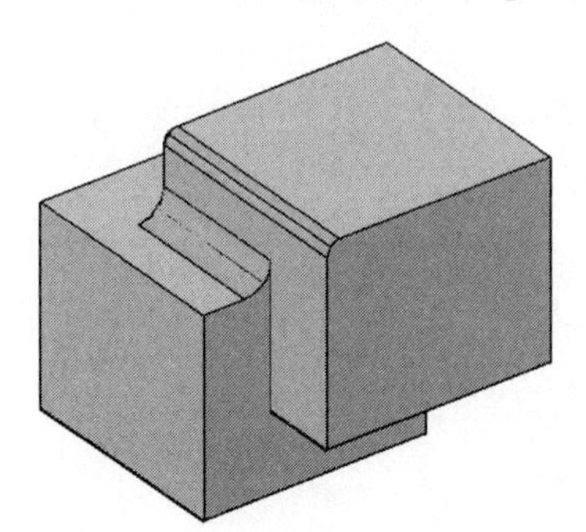

'Filter

Rel.12 Creates filter lists that can be applied when creating selection sets.

Command	Alias	Ctrl+	F-key	Alt+	Menu Bar	Tablet
filter	fi	...	...	...	...	...

Command: filter

Displays dialog box:

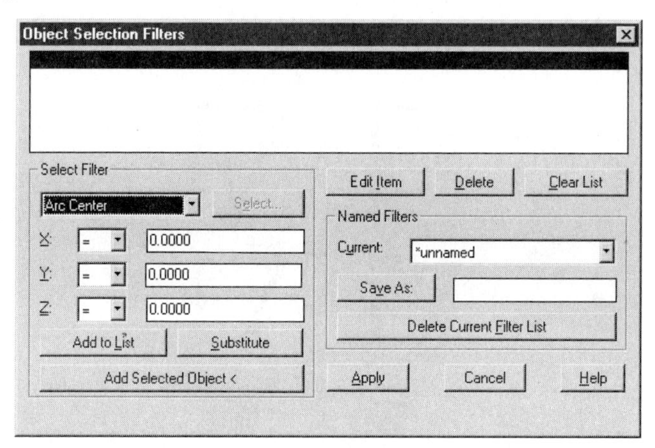

*After you click **Apply**, AutoCAD continues in the command area:*

Applying filter to selection.

Select objects: *(Select one or more objects.)*

Select objects: *(Press Enter to end object selection.)*

Exiting filtered selection.

AutoCAD highlights filtered objects with grips.

OBJECT SELECTION FILTERS DIALOG BOX

Select Filter

Select displays all items of the specified type in the drawing.

X, Y, Z specifies the object's coordinates.

Add to List adds the current select-filter option to the filter list.

Substitute replaces a highlighted filter with the selected filter.

Add Selected Object selects the object to be added from the drawing.

Edit Item edits the highlighted filter item.

Delete deletes the highlighted filter item.

Clear list clears the entire filter list.

Named Filter

Current selects the named filter from the list.

Save As saves the filter list with a name and the *.nfl* extension.

Delete Current Filter List deletes the named filter.

Apply closes the dialog box, and applies the filter operation.

COMMAND LINE OPTION

Select options selects the objects to be filtered; use the **All** option to select all non-frozen objects in the drawing.

RELATED COMMANDS

Any AutoCAD command with a 'Select objects' prompt.

QSelect creates a selection set quickly via a dialog box.

Select creates a selection set via the command line.

RELATED FILE

***.nfl** is a named filter list.

TIPS

- The selection set created by Filter is accessed via the **P** (previous) selection option.
- Alternatively, **'Filter** is used transparently at the 'Select objects:' prompt.
- Filter uses the following grouping operators:

**Begin OR	*with*	**End OR
**Begin AND	*with*	**End AND
**Begin XOR	*with*	**End XOR
**Begin NOT	*with*	**End NOT

- Filter uses the following relational operators:

Operator	Meaning
<	Less than.
<=	Less than or equal to.
=	Equal to.
!=	Not equal to.
>	Greater than.
>=	Greater than or equal to.
*	All values.

- Save selection sets by name to an *.nfl* (short for *named filter*) file on disk for use in other drawings or editing sessions.

Quick Start Tutorial

Filtering Selection Sets

In this tutorial, you erase all construction lines (xlines) from a drawing:

Step 1

1. Start the Erase command, and then invoke the Filter command transparently:

   Command: erase
   Select objects: 'filter

   Notice that AutoCAD displays the Object Selection Filters dialog box:

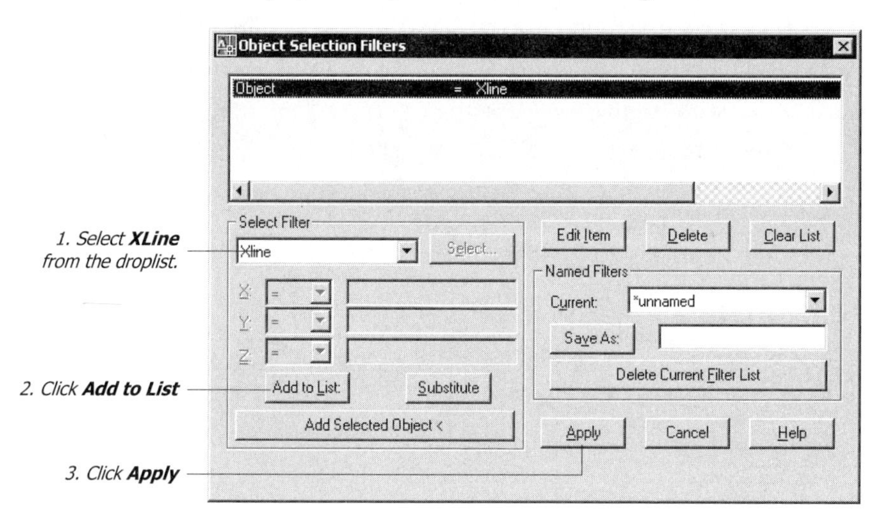

Step 2

In the **Select Filter** section:

1. Select **Xline** from the drop list.
2. Click the **Add to List** button.
3. Click **Apply**. Notice that the Filter command continues by displaying prompts on the command line, 'Applying filter to selection.'
4. To specify that Filter should search the entire drawing, enter the **All** option:

   Select objects: all
   n **found** *n* **were filtered out.**

Step 3

1. Exit the Filter command by pressing Enter:

   Select objects: *(Press* **ENTER** *to end object selection.)*
   Exiting filtered selection. <Selection set: 7>
   n **found**

2. AutoCAD resumes the **Erase** command. Press **Enter** to end it:

   Select objects: *(Press* **ENTER** *to end the command.)*

Notice that AutoCAD uses the selection set created by the Filter command to erase the xlines.

Find

2000 Finds, and optionally replaces, text in drawings.

Command	Alias	Ctrl+	F-key	Alt+	Menu Bar	Tablet
find	...	...	...	**EF**	**Edit** ↳**Find**	**X10**

Command: find

Displays dialog box:

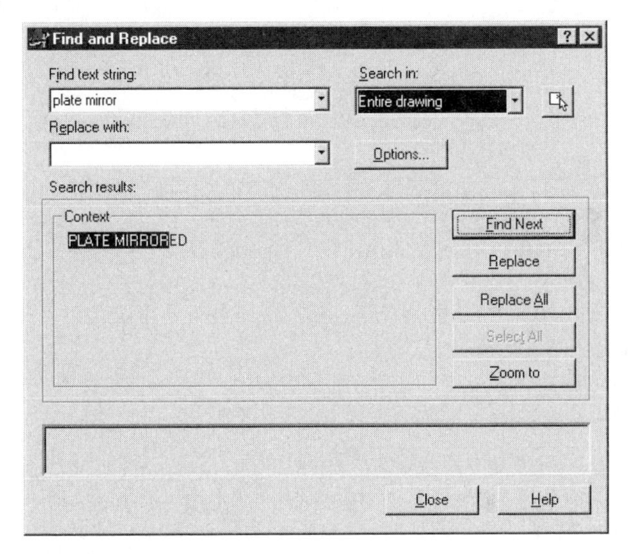

FIND AND REPLACE DIALOG BOX

Find text string specifies the text to find; enter text or click the down arrow to select one of the most recent lines of text searched for.

Replace with specifies the text to be replaced (only if replacing found text); enter text or click the down arrow to select one of the most recent lines of text searched for.

Search in:

- **Current Selection** searches the current selection set for text; click the **Select objects** button to create a selection set, if necessary.
- **Entire Drawing** searches the entire drawing.

Select objects allows you to select objects in the drawing; press ENTER to return to dialog box.

Options displays the Find and Replace Options dialog box.

Search Results

Context displays the found text in its context.

Find Next finds the text entered in the **Find text string** field.

Replace replaces a single found text with the text entered in the **Replace with field**.

Replace All replaces all instances of the found text.

Select All selects all objects containing the text entered in the **Find text string** field; AutoCAD displays the message "AutoCAD found and selected *n* objects that contain "..."."

Zoom to zooms in to the area of the current drawing containing the found text.

FIND AND REPLACE OPTIONS DIALOG BOX

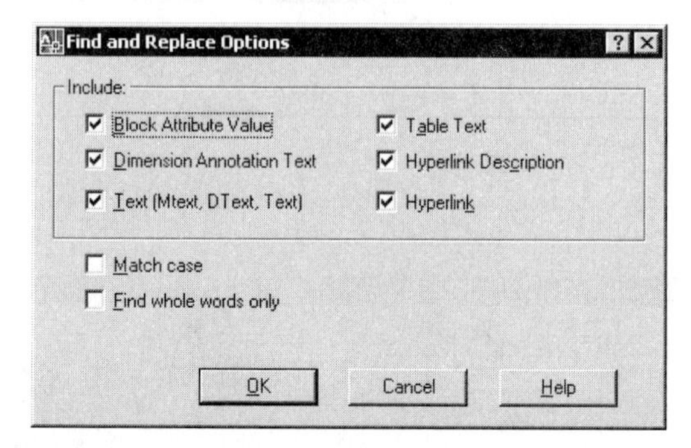

Include

☑ **Block Attribute Value** finds text in attributes.

☑ **Dimension Annotation Text** finds text in dimensions.

☑ **Text (Mtext, DText, Text)** finds text in paragraph text placed by the **MText** command, single-line text placed by the **Text** command, and field text placed by the **Field** command.

☑ **Table Text** finds text in tables.

☑ **Hyperlink Description** finds text in the description of a hyperlink.

☑ **Hyperlink** finds text in a hyperlink.

☐ **Match case** finds text that exactly matches the uppercase and lowercase pattern; for example, when searching for "Quick Reference," AutoCAD would find "Quick Reference" but not "quick reference."

☐ **Find whole words only** finds text that exactly matches whole words; for example, when searching for "Quick Reference," AutoCAD would find "Quick Reference" but not "Quickly Reference."

RELATED COMMANDS

AttDef creates attribute text.

DdEdit edits text.

Dim*xxx* creates dimension text.

Field places updatable text.

Hyperlink creates hyperlinks and hyperlink descriptions.

MText creates paragraph text.

Properties edits selected text.

QSelect finds text objects.

Text creates single-line text.

TIPS

- To find database links, use the dbConnect command.
- The QSelect command places text in a selection set.

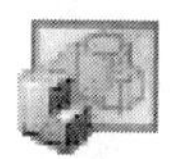

FlatShot

2004 Creates 2D blocks flattened from 3D models in the current viewport; short for "FLATten snapSHOT"; formerly an Express Tool.

Command	Alias	Ctrl+	F-key	Alt+	Menu Bar	Tablet
flatshot	**fshot**	...	...	...	...	...

Command: flatshot

Displays dialog box:

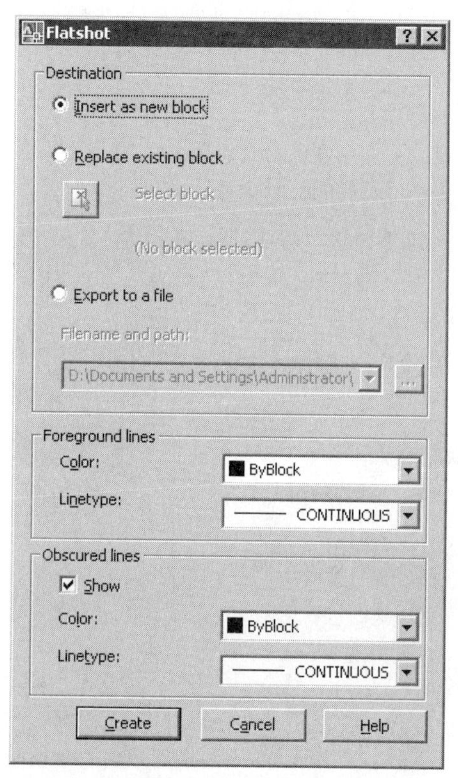

*Click **Create**. The command continues with **-Insert** command-like prompts:*

Units: Unitless Conversion: 1.0000

Specify insertion point or [Basepoint/Scale/X/Y/Z/Rotate]: *(Pick a point, or enter an option.)*

Enter X scale factor, specify opposite corner, or [Corner/XYZ] <1>: *(Press Enter, or enter an option.)*

Enter Y scale factor <use X scale factor>: *(Press Enter, or enter a scale factor.)*

Specify rotation angle <0>: *(Press Enter, or enter a rotation angle.)*

DIALOG BOX OPTIONS

Destination options

Insert As New Block inserts the flattened model as a block.

Replace Existing Block replaces the existing block, if it already exists.

Select Block selects the block to replace; press Enter to return to the dialog box.

Export to a File saves the flattened model to a *.dwg* file on disk.

Foreground Lines options

Color specifies the color of non-hidden (non-obscured) lines; click **Select Color** to select other colors.

Linetype specifies the linetype of non-hidden lines; click **Select Linetype** to load additional linetypes into the drawing.

Obscured Lines options

Show toggles the display of hidden-lines.

Color specifies the color of hidden (obscured) lines.

Linetype specifies the linetype of hidden lines.

Create closes the dialog box, and then creates the flattened view; the original model is retained.

TIPS

- This command works with solids and surfaces only; it does not work with other kinds of 3D models, such as those made of faces and thickened 2D objects.

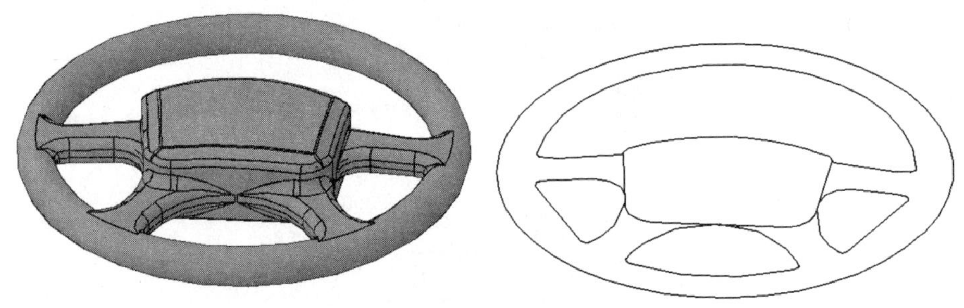

***Left**: Original 3D model.*
***Right**: 2D block created by the FlatShot command.*

- The command creates a block from the 3D geometry; the block can be exploded with the Explode command for further editing.
- This command flattens the current screen view onto the current UCS. For the best result, set the UCS to View before using this command. This ensures the flattened view looks like the screen view. Finally, insert the new block into the WCS.

Removed Commands

Fog was removed from AutoCAD 2007; now an alias for the **RenderEnvironment** command.

GifIn was removed from Release 14. In its place, use **Image**.

GeographicLocation

2007 Positions the sun (light) according to the selected location.

Command	Alias	Ctrl+	F-key	Alt+	Menu Bar	Tablet
geographiclocation	**geo**	...	...	**VELG**	**View**	...
	north				↳**Render**	
	northdir				↳**Light**	
					↳**Geographic Location**	

Command: geographiclocation

Displays dialog box:

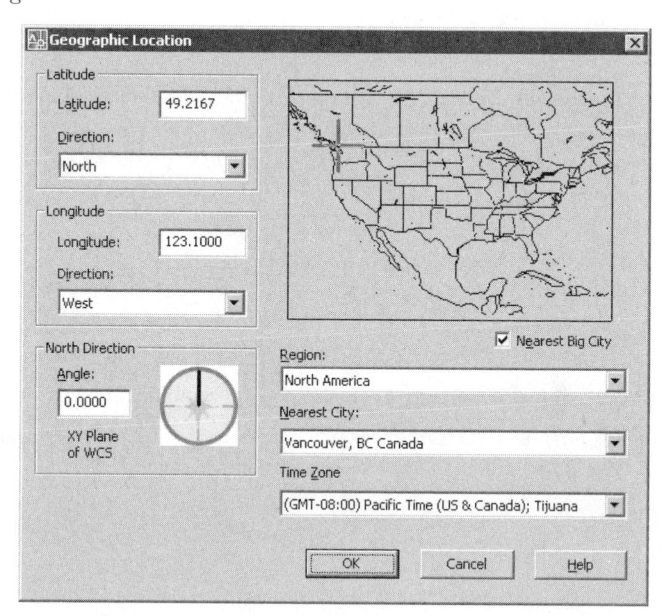

DIALOG BOX OPTIONS

Latitude sets the latitude from 0 to 90 degrees; when a city is selected from the map, its latitude is shown here.

Direction selects North or South of the equator.

Longitude sets the longitude from 0 to 180.

Direction selects East or West from the Prime Meridian.

Angle specifies the direction of North; default is the positive y axis.

Map options

You can click the map to locate the latitude and longitude; the red cross shows the current location.

☑ **Nearest Big City** displays the name of the nearest city to the latitude and longitude values entered.

Region selects a region or the entire world.

Nearest City selects a city.

Time Zone specifies the time zone.

RELATED COMMAND

SunProperties opens the Sun Properties palette.

RELATED SYSTEM VARIABLES

Latitude stores the current latitude.

Longitude stores the current longitude.

NorthDirection stores the angle from the positive y axis.

Timezone stores the current time zone.

TIPS

- You can access the sunlight through the Light control panel on the Dashboard:

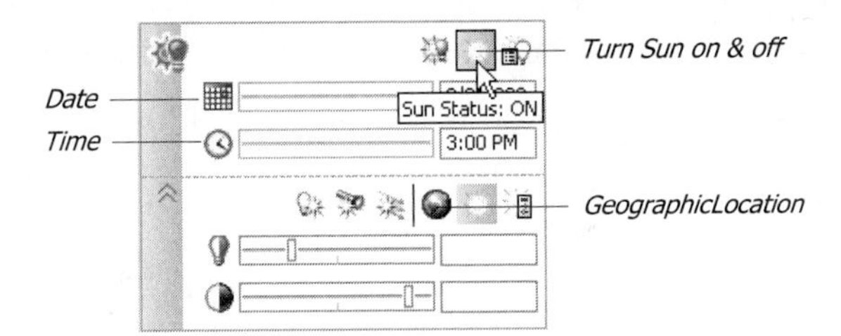

- In addition to GeographicLocation, you can also control the sunlight by date and time.
- This command uses names and coordinates of cities found in the *.map* files found in *\AutoCAD 2007\Support*, which are no longer user-editable.

'GotoUrl

2000 Goes to hyperlinks contained by objects.

Command	Alias	Ctrl+	F-key	Alt+	Menu Bar	Tablet
gotourl	...	...	...	...	...	...

Command: gotourl
Select objects: *(Select one or more objects.)*
Select objects: *(Press Enter to end object selection.)*
browser Enter Web location (URL) <http://www.autodesk.com>: http://www.upfrontezine.com
AutoCAD launches your computer's default Web browser, and attempts to access the URL.

COMMAND LINE OPTION

Select objects selects one or more objects containing a hyperlink.

RELATED COMMANDS.

Browser launches the Web browser.

Hyperlink attaches, edits, and removes hyperlinks from objects.

TIPS

- This command is meant for use by macros and menus.
- AutoCAD uses this command for the shortcut menu's Hyperlink | Open option.

Gradient

2006 Floods areas with gradient fills.

Command	Alias	Ctrl+	F-key	Alt+	Menu Bar	Tablet
gradient	**gd**	...	...	**DG**	**Draw** ↳**H**	...

Command: gradient

Displays the Gradient tab of the dialog box:

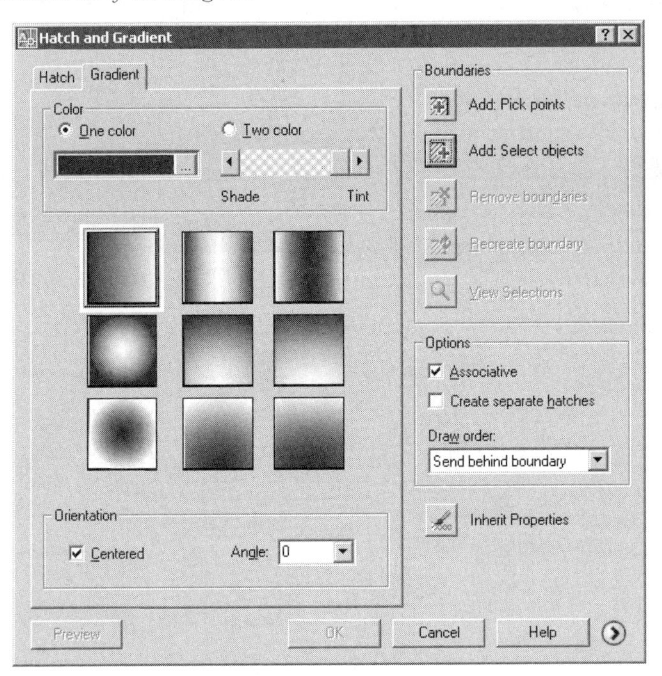

HATCH AND GRADIENT DIALOG BOX

Color

⊙**One color** produces color-shade gradients.

Shade - Tint slider varies the second color between white (shade) and black (tint).

○**Two color** produces two-color gradients.

Color 1 selects the first color.

Color 2 selects the second color.

... displays the Select Color dialog box.

Orientation

☑ **Centered** centers the gradient in the hatch area.

☐ **Centered** moves gradient up and to the left.

Angle rotates the gradient.

*See the **BHatch** command for other options.*

RELATED SYSTEM VARIABLES

GfAng specifies the angle of the gradient fill; ranges from 0 to 360 degrees.

GfClr1 specifies the first gradient fill color in RGB format, such as "RGB 000, 128, 255."

GfClr2 specifies the second gradient fill color in RGB format.

GfClrLum specifies the luminescence of one-color gradient fills, from 0.0 (black) to 1.0 (white).

GfClrState specifies whether the gradient fill is one-color or two-color.

GfName specifies the gradient fill pattern:

1 — Linear.

2 — Cylindrical.

3 — Inverted cylindrical.

4 — Spherical.

5 — Inverted spherical.

6 — Hemispherical.

7 — Inverted hemispherical.

8 — Curved.

9 — Inverted curved.

GfShift specifies whether the gradient fill is centered or shifted to the upper-left.

TIP

- The Gradient command simply displays the Gradient tab of the Hatch and Fill dialog box.

'GraphScr

V. 2.1 Places the graphics window in front of the text window.

Command	Alias	Ctrl+	F-key	Alt+	Menu Bar	Tablet
graphscr	...	...	F2	...	...	...

Command: graphscr

AutoCAD displays the drawing window:

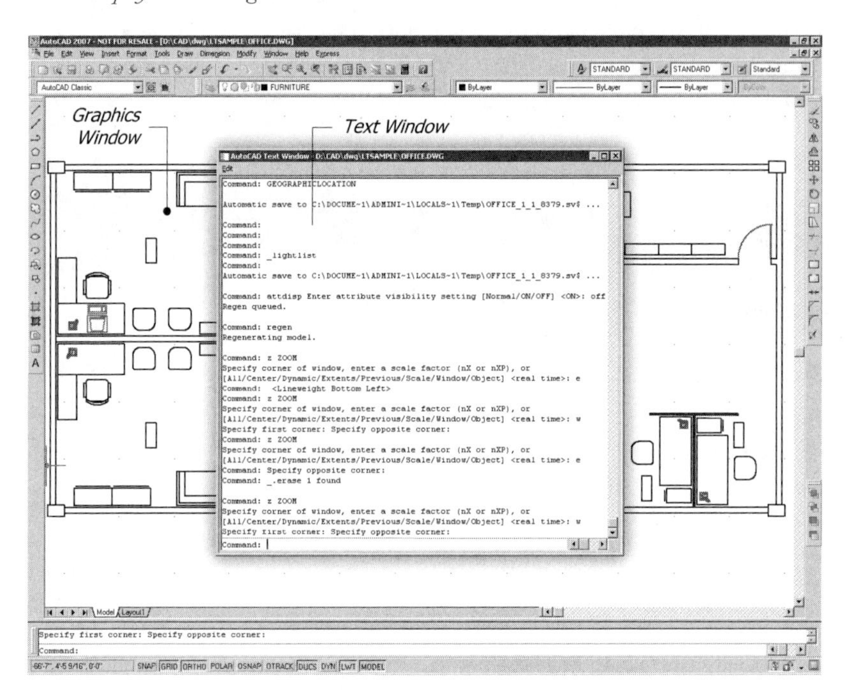

RELATED COMMANDS

CopyHist copies text from the Text window to the Clipboard.

TextScr switches from the graphics window to the Text window.

RELATED SYSTEM VARIABLE

ScreenMode indicates whether the current screen is in text or graphics mode. Since the advent of Windows, only holds the value of 3: dual screen, displaying both text and graphics.

TIP

- The Text window appears frozen when a dialog box is active. Click the dialog box's **OK** or **Cancel** button to regain access to the Text window.

'Grid

V. 1.0 Displays a grid of reference dots within the current drawing limits.

Command	Alias	Ctrl+	F-key	Alt+	Status Bar	Tablet
grid	...	G	F7	...	GRID	...

Command: grid

This command has changed significantly in AutoCAD 2007:

Specify grid spacing(X) or [ON/OFF/Snap/Major/aDaptive/Limits/Follow/Aspect] <5'-0">: *(Enter a value, or enter an option.)*

When the visual style is set to 2D wireframe, the grid is displayed as dots.

In all other visual styles, the grid is displayed as lines.

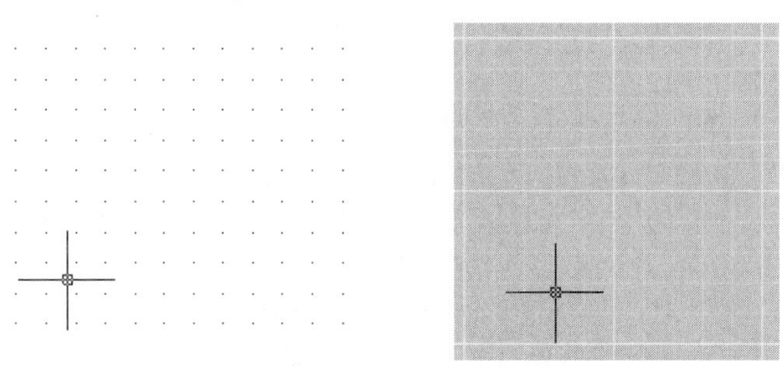

***Left:** Grid of dots displayed by 2D wireframe visual style.*
***Right:** Grid of lines displayed by other visual styles.*

COMMAND LINE OPTIONS

Specify grid spacing(X) sets the x and y direction spacing; an **x** following the value sets the grid spacing to a multiple of the current snap setting, such as 2x.

ON turns on grid markings.

OFF turns off grid markings.

Snap makes the grid spacing the same as the snap spacing.

Major specifies the number of minor grid lines per major line; default = 5 (*new to AutoCAD 2007*).

aDaptive displays fewer grid lines when zoomed out (*new to AutoCAD 2007*).

Limits determines whether the grid is limited to the rectangle defined by the Limits command (*new to AutoCAD 2007*).

Follow matches the grid plane to the dynamic UCS (*new to AutoCAD 2007*).

Aspect specifies different spacing for grid lines/dots in the x and y directions.

RELATED COMMANDS

DSsettings sets the grid via a dialog box.

Options sets the colors for grid lines.

Limits sets the limits of the grid in WCS.

Snap sets the snap spacing.

VsCurrent changes grid from dots (2D wireframe) to lines (other modes).

RELATED SYSTEM VARIABLES

GridDisplay determines how the grid is displayed (*new to AutoCAD 2007*):

- **0** — grid restricted to the area specified by the **Limits** command.
- **1** — grid goes beyond the **Limits** command.
- **2** — adaptive display limits grid density when drawing is zoomed out.
- **4** — grid lines are more closely spaced when zoomed in; requires 2 turned on.
- **8** — grid plane follows the x,y-plane of the dynamic UCS.

GridMajor specifies the number of minor grid lines per major lines (*new to AutoCAD 2007*).

GridMode toggles the grid in the current viewport.

GridUnit specifies the current grid x,y-spacing.

LimMin specifies the x,y coordinates of the lower-left corner of the grid display.

LimMax specifies the x,y coordinates of the upper-right corner of the grid display.

SnapStyl displays a normal or isometric grid.

TIPS

- The grid is most useful when set to the snap spacing, or to a multiple of the snap spacing.
- When the grid spacing is set to 0, it matches the snap spacing.
- The **Options** command sets the color of grid lines. The grid line along the x axis is colored red, while the y axis line is green (*new to AutoCAD 2007*).

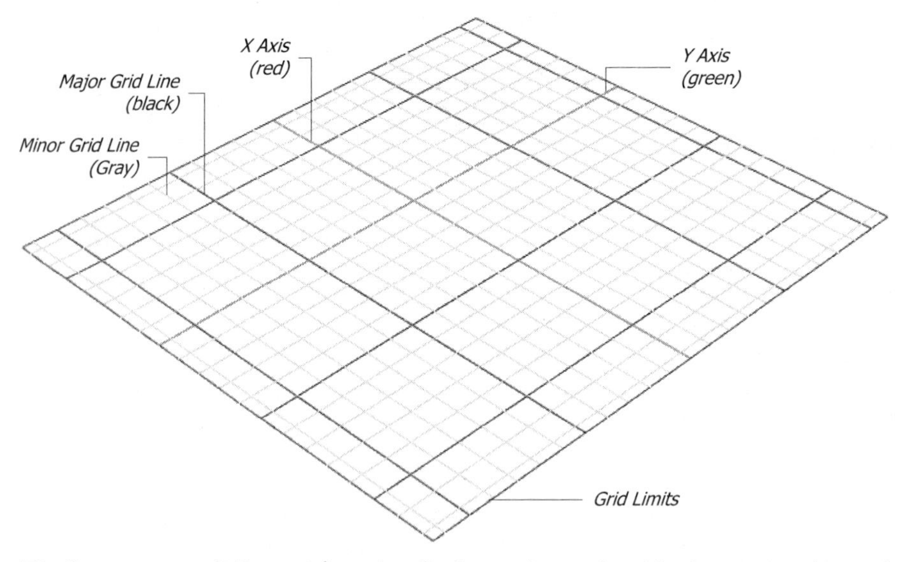

- The **Snap** command's **Isometric** option displays an isometric grid; when perspective mode is turned on, the grid is also displayed in perspective.
- You can set a different grid spacing in each viewport, and a different grid spacing in the x and y directions.
- Rotate the grid with the **Snap** command's **Rotate** option.
- Grid markings are not plotted; to create a plotted grid, use the **Array** command to place an array of points or lines.

Group

Rel.13 Creates named selection sets of objects.

Commands	Aliases	Ctrl	F-key	Alt+	Menu Bar	Tablet
group	**g**	...	...	...	...	**X8**
-group	**-g**					

Command: group

Displays dialog box:

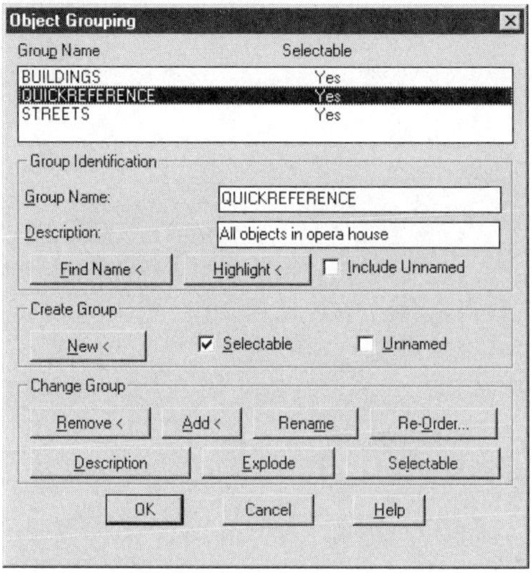

OBJECT GROUPING DIALOG BOX

Group Name lists the names of groups in the drawing.

Group Identification

Group Name displays the name of the current group.

Description describes the group; may be up to 64 characters long.

Find Name lists the name(s) of group(s) that a selected object belongs to.

Highlight highlights the objects included in the current group.

☐ **Include Unnamed** lists unnamed groups in the dialog box.

Create Group

New selects objects for the new group.

☑ **Selectable** toggles selectability: picking one object picks the entire group.

☐ **Unnamed** creates an unnamed group; AutoCAD gives the name ***A***n*, where *n* is a number that increases with each group.

Change Group

Remove removes objects from the current group.

Add adds objects to the current group.

Rename renames the group.

Re-order changes the order of objects in the group; displays Order Group dialog box.

Description changes the description of the group.

Explode removes the group description; does not erase group members.

Selectable toggles selectability.

ORDER GROUP DIALOG BOX

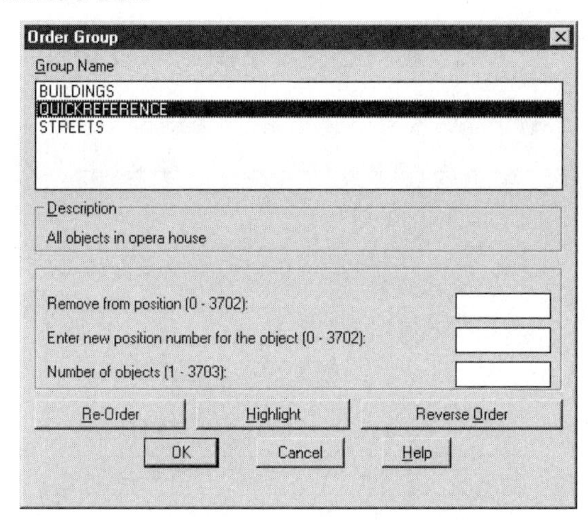

Group Name lists the names of groups in the current drawing.

Description describes the selected group.

Remove from position (0 - *n***)** selects the object to move.

Replace at position (0 - *n***)** moves the group name to a new position.

Number of objects (1 - *n***)** lists the number of objects to reorder.

Re-Order applies the order changes.

Highlight highlights the objects in the current group.

Reverse Order reverses the order of the groups.

-GROUP Command

Command: -group

Enter a group option
[?/Order/Add/Remove/Explode/REName/Selectable/Create] <Create>:
(Enter an option.)

COMMAND LINE OPTIONS

? lists the names and descriptions of currently-defined groups.

Order changes the order of objects within the group.

Add adds objects to the group.

Remove removes objects from the group.

Explode removes the group definition from the drawing.

RENname renames the group.

Selectable toggles whether the group is selectable.

Create creates a newly-named group from the objects selected.

RELATED COMMANDS

Block creates named symbols from a group of objects.

Select creates selection sets.

RELATED SYSTEM VARIABLE

PickStyle toggles whether groups are selected by the usual selection process:

- **0** — Groups and associative hatches are not selected.
- **1** — Groups are included in selection sets (default).
- **2** — Associative hatches are included in selection sets.
- **3** — Both are selected.

TIPS

- You can toggle groups on and off with the CTRL+SHIFT+A shortcut keystroke.
- Use CTRL+H to toggle between selecting the entire group, and objects within the group.
- Consider a group as a named selection set; unlike a regular selection set, a group is not "lost" when the next group is created.
- Group descriptions can be up to 64 characters long.
- Anonymous groups are unnamed; AutoCAD refers to them as ***A***n.

Gsb...

2007 Runs a series of benchmarks that test the graphics display speed (short for "Graphic Speed Benchmark"; undocumented commands).

Command	Alias	Ctrl+	F-key	Alt+	Status Bar	Tablet
gsb1	...	...	...	...	...	...

Command: gsb1

This command rotates, pans, and zooms the current drawing, and then displays the result:

```
Wireframe       : 31.230437 seconds, 23.086453 fps
Gouraud Shaded  : 23.979216 seconds, 30.067705 fps
```

Other graphics benchmarking commands:

GsAutoOrbit rotates 3D models about the x and y axes.

GsAutoZoomPan zooms 3D models in and out.

Gsb2, **Gsb3**, and **Gsb4** run the wireframe and gouraud-shaded benchmarks at different speeds.

Gsb5 runs the benchmark in wireframe, hidden-line, flat-shaded, and Gouraud-shaded modes.

GsbXy, **GsbxyAutomated**, **GsbXyFlat**, **GsbXyG3**, **GsbXyGouraud**, **GsbXyHidden**, and **GsbXyWireframe** rotate 3D models in a variety of rendered modes about the x and y axes.

GsbXy rotates 3D model in all four rendering modes, as listed above.

GsbXyAutomated seems to crash AutoCAD.

GsbXyFlat runs benchmark in flat-shaded mode.

GsbXyG3 runs benchmark in a faster Gouraud-shaded mode.

GsbXyGouraud runs benchmark in Gouraud-shaded mode.

GsbXyHidden runs benchmark in hidden-line removal mode.

GsbXyWireframe runs benchmark in wireframe mode.

GsDolly moves the viewpoint along an axis:

Command: gsdolly
Enter Camera Dolly Distance: *(Enter distance.)*
Enter Active Axis: *(Specify x, y, or z axis.)*

GsClipBack and **GsFrontClip** and specify back and front clipping planes,

Command: gsfrontclip
Enter Front Clipping Plane Position: *(Enter distance.)*

GlClipBackOn and **GsClipFrontOn** toggle the clipping planes:

Command: gsclipfronton
Front Clip On: *(Enter on or off.)*

GsOrbit rotates model by a given angle about a specified axis:

Command: gsorbit
Enter Orbit Angle (degrees): *(Enter orbit angle.)*
Enter Active Axis: *(Specify x, y, or z axis.)*

GsPan pans model a given distance and along a specified axis:

Command: gspan
Enter Camera Pan Distance: *(Enter pan distance.)*
Enter Active Axis: *(Specify x, y, or z axis.)*

GsTestBenchmark processes 3D model with a variety of movements and shade modes; reports results.

GsTestRegress rotates and zooms model in 3D space with different shade modes.

GsZoom prompts for a zoom factor:

Command: gszoom
Enter Camera Zoom Factor: *(Enter a value.)*

RELATED COMMANDS

3dConfig configures the capabilities of graphics boards.

Options selects the display driver for graphics boards.

TIPS

- The g*stext.arx* application must be loaded first with the AppLoad command.
- To record the results to a log file, first execute the LogfileOn command.
- The results reported by the benchmarking commands depend on the speed of the graphics board and the complexity of the drawing.
- These commands can be used to compare the speed of different graphics boards.

Changed Command

As of AutoCAD 2006, the **Hatch** command acts like the **BHatch** command, while the **-Hatch** command is like the **-BHatch** command.

Hatch

Rel.12 Applies associative hatch pattern to objects within boundaries.

Commands	Aliases	Ctrl+	F-key	Alt+	Menu Bar	Tablet
hatch	**bh**	...	...	**DH**	**Draw**	**P9**
	bhatch				**↳Hatch**	
	h					
-hatch	**-bhatch**					

Command: hatch

Displays dialog box:

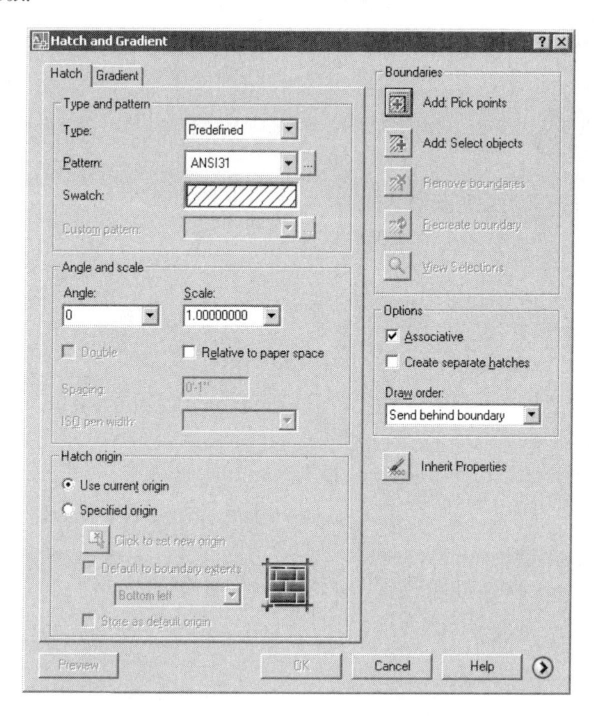

HATCH AND GRADIENT DIALOG BOX

Type and Pattern

Type selects the pattern type:

Type	Meaning
Predefined	Hatches stored in *acad.pat* and *acadiso.pat* files.
User Defined	Parallel-line hatches with spacing defined by you.
Custom	Hatches defined by *.pat* files added to AutoCAD's search path.

Pattern selects hatch pattern.

... displays Hatch Pattern Palette dialog box, showing sample pattern types.

Swatch displays a non-scaled preview of the hatch pattern; click to display the Hatch Pattern Palette dialog box.

Custom Pattern lists the custom patterns, if available.

Angle and Scale

Angle specifies the hatch pattern rotation; default = 0 degrees.

Scale specifies the hatch pattern scale; default = 1.0.

Double (available when User-defined type is selected):

☑ User-defined hatch is applied a second time at 90 degrees to the first pattern.

☐ User-defined hatch is applied once.

Relative to Paper Space (available when hatch is applied through layout viewport):

☑ Hatch patterns scale relative to layout scale.

☐ Hatch patterns scale independent of layout scale.

Spacing specifies the spacing between the lines of a user-defined hatch pattern.

ISO Pen Width patterns scale according to pen width.

Hatch Origin

⊙ **Use Current Origin** uses x,y coordinates stored in the HpOrigin system variable.

○ **Specified Origin** specifies new hatch origins:

Click to Set New Origin prompts "Specify origin point:" to locate new origin in drawing.

☑ **Default to Boundary Extents** places the new origin at one of the four corners of the extents or at the center of the rectangular *extents* of the hatch.

☐ **Store as Default Origin** stores the coordinates in the HpOrigin system variable.

Boundaries

Add: Pick Points prompts, "Pick internal point or [Select objects/remove Boundaries]:". Pick a point inside an area; AutoCAD automatically detects the boundary, which will be filled with the hatch pattern.

Add: Select Objects prompts, "Select objects or [picK internal point/remove Boundaries]:". Pick one or more objects, which will be filled with the hatch pattern.

Remove Boundaries prompts "Select objects or [Add boundaries]:". Pick one or more boundaries to be removed from the hatch pattern selection set.

Recreate Boundaries places polylines or regions around selected hatches.

View Selections views hatch pattern selection set.

Options

Associative:

☑ Automatically updates hatches when boundary or properties are modified; pattern is created as a hatch object.

☐ Hatch is static; pattern is created as a block.

Create separate hatches:

☑ Each hatch pattern is independent.

☐ Patterns are a single pattern, when multiple hatches are created at once.

Draw Order:

- **Do not assign** places hatch normally.
- **Send to back** places hatch behind all other overlapping objects in the drawings.
- **Bring to front** places hatch in front of all other overlapping objects.
- **Send behind boundary** places hatch behind its boundary.
- **Bring in front of boundary** places hatch in front of its boundary.

Inherit Properties sets the hatch pattern parameters from an existing hatch pattern.
Preview previews the hatch pattern.

More Options

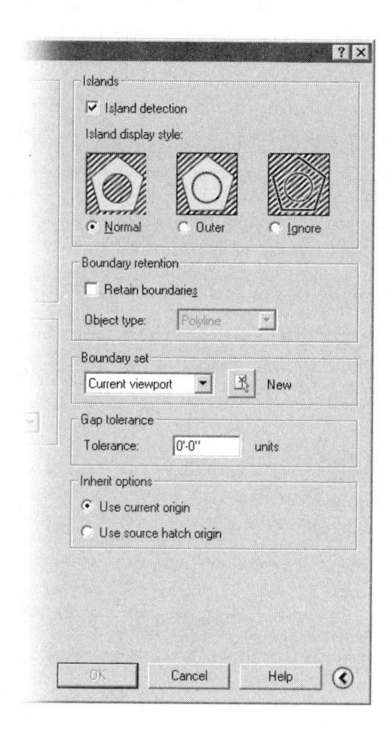

Islands

Island detection:

☑ Island detection is turned on.

☐ Island detection is turned off.

Island display style:

⊙ **Normal** turns hatching off and on each time it crosses a boundary; text is not hatched.

○ **Outer** hatches only the outermost areas; text is not hatched.

○ **Ignore** hatches everything within the boundary; text is hatched.

Boundary Retention

Retain Boundaries:

☑ Boundary polylines or regions (created during the boundary hatching process) are kept after Hatch finishes.

☐ Boundaries are discarded after hatching.

Object Type constructs the boundary from:

- **Polylines** forms the boundary from a polyline.
- **Region objects** forms the boundary from a 2D region.

Boundary Set

Boundary Set defines how objects are analyzed for defining boundaries; not available when **Select Objects** is used to define the boundary (default = current viewport).

New creates new boundary sets; AutoCAD dismisses the dialog box, and then prompts you to select objects.

Gap Tolerance

Tolerance specifies the maximum gap between boundary objects. (Objects don't need to touch to form a valid hatching boundary.) Range is 0 to 5000 units.

Gradient tab

See Gradient command.

-HATCH Command

Command: -hatch

Current hatch pattern: ANSI31

Specify internal point or [Properties/Select/draW boundaries/remove Boundaries/Advanced/DRaw order/Origin]: *(Pick a point, or specify an option.)*

COMMAND LINE OPTIONS

Internal Point option

Pick a point creates a boundary of the area surrounding the pick point.

Property options

Enter a pattern name or [?/Solid/User defined] <ANSI31>: *(Enter a name, or select an option.)*

Specify a scale for the pattern <1.0000>: *(Enter a scale factor.)*

Specify an angle for the pattern <0>: *(Enter an angle.)*

Enter a pattern name allows you to name the hatch pattern.

? lists the names of available hatch patterns.

Solid floods the area with a solid fill in the current color.

User defined creates a simple, user-defined hatch pattern.

Specify a scale specifies the hatch pattern angle (default = 0 degrees).

Specify an angle specifies the hatch pattern scale (default: = 1.0).

Select options

Select objects: *(Select one or more objects.)*

Select objects: *(Press Enter.)*

Select objects selects one or more objects to fill with hatch pattern.

Draw Boundary options

Retain polyline boundary? [Yes/No] <N>: *(Type **Y** or **N**.)*

Specify start point: *(Pick point.)*

Specify next point or [Arc/Length/Undo]: *(Pick point, or enter an option.)*

Specify next point or [Arc/Close/Length/Undo]: *(Pick point or enter an option.)*
Specify start point for new boundary or <Accept>: *(Press Enter to end option.)*

Retain polyline boundary:

- **On** keeps the boundary created during the hatching process after -Hatch finishes.
- **Off** discards the boundary.

Specify start point / next point prompts to draw a closed polyline.

Remove Boundaries options
Select objects or [Add boundaries]: *(Select one or more boundary objects.)*
Select objects or [Add boundaries/Undo]: *(Press* **ENTER**, *or type* **U**.*)*

Select objects selects hatch boundaries to remove.

Undo adds the removed island.

Advanced options
Enter an option [Boundary set/Retain boundary/Island detection/Style/Associativity/Gap tolerance]: *(Specify an option.)*

Boundary set defines the objects analyzed when a boundary is defined by a specified pick point.

Retain boundary:

- **On** the boundary created during the hatching process is kept after -Hatch finishes.
- **Off** the boundary is discarded.

Island detection:

- **On** objects within the outermost boundary are used as boundary objects.
- **Off** all objects within the outermost boundary are filled.

Style selects the hatching style: ignore, outer, or normal.

Associativity:

- **On** hatch pattern is associative.
- **Off** hatch pattern is not associative.

Gap tolerance specifies the largest gap permitted in the boundary:
Specify a boundary gap tolerance value <0>: *(Enter a value between 0 and 5000.)*

Draw Order options
Enter draw order [do Not assign/send to Back/bring to Front/send beHind boundary/bring in front of bounDary] <send beHind boundary>: *(Specify an option.)*

do Not assign places hatches normally.

send to Back places hatches behind all other overlapping objects in the drawing.

bring to Front places hatches in front of all other overlapping objects.

send beHind boundary places hatches behind its boundary.

bring in front of bounDary places hatch in front of the boundary.

Origin options
[Use current origin/Set new origin/Default to boundary extents] <Use current origin>: *(Enter an option.)*

Use current origin uses the x,y coordinates stored in the **HpOrigin** system variable.

Set new origin prompts to pick a point in the drawing as the new origin.

Default to boundary extents places the new origin at one of the four corners of the extents or at the center of the rectangular *extents* of the hatch.

Hatch Pattern Palette dialog box

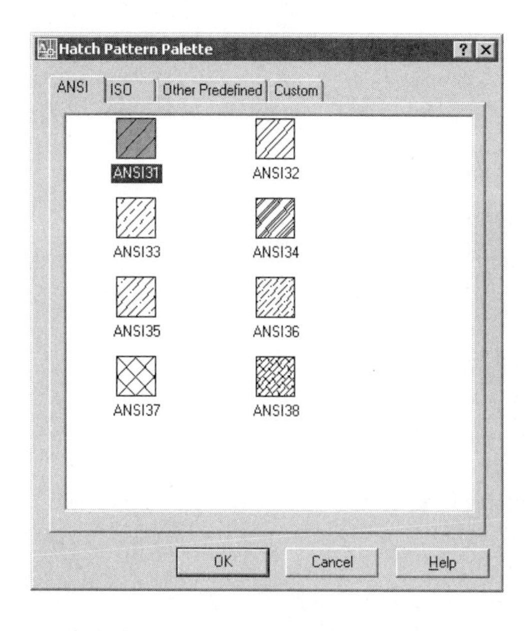

ANSI Patterns:

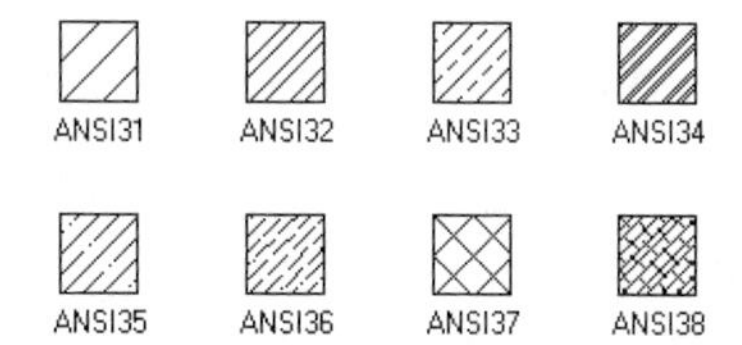

ISO Patterns:

Other Predefined Patterns:

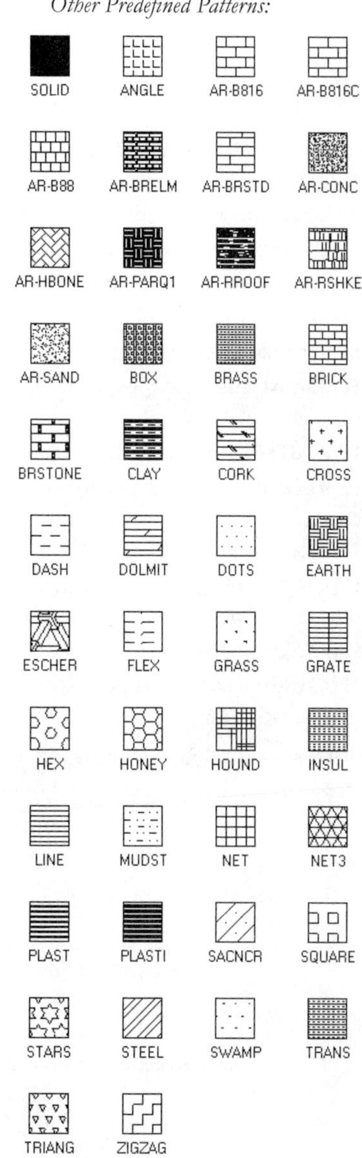

RELATED COMMANDS

AdCenter places hatch patterns from other drawings.

Boundary traces polylines automatically around closed boundaries.

Convert converts Release 13 (and earlier) hatch patterns into Release 14-2007 format.

Hatch places nonassociative hatch patterns.

HatchEdit edits hatch patterns.

PsFill floods closed polylines with PostScript fill patterns.

ToolPalette stores and places selected hatch patterns and fill colors.

RELATED SYSTEM VARIABLES

DelObj toggles whether boundary is erased after hatch is placed.

FillMode determines whether hatch patterns are displayed.

HpAng specifies the current hatch pattern angle; default = 0.

HpAssoc determines whether or not hatches are associative.

HpBound specifies the hatch boundary object: polyline or region.

HpDouble specifies single or double hatching.

HpDrawOrder controls the display order of the hatch pattern relative to other overlapping objects.

HpGalTol reports the current gap tolerance.

HpInherit determines whether the MatchProp command copies hatch origins from the source objects.

HpName names the current hatch pattern.

HpObjWarning specifies the maximum number of hatch boundary objects that can be selected before AutoCAD sounds the alert; default = 10000.

HpOrigin specifies the x,y coordinates for the origin of new hatches.

HpOriginMode determines the default hatch origin point.

- **0** — Uses HpOrigin.
- **1** — Bottom left corner of boundary's rectangular extents.
- **2** — Bottom right corner of boundary's rectangular extents.
- **3** — Top right corner of boundary's rectangular extents.
- **4** — Top left corner of boundary's rectangular extents.
- **5** — Center of boundary's rectangular extents.

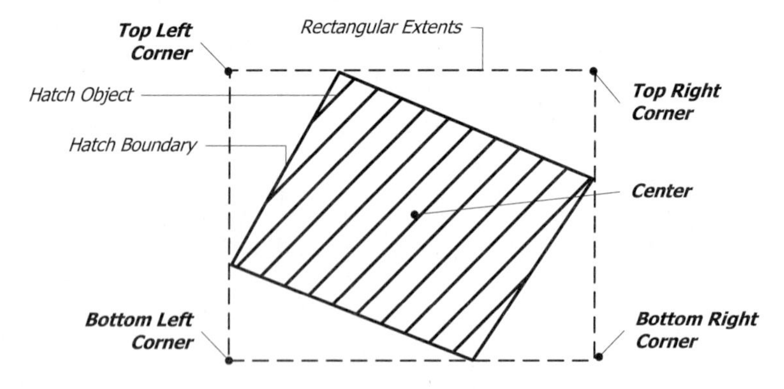

HpScale specifies the current hatch pattern scale factor; default = 1.

HpSeparate determines whether a single hatch object or separate hatch objects are created when operating on several closed boundaries.

HpSpace specifies the current hatch pattern spacing factor; default = 1.

OsnapHatch determines whether object snap snaps to hatch patterns.

PickStyle controls the selection of hatch patterns:

0 — Neither groups nor hatches selected.

1 — Groups selected (default).

2 — Associative hatches selected.

3 — Both selected.

SnapBase specifies the starting coordinates of hatch pattern (default = 0,0).

RELATED FILES

acad.pat contains the ANSI and other hatch pattern definitions.

acadiso.pat contains the ISO hatch pattern definitions.

TIPS

- This command first generates a boundary, and then hatches the inside area.
- Use the Boundary command to create just the boundary.
- Hatch stores hatching parameters in the pattern's extended object data.
- Bringing the pattern to the front (through the Draw Order option) makes it easier to edit the hatch.
- All ANSI patterns are defined as 45° lines. To keep at 45°, the angle should be specified as 0.
- Set the OsnapHatch system variable to 0 to turn off object snapping to hatch and fill patterns.
- Hatches, solid fills, and gradient fills can be dragged into the Tool Palette window.
- The Trim command trims hatches.
- As of AutoCAD 2006, the Gradient command displays the Gradient tab of the Hatch and Gradient dialog box.

HatchEdit

Rel.13 Edits associative hatch objects.

Commands	Alias	Ctrl+	F-key	Alt+	Menu Bar	Tablet
hatchedit	**he**	...	...	**MOH**	**Modify** ↳**Object** ↳**Hatch**	**Y16**
-hatchedit						

Command: hatchedit

Select associative hatch object: *(Select one hatch object.)*

Displays dialog box:

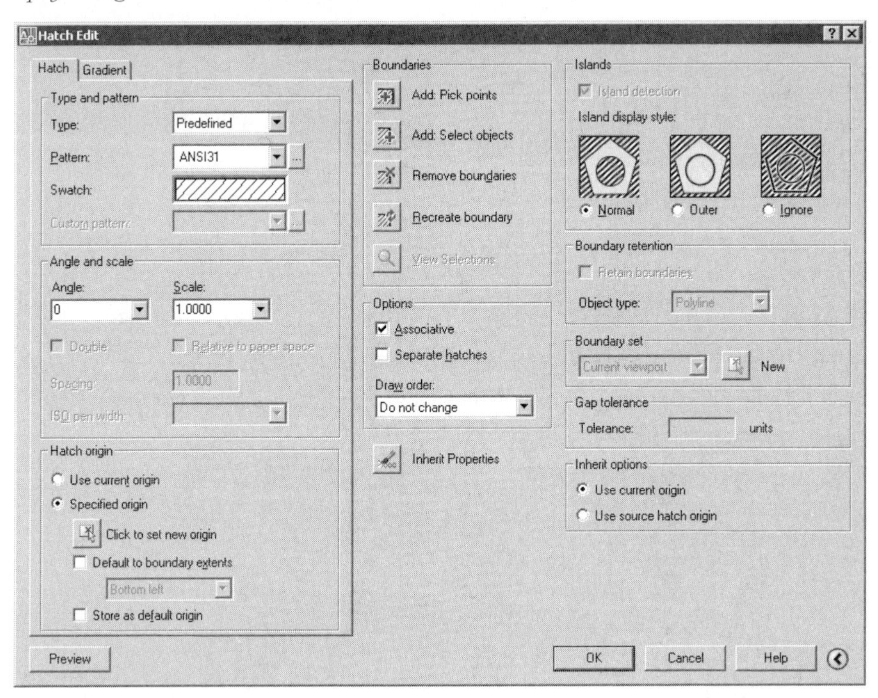

HATCH EDIT DIALOG BOX

*See the **BHatch** and **Gradient** commands for options.*

-HATCHEDIT Command

Command: -hatchedit

Select associative hatch object: *(Select one object.)*

Enter hatch option [DIsassociate/Style/Properties/DRaw order/ADd boundaries/Remove boundaries/recreate Boundary/ASsociate/separate Hatches/Origin] <Properties>: *(Enter an option.)*

COMMAND LINE OPTIONS

Select selects a single associative hatch pattern.

DIsassociate removes associativity from hatch pattern.

Style options

Enter hatching style [Ignore/Outer/Normal] <Normal>: *(Enter an option.)*

Normal hatches alternate boundaries (default).

Outer hatches only outermost boundary.

Ignore hatches everything within outermost boundary.

Properties options

Enter a pattern name or [?/Solid/User defined] <ANSI31>: *(Enter name or type an option.)*

Specify a scale for the pattern <1.0000>: *(Enter scale factor.)*

Specify an angle for the pattern <0>: *(Enter pattern angle.)*

Pattern name specifies the name of a valid hatch pattern.

? lists the available hatch pattern names.

Solid replaces the hatch pattern with solid fill.

User defined creates a new on-the-fly hatch pattern.

Scale changes the hatch pattern scale.

Angle changes the hatch pattern angle.

Draw Order options

Enter draw order [do Not assign/send to Back/bring to Front/send beHind boundary/bring in front of bounDary] <send beHind boundary>: *(Specify an option.)*

do Not assign places hatch normally.

send to Back places hatch behind all other overlapping objects in the drawings.

bring to Front places hatch in front of all other overlapping objects.

send beHind boundary places hatch behind its boundary.

bring in front of bounDary places hatch in front of its boundary.

Add Boundaries options

Specify internal point or [Select objects]: *(Pick a point or select objects.)*

Specify internal point selects points within closed objects.

Select objects selects one or more objects to add to the boundary collection.

Remove Boundaries options

Select objects or [Add boundaries]:*(Pick a point or select objects.)*

Select objects selects one or more objects to remove from the boundary collection.

Add boundaries switches to add boundary mode.

Recreate Boundaries options

Enter type of boundary object [Region/Polyline] <Polyline>: *(Type* **R** *or* **P***.)*

Reassociate hatch with new boundary? [Yes/No] <N>: *(Type* **Y** *or* **N***.)*

Enter type of boundary object:

- **Region** switches boundary to region object.
- **Polyline** switches boundary to polyline object.

Reassociate hatch with new boundary?

- **Yes** associates hatch with changed boundary type.
- **No** does not associate hatch with changed boundary type.

Associate options

Associates hatch with the selected boundary objects.

Specify internal point or [Select objects]: *(Pick a point or select objects.)*

Specify internal point selects points within closed objects.

Select objects selects one or more boundary objects to associate with hatch.

Origin options

Associates hatch with the selected boundary objects.

[Use current origin/Set new origin/Default to boundary extents] <Use current origin>: *(Enter an option.)*

Use current origin retains the current origin.

Set new origin prompts for a new origin point, and then asks whether to store it as the new default origin.

Default to boundary extents specifies one of bottom Left (default), bottom Right, top rIght, top lEft, or Center.

Separate Hatches separates a single object consisting of multiple hatch areas into independent hatch objects.

RELATED COMMANDS

BHatch applies associative hatch patterns.

Explode explodes a hatch pattern block into lines.

Gradient applies associative fills to areas.

MatchProp matches the properties of one hatch to other hatches.

Properties changes properties of hatch patterns.

TIPS

- HatchEdit works with associative and non-associative hatch objects.
- AutoCAD cannot change a non-associative hatch to associative; uncheck the **Associative** option to convert a non-associative hatch to associative. Also, you can use the Explode command to reduce hatches to lines.
- As an alternative to entering this command, double-click hatch patterns to display the Hatch Edit dialog box.
- To select a solid fill, click the outer edge of the hatch pattern, or use a crossing window selection on top of the solid fill.
- Even though the Hatch Edit dialog box looks identical to the BHatch dialog box, many options are not available in the Hatch Edit dialog box.

Helix

2007 Draws helixes.

Commands	Alias	Ctrl+	F-key	Alt+	Menu Bar	Tablet
helix	...	...	...	DX	Draw ↳Helix	...

Command: helix

Number of turns = 3.0000 Twist=CCW

Specify center point of base: *(Pick point 1.)*

Specify base radius or [Diameter] <1.0>: *(Enter radius 2, or type D.)*

Specify top radius or [Diameter] <1.5>: *(Enter radius 3, or type D.)*

Specify helix height or [Axis endpoint/Turns/turn Height/tWist] <1.0>: *(Enter height 4, or enter an option.)*

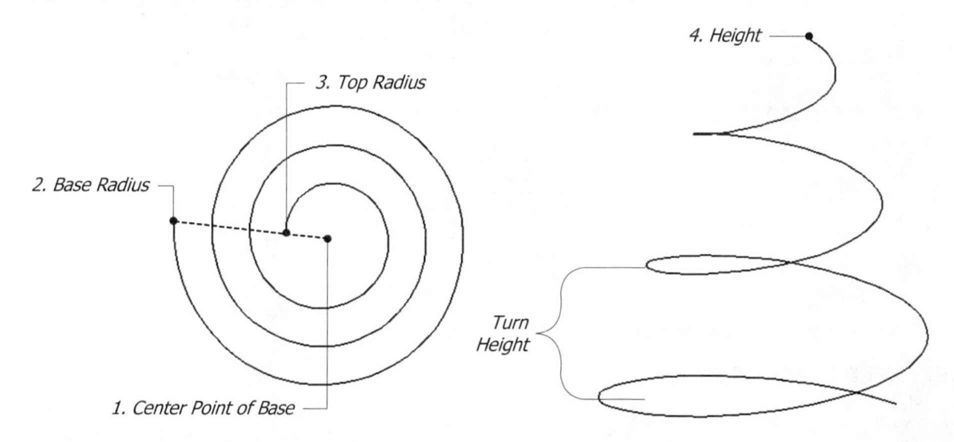

COMMAND OPTIONS

Center point of base specifies the coordinates of the helix's base; when z is not specified, the helix is drawn at the current elevation.

Base radius specifies the radius of the base.

Diameter specifies the diameter of the base.

Top radius specifies the radius of the top; value must be positive.

Helix height specifies the height of the helix; if negative, the helix is drawn downwards from its base.

Axis endpoint specifies a second point (from the base center point) to draw the helix at an angle.

Turns specifies the number of turns the helix makes; default = 3.

turn Height specifies the height of one turn.

tWist switches between drawing the helix clockwise and counterclockwise (default).

TIP

- To draw a spring, make the top and bottom radii the same.

'Help

V. 1.0 Lists information for using AutoCAD's commands.

Command	Alias	Ctrl+	F-key	Alt+	Menu Bar	Tablet
help	'?	...	F1	HH	Help ↳Help	Y7

Command: help

Displays window:

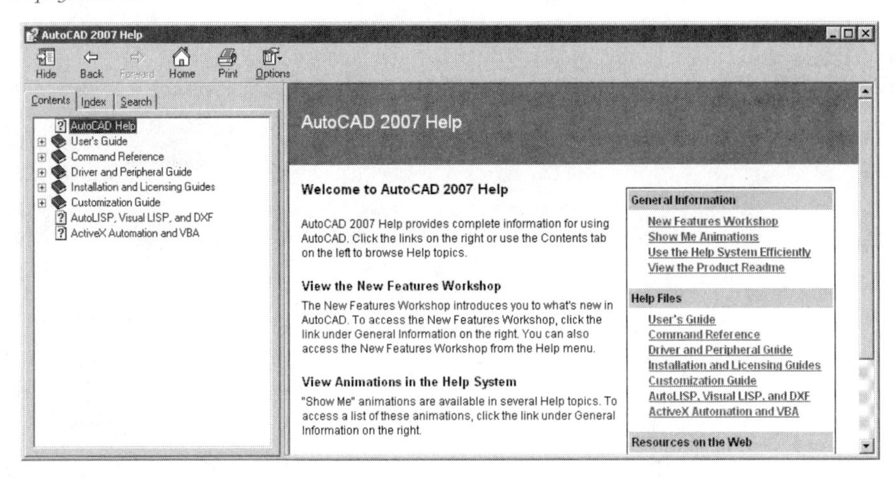

RELATED COMMANDS

All.

RELATED FILE

acad170.chm is AutoCAD 2007's primary help file.

TIPS

- Because 'Help, '?, and **F1** are transparent commands, you can use them during another command to get help with the command's options.
- Pressing **F1** during a command displays context-sensitive help.

Removed Commands

Hide now operates only in 2D wireframe mode; as of AutoCAD 2007 it is an alias for the **VsCurrent** command in all other modes, such as 3D wireframe and Realistic.

HlSettings was removed from AutoCAD 2007; it was replaced by the **VisualStyles** command.

HpConfig was removed from AutoCAD 2000; it was replaced by the **Plot** command.

Hyperlink

2000 Attaches hyperlinks (URLs) to objects in drawings.

Commands	Alias	Ctrl+	F-key	Alt+	Menu Bar	Tablet
hyperlink	...	k	...	IH	Insert ⇘Hyperlink	...
-hyperlink						

Command: hyperlink
Select objects: *(Select one or more objects.)*
Select objects: *(Press Enter to end object selection.)*

Displays dialog box:

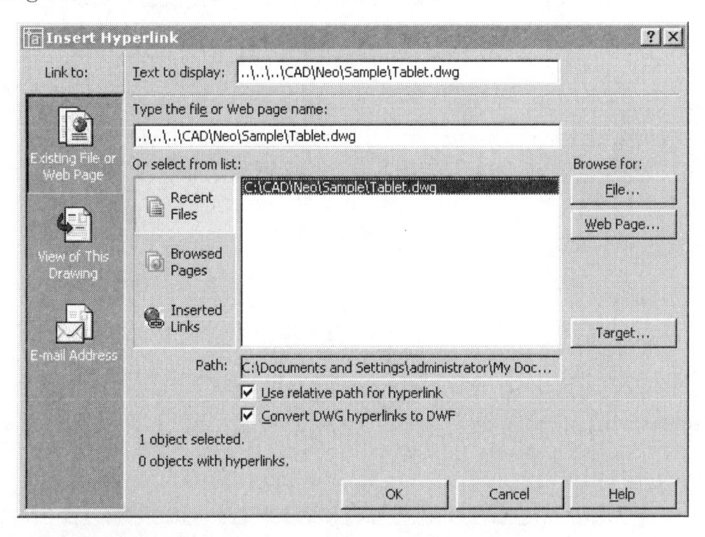

COMMAND LINE OPTION

Select object selects the objects to which to attach the hyperlink.

INSERT HYPERLINK DIALOG BOX

Existing File or Web Page page

Text to display describes the hyperlink; the description is displayed by the tooltip; when blank, the tooltip displays the URL.

Type the file or Web page name specifies the hyperlink (URL) to associate with the selected objects; the hyperlink may be any file on your computer, on any computer you can access on your local network, or on the Internet.

Or select from list:

- **Recent files** lists drawings recently opened in AutoCAD.
- **Browsed pages** lists pages recently viewed with Web browsers, and other software.
- **Inserted links** lists URLs recently entered in Web browsers.

File opens the Browse the Internet - Select Hyperlink dialog box.

Web Page starts up a simple Web browser.

Target specifies a location in the file, such as a target in an HTML file, a named view in AutoCAD, or a page in a spreadsheet document. Displays Select Place in Document dialog box.

Path displays the full path and filename to the hyperlink; only the file name appears when **Use relative path for hyperlink** is checked.

☑ **Use relative path for hyperlink** toggles use of the path for relative hyperlinks in the drawing; when "" (null), the drawing paths stored in **AcadPrefix** are used.

☑ **Convert DWG hyperlinks to DWF** preserves hyperlinks when drawings are exported in DWF format.

Remove link removes the hyperlink from the object; this button appears only if you select an object that already has a hyperlink.

View of This Drawing Page

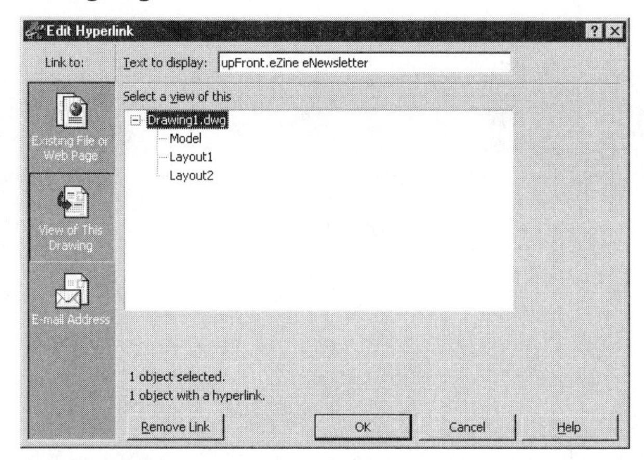

Select a view of this selects a layout, named view, or named plot.

Email Address Page

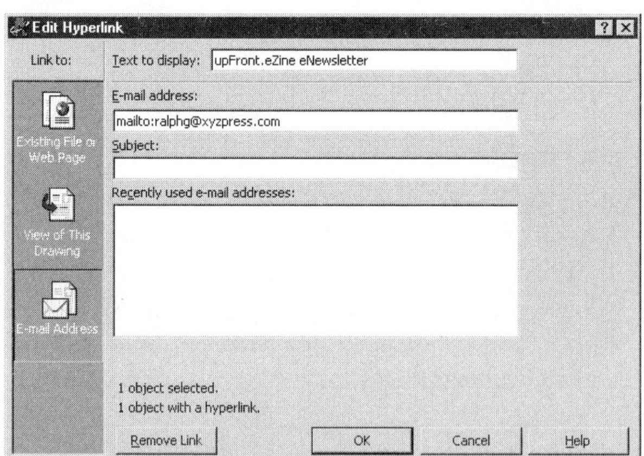

Email address specifies the email address; the *mailto:* prefix is added automatically.

Subject specifies the text that will be added to the Subject line.

Recently used e-mail addresses lists the email addresses recently entered.

-HYPERLINK Command

Command: -hyperlink
Enter an option [Remove/Insert] <Insert>: *(Enter an option.)*
Enter hyperlink insert option [Area/Object] <Object>: *(Enter an option.)*
Select objects: *(Select one or more objects.)*
Select objects: *(Press* **Enter** *to end object selection.)*
Enter hyperlink <current drawing>: *(Enter hyperlink address.)*
Enter named location <none>: *(Optional: Enter a bookmark.)*
Enter description <none>: *(Optional: Enter a description.)*

COMMAND LINE OPTIONS

Remove removes a hyperlink from selected objects or areas.

Insert adds a hyperlink to selected objects or areas.

Select objects selects the object to which the hyperlink will be added.

Enter hyperlink specifies the filename or hyperlink address.

Enter named location (*optional*) specifies a location within the file or hyperlink.

Enter description (*optional*) describes the hyperlink.

RELATED COMMANDS

HyperlinkOptions toggles the display of the hyperlink cursor, shortcut menu, and tooltip.

HyperlinkOpen opens hyperlinks (URLs) via the command line.

HyperlinkBack returns to the previous URL.

HyperlinkFwd moves forward to the next URL; works only when the HyperlinkBack command was used, otherwise AutoCAD complains, "** No hyperlink to navigate to **".

HyperlinkStop stops the display of the current hyperlink.

GoToUrl displays specific Web pages.

PasteAsHyperlink pastes hyperlinks to selected objects.

RELATED SYSTEM VARIABLE

HyperlinkBase specifies the path for relative hyperlinks in the drawing; when "" (null), the drawing paths stored in AcadPrefix are used.

TIPS

- If the drawing has never been saved, AutoCAD is unable to determine the default *relative folder*. For this reason, the Hyperlink command prompts you to save the drawing.
- By using hyperlinks, you can create a *project document* consisting of drawings, contacts (word processing documents), project timelines, cost estimates (spreadsheet pages), and architectural renderings. To do so, create a "title page" of an AutoCAD drawing with hyperlinks to the other documents.
- To edit a hyperlink with Hyperlink, select object(s), make editing changes, and click OK.
- To remove a hyperlink with Hyperlink, select the object, and then click Remove Link.
- An object can have just one hyperlink attached to it; more than one object, however, can share the same hyperlink.
- The alternate term for "hyperlink" is *URL*, which is short for "universal resource locator," the universal file naming system used by the Internet.

'HyperlinkOpen/Back/Fwd/Stop

2000 Controls the display of hyperlinked pages (undocumented commands).

Commands	Alias	Ctrl+	F-key	Alt+	Menu Bar	Tablet
hyperlinkopen	...	...	...	...	...	...
hyperlinkback						
hyperlinkfwd						
hyperlinkstop						

Command: hyperlinkopen
Enter hyperlink <current drawing>: *(Enter a hyperlink option.)*
Enter named location <none>: *(Optional: Enter a bookmark.)*
Displays the specified Web page or file, if possible.

Command: hyperlinkback
Returns to the previous hyperlinked page.

Command: hyperlinkstop
Stops displaying the Web page.

Command: hyperlinkfwd
*Hyperlinks to the next page; can be used only after the **HyperLinkBack** command.*

COMMAND LINE OPTIONS

Enter hyperlink enters a URL (uniform resource locator) or a filename.

Enter named location enters a named view or other valid target.

RELATED COMMANDS

Hyperlink attaches a hyperlink to objects.

HyperlinkOptions specifies the options for hyperlinks.

GoToUrl displays a specific Web page.

RELATED TOOLBAR ICONS

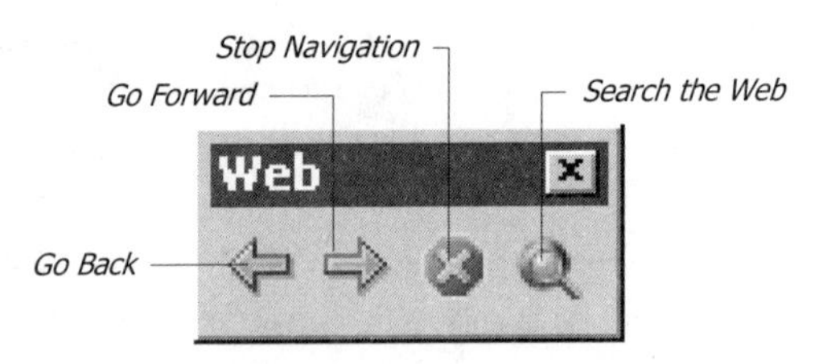

Go Back goes to the previous hyperlink; executes the HyperlinkBack command.

Go Forward goes to the next hyperlink; executes the HyperlinkFwd command.

Stop Navigation stops loading the current hyperlink file; executes the HyperlinkStop command.

Browse the Web displays the Web browser; executes the Browser command.

'HyperlinkOptions

2000 Toggles the display of the hyperlink cursor, shortcut menu, and tooltip.

Command	Alias	Ctrl+	F-key	Alt+	Menu Bar	Tablet
hyperlinkoptions	...	...	...	...	...	

Command: hyperlinkoptions

Display hyperlink cursor tooltip and shortcut menu? [Yes/No] <Yes>: *(Type* **Y** *or* **N***.)*

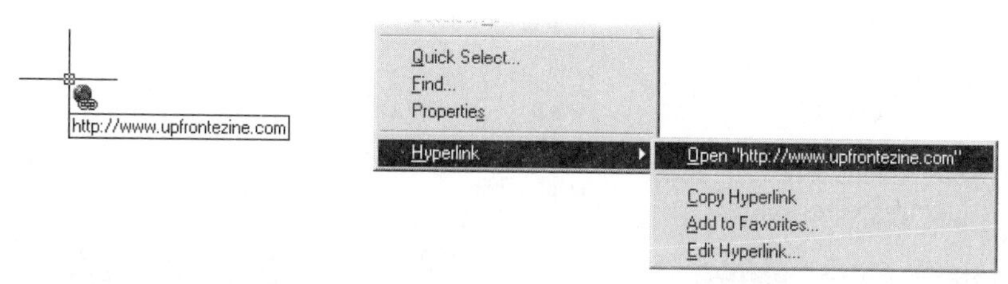

***Left**: Hyperlink cursor and tooltip.*
***Right**: Hyperlink shortcut menu.*

COMMAND LINE OPTIONS

Display hyperlink cursor and shortcut menu toggles the display of the hyperlink cursor and shortcut menu.

Display hyperlink tooltip toggles the display of the hyperlink tooltip.

SHORTCUT MENU

Select an object containing a hyperlink; then right-click to display the cursor menu.

***Hyperlink** options:*

Open "url" launches the appropriate applications, and loads the file referenced by the URL.

Copy Hyperlink copies hyperlink data to the Clipboard; use the PasteAsHyperlink command to paste the hyperlink to selected objects.

Add to Favorites adds the hyperlink to a favorites list.

Edit Hyperlink displays the Edit Hyperlink dialog box; see the Hyperlink command.

RELATED COMMANDS

Hyperlink attaches a hyperlink to objects.

Options determines options for most other aspects of AutoCAD.

RELATED SYSTEM VARIABLE

HyperlinkBase specifies the path for relative hyperlinks in the drawing; when "" (null), the drawing paths stored in AcadPrefix are used.

TIP

- Answering "N" to this command makes hyperlinks unavailable.

'Id

V. 1.0 Identifies the 3D coordinates of specified points (short for IDentify).

Command	Alias	Ctrl+	F-key	Alt+	Menu Bar	Tablet
id	...	...	...	TQI	Tools ↳Inquiry ↳Id Point	U9

Command: id
Specify point: *(Pick a point.)*

Sample output:

X = 1278.0018 Y = 1541.5993 Z = 0.0000

COMMAND LINE OPTION

Specify point picks a point.

RELATED COMMANDS

List lists information about picked objects.

Point draws points.

RELATED SYSTEM VARIABLE

LastPoint contains the 3D coordinates of the last picked point.

TIPS

- The Id command stores the picked point in the LastPoint system variable. Access that value by entering @ at the next prompt for a point value.
- Invoke the Id command to set the value of the LastPoint system variable, which can be used as relative coordinates in another command.
- If a 2D point is specified, the z-coordinate displayed by Id is the current elevation setting; otherwise, the z-coordinate is that of the specified point.
- When you use Id with an object snap, the z-coordinate is the object-snapped value.

Replaced Command

Image was removed from AutoCAD 2007; it was replaced by the **ExternalReferences** command.

-Image

Rel.14 Controls the attachment of raster images at the command line.

Commands	Alias	Ctrl+	F-key	Alt+	Menu Bar	Tablet
-image	-im	...	...	...	...	...

Command: -image

Enter image option [?/Detach/Path/Reload/Unload/Attach] <Attach>: *(Enter an option.)*

COMMAND LINE OPTIONS

? lists currently-attached image files.

Detach erases images from drawings.

Path lists the names of images in the drawing.

Reload reloads image files into drawings.

Unload removes images from memory without erasing them.

Attach displays the Attach Image File dialog box; see the ImageAttach command.

RELATED COMMANDS

ExternalReferences replaces the dialog box version of the Image command.

ImageAdjust controls the brightness, contrast, and fading of images.

ImageAttach attaches images to the current drawing.

ImageClip creates clipping boundaries around images.

ImageFrame toggles the display of image frames.

ImageQuality toggles between draft and high-quality mode.

RELATED SYSTEM VARIABLE

ImageHlt toggles whether the entire image is highlighted.

TIPS

- The -Image command loads files with the following extensions: *.bil*, *.bmp*, *.cal*, *.cg4*, *.dib*, *.flc*, *.fli*, *.gif*, *.gp4*, *.igs*, *.jpg*, *.mil*, *.pct*, *.pcx*, *.png*, *.rlc*, *.rle*, *.rst*, *.tga*, and *.tif*. Note that *.igs* is not IGES, but a raster format used for mapping. Autodesk's RasterDesign software permits viewing of additional raster formats, such as *.sid* and *.ecw*. This command handles raster images of these color depths:

  Bitonal — black and white (monochrome).

  8-bit gray — 256 shades of gray.

  8-bit color — 256 colors.

  24-bit color — 16.7 million colors.

- AutoCAD can display one or more images in any viewport. There is no theoretical limit to the number and size of images.
- Images do not display in shaded visual modes.
- The DrawOrder command can place images "behind" other objects.
- Images are always xrefs; the source image file must be present when the drawing is opened.
- The Image command was replaced by the ExternalReferences command AutoCAD 2007.

ImageAdjust

Rel.14 Controls brightness, contrast, and fade of attached raster images.

Commands	Alias	Ctrl+	F-key	Alt+	Menu Bar	Tablet
imageadjust	**iad**	...	...	**MOIA**	**Modify** ↳**Object** ↳**Image** ↳**Adjust**	**X20**
-imageadjust						

Command: imageadjust
Select image(s): *(Select one or more image objects.)*
Select image(s): *(Press Enter to end object selection.)*

Displays dialog box:

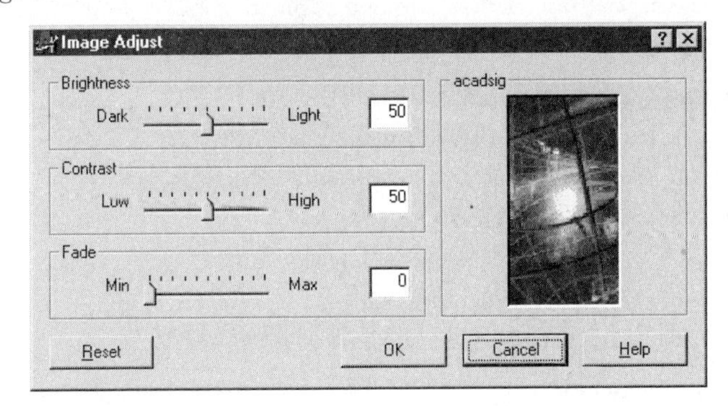

IMAGE ADJUST DIALOG BOX

Brightness options

Dark reduces the brightness of the image as values approach 0.

Light increases the brightness of the image as values approach 100.

Contrast

Low reduces the image contrast as values approach 0.

High increases the image contrast as values approach 100.

Fade

Min reduces the image fade as values approach 0.

Max increases the image fade as values approach 100.

Reset resets the image to its original parameters; default values are:

Brightness — 50

Contrast — 50

Fade — 0

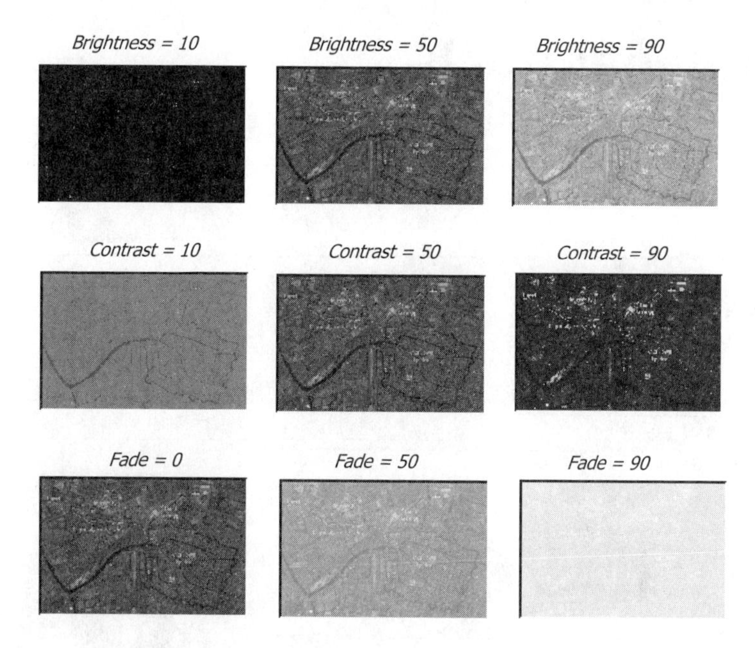

-IMAGEADJUST Command

Command: -imageadjust

Select image(s): *(Select one or more images.)*

Select image(s): *(Press Enter to end object selection.)*

Enter image option [Contrast/Fade/Brightness] <Brightness>: *(Enter an option.)*

COMMAND LINE OPTIONS

Contrast option

Enter contrast value (0-100) <50>: *(Enter a value.)*

Enter contract value adjusts the contrast between 0% contrast and 100%; default = 50.

Fade Ooption

Enter fade value (0-100) <0>: *(Enter a value.)*

Enter fade option adjusts the fading between 0% faded and 100%; default = 0.

Brightness option

Enter brightness value (0-100) <50>: *(Enter a value.)*

Enter brightness value adjusts the brightness between 0% brightness and 100%; default = 50.

RELATED COMMANDS

ExternalReferences controls the loading of raster image files in drawings.

ImageAttach attaches images to the current drawing.

ImageClip creates clipping boundaries on images.

ImageFrame toggles display of image frames.

ImageQuality toggles display between draft and high-quality mode.

ImageAttach

Rel.14 Selects raster files to attach to drawings.

Command	Alias	Ctrl+	F-key	Alt+	Menu Bar	Tablet
imageattach	iat	...	...	II	Insert ↳Raster Image	...

Command: imageattach

Displays file dialog box:

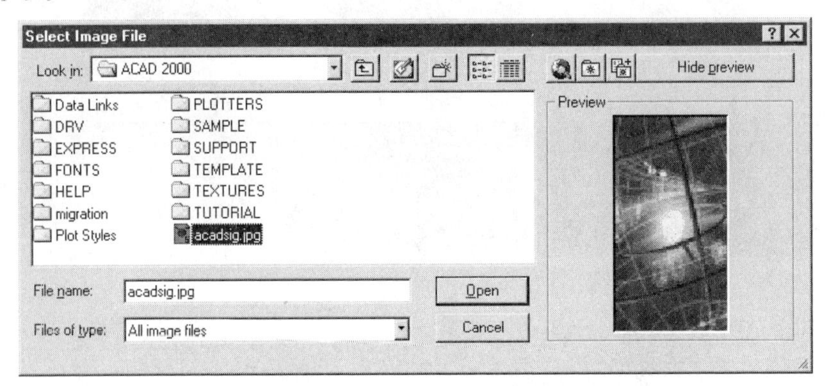

Select an image file, and then click ***Open.***

Displays dialog box:

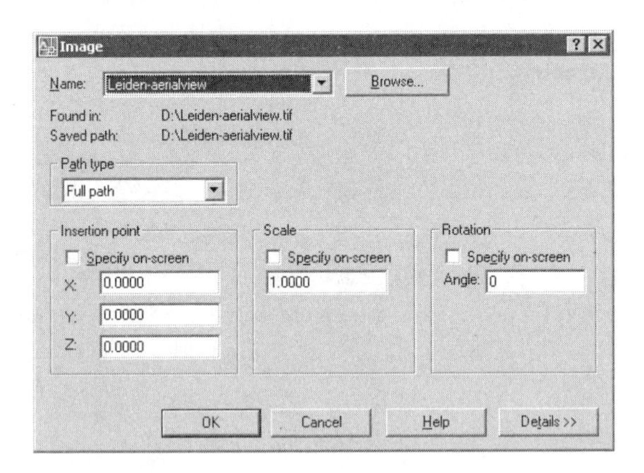

DIALOG BOX OPTIONS

Name selects names from a list of previously-attached images.

Browse selects files; displays Select Image File dialog box.

Path Type

Full Path saves the path to the image file.

Relative Path saves the path relative to the current drawing file.

No Path does not save the path.

Insertion Point

☑ **Specify On-Screen** specifies the insertion point of the image in the drawing, after the dialog box is dismissed.

X, Y, Z specifies the x, y, z coordinates of the lower-left corner of the image.

Scale

☑ **Specify on-screen** specifies the scale of the image (relative to the lower-left corner) in the drawing after the dialog box is dismissed.

Scale specifies the scale of the image; a positive value enlarges the image, while a negative value reduces it.

Rotation

☐ **Specify on-screen** specifies the rotation angle of the image about the lower-left corner in the drawing, after the dialog box is dismissed.

Angle specifies the angle to rotate the image; positive angles rotate the image counterclockwise.

Details expands the dialog box to display information about the image:

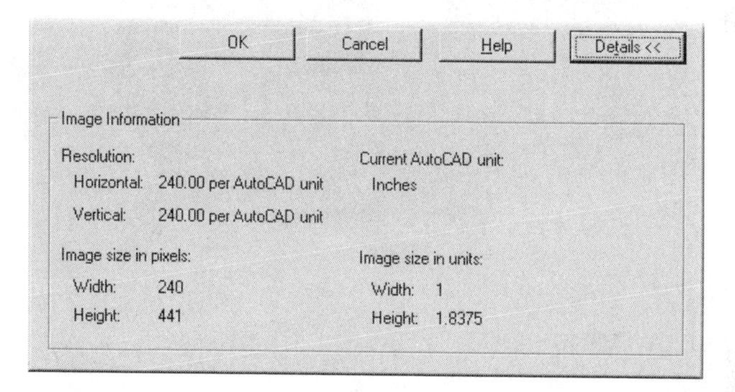

RELATED COMMAND

ExternalReferences controls the loading of raster image files in drawings.

RELATED SYSTEM VARIABLE

InsUnits specifies the drawing units for the inserted image.

TIPS

- For a command-line version of the ImageAttach command, use the -Image command's **Attach** option.
- This dialog box no longer selects units from the Current AutoCAD Unit list box; as of AutoCAD 2000, use the InsUnits system variable.

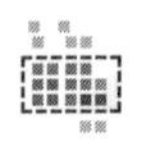

ImageClip

Rel.14 Clips raster images.

Command	Alias	Ctrl+	F-key	Alt+	Menu Bar	Tablet
imageclip	**icl**	...	...	**MCI**	**Modify** **↳Clip** **↳Image**	**X22**

Command: imageclip

Select image to clip: *(Select one image object.)*

Enter image clipping option [ON/OFF/Delete/New boundary] <New>: *(Enter an option.)*

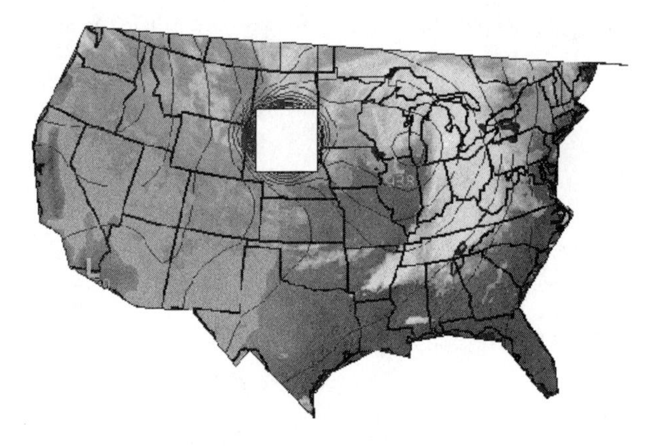

COMMAND LINE OPTIONS

Select image selects one image to clip.

ON turns on a previous clipping boundary.

OFF turns off the clipping boundary.

Delete erases the clipping boundary.

New Boundary Options

Enter clipping type [Polygonal/Rectangular] <Rectangular>: *(Enter an option.)*

Polygonal creates a polygonal clipping path.

Rectangular creates a rectangular clipping boundary.

Polygonal Options

Specify first point: *(Pick a point.)*

Specify next point or [Undo]: *(Pick a point, or type* **U***.)*

Specify next point or [Undo]: *(Pick a point, or type* **U***.)*

Specify next point or [Close/Undo]: *(Pick a point, or enter an option.)*

Specify next point or [Close/Undo]: *(Enter* **C** *to close the polygon.)*

Specify first point specifies the start of the first segment of the polygonal clipping path.

Specify next point specifies the next vertex.

Undo undoes the last vertex.

Close closes the polygon clipping path.

Rectangular Options

Specify first corner point: *(Pick a point.)*

Specify opposite corner point: *(Pick a point.)*

Specify first corner point specifies one corner of the rectangular clip.

Specify opposite corner point specifies the second corner.

When you select an image with a clipped boundary, AutoCAD prompts:

Delete old boundary? [No/Yes] <Yes>: *(Type* **N** *or* **Y***.)*

Delete old boundary?

Yes removes the previously-applied clipping path.

No exits the command.

RELATED COMMANDS

ExternalReferences controls the loading of raster image files in the drawing.

ImageAdjust controls the brightness, contrast, and fading of the image.

ImageAttach attaches an image in the current drawing.

ImageFrame toggles the display of the image's frame.

ImageQuality toggles the display between draft and high-quality mode.

XrefClip clips *.dwg* drawing files attached as externally-referenced files.

TIPS

- You can use object snap modes on the image's frame, but not on the image itself.
- To clip a hole in the image, create the hole, and then double back on the same path:

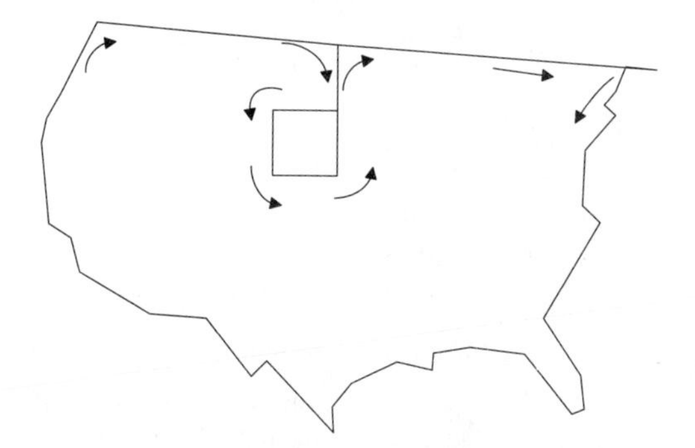

- For a rounded clipping path, apply the **PEdit** command.

ImageFrame

Rel.14 Toggles the frame around raster images for display and plotting.

Command	Alias	Ctrl+	F-key	Alt+	Menu Bar	Tablet
imageframe	...	...	...	**MOIF**	**Modify** ↳**Object** ↳**Image** ↳**Frame**	...

Command: imageframe
Enter image frame setting [0, 1, 2] <1>: *(Enter a number.)*

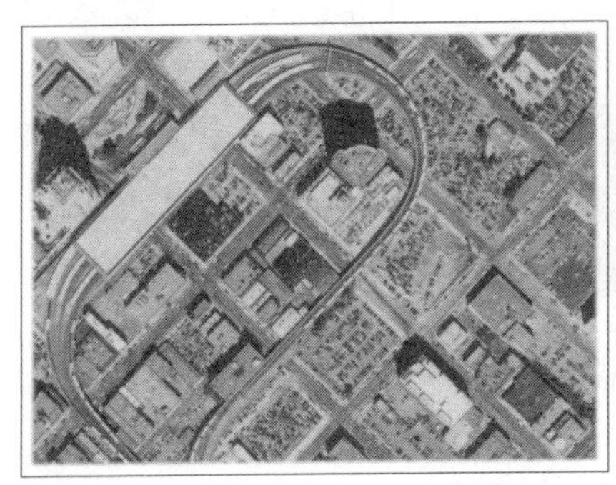

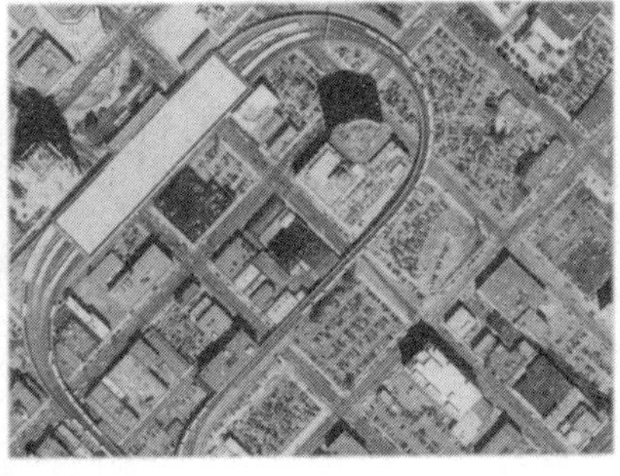

Left: *Image frame turned on*
Right: *Turned off.*

COMMAND LINE OPTIONS

0 hides image frames.

1 displays and plots image frames.

2 displays, but does not plot, image frames.

RELATED COMMANDS

ExternalReferences controls the loading of raster image files in the drawing.

ImageAdjust controls the brightness, contrast, and fading of the image.

ImageAttach attaches an image in the current drawing.

ImageClip creates a clipping boundary of the image.

ImageQuality toggles the display between draft and high-quality mode.

TIPS

- *Warning!* When ImageFrame is turned off, you cannot select the image.
- Frames are turned on (or off) in all viewports.
- This command applies to all image frames; they cannot be selectively toggled.

ImageQuality

Rel.14 Toggles the quality of raster images.

Command	Alias	Ctrl+	F-key	Alt+	Menu Bar	Tablet
imagequality	...	...	...	MOIQ	Modify ↳Object ↳Image ↳Quality	...

Command: imagequality

Enter image quality setting [High/Draft] <High>: *(Type* **H** *or* **D***.)*

COMMAND LINE OPTIONS

High displays images at a higher quality.

Draft displays images at a lower quality.

RELATED COMMANDS

ExternalReferences controls the loading of raster image files in the drawing.

ImageAdjust controls the brightness, contrast, and fading of the image.

ImageAttach attaches an image in the current drawing.

ImageClip creates a clipping boundary on the image.

ImageFrame toggles the display of the image's frame.

TIPS

- High quality displays the image more slowly; draft quality displays the image more quickly.
- I find that *draft* quality looks better (crisper) than high quality (blurred), but technical editor Bill Fane finds the quality varies with the type of image — lines versus shades, and insertion scale versus image file resolution.
- This command affects the display only; AutoCAD always plots images in high quality.

Import

Rel.13 Imports vector files into drawings.

Command	Alias	Ctrl+	F-key	Alt+	Menu Bar	Tablet
import	imp	...	...	...	...	T2

Command: import

Displays dialog box:

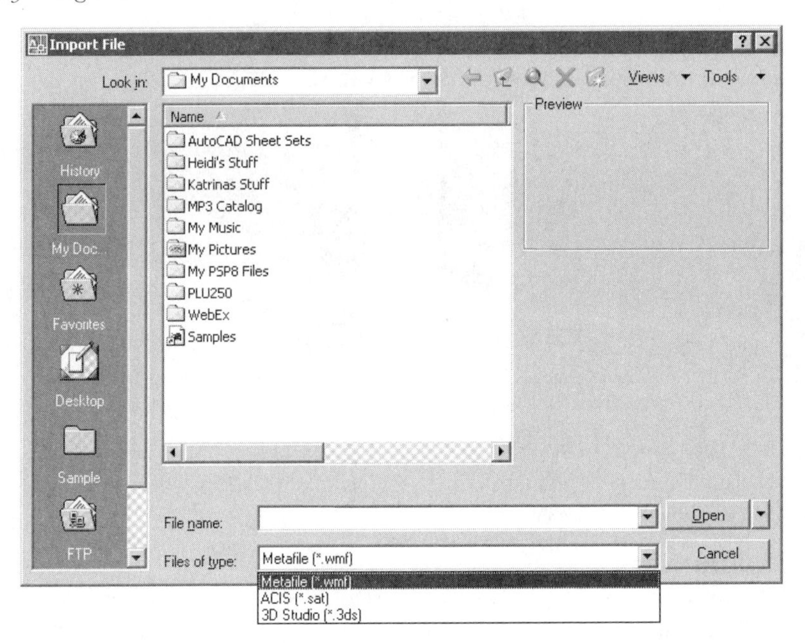

IMPORT FILE DIALOG BOX

Look in selects the folder (subdirectory) and drive from which to import the file.

File name specifies the name of the file, or accepts the default.

File of type selects the file format in which to import the file:

- Metafile (*.wmf).
- ACIS (*.sat).
- 3D Studio (*.3ds).

Open imports the file.

Cancel dismisses the dialog box, and returns to AutoCAD.

RELATED COMMANDS

AppLoad loads AutoLISP, VBA, and ObjectARX routines.

DxbIn imports a DXB file.

ExternalReferences displays *.dwg*, image, and *.dwf* files in the current drawing.

Export exports the drawing in several vector and raster formats.

Load imports SHX shape objects.

Insert places another drawing in the current drawing as a block.

InsertObj places an OLE object in the drawing via the Clipboard.

LsLib imports landscape objects.

MatLib imports rendering material definitions.

MenuLoad loads customization and menu files into AutoCAD.

Open opens AutoCAD (any version) *.dwg* and *.dxf* files.

PasteClip pastes objects from the Clipboard.

PasteSpec pastes or links objects from the Clipboard.

Replay displays renderings in TIFF, Targa, or GIF formats.

VSlide displays *.sld* slide files.

XBind imports named objects from another *.dwg* file.

TIPS

- The Import command acts as a "shell" command; it launches other AutoCAD commands that perform the actual import function. Other options may be available with the actual command, such as insertion point and scale.

Format	Meaning	Related Command
Metafile	Windows metafile WMF	**WmfIn**
ACIS	ASCII SAT	**AcisIn**
3D Studio	3D Studio 3DS format	**3dsIn**

- To import *.dxf* files, use the Open command.
- AutoCAD no longer imports PostScript and EPS files.

Removed Commands

INetCfg was removed from AutoCAD 2000; it was replaced by Windows' Internet configuration.

INetHelp was removed from AutoCAD 2000; it was replaced by AutoCAD's standard online help.

Imprint

2007 Imprints objects on the faces of 3D solids; updated from the SolidEdit command's Body/Imprint option.

Command	Alias	Ctrl+	F-key	Alt+	Menu Bar	Tablet
imprint	...	...	...	**MNI**	**Modify** **↳Solid Editing** **↳Imprint Edges**	...

Command: imprint

Select a 3D solid: *(Select one 3D solid.)*

Select an object to imprint: *(Select another 2D, 3D, surface, or solid object.)*

Delete the source object [Yes/No] <N>: *(Type **Y** or **N**.)*

Select an object to imprint: *(Press **Enter** to exit the command.)*

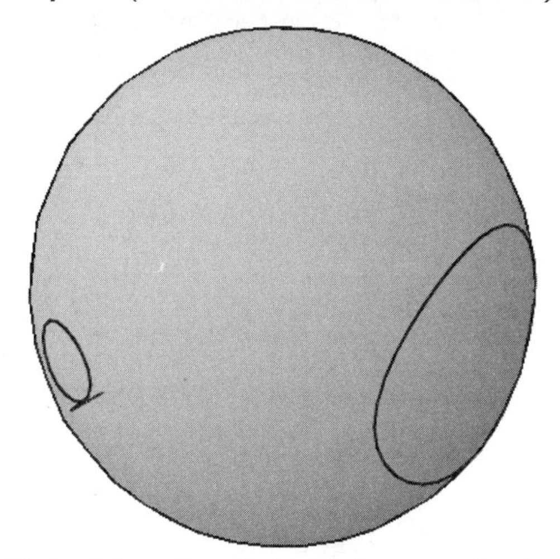

The two circles are the result of imprinting a cone on the sphere.

COMMAND OPTIONS

Select a 3D solid selects the 3D solid to be imprinted.

Select an object selects the object that will do the imprinting; can be an arc, circle, line, 2D polyline, ellipse, spline, region, and 3D polyline, or 3D solid. Also called the "source" object.

Delete the source object toggles deletion of the source object.

RELATED COMMAND

SolidEdit performs the same function as its Body/Imprint option.

Ink...

2006 Group of commands that enable ink and gestures on TabletPC-supported versions of AutoCAD (undocumented commands).

Commands	Aliases	Ctrl+	F-key	Alt+	Menu Bar	Tablet
Ink...	...	...	...	...	...	...

Warning! These commands do not operate unless AutoCAD is running on a TabletPC or on a regular computer loaded with the TabletPC SDK, and the asdkTPCtest.arx application is loaded.

INKBORDERDISPLAY Command

Toggles the display of a border around the ink objects.

Command: inkborderdisplay

Enter Border Display (0 = None, 1 = Bounding Box, 2 = Spine): *(Enter a number.)*

0 suppresses the display of the bounding box.

1 surrounds the ink objects with a bounding box.

2 indicates ink objects with a vertical line.

INKCOLOR Command

Specifies the color of the ink.

Command: inkcolor

Enter Red component: *(Enter a number between 0 and 255.)*

Enter Green component: *(Enter a number between 0 and 255.)*

Enter Blue component: *(Enter a number between 0 and 255.)*

Red specifies the amount of red for the ink color; 0 = no red, 255 = full red.

Green specifies the amount of green for the ink color.

Blue specifies the amount of blue for the ink color.

INKGESTURE Command

Enables ink gesture mode in AutoCAD.

Command: inkgesture

Enable/Disable gestures <Enable>: *(Type E or D.)*

Enable enables gestures. *Warning!* This option causes AutoCAD to crash on computers without TabletPC support.

Disable disables gesture mode.

INKHILITE Command

Toggles highlight mode.

Command: inkhilite

INKOLECREATE Command

Creates OLE (object linking and embedding) objects of ink.

Command: inkolecreate

Enter CreateOle (0 = No 1 = yes): *(Type* **0** *or* **1**.*)*

0 turns off Create OLE mode.

1 turns on Create OLE mode.

INKPENWIDTH Command

Specifies the width of ink.

Command: inkpenwidth

Enter pen width (HIMETRIC): *(Enter a value.)*

Enter pen width specifies the width of the ink lines.

INKRECO Command

Converts ink to text by attempting to recognize the handwriting.

Command: inkreco

INKRLINE Command

Toggles redline mode.

Command: inkrline

INKTRANSPARENCY Command

Toggles redline mode.

Command: inktransparency

Enter transparency ([0, 255], 0 = opaque): *(Enter a value.)*

Enter transparency specifies the transparency of the ink lines.

RELATED COMMANDS

Sketch draws freehand lines and polylines.

Tablet provides support for tablet input devices.

TIPS

- Before using these command, you must load the *asdkTPCtest.arx* application into AutoCAD with the AppLoad command.
- These commands are for demonstration purposes, to show what is possible using ink and gestures with AutoCAD. These commands work only in 2D; they do not work in USCs and views other than the WCS.
- Some of these commands will crash AutoCAD if it is not running on a TabletPC or on a regular computer with the TabletPC SDK installed.
- The TabletPC SDK (software development kit) is available from www.microsoft.com/downloads/details.aspx?FamilyID=4b14b74a-27e4-42c4-862f-273f6302ea4f&displaylang=en. More information on ink is available at msdn.microsoft.com/library/en-us/dntab101/html/Tab101C03.asp.
- Autodesk recommends that drawings with ink objects not be saved.

Insert

V. 1.0 Inserts previously-defined blocks into drawings.

Commands	Aliases	Ctrl+	F-key	Alt+	Menu Bar	Tablet
insert	i	...	...	IB	Insert	T5
	inserturl				↳Block	
-insert	-i					

Command: insert

Displays dialog box:

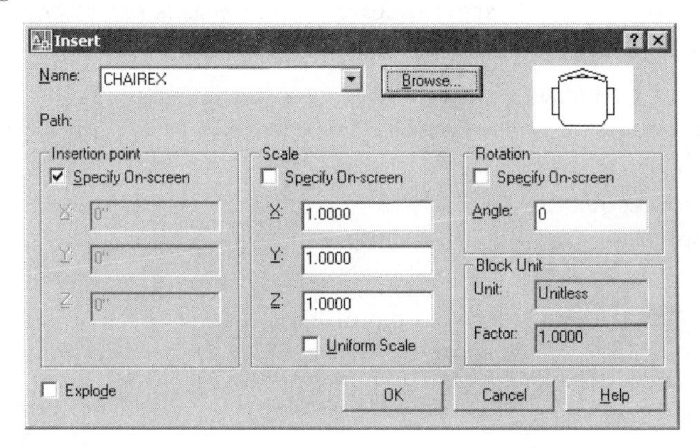

INSERT DIALOG BOX

Name selects the name from a list of previously-inserted blocks.

Browse displays the Select Drawing File dialog box; select a block in either file format:

- **Drawing (*.dwg)** AutoCAD drawing file.
- **DXF (*.dxf)** drawing interchange file.

Insertion point

☑ **Specify On-Screen** specifies the insertion point in the drawing, after closing dialog box.

X, Y, Z specifies the x, y, z coordinates of the lower-left corner of the block.

Scale

☑ **Specify on-screen** specifies the scale of the block (relative to the lower-left corner) in the drawing, after the dialog box is dismissed.

X, Y, Z specifies the x,y,z scale of the block; positive values enlarge the block, while negative values reduce it.

☐ **Uniform Scale** forces the y and z scale factors to be the same as the x scale factor.

Rotation

☐ **Specify on-screen** specifies the rotation angle of the block (about the lower-left corner) in the drawing after the dialog box is dismissed.

Angle specifies the angle to rotate the block; positive angles rotate the block counterclockwise.

Unit Block

Unit specifies the units for the block based on the value stored in the InsUnit variable.

Factor reports the scale factor based on the InsUnit variable.

☐ **Explode** explodes the block upon insertion.

-INSERT Command

Enter block name or [?]: *(Enter name, or* **?***.)*

Specify insertion point or [Basepoint/Scale/X/Y/Z/Rotate/PScale/PX/PY/PZ/PRotate]: *(Pick a point, or enter an option.)*

Enter X scale factor, specify opposite corner, or [Corner/XYZ] <1>: *(Enter a scale factor, or enter an option.)*

Enter Y scale factor <use X scale factor>: *(Enter a scale factor, or press Enter.)*

Specify rotation angle <0>: *(Enter a rotation angle.)*

COMMAND LINE OPTIONS

Block name specifies the name of the block to be inserted.

? lists the names of blocks stored in the drawing.

Specify insertion point specifies the lower-left corner of the block's insertion point.

Basepoint places the block temporarily to allow you to specify a new base point. AutoCAD prompts:

Specify base point: *(Pick a point.)*

Scale specifies the scale for the x, y, and z factors equally. AutoCAD prompts:

Specify scale factor for XYZ axes: *(Enter a scale factor.)*

X indicates the x scale factor.

Y indicates the y scale factor.

Z indicates the z scale factor.

Rotate specifies the rotation angle.

PScale supplies a predefined x, y, and z scale factor.

PX supplies a predefined x scale factor.

PY supplies a predefined y scale factor.

PZ supplies a predefined z scale factor.

PRotate supplies a predefined rotation angle.

Corner indicates the x and y scale factors by pointing on the screen.

XYZ displays the x, y, and z scale submenu.

INPUT OPTIONS

In response to the 'Block Name' prompt, you can enter:

~ — Display a dialog box of drawings stored on disk: `Block name: ~`

* — Insert block exploded: `Block name: *filename`

= — Redefine existing block with a new block: `Block name: oldname=newname`

In response to the 'Insertion point' prompt, you can enter:

Scale — Specify x, y, and z scale factors.
PScale — Preset x, y, and z scale factors.
XScale — Specify x scale factor.
PxScale — Preset x scale factor.
YScale — Specify y scale factor.
PyScale — Preset y scale factor.
ZScale — Specify z scale factor.
PzScale — Preset z scale factor.
Rotate — Specify rotation angle.
PRotate — Preset rotation angle.

RELATED COMMANDS

Block creates a block out of a group of objects.

Explode reduces inserted blocks to their constituent objects.

ExternalReferences displays drawings stored on disk in the drawing.

MInsert inserts a block as a blocked rectangular array.

Rename renames blocks.

WBlock writes blocks to disk.

RELATED SYSTEM VARIABLES

ExplMode toggles whether non-uniformly scaled blocks can be exploded.

InsBase specifies the name of the most-recently inserted block.

InsUnits specifies the drawing units for the inserted block.

TIPS

- You can insert any other AutoCAD drawing into the current drawing.
- A *preset* scale factor or rotation means the dragged image is shown at that scale, but you can enter a new scale when inserting.
- Drawings are normally inserted as a block; prefix the filename with an * (*asterisk*) to insert the drawing as separate objects.
- Redefine all blocks of the same name in the current drawing by adding the = (*equal*) suffix after its name at the 'Block name' prompt.
- Insert a mirrored block by supplying a negative x or y scale factor, such as:

  X scale factor: -1

- AutoCAD converts a negative z scale factor into its absolute value, which always makes it positive.
- As of AutoCAD Release 13, you can explode a mirrored block and a block inserted with different scale factors, when the system variable ExplMode is turned on.
- As of AutoCAD 2004, the Insert command no longer imports drawings in XML format (designXML).
- When inserting blocks with attached xrefs, the xrefs are retained.

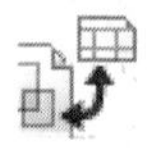

InsertObj

Rel.13 Places OLE objects as linked or embedded objects in drawings (short for INSERT OBJect).

Command	Alias	Ctrl+	F-key	Alt+	Menu Bar	Tablet
insertobj	**io**	...	...	**IO**	**Insert** ⤷**OLE Object**	**T1**

Command: insertobj

Displays dialog box:

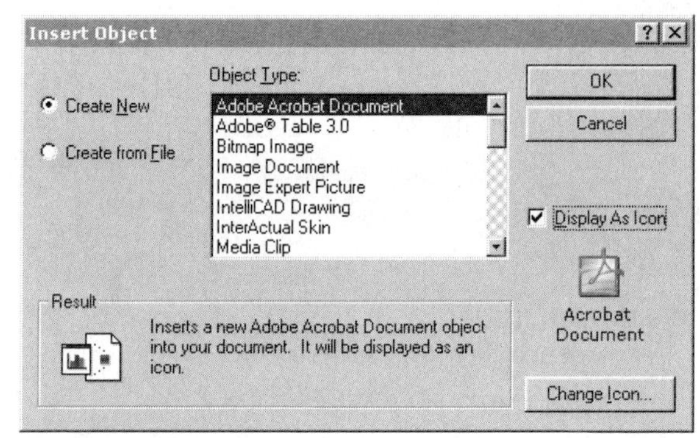

INSERT OBJECT DIALOG BOX

⊙**Create New** creates new objects in other applications, and then embeds them in the drawing.

○**Create from File** selects a file to embed in or link to the current drawing.

Object Type selects an object type from the list; the related application automatically launches if you select the **Create New** option.

☑ **Display As Icon** displays the object as an icon, rather than as itself.

Change Icon selects another icon.

RELATED COMMANDS

OleLinks controls the OLE links.

PasteSpec places an object from the Clipboard in the drawing as a linked object.

RELATED SYSTEM VARIABLES

MsOleScale determines the scale of OLE objects placed in model space.

OleHide toggles the display of OLE objects.

OleQuality determines the plot quality of OLE objects.

OleStartup specifies whether the source apps of embedded OLE objects load for plotting.

RELATED WINDOWS COMMANDS

Edit | Copy copies an object to the Clipboard for use in other Windows applications.

File | Update updates an OLE object from another application.

Renamed Command

InsertUrl was removed from AutoCAD 2000; it was replaced by the **Insert** command's **Browse | Search the Web** option.

Interfere

Rel.11 Determines the interference of two or more 3D solid objects; optionally creates a 3D solid body of the common volumes.

Command	Alias	Ctrl+	F-key	Alt+	Menu Bar	Tablet
interfere	**inf**	...	...	**DM3I**	**Draw** ↳**3D Operations** ↳**Interference Checking**	...
-interfere						

Command: interfere

This command has changed substantially in AutoCAD 2007:

Select first set of objects or [Nested selection/Settings]: *(Select one or more solid objects, or enter an option.)*

Select second set of objects or [Nested selection/checK first set] <checK>: *(Press Enter to check, or enter an option.)*

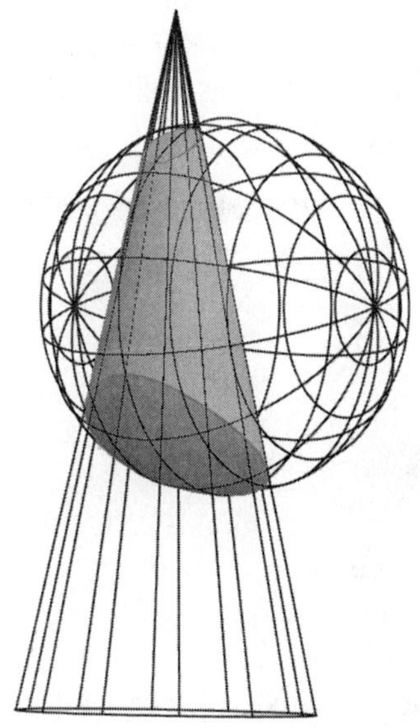

An interfering solid (shaded at center).

COMMAND LINE OPTIONS

Select first set of objects checks solids in a single selection set for interference with one another.

Nested selection selects solids nested in blocks and xrefs.

Settings displays the Interference Settings dialog box.

checK displays the Interference Checking dialog box.

DIALOG BOX OPTIONS

Interference Settings dialog box

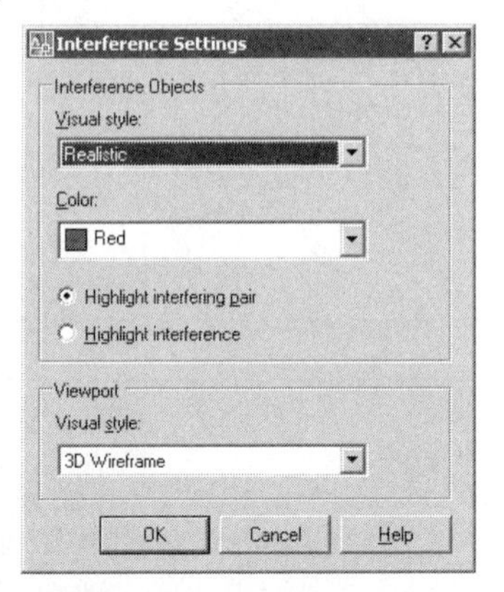

Visual Style specifies the visual style of interference objects.

Color specifies the color of interference objects.

⊙ **Highlight Interfering Pair** highlights the pair of solids that interfere with each other.

○ **Highlight Interference** highlights the interference object(s) created from the interfering pair.

Visual Style specifies the visual style while checking for interference.

Interference Checking dialog box

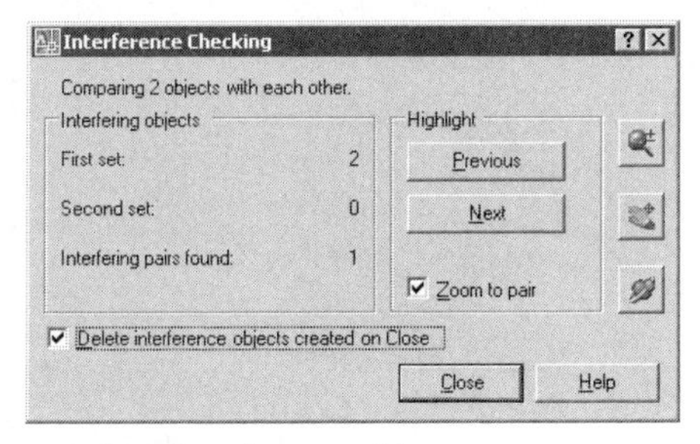

Previous highlights the previous interference object.

Next highlights the next interference object.

☑ **Zoom to Pair** zooms to interference object(s) when clicking **Previous** and Next**.**

☑ **Delete Interference Objects Created on Close** deletes the interference object(s) after closing the dialog box.

Zoom temporarily dismisses the dialog box, and then starts the Zoom command.

Pan temporarily dismisses the dialog box, and then starts the Pan command.

3D Orbit temporarily dismisses the dialog box, and then starts the 3dOrbit command.

-Interfere Command

Command: -interfere

Select first set of objects or [Nested selection]: *(Select one or more solid objects, or enter an option.)*

Select second set of objects or [Nested selection/checK first set] <checK>: *(Press Enter to check, or enter an option.)*

Comparing 2 objects with each other.

Interfering objects (first set): 2

(second set): 0

Interfering pairs: 1

Create interference objects? [Yes/No] <No>: *(Type **Y** or **N**.)*

RELATED COMMANDS

Intersect creates a new volume from the intersection of two volumes.

Section creates a 2D region from a 3D solid.

Slice slices a 3D solid with a plane.

RELATED SYSTEM VARIABLES

InterfereColor specifies the color of interference objects (*new to AutoCAD 2007*).

InterfereObjVs specifies the visual style of interference objects (*new to AutoCAD 2007*).

InterferenceVpSs specifies the visual style for the current viewport during interference checking (*new to AutoCAD 2007*).

Intersect

Rel.11 Creates 3D solids of 2D regions through the Boolean intersection of two or more solids or regions.

Command	Alias	Ctrl+	F-key	Alt+	Menu Bar	Tablet
intersect	**in**	...	...	**MNI**	**Modify** **↳Solids Editing** **↳Intersect**	**X17**

Command: intersect
Select objects: *(Select one or more solid objects.)*
Select objects: *(Press Enter to end object selection.)*

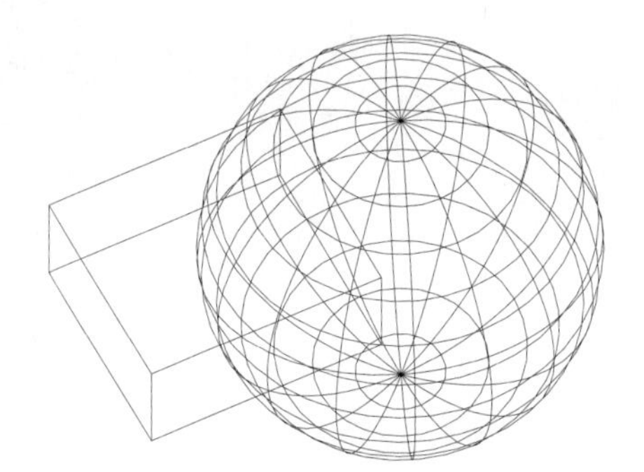

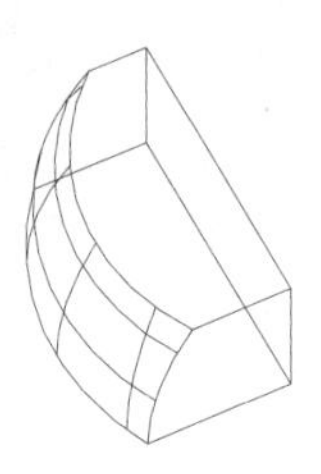

Left: *Two intersecting solids.*
Right: *The result of applying the Intersect command.*

COMMAND LINE OPTION

Select objects selects two or more objects to intersect.

RELATED COMMANDS

Interfere creates a new volume from the interference of two or more volumes.

Subtract subtracts one 3D solid from another.

Union joins 3D solids into a single body.

RELATED SYSTEM VARIABLES

ShowHist toggles display of history in solids (*new to AutoCAD 2007*).

SolidHist toggles the retention of history in solids (*new to AutoCAD 2007*).

TIPS

- You can use this command on 2D regions and 3D solids.
- The **Interference** and **Intersect** commands may seem similar. Here is the difference between the two:

  Intersect *erases* all of the 3D solid parts that do not intersect.

  Interfere *creates a new object* from the intersection; it does not erase the original objects.

'Isoplane

V. 2.0 Changes the crosshair orientation and grid pattern among the three isometric drawing planes.

Command	Alias	Ctrl+	F-key	Alt+	Menu Bar	Tablet
isoplane	...	E	F5	...	...	...

Command: isoplane
Enter isometric plane setting [Left/Top/Right] <Top>: *(Enter an option, or press Enter.)*

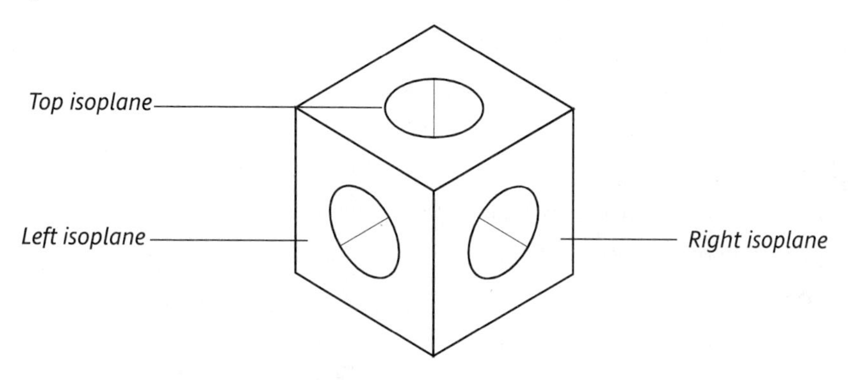

COMMAND LINE OPTIONS

Left switches to the left isometric plane.

Top switches to the top isometric plane.

Right switches to the right isometric plane.

ENTER switches to the next isometric plane in the following order: left, top, right.

RELATED COMMANDS

Options displays a dialog box for setting isometric mode and planes.

Ellipse draws isocircles.

Snap turns on isometric drawing mode.

RELATED SYSTEM VARIABLES

SnapIsoPair contains the current isometric plane.

GridMode toggles grid visibility.

GridUnit specifies the current grid x,y spacing.

LimMin specifies the x,y coordinates of the lower-left corner of the grid display.

LimMax holds the x,y coordinates of the upper-right corner of the grid display.

SnapStyl specifies a normal or isometric grid:

0 — Normal (default).
1 — Isometric grid.

Quick Start Tutorial

Creating Isometric Dimensions

AutoCAD's dimensions must be modified for isometric drawings, so that the dimension text looks "correct" in isometric mode. This involves two steps — (1) creating isometric text styles, and (2) changing dimension variables — repeated three times, once for each isoplane.

Step 1

To create isometric text styles:

1. From the **Format** menu, select **Text Style**.
2. When the Text Style dialog box appears, click **New**.
3. Enter **isotop** for the name of the new text style, which is used for text in the top isoplane.
4. Click **OK**.
5. When the Text Style dialog box reappears, select *Simplex.Shx* from **Font Name**.
6. Change the **Oblique Angle** to **-30**.
7. Click **Apply**.
8. Create text styles for the other two isoplanes:

Style Name	Font Name	Oblique Angle
IsoTop	*simplex.shx*	-30
IsoRight	*simplex.shx*	30
IsoLeft	*simpelx.shx*	30

Enter these values into the Text Style dialog box, click **Apply**, and then **Close**.

Step 2

To create isometric dimension styles:

1. Create the dimension styles for the three isoplanes by selecting **Dimension Styles** from the **Format** menu.
2. Create a new dimension style:

   Click **New**.

   Enter **IsoLeft** in the **New Style Name** field.

   Click **Continue**.

3. Force dimension text to align with the dimension line:

   Select the **Text** tab.

   Select **Aligned with dimension line** in the **Text Alignment** section.

4. Specify text style for dimension text:

   Select **ISOLEFT** from the **Text Style** list box.

   Click **OK**.

5. One of the three needed dimension styles has been created. Create dimstyles for all isoplanes using these parameters:

Dimstyle Name	Text Style
Isotop	IsoTop
Isoright	IsoRight
Isoleft	IsoLeft

6. Click **Close** to exit the Dimension Style Manager dialog box.

Step 3

To place linear dimensions in an isometric drawing, you must use the DimAligned command, because it aligns the dimension along the isometric axes: place all dimensions in one isoplane, and then switch to the next isoplane.

1. Press **F5** to switch to the appropriate isoplane, such as **Top**.
2. Use the DimStyle command to select the associated dimension style, such as **IsoTop**.
3. Place the dimension with the DimAligned command; it is helpful to use **INTersection** object snaps.
4. Use the DimEdit command's **Oblique** option to skew the dimension by 30 or -30 degrees, as follows:

IsoPlane	DimStyle	Oblique Angle
Top	IsoTop	30
Left	IsoLeft	30
Right	IsoRight	-30

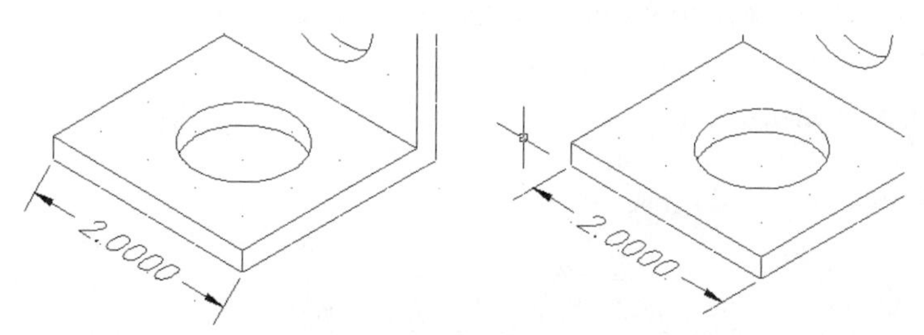

***Left**: Aligned dimension text before...*
***Right**: ...and after applying DimEdit's Oblique option.*

5. To place a leader, use the **Standard** dimstyle and **Standard** text style.

JogSection

2007 Adds 90-degree jogs to section planes.

Command	Alias	Ctrl+	F-key	Alt+	Menu Bar	Tablet
jogsection	...	...	...	...	...	...

Command: jogsection

This command requires that at last one section be placed in the drawing with the ***SectionPlane*** *command.*

Select section object: *(Select one section objects.)*

Specify a point on the section line to add jog: *(Pick a point along the section line.)*

Specify a point on the section line to add jog: *(Press Enter to exit the command.)*

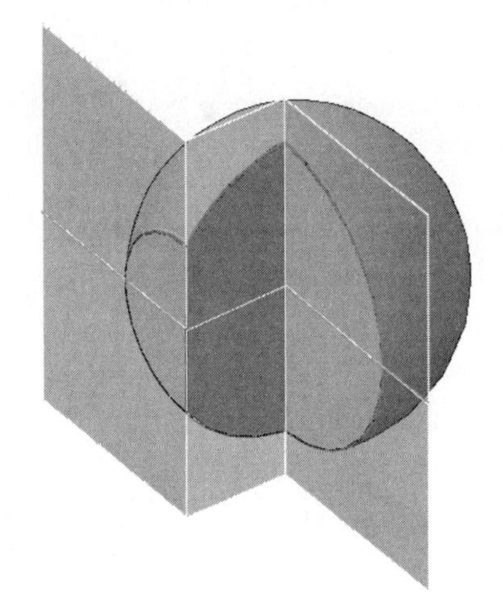

COMMAND LINE OPTIONS

Select section object selects the section to be jogged.

Specify a point on the section line to add jog picks the jog location; the jog angle is always 90 degrees.

RELATED COMMANDS

LiveSection activates section planes.

SectionPlane places section planes in drawings.

TIPS

- Use osnap modes to help position the jog more accurately.
- To create the jog section illustrated above, use the commands in this order:
  1. **SectionPlane** command to place the section plane.
  2. **LiveSection** command to edit the section plane.
  3. **JogSection** command to add a jog to the section plane.
- After the jog is placed, you can use grips to modify it.
- Jogs can be placed while placing the section plane during the SectionPlane command; use this command to place jogs after the section plane in place.

Join

2006 Joins open objects to make one object or closed objects.

Command	Alias	Ctrl+	F-key	Alt+	Menu Bar	Tablet
join	j	...	...	MJ	Modify Join	...

Command: join

Select source object: *(Select an object.)*

Select *objects* **to join to source or [cLose]:** *(Select other objects, or enter* **L** *to close open object.)*

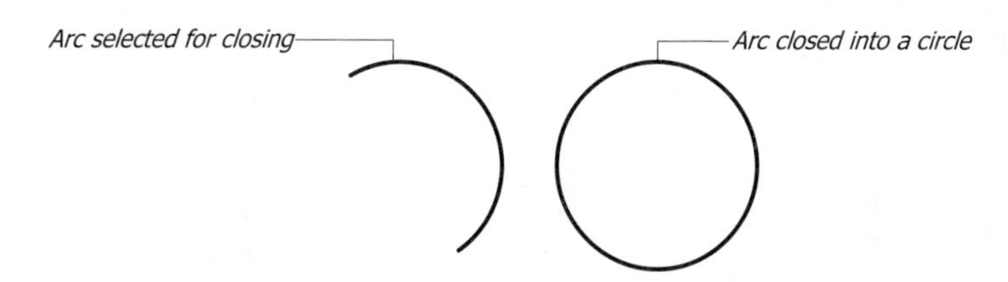

COMMAND LINE OPTIONS

Select source object selects object to be joined or closed.

Select objects to join joins similar objects. The following objects can be joined, even when there is a gap between them:

- Colinear lines — joined into a single line.
- Arcs in the same imaginary circle — joined into a single arc.
- Elliptical arcs in the same imaginary ellipse — joined into a single elliptical arc.

These objects can only be joined when no gap exists between them:

- Polylines, lines, and arcs — joined into a single polyline.
- Splines — joined into a single spline.

cLose closes arcs and elliptical arcs:

- Arcs become circles.
- Elliptical arcs become ellipses.

RELATED COMMANDS

Break turns circles and ellipses into arcs.

PEdit joins lines, arcs, and polylines with its **Join** option.

TIP

- If arcs have been dimensioned with the DimArc command, the closed circles are disassociated from the dimension.

JpgOut

2004 Exports drawings in JPEG format.

Command	Alias	Ctrl+	F-key	Alt+	Menu Bar	Tablet
jpgout	...	...	...	...	...	...

Command: jpgout

*Displays Create Raster File dialog box. Specify a filename, and then click **Save**.*

Select objects or <all objects and viewports>: *(Select objects, or press Enter to select all objects and viewports.)*

COMMAND LINE OPTIONS

Select objects selects specific objects.

All objects and viewports selects all objects and all viewports, whether in model space or in layout mode.

RELATED COMMANDS

BmpOut exports drawings in BMP (bitmap) format.

Image places raster images in the drawing.

Plot outputs drawings in JPEG and other raster formats.

PngOut exports drawings in PNG (portable network graphics) format.

TifOut exports drawings in TIFF (tagged image file format) format.

TIPS

- JPEG files are often used by digital cameras and Web pages.
- For more control over the resulting *.jpg* file, use the Plot command's PublishToWebJpg driver. For example, it can specify a high resolution such as 2550x3300, whereas the JpgOut command is limited to the resolution of your computer's monitor (1280x1024, typically).
- When using the JpgOut command to save a rendering, the rendering effects of the VisualStyles command are preserved, but not those of the Render command.
- The drawback to saving drawings in JPEG format is that the image is less clear (due to artifacts) than in other formats; the advantage is that JPEG files are highly compressed.

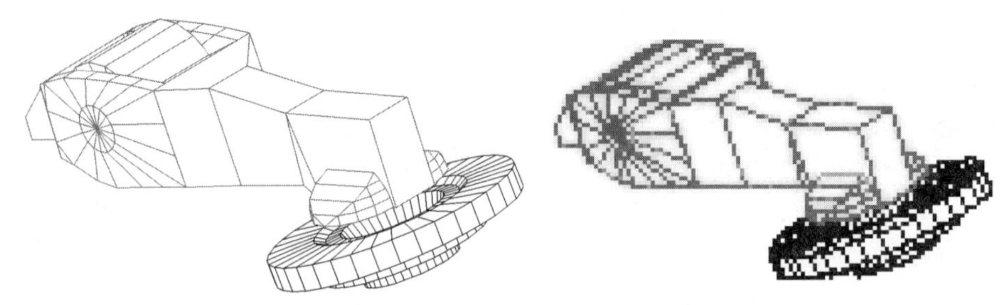

***Left**: Original AutoCAD image zoomed-in.*
***Right**: Enlarged JPEG image with artifacts.*

- This command provides no options for specifying the level of compression.
- JPEG is short for "joint photographic expert group."

JustifyText

2001 Changes the justification of text.

Command	Alias	Ctrl+	F-key	Alt+	Menu Bar	Tablet
justifytext	...	...	...	**MOTJ**	**Modify** **↳Object** **↳Text** **↳Justify**	...

Command: justifytext
Select objects: *(Select one or more text objects.)*
Select objects: *(Press Enter to end object selection.)*
Enter a justification option
[Left/Align/Fit/Center/Middle/Right/TL/TC/TR/ML/MC/MR/BL/BC/BR]
<Left>: *(Enter an option, or press* ENTER*.)*

COMMAND LINE OPTIONS

Select objects selects one or more text objects in the drawing.

Align aligns the text between two points with adjusted text height.

Fit fits the text between two points with fixed text height.

Center centers the text along the baseline.

Middle centers the text horizontally and vertically.

Right right-justifies the text.

TL justifies to top-left.

TC justifies to top-center.

TR justifies to top-right.

ML justifies to middle-left.

MC justifies to middle-center.

MR justifies to middle-right.

BL justifies to bottom-left.

BC justifies to bottom-center.

BR justifies to bottom-right.

RELATED COMMANDS

Text places text in the drawing.

DdEdit edits text.

ScaleText changes the size of text.

Style defines text styles.

Properties changes the justification, but moves text.

TIPS

- This command works with text, mtext, leader text, and attribute text.
- When the justification is changed, the text does not move to reflect the changed insertion point.

Layer Tools

2007 Manipulates layers; a group of commands formerly in Express Tools.

Command	Alias	Ctrl+	F-key	Alt+	Menu Bar	Tablet
...	...	...	...	FA	Format ↳Layer Tools	...

This group of commands perform very focused operations on layers and objects. Many commands are paired, one undoing the action of another.

LayOff turns off the layer of the selected object.
LayOn turns on all layers, except frozen layers.

LayIso turns off all layers except the ones holding selected objects.
LayVpi isolates the selected object's layer in the current viewport
LayWalk displays objects on selected layers, turning off all other layers.
LayUnIso turns on layers that were turned off with the LayIso command

LayFrz freezes the layers of the selected objects.
LayThw thaws all layers.

LayLck locks the layer of the selected object.
LayULk unlocks the layer of a selected object.

LayCur changes the layer of selected objects to the current layer.
LayMch changes the layers of selected objects to that of another selected object.
LayMCur makes the selected object's layer current (similar to Ai_Molc command).

CopyToLayer copies objects to another layer; see CopyToLayer command.
LayMrg moves objects to another layer.

LayDel erases all objects from the specified layer, and then purges the layer from drawing.

LayCur

LayCur changes the layer of selected objects to the current layer (short for "LAYer CURrent").

Command: laycur
Select objects to be changed to the current layer: *(Select one or more objects.)*
Select objects to be changed to the current layer: *(Press Enter.)*
One object changed to layer *layername* **(the current layer).**

LayDel and -LayDel

LayDel and -LayDel erase all objects from the specified layer, and then purge the layer from drawing (short for "LAYer DELete"). The current layer cannot be deleted. If you make a mistake, use the U command to restore the purged layer name and its erased objects.

Command: laydel
Select object on layer to delete or [Name]: *(Select an object, or type* **N***.)*

Selected layers: *layername*.

Select object on layer to delete or [Name/Undo]:

********** WARNING **********

You are about to delete layer "*layername***" from this drawing.**

Do you wish to continue? [Yes/No] <No>: *(Type* **Y** *or* **N***.)*

Deleting layer "*layername***".**

1 layer deleted.

COMMAND OPTIONS

Name displays the Delete Layers dialog box; select one or more layers to delete and purge.

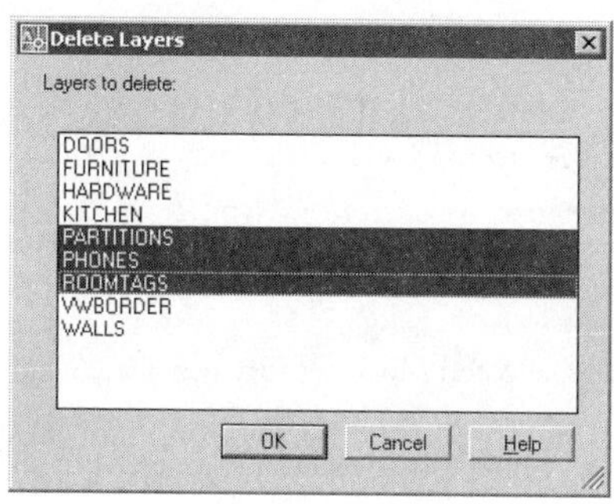

Undo undoes the last layer delete.

LayFrz

LayFrz freezes the layers of the selected objects (short for "LAYer FReeZe").

Command: layfrz

Current settings: Viewports=Vpfreeze, Block nesting level=Block

Select an object on the layer to be frozen or [Settings/Undo]: *Select an object, or enter an option.)*

Layer "*layername***" has been frozen.**

Select an object on the layer to be frozen or [Settings/Undo]: *(Press Enter, or enter an option.)*

COMMAND OPTIONS

Settings offers these options:

Enter setting type for [Viewports/Block selection] selects options for viewports or blocks.

In paper space viewport use determines how to freeze layers in viewports:

- **Freeze** freezes the layers in all viewports.
- **Vpfreeze** freezes just the current viewport.

Enter Block Selection nesting lever: determines how to freeze layers in blocks and xrefs:

- **Block** freezes the entire block's layer, but freezes the object's layer in xrefs.
- **Entity** freezes the object's layer
- **None** freezes neither the block nor xref.

Undo undoes the last freeze action.

LayIso

LayIso turns off all layers except the ones holding selected objects (short for "LAYer ISOlate").

Command: layiso

Current setting: Viewports=Vpfreeze

Select objects on the layer(s) to be isolated or [Settings]: *(Select an object, or type* **S***).*

Select objects on the layer(s) to be isolated or [Settings]: *(Press Enter.)*

Layer *layername* **has been isolated.**

COMMAND OPTION

Settings offers this option:

In paper space viewport use determines how to freeze layers in viewports:

Freeze freezes the layers in all viewports.

Vpfreeze freezes just the current viewport.

LayLck

LayLck locks the layer of the selected object (short for "LAYer LoCK").

Command: laylck

Select an object on the layer to be locked: *(Select an object.)*

Layer "*layername***" has been locked.**

LayMch and -LayMch

LayMch and -LayMch change the layers of selected objects to that of a selected object (short for "LAYer MatCH").

Command: laymch

Select objects to be changed: *(Pick one or more objects.)*

Select objects: *(Press Enter.)*

Select object on destination layer or [Name]: *(Pick an object on another layer.)*

One object changed to layer "*layername***"**

COMMAND OPTION

Name displays the Change to Layer dialog box; see LayDel command.

In the -LayMch command, the ***Name*** *option displays the following prompt:*

Enter layer name: *(Type the valid name of a layer.)*

LayMCur

LayMCur makes the selected object's layer current, just like the Ai_Molc command (short for "LAYer Make CURrent").

Command: laycur

Select object whose layer will become current: *(Pick an object.)*

layername **is now the current layer.**

LayMrg and -LayMrg

LayMrg and -LayMrg move objects to another layer (short for "LAYer MeRGe").

Command: laymrg

Select object on layer to merge or [Name]: *(Pick an object.)*

Selected layers: *layername*.

Select object on layer to merge or [Name/Undo]: *(Pick another object.)*

Selected layers: *layername, anothername*.

Select object on layer to merge or [Name/Undo]: *(Press Enter.)*

Select object on target layer or [Name]: *(Pick an object.)*

COMMAND OPTION

Name displays the Merge to Layer dialog box; see LayDel command.

*In the -LayMrg command, the **Name** option displays the following prompt:*

Enter layer name: *(Type the name of a layer.)*

LayOff

LayOff turns off the layer of the selected object.

Command: layoff

Current settings: Viewports=Vpfreeze, Block nesting level=Block

Select an object on the layer to be turned off or [Settings/Undo]: *(Select an object, or enter an option.)*

Layer "0" has been turned off.

See LayFrz for the meaning of the Setting and Undo options.

LayOn

LayOn turns on all layers, except frozen ones.

Command: layon

Warning: layer "*layername***" is frozen. Will not display until thawed.**

All layers have been turned on.

LayThw

LayThw thaws all layers (short for "LAYer THaW").

Command: laythw

All layers have been thawed.

LayULk

LayULk unlocks the layer of a selected object (short for "LAYer UnLocK").

Command: layulk

Select an object on the layer to be unlocked: *(Pick an object on a locked layer.)*

Layer "*layername***" has been unlocked.**

LayUnIso

LayUnIso turns on layers that were turned off with the last LayIso command (short for "LAYer UNISOlate").

Command: layuniso

Layers isolated by LAYISO command have been restored.

LayVpi

LayVpi isolates the selected object's layer in the current viewport (short for "LAYer ViewPort Isolate"). This command works only in paper space with two or more viewports.

Command: layvpi

Current settings: Viewports=Vpfreeze, Block nesting level=Block

Select objects on the layer to be isolated in viewport or [Settings/Undo]:
(Select one or more objects, or enter an option.)

Layer *layername* **has been frozen in all viewports but the current one.**

Select an object on the layer to be isolated in viewport or [Settings/Undo]:
(Press Enter to exit the command.)

See LayFrz for the meaning of the Setting and Undo options.

LayWalk

LayWalk displays objects on selected layers.

Command: laywalk

Displays dialog box:

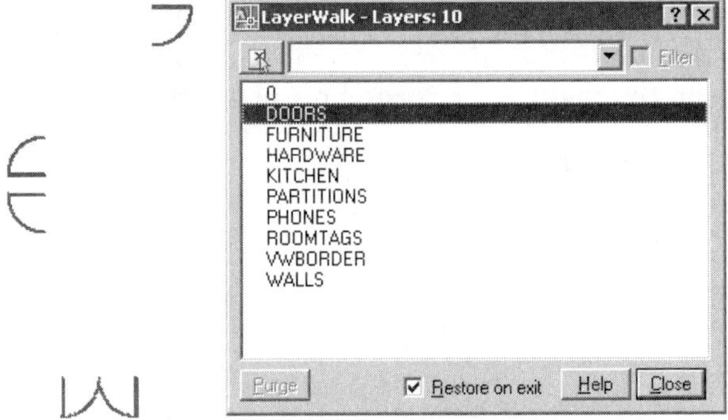

Left: *Doors are displayed because layer Doors is selected.*
Right: *LayerWalk dialog box.*

DIALOG BOX OPTIONS

Select objects selects one or more objects in the drawing, and reports their name(s).

☐ **Filter** filters the layer list according to the rules of the selected filter.

Purge removes the selected layer, if it contains no objects.

☑ **Restore on Exit** restores layer display upon exiting the dialog box.

Close closes the dialog box and exits the command.

SHORTCUT MENU OPTIONS

Right-click the LayerWalk dialog box to access the following shortcut menu:

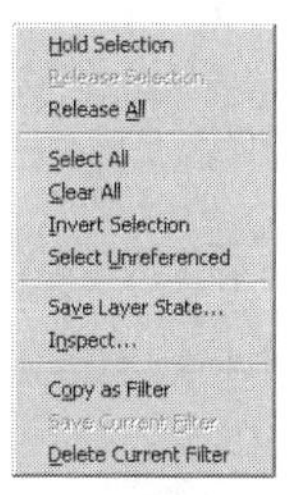

Hold Selection adds an asterisk (*) to the layer name, and always displays it.

Release Selection removes the asterisk from the selected layers.

Release All removes all asterisks, and then turns off the **Always Show** option.

Select All selects all layers, and displays them.

Clear All unselects all layers, hiding them.

Invert Selection selects unselected layers, and vice versa.

Select Unreferenced selects all layers with no objects; the **Purge** button removes them from the drawing.

Save Layer State saves the selected layers for use by the Layer States Manager.

Inspect displays the Inspect dialog box, which reports on the total number of layers, number of selected layers, and the total number of objects on the selected layers.

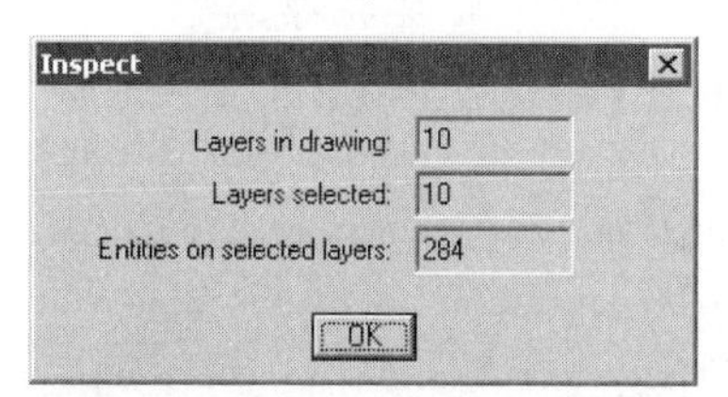

Copy as Filter adds the selected layer's name to the Filter text box; edit the name with the **?** and * wildcard characters.

Save Current Filter saves the filter for reuse.

Delete Current Filter erases the filter from the filter list.

RELATED COMMANDS

Layer controls layer creation and properties.

CopyToLayer copies objects from one layer to another.

RELATED SYSTEM VARIABLE

CLayer is the name of the current layer.

TIPS

- In the LayerWalk dialog box, double-click a layer name to keep it on always; an asterisk appears beside its name.
- Also in the LayerWalk dialog box, hold down the **Ctrl** key to select more than one layer at a time.

'Layer

V. 1.0 Controls the creation, status, and visibility of layers.

Commands	Aliases	Ctrl+	F-key	Alt+	Menu Bar	Tablet
layer	**la**	...	...	**OL**	**Format**	**U5**
	ddlmodes				**↳Layer**	
-layer	**-la**					

Command: layer

Displays dialog box:

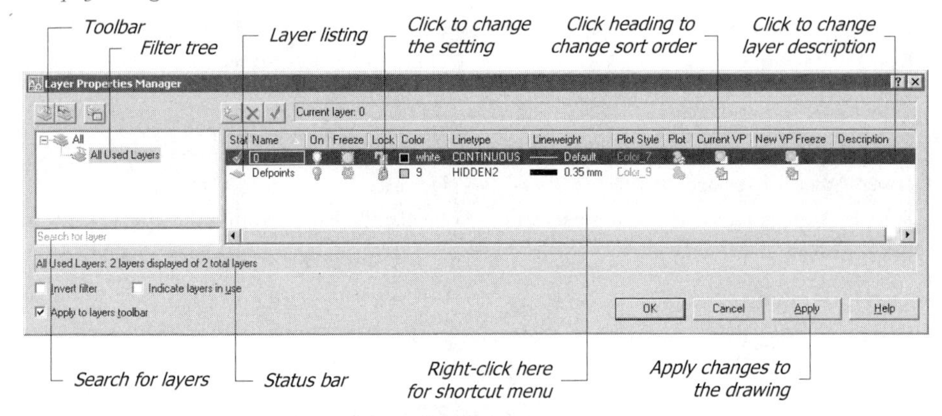

LAYER PROPERTIES MANAGER DIALOG BOX

Click a header name to sort alphabetically (A-Z); click a second time for reverse-alphabetical sort (Z-A).

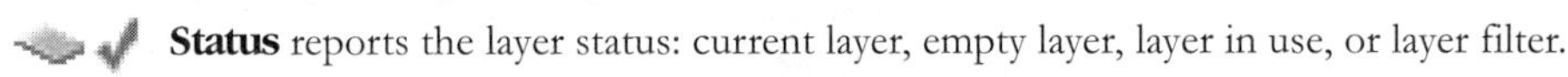

Status reports the layer status: current layer, empty layer, layer in use, or layer filter.

Name lists the names of layers in the current drawing.

 **On** toggles layers between on and off.

 **Freeze** toggles layers between thawed and frozen in all viewports.

 **Lock** toggles layers between unlocked and locked.

■ **Color** specifies the color of objects on layers.

Linetype specifies the linetype for objects on layers.

Lineweight specifies the lineweight for objects on layers.

Plot Style specifies the plot style for objects on layers.

 **Plot** toggles layers between plot and no-plot.

The following two options appear only in layout mode (paper space):

 **Current VP Freeze** specifies whether the layer is frozen in the current viewport.

 **NewVP Freeze** specifies whether the layer is frozen in new viewports.

Description provides space to describe the layer.

☐ **Invert filter** inverts the display of layer names; for example, when **Show all used layers** is selected, the **Invert filter** option causes all layers with no content to be displayed; alternatively, press **ALT+I**.

☑ **Apply to layers toolbar** applies the filter to the layer names displayed by the Layers toolbar; alternatively, press **ALT+T**.

Apply applies the changes to the layers without exiting the dialog box.

OK exits the dialog box, and applies changes to the layers.

Cancel exits the dialog box, and leaves layers unchanged.

Dialog Box toolbar

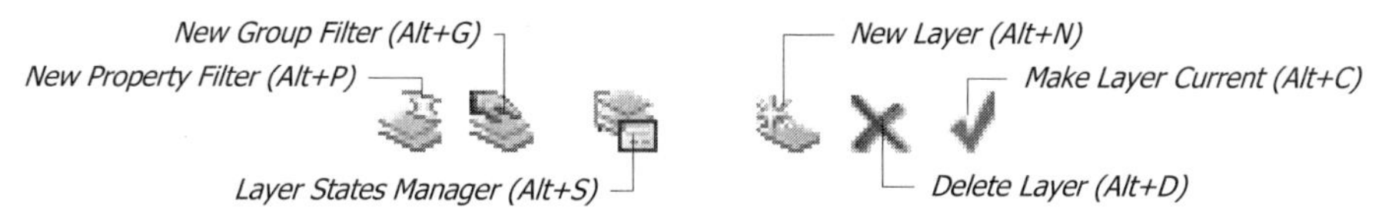

New Property Filter displays New Filter Properties dialog box.

New Group Filter adds an item to the group filter list, initially named "Group Filter 1."

Layer States Manager displays Layer States Manager dialog box.

New Layer adds layers to the drawing, initially named "Layer1."

Delete Layer purges the selected layer; some layers cannot be deleted, as described by the warning dialog box:

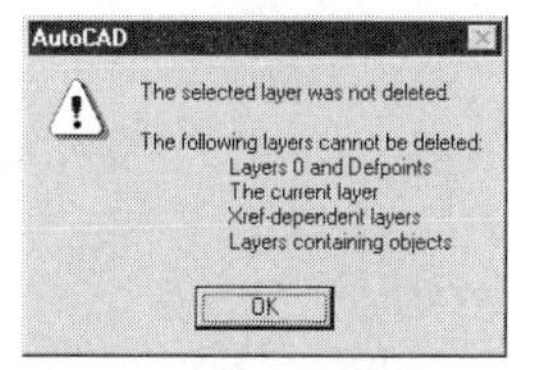

Current Layer sets the selected layer as current.

Shortcut Menus

Most layer options are found in right-click menus.

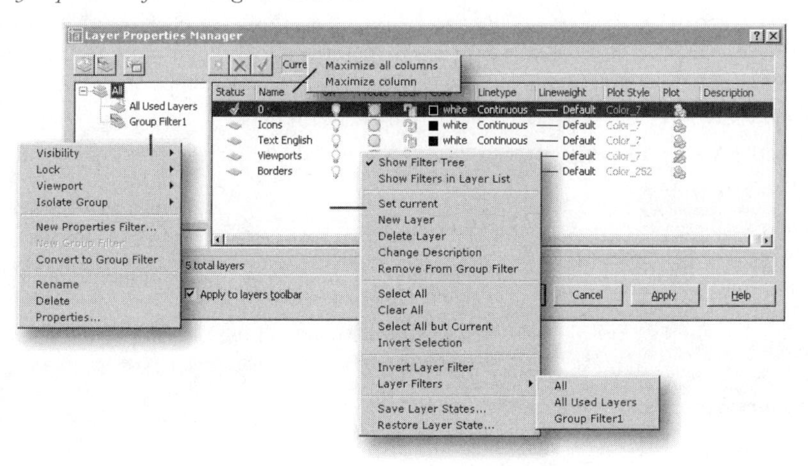

Column Headers

Maximize Column changes the selected column width to display all column content.

Maximize All Columns changes the width of all columns to display content of all columns. To display complete headers, you must resize the column by dragging the header separators. You can resize the dialog box.

Filter Tree

Depending on where you right-click in the filter tree area, some menu options may be grayed-out, indicating they are unavailable.

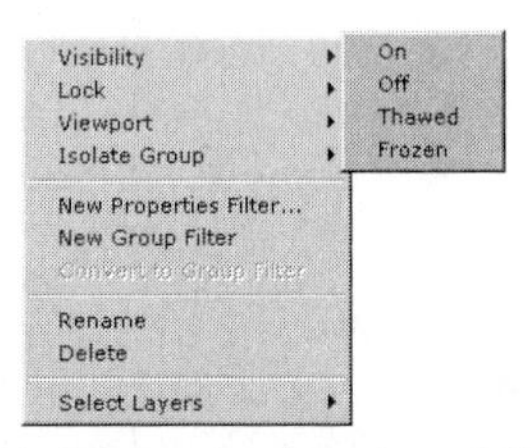

Visibility toggles the visibility of all layers in the selected filter or group:

- **On** displays, plots, and regenerates objects; includes objects during hidden-line removal.
- **Off** does not display or plot objects; does not include objects during hidden-line removal, and the drawing is not regenerated when the layer is turned on.
- **Thawed** reverses the **Frozen** option.
- **Frozen** does not display or plot objects; includes objects during hidden-line removal, and the drawing is regenerated when the layer is thawed.

Lock determines whether objects can be edited:

- **Lock** prevents objects from being edited; all other operations that don't involve editing are permitted, such as object snaps.
- **Unlock** allows objects to be edited.

Viewport specifies whether layers are frozen in layout mode (unavailable in model space):

- **Freeze** applies Current VP Freeze to all layers (in the filter or group) of the current viewport.
- **Thaw** turns off Current VP Freeze for layers in the filter or group.

Isolate Group turns off all layers not part of the filter or group; only layers that are part of the filter or group are visible in the drawing. In model space, this option applies to all layers; in layout mode, this option applies selectively, depending on the suboption selected:

- **All Viewports** freezes all layers not in the filter or group.
- **Active Viewport Only** freezes all layers (not in the filter or group) in the current viewport only.

New Properties Filter displays the Layer Filter Properties dialog box.

New Group Filter creates a new layer group filter named "Group Filter 1," which you can rename to something more meaningful. To add layers to groups:

1. Hold down the CTRL key, and then select layer names in the layer list (right-hand pane).
2. Drag the layers onto the group filter name (left-hand pane).

As an alternative, follow these steps:

1. Right-click the group filter name.
2. From the shortcut menu, select **Select Layers | Add**.
3. In the drawing, click on objects whose layer you want to add to the group.
4. Press **Enter** when done selecting objects.

Convert to Group Filter converts the selected filter to a group; the name does not change, but the icon changes to indicate a group filter.

Rename changes the name of the filter or group. As an alternative, click the name twice, slowly, and then enter a new name.

Delete erases the filter or group; does not erase the layers in the filter or group. The All, All Used Layers, and Xref filters cannot be erased.

Properties displays the Layer Filter Properties dialog box (available only when a filter is selected).

Select Layers allows you to add and remove layers by selecting objects in the drawing (available only when a group is selected):

- **Add** allows you to select objects in the drawing; press ENTER to return to the layer dialog box.
- **Replace** removes existing layers from the group, and adds the newly-selected layers.

Layer List

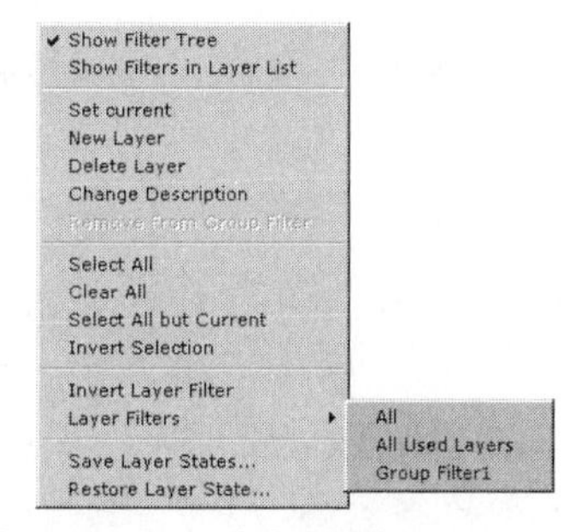

Show Filter Tree toggles filter tree view (left-hand pane).

Show Filters in Layer List toggles the display of filters in the layer list view (right-hand pane); when off (no check mark), only layers are shown.

Set Current sets the selected layer as current. Alternatively, press **Alt+C** or click the **Set Current** button on the dialog box's toolbar.

New Layer creates a new layer, names it "Layer1," and gives it the properties of the currently-selected layer. As alternatives, press ALT+N or click the **New Layer** button on the dialog box's toolbar.

Delete Layer erases the selected layers from the drawing; layers 0, Defpoints, the current layer, xref-dependent layers, and those containing objects cannot be erased. As alternatives, press ALT+D or click the **Delete Layer** button on the dialog box's toolbar.

Change Description adds and changes layer description text. As an alternative, click the description twice, and then edit the text.

Remove from Group Filter removes the selected layers from the select group.

Select All selects all layers; alternatively, press CTRL+A.

Clear All unselects all layers; alternatively, click a single layer.

Select All but Current selects all layers, except the current layer.

Invert Selection unselects selected layers, and selects all other layers.

Invert Layer Filter displays all layers not in the selected filter; alternatively, press ALT+I.

Layer Filters displays a submenu listing the names of filters; alternatively, turn on the filter tree view (left-hand pane). Select a filter name to apply it to the layer list.

Save Layer States displays the New Layer State to Save dialog box.

Restore Layer State displays the Layer States Manager dialog box; alternatively, press ALT+S.

LAYER FILTERS PROPERTIES DIALOG BOX

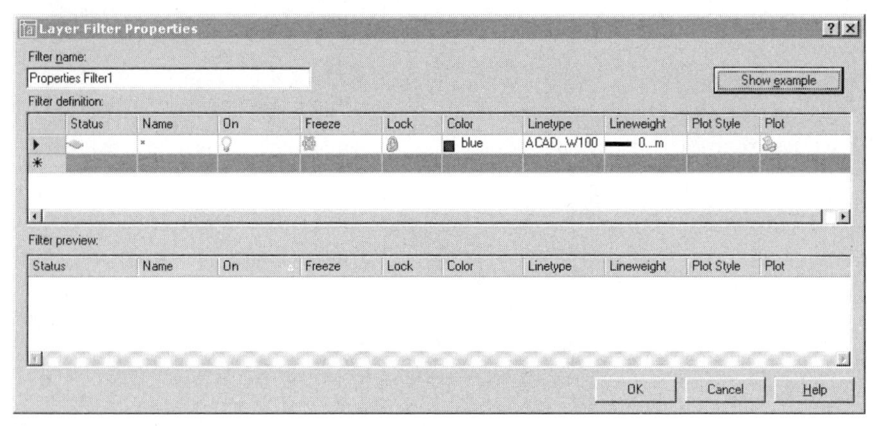

Filter name names the filter.

Show Example displays the filter result in the Filter Preview area (lower half).

Filter definition defines the parameters of the filter:

- **Status** specifies the status of the layer.
- **Layer name** specifies the names of layers to filter.
- **On** selects the layers that are on, off, or both; leave blank for both.
- **Freeze** selects the layers that are frozen, thawed, or both.
- **Lock** selects the layers that are locked, unlocked, or both.
- **Colors** selects the layers of a specific color.
- **Lineweight** selects the layers with a specific lineweight.
- **Linetype** selects the layers with a specific linetype.
- **Plot style** selects the layers with a specific plot style.
- **Plot** selects the layers that plot, do not plot, or both.
- **Current VP Freeze** selects the layers that are frozen, thawed, or both, in the current viewport.
- **New VP Freeze** selects the layers that are frozen, thawed, or both, in the new viewport.

Shortcut Menus

Right-click filter headers:

Maximize all columns
Maximize column

Maximize Column changes the selected column width to display all column content.

Maximize All Columns changes the width of all columns to display content of all columns.

Right-click filter definition row:

Duplicate Row
Delete Row

Duplicate Row duplicates filter definitions.

Delete Row erases the row.

NEW LAYER STATE TO SAVE DIALOG BOX

*To access this dialog box, right-click in the Layer List, and select **Save Layer States** from the shortcut menu.*

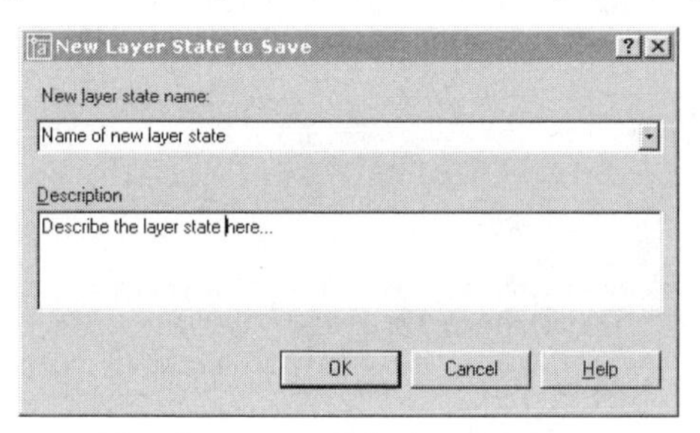

New layer state name specifies the name of the layer state.

Description provides an optional description of the layer state.

OK displays the Layer States Manager dialog box (described below).

LAYER STATES MANAGER DIALOG BOX

*To access this dialog box, right-click in the Layer List, and select **Restore Layer States** from the shortcut menu. Alternatively, click the Layer States Manager button, or press* ALT+S.

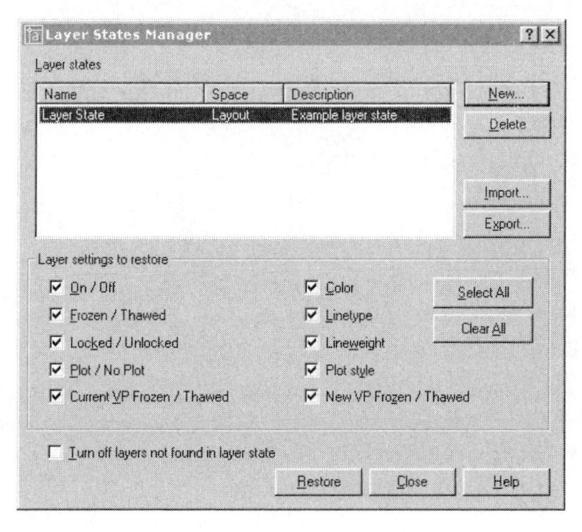

To rename a layer state, click its name twice, and then edit the name.

To edit the description, click the description twice, and then edit the text.

To sort layer states, click the headers – Name, Space, and Description.

Layer states lists the names of layer states that can be restored.

New displays the New Layer State to Save dialog box.

Delete deletes the selected layer state.

Import imports *.las* (layer state) files; layer states can be shared among drawings.

Export exports the selected layer state to *.las* files.

Layer settings to restore selects the settings to restore.

Select All selects all settings.

Clear All unselects all settings (turns them off).

☐ **Turn off layers not found in layer state** turns off layers with unsaved settings.

Restore closes the dialog box, and restores the layer state.

-LAYER Command

Command: -layer

Current layer: "0"

Enter an option [?/Make/Set/New/ON/OFF/Color/Ltype/LWeight/Plot/PStyle/Freeze/Thaw/LOck/Unlock/stAte]: *(Enter an option.)*

COMMAND LINE OPTIONS

Groups and filters cannot be created by this command.

Color indicates the color for all objects drawn on the layer.

Freeze disables the display of the layer.

LOck locks the layer.

Ltype indicates the linetype for all objects drawn on the layer.

LWeight specifies the lineweight.

Make creates a new layer, and makes it current.

New creates a new layer.

OFF turns off the layer.

ON turns on the layer.

Plot determines whether the layer is plotted.

PStyle specifies the plot style (available only when plot styles are attached to the drawing).

Set makes the layer current.

stAte sets and saves layer states.

Thaw unfreezes the layer.

Unlock unlocks the layer.

? lists the names of layers in the drawing.

stAte Options

Enter an option [?/Save/Restore/Edit/Name/Delete/Import/EXport]: *(Enter an option).*

? lists the names of layer states in the drawing.

Save saves the layer state and properties by name; properties include on, frozen, lock, plot, newvpfreeze, color, linetype, lineweight, and plot style.

Restore restores a named state.

Edit changes the settings of named states.

Name renames named states.

Delete erases named states from the drawing.

Import opens layer states *.las* files.

Export saves a selected named state to a *.las* file.

RELATED COMMANDS

LayerGroup manipulates layers; a group of commands.

LayerP returns to the previous layer.

Change moves objects to different layers via the command line.

Properties moves objects to different layers via a dialog box.

LayTrans translates layer names.

Purge removes unused layers from drawings.

Rename renames layers.

View controls visibility of layers with named views.

VpLayer controls the visibility of layers in paper space viewports.

RELATED SYSTEM VARIABLES

CLayer contains the name of the current layer.

LayerFilterAlert removes extraneous filters.

ShowLayerUsage toggles the layer usage icons.

RELATED FILE

***.las** are layer state files, which use the DXF format.

TIPS

- A *frozen* layer cannot be seen or edited.
- A *locked* layer can be seen, but not edited.
- Layer "Defpoints" never plots, no matter its setting.
- To create more than one new layer at a time, use commas to separate layer names in -Layer.
- For new layers to take on properties of an existing layer, select the layer before clicking New.
- If layer names appear to be missing, they may have been filtered from the list.
- Turn off Plot on several random layers, and then drag the header until the Plot column is no longer shown. According to Lynn Allen, this is guaranteed to drive coworkers nuts!

LayerP, LayerPMode

2002 Undoes changes made to layer settings (short for LAYER Previous); toggles layer-previous mode (short for LAYER Previous MODE).

Command	Alias	Ctrl+	F-key	Alt+	Menu Bar	Tablet
layerp	...	...	...	...	...	...

Command: layerp
Restored previous layer status

LAYERPMODE Command

Command: layerpmode
Enter LAYERP mode [ON/OFF] <ON>: *(Type* **ON** *or* **OFF**.*)*

COMMAND LINE OPTIONS

On turns on layer previous mode (default).

Off turns off layer previous mode.

RELATED COMMANDS

Layer creates and sets layers and modes.

Layer Group manipulates layers; a group of commands.

View stores layer settings with named views.

RELATED SYSTEM VARIABLE

CLayer specifies the name of the current layer.

TIPS

- This command acts like an "undo" command for changes made to layers only, for example, the layer's color or lineweight.
- LayerPMode command must be turned on for the LayerP command to work.
- The LayerP command does not undo the renaming, deleting, purging, or creating of layers.
- When layer previous mode is on, AutoCAD tracks changes to layers.

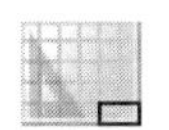

Layout

2000 Creates and deletes paper space layouts on the command line.

Commands	Alias	Ctrl+	F-key	Alt+	Menu Bar	Tablet
layout	...	...	...	**IL**	**Insert ↳Layout**	...
-layout	**lo**					

Command: layout *or* -layout
Enter layout option [Copy/Delete/New/Template/Rename/SAveas/Set/?] <set>: *(Enter an option.)*

COMMAND LINE OPTIONS

Copy copies a layout to create a new layout.

Delete deletes a layout; the Model tab cannot be deleted.

New creates a new layout tab, automatically generating the name for the layout (default = first unused tab in the form Layout*n*), which you may override.

Template displays the Select File dialog box, which allows you to select a *.dwg* drawing or *.dwt* template file to use as a template for a new layout. If the file has layouts, it displays the Insert Layout(s) dialog box.

Rename renames a layout.

SAveas saves the layouts in a drawing template (*.dwt*) file. The last current layout is used as the default for the layout to save.

Set makes a layout current.

? lists the layouts in the drawing in a format similar to the following:

```
Active Layouts:
Layout: Layout1        Block name: *PAPER_SPACE.
Layout: Layout2        Block name: *Paper_Space5.
```

Shortcut Menu

Right-click any layout tab:

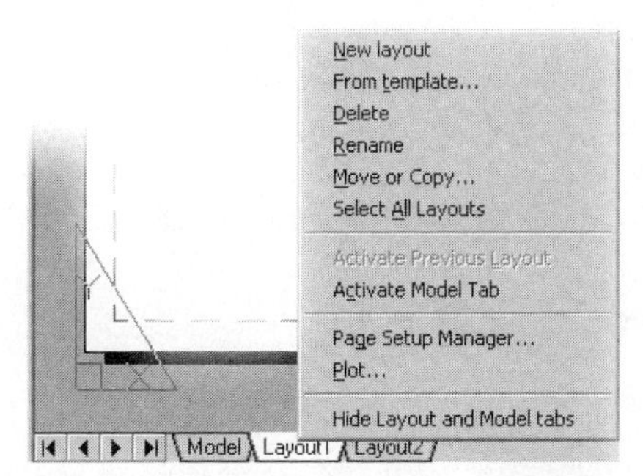

New layout creates a new layout with the default name of **Layout***n*.

From template displays the Select File and Insert Layout dialog boxes.

Delete deletes the selected layout; displays a warning dialog box:

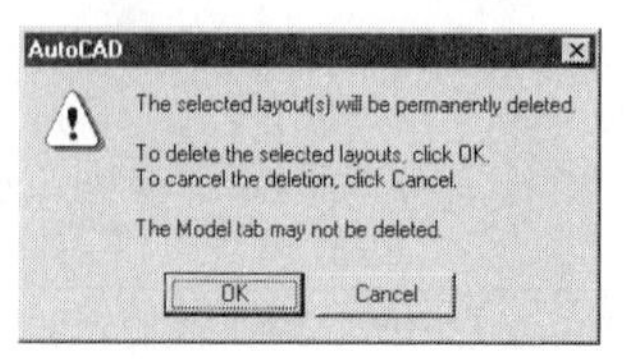

Rename displays the Rename Layout dialog box:

Move or Copy displays the Move or Copy dialog box.

Select All Layouts selects all layouts.

Activate Previous Layout returns to the previously-accessed layout, useful when a drawing has many layouts.

Activate Model Tab returns to the Model tab.

Page Setup displays the Page Setup dialog box; see the PageSetup command.

Plot displays the Plot dialog box; see the Plot command.

Hide Layout and Model tabs moves the tabs to the status bar (*new to AutoCAD 2007*).

INSERT LAYOUT(S) DIALOG BOX

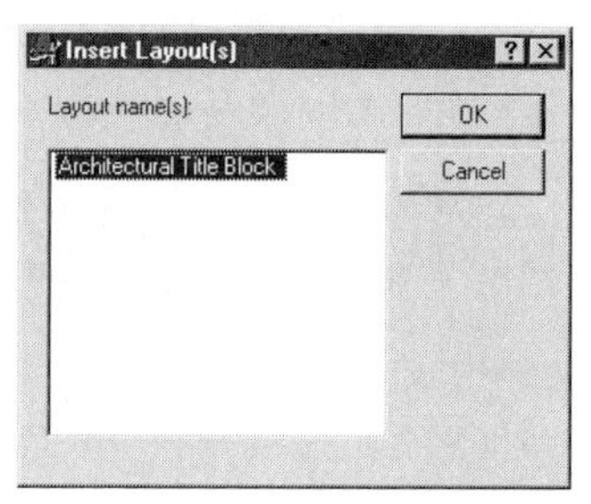

Layout names(s) lists the names of layouts found in the selected template or drawing; you may select more than one layout at a time by holding down the **Ctrl** key.

OK adds the selected layouts to the current drawing.

Cancel dismisses the dialog box and cancels the command.

Move or Copy dialog box

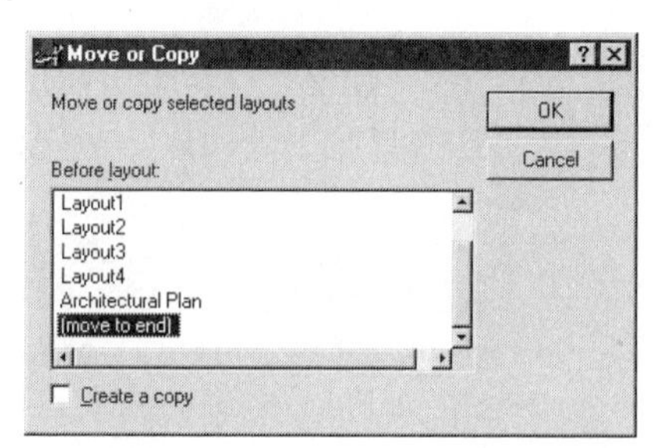

Before layout lists the names of layouts; select one to appear before the current layout.

(move to end) moves the current layout to the end of layouts.

☐ **Create a copy** copies selected layouts.

RELATED COMMAND

LayoutWizard creates and deletes paper space layouts via a wizard.

STARTUP SWITCH

/layout specifies the layout tab to show when AutoCAD starts up.

TIPS

- "Layout" is the name for paper space as of AutoCAD 2000.
- A layout name can be up to 255 characters long; the first 31 characters are displayed in the tab. Drawings can have up to 255 layouts each.
- To switch between layouts, click tabs located below the drawing.

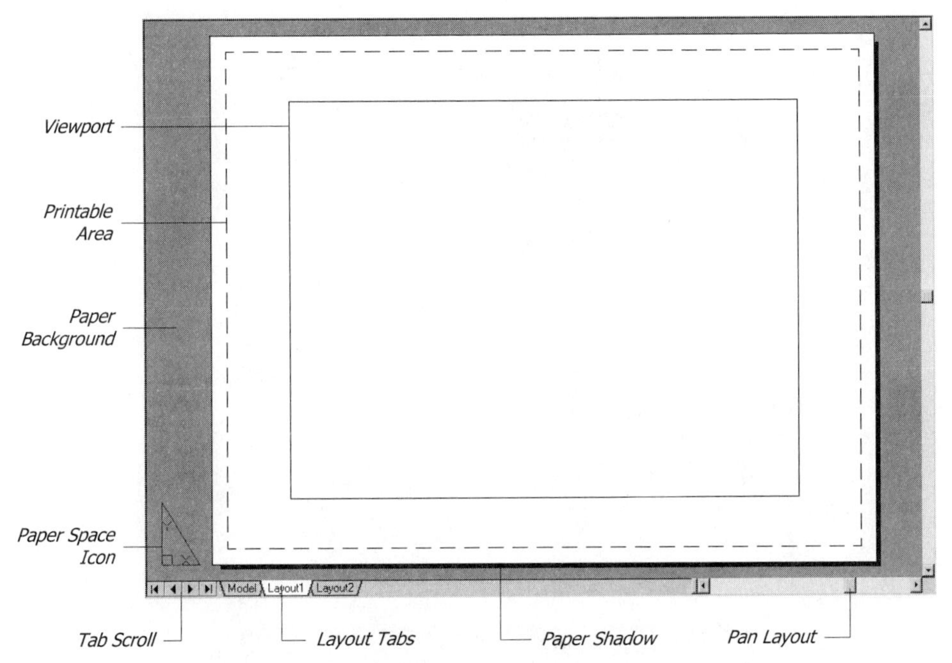

- The Model tab cannot be deleted, renamed, moved, or copied.
- Layout tabs can be moved to the status bar: right-click a tab, and then select **Hide Layout and Model tabs** (*new to AutoCAD 2007*):

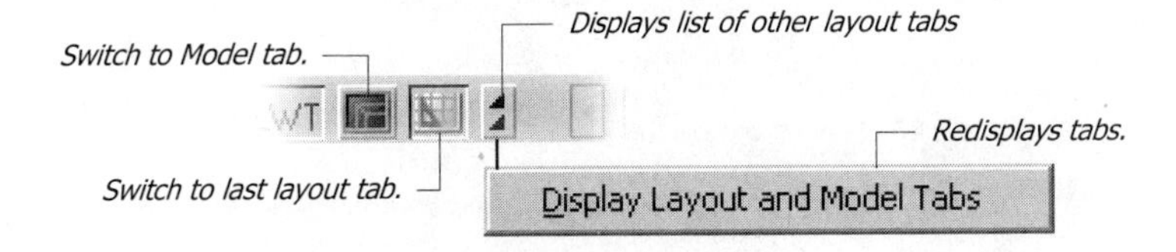

LayoutWizard

2000 Creates and deletes paper space layouts via a wizard.

Command	Alias	Ctrl+	F-key	Alt+	Menu Bar	Tablet
layoutwizard	...	...	...	**ILW**	**Insert** ↳**Layout** ↳**Create Layout Wizard**	...
				TZC	**Tool** ↳**Wizards** ↳**Create Layout**	

Command: layoutwizard

Displays dialog box:

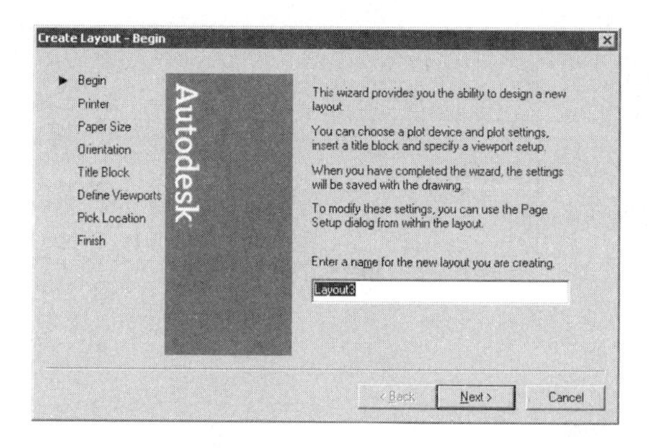

CREATE LAYOUT DIALOG BOX

Enter a name specifies the name for the layout.

Back displays the previous dialog box.

Next displays the next dialog box.

Cancel cancels the command.

Printer page

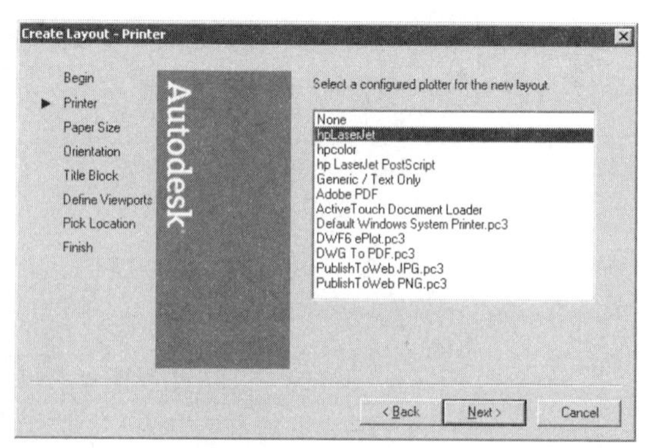

Select a configured plotter selects a printer or plotter to output the layout.

Paper Size page

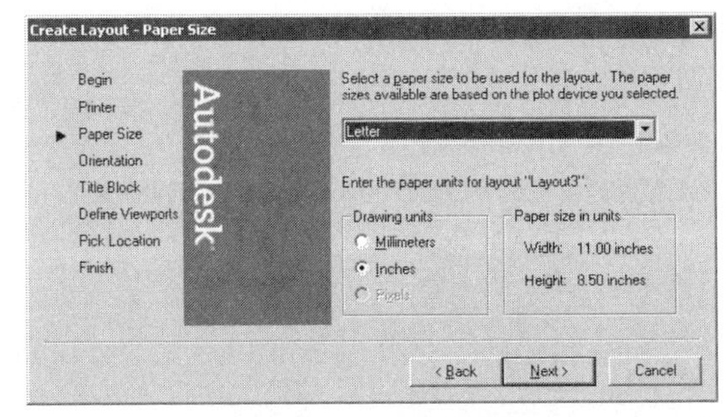

Select a paper size... selects a size of paper supported by the output device.

Enter the paper units

⊙ **Millimeters** measures paper size in metric units.

○ **Inches** measures paper size in Imperial units.

○ **Pixels** measures paper size in dots per inch.

Orientation page

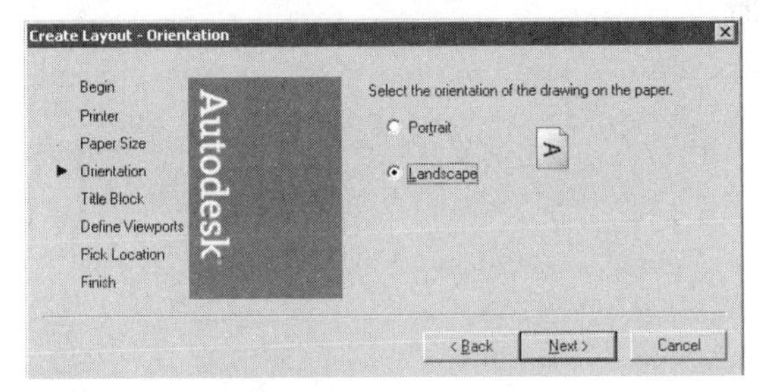

Select the orientation

- **Portrait** plots the drawing vertically.
- **Landscape** plots the drawing horizontally.

Title Block page

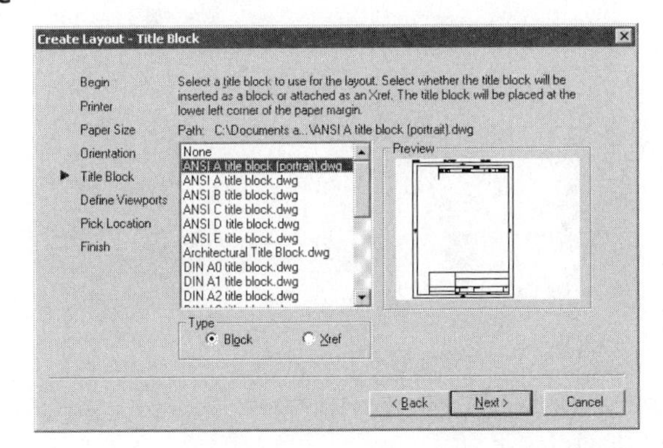

Select a title block... specifies a title border for the drawing as a block or an xref:

⊙**Block** inserts the title border drawing.

○**Xref** references the title border drawing.

Define Viewports page

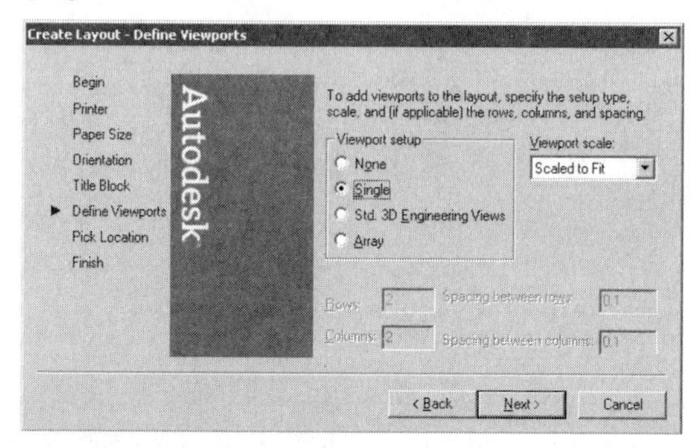

Viewport Setup

○**None** creates no viewport.

⊙**Single** creates a single paper space viewport.

○**Std. 3D Engineering Views** creates top, front, side, and isometric views.

○**Array** creates a rectangular array of viewports.

Viewport scale

- **Scaled to Fit** fits the model to the viewport.
- *mm:nn* specifies a scale factor, ranging from 100:1 to 1/128":1'0".

Rows specifies the number of rows for arrayed viewports.

Columns specifies the number of columns for arrayed viewports.

Spacing between rows specifies the vertical distance between viewports.

Spacing between columns specifies the horizontal distance between viewports.

Pick Location page

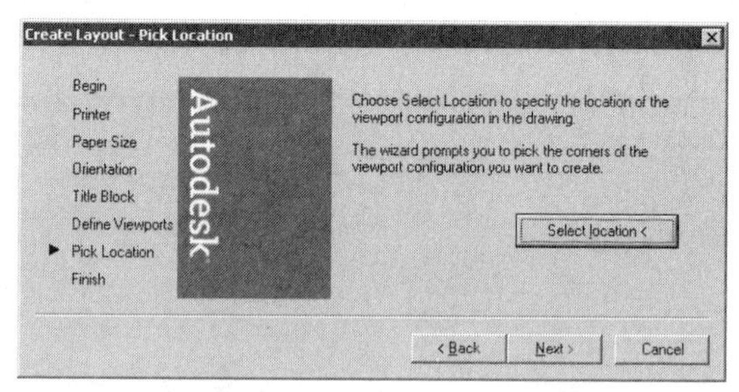

Select location specifies the corners of a rectangle holding the viewports. AutoCAD prompts:

Regenerating layout.
Specify first corner: *(Pick a point.)*
Specify opposite corner: *(Pick a point.)*
Regenerating model.

AutoCAD returns to the Create Layout dialog box.

Finish page

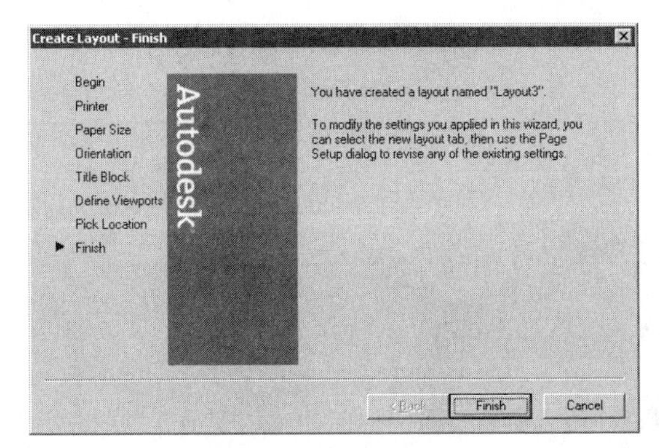

Finish exits the dialog box and creates the layout.

RELATED COMMANDS

Layout creates a layout on the command line.

RELATED SYSTEM VARIABLES

CTab contains the name of the current tab.

LayTrans

2002 Translates layer names (short for LAYer TRANSlation).

Command	Alias	Ctrl+	F-key	Alt+	Menu Bar	Tablet
laytrans	...	...	...	TSSL	Tools ↳CAD Standards ↳Layer Translator	...

Command: laytrans

Displays dialog box:

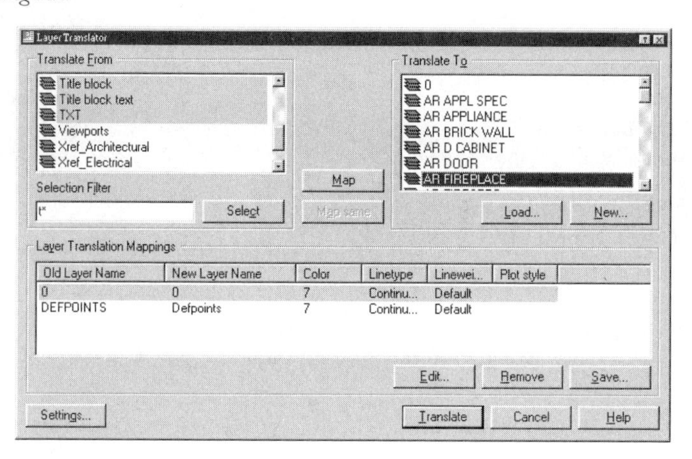

LAYER TRANSLATOR DIALOG BOX

Translate From

Translate From lists the names of layers in the current drawing; icons indicate whether the layer is being used (referenced):

Layer contains at least one object.

Layer contains no objects, and can be purged.

Selection Filter specifies a subset of layer names through wildcards.

Select highlights the layer names that match the selection filter.

Map maps the selected layer(s) in the Translate From column to the selected layer in the Translate To column.

Map same maps layers automatically with the same name.

Translate To

Translate To lists layer names in the drawing opened with the Load button.

Load accesses the layer names in another drawing via the Select Drawing File dialog box.

New creates a new layer via the New Layer dialog box.

Layer Translation Mappings

Edit edits the linetype, color, lineweight, and plot style settings via the Edit Layer dialog box; identical to the New Layer dialog box.

Remove removes the selected layer from the list.

Save saves the matching table to a *.dws* (drawing standard) file.

Settings specifies translation options via the Settings dialog box.

Translate changes the names of layers, as specified by the Layer Translation Mappings list.

NEW LAYER DIALOG BOX

Identical to the Edit Layer dialog box.

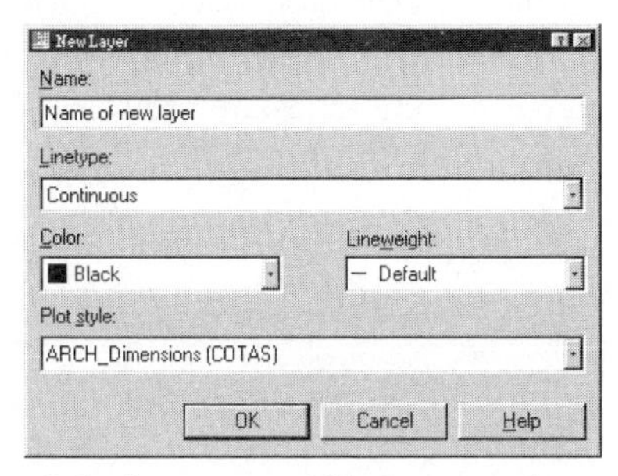

Name specifies the name of the layer, up to 255 characters long.

Linetype selects a linetype from those available in the drawing.

Color selects a color; select **Other** for the Select Color dialog box.

Lineweight selects a lineweight.

Plot style selects a plot style from those available in the drawing; this option is not available if plot styles have not been enabled in the drawing.

SETTINGS DIALOG BOX

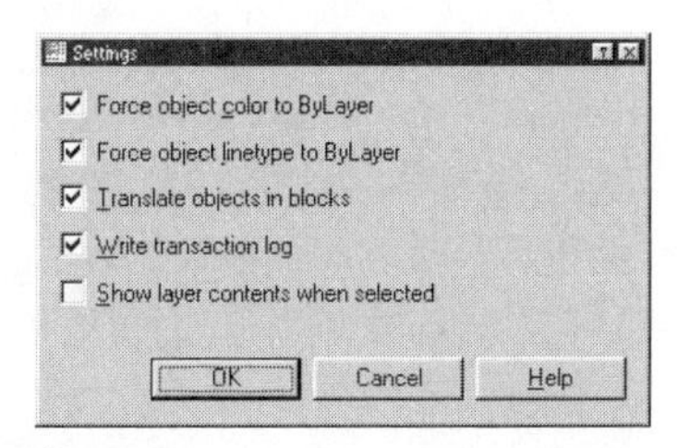

☑ **Force object color to Bylayer** forces every translated layer to take on color Bylayer.

☑ **Force object linetype to Bylayer** forces every translated layer to take on linetype Bylayer.

☑ **Translate objects in blocks** forces objects in blocks to take on new layer assignments.

☑ **Write transaction log** writes the results of the translation to a *.log* file, using the same filename as the drawing. When command is complete, AutoCAD reports:

Writing transaction log to *filename*.log.

☐ **Show layer contents when selected** lists the names of selected layers only in the Translate From list.

RELATED COMMANDS

Standards creates the standards for checking drawings.

CheckStandards checks the current drawing against a list of standards.

Layer creates and sets layers and modes.

RELATED FILES

.*dws drawing standard file; saved in DWG format.

.*log log file recording layer translation; saved in ASCII format.

TIPS

- You can purge unused layers (those prefixed by a white icon) within the Layer Translator dialog box:

  1. Right click any layer name in the **Translate From** list.
  2. Select **Purge Layers**. The layers are removed from the drawing.

- You can load layers from more than one drawing file; duplicate layer names are ignored.

Leader

Rel.13 Draws leader lines with one or more lines of text.

Command	Alias	Ctrl+	F-key	Alt+	Menu Bar	Tablet
leader	lead	...	...	...	...	R7

Command: leader

Specify leader start point: *(Pick a starting point, such as at 1 illustrated below.)*

Specify next point: *(Pick the shoulder point — point 2.)*

Specify next point or [Annotation/Format/Undo] <Annotation>: *(Press Enter to start the annotation, or pick another point — point 3 — or enter an option.)*

Specify next point or [Annotation/Format/Undo] <Annotation>: *(Press Enter to specify the text — 4.)*

Enter first line of annotation text or <options>: *(Enter text — 5 — or press Enter for options.)*

Enter next line of annotation text: *(Press Enter.)*

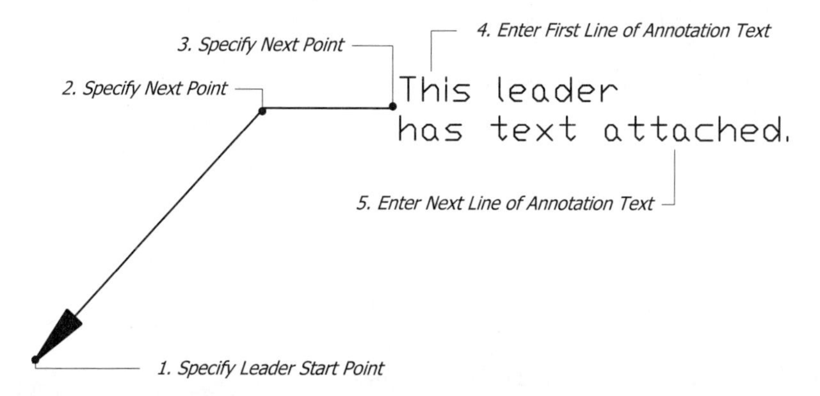

COMMAND LINE OPTIONS

Specify start point specifies the location of the arrowhead.

Specify next point positions the leader line's vertex.

Undo undoes the leader line to the previous vertex.

Format options

Enter leader format option [Spline/STraight/Arrow/None] <Exit>: *(Enter an option.)*

Spline draws the leader line as a NURBS (short for non uniform rational Bezier spline) curve; see Spline command.

STraight draws a straight leader line (default).

Arrow draws the leader with an arrowhead (default).

None draws the leader with no arrowhead.

Annotation options

Enter first line of annotation text or <options>: *(Press* **ENTER**.*)*

Enter an annotation option [Tolerance/Copy/Block/None/Mtext] <Mtext>: *(Enter an option.)*

Enter first line of annotation text specifies the leader text.

Tolerance places one or more tolerance symbols; see the Tolerance command.

Copy copies text from another part of the drawing.

Block places a block; see the -Insert command.

None specifies no annotation.

MText displays the Text Formatting toolbar; see the MText command.

RELATED DIM VARIABLES

DimAsz specifies the size of the arrowhead and the hookline.

DimBlk specifies the type of arrowhead.

DimClrd specifies the color of the leader line and the arrowhead.

DimGap specifies the gap between hookline and annotation (gap between box and text).

DimScale specifies the overall scale of the leader.

TIPS

- This command draws several types of leader:

- The text in a leader is an mtext (multiline text) object.
- Use the \P metacharacter to create line breaks in leader text.
- Autodesk recommends using the QLeader command, which has a dialog box for settings and more options.

Lengthen

Rel.13 Lengthens and shortens open objects by several methods.

Command	Alias	Ctrl+	F-key	Alt+	Menu Bar	Tablet
lengthen	len	...	...	MG	Modify ↳Lengthen	W14

Command: lengthen
Select an object or [DElta/Percent/Total/DYnamic]: *(Select an open object.)*
Current length: ***n.nnnn***

COMMAND LINE OPTIONS

Select an object displays length; does not change the object.

DElta options
Enter delta length or [Angle] <0.0000>: *(Enter a value, or type* ***A****.)*
Specify second point: *(Pick a point.)*
Select an object to change or [Undo]: *(Select an open object, or type* ***U****.)*

Enter delta length changes the length by the specified amount.

Angle changes the angle by the specified value.

Undo undoes the most recent lengthening operation.

Percent options
Enter percentage length <100.0000>: *(Enter a value.)*
Select an object to change or [Undo]: *(Select an open object, or type* ***U****.)*

Enter percent length changes the length to a percentage of the current length.

Undo undoes the most recent lengthening operation.

Total options
Specify total length or [Angle] <1.0000)>: *(Enter a value, or type* ***A****.)*
Select an object to change or [Undo]: *(Select an open object, or type* ***U****.)*

Specify total length changes the length by an absolute value.

Angle changes the angle by the specified value.

Undo undoes the most recent lengthening operation.

DYnamic options
Select an object to change or [Undo]: *(Select an open object, or type* ***U****.)*
Specify new end point: *(Pick a point.)*

Specify new end point dynamically changes the length by dragging.

Undo undoes the most-recent lengthening operation.

TIPS

- **Lengthen** command only works with open objects, such as lines, arcs, and polylines; it does not work with closed objects, such as circles, polygons, and regions.
- **DElta** changes lengths using these measurements: (1) distance from endpoint of selected object to pick point; or (2) incremental length measured from endpoint of angle.
- **Angle** option applies only to arcs.

Light

Rel.12 Places point, spot, and distant lights in drawings for use by Render.

Command	Aliases	F-key	Alt+	Menu Bar	Tablet
light	**pointlight**	**...**	**VEL**	**View**	**O1**
	distantlight			**↳Render**	
	spotlight			**↳Light**	

Command: light

This command is substantially changed in AutoCAD 2007.

Enter light type [Point/Spot/Distant] <Spot>: *(Enter an option.)*

COMMAND LINE OPTIONS

Point places point lights, which radiate light in all directions; you can also use the PointLight alias.

Spot places spot lights, which direct light at a target point; you can also use the SpotLight alias.

Distant places distant lights, which do not attenuate, like the sun; you can also use the DistantLight alias.

 Point light options

Specify source location <0,0,0>: *(Pick a point, or enter x,y,z coordinates.)*
Enter an option to change [Name/Intensity/Status/shadoW/Attenuation/Color/eXit] <eXit>: *(Press Enter, or enter an option.)*

Name names the light; if you do not give it a name, Autodesk assigns generic names, such as "Pointlight1."

Intensity specifies the light's brightness, ranging from 0 to the largest real number.

Status toggles the light on and off.

Shadow determines how the light casts shadows:

- **Off** — shadows are turned off.
- **Sharp** — shadows have sharp edges.
- **Soft** — shadows have soft edges; most realistic.

Attenuation specifies how the light diminishes or decays with distance:

- **None** — light does not diminish.
- **Inverse Linear** — light intensity diminishes by the inverse of the linear distance.
- **Inverse Squared** — light intensity diminishes by the inverse of the square of the distance.
- **Attenuation Start Limit** — specifies where light starts (offset from the light's center); of the light. *Warning*! Attenuation limit options are not supported by OpenGL display drivers.
- **Attenuation End Limit** — specifies where light ends; no light is cast beyond this point.

Color specifies the light's color.

eXit exits the command and places the light glyph in the drawing.

Spot light options

Specify source location <0,0,0>: *(Pick a point.)*
Specify target location <0,0,-10>: *(Pick another point.)*
Enter an option to change [Name/Intensity/Status/Hotspot/Falloff/shadoW/ Attenuation/Color/eXit] <eXit>: *(Press Enter, or enter an option.)*

Name names the light; if you do not give it a name, Autodesk assigns generic names, such as "Spottlight1."

Intensity specifies the light's brightness, ranging from 0 to the largest real number.

Status toggles the light on and off.

Hotspot defines the angle of the brightest (inner) cone of light (a.k.a. beam angle); ranges from 0 to 160 degrees.

Falloff defines the angle of the full (outer) cone of light (a.k.a. field angle); ranges from 0 to 160 degrees.

Shadow determines how the light casts shadows:

- **Off** — shadows are turned off.
- **Sharp** — shadows have sharp edges.
- **Soft** — shadows have soft edges; most realistic.

Attenuation specifies how the light diminishes or decays with distance:

- **None** — light does not diminish.
- **Inverse Linear** — light intensity diminishes by the inverse of the linear distance.
- **Inverse Squared** — light intensity diminishes by the inverse of the square of the distance.
- **Attenuation Start Limit** — specifies where light starts (offset from the light's center); of the light.
- **Attenuation End Limit** — specifies where light ends; no light is cast beyond this point.

Color specifies the light's color.

eXit exits the command and places the light glyph in the drawing.

Distant light options

Specify light direction FROM <0,0,0> or [Vector]: *(Pick a point, or type* **V***.)*
Specify light direction TO <1,1,1>: *(Pick another point.)*
Enter an option to change [Name/Intensity/Status/shadoW/Color/eXit] <eXit>: *(Press Enter, or enter an option.)*

From specifies the location of the distant light.

Vector prompts you to enter a vector direction, such as 0,-1,1.

To specifies the direction of the light.

Name names the light; if unnamed by you, Autodesk assigns generic names, such as Pointlight1.

Intensity specifies the light's brightness, ranging from 0 to the largest real number.

Status toggles the light on and off.

shadoW determines how the light casts shadows:

- **Off** — shadows are turned off.
- **Sharp** — shadows have sharp edges.

- **Soft** — shadows have soft edges; most realistic.

Color specifies the light's color.

eXit exits the command and places the light in the drawing; distant lights show no glyphs.

EDITING LIGHTS

Lights can be edited through the Properties palette, the Light List palette (see the LightList command), or through grips editing (point and spot lights only).

Point lights

Point light glyphs can be relocated with its Position grip:

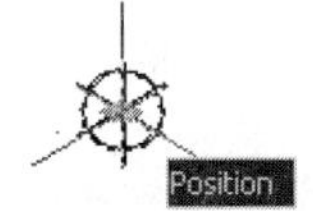

Spot lights

Spot light glyphs can be extensively edited using grips:

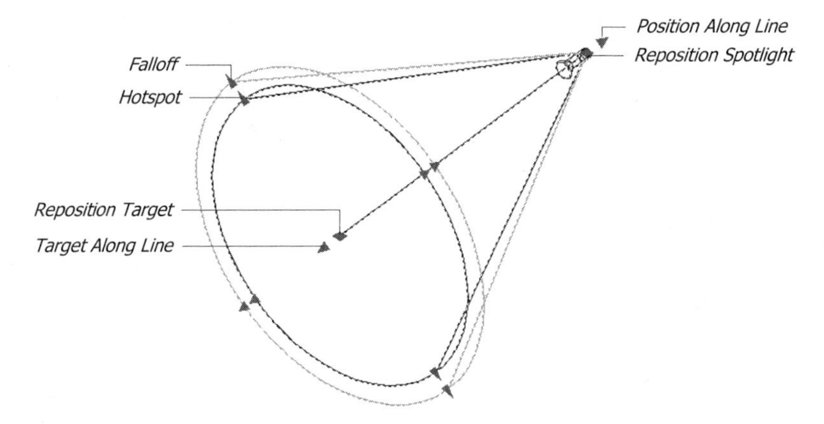

REALTED COMMANDS

LightList displays a list of lights defined in the current drawing.

GeographicLocation determines the position of the sun based on longitude and latitude.

SunProperties specifies the properties of sunlight.

Dashboard provides additional controls for lights, sun position, and shadows.

Options changes the colors of light glyphs.

RELATED SYSTEM VARIABLES

LightGlyphDisplay toggles the display of light glyphs.

ShadowPlaneLocation locates a ground plane onto which shadows can be cast.

TIPS

- This command works in model space only.
- As of AutoCAD 2007, light glyphs replace blocks that represented lights.
- When the drawing has no lights defined, AutoCAD assumes ambient light, which ensures every object in the scene has illumination; ambient light is an omnipresent light source.
- Use the Dashboard's Light control panel to turn on the sun light.
- Use point lights to simulate light bulbs.
- Light can be named using upper- and lowercase letters, numbers, spaces, hyphens, and underscores to a maximum of 256 characters.
- Turning on shadows slows down AutoCAD's display speed.
- Attenuation start limits and end limits are not supported by graphics boards using the OpenGL driver (*wopengl9.hdi*). Use the 3dConfig command's View Tune Log to identify the driver your computer's graphics board uses.
- Placement starts with a bright light over your left shoulder and a dimmer one just over your right one.

Shadows cast onto the shadow plane by the VsShadows system variable.

DEFINITIONS

Constant light — attenuation is 0; default intensity is 1.0.

Inverse linear light — light strength decreases to ½-strength two units of distance away, and ¼-strength four units away; default intensity is ½ extents distance.

Inverse square light — light strength decreases to ¼-strength two units away, and 1/8-strength four units away; default intensity is ½ the square of the extents distance.

Extents distance — distance from minimum lower-left coordinate to the maximum upper-right.

RGB color — three primary colors (red, green, blue) shaded from black to white.

HLS color — changes colors by hue (color), lightness, and saturation (less gray).

Hotspot — brightest cone of light; beam angle ranges from 0 to 160 degrees (default: 45 degrees).

Falloff — angle of the full light cone; field angle ranges from 0 to 160 degrees (default: 45 degrees).

 LightList, LightListClose

2007 Controls the properties of lights placed in drawings.

Command	Alias	Ctrl+	F-key	Alt+	Menu Bar	Tablet
lightlist	...	...	...	VELL	**View** ↳**Render** ↳**Light** ↳**Light List**	...
lightlistclose						

This command is meant to edit light properties, after lights have been placed in the drawing.

Command: lightlist

Displays palette (the sun light does not appear here; see the SunProperties command):

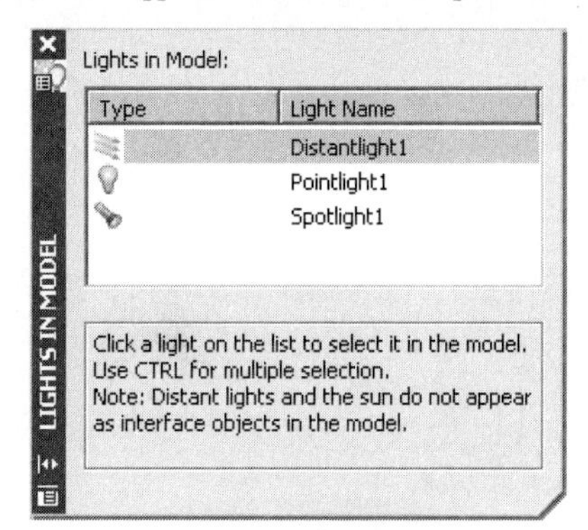

Command: lightlistclose

Closes the palette.

RIGHT-CLICK MENU

Right-click a light name to display the shortcut menu:

Delete Light erases the light from the drawing.

Properties displays the Properties palette.

DISTANT LIGHT PROPERTIES

General properties

Name names the light; if you did not name it, , AutoCAD assigns a generic name, such as "Distantlight1."

On/Off Status toggles the light on and off.

Shadows toggles the display of shadows.

Intensity Factor specifies the light's brightness, ranging from 0 to the largest real number.

Color specifies the light's color.

Geometry properties

From Vector X, Y, Z specifies the x, y, z coordinates of the distant light.

To Vector X, Y, Z specifies the x, y, z coordinates of the light's target.

Source Vector X, Y, Z specifies the x, y, z coordinates of the light source based on altitude and azimuth settings.

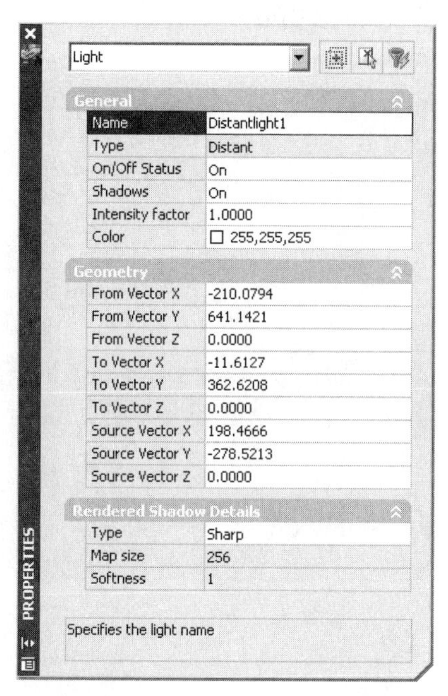

Rendered Shadow Details properties

Type determines how the light casts shadows:

- **Off** — shadows are turned off.
- **Sharp** — shadows have sharp edges.
- **Soft** — shadows have soft edges; most realistic.

Map size specifies the size of the shadow map; range is 64x64 to 4096x4096.

Softness determines the softness of the shadow's edges.

POINT LIGHT PROPERTIES

General Properties

Name names the light

Type toggles between point and spot lights.

On/Off Status toggles the light on and off.

Shadows toggles the display of shadows.

Intensity Factor specifies the light's brightness, ranging from 0 to the largest real number.

Color specifies the light's color.

Plot Glyph determines whether the light glyph is plotted.

Geometry properties

Position X, Y, Z specifies the x, y, z coordinates of the light.

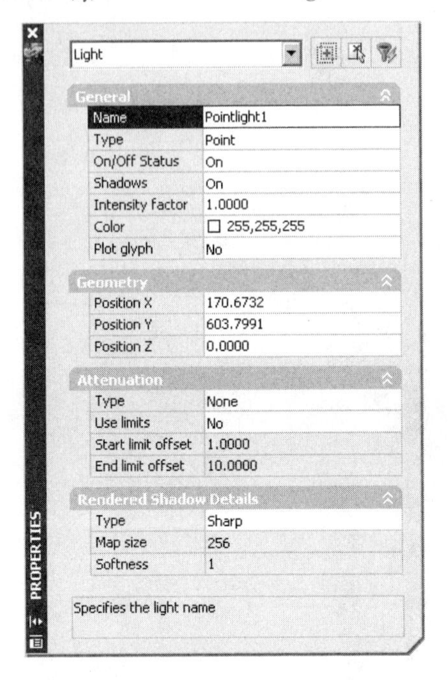

Attenuation

Type specifies how the light diminishes or decays with distance:

- **None** — light does not diminish.
- **Inverse Linear** — light intensity diminishes by the inverse of the linear distance.
- **Inverse Squared** — light intensity diminishes by the inverse of the square of the distance.

Use Limits toggles whether attenuation limits are employed:

Start Limit Offset — specifies where the light beam starts (offset from the light's center).

End Limit Offset — specifies where the light beam ends; no light is cast beyond this point.

Rendered Shadow Details properties

Type determines how the light casts shadows:

- **Off** — shadows are turned off.
- **Sharp** — shadows have sharp edges.
- **Soft** — shadows have soft edges; most realistic.

Map size specifies the size of the shadow map; range is 64x64 to 4096x4096.

Softness determines the softness of the shadow's edges.

SPOT LIGHT PROPERTIES

General Properties

Name names the light.

Type toggles between spot and point lights.

On/Off Status toggles the light on and off.

Shadows toggles the display of shadows.

Hotspot Angle defines the angle of the brighter, inner cone of light; ranges from 0 to 160 degrees.

Falloff Angle defines the angle of the full, outer cone of light; ranges from 0 to 160 degrees.

Intensity Factor specifies the light's brightness, ranging from 0 to the largest real number.

Color specifies the light's color.

Plot Glyph determines whether the light glyph is plotted.

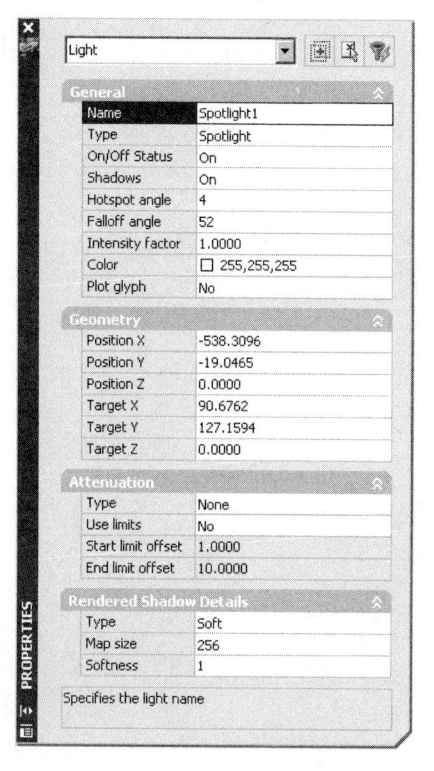

Geometry properties

Position X, Y, Z specifies the x, y, z coordinates of the light.

Target X, Y, Z specifies the x, y, z coordinates of the target point.

Attenuation

Type specifies how the light diminishes or decays with distance:

- **None** — light does not diminish.
- **Inverse Linear** — light intensity diminishes by the inverse of the linear distance.
- **Inverse Squared** — light intensity diminishes by the inverse of the square of the distance.

Use Limits toggles whether attenuation limits are employed:

Start Limit Offset — specifies where the light beam starts (offset from the light's center).

End Limit Offset — specifies where the light beam ends; no light is cast beyond this point.

Rendered Shadow Details properties

Type determines how the light casts shadows:

- **Off** — shadows are turned off.
- **Sharp** — shadows have sharp edges.
- **Soft** — shadows have soft edges; most realistic.

Map size specifies the size of the shadow map; range is 64x64 to 4096x4096.

Softness determines the softness of the shadow's edges.

RELATED COMMANDS

Light places lights in the current drawing.

Dashboard provides additional controls for lights.

RELATED SYSTEM VARIABLE

LightGlyphDisplay toggles the display of light glyphs.

'Limits

<u>**V. 1.0**</u> Defines the 2D limits in the WCS for the grid markings and the Zoom All command; optionally prevents specifying points outside of limits.

Command	Alias	Ctrl+	F-key	Alt+	Menu Bar	Tablet
limits	...	...	...	OA	Format ↳Drawing Limits	V2

Command: limits

Reset Model *(or* **Paper***)* **space limits:**

Specify lower left corner or [ON/OFF] <0.0,0.0>: *(Pick a point, or type* **ON** *or* **OFF***.)*

Specify upper right corner <12.0,9.0>: *(Pick another point.)*

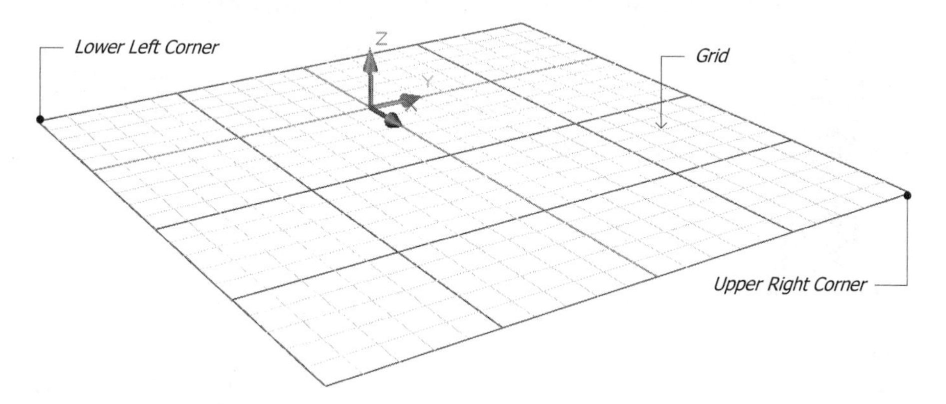

COMMAND LINE OPTIONS

OFF turns off limits checking.

ON turns on limits checking.

ENTER retains limits values.

RELATED COMMANDS

Grid displays grid dots, which are bounded by limits.

Zoom displays the drawing's extents or limits (whichever is larger) with the **All** option.

RELATED SYSTEM VARIABLES

GridDisplay determines if the grid is constrained by the limits.

LimCheck toggles the limit's drawing check.

LimMin specifies the lower-right 2D coordinates of current limits.

LimMax specifies the upper-left 2D coordinates of current limits.

TIP

- As of AutoCAD 2007, the grid can ignore the limits set by this command; see DSettings and Grid commands.

Line

V. 1.0 Draws straight 2D and 3D lines.

Command	Alias	Ctrl+	F-key	Alt+	Menu Bar	Tablet
line	**l**	...	...	**DL**	**Draw** ↳**Line**	**J10**

Command: line
Specify first point: *(Pick a starting point.)*
Specify next point or [Undo]: *(Pick another point, or type* **U**.*)*
Specify next point or [Undo]: *(Pick another point, or type* **U**.*)*
Specify next point or [Close/Undo]: *(Pick another point, or enter an option.)*
Specify next point or [Close/Undo]: *(Press Enter to end the command.)*

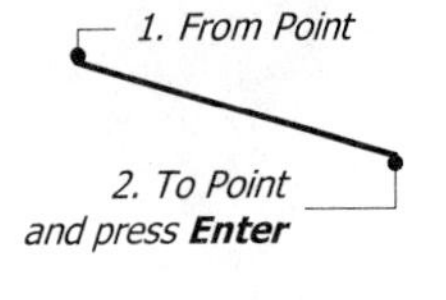

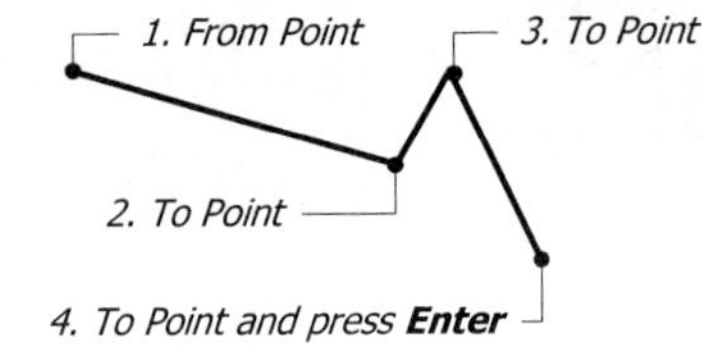

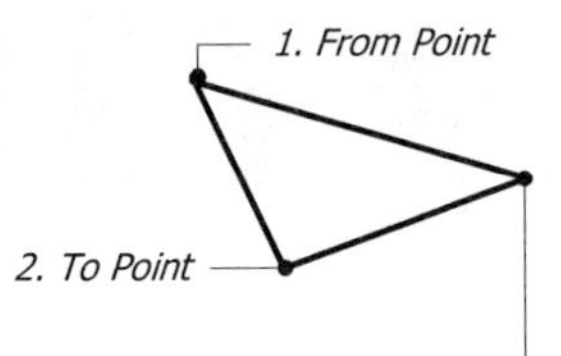

COMMAND LINE OPTIONS

Close closes the line from the current point to the starting point.

Undo undoes the last line segment drawn.

Enter continues the line from the last endpoint at the 'From point' prompt; terminates the Line command at the 'To point' prompt.

RELATED COMMANDS

MLine draws up to 16 parallel lines.

PLine draws polylines and polyline arcs.

Trace draws lines with width.

Ray creates semi-infinite construction lines.

Sketch draws continuous line segments, with up-down control.

XLine creates infinite construction lines.

RELATED SYSTEM VARIABLES

Lastpoint specifies the last-entered coordinate triple (x,y,z-coordinate).

Thickness determines the thickness of the line.

TIPS

- To draw 2D lines, enter x,y coordinate pairs; the z coordinate takes on the value of the Elevation system variable.
- To draw 3D lines, enter x,y,z coordinate triples.
- When system variable Thickness is not zero, the line has thickness, which makes it a plane perpendicular to the current UCS.

'Linetype

V. 2.0 Loads linetype definitions into the drawing, creates new linetypes, and sets the working linetype.

Commands	Aliases	Ctrl+	F-key	Alt+	Menu Bar	Tablet
linetype	lt ltype ddltype	...	...	ON	Format ↳Linetype	U3
-linetype	-lt -ltype					

Command: linetype

Displays dialog box:

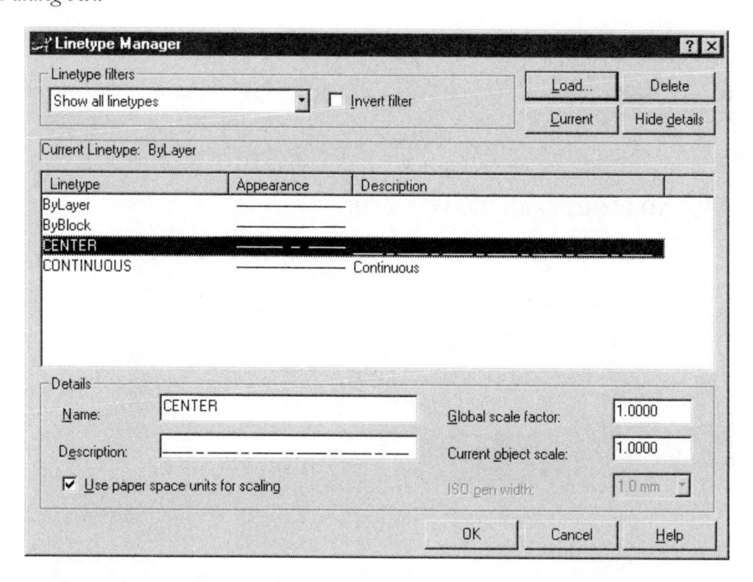

LINETYPE MANAGER DIALOG BOX

Linetype Filters

- **Show all linetypes** displays all linetypes defined in the current drawing.
- **Show all used linetypes** displays all linetypes being used.
- **All xref dependent linetypes** displays linetypes in externally-referenced drawings.

☐ **Invert filter** inverts the display of layer names; for example, when **Show all used linetypes** is selected, the **Invert filter** option displays all linetypes not used in the drawing.

Loads displays the Load or Reload Linetypes dialog box.

Current sets the selected layer as current.

Delete purges the selected linetypes; some linetypes cannot be deleted, as described by the warning dialog box:

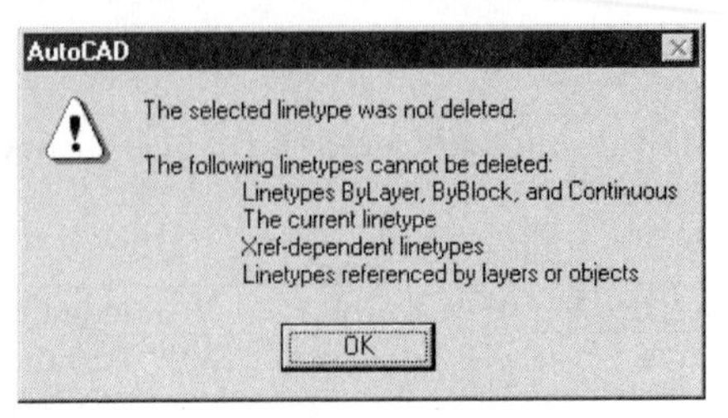

Show/Hide Details toggles the display of the Details portion of the Linetype Properties Manager dialog box.

Details

Name names the selected linetype.

Description describes the linetype.

☑ **Use paper space units for scaling** specifies that paper space linetype scaling is used, even in model space.

Global scale factor specifies the scale factor for all linetypes in the drawing.

Current object scale specifies the individual object scale factor for all subsequently-drawn linetypes, multiplied by the global scale factor.

ISO pen width applies standard scale factors to ISO (international standards) linetypes.

LOAD OR RELOAD LINETYPES DIALOG BOX

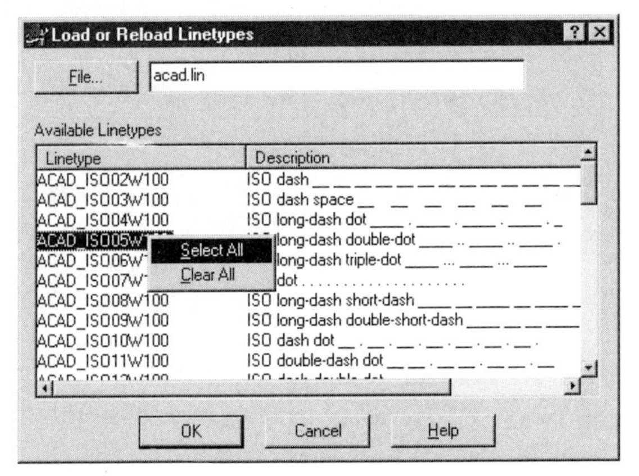

File names the *.lin* linetype definition file.

SHORTCUT MENU

Right-click any linetype name in the Linetype Manager dialog box:

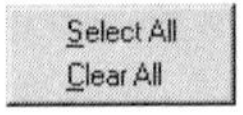

Select All selects all linetypes.

Clear All selects no linetypes.

-LINETYPE Command

Command: -linetype

Enter an option [?/Create/Load/Set]: *(Enter an option.)*

COMMAND LINE OPTIONS

Create creates new user-defined linetypes.

Load loads linetypes from linetype definition (*.lin*) files.

Set sets the working linetype.

? lists the linetypes loaded into the drawing.

RELATED COMMANDS

Change changes objects to a new linetype; changes linetype scale.

ChProp changes objects to a new linetype.

LtScale sets the scale of the linetype.

Rename changes the name of the linetype.

RELATED SYSTEM VARIABLES

CeLtype specifies the current linetype setting.

LtScale specifies the current linetype scale.

PsLtScale specifies the linetype scale relative to paper scale.

PlineGen controls how linetypes are generated for polylines.

TIPS

- The only linetypes defined initially in a new AutoCAD drawing are:

  Continuous draws unbroken lines.

  Bylayer specifies linetype by the layer setting.

  Byblock specifies linetypes by the block definition.

- Linetypes must be loaded from *.lin* definition files before being used in a drawing.
- When loading one or more linetypes, it is faster to load all linetypes, and then use the Purge command to remove the linetype definitions that have not been used in the drawing.
- As of AutoCAD Release 13, objects can have independent linetype scales.
- Express Tools has the MkLType command for creating custom linetypes.

RELATED FILE

The following standard linetypes are in *\autocad 2007\support\acad.lin*:

ACAD_ISO02W100	ISO dash __ __ __ __ __ __ __ __ __ __ __ __ __
ACAD_ISO03W100	ISO dash space __ __ __ __ __ __
ACAD_ISO04W100	ISO long-dash dot ____ . ____ . ____ . ____ . _
ACAD_ISO05W100	ISO long-dash double-dot ____ .. ____ .. ____ .
ACAD_ISO06W100	ISO long-dash triple-dot ____ ... ____ ... ____
ACAD_ISO07W100	ISO dot . . . .
ACAD_ISO08W100	ISO long-dash short-dash ____ __ ____ __ ____ _
ACAD_ISO09W100	ISO long-dash double-short-dash ____ __ __ ____
ACAD_ISO10W100	ISO dash dot __ . __ . __ . __ . __ . __ .
ACAD_ISO11W100	ISO double-dash dot __ __ . __ __ . __ __ . __
ACAD_ISO12W100	ISO dash double-dot __ . . __ . . __ . . __ . .
ACAD_ISO13W100	ISO double-dash double-dot __ __ . . __ __ . .
ACAD_ISO14W100	ISO dash triple-dot __ . . . __ . . . __ . . .
ACAD_ISO15W100	ISO double-dash triple-dot __ __ . . . __ __ .
BATTING	Batting SSSSSSSSSSSSSSSSSSSSSSSSSSSSSSSSSSSSSSS
BORDER	Border __ __ . __ __ . __ __ . __ __ . __ __ .
BORDER2	Border (.5x) __.__.__.__.__.__.__.__.__.__.__.
BORDERX2	Border (2x) ____ ____ . ____ ____ . ___
CENTER	Center ____ _ ____ _ ____ _ ____ _ ____ _ ____
CENTER2	Center (.5x) ___ _ ___ _ ___ _ ___ _ ___ _ ___
CENTERX2	Center (2x) ________ __ ________ __ _____
DASHDOT	Dash dot __ . __ . __ . __ . __ . __ . __
DASHDOT2	Dash dot (.5x) _._._._._._._._._._._._._._.
DASHDOTX2	Dash dot (2x) ____ . ____ . ____ . ___
DASHED	Dashed __ __ __ __ __ __ __ __ __ __ __ __ __ _
DASHED2	Dashed (.5x) _ _ _ _ _ _ _ _ _ _ _ _ _ _ _ _ _
DASHEDX2	Dashed (2x) ____ ____ ____ ____ ____ ___
DIVIDE	Divide ____ . . ____ . . ____ . . ____ . . ____
DIVIDE2	Divide (.5x) __..__..__..__..__..__..__..__.._
DIVIDEX2	Divide (2x) ________ . . ________ . . _
DOT	Dot . . . .
DOT2	Dot (.5x) ....
DOTX2	Dot (2x) . . . .
FENCELINE1	Fenceline circle ----0-----0----0-----0----0---
FENCELINE2	Fenceline square ----[]-----[]----[]-----[]----
GAS_LINE	Gas line ----GAS----GAS----GAS----GAS----GAS---
HIDDEN	Hidden __ __ __ __ __ __ __ __ __ __ __ __ __ __
HIDDEN2	Hidden (.5x) _ _ _ _ _ _ _ _ _ _ _ _ _ _ _ _ _ _
HIDDENX2	Hidden (2x) ____ ____ ____ ____ ____ ____ ____
HOT_WATER_SUPPLY	Hot water supply ---- HW ---- HW ---- HW ----
PHANTOM	Phantom ______ __ __ ______ __ __ ______
PHANTOM2	Phantom (.5x) ___ _ _ ___ _ _ ___ _ _ ___ _ _
PHANTOMX2	Phantom (2x) ____________ ____ ____ _
TRACKS	Tracks -|-|-|-|-|-|-|-|-|-|-|-|-|-|-|-|-|-|-|-
ZIGZAG	Zig zag /\/

List

V. 1.0 Lists information about selected objects in the drawing.

Command	Alias	Ctrl+	F-key	Alt+	Menu Bar	Tablet
list	li ls showmat	...	...	TQL	Tools ↳Inquiry ↳List	U8

Command: list
Select objects: *(Select one or more objects.)*
Select objects: *(Press Enter to end object selection.)*

Sample output:

```
        3DSOLID   Layer: "0"
                          Space: Model space
                 Color: 3 (green)    Linetype: "BYLAYER"
                 Handle = 2e7
       History = None
  Show History = No
 Bounding Box: Lower Bound X = 6.2210    , Y = 6.6221    , Z = -1.9859
               Upper Bound X = 10.1928   , Y = 10.5939   , Z = 1.9859
```

COMMAND LINE OPTIONS

Enter continues the display.

Esc cancels the display.

F2 returns to graphics screen.

RELATED COMMANDS

Area calculates the area and perimeter of selected objects.

DbList lists information about *all* objects in the drawing.

MassProp calculates the properties of 2D regions and 3D solids.

TIPS

- This command lists the following information only under certain conditions:

  Color — when not set BYLAYER.
  Linetype — when not set BYLAYER.
  Thickness — when not 0.
  Elevation — when z coordinate is not 0.
  Extrusion direction — when z axis differs from current UCS.

- Object handles are described by hexadecimal numbers.
- Alternatives to this command include Properties, Dist, and Id.

Removed Command

ListURL was removed from AutoCAD 2000; it was replaced by **-Hyperlink**.

LiveSection

2007 Activates section planes to show the interior of 3D models.

Command	Alias	Ctrl+	F-key	Alt+	Menu Bar	Tablet
livesection	...	...	...	...	...	...

As an alternatively, you can double-click a section plane to make it live.

Command: livesection

Select section object: *(Select one selection plane.)*

Selecting the live section activate its dynamic grips illustrated below.

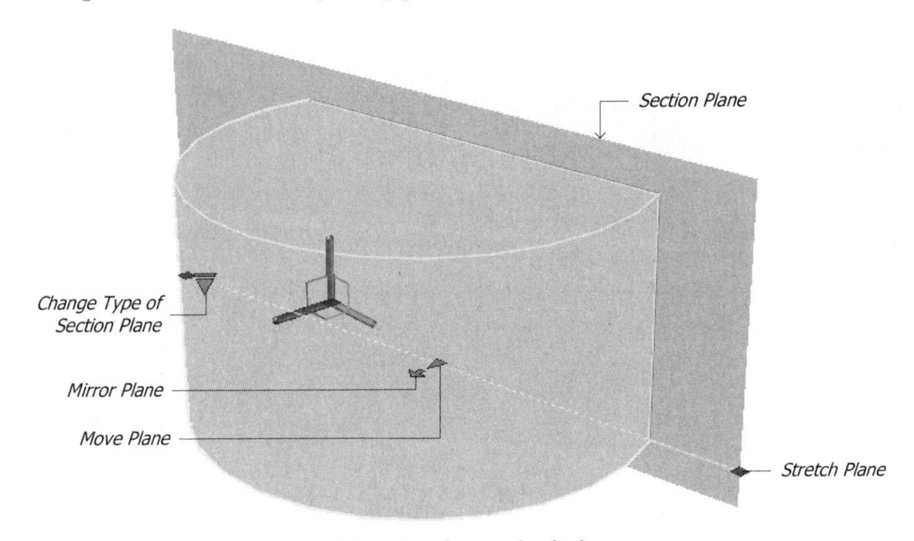

Live section planes can be edited.

COMMAND LINE OPTION

Select section object selects the section plane to make live.

SHORTCUT MENU OPTIONS

Right-click the live section plane for these additional options:

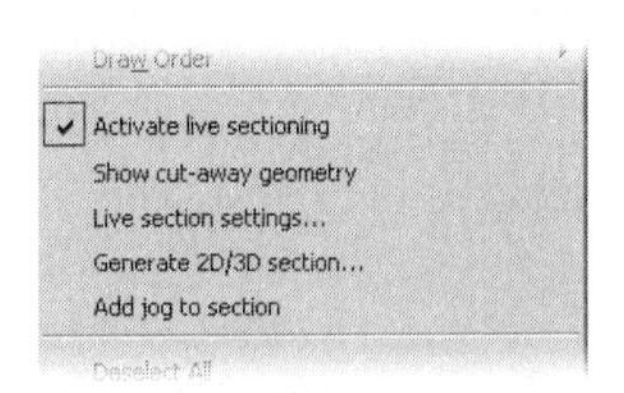

Activate live sectioning toggles live sectioning on and off.

Show cut-away geometry displays the hidden portions in another color.

Live section settings displays the Section Settings dialog box.

Generate 2D/3D section displays the Generate Section/Elevation dialog box, and then creates a second model of the 2D section or 3D elevation.

Add jog to section adds a 90-degree jog to the section plane; see the JogSection command.

DYNAMIC GRIP MENU

Click the lookup grip:

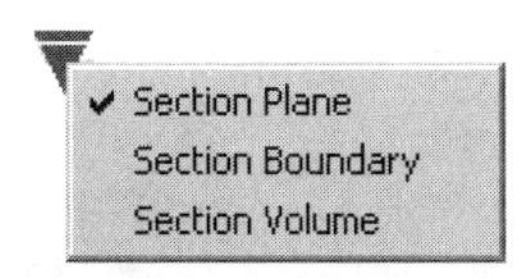

Section Plane displays the section line and cutting plane; the plane extends to "infinity" on all four sides.

Section Boundary displays a 2D box limiting the x,y extents of the cutting plane.

Section Volume displays a 3D box limiting the extent of the cutting plane in all directions.

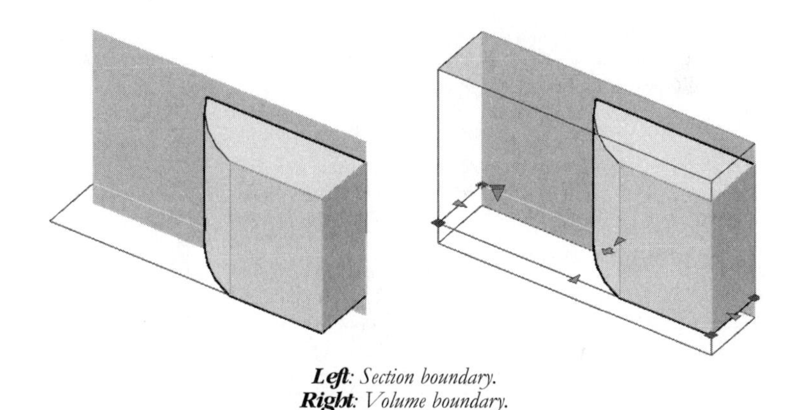

***Left**: Section boundary.*
***Right**: Volume boundary.*

SECTION SETTINGS DIALOG BOX

☑ **Activate Live Section** turns on live sectioning for the selected section.

Intersection Boundary options

Color specifies the color of the intersection boundary.

Linetype specifies the linetype.

Linetype Scale specifies the linetype scale.

Lineweight specifies the lineweight.

Intersection Fill options

Show toggles the display of the intersection fill.

Face Hatch determines if the fill is solid, a hatch pattern, or none.

Angle specifies the angle of the hatch pattern.

Hatch Scale specifies the hatch pattern scale factor.

Hatch Spacing specifies the spacing between hatch lines.

Color specifies the color of the intersection fill.

Linetype specifies the linetype.

Linetype Scale specifies the linetype scale.

Lineweight specifies the lineweight.

Surface Transparency changes the translucency of the intersection fill; ranges from 0 to 100.

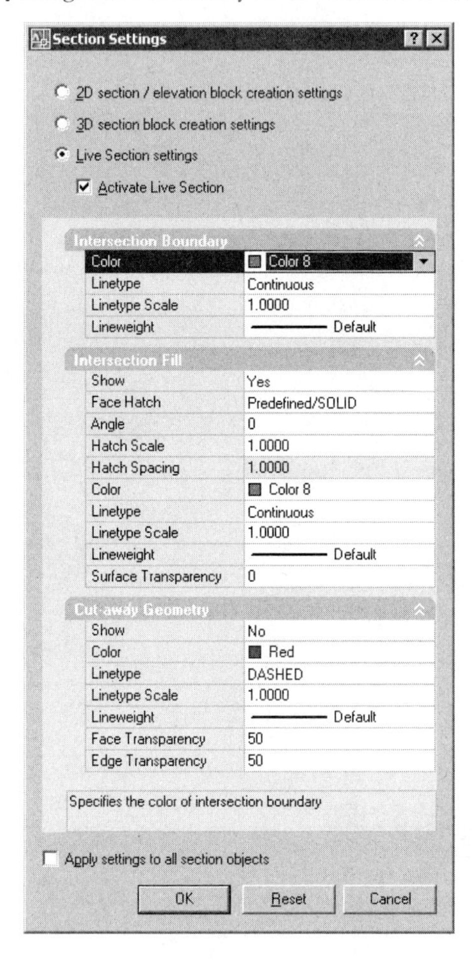

Cut-away Geometry options

Show toggles the display of cut-away geometry (removed by the live section plane).

Color specifies the color of the intersection fill.

Linetype specifies the linetype.

Linetype Scale specifies the linetype scale.

Lineweight specifies the lineweight.

Face Transparency changes the translucency of the cutaway faces.

Edge Transparency changes the translucency of the cutaway edges.

☑ **Apply Settings to All Section Objects** applies settings to future sectionings.

Reset reverts settings to default values.

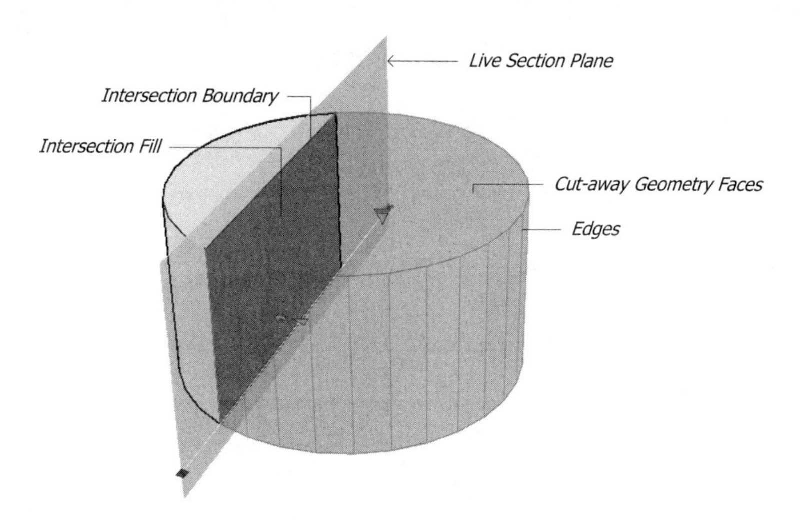

GENERATE SECTION/ELEVATION DIALOG BOX

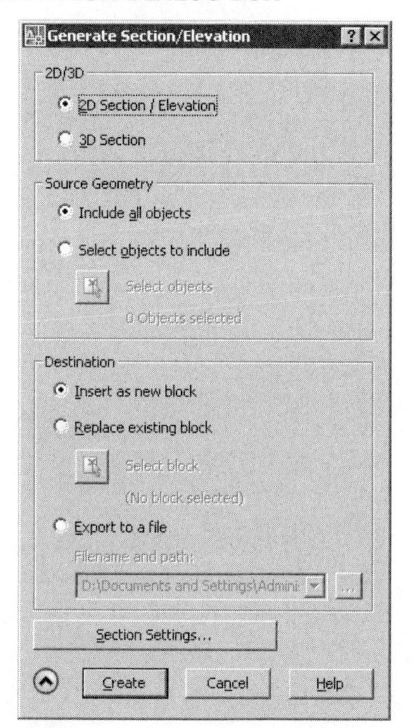

2D/3D options

⊙ **2D Section/Elevation** generates 2D sections (a.k.a. elevations).

○ **3D Section** generates 3D sections.

The following options are visible only when the More button is clicked:

Source Geometry options

⊙ **Include All Objects** sections all 3D solids, surfaces, and regions in the drawing, blocks, and xrefs.

○ **Select Objects to Include** selects manually the 3D solids, surfaces, and regions to be sectioned.

Select Objects allows you to pick the objects to section with a cursor.

Destination options

⊙ **Insert as New Block** inserts the section as a block in the drawing.

○ **Replace Existing Block** replaces an existing section block.

Select Block selects the block to replace.

○ **Export to a File** saves the section as an AutoCAD drawing on disc.

Filename and Path specifies the *.dwg* file name and path by which the section is saved.

Section Settings displays the Section Settings dialog box.

Create creates the section.

RELATED AUTOCAD COMMANDS

SectionPlane places section planes in 3D objects.

JogSection adds 90-degree jogs to section planes.

TIPS

- Once live, section planes can be edited with dynamic grips, as illustrated above.
- Live sectioning works on 3D objects and regions in model space only.
- Only one section plane can be live at a time.

Load

V. 1.0 Loads SHX-format shape files into drawings.

Command	Alias	Ctrl+	F-key	Alt+	Menu Bar	Tablet
load	...	...	...	...	...	...

Command: load

Displays Load Shape File dialog box. Select an .shx file, and the click ***Open.***

COMMAND LINE Options

None.

RELATED AUTOCAD COMMAND

Shape inserts shapes into the current drawing.

RELATED FILES

* ***.shp*** are source code files for shapes.

* ***.shx*** are compiled shape files.

In \autocad 2007\support folder:

gdt.shx and ***gdt.shp*** are geometric tolerance shapes used by the Tolerance command.

ltypeshp.shx and ***ltypeshp.shp*** are linetype shapes used by the Linetype command.

TIPS

- Shapes are more efficient than blocks, but are harder to create.
- The Load command cannot load *.shx* files meant for fonts. For example, AutoCAD complains, "gdt.shx is a normal text font file, not a shape file."
- Do not confuse this command with the AutoLISP **(load)** function, which loads *.lsp* files.

Loft

2007 Creates 3D surface or solid lofts between two or more objects.

Command	Alias	Ctrl+	F-key	Alt+	Menu Bar	Tablet
loft	...	...	...	**DL**	**Draw** **↳Modeling** **↳Loft**	...

Command: loft
Select cross-sections in lofting order: *(Pick two or more cross-section objects.)*
Select cross-sections in lofting order: *(Press Enter.)*
Enter an option [Guides/Path/Cross-sections only] <Cross-sections only>: *(Press Enter, or enter an option.)*

AutoCAD creates an initial lofting, and then displays the Loft Settings dialog box:

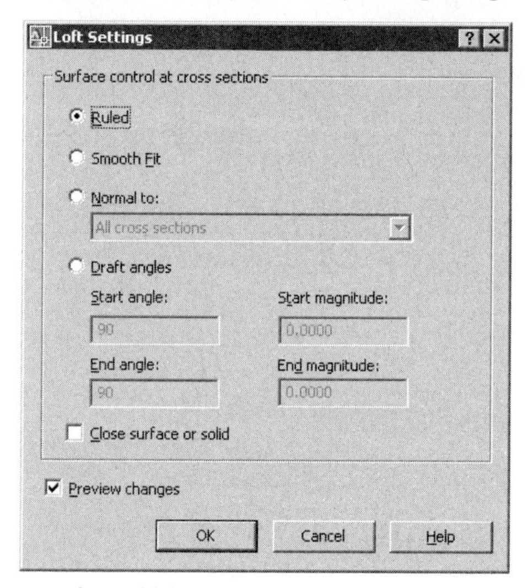

Click ***OK****. AutoCAD creates the final lofting:*

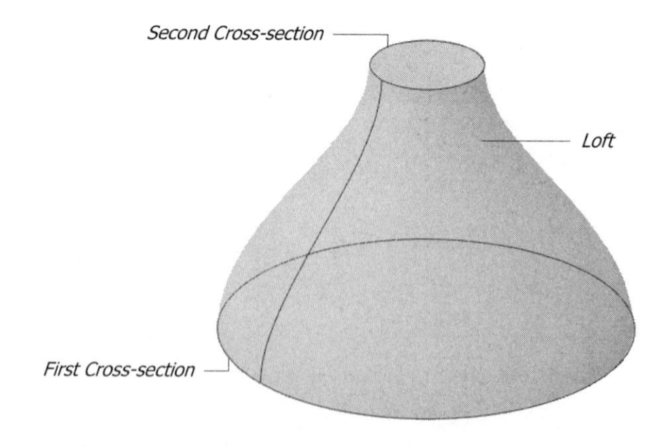

COMMAND LINE OPTIONS

Select cross-sections selects the objects that define the lofting surface: lines, arcs, elliptical arcs, circles, ellipses, polylines, and points (for the first and last cross-sections only).

Cross-sections only uses the cross-section objects selected earlier.

Path selects a single path object that guides the loft: a line, arc, elliptical arc, circle, ellipse, spline, helix, polyline, or 3D polyline.

Guides selects two or more guide-curve objects that define the loft; guides must start on the first cross-section, intersect each cross-section, and end at the last cross-section. Guide objects can be lines, arcs, elliptical arcs, 2D and 3D polylines, and splines.

DIALOG BOX OPTIONS

⊙ **Ruled** creates smoothly-ruled lofts between cross sections, but sharp-edged rules at cross sections.

○ **Smooth Fit** creates smoothly ruled lofts between cross sections, but sharp edges at the starting and ending cross sections.

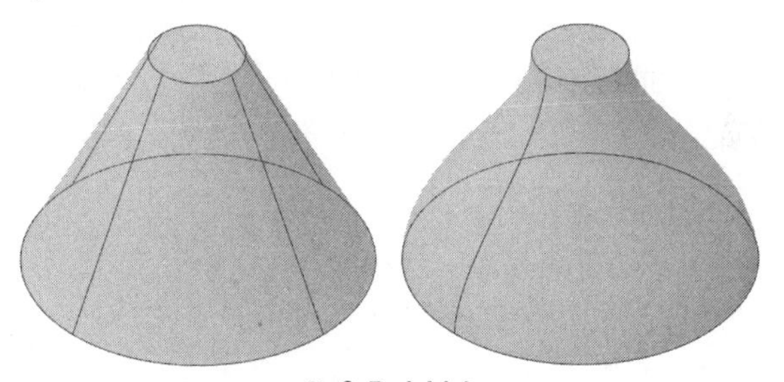

***Left**: Ruled loft.*
***Right**: Normal to start and end cross-sections.*

○ **Normal to** controls the surface normal where the loft passes through the cross-sections:

- **All Cross Sections** — surface normal is normal to all cross-sections.
- **Start Cross Section** — surface normal is normal to the starting cross-section.
- **End Cross Section** — surface normal is normal to the ending cross-section.
- **Start and End Cross Sections** — surface normal is normal to both starting and ending cross-sections.

○ **Draft Angles** controls the draft angle and magnitude of the first and last cross-sections of the lofted solid or surface.

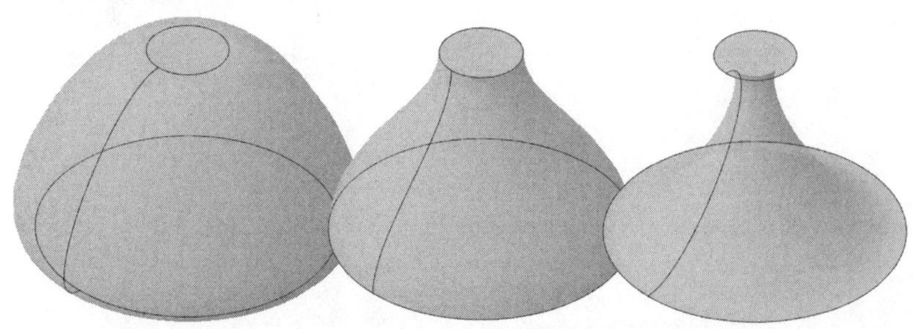

***Left to right:** Loft angles of 0, 80 and 180 degrees.*

Start Angle specifies the draft angle of the starting cross-section.

Start Magnitude controls the distance between the starting cross-section in the direction of the draft angle but before the loft bends toward the next cross-section.

End Angle specifies the draft angle of the starting cross-section.

End Magnitude controls the distance between the ending cross-section in the direction of the draft angle but before the loft bends toward the previous cross-section.

☐ **Close Surface or Solid** (when off) creates open lofts; when on, closes the loft.

☑ **Preview Changes** shows changes to the loft as options are changed in this dialog box.

RELATED AUTOCAD COMMAND

Shape inserts shapes into the current drawing.

RELATED SYSTEM VARIABLES

DelObj toggles whether the cross-section geometry is erased after the loft is created.

LoftAng1 stores the default for the first lofting angle.

LoftAng2 stores the default for the last lofting angle.

LoftMag1 specifies the magnitude of the first lofting angle.

LoftMag2 specifies the magnitude of the last lofting angle.

LoftNormals specifies where lofting normals are placed.

LoftParam specifies the loft shape.

SurfU specifies the number of m-direction isolines on lofts.

SurfV specifies the number of n-direction isolines on lofts.

TIPS

- The cross-section curves must be all open or all closed; open ones create surfaces, while closed ones create solids.
- After the loft is created, its shape can be edited with grips, like spline curves.
- Use this command to create surfaces from contour maps and boat hulls.

LogFileOn, LogFileOff

Rel.13 Turns on and off the command logging to *.log* files.

Command	Alias	Ctrl+	F-key	Alt+	Menu Bar	Tablet
logfileon	...	...	...	...	...	...
logfileoff	...	...	...	...	...	...

Command: logfileon

AutoCAD begins recording command-line text to the log file.

Command: logfileoff

AutoCAD stops recording command-line text, and closes the log file.

COMMAND LINE OPTIONS

None.

RELATED AUTOCAD COMMAND

CopyHist copies all command text from the Text window to the Clipboard.

RELATED SYSTEM VARIABLES

LogFileName specifies the name of the log file.

LogFilePath specifies the path for the log files for all drawings in a session.

LogFileMode toggles whether text window is written to log file:

0 — text not written to file (default).

1 — text written to file.

TIPS

- AutoCAD places a dashed line at the end of each log file session.
- If log file recording is left on, it resumes when AutoCAD is next loaded, which can result in very large log files.
- The default log file name is the same as the drawing name, and is stored in folder *C:\Documents and Settings\username\Local Settings\Application Data\Autodesk\AutoCAD 2007\R16.1\enu*. You can give the file a different folder and name with the Options command's Files tab, or with system variables LogFileName and LogFilePath.
- *Historical note:* In some early versions of AutoCAD, **CTRL+Q** meant "quick screen print," which output the current screen display to the printer. **CTRL+Q** reappeared in AutoCAD Release 14 to record command text to a file. As of AutoCAD 2004, the **CTRL+Q** shortcut quits AutoCAD, instead of toggling the log file — curious, given that there is already a keyboard shortcut, **ALT+F4**, that quits AutoCAD.

Removed Commands

The **LsEdit**, **LsLib**, and **LsNew** commands were removed from AutoCAD 2007.

'LtScale

V. 2.0 Sets the global scale factor of linetypes (short for Line Type SCALE).

Command	Alias	Ctrl+	F-key	Alt+	Menu Bar	Tablet
ltscale	lts	...	...	...	...	...

Command: ltscale
Enter new linetype scale factor <1.0000>: *(Enter a scale factor.)*
Regenerating drawing.

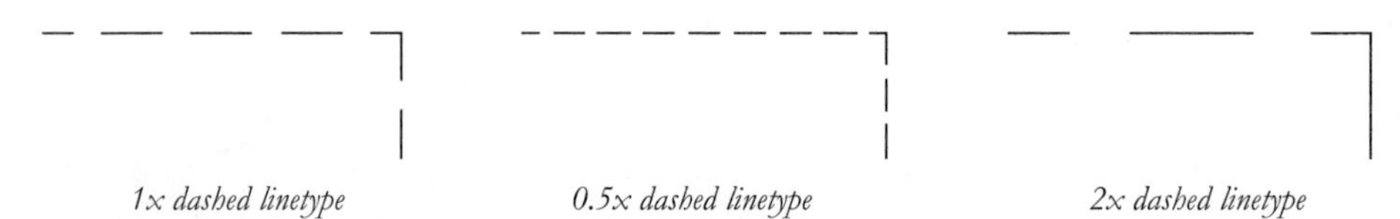

1x dashed linetype *0.5x dashed linetype* *2x dashed linetype*

COMMAND LINE OPTION

Enter new linetype scale factor changes the global scale factor of all linetypes in drawings.

RELATED COMMANDS

ChProp changes the linetype scale of one or more objects.

Properties changes the linetype scale of objects.

Linetype loads, creates, and sets the working linetype.

RELATED SYSTEM VARIABLES

LtScale contains the global linetype scale factor.

CeLtScale specifies the current object linetype scale factor relative to the global scale.

PlineGen controls how linetypes are generated for polylines.

PsLtScale specifies that the linetype scale is relative to paper space.

TIPS

- If the linetype scale is too large, the linetype appears solid.
- If the linetype scale is too small, the linetype appears as a solid line that redraws very slowly.
- In addition to setting the scale with the LtScale command, the *acad.lin* file contains each linetype in three scales: normal, half-size, and double-size.
- You can change the linetype scaling of individual objects, which is then multiplied by the global scale factor specified by the LtScale command.

'LWeight

2000 Sets the current lineweight (display width) of objects.

Commands	Aliases	Ctrl+	Status Bar	Alt+	Menu Bar	Tablet
lweight	**lw**	...	**LWT**	**OW**	**Format**	**W14**
	lineweight				**↳Lineweight**	
-lweight						

Command: lweight

Displays dialog box:

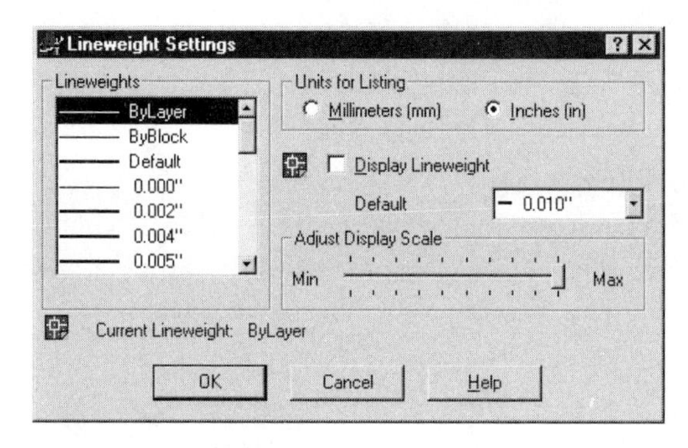

LINEWEIGHT SETTINGS DIALOG BOX

Lineweights lists lineweight values.

Units for Listing

⊙ **Millimeters (mm)** specifies lineweight values in millimeters.

○ **Inches (in)** specifies lineweight values in inches.

☐ **Display Lineweight** toggles the display of lineweights; when checked, lineweights are displayed.

Default specifies the default lineweight for layers (default = 0.01" or 0.25 mm).

Adjust Display Scale controls the scale of lineweights in the Model tab, which displays lineweights in pixels.

SHORTCUT MENU

*Right-click **LWT** on status bar to display shortcut menu:*

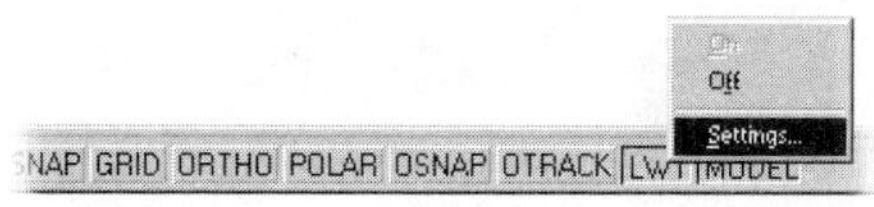

On turns on lineweight display.

Off turns off lineweight display.

Settings displays Lineweight Settings dialog box.

-LWEIGHT Command

Command: -lweight

Enter default lineweight for new objects or [?]: *(Enter a value, or type **?**.)*

COMMAND LINE OPTIONS

Enter default lineweight specifies the current lineweight; valid values include Bylayer, Byblock, and Default.

? lists the valid values for lineweights:

```
ByLayer ByBlock Default
 0.000" 0.002"  0.004"  0.005" 0.006"  0.007"
 0.008" 0.010"  0.012"  0.014" 0.016"  0.020"
 0.021" 0.024"  0.028"  0.031" 0.035"  0.039"
 0.042" 0.047"  0.055"  0.062" 0.079"  0.083"
```

ByLayer
ByBlock
Default
0.00 mm
0.05 mm
0.09 mm
0.13 mm
0.15 mm
0.18 mm
0.20 mm
0.25 mm
0.30 mm
0.35 mm
0.40 mm
0.50 mm
0.53 mm
0.60 mm
0.70 mm
0.80 mm
0.90 mm
1.00 mm
1.06 mm
1.20 mm
1.40 mm
1.58 mm
2.00 mm
2.11 mm

RELATED COMMAND

Properties changes lineweights of selected objects.

RELATED SYSTEM VARIABLES

LwDefault specifies the default linewidth; default = 0.01" or 0.25 mm.

LwDisplay toggles the display of lineweights in the drawing.

LwUnits determines whether the lineweight is measured in inches or millimeters.

TIPS

- To create custom lineweights for plotting, use the Plot Style Table Editor.
- A lineweight of 0 plots the lines at the thinnest width of which the plotter is capable, usually one pixel or one dot wide.
- Linewieghts are "hardwired" into AutoCAD, and cannot be customized.
- Linewieghts can be set through layers and the Properties toolbar.

Replaced Command

MakePreview was removed from AutoCAD Release 14; it was replaced by the **RasterPreview** system variable, which controls the creation of previews when drawings are saved.

Markup, MarkupClose

2005 Opens and closes the Markup Set Manager palette.

Command	Alias	Ctrl+	F-key	Alt+	Menu Bar	Tablet
markup	**msm**	**7**	**...**	**TPM**	**Tools** ↳**Palettes** ↳**Markup Set Manager**	**...**
markupclose		**7**				

Command: markup

Displays palette:

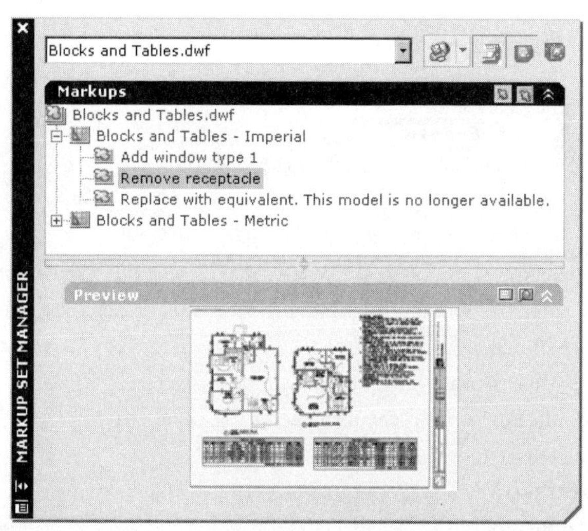

To open a markup set, select Open. AutoCAD displays the Open Markup DWF dialog box. Select a .dwf file, and then click Open.

If the file contains no markup data, AutoCAD complains, 'Filename.dwf does not contain any markup data. Would you like to open this DWF file in the viewer?' Click Yes to open the file in the DWF Viewer; click No to cancel.

Command: markupclose

Closes the palette.

MARKUP SET MANAGER ICONS

Close Window closes the Markup Set Manager palette; reopen with CTRL+7.

Show All Sheets shows all sheets with markups.

Hide Non-Markup Sheets hides all sheets without markups.

Collapse collapses the preview/details area.

Details displays details about the markup sheet.

Preview displays a bitmap image of the sheet.

Palette toolbar

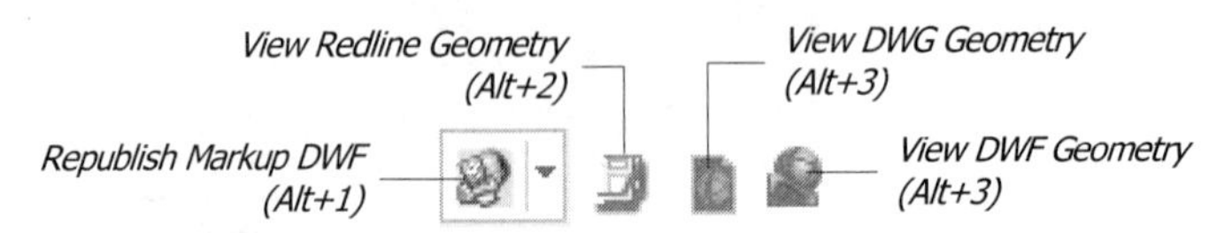

Republish Markup DWF displays menu (**ALT+1**):

- **Republish All Sheets** republishes all sheets as a background job using **Publish**.
- **Republish Markup Sheets** republishes only sheets with markups.

View Redline Geometry toggles the display of redline markup objects (**ALT+2**).

View DWG Geometry toggles the display of *.dwg* drawing files (**ALT+3**).

View DWF Geometry toggles the display of *.dwf* files (**ALT+4**).

SHORTCUT MENU

Right-click a markup file:

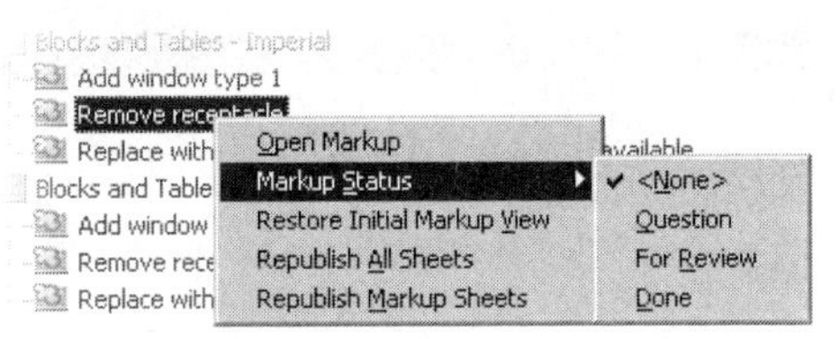

Open Markup opens the markup data in AutoCAD; alternatively, double-click the markup.

Markup Status displays a submenu for changing the status of a markup sheet:

- **None** indicates no change in status.
- **Question** indicates markup has questions to be answered.
- **For Review** indicates markup changes need to be reviewed.
- **Done** indicates markup is done.

Restore Initial Markup View resets all markup views to their original state.

Republish All Sheets republishes all sheets as a background job using Publish.

Republish Markup Sheets republishes only sheets with markups.

RELATED COMMANDS

OpenDwfMarkup opens *.dwf* files containing markup data.

DwfAttach attaches *.dwf* files as underlays.

3dDwfOut exports 3D drawings in DWF format.

RELATED SYSTEM VARIABLE

MsmState reports whether the Markup Set Manager palette is open.

TIPS

- MSM is short for "markup set manager."
- This command reads only *.dwf* files marked-up with Composer; it does not read other *.dwf* files.

 MassProp

Rel.11 Reports the mass properties of 3D solid models, bodies, and 2D regions (short for MASS PROPerties).

Command	Alias	Ctrl+	F-key	Alt+	Menu Bar	Tablet
massprop	...	...	...	**TQM**	**Tools** ↳**Inquiry** ↳**Region/Mass Properties**	**U7**

Command: massprop
Select objects: *(Select one or more regions and/or solid model objects.)*
Select objects: *(Press Enter.)*
Write analysis to a file? [Yes/No] <N>: *(Type **Y** or **N**.)*

Example output of a solid sphere:

```
 ----------------    SOLIDS    ----------------

Mass:                    9.9378
Volume:                  9.9378
Bounding box:         X: 4.7662  --  7.1926
                      Y: 4.9973  --  8.1426
                      Z: -3.8483  --  1.6878
Centroid:             X: 5.8455
                      Y: 6.6391
                      Z: -0.2218
Moments of inertia:   X: 455.3659
                      Y: 354.4693
                      Z: 784.7290
Products of inertia: XY: 384.7960
                     YZ: -10.0200
                     ZX: -14.3752
Radii of gyration:    X: 6.7692
                      Y: 5.9723
                      Z: 8.8862
Principal moments and X-Y-Z directions about centroid:
                      I: 17.1241 along [0.9515 0.3078 0.0000]
                      J: 16.6049 along [-0.2741 0.8472 -0.4552]
                      K: 4.6438 along [-0.1401 0.4331 0.8904]

Write analysis to a file? [Yes/No] <N>:
```

COMMAND LINE OPTIONS

Select objects selects the solid model objects (2D regions, 3D solids, and bodies) to analyze.

Write analysis to a file:

Yes writes mass property reports to *.mpr* files.

No does not write reports to file.

RELATED COMMANDS

Area calculates the area and perimeter of non-solid objects.

Properties displays information about all objects.

RELATED FILE

*.*mpr* is the file to which MassProp writes its results (mass properties report).

TIPS

- This command can be used with 2D regions as well as 3D solids; it cannot be used with 3D surface models or 2D non-region objects.
- Distances are relative to the current UCS. For them to be meaningful, you should orient the UCS to a significant feature on the object.
- Mass properties are computed as if the selected regions were unioned, and as if the selected solids were unioned.
- As of Release 13, AutoCAD's solid modeling no longer allows you to apply a material density to a solid model. All solids and bodies have a density of 1.
- AutoCAD only analyzes regions coplanar (lying in the same plane) with the first region selected.

DEFINITIONS

Bounding Box

– the lower-right and upper-left coordinates of a rectangle enclosing 2D regions.

– the x,y,z coordinate triple of a 3D box enclosing 3D solids or bodies.

Centroid

– the x,y,z coordinates of the center of 2D regions.

– the center of mass of 3D solids and bodies.

Mass

– equal to the volume, because density = 1; not calculated for regions.

Moment of Inertia

– for 2D regions = Area * Radius^2

– for 3D bodies = Mass * Radius^2.

Perimeter

– total length of inside and outside loops of 2D regions (not calculated for 3D solids).

Product of Inertia

– for 2D regions = Mass * Distance (of centroid to y,z axis) * Distance (of centroid to x,z axis).

– for 3D bodies = Mass * Distance (of centroid to y,z axis) * Distance (of centroid to x,z axis)

Radius of Gyration

– for 2D regions and 3D solids = $(\text{MomentOfInertia} / \text{Mass})^{1/2}$

Volume

– 3D space occupied by a 3D solid or body (not calculated for regions).

MatchCell

2005 Matches the properties of table cells.

Command	Alias	Ctrl+	F-key	Alt+	Menu Bar	Tablet
matchcell	...	...	...	...	...	...

Command: matchcell
Select source cell: *(Select a cell in a table.)*
Select destination cell: *(Select one or more cells.)*

COMMAND LINE OPTIONS

Select source cell gets property settings from the source cell.

Select destination cell passes property settings to the destination cells.

RELATED COMMANDS

Table creates new tables in drawings.

TableStyle defines table styles.

MatchProp copies properties between objects other than cells.

RELATED SYSTEM VARIABLE

CTable Style specifies the name of the current table style.

TIPS

- Use this command to copy formatting from one cell to another.
- This command can also be launched by right-clicking a selected cell, and then choosing Match Cell from the shortcut menu.
- Use the MatchProp command to copy properties from one table to another.

'MatchProp

Rel.14 Matches the properties between selected objects (short for MATCH PROPerties).

Command	Alias	Ctrl+	F-key	Alt+	Menu Bar	Tablet
matchprop	**ma**	**...**	**...**	**MM**	**Modify**	**Y14**
	painter				**⇘Match Properties**	

Command: matchprop

Select source object: *(Select a single object.)*

Current active settings: Color Layer Ltype Ltscale Lineweight Thickness PlotStyle Dim Text Hatch Polyline Viewport Table Material Shadow display

Select destination object(s) or [Settings]: *(Pick one or more objects, or type* **S***.)*

Select destination object(s) or [Settings]: *(Press Enter to exit command.)*

COMMAND LINE OPTIONS

Select source object gets property settings from the source object.

Select destination object(s) passes property settings to the destination objects.

Settings displays dialog box:

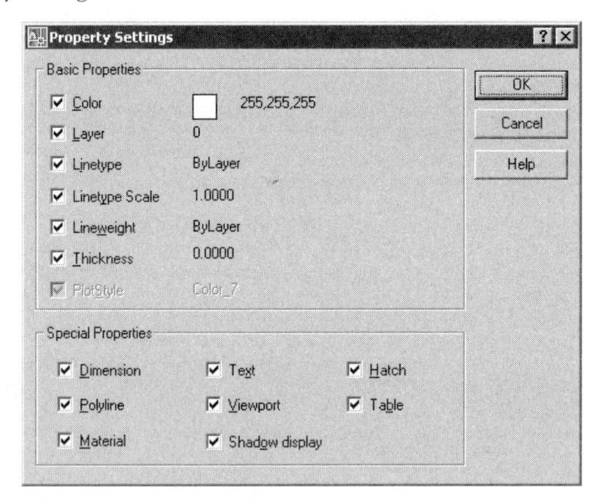

PROPERTY SETTINGS DIALOG BOX

Basic Properties

☑ **Color** specifies the color for destination objects; not available when OLE objects are selected.

☑ **Layer** specifies the layer name for destination objects; not available when OLE objects are selected.

☑ **Linetype** specifies the linetype for destination objects; not available when attributes, hatch patterns, mtext, OLE objects, points, or viewports are selected.

☑ **Linetype Scale** specifies the linetype scale for destination objects; not available when attributes, hatch patterns, mtext, OLE objects, points, or viewports are selected.

☑ **Lineweight** specifies the lineweight for destination objects.

☑ **Thickness** specifies the thickness for destination objects; available only for objects that can have thickness: arcs, attributes, circles, lines, mtext, points, 2D polylines, regions, text, and traces.

☑ **Plot Style** specifies the plot style; not available when PStylePolicy = 1 (color-dependent plot style mode), or when OLE objects are selected.

Special Properties

☑ **Dimension** copies the dimension styles of dimensions, leaders, and tolerance objects.

☑ **Text** copies the text style of text and mtext objects.

☑ **Hatch** copies the hatch pattern of hatched objects.

☑ **Polyline** copies the width and linetype generation of polylines; curve fit, elevation, and variable width properties are not copied.

☑ **Viewport** copies all properties of viewport objects, except clipping, UCS-per-viewport, freeze-thaw settings, and viewprot scale.

☑ **Table** copies style of table objects.

☑ **Material** copies the material name *(new to AutoCAD 2007).*

☑ **Shadow display** copies type of shadow casting *(new to AutoCAD 2007).*

RELATED COMMAND

Properties changes most aspects of one selected object.

RELATED SYSTEM VARIABLE

PStylePolicy determines whether the PlotStyle option is available.

TIPS

- In other Windows applications, this command is known as Format Painter.
- Use this command to make pattern properties the same among multiple hatches.
- Use the MatchCell command to copy properties from one table cell to another.

Removed Command

The **MatLib** command was removed from AutoCAD 2007; it is replaced by the Materials group of the **ToolPalette** command.

MaterialAttach

2007 Attaches rendering materials to layers.

Command	Alias	Ctrl+	F-key	Alt+	Menu Bar	Tablet
materialattach	...	...	...	...	...	...

Command: materialattach

Before using this command, add materials to the drawing through the Materials group of the Tool palette.

Displays dialog box:

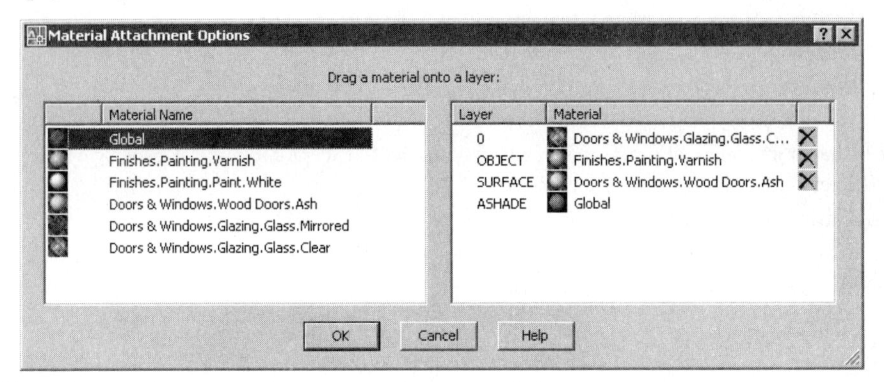

To attach a material to a layer, drag the material name onto the layer name.

DIALOG BOX OPTIONS

Material Name lists the names of materials defined in the drawing.

Layer lists the names of layers in the current drawing.

Material lists the material attached to the layer.

X (*remove*) removes the material from the layer; the layer reverts to the Global material.

RELATED COMMANDS

Materials creates and edits materials.

Render displays the materials applied to layers.

RELATED SYSTEM VARIABLE

CMaterial specifies the default material for newly-created objects; default = bylayer.

TIPS

- This command is limited to attaching materials to layers; to apply materials to objects, drag them from the Materials group of the Tool palette onto objects in the drawing.
- This command can also be launched from the Dashboard.
- Materials can be matched between objects using the MatchProp command.
- Layers without a user-attached material are automatically assigned the Global material.
- Materials can be removed from drawings with the Purge command.

MaterialMap

2007 Maps materials to the faces of objects, and adjusts the mapping.

Command	Alias	Ctrl+	F-key	Alt+	Menu Bar	Tablet
materialmap	setuv	...	...	VEA	View ↳Render ↳Mapping	...

To see the effect of this command, change the visual style to Realistic with the ***vs*** *alias.*

Command: materialmap

Select an option [Box/Planar/Spherical/Cylindrical/copY mapping to/Reset mapping]<Box>: *(Enter an option.)*

Select faces or objects: *(Pick one or more objects.)*

Select faces or objects: *(Press Enter.)*

Accept the mapping or [Move/Rotate/reseT/sWitch mapping mode]: *(Press Enter, or enter an option.)*

COMMAND LINE OPTIONS

Planar maps the material once over the entire object, scaling it vertically and horizontally; meant for use on faces.

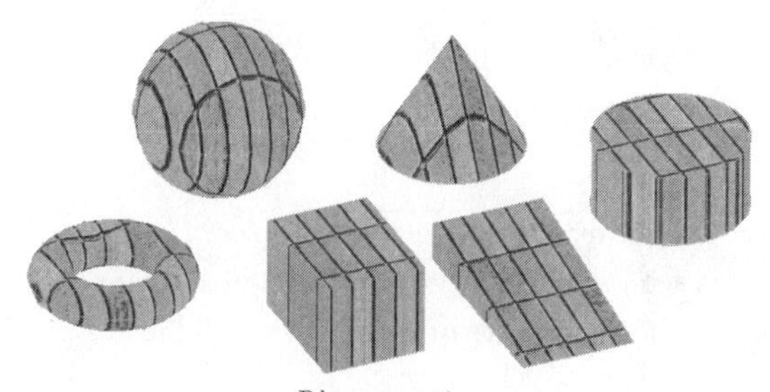

Planar mapping.

Box repeats the mapping on each face of objects; meant for rectangular objects; default mapping for new objects.

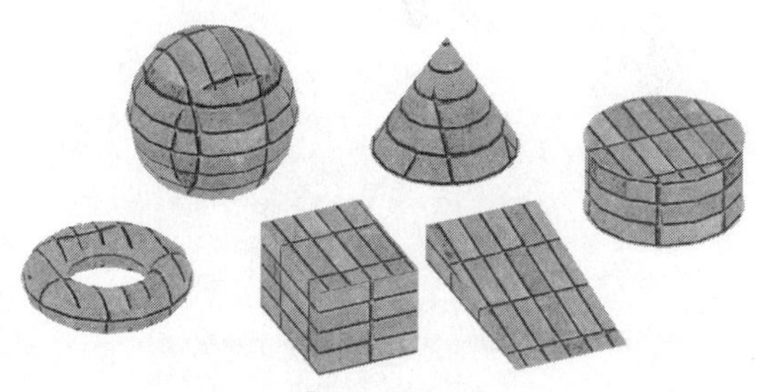

Box mapping.

Cylindrical warps the material horizontally, and scales the material vertically; meant for use on cylindrical objects.

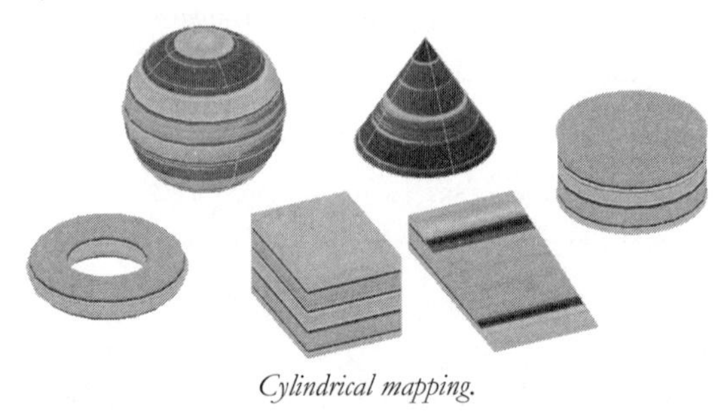

Cylindrical mapping.

Spherical warps the material horizontally and vertically; meant for use on round objects.

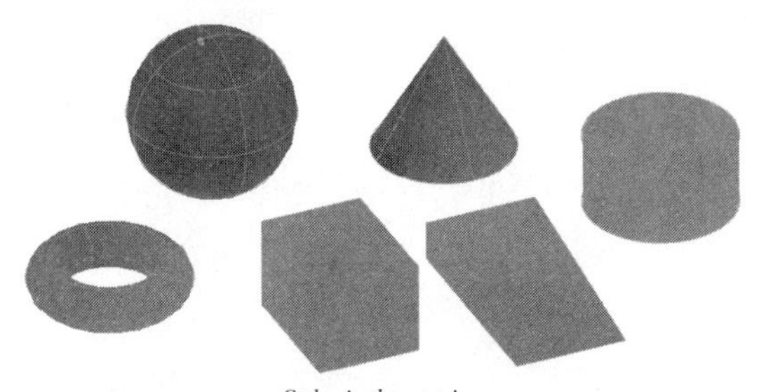

Spherical mapping.

copY mapping to copies the mapping style to other objects.

Reset mapping (and **reseT**) resets the u,v-mapping coordinates to their default values.

Select faces or objects selects the faces or objects to which the mapping style should be applied; hold down the **Ctrl** key to select faces.

Accept the mapping exits the command.

Move moves the material mapping interactively; see the Move3D command. Use the triangular grips to change the size of the map in the x, y, and z directions.

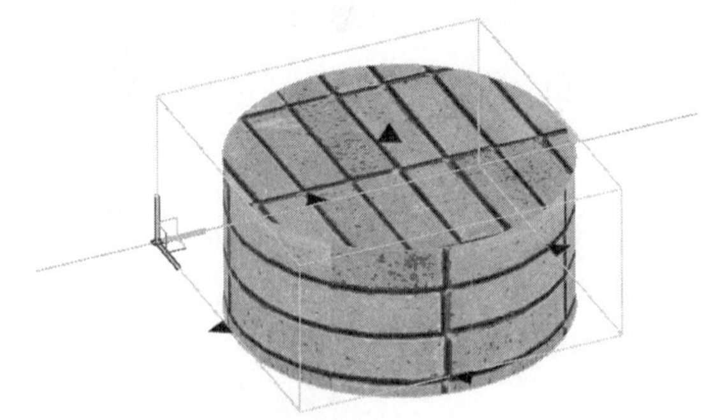

Rotating the material map interactively.

Rotate rotates the material mapping interactively; see the Rotate3D command.

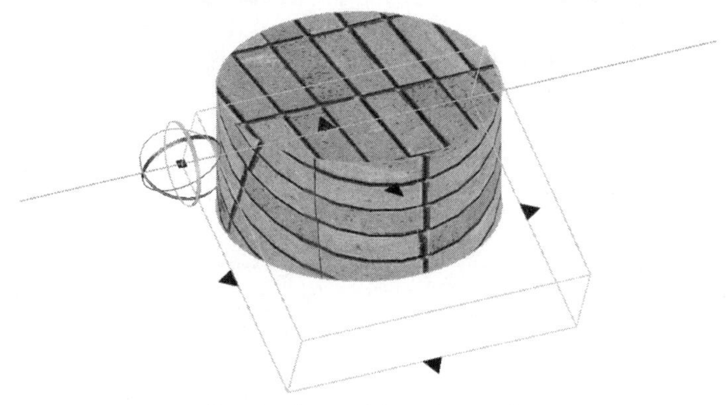

Rotating the material map interactively.

sWitch mapping mode switches to one of the other mapping modes: box, planar, spherical, or cylindrical.

RELATED COMMANDS

Materials creates and edits materials.

Render displays the materials applied to layers.

RELATED SYSTEM VARIABLE

CMaterial specifies the default material for newly-created objects; default = bylayer.

TIP

- Materials can be mapped only to 3D solids, 3D surfaces, faces, and 2D objects with thickness.

Materials, MaterialsClose

2007 Creates and edits materials for use by the Render command and the Realistic visual style.

Command	Aliases	Ctrl+	F-key	Alt+	Menu Bar	Tablet
materials	**mat**	...	...	**VEM**	**View**	**Q1**
	rmat				**↳Render**	
	finish				**↳Materials**	
materialsclose						

Command: materials

Displays palette:

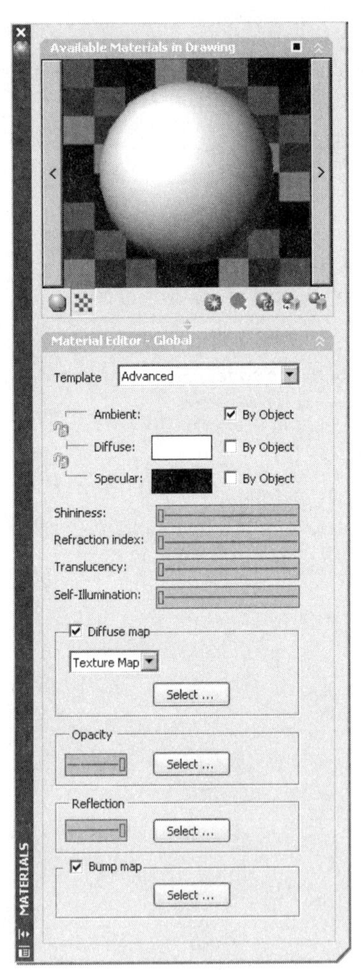

Command: materialsclose

Closes the palette.

MATERIALS PALETTE

Template specifies the generic type of material:

- **Realistic** [*noted by* R *below*] provides a basic set of parameters.
- **Realistic Metal** [*noted by* RM *below*] provides realistic parameters suitable for metal; leaves out Refraction Index, for example.
- **Advanced** [*noted by* A *below*] provides more options than Realistic; includes Reflection, for example.
- **Advanced Metal** [*noted by* AM *below*] provides advanced parameters suitable for metal.

[A | AM] **Ambient** specifies color for faces lighted only by ambient light.

[R | RM | A | AM] **Diffuse** specifies the main color of the material.

[A] **Specular** specifies the color of highlights on shiny materials.

[R | A | AM] **By Object** toggles whether the material color is based on the object's color.

[A | AM] **Ambient-Diffuse Lock** sets the ambient color the same as the diffuse color.

[A] **Diffuse-Specular Lock** sets the specular color the same as the diffuse color.

[R | RM | A | AM] **Shininess** adjusts the amount of shininess by changing the size of the highlight; very shiny objects have smaller and brighter shine spots. Range is 0 to 100.

[R] **Refraction Index** adjusts the refraction index (how light bends going through translucent objects):

- 1.0 — no distortion.
- 3.0 — increased distortion.

[R | A] **Translucency** adjusts translucency (see-through-ness) and light scattering within the material:

- 0 — opaque.
- 100 — fully translucent.

[R | RM | A | AM] **Self-illumination** causes objects to give off light. Range is 0 to 100; when 0, there is no no self-illumination.

[R | RM | A | AM] ☑ **Diffuse Map** toggles diffuse mapping:

- **Texture map** — applies an image file: *.tga*, *.bmp*, *.rle*, *.dib*, *.png*, *.jpg*, *.jpeg*, *.tif*, *.gif*, and *.pcx*. AutoCAD stores 340 JPEG texture files in *C:\Documents and Settings\All Users\Application Data\Autodesk\AutoCAD 2007\R17.0\enu\Textures.*
- **Wood** — displays parameters suitable for simulating wood grain.
- **Marble** — displays parameters suitable for simulating marble veins.

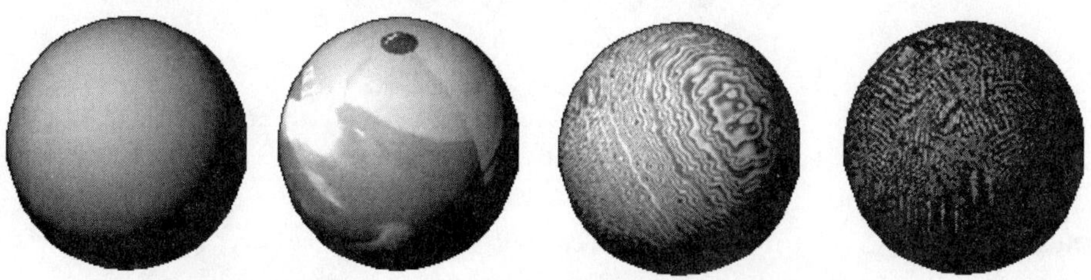

From left to right: *None; texture map applied; wood grain; and marble vein.*

After a texture map is opened, these buttons appear:

Adjust Bitmap Button changes the scale, tiling, offset, and rotation of the image; displays the Adjust Bitmap dialog box.

Delete Map Information removes diffuse map from the material.

After Wood diffuse map is selected, this button appears:

Edit Map displays the Wood window for editing the wood grain.

Slider adjusts the intensity of the wood grain.

After Marble diffuse map is selected, this button appears:

Edit Map displays the Marble window for editing the marble vein.

Slider adjusts the intensity of the marble vein.

Opacity, Reflection, and Bump Map options

  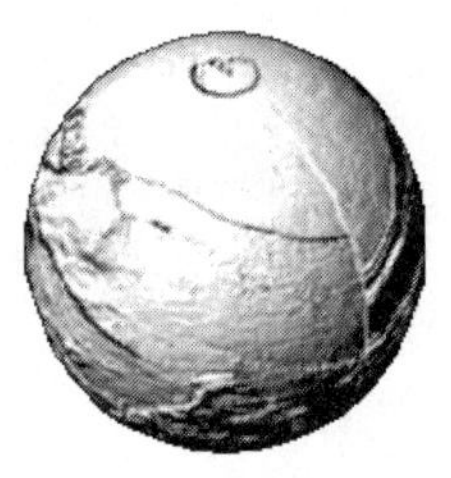

From left to right: *Opacity (transparency); reflection; and bump mapping.*

[R | RM | A | AM] **Opacity** specifies areas of transparency. Select a file that defines transparency:

- **Black** — opaque areas.
- **Gray** — translucent areas.
- **White** — transparent areas.

Slider adjust the amount of overall opacity.

Select Image selects the image that defines the areas of transparency; displays the Select Image File dialog box. AutoCAD stores a single JPEG opacity (cutout) file in *C:\Documents and Settings\All Users\Application Data\Autodesk\AutoCAD 2007\R17.0 \enu\Textures\Cutout*

Opacity (cutout) image for trellis.

[RM | A | AM] **Reflection** adjusts the amount of scene reflection metal surfaces. Range is 0 to 100; the higher the value, the more reflective the material.

Slider adjust the amount of reflectivity.

Select Image selects the image to be reflected off the material's surface; displays the Select Image File dialog box.

[R | RM | A | AM] **Bump Map** adds surface bumpiness.

Slider adjusts the amount of bump texture. Range is -1000 (inverts the bump map) to 1000.

Select Image selects the image to be applied to the material's surface; displays the Select Image File dialog box. AutoCAD stores 115 JPEG bump files in *C:\Documents and Settings\All Users\ Application Data\Autodesk\AutoCAD 2007\R17.0\enu\Textures\Bump.*

Bump map image for woven basket.

ADJUST BITMAP DIALOG BOX

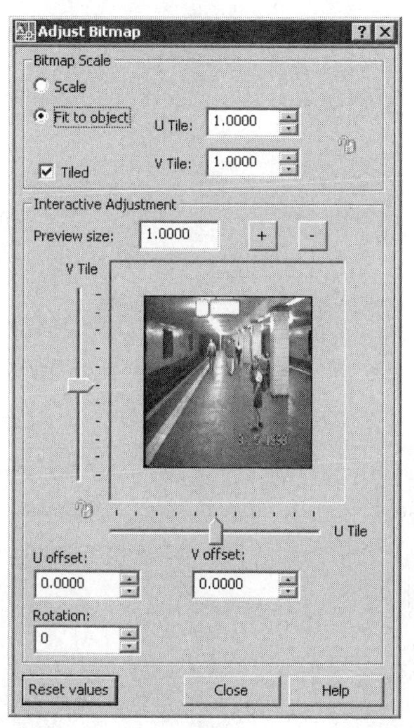

Scale size the image (bitmap) to the object.

Units specifies the units for scaling, ranging from None to Kilometers.

Width and **Height** specify the width (and height) of the image.

Fit to Object sizes the image to fit the object.

U Tile and **V Tile** repeat the image in the x (and y) direction; range is from 0.02 (magnified) to 50 (reduced).

Tiled arrays the image to fit the object.

Lock Aspect Ratio makes the height and width (U and V) sizes the same.

belo

Interactive Adjustment options

Preview Size zooms the preview image; range is from 0.0001 (zoomed in, large image) to 495176003 (zoom out, small image). You can drag the preview image to change its offsets.

U Tile and **V Tile** sliders change the amount of tiling along the U and V axes.

Lock preserves the image's aspect ratio.

U Offset and **V Offset** relocate the image's starting point (lower left corner) along the U and V axes.

Rotation rotates the image around the W axis (equivalent of the z axis). This does not work for spherical and cylindrical mapping; use the MaterialMap command for them instead.

Reset Values returns the original values.

WOOD WINDOW

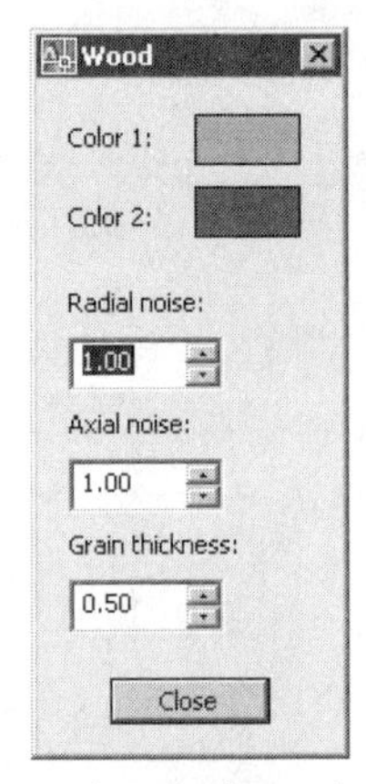

Color 1 and **Color 2** specify the first and second wood grain colors.

Radial Noise adjusts randomness of the pattern perpendicular to the wood grain; range is 0 (least noise) to 100 (most noise).

From left to right: *Radial noise ranging from 1, to 10, to 100.*

Axial Noise adjusts randomness of the pattern parallel for the grain; range is 0 (least) to 100 (most).

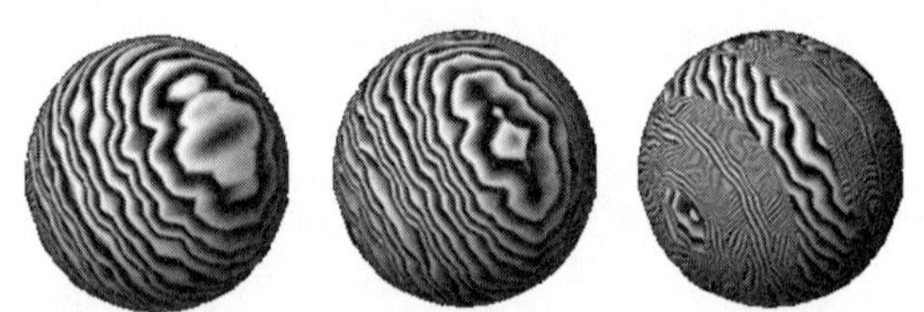

From left to right: *Axial noise ranging from 0, to 10, to 100.*

Grain Thickness adjusts the thickness of the grain; range is 0 (very thin) to 100 (very wide).

From left to right: *Grain thickness ranging from 0.5, to 1, to 10.*

MARBLE WINDOW

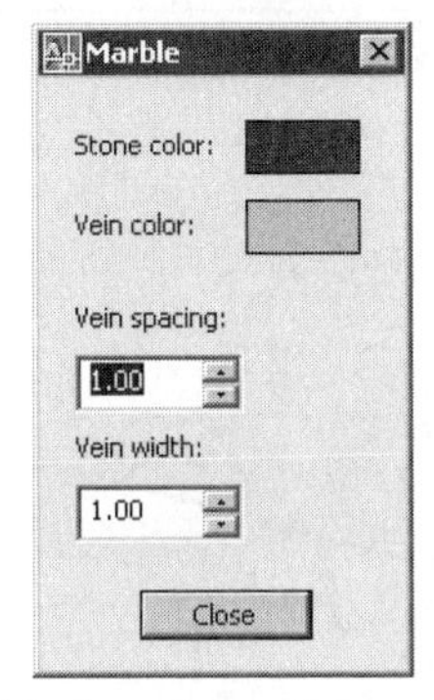

Stone Color and **Vein Color** specify the colors of the stone and vein.

Vein Spacing and **Vein Width** set the spacing between veins and of the veins. Range is 0 (narrow) to 100 (wide).

RELATED COMMANDS

MaterialAttach attaches materials by layer.

MaterialMap adjusts mapping on surfaces.

RELATED SYSTEM VARIABLE

MatState reports whether the palette is open.

TIPS

- A *material* defines the look of a rendered object: coloring, reflection or shine, roughness, and ambient reflection.
- Materials appear when visual style is set to Realistic and when the Render command is used.
- By default, a drawing contains a single material definition, called *GLOBAL*, with the default parameters for color, reflection, roughness, and ambience.
- Materials do not define the density of 3D solids and bodies.

Measure

V. 2.5 Places points or blocks at constant intervals along lines, arcs, circles, and polylines.

Command	Alias	Ctrl+	F-key	Alt+	Menu Bar	Tablet
measure	**me**	...	...	**DOM**	**Draw** ↳**Point** ↳**Measure**	**V12**

Command: measure

Select object to measure: *(Pick a single object.)*

Specify length of segment or [Block]: *(Enter a value, or type* **B***.)*

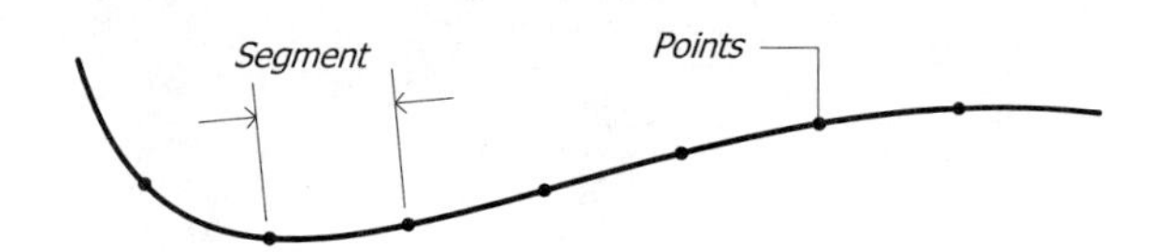

COMMAND LINE OPTIONS

Select object selects a single object for measurement.

Specify length of segment indicates the distance between markers.

Block Options

Enter name of block to insert: *(Enter name.)*

Align block with object? [Yes/No] <Y>: *(Enter* **Y** *or* **N***.)*

Specify length of segment: *(Enter a value.)*

Enter name of block indicates the name of the block to use as a marker; the block must already exist in the drawing.

Align block with object? aligns the block's x axis with the object.

RELATED COMMANDS

Block creates blocks that can be used with the **Measure** command.

Divide divides objects into a specific number of segments.

RELATED SYSTEM VARIABLES

PdMode controls the shape of a point.

PdSize controls the size of a point.

TIPS

- You must define the block before it can be used with this command.
- The **Measure** command does not place points (or blocks) at the beginning and end of measured objects.

Removed Command

MeetNow was removed from AutoCAD 2004.

Menu

V. 1.0 Loads *.cui*, *.mns*, and *.mnu* menu files.

Command	Alias	Ctrl+	F-key	Alt+	Menu Bar	Tablet
menu	...	...	...	...	...	...

Command: menu

Displays the Select Menu File dialog box.

Select a .cui, .mns, or .mnu file, and then click **Open**.

COMMAND LINE OPTIONS

None.

RELATED COMMANDS

Cui controls most aspects of customization.

CuiLoad loads a partial menu file.

Tablet configures digitizing tablet for use with overlay menus.

RELATED SYSTEM VARIABLES

MenuName specifies the name of the currently-loaded menu file.

MenuCtl determines whether sidescreen menu pages switch in parallel with commands entered at the keyboard.

MenuEcho suppresses menu echoing.

ScreenBoxes specifies the number of menu lines displayed on the side menu.

RELATED FILES

****.cui*** source file for menus, toobars, and so on; stored in XML format.

****.mns*** source for obsolete menu files; stored in ASCII format.

****.mnu*** source for obsolete menu template files; stored in ASCII format.

TIPS

- AutoCAD automatically converts *.mns* and *.mns* files into *.cui* files.
- The *.cui* file defines the function of the screen menu, menu bar, cursor menu, icon menus, digitizing tablet menus, pointing device buttons, toolbars, help strings, and the AUX: device.
- AutoCAD includes the *custom.cui* file for customizing menus independently of *acad.cui*.

Replaced Commands

The **MenuLoad** and **MenuUnload** commands became aliases for the **CuiLoad** and **CuiUnload** commands in AutoCAD 2006.

MInsert

V. 2.5 Inserts an array of blocks as a single block (short for Multiple INSERT).

Command	Alias	Ctrl+	F-key	Alt+	Menu Bar	Tablet
minsert	...	...	...	...	...	...

Command: minsert

Enter block name or [?]: *(Enter a name, type* **?***, or enter* **~** *to select a .dwg file.)*

Specify insertion point or [Scale/X/Y/Z/Rotate/PScale/PX/PY/PZ/PRotate]: *(Pick a point, or enter an option.)*

Enter X scale factor, specify opposite corner, or [Corner/XYZ] <1>: *(Enter a value, pick a point, or enter an option.)*

Enter Y scale factor <use X scale factor>: *(Enter a value, or press Enter.)*

Specify rotation angle <0>: *(Enter a value, or press Enter.)*

Enter number of rows (---) <1>: *(Enter a value.)*

Enter number of columns (|||) <1>: *(Enter a value.)*

Enter distance between rows or specify unit cell (---): *(Enter a value.)*

Specify distance between columns (|||): *(Enter a value.)*

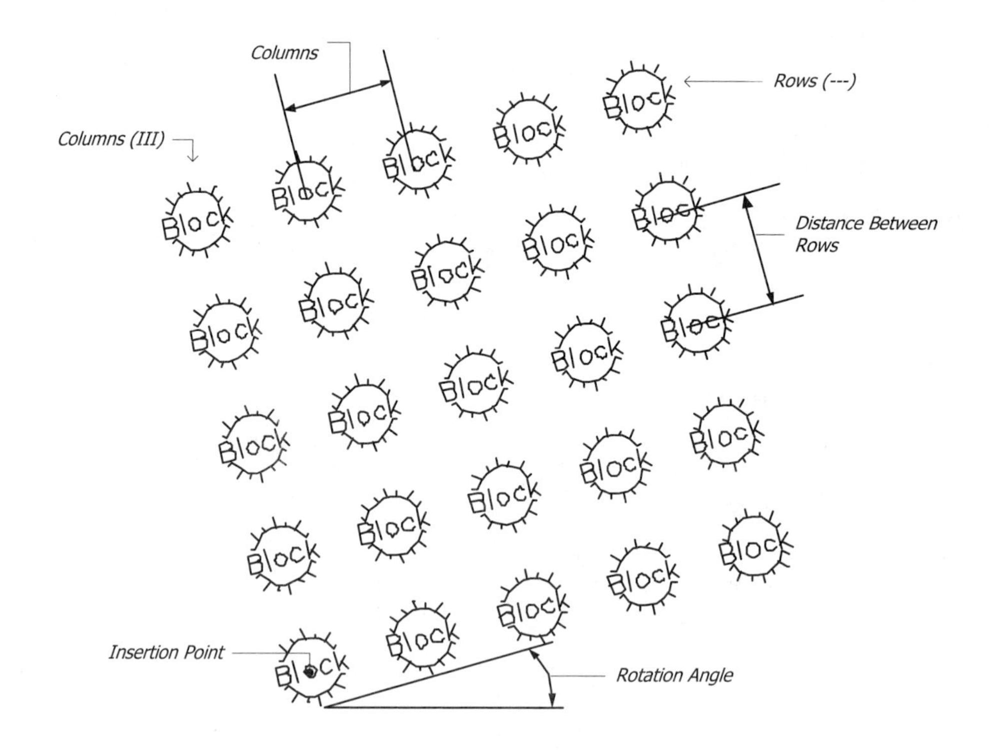

COMMAND LINE OPTIONS

Enter block name names the block to be inserted; the block must already exist in the drawing.

? lists the names of blocks stored in the drawing.

Specify insertion point specifies the x,y coordinates of the first block.

P supplies predefined scale and rotation values.

X scale factor indicates the x scale factor.

Specify opposite corner specifies a second point that indicates the x,y-scale factor.

Corner indicates the x and y scale factors by picking two points on the screen.

XYZ specifies x, y, and z scaling.

Specify rotation angle specifies the angle of the array.

Number of rows specifies the number of horizontal rows.

Number of columns specifies the number of vertical columns.

Distance between rows specifies the distance between rows.

Specify unit cell shows the cell distance by picking two points on the screen.

Distance between columns specifies the distance between columns.

RELATED COMMANDS

3dArray creates 3D rectangular and polar arrays.

Array creates 2D rectangular and polar arrays.

Block creates a block.

Insert inserts single blocks.

TIPS

- The array placed by this command is a single block.
- You *cannot* explode blocks created by this command.
- You may redefine blocks created by the MInsert command.

Mirror

V. 2.0 Creates a mirror copy of a group of objects in 2D space.

Command	Alias	Ctrl+	F-key	Alt+	Menu Bar	Tablet
mirror	mi	...	...	MI	Modify ⤷Mirror	V16

Command: mirror
Select objects: *(Pick one or more objects.)*
Select objects: *(Press Enter to end object selection.)*
Specify first point of mirror line: *(Pick a point.)*
Specify second point of mirror line: *(Pick another point.)*
Delete source objects? [Yes/No] <N>: *(Type **Y** or **N**.)*

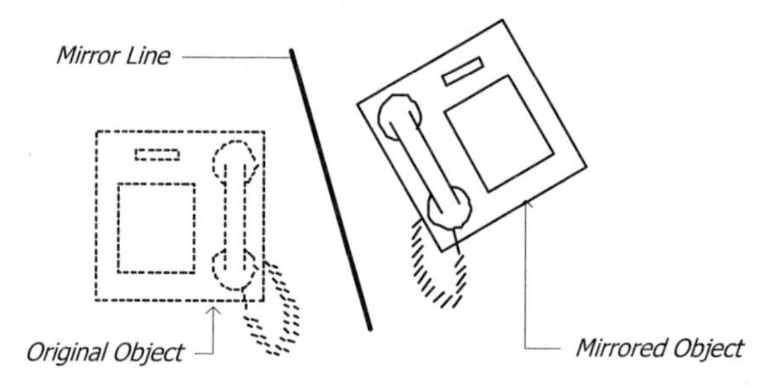

COMMAND LINE OPTIONS

Select objects selects the objects to mirror.

First point specifies the starting point of the mirror line.

Second point specifies the endpoint of the mirror line.

Delete source objects deletes selected objects.

RELATED COMMANDS

Copy creates non-mirrored copies of objects.

Mirror3d mirrors objects in 3D-space.

RELATED SYSTEM VARIABLE

MirrText determines whether text is mirrored by this command. Since AutoCAD 2005, this variable defaults to 0 (text is not mirrored).

TIPS

- This command is excellent for cutting drafting work by half for symmetrical objects. For doubly-symmetrical objects, use Mirror twice.
- Although you can mirror a viewport in paper space, this does not mirror the model space objects inside the viewport.
- Turn on Ortho mode to ensure that the mirror is perfectly horizontal or vertical.
- The mirror line becomes a mirror plane in 3D; it is perpendicular to the x,y plane of the UCS containing the mirror line.

Mirror3d

Rel.11 Mirrors objects about a plane in 3D space.

Command	Alias	F-key	Alt+	Menu Bar	Tablet
mirror3d	**3dmirror**	...	**M3D**	**Modify** ↳**3D Operation** ↳**3D Mirror**	**W21**

Command: mirror3d
Select objects: *(Pick one or more objects.)*
Select objects: *(Press Enter to end object selection.)*
Specify first point of mirror plane (3 points) or
[Object/Last/Zaxis/View/XY/YZ/ZX/3points] <3points>: *(Pick a point, or enter an option.)*
Delete old objects? <N>: *(Type **Y** or **N**.)*

COMMAND LINE OPTIONS

Select objects selects the objects to be mirrored in space.

Specify first point specifies the first point of the mirror plane.

Object selects a circle, arc or 2D polyline segment as the mirror plane.

Last selects the last-picked mirror plane.

View specifies that the current view plane is the mirror plane.

XY specifies that the x,y plane is the mirror plane.

YZ specifies that the y,z plane is the mirror plane.

ZX specifies that the z,x plane is the mirror plane.

Zaxis defines the mirror plane by a point on the plane and the normal to the plane, i.e., the z axis.

3points defines three points on the mirror plane.

RELATED COMMANDS

Align translates and rotates objects in 2D planes and 3D space.

Mirror mirrors objects in 2D space.

Rotate3d rotates objects in 3D space.

RELATED SYSTEM VARIABLE

MirrText determines whether text is mirrored by the Mirror and Mirror3D commands:

0 — Text is not mirrored about the horizontal axis (*default*).

1 — Text is mirrored.

MlEdit

Rel.13 Edits multilines (*short for MultiLine EDITor*).

Command	Alias	Ctrl+	F-key	Alt+	Menu Bar	Tablet
mledit	...	...	...	**MOM**	**Modify** ↳**Object** ↳**Multiline**	**Y19**
-mledit						

Command: mledit

Displays dialog box:

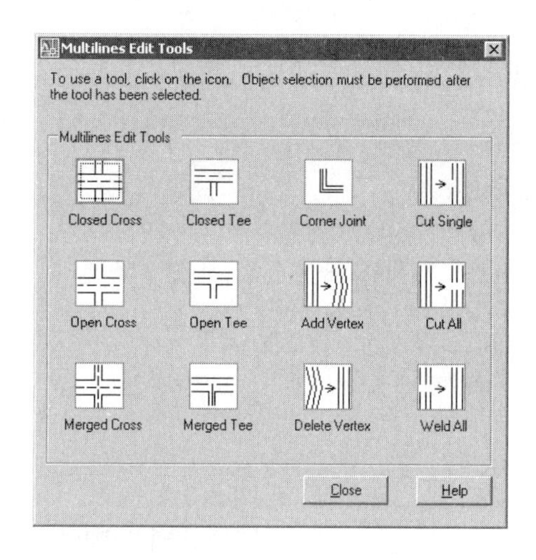

MULTILINE EDIT TOOLS DIALOG BOX

Closed Cross closes the intersection of two multilines.

Open Cross opens the intersection of two multilines.

Merged Cross merges a pair of multilines: opens exterior lines; closes interior lines.

Closed Tee closes T-intersections.

Open Tee opens T-intersections.

Merged Tee merges T-intersection by opening exterior lines and closing interior lines.

Corner Joint creates corner joints with pairs of intersecting multilines.

Add Vertex adds vertcies (*joints*) to multiline segments.

Delete Vertex removes vertices from multiline segments.

Cut Single places gaps in a single line of multilines.

Cut All places gaps in all lines of multilines.

Weld All removes gaps from multilines.

-MLEDIT Command

Command: -mledit

Enter mline editing option [CC/OC/MC/CT/OT/MT/CJ/AV/DV/CS/CA/WA]: *(Enter an option.)*

COMMAND LINE OPTIONS

AV adds vertices.

DV deletes vertices.

CC closes crossings.

OC opens crossings.

MC merges crossings.

CT closes tees.

OT opens tees.

MT merges tees.

CJ creates corner joints.

CS cuts a single line.

CA cuts all lines.

WA welds all lines.

U undoes the most-recent multiline edit.

RELATED COMMANDS

MLine draws up to 16 parallel lines.

MlStyle defines the properties of a multiline.

RELATED SYSTEM VARIABLES

CMlJust specifies the current multiline justification mode.

CMlScale specifies the current multiline scale factor (default = 1.0).

CMlStyle specifies the current multiline style name (default = " ").

RELATED FILE

***.mln** is the multiline style definition file.

TIPS

- Use the Cut All option to open up a gap before placing door and window symbols in a multiline wall.
- Use the Weld All option to close up a gap after removing the door or window symbol in a multiline.
- Use the Stretch command to move a door or window symbol in a multiline wall.
- When you open a gap in a multiline, AutoCAD does not cap the sides of the gap. You may need to add the end caps with the Line command.

MLine

Rel.13 Draws up to 16 parallel lines (short for Multiple LINE).

Command	Alias	Ctrl+	F-key	Alt+	Menu Bar	Tablet
mline	**ml**	...	...	**DM**	**Draw** **↳Multiline**	**M10**

Command: mline

Current settings: Justification = Top, Scale = 1.00, Style = STANDARD

Specify start point or [Justification/Scale/STyle]: *(Pick a point, or enter an option.)*

Specify next point: *(Pick a point.)*

Specify next point or [Undo]: *(Pick a point, or type* **U***.)*

Specify next point or [Close/Undo]: *(Pick a point, or else enter an option.)*

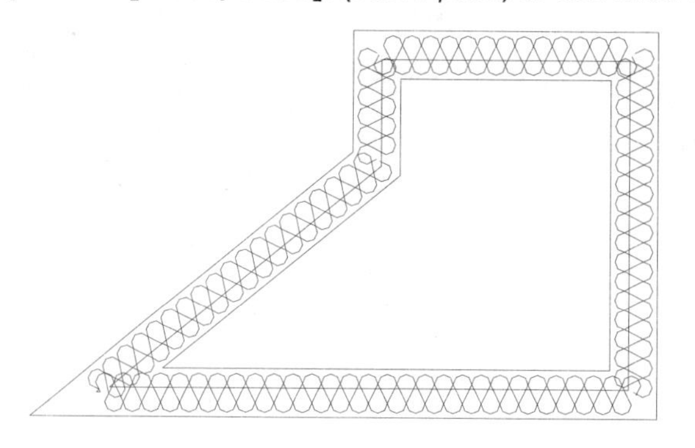

COMMAND LINE OPTIONS

Specify start point indicates the start of the multiline.

Specify next point indicates the next vertex.

Undo removes the most recently-added segment.

Close closes the multiline to its start point.

Justification options

Enter justification type [Top/Zero/Bottom] <top>: *(Enter an option.)*

Top draws the top line of the multiline at the cursor; remainder of multiline is "below" the cursor.

Zero draws the center (*zero offset point*) of the multiline at the cursor.

Bottom draws the bottom of the multiline at the cursor; remainder of the multiline is "above" the cursor.

Scale options

Enter mline scale <1.00>: *(Enter a value.)*

Enter mline scale specifies the scale of the width of the multiline; see Tips for examples.

STyle options

Enter mline style name or [?]: *(Enter style name, or type* **?***.)*

Enter mline style name specifies the name of the multiline style.

? lists the names of the multiline styles defined in drawing.

RELATED COMMANDS

MlEdit edits multilines.

MlProp defines the properties of a multiline.

RELATED SYSTEM VARIABLES

CMlJust specifies the current multiline justification:

- **0** — Top (default).
- **1** — Middle.
- **2** — Bottom.

CMlScale specifies the current multiline scale factor (default = 1.0).

CMlStyle specifies the current multiline style name (default = "").

RELATED FILE

.mln is the multiline style definition file.

TIPS

- Examples of scale factors:

Scale	Meaning
1.0	Default scale factor.
2.0	Draws multiline twice as wide.
0.5	Draws multiline half as wide.
-1.0	Flips multiline.
0	Collapses multiline to a single line.

- Multiline styles are stored in *.mln* files in DXF-like format.

MlStyle

Rel.13 Defines the characteristics of multilines (short for MultiLine STYLE).

Command	Alias	Ctrl+	F-key	Alt+	Menu Bar	Tablet
mlstyle	...	...	...	**OM**	**Format** ↳**Multiline Style**	**V5**

Command: mlstyle

Displays dialog box:

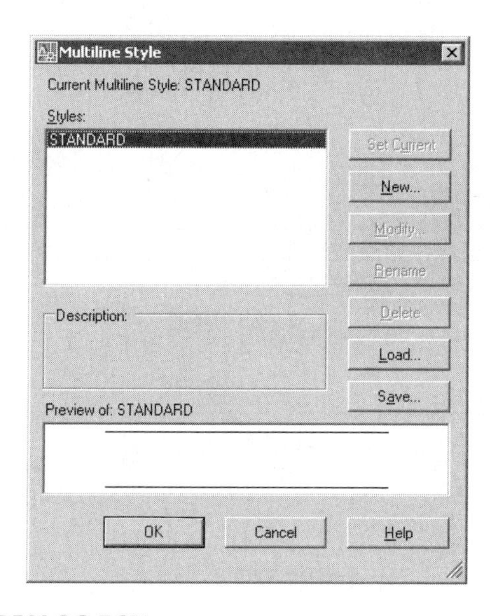

MUTLILINE STYLE DIALOG BOX

Set Current sets the selected multiline style name as the working style.

New creates new multiline styles; displays the Create New Multiline Style dialog box.

Modify changes the properties of existing multiline styles.

Rename changes the name of the selected multiline style.

Delete removes the multiline style from the **Styles** list.

Load loads styles from *.mln* files; displays the Load Multiline Styles dialog box.

Save saves a multiline style or renames a style; displays Save Multiline Style dialog box.

CREATE NEW MULTILINE STYLE DIALOG BOX

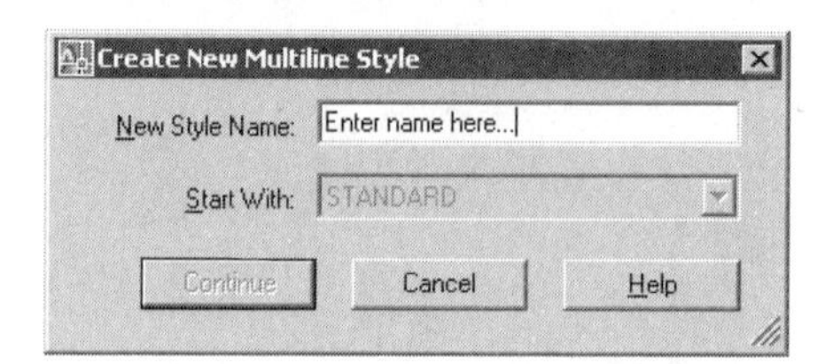

New Style Name specifies the name of the new multiline style. *Warning!* The name must contain no spaces.

Start With copies the multiline style from an existing style.

Continue displays the New Multiline Style dialog box.

NEW/MODIFY MULTILINE STYLE DIALOG BOX

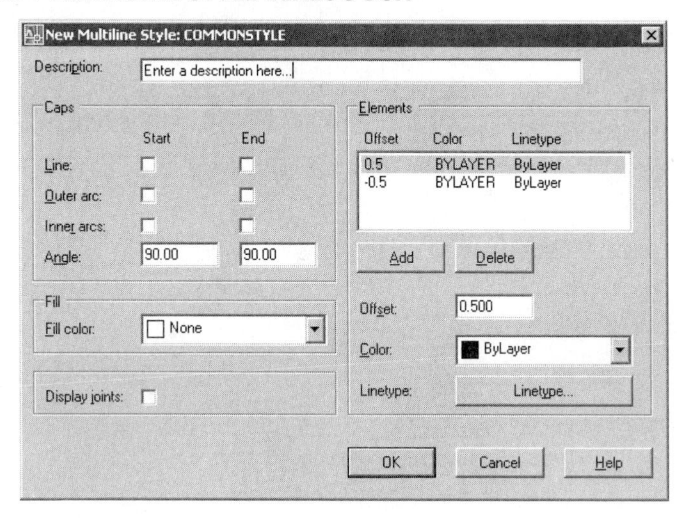

Caps options

☐ **Line** draws a straight line start and/or end cap.

☐ **Outer Arc** draws an arc to cap the outermost pair of lines.

☐ **Inner Arcs** draws an arc to cap all inner pairs of lines.

Angle specifies the angle for straight line caps.

Fill options

Fill Color lists common colors, and displays the Select Color dialog box.

☐ **Display Joints** toggles the display of joints (miters) at vertices; affects all multiline segments.

Elements options

Add adds an element (line).

Delete deletes an element.

Offset specifies the distance from origin to element; the origin is often the center line.

Color specifies the element color; displays Select Color dialog box.

Linetype specifies the element linetype; displays Select Linetype dialog box.

LOAD MULTILINE STYLES DIALOG BOX

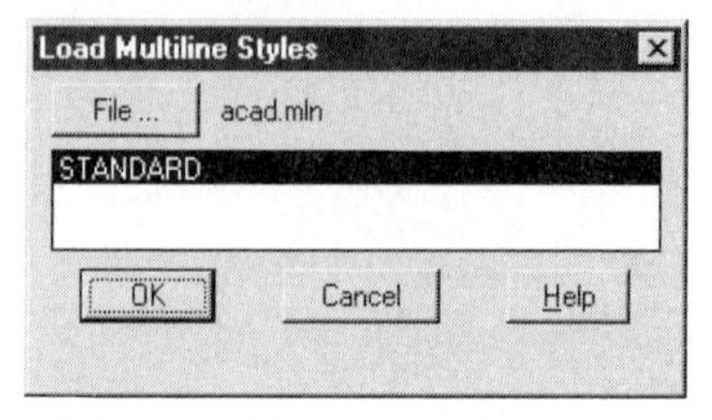

File selects an *.mln* multiline definition file.

RELATED COMMANDS

MlEdit edits multilines.

MLine draws up to 16 parallel lines.

RELATED SYSTEM VARIABLES

CMlJust specifies the current multiline justification:

0 — Top (default).

1 — Middle.

2 — Bottom.

CMlScale specifies the current multiline scale factor (default = 1.0).

CMlStyle specifies the current multiline style name (default = " ").

RELATED FILE

acad.mln is a multiline style definition file.

TIPS

- Use the **MlEdit** command to create (or close up) gaps for door and window symbols in multiline walls.
- The multiline scale factor has the following effect on the look of a multiline:

Scale	Meaning
1.0	Specifies the default scale factor (normal size).
0.5	Draws multiline half as wide.
2.0	Draws multiline twice as wide, not twice as long.
-1.0	Flips multiline about its origin.
0.0	Collapses multiline to a single line.

- The *.mln* file describes multiline styles in a DXF-like format.
- You cannot change the properties of the Standard style.
- Once multilines are placed in drawings, you cannot change their properties or their style. Once drawn, that's it! The workaround is to erase, change styles, and then reapply the **MLine** command.

Model

2000 Switches to Model tab.

Command	Alias	Ctrl+	F-key	Alt+	Menu Bar	Tablet
model	...	...	...	...	...	...

Command: model

Switches to the model tab.

COMMAND LINE OPTIONS

None.

RELATED COMMANDS

Layout creates layouts.

MSpace switches to model space.

RELATED SYSTEM VARIABLE

Tilemode switches between Model and Layout tabs.

TIPS

- This command automatically sets TileMode to 1.
- As an alternative to this command, you can select the Model tab:

  ...or click the button or icon (*new to AutoCAD 2007*) on the status bar :

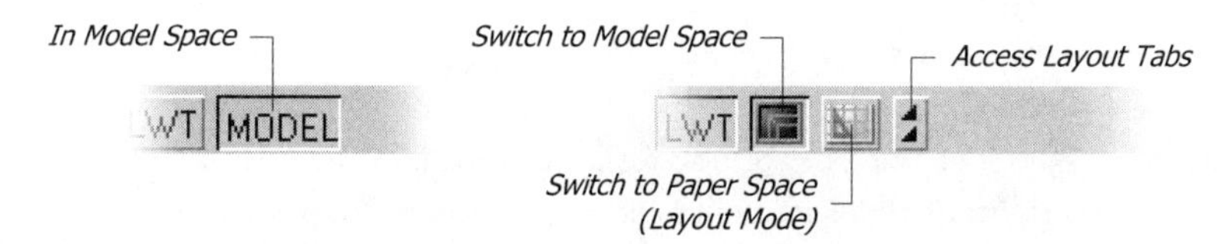

- The Model tab replaces the TILE button on the status bar of AutoCAD Release 13 and 14.
- *Historical Note:* The system variable is named "tile"mode, because model space can only display *tiled* viewports. (Paper space, or layout mode, can display overlapping viewports.) Turning off tiled-viewport mode meant AutoCAD was switching to paper space, where viewports no longer had to be tiled.

  Going back further, it was a graphic board manufacturer, Control Systems, that first figured out how to make AutoCAD display four tiled viewports at once. Autodesk added the feature to AutoCAD Release 10.

  All of which leads to a question I cannot answer: Why can't viewports be tiled in model space? Technical editor Bill Fane attempts an answer: "Because that's the way we've always done it!" Seriously, though...

  The idea is that the model resides in model space, and then is looked at through viewports in paper space. Because multi-view drawings can be created, scaled, and plotted more easily in paper space, there is no need for floating viewports in model space.

Move

V. 1.0 Moves one or more objects.

Command	Alias	Ctrl+	F-key	Alt+	Menu Bar	Tablet
move	m	...	...	MV	Modify ↳Move	V19

Command: move

Select objects: *(Select one or more objects.)*

Select objects: *(Press Enter to end object selection.)*

Specify base point or [Displacement] <Displacement>: *(Pick a point.)*

Specify second point of displacement or <use first point as displacement>:
(Pick a point, or press Enter.)

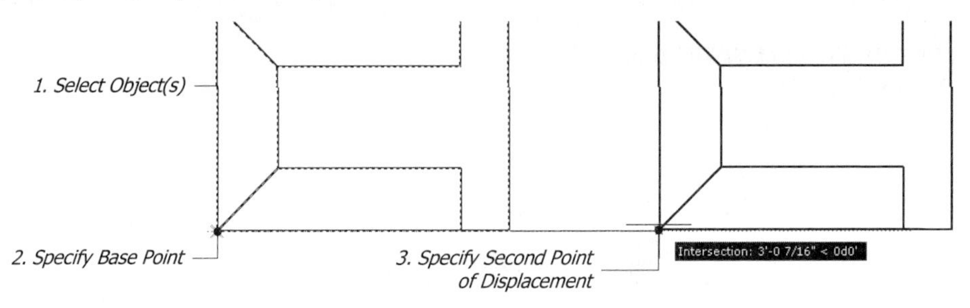

COMMAND LINE OPTIONS

Select objects selects the objects to copy.

Specify base point indicates the starting point for the move.

Displacement specifies relative x,y,z displacement when you press Enter at the next prompt.

Specify second point of displacement indicates the distance to move; you can use absolute (x,y), relative (@x,y), and polar (x<angle) coordinates.

RELATED COMMANDS

Copy copies the selected objects.

MlEdit moves the vertices of a multiline.

PEdit moves the vertices of a polyline.

TIP

- When you press Enter at the 'Specify Second Point of Displacement' prompt, AutoCAD uses the first point as the "displacement."

  For example, when you enter **4,3** at the 'Specify base point' prompt, and press Enter at the second prompt, AutoCAD moves the objects 4 units in the x direction and 3 units in the y.

MRedo

2004 Reverses the effect of the Undo command (short for Multiple REDO).

Command	Alias	Ctrl+	F-key	Alt+	Menu Bar	Tablet
mredo	...	Y	...	...	...	...

Command: mredo
Enter number of actions or [All/Last]: *(Enter an option.)*

COMMAND LINE OPTIONS

Enter number of actions redoes the specified number of steps.

All redoes all commands undone.

Last redoes the last command.

RELATED COMMANDS

Redo redoes a single undo.

U undoes a single command.

Undo undoes one or more commands.

TIPS

- The MRedo button on the toolbar lists the redoable actions:

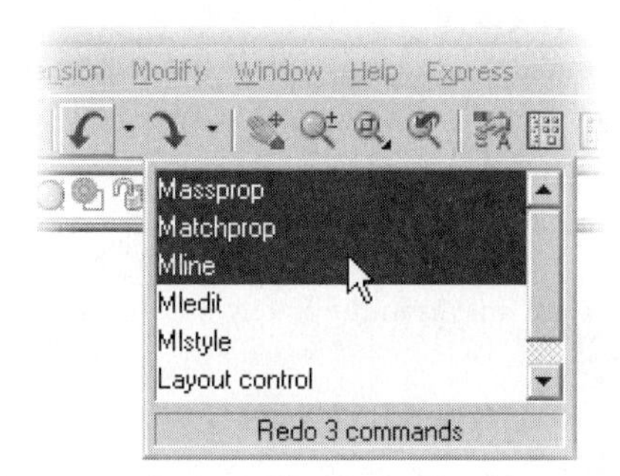

- This command allows you to undo several undoes, but does not allow you to skip over actions.

MSlide

Ver.2.0 Saves the current viewport as a *.sld* slide file on disk (short for Make SLIDE).

Command	Alias	Ctrl+	F-key	Alt+	Menu Bar	Tablet
mslide	...	...	...	...	...	...

Command: mslide

Displays Create Slide File dialog box. Specify a file name, and then click ***Save.***

COMMAND LINE OPTIONS

None.

RELATED COMMANDS

Save saves the current drawing as a DWG-format drawing file.

SaveImg saves the current view as a TIFF, Targa, or GIF-format raster file.

VSlide displays an SLD-format slide file in AutoCAD.

RELATED FILES

.sld files store slides created by this command.

.slb files store libraries of slides.

RELATED AUTODESK PROGRAM

SlideLib.exe compiles a group of slides into an SLB-format slide library file.

TIPS

- You view slides with the VSlide command.
- The slide was a predecessor to viewing raster and vector images inside AutoCAD.
- Slide files are used to create the images in the (rarely used) palette dialog boxes.

MSpace

Rel.11 Switches the drawing from paper space to model space (short for Model SPACE).

Command	Alias	Ctrl+	F-key	Alt+	Menu Bar	Tablet
mspace	ms	...	...	...	...	L4

Command: mspace

*In model tab, AutoCAD complains, '** Command not allowed in Model Tab **'.*

In model space of a layout tab, AutoCAD complains, 'Already in model space.'

In a layout tab (paper space), AutoCAD switches to model space in layout mode, and highlights a viewport:

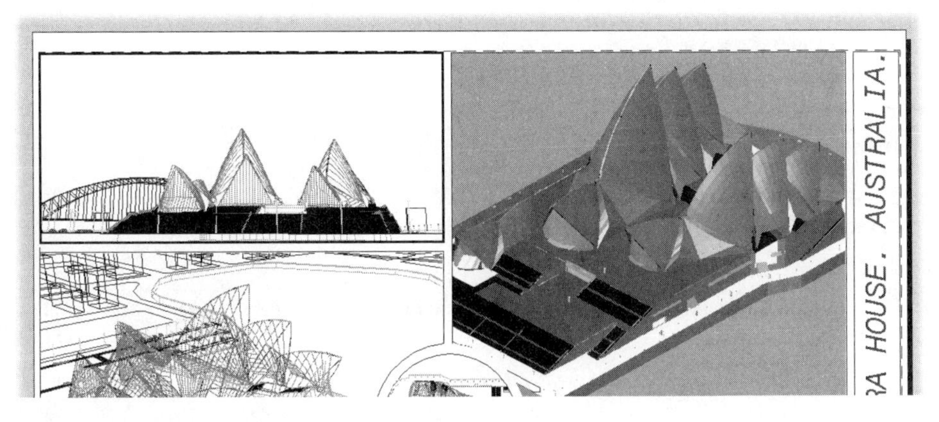

COMMAND LINE OPTIONS

None.

RELATED COMMANDS

PSpace switches from model space to paper space.

Model switches from layout mode to model mode.

Layout switches from model mode to layout mode.

RELATED SYSTEM VARIABLES

MaxActVp specifies the maximum number of viewports with visible objects; default=64.

TileMode specifies the current setting of tiled viewports.

TIPS

- To switch quickly between paper space and model space, click the Model and Paper icons on the status bar:

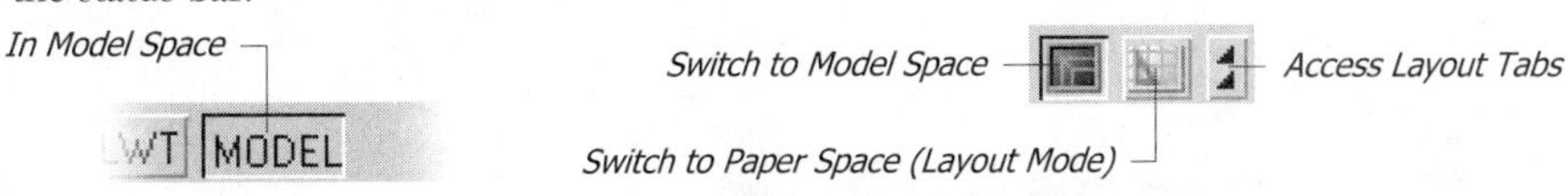

- AutoCAD clears the selection set when moving between paper space and model space.
- Note that there is a difference between model/layout *tabs* and the model/paper *button.*

MTEdit

Rel.13 Edits mtext objects (short for Multiline Text EDITor).

Command	Alias	Ctrl+	F-key	Alt+	Menu Bar	Tablet
mtedit	...	...	...	...	...	...

Command: mtedit

Select an MTEXT object: *(Pick an mtext object.)*

*Displays Text Formatting toolbar; see **MText** command.*

COMMAND LINE OPTION

Select an MTEXT object selects one paragraph text object for editing.

RELATED COMMANDS

DdEdit displays the text editor appropriate for the text object.

Properties changes the properties of an mtext object.

MtProp specifies the properties of an mtext object.

RELATED SYSTEM VARIABLE

MTextEd specifies the name of the external text editor to place and edit multiline text.

TIPS

- This command displays the same dialog box as the DdEdit command when an mtext object is selected.
- You can also invoke the mtext editor by double-clicking mtext.

A MText

Rel.13 Creates multiline, or paragraph, text objects that fit the width defined by the boundary box (short for Multline TEXT).

Command	Alias	Ctrl+	F-key	Alt+	Menu Bar	Tablet
mtext	**t**	...	...	**DXXM**	**Draw**	**J8**
	mt				**↳Text**	
					↳Multiline Text	
-mtext	**-t**					

Command: mtext

Current text style: "Standard" Text height: 0.20

Specify first corner: *(Pick a point, 1.)*

AutoCAD displays the mtext bounding box:

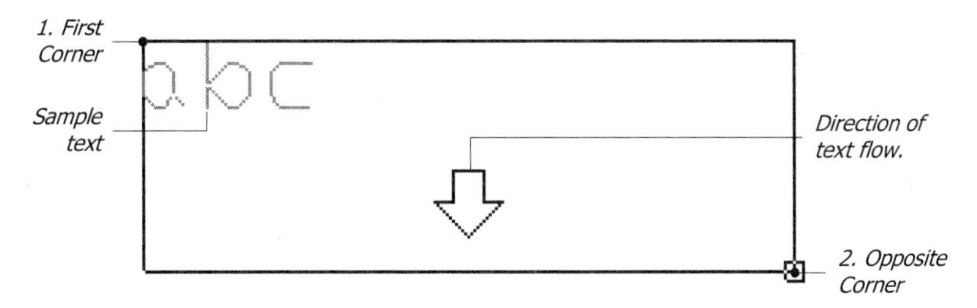

Specify opposite corner or [Height/Justify/Line spacing/Rotation/Style/ Width]: *(Pick another point, 2, or enter an option.)*

Displays toolbars and text editing area:

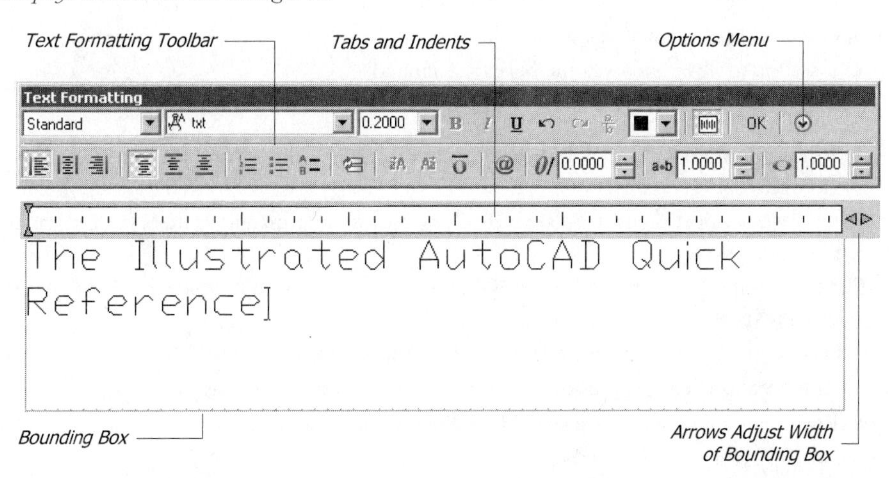

TEXT FORMATTING TOOLBAR

Toolbar redesigned with AutoCAD 2006:

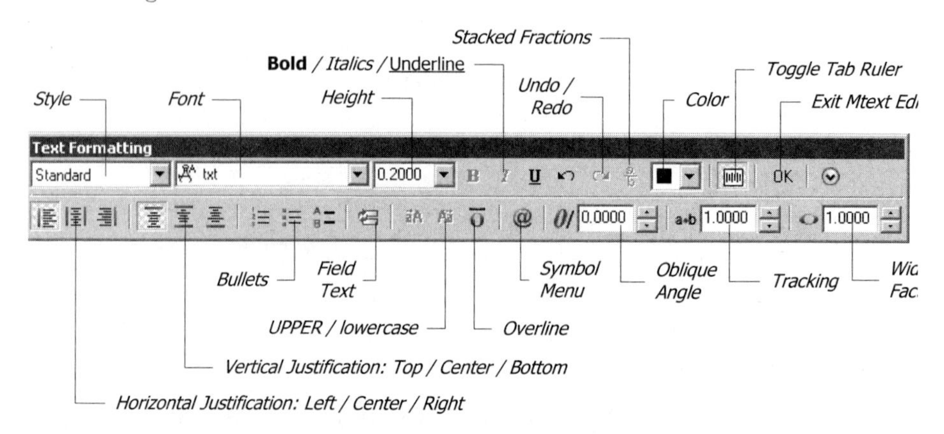

Style selects a predefined text style; see the Style command.

Font selects a TrueType (*.ttf*) or AutoCAD (*.shx*) font name (default=TXT).

Height specifies the height of the text in units (default = 0.2 units for Imperial drawings, 2.5 units for metric drawings).

Bold boldfaces the text, if permitted by the font.

Italic *italicizes* the text, if permitted by the font.

Underline underlines the text.

Undo undoes the last action.

Redo undoes the last undo.

Stack Fraction stacks a pair of characters separated by slash.

Color selects color for text; choose **Other Color** to display Select Color dialog box.

Ruler toggles the display of the tab and indent ruler.

OK closes the toolbar, and exits the MText command.

Menu displays the shortcut menu of options.

Horizontal Justification aligns the text horizontally against the left margin, centers it, or aligns it to the right margin.

Vertical Justification aligns the text vertically against the top margin, centers it, or aligns it to the bottom margin.

Bullets automatically applies bullet symbols, numbers (1., 2., 3., etc), or letters (A., B., C., etc) to selected text.

Field inserts field text from the Field dialog box; see the Field command.

Upper/ lowercase changes the selected text to all upper or all lowercase.

Overline draws a line over the selected text, the inverse of underline.

Symbol displays a menu of symbols.

Oblique Angle slants the text.

Tracking changes the spacing between characters, tighter or looser.

Width Factor makes characters wider and narrower.

Tab Ruler

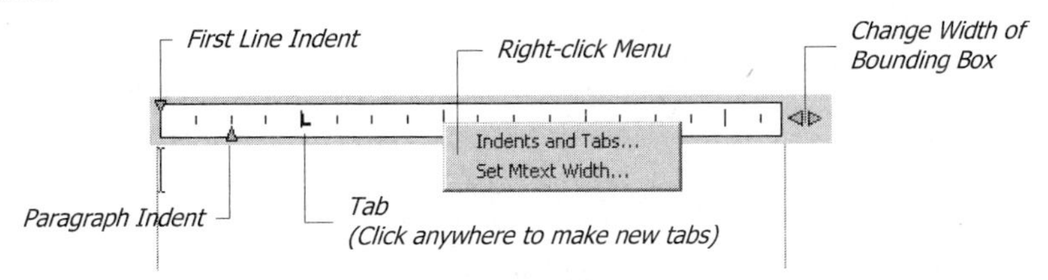

Indents and Tabs displays Indents and Tabs dialog box.

Set MText Width displays Set MText Width dialog box.

INDENTS AND TABS DIALOG BOX

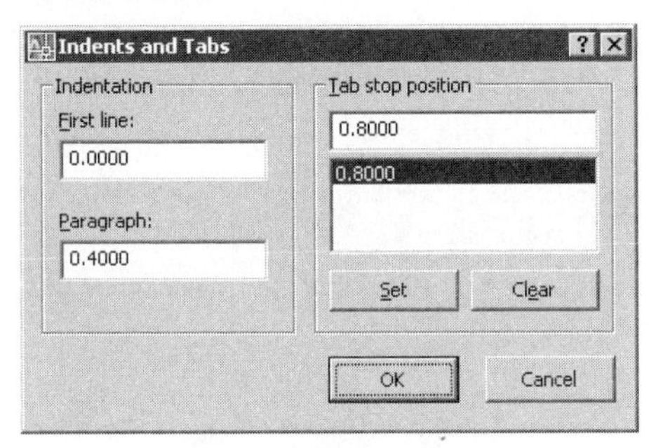

Indentation options

First line specifies the indent distance for the first line of a paragraph; enter a negative number to create a hanging indent.

Paragraph specifies the indent distance for the entire paragraph.

Tab Stop Position options

Set adds the tab position.

Clear removes the selected tab position.

SET MTEXT WIDTH DIALOG BOX

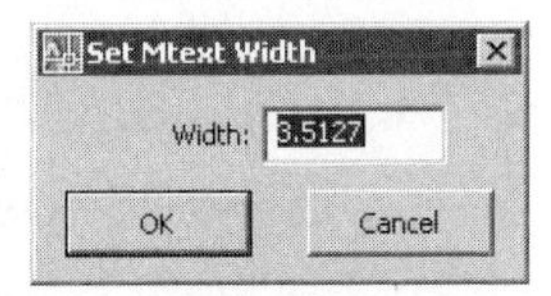

Width changes the width of the mtext bounding box; as an alternative, change the boundary box's size by dragging its right and bottom borders.

SHORTCUT MENUS

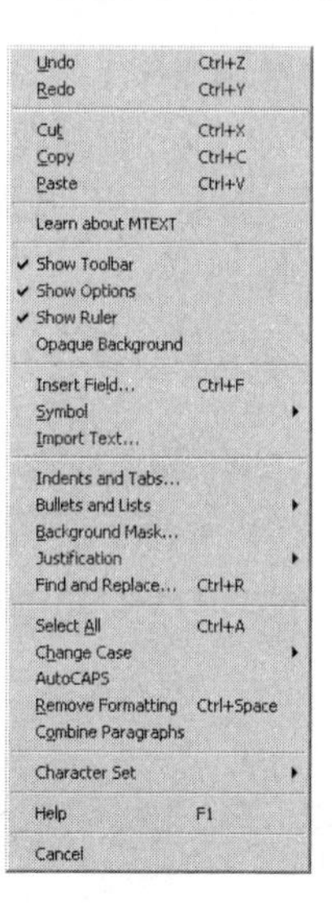

Right-click text to display menu, or enter shortcut keystrokes:

Undo (**Ctrl+Z**) undoes the last action.

Redo (**Ctrl+Y**) undoes the last undo.

Cut (**Ctrl+X**) removes selected text, and places it on the Clipboard.

Copy (**Ctrl+C**) copies selected text, and places it on the Clipboard.

Paste (**Ctrl+V**) inserts text from the Clipboard.

Learn about MTEXT opens the New Features Workshop.

✓ **Show Toolbar** toggles the display of the Text Formatting toolbar.

✓ **Show Options** toggles the display of the toolbar's lower half.

✓ **Show Ruler** toggles the display of the tab ruler.

Opaque Background toggles the opacity of the bounding box.

Insert Field (**Ctrl+F**) displays the Field dialog box; select a field, and then click **OK**. See the Field command.

Symbol inserts symbols in the text (click **Other** for Character Map dialog box):

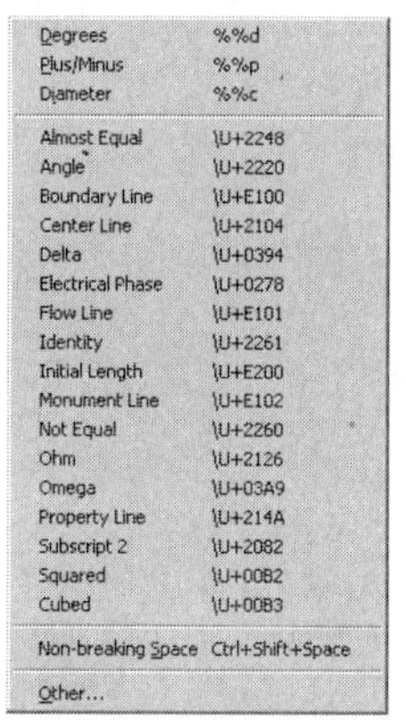

Import Text imports text from ASCII (*.txt*) and RTF format files; displays the Open dialog box. *Warning!* The maximum size of text file is limited to 32KB.

Indents and Tabs displays the Indents and Tabs dialog box.

Bullets and Lists displays a submenu of options:

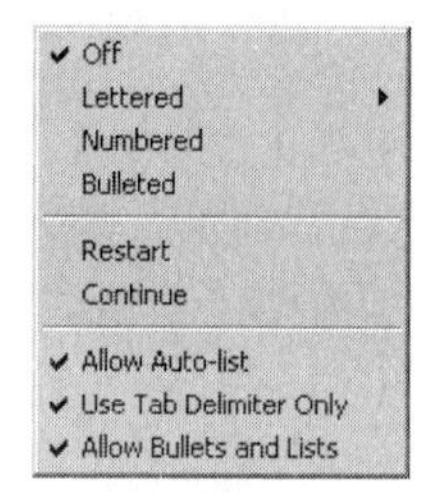

Background Mask displays the Background Mask dialog box.

Justification selects a justification mode:

TL — top left (default).
TC — top center.
TR — top right.
ML — middle left.
MC — middle center.
MR — middle right.
BL — bottom left.
BC — bottom center.
BR — bottom right.

Find and Replace (**Ctrl+R**) displays the Find and Replace dialog box.

Select All (**Ctrl+A**) selects all the text in the bounding box.

Change Case changes the case of the text:

- **UPPERCASE** (**Ctrl+Shift+U**) changes selected characters to uppercase.
- **lowercase** (**Ctrl+Shift+L**) changes selected characters to lowercase.

AutoCAPS places text as uppercase as it is typed.

Remove Formatting removes bold, italic, and underlined formatting from selected text.

Combine Paragraphs combines selected text into a single paragraph.

Stack stacks text on either side of the / , ^ , or # symbol.

Character Set selects alternate character sets, such as Western, Hebrew, and Thai.

Help displays online help, just like pressing **F1**.

Cancel exits the mtext editor, like pressing ESC.

BACKGROUND MASK DIALOG BOX

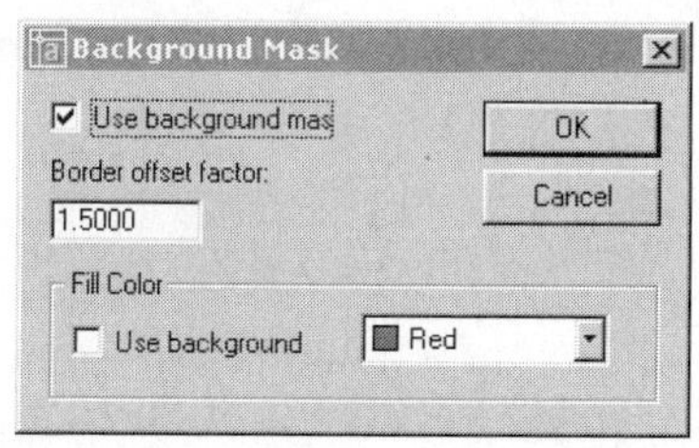

☑ **Use background mask** toggles the display of the background mask. Note that the background mask applies to the full width of the mtext block, rather than the width of text.

Border offset factor determines the distance that the "margin" extends beyond the text. The *factor* is based on the text height: 1.0 means there is no offset; 1.5 means the offset distance is 1.5 times the text height. Maximum value = 5.0; minimum value = 1.0.

Fill Color options

Use background:

☑ Uses the background color (usually white or black).

☐ Uses the specified color; for more colors, chose **Select Color**.

FIND AND REPLACE DIALOG BOX

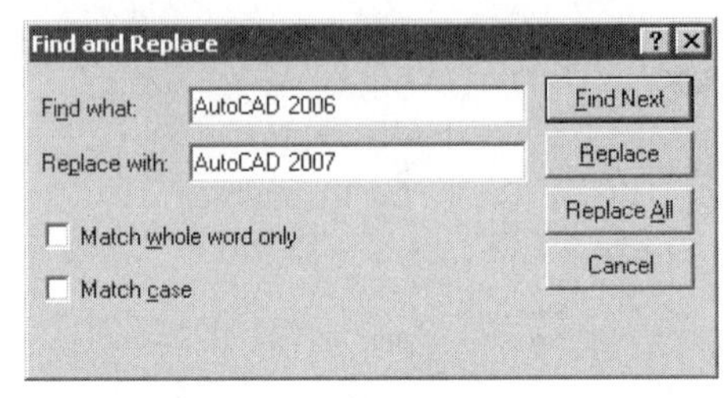

Find what specifies the text to search for. This dialog box searches only text within the bounding box; to search for text in the entire drawing, use the **Find** command.

Replace with specifies the replacement text; leave blank to search only.

Match whole word only

☑ Matches the entire word(s).

☐ Matches parts of the word(s).

Match case

☑ Matches the case of the words.

☐ Ignores the word case.

Find Next finds the next occurrence of the word(s).

Replace replaces the found occurrence .

Replace All replaces all occurrence .

Cancel dismisses the dialog box.

-MTEXT Command

Command: -mtext

Current text style: STANDARD. Text height: 0.2000

Specify first corner: *(Pick a point.)*

Specify opposite corner or [Height/Justify/Line spacing/Rotation/Style/ Width]: *(Pick another point.)*

MText: *(Enter text.)*

MText: *(Press Enter to end the command.)*

COMMAND LINE OPTIONS

Height specifies the height of UPPERCASE text (*default = 0.2 units*).

Justify specifies a justification mode.

Line spacing specifies the distance between lines of text. AutoCAD prompts, "Enter line spacing type [At least/Exactly] <At least>:" and then "Enter line spacing factor or distance <1x>:".

Rotation specifies the rotation angle of the boundary box.

Style selects the text style for multiline text (default = STANDARD).

Width sets the width of the boundary box; a width of 0 eliminates the boundary box.

RELATED COMMANDS

DdEdit edits mtext; alternatively, double-click the mtext to edit.

Properties changes all aspects of mtext.

MtProp changes properties of multiline text.

PasteSpec pastes formatted text from the Clipboard into the drawing.

Style creates a named text style from a font file.

RELATED SYSTEM VARIABLES

CenterMt determines how the bounding box is stretched.

MTextEd names the external text editor for placing and editing multiline text.

MTJigString specifies the sample text displayed when the bounding box is created; default = "abc".

TIPS

- Use the MTextEd system variable to define a different text editor. To return to the old interface, enter the following system variables:

  MTextEd=oldeditor
  MTExtFixed=1

- The Import Text option is limited to ASCII (unformatted) and RTF (rich text format) text files no more than 32KB in size.
- To import Word documents, copy the text to the Clipboard, and then press **Ctrl+V** in the mtext editor. Most, but not all, formatting is retained.
- To import formatted text, copy text from the word processor to the Clipboard, and then use AutoCAD's PasteSpec command.
- To link text in the drawing with a word processor, use the InsertObj command. When the word processor updates, the linked text is updated in the drawing.
- The mtext editor displays the diameter symbol as "%%c" and nonbreaking spaces as hollow rectangles, but these are displayed correctly in drawings.
- Stacked text can be created on either side of the following symbols:
  - **Carat** (^) stacks text as left-justified tolerance values.
  - **Forward slash** (/) stacks text as center-justified fractional-style values; the slash is converted to a horizontal bar.
  - **Pound sign** (#) stacks text with a tall diagonal bar.

  Use the stack tool a second time to unstack stacked text.

MtProp

Rel.13 . Changes the properties of multiline text (short for Multline Text PROPerties; an undocumented command).

Command	Alias	Ctrl+	F-key	Alt+	Menu Bar	Tablet
mtprop	...	...	...	...	...	...

Command: mtprop

Select an MText object: *(Pick an mtext object.)*

*Displays the Text Formatting toolbar; see the **MText** command.*

COMMAND LINE OPTIONS

*See the **MText** command.*

RELATED COMMANDS

DdEdit edits multiline text.

MText places multiline text.

Style creates a named text style from a font file.

TIP

- You can also invoke the mtext editor by double-clicking mtext.

Multiple

V. 2.5 Repeats commands automatically that do not repeat on their own.

Command	Alias	Ctrl+	F-key	Alt+	Menu Bar	Tablet
multiple	...	...	...	...	...	...

Command: multiple
Enter command name to repeat: *(Enter command name.)*

This command can also be used as a command modifier:
Command: multiple circle
3P/2P/TTR/<Center point>: *(Pick a point, or enter an option.)*
Diameter/<Radius>: *(Enter an option.)*
circle 3P3P/2P/TTR/<Center point>: *(Pick a point, or enter an option.)*
Diameter/<Radius>: *(Enter an option.)*
circle 3P3P/2P/TTR/<Center point>: *(Press ESC to end command.)*

COMMAND LINE OPTIONS

Enter command name to repeat specifies the name of the command to repeat.

ESC stops the command from automatically repeating itself.

COMMAND INPUT OPTIONS

SPACEBAR repeats the previous command.

CLICK repeats a command by clicking on any blank spot of the tablet menu.

RELATED COMMANDS

Redo undoes an undo.

U undoes the previous command; undoes one multiple command at a time.

RELATED COMMAND MODIFIERS

' *(apostrophe)* allows the use of some commands within another command.

. *(period)* forces the use of undefined commands.

- *(dash)* forces the display of prompts on the command line for some commands.

+ *(plus)* prompts for the tab number of tabbed dialog boxes.

_ *(underscore)* uses the English command in international versions of AutoCAD.

(*(open parenthesis)* executes AutoLISP functions on the command line.

$(*(dollar and parenthesis)* executes Diesel functions on the command line.

TIPS

- Use the **Multiple** command to repeat commands that do not repeat on their own. This command does not cause options to repeat.
- *Warning*! Multiple **U** will undo all edits in the drawing.

MView

Rel.11 Creates and manipulates overlapping viewports (short for Make VIEWports).

Command	Alias	Ctrl+	F-key	Alt+	Menu Bar	Tablet
mview	mv	...	...	...	...	M4

Command: mview

*In Model tab, AutoCAD complains, '** Command not allowed in Model Tab **'.*

In a layout tab, AutoCAD prompts:

Specify corner of viewport or
[ON/OFF/Fit/Shadeplot/Lock/Object/Polygonal/Restore/2/3/4]<Fit>: *(Pick a point, or enter an option.)*

Specify opposite corner: *(Pick a point.)*

Regenerating drawing.

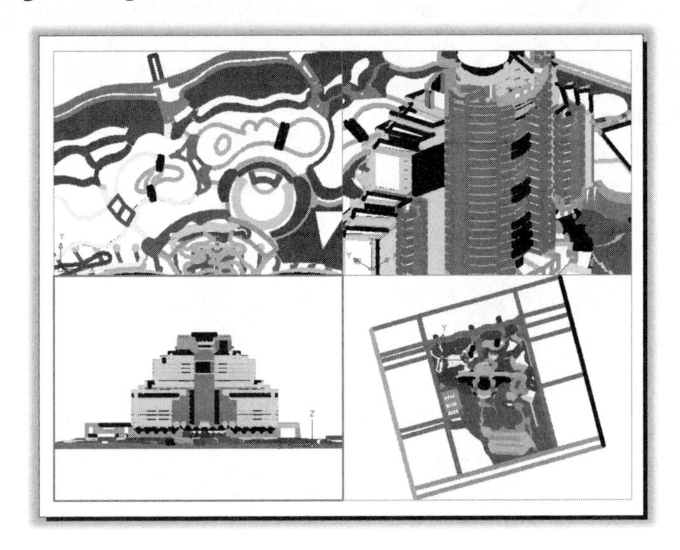

COMMAND LINE OPTIONS

Specify corner of viewport indicates the first point of a single viewport (default).

Fit creates a single viewport that fits the screen.

Shadeplot creates hidden-line or shaded views during plotting and printing.

Hideplot performs hidden-line removal during plotting and printing (*undocumented option*).

Lock locks the selected viewport's scale relative to model space.

Object converts a circle, closed polyline, ellipse, spline, or region into a viewport.

OFF turns off a viewport.

ON turns on a viewport.

Polygonal creates a multisided viewport of straight lines and arcs.

Restore restores a saved viewport configuration.

2 (Two Viewports) options

Enter viewport arrangement [Horizontal/Vertical] <Vertical>: *(Enter an option.)*

Specify first corner or [Fit] <Fit>: *(Pick a point, or enter an option.)*

Horizontal stacks two viewports.

Vertical places two viewports side-by-side (default).

3 (Three Viewports) options

[Horizontal/Vertical/Above/Below/Left/Right]<Right>: *(Enter an option.)*

Specify first corner or [Fit] <Fit>: *(Pick a point, or enter an option.)*

Horizontal stacks the three viewports.

Vertical places three side-by-side viewports.

Above places two viewports above the third.

Below places two viewports below the third.

Left places two viewports to the left of the third.

Right places two viewports to the right of the third *(default)*.

4 (Four Viewports) options

Specify first corner or [Fit] <Fit>: *(Pick a point, or enter an option.)*

Fit creates four identical viewports that fit the viewport.

First Point indicates the area of the four viewports (default).

Shadeplot options

Shade plot? [As displayed/Wireframe/Hidden/Visual style/Rendered] <As displayed>: *(Enter an option.)*

As displayed plots the drawing as displayed.

Wireframe plots the drawing as a wireframe.

Hidden plots the drawing with hidden lines removed.

Visual style plots the drawing with specified visual style *(new to AutoCAD 2007).*

Rendered plots the drawing rendered.

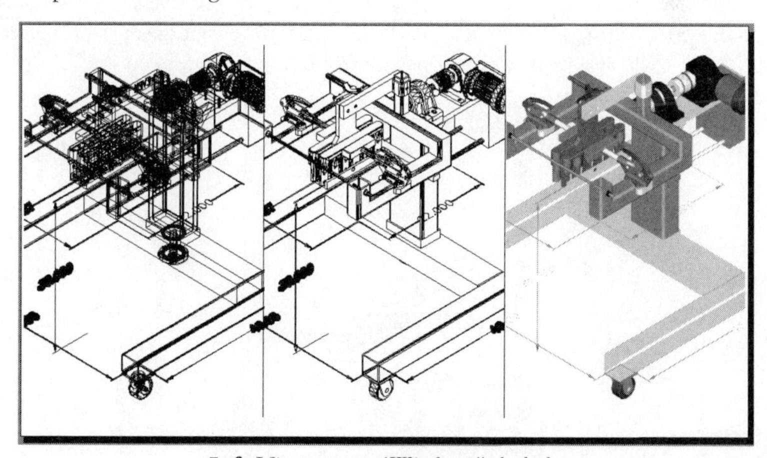

Left: *Viewport set to "Wireframe" shadeplot.*
Center: *Viewport set to "Hidden" shadeplot.*
Right: *Viewport set to "Rendered" shadeplot.*

Hideplot Options *(undocumented)*
Hidden line removal for plotting [ON/OFF]: on
Select objects: *(Select one or more viewports.)*

Hidden line removal for plotting toggles the removal of hidden lines for plotted output.

RELATED COMMANDS

Layout creates new layouts.

MSpace switches to model space.

PSpace switches to paper space before creating viewports.

RedrawAll redraws all viewports.

RegenAll regenerates all viewports.

VpLayer controls the visibility of layers in each viewport.

VPorts creates tiled viewports in model space.

Zoom zooms a viewport relative to paper space via the XP option.

RELATED SYSTEM VARIABLES

CvPort specifies the number of the current viewport.

MaxActVp controls the maximum number of active viewports:

- **2** — Minimum.
- **64** — Default and maximum.

PsLtScale specifies linetype scaling in paper space.

TileMode controls the availability of overlapping viewports.

TIPS

- Drawings can have up to 32,767 viewports, but only a maximum of 64 that show content.
- Although the system variable MaxActVp limits the number of simultaneously-visible viewports, the Plot command plots all viewports.
- Snap, Grid, Hide, Shade, Ucs, and so on, can be set separately in each viewport.
- Some of this command's options are also available from the Properties palette: select a viewport, right-click, and then select Properties from the shortcut menu.

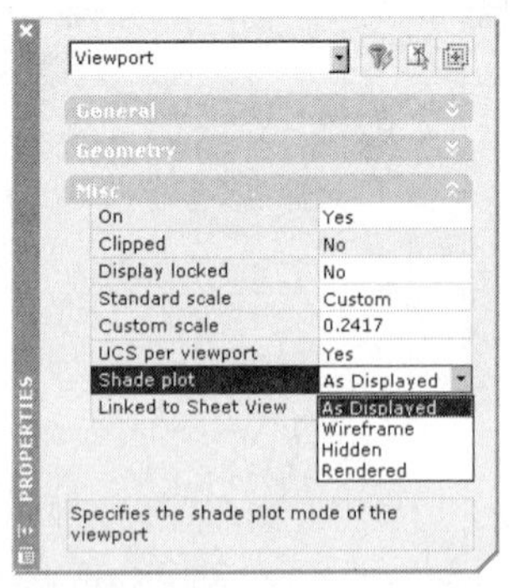

- Press **Ctrl+R** to switch between viewports.
- Only rectangular viewports can be locked.

- The preset viewports created by the **MView** command have these shapes:

Fit Viewport

Fit	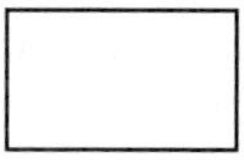 Creates a single viewport.

2 Viewports

Horizontal	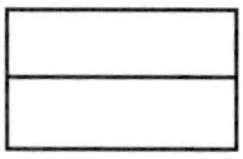 Creates one viewport over another viewport.
Vertical	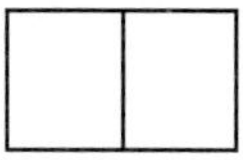 Creates one viewport beside another (default).

3 Viewports

Horizontal	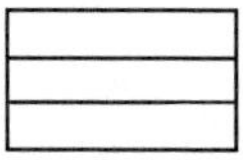 Creates three viewports over each other.
Vertical	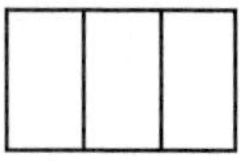 Creates three viewports side-by-side.
Above	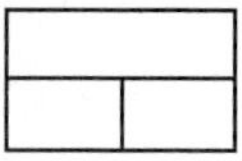 Creates one viewport over two viewports.
Below	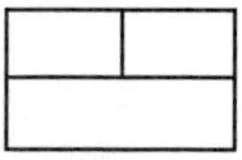 Creates one viewport below two viewports.
Left	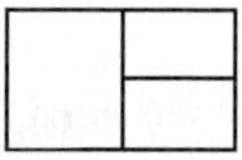 Creates one viewport left of two viewports.
Right	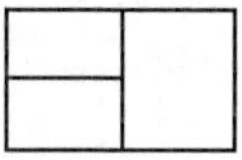 Creates one viewport right of two viewports (default).

Four Viewports

4	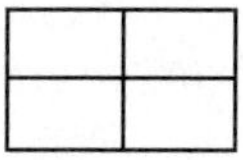 Splits the current viewport into four viewports.

'MvSetup

Rel. 11 Sets up a drawing quickly, complete with a predrawn border. Optionally sets up multiple viewports, sets the scale, and aligns views in each viewport (short for Model View SETUP).

Command	Alias	Ctrl+	F-key	Alt+	Menu Bar	Tablet
mvsetup	...	...	...	...	...	...

Command: mvsetup

When in model space:

Enable paper space? [No/Yes] <Y>: *(Type* **Y** *or* **N**.*)*

Command prompts in model tab (not paper space):

Enter units type [Scientific/Decimal/Engineering/Architectural/Metric]: *(Enter an option.)*

Enter the scale factor: *(Specify a distance.)*

Enter the paper width: *(Specify a distance.)*

Enter the paper height: *(Specify a distance.)*

Command prompts in layout mode (paper space):

Enter an option [Align/Create/Scale viewports/Options/Title block/Undo]: *(Enter an option.)*

COMMAND LINE OPTIONS

Align options

Pans the view to align a base point with another viewport.

Enter an option [Angled/Horizontal/Vertical alignment/Rotate view/Undo]: *(Enter an option.)*

Angled specifies the distance and angle from a base point to a second point.

Horizontal aligns views horizontally with a base point in another viewport.

Vertical alignment aligns views vertically with a base point in another viewport.

Rotate view rotates the view about a base point.

Undo undoes the last action.

Create options

Enter option [Delete objects/Create viewports/Undo] <Create>: *(Enter an option.)*

Delete objects erases existing viewports.

Create viewports creates viewports in these configurations:

Layout	Meaning
0	No layout.
1	Single viewport.
2	Standard engineering layout.
3	Array viewports along x and y axes.

Undo undoes the last action.

Scale Viewports options

Select the viewports to scale...

Select objects: *(Pick a viewport.)*

Select objects: *(Press* **ENTER** *to end object selection.)*

Set the ratio of paper space units to model space units...

Enter the number of paper space units <1.0>: *(Enter a value.)*

Enter the number of model space units <1.0>: *(Enter a value.)*

Select objects selects one or more viewports.

Enter the number of paper space units scales the objects in the viewport with respect to drawing objects.

Enter the number of model space units scales the objects in the viewport with respect to drawing objects.

Options options

Enter an option [Layer/LImits/Units/Xref] <exit>: *(Enter an option.)*

Layer specifies the layer name for the title block.

Limits specifies whether to reset limits after title block insertion.

Units specifies inch or millimeter paper units.

Xref specifies whether title is inserted as a block or as an external reference.

Title Block options

Enter title block option [Delete objects/Origin/Undo/Insert] <Insert>: *(Enter an option.)*

Delete objects erases an existing title block from the drawing.

Origin relocates the origin.

Undo undoes the last action.

Insert displays the available title blocks.

RELATED SYSTEM VARIABLE

TileMode specifies the current setting of TileMode.

RELATED FILES

mvsetup.dfs is the default settings file for this command.

acadiso.dwg is a template drawing with ISO (international standards) defaults.

All *.dwt* template drawings can also be used as templates.

RELATED COMMANDS

LayoutWizard sets up the viewports via a “wizard.”

TIPS

- When option **2 (Std. Engineering)** is selected at the **Create** option, the following views are created (counterclockwise from upper left):

  Top view.

  Isometric view.

  Front view.

  Right view.

- To create the title block, MvSetup searches the path specified by the AcadPrefix variable. If the appropriate drawing cannot be found, MvSetup creates the default border.
- MvSetup makes use the following predefined title blocks:

0:	None
1:	ISO A4 Size(mm)
2:	ISO A3 Size(mm)
3:	ISO A2 Size(mm)
4:	ISO A1 Size(mm)
5:	ISO A0 Size(mm)
6:	ANSI-V Size(in)
7:	ANSI-A Size(in)
8:	ANSI-B Size(in)
9:	ANSI-C Size(in)
10:	ANSI-D Size(in)
11:	ANSI-E Size(in)
12:	Arch/Engineering (24 x 36in)
13:	Generic D size Sheet (24 x 36in)

- The metric A0 size is similar to the Imperial E-size, while the metric A4 size is similar to A-size.
- This command provides the following preset scales (scale factor shown in parentheses):

Architectural Scales	Scientific Scales	Decimal Scales	Engineering Scales	Metric Scales
(480) 1/40"=1'	**(4.0) 4 TIMES**	**(4.0) 4 TIMES**	**(120) 1"=10'**	**(5000) 1:5000**
(240) 1/20"=1'	**(2.0) 2 TIMES**	**(2.0) 2 TIMES**	**(240) 1"=20'**	**(2000) 1:2000**
(192) 1/16"=1'	**(1.0) FULL**	**(1.0) FULL**	**(360) 1"=30'**	**(1000) 1:1000**
(96) 1/8"=1'	**(0.5) HALF**	**(0.5) HALF**	**(480) 1"=40'**	**(500) 1:500**
(48) 1/4"=1'	**(0.25) QUARTER**	**(0.25) QUARTER**	**(600) 1"=50'**	**(200) 1:200**
(24) 1/2"=1'			**(720) 1"=60'**	**(100) 1:100**
(16) 3/4"=1'			**(960) 1"=80'**	**(75) 1:75**
(12) 1"=1'			**(1200) 1"=100'**	**(50) 1:50**
(4) 3"=1'				**(20) 1:20**
(2) 6"=1'				**(10) 1:10**
(1) FULL				**(5) 1:5**
				(1) FULL

- You can add your own title block with the **Add** option. Before doing so, create the title block as an AutoCAD drawing.

Quick Start Tutorial

Using MvSetup

MvSetup has many options, but does not present them in a logical fashion. To set up a drawing with MvSetup, follow these basic steps:

Start the **MvSetup** command:

Command: mvsetup

Model Space options

Enable paper space? [Yes/No]: n

1. Select the units and scale factor:

   Enter units type [Scientific/Decimal/Engineering/Architectural/Metric]: *(Enter an option.)*
   Enter the scale factor: *(Enter a number from the list displayed in the Text window.)*

2. Specify the size of the paper:

   Enter the paper width: *(Enter the width.)*
   Enter the paper height: *(Enter the height.)*

Notice that AutoCAD draws a rectangle the shape of the paper, scaled to the required size.

Paper Space options

Enable paper space? [Yes/No]: y

1. Select a title block with the **Title block** option, and then insert it with the **Insert** option:

   Enter an option [Align/... /Title block/Undo]: *(Type* **T***.)*
   Enter title block option [Delete objects/... /Insert] <Insert>: *(Type* **I***.)*

2. Create the viewports with the **Create** option:

   Enter an option [Align/Create/Scale viewports/ ... /Undo]: c
   Enter option [Delete objects/Create viewports/Undo] <Create>: c
   Enter layout number to load or [Redisplay]: *(Enter a number from the list.)*

3. Place the viewport(s):

   Specify first corner of bounding area for viewport(s): *(Pick a point.)*
   Specify opposite corner: *(Pick another point.)*
   Specify distance between viewports in X direction <0.0>: *(Enter a distance.)*
   Specify distance between viewports in Y direction <0.25>:

4. Align the model in the viewports with the **Align** option:

   Enter an option [Align/Create/ ... /Undo]: a
   Enter an option [Angled/Horizontal/Vertical alignment/Rotate view/ Undo]: *(Enter an option.)*

5. Scale the viewports. Make the object the same size in all four viewports with the **Scale viewports** option. When you are prompted to 'Select objects', select the four *viewports*, not the objects in the viewports.

6. Save your work when done!

NetLoad

2005 Loads *.dll* files written with Microsoft's .Net programming interface.

Command	Alias	Ctrl+	F-key	Alt+	Menu Bar	Tablet
netload	...	...	...	...	...	...

Command: netload

Displays the Choose .NET Assembly dialog box.

Select a .dll file, and then click ***Open.***

COMMAND LINE OPTIONS

None.

TIP

- Some parts of AutoCAD written in .Net include the new Layer dialog box and the migration utility.

New

Rel.12 Starts new drawings from template drawings, from scratch, or through step-by-step drawing setup "wizards."

Command	Alias	Ctrl+	F-key	Alt+	Menu Bar	Tablet
new	**...**	**N**	**...**	**FN**	**File** **↳New**	**T24**

Command: new

*AutoCAD displays one of three interfaces, depending on the settings of the **FileDia** and **Startup** variables.*

FileDia	Startup	New
1	1	Displays Startup wizard.
1	0	Displays Select Template dialog box.
0	1 or 0	Prompts for *.dwt* file at command line.

SELECT TEMPLATE DIALOG BOX

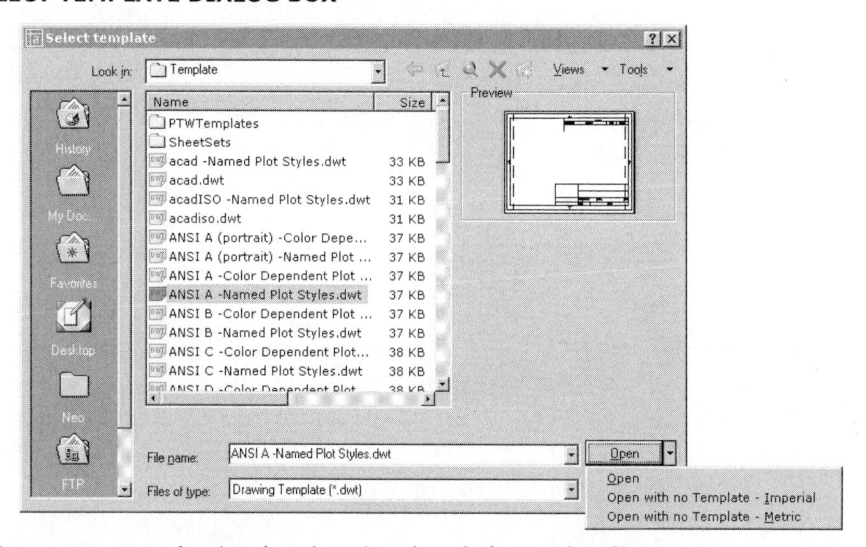

Open creates new drawings based on the selected *.dwt* template file.

Open with no template - Imperial creates new drawings with default Imperial values.

Open with no template - Metric creates new drawings with default metric values.

STARTUP DIALOG BOX

The Startup wizard is displayed when ***Startup*** *is set to 1 and when AutoCAD first starts.*

The Create New Drawing wizard is similar, and is displayed when subsequent new drawings are opened.

Open a Drawing page

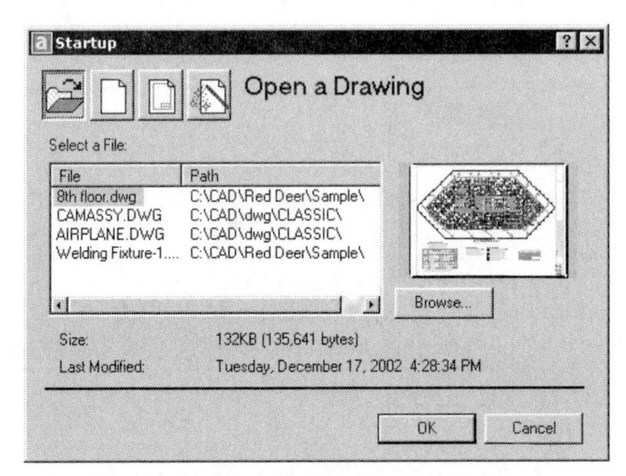

(This view is not found in the Create New Drawing wizard.)

Select a File selects one of the four drawings listed.

Browse displays the Select File dialog box; see the **Open** command.

Start from Scratch page

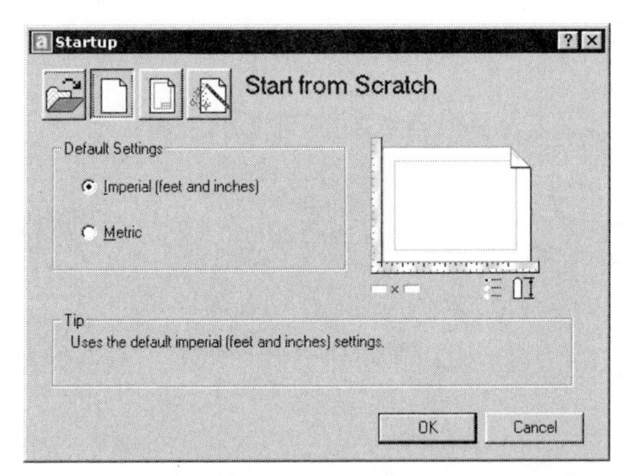

⊙**English** creates a new drawing with AutoCAD's default values and **InsUnits**=1 (inches).

○**Metric** creates a new drawing with AutoCAD's default values and **InsUnits**=4 (millimeters).

Use a Template page

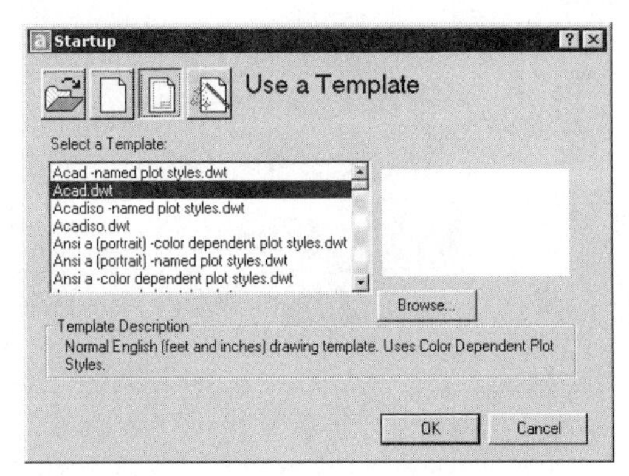

Select a Template creates a new drawing based on the selected *.dwt* template file.

Browse displays the Select a Template File dialog box.

Use a Wizard page

Select a Wizard

- **Advanced Setup** sets up a new drawing in several steps.
- **Quick Setup** sets up a new drawing in two steps.

QUICK SETUP WIZARD

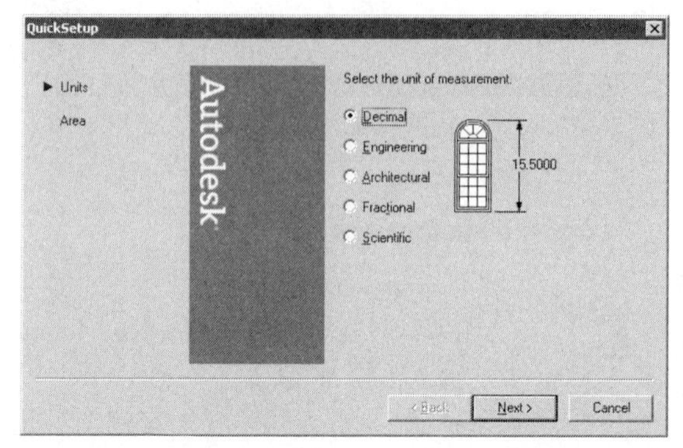

⊙**Decimal** displays units in decimal (or "metric") notation (default): 123.5000.

○**Engineering** displays units in feet and decimal inches: 10'-3.5000".

○**Architectural** displays units in feet, inches, and fractional inches: 10' 3-1/2".

○**Fractional** displays units in inches and fractions: 123 1/2.

○**Scientific** displays units in scientific notation: 1.235E+02.

Cancel cancels the wizard, and returns to the previous drawing.

Back moves back one step.

Next moves forward one step.

Area page

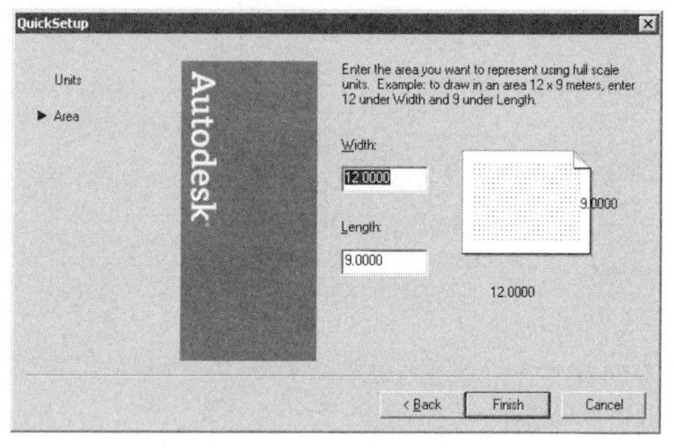

Width specifies the width of the drawing in real-world (not scaled) units; default = 12 units.

Length specifies the length or depth of the drawing in real-world units; default = 9 units.

ADVANCED SETUP WIZARD

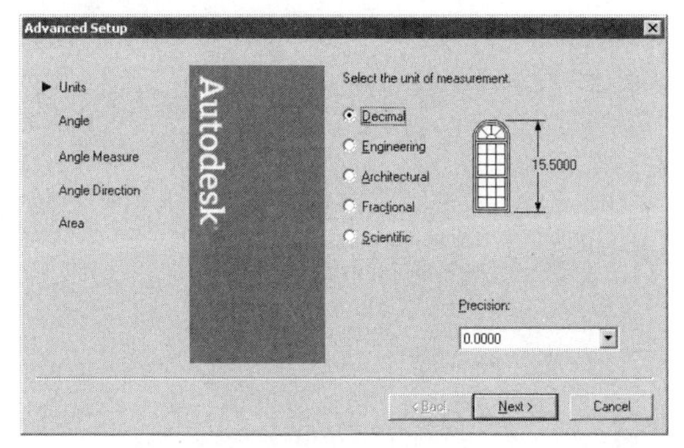

⊙**Decimal** displays units in decimal (or "metric") notation (default): 123.5000.

O**Engineering** displays units in feet and decimal inches: 10'-3.5000".

O**Architectural** displays units in feet, inches, and fractional inches: 10' 3-1/2".

O**Fractional** displays units in inches and fractions: 123 1/2.

O**Scientific** displays units in scientific notation: 1.235E+02.

O**Precision** selects the precision of display up to 8 decimal places or 1/256.

Angle page

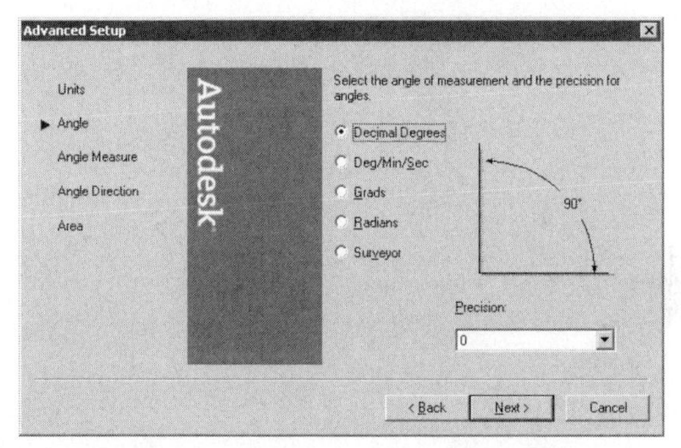

⊙**Decimal Degrees** displays decimal degrees (default): 22.5000.

O**Deg/Min/Sec** displays degrees, minutes, and seconds: 22 30.

O**Grads** displays grads: 25g.

O**Radians** displays radians: 25r.

O**Surveyor** displays surveyor units: N 25d0'0" E.

Precision selects a precision up to 8 decimal places.

Angle Measure page

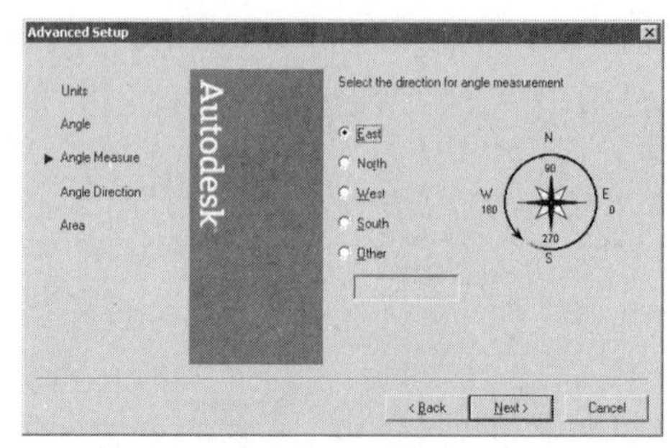

⊙**East** specifies that zero degrees points East (default).

○**North** specifies that zero degrees points North.

○**West** specifies that zero degrees points West.

○**South** specifies that zero degrees points South.

○**Other** specifies any of the 360 degrees as zero degrees.

Angle Direction page

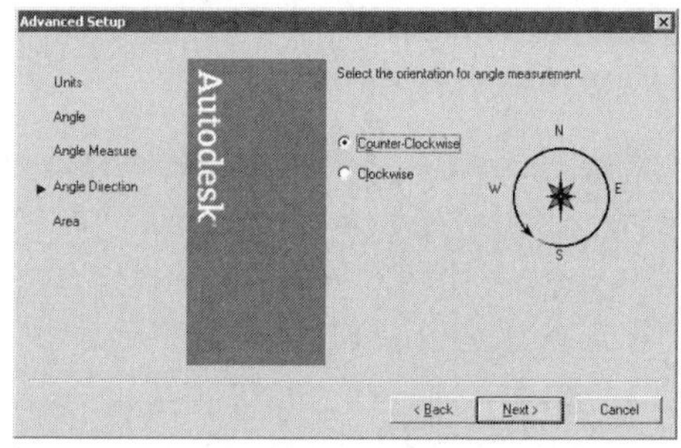

Counter-Clockwise measures positive angles counterclockwise from 0 degrees (default).

Clockwise measures positive angles clockwise from 0 degrees.

Area page

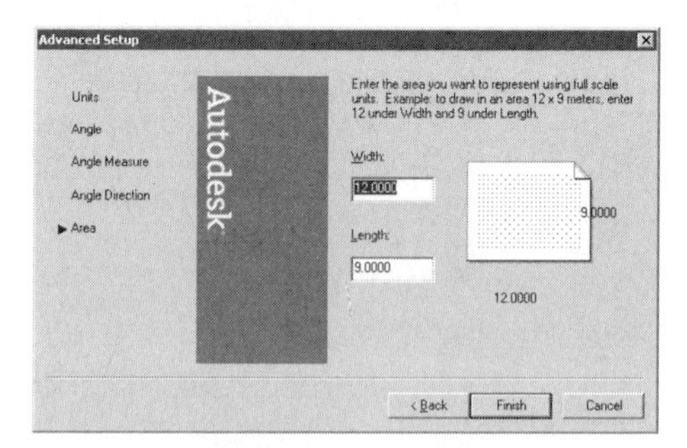

Width specifies the width of the drawing in real-world (not scaled) units; default = 12 units.

Length specifies the length or depth of the drawing in real-world units; default = 9 units.

COMMAND LINE SWITCHES

Switches used by Target field on Shortcut tab in the AutoCAD desktop icon's Properties dialog box:

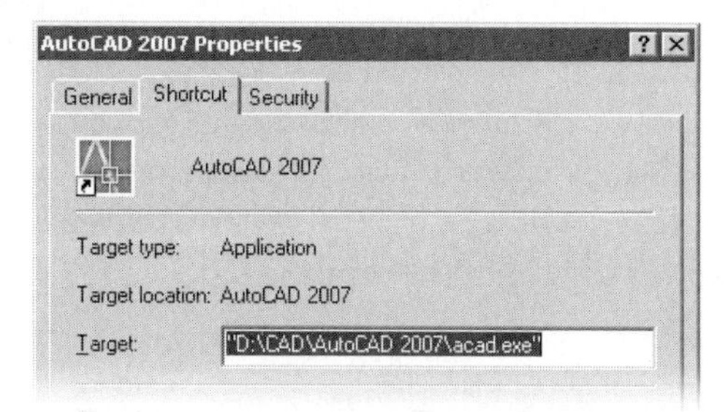

/b runs a script file after AutoCAD starts; uses the following format:

acad.exe "\acad 2007\drawing.dwg" /b "file name.scr"

/c specifies the path for alternative hardware configuration file; default = *acad2007.cfg*.

/layout specifies the layout to display.

/ld loads the specified ARx and DBx applications.

/nologo suppresses the display of the AutoCAD logo screen.

/nossm prevents Sheet Set Manager palette from loading.

/p specifies a user-defined profile to customize AutoCAD's user interface.

/pl publishes a set of drawings defined by a *.dsd* file in the background.

/r restores the default pointing device.

/s specifies additional support folders; maximum is 15 folders, with each folder name separated by a semicolon.

/set specifies the *.dst* sheet set file to load.

/t specifies the *.dwt* template drawing to use.

/v specifies the named view to display upon startup of AutoCAD.

/w specifies the workspace with which to first display (*new to AutoCAD 2007*).

Command Line Options

When ***FileDia*** *= 0, AutoCAD prompts you at the command line:*

Command: new

Enter template file name or [. (for none)] <*default .dwt file path name*>: *(Enter the path and name of a .dwt , .dwg, or .dws file.)*

Alternatively, enter the following options:

ENTER accepts the default template drawing file.

. (period) eliminates use of a template; AutoCAD uses either *acad.dwt* or *acadiso.dwt*, depending on the setting of the **MeasureInit** system variable.

~ (tilde) forces the display of the Select Template dialog box.

RELATED COMMANDS

QNew starts a new drawing based on a predetermined template file.

SaveAs saves the drawing in *.dwg* or *.dwt* formats; creates template files.

RELATED SYSTEM VARIABLES

DbMod indicates whether the drawing has changed since being loaded.

DwgPrefix indicates the path to the drawing.

DwgName indicates the name of the current drawing.

FileDia displays prompts at the 'Command:' prompt.

InsUnits determines whether blocks are inserted in inches, millimeters, or other units.

Startup determines whether the dialog box or the wizard is displayed.

MeasureInit determines whether the units are imperial or metric.

RELATED FILES

wizard.ini names and describes template files.

****.dwt*** are template files stored in *.dwg* format.

TIPS

- Until you give the drawing a name, AutoCAD names it *drawing1.dwg* (and subsequent drawings created in the same session as *drawing2.dwg*, *drawing3.dwg*, and so on).
- The default template drawing is the last one used, and not necessarily *acad.dwg*.
- Edit and save *.dwt* template drawings to change the defaults for new drawings.
- When you press CTRL+N, AutoCAD's behavior differs from Microsoft Office programs: Office programs display a new document that takes on the properties of the current document.
- Turning off **FileDia** (set to 0) is meant for use with macros and programs.
- The **Startup** system variable can be set in the Options dialog box: from the Tools menu, select Options; click the System tab (not the Open and Save tab!). In the General Options section, the Startup option offers two settings:

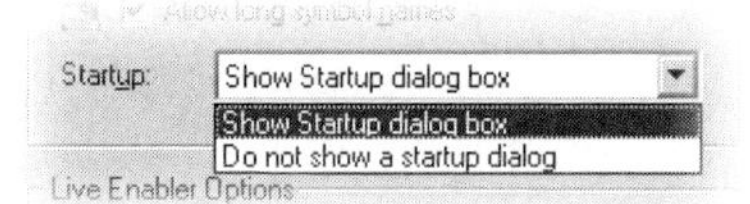

  - Do not show a startup dialog displays the Select Template dialog box.
  - Show Startup dialog displays the Startup and Create New Drawing wizards.
- The Today window was removed in AutoCAD 2004.

NewSheetset

2005 Runs the New Sheet Set wizard.

Command	Alias	Ctrl+	F-key	Alt+	Menu Bar	Tablet
newsheetset	...	...	...	**TSNN**	**Tools** ↳**Wizards** ↳**New Sheet Set**	...

Command: newsheetset

Displays the Create Sheet Set dialog box.

CREATE SHEET SET DIALOG BOX

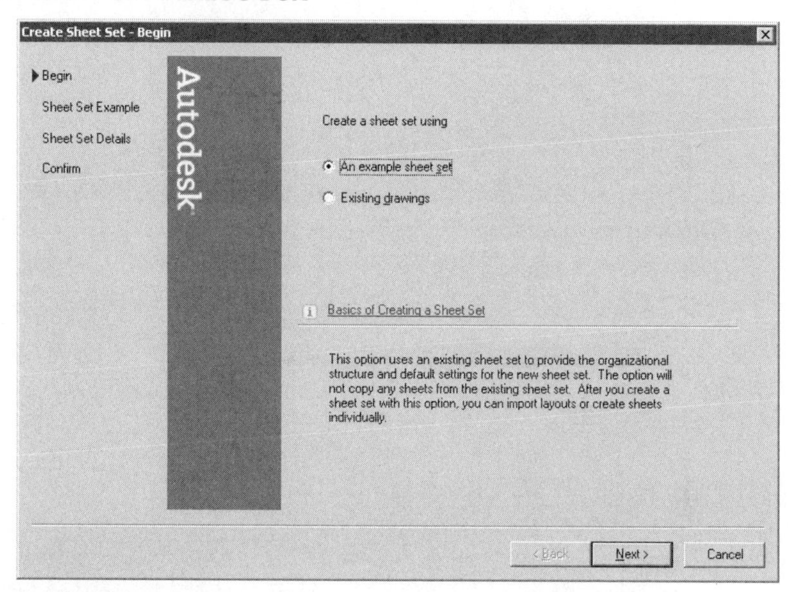

Create a sheet set using:

⊙**An example sheet set** creates a new sheet set based on "templates."

○**Existing drawings** selects one or more folders holding drawings, which are imported into the sheet set.

Click **Next** to continue.

Sheet Set Example Page

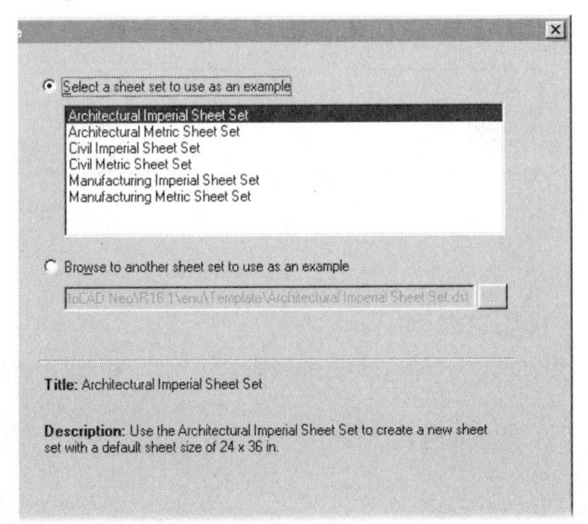

⊙ **Select a sheet set to use as an example** selects one of the sheet sets provided by Autodesk.

○ **Browse to another sheet set to use as an example** opens a *.dst* sheet set data file.

... Opens the Browse for Sheet Set dialog box; select a *.dst* file, and then click **Open**.

Sheet Set Details Page

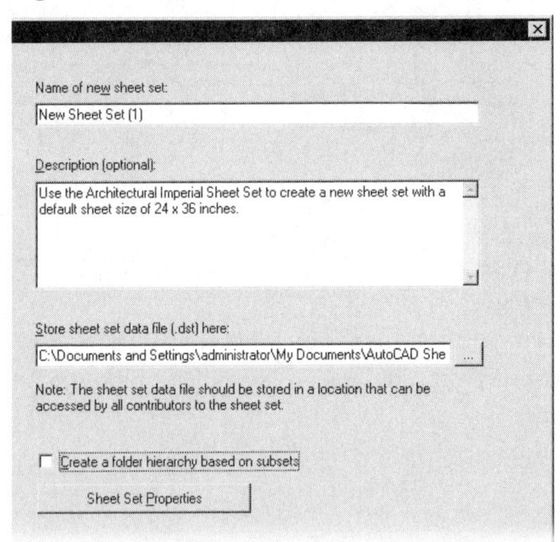

Name of new sheet set names the sheet set.

Description describes the sheet set.

Store sheet set data file here specifies the location of the *.dst* sheet set data file.

... displays the Browse for Sheet Set Folder dialog box; select a folder, and then click **Open**.

Create a folder hierarchy based on subsets toggles creation of one or more folders to hold subset files.

Sheet Set Propeties displays the Sheet Set Properties dialog box.

Finishes with the Confirm page, illustrated below.

Existing Drawings

Sheet Set Details page

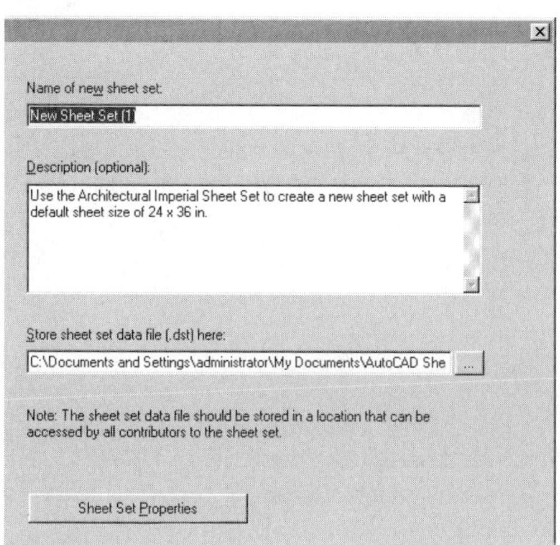

Name of new sheet set names the sheet set.

Description describes the sheet set.

Store sheet set data file here locates the sheet set data file.

... displays the Browse for Sheet Set Folder dialog box; select a folder, and then click **Open**.

Sheet Set Properties displays the Sheet Set Properties dialog box.

Choose Layouts page

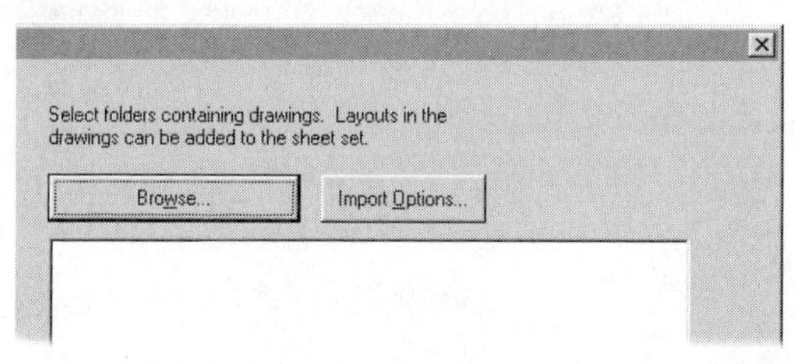

Browse displays Browse for Folder dialog box; select a folder, and then click **OK**.

Import Options displays the Import Options dialog box.

Confirm page

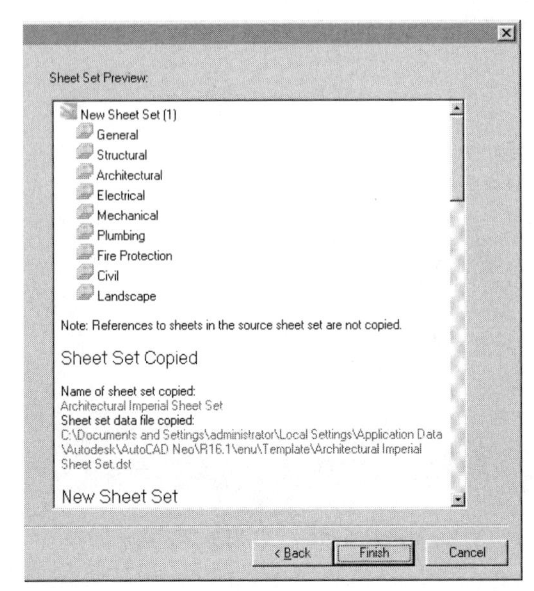

Press **CTRL+C** to copy the settings, after selecting all text. (In a text editor, press **CTRL+V** to paste the text in a document.)

Click **Back** to change and correct settings.

Click **Finish** to create the new sheet set.

SHEET SET PROPERTIES DIALOG BOX

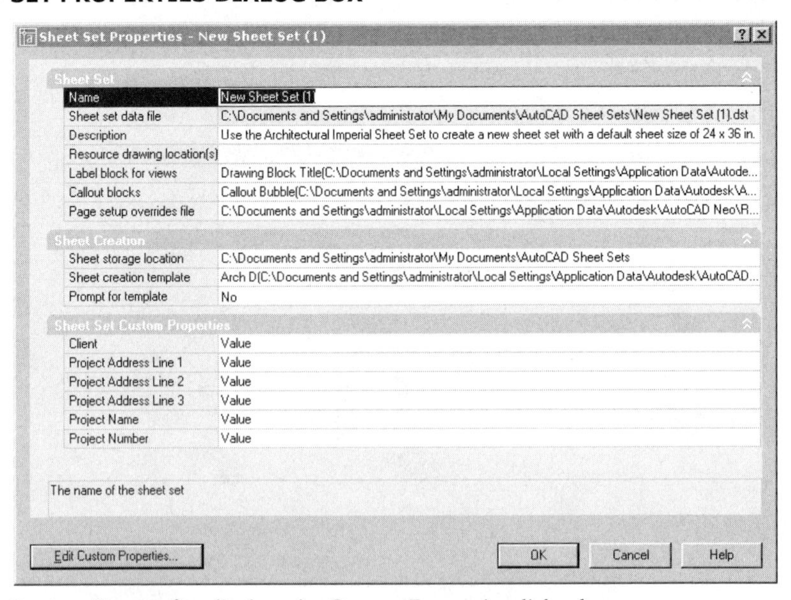

Edit Custom Properties displays the Custom Properties dialog box.

CUSTOM PROPERTIES DIALOG BOX

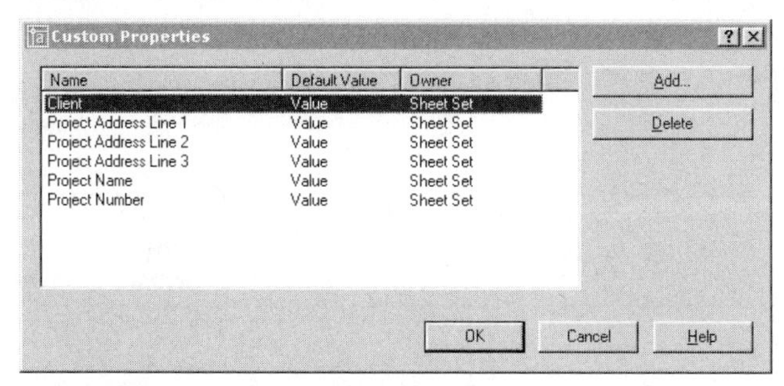

Add displays the Add Custom Property dialog box.

Delete removes the selected custom property without warning; click **Cancel** to undo erasure.

ADD CUSTOM PROPERTY DIALOG BOX

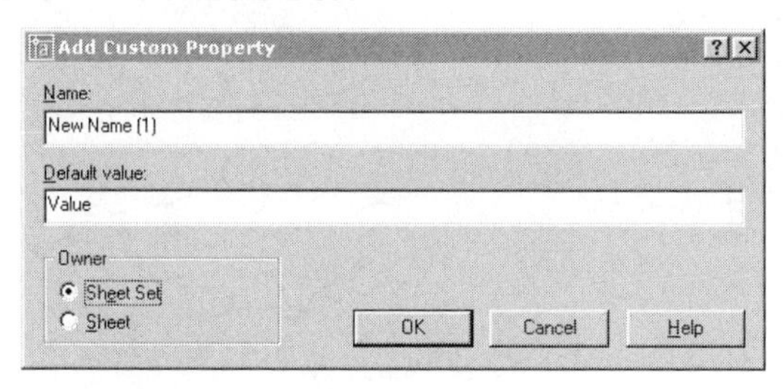

Name names the custom property.

Default value specifies the default value.

Owner:

⊙ **Sheet set** indicates that the custom property belongs to the sheet set.

○ **Sheet** indicates that the custom property belongs to the sheet.

IMPORT OPTIONS DIALOG BOX

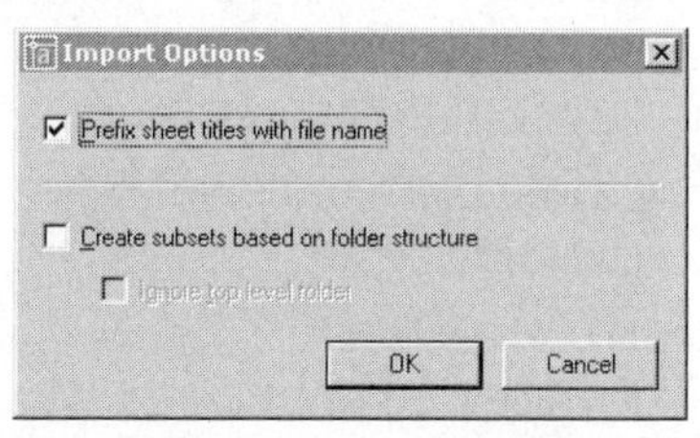

☑ **Prefix sheet titles with file name** tags sheet set names with file names.

☐ **Create subsheets based on folder structure** generates sheets based on folder names.

☐ **Ignore top level folder** ignores the topmost folder in sheet name generation.

RELATED COMMANDS

OpenSheetset opens existing sheet sets.

Sheetset displays the Sheet Set Manager palette.

Offset

V. 2.5 Draws parallel lines, arcs, circles and polylines; repeats automatically until canceled.

Command	Alias	Ctrl+	F-key	Alt+	Menu Bar	Tablet
offset	o	...	...	MS	Modify ↳Offset	V17

Command: offset

Specify offset distance or [Through/Erase/Layer] <1.0000>: *(Enter a distance or an option.)*

Select object to offset or [Exit/Undo]: *(Select an object.)*

Specify point on side to offset or[Exit/Multiple/Undo]: *(Pick a point.)*

Select object to offset or [Exit/Undo]: *(Select another object, or press* ESC *to end the command.)*

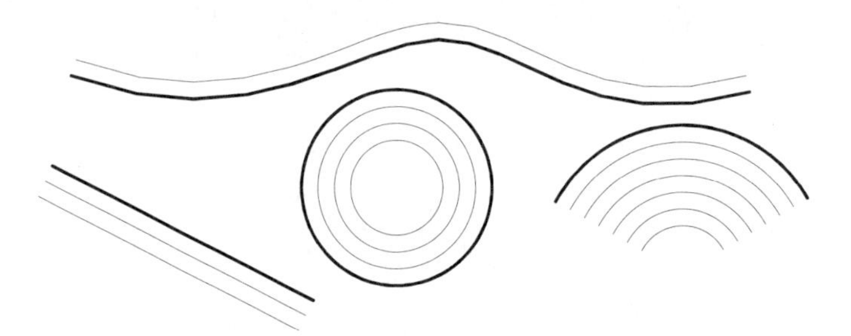

Offset objects, with original objects shown bold.

COMMAND LINE OPTIONS

Offset distance specifies the perpendicular distance to offset.

Through indicates the offset distance.

Erase erases the original object.

Layer specifies the layer on which to place offset objects.

Exit exits the command, as does ESC.

Undo undoes the last offset operation.

Multiple repeats the offset from the last object.

RELATED COMMANDS

Copy creates one or more copies of a group of objects.

MLine draws up to 16 parallel lines.

RELATED SYSTEM VARIABLES

OffsetDist specifies the current offset distance.

OffsetGapType determines how to close gaps created by offset polylines.

TIPS

- Offsets of curved objects change their radii.
- AutoCAD complains "Unable to offset" when curved objects are smaller than the offset distance.

OleConvert

2000 Converts OLE objects, if possible (an undocumented command).

Command	Alias	Ctrl+	F-key	Alt+	Menu Bar	Tablet
oleconvert	...	...	...	...	...	...

Select an OLE object before entering the command.

Command: oleconvert

Displays dialog box:

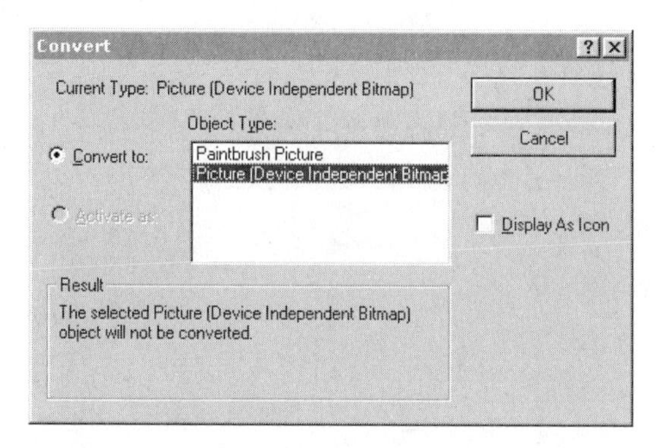

CONVERT DIALOG BOX

Object Type selects the type of object to convert to; list of types varies, depending on the object.

⊙ **Convert to** converts the OLE object to another type.

○ **Activate as** opens the OLE object with the source application.

☐ **Display as Icon** displays the OLE object as an icon.

RELATED COMMANDS

InsertObj places an OLE object in the drawing.

PasteSpec pastes objects from the Clipboard as linked objects in the drawing.

OleReset returns the OLE object to its original form.

RELATED SYSTEM VARIABLES

OleHide toggles the display of OLE objects.

OleQuality determines the plot quality of OLE objects.

OleStartup loads the source application for embedded OLE objects before plotting.

TIPS

- In many cases, this command is unable to convert the OLE objects.
- You can also access this command by: (1) selecting OLE object, (2) right-clicking, and (3) selecting **OLE | Convert** from the shortcut menu.
- OLE is short for "object linking and embedding." It is technology invented by IBM for its OS/2 operating system. OLE allows documents and objects from other applications to exist in "foreign" documents — documents that would not otherwise accept unknown objects.

OleLinks

Rel.13 Changes, updates, and cancels OLE links between the drawing and other Windows applications (short for Object Linking and Embedding LINKS).

Command	Alias	Ctrl+	F-key	Alt+	Menu Bar	Tablet
olelinks	...	...	...	EO	Edit ↳OLE Links	...

Command: olelinks

When no OLE links are in the drawing, the command does nothing.

When at least one OLE object is in the drawing, displays dialog box:

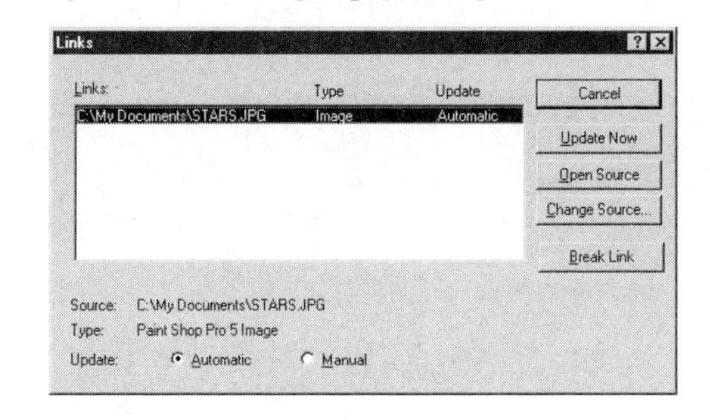

LINKS DIALOG BOX

Links displays a list of linked objects: source filename, type of file, and update mode — automatic or manual.

Update selects either automatic or manual updates.

Update Now updates selected links.

Open Source starts the source application program.

Break Link cancels the OLE link; keeps the object in place.

Change Source displays the Change Source dialog box.

RELATED COMMANDS

InsertObj places an OLE object in the drawing.

PasteSpec pastes objects from the Clipboard as linked objects in the drawing.

OleScale specifies the size, scale, and properties of selected OLE objects.

RELATED SYSTEM VARIABLES

MsOleScale specifies the scale of OLE objects in model space.

OleHide toggles the display of OLE objects.

OleQuality determines the plot quality of OLE objects.

OleStartup loads the source applications for embedded OLE objects before plotting.

RELATED WINDOWS COMMANDS

Edit | Copy copies objects from the source application to the Clipboard.

File | Update updates the linked object in the source application.

OleOpen, OleReset

2000 Opens OLE objects in their source applications; resets OLE objects (undocumented commands).

Command	Alias	Ctrl+	F-key	Alt+	Menu Bar	Tablet
oleopen	...	...	...	...	...	...
olereset	...	...	...	...	...	...

Select an OLE object before entering the command.

Command: oleopen

Opens the OLE object in its source application. For example, selecting a text document opens it in the default word processor.

Command: olereset

Returns the OLE object to it original form.

RELATED COMMANDS

InsertObj places an OLE object in the drawing.

PasteSpec pastes objects from the Clipboard as linked objects in the drawing.

RELATED SYSTEM VARIABLES

MsOleScale specifies the scale of OLE objects in model space.

OleHide toggles the display of OLE objects.

OleQuality determines the plot quality of OLE objects.

OleStartup loads the source applications for embedded OLE objects before plotting.

TIPS

- You can also access these commands by: (1) selecting OLE objects, (2) right-clicking, and (3) selecting **OLE | Open** from the shortcut menu (or selecting **OLE | Reset**).
- If you do not first select an OLE object, AutoCAD complains, 'Unable to find OLE object. Object must be selected before entering the command.'

OleScale

2000 Modifies the properties of OLE objects.

Command	Alias	Ctrl+	F-key	Alt+	Shortcut Menu	Tablet
olescale	...	...	...	...	...	...

Command: olescale

Displays dialog box:

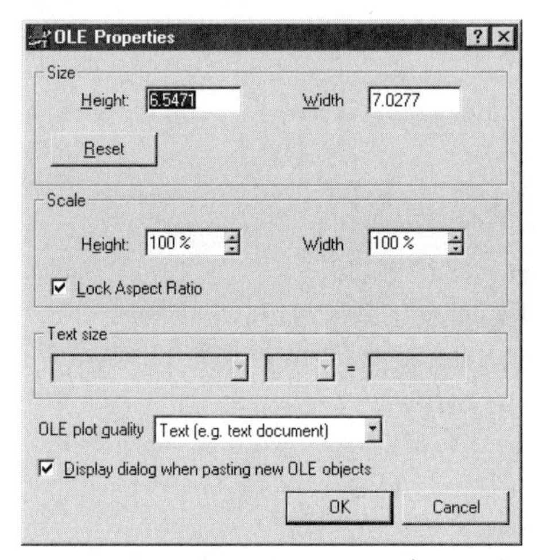

Alternatively, right-click a selected OLE object, and choose ***OLE | Text Size*** *from the shortcut menu. AutoCAD displays an abbreviated dialog box:*

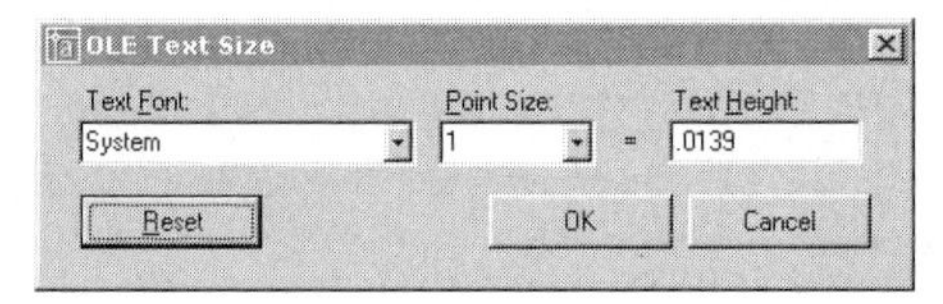

OLE PROPERTIES DIALOG BOX

The following options apply to text placed as OLE objects in drawings.

Size

Height changes the height of the OLE object; displays the current height.

Width changes the width of the OLE object; displays the current width.

Reset resets the OLE object to its size when first inserted into the drawing.

Scale

Height changes the height of the OLE object by a percentage of the original height.

Width changes the width of the OLE object by a percentage of the original width.

☑ **Lock Aspect Ratio** changes the Width size and ratio to match the Height, and vice versa.

Text Size

Font displays the fonts used by the OLE object, if any.

Point Size displays the text height in point sizes; limited to the point sizes available for the selected font, if any (1 point = $^1/_{72}$ inch).

Text Height specifies the text height in drawing units.

OLE Plot Quality determines the quality of the pasted object when plotted:

Line Art — Text is plotted as text; no colors or shading are preserved; some graphical images are not plotted, while others are plotted as monochrome images (black and white only, no shades of gray).

Text — All text formatting is preserved; text is plotted as graphics, which plots less cleanly than the **Line Art** setting; graphics are plotted less cleanly and at reduced colors than **Graphics** and **Photograph** settings.

Graphics — Graphics are plotted at a reduced number of colors (fewer shades of gray or "posterization"); all text formatting is preserved; text is plotted as graphics, but more cleanly than with **Text** and **Photograph** settings.

Photograph — Graphics are plotted at reduced resolution and colors; all text formatting is preserved; text is plotted as graphics.

High Quality — Graphics are plotted at full resolution and colors.

Photograph — All text formatting preserved; text plotted more cleanly than with **Text** and **Graphics** settings.

Display dialog when pasting new OLE object:

☑ Displays the OLE Properties dialog box automatically when inserting an OLE object.

☐ Does not display the dialog box.

RELATED SYSTEM VARIABLES

MsOleScale specifies the scale of OLE objects in model space.

OleHide toggles the display of OLE objects in the drawing and in plots.

OleQuality specifies the quality of displaying and plotting OLE objects.

OleStartup loads the source application of embedded OLE objects for plots.

RELATED COMMANDS

MsOleScale sets the scale for text in OLE objects.

InsertObj inserts an OLE object into the drawing.

OleLinks modifies the link between the object and its source.

PasteSpec allows you to paste an object with a link.

TIPS

- Change **OleStartup** to 1 to load the OLE source application, which may help improve the plot quality of OLE objects.
- The **OLE Plot Quality** list box determines the quality of the pasted object when plotted. I recommend the **Line Art** setting for text, unless the text contains shading and other graphical effects.

Oops

V. 1.0 Restores the last-erased group of objects; restores objects removed by the **Block** and **-Block** commands.

Command	Alias	Ctrl+	F-key	Alt+	Menu Bar	Tablet
oops	...	...	...	...	...	...

Command: oops

COMMAND LINE OPTIONS

None.

RELATED COMMANDS

Block, **Erase**, and **WBlock** can use this command to return erased objects.

U undoes the most recent command.

TIP

- This command restores only the most-recently erased object; use the Undo command to restore earlier objects.

Open

Rel.12 Loads one or more drawings into AutoCAD.

Command	Alias	Ctrl+	F-key	Alt+	Menu Bar	Tablet
open	**openurl**	**O**	**...**	**FO**	**File** **↳Open**	**T25**

Command: open

Displays dialog box:

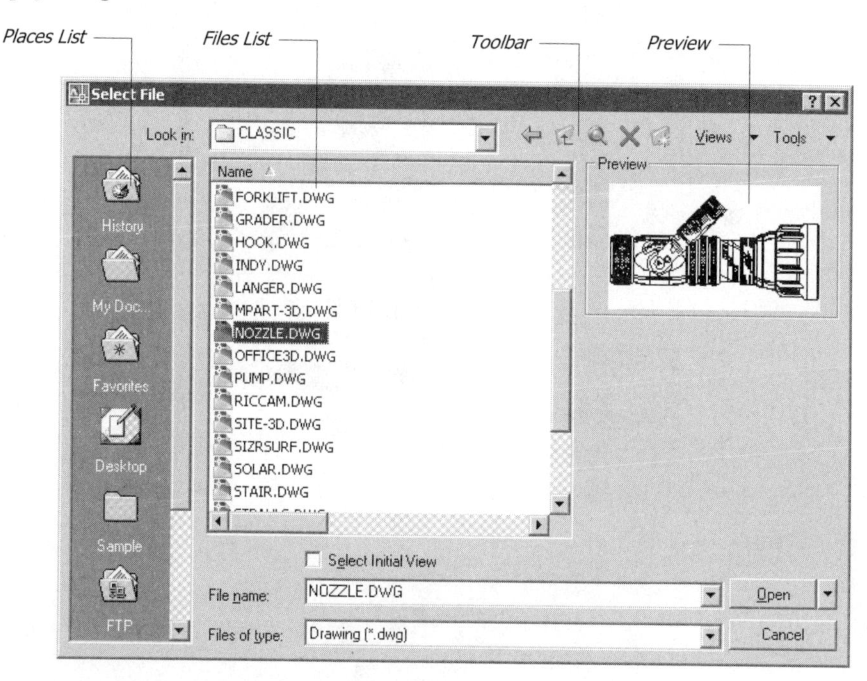

SELECT FILE DIALOG BOX

Look in selects the network drive, hard drive, or folder (subdirectory).

Preview displays a preview image of AutoCAD drawings.

☐ **Select initial view** selects a named view if the drawing has saved views. (After drawing is opened, AutoCAD displays the Select Initial View dialog box listing named views.)

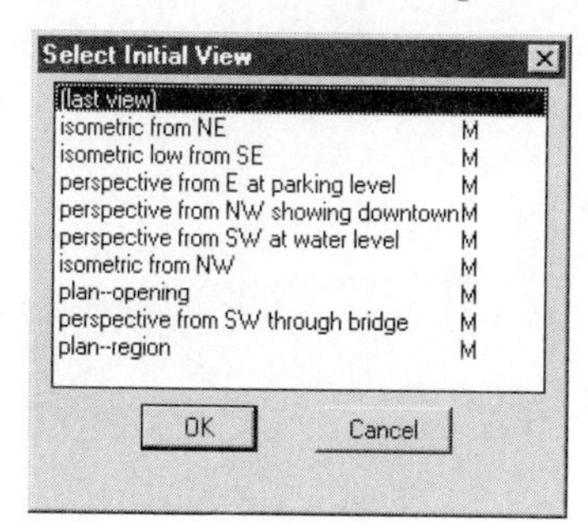

M indicates a view created in model space.

P indicates a view created in paper space (layout mode).

File name specifies the name of the drawing.

Files of type specifies the type of file:

- **Drawing (*.dwg)** AutoCAD drawing file.
- **Standard (*.dws)** drawing standards file; see the **Standards** command.
- **DXF (*.dxf)** drawing interchange file; see the **DxfIn** command.
- **Drawing Template File (*.dwt)** template drawing file; see the **New** command.

Open opens the selected drawing file(s); to open more than one drawing at a time:

- In the files list, hold down the SHIFT key to select a contiguous range of files:

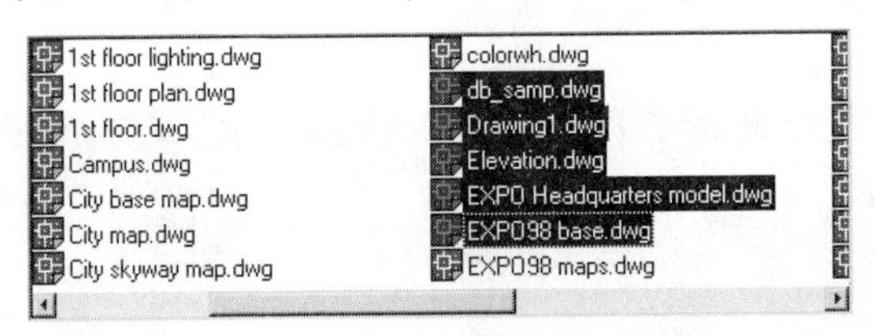

- Hold down the CTRL key to select two or more non-contiguous files:

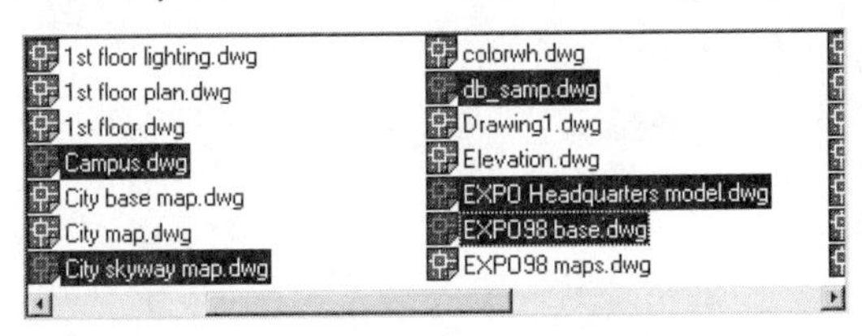

Cancel dismisses the dialog box without opening a file.

Open Button

Open opens the drawing.

Open as read-only loads the drawing, but you cannot save changes to the drawing except under another file name. AutoCAD displays "Read Only" on the title bar.

Partial Open loads selected layers or named views; displays the Partial Open dialog box; not available for *.dxf* and template files.

Partial Open Read-Only partially loads the drawing in read-only mode.

PARTIAL OPEN DIALOG BOX

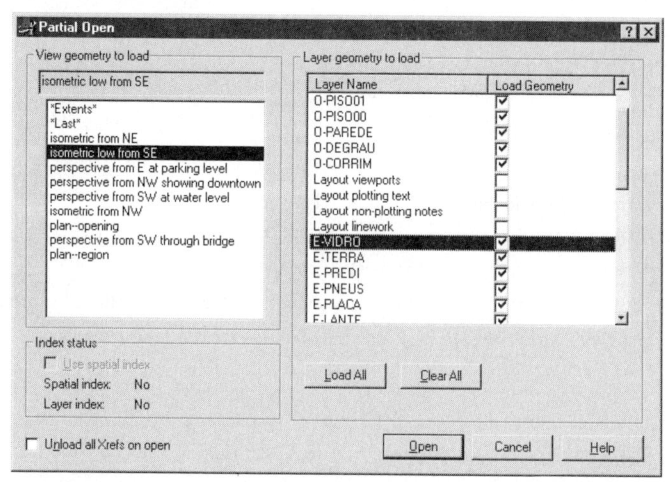

View geometry to load selects the model space views to load; paper space views are not available for partial loading.

Layer Geometry to Load

Load Geometry selects the layers to load.

Load All loads all layers.

Clear All deselects all layer names.

Index Status

Available only when the drawing was saved with spatial indices.

Use spatial index determines whether to use the spatial index for loading, if available.

Spatial Index indicates whether the drawing contains the spatial index.

Layer Index indicates whether the drawing contains the layer index.

☐ **Unload all xrefs on open** loads externally-referenced drawings when opening the drawing.

Open opens the drawing, and then partially loads the geometry.

DIALOG BOX TOOLBAR

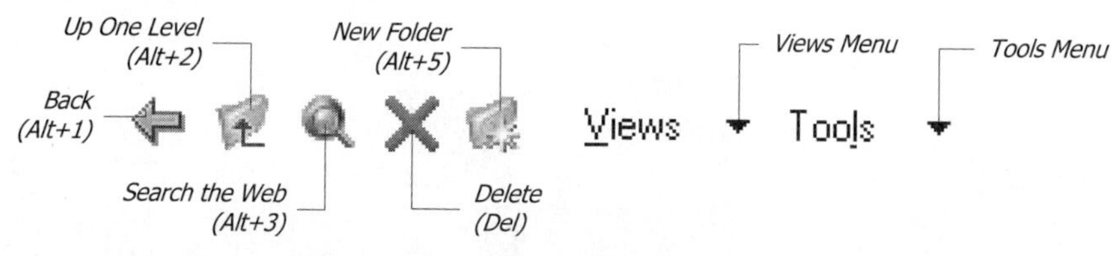

Back returns to the previous folder (keyboard shortcut **ALT+1**).

Up moves up one level to the next folder or drive (**ALT+2**).

Search the Web displays the Browse the Web window (**ALT+3**); see Browser command.

Delete removes the selected file(s); does not delete folders or drives (**DEL**).

Create New Folder creates new folders (**ALT+5**).

Views provides display options:

- **List** displays the file and folder names only.
- **Details** displays file and folder names, type, size, and date.
- **Thumbnails** displays thumbnail images of *.dwg* files.
- **Preview** toggles display of the preview window.

Tools provides file-oriented tools:

- **Find** displays the Find dialog box for searching files.
- **Locate** searches for the file along AutoCAD's search paths.
- **Add/Modify FTP Locations** displays a dialog box for storing the logon names and passwords for FTP (file transfer protocol) sites.
- **Add Current Folder to Places** adds the selected folders to the Places sidebar.
- **Add to Favorites** adds the selected files and folders to the Favorites list.

PLACES LIST

History displays files opened by AutoCAD during the last four weeks.

My Documents displays files and folders in the *my documents* folder.

Favorites displays files and folders in the *favorites* folder.

Desktop displays the contents of the *desktop* folder.

FTP displays the FTP Locations list.

Buzzsaw.com goes to the www.autodesk.com/buzzsaw Web site.
(The Point A and RedSpark destinations were removed from AutoCAD 2004.)

SHORTCUT MENUS

Places List menu

Right-click icons in the Places list:

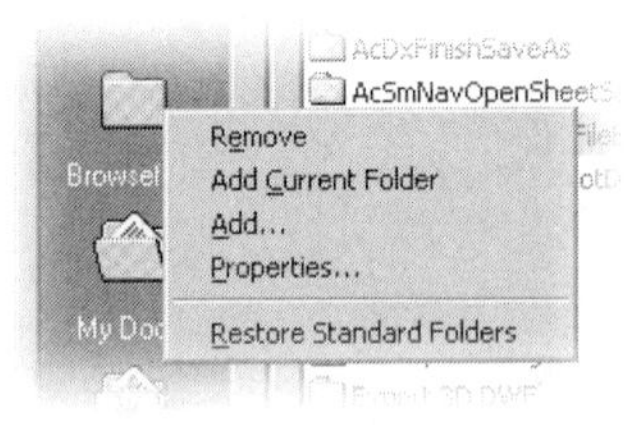

Remove removes a folder from the list.

Add Current Folder adds the selected folder to the list; you can also drag a folder from the file list into the Places list.

Add displays the Add Places Item dialog box. (It's much easier simply to drag a folder onto the Places list.)

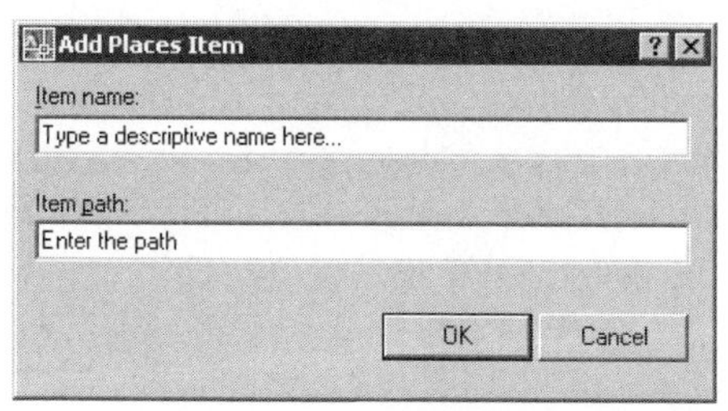

Properties displays the Places Item Properties dialog box, which is identical to the Add Places Item dialog box, and is available only for items you've added.

Restore Standard Folders restores the folders shown above.

File List

Right-click the File list without selecting a file or folder:

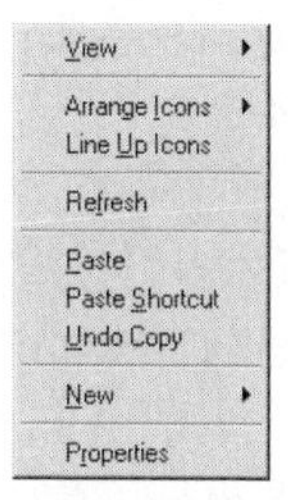

View switches between filename views: large icons, small icons, list, thumbnails, and details.

Arrange Icons arranges icons by name, type, size, and date.

Line Up Icons places icons in an orderly pattern.

Refresh updates the folder listing.

Paste pastes a file from the Clipboard.

Paste Shortcut pastes a file from the Clipboard as a shortcut.

Undo Copy undoes the copy-paste operation; available only after a copy or paste.

New creates a new folder (subdirectory) or shortcut.

Properties displays the Properties dialog box of the folder selected by the Look in list.

Right-click a file or folder name:

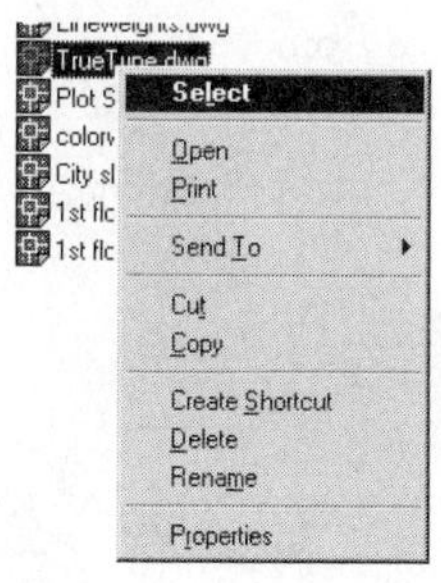

Select opens the drawing in AutoCAD.

Open opens the drawing in AutoCAD also.

Print does not work, even though it is on the menu.

Send To copies the file to another drive; may not work with some software.

Cut cuts the file to the Clipboard.

Copy copies the file to the Clipboard.

Create Shortcut creates a shortcut icon for the selected file.

Delete erases the file; displays a warning dialog box.

Rename renames the file; displays a warning dialog box if you change the extension.

Properties displays the Properties dialog box in read-only mode; use the **DwgProps** command within AutoCAD to change the settings.

Command Line Options

When FileDia = 0:

Command: open

Enter name of drawing to open <.>: *(Enter name of drawing file.)*

Opening an AutoCAD format file

Enter name specifies the (optional) path and name of the *.dwg*, *.dxf*, or *.dws* file.

If you leave out the extension, AutoCAD assumes a *.dwg* file.

Enter tilde (~) to display the Select File dialog box.

RELATED SYSTEM VARIABLES

TaskBar displays a button on the Windows taskbar for every open drawing.

DbMod indicates whether the drawing has been modified.

DwgCheck checks if the drawing was last edited by AutoCAD.

DwgName contains the drawing's filename.

DwgPrefix contains the drive and folder of the drawing.

DwgTitled indicates whether the drawing has a name other than *drawing1.dwg.*

FileDia toggles the interface between dialog boxes and the command-line.

FullOpen indicates whether the drawing is fully or partially opened.

RELATED COMMANDS

FileOpen opens drawings without the dialog box.

SaveAs saves drawings with new names.

PartialLoad loads additional portions of partially-opened drawings.

TIPS

- When drawing filenames are dragged from Windows Explorer into open drawings, they are inserted as blocks; when dragged to AutoCAD's title bar, they are opened as drawings.
- You can double-click *.dwg* files in Windows Explorer to open them in AutoCAD. If AutoCAD is not yet running, it starts with the last-used profile and vertical application.
- DXF and template files cannot be partially opened.
- After a drawing is partially opened, use PartiaLoad to load additional parts of the drawing.
- When a partially-opened drawing contains a bound xref, only the portion of the xref defined by the selected view is bound to the partially-open drawing.
- OpenUrl opened drawings over the Internet in earlier releases of AutoCAD.

Alaised Command

OpenUrl was removed from AutoCAD 2000; it now activates the **Open** command.

OpenDwfMarkup

2005 Opens *.dwf* markup files in AutoCAD, and then loads the Markup Set Manager palette.

Command	Alias	Ctrl+	F-key	Alt+	Menu Bar	Tablet
opendwfmarkup	...	...	...	**FK**	**File** **↳Load Markup Set**	...

Command: opendwfmarkup

Displays the Open Markup DWF dialog box. Select a .dwf file, and then click ***Open****.*

If the .dwf file contains no markup data, AutoCAD complains, 'Filename.dwf does not contain any markup data. Would you like to open this DWF file in the viewer?' Click ***Yes*** *to open the file in Composer; click* ***No*** *to cancel.*

COMMAND OPTIONS

See ***Markup*** *command.*

RELATED COMMANDS

Markup displays the Markup Set Manager, and also opens *.dwf* files containing markup data.

MarkupClose closes the Markup Set Manager palette.

RmlIn imports *.rml* redline markup files created by Volo View.

RELATED SYSTEM VARIABLES

MsmState reports whether the Markup Set Manager palette is open.

FileDia determines whether this command displays a dialog box or command-line prompts.

TIP

- MSM is short for "markup set manager." DWF is short for "drawing Web format."

OpenSheetset

2005 Loads sheet sets into the current drawing.

Command	Alias	Ctrl+	F-key	Alt+	Menu Bar	Tablet
opensheetset	...	...	...	**FE**	**File** ↳**Open Sheet Set**	...
-opensheetset						

Command: opensheetset

Displays Open Sheet Set dialog box. Select a .dst file, and then click ***Open.***

AutoCAD displays the Sheet Set Manager. See ***Sheetset*** *command.*

-OPENSHEETSET Command

Command: -opensheetset

Enter name of sheet set to open: *(Enter .dst file name.)*

Enter the path and name of a *.dst* file, or enter tilde (~) to display the file dialog box.

RELATED COMMANDS

NewSheetset creates new sheet sets.

Sheetset opens the Sheet Set Manager palette.

SheetsetClose closes the Sheet Set Manager palette.

RELATED SYSTEM VARIABLES

SsFound records path and file name of the sheet set.

SsLocate toggles whether AutoCAD opens sheet sets associated with the drawing.

SsmAutoOpen toggles whether AutoCAD displays the Sheet Set Manager palette when the drawing is opened.

SsmState reports whether the Sheet Set Manager is active.

FileDia determines whether this command displays a dialog box or command-line prompts.

TIPS

- The **Sheetset** command automatically opens the previously-opened sheet set file.
- SSM is short for "sheet set manager."

Options

2000 Sets system and user preferences.

Commands	Aliases	Ctrl+	F-key	Alt+	Menu Bar	Tablet
options	**op**	...	...	**TN**	**Tools**	**Y10**
	gr, ddgrips				**↳Options**	
	preferences					
	ddselect					
+options						

Command: options

Displays dialog box.

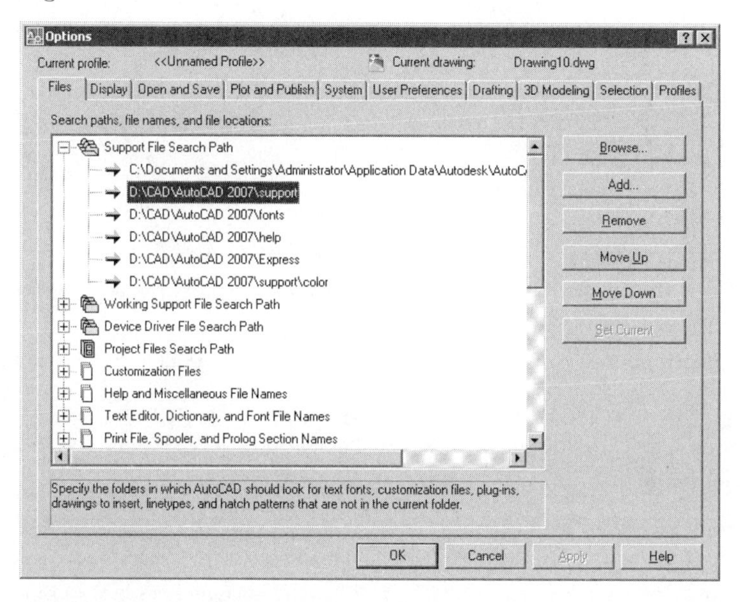

OPTIONS DIALOG BOX

OK applies the changes, and closes the dialog box.

Cancel cancels the changes, and closes the dialog box.

Apply applies the changes, and keeps the dialog box open.

Files Tab

Search paths, file names, and file locations specifies folders and files used by AutoCAD.

Browse displays the Browse for Folder dialog box for selecting folders, and the Select a File dialog box for selecting files.

Add adds an item below the selected path or file name.

Remove removes the selected item without warning; click CANCEL to undo the removal.

Move Up moves the selected item above (or before) the preceding item; applies to search paths only.

Move Down moves the selected item below the following item; applies to search paths only.

Set Current makes current the selected project names and spelling dictionaries only.

Display tab

 — *indicates setting is saved to a related system variable.*

Window Elements

☐ **Display scroll bars in drawing window** toggles the presence of the horizontal and vertical scroll bars.

☐ **Display screen menu** toggles the presence of the screen menu.

☐ **Use large buttons for Toolbars** switches toolbar buttons between normal and large size.

☑ **Show Tooltips** toggles the display of tooltips.

☑ **Show shortcut keys in Tooltips** toggles the display of shortcut keystrokes in tooltips.

Colors specifies the colors for the graphics and text windows; displays Color Options dialog box.

Fonts specifies the font for text on the command line; displays the Command Line Window Font dialog box.

Layout Elements

☑ **Display Layout and Model tabs** toggles the presence of the Model and Layout tabs.

☑ **Display printable area** toggles the display of dashed margin lines in layout modes.

☑ **Display paper background** toggles the presence of the page in layout modes.

☑ **Display paper shadow** toggles the presence of the drop shadow under the page in layout modes.

☐ **Show Page Setup Manager for new layouts** specifies whether the Page Setup dialog box is displayed when you create a new layout. Use this dialog box to set options related to paper and plot settings.

☑ **Create viewport in new layouts** toggles the automatic creation of a single viewport for new layouts.

Crosshair size specifies the size of the crosshair cursor; range is 1% to 100% of the viewport (stored in system variable CursorSize); default = 5%.

Display Resolution

Arc and circle smoothness controls the displayed smoothness of circles, arcs, and other curves; range is 1 to 20000 (ViewRes); default = 1000.

Segments in a polyline curve specifies the number of line segments used to display polyline curves; range is -32767 to 32767 (SplineSegs); default = 8.

Rendered object smoothness controls the displayed smoothness of shaded and rendered curves; range is 0.01 to 10 (FaceTRes) default = 0.5.

Contour lines per surface specifies the number of contour lines on solid 3D objects; range is 0 to 2047 (IsoLines); default = 4.

Display Performance

☐ **Pan and zoom with raster and OLE** toggles the display of raster and OLE images during
realtime pan and zoom (RtDisplay).

☑ **Highlight raster image frame only** highlights only the frame, and not the entire raster image, when on (ImageHlt).

☑ **Apply solid fill** toggles the display of solid fills in multilines, traces, solids, solid fills, and wide polylines; this option does not come into effect until you click **OK**, and then use the **Regen** command (FillMode).

☐ **Show text boundary frame only** toggles the display of rectangles in place of text; this option does not come into effect until you click OK, and then use the Regen command (QTextMode).

☐ **Show silhouettes in wireframe** toggles the display of silhouette curves for 3D solid objects; when off, tesselation lines are drawn when hidden-line removal is applied to the 3D object (DispSilh).

Reference fading intensity specifies the amount of fading during in-place reference editing; range is 0% to 90% (XFadeCtl); default = 50%.

Open and Save tab

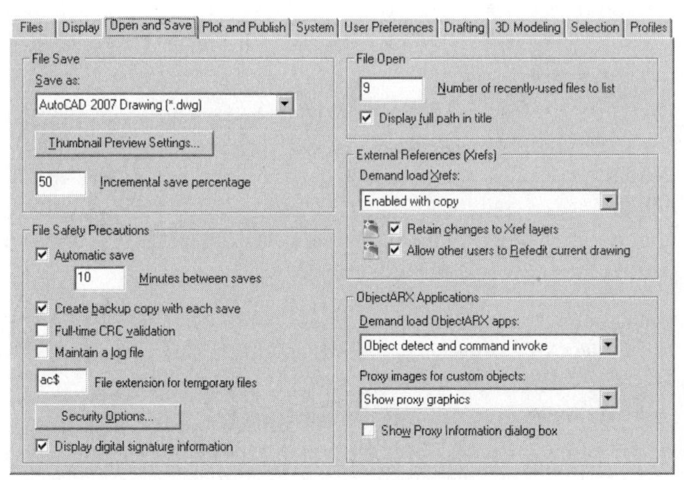

File Save

Save as specifies the default file format used by the Save and SaveAs commands; default = "AutoCAD 2007 Drawing (*.dwg)."

Thumbnail Preview Settings displays Thumbnail Preview Settings dialog box.

Incremental save percentage indicates the percentage of wasted space allowed in a drawing file before a full save is performed; range is 0% to 100% (ISavePercent); default = 50.

File Safety Precautions

☑ **Automatic save** saves the drawing automatically at prescribed time intervals (SaveFile and SaveFilePath).

Minutes between saves specifies the duration between automatic saves (SaveTime); default = 10 minutes.

☑ **Create backup copy with each save** creates backup copies when drawings are saved; (**ISavBak**).

☐ **Full-time CRC validation** performs cyclic redundancy check (*CRC*) error-checking each time an object is read into the drawing.

☐ **Maintain a log file** saves the Text window text to a log file (LogFileMode); default = off.

File extension for temporary files specifies the filename extension for temporary files created by AutoCAD (NodeName); default = *.ac$*.

Security Options displays the Security Options dialog box; see the SecurityOptions command.

☑ **Display digital signature information** displays digital signature information when opening files with valid digital signatures (SigWarn).

File Open

Number of recently-used files to list specifies the number of recently-opened filenames to list in the Files menu; minimum = 0; default and maximum = 9.

☑ **Display full path in title** displays the drawing file's path in AutoCAD's titlebar.

External References (Xrefs)

Demand load Xrefs specifies the style of demand loading of externally-referenced drawings a.k.a. xrefs (XLoadCtl):

- **Disabled** turns off demand loading.
- **Enabled** turns on demand loading to improve performance, but the drawing cannot be edited by another user; default.
- **Enabled with copy** turns on demand loading; loads a copy of the drawing so that another user can edit the original.

☐ **Retain changes to Xref layers** saves changes to properties for xref-dependent layers (VisRetain).

☑ **Allow other users to Refedit current drawing** allows another user to edit the current drawing when referenced by another drawing (XEdit).

ObjectARX Applications

Demand load ObjectARX apps demand-loads an ObjectARx application when the drawing contains proxy objects (DemandLoad):

- **Disable load on demand** turns off demand loading.
- **Custom object detect** demand-loads applications when drawings contain proxy objects.
- **Command invoke** demand-loads applications when commands (of the applications) are invoked.
- **Object detect and command invoke** demand-loads the application when the drawing contains proxy objects, or when one of the application's commands is invoked; default.

Proxy images for custom objects specifies how proxy objects are displayed:

- **Do not show proxy graphics** does not display proxy objects.
- **Show proxy graphics** displays proxy objects.
- **Show proxy bounding box** displays a rectangle instead of the proxy object.

☑ **Show Proxy Information dialog box** displays a warning dialog box when a drawing contains proxy objects (ProxyNotice).

Plot and Publish tab

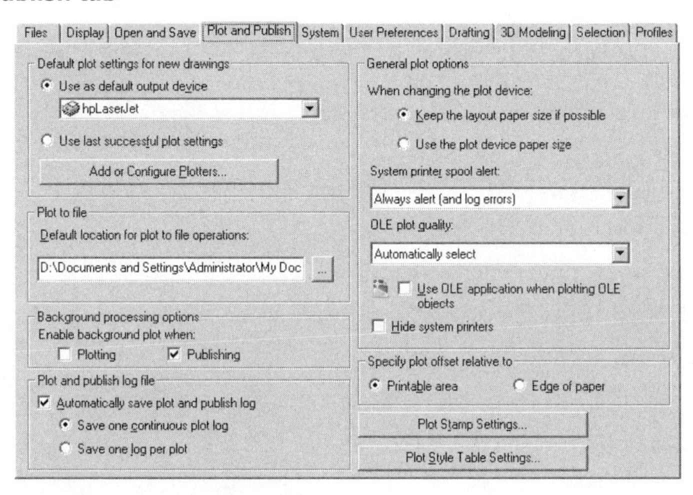

Default Plot Settings for New Drawings

⊙**Use as default output device** selects the default output device.

○**Use last successful plot settings** reuses the plot settings from the last successful plot.

Add or configure plotters displays Plotter Manager window; see PlotterManager command.

Plot to File

Default location for plot to file operations specifies the folder in which to store plot files; default=*documents and settings**username**my documents*. (**...** displays dialog box for selecting folder.)

Background Processing

☐ **Plotting** executes the Plot command in the background (BackgroundPlot).

☑ **Publishing** executes the Publish command in the background (BackgroundPlot).

Plot and Publish Log File

☑ **Automatically save plot and publish log file:**

⊙ **Save one continuous plot log** saves all log data in a single file.

○ **Save one log per plot** saves log data in separate files.

General Plot Options

When changing the plot device:

⊙ **Keep the layout paper size if possible** uses the paper size specified by the Page Setup dialog box's Layout Settings tab, provided the output device can handle the paper size (PaperUpdate); default = on.

○ **Use the plot device paper size** uses the paper size specified by the *.pc3* plotter configuration file (PaperUpdate); default = off.

System printer spool alert displays an alert when a spooled drawing has a conflict:

- **Always alert (and log errors)** displays alert, and logs the error message.
- **Alert first time only (and log errors)** displays the alert once, but logs all error messages.
- **Never alert (and log first error)** does not display an alert, but logs the first error message.
- **Never alert (do not log errors)** neither displays an alert, nor logs any error messages.

OLE plot quality determines the quality of OLE objects when plotted (OleQuality); default = Automatically Select; see the OleScale command.

☐ **Use OLE application when plotting OLE objects** launches the application that created the OLE object when plotting a drawing with an OLE object (OleStarup).

☐ **Hide system printers** hides the names of Windows system printers not specific to CAD.

Specify Plot Offset Relative To

⊙ **Printable area** measures offsets to plotter margins.

○ **Edge of paper** measures offsets to paper edges.

Plot Stamp Settings displays the Plot Stamp dialog box; see PlotStamp command.

Plot Style Table Settings displays the Plot Style Table Settings dialog box.

System tab

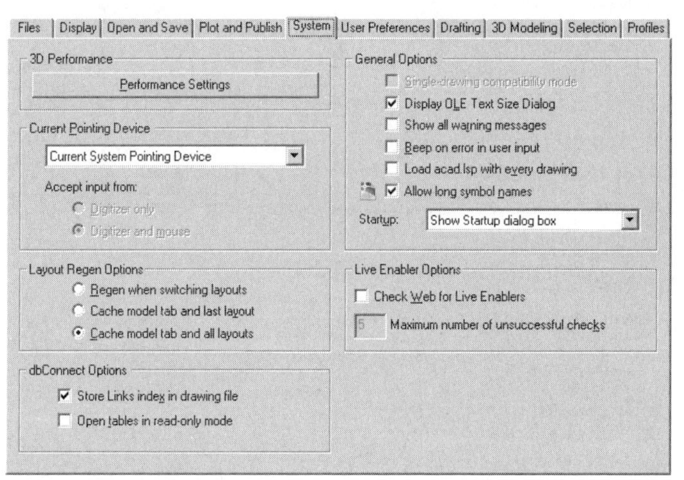

3D Performance

Performance Settings displays the Adaptive Degradation and Performance Tuning dialog box; see the 3dConfig command.

Current Pointing Device

Current Pointing Device selects the pointing device driver:

- **Current System Pointing Device** selects the pointing device used by Windows.
- **Wintab Compatible Digitizer** selects a Wintab-compatible digitizer driver.

Accept input from *(available only when a digitizing tablet is attached):*

○ **Digitizer only** reads input from the digitizer, and ignores the mouse.

⊙ **Digitizer and mouse** reads input from the digitizer and the mouse.

Layout Regen

○ **Regen when switching layouts** regenerates the drawing each time layouts are switched.

○ **Cache model tab and last layout** saves display lists of the model tab and last layout accessed.

⊙ **Cache model tab and all layouts** saves the display list of the model tab and all layouts.

dbConnect Options

☑ **Store links index in drawing file** stores the database index in the drawing file.

☐ **Open tables in read-only mode** opens database tables in read-only mode.

General Options

☐ **Single-drawing compatibility mode** forces the Single-drawing Interface (SDI), which limits AutoCAD to opening a single drawing at a time; this may be required for compatibility with some third-party applications (SDI).

☑ **Display OLE Text Size dialog** displays the OLE Properties dialog box after an OLE object is inserted in the drawing; see the OleScale command.

☐ **Show all warning messages** displays all dialog boxes with the Don't Display This Warning Again option.

☐ **Beep on error in user input** beeps the computer when AutoCAD detects a user error.

☐ **Load acad.lsp with every drawing** loads the *acad.lsp* file with every drawing (AcadLspAsDoc).

☑ **Allow long symbol names** allows symbol names — such as layers, dimension styles, and blocks, — to be up to 255 characters long, and to include letters, numbers, blank spaces, and most punctuation marks; when off, names are limited to 31 characters, and spaces may not be used (ExtNames).

Startup:

- **Show Startup dialog box** displays Startup dialog box when AutoCAD launches or Create New Drawing dialog box when AutoCAD is already running.
- **Do not show a startup dialog** displays the Select Template File dialog box.

Live Enabler Options

☐ **Check Web for Live Enablers** checks whenever an Internet connection is present.

Maximum number of unsuccessful checks limits how often AutoCAD attempts to call home; default = 5.

User Preferences tab

Windows Standard Behavior

☑ **Double click editing** toggles editing objects by double-clicking them; see Cui command (*new to AutoCAD 2007*).

☑ **Shortcut menus in drawing area:** toggles the effect of right-clicking in the drawing area: when on, displays a shortcut menu; when off, is equivalent to pressing the Enter key (ShortCutMenu).

Right-click Customization displays the Right-Click Customization dialog box (ShortCutMenu).

Insertion Scale

Source content units specifies the default units when an object is inserted into the drawing from AutoCAD DesignCenter; Unspecified-Unitless means the object is not scaled when inserted (InsUnitsDefTarget); default = inches or mm.

Target drawing units specifies the default units when "insert units" are not specified by the InsUnits system variable (InsUnitsDefTarget); default = inches or mm.

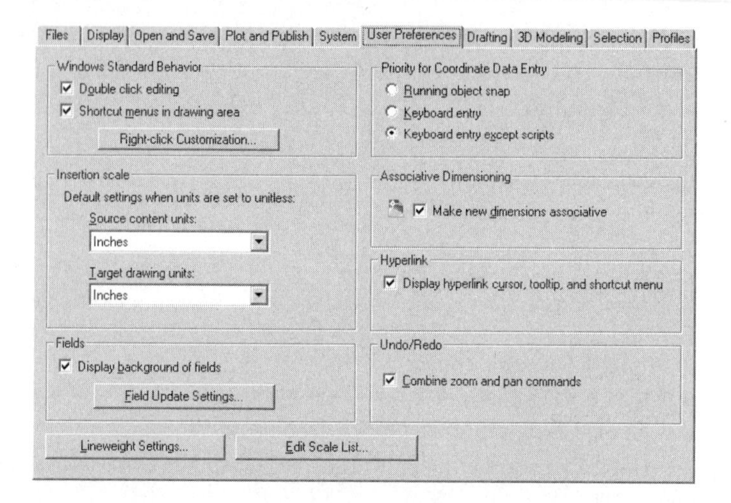

Field

☑ **Display background of fields** displays gray behind field text (FieldDisplay).

Field Update Settings displays the Field Update Settings dialog box; see UpdateField command.

The Hidden Line Settings button was removed from AutoCAD 2007.

Lineweight Settings displays the Lineweight Settings dialog box; see Lineweight command.

Edit Scale List displays the Edit Scale List dialog box; see ScaleListEdit command.

Priority for Coordinate Data Entry

○ **Running object snap** means that osnap overrides coordinates entered at the keyboard (OSnapCoord).

○ **Keyboard entry** means that coordinates entered at the keyboard override osnaps (OSnapCoord).

⊙ **Keyboard entry except scripts** means that coordinates entered at the keyboard override running object snaps, except when coordinates are provided by a script (OSnapCoord).

Associative Dimensioning

☑ **Make new dimension associative** means that dimensions are associated with objects; when off, dimensions are associated with defpoints (DimAssoc).

Hyperlink

☑ **Display hyperlink cursor and shortcut menu** displays hyperlink cursor (looks like chain links and the planet earth, when the cursor passes over objects containing hyperlinks), and adds the Hyperlink option to shortcut menus when right-clicking objects containing hyperlinks. See the HyperlinkOptions command.

Undo/Reo

☑ **Combine zoom and pan commands** combines all sequential zooms and pans into a single undo or redo.

Drafting tab

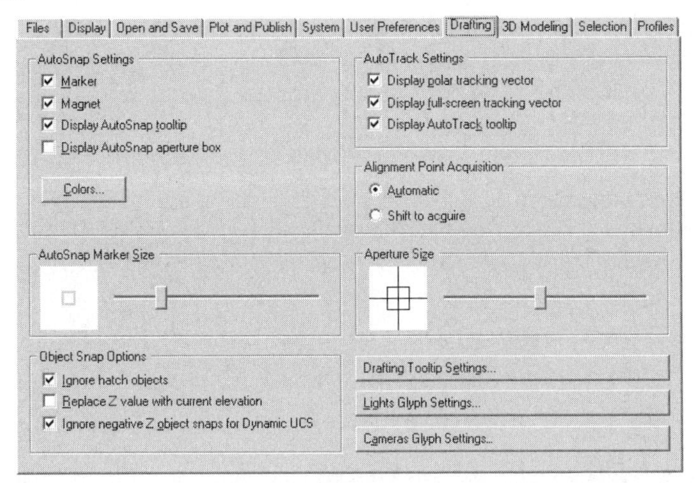

AutoSnap Settings

☑ **Marker** displays the object snap icon (AutoSnap).

☑ **Magnet** turns on the AutoSnap magnet (AutoSnap).

☑ **Display AutoSnap tooltip** displays the AutoSnap tooltip (AutoSnap).

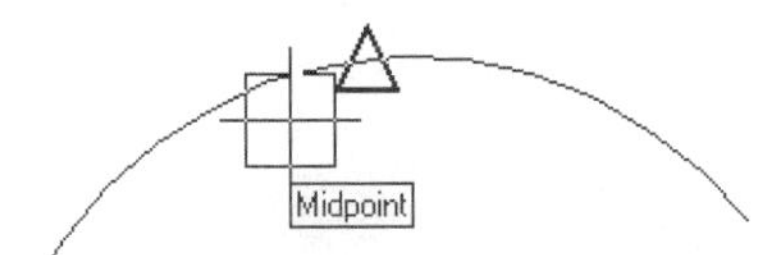

☐ **Display AutoSnap aperture box** displays the AutoSnap aperture box (ApBox).

Colors specifies the color of the AutoSnap icons; default = color #31. Displays the Drawing Window Colors dialog box.

AutoSnap marker size sets the size for the AutoSnap icon; range is 1 to 20 pixels.

Object Snap Options

☑ **Ignore hatch objects** prevents AutoCAD from snapping to hatch objects.

☐ **Replace Z value with current elevation**

☑ **Ignore negative Z object snaps for Dynamic UCS** (OsOptions; *new to AutoCAD 2007*).

AutoTrack Settings

☑ **Display polar tracking vector** displays the Polar Tracking vectors at specific angles (TrackPath).

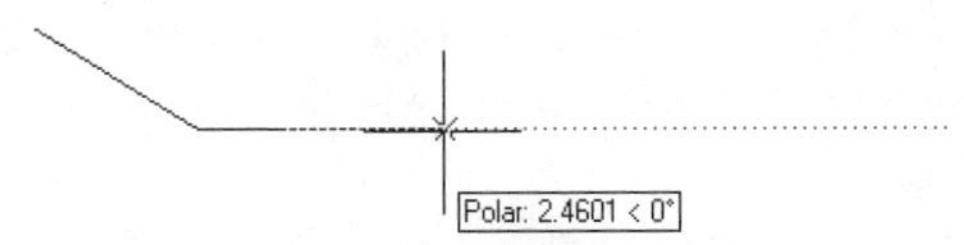

Polar tracking vector (dotted line) and AutoTrack tooltip.

☑ **Display full-screen tracking vector** displays the tracking vectors (TrackPath).

☑ **Display AutoTrack tooltip** displays the AutoTrack tooltip (AutoSnap).

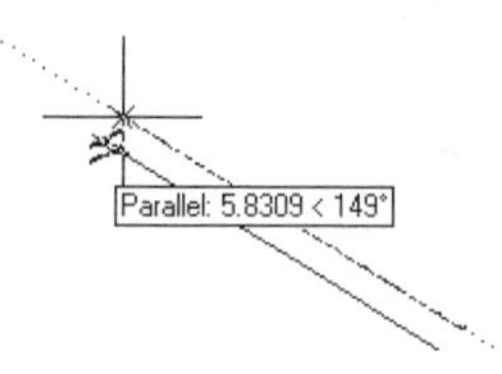

Full-screen tracking vector and AutoTrack tooltip.

Alignment Point Acquisition

⊙ **Automatic** displays tracking vectors automatically when aperture moves over an object snap.

○ **Shift to acquire** displays tracking vectors when pressing Shift and moving the aperture over an object snap.

Aperture Size

Aperture Size sets the size for the aperture; range is 1 to 50 pixels (Aperture); default = 10.

Drafing Tooltip Settings displays the Tooltip Appearance dialog box; see DSettings command.

Light Glyphs Settings displays the Light Glyphs Appearance dialog box (*new to AutoCAD 2007*).

Camera Glyphs Settings displays the Camera Glyphs Appearance dialog box (*new to AutoCAD 2007*).

3D Modeling tab

(New to AutoCAD 2007.)

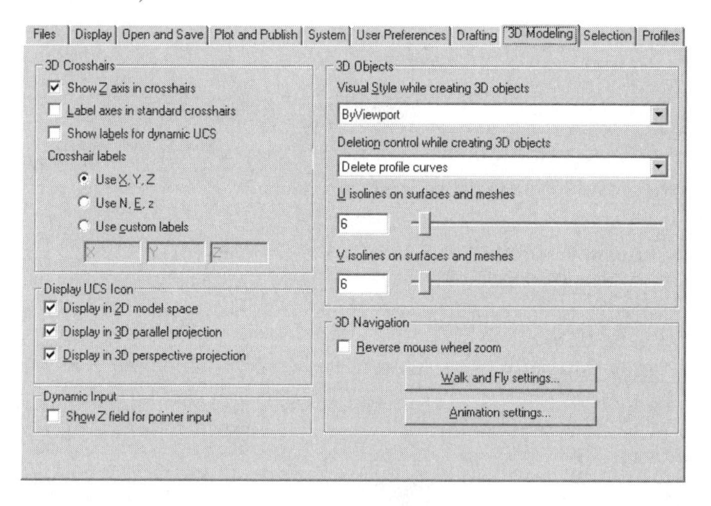

3D Crosshairs

☑ **Show Z Axis in Crosshairs** toggles display of z axis in crosshair cursor.

☐ **Label Axes in Standard Crosshairs** toggles whether labels (X, Y, and Z) are displayed with the crosshair cursor.

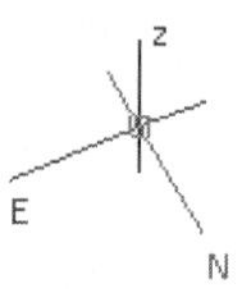

☐ **Show Labels for Dynamic UCS** displays labels during dynamic UCS, even when the axis labels are off.

☐ **Crosshair Labels**:

⊙ **Use X, Y, Z** labels the axes with X, Y, and Z.

○ **Use N, E, z** labels the axes with N (north), E (east), and z.

○ **Use Custom Labels** labels each axis with up to eight user-defined characters.

Display UCS Icon

☑ **Display in 2D Model Space** toggles display of the UCS icon in 2D Wireframe visual style.

☑ **Display in 3D Parallel Projection** toggles display of the UCS icon in parallel projection with non-2D visual styles.

☑ **Display in 3D Perspective Projection** toggles display of the UCS icon in perspective projection with non-2D visual styles.

Dynamic Input

☐ **Show Z Field for Pointer Input** toggles display of the z coordinate during dynamic input.

3D Objects

Visual Style While Creating 3D Objects while dragging 3D primitives into shape (DragVs system variable):

- **By Viewport** uses the viewport's setting (see the MView command).
- 2D Wireframe.
- 3D Hidden.
- 3D Wireframe.
- Conceptual.
- Realistic.

Deletion Control While Creating 3D Objects (DelObj):

- Retain defining geometry.
- Delete profile curves.
- Delete profile and path curves.
- Prompt to delete profile curves.
- Prompt to delete profile and path curves.

U and **V Isolines on Surfaces and Meshes** specifies the number of isolines in the u and v directions (SurfU and SurvV).

3D Navigation

☐ **Reverse Mouse Wheel Zoom** switches the direction of zoom when the mouse wheel is rolled (ZoomWheel).

Walk and Fly Settings displays the Walk and Fly Settings dialog box; see the WalkFlySettings command.

Animation Settings displays the Animation Settings dialog box; see the AniPath command.

Selection tab

This tab is displayed when the DdGrips command is entered.

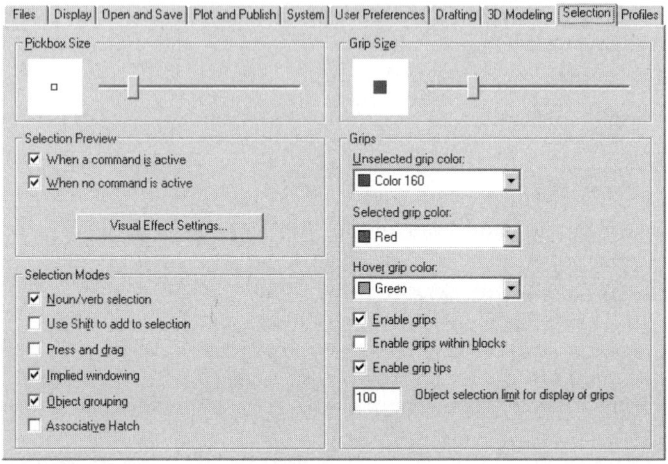

Pickbox Size specifies the size of the pickbox; range is 1 to 20 pixels (PickBox); default = 3.

Grip size specifies the size of grips; range is from 1 to 20 pixels (GripSize); default = 5.

Selection Preview

☑ **When a command is active** previews object selection during commands.

☑ **When no command is active** previews object selection as cursor passes over objects.

Visual Effect Settings displays the Visual Effects Settings dialog box.

Selection Modes

☑ **Noun/verb selection** allows you to select an object before executing an editing command (PickFirst).

☐ **Use Shift to add to selection** requires you to press Shift to add or remove objects from the selection set, a la Windows (PickAdd).

☐ **Press and drag** allows you to create the selection window by dragging (PickDrag).

☑ **Implied windowing** creates a selection window when you pick a point in the drawing that does not pick an object (PickAuto).

☑ **Object grouping** selects the entire group when an object in the group is selected (PickStyle).

☐ **Associative hatch** selects boundary objects, along with the associative hatch patterns (PickStyle).

Grips

Unselected grip color specifies color of unselected (cold) grips (GripColor); default = blue.

Selected grip color specifies the color of selected (hot) grip (GripHot); default = red.

Hover grip color specifies the color of the grip when the cursor hovers over it; default = green.

☑ **Enable grips** displays grips on selected objects (Grips).

☐ **Enable grips within blocks** displays all grips for every object in the selected block; when off, a single grip at the block's insertion point is displayed (GripBlock).

☑ **Enable grip tips** toggles the display of grip tips on custom objects (GripTips).

Object selection limit for display of grips limits the number of selected objects that display grips (GripObjLimit); default = 100.

Profiles tab

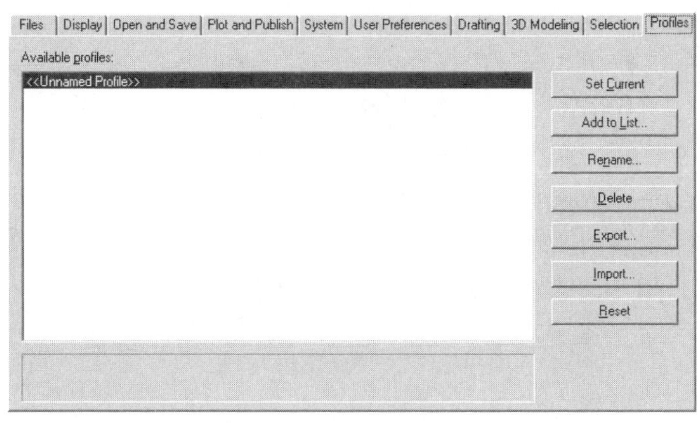

Set Current sets the selected profile as the current profile.

Add to List displays the Add Profile dialog box; allows you to enter a name and description for the new profile.

Rename displays the Change Profile dialog box; allows you to change the name and the description of the selected profile.

Delete erases the selected profile; the current profile cannot be erased.

Export exports profiles as *.arg* files.

Import imports *.arg* profiles into AutoCAD.

Reset resets the selected profile to AutoCAD's default settings.

COLOR OPTIONS DIALOG BOX

In the Display tab, click the Colors button:

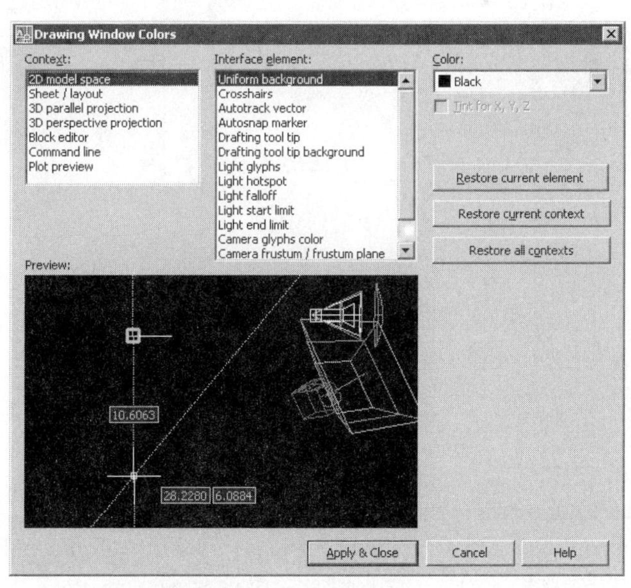

Context selects the user interface.

Interface element selects the individual interface item.

Color selects the color for the interface element.

☐ **Tint for X, Y, Z** selects colors for the crosshair cursor and 3D UCS icon's x, y, z axes.

Restore current element sets the selected item back to its default color.

Restore current context sets the selected UI back to its default colors.

Restore all contexts sets all UIs back to their default colors.

COMMAND LINE WINDOW FONT DIALOG BOX

In the Display tab, click the Font button:

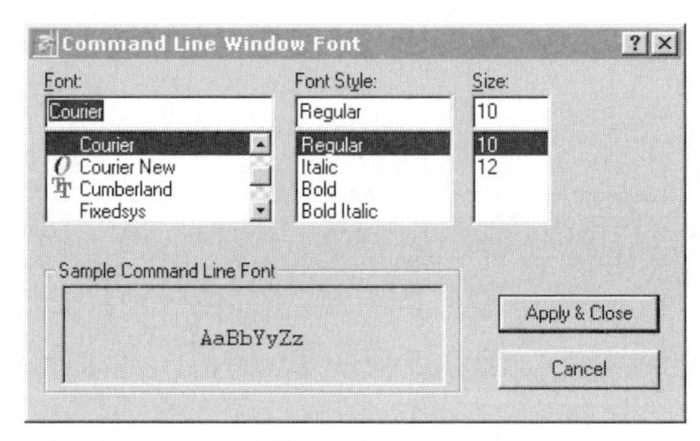

Font selects the font for the command line palette.

Font Style changes the style.

Size specifies the size of the text.

VISUAL EFFECT SETTINGS DIALOG BOX

In the Selection tab, click the Visual Effect Settings button:

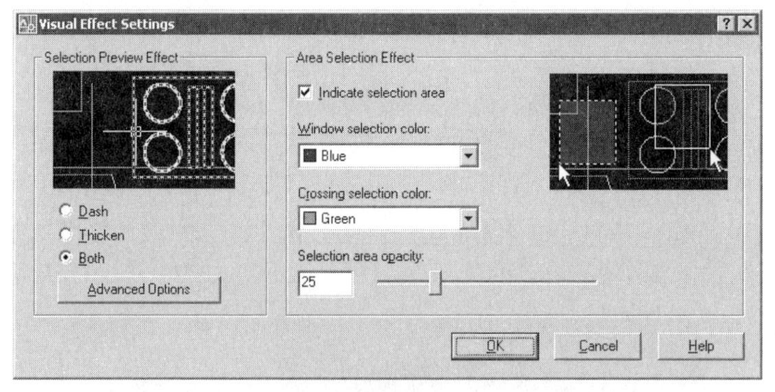

○ **Dash** changes selected object lines to dashes.

○ **Thicken** thickens selected object lines.

⊙ **Both** changes selected object lines to thick dashes.

Advanced Options displays the Advanced Options dialog box.

☑ **Indicated selection area** toggles filled selection areas.

Window selection color selects the color for Window selection mode.

Crossing selection color selects the color for Crossing selection mode.

Selection area opacity specifies the "see thru ness" of the selection area; 100=opaque.

Advanced Options dialog box

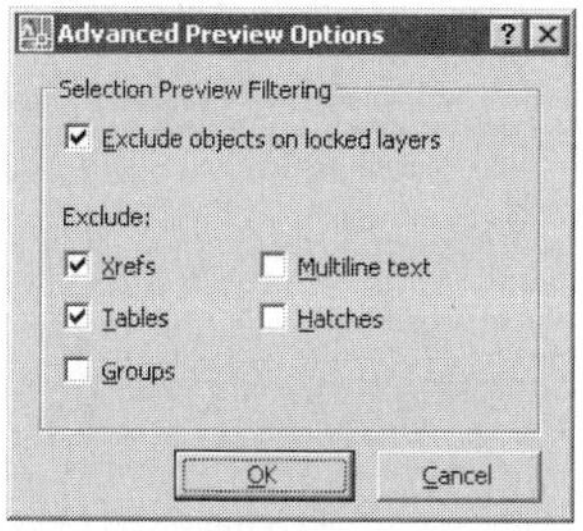

☑ **Exclude objects on locked layers** ignores objects on locked layers, which cannot be edited.

Exclude excludes specific object types.

PLOT STYLE TABLE SETTINGS DIALOG BOX

In the Plot and Publish tab, click the Plot Style Table Settings button:

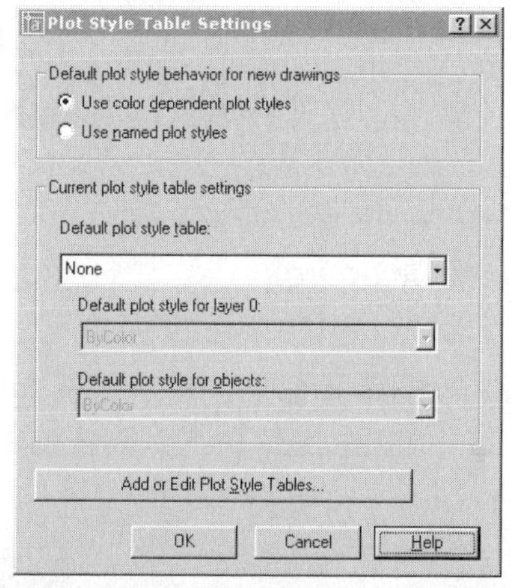

Default Plot Style Behavior for New Drawings (PStylePolicy).

⊙ **Use Color Dependent Plot Styles** uses color-dependent plot styles in new drawings; plotted colors are defined by a number between 1 and 255.

○ **Use Named Plot Styles** uses named plot styles in new drawings; objects are plotted based on plot style definitions.

Current Plot Style Table Settings

Default Plot Style Table selects the plot style table to attach to new drawings.

Default Plot Style for Layer 0 specifies the default plot style for Layer 0 in new drawings (DefLPlStyle).

Default Plot Style for Objects specifies the default plot style assigned to new objects (DefPlStyle). The list displays BYLAYER, BYBLOCK, Normal styles, and any plot styles defined in the currently-loaded plot style table.

Add or Edit Plot Style Tables displays the Plot Style Manager window; see the StylesManager command.

THUMBNAIL PREVIEW SETTINGS DIALOG BOX

In the Open and Save tab, click the Thumbnail Preview Settings button:

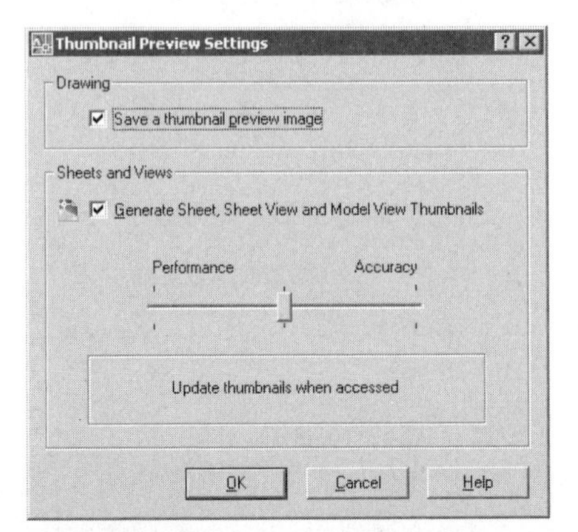

☑ **Save a Thumbnail Preview Image** saves the current view of the drawing as a bitmap preview image in the *.dwg* file (RasterPreview).

☑ **Generate Sheet, Sheet View and Model View Thumbnails** updates thumbnails.

Performance updates thumbnail previews only when UpdateThumbsNow is executed.

Accuracy updates thumbnail previews with every drawing save.

VISUAL EFFECT SETTINGS DIALOG BOX

In the User Preferences tab, click the Right Click Customization button:

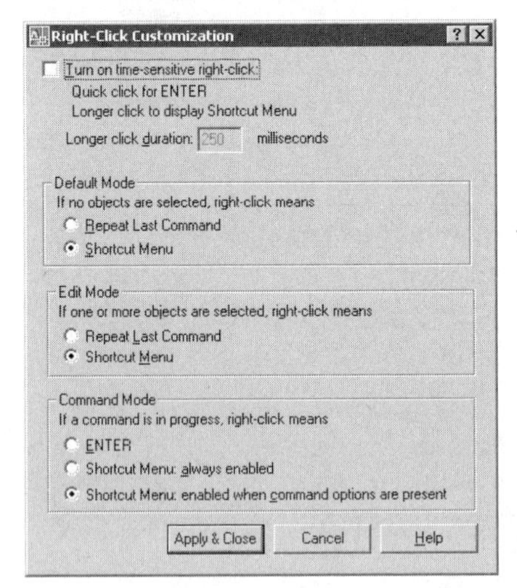

☐ **Turn on Time-Sensitive Right-Click** toggles Enter-or-shortcut menu.

Default Mode:

○ **Repeat Last Command** repeats the previous command, like pressing Enter.

⊙ **Shortcut Menu** displays the appropriate shortcut menu.

Edit Mode:

○ **Repeat Last Command** repeats the previous command, like pressing Enter.

⊙ **Shortcut Menu** displays the appropriate shortcut menu.

Command Mode during a command:

○ **ENTER** disables the shortcut menu, and executes Enter.

○ **Shortcut Menu: Always Enabled** enables the shortcut menu, and disables Enter.

⊙ **Shortcut Menu: enabled when command options are present** and if no options available, executes Enter.

FIELD UPDATE SETTINGS DIALOG BOX

In the User Preferences tab, click the Field Update Settings button:

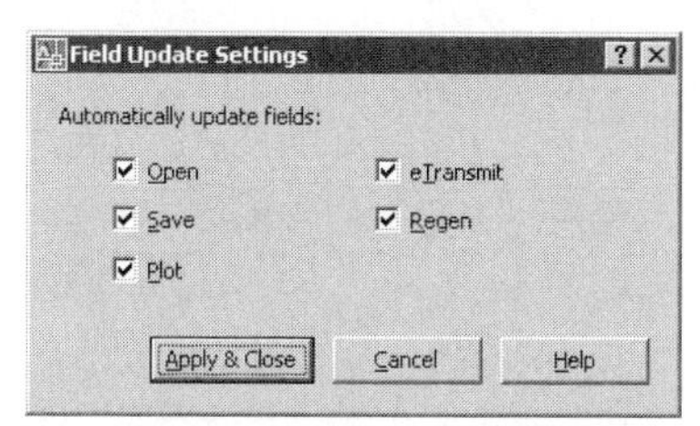

☑ Specifies which commands force AutoCAD to update field text.

LIGHT AND CAMERA GLYPH APPEARANCE DIALOG BOXES

In the Drafting tab, click the Light Glyph Settings or Camera Glyph Settings button:

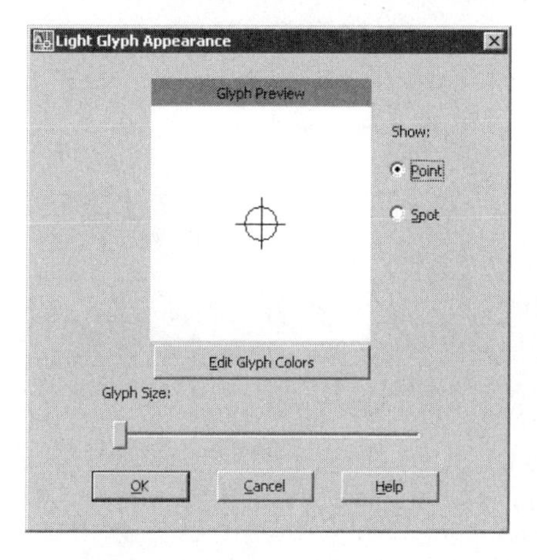

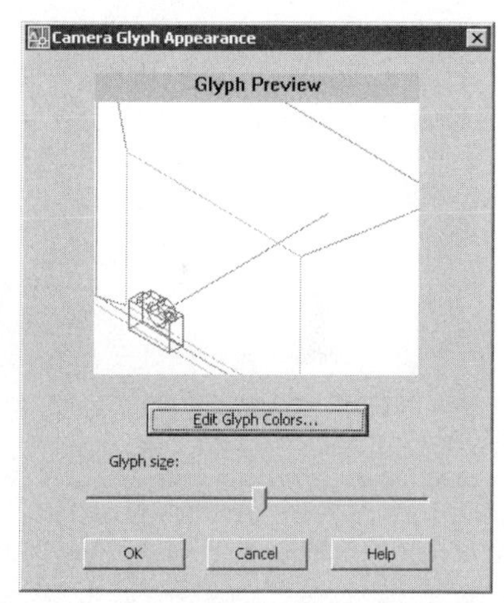

Edit Glyph Colors changes the color(s) of the glyphs; displays the Drawing Windows Color dialog box.

Glyphy size changes the size of the glyph.

Light Glyph Appearance

Show:

⊙ **Point** glyph for point lights (PointLight command).

○ **Spot** glyph for point lights (SpotLight command).

+OPTIONS COMMAND

Command: +options

Tab index <0>: *(Enter a digit between 0 and 9.)*

COMMAND LINE OPTION

Tab index specifies the tab to display:

0 — Files tab.

1 — Display tab.

2 — Open and Save tab.

3 — Plot and Publish tab.

4 — System tab.

5 — User Preferences tab.

6 — Drafting tab.

7 — 3D Modeling tab.

8 — Selection tab.

9 — Profiles tab.

TIPS

- *Grips* are small squares that appear on an object when the object is selected at the 'Command' prompt. In other Windows applications, grips are known as *handles*.

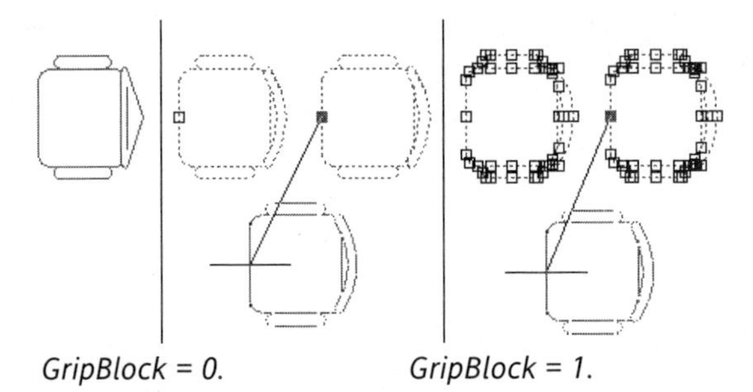

GripBlock = 0. *GripBlock = 1.*

- When an object is first selected, the grips are blue, and are called *unselected* or *cold* grips. When a grip is selected, it turns into a solid red square; this is called a *hot* grip. Press **ESC** to turn off unselected grips; press ESC twice to turn off hot grips.
- A larger pickbox makes it easier to select object generally, but also easier to select objects accidentally.
- The first time you use the DrawOrder command, it turns on all object sort method options.
- It's best to choose a crosshair color that is different from your drawing colors.

'Ortho

V. 1.0 Constrains drawing and editing commands to the vertical and horizontal directions only (short for ORTHOgraphic).

Command	Alias	Ctrl+	F-key	Alt+	Status Bar	Tablet
ortho	...	L	F8	...	ORTHO	...

Command: ortho
Enter mode [ON/OFF] <OFF>: *(Enter* **ON** *or* **OFF**.*)*

COMMAND LINE OPTIONS

OFF turns off ortho mode.

ON turns on ortho mode.

STATUS BAR OPTIONS

Ortho mode can be toggled on and off by clicking ORTHO on the status bar:

10.1046, 0.9275 , 0.0000 SNAP GRID ORTHO POLAR OSNAP

RELATED COMMANDS

DSettings toggles ortho mode via a dialog box.

Snap rotates the ortho angle.

RELATED SYSTEM VARIABLES

OrthoMode stores the current ortho modes.

PolarMode specifies polar and object snap tracking.

SnapAng specifies the rotation angle of the ortho cursor.

TIPS

- Use ortho mode when you want to constrain your drawing and editing to right angles.
- This command has been effectively replaced by the Polar command, which is not restricted to 90-degree increments.
- Rotate the angle of ortho with the Snap command's Rotate option.
- In isoplane mode, ortho mode constrains the cursor to the current isoplane.
- AutoCAD ignores ortho mode when you enter coordinates from the keyboard and in perspective mode; ortho is also ignored by object snap modes.
- Ortho is not necessarily horizontal or vertical; its orientation is determined by the current UCS and snap alignment.

'-OSnap

V. 2.0 Sets and turns on and off object snap modes at the command line (short for Object SNAP.

Command	Alias	Ctrl+	F-key	Alt+	Status Bar	Tablet
-osnap	-os	...	F3	...	OSNAP	T15 - U22

The OSnap command displays the Drafting Settings dialog box; see the DSettings command.

Command: -osnap
Current osnap modes: Ext
Enter list of object snap modes: *(Enter one or more modes separated by commas.)*

COMMAND LINE OPTIONS

You only need enter the first three letters as the abbreviation for each option:

APParent snaps to the intersection of two objects that don't physically cross, but appear to intersect on the screen, or would intersect if extended.

CENter snaps to the center point of arcs and circles.

ENDpoint snaps to the endpoint of lines, polylines, traces, and arcs.

EXTension snaps to the extension path of objects.

FROm extends from a point by a given distance.

INSertion snaps to the insertion point of blocks, shapes, and text.

INTersection snaps to the intersection of two objects, or to a self-crossing object, or to objects that would intersect if extended.

MIDpoint snaps to the middle point of lines and arcs.

NEArest snaps to the object nearest the crosshair cursor.

NODe snaps to a point object.

NONe turns off all object snap modes temporarily.

OFF turns off all object snap modes.

PARallel snaps to a parallel offset.

PERpendicular snaps perpendicularly to objects.

QUAdrant snaps to the quadrant points of circles and arcs.

QUIck snaps to the first object found in the database.

TANgent snaps to the tangent of arcs and circles.

STATUS BAR OPTIONS

Right-click ***OSNAP*** *on the status bar:*

On turns on previously-set object snap modes.

Off turns off all running object snaps.

Settings displays the Object Snap tab of the Drafting Settings dialog box.

SHORTCUT MENU

Hold down the Ctrl key, and then right-click anywhere in the drawing:

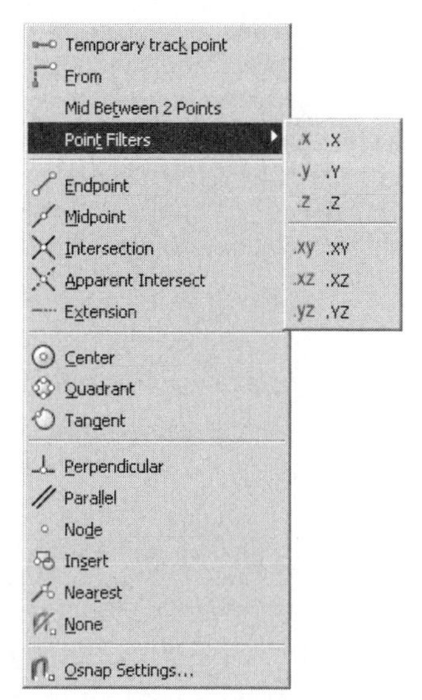

Temporary track point invokes Tracking mode; see the Tracking command.

From locates temporary points during drawing and editing commands. AutoCAD prompts:

From point: from
Base point: *(Pick a point.)*
<Offset>: *(Pick another point.)*

Mid Between 2 Points locates a point between two pick points.

From point: m2p
First point of mid: *(Pick a point.)*
Second point of mid: *(Pick another point.)*

Point filters invokes point filter modes:

.x — **need YZ:** *(Enter value for y and z.)*
.y — **need XZ:** *(Enter value for x and z.)*
.z — **need XY:** *(Enter value for x and y.)*
.xy — **need Z:** *(Enter value for z.)*
.xz — **need Y:** *(Enter value for y.)*
.yz — **need X:** *(Enter value for x.)*

Osnap Settings displays the Object Snap tab of the Drafting Settings dialog box; see **DSettings** command.

RELATED SYSTEM VARIABLES

Aperture specifies the size of the object snap aperture in pixels.

AutoSnap controls the display of AutoSnap (default = 63).

OsnapNodeLegacy determines whether the NODe object snap snaps to text insertion points.

OsOptions toggles between using the z coordinate and the current elevation setting (*new to AutoCAD 2007*).

OsnapCoord overrides object snaps when entering coordinates at 'Command' prompt.

0 — Object snap overrides keyboard.
1 — Keyboard overrides object snap settings.
2 — Keyboard overrides object snap settings, except during a script.

OsnapZ determines whether osnaps uses the z coordinate or the elevation.

OsMode stores the current object snap mode(s):

0 — NONe (default).
1 — ENDpoint.
2 — MIDpoint.
4 — CENter.
8 — NODe.
16 — QUAdrant.
32 — INTersection.
64 — INSertion.
128 — PERpendicular.
256 — TANgent.
512 — NEArest.
1024 — QUIck.
2048 — APParent intersection.
4096 — EXTension.
8192 — PARallel.

TempOverrides toggles temporary override keys.

TEMPORARY OVERRIDE KEYS

These keystrokes temporarily override object snaps during commands:

Override	Left Side	Right Side
Disables all osnaps and tracking	SHIFT+D	SHIFT+L
Enables osnap enforcement	SHIFT+S	SHIFT+;
Overrides CENter osnap	SHIFT+C	SHIFT+,
Overrides ENDpoint osnap	SHIFT+E	SHIFT+P
Overrides MIDpoint osnap	SHIFT+V	SHIFT+M
Toggles osnap mode	SHIFT+A	SHIFT+'

TIPS

- The Aperture command controls the drawing area through which AutoCAD searches.
- If AutoCAD finds no snap matching the current modes, then the pick point is selected.
- The **m2p** modifier can also be entered as **mtp**.
- The APPint and INT object snap modes should not be used together.
- To turn on more than one object snap at a time, use a comma to separate mode names:

Enter list of object snap modes: int,end,qua

- The location of object snaps:

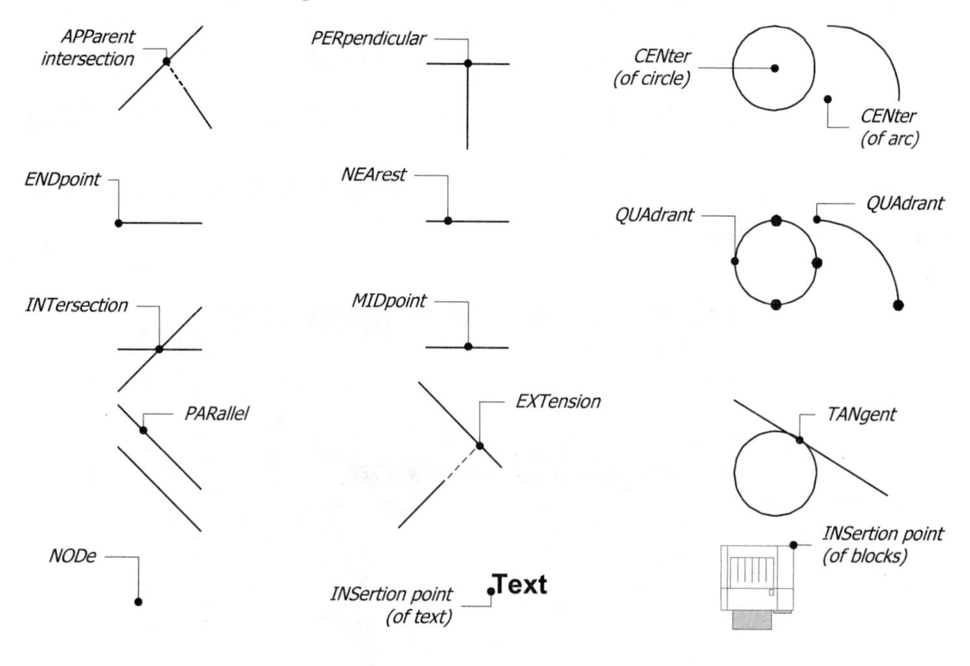

PageSetup

2000 Sets up model and layout views in preparation for plotting drawings.

Command	Alias	Ctrl+	F-key	Alt+	Menu Bar	Tablet
pagesetup	...	...	...	**FG**	**File** **↳Page Setup**	**V25**

Command: pagesetup

Displays dialog box:

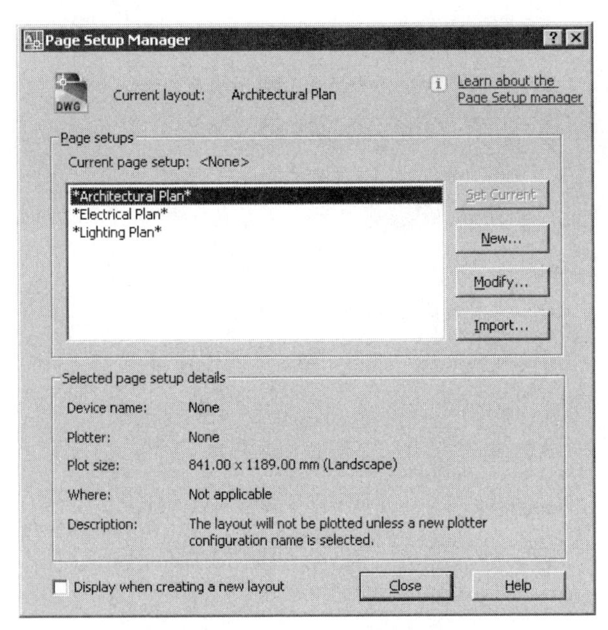

PAGE SETUP MANAGER DIALOG BOX

Set Current selects the page setup to be used by the drawing.

New displays the New Page Setup dialog box.

Modify displays the Page Setup dialog box.

Import displays the Select Page Setup From File dialog box; select a *.dwg*, *.dwt*, or *.dxf* file, and then click Open; displays the Import Page Setups dialog box.

☐ **Display when creating a new layout** displays this dialog box the first time you click a layout tab in new drawings.

NEW PAGE SETUP DIALOG BOX

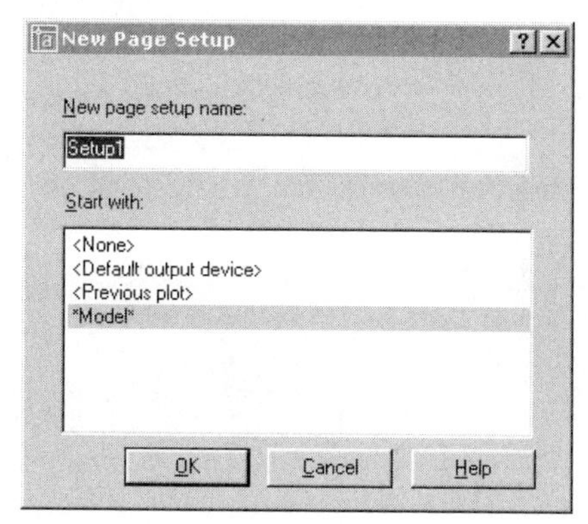

New Page Setup Name specifies the name of the setup.

Start With selects the default settings:

> **<None>** — no page setup is used as the template.
>
> **<Default Output Device>** — selects the plotter specified as the default output device in the Options dialog box.
>
> **<Previous Plot>** — uses settings from a previous plot job.
>
> * *Layout Name** — selects an existing layout.

OK displays the Page Setup dialog box.

PAGE SETUP DIALOG BOX

See Plot command.

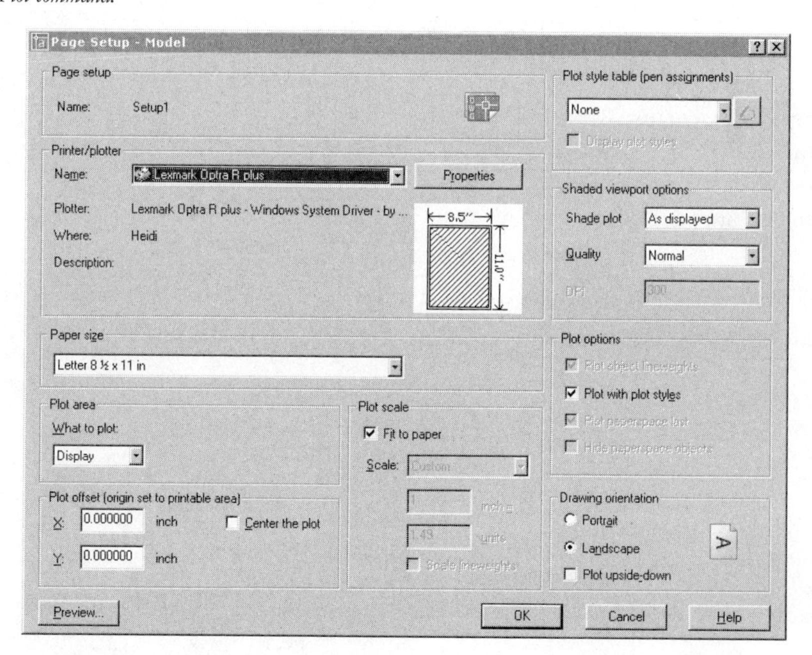

IMPORT PAGE SETUPS DIALOG BOX

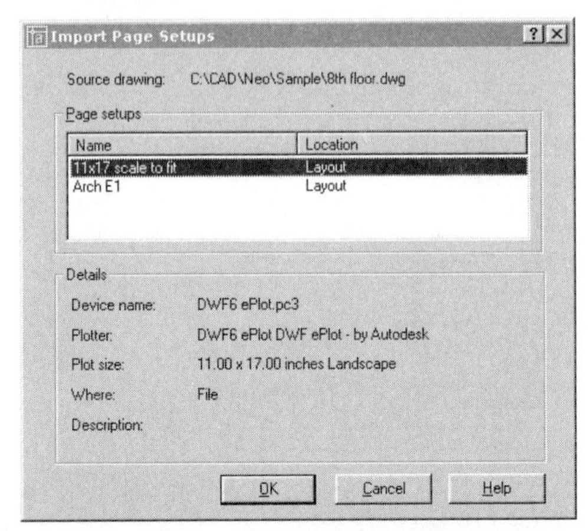

Name lists the names of page setups found in the drawing, if any.

Location indicates the location of the page setups: model or layout.

OK returns to the Page Setup Manager dialog box.

SHORTCUT MENU

Right-click a page setup name:

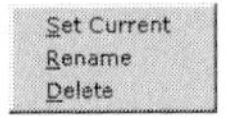

Set Current selects the page setup to use for the drawing.

Rename changes the name of the page setup.

Delete erases the page setup from the drawing, without warning.

RELATED COMMANDS

Layout creates new layouts.

LayoutWizard creates layouts and page setups.

Plot plots drawings based on the settings of the PageSetup command.

TIPS

- The sheet set icon shows when this command is activated from the Sheet Set Manager palette.
- Right-click **Model** or **Layout** tabs, and then select **Page Setup Manager**.

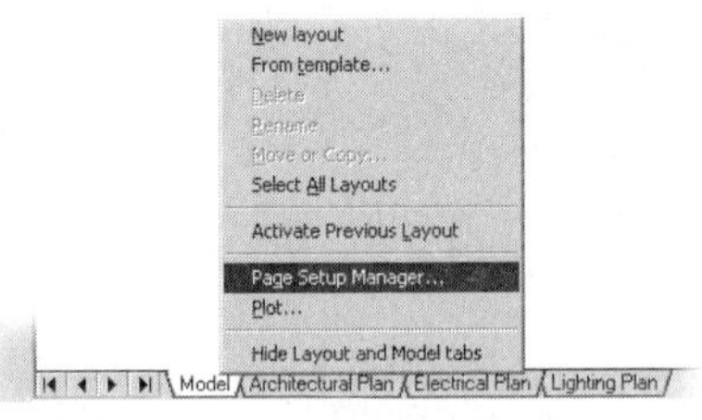

- Page setups cannot be applied to entire sheet sets after they are created.

'Pan

V. 1.0 Moves the view in the current viewport.

Commands	Aliases	Ctrl+	F-key	Alt+	Menu Bar	Tablet
pan	p rtpan 3dpantransparent	...	...	VPT	View ↳Pan ↳Realtime	N11- P11
-pan	-p			VPP	View ↳Pan ↳Point	

Command: pan

Press Esc or Enter to exit, or right-click to display shortcut menu. *(Move cursor to pan, and then press* ESC *to exit command.)*

Enters real-time panning mode, and displays hand cursor:

Drag the hand cursor to pan the drawing in the viewport.

Press Enter or ESC *to return to the 'Command' prompt.*

SHORTCUT MENU OPTIONS

During real-time pan mode, right-click the drawing to display shortcut menu:

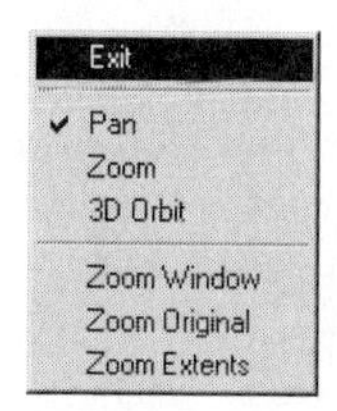

Exit exits real-time pan mode; returns to the 'Command' prompt.

Pan switches to real-time pan mode, when in real-time zoom mode.
Zoom switches to real-time zoom mode; see the Zoom command.
3D Orbit switches to 3D orbit mode; see the 3dOrbit command.

Zoom Window prompts you, 'Press pick button and drag to specify zoom window.'
Zoom Original returns to the view when you first started the Pan command.
Zoom Extents displays the entire drawing.

-PAN Command

Command: -pan
Displacement: *(Pick a point, or enter x,y-coordinates.)*
Second point: *(Pick another point, or enter x,y-coordinates.)*

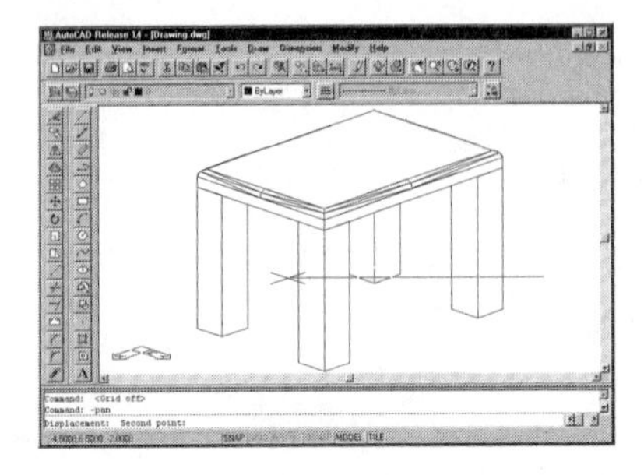

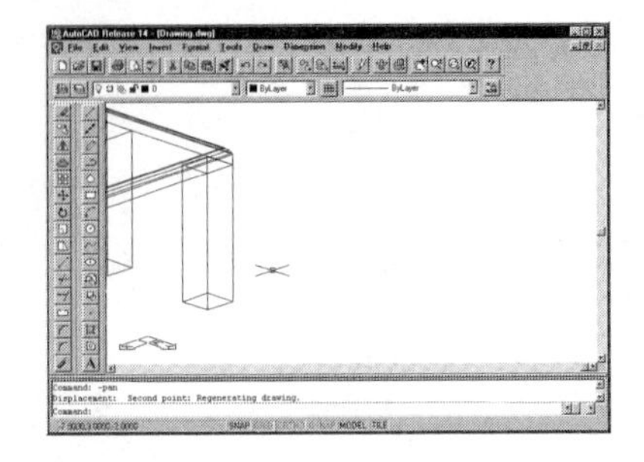

Left: *Before panning to the left.*
Right: *After panning to the left:*

COMMAND LINE OPTIONS

ENTER or **ESC** exits real-time panning mode.

Displacement specifies the distance and direction to pan the view.

Second point pans to this point.

RELATED COMMANDS

DsViewer displays the Aerial View palette, which pans in an independent palette.

RegenAuto determines how regenerations are handled.

View saves and restores named views.

Zoom pans with the **Dynamic** option.

3dOrbit pans during perspective mode.

RELATED SYSTEM VARIABLES

MButtonPan determines the action of a mouse's third button or wheel.

RtDisplay toggles the display of raster and OLE images during realtime panning.

VtDuration specifies the duration of view transitions.

VtEnable turns on smooth view transitions.

VtFps specifies the speed of view transitions.

TIPS

- You can pan each viewport independently.
- Use **'Pan** to start drawing objects in one area of the drawing, pan over, and then continue working in another area of the drawing.
- You cannot use transparent pan under the following conditions: perspective mode, VPoint, DView, another Pan command, View, Zoom, or 3dOrbit.
- When the drawing no longer moves during real-time panning, you have reached the panning limit; AutoCAD changes the hand icon to show the limit. (Use Regen to increase the limit.)

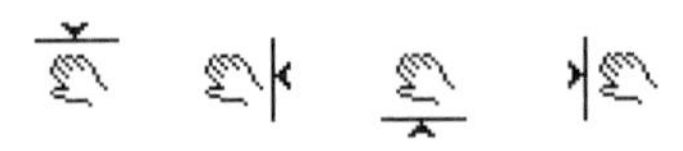

- As an alternative, use the horizontal and vertical scroll bars to pan the drawing.
- With MButtonPan set to 1, pan by holding down the wheelbutton and moving the mouse.

PartialCui

2006 Loads *.cui* files into AutoCAD.

Command	Alias	Ctrl+	F-key	Alt+	Menu Bar	Tablet
partialcui	...	...	...	TC	**Tools** ↳**Customize** ↳**Import Customizations**	...

Command: partialcui

Displays Select Customization File dialog box.

Select a .cui, .mnu, or .mns file, and then click Open.

TIPS

- The **U** command reverses the effect of this command.
- This command is meant to replace the MenuLoad command.

PartiaLoad

2000 Loads additional views and layers of a partially-loaded drawing; this command works only with drawings that have been partially loaded.

Commands	Alias	Ctrl+	F-key	Alt+	Menu Bar	Tablet
partialload	...	...	...	**FL**	**File** **↳Partial Load**	
-partialload						

Command: partialload

If the current drawing has not been partially opened through the Open command's Partial Open option, AutoCAD complains, 'Command not allowed unless the drawing has been partially opened.'

Otherwise, displays dialog box:

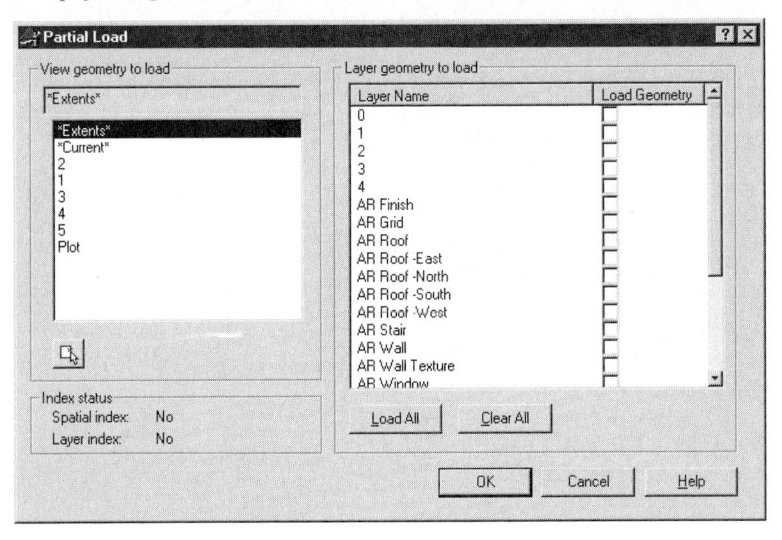

PARTIAL LOAD DIALOG BOX

View geometry to load selects the model space views to load. (Paper space views are not available for partial loading.)

Pick a Window button dismisses the dialog box temporarily to allow you to specify an area that becomes the view to load. AutoCAD prompts you:

Specify first corner: *(Pick a point.)*

Specify opposite corner: *(Pick a point.)*

When the Partial Load dialog box returns, *New View* is listed by the View Geometry to Load list.

Layer Geometry to Load

Load Geometry selects the layers to load.

Load All loads all layers.

Clear All deselects all layer names.

-PARTIALOAD Command

Command: -partiaload

Specify first corner or [View]: *(Pick a point, or type **V**.)*

Enter view to load or [?] <*Extents*>: *(Enter a view name, or type **?**.)*

Enter layers to load or [?] <none>: *(Enter a layer name, press **ENTER** for no layers, or type **?**.)*

COMMAND LINE OPTIONS

Specify first corner specifies a corner to create a new view.

Specify opposite corner specifies the second corner; causes geometry in the new view to be loaded into the drawing.

View options

Enter view to load loads the geometry found in the view into the drawing.

? lists the names of views in the drawing:

Enter view name(s) to list <*>: *(Enter a view name, or type *****.)*

Sample display:

Saved model space views:
View name"1"
"Plot"

Extents loads all geometry into the drawing.

Enter layers to load loads geometry on the layers into the drawing.

? lists the names of layers in the drawing:

Enter layers to list <*>: *(Press Enter.)*

Sample display:

Layer names:
"0"
"3"

None loads no layers.

RELATED COMMANDS

Open partially opens a drawing via a dialog box.

PartialOpen partially opens a drawing via the command line.

RELATED SYSTEM VARIABLE

FullOpen indicates whether the drawing is fully or partially opened.

-PartialOpen

2000 Opens a drawing and loads selected layers and views.

Command	Alias	F-key	Alt+	Menu Bar	Tablet
-partialopen	partialopen	...	...	...	...

Command: -partialopen
Enter name of drawing to open <filename.dwg>: *(Enter a file name.)*
Enter view to load or [?]<*Extents*>: *(Enter a view name, or type* **?***.)*
Enter layers to load or [?]<none>: *(Enter a layer name, or type* **?***.)*
Unload all Xrefs on open? [Yes/No] <N>: *(Type* **Y** *or* **N***.)*

COMMAND LINE OPTIONS

Enter name of drawing to open specifies the name of the DWG file.

Enter view to load loads the geometry found in the view into the drawing.

? lists the names of the views in the drawing.

Extents loads all geometry into the drawing.

Enter layers to load loads geometry on the layers into the drawing.

? lists the names of layers in the drawing.

None loads no layers.

Unload all Xrefs on open:

Yes does not load externally-referenced drawings.

No loads all externally-referenced drawings.

RELATED COMMANDS

Open displays the Select Drawing dialog box; includes the **Partial Open** option.

PartialLoad loads additional views or layers of a partially-opened drawing.

RELATED SYSTEM VARIABLES

FullOpen indicates whether the drawing is fully or partially opened.

TIPS

- **PartialOpen** is an alias for the **-PartialOpen** command.
- As an alternative, you may use the **Partial Open** option of the **Open** command's Select File dialog box.

PasteAsHyperlink

2000 Pastes object as hyperlinks in drawings; works only if the Clipboard contains appropriate data.

Command	Alias	Ctrl+	F-key	Alt+	Menu Bar	Tablet
pasteashyperlink	...	...	...	EH	**Edit** ↳**Paste as Hyperlink**	...

Command: pasteashyperlink
Select objects: *(Select one or more objects.)*
Select objects: *(Press Enter.)*

COMMAND LINE OPTION

Select objects selects the objects to which the hyperlink will be pasted.

RELATED COMMANDS

CopyClip copies a hyperlink to the Clipboard.

Hyperlink adds hyperlinks to selected objects.

SelectURL highlights all objects with hyperlinks.

RELATED SYSTEM VARIABLES

None.

PasteBlock

2000 Pastes objects as blocks in drawings; works only when the Clipboard contains AutoCAD objects.

Command	Alias	Ctrl+	F-key	Alt+	Menu Bar	Tablet
pasteblock	...	Shift+V	...	EK	Edit ↳Paste as Block	...

Command: pasteblock
Specify insertion point: *(Pick a point.)*

COMMAND LINE OPTION

Specify insertion point specifies the position for the block.

RELATED COMMANDS

Insert inserts blocks in the drawing.

CopyClip copies a block to the Clipboard.

RELATED SYSTEM VARIABLES

None.

TIPS

- This command does not work when the Clipboard contains data that cannot be pasted as a block.
- This command pastes any AutoCAD object as a block, generating a generic block name similar to "A$C65D94228."
- If the Clipboard contains a block, the block is nested by this command.
- As of AutoCAD 2004, you can use the CTRL+SHIFT+V shortcut.

PasteClip

Rel.13 Places objects from the Clipboard in drawings (short for PASTE CLIPboard).

Command	Alias	Ctrl+	F-key	Alt+	Menu Bar	Tablet
pasteclip	...	V	...	EP	**Edit** ⇨**Paste**	U13

Command: pasteclip

When the Clipboard contains AutoCAD objects, the following prompt is displayed:

Specify insertion point: *(Pick a point.)*

COMMAND LINE OPTION

Specify insertion point specifies the lower-left corner of the pasted object.

SHORTCUT MENU

Right-click pasted objects to display shortcut menu. Not all options appear with all pasted objects.

OLE displays submenu:

Open opens the object in its source application; does not work with static ActiveX objects.

Reset resets the object to its original settings.

Text Size displays the Text Size dialog box; available only when the object contains text.

Convert converts (or, in most cases, does not convert) the object to another format; displays dialog box:

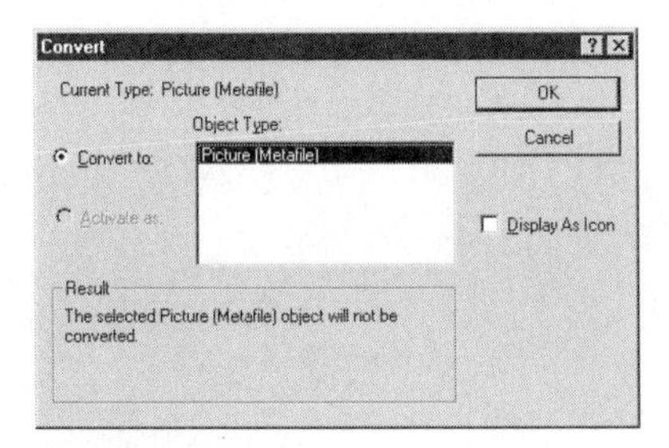

RELATED COMMANDS

CopyClip copies selected objects to the Clipboard.

Insert inserts an AutoCAD drawing in the drawing.

InsertObj inserts an OLE object in the drawing.

PasteSpec places the Clipboard object as pasted or linked objects.

RELATED SYSTEM VARIABLE

OleHide toggles the display of the OLE object (1 = off).

TIPS

- Graphical objects are placed in the drawing as OLE objects.
- Text is usually — but not always — placed in the drawing as Mtext objects.
- Use the **PasteBlock** command to paste objects as an AutoCAD block.

PasteOrig

2000 Pastes AutoCAD objects from the Clipboard into the current drawing at the objects' original locations (short for PASTE ORIGinal).

Command	Alias	Ctrl+	F-key	Alt+	Menu Bar	Tablet
pasteorig	...	...	...	**ED**	**Edit** **↳Paste to Original Coordinates**	...

Command: pasteorig

COMMAND LINE OPTIONS

None.

RELATED COMMANDS

CopyClip copies a drawing to the Clipboard.

Insert inserts an AutoCAD drawing in the drawing.

PasteBlock pastes AutoCAD objects as a block at a user-specified insertion point.

RELATED SYSTEM VARIABLES

None.

TIPS

- Use this command to copy objects from one drawing to another.
- This command cannot be used to paste objects into the drawing from which they originate.

PasteSpec

Rel.13 Pastes Clipboard objects in the current drawing as embedded, linked, pasted, or AutoCAD objects (short for PASTE SPECial).

Command	Alias	Ctrl+	F-key	Alt+	Menu Bar	Tablet
pastespec	**pa**	...	...	**ES**	**Edit** ↳**Paste Special**	...

Command: pastespec

Displays dialog box:

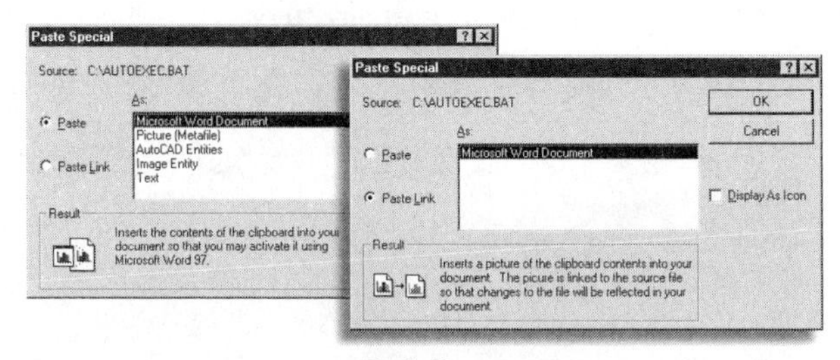

Left: *Paste.*
Right: *Paste Link.*

PASTE SPECIAL DIALOG BOX

⊙**Paste** pastes the object as an embedded object.

○**Paste Link** pastes the object as a linked object.

☐ **Display as Icon** displays the object as an icon from the originating application.

Change Icon allows you to select the icon; displays dialog box:

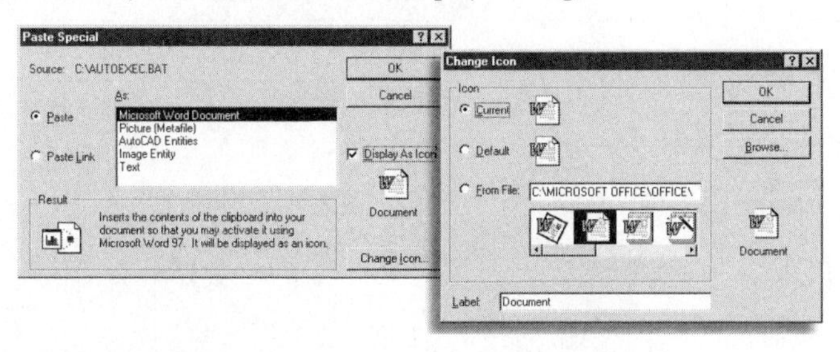

RELATED COMMANDS

CopyClip copies the drawing to the Clipboard.

InsertObj inserts an OLE object in the drawing.

OleLinks edits the OLE link data.

PasteClip places the Clipboard object as a pasted object.

RELATED SYSTEM VARIABLE

OleHide toggles the display of OLE objects (1 = off).

PcInWizard

2000 Converts PCP and PC2 plot configurations to PC3 format.

Command	Alias	Ctrl+	F-key	Alt+	Menu Bar	Tablet
pcinwizard	...	...	...	TZI	Tools ↳Wizard ↳Import Plot Settings	...

Command: pcinwizard

Displays dialog box.

IMPORT PCP OR PC2 PLOT SETTINGS DIALOG BOX

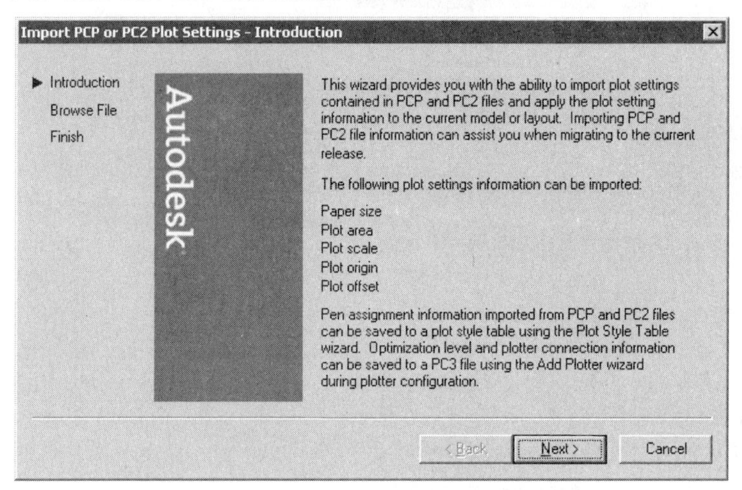

Back returns to the previous step.

Next continues to the next step.

Cancel cancels the wizard.

Browse File page

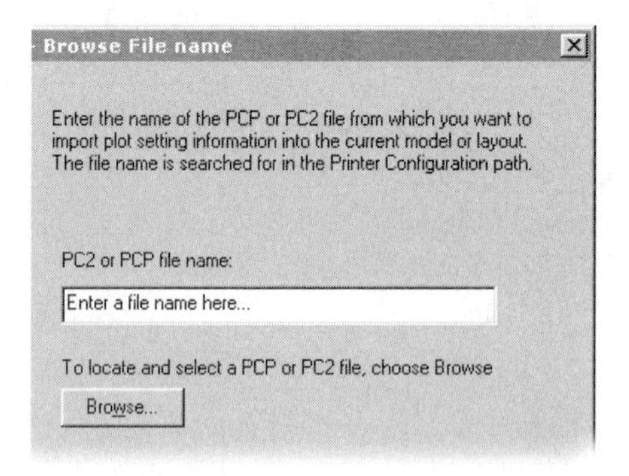

PCP or PC2 filename specifies the name of the *.pcp* or *pc2* file.

Browse displays the Import dialog box.

Finish page

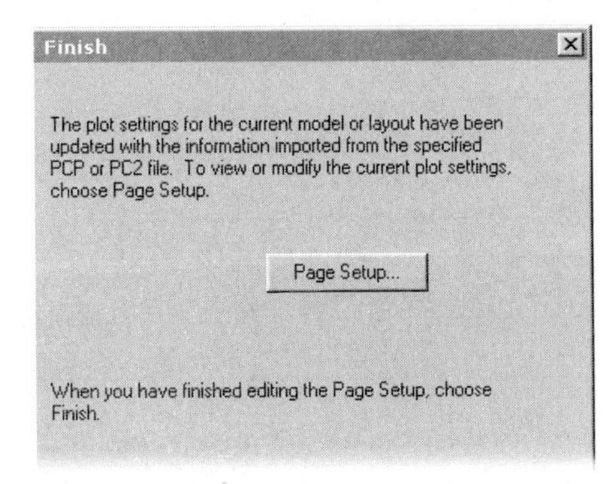

Page Setup displays the Page Setup dialog box; see the PageSetup command.

Finish completes the importation.

RELATED COMMANDS

PageSetup creates and modifies page setups.

Plot uses PC3 files to control the plotter configuration.

RELATED SYSTEM VARIABLES

None.

TIPS

- This command imports *.pcp* and *.pc2* files, and applies them to the current layout or model tab.
- Only use this command when have you created *.pcp* or *.pc2* files with earlier versions of AutoCAD.
- PCP is short for "plotter configuration parameters":
  - *.pcp* files are used by AutoCAD Release 13.
  - *.pc2* files are used by AutoCAD Release 14.
  - *.pc3* files are used by AutoCAD 2000 to now.
- AutoCAD imports the following information from *.pcp* and *.pc2* files: paper size, plot area, plot scale, plot origin, and plot offset.
- To import color-pen mapping, use the Plot Style Table wizard, run by the *StyShWiz.Exe* program.
- To import the optimization level and plotter connection, use the Add-a-Plotter wizard, run by the *AddPlWiz.Exe* program.

PEdit

V. 2.1 Edits 2D polylines, 3D polylines, or 3D meshes — depending on the selected object (short for Polyline EDIT).

Command	Alias	Ctrl+	F-key	Alt+	Menu Bar	Tablet
pedit	**pe**	...	...	**MOP**	**Modify** ↳**Object** ↳**Polyline**	**Y17**

Command: pedit

Options vary, depending on whether a 2D polyline, 3D polyline, or polymesh is picked.

2D Polyline Operations

Select polyline or [Multiple]: *(Pick a 2D polyline, or type* **M.***)*

Enter an option [Open/Join/Width/Edit vertex/Fit/Spline/Decurve/Ltype gen/Undo]: *(Enter an option.)*

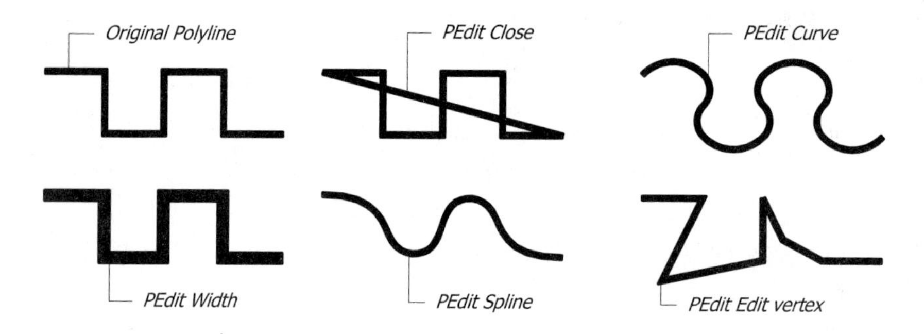

COMMAND LINE OPTIONS

Multiple selects more than one polyline to edit.

Close closes an open polyline by joining the two endpoints with a single segment.

Decurve reverses the effects of **Fit** and **Spline** operations.

Edit vertex options are listed below.

Fit fits a curve to the tangent points of each vertex.

Ltype gen specifies the linetype generation style.

Join adds other polylines to the current polyline.

Open opens a closed polyline by removing the last segment.

Spline fits a splined curve along the polyline.

Undo undoes the most recent PEdit operation.

Width changes the width of the entire polyline.

eXit exits the PEdit command.

Edit Vertex options
Enter a vertex editing option [Next/Previous/Break/Insert/Move/Regen/Straighten/Tangent/Width/eXit] <N>: *(Enter an option.)*

*Note: It is much easier to use grips to edit polyline vertices than this **Edit vertex** option:*

Break removes a segment or breaks the polyline at a vertex:

Next moves the x-marker to the next vertex.

Previous moves the x-marker to the previous vertex.

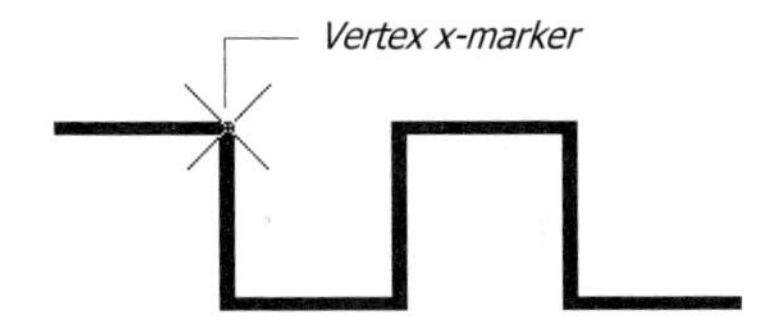

Go performs the break.

eXit exits the **Break** sub-submenu.

Insert inserts another vertex.

Move relocates a vertex; more easily accomplished with grips editing.

Next moves the x-marker to the next vertex.

Previous moves the x-marker to the previous vertex.

Regen regenerates the screen to show the effect of the PEdit command.

Straighten draws a straight segment between two vertices:

Next moves the x-marker to the next vertex.

Previous moves the x-marker to the previous vertex.

Go performs the straightening.

eXit exits the **Straighten** sub-submenu.

Tangent shows the tangent to the current vertex.

Width changes the width of a segment.

eXit exits the **Edit vertex** submenu.

3D Polyline Operations
Command: pedit
Select polyline or [Multiple]: *(Pick a 3D polyline, or type **M**.)*
Enter an option [Open/Edit vertex/Spline curve/Decurve/Undo]: *(Enter an option.)*

COMMAND LINE OPTIONS

Multiple selects more than one 3D polyline to edit.

Close closes an open polyline.

Decurve reverses the effects of the **Fit curve** and **Spline curve** operations.

Open removes the last segment of a closed polyline.

Spline curve fits a splined curve along the polyline.

Undo undoes the most recent PEdit operation.

eXit exits the PEdit command.

Edit Vertex options

Enter a vertex editing option
[Next/Previous/Break/Insert/Move/Regen/Straighten/eXit] <N>: *(Enter an option.)*

Break removes a segment or breaks the polyline at a vertex:

- **Next** moves the x-marker to the next vertex.
- **Previous** moves the x-marker to the previous vertex.
- **Go** performs the break.
- **eXit** exits the **Break** sub-submenu.

Insert inserts another vertex.

Move relocates a vertex; more easily accomplished with grips editing.

Next moves the x-marker to the next vertex.

Previous moves the x-marker to the previous vertex.

Regen regenerates the screen to show the effect of PEdit options.

Straighten draws a straight segment between two vertices:

- **Next** moves the x-marker to the next vertex.
- **Previous** moves the x-marker to the previous vertex.
- **Go** performs the straightening.
- **eXit** exits the **Straighten** sub-submenu.

eXit exits the **Edit vertex** submenu.

3D Mesh Operations

Select polyline or [Multiple]: *(Pick a 3D mesh, or type* **M.***)*
Enter an option [Edit vertex/Smooth surface/Desmooth/Mclose/Nclose/Undo]: *(Enter an option.)*

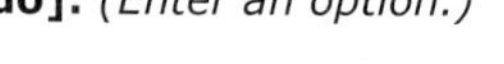

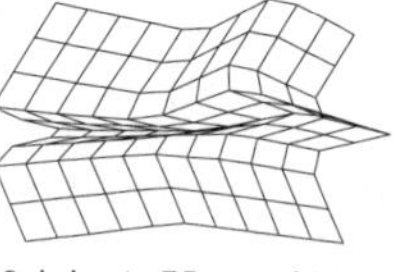

Original 3D surface

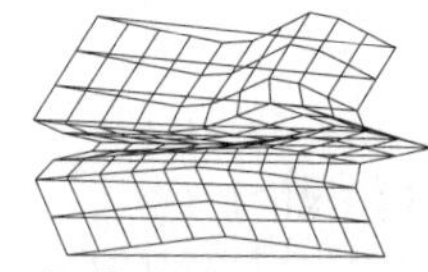

Pedit Mclose

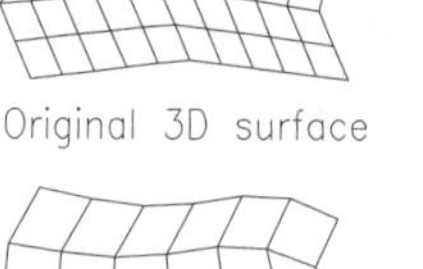

Pedit Smooth surface

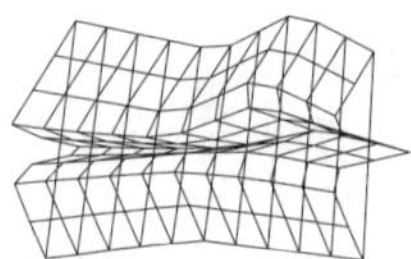

Pedit Nclose

COMMAND LINE OPTIONS

Multiple selects more than one 3D mesh to edit.

Desmooth reverses the effect of the **Smooth** surface options.

Mclose closes the mesh in the m-direction.

Mopen opens the mesh in the m-direction.

Nclose closes the mesh in the n-direction.

Nopen opens the mesh in the n-direction.

Smooth surface smooths the mesh with a Bezier-spline.

Undo undoes the most recent PEdit operation.

eXit exits the PEdit command.

Edit Vertex Options

Current vertex (0,0).

Enter an option [Next/Previous/Left/Right/Up/Down/Move/REgen/eXit] <N>: *(Enter an option.)*

Down moves the x-marker down the mesh by one vertex.

Left moves the x-marker along the mesh by one vertex left.

Move relocates the vertex to a new position; more easily accomplished with grips editing.

Next moves the x-marker along the mesh to the next vertex.

Previous moves the x-marker along the mesh to the previous vertex.

REgen regenerates the drawing to show the effects of the PEdit command.

Right moves the x-marker along the mesh by one vertex right.

Up moves the x-marker up the mesh by one vertex.

eXit exits the **Edit vertex** submenu.

RELATED COMMANDS

Break breaks 2D polylines at any position.

Chamfer chamfers all vertices of 2D polylines.

Convert converts older polylines to the new lwpolyline format.

EdgeSurf draws 3D meshes.

Fillet fillets all vertices of 2D polylines.

Join joins like objects, as well as lines, arcs, and polylines into a single polyline.

PLine draws 2D polylines.

RevSurf draws 3D meshes of revolution.

RuleSurf draws 3D ruled meshes.

TabSurf draws 3D tabulated meshes.

3D draws 3D surface objects.

3dPoly draws 3D polylines.

RELATED SYSTEM VARIABLES

SplFrame determines the visibility of a polyline spline frame:

0 — Does not display control frame (default).
1 — Displays control frame.

SplineSegs specifies the number (-32768 to 32767) of lines or arcs used to draw a splined polyline (default = 8); when the number is negative, arc segments are used, when positive, line segments.

SplineType determines the Bezier-spline smoothing for 2D and 3D polylines:

5 — Quadratic Bezier spline.
6 — Cubic Bezier spline.

SurfType determines the smoothing using the **Smooth** option:

5 — Quadratic Bezier spline.
6 — Cubic Bezier spline.
7 — Bezier surface.

TIPS

- During vertex editing, move the x-marker by pressing the spacebar, the **ENTER** key, right-clicking, or pressing mouse button #2. The marker is moved to the next or previous vertex, whichever was used last.
- It's easier to edit the position of vertices with grips, than with this command.

PFace

Rel.11 Draws multisided 3D meshes; intended for use by AutoLISP and ARx programs (short for Poly FACE).

Command	Alias	Ctrl+	F-key	Alt+	Menu Bar	Tablet
pface	...	...	...	...	...	...

Command: pface
Specify location for vertex 1: *(Pick point 1.)*
Specify location for vertex 2 or <define faces>: *(Pick point 2, or press Enter.)*
Face 1, vertex 1:
Enter a vertex number or [Color/Layer]: *(Type a number, or enter an option.)*
Face 1, vertex 2:
Enter a vertex number or [Color/Layer] <next face>: *(Type a number, or enter an option.)*

...and so on.

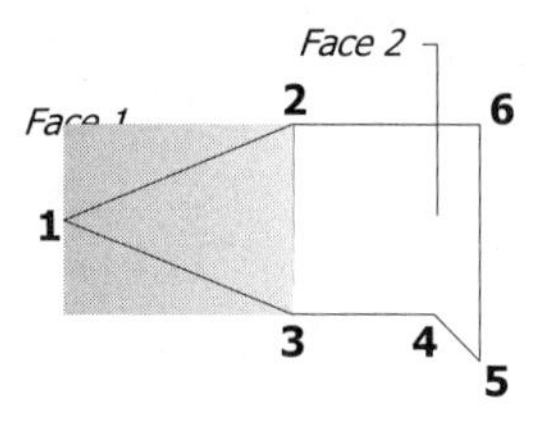

COMMAND LINE OPTIONS

Vertex defines the location of a vertex.

Face defines the faces, based on vertices.

Color gives the face a different color.

Layer places the face on a different layer.

RELATED COMMANDS

3dFace draws three- and four-sided 3D meshes.

3dMesh draws a 3D mesh with polyfaces.

RELATED SYSTEM VARIABLES

PFaceVMax specifies the maximum number of vertices per polyface (default = 4).

SplFrame controls the display of invisible faces (default = 0, not displayed).

TIPS

- The 3dFace command creates 3- and 4-sided meshes, while this command creates meshes of an arbitrary number of sides, and allows you to control the layer and the color of each face.
- The maximum number of vertices in the m- and n-direction is 256 when entered from the keyboard, and 32767 when entered from DXF files or from programs.
- Make an edge invisible by entering a negative number for the beginning vertex of the edge.
- This command is meant for programmers. To draw 3D mesh objects, use these commands instead: 3D, 3dMesh, RevSurf, RuleSurf, EdgeSurf, or TabSurf.

Plan

Rel.10 Displays the plan view of the WCS, the current UCS, or a named UCS.

Command	Alias	Ctrl+	F-key	Alt+	Menu Bar	Tablet
plan	...	...	...	**V3P**	**View** ↳**3D Views** ↳**Plan View**	**N3**

Command: plan

Enter an option [Current ucs/Ucs/World] <Current>: *(Enter an option.)*

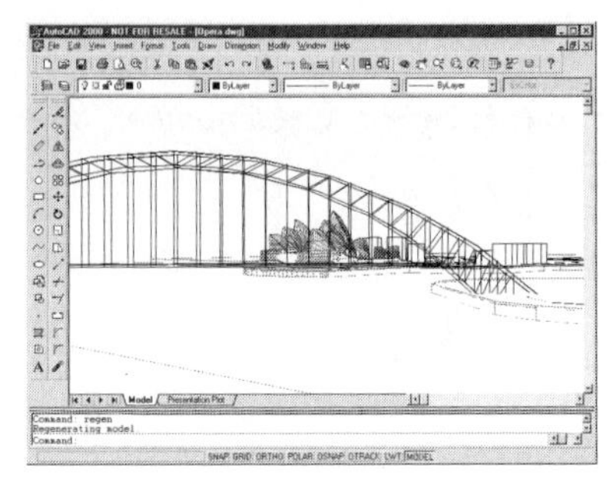

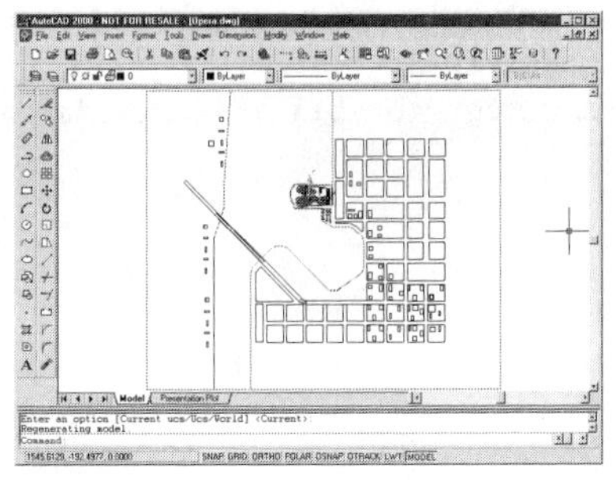

Left: *Example of a 3D view.*
Right: *After using the Plan World command.*

COMMAND LINE OPTIONS

Current UCS shows the plan view of the current UCS.

Ucs shows the plan view of the named UCS.

World shows the plan view of the WCS.

RELATED COMMANDS

UCS creates new UCS views.

VPoint changes the viewpoint of 3D drawings.

RELATED SYSTEM VARIABLES

UcsFollow displays the plan view for UCS or WCS automatically.

ViewDir contains the x,y,z coordinates of the current view direction.

TIPS

- Entering **VPoint 0,0,0** is an alternative to using this command.
- This command turns off perspective mode and clipping planes.
- Plan does not work in paper space; AutoCAD complains, "** Command only valid in Model space **."
- This command is an excellent method for turning off perspective mode.

PlaneSurf

2007 Draws flat rectangular surfaces, or converts 2D closed objects into surfaces (short for PLANar SURFace).

Command	Alias	Ctrl+	F-key	Alt+	Menu Bar	Tablet
planesurf	...	...	...	**DMF**	**Draw** ↳**Modeling** ↳**Planar Surface**	...

Command: planesurf
Specify first corner or [Object] <Object>: *(Pick a point, or type* **O***.)*
Specify other corner: *(Pick another point.)*

COMMAND LINE OPTIONS

Specify first corner picks the first corner to make a rectangular surface.

Object converts the selected closed 2D object into a surface; choose a line, circle, arc, ellipse, elliptical arc, 2D polyline, planar 3D polyline, or spline.

Specify other corner specifies the other corner of the rectangular surface.

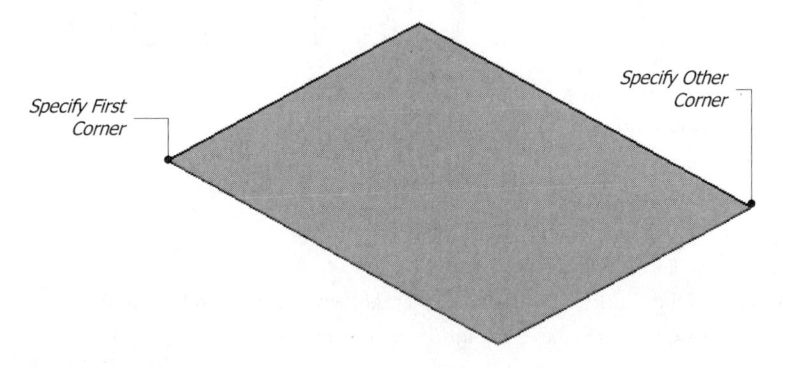

RELATED COMMAND

ConvertToSurface converts non-planar objects to surfaces.

TIP

- Use this command to create "floors."

PLine

V. 1.4 Draws complex 2D lines made of straight and curved sections of constant and variable width; treated as a single object (short for Poly LINE).

Command	Alias	Ctrl+	F-key	Alt+	Menu Bar	Tablet
pline	pl	...	...	DP	Draw ⇨Polyline	N10

Command: pline
Specify start point: *(Pick a point.)*
Current line-width is 0.0000
Specify next point or [Arc/Halfwidth/Length/Undo/Width]: *(Pick a point, or enter an option.)*
Specify next point or [Arc/Close/Halfwidth/Length/Undo/Width]: *(Pick a point, or enter an option.)*

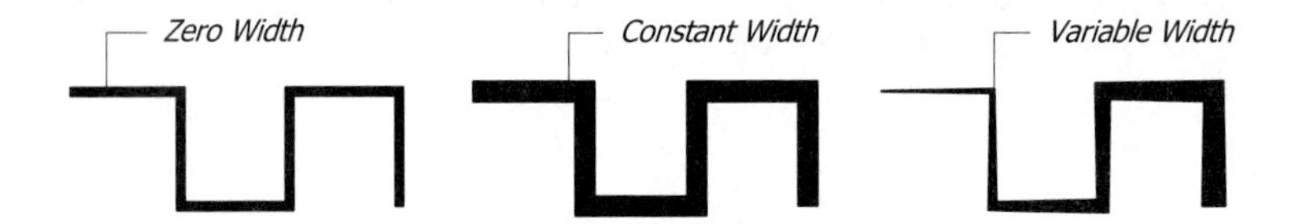

COMMAND LINE OPTIONS

Specify start point indicates the start of the polyline.

Close closes the polyline with a line segment.

Halfwidth indicates the half-width of the polyline.

Length draws a polyline tangent to the last segment.

Undo erases the last-drawn segment.

Width indicates the width of the polyline; start and end widths can be different.

Endpoint of line indicates the endpoint of the polyline.

Arc options
Specify endpoint of arc or [Angle/CEnter/CLose/Direction/Halfwidth/Line/Radius/Second pt/Undo/Width]: *(Enter an option.)*

Endpoint of arc indicates the arc's endpoint.

Angle indicates the arc's included angle.

CEnter indicates the arc's center point.

CLose uses an arc to close a polyline.

Direction indicates the arc's starting direction.

Halfwidth indicates the arc's half width.

Line switches back to the menu for drawing lines.

Radius indicates the arc's radius.

Second pt draws a three-point arc.

Undo erases the last drawn arc segment.

Width indicates the width of the arc.

RELATED COMMANDS

Boundary draws a polyline boundary.

Donut draws solid-filled circles as polyline arcs.

Ellipse draws ellipses as polyline arcs when PEllipse = 1.

Explode reduces a polyline to lines and arcs with zero width.

Fillet fillets polyline vertices with a radius.

PEdit edits the polyline's vertices, widths, and smoothness.

Polygon draws polygons as polylines of up to 1024 sides.

Rectang draws a rectangle out of a polyline.

Sketch draws polyline sketches, when SkPoly = 1.

Xplode explodes a group of polylines into lines and arcs of zero width.

3dPoly draws 3D polylines.

RELATED SYSTEM VARIABLES

PLineGen specifies the style of linetype generation:

- **0** — Vertex to vertex (default).
- **1** — End to end (compatible with Release 12).

PLineType controls the conversion of old (pre-R14) polylines and the creation of lwpolyline objects by the PLine command:

- **0** — Old polylines not converted; old-format polylines created.
- **1** — Not converted; lwpolylines created.
- **2** — Polylines in pre-R14 drawings converted on open;

PLine command creates lwpolyline objects (default).

PLineWid sets the current width of polyline (default = 0.0).

TIPS

- **Boundary** uses a polyline to outline a region automatically; use the List command to find its area.
- If you cannot see a linetype on a polyline, change system variable PlineGen to 1; this regenerates the linetype from one end of the polyline to the other.
- If the angle between a joined polyline and polyarc is less than 28 degrees, the vertex is chamfered; at greater than 28 degrees, the vertex is not chamfered.
- Use the object snap modes INTersection or ENDpoint to snap to the vertices of a polyline.

Plot

V.1.0 Creates hard copies of drawings on vector, raster, and PostScript plotters; also plots to files on disk.

Commands	Alias	Ctrl+	F-key	Alt+	Menu Bar	Tablet
plot	print	P	...	FP	File	W25
	dwfout				↳Plot	
-plot						

Command: plot

Displays dialog box:

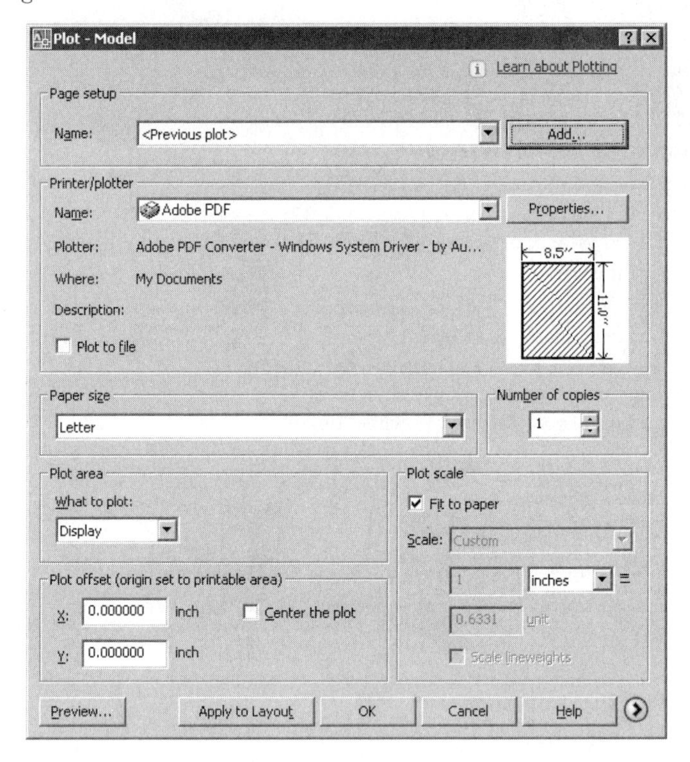

PLOT DIALOG BOX

Page Setup

Name selects or imports a previous plot setup.

Add prompts for the name of a new plot setup; displays the Add Page Setup dialog box.

Printer/Plotter Configuration

Name selects a system printer, a *.pc3* named plotter configuration, or previous plot setting.

Properties displays the Plotter Configuration Editor dialog box; see PlotterManager command.

☐ **Plot to file** plots the drawing to a *.plt* file.

Paper Size

Paper size displays paper sizes supported by the selected output device.

Plot Area

What to plot:

- **Layout** plots all parts of the drawing within the margins of the specified paper size; the origin is calculated from 0,0 in the layout (not available in model space).
- **Display** plots all parts of the drawing within the current viewport.
- **Extents** plots the entire drawing.
- **Limits** plots the drawing in the Model tab within a rectangle defined by the Limits command.
- **View** plots a view saved with the View command (not available when no views are named).
- **Window** plots all parts of the drawing within a picked rectangle.

Plot Offset

Center the plot centers the plot on the paper.

X specifies the plot origin offset from x = 0; the origin varies, depending on the plotter model.

Y specifies the plot origin offset from y = 0.

☑ **Center the plot** centers the plot between the page edge or printer margins (not available when the Layout option is selected, above).

Number of Copies

Number of copies specifies the number of copies to plot.

Plot Scale

☑ **Fit to paper** scales the drawing to fit within the plotter's margins (not available when the **Layout** option selected).

Scale specifies the plotting scale; ignored by layouts.

- **Custom** specifies user-defined scale; these can be edited by the ScaleListEdit command.

Inches / mm specifies the number of inches (or millimeters) on the plotted page.

Unit specifies number of units in drawing.

☐ **Scale lineweights** scales lineweights proportionately to the plot scale.

Preview displays the preview window; see the **Preview** command.

Apply to layout applies setting to layout.

OK starts the plot.

EXPANDED DIALOG BOX

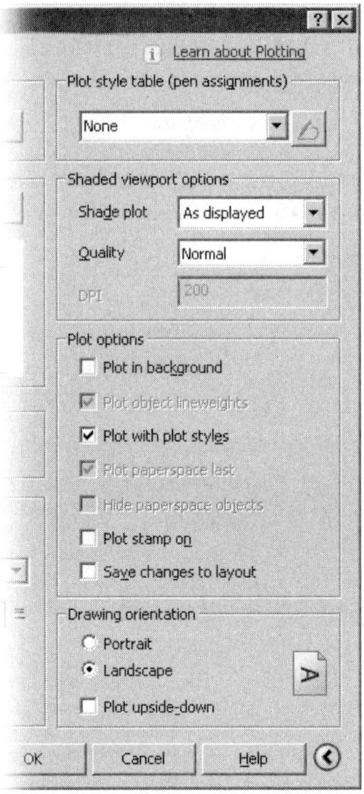

Plot Style Table (Pen Assignments)

Drop list selects a .ctb plot style table file; see the PlotStyle command.

New displays the Add Color-Dependent Plot Style Table wizard; see the StylesManager command.

Edit button displays the Plot Style Table Editor dialog box; see the StylesManager command.

Shaded Viewport Options

Available only in model space; in layouts, right-click the viewport border, and then select Properties.

Shade Plot plots the viewport as-displayed, in a visual style, or rendered.

Quality sets print quality as draft (wireframe), preview (quarter-resolution, up to 150dpi), normal (half resolution, up to 300dpi), presentation (maximum printer resolution up to 600dpi), maximum (maximum printer resolution), or custom (as specified by DPI text box).

DPI specifies the resolution for the output device; DPI is short for "dots for inch."

Plot Options

☐ **Plot in background** plots drawings as a background process.

☑ **Plot object lineweights** plots objects with lineweights (available only when plot style table turned off).

☑ **Plot with plot styles** plots objects using the plot styles.

☑ **Plot paperspace last** plots the model space objects first, followed by paper space objects (available in layouts only).

☐ **Hide paperspace objects** removes hidden lines before plotting.

☐ **Plot Stamp On** stamps the plot; when on, the Plot Stamp Settings button appears. See PlotStamp command.

☐ **Save changes to layouts** saves changes made in this dialog box to the layout.

Draw Orientation

⊙ **Portrait** positions the paper vertically.

○ **Landscape** positions the paper horizontally.

☐ **Plot Upside Down** plots the drawing upside down.

-PLOT Command

Command: -plot

Detailed plot configuration? [Yes/No] <No>: *(Type* ***Y*** *or* ***N****.)*

Brief Plot Configuration Options

Enter a layout name or [?] <Model>: *(Enter a layout name, or type* **?***.)*

Enter a page setup name <>: *(Enter a page setup name, or press* ENTER *for none.)*

Enter an output device name or [?] <default printer>: *(Enter a printer name, or type* **?** *for a list of printers.)*

Write the plot to a file [Yes/No] <N>: *(Type* **Y** *or* **N***.)*

Save changes to page setup [Yes/No]? <N> *(Type* **Y** *or* **N***.)*

Proceed with plot [Yes/No] <Y>: *(Type* **Y** *or* **N***.)*

Detailed Plot Configuration Options

Enter a layout name or [?] <Model>: *(Enter a layout name, or type* **?***.)*

Enter an output device name or [?] <default printer>: *(Enter a printer name, or type* **?** *for a list of printers.)*

Enter paper size or [?] <Letter 8 ½ x 11 in>: *(Enter a paper size, or type* **?***.)*

Enter paper units [Inches/Millimeters] <Inches>: *(Type* **I** *or* **M***.)*

Enter drawing orientation [Portrait/Landscape] <Landscape>: *(Type* **P** *or* **L***.)*

Plot upside down? [Yes/No] <No>: *(Type* **Y** *or* **N***.)*

Enter plot area [Display/Extents/Limits/View/Window] <Display>: *(Enter an option.)*

Enter plot scale (Plotted Inches=Drawing Units) or [Fit] <Fit>: *(Enter a scale factor, or type* **F***.)*

Enter plot offset (x,y) or [Center] <0.00,0.00>: *(Enter an offset, or type* **C***.)*

Plot with plot styles? [Yes/No] <Yes>: *(Type* **Y** *or* **N***.)*

Enter plot style table name or [?] (enter . for none) <>: *(Enter a name, or enter an option.)*

Plot with lineweights? [Yes/No] <Yes>: *(Type* **Y** *or* **N***.)*

Enter shade plot setting [As displayed/Wireframe/Hidden/Visual styles/ Rendered] <As displayed>: *(Enter an option.)*

Write the plot to a file [Yes/No] <N>: *(Type* **Y** *or* **N***.)*

Save changes to page setup? Or set shade plot quality? [Yes/No/Quality] <N>: *(Enter an option.)*

Proceed with plot [Yes/No] <Y>: *(Type* **Y** *or* **N***.)*

COMMAND OPTIONS

Enter a layout name specifies the name of a layout; enter Model for model view or **?** for a list of layout names.

Enter an output device name specifies the name of a printer, or enter **?** for a list of print devices.

Enter paper size specifies the name of a paper size, or enter **?** for a list of paper sizes supported by the print device.

Enter paper units specifies Inches (imperial) or Millimeters (metric) units.

Enter drawing orientation specifies Portrait (vertical) or Landscape (horizontal) orientation.

Plot upside down:

- **Yes** plots the drawing upside down.
- **No** plots the drawing normally.

Enter plot area specifies that the current Display, Extents of the drawing, Limits defined by the **Limits** command, View name, or window area be plotted.

Enter plot scale specifies the scale using the Plotted Inches=Drawing Units format; or enter F to fit the drawing to the page.

Enter plot offset specifies the distance between the lower left corner of the paper and the drawing; or enter **C** to center the drawing on the page.

Plot with plot styles:

- **Yes** prompts you to specify the name of a plot style.
- **No** ignores plot styles, and uses color-dependent styles.

Plot with lineweights:

- **Yes** uses lineweight settings to draw thicker lines.
- **No** ignores lineweights.

Enter shade plot setting specifies whether the plot should be rendered As displayed, Wireframe, Hidden lines removed, named Visual style, or Rendered.

Write the plot to a file:

- **Yes** sends the plot to a file.
- **No** sends the plot to the plot device.

Save changes to page setup:

- **Yes** saves the settings you specified during the previous set of questions.
- **No** does not save the settings.
- **Quality** specifies the quality of the plot.

Proceed with plot:

- **Yes** plots the drawing.
- **No** cancels the plot.

Quality options

Enter shade plot quality [Draft/Preview/Normal/pResentation/Maximum/Custom] <Normal>: *(Select an option.)*

Draft, Preview, Normal, pResentation, and Maximum apply preset dpi (dots per inch) settings to the plot. The higher the dpi, the better the quality, but the slower the plot.

Custom allows you to specify the dpi setting; default = 150dpi.

RELATED COMMANDS

PageSetup selects one or more plotter devices.

Preview goes directly to the plot preview screen.

Publish plots multi-page drawings.

PublishToWeb plots drawings to HTML format.

PsOut saves drawings in *.eps* format.

ViewPlotDetails reports on succussful and unsuccessful plots.

RELATED SYSTEM VARIABLES

BackgroundPlot toggles background plotting.

BgrndPlotTimeout specifies the time AutoCAD waits for plotter to respond.

FullPlotPath sends either the full path or just the drawing name to the plot spooler software.

RasterDpi specifies the scaling for raster devices.

TextFill toggles the filling of TrueType fonts.

TextQlty specifies the quality of TrueType fonts.

RELATED FILES

*.***pc3*** holds plotter configuration parameter files.

*.***plt*** holds plot files created with this command.

TIPS

- AutoCAD R12 replaced PrPlot with the Plot command. R13 removed -p freeplot (plotting without using up a network license). AutoCAD 2005 removed the partial preview.
- AutoCAD cannot plot perspective views to-scale, only to-fit.
- Background plotting displays an icon in the tray; click it for options.

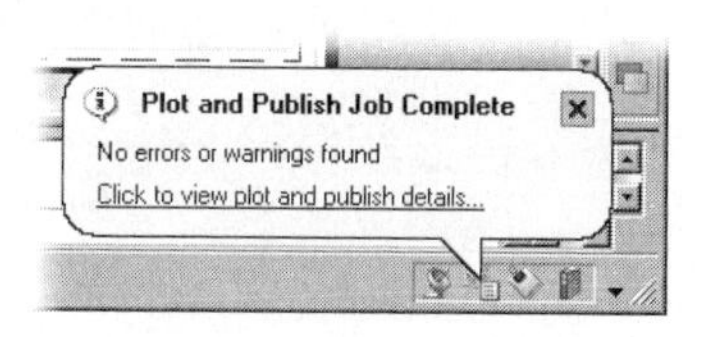

- You must wait for one background plot to complete before beginning the next one.
- Turn off BackgroundPlot when plotting drawings with scripts.

'PlotStamp

2000i Adds information about drawings to hard copy plots.

Commands	Alias	F-key	Alt+	Menu Bar	Tablet
plotstamp	**ddplotstamp**	...	...	...	...
...					
-plotstamp					

Command: plotstamp

Displays dialog box:

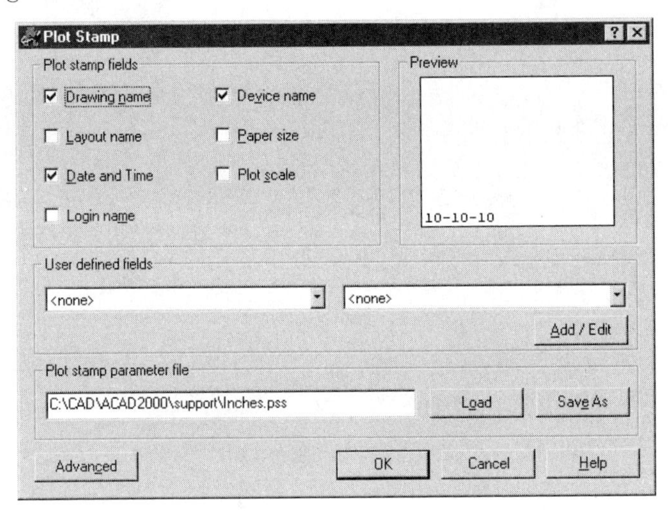

PLOT STAMP DIALOG BOX

Plot Stamp Fields

☑ **Drawing name** adds the path and name of the drawing.

☐ **Layout name** adds the layout name.

☑ **Date and time** adds the date (short format) and time of the plot.

☐ **Login name** adds the Windows login name, as stored in the LogInName system variable.

☑ **Device name** adds the plotting device's name.

☐ **Paper size** adds the paper size, as currently configured.

☐ **Plot scale** adds the plot scale.

User Defined Fields

Add/Edit displays the User Defined Fields dialog box.

Plot Stamp Parameter File

Load/Save displays the Plotstamp Parameter File Name dialog box for opening (or saving) *.pss* files.

Advanced displays the Advanced Options dialog box.

ADVANCED OPTIONS DIALOG BOX

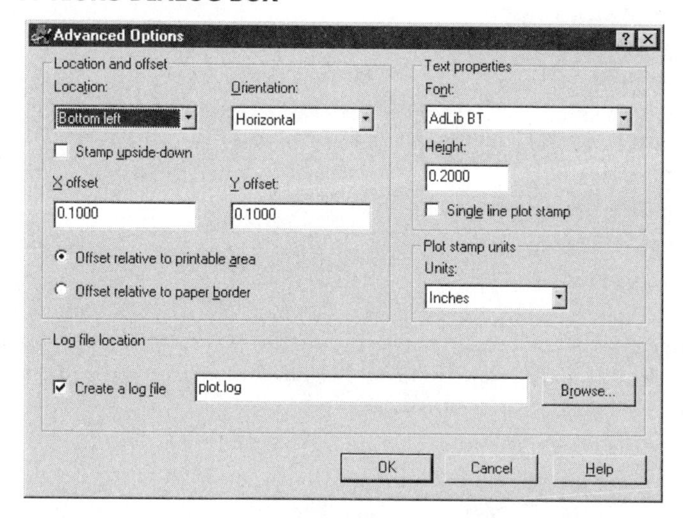

Location and Offset

Location specifies the location of the plot stamp: Top Left, Bottom Left (default), Bottom Right, or Top Right — relative to the orientation of the drawing on the plotted page.

Stamp upside down:

☑ Draws the plot stamp upside down.

☐ Draws the plot stamp rightside up.

Orientation rotates the plot stamp relative to the page: Horizontal or Vertical.

X offset specifies the distance from the corner of the printable area or page; default = 0.1.

Y offset specifies the distance from the corner of the printable area or page; default = 0.1

⊙ **Offset relative to printable area.**

○ **Offset relative to paper border.**

Text Properties

Font selects the font for the plot stamp text.

Height specifies the height of the plot stamp text.

Single line plot stamp:

☑ Plotstamp text is placed on a single line.

☐ Plotstamp text is placed on two lines.

Plot Stamp Units

Units selects either inches, millimeters, or pixels.

Log File Location

☑ **Create a log file** saves the plot stamp text to a text file.

Browse displays the Log File Name dialog box.

-PLOTSTAMP Command

Command: -plotstamp

Current plot stamp settings:

Displays the current setting of nearly 20 plot stamp settings.

Enter an option [On/OFF/Fields/User Fields/Log file/LOCation/Text properties/UNits]: *(Enter an option.)*

COMMAND LINE OPTIONS

On turns on plot stamping.

OFF turns off plot stamping.

Fields specifies plot stamp data: drawing name, layout name, date and time, login name, plot device name, paper size, plot scale, comment, write to log file, log file path, location, orientation, offset, offset relative to, units, font, text height, and stamp on single line.

User fields specifies two user-defined fields.

Log file writes the plotstamp data to a file instead of the drawing.

LOCation specifies the location and orientation of the plotstamp.

Text properties specifies the font name and height.

UNits specifies the units of measurement: inches, millimeters, or pixels.

RELATED COMMAND

Plot plots the drawing.

RELATED FILES

__.log__ is the plotstamp log file; stored in ASCII text format.

__.pss__ holds the plotstamp parameters; stored in binary format.

TIPS

- When the options of the Plot Stamp dialog box are grayed out, or when the -Plotstamp command reports 'Current plot stamp file or directory is read only,' this means that the *Inches.pss* or *Mm.pss* file in the *Support* folder is read-only. To change, in Explorer: (1) right-click the file, (2) select **Properties**, (3) uncheck **Read-only**, and (4) click **OK**.
- You can access the Plot Stamp dialog box via the Plot command's More Options dialog box:

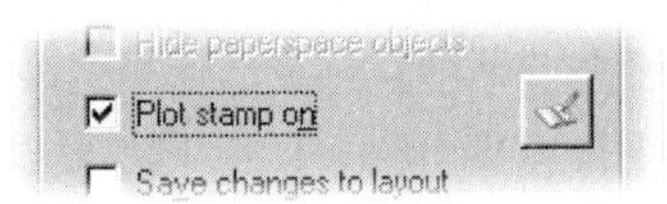

- *Warning!* Too large of an offset value positions the plotstamp text beyond the plotter's printable area, which causes the text to be cut off. To prevent this, use the **Offset relative to printable area** option.

PlotStyle

2000 Selects and assigns plot styles to objects.

Commands	Alias	Ctrl+	F-key	Alt+	Menu Bar	Tablet
plotstyle	...	...	...	**OY**	**Format** ↳**Plot Style**	...
-plotstyle						

Command: plotstyle

When no objects are selected, displays the Current Plot Style dialog box.

When one or more objects are selected, displays the Select Plot Style dialog box.

SELECT PLOT STYLE DIALOG BOX

Displayed when one or more objects are selected:

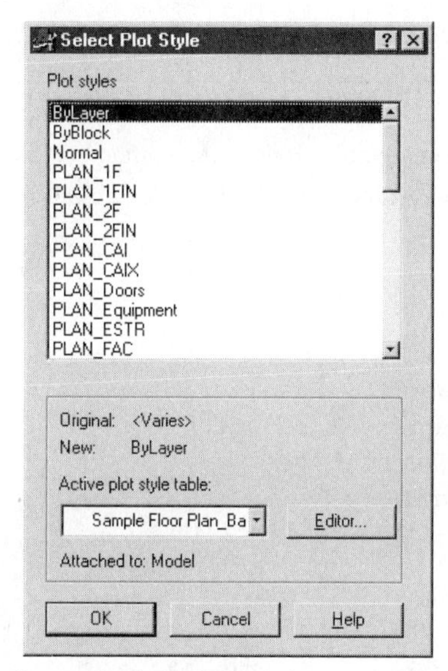

Plot styles lists the plot styles available in the drawing.

Active plot style table selects the plot style table to attach to the current drawing.

Editor displays the Plot Style Table Editor dialog box; see the StylesManager command.

CURRENT PLOT STYLE DIALOG BOX

Displayed when no objects are selected:

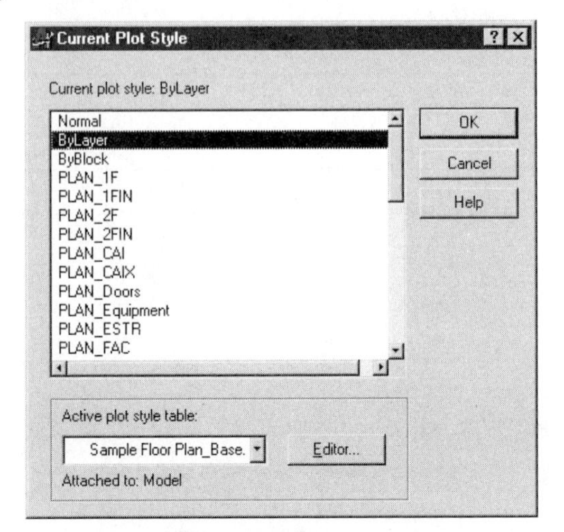

Plot styles lists the plot styles available in the drawing.

Active plot style table selects the plot style table to attach to the current drawing.

Editor displays the Plot Style Table Editor dialog box; see the StylesManager command.

-PLOTSTYLE Command

Command: -plotstyle
Current plot style is "Default"
Enter an option [?/Current] : *(Type* **?** *or* **C***.)*
Set current plot style : *(Enter a name.)*
Current plot style is "*plotstylename*"
Enter an option [?/Current] : *(Press Enter.)*

COMMAND LINE OPTIONS

Current prompts you to change plot styles.

Set current plot style specifies the name of the plot style.

ENTER exits the command.

? displays plot style names; sample output:

```
Plot Styles:
-----------------
ByLayer
ByBlock
Normal
ARCH_Dimensions for stairway
Current plot style is "PLAN_ESTR"
```

RELATED COMMANDS

Plot plots drawings with plot styles.

PageSetup attaches plot style tables to layouts.

StylesManager modifies plot style tables.

Layer specifies plot styles for layers.

RELATED SYSTEM VARIABLES

CPlotStyle specifies the plot style for new objects; defined values include "ByLayer," "ByBlock," "Normal," and "User Defined."

DefLPlStyle specifies the plot style name for layer 0.

DefPlStyle specifies the default plot style for new objects.

PStyleMode indicates whether the drawing is in Color-Dependent or Named Plot Style mode.

PStylePolicy determines whether object color properties are associated with plot style.

TIPS

- This command does not operate until a plot style table has been created for the drawing; the **PlotStyle** command displays this message box:

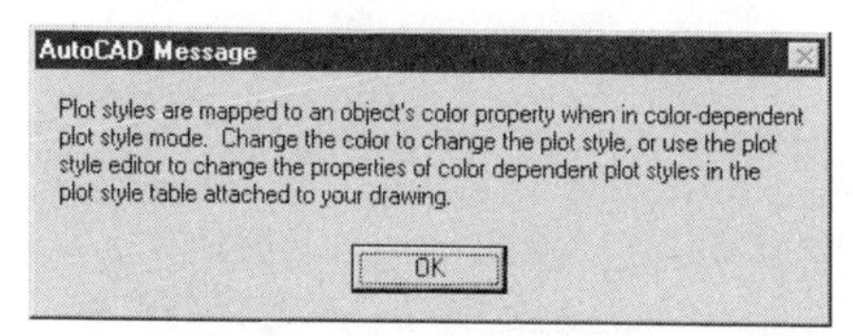

- A plot style can be assigned to any object and to any layer.
- A plot style can override the following plot settings: color, dithering, gray scale, pen assignment, screening, linetype, lineweight, end style, join style, and fill style.
- Plot styles are useful for plotting the same layout in different ways, such as emphasizing objects using different lineweights or colors in each plot.
- Plot style tables can be attached to the Model tab and layout tabs, and attach different plot style tables to layouts, to create different looks for plots.
- By default, drawings do not use plot styles. To turn on plot styles, follow the tutorial.

Applying Plot Styles

Before you can apply plot styles to a new drawing, you must turn on plot styles; notice that Plot Style Control in the Object Properties toolbar is grayed out.

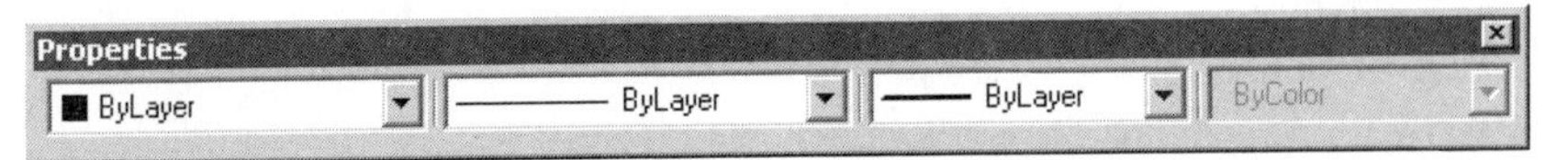

Follow these steps to turn on plot style:

1. From the **Tools** menu, select **Options**.
2. In the Options dialog box, select the **Plot and Publish** tab, and then click the **Plot Style Table Settings** button.
3. In the Plot Style Table Settings dialog box, select the **Use Named Plot Styles** option.

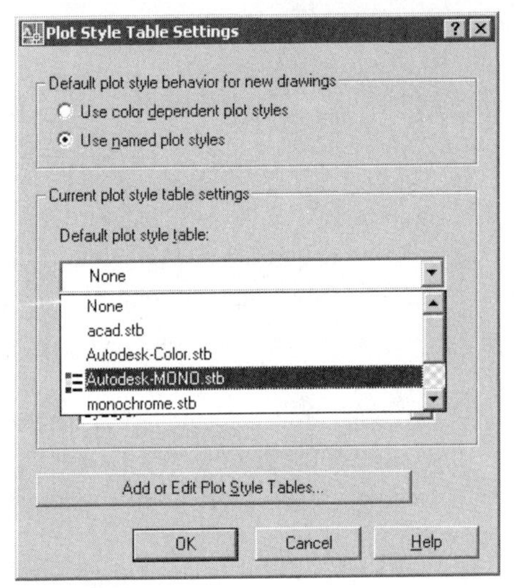

4. From the **Default plot style** list, select any *.stb* plot style.
5. Click **OK** to dismiss both dialog boxes.
6. The Plot Style Control in the Properties toolbar and Layer Properties Manager dialog box is not available until the *next* new drawing you start, curiously enough.

PlotterManager

2000 Displays the Plotters window, the Add-A-Plotter wizard, and PC3 configuration Editor.

Command	Alias	Ctrl+	F-key	Alt+	Menu Bar	Tablet
plottermanager	...	...	...	**FM**	**File** ↳**Plotter Manager**	**Y24**

Command: plottermanager

Displays window:

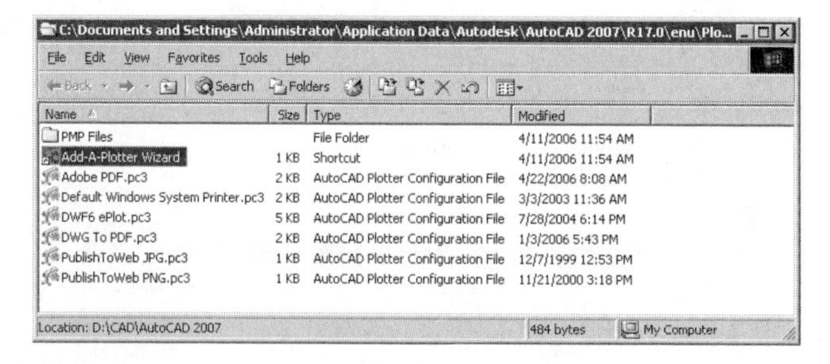

PLOTTERS WINDOW

Add-a-Plotter Wizard adds a plotter configuration; double-click to run the Add Plotter wizard.

***.pc3** specifies parameters for creating plotted output; double-click to display the Plotter Configuration Editor; see the StylesManager command.

ADD PLOTTER WIZARD

The following steps create PC3 plotter configurations for plotting drawings to generic HP plotters; the steps are similar to creating .pc3 files for other types of plotters:

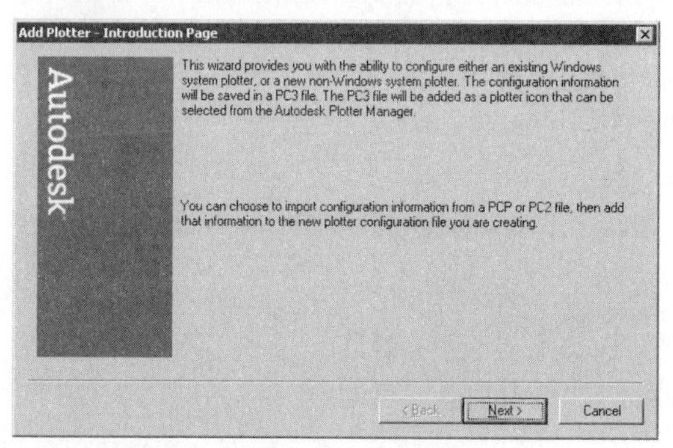

Next displays the next dialog box.

Cancel dismisses the dialog box.

Begin page

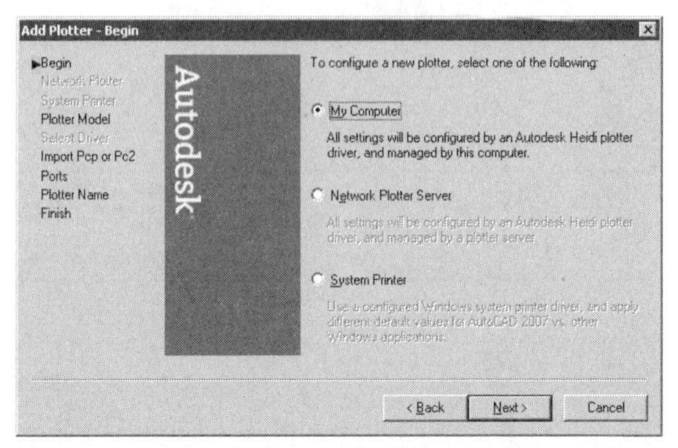

My Computer configures printers and plotters connected to your computer.

Network Plotter Server configures printers and plotters connected to other computers on networks.

System Printer configures the default Windows printer.

Plotter Model page

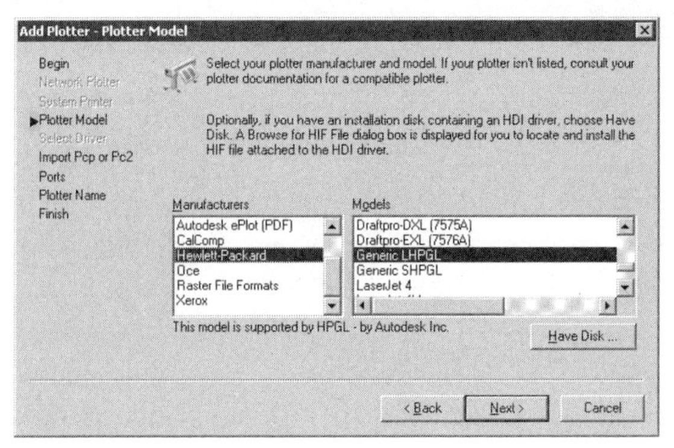

Manufacturer selects a brand name of plotter.

Model selects a specific model number.

Have Disk selects plotter drivers provided by the manufacturer; displays the Open dialog box.

Import Pcp or Pc2 page

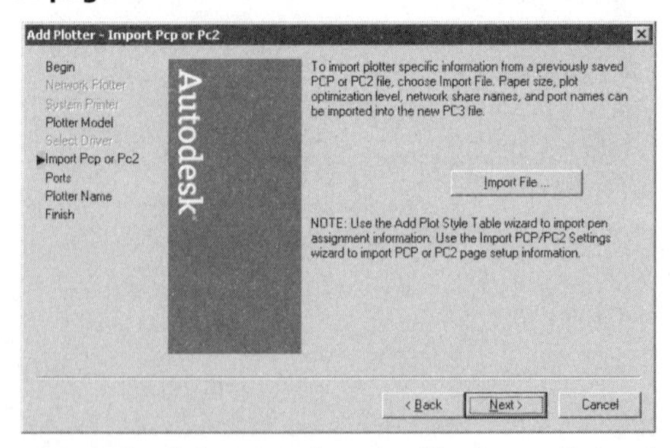

Import File imports *.pcp* and *.pc2* plotter configuration files from earlier versions of AutoCAD.

Ports page

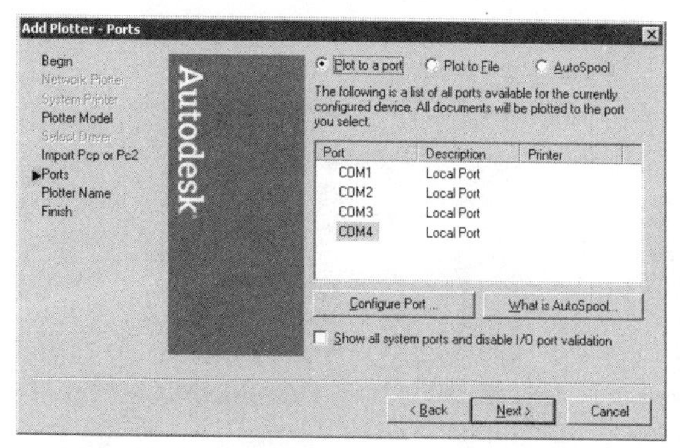

⊙ **Plot to a port** sends the plot to an output port.

○ **Plot to File** plots the drawing to a file with user-defined file names; default file name is the same as the drawing name with the *.plt* extension; PostScript plot files are given the *.eps* extension.

○ **AutoSpool** plots the drawing to files with the file name generated by AutoCAD, and then executes the command specified in the Option dialog box's Files tab. Enter the name of the program that should process the AutoSpool file in **Print Spool Executable**. You may include these DOS command-line arguments:

Argument	Meaning
%s	Substitutes path, spool filename, and extension.
%d	Specifies the path, AutoCAD drawing name, and extension.
%c	Describes the device.
%m	Returns the plotter model.
%n	Specifies the plotter name.
%p	Specifies the plotter number.
%h	Returns the height of the plot area in plot units.
%w	Returns the width of the plot area in plot units.
%i	Specifies the first letter of the plot units.
%l	Specifies the login name (***LogInName*** *system variable*).
%u	Specifies the user name.
%e	Specifies the equal sign (=).
%%	Specifies the percent sign (%).

Port options

Port lists the virtual ports defined by the Windows operating system:

Port	Meaning
USB	Universal serial bus.
COM	Serial port.
LPT	Parallel port.
HDI	Autodesk's Heidi Device Interface.

Description describes the type of port:

Description	Meaning
Local Port	Printer is connected to your computer.
Network Port	Printer is accessible through the network.

Printers describes the brand name of the printer connected to the port.

Configure Port displays the Configure Port dialog box; allows you to specify parameters specific to the port, such as timeout and protocol.

What is AutoSpool? displays an explanatory dialog box:

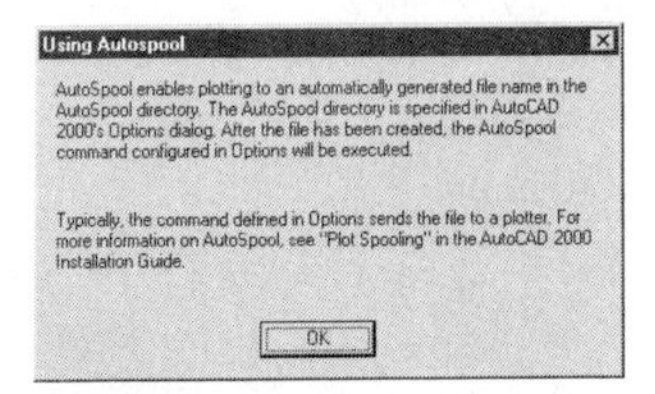

☐ **Show all system ports and disable I/O port validation** prevents AutoCAD from checking whether the port is valid.

Plotter Name page

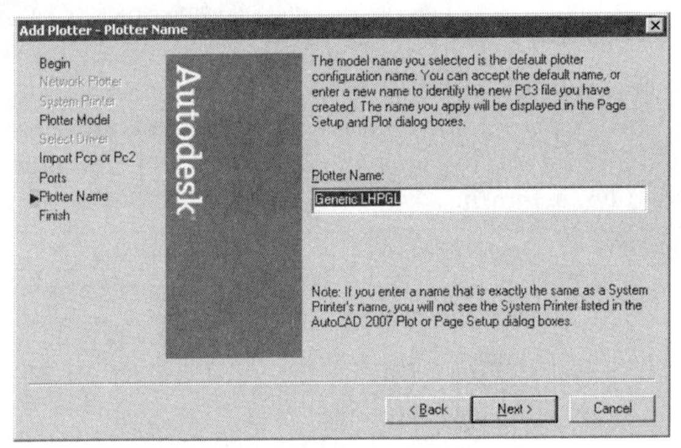

Plotter name specifies a user-defined name for the plotter configuration; you may have many different configurations for a single plotter.

Finish Page

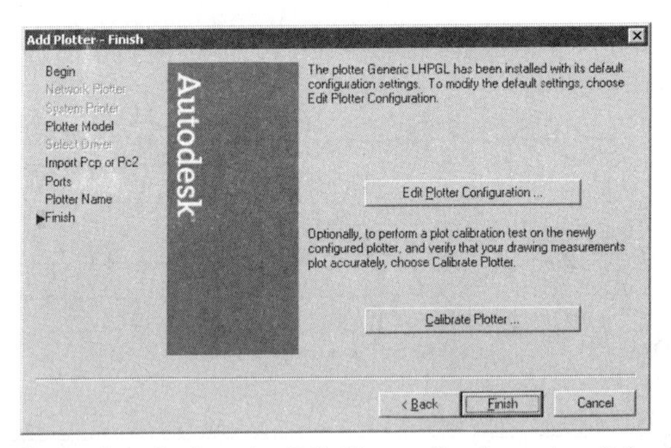

Edit Plotter Configuration displays the Edit Plotter Configuration dialog box.

Calibrate Plotter displays the Calibrate Plotter wizard.

PLOTTER CONFIGURATION EDITOR DIALOG BOXES

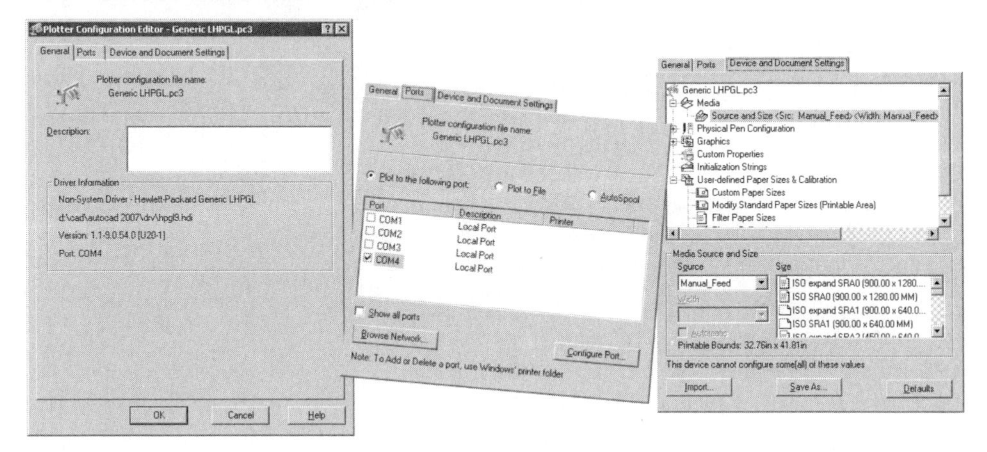

General tab

Description allows you to provide a detailed description of the plotter configuration.

Ports tab

⊙ **Plot to the following port** sends the plot to an output port.

○ **Plot to File** plots the drawing to a file with a user-defined file name; default file name is the same as the drawing name, with a *.plt* extension; PostScript plot files are given the *.eps* extension.

○ **AutoSpool** plots the drawing to a file with a filename generated by AutoCAD, and then executes the command specified in the Option dialog box's Files tab.

☐ **Show all ports** lists all ports on the computer.

Browse Network displays the Browse for Printer dialog box; selects a printer on the network.

Configure Port displays the Configure Port dialog box; allows you to specify the parameters specific to the port, such as timeout and protocol.

Device and Document Settings tab

Media specifies the paper source, paper size, type, and destination.

Physical Pen Configuration specifies the physical pens in the pen plotter (for pen plotters only).

Graphics specifies settings for plotting vector and raster graphics and TrueType fonts.

Custom Properties specifies settings specific to the plotter, printer, or other output device.

Initialization Strings specifies the control codes for pre-initialization, post-initialization, and termination (for non-system plotters only).

Calibration Files and Paper Sizes calibrates the plotter by specifying the *.pmp* file; adds and modifies custom paper sizes; see the Calibrate Plotter wizard.

Import imports *.pcp* and *.pc2* plotter configuration files from earlier versions of AutoCAD.

Save As saves the plotter configuration data to a *.pc3* file.

Defaults resets the plotter configuration settings to the previously-saved values.

RELATED COMMANDS

Plot plots drawings with plot styles.

PageSetup attaches plot style tables to layouts.

StylesManager modifies plot style tables.

AddPlWiz.Exe runs the Add Plotter wizard.

RELATED SYSTEM VARIABLES

PaperUpdate toggles the display of a warning before AutoCAD plots a layout with a paper size different from the size specified by the plotter configuration file:

0 — Displays warning dialog box (default).

1 — Changes paper size to match the size specified by the plotter configuration file.

PlotRotMode controls the orientation of plots:

0 — Aligns rotation icon with media at the lower left for 0 degrees; calculates x and y origin offsets relative to lower-left corner.

1 — Aligns the lower-left corner of plotting area with lower-left corner of the paper.

2 — Same as 0, except x and y origin offsets relative to the rotated origin position.

TIPS

- You can create and edit *.pc3* plotter configuration files without AutoCAD. From the Start button on the Windows toolbar, select **Settings | Control Panel | Autodesk Plotter Manager**.

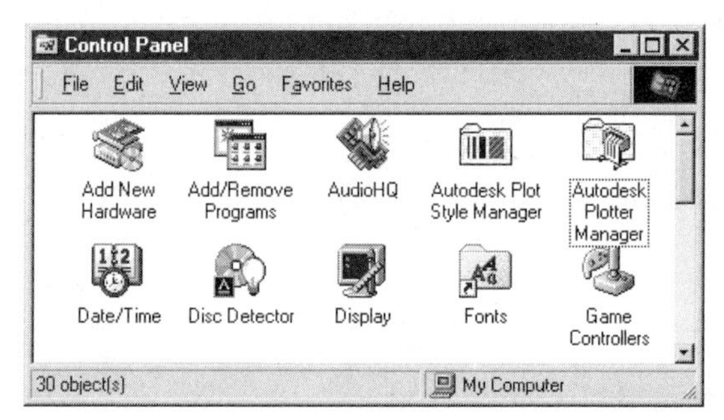

- The *dwf classic.pc3* file specifies parameters for creating *.dwf* files via the Release 14-compatible DwfOut command.
- The *dwf eplot.pc3* file specifies parameters for creating *.dwf* files via the Plot command.

PngOut

2004 Exports the current view in PNG raster format.

Command	Alias	Ctrl+	F-key	Alt+	Menu Bar	Tablet
pngout	...	...	...	...	...	...

Command: pngout

*Displays Create Raster File dialog box. Specify a filename, and then click **Save**.*

Select objects or <all objects and viewports>: *(Select objects, or press Enter to select all objects and viewports.)*

COMMAND LINE OPTIONS

Select objects selects specific objects.

All objects and viewports selects all objects and all viewports, whether in model space or in layout mode.

RELATED COMMANDS

BmpOut exports drawings in BMP (bitmap) format.

Image places raster images in the drawing.

JpgOut exports drawings in JPEG (joint photographic expert group) format.

Plot exports drawings in PNG and other raster formats.

TifOut exports drawings in TIFF (tagged image file format) format.

TIPS

- The rendering effects of the VsCurrent command are preserved, but not those of the Render command.
- PNG files are used as a royalty-free alternative to JPEG files.
- PNG is short for "portable network graphics."

Point

V. 1.0 Draws a 3D point.

Command	Alias	Ctrl+	F-key	Alt+	Menu Bar	Tablet
point	po	...	...	DO	Draw ↳Point	O9

Command: point
Current point modes: PDMODE=0 PDSIZE=0.0000
Specify a point: *(Pick a point.)*

COMMAND LINE OPTION

Point positions a point, or enters a 2D or 3D coordinate.

RELATED COMMANDS

DdPType displays a dialog box for selecting PsMode and PdSize.

Regen regenerates the display to see the new point mode or size.

RELATED SYSTEM VARIABLES

PDMode determines the appearance of a point:

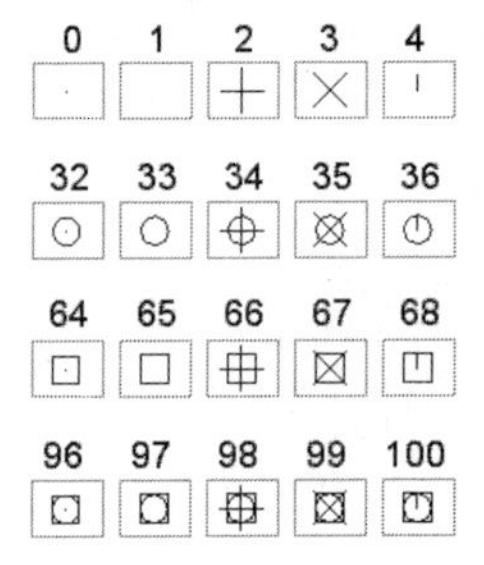

PDSize determines the size of a point:

0 — 5% of height of the **ScreenSize** system variable (default).

1 — No display.

-10 — *(Negative)* Ten percent of the viewport size.

10 — (*Positive*) Ten pixels in size.

TIPS

- The size and shape of points are determined by PdSize and PdMode; changing these changes the appearance and size of *all* points in the drawing with the next regeneration (Regen).
- Entering only x,y coordinates places the point at the z coordinate of the current elevation; setting Thickness to a value draws points as lines in 3D space.
- Prefix coordinates with * (such as *1,2,3) to place points in the WCS, rather than in the UCS. Use the object snap mode NODe to snap to points.
- Points plot according to the setting of PdMode.

Aliased Command

The **SpotLight** command's options are accessed from the **Light** command.

Polygon

V. 2.5 Draws 2D polygons of between three 3 and 1024 sides.

Command	Alias	Ctrl+	F-key	Alt+	Menu Bar	Tablet
polygon	**pol**	...	...	**DY**	**Draw** ↳**Polygon**	**P10**

Command: polygon
Enter number of sides <4>: *(Enter a number.)*
Specify center of polygon or [Edge]: *(Pick a point, or type* **E**.*)*
Enter an option [Inscribed in circle/Circumscribed about circle] <I>: *(Type* **I** *or* **C**.*)*
Specify radius of circle: *(Enter a value, or pick two points.)*

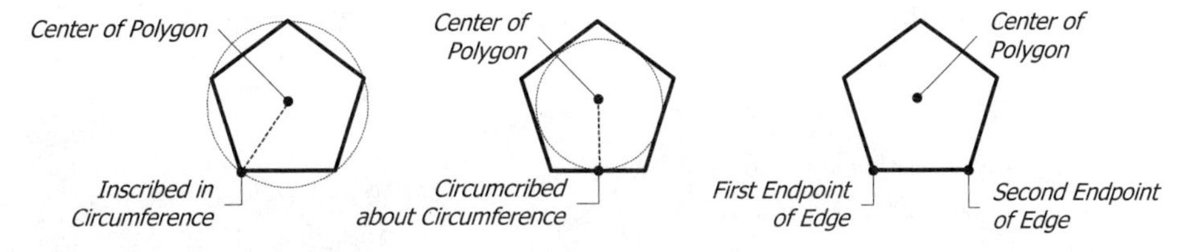

COMMAND LINE OPTIONS

Center of polygon indicates the center point of the polygon; then:

C (*Circumscribed*) fits the polygon outside an indicated circumference.

I (*Inscribed*) fits the polygon inside an indicated circumference.

Edge draws the polygon based on the length of one edge.

RELATED COMMANDS

PEdit edits polylines, including polygons.

Rectang draws squares and rectangles.

RELATED SYSTEM VARIABLE

PolySides specifies the most recently entered number of sides (default = 4).

TIPS

- Polygons are drawn as a polyline; use PEdit to edit the polygon, such as its width.
- The pick point determines the polygon's first vertex; polygons are drawn counterclockwise.
- Use the system variable PolySides to preset the default number of polygon sides.
- Use the Snap command to place the polygon precisely; use INTersection or ENDpoint object snap modes to snap to the polygon's vertices.

Aliased Command

Preferences was removed from AutoCAD 2000; it now is an alias for the **Options** command.

PolySolid

2007 Draws 3D solid "walls" made of straight and curved sections in a manner similar to polylines.

Command	Alias	Ctrl+	F-key	Alt+	Menu Bar	Tablet
polysolid	**psolid**	...	...	**DMP**	**Draw** ↳**Modeling** ↳**Polysolid**	...

Command: polysolid

Specify start point or [Object/Height/Width/Justify] <Object>: *(Pick a point, or enter an option.)*

Specify next point or [Arc/Undo]: *(Pick a point, press Enter, or enter an option.)*

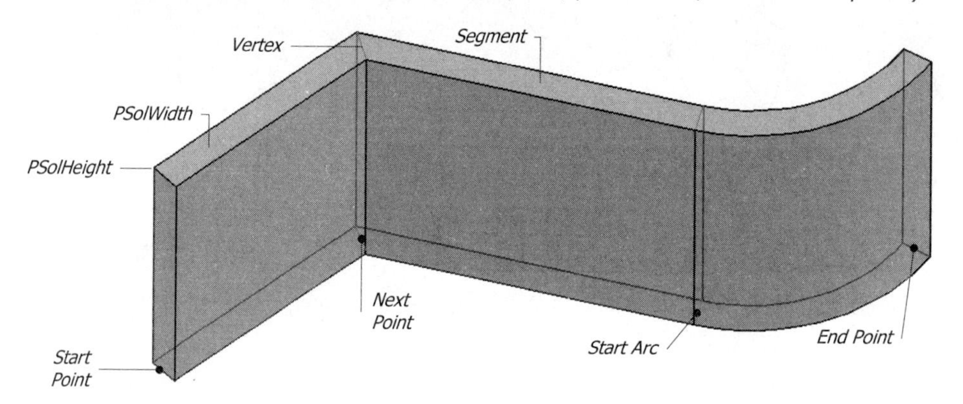

COMMAND LINE OPTIONS

Specify start point indicates the starting point by picking a point or entering x,y coordinates.

Object selects the line, arc, circle, or 2D polyline to convert into a solid.

Height specifies the height of the polysolid; default value is stored in the PSolHeight system variable.

Width specifies the width; default value is found in the PSolWidth system variable.

Justify determines whether the base of the polysolid is drawn to the left, on the center, or to the right of the pick points.

Specify the next point indicates the vertex of the wall; press Enter to stop drawing the polysolid.

Arc switches to arc-drawing mode with prompts similar to that of the PLine command.

Undo undraws the last segment.

Close closes the polysolid by joining the endpoint with the start point.

RELATED SYSTEM VARIABLES

PSolHeight specifies the default height of polysolids.

PSolWidth specifies the default width of polysolids.

PressPull

2007 Extends and shortens the faces of 3D solids.

Command	Alias	Ctrl+	F-key	Alt+	Menu Bar	Tablet
presspull	...	...	...	...	...	...

Command: presspull

Click inside bounded areas to press or pull. *(Pick a face, and then drag.)*

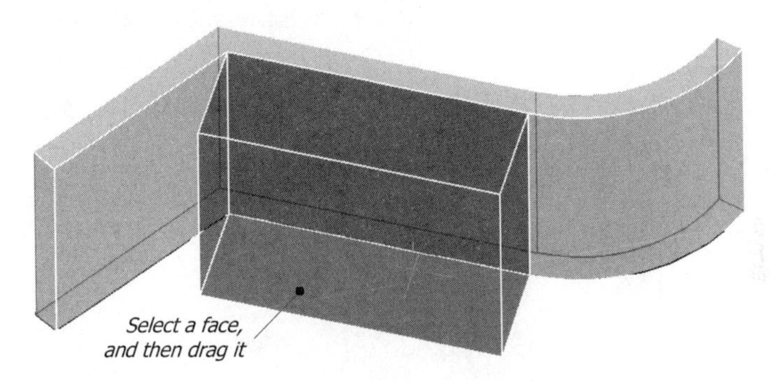

Select a face,
and then drag it

COMMAND LINE OPTION

Click inside bounded areas lets you drag the following faces:

- Areas that can be hatched by picking a point with zero gap tolerance, such as circles.
- Areas enclosed by crossing coplanar linear geometry — includes edges and geometry in blocks.
- Closed polylines, regions, 3D faces, and 2D solids with coplanar vertices.
- Areas created by geometry (including edges on faces) drawn coplanar to any face of 3D solids.

RELATED COMMANDS

Extrude converts 2D areas into 3D solids with optional tapered sides.

Thicken converts surfaces to 3D solids.

TIPS

- This command can be used to lengthen and shorten the faces of 3D solids.
- When used on 2D objects, like circles, the circle is turned into a 3D solid.
- When pushpulling "holes" (cylinders subtracted from solids), the result may be unexpected.
- I recommend using this command only for creating solids from closed boundaries and regions.
- Don't use this command for changing existing solids, because it then creates a new solid based on the perimeter of the selected face. Instead, use CTRL+**Pick** to select a face, and then use grips editing.

Preview

Rel.13 Displays plot preview; bypasses the Plot command.

Command	Alias	Ctrl+	F-key	Alt+	Menu Bar	Tablet
preview	pre	...	...	FV	File ⇘Plot Preview	X24

Command: preview

Press ESC or ENTER to exit, or right-click to display shortcut menu.

Displays preview screen:

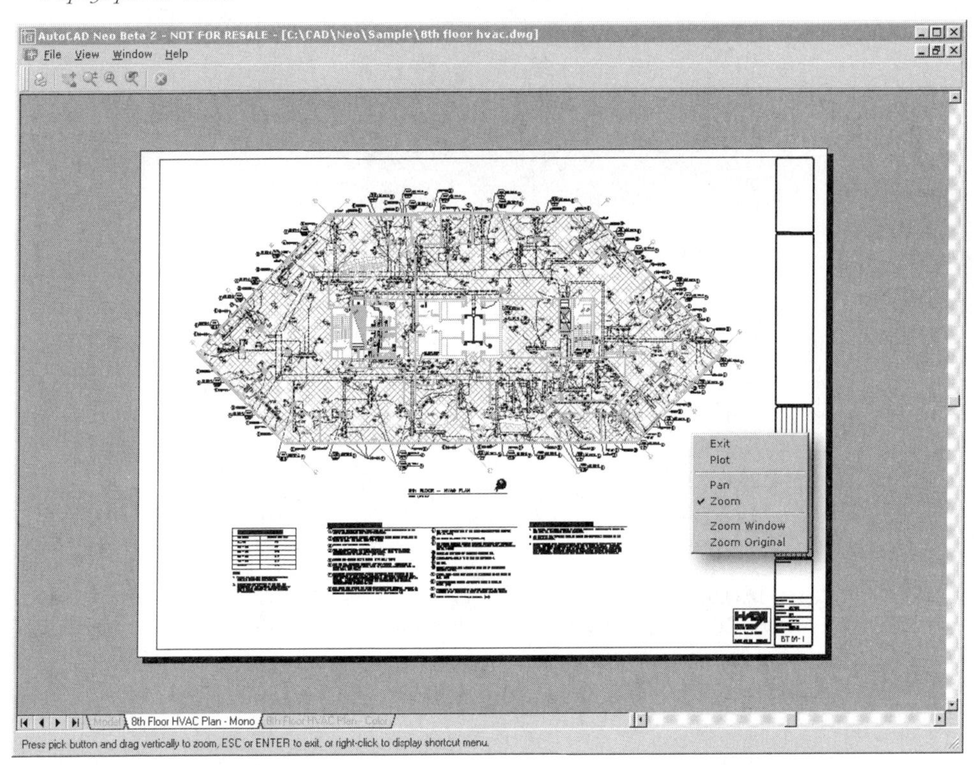

COMMAND LINE OPTION

ESC returns to the drawing window.

Preview Toolbar

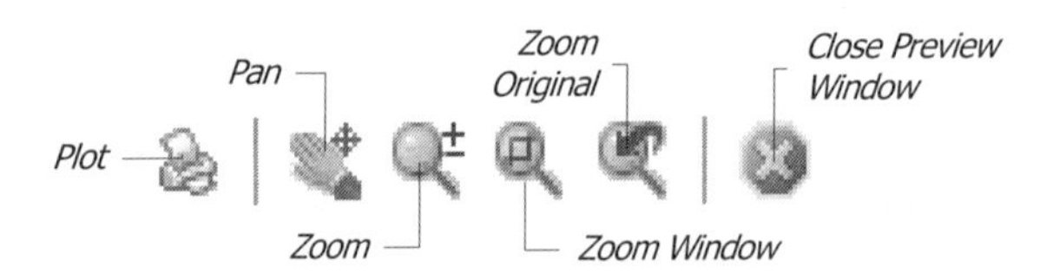

SHORTCUT MENU

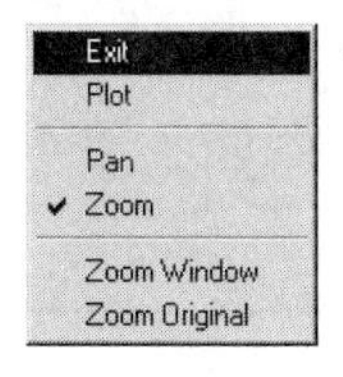

Exit exits preview mode.

Plot plots the drawing; goes immediately to plotting, bypassing the plot dialog box.

Pan pans the preview image in real time.

Zoom enlarges and shrinks the preview image in real time.

Zoom Window zooms into a windowed area.

Zoom Original returns to the orignal size.

RELATED COMMANDS

Plot plots the drawing.

PageSetup enables plot preview once a plotter is assigned to the layout.

TIPS

- This command does not operate when no plotter is assigned; use the PageSetup command to assign a plotter to the Model and layout tabs.
- Press ESC to exit preview mode.
- Partial preview mode was removed with AutoCAD 2005.
- To zoom or pan, select the mode, and then hold down the left mouse button. The image zooms or pans as you move the mouse.
- To change the background color of the preview screen, go to **Tools | Options | Display | Colors**, and then select "Plot Preview Background" from the Window Element droplist.

'Properties, 'PropertiesClose

2000 Opens and closes the Properties palette for examining and modifying properties of selected objects.

Command	Aliases	Ctrl+	F-key	Alt+	Menu Bar	Tablet
properties	**ch**	**1**	**...**	**TP**	**Tools**	**Y12-Y13**
	mo				**↳Palettes**	
	pr, props				**↳Properties**	
	ddchprop			**MP**	**Modify**	
	ddmodify				**↳Properties**	
propertiesclose	**prclose**	**1**	**...**	**TP**	**Tools**	**...**
					↳Properties	

Alternatively, you can right-click any object and select Properties from the shortcut menu.

Command: properties

Displays palettte with different content, depending on the objects selected. When no objects are selected:

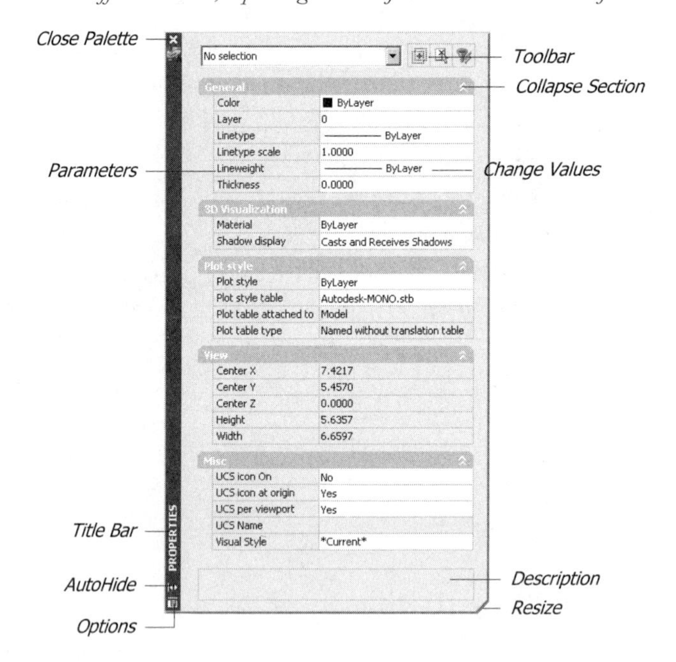

Command: propertiesclose

Closes palette.

Palette toolbar

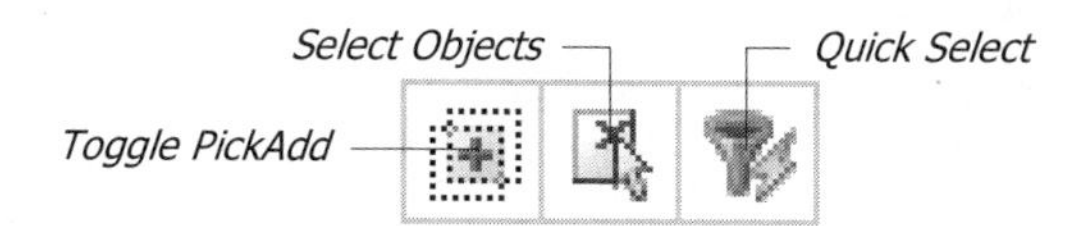

Toggle Value of PickAdd Variable controls selections when the Shift key is held down:

- **Off** adds objects to the selection set.
- **On** replaces objects in the selection set.

Select Objects prompts, 'Select objects:'; you can also just pick objects without this button.

Quick Select displays the Quick Select dialog box; see the QSelect command.

RELATED COMMANDS

ChProp changes an object's color, layer, linetype, and thickness.

Style creates and changes text styles.

RELATED SYSTEM VARIABLES

OpmState reports whether the Properties palette is open.

CeColor specifies the current color.

CeLtScale specifies the current linetype scale factor.

CeLtype specifies the current linetype.

CLayer specifies the current layer.

Elevation specifies the current elevation in the z-direction.

Thickness specifies the current thickness in the z-direction.

TIPS

- To count objects in the drawing, press Ctrl+A to select all objects, and then click the Selection list:

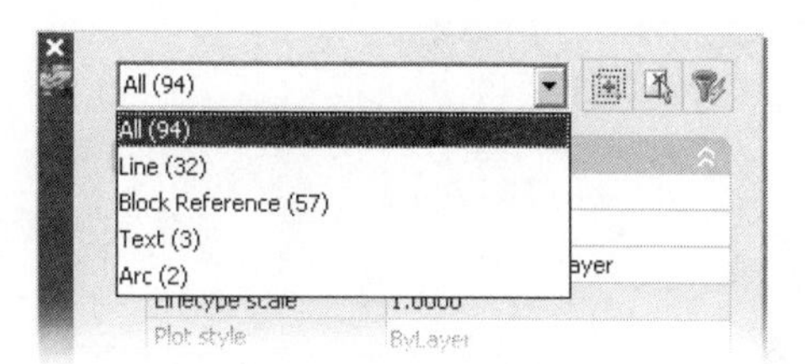

- As an alternative to entering the PropertiesClose command, you can click the **x** button on the palette's title bar.
- Double-click the title bar to dock and undock the palette within the AutoCAD window.
- The Properties palette can be dragged larger and smaller by its edges.
- When an item is displayed by gray text, it cannot be modified.
- This command can only be used transparently after it has been used at least once non-transparently.

Removed Command

The **PsDrag** command was discontinued with AutoCAD 2000i. It has no replacement.

PSetupIn

2000 Imports user-defined page setups into the current drawing layout (short for Page SETUP IN).

Commands	Alias	Ctrl+	F-key	Alt+	Menu Bar	Tablet
psetupin	...	...	...	...	...	...
-psetupin						

Command: psetupin

Displays the Select File dialog box. Select a .dwg, .dwt, or .dxf file, and then click ***Open.***

AutoCAD displays dialog box:

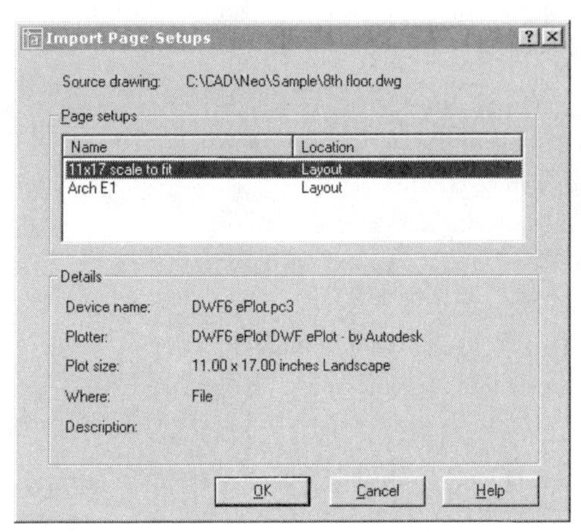

IMPORT PAGE SETTINGS DIALOG BOX

Name lists the names of page setups.

Location lists the location of the setups.

OK closes the dialog box, and loads the page setup.

-PSETUPIN Command

Command: -psetupin

Enter filename: *(Enter .dwg file name.)*

Enter user defined page setup to import or [?]: *(Enter a name, or type* **?***.)*

COMMAND LINE OPTIONS

Enter filename enters the name of a drawing file.

Enter user defined page setup to import enters the name of a page setup.

? lists the names of page setups in the drawing.

RELATED COMMANDS

PageSetup creates a new page setup configuration.

Plot plots the drawing.

PsFill

Rel.12 Fills 2D polyline outlines with raster PostScript patterns (short for PostScript FILL; an undocumented command).

Command	Alias	Ctrl+	F-key	Alt+	Menu Bar	Tablet
psfill	...	...	...	...	...	...

Command: psfill
Select polyline: *(Pick an outline.)*
Enter PostScript fill pattern name (. = none) or [?] <.>: *(Enter pattern name, or type* **?***.)*

COMMAND LINE OPTIONS

Select polyline selects the closed polyline to fill.

PostScript pattern specifies the name of the fill pattern.

. (dot) selects no fill pattern.

? lists the available fill patterns.

* specifies that the pattern should not be outlined with a polyline.

RELATED COMMANDS

BHatch fills an area with vector hatch patterns and solid colors.

PsOut exports drawings as PostScript files.

RELATED SYSTEM VARIABLE

PSQuality specifies the display options for PostScript images:

75 — (*Positive*) Displays filled image at 75dpi (default).
0 — Displays bounding box and filename; no image.
-75 — (*Negative*) Displays image outline at 75dpi; no fill.

TIPS

- These fill patterns are available:

Grayscale	RGBcolor
AIlogo	Lineargray
Radialgray	Square
Waffle	Zigzag
Stars	Brick
Specks	

- You cannot see fill patterns in AutoCAD drawings until they are exported with the **PsOut** command.

Removed Command

PsIn command was removed with AutoCAD 2000i; there is no replacement.

PsOut

Rel.12 Exports the current drawing as an encapsulated PostScript file (an undocumented command).

Command	Alias	Ctrl+	F-key	Alt+	Menu Bar	Tablet
psout	...	...	...	**FE**	**File** ⇨**Export**	...

Command: psout

Displays the Create PostScript File dialog box.

Select ***Tools | Options*** *to display dialog box of options:*

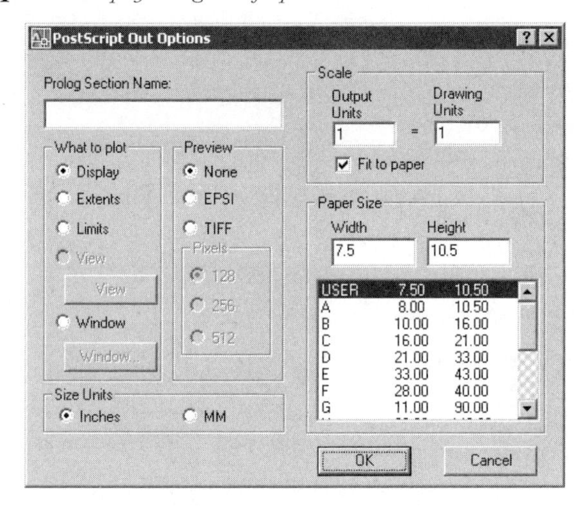

POSTSCRIPT OUT OPTIONS DIALOG BOX

Prolog Section Name specifies the name of the optional prolog section read from the *acad.psf* file. It customizes the PostScript output.

What to Plot

⊙**Display** selects the display in the current viewport.

○**Extents** selects the drawing extents.

○**Limits** selects the drawing limits.

○**View** selects a named view.

○**Window** picks two corners of a window.

Preview

⊙**None** specifies no preview image (default).

○**EPSI** specifies Macintosh preview image format.

○**TIFF** specifies Tagged Image File Format (recommended).

Pixels

128 specifies a preview image size of 128x128 pixels (default).

256 specifies a preview image size of 256x256 pixels.

512 specifies a preview image size of 512x512 pixels (recommended).

Size Units

Inches specifies the plot parameters in inches.

MM specifies the plot parameters in millimeters.

Scale

Output Units scales the output units.

Drawing Units specifies the drawing units.

Fit to Paper forces the drawing to fit paper size.

Paper Size

Width enters a width for the output size.

Height enters a height for the output size.

RELATED COMMAND

Plot exports the drawing in a variety of formats, including raster EPS and PDF.

RELATED SYSTEM VARIABLE

PSProlog specifies the PostScript prolog information.

RELATED FILE

.eps extension of file produced by PsOut.

TIPS

- The "screen preview image" is only used for screen display purposes, since graphics software generally cannot display PostScript graphic files.
- When you select the **Window** option, AutoCAD prompts you for the window corners *after* you finish selecting options.
- Although Autodesk recommends using the smallest screen preview image size (128x128), even the largest preview image (512x512) has a minimal effect on file size and screen display time. Some software programs, such as those from Microsoft, might reject an *.eps* file when the preview image is larger than 128x128. The screen preview image size has no effect on the quality of the PostScript output.
- If you're not sure which screen preview format to use, select TIFF.
- AutoCAD no longer imports PostScript files, but can generate PDF files with the Plot command.

PSpace

Rel.11 Switches from model space to paper space in layout mode (short for Paper SPACE).

Command	Alias	Ctrl+	F-key	Alt+	Menu Bar	Tablet
pspace	ps	...	...	...	...	L5

This command does not operate in model space.

Command: pspace

COMMAND LINE OPTIONS

None.

RELATED COMMANDS

MSpace switches from paper space to model space.

MView creates viewports in paper space.

UcsIcon toggles the display of the paper space icon.

Zoom scales paper space relative to model space with the **XP** option.

RELATED SYSTEM VARIABLES

MaxActVP specifies the maximum number of viewports displaying an image.

PsLtScale specifies the linetype scale relative to paper space.

TileMode allows paper space when set to 0.

TIPS

- Use paper space to layout multiple views of a single drawing.
- In layout mode, you can switch to paper space by clicking the word **MODEL** on the status bar; switch to a layout's model space by clicking the word **PAPER**.

4.0679, 0.4813, 0.0000 SNAP GRID ORTHO POLAR OSNAP OTRACK LWT MODEL

4.0679, 0.4813, 0.0000 SNAP GRID ORTHO POLAR OSNAP OTRACK LWT PAPER

- When a drawing is in paper space, AutoCAD displays **PAPER** on the status line and the paper space icon:

W X

- Paper space is known as "drawing composition" in some other CAD packages.

'Publish

2004 Outputs multiple layout sheets from one or more drawings as a single, multi-page *.dwf* file or hardcopy plot.

Command	Alias	Ctrl+	F-key	Alt+	Menu Bar	Tablet
publish	...	...	...	**FH**	**File** ↳**Publish**	...
-publish						

Command: publish

Displays dialog box:

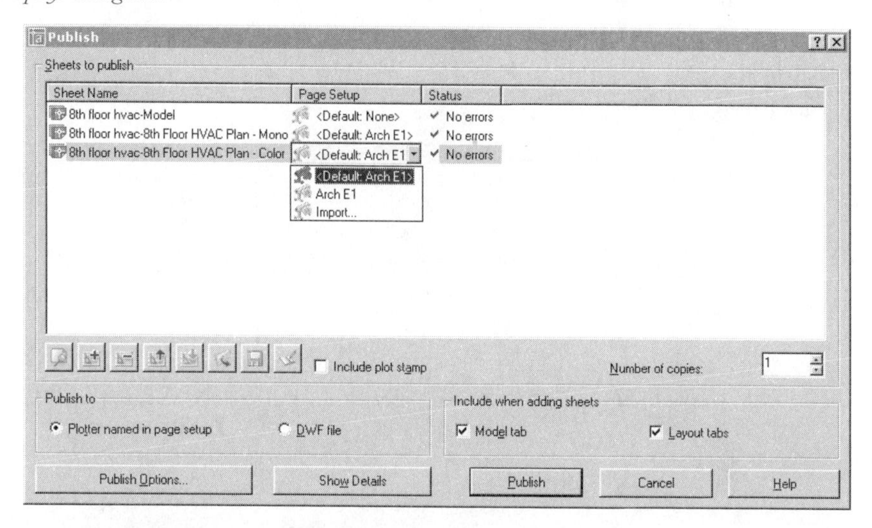

PUBLISH DIALOG BOX

List of Drawing Sheets

Sheet Name concatenates the drawing name and the layout name with a dash (-); edit sheet names with the **Rename** option.

Page Setup lists the named page setup for each layout; click to select other setups.

Status displays a message as layouts are published.

☐ **Include plot stamp** adds plot stamp data to the edge of each plot; see the PlotStamp command. (Click the **Plot Stamp Settings** button to access the Plot Stamp dialog box.)

Number of copies allows you make multi-copy prints (available only when plotting).

Publish To

⊙**Plotters named in page setups** generates hardcopy plots, or to-file plots.

○**DWF file** generates multi-sheet *.dwf* files.

Include when Adding Sheets

Model Tab determines whether the model layout is included with the list of sheets.

Layout Tabs includes all layouts.

Publish Options displays the Publish Options dialog box.

Show Details expands the dialog box to provide added information about the selected sheet.

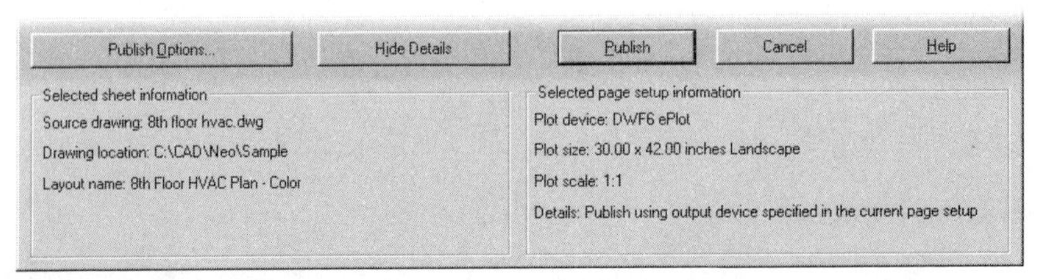

Publish generates the *.dwf* file or plot; displays Now Plotting dialog box.

PUBLISH OPTIONS DIALOG BOX

Redesigned in AutoCAD 2006.

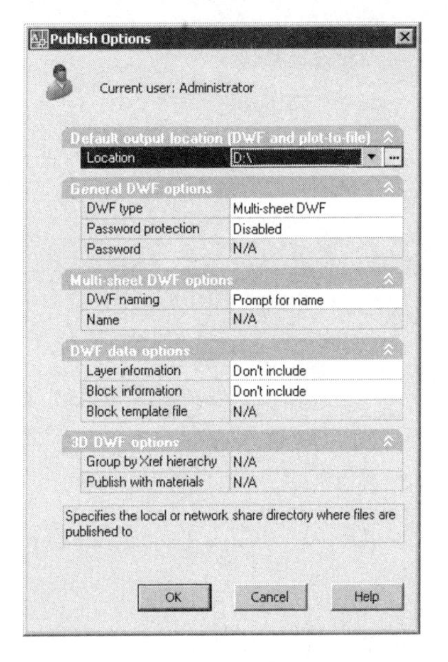

Default Output Location

Location specifies the folder in which to save the *.dwf* and *.plt* plot files.

... displays the Select a Folder for Generated Files dialog box; select a folder, and click **Select**.

General DWF

DWF Type:

- **Single-Sheet DWF** generates one *.dwf* file for each sheet.
- **Multi-Sheet DWF** generates a single *.dwf* file for all sheets.

Password Protection:

- **Disabled** leaves out the password.
- **Prompt for Password** prompts you for the password during publishing. Passwords are case-sensitive, and may consist of letters, numbers, punctuation, and non-ASCII characters. *Warning!* Passwords cannot be recovered.
- **Specify Password** specifies the password.

Password specifies the password.

Multi-Sheet DWF

DWF Naming:

- **Prompt for Name** prompts you for the file and folder names during publishing.
- **Specify Name** specifies file and folder names for the multisheet *.dwf* file.

Name:

... displays the Select DWF File dialog box; select a file name, and then click **Select**.

DWF Data

Layer information toggles whether layers are included in *.dwf* files.

Block information toggles whether blocks are included in *.dwf* files.

Block template file determines whether the *.blk* file is created or edited.

... displays the Publish Block Template dialog box; specifies which blocks and attributes to include in *.dwf* files.

3D DWF

Grouped by Xref Hierarchy displays objects by xref in the DWF Viewer.

Publish with Materials displays objects with materials in the DWF Viewer.

SHORTCUT MENU

Right-click in a sheet name the Sheets to Publish area:

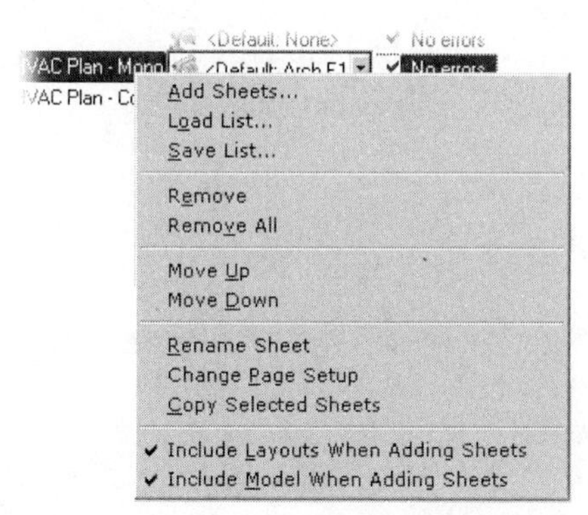

Add Sheets displays the Select Drawings dialog box; choose one or more drawings, and then click **Select**.

Load List displays the Load List of Sheets dialog box; select a *.dsd* drawing set description or *.bp3* batch plot list file, and then click **Load**.

Save List displays the Save List As dialog box; enter a file name, and then click **Save**.

Remove removes selected sheets from the list, wtihout warning.

Remove All removes all sheets.

Move Up moves the selected sheets up the list.

Move Down moves the selected sheets down the list.

Rename Sheet renames the selected sheet.

Change Page Setup displays the Page Setup drop list.

Copy Selected Sheets copies selected sheets, and adds the "-Copy*n*"suffix to the sheet name.

Include Layouts when Adding Sheets includes all layouts in the sheet set.

Include Model when Adding Sheets includes model space with the layouts.

Dialog Box Toolbar

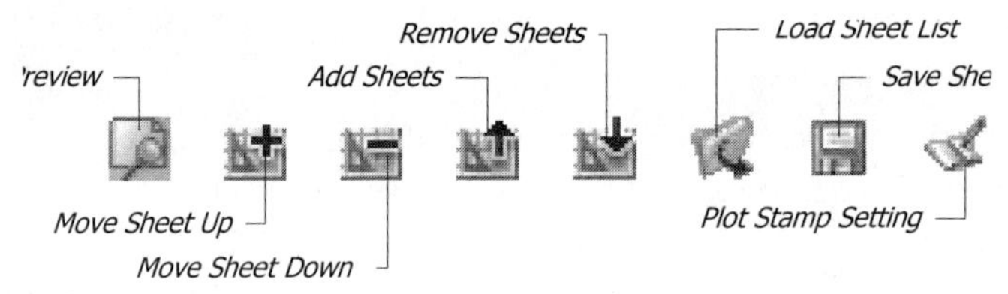

Preview displays the plot preview window; see the Preview command.

Add Sheets displays the Select Drawings dialog box; choose one or more drawings, and then click **Select**.

Remove Sheets removes selected sheets from the list, without warning.

Move Sheet Up moves the selected sheets up the list.

Move Sheet Down moves the selected sheets down the list.

Load Sheet List displays the Load List of Sheets dialog box; select a *.dsd* drawing set description or *.bp3* batch plot list file, and then click **Load**.

Save Sheet List displays the Save List As dialog box; enter a file name, and then click **Save**.

Plot Stamp Settings displays the Plot Stamp dialog box; see PlotStamp command.

-PUBLISH Command

Command: -publish

If drawing is not saved, displays Drawing Modified - Save Changes dialog box. Click ***OK***.

Displays Select List of Sheets dialog box. Select a .dsd file, and then click ***Select***.

Immediately plots the drawing set, and generates a log file.

RELATED COMMANDS

Plot outputs drawings as *.dwf* files.

PublishToWeb coverts drawings to Web pages.

3dDwfPublish exports model space drawings in 3D DWF format.

STARTUP SWITCH

/ pl specifies the *.dsd* file to publish in the background when AutoCAD starts up.

TIPS

- The Model tab is included only when the **Include Model When Adding Sheets** setting is turned on.
- DWF passwords are case sensitive. If the password is lost, it cannot be recovered, and the *.dwf* file cannot be opened. Erase the *.dwf* file, and create a new set.
- The -Publish command is good for generated sheets when a *.dsd* (drawing set description) file exists.
- To have AutoCAD publish a set of drawings unattended, follow these steps:
  1. Open the drawings in AutoCAD, and then use the Publish command to create a *.dsd* file. (Click the **Save Sheet list** button.)
  2. Exit AutoCAD.
  3. Modify the Target of AutoCAD's desktop item's Properties:

     "D:\CAD\AutoCAD 2007\acad.exe" **/pl "C:\path name\file name.dst"**
  4. Close the Properties dialog box.
  5. Double-click the icon, and AutoCAD starts plotting.
- This command can only be used transparently after it has been used at least once non-transparently.

'PublishToWeb

2000i Exports drawings as DWF, JPG, or PNG images embedded in pre-formatted Web pages.

Command	Alias	Ctrl+	F-key	Alt+	Menu Bar	Tablet
publishtoweb	...	...	...	**FW**	**File** **↳Publish to Web**	**X25**

Before starting this command, save drawings.

Command: publishtoweb

Displays dialog box:

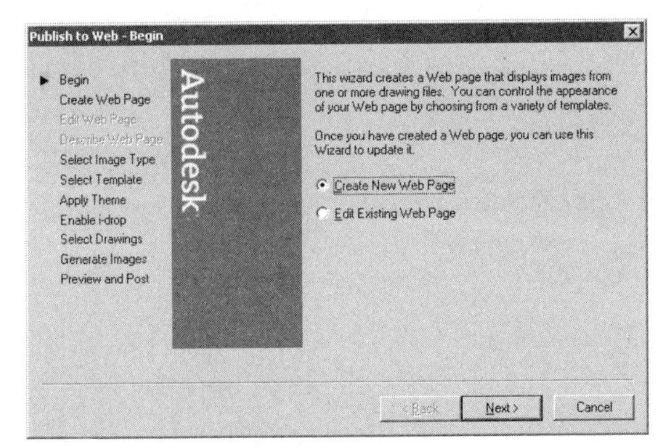

⊙ **Create New Web Page** guides you through the steps to create new Web pages from drawings.

○ **Edit Existing Web Page** guides you through the steps to edit existing Web pages created earlier by this command.

Create Web Page page

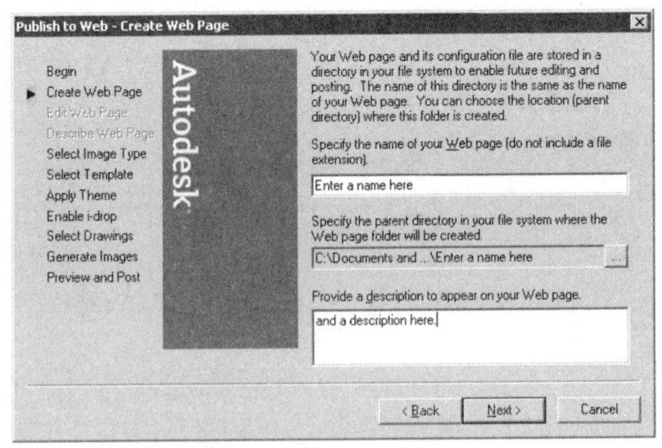

Specify the name of your Web page requires you to enter a file name. (AutoCAD uses the name for the files making up the Web page, which allows you later to edit the Web page.) The name also appears at the top of the Web page.

Provide a description to appear on your Web page provides a description appearing below the name on the Web page.

Select Image Type page

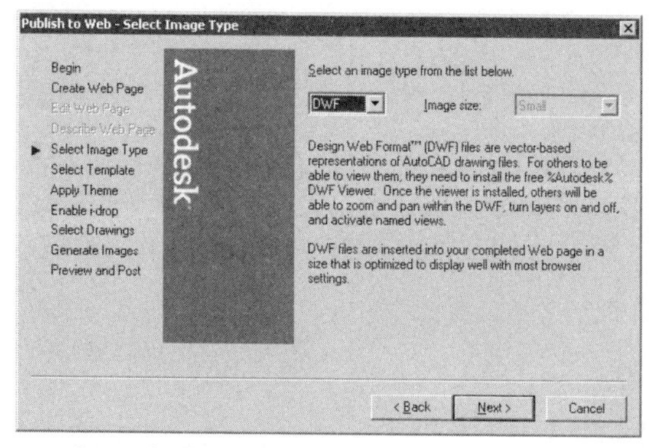

Select an image type from the list below:

- **DWF** (drawing Web format) is a vector format that displays cleanly, and can be zoomed and panned; not all Web browsers can display DWF.
- **JPEG** (joint photographic experts group) is a raster format that all Web browsers display; may create *artifacts* (details that don't exist).
- **PNG** (portable network graphics) is a raster format that does not suffer the artifact problem, but some Web browsers do not display PNG.

Image size selects a size of raster image (available for JPEG and PNG only):

- **Small**— 789x610 resolution.
- **Medium** — 1009x780 resolution.
- **Large** — 1302x1006 resolution.
- **Extra Large** — 1576x1218 resolution.

Select Template page

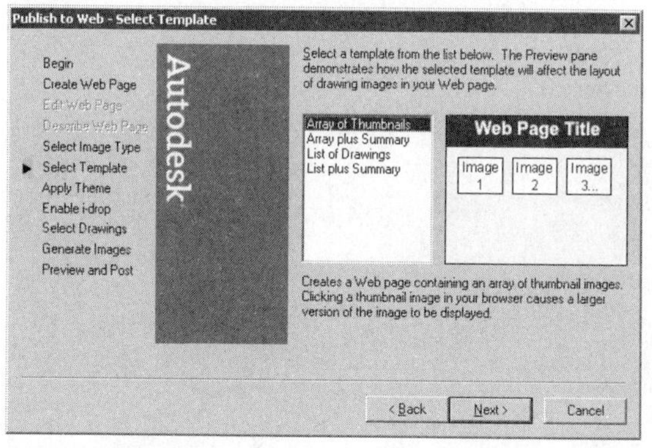

Select a template from the list below selects one of the pre-designed formats for the Web page.

Apply Theme page

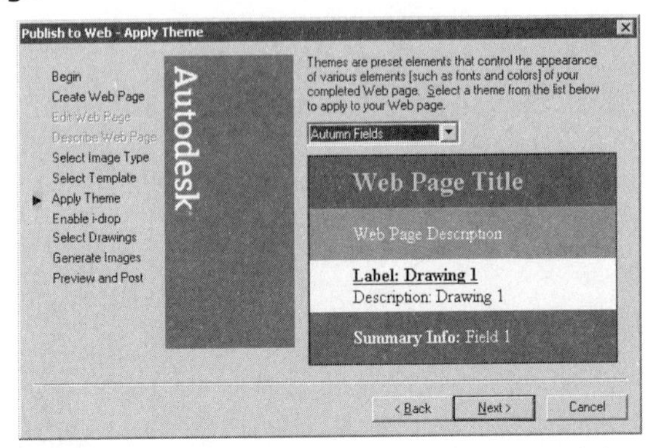

Select a theme from the list below selects one of the pre-designed themes (colors and fonts) for the Web page.

Enable i-drop page

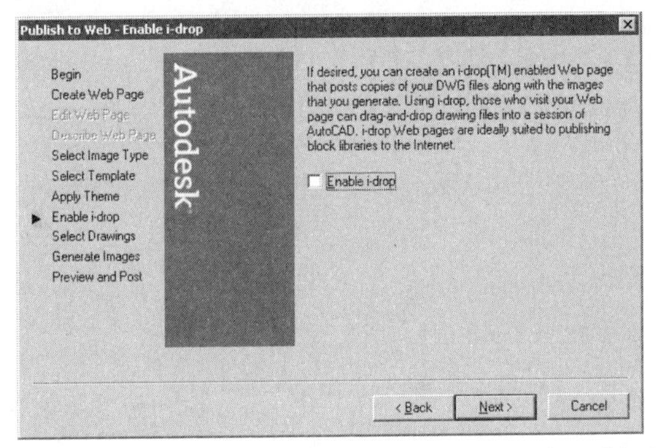

☐ **Enable i-drop** adds i-drop capability to the Web page. This allows blocks to be dragged from the Web page directly into AutoCAD and other Autodesk software.

Select Drawings page

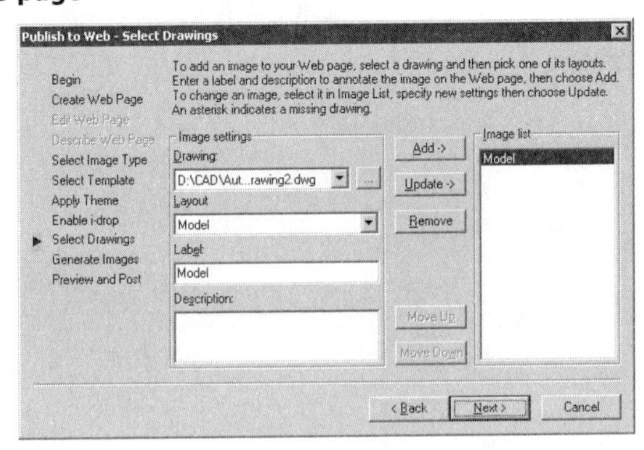

Image Settings

Drawing selects the drawing; the current drawing is the default.

Layout selects the name of a layout, or Model space.

Label names the image.

Description provides a description that appears with the drawing on the Web page.

Add adds the image setting to the image list.

Update changes the image setting in the image list.

Remove removes the image setting from the image list.

Move up moves the image setting up the image list.

Move down moves the image setting down the image list.

Generate Images page

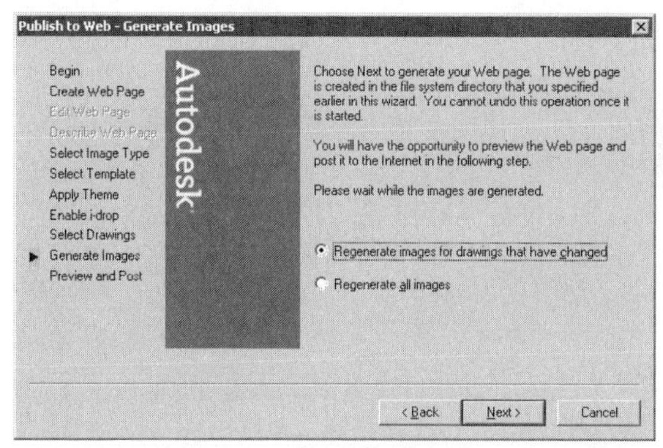

⊙**Regenerate images for drawings that have changed** updates images for those drawings that have been edited.

○**Regenerate all images** regenerates all images from all drawings; ensures all are up to date.

Preview and Post page

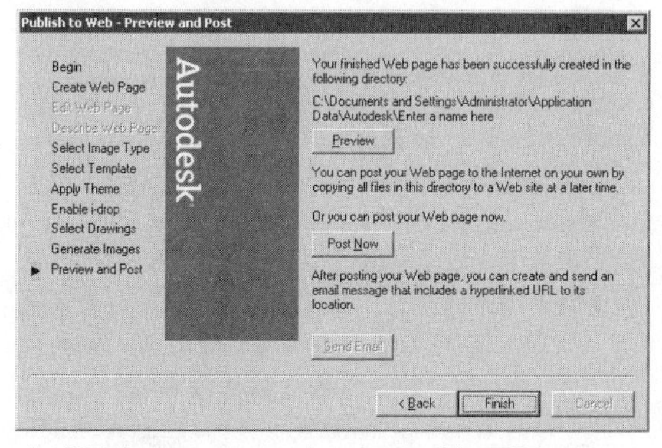

Preview launches the Web browser to preview the resulting Web page.

Post Now uploads the files (HTML, JPEG, PNG, DWF, and so on) to the Web site.

Send Email sends an email to alert others of the posted Web page.

Examples of resulting Web pages:

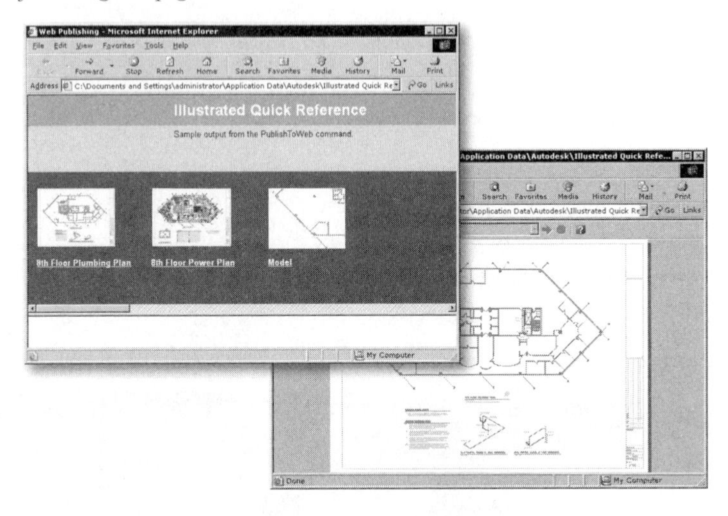

RELATED COMMANDS

Publish exports drawings as multi-sheet *.dwf* files.

Plot exports drawings as *.dwf* files via the **ePlot** option.

Hyperlink places hyperlinks in drawings.

RELATED FILES

*** .ptw** are PublishToWeb parameter files, stored in tab-delimited ASCII file.

*** .js** are JavaScript files.

*** .jpg** are joint photographic experts group (raster image) files.

*** .png** are portable network graphics (raster image) files.

*** .dwf** are drawing Web format (vector image) files.

TIPS

- Use the **Regenerate all images** option, unless you have an exceptionally slow computer or a large number of drawings to process. The Generate Images step can take a long time.
- After you click **Preview** to view the Web page (and after AutoCAD launches the Web browser), click the **Back** button to make changes, if the result is not to your liking.
- The **Post Now** option works only if you have correctly set up the FTP (file transfer protocol) parameters. If so, you can have AutoCAD upload the HTML files directly to a Web site. If not, use a separate FTP program to upload the files from the *\windows\applications data\autodesk* folder.
- You can customize the themes and templates by editing the *acwebpublish.css* (themes) and *acwebpublish.xml* (templates) files.
- This command can only be used transparently after it has been used at least once non-transparently.
- This command fails if you use the Options command to redirect templates to a folder other than *\template,* because AutoCAD expects to find the *\ptwtemplate* folder in *\template.*

Purge

V. 2.1 Removes unused, named objects from the drawing: blocks, dimension styles, layers, linetypes, plot styles, shapes, text styles, table styles, application ID tables, and multiline styles.

Commands	Alias	Ctrl+	F-key	Alt+	Menu Bar	Tablet
purge	**pu**	...	...	**FUP**	**File** ↳**Drawing Utilities** ↳**Purge**	**X25**
-purge						

Command: purge

AutoCAD 2007 add Materials and Visual Styles. Displays dialog box:

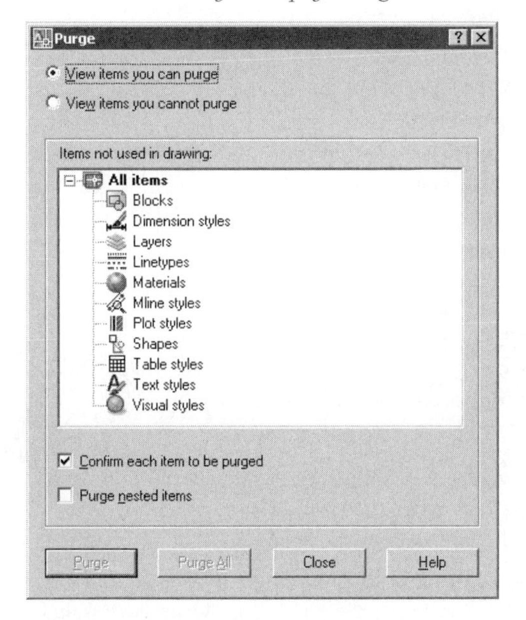

PURGE DIALOG BOX

⊙**View items you can purge** lists objects that can be purged from the drawing.

○**View items you cannot purge** lists objects that cannot be purged from the drawing.

☑ **Confirm each item to be purged** displays a confirmation dialog box for each object being purged.

☐ **Purge nested items** purges nested objects, such as unused blocks within unused blocks.

-PURGE Command

Command: -purge

Enter type of unused objects to purge
[Blocks/Dimstyles/LAyers/LTypes/MAterials/Plotstyles/SHapes/textSTyles/Mlinestyles/Tablestyles/Visualstyles/Regapps/All]: *(Enter an option.)*

Sample response:

Purge linetype CENTER? <N> **y**
Purge linetype CENTER2? <N> **y**

COMMAND LINE OPTIONS

Blocks purges named but unused (not inserted) blocks.

Dimstyles purges unused dimension styles.

LAyers purges unused layers.

LTypes purges unused linetypes.

MAterials purges unused linetypes (*new to AutoCAD 2007*).

Plotstyle purges unused plot styles.

SHapes purges unused shape files.

textSTyles purges unused text styles.

Mlinestyles purges unused multiline styles.

Tablestyles purges unused table styles.

Visualstyles purges unused visual styles (*new to AutoCAD 2007*).

Regapps purges unused application ID tables of registered applications.

All purges dall named objects, if possible.

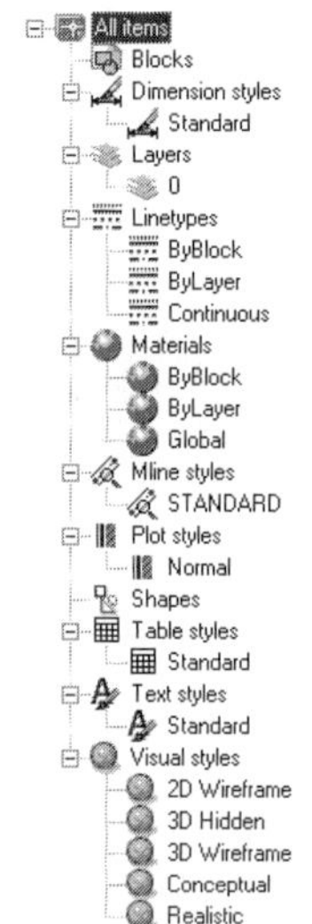

RELATED COMMAND

-WBlock writes the current drawing to disk with the * option, and removes spurious information from the drawing.

TIPS

- It may be necessary to use this command several times; follow each purge with the Close command, then open the drawing, and purge again. Repeat until the Purge command reports nothing to purge.
- The **View items you cannot purge** option lists items being used, as well as system items that can never be purged, as illustrated at right.

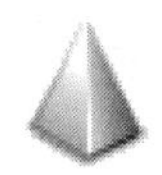

Pyramid

2007 Draws 3D solid pyramids with pointy or flat tops, and 3 to 32 sides.

Command	Alias	Ctrl+	F-key	Alt+	Menu Bar	Tablet
pyramid	**pyr**	...	...	**DMY**	**Draw** **↳Modeling** **↳Pyramid**	...

Command: pyramid
4 sides Circumscribed
Specify center point of base or [Edge/Sides]: *(Pick a point, or enter an option.)*
Specify base radius or [Inscribed]: *(Pick a point, or type* **I***.)*
Specify height or [2Point/Axis endpoint/Top radius]: *(Pick a point, or enter an option.)*

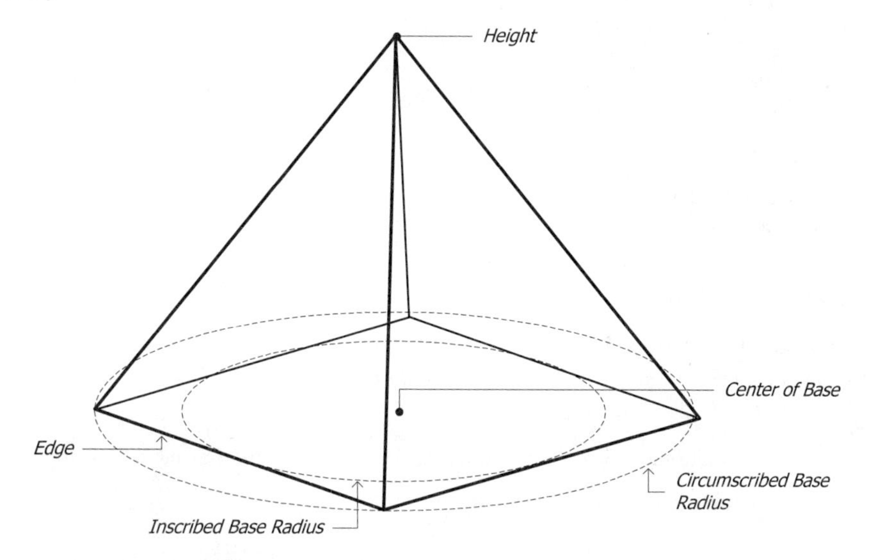

COMMAND LINE OPTIONS

The base is defined by a center point or edge length, and the number of sides:

Specify center point of base locates the center of the pyramid's base by x,y,z coordinates or by picking a point in the drawing.

Edge specifies the size of the base by the length of one edge; the base is always made from equilateral sides.

Sides specifies the number of sides; the range is 3 to 32, and the default is 4.

The base is drawn like a polygon: circumscribed or inscribed:

Specify base radius specifies the radius of a "circle" that circumscribes the base.

Inscribed specifies the radius of a "circle" inscribed within the base.

Specify height indicates the height by x,y,z coordinates or by picking a point in the drawing.

2Point specifies the height by picking two points.

Axis endpoint specifies the location and angle of the top point; this allows the pyramid to be drawn at an angle; the top and bottom are perpendicular to the axis.

Top radius specifies the size of the flat top to the pyramid.

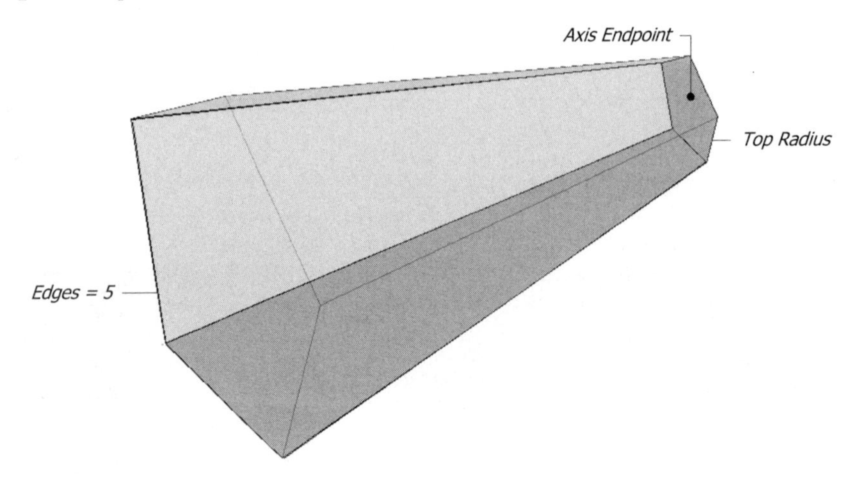

RELATED COMMANDS

Ai_Pyramid draws 3D mesh pyramids.

Cone draws 3D solid cones.

Extrude creates 3D objects with optional tapered sides.

RELATED SYSTEM VARIABLE

DragVs specifies the visual style while creating 3D solids.

TIPS

- To create a pyramid-like object with a non-equilateral base, use the Extrude command.
- After the pyramid is drawn, it can be edited with grips.

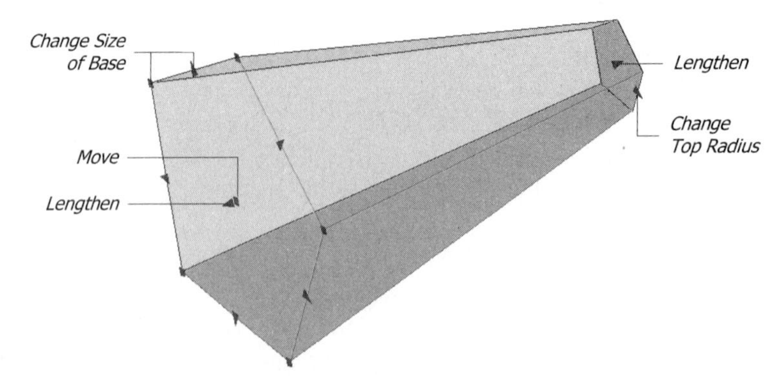

- When the top radius equals the base radius, this command draws a prism.

QDim

2000 Dimensions entire objects (with continuous, baseline, ordinate, radius, diameter, and staggered dimensions) using just three picks (short for Quick DIMensioning).

Command	Alias	Ctrl+	F-key	Alt+	Menu Bar	Tablet
qdim	...	...	...	NQ	Dimension ↳Quick Dimenion	W1

Command: qdim
Select geometry to dimension: *(Select one or more objects; press* ***CTRL+A*** *to select the entire drawing.)*
Select geometry to dimension: *(Press Enter to end object selection.)*
Specify dimension line position, or
[Continuous/Staggered/Baseline/Ordinate/Radius/Diameter/datumPoint/ Edit/seTtings] <Continuous>: *(Enter an option.)*

COMMAND LINE OPTIONS

Select geometry to dimension selects objects to dimension.

Specify dimension line position specifies the location of the dimension line.

Continuous draws continuous dimensions.

Staggered draws staggered dimensions.

Baseline draws baseline dimensions.

Ordinate draws ordinate dimensions relative to the UCS origin.

Radius draws radial dimensions; prompts 'Specify dimension line position:'.

Diameter draws diameter dimensions; prompts 'Specify dimension line position:'.

datamPoint sets a new datum point for ordinate and baseline dimensions; prompts 'Select new datum point:'.

Edit options
Indicate dimension point to remove, or [Add/eXit] <eXit>: *(Select a dimension, point, or enter an option.)*

Indicate dimension to remove selects the dimension point to remove from the dimension.

Add adds a dimension point to the dimension.

eXit returns to dimension drawing mode.

seTtings options
Associative dimension priority [Endpoint/Intersection] <Endpoint>: *(Enter an option.)*

Endpoint applies associative dimensions to endpoints over intersections.

Intersection applies associative dimensions to intersections over endpoints.

RELATED COMMANDS

DimStyle creates dimension styles, which specify the look of a dimension.

Dim*xxx* draws other kinds of dimensions.

QLeader draws leaders.

RELATED SYSTEM VARIABLE

DimStyle specifies the current dimension style.

TIPS

- Example of continuous dimensions:

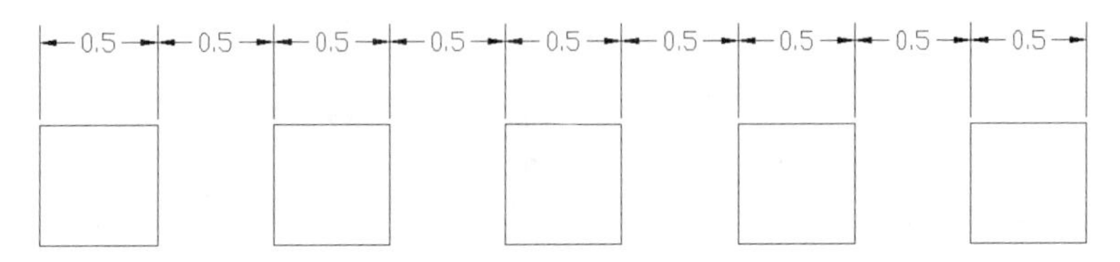

- Example of staggered dimensions:

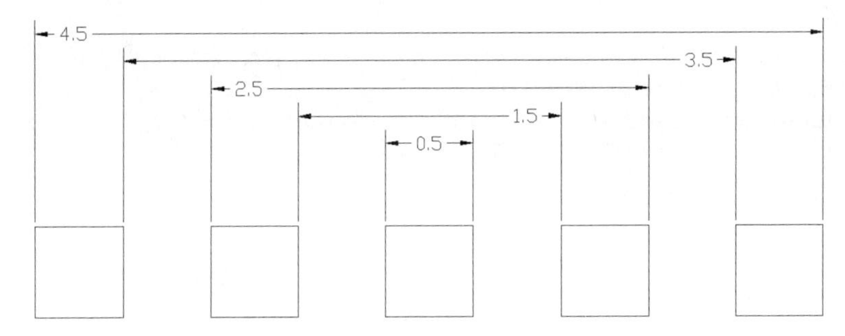

- Example of ordinate dimensions:

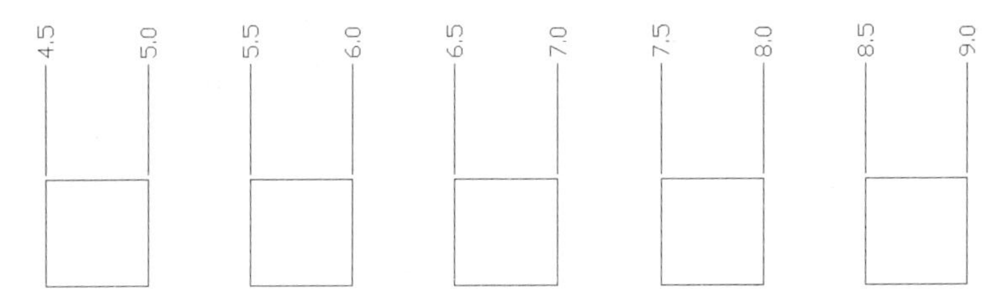

- As of AutoCAD 2004, dimensions created with QDim are fully associative.
- This command sometimes fails. AutoCAD complains, "Invalid number of dimension points found."

QLeader

Rel.14 Draws leaders; dialog box specifies options for custom leaders and annotations (short for Quick LEADER).

Command	Alias	Ctrl+	F-key	Alt+	Menu Bar	Tablet
qleader	**le**	...	...	**NE**	**Dimension ↳Leader**	**W2**

Command: qleader

Specify first leader point, or [Settings] <Settings>: *(Pick point 1 for the arrowhead, or type* **S***.)*

When ***S*** *is entered, displays dialog box:*

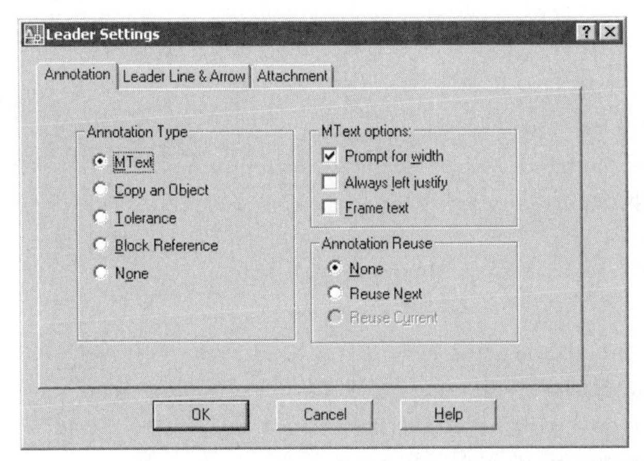

Click ***OK*** *to continue with the command; the prompts vary, depending on the options selected in the dialog box:*

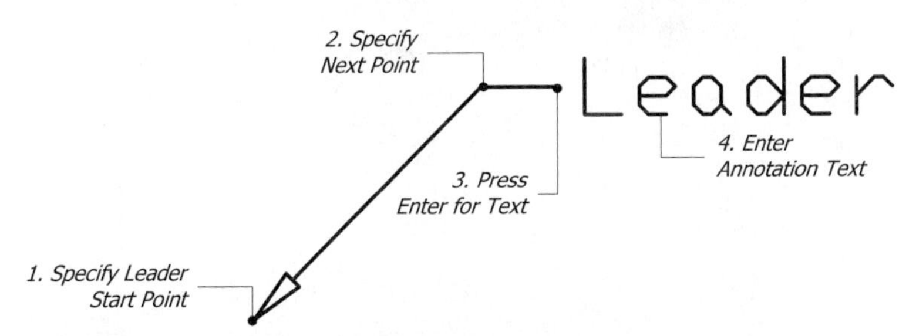

Specify next point: *(Pick point 2 for the leader shoulder.)*

Specify next point: *(Pick point, or press Enter for text options.)*

Specify text width <0.0000>: *(Enter a value, or press Enter.)*

Enter first line of annotation text <Mtext>: *(Enter text,or press Enter for mtext editor.)*

Enter next line of annotation text: *(Press Enter to end command.)*

COMMAND LINE OPTIONS

Specify first leader point picks the location for the leader's arrowhead; press Enter to display tabbed dialog box.

Specify next point picks the vertices of the leader; press Enter to end the leader line.

Specify text width specifies the width of the bounding box for the leader text.

Enter first line of annotation text specifies the text for leader annotation; press Enter twice to end.

Enter next line of annotation text specifies more text; press Enter once to end.

LEADER SETTINGS DIALOG BOX

Annotation Type

⊙**MText** prompts you to enter text for the annotation.

○**Copy an Object** attaches any object in the drawing as an annotation.

○**Tolerance** prompts you to select tolerance symbols for the annotation.

○**Block Reference** prompts you to select a block for the annotation.

○**None** attaches no annotation.

MText Options

☑ **Prompt for width** displays the 'Specify text width' prompt.

☐ **Always left justify** left justifies the text, even when the leader is drawn to the right.

☐ **Frame text** places a rectangle around the text.

Annotation Reuse

⊙**None** does not retain annotation for next leader.

○**Reuse Next** remembers the current annotation for the next leader.

○**Reuse Current** uses the last annotation for the current leader.

Leader Line & Arrow tab

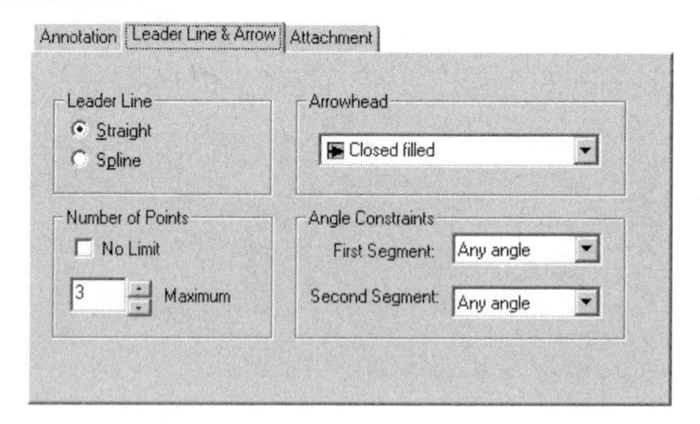

Leader Line

⊙**Straight** draws the leader with straight lines.

○**Spline** draws the leader as a spline curve.

Number of Points

☐ **No limit** keeps prompting for leader vertex points until you press Enter.

Maximum stops the command prompting for leader vertex points; default=3.

Arrowhead

Arrowhead selects the type of arrowhead, including Closed filled (default), None, and User Arrow.

Angle Constraints

First Segment selects from Any angle (user-specified), Horizontal (0 degrees), 90, 45, 30, or 15-degree leader line, first segment.

Second Segment selects from Any angle (user-specified), Horizontal, 90, 45, 30, or 15-degree leader line, second segment.

Attachment tab

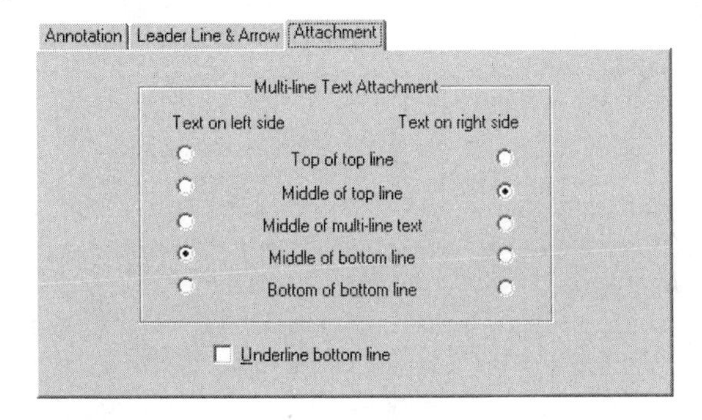

Multiline Text Attachment

Left Side positions the annotation at one of several locations relative to the last leader segment, when the annotation is located to the left of the leader:

O Top of top line.

O Middle of top line.

⊙ Middle of multi-line text.

O Middle of bottom line (default).

O Bottom of bottom line.

Right Side positions the annotation at one of several locations relative to the last leader segment, when the annotation is located to the right of the leader.

☐ **Underline bottom line** underlines the last line of leader text.

RELATED COMMANDS

DdEdit edits leader text; see the MText command.

DimStyle sets dimension variables, including leaders.

Leader draws leaders without dialog boxes.

TIPS

- This command draws leaders, just like the Leader command in AutoCAD Release 13 and 14. The difference is that it brings up a triple-tab dialog box for setting the leader options.
- Some options have interesting possibilities, such as using a mtext, text, block reference, or tolerance in place of the leader text.

QNew

2004 Starts new drawings based on a default template file *(short for Quick NEW)*.

Command	Alias	Ctrl+	F-key	Alt+	Menu Bar	Tablet
qnew	...	...	...	...	...	...

Command: qnew

Display depends on the ***Startup*** *option in the General Options section of the System tab (Options dialog box). See the* ***New*** *command.*

RELATED COMMANDS

New starts new drawings.

SaveAs saves drawings in *.dwg* or *.dwt* format; creates template files.

RELATED SYSTEM VARIABLES

DbMod indicates whether the current drawing has changed since being loaded.

DwgPrefix indicates the path to the current drawing.

DwgName indicates the name of the current drawing.

FileDia displays file prompts at the 'Command' prompt.

RELATED FILES

uizard ini holds the names and descriptions of template files.

***.dut** are template files stored in *.dwg* format.

TIPS

- The New and QNew commands operate in an identical manner, except when a template drawing is specified by the Option command's Files tab:

- The toolbar icon executes the QNew command, while the **File | New** menu pick executes the New command.

QSave

Rel.12 Saves the current drawing without requesting a file name (short for Quick SAVE).

Command	Alias	Ctrl+	F-key	Alt+	Menu Bar	Tablet
qsave	...	S	...	FS	File	U24-
					↳Save	U25

Command: qsave

If the drawing has never been saved, displays the Save Drawing As dialog box.

COMMAND LINE OPTIONS

None.

RELATED COMMANDS

Quit ends AutoCAD, with or without saving the drawing.

Save saves the drawing, after requesting the filename.

RELATED SYSTEM VARIABLES

DBMod indicates whether the drawing has changed since it was loaded.

DwgName specifies the current drawing filename (default = *drawing1*).

DwgTitled specifies the status of drawing's filename:

0 — Drawing is named *drawing1* (default).

1 — Drawing was given another name.

TIPS

- When the drawing is unnamed, the QSave command displays the Save Drawing As dialog box to request a file name; see the SaveAs command.
- When the drawing file, its folder, or drive (such as a CD-ROM drive) is marked read-only, use the SaveAs command to save the drawing by another file name, or to another folder or drive.

QSelect

2000 Creates selection sets of objects based on their properties (short for Quick SELECT).

Command	Alias	Ctrl+	F-key	Alt+	Menu Bar	Tablet
qselect	...	...	...	TK	Tools ↳Quick Select	X9

Command: qselect

Displays dialog box:

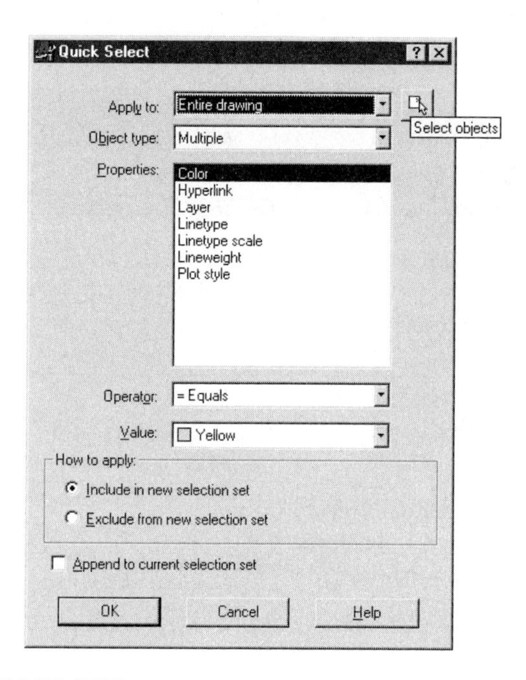

QUICK SELECT DIALOG BOX

Apply to applies the selection criteria to the entire drawing or current selection set; to create a selection set, click the **Select Objects** button.

Select Objects allows you to select objects; AutoCAD prompts: 'Select objects: '. Right-click or press ENTER to return to this dialog box.

Object type lists the objects in the selection set; allows you to narrow the selection criteria to specific types of objects (default = Multiple).

Properties lists the properties valid for the selected object types; when you select more than one object type, only the properties in common are listed.

Operator lists logical operators available for the selected property; operators include:

Operator		Meaning
=	**Equals**	Selects objects equal to the property.
<>	**Not Equal To**	Selects objects different from the property.
>	**Greater Than**	Selects objects greater than the property.
<	**Less Than**	Selects objects less than the property.
*	**Wildcard Match**	Selects objects with matching text.

Value specifies the property value for the filter. If known values for the selected property are available, Value becomes a list from which you can choose a value. Otherwise, enter a value.

How to Apply

⊙ **Include in new selection set** creates a new selection set.

○ **Exclude from new selection set** inverts the selection set, excluding all objects that match the selection criteria.

Append to current selection set

☑ adds to the current selection set.

☐ replaces the current selection set.

RELATED COMMANDS

Select selects objects on the command line.

Filter runs a more sophisticated version of the QSelect command.

RELATED SYSTEM VARIABLES

None.

TIPS

- Use this command to select objects based on their properties; use the Select command to select objects based on their location in the drawing.
- You may select objects before entering the QSelect command, and then add or remove objects from the selection set with the Quick Select dialog box's options.
- Since this command is not transparent, you cannot use it within other commands; instead, use the 'Filter command.
- As of AutoCAD 2005, this command also selects block insertions and tables.
- This command works with the properties of proxy objects created by ObjectARX applications.

'QText

V. 2.0 Displays lines of text as rectangular boxes (short for Quick TEXT).

Command	Alias	Ctrl+	F-key	Alt+	Menu Bar	Tablet
qtext	...	...	...	...	...	...

Command: qtext
Enter mode [ON/OFF] <OFF>: *(Enter* **ON** *or* **OFF**.*)*

ILLUSTRATED
AUTOCAD
QUICK REFERENCE

Left: *Normally displayed text.*
Right: *Quick text after regeneration.*

COMMAND LINE OPTIONS

ON turns on quick text after the next RegenAll command.

OFF turns off quick text after the next RegenAll command.

RELATED COMMAND

RegenAll regenerates the screen, which makes quick text take effect.

RELATED SYSTEM VARIABLE

QTextMode holds the current state of quick text mode.

TIPS

- A regeneration is required before AutoCAD displays text in quick outline form:

  Command: regenall
  Regenerating model.

- To reduce regen time, use QText to turn lines of text into rectangles, which redraw faster.
- The length of a qtext box does not necessarily match the actual length of text.
- Turning on QText affects text during plotting; qtext blocks are plotted as rectangles.
- To find invisible text (such as text made of spaces), turn on QText, thaw all layers, zoom to extents, and use the RegenAll command.
- The rectangles displayed by this command are affected by lineweights.

'QuickCalc, 'QcClose

2006 Opens and closes the QuickCalc palette.

Command	Alias	Ctrl+	F-key	Alt+	Menu Bar	Tablet
quickcalc	**qc**	**8**	...	**TPQ**	**Tools** **↳Palettes** **↳Quick Calc**	...
qcclose	...	**8**	...			

Command: quickcalc

Displays palette:

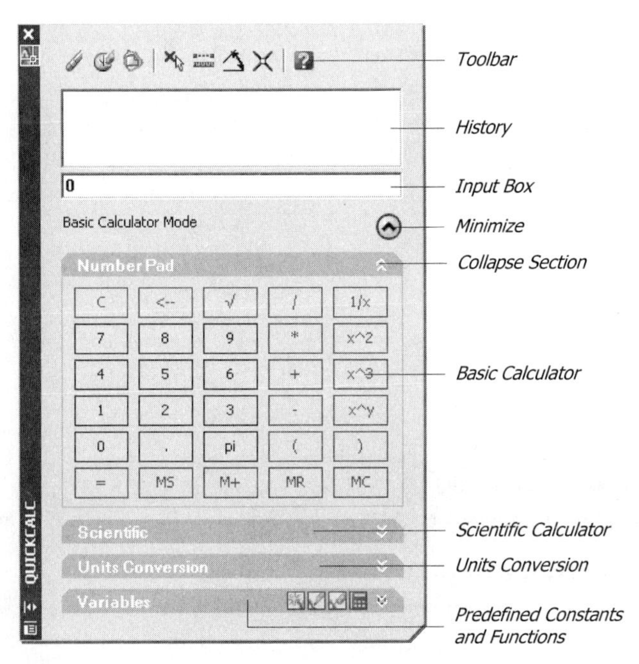

Command: qcclose

Closes the palette.

QUICKCALC PALETTE

Basic Calculator performs arithmetic and algebraic functions.

Scientific Calculator adds logarithmic and trigonometric functions (see below, left).

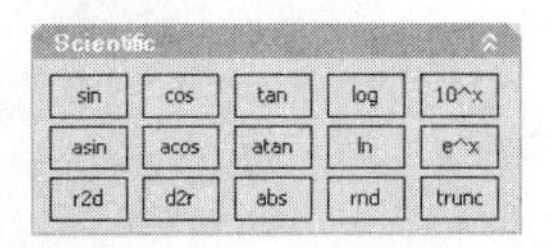

Units Conversion	
Units type	Length
Convert from	Meters
Convert to	Meters
Value to convert	0
Converted value	

Units Conversion converts between units based on the *acad.unt* file, although does not include all conversion provided by the file, such as fortnights (see above, right).

Predefined Constants and Functions defines the values of variables.

Window Toolbars

QuickCalc Toolbar

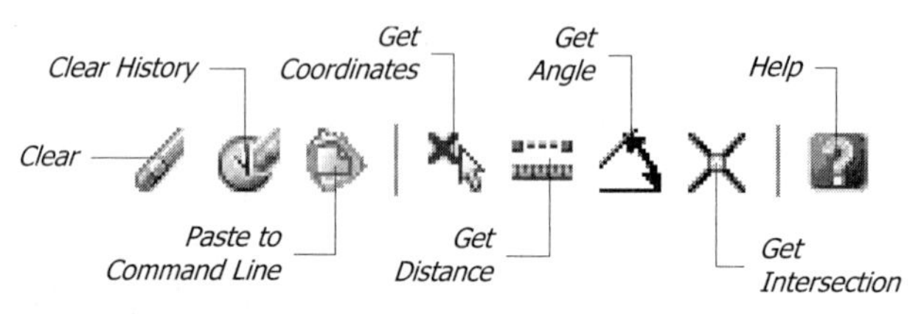

Clear clears the input area.

Clear History clears the history list.

Paste to Command Line places the value on the command line.

Get Coordinates returns the x,y,z coordinates of the pick point.

Get Distance returns the distance between two points picked in the drawing.

Get Angle returns the angle between two points relative to the x axis.

Get Intersection returns the x,y,z coordinates of the intersection of two lines defined by four pick points.

Help displays online help.

Variables Toolbar

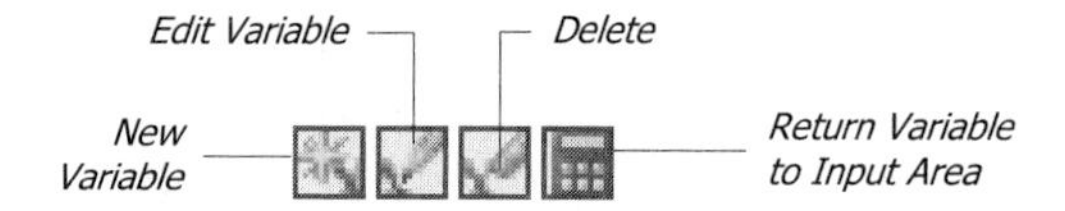

New Variable displays the Variable Definition dialog box.

Edit Variable also opens the Variable Definition dialog box.

Delete erases the selected variable.

Return Variable to Input Area places variable on the input line.

VARIABLE DEFINITION DIALOG BOX

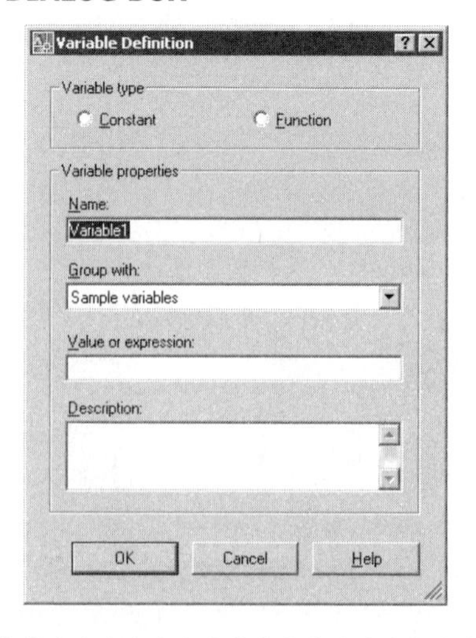

Variable Type:

Constant stores variables as constant values.

Function stores variables as functions.

Variable Properties:

Name specifies the name of the variable.

Group With stores the variable with a user-defined group.

Value or Expression defines the value of the constant, or expression for the variable.

Description describes the constant or variable.

RELATED COMMAND

Cal displays the geometry calculator function on the command line.

RELATED SYSTEM VARIABLES

CalcInput determines whether mathematical expressions are evaluated in dialog boxes.

QcState reports whether the QuickCalc palette is open or not.

RELATED FILE

acad.unt stores conversion units.

TIPS

- This command can be used during other commands to provide numeric input.
- You can enter expressions in fields of dialog boxes that expect numbers, such as the one illustrated below. Use the following syntax:

=expression (Press ***ALT****+Enter)*

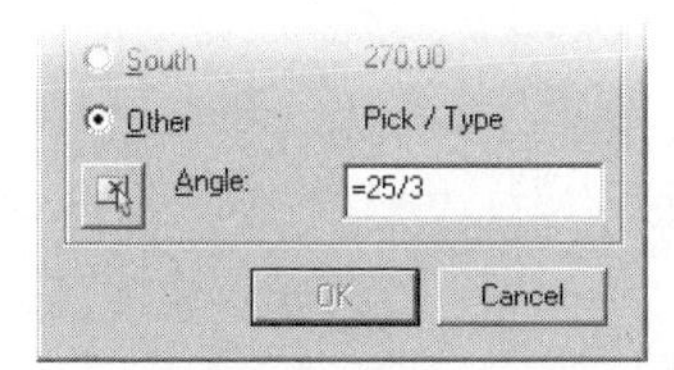

- Unlike most calculators, QuickCalc does not calculate answers when you click a function; instead, it builds *expressions*. It expects you to compose expressions: enter the expression, edit it, and then click **=** or press Enter.
- Use the history area to modify expressions and recalculate them.
- You can also use AutoLISP to perform calculations at the command line.

Quit

V. 1.0 Exits AutoCAD without saving changes made to the drawing after the most recent QSave or SaveAs command.

Command	Alias	Ctrl+	F-key	Alt+	Menu Bar	Tablet
quit	**exit**	**Q**	**Alt+F4**	**FX**	**File ↳Exit**	**Y25**

Command: quit

Displays dialog box:

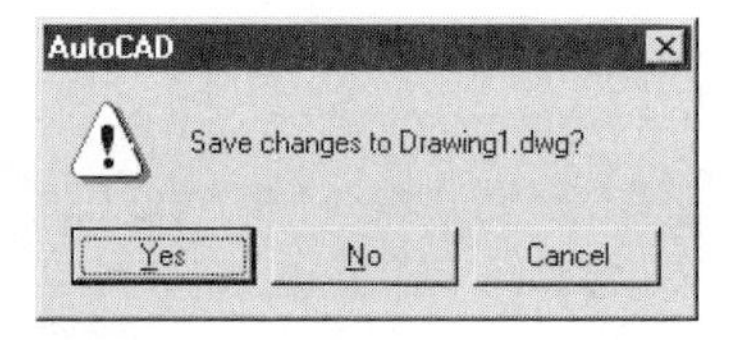

AUTOCAD DIALOG BOX

Yes saves changes before leaving AutoCAD.

No does not save changes.

Cancel does not quit AutoCAD; returns to drawing.

RELATED COMMANDS

Close closes the current drawing.

SaveAs saves the drawing by another name and to another folder or drive.

RELATED SYSTEM VARIABLE

DBMod indicates whether the drawing has changed since it was loaded.

RELATED FILES

*****.dwg*** are AutoCAD drawing files.

*****.bak*** are backup files.

*****.bk*n*** are additional backup files.

TIPS

- You can change a drawing, yet preserve its original format:
  1. Use the SaveAs command to save the drawing by another name.
  2. Use the Quit command to preserve the drawing in its most recently-saved state.
- Even if you accidentally save over a drawing, you can recover the previous version — when AutoCAD is set to save backup files automatically. (See the Options command). Follow these steps:
  1. Use Windows Explorer to rename the drawing file.
  2. Use Windows Explorer to rename the *.bak* (backup) extension to *.dwg*.
- You cannot save changes to a read-only drawing with the Quit command; use the SaveAs command instead.

R14PenWizard

2000 Helps create color-dependent plot style tables (an undocumented command).

Command	Alias	Ctrl+	F-key	Alt+	Menu Bar	Tablet
r14penwizard	...	...	...	**TZD**	**Tools** **↳Wizard** **↳Add Color-Dependent Plot Style Table**	...

You can also access this command through the Windows Control Panel's ***Autodesk Plot Style Manager****.*
(This command makes the Plot command compatible with versions of AutoCAD prior to 2000.)

Command: r14penwizard

Displays Add Color-Dependent Plot Style Table wizard.

ADD COLOR-DEPENDENT PLOT STYLE TABLE DIALOG BOX

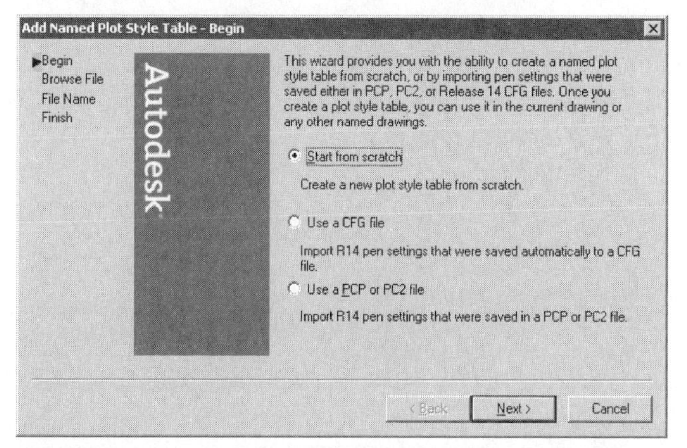

⊙ **Start from scratch** creates a new color-dependent plot style table file.

○ **Use a CFG file** converts the plotter pen settings stored in AutoCAD Release 14 *acad.cfg* files.

○ **Use a PCP or PC2 file** converts the plotter pen settings stored in the plotter configuration parameter *.pcp* and *.pc2* files of earlier versions of AutoCAD.

Back goes back to the previous dialog box.

Next moves to the next dialog box.

Cancel exits the wizard.

File Name page

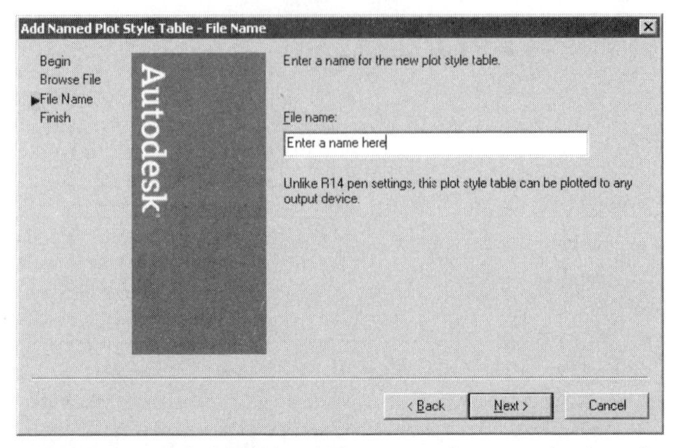

File name specifies the name of the file in which to store the new color-dependent plot style table.

Finish page

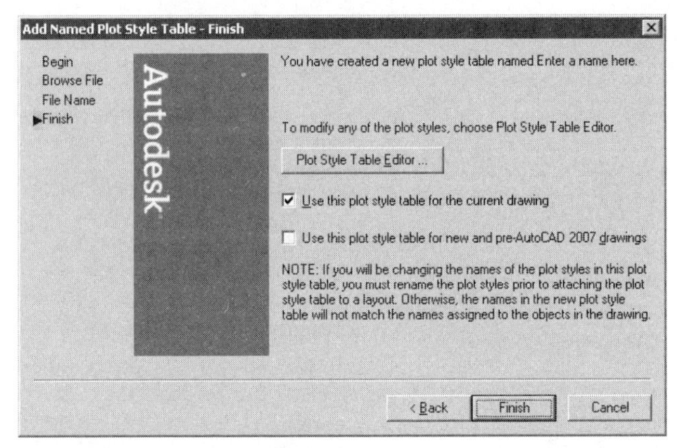

Plot Style Table Editor displays the Plot Style Table Editor dialog box; see StylesManager command.

☑ **Use this plot style table for the current drawing** applies the plot style table to the current drawing.

☐ **Use this plot style table for new and pre-AutoCAD 2004 drawings** applies the plot style table to all new drawings and drawings created by versions of AutoCAD prior to 2004.

RELATED COMMANDS

PcInWizard imports *.pcp* and *.pc2* configuration plot files into the current layout.

PlotterManager accesses the Add Plotter wizard and Assign Plot Style wizard.

Plot plots the drawing.

StylesManager displays the Plot Style Table Editor dialog box.

Ray

Rel.13 Draws semi-infinite construction lines.

Command	Alias	Ctrl+	F-key	Alt+	Menu Bar	Tablet
ray	...	...	...	DR	Draw ↳Ray	K10

Command: ray
Specify start point: *(Pick a point.)*
Specify through point: : *(Pick another point.)*
Specify through point: : *(Press Enter to end the command.)*

Start point.

COMMAND LINE OPTIONS

Start point specifies the starting point of the ray.

Through point specifies the point through which the ray passes.

RELATED COMMANDS

Properties modifies rays.

Line draws finite line segments.

XLine draws infinite construction lines.

TIPS

- Ray objects have an endpoint at one end and are "infinite" in length at the other end.
- Rays display and plot, but do not affect the extents of the drawing.
- A ray has all the properties of a line (including color, layer, and linetype), and can be used as a cutting edge for the Trim command.
- The technical editor defines infinite: "You can type in coordinates for the ray's start and through points with restrictions. Scientific notation allows a maximum of e99. Maximum command line input is 255 characters. Thus, the start point can be -999...999E99, -999...999E99 and through point can be -999...999E99, -999...999E99 — where ... consists of enough 9s to total 255 characters."

Removed Command

RConfig — render configuration — was removed from Release 14. It is replaced by **Render**.

Recover

Rel.12 Recovers damaged drawings without user intervention.

Command	Alias	Ctrl+	F-key	Alt+	Menu Bar	Tablet
recover	...	...	...	**FUR**	**File** ↳**Drawing Utilities** ↳**Recover**	...

Command: recover

Displays the Select File dialog box. Select a .dwg, .dxf, dws, or .dwt file, and then click ***Open.***

Sample output:

```
Drawing recovery.
Drawing recovery log.
Scanning completed.

Validating objects in the handle table.
Valid objects 14526  Invalid objects 0
Validating objects completed.
16 error opening *Model_Space's layout.
Setting layout id to null.
16 error opening *Paper_Space's layout.
Setting layout id to null.

    Salvaged database from drawing.
A vertex was added to a 3D pline (3A18) which had only one vertex.
Auditing Header
Auditing Tables
Auditing Entities Pass 1
Pass 1 14500   objects audited
Auditing Entities Pass 2
Pass 2 14500   objects audited
Auditing Blocks
 10      Blocks audited
Total errors found 2 fixed 2
Erased 0 objects
```

RELATED COMMANDS

Audit checks a drawing for integrity.

DrawingRecovery displays a window listing recoverable drawings.

TIPS

- The Recover command does not ask permission to repair damaged parts of the drawing file; use the Audit command if you want to control the repair process.
- The Quit command discards changes made by the Recover command.
- If the Recover and Audit commands do not fix the problem, try using the DxfOut command, followed by the DxfIn command — assuming the file is not so damaged it cannot be opened at all.

Rectang

Rel.12 Draws squares and rectangles with a variety of options.

Command	Aliases	Ctrl+	F-key	Alt+	Menu Bar	Tablet
rectang	**rec**	...	...	**DG**	**Draw**	**Q10**
	rectangle				**↳Rectangle**	

Command: rectang

Specify first corner point or [Chamfer/Elevation/Fillet/Thickness/Width]: : *(Pick a point, or enter an option)*

Specify other corner point or [Area/Dimensions/Rotation]: *(Pick another point.)*

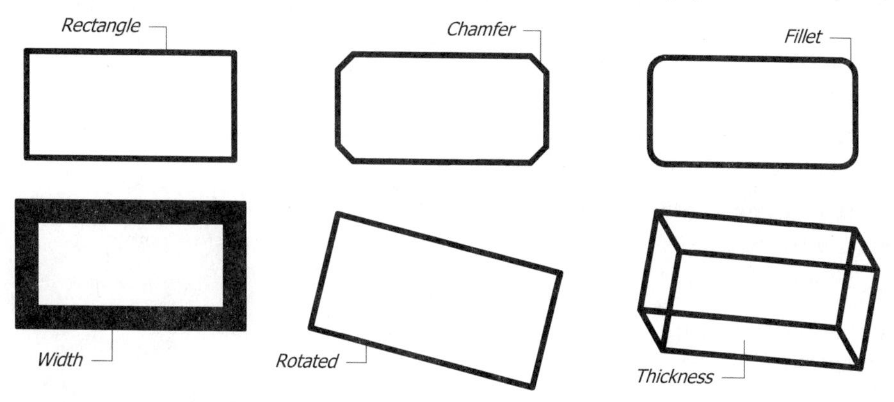

COMMAND LINE OPTIONS

Specify first corner point picks the first corner of the rectangle.

Specify other corner point picks the opposite corner of the rectangle.

Chamfer options

Specify first chamfer distance for rectangles <0.0>: : *(Enter a distance value.)*

Specify second chamfer distance for rectangles <0.0>: *(Enter another distance value.)*

First chamfer distance for rectangles sets the first chamfer distance for all four corners.

Second chamfer distance for rectangles sets the second chamfer distance for all four corners.

Elevation option

Specify the elevation for rectangles <0.0>: *(Enter an elevation value.)*

Elevation for rectangles sets the elevation (*height of the rectangle in the z direction*).

Fillet option

Specify fillet radius for rectangles <1.0>: *(Enter a radius value.)*

Fillet radius for rectangles sets the fillet radius for all four corners of the rectangle.

Thickness option

Specify thickness for rectangles <0.0>: *(Enter a thickness value.)*

Thickness for rectangles sets the thickness of the rectangle's sides in the z direction.

Width option

Specify line width for rectangles <0.0>: *(Enter a width value.)*

Width for rectangles sets the width of all segments of the rectangle's four sides.

Area option

Enter area of rectangle in current units <100.0>: *(Enter an area value.)*

Calculate rectangle dimensions based on [Length/Width] <Length>: *(Type* **L** *or* **W**.*)*

Enter rectangle length <10.0>:

Length specifies the length of the rectangle; AutoCAD determines the width.

Width specifies the width of the rectangle; AutoCAD determines the length.

Dimensions options

Specify length for rectangles <0.0>: *(Enter a length value.)*

Specify width for rectangles <0.0>: *(Enter a width value.)*

Specify other corner point or [Area/Dimension/Rotation]: *(Pick a point.)*

Specify length specifies the length along the x axis.

Specify width specifies the length along the y axis.

Specify other corner point specifies the orientation of the rectangle.

Rotation options

Specify rotation angle or [Pick points] <0>: *(Enter an angle.)*

Rotation angle specifies the angle from the x axis to rotate about the first corner.

Pick points prompts to pick two points that show the angle.

RELATED COMMANDS

Donut draws solid-filled circles with polylines.

Ellipse draws ellipses with polylines, when PEllipse = 1.

PEdit edits polylines, including rectangles.

PLine draws polylines and polyline arcs.

Polygon draws polygons — 3 to 1024 sides — from polylines.

RELATED SYSTEM VARIABLE

PlineWid saves the current polyline width.

TIPS

- Rectangles are drawn from polylines; use the PEdit command to change the rectangle, such as the width of the polyline.
- The values you set for the Chamfer, Elevation, Fillet, Thickness, and Width options become the default for the next execution of the Rectangle command.
- The pick point determines the location of the rectangle's first vertex; rectangles are drawn counterclockwise.
- Use the Snap command and object snap modes to place the rectangle precisely.
- Use object snap modes ENDpoint or INTersection to snap to the rectangle's vertices.
- This command ignores the settings in the ChamferA, ChamferB, Elevation, FilletRad, PLineWid, and Thickness system variables.

Redefine

Rel. 9 Restores the meaning of AutoCAD commands disabled by the **Undefine** command.

Command	Alias	Ctrl+	F-key	Alt+	Menu Bar	Tablet
redefine	...	...	...	...	...	...

Command: redefine
Enter command name: *(Enter command name.)*

COMMAND LINE OPTION

Enter command name specifies the name of the AutoCAD command to redefine.

RELATED COMMANDS

All AutoCAD commands can be redefined.

Undefine disables the meaning of AutoCAD commands.

TIPS

- Prefix any command with a . (*period*) to redefine the undefinition temporarily, as in:

  Command: .line

- Prefix any command with an _ (*underscore*) to make an English-language command work in any linguistic version of AutoCAD, as in:

  Command: _line

- You must undefine a command with the Undefine command before using the Redefine command.
- *Warning!* If you undefine the Redefine command, it cannot be redefined with the Redefine command.

Redo

V. 2.5 Reverses the effect of the most-recent U and Undo commands.

Command	Alias	Ctrl+	F-key	Alt+	Menu Bar	Tablet
redo	...	...	...	**ER**	**Edit** ↳**Redo**	**U12**

Command: redo

COMMAND LINE OPTIONS

None.

RELATED COMMANDS

MRedo redoes more than one undo.

Oops un-erases the most recently-erased objects.

U undoes the most recent AutoCAD command.

Undo undoes the most recent series of AutoCAD commands.

TIPS

- The **Redo** command is limited to reversing a single undo, while the **Undo** and **U** commands can undo operations all the way back to the beginning of the editing session.
- The **Redo** command must be used immediately following the **U** or **Undo** command.

'Redraw, 'RedrawAll

V. 1.0 Redraws viewports to clean up the screen.

Command	Alias	Ctrl+	F-key	Alt+	Menu Bar	Tablet
redraw	**r**	...	...	...	...	...
redrawall	**ra**	...	...	**VR**	**View** ↳**Redraw**	**Q11-R11**

Command: redraw

Reraws the current viewport.

Command: redrawall

Redraws all viewports.

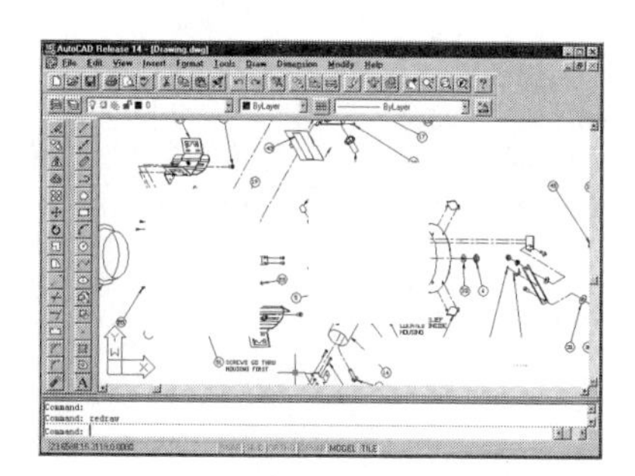

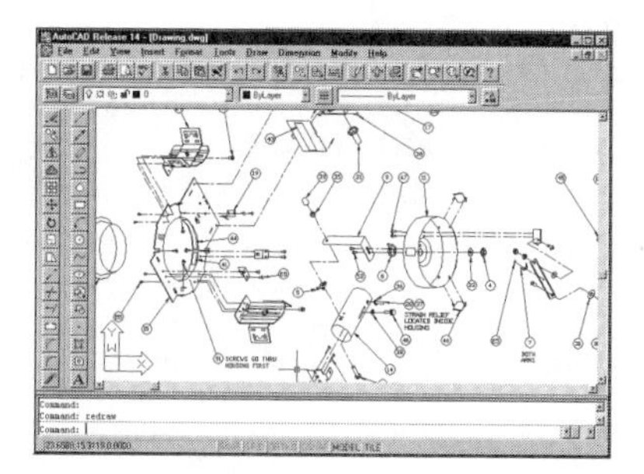

Left: *Before redraw, portions of the drawing are "missing."*
Right: *The drawing is clean after the redraw.*

COMMAND LINE OPTION

ESC stops the redraw.

RELATED COMMAND

Regen regenerates the current viewport.

RELATED SYSTEM VARIABLE

SortEnts controls the order of redrawing objects:

- **0** — Sorts by the order in the drawing database.
- **1** — Sorts for object selection.
- **2** — Sorts for object snap.
- **4** — Sorts for redraw.
- **8** — Sorts for creating slides.
- **16** — Sorts for regenerations.
- **32** — Sorts for plotting.
- **64** — Sorts for PostScript plotting.

TIPS

- Use this command to clean up the screen after a lot of editing; some commands automatically redraw the screen when they are done.
- **Redraw** does not affect objects on frozen layers.

RefClose

2000 Saves or discards changes made to reference objects (blocks and xrefs) edited in-place (short for REFerence CLOSE).

Command	Alias	Ctrl+	F-key	Alt+	Menu Bar	Tablet
refclose	...	...	...	...	**Tools** **⇘Xref and Block In-place Editing** **⇘ Close Reference**	...
				...	**Tools** **⇘Xref and Block In-place Editing** **⇘Save Reference Edits**	

Command: refclose
Enter option [Save/Discard reference changes] <Save>: *(Enter an option.)*

COMMAND LINE OPTIONS

Save saves the editing changes made to the block or externally-referenced file.

Discard discards the changes.

RELATED COMMANDS

RefEdit edits blocks and externally-referenced files attached to the current drawing.

Insert inserts a block in the drawing.

XAttach attaches an externally-referenced drawing.

RELATED SYSTEM VARIABLE

RefEditName specifies the filename of the referenced file being edited.

TIPS

- AutoCAD prompts you with a warning dialog box to ensure you really want to discard or save the changes made to the reference:

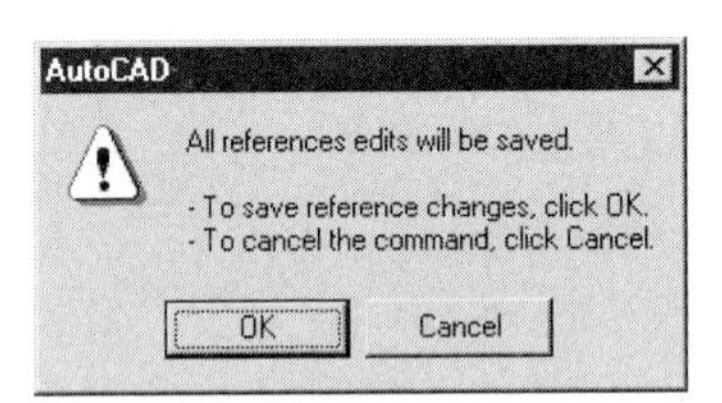

- You can use this command only after the RefEdit command; otherwise AutoCAD reports, "** Command not allowed unless a reference is checked out with RefEdit command **."

RefEdit

2000 Edits-in-place blocks and externally-referenced files attached to the current drawing (short for REFerence EDIT).

Commands	Alias	Ctrl+	F-key	Alt+	Menu Bar	Tablet
refedit	...	...	...	...	**Tools** ⇘**Xref and Block In-place Editing** ⇘**Edit Reference In-Place**	...
-refedit						

Command: refedit

Select reference: *(Pick an externally-referenced drawing or block.)*

Displays dialog box:

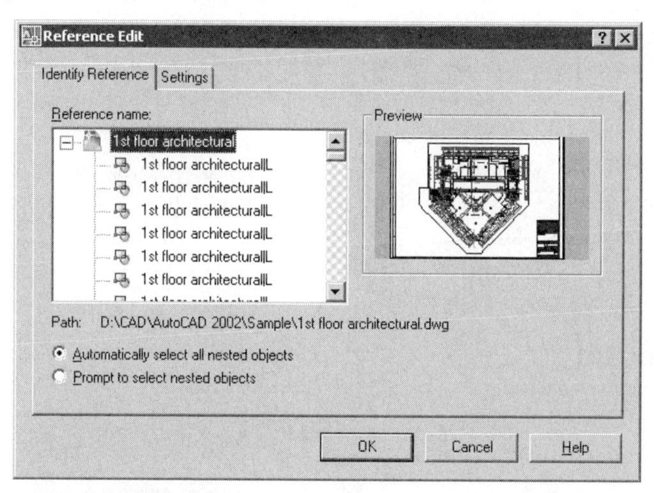

Click ***OK*** *to continue with the command:*

Select nested objects: *(Pick one or more objects.)*

Use REFCLOSE or the Refedit toolbar to end reference editing session.

Displays RefEdit toolbar.

COMMAND LINE OPTIONS

Select reference selects an externally-referenced drawing or an inserted block for editing.

Select nested objects selects objects within the reference — this becomes the selection set of objects that you may edit; you may select all nested objects with the All option (with the exception of OLE objects and objects inserted with the MInsert command, which cannot be refedited).

REFERENCE EDIT DIALOG BOX

Reference name lists a tree of the selected reference object and its nested references; a single reference can be edited at a time.

Preview displays a preview image of the selected reference.

⊙ **Automatically select all nested objects** selects all nested objects.

○ **Prompt to select nested objects** prompts you. 'Select nested objects.'

Settings Tab

Create unique layer, style, block names

- ☑ Prefixes layer and symbol names of extracted objects with **$*n*$**.
- ☐ Retains the names of layers and symbols, as in the reference.

Display attribute definitions for editing (available only when an xref contains attributes):

- ☑ Makes non-constant attributes invisible; attribute definitions can be edited.
- ☐ Attribute definitions cannot be edited.

Note: When edited attributes are saved back to the block reference, use the AttRedef or BAttMan commands to update attributes of the original references.

Lock objects not in working set options:

- ☑ Locks unselected objects in a manner similar to locking layers.
- ☐ Does not lock objects.

REFEDIT TOOLBAR

This toolbar appears automatically after you select nested objects to edit:

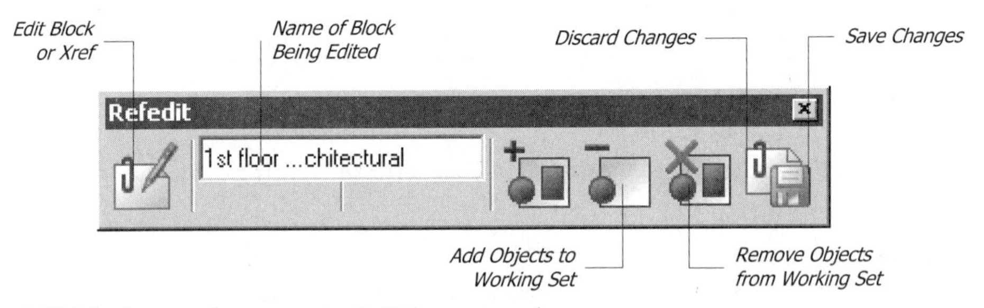

Edit block or xref executes the RefEdit command.

Add objects to working set executes the RefSet Add command.

Remove objects from working set executes the RefSet Remove command.

Discard changes to reference executes the RefClose Discard command.

Save back changes to reference executes the **RefClose Save** command.

-REFEDIT Command

Command: -refedit

Select reference: *(Pick an externally-referenced drawing, or a block.)*

Select nesting level [Ok/Next] <Next>: *(Type **O** or **N**.)*

Enter object selection method [All/Nested] <All>: *(Type **A** or **N**.)*

Display attribute definitions [Yes/No] <No>: *(Type **Y** or **N**.)*

Use REFCLOSE or the Refedit toolbar to end reference editing session.

*Displays **Refedit** toolbar.*

COMMAND LINE OPTIONS

Select reference selects an externally-referenced drawing or an inserted block for editing.

Select nesting level selects objects within the reference — this becomes the selection set of objects that you may edit; you may select all nested objects with **All** (with the exception of OLE objects and objects inserted with the MInsert command, which cannot be refedited).

Enter object selection method:

All selects all objects.

Nested selects only nested objects.

Display attribute definitions:

Yes makes non-constant attributes invisible; attribute definitions can be edited.

No means attribute definitions cannot be edited.

RELATED COMMANDS

RefSet adds and removes objects from a working set.

RefClose saves or discards editing changes to the reference.

AttRedef redefines blocks and updates attributes.

BAttMan controls all aspects of attributes in blocks.

BEdit edits blocks.

RELATED SYSTEM VARIABLES

RefEditName stores the name of the externally-referenced file or block being edited.

XEdit determines whether the current drawing can be edited while being referenced by another drawing.

XFadeCtl specifies the amount of fading for objects not being edited in place.

TIPS

- You may find it more convenient to use the XEdit or BEdit commands.
- In layouts, the drawing must be in model mode for you to select an xref.
- OLE objects and objects inserted with the MInsert command cannot be refedited.
- AutoCAD identifies the "working set" as those objects that you have selected to edit in-place.
- Objects *not* in a working set cannot be selected.

RefSet

2000 Adds and removes objects from working sets (short for REFerence SET).

Command	Alias	Ctrl+	F-key	Alt+	Menu Bar	Tablet
refset	...	...	...	...	**Tools**	...
					↳Xref and Block In-place Editing	
					↳Add to Working Set	
				...	**Tools**	
					↳Xref and Block In-place Editing	
					↳Remove from Working Set	

Command: refset
Transfer objects between the Refedit working set and host drawing...
Enter an option [Add/Remove] <Add>: *(Type* **A** *or* **R***.)*
Select objects: *(Pick one or more objects.)*

COMMAND LINE OPTIONS

Add prompts you to select objects to add to the working set.

Remove prompts you to select objects to remove from the working set.

Select objects selects the objects to be added or removed.

RELATED COMMANDS

RefEdit edits reference objects in place.

RefClose saves or discards editing changes to the reference.

RELATED SYSTEM VARIABLES

RefEditName stores the name of the xref or block being edited.

XEdit determines whether the current drawing can be edited while being referenced by another drawing.

XFadeCtl specifies the amount of fading for objects not being edited in place.

TIPS

- The purpose of this command is to add objects to — or remove them from — the "working set" of objects, while you are performing in-place editing of a block or an externally-referenced drawing.
- You cannot add or remove objects on locked layers. AutoCAD complains, '** *n* selected objects are on a locked layer.'

Regen, RegenAll

V. 1.0 Regenerates viewports to update the drawing.

Command	Alias	Ctrl+	F-key	Alt+	Menu Bar	Tablet
regen	re	...	...	VG	View ↳Regen	J1
regenall	rea	...	...	VA	View ↳Regen All	K1

Command: regen
Regenerating model.

Regenerates the current viewport.

Command: regenall
Regenerating model.

Regenerates all viewports.

COMMAND LINE OPTION

ESC cancels the regeneration.

RELATED COMMANDS

Redraw cleans up the current viewport quickly.

RegenAuto checks with you before doing most regenerations.

ViewRes controls whether zooms and pans are performed at redraw speed.

RELATED SYSTEM VARIABLES

RegenMode toggles automatic regeneration.

WhipArc determines how circles and arcs are displayed:

0 — Circles and arcs displayed as vectors (default).

1 — Circles and arcs displayed as true circles and arcs.

TIPS

- Some commands automatically force a regeneration of the screen; other commands queue the regeneration.
- The Regen command reindexes the drawing database for better display and object selection performance.
- To save on regeneration time, freeze layers you are not working with, apply QText to turn text into rectangles, and place hatching on its own layer.
- Use the RegenAll command to regenerate all viewports.

'RegenAuto

V. 1.2 Prompts before performing regenerations, when turned off (short for REGENeration AUTOmatic).

Command	Alias	Ctrl+	F-key	Alt+	Menu Bar	Tablet
regenauto	...	...	...	...	...	...

Command: regenauto
Enter mode [ON/OFF] <ON>: *(Enter* **ON** *or* **OFF***.)*

COMMAND LINE OPTIONS

OFF turns on "About to regen, proceed?" message.

ON turns off "About to regen, proceed?" message.

RELATED COMMANDS

Regen forces a regeneration in the current viewport.

RegenAll forces a regeneration in all viewports.

RELATED SYSTEM VARIABLES

Expert suppresses "About to regen, proceed?" message when value is greater than 0.

RegenMode specifies the current setting of automatic regeneration:

0 — Off.

1— On (default).

TIPS

- If a regeneration is caused by a transparent command, AutoCAD delays it and responds with the message, "Regen queued."
- When RegenAuto is set to off, the following prompt is displayed with every command that causes a regeneration:

  Command: regen
  About to regen, proceed? <Y>: *(Press* **ENTER***.)*
- AutoCAD Release 12 reduces the number of regenerations by expanding the virtual screen from 16 bits to 32 bits.
- This command was useful when computers were so slow that a regeneration could take several minutes.

Region

Rel.11 Creates 2D regions from closed objects.

Command	Alias	Ctrl+	F-key	Alt+	Menu Bar	Tablet
region	reg	...	...	DN	Draw ⇨Region	R9

Command: region
Select objects: *(Select one or more closed objects.)*
Select objects: *(Press Enter to end object selection.)*
1 loop extracted.
1 region created.

COMMAND LINE OPTION

Select objects selects objects to convert to a region; AutoCAD discards unsuitable objects.

RELATED COMMANDS

All drawing commands.

RELATED SYSTEM VARIABLE

DelObj toggles whether objects are deleted during conversion by the Region command.

TIPS

- This command converts closed line sets, closed 2D and planar 3D polylines, and closed curves.
- This command rejects open objects, intersections, and self-intersecting curves.
- The resulting region is unpredictable when more than two curves share an endpoint.
- Polylines with width lose their width when converted to a region.
- Island are "holes" in regions.

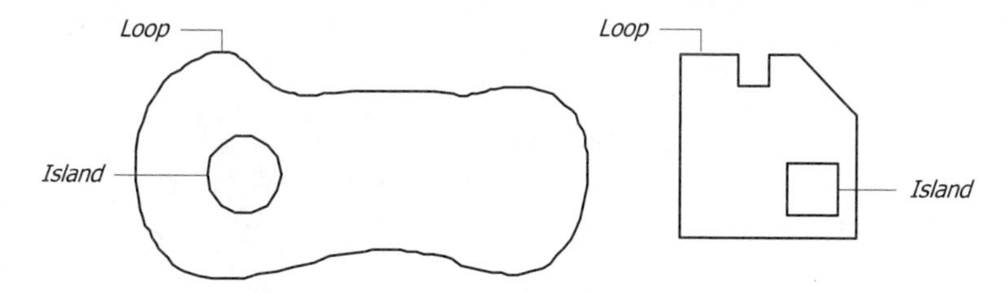

DEFINITIONS

Curve — an object made of circles, ellipses, splines, and joined circular and elliptical arcs.

Island — a closed shape fully within (not touching or intersecting) another closed shape.

Loop — a closed shape made of closed polylines, closed lines, and curves.

Region — a 2D closed area defined as a ShapeManager object.

Reinit

Rel.12 Reinitializes digitizers and input-output ports, and reloads the *acad.pgp* file (short for REINITialize).

Command	Alias	Ctrl+	F-key	Alt+	Menu Bar	Tablet
reinit	...	...	...	...	...	...

Command: reinit

Displays dialog box:

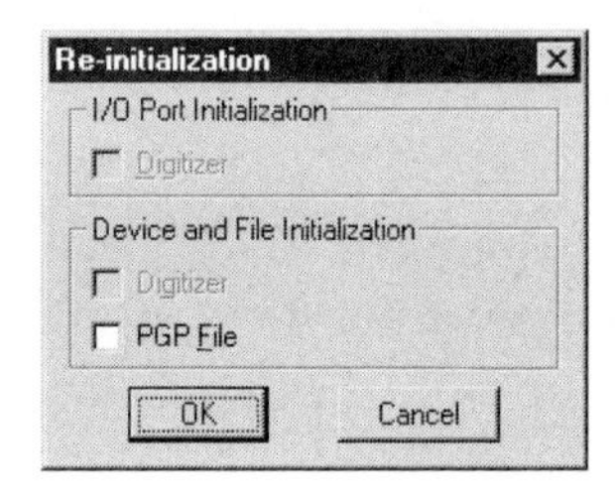

RE-INITIALIZATION DIALOG BOX

I/O Port Initialization

☐ **Digitizer** reinitializes ports connected to the digitizer; grayed out if no digitizer is configured.

Device and File Initialization

☐ **Digitizer** reinitializes the digitizer driver; grayed out if no digitizer is configured.

☐ **PGP File** reloads the *acad.pgp* file.

RELATED COMMAND

CuiLoad reloads menu files.

RELATED SYSTEM VARIABLE

Re-Init reinitializes via system variable settings.

RELATED FILES

acad.pgp is the program parameters file in the hidden *\documents and settings\<username>\application data\autodesk\autocad 2007\r16.2\enu\support* folder.

.hdi are device drivers specific to AutoCAD.

TIPS

- AutoCAD allows you to connect both the digitizer and the plotter to the same port, since you do not need the digitizer during plotting; use this command to reinitialize the digitizer after plotting.
- AutoCAD automatically reinitializes all ports and reloads the *acad.pgp* file each time a drawing is loaded.

Rename

V. 2.1 Changes the names of blocks, dimension styles, layers, linetypes, plot styles, text styles, table styles, UCS names, views, and viewports.

Commands	Aliases	Ctrl+	F-key	Alt+	Menu Bar	Tablet
rename	**ren**	...	...	**OR**	**Format ↳Rename**	**V1**
-rename	**-ren**					

Command: rename

Displays dialog box:

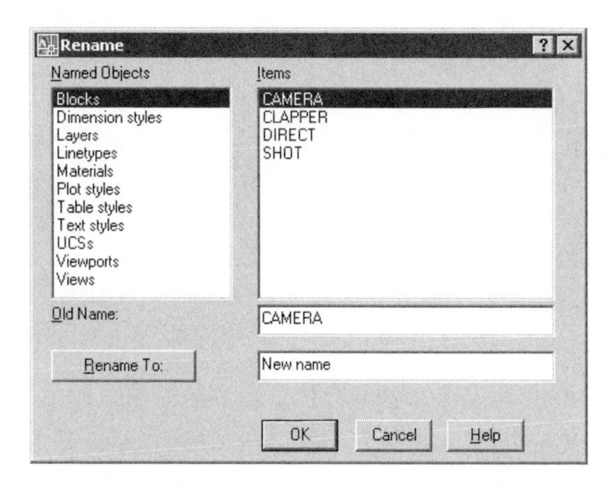

RENAME DIALOG BOX

Named Objects lists named objects in the drawing.

Items lists the names of named objects in the current drawing.

Old Name displays the current name of an object to be renamed.

Rename to allows you to enter a new name for the object.

-RENAME Command

Command: -rename

Enter object type to rename

[Block/Dimstyle/LAyer/LType/Material/Plotstyle/textStyle/Tablestyle/Ucs/VIew/VPort]: *(Enter an option.)*

Example usage:

[Block/Dimstyle/LAyer/LType/Material/Plotstyle/textStyle/Tablestyle/Ucs/VIew/VPort]: b

Enter old block name: diode-2

Enter new block name: diode-02

COMMAND LINE OPTIONS

Block changes the names of blocks.

Dimstyle changes the names of dimension styles.

LAyer changes the names of layers.

LType changes the names of linetypes.

Material changes the names of materials (*new to AutoCAD 2007*).

Plotstyle changes the names of plot styles.

Tablestyle changes the names of table styles.

textStyle changes the names of text styles.

Ucs changes the names of UCS configurations.

VIew changes the names of view configurations.

VPort changes the names of viewport configurations.

RELATED SYSTEM VARIABLES

CeLType names the current linetype.

CLayer names the current layer.

CMaterial names the current material (*new to AutoCAD 2007*).

CTableStyle names the current table style.

DimStyle names the current dimension style.

InsName names the current block.

TextStyle names the current text style.

UcsName names the current UCS view.

TIPS

- You cannot rename layer "0", anonymous blocks, groups, or linetype "Continuous."
- To rename a group of similar names, use * (the wildcard for "all") and ? (the wildcard for a single character).
- Names can be up to 255 characters in length.
- The Properties command does *not* allow you to rename blocks.
- The PlotStyle option is available only when plot styles are attached to the drawing.
- You cannot use this command during RefEdit.

Render

Rel.12 Renders 3D objects in an independent window.

Command	Alias	Ctrl+	F-key	Alt+	Menu Bar	Tablet
render	**rr**	**...**	**...**	**VER**	**View** ↳**Render** ↳**Render**	**M1**

This command is substantially changed in AutoCAD 2007.

Command: render

Displays window (illustrated below), or outputs the rendering in the current viewport, of a windowed selection, or of selected objects — depending on the setting in the RPref command.

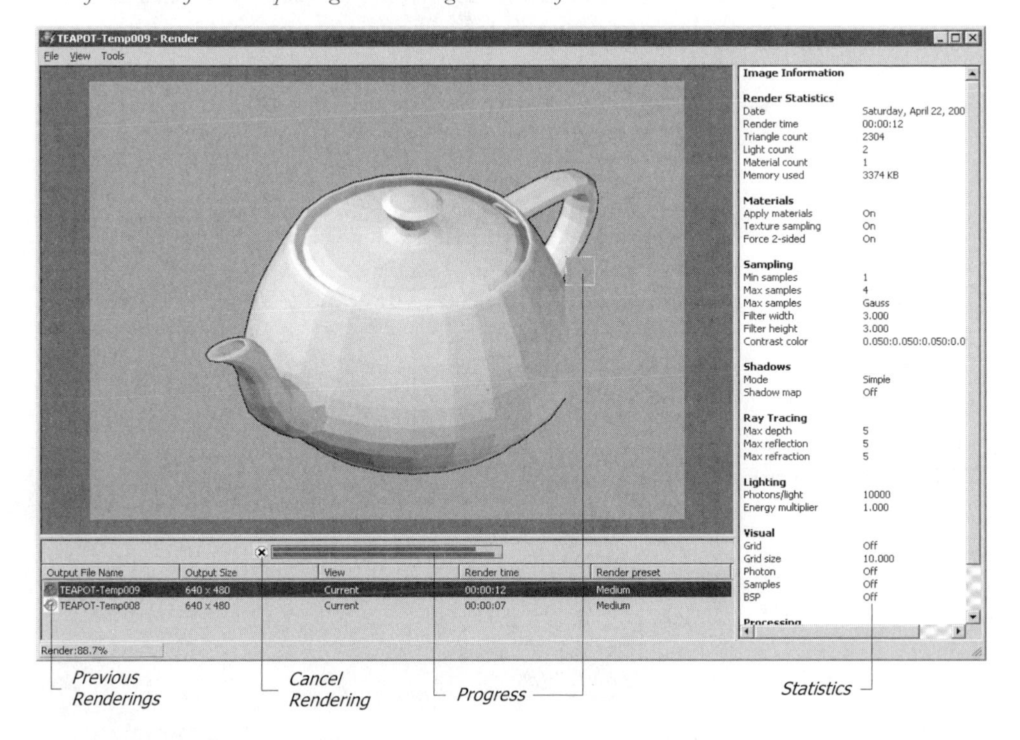

RENDER WINDOW MENU BAR

File

Save saves the rendered image to a file in BMP, TIFF, PNG, JPEG, Targa, or PCX formats. While the SaveImg command does not apply to this window, it has the same options as this command.

Save Copy saves a copy of the image by another name or to another folder.

Exit closes the Render window; reopen with the RenderWin command.

View

Status Bar toggles the display of the status bar.

Statistics Pane toggles the display of the statistics pane.

Tools

Zoom + makes the image up to 64x larger.

Zoom - makes the image up to 64x smaller.

Use the scroll bars or mouse wheel to pan the image. Hold down the Ctrl key and use the mouse wheel to zoom.

PROGRESS METERS

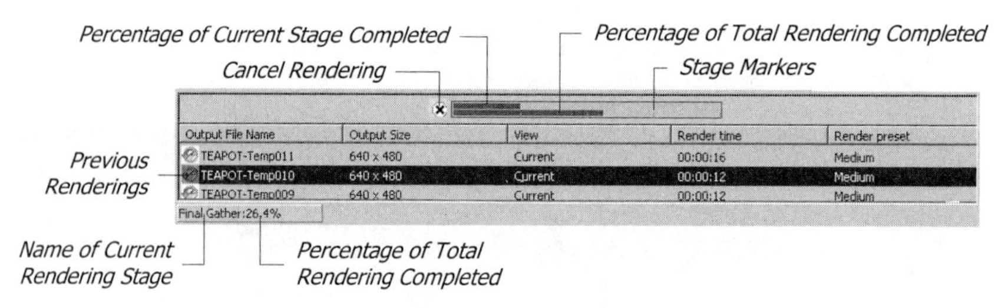

x cancels the rendering, as does pressing the **ESC** key.

Upper bar shows the progress within the current phase: translation, optional photon emission, optional final gather, and render.

Lower bar shows the progress of the entire rendering; also displayed by the status bar. The history pane reports the time each rendering took.

Status bar reports the name of the current rendering stage, and the current percentage of the total rendering.

SHORTCUT MENU

Right-click the history pane:

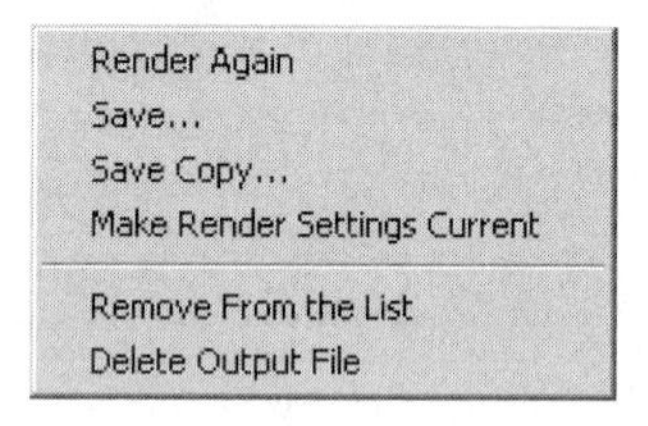

Render Again re-renders the selected history item.

Save saves the rendering to an image file.

Save Copy saves the rendering to a different folder.

Make Render Settings Current loads the rendering style associated with the history item.

Remove From The List removes the history item from the list.

Delete Output File erases the rendering from the image pane

RELATED COMMANDS

RenderEnvironment sets up fog effects.

RenderPresets sets up rendering defaults for use with this command; displays the Render Presets Manager dialog box.

RPref sets up rendering styles; displays the Advanced Render Settings palette.

RenderWin acceses the Render window.

TIPS

- Use the scoll bars or mouse wheel to pan the image. Hold down the **CTRL** key, and use the mouse wheel to zoom.
- Rendering is performed in four steps: translation, photon emission, final gather, and render. The photon emission and final gather steps occur only if specified in the render preferences.

RenderCrop

2007 Renders a windowed area of the current viewport.

Command	Alias	Ctrl+	F-key	Alt+	Menu Bar	Tablet
rendercrop	rc	...	...	...	...	...

Command: rendercrop
Pick crop window to render: *(Pick a point, or enter x,y coordinates.)*
please enter the second point: *(Pick another point.)*
Renders the windowed area:

COMMAND OPTIONS

Pick crop window to render picks the first point of the area to be rendered.

please enter the second point picks the other corner of the rectangle. (Note the unusual wording of the prompt.)

RELATED COMMANDS

RenderWin renders the drawing in an independent window.

RPref sets up rendering preferences.

TIPS

- Rendering a windowed area is faster than rendering the entire viewport.
- This command does not allow you to specify rendering preferences; before using this command, run the RPref command to set up preferences.

RenderEnvironment

2007 Sets the parameters for fog effects in renderings; the former Fog command.

Command	Alias	Ctrl+	F-key	Alt+	Menu Bar	Tablet
renderenvironment	fog	...	...	VEE	View Render ↳Render Environment	P2

Command: renderenvironment

Displays dialog box:

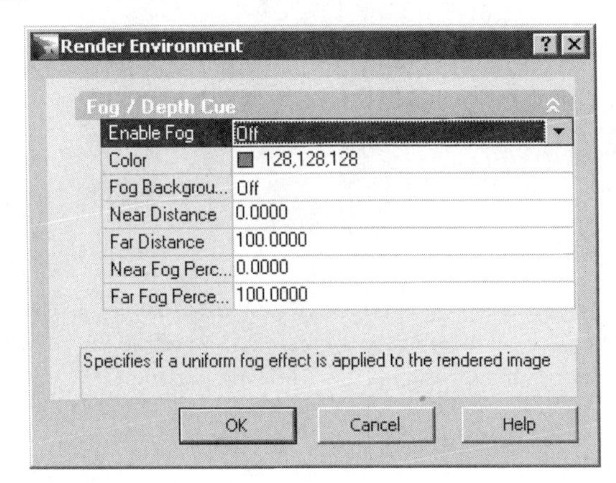

DIALOG BOX OPTIONS

Enable Fog toggles fog settings.

Color specifies the color of the fog; displays the Select Color dialog box.

Fog Background applies the fog effect to the background; see the View command.

Near Distance indicates the start of the fog effect, measured as a decimal (0.0 - 1.0) from the camera.

Far Distance indicates the distance from the camera where the fog effect ends.

Near Fog Percentage specifies the near intensity of the fog effect.

Far Fog Percentage specifies the far intensity of the fog effect.

RELATED COMMAND

Render creates renderings with the fog effect.

TIPS

- First set up the RenderEnvironment command, and then use Render to see the effect.
- The fog can be any color; here are some suggestions:

Color	RGB	HLS	Effect
White	1,1,1	0,1,0	Fog.
Black	0,0,0	0,0,0	Distance.
Green	0,1,0	0.33,0.5,0	Alien mist.

- The effect of using white as the fog color:

- It can be tricky getting the fog effect to work. Use the 3dClip command to set the back clipping plane at the back of the model, or where you want the fog to have its full effect. Then use RenderEnvironment to set up the following parameters:
  - Enable fog: On
  - Color: White
  - Fog Background: On or Off
  - Near distance: 0.70
  - Far distance: 1.00
  - Near fog percentage: 0.00
  - Far fog percentage: 100

RenderPresets

2007 Creates rendering styles; displays AutoCAD's longest dialog box.

Command	Alias	Ctrl+	F-key	Alt+	Menu Bar	Tablet
renderpresets	**rp**	...	...	...	...	...
	rfileopt					

Command: renderpresets

Displays dialog box:

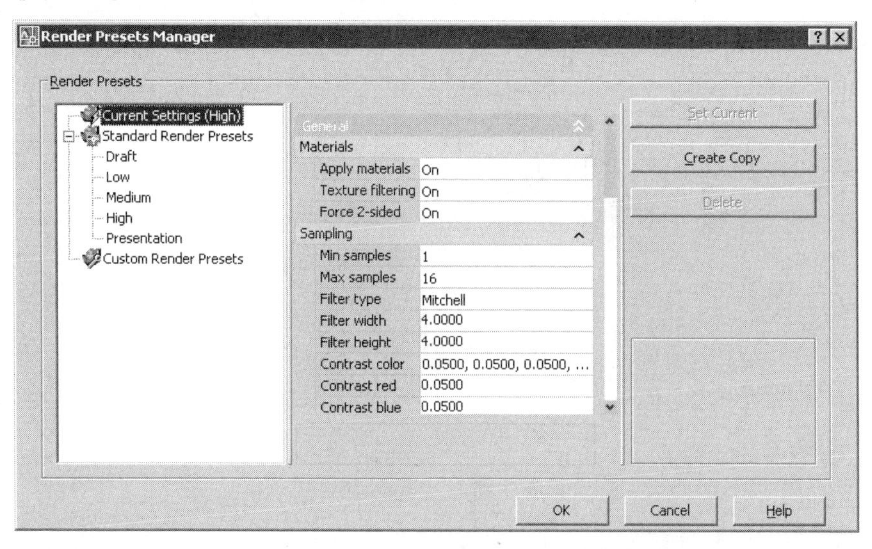

DIALOG BOX OPTIONS

See the RPref command for details of the parameters.

Set Current sets the selected rendering style as the default; the Render and RenderCrop commands use this rendering style.

Create Copy creates a copy of the parameters so that you can modify them for your needs.

Delete removes the named rendering style; you cannot delete the system styles.

TIPS

- AutoCAD comes preset with five styles: Draft, Low, Medium, High, and Presentation:
    - **Draft** rendering is very fast, but inaccurate and very coarse. Time to render the *teapot.dwg*: 3.5 seconds.
    - **Low** rendering is fast but inaccurate; time to render: 4.5 seconds.
    - **Medium** rendering is slightly slower but much more accurate; time to render: 10 seconds.
    - **High** rendering is slower and more accurate; time to render: 15 seconds.
    - **Presentation** rendering is very slow and very accurate; time to render: 55 seconds.

***Left**: Teapot.dwg rendered in draft mode; notice the display errors.*
***Right**: Drawing rendered in presentation mode; notice the material.*

- The complete contents of the Render Presets Manager look like this (see the RPref command for details):

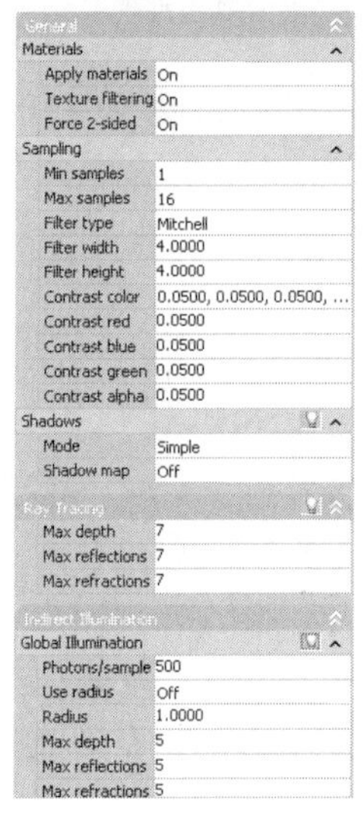

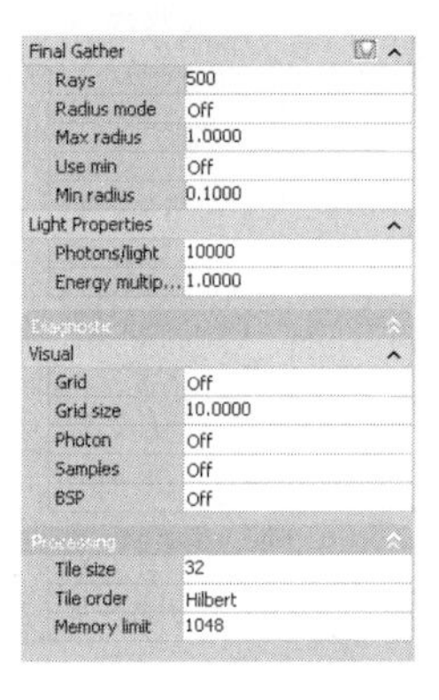

- The teapot was used in the early days of computer rendering software as a complex test case to ensure the rendering was completed accurately.

Removed Command

RenderUnload was removed from Release 14.

RenderWin

2007 Displays the Render window with the most recent rendering; formerly the RendScr command.

Command	Alias	F-key	Alt+	Menu Bar	Tablet
renderwin	rendscr	...	...	...	...
	rw				

Command: renderwin

Displays the Render window with the most recent rendering.

See Render command for details.

RELATED COMMANDS

Render generates renderings in the Render window.

RenderCrop generates renderings in a windowed area of the viewport.

RenderPresets sets the rendering parameters.

TIPS

- If no rendering has occurred in this drawing, AutoCAD complains, "RENDERWIN Drawing contains no render history," and does not display the Render window.
- If this command fails to display the Render window, it's because the window is already open, but covered up by the AutoCAD window. Look for its button on the taskbar.

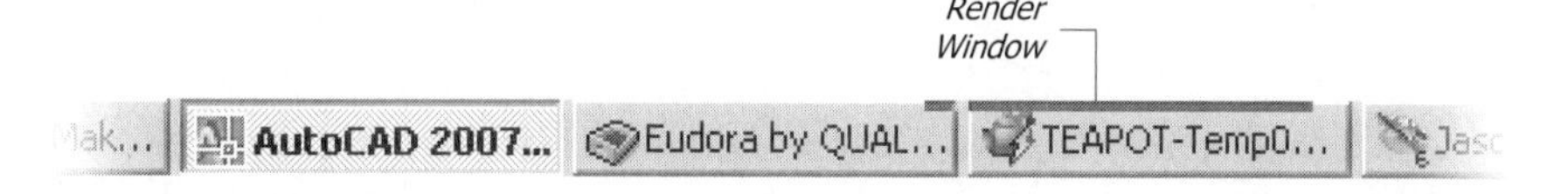

Removed Commands

The **Replay** command was removed from AutoCAD 2007.

The **RendScr** command is an alias of the **RenderWin** command as of AutoCAD 2007.

ResetBlock

2006 Resets dynamic block references to their default values.

Command	Alias	Ctrl+	F-key	Alt+	Menu Bar	Tablet
resetblock	...	...	...	...	...	...

Command: resetblock
Select objects: *(Select one or more blocks.)*
Select objects: *(Press Enter to end object selection.)*
***n* blocks reset.**

COMMAND LINE OPTION

Select objects selects the blocks to be reset.

RELATED COMMAND

BEditor creates and edits dynamic blocks.

TIP

- Dynamic blocks can be changed in many ways, according to the actions defined in the Block Editor. Use this command to reset blocks to their original definition.

'Resume

V. 2.0 Resumes script files paused by an error, or by pressing the Backspace or ESC keys.

Command	Alias	Ctrl+	F-key	Alt+	Menu Bar	Tablet
resume	...	...	...	...	...	...

Command: resume

COMMAND LINE OPTIONS

BACKSPACE pauses the script file.

ESC stops the script file.

RELATED COMMANDS

RScript reruns the current script file.

Script loads and runs a script file.

RevCloud

2004 Draws revision clouds, and converts objects into revision clouds.

Command	Alias	Ctrl+	F-key	Alt+	Menu Bar	Tablet
revcloud	...	...	...	**DU**	**Draw** ↳**Revision Cloud**	...

Command: revcloud
Minimum arc length: 0.5000 Maximum arc length: 0.5000 Style: Normal
Specify start point or [Arc length/Object/Style] <Object>: s
Select arc style [Normal/Calligraphy] <Normal>: c
Arc style = Calligraphy
Specify start point or [Arc length/Object/Style] <Object>: *(Pick a point, or enter an option.)*
Guide crosshairs along cloud path... *(Move cursor to create cloud.)*
Revision cloud finished.

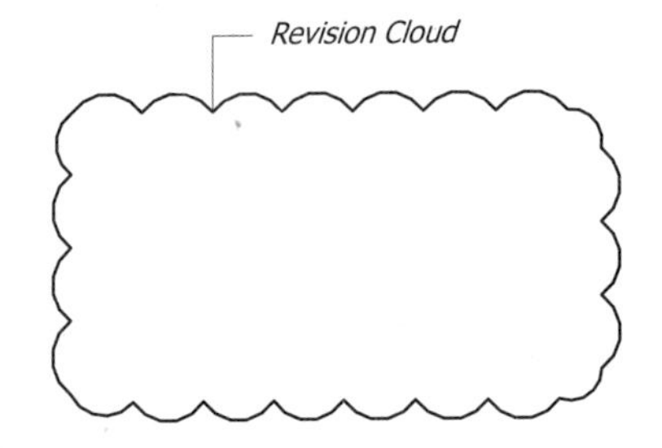

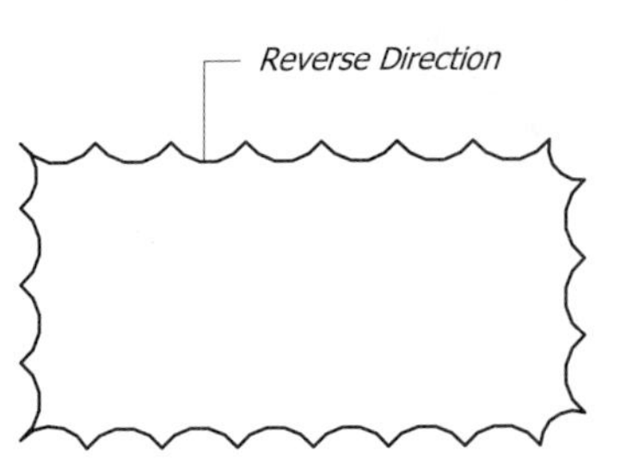

COMMAND LINE OPTIONS

Specify start point specifies the starting point of the cloud.

Guide crosshairs along cloud path outlines the cloud.

Arc Length Options

Specify minimum length of arc <0.5000>: *(Enter a minimum value.)*
Specify maximum length of arc <1.0000>: *(Enter a maximum value.)*

Specify minimum length of arc specifies the minimum arc length.

Specify maximum length of arc specifies the maximum arc length.

Object Options

Select object: *(Select one object to convert to a revision cloud.)*
Reverse direction [Yes/No] <No>: *(Type **Y** or **N**.)*

Select object selects the object to convert into a revision cloud.

Reverse direction turns the revision cloud inside-out.

Style Options

Select arc style [Normal/Calligraphy] <Normal>: *(Type* **N** *or* **C**.*)*

Normal draws clouds with uniform-width arcs.

Calligraphy draws clouds from variable-width polyarcs.

RELATED SYSTEM VARIABLE

DimScale affects the size and width of the arcs.

TIPS

- When the cursor reaches the start point, the revision cloud closes automatically.
- The revision cloud is drawn as a polyline.
- To edit the revision cloud, select it, and then move the grips.
- The arc length is not available from a system variable, because it is stored in the Windows registry.
- The arc length is multiplied by the value stored in the DimScale system variable.

Revolve

Rel.11 Creates 3D solid objects by revolving closed 2D objects about an axis.

Command	Alias	Ctrl+	F-key	Alt+	Menu Bar	Tablet
revolve	**rev**	...	...	**DMR**	**Draw** **↳Modeling** **↳Revolve**	**Q7**

Command: revolve

This command has revised option prompts in AutoCAD 2007:

Current wire frame density: ISOLINES=4

Select objects to revolve: *(Select one or more closed objects.)*

Select objects to revolve: *(Press Enter to end object selection.)*

Specify axis start point or define axis by [Object/X/Y/Z] <Object>: *(Pick a point, or enter an option.)*

Specify angle of revolution or [STart angle] <360>: *(Enter a value.)*

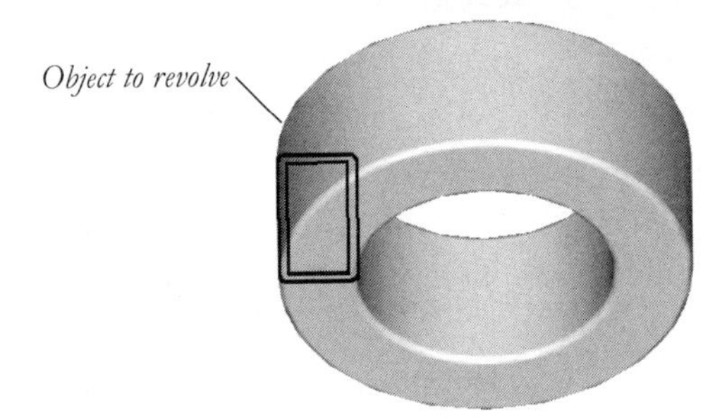

The filleted rectangle was used to create the revolved object.

COMMAND LINE OPTIONS

Select objects to revolve selects a closed object to revolve: closed polyline, circle, ellipse, donut, polygon, closed spline, or a region.

Specify axis start point indicates the axis of revolution; you must specify the endpoint.

Object selects the object that determines the axis of revolution.

X uses the positive x axis as the axis of revolution.

Y uses the positive y axis as the axis of revolution.

Specify angle of rotation specifies the amount of rotation; full circle = 360 degrees.

STart angle specifies the starting angle.

RELATED COMMANDS

Extrude extrudes 2D objects into a 3D solid model.

RevSurf rotates open and closed objects, forming a 3D surface.

TIPS

- Revolving an open object creates surfaces; closed objects, solids.
- This command does not work with open objects or self-intersecting polylines.

RevSurf

Rel.10 Generates 3D meshes of revolution defined by a path curve and an axis (short for REVolved SURFace); does not draw surfaces.

Command	Alias	Ctrl+	F-key	Alt+	Menu Bar	Tablet
revsurf	...	...	...	DMMM	Draw ↳Modeling ↳Meshes ↳Revolved Surface	O8

Command: revsurf

Current wire frame density: SURFTAB1=6 SURFTAB2=6

Select object to revolve: *(Select an object.)*

Select object that defines the axis of revolution: *(Select an object.)*

Specify start angle <0>: *(Enter a value.)*

Specify included angle (+=ccw, -=cw) <360>: *(Enter a value.)*

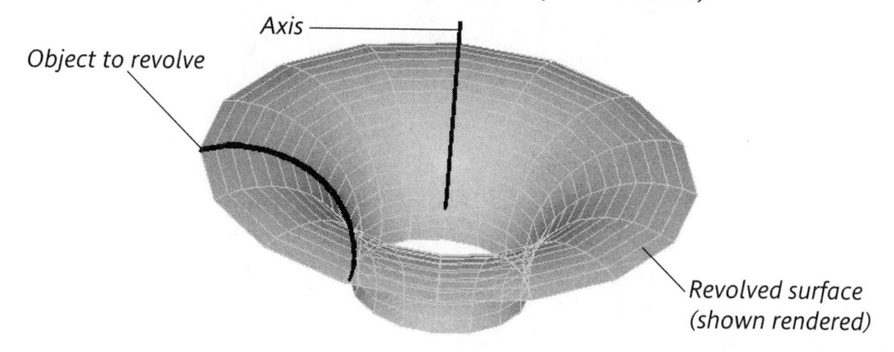

COMMAND LINE OPTIONS

Select object to revolve selects the single object that will be revolved about an axis.

Select object that defines the axis of revolution selects the axis object.

Start angle specifies the starting angle.

Included angle specifies the angle to rotate about the axis.

RELATED COMMANDS

PEdit edits revolved meshes.

Revolve revolves a 2D closed object into a 3D solid.

RELATED SYSTEM VARIABLES

SurfTab1 specifies the mesh density in the m-direction (default = 6).

SurfTab2 specifies the mesh density in the n-direction (default = 6).

TIPS

- This command works with open and closed objects.
- If a multi-segment polyline is the axis of revolution, the rotation axis is defined as the vector pointing from the first vertex to the last vertex, ignoring the intermediate vertices.

Rotate

V. 2.5 Rotates objects about an axis perpendicular to the current UCS.

Command	Alias	Ctrl+	F-key	Alt+	Menu Bar	Tablet
rotate	ro	...	...	MR	Modify ↳Rotate	V20

Command: rotate
Current positive angle in UCS: ANGDIR=counterclockwise ANGBASE=0
Select objects: *(Select one or more objects.)*
Select objects: *(Press Enter to end object selection.)*
Specify base point: *(Pick a point.)*
Specify rotation angle or [Copy/Reference]: *(Enter a value, or type **R**.)*

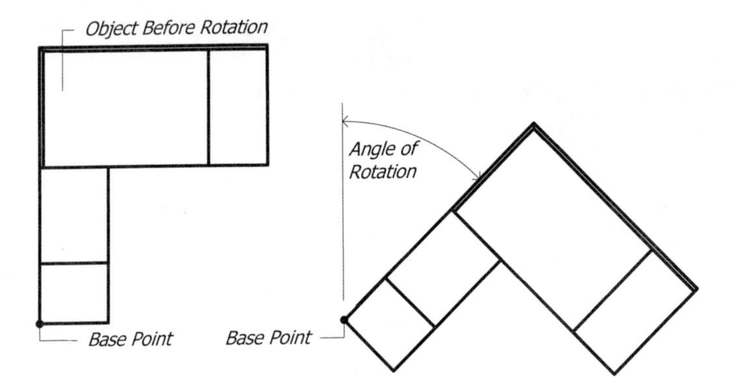

COMMAND LINE OPTIONS

Select objects selects the objects to be rotated.

Specify base point picks the point about which the objects will be rotated.

Specify rotation angle specifies the angle by which the objects will be rotated.

Reference allows you to specify the current rotation angle and new rotation angle.

Copy rotates a copy of the object, leaving the original in place.

RELATED COMMANDS

3dRotate rotates objects in 3D space using a 3D grip tool.

UCS rotates the coordinate system.

Snap command's Rotate option rotates the cursor.

TIPS

- AutoCAD rotates the selected object(s) about the base point.
- At the 'Specify rotation angle' prompt, you can show the rotation by moving the cursor. AutoCAD dynamically displays the new rotated position as you move the cursor.
- Use object snap modes, such as INTersection, to position the base point and rotation angle(s) accurately.

Rotate3D

Rel.11 Rotates objects about any axis in 3D space with a user interface that's different from the 3dRotate command.

Command	Alias	Ctrl+	F-key	Alt+	Menu Bar	Tablet
rotate3d	...	...	...	M3R	...	W22

Command: rotate3d
Current positive angle: ANGDIR=counterclockwise ANGBASE=0
Select objects: *(Select one or more objects.)*
Select objects: *(Press Enter to end object selection.)*
Specify first point on axis or define axis by
[Object/Last/View/Xaxis/Yaxis/Zaxis/2points]: *(Pick a point, or enter an option.)*
Specify second point on axis: *(Pick another point.)*
Specify rotation angle or [Reference]: *(Enter a value, or type* **R**.*)*

COMMAND LINE OPTIONS

Select objects selects the objects to be rotated.

Specify rotation angle rotates objects by a specified angle: relative rotation.

Reference specifies the starting and ending angle: absolute rotation.

Define Axis By Options

Object selects object to specify the rotation axis.

Last selects the previous axis.

View makes the current view direction the rotation axis.

Xaxis makes the x axis the rotation axis.

Yaxis makes the y axis the rotation axis.

Zaxis makes the z axis the rotation axis.

2 points defines two points on the rotation axis.

RELATED COMMANDS

3dRotate rotates objects in 3D space using a 3D grip tool.

Rotate rotates objects in 2D space.

Align rotates, moves, and scales objects in 3D space.

Mirror3d mirrors objects in 3D space.

Removed Commands

The **RmlIn** command was removed from AutoCAD 2006; its replacement is **Markup**.

The **RMat** command was removed from AutoCAD 2007; its replacement is **MaterialAttach**.

RPref, RPrefClose

Rel.12 Specifies options for the Render and RenderCrop commands (short for Render PREFerences).

Command	Alias	Ctrl+	F-key	Alt+	Menu Bar	Tablet
rpref	**rpr**	...	...	**VED**	**View** ↳**Render** ↳**Advanced Render Settings**	**R2**
rprefclose						

Command: rpref

Displays palette:

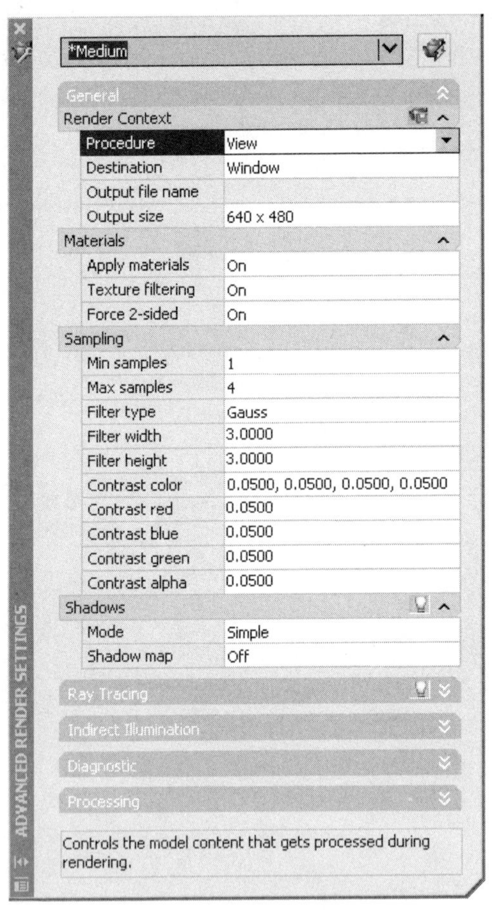

Command: rprefclose

Clsoes the palette.

Options are identical to the Render Presets Manager dialog box (RenderPresets command).

GENERAL

performs the render operation.

Render Context

toggles the Output File Name parameter for the Render command's Save option.

Procedure determines which part of the drawing is rendered:

- **View** — renders the entire viewport.
- **Crop** — prompts for you to pick two points, like the RenderCrop command.
- **Selected** — prompts you to select the objects to be rendered.

Destination determines where the model will be rendered:

- **Viewport** — model is rendered in the current viewport.
- **Window** — model is rendered in the Render window.

Output file name specifies the name and format of images saved to disc; see SaveImg command.

Output size specifies the size of the rendered image; smaller views render faster.

Materials

Apply Materials toggles rendering of materials, if applied to the model. If not applied, the model takes on the look defined by the Global material. See Materials command.

Texture Filtering toggles the rendering of texture maps.

Force 2-Sided toggles the rendering of both sides of faces; single-sided is faster.

Sampling

Min Samples specifies the minimum sample rate (number of samples per pixel). Range is 1/64 (one out of every 64 pixels is sampled) to 1024 (1,024 samples per pixel). More samples result in higher quality but slower rendering speed.

Max Samples specifies the maximum sample rate used when neighboring samples exceed the contrast limit (defined below).

Filter Type specifies how multiple samples are combined into one:

- **Box** — sums all samples in the filter area with equal weight (fastest).
- **Gauss** — weights samples using a bell curve.
- **Triangle** — weights the samples using a pyramid.
- **Mitchell** — weights the samples using a steeper bell curve.
- **Lanczos** — weights the samples using a steep bell curve that diminishes the effect of samples at the edge of the filter area.

Filter Width and **Filter Height** specify the size of the filtered area; range is 0 to 8. Larger values soften the image, but increase rendering time.

Contrast Color selects the threshold RGBA (red, green, blue, alpha) values; displays the Select Color dialog box. Range is from 0 to 1, where 0 is black and 1 is white.

Contrast Red, **Contrast Blue**, and **Contrast Green** indicate threshold values for the red, blue, and green components of samples.

Contrast Alpha indicates the threshold value for the alpha component, typically used for transparency.

Shadows

toggles shadow-casting.

Mode specifies the type of shadow casting:

- **Simple** — casts random-order shadow.
- **Sort** — casts shadows from objects to lights.
- **Segments** — casts shadows along the light ray.

Shadow Map Controls toggles shadow mapping:

On — shadow-mapped shadows.

Off — ray-traced shadows.

RAY TRACING

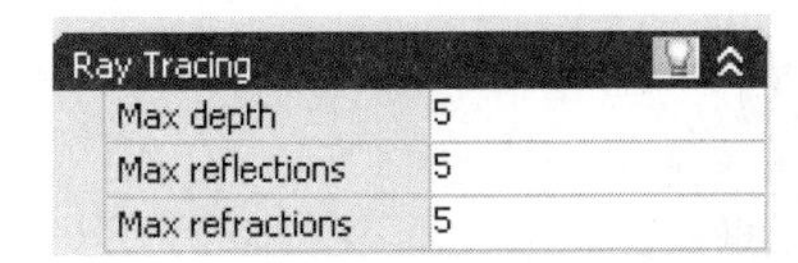

toggles ray tracing.

Max Depth stops raytracing when the number of reflections reaches this number. If rays refract twice and reflect three times, the ray tracing is stopped when this value is set to 5.

Max Reflections stops ray tracing after this many *reflections* (light bounced off a surface). Turns off reflections when set to 0.

Max Refractions stops ray tracing after this many *refractions* (light distorted going through an object). Turns off refractions when set to 0.

INDIRECT ILLUMINATION

Indirect Illumination	
Global Illumination	
Photons/sample	500
Use radius	Off
Radius	1.0000
Max depth	5
Max reflections	5
Max refractions	5
Final Gather	
Rays	200
Radius mode	Off
Max radius	1.0000
Use min	Off
Min radius	0.1000
Light Properties	
Photons/light	10000
Energy multiplier	1.0000

Global Illumination

toggles global illumination.

Photons/Samples specifies the number of photons used to compute the intensity of global illumination. Range is 1 to 2147483647; more photons make for less noise, more blur, and longer renderings.

Use Radius toggles the use of photon sizing. When off, photons are determined to be 1/10 the radius of the viewport.

Radius specifies the area in which photons are used.

Max Depth, Max Reflections, and **Max Refractions** settings are the same as for ray tracing.

Final Gather

toggles gathering (additional global illumination calculations).

Rays specifies the number of rays used to compute indirect illumination. More rays make for less noise but longer renderings.

Radius Mode specifies the radius mode:

- **On** — Max Radius setting is used.
- **Off** — maximum radius is 10 percent of the maximum model radius.
- **View** — Max Radius setting in pixels.

Max Radius specifies the maximum radius used for final gathering. Lower values improve quality but increase rendering time.

Use Min Controls toggles the minimum radius setting.

Min Radius specifies the minimum radius used for final gathering. Higher values improve quality but increase rendering time.

Light Properties

Affects how lights behave when calculating indirect illumination.

Photons/Light specifies the number of photons emitted by each light. Larger values increase accuracy but also increase the rendering time and RAM usage.

Energy Multiplier multiplies the global illumination intensity.

DIAGNOSTIC

Helps track down rendering peculiarities.

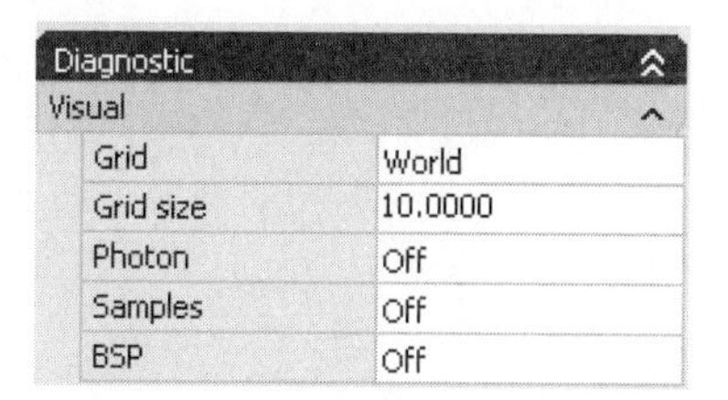

Visual

Grid shows the coordinate space of:

- **Object** — local u,v,w coordinates for each object.
- **World** — world x,y,z coordinates for all objects.
- **Camera** — imposes a rectangular grid on the view.

Grid Size sets the size of the grid.

Photon renders the effect of a photon map, if present:

- **Density** — renders the photon map; red = high density; blue = low density.
- **Irradiance** — shades photons based on their irradiance; red = maximum irradiance; blue = low irradiance.

BSP (binary space partition) helps track down very slow renderings by visualizing the BSP raytrace parameters:

- **Depth** — illustrates the tree depth; red = top faces; blue = deep faces.
- **Size** — colors the leaf size in the tree.

Processing

Tile Size specifies the rendering tile size; smaller values result in longer rendering times.

Tile Order specifies the tile render order:

- **Hilbert** — renders the next tile based on the difficulty of switching to the next one.
- **Spiral** — starts at the center and then spirals outward.
- **Left to Right** — bottom to top, left column to right.
- **Right to Left** — bottom to top, right column to left.
- **Top to Bottom** — right to left, top row to bottom.
- **Bottom to Top** — right to left, bottom row to top.

Memory Limit discards geometry to allocate memory for other objects, if this memory limit is reached.

RELATED COMMANDS

Render and **RenderCrop** use these settings to render 3D models.

RenderPresets creates rendering styles.

RScript

V. 2.0 Repeats script files (short for Repeat SCRIPT).

Command	Alias	Ctrl+	F-key	Alt+	Menu Bar	Tablet
rscript	...	...	...	...	...	...

Command: rscript

COMMAND LINE OPTIONS

None.

RELATED COMMANDS

Resume resumes a script file after being interrupted.

Script loads and runs a script file.

TIP

- Placed at the end of a script file, this command causes the script file to run repeatedly until canceled with **BACKSPACE** or **ESC**.

RuleSurf

Rel.10 Draws 3D ruled meshes between two objects (short for RULEd SURFace); does not draw surfaces.

Command	Alias	Ctrl+	F-key	Alt+	Menu Bar	Tablet
rulesurf	...	...	...	**DMMR**	**Draw** ↳**Modeling** ↳**Meshes** ↳**Ruled Surf**	**Q8**

Command: rulesurf
Select first defining curve: *(Pick an object.)*
Select second defining curve: *(Pick another object.)*

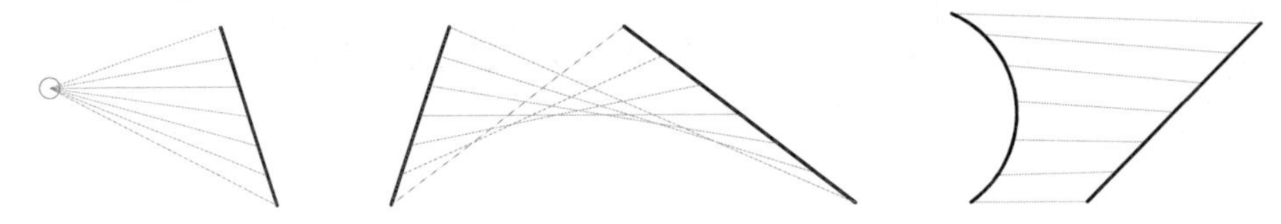

Left to right: *Ruled meshes created between point and line; line and line; and arc and line.*

COMMAND LINE OPTIONS

Select first defining curve selects the first object for the ruled mesh.

Select second defining curve selects the second object for the ruled mesh.

RELATED COMMANDS

EdgeSurf draws a 3D mesh bounded by four edges.

RevSurf draws a 3D mesh of revolution.

TabSurf draws a 3D tabulated mesh.

RELATED SYSTEM VARIABLE

SurfTab1 determines the number of faces drawn.

TIPS

- This command uses these objects as the boundary curve: points, lines, arcs, circles, polylines, and 3D polylines.
- Pick order is important: the ruled mesh is drawn starting at the endpoint nearest the pick point. This can result in a twisted surface, as illustrated by the line-line example above.
- Both boundaries must either be closed or open; the exception is the point object.
- This command begins drawing its mesh as follows:

  Open object — from the object's endpoint closest to your pick.
  Circle — from the zero-degree quadrant.
  Closed polyline — from the last vertex.

- RuleSurf creates meshes with circles in the opposite direction from that of closed polylines.

Save, SaveAs

V. 2.0 Saves the current drawing to disk, after prompting for a file name.

Commands	Alias	Ctrl+	F-key	Alt+	Menu Bar	Tablet
save	**saveurl**	...	...	...	...	...
saveas	...	**Shift+S**	...	**FA**	**File** ↳**Save As**	**V24**

Command: save *or* saveas

Displays dialog box.

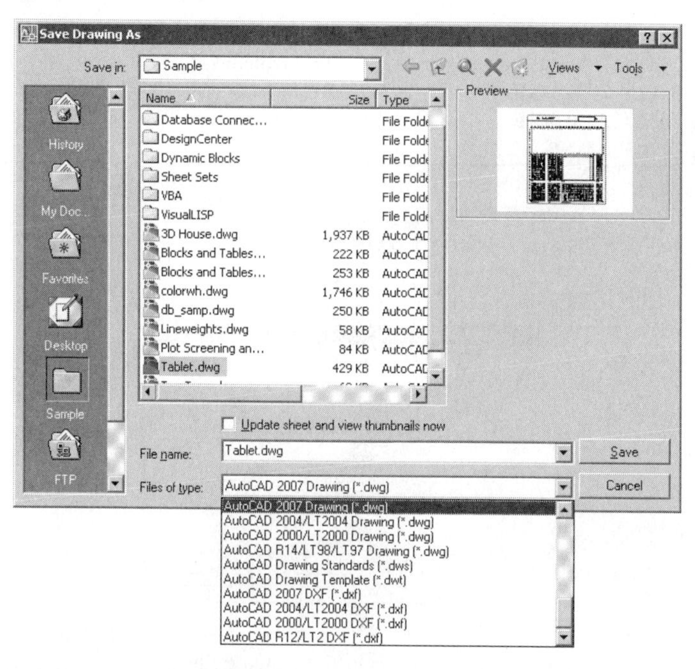

SAVING DRAWING AS DIALOG BOX

Save in selects the folder, hard drive, or network drive in which to save the drawing.

☑ **Update sheet and view thumbnails now**.

File name names the drawing; maximum = 255 characters; default=*drawing1.dwg*.

Save saves the drawing, and then returns to AutoCAD.

Cancel does not save the drawing.

When a drawing of the same name already exists in the same folder, displays dialog box:

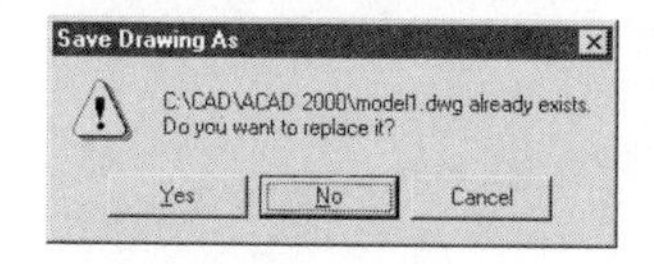

Files of type saves the drawing in a variety of formats:

Save as type	Saves drawings in...
DWG formats:	
AutoCAD 2007 DWG	Native format.
AutoCAD 2004/LT 2004 Drawing	Format compatible with AutoCAD 2004/5/6.
AutoCAD 2000/LT 2000 Drawing	Format compatible with AutoCAD 2000/i/2.
AutoCAD R14/LT98/LT97 Drawing	Format compatible with AutoCAD Release 14.
DWS and DWT formats:	
Drawing Standards	AutoCAD standards format.
Drawing Template	Folder named *template*.
DXF formats:	
AutoCAD 2007 DXF	AC17 DXF format.
AutoCAD 2004/LT 2004 DXF	AC16 DXF format .
AutoCAD 2000/LT 2000 DXF	AC15 DXF format.
AutoCAD R12/ LT 2 DXF	AC12 DXF format.

SAVEAS OPTIONS DIALOG BOX

From the dialog box's toolbar, select ***Tools | Options***:

DWG Options tab

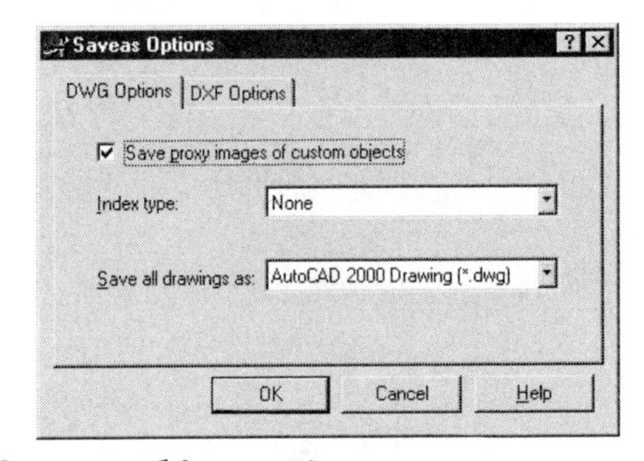

Save proxy images of custom objects option:

☑ Saves an image of the custom objects in the drawing file.

☐ Saves a frame around each custom object, instead of an image.

Index type saves indices with drawings; useful only for xrefed and partially-loaded drawings:

- **None** — creates no indices (default).
- **Layer** — loads layers that are on and thawed.
- **Spatial** — loads only the visible portion of clipped xrefs.
- **Layer and Spatial** — combines the above two options.

Save all drawings as selects the default format for saving drawings.

DXF Options tab

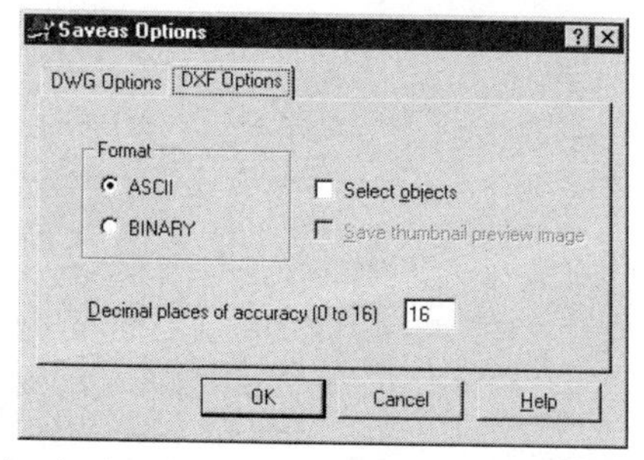

⊙**ASCII** saves the drawing in ASCII format, which can be read by humans – as well as by common text editors – but takes up more disk space.

○**BINARY** saves the drawing in binary format, which takes up less disk space, but cannot be read by some software programs.

☐ **Select objects** allows you to save selected objects to the DXF file; AutoCAD prompts you, 'Select objects:'.

☐ **Save thumbnail preview image** saves a thumbnail image of the drawing.

Decimal places of accuracy specifies the number of decimal places for real numbers: 0 to 16.

TEMPLATE DESCRIPTION DIALOG BOX

Displayed when drawing is saved as .dwt template file:

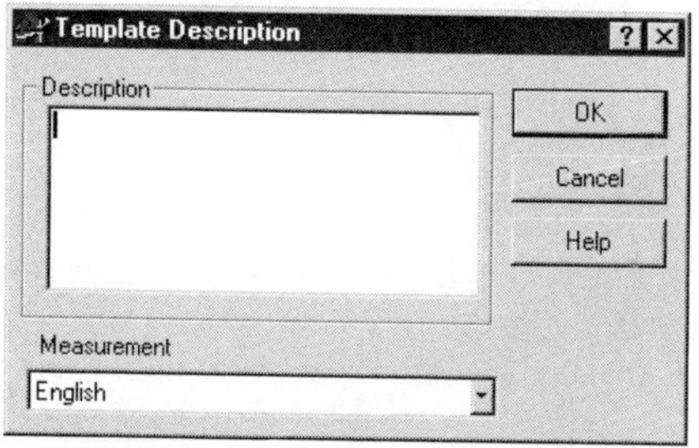

Description describes the template file; stored in *wizard.ini.*

Measurement selects English or metric as the default measurement system.

RELATED COMMANDS

Export saves the drawing in a variety of raster and vector formats.

Plot saves the drawing in even more raster and vector formats.

Quit exits AutoCAD without saving the drawing.

QSave saves the drawing without prompting for a name.

RELATED SYSTEM VARIABLES

DBMod indicates that the drawing was modified since being opened.

DwgName specifies the name of the drawing; *drawing1.dwg* when unnamed.

TIPS

- The Save and SaveAs commands perform exactly the same function.
- Use these commands to save the drawing by another name.
- When a drawing is opened as read-only, use these commands to save the drawing to another filename.
- To save drawings without seeing the Save Drawing As dialog box, use the QSave command.
- All newer releases of AutoCAD can read drawing files produced by older releases, but AutoCAD 2004/5/6 and earlier cannot open drawings created by AutoCAD 2007. This command can translate *.dwg* files to make them compatible with earlier releases. The problem is that incompatible objects are modified:

    - The Materials and ShadowDisplay properties are ignored and preserved by AutoCAD 2004/5/6. For example, apply a material to an object in AutoCAD 2007, export in 2004 DWG format. In AutoCAD 2006, the material is not displayed; you can modify the object, and when reopened in 2007, the original material definition is preserved.
    - New 3D solid objects, such as pyramids, are translated correctly to earlier releases.
    - The helix is new to AutoCAD 2007. When translated to DWG 2004/5/6 format, it appears as a proxy entity. (Proxy entities are visual representations of ARx-defined objects or "custom objects" defined in other AutoCAD-based applications. Because AutoCAD does not have access to the ARX code, Autodesk designed a metaformat to represent custom objects — proxy entities.) You can change properties of proxies, such as color and layer, and perform simple editing tasks, such as moving and rotating. But the object itself cannot be edited, such as changing the height and radius of a helix. The Explode command changes the helix to a series of arced polylines. A better solution is to use AutoCAD 2007's Sweep command to convert the helix into a coil. Because the coil is a 3D solid entity, AutoCAD 2004/5/6 represents it accurately, and you can edit it like any other solid.
    - AutoCAD 2007 has extruded, lofted, planar, revolved, and swept surfaces. When translated to earlier releases, these surfaces become proxy entities. When exploded, they become arcs, polylines, and circles. Use AutoCAD 2007's ConvertToSolid command to convert surfaces to solids, which are acceptable to AutoCAD 2004/5/6.
    - When drawings containing the new SectionObject object are exported to DWG 2004 format, the section plane becomes a proxy entity. When exploded, it becomes lines and polylines.
    - Cameras and lights are represented by a new 3D icon called "glyphs." When exported to DWG 2004 format, glyphs do not appear. Glyphs are ignored and preserved by AutoCAD 2004/5/6.

Removed Command

SaveAsR12 was removed from AutoCAD Release 14. Use **SaveAs** instead, which saves drawings in earlier formats.

SaveImg

Rel.12 Saves a rendering of the current viewport image as raster files on disk (short for SAVE IMAge).

Command	Alias	Ctrl+	F-key	Alt+	Menu Bar	Tablet
saveimg	...	...	...	TYS	Tools ↳Display Image ↳Save	V7

Command: saveimg

This command has changed in AutoCAD 2007.

Displays the Render Output File dialog box.

Provide a file name, select a format, and then click ***Save****.*

One of the following dialog boxes appears:

IMAGE OPTIONS DIALOG BOXES

BMP Image Options dialog box

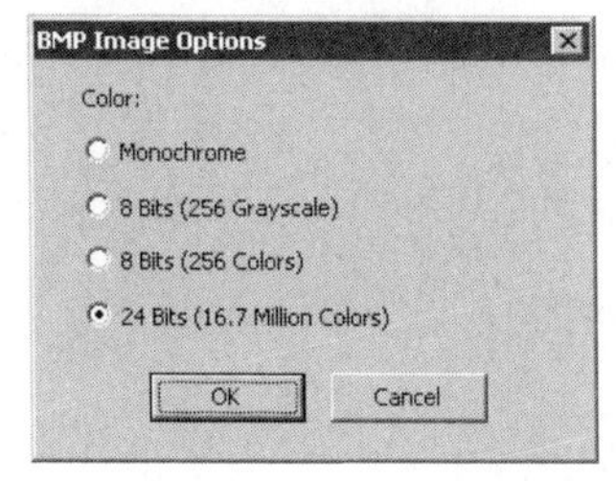

○ **Monochrome** converts all lines to black.

○ **8 Bits (256 Grayscale)** reduces the image to shades of gray.

○ **8 Bits (256 Colors)** reduces the image to 256 colors.

⊙ **24 Bits (16.7 Million Colors)** maintains all colors in the image.

PCX Image Options dialog box

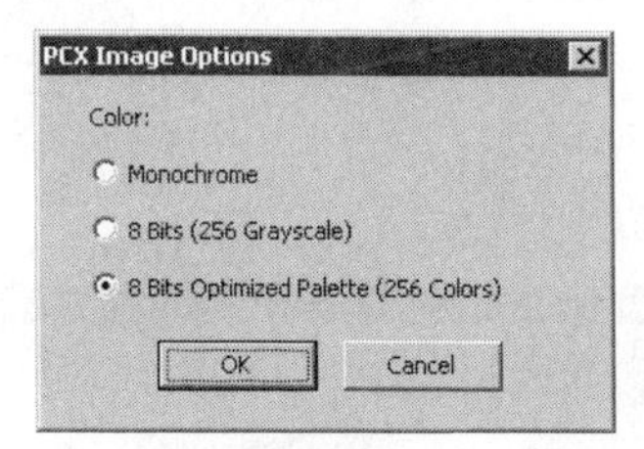

○ **Monochrome** converts all lines to black.

○ **8 Bits (256 Grayscale)** reduces the image to shades of gray.

⊙ **8 Bits Optimized (256 Colors)** reduces the image to the best 256 colors possible.

Targa Image Options dialog box

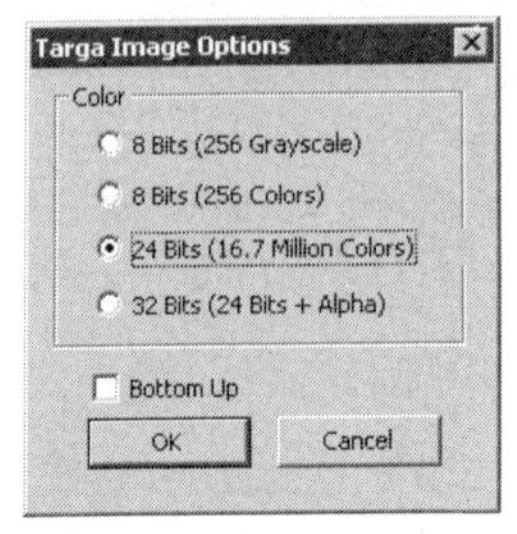

○ **8 Bits (256 Grayscale)** reduces the image to shades of gray.

○ **8 Bits (256 Colors)** reduces the image to 256 colors.

⊙ **24 Bits (16.7 Million Colors)** maintains all colors in the image.

○ **32 Bits (23 Bits + Alpha)** maintains all colors, plus transparency information (alpha).

☐ **Bottom Up** stores data in the file from the bottom to the top of the image.

TIFF Image Options dialog box

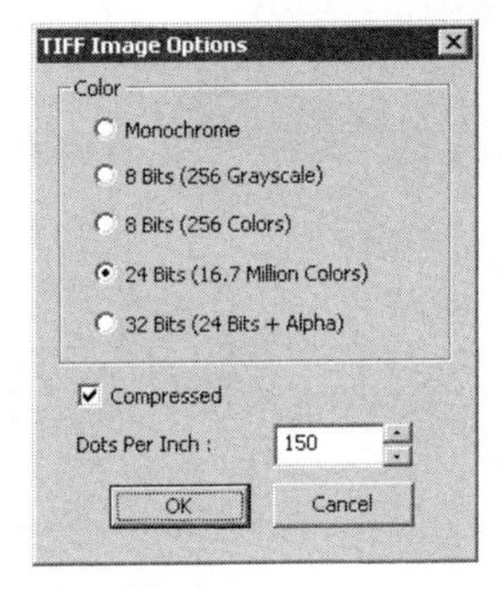

○ **Monochrome** converts all lines to black.

○ **8 Bits (256 Grayscale)** reduces the image to shades of gray.

○ **8 Bits (256 Colors)** reduces the image to 256 colors.

⊙ **24 Bits (16.7 Million Colors)** maintains all colors in the image.

○ **32 Bits (23 Bits + Alpha)** maintains all colors, plus transparency information (alpha).

☑ **Compressed** reduces the size of the file by compressing the data.

Dots Per Inch specifies the resolution of the image; range is 0 to 1000.

JPEG Image Options dialog box

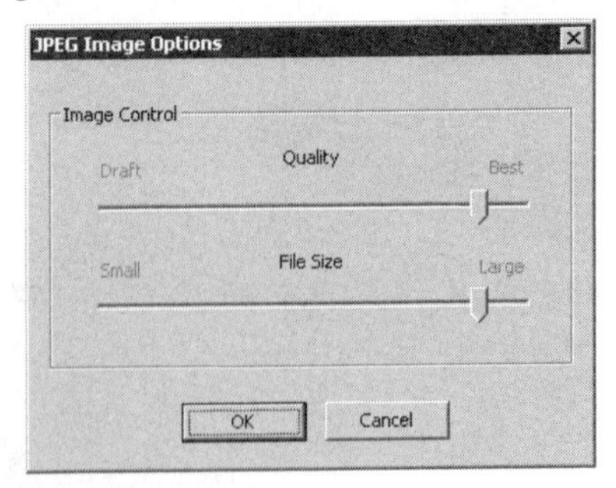

Quality adjust the amount of compression.

File size also adjusts the amount of compression; increase the compression for smaller file sizes. The sliders are linked, and cannot be set independently

PNG Image Options dialog box

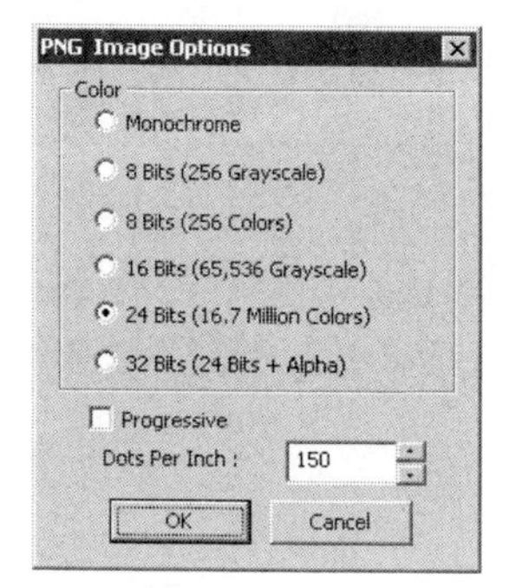

○ **Monochrome** converts all lines to black.

○ **8 Bits (256 Grayscale)** reduces the image to shades of gray.

○ **8 Bits (256 Colors)** reduces the image to 256 colors.

⊙ **24 Bits (16.7 Million Colors)** maintains all colors in the image.

○ **32 Bits (23 Bits + Alpha)** maintains all colors, plus transparency information (alpha).

☑ **Progressive** displays the image more quickly by showing more of it, until all details are filled in.

Dots Per Inch specifies the resolution of the image; range is 0 to 1000.

RELATED COMMANDS

ImageAttach attaches raster images to drawings.

JpgOut saves selected objects in JPEG raster format.

PngOut saves selected objects in PNG raster format.

TifOut saves selected objects in TIFF raster format.

WmfOut saves selected objects in WMF vector format.

CopyClip copies selected objects to the Clipboard.

PRT SCR saves the entire screen to the Clipboard.

TIPS

- This command saves the entire viewport; in contrast, the JpgOut, PngOut, TifOut, and WmfOut commands prompt you to select objects.
- The GIF format, which is one of the two most common formats used to display images on the Internet (and was found in AutoCAD Release 12 and 13), was replaced by the BMP format in AutoCAD Release 14.

Aliased Command

SaveUrl was removed from AutoCAD 2000; it now activates the **Save**.

Scale

V. 2.5 Makes selected objects smaller or larger.

Command	Alias	Ctrl+	F-key	Alt+	Menu Bar	Tablet
scale	**sc**	...	...	**ML**	**Modify ↳Scale**	**V21**

Command: scale
Select objects: *(Select one or more objects.)*
Select objects: *(Press Enter to end object selection.)*
Specify base point: *(Pick a point.)*
Specify a scale factor or [Copy/Reference]: *(Enter scale factor, or type* **C** *or* **R**.*)*

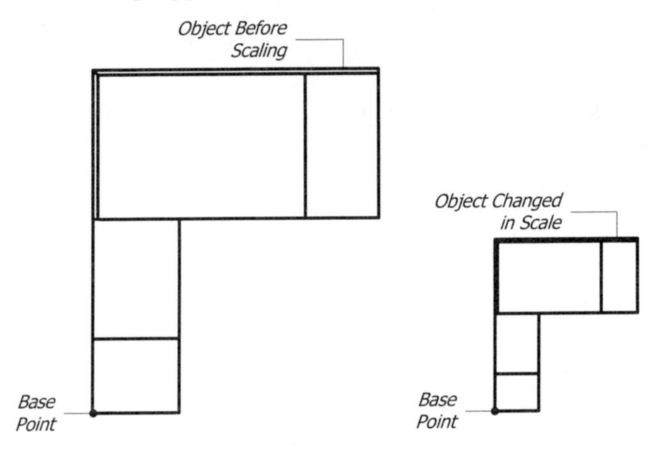

COMMAND LINE OPTIONS

Base point specifies the point from which scaling takes place.

Reference requests a distance, followed by a new distance.

Copy scales a copy of the selected objects, leaving the originals in place.

Scale factor indicates the scale factor, which applies equally in the x and y directions.

Scale Factor	Meaning
> **1.0**	Enlarges object(s).
= **1.0**	Makes no change.
> **0.0** *and* < **1.0**	Reduces object(s).
= **0.0** *or negative*	Illegal values.

Reference options
Specify reference length <1>: *(Enter a base length.)*
Specify new length: *(Enter a new length.)*

Specify reference length specifies a baseline length; you can enter a value, or pick two points.

Specify new length specifies a new length that determines the scale factor.

RELATED COMMANDS

Align scales an object in 3D space.

Insert allows a block to be scaled independently in the x, y, and z directions.

Plot allows a drawing to be plotted at any scale.

TIPS

- You can interactively scale the object by moving the cursor to make the object larger and smaller.
- The objects scale from the base point.
- Points are the only object that cannot be scaled.
- Scale factors larger than 1.0 grow the object; smaller than 1.0 shrink the object.
- This command changes the size in all dimensions (x,y,z) equally; to change an object in just one dimension, use the Stretch command.
- The **Reference** option is useful for:
  - Scaling raster images to match the size of a drawing.
  - Making one object the same size as another.
  - Changing the units in a drawing. For example, use 1:12 to change decimal feet into inches, or 2.54:1 to change cm to inches.

'ScaleListEdit

2006 Edits lists of scale factors.

Command	Alias	Ctrl+	F-key	Alt+	Menu Bar	Tablet
scalelistedit	...	...	...	OE	Format ↳Scale List	...
-scalelistedit						

Command: scalelistedit

Displays dialog box:

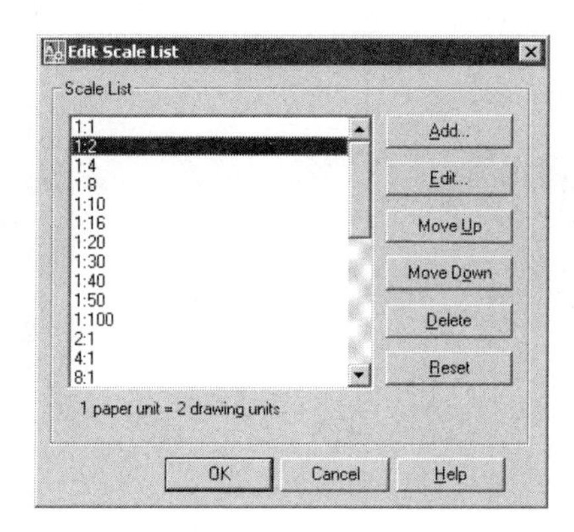

DIALOG BOX OPTIONS

Add adds scale factors to the list; displays the Add Scale Factor dialog box.

Edit changes the selected scale factor; displays the Edit Scale Factor dialog box.

Move Up moves the selected scale factor up the list.

Move Down moves the selected scale factor down the list.

Delete erases the selected scale factors from the list without warning; click **Cancel** or **Restore** to recover accidentally-erased scale factors.

Reset restores the list to its original status; AutoCAD warns custom scale factors will be lost.

Add Scale Factor dialog box

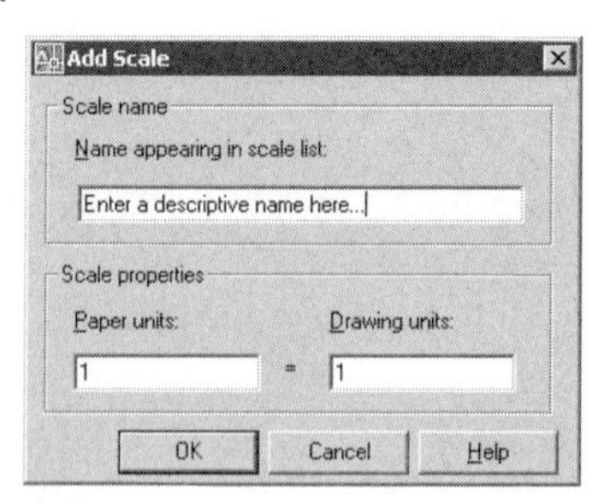

Name Appearing in Scale List provides descriptive names that appear in place of scale factors. These may be an actual scale factor, such as 1:10.

Paper Units specifies the size of objects on paper. This value is usually 1 for drawings larger than the paper.

Drawing Units specifies the size of objects in the drawing. This value is usually 1 for drawings smaller than the paper.

-ScaleListEdit Command

Command: -scalelistedit

Enter option [?/Add/Delete/Reset/Exit] <Add>: *(Enter an option.)*

Enter name for new scale: *(Type a name.)*

Enter scale ratio: *(Type the scale ratio, such as 1:1.5.)*

COMMAND LINE OPTIONS

? lists the scale factors in the Text window.

	Scale Name	Paper Units	Drawing Units	Effective Scale
1:	1:1	1.0000	1.0000	1.0000
2:	1:2	1.0000	2.0000	0.5000
3:	1:4	1.0000	4.0000	0.2500
4:	1:8	1.0000	8.0000	0.1250
5:	1:10	1.0000	10.0000	0.1000
6:	1:16	1.0000	16.0000	0.0625
7:	1:20	1.0000	20.0000	0.0500
8:	1:30	1.0000	30.0000	0.0333
9:	1:40	1.0000	40.0000	0.0250
10:	1:50	1.0000	50.0000	0.0200
11:	1:100	1.0000	100.0000	0.0100
12:	2:1	2.0000	1.0000	2.0000
13:	4:1	4.0000	1.0000	4.0000
14:	8:1	8.0000	1.0000	8.0000
15:	10:1	10.0000	1.0000	10.0000
16:	100:1	100.0000	1.0000	100.0000
17:	1/128" = 1'-0"	0.0078	12.0000	0.0007
18:	1/64" = 1'-0"	0.0156	12.0000	0.0013
19:	1/32" = 1'-0"	0.0313	12.0000	0.0026
20:	1/16" = 1'-0"	0.0625	12.0000	0.0052
21:	3/32" = 1'-0"	0.0938	12.0000	0.0078
22:	1/8" = 1'-0"	0.1250	12.0000	0.0104
23:	3/16" = 1'-0"	0.1875	12.0000	0.0156
24:	1/4" = 1'-0"	0.2500	12.0000	0.0208
25:	3/8" = 1'-0"	0.3750	12.0000	0.0313
26:	1/2" = 1'-0"	0.5000	12.0000	0.0417
27:	3/4" = 1'-0"	0.7500	12.0000	0.0625
28:	1" = 1'-0"	1.0000	12.0000	0.0833
29:	1-1/2" = 1'-0"	1.5000	12.0000	0.1250
30:	3" = 1'-0"	3.0000	12.0000	0.2500
31:	6" = 1'-0"	6.0000	12.0000	0.5000
32:	1'-0" = 1'-0"	12.0000	12.0000	1.0000

Add adds scale factors.

Delete erases scale factors.

Reset restores the list of scale factors to the original factory setting.

Exit exits the command.

TIPS

- This list of scale factors is displayed by the Plot command and the Properties command for viewports.
- Use this command to shorten the list of unused scale factors, and to add custom scales.

ScaleText

2002 Changes the height of text objects relative to their insertion points.

Command	Alias	Ctrl+	F-key	Alt+	Menu Bar	Tablet
scaletext	...	...	...	MOTS	Modify ↳Object ↳Text ↳Scale	...

Command: scaletext
Select objects: *(Select one or more objects.)*
Select objects: *(Press Enter to end object selection.)*
Enter a base point option for scaling [Existing/Left/Center/Middle/Right/TL/TC/TR/ML/MC/MR/BL/BC/BR] <Existing>: *(Enter an option, or press* ENTER *to keep insertion point as is.)*
Specify new height or [Match object/Scale factor] <0.2>: *(Enter a value or option.)*

COMMAND LINE OPTIONS

Enter a base point option for scaling specifies the point from which scaling takes place.

Existing uses the existing insertion point for each text object.

Specify new height specifies the new height of the text.

Match object matches the height of another text object.

Scale factor scales the text by a factor:

Scale Factor	Meaning
> **1.0**	Enlarges text.
1.0	Makes no change in size.
> **0.0** *and* < **1.0**	Reduces text.
0.0 *or negative*	Illegal values.

Scale Factor Options

Specify scale factor or [Reference] <2.0000>: *(Enter a value, or type* **R***.)*

Specify scale factor scales the text by a factor; see **Scale** command.

Reference supplies a reference value, followed by a new value.

RELATED COMMANDS

Justify changes the justification of text.

Properties changes all aspects of the text.

Style creates text styles.

TIP

- *Warning!* Scaling text larger may make it overlap nearby text.

Removed Command

The **Scene** command was removed from AutoCAD 2007.

'Script

V. 1.4 Runs ASCII files containing sequences of AutoCAD commands and options.

Command	Alias	Ctrl+	F-key	Alt+	Menu Bar	Tablet
script	scr	...	...	TR	Tools ↳Run Script	V9

Command: script

Displays the Select Script File dialog box. Select an .scr file, and then click ***Open.***

Script file begins running as soon as it is loaded.

COMMAND LINE OPTIONS

BACKSPACE interrupts the script.

ESC stops the script.

RELATED COMMANDS

Delay specifies the delay in milliseconds; pauses the script before executing the next command.

Resume resumes a script after it is interrupted.

RScript repeats script files.

STARTUP SWITCH

/b specifies the script file name to run when AutoCAD starts up.

TIPS

- Since the Script command is transparent, it can be used during another command.
- Use the **/s** command-line switch to run a script when AutoCAD starts.
- Prefix the VSlide command with * to preload it; this results in a faster slide show:

  ```
  *vslide
  ```

- You can make a script file more flexible — such as pausing for user input, or branching with conditionals — by inserting AutoLISP functions.

Section

Rel.11 Creates 2D region objects from the intersection of a plane and a 3D solid.

Command	Alias	Ctrl+	F-key	Alt+	Menu Bar	Tablet
section	sec	...	...	DIE	Draw ↳Solid ↳Section	...

Command: section
Select objects: *(Select one or more objects.)*
Select objects: *(Press Enter to end object selection.)*
Specify first point on section plane by [Object/Zaxis/View/XY/YZ/ZX/ 3points]: *(Pick a point, or enter an option.)*

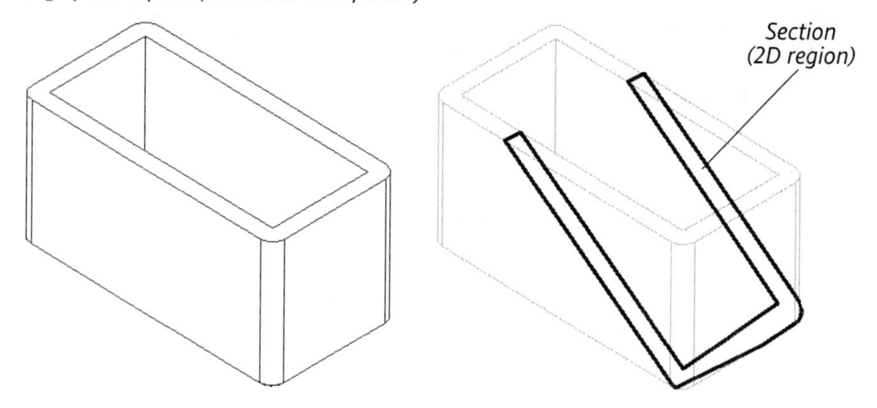

COMMAND LINE OPTIONS

Select objects selects the 3D solid objects to be sectioned.

Object aligns the section plane with a selected object: circle, ellipse, arc, elliptical arc, 2D spline, or 2D polyline.

Zaxis specifies the normal (*z axis*) to the section plane.

View uses the current view plane as the section plane.

XY uses the x,y plane of the current view.

YZ uses the y,z plane of the current view.

ZX uses the z,x plane of the current view.

3 points picks three points to specify the section plane.

RELATED COMMAND

Slice cuts a slice out of a solid model, creating another 3D solid.

TIPS

- Section regions are placed on the current layer, not the object's layer.
- Regions are ignored.
- One cutting plane can section multiple solids at a time.

SectionPlane

2007 Places "section plane" objects that visually cut 3D objects.

Command	Alias	Ctrl+	F-key	Alt+	Menu Bar	Tablet
sectionplane	splane	...	...	DME	Draw ↳Modeling ↳Section Plane	...

Command: sectionplane

Select face or any point to locate section line or [Draw section/Orthographic]: *(Pick a face or a point, or enter an option.)*

Specify through point: *(Pick another point.)*

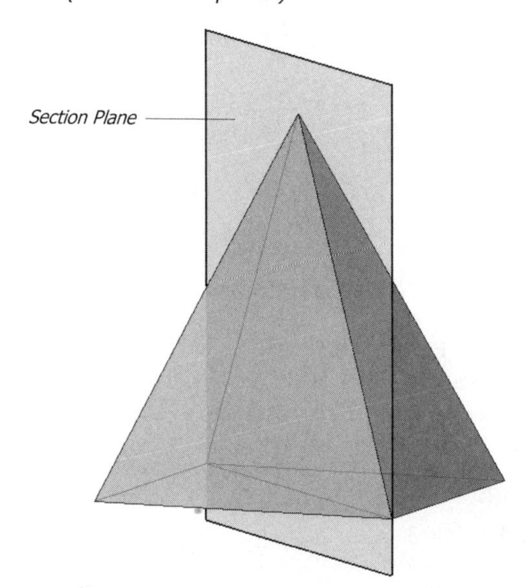

After the section plane is placed, double-click it to make it "live" (active). See the LiveSection command.

COMMAND LINE OPTIONS

Select face aligns the section plane with the face; you can pick a region or a face on a 3D solid. (There is no need to hold down the CTRL key to select faces.)

Select any point specifies the starting point of the section plane; if this command insists on aligning the section plane with a face, override it with the **Draw section** option.

Draw section places the section plane by picking points.

Orthographic places the section plane parallel to standard views: front, back, top, bottom, left, or right view.

Specify through point positions the section plane.

Draw Section options

It can be helpful to use ENDpoint and QUAdrant object snaps with these prompts.

Specify start point: *(Pick a point, 1.)*

Specify next point: *(Pick another point, 2.)*

Specify next point or ENTER to complete: *(Press Enter to skip to the next prompt, or pick a third point, 3.)*

Specify next point in direction of section view: *(Pick a point to align the section plane, 4.)*

Specify start point specifies the starting point of the section plane.

Specify next point allows you to draw a multi-segment section plane, as illustrated below.

Direction of section view specifies the cut-away portion.

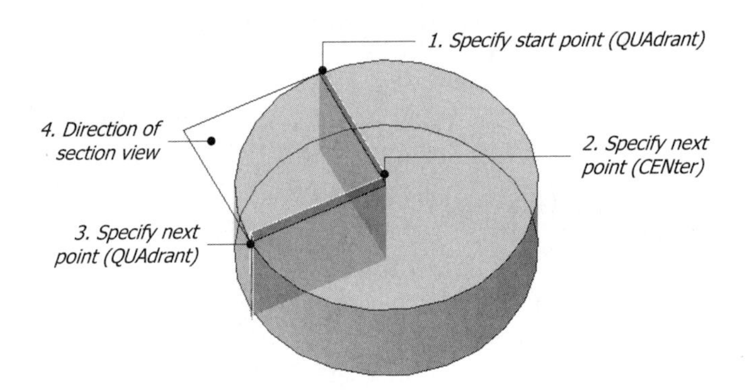

Orthographic option

Align section to: [Front/Back/Top/Bottom/Left/Right]: *(Enter an option.)*

Front, Back, Top, Bottom, Left, or **Right** aligns the section plane with current UCS. Live sectioning is turned on automatically (the forward portion of the sectioned objects being removed from view). See the LiveSection command.

RELATED COMMANDS

LiveSection edits section planes.

Section creates region slices from 3D models.

Interfere creates new objects from the interference of two or more 3D solids.

TIPS

- This is a two-part command:
    1. Creating the section plane with the SectionPlane command; and then
    2. Editing the section plane with the LiveSection command.
- "Sectionplane" is an distinct object type introduced with AutoCAD 2007.

SecurityOptions

2004 Sets passwords and digital signatures for securing drawings against unauthorized access.

Command	Alias	Ctrl+	F-key	Alt+	Menu Bar	Tablet
securityoptions	...	...	...	...	...	...

Command: securityoptions

Displays dialog box:

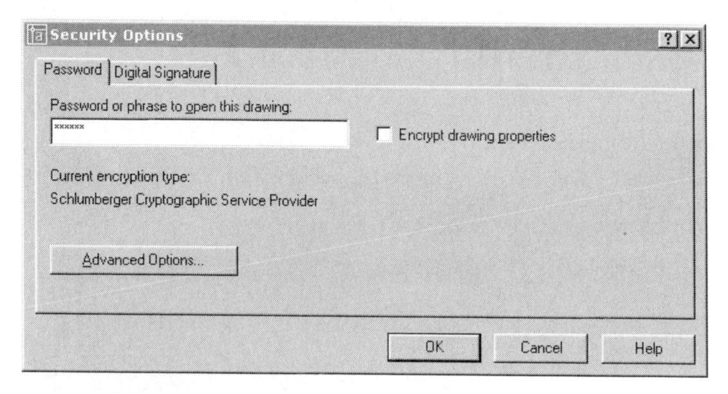

SECURITY OPTIONS DIALOG BOX

Password or phrase to open this drawing assigns a password to lock the drawing.

☐ **Encrypt drawing properties** encrypts drawing properties data; see the DwgProps command.

Advanced Options displays the Advanced Options dialog box; available only after a password is entered.

OK displays the Confirm Password dialog box:

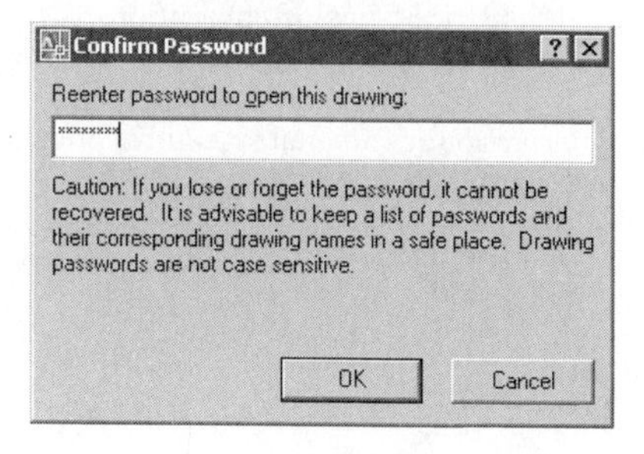

Digital Signature tab

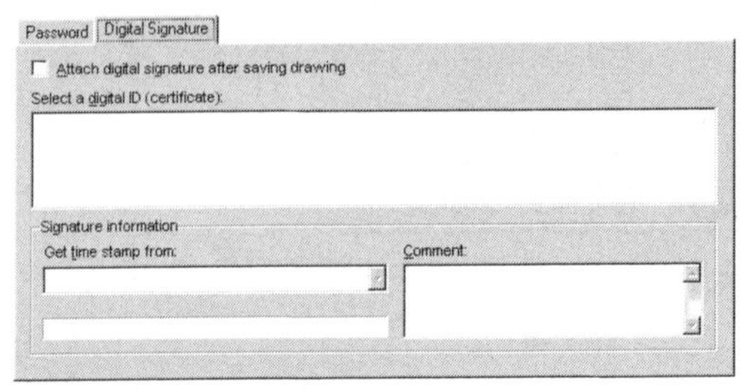

☐ **Attach digital signature after saving drawing** attaches the digital signature the next time the drawing is saved.

Select a digital ID (certificate) lists digital signature services.

Signature information

Get time stamp from lists time servers.

Comment provides additional comments to be included with the digital signature.

ADVANCED OPTIONS DIALOG BOX

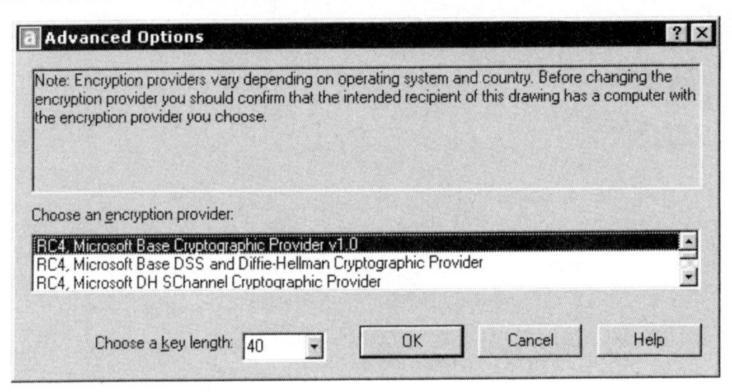

Choose an encryption provider selects from several encryption types.

Choose a key length selects 40, 48, or 56 bits of encryption.

RELATED COMMAND

SigValidate displays information about the digital signatures attached to drawings.

RELATED SYSTEM VARIABLE

SigWarn determines whether a warning alerts that drawings are opened with digital signatures.

TIPS

- When opening drawings protected by passwords, AutoCAD displays the **Password** dialog box for entering the password:

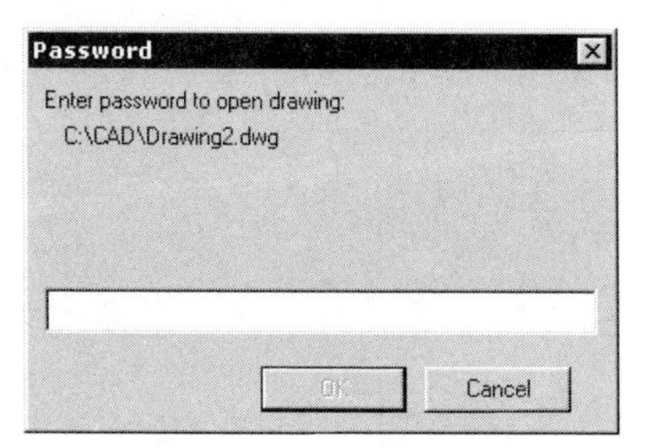

- When the password entered is correct, AutoCAD opens the drawing.
- When the password entered is incorrect, AutoCAD complains:

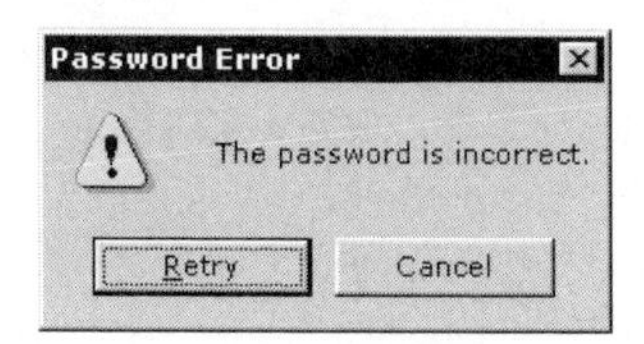

- To remove the password, in the **Security Options** dialog box, remove the password, and then click **OK**.
- The first time you assign a digital signature to the drawing, you are prompted to obtain one:

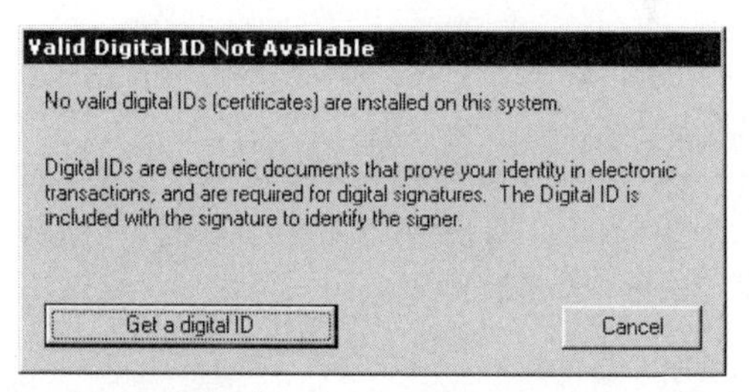

- *Warning!* If the password is lost, the drawing is not recoverable. Autodesk recommends that you (1) save a backup copy of the drawing not protected with any password; and (2) maintain a list of password names stored in a secure location.
- *Warning!* The encryption provider must be the same on the receiving computers; otherwise the drawings cannot be opened.
- To remove this feature from AutoCAD, install with the **Custom** option, and then unselect **Drawing Encryption**.

Select

V. 2.5 Creates a selection set of objects based on their location in the drawing.

Command	Alias	Ctrl+	F-key	Alt+	Menu Bar	Tablet
select	...	...	...	...	...	...

Command: select
Select objects: *(Select one or more objects, or enter an option.)*
Select objects: *(Press Enter to end object selection.)*

COMMAND LINE

pick selects a single object.

AU switches from pick mode to **C** or **W** mode, depending on whether an object is found at the initial pick point (short for AUtomatic).

ALL selects all objects in the drawing; press CTRL+A to select all objects.

BOX goes into **C** or **W** mode, depending on how the cursor moves.

C selects objects in and crossing the selection box (Crossing).

CP selects all objects inside and crossing the selection polygon (Crossing Polygon).

F selects all objects crossing a selection polyline (Fence).

G selects objects contained in a named group (Group).

M makes multiple selections before AutoCAD scans the drawing; saves time in a large drawing (Multiple).

SI selects only a single set of objects before terminating the Select command (SIngle).

W selects all objects inside the selection box (Window).

WP selects objects inside the selection polygon (Windowed Polygon).

L selects the last-drawn object still visible on the screen (Last).

P selects previously-selected objects (Previous).

R removes objects from the selection set (Remove).

U removes the selected objects most-recently added (Undo).

A continues to add objects after using the **R** option (Add).

ENTER completes the **Select** command.

ESC abandons the **Select** command.

RELATED COMMANDS

All commands that prompt 'Select objects'.

Filter specifies objects to be added to the selection set.

QSelect selects objects via a dialog box interface based on their properties.

RELATED SYSTEM VARIABLES

PickAdd controls how objects are added to a selection set.

PickAuto controls automatic windowing at the 'Select objects' prompt.

PickDrag controls the method of creating a selection box.

PickFirst controls the command-object selection order.

TIPS

- Objects selected by the Select command become the Previous selection set and may be selected by any subsequent 'Select objects:' prompt by responding **P** for "previous."
- To view a list of all selection options, enter any non-valid text at the prompt, such as "asdf":

  Command: select
  Select objects: asdf
  Invalid selection
  Expects a point or
  Window/Last/Crossing/BOX/ALL/Fence/WPolygon/CPolygon/Group/ Add/Remove/Multiple/Previous/Undo/AUto/SIngle
  Select objects:

- Pressing CTRL+A selects all objects in the drawing without needing the Select command.
- **All** selects all objects in the drawing, except those on frozen and locked layers.
- **Crossing** selects objects within and crossing the selection rectangle.
- **Window** selects objects within the selection rectangle.
- **Fence** selects objects crossing the selection polyline.
- **CPolygon** selects objects within and crossing the selection polygon.
- **WPolygon** selects objects within the selection polygon.
- Selection preview highlights objects as the cursor passes over them. This can be useful for determining which objects belong together; for example, lines and arcs vs polylines, or blocks. However, it slows down the display speed with large, complex objects. In the figure below, the chair block is highlighted by the cursor:

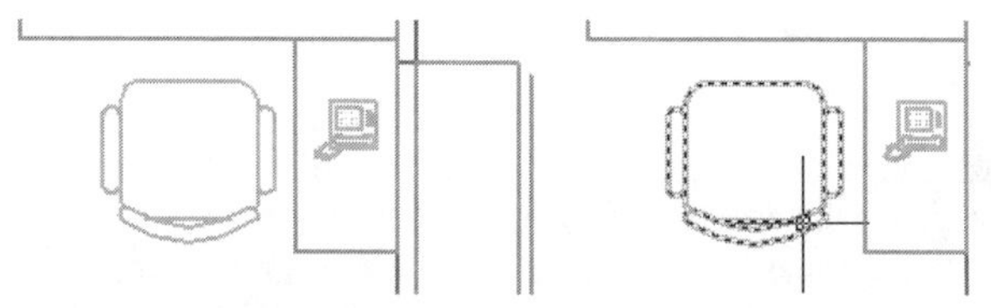

  You can change preview highlighting with the Options command: chose the **Selection** tab, and then click the **Visual Effect Settings** button.

SelectURL

Rel.14 Highlights all objects and areas that have hyperlinks attached to them (an undocumented command).

Command	Alias	Ctrl+	F-key	Alt+	Menu Bar	Tablet
selecturl	...	...	...	...	...	...

Command: selecturl

Highlights hyperlinked objects and areas with grips.

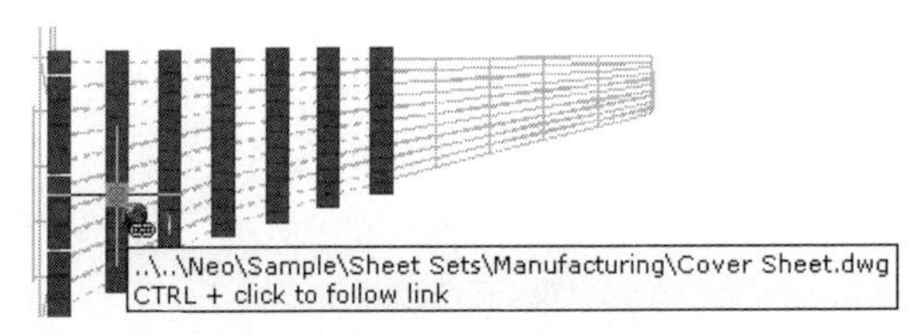

If there are no hyperlinks in the drawing, AutoCAD reports:

No objects with hyperlinks found.

COMMAND LINE OPTIONS

None.

RELATED COMMANDS

AttachURL attaches URLs to objects and areas.

Hyperlink attaches URLs to objects via a dialog box.

TIPS

- Examples of URLs include:

http://www.autodeskpress.com	Autodesk Press Web site.
news://adesknews.autodesk.com	Autodesk news server.
ftp://ftp.autodesk.com	Autodesk FTP server.
http://www.upfrontezine.com	Author Ralph Grabowski's Web site.

- Do not delete layer URLLAYER.
- The URL is stored as follows:

Attachment	URL
One object	Stored as xdata (extended entity data).
Multiple objects	Stored as xdata in each object.
Area	Stored as xdata in a rectangle object on layer URLLAYER.

DEFINITIONS

DWF — short for "drawing Web format," Autodesk's file format for displaying drawings on the Internet.

URL — short for "uniform resource locator," the universal file naming convention.

'SetIDropHandler

2004 Specifies how i-drop objects should be handed in drawings.

Command	Alias	Ctrl+	F-key	Alt+	Menu Bar	Tablet
'setidrophandler	...	...	...	...	...	...

Command: setidrophandler

Displays dialog box:

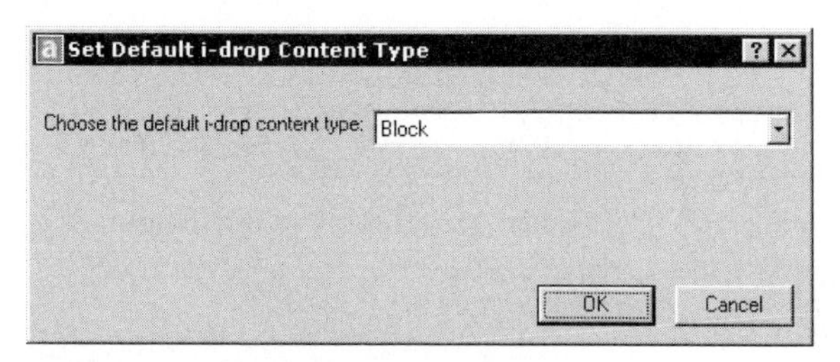

SET DEFAULT I-DROP CONTENT TYPE DIALOG BOX

Choose the default i-drop content type specifies how i-drop objects are handled when placed in drawings.

TIPS

- When i-drop content is dragged from Web pages into drawings, AutoCAD treats the objects as blocks.
- Ever since this command was introduced, its only option has been "Block", so this command serves no purpose.
- For more information on i-drop (short for "Internet drag'n drop") technology, see www.autodesk.com/developidrop.

Removed Command

The **SetUV** command was removed from AutoCAD 2007; its replacement is **MaterialMap**.

'SetVar

<u>**V. 2.5**</u> Lists the settings of system variables, and allows you to change variables that are not read-only (short for SET VARiable).

Command	Alias	Ctrl+	F-key	Alt+	Menu Bar	Tablet
setvar	**set**	...	...	**TQV**	**Tools** **↳Inquiry** **↳Set Variable**	**U10**

Command: setvar

Enter variable name or [?]: *(Enter a name, or type **?**.)*

COMMAND LINE OPTIONS

Enter variable name names the system variable you want to access.

? lists the names and settings of system variables.

TIPS

- See Appendix C for a complete list of all system variables found in AutoCAD.
- Example usage of this command:

  Command: setvar

  Enter variable name or [?]: visretain

  Enter new value for VISRETAIN <0>: 1
- Almost all system variables can be entered without the SetVar command. For example:

  Command: visretain

  New value for VISRETAIN <0>: 1
- System variables marked "read only" cannot be changed:

  Command: _pkser

  _PKSER = "341-35000000" (read only)

Replaced Commands

Shade was replaced in AutoCAD 2000 by the **ShadeMode** command's **Flat+edges** option.

ShadeMode was replaced by the **-ShadeMode** command in AutoCAD 2007; **ShadeMode** is now an alias for the **VsCurrent** command.

-ShadeMode

2000 Generates renderings of 3D models quickly in a variety of modes (replaces the Shade and ShadeMode commands).

Command	Alias	Ctrl+	F-key	Alt+	Menu Bar	Tablet
-shademode	shade	...	...	VS	View ⇨Shade	N2

Command: -shademode
Current mode: 2D wireframe
Enter option [2D wireframe/3D wireframe/Hidden/Flat/Gouraud/ fLat+edges/gOuraud+edges] <2D wireframe>: *(Enter an option.)*

COMMAND LINE OPTIONS

2D wireframe displays wireframe models in 2D space.

3D wireframe displays wireframe models in 3D space.

Hidden removes hidden faces.

Flat displays flat shaded faces.

fLat+edges displays flat shaded faces, with outlined faces of the background color.

Gouraud displays smooth shaded faces.

gOuraud+edges displays smooth shaded faces, with outlined faces of the background color.

RELATED COMMANDS

VsCurrent displays viewports in a variety of visual styles.

MView specifies hidden-line removal of individual viewports for plotting.

Render performs more realistic renderings than does -ShadeMode.

3dOrbit removes hidden lines and shades perspective views in real-time.

TIPS

- As an alternative to the -ShadeMode command, the Render command creates high-quality renderings of 3D drawings, but takes somewhat longer to do so.
- The smaller the viewport, the faster the shading.
- Drawing (at left, below) displayed in -ShadeMode's default **2D wireframe** mode; note the standard 2D UCS icon. Wireframe colors are based on layer and object colors.

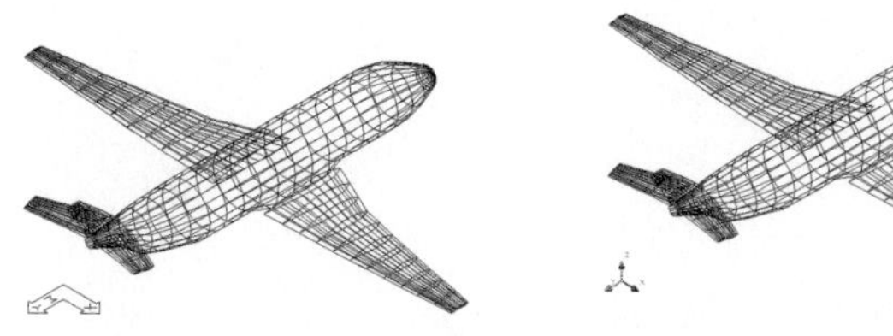

- Drawing (at right, above) displayed in **3D wireframe** mode; note the 3D UCS icon. Wireframe colors are based on the color of the rendering materials, if any.

- Drawing (at left, below) displayed in **Hidden** mode; faces hidden by other faces are not displayed.

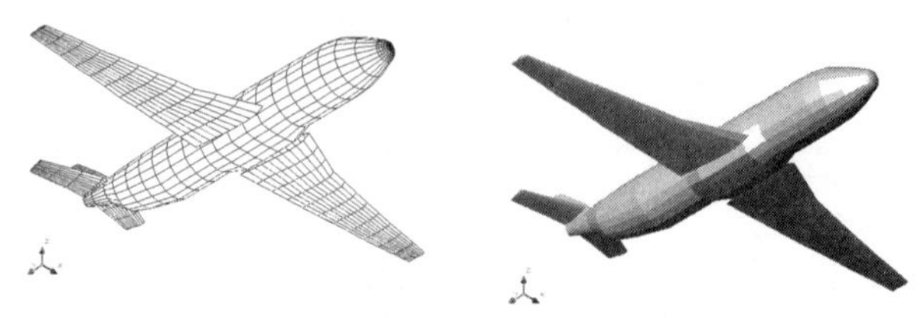

- Drawing (at right, above) displayed in **Flat** mode; each face is filled with shaded grays.
- Drawing (at left, below) displayed in **Gouraud** mode; faces are smoothed.

- Drawing (at right, above) displayed in **fLat+edges** mode; faces are outlined by the background color (*white*).
- Drawing (below) displayed in **gOuraud+edges** mode; each face is outlined.

Shape

V. 1.0 Inserts shapes into drawings.

Command	Alias	Ctrl+	F-key	Alt+	Menu Bar	Tablet
shape	...	...	...	...	...	...

Command: shape
Enter shape name or [?]: *(Enter the name, or type* ***?****.)*
Specify insertion point: *(Pick a point.)*
Specify height <1>: *(Specify the height.)*
Specify rotation angle <0>: *(Specify the angle.)*

COMMAND LINE OPTIONS

Enter shape name indicates the name of the shape to insert.

? lists the names of currently-loaded shapes.

Specify insertion point indicates the insertion point of the shape.

Specify height specifies the height of the shape.

Specify rotation angle specifies the rotation angle of the shape.

RELATED COMMANDS

Load loads SHX-format shape files into the drawings.

Insert inserts blocks into drawings.

Style loads *.shx* font files into drawings.

RELATED SYSTEM VARIABLE

ShpName specifies the default *.shp* file name.

TIPS

- Shapes are used to define the text and symbols found in complex linetypes.

Some electronic shapes.

- Shapes were an early alternative to blocks, but now are used primarily with complex linetypes. They take up extremely small amounts of memory, but are difficult to create.
- Shapes are defined by *.shp* files, which must be compiled into *.shx* files before they can be loaded by the Load command.
- Shapes must be loaded by the Load command before they can be inserted with the Shape command.
- Compile an *.shx* file into an *.shp* file with the Compile command.

Sheetset, SheetsetHide

2005 Displays and hides the Sheetset Manager palette.

Command	Alias	Ctrl+	F-key	Alt+	Menu Bar	Tablet
sheetset	**ssm**	**4**	...	...	**Tools** ↳**Palettes** ↳**Sheetset Manager**	...
sheetsethide	...	**4**	...	...	**Tools** ↳**Palettes** ↳✓**Sheetset Manager**	...

Command: sheetset

Displays palette:

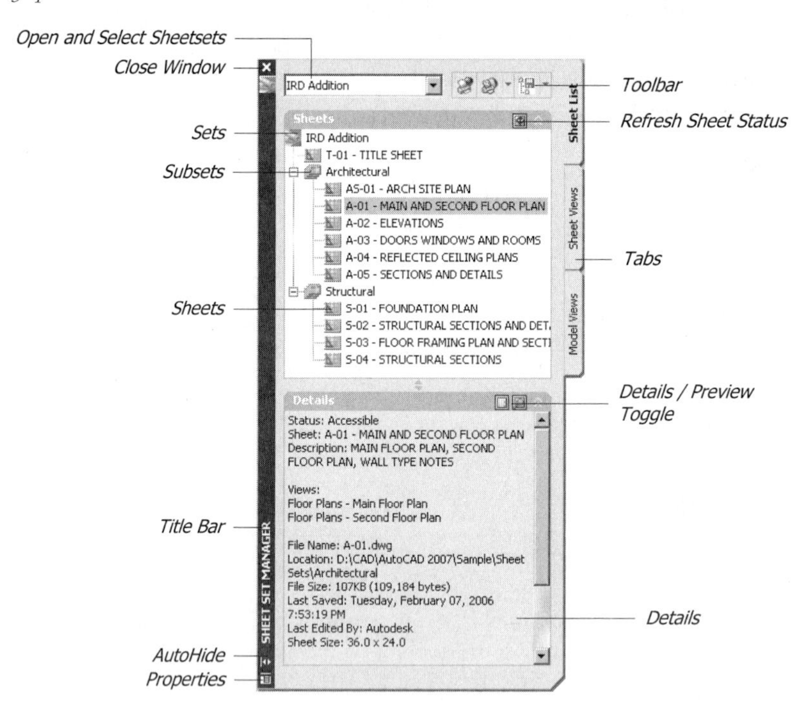

Command: sheetsethide

Closes the palette:

SHEETSET MANAGER PALETTE

x closes the palette; alternatively, use the SheetsetClose command.

Toolbar changes, depending on the tab selected.

Title bar docks against the side of AutoCAD, or can be dragged around.

Tabs change between Sheet List, View List, and Resource Drawings views.

Views lists sheets and sheet set names; right-click for shortcut menus.

- **Sheet List** displays names of sheets, which are organized into subsets.
- **Sheet Views** displays views for each sheet set; views can be organized into categories.
- **Model Views** displays folders and files used by the sheet set.

Open and select sheet sets displays a droplist of recently-opened sheet sets.

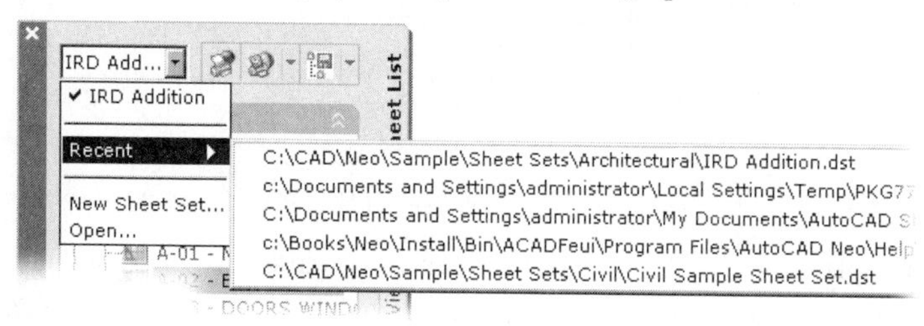

- **Recent** lists recently-opened *.dst* sheet set data files.
- **New Sheet Set** starts the Create Sheet Set wizard; see NewSheetset command.
- **Open** displays the Open Sheetset dialog box; select a *.dst* file, and then click **Open**. See the **NewSheetset** command.

Refresh Sheet Status (**F5**) updates the list of files and information stored in sheet set data *.dst* files, checks resource folders for new and removed drawings, and checks all drawings for new and removed model space views.

Details lists information about selected sheets and sheet sets.

Sheet Preview shows preview images of selected sheets.

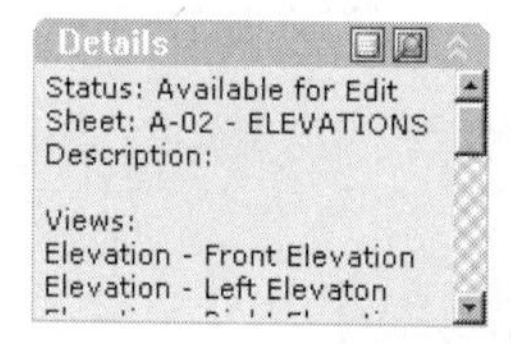

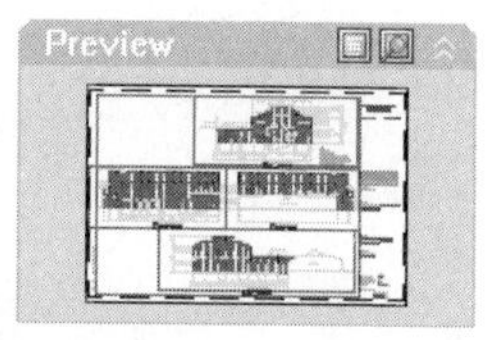

AutoHide reduces palette to titlebar when cursor not over the palette.

Properties displays a shortcut menu.

Sheet List toolbar

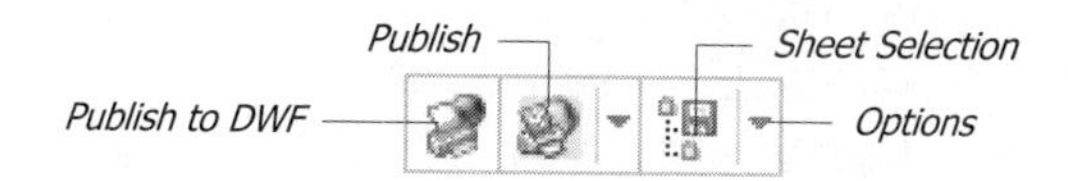

SHORTCUT MENUS

 *Right-click a sheet set name:*

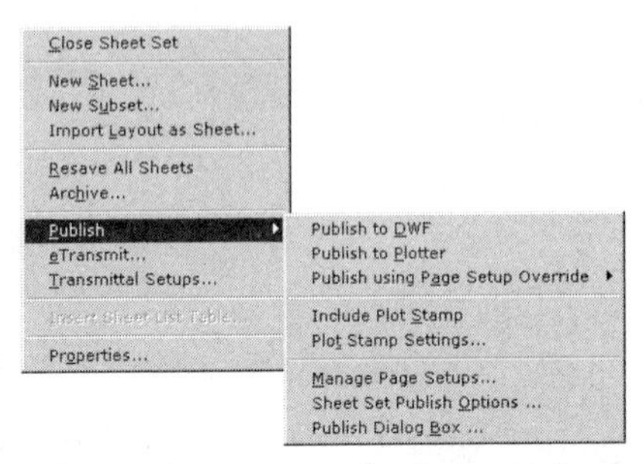

Close Sheet Set closes the entire sheet set; use the **Open** option to open another sheet set.

New Sheet displays the New Sheet dialog box; enter a number and name, and then click **OK**.
New Subset displays the Subsheet Properties dialog box; enter a name, and then click **OK**.
Import Layout as Sheet displays the Import Layout as Sheet dialog box; select a drawing and layouts, and then click **OK**.

Resave All Sheets saves sheet set information stored in each drawing file.
Archive displays the Archive as Sheet Set dialog box; see the Archive command.

Publish plots the sheet set as *.dwf* files or to a plotter; see the Publish command.
eTransmit collates the sheet set for transmittal by CD or email; see the eTransmit command.
Transmittal Setups displays the Transmittal Setup dialog box; see the eTransmit command.

Insert Sheet List Table creates a table listing all sheets in the sheet set, and adds it to a sheet.

Properties displays the Sheet Set Properties dialog box; make changes, and then click **OK**.

Subset

 *Right-click a subset name:*

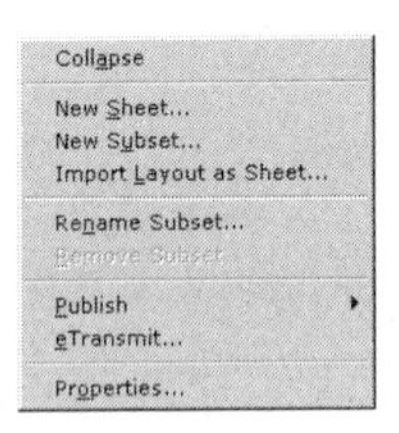

Collapse hides the names of sheets.
Rename Subset changes the name of the subset.
Remove Subset erases the subset; available only when subset contains no sheets.
Properties displays the Subset Properties dialog box; make changes, and then click **OK**.

Sheet

 *Right-click a sheet name:*

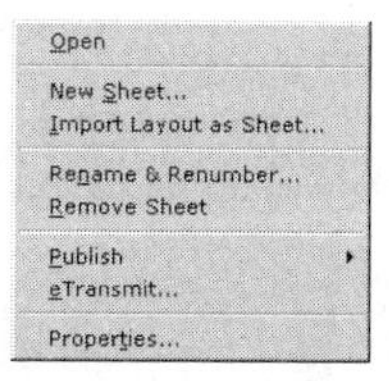

Open opens the drawing in a new window.

Rename & Renumber displays the Rename & Renumber Sheet dialog box.

Remove Sheet removes the sheet from the set; does *not* erase the drawing or layout.

Properties displays the Sheet Properties dialog box; make changes, and then click **OK**.

Sheet Views tab

This tab was formerly called "View List."

 *Right-click a sheet set name:*

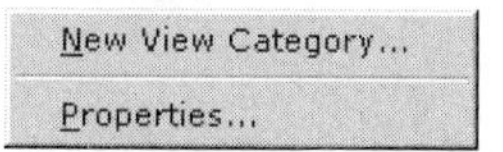

New View Category displays the View Category dialog box; enter info, and then click **OK**.

Properties displays the Sheetset Properties dialog box; make changes, and then click **OK**.

Category

 *Right-click a category name:*

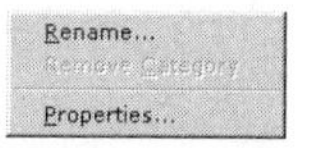

Rename renames the category.

Remove Category erases the category; available only when category is empty.

Properties displays the View Category dialog box; make changes, and then click **OK**.

View

 *Right-click a view name:*

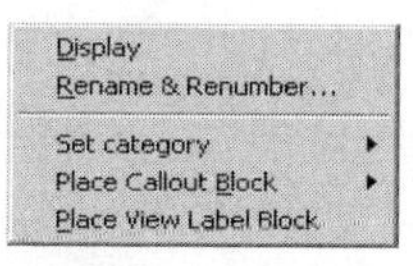

Open displays the drawing in a new window.

Rename & Renumber displays the Rename & Renumber View dialog box.

Set Category selects a view category, if defined. See the View command.

Place Callout Block:

- **Callout Bubble** places "Callout Bubble" block in drawings, with view number (1) and sheet number (A-02) field text filled in; see the **Insert** command.

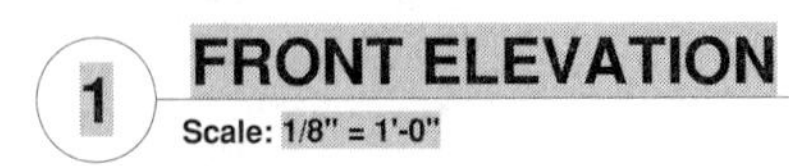

Place View Label Block places "Drawing Name" block in drawings, with view number (1), view name (Front Elevation), and viewport scale filled in as field text. (Gray background indicates field text; see the **Field** command.)

Model Views tab

This tab was formerly called "Drawing Resources."

Path

Right-click a path name:

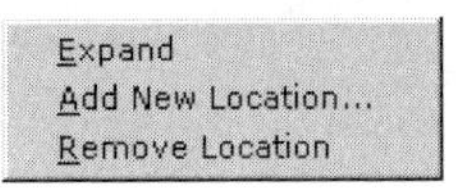

Expand displays folders, drawings, and views.

Add New Location displays the Browse for Folder dialog box; select a folder, and then click **Open**.

Remove Location deletes a location, with no warning.

Drawing

 *Right-click a drawing name:*

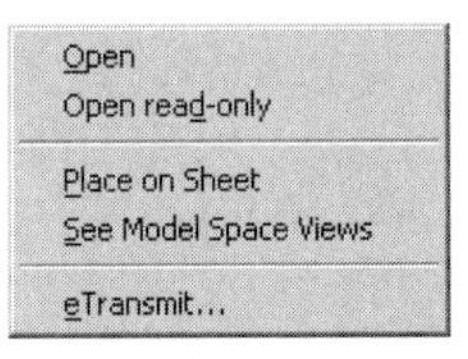

Open opens the drawing in a new window.

Open read-only opens the drawing read-only, so that changes cannot be saved.

Place on Sheet places the drawing in the current sheet; works only in layout mode.

See Model Space Views displays model space view icons, if any.

eTransmit packages drawings with support files; see the eTransmit command.

View

 *Right-click a view name:*

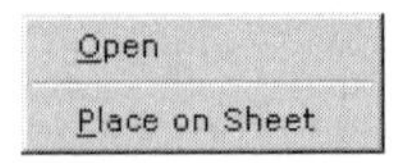

Open opens the view in a new window.

Place on Sheet places the drawing in the current sheet; works only in layout mode.

NEW SHEET / SUBSET DIALOG BOX

Similar dialog boxes are used for properties of subsets and new sheets.

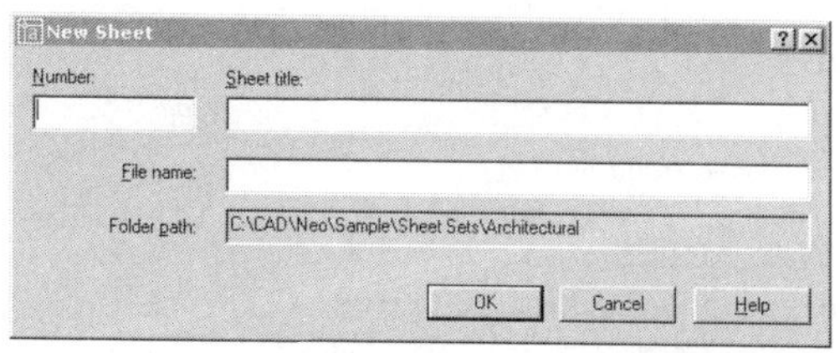

Number specifies the order number of the sheet; allows sheets to be placed in correct order.

Sheet Title names the sheet, usually with the name of the related layout tab.

File Name names the sheet file, usually consisting of the sheet number and sheet title.

SUBSET SHEET PROPERTIES DIALOG BOX

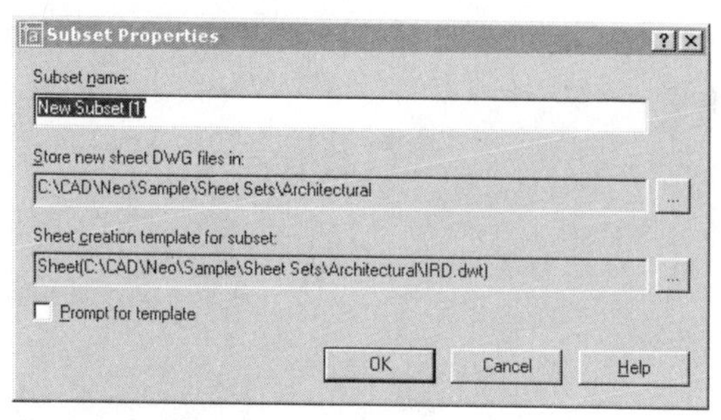

Subset name specifies the name of the subset.

Store new sheet DWG files in selects the folder for storing drawing files.

... displays Browser for Folder dialog box; select a folder, and then click **Open**.

Sheet Creation Template for Subset specifies the name of the *.dwt* template file to use.

... displays Browser for Folder dialog box; select a folder, and then click **Open**.

☐**Prompt for template** prompts for the template name later.

IMPORT LAYOUTS AS SHEETS DIALOG BOX

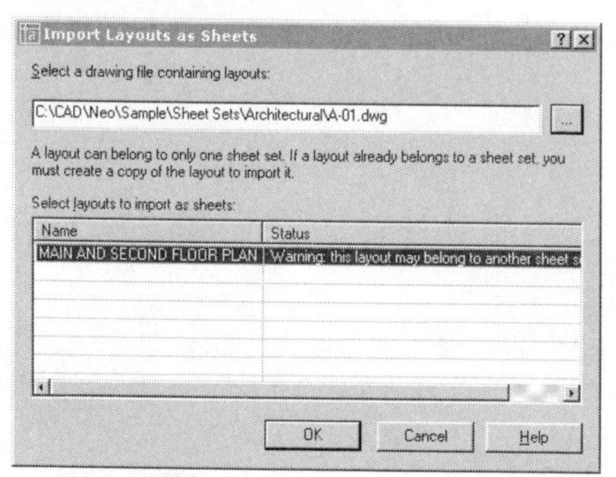

Select a drawing file containing layouts specifies the name of a drawing file.

... displays Browser for Folder dialog box; select a folder, and then click **Open**.

Select layouts to import as sheets lists the names of layouts found in the drawing file.

RELATED COMMANDS

NewSheetset creates new sheet sets.

OpenSheetset opens a specific sheet set.

RELATED SYSTEM VARIABLES

SsFound records the path and file name of current sheet set.

SsLocate toggles whether AutoCAD opens the sheet sets associated with drawings.

SsmAutoOpen toggles whether AutoCAD displays the Sheet Set Manager palette when drawings are opened.

SsmPollTime determines how often AutoCAD updates the Sheet Set Manager.

SsmSheetStatus specifies how data in sheetsets are updated.

SsmState reports whether the Sheet Set Manager palette is open.

STARTUP SWITCHES

/nossm (no Sheet Set Manager) prevents the Sheet Set Manager palette from opening at startup.

/set loads a specified *.dst* sheet set data file at startup.

RELATED FILES

.dst are sheet set data files.

TIPS

- The purpose of sheet sets is to create a hierarchy of drawings, arranged in order; as well, drawings can be organized into subsets (sheets) and categories (views).
- *SSM* is short for "sheet set manager."
- Sheet sets can be created with the NewSheetset command, based either on existing drawings or on another sheet set.
- Sheet sets, subsets, and sheets have properties, such as titles, descriptions, file paths, and custom properties.

Shell

V. 2.5 Starts a new instance of the Windows command interpreter; runs external commands defined in *acad.pgp*.

Command	Alias	Ctrl+	F-key	Alt+	Menu Bar	Tablet
shell	sh	...	...	...	...	...

Command: shell

OS Command: *(Enter a command.)*

Opens window:

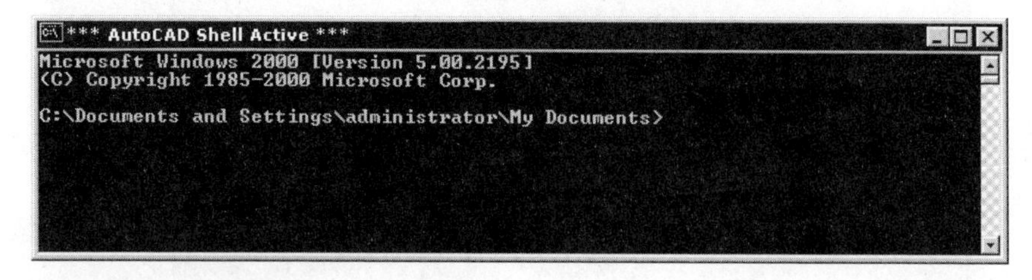

COMMAND LINE OPTIONS

OS Command specifies an operating system command, and then closes the new instance of the Windows command interpreter.

ENTER remains in the operating system's command mode for more than one command.

Exit returns to AutoCAD from the OS Command mode.

RELATED COMMAND

Quit exits AutoCAD.

RELATED FILE

acad.pgp is the external command definition file.

TIP

- This command was an early solution to the problem of no multi-tasking or task-switching in DOS, before Windows. It is rarely used today.

Aliased Command

ShowMat was removed from AutoCAD 2007; it now activates the **List** command.

SigValidate

2004 Displays digital signature information stored in drawings.

Command	Alias	Ctrl+	F-key	Alt+	Menu Bar	Tablet
sigvalidate	...	...	...	...	...	...

Command: sigvalidate

Displays dialog box:

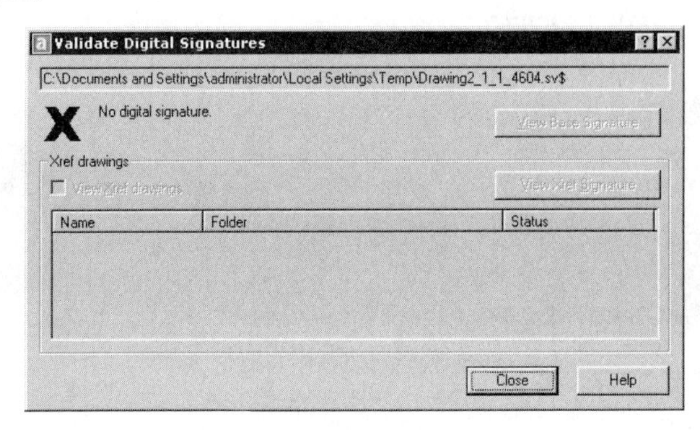

VALIDATE DIGITAL SIGNATURES DIALOG BOX

View Base Signature displays the Digital Signature Contents dialog box.

Xref Drawings

View Xref Drawings

☑ Lists xref drawings attached to the drawing.

☐ Does not list xrefs.

View Xref Signature displays the Digital Signatures Contents dialog box for the selected xref; available only when the selected xref has a digital signature.

RELATED COMMAND

SecurityOptions attaches digital signatures to drawings.

TIPS

- Digital signatures validate the authenticity of drawings.
- Digital signatures indicate whether the drawing changed since being signed.
- Autodesk notes that digital signatures become invalid for these reasons: the drawing was corrupted when the digital signature was attached; the drawing was corrupted in transit; or the digital ID is no longer valid.
- This command requires that the drawing be saved first; if not, AutoCAD reports, "This drawing has been modified. The last saved version will be validated."
- If this command does not work, it was removed from AutoCAD; re-install with the **Custom** option, and then select Drawing Encryption.

Sketch

V. 1.4 Allows freehand drawing as a series of lines or polylines.

Command	Alias	Ctrl+	F-key	Alt+	Menu Bar	Tablet
sketch	...	...	...	...	...	...

Command: sketch
Record increment <0.1000>: *(Enter a value.)*
Sketch. Pen eXit Quit Record Erase Connect .
Click pick button to begin drawing:
<Pen down>
Click pick button again to stop drawing:
<Pen up>
Press Enter to record sketching and exit ***Sketch:***
nnn **lines recorded.**

COMMAND LINE OPTIONS

Commands can be invoked by mouse and digitizer buttons:

Connect connects to the last drawing segment (*as an alternative, press button #6).*

Erase erases temporary segments as the cursor moves over them (*button #5*).

eXit records the temporary segments, and exits the Sketch command (*button #3 or Spacebar or Enter*).

Pen lifts and lowers the pen (*pick button #1*).

Quit discards temporary segments, and exits the Sketch command (*button #4 or* ESC).

Record records the temporary segments as permanent (*button #2*).

. (*Period*) connects the last segment to the current point (*button #1*).

RELATED COMMANDS

Line draws line segments.

PLine draws polyline and polyline arc segments.

RELATED SYSTEM VARIABLES

SketchInc specifies the current recording increment for the Sketch command (default = 0.1).

SKPoly controls the type of sketches recorded:

0 — Record sketches as lines (default).

1 — Record sketches as polylines.

TIPS

- During the **Sketch** command, the definitions of the pointing device's buttons change to:

Button	Meaning	Keystroke
0	Raises and lowers the ***p***en	P
1	Draws line to the current ***point***	.
2	***R***ecords the sketch	R
3	Records the sketch and e***X***its	X *or* **Enter** *or* **Spacebar**
4	Discards the sketch and ***Q***uits	Q *or* **ESC**
5	***E***rases the sketch	E
6	***C***onnects to last-drawn segment	C

- Only the first two or three button commands are available on a mouse.
- Pull-down menus are unavailable during the Sketch command.

Slice

Rel.11 Cuts 3D solids with planes, creating one or two 3D solids.

Command	Alias	Ctrl+	F-key	Alt+	Menu Bar	Tablet
slice	sl	...	...	M3S	Draw ↳3D Operations ↳Slice	...

Command: slice
Select objects: *(Select one or more solid objects.)*
Select objects: *(Press Enter to end object selection.)*
Specify first point on slicing plane by [Object/Zaxis/View/XY/YZ/ZX/3points] <3points>: *(Pick a point, or enter an option.)*
Specify a point on desired side of the plane or [keep Both sides]: *(Pick a point, or type **B**.)*

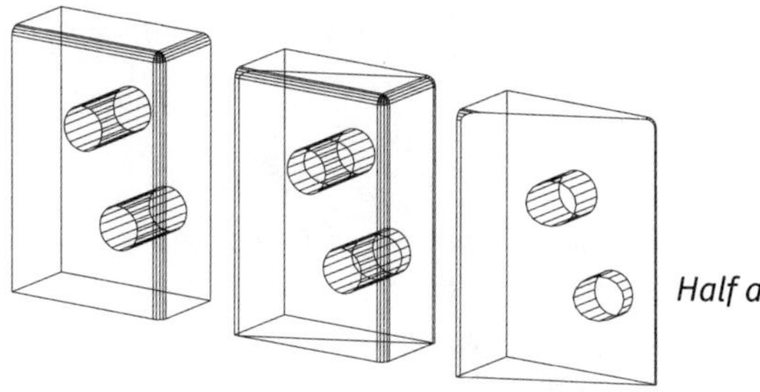

COMMAND LINE OPTIONS

Select objects selects the 3D solid model to slice.

Slicing Plane Options

Object aligns the cutting plane with a circle, ellipse, arc, elliptical arc, 2D spline, or 2D polyline.
View aligns the cutting plane with the viewing plane.
XY aligns the cutting plane with the x,y plane of the current UCS.
YZ aligns the cutting plane with the y,x plane of the current UCS.
Zaxis aligns the cutting plane with two normal points.
ZX aligns the cutting plane with the z,x plane of the current UCS.
3 points aligns the cutting plane with three points.

keep Both sides retains both halves of the cut solid model.
Specify a point on the desired side of the plane retains either half of the cut solid model.

TIPS

- This command cannot slice a 2D region, a 3D wireframe model, or other 2D shapes.
- It is helpful to use object snap when specifying points.

'Snap

V. 1.0 Sets the drawing "resolution," the origin for the grid and hatches, isometric mode, and the angle of the grid, hatches, and orthographic mode.

Command	Alias	Ctrl+	F-key	Alt+	Status Bar	Tablet
snap	sn	B	F9	...	SNAP	...

Command: snap

Specify snap spacing or [ON/OFF/Aspect/Style/Type] <0.5>: *(Enter a value, or an option.)*

COMMAND LINE OPTIONS

Snap spacing sets the snap increment.

Aspect sets separate x and y increments.

OFF turns off snap.

ON turns on snap.

Rotate rotates the crosshairs for snap and grid; a hidden option in AutoCAD 2007.

Style switches between standard and isometric style.

Type switches between grid or polar snap.

Aspect options

Specify horizontal spacing <0.5>: *(Enter a value.)*

Specify vertical spacing <0.5>: *(Enter a value.)*

Specify horizontal spacing specifies the spacing between snap points along the x-direction.

Specify vertical spacing specifies the spacing between snap points along the y-direction.

Rotate options

Specify base point <0.0,0.0>: *(Enter a value.)*

Specify rotation angle <0>: *(Enter a value.)*

Specify base point specifies the point from which snap increments are determined.

Specify rotation angle rotates the snap about the base point.

Style options

Enter snap grid style [Standard/Isometric] <S>: *(Type **S** or **I**.)*

Specify vertical spacing <0.5>: *(Enter a value.)*

Standard specifies rectangular snap.

Isometric specifies isometric snap.

Specify vertical spacing specifies the spacing between snap points along the y-direction.

Type Options

Enter snap type [Polar/Grid] <Grid>: *(Type **P** or **G**.)*

Polar specifies polar snap.

Grid specifies rectangular snap.

SHORTCUT MENU

Click **SNAP** *to turn snap on and off; right-click for shortcut menu:*

Polar Snap On turns on polar snap.

Grid Snap On turns on snap.

Off turns off snap.

Settings displays the Snap and Grid tab of the Drafting Settings dialog box.

RELATED COMMANDS

DSettings sets snap values via a dialog box.

Grid turns on the grid.

Isoplane switches to a different isometric drawing plane.

RELATED SYSTEM VARIABLES

SnapAng specifies the current angle of the snap rotation.

SnapBase specifies the base point of the snap rotation.

SnapIsoPair specifies the current isometric plane setting.

SnapMode determines whether snap is on.

SnapStyl determines the style of snap.

SnapUnit specifies the current snap increment in the x and y directions.

TIPS

- Setting the snap is setting the cursor resolution. For example, setting a snap distance of 0.1 means that when you move the cursor, it jumps in 0.1 increments. You can, however, still type in numerical values of greater resolution, such as 0.1234.
- There is no snap distance in the z-direction.
- The **Aspect** option is not available when the **Style** option is set to **Isometric**; you may, however, rotate the isometric grid.
- The **Rotate** option was removed from view in AutoCAD 2007, but still works.
- The options of the **Snap** command affect several other commands:

Snap Option	Command	Effect
Style	Ellipse	Adds **Isocircle** option to the Ellipse command.
SnapBase	Hatch	Relocates the origin of the hatch.
Rotate	Hatch	Rotates the hatching angle.
Rotate	Grid	Rotates the grid display.
Rotate	Ortho	Rotates the ortho angle.

- You can toggle snap mode by clicking the word **SNAP** on the status bar.

SolDraw

Rel.13 Creates profiles and sections in viewports created with the **SolView** command (short for SOLids DRAWing).

Command	Alias	Ctrl+	F-key	Alt+	Menu Bar	Tablet
soldraw	...	...	...	DMUD	Draw ↳Modeling ↳Setup ↳Drawing	...

Command: soldraw
Select viewports to draw ..
Select objects: *(Select one or more viewports.)*
Select objects: *(Press Enter to end object selection.)*

COMMAND LINE OPTION

Select objects selects a viewport; must be a floating viewport in paper space.

RELATED COMMANDS

SolProf creates profile images of 3D solids.

SolView creates floating viewports.

TIPS

- The SolView command must be used before this command.
- This command performs the following actions:
  1. Creates visible and hidden lines representing the silhouette and edges of solids in the viewport.
  2. Projects to a plane perpendicular to the viewing direction.
  3. Generates silhouettes and edges for all 3D solids and portions of solids behind the cutting plane.
  4. Crosshatches sectional views.
- Existing profiles and sections in the selected viewport are erased.
- All layers — except the ones needed to display the profile or section — are frozen in each viewport.
- The following layers are used by SolDraw, SolProf, and SolView: *viewname*-**VIS**, *viewname*-**HID**, and *viewname*-**HAT**.
- Hatching uses the values set in system variables HpName, HpScale, and HpAng.

Solid

V. 1.0 Draws solid-filled triangles and quadrilaterals; does *not* create 3D solids.

Command	Alias	Ctrl+	F-key	Alt+	Menu Bar	Tablet
solid	**so**	...	...	**DMM2**	**Draw** ↳**Modeling** ↳**Meshes** ↳**2D Solid**	**L8**

Command: solid
Specify first point: *(Pick a point.)*
Specify second point: *(Pick a point.)*
Specify third point: *(Pick a point.)*
Specify fourth point or <exit>: *(Pick a point, or press Enter to draw triangle.)*

Three-point solid.

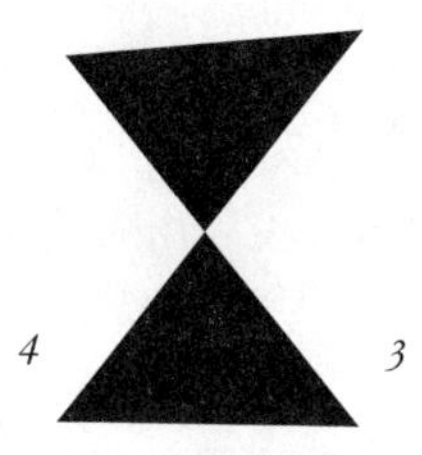

Pick order makes a difference for four-point solids.

COMMAND LINE OPTIONS

First point picks the first corner.

Second point picks the second corner.

Third point picks the third corner.

Fourth point picks the fourth corner; alternatively, press ENTER to draw triangle.

ENTER draws quadilaterials; ends the **Solid** command.

RELATED COMMANDS

Fill turns object fill off and on.

Hatch fills any shape with a solid fill pattern.

Trace draws lines with width.

PLine draws polylines and polyline arcs with width.

RELATED SYSTEM VARIABLE

FillMode determines whether solids are displayed filled or outlined.

SolidEdit

2000 Edits the faces and edges of 3D solids (short for SOLids EDITor).

Command	Alias	Ctrl+	F-key	Alt+	Menu Bar	Tablet
solidedit	...	...	...	MN	Modify ⇘Solid Editing	...

Command: solidedit
Solids editing automatic checking: SOLIDCHECK=1
Enter a solids editing option [Face/Edge/Body/Undo/eXit] <eXit>: *(Enter an option.)*

COMMAND LINE OPTIONS

Undo undoes the last editing actions, one at a time, up to the start of this command.

eXit exits body mode.

Face options

Enter a face editing option [Extrude/Move/Rotate/Offset/Taper/Delete/Copy/coLor/Undo/eXit] <eXit>: *(Enter an option.)*

Extrude extrudes one or more faces to the specified distance, or along a path; a positive value extrudes the face in the direction of its normal.

Move moves one or more faces the specified distance.

Rotate rotates one or more faces about an axis by a specified angle.

Offset offsets one or more faces by the specified distance, or through a specified point; a positive value increases the size of the solid, while a negative value decreases it.

Taper tapers one or more faces by a specified angle; a positive angle tapers in, while a negative angle tapers out.

Delete removes the selected faces; also removes attached chamfers and fillets.

Copy copies the selected faces as a 3D region or a 3D body object.

Color changes the color of the selected faces.

Edge options

Enter an edge editing option [Copy/coLor/Undo/eXit] <eXit>: *(Enter an option.)*

Copy copies the selected 3D edges as a line, arc, circle, ellipse, or spline.

coLor changes the color of the selected edges.

Body options

Enter a body editing option [Imprint/seParate solids/Shell/cLean/Check/Undo/eXit] <eXit>: *(Enter an option.)*

Imprint imprints a selection set of arcs, circles, lines, 2D and 3D polylines, ellipses, splines, regions, bodies, and 3D solids on the face of a 3D solid. This function is also available through the Imprint command (*new to AutoCAD 2007*).

seParate solids separates 3D solids into independent 3D solid objects; the solid objects must have disjointed volumes; this option does *not* separate 3D solids that were joined by a Boolean editing commands, such as Intersect, Subtract, and Union.

Shell creates a hollow, thin-walled solid of specified thickness; a positive thickness creates a shell on the inside of the solid, while a negative value creates the shell on the outside of the solid.

cLean removes redundant edges and vertices, imprints, unused geometry, shared edges, and shared vertices.

Check checks whether the object is a 3D solid; duplicates the function of the SolidCheck system variable.

RELATED SYSTEM VARIABLE

SolidCheck toggles solid validation on and off (default = on).

RELATED COMMAND

Imprint imprints the edges of objects onto 3D solid models (*new to AutoCAD 2007*).

TIP

- When working with this command, you can select a face, an edge, or an internal point on a face. Alternatively, use the **CP** (*crossing polygon*), **CW** (*crossing window*), or **F** (*fence*) object selection options.

SolProf

Rel.13 Creates profile images of 3D solids (short for SOLid PROFile).

Command	Alias	Ctrl+	F-key	Alt+	Menu Bar	Tablet
solprof	...	...	...	DMUP	Draw ↳Modeling ↳Setup ↳Profile	...

Command: solprof
Select objects: *(Select one or more objects.)*
Select objects: *(Press* **ENTER** *to end object selection.)*
Display hidden profile lines on separate layer? [Yes/No] <Y>: *(Type* **Y** *or* **N***.)*
Project profile lines onto a plane? [Yes/No] <Y>: *(Type* **Y** *or* **N***.)*
Delete tangential edges? [Yes/No] <Y>: *(Type* **Y** *or* **N***.)*
n **solids selected.**

COMMAND LINE OPTIONS

Select objects selects the objects to profile.

Display hidden profile lines on separate layer?

No specifies that all profile lines are visible; a block is created for the profile lines of every selected solid.

Yes generates just two blocks: one for visible lines and one for hidden lines.

Project profile lines onto a plane?

No creates profile lines with 3D objects.

Yes creates profile lines with 2D objects.

Delete tangential edges?

No does not display *tangential edges*, the transition line between two tangent faces.

Yes displays tangential edges.

RELATED COMMANDS

SolDraw creates profiles and sections in viewports.

SolView creates floating viewports.

TIPS

- The **SolView** command must be used before the SolProf command.
- Solids that share a common volume can produce dangling edges, if you generate profiles with hidden lines. To avoid this, first use the Union command.
- AutoCAD must display a layout, and a model space viewport must be active, before you can use the SolProf command.

SolView

Rel.13 Creates floating viewports in preparation for the SolDraw and SolProf commands (short for SOLid VIEWs).

Command	Alias	Ctrl+	F-key	Alt+	Menu Bar	Tablet
solview	**...**	**...**	**...**	**DMUV**	**Draw** ↳**Modeling** ↳**Setup** ↳**View**	**...**

Command: solview

Enter an option [Ucs/Ortho/Auxiliary/Section]: *(Enter an option.)*

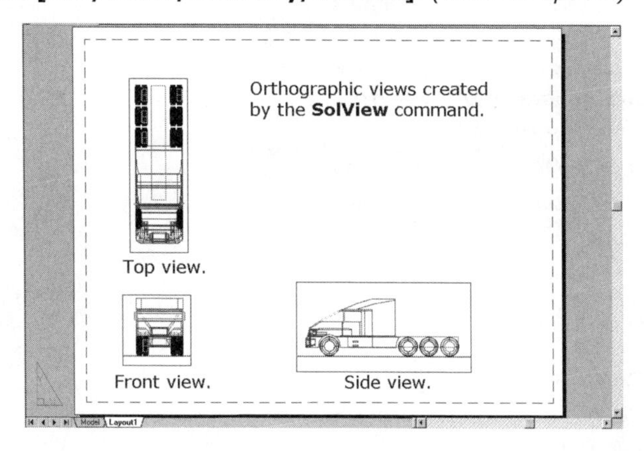

COMMAND LINE OPTIONS

Ucs options

Enter an option [Named/World/?/Current] <Current>: *(Enter an option.)*

Named creates a profile view using the x,y plane of a named UCS.

World creates a profile view using the x,y plane of the WCS.

? lists the names of existing UCSs.

Current creates a profile view using the x,y-plane of the current UCS.

Ortho options

Specify side of viewport to project: *(Pick a point.)*

Specify view center: *(Pick a point.)*

Enter view name: *(Enter a name.)*

Pick side of viewport to project selects the edge of one viewport.

View center picks the center of the view.

Clip picks two corners for a clipped view.

View name names the view.

Auxiliary options

Specify first point of inclined plane: *(Pick a point.)*

Specify second point of inclined plane: *(Pick a point.)*

Specify side to view from: *(Pick a point.)*

Specify view center: *(Pick a point.)*

Enter view name: *(Enter a name.)*

Inclined plane's 1st point picks the first point.

Inclined plane's 2nd point picks the second point.

Side to view from determines the view side.

Section options

Specify first point of cutting plane: *(Pick a point.)*

Specify second point of cutting plane: *(Pick a point.)*

Specify side to view from: *(Pick a point.)*

Enter view scale <5.9759>: *(Pick a point.)*

Specify view center <specify viewport>: *(Pick a point.)*

Specify first corner of viewport: *(Pick a point.)*

Specify opposite corner of viewport: *(Pick a point.)*

Enter view name: *(Enter a name.)*

Cutting plane 1st point picks the first point.

Cutting plane 2nd point picks the second point.

Side to view from determines the view side.

Viewscale specifies the scale of the new view.

eXit exits the command.

RELATED COMMANDS

SolDraw creates profiles and sections in viewports.

SolProf creates profile images of 3D solids.

TIPS

- This command creates orthographic, auxiliary, and sectional views.
- This command must be used before the SolDraw command, because it creates the layers required by SolDraw.
- This command is useful for creating layouts for the SolProf command.
- The layers created by this command have the following prefixes:

Layer Name	View
viewname-**VIS**	Visible lines.
viewname-**HID**	Hidden lines.
viewname-**DIM**	Dimensions.
viewname-**HAT**	Hatch patterns for sectional views.

- *Warning!* Autodesk warns that "The information stored on these layers is deleted and updated when you run SolDraw. Do not place permanent drawing information on these layers."

'SpaceTrans

2002 Converts distances between model and space units (short for SPACE TRANSlation).

Command	Alias	Ctrl+	F-key	Alt+	Menu Bar	Tablet
spacetrans	...	...	...	...	...	...

Command: spacetrans

This command does not operate in Model tab. In model view of a layout tab:

Specify paper space distance <1.000>: *(Enter a value.)*

In paper space of a layout tab:

Select a viewport: *(Pick a point.)*

Specify model space distance <1.000>: *(Enter a value.)*

COMMAND LINE OPTIONS

Specify paper space distance specifies the paper space length to be converted to its model space equivalent, usually the scale factor.

Select a viewport selects a paper space viewport.

Specify model space distance specifies the model space length to be converted to its paper space equivalent, usually the scale factor.

RELATED COMMANDS

Text places text in the drawing.

SolProf creates profile images of 3D solids.

TIPS

- This command is meant to be used transparently during other commands, and not at the 'Command:' prompt. It does not work in model tab.
- The purpose of this command is to convert lengths between model and paper space.
- Autodesk recommends using this command for converting model space text heights into paper space text heights.

'Spell

Rel.13 Checks the spelling of text in the drawing.

Command	Alias	Ctrl+	F-key	Alt+	Menu Bar	Tablet
spell	**sp**	...	...	**TE**	**Tools** ↳**Spelling**	**T10**

Command: spell
Select objects: *(Select one or more text objects.)*
Select objects: *(Press Enter to end object selection.)*

When unrecognized text is found, displays dialog box:

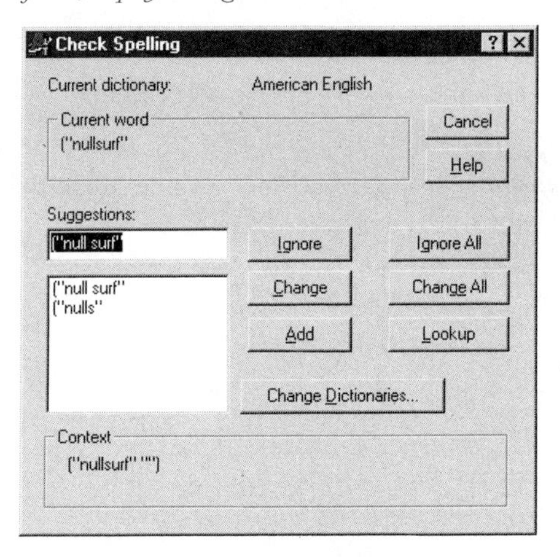

When selected text is recognized, or when spelling check is complete, displays dialog box:

CHECK SPELLING DIALOG BOX

Ignore ignores the word, and goes to the next word.

Ignore All ignores all words with this spelling.

Change changes the word to the suggested spelling.

Change All changes all words with this spelling.

Add adds the word to the user (custom) dictionary.

Lookup checks the spelling of the word in the Suggestions box.

Change dictionaries selects a different dictionary; displays the Change Dictionaries dialog box.

CHANGE DICTIONARIES DIALOG BOX

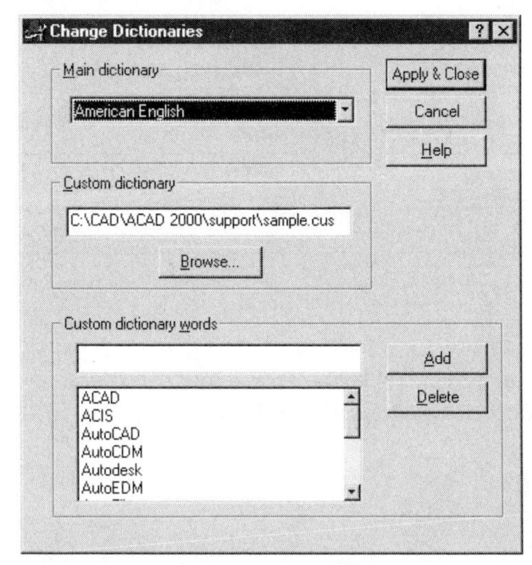

Main dictionary selects a language for the dictionary.

Custom Dictionary

Directory specifies the drive, folder, and filename of the custom dictionary.

Browse displays the **Select Custom Dictionary** dialog box.

Custom Dictionary Words

Add adds a word.

Delete removes a custom word from the dictionary.

RELATED COMMANDS

DdEdit edits text.

MText places paragraph text.

Text places lines of text in the drawing.

RELATED SYSTEM VARIABLES

DctCust names the custom spelling dictionary.

DctMain names the main spelling dictionary.

RELATED FILES

enu.dct is the dictionary word file.

.cus are the custom dictionary files.

TIPS

- A spell checker does *not* check your spelling; words that are spelled correctly but used incorrectly (such as *its* and *it's*) are not flagged. Rather, a spell checker looks for words that it does not recognize, namely words not in its dictionary file.
- As of AutoCAD 2002, the Spell command also checks words in blocks.

Sphere

Rel.11 Draws 3D spheres as solid models.

Command	Alias	Ctrl+	F-key	Alt+	Menu Bar	Tablet
sphere	...	...	...	DMS	Draw ↳Modeling ↳Sphere	K7

Command: sphere

This command's prompts have changed in AutoCAD 2007:

Specify center point or [3P/2P/Ttr]: *(Pick a point, or enter an option.)*

Specify radius or [Diameter] <1.0>: *(Enter a value, or type* **D***.)*

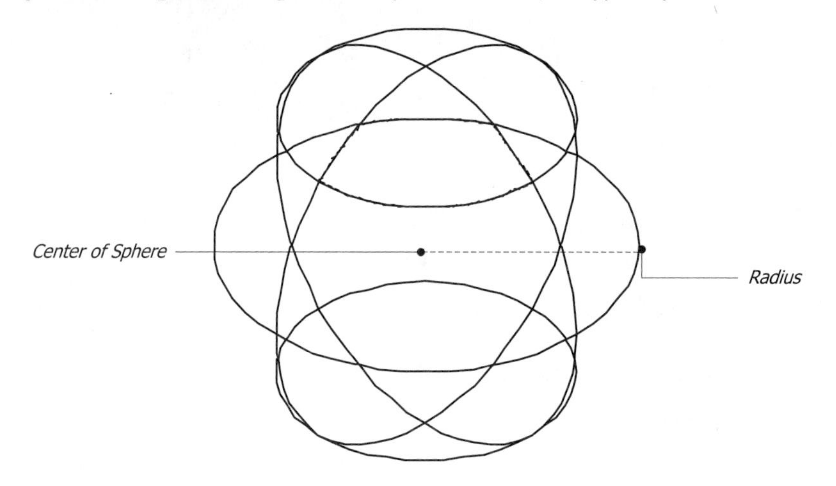

COMMAND LINE OPTIONS

Specify center point locates the center point of the sphere.

3P specifies three points on the outer circumference (equator) of the sphere (*new to AutoCAD 2007*).

2P specifies two points on the outer diameter of the sphere (*new to AutoCAD 2007*).

Ttr specifies two tangent points plus the radius of the outer circumference of the sphere (*new to AutoCAD 2007*).

Radius specifies the radius of the sphere.

Diameter specifies the diameter of the sphere.

RELATED COMMANDS

Cone draws solid cones.

Cylinder draws solid cylinders.

Torus draws solid tori.

Ai_Sphere draws surface meshed spheres.

RELATED SYSTEM VARIABLES

DragVs specifies the visual style during creation of solids.

DispSilh toggles the silhouette display of 3D solids.

IsoLines specifies the number of tessellation lines that define the surface of curved 3D solids; range is from 0 to 2047.

TIPS

- This command places the sphere's central axis parallel to the z axis of the current UCS, with the latitudinal isolines parallel to the x,y plane.
- Notice the effect of the DispSilh system variable, which toggles the silhouette display of 3D solids, after executing the Hide command:

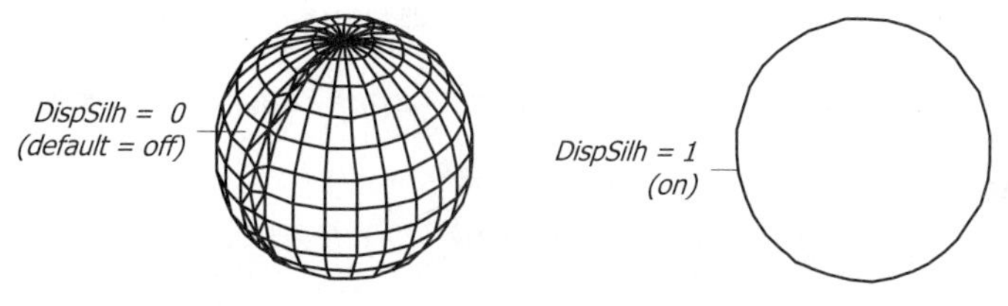

- You must use the Regen command after changing the DispSilh and IsoLines system variables.
- Notice the effect of the IsoLines system variable, which controls the number of tessellation lines used to define the surface of a 3D solid:

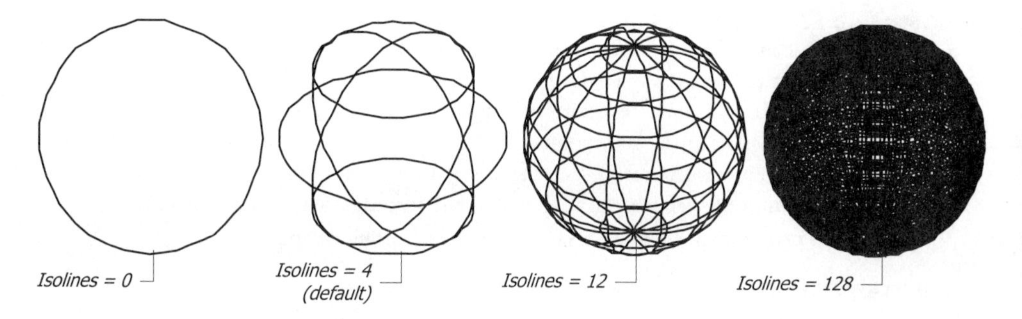

- Spheres can be edited using grips:

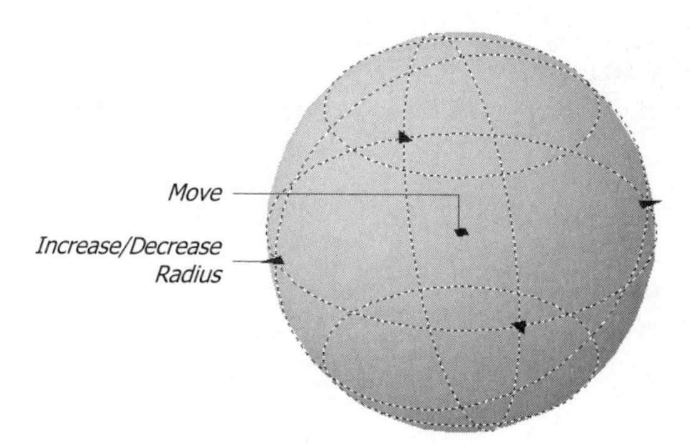

Spline

Rel.13 Draws NURBS (Non-Uniform Rational Bezier Spline) curves.

Command	Alias	Ctrl+	F-key	Alt+	Menu Bar	Tablet
spline	spl	...	...	DS	Draw ↳Spline	L9

Command: spline
Specify first point or [Object]: *(Pick a point, or type* **O***.)*
Specify next point: *(Pick a point.)*
Specify next point or [Close/Fit tolerance] <start tangent>: *(Pick a point, or enter an option.)*
Specify next point or [Close/Fit tolerance] <start tangent>: *(Press Enter to define tangency points.)*
Specify start tangent: *(Pick a point, or press Enter .)*
Specify end tangent: *(Pick a point, or press Enter ..)*

Open spline

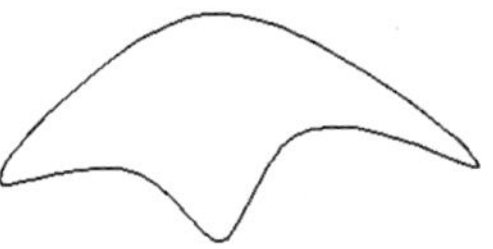

Closed spline

COMMAND LINE OPTIONS

Specify first point picks the starting point of the spline.
Object converts 2D and 3D splined polylines into a NURBS spline.
Specify next point picks the next tangent point.
Close closes the spline at the start point.
Fit changes the spline tolerance; 0 = curve passes through fit points.
Specify start tangent specifies the tangency of the starting point of the spline.
Specify end tangent specifies the tangency of the endpoint of the spline.

RELATED COMMANDS

PEdit edits splined polylines.
PLine draws splined polylines.
SplinEdit edits NURBS splines.

RELATED SYSTEM VARIABLE

DelObj toggles whether the original polyline is deleted with the **Object** option.

TIP

- A closed spline has the same start and end tangent.

SplinEdit

Rel.13 Edits NURBS splines.

Command	Alias	Ctrl+	F-key	Alt+	Menu Bar	Tablet
splinedit	spe	...	...	MOS	Modify ↳Object ↳Spline	Y18

Command: splinedit

Select spline: *(Select one spline object.)*

Enter an option [Fit data/Close/Move vertex/Refine/rEverse/Undo]: *(Enter an option.)*

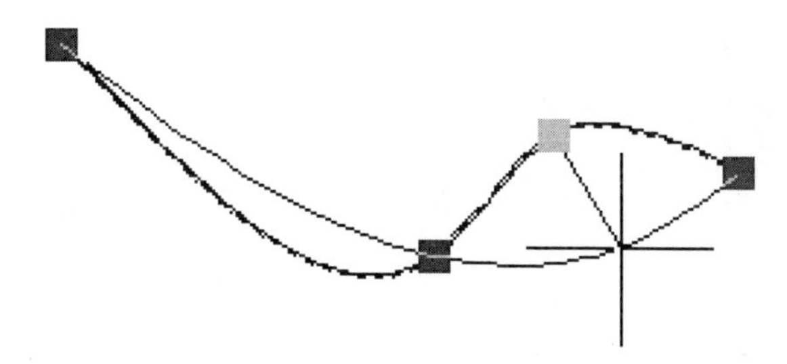

COMMAND LINE OPTIONS

Close closes the spline, if open.

Move vertex moves a control vertex.

Open opens the spline, if closed.

Refine adds a control point; changes the spline's order or weight.

rEverse reverses the spline's direction.

Undo undoes the most-recent edit change.

eXit exits the command.

Fit Data Options

Enter a fit data option [Add/Close/Delete/Move/Purge/Tangents/toLerance/eXit] <eXit>: *(Enter an option.)*

Add adds fit points.

Close closes the spline, if open.

Delete removes fit points.

Move moves fit points.

Open opens the spline, if closed.

Purge removes fit point data from the drawing.

Tangents edits the start and end tangents.

toLerance refits the spline with the new tolerance value.

eXit exits suboptions.

RELATED COMMANDS

PEdit edits a splined polyline.

PLine creates polylines, including splined polylines.

Spline draws a NURBS spline.

TIPS

- The spline loses its fit data when you use these SplinEdit options:
  - **Refine.**
  - **Fit Purge.**
  - **Fit Tolerance** followed by **Fit Move.**
  - **Fit Tolerance** followed by **Fit Open** or **Fit Close**.
- The maximum order for a spline is 26; once the order has been elevated, it cannot be reduced.
- The larger the weight, the closer the spline is to the control point.
- This command automatically converts a spline-fit polyline into a spline object, even if you do not edit the polyline.

Aliased Command

Spotlight is also handled by the **Light** command.

Standards

2002 Loads standards into the current drawing.

Command	Alias	Ctrl+	F-key	Alt+	Menu Bar	Tablet
standards	**sta**	...	...	**TSC**	**Tools** **↳CAD Standards** **↳Configure**	...

Command: standards

Displays dialog box:

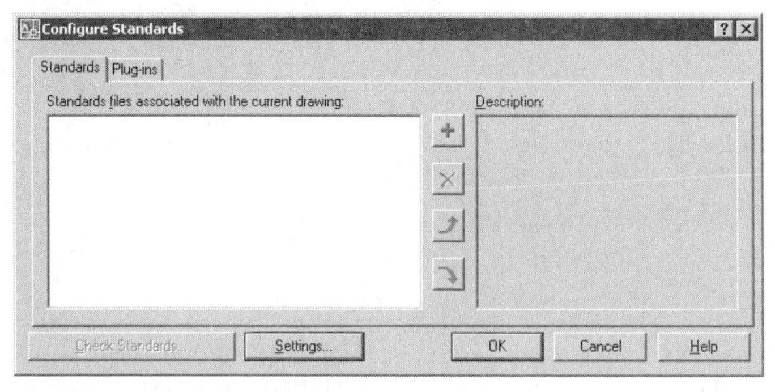

To add standards to the drawing, click the + icon, and then select a .dws file.

CONFIGURE STANDARDS DIALOG BOX

+ **Add** attaches a *.dws* standards file to the drawing (**F3**); displays the Select Standards File dialog box; more than one *.dws* file can be attached.

× **Remove Standards** removes the selected standards file (DEL).

Move up moves the selected *.dws* standards file higher in the list; available only when two or more standards files are loaded.

Move down moves the selected standards file lower in the list.

Check Standards displays the Check Standards dialog box; see the CheckStandards command.

Settings displays CAD Standards Settings dialog box. See CheckStandards.

Once a standard has been attached to the drawing, the dialog box looks like this:.

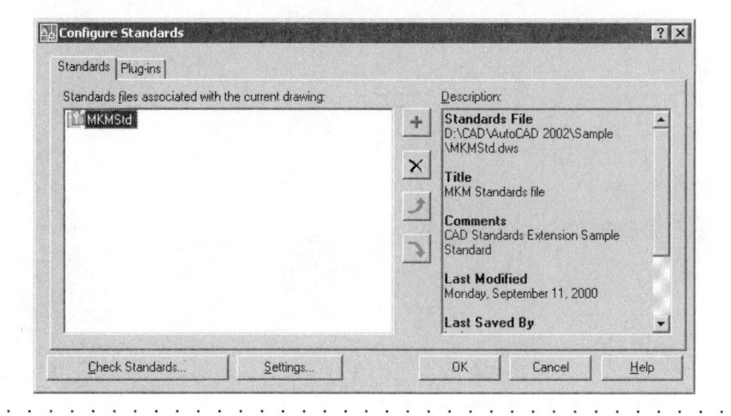

Plug-ins tab

Displays information about each item in the CAD standard:

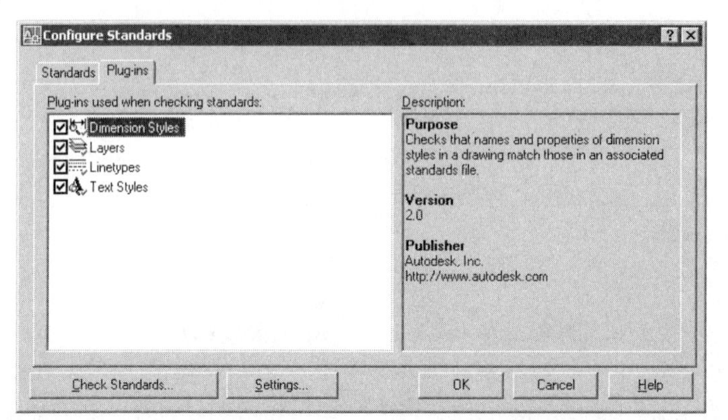

☑ **Dimension Styles** checks if the drawing's dimension styles conform; click to turn off checking.

☑ **Layers** checks if the drawing's layer properties styles conform.

☑ **Linetypes** checks if the drawing's linetype definitions styles conform.

☑ **Text Styles** checks if the drawing's text styles conform.

RELATED COMMAND

CheckStandards checks the drawing against standards loaded by this command.

RELATED FILE

* ***.dws*** are AutoCAD standards files stored in DWG format.

Removed Command

Stats was removed from AutoCAD 2007.

'Status

V. 1.0 Displays information about the current drawing and environment.

Command	Alias	Ctrl+	F-key	Alt+	Menu Bar	Tablet
status	...	...	...	**TQS**	**Tools** **↳Inquiry** **↳Status**	...

Command: status

Example output:

```
223 objects in Drawing6.dwg
Model space limits are X:     0.0000   Y:     0.0000  (Off)
                       X:    12.0000   Y:     9.0000
Model space uses       X:    17.7037   Y:    20.3182
                       X:    38.1227   Y:    31.1948 **Over
Display shows          X:    16.4004   Y:    17.8401
                       X:    39.5708   Y:    34.9249
Insertion base is      X:     0.0000   Y:     0.0000   Z:     0.0000
Snap resolution is     X:     0.5000   Y:     0.5000
Grid spacing is        X:     0.5000   Y:     0.5000

Current space:         Model space
Current layout:        Model
Current layer:         "0"
Current color:         BYLAYER -- 7 (white)
Current linetype:      BYLAYER -- "Continuous"
Current material:      BYLAYER -- "Global"
Current lineweight:    BYLAYER
Current elevation:     0.0000  thickness:     0.0000
Fill on  Grid off  Ortho on  Qtext off  Snap off  Tablet off
Object snap modes:     Center, Endpoint, Insert, Intersection, Midpoint,
                       Nearest, Node, Perpendicular, Quadrant, Tangent,
                       Appint, Extension, Parallel
Free dwg disk (D:) space: 2344.3 MBytes
Free temp disk (C:) space: 3130.4 MBytes
Press ENTER to continue:
Free physical memory: 209.4 Mbytes (out of 1023.5M).
Free swap file space: 857.8 Mbytes (out of 1694.3M).
```

COMMAND LINE OPTIONS

Enter continues the listing.

F2 returns to the graphics screen.

RELATED COMMANDS

DbList lists information about all the objects in the drawing.

List lists information about the selected objects.

Properties lists information about selected objects in a palette.

DEFINITIONS

Model Space limits, Paper Space limits – the x,y coordinates stored in the LimMin and LimMax system variables; 'Off' indicates limits checking is turned off (LimCheck).

Model Space uses, Paper Space uses – the x,y coordinates of the lower-left and upper-right extents of objects in the drawing; 'Over' indicates drawing extents exceed the drawing limits.

Display shows -- the x,y coordinates of the lower-left and upper-right corners of the current display.

Insertion base is – the x,y,z coordinates, as stored in system variable InsBase.

Snap resolution is – the snap settings, as stored in the SnapUnit system variable.

Grid spacing is – the grid settings, as stored in the GridUnit system variable.

Current space – the indication of whether model space or paper space is current.

Current layout – the name of the current layout.

Current layer, Current color, Current linetype, Current lineweight, Current plot style, Current elevation, Material Thickness – the current values for the layer name, color, linetype name, elevation, and thickness, as stored in system variables CLayer, CeColor, CeLType, CeLweight, CMaterial, CPlotSytle, Elevation, and Thickness.

Fill, Grid, Ortho, Qtext, Snap, Tablet – the current settings for the fill, grid, ortho, qtext, snap, and tablet modes, as stored in the system variables FillMode, GridMode, OrthoMode, TextMode, SnapMode, and TabMode.

Object Snap modes – the currently-set object modes, as stored in the system variable OsMode.

Free disk (dwg + temp = C) – the amount of free disk space on the drive storing AutoCAD's temporary files, as held by system variable TempPrefix.

Free physical memory – the amount of free RAM.

Free swap file space – the amount of free space in AutoCAD's swap file on disk.

StlOut

Rel.12 Exports 3D solids and bodies in binary or ASCII SLA format (short for STereoLithography OUTput).

Command	Alias	Ctrl+	F-key	Alt+	Menu Bar	Tablet
stlout	...	...	...	FE ↳STL	File ↳Export ↳Lithography	...

Command: stlout
Select a single solid for STL output...
Select objects: *(Select one or more solid objects.)*
Select objects: *(Press Enter to end object selection.)*
Create a binary STL file? [Yes/No] <Y>: *(Type* **Y** *or* **N***.)*

Displays the Create STL File dialog box; enter a name, and then click ***Save***.

COMMAND LINE OPTIONS

Select objects selects a single 3D solid object.

Y creates a binary-format *.sla* file.

N creates an ASCII-format *.sla* file.

RELATED COMMANDS

All solid modeling commands.

AcisOut exports 3D solid models to an ASCII SAT-format ACIS file.

AmeConvert converts AME v2.x solid models into ACIS models.

RELATED SYSTEM VARIABLE

FaceTRes determines the resolution of triangulating solid models.

RELATED FILE

***.stl** is the SLA-compatible file with STL extension created by this command.

TIPS

- The solid model must lie in the positive x,y,z-octant of the WCS.
- This command exports a single 3D solid; it does not export regions or any other AutoCAD object.
- Even though this command prompts you twice to 'Select objects', selecting more than one solid causes AutoCAD to complain, "Only one solid per file permitted."
- The resulting *.stl* file cannot be imported back into AutoCAD.

DEFINITIONS

STL — STereoLithography data file, a faceted representation of the model.

SLA — StereoLithography Apparatus.

 Stretch

V. 2.5 Lengthens, shortens, or distorts objects.

Command	Alias	Ctrl+	F-key	Alt+	Menu Bar	Tablet
stretch	**s**	...	...	**MH**	**Modify ↳Stretch**	**V22**

Command: stretch
Select objects to stretch by crossing-window or crossing-polygon...
Select objects: *(Type* **C** *or* **CP***.)*
Specify first corner: *(Pick a point.)*
Specify opposite corner: *(Pick a point.)*
Select objects: *(Press Enter to end object selection.)*
Specify base point or displacement: *(Pick a point.)*
Specify second point of displacement or <use first point as displacement>:
(Pick a point.)

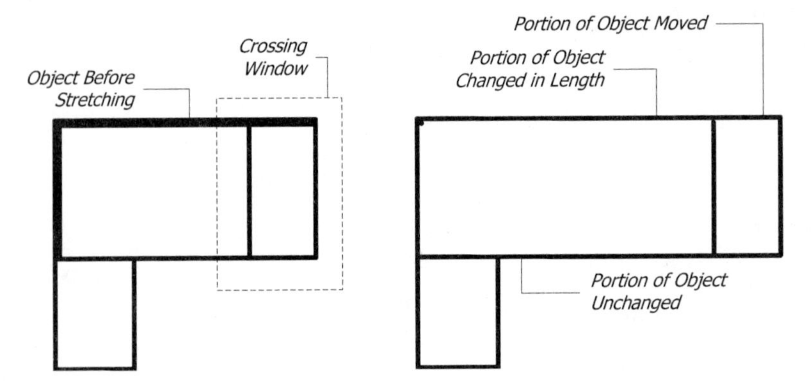

COMMAND LINE OPTIONS

First corner selects object; must be **CPolygon** or **Crossing** object selection.

Select objects selects other objects using any selection mode.

Base point indicates the starting point for stretching.

Second point makes the object larger or smaller.

RELATED COMMANDS

Lengthen changes the size of open objects.

Scale increases or decreases the size of objects.

TIPS

- The effect of Stretch on objects is not always obvious; be prepared to use the U command.
- Objects entirely within the selection window are moved; objects crossing the selection window are stretched.
- This command will not move a hatch pattern unless the hatch's origin is included in the selection set. This command does not stretch 3D solids, nor the width and tangents of polylines — although it does move them.

'Style

V. 2.0 Creates and modifies text styles, which define the properties of fonts.

Commands	Aliases	Ctrl+	F-key	Alt+	Menu Bar	Tablet
style	st	...	...	OS	Format	U2
	ddstyle				↳Text Style	
-style						

Command: style

Displays dialog box:

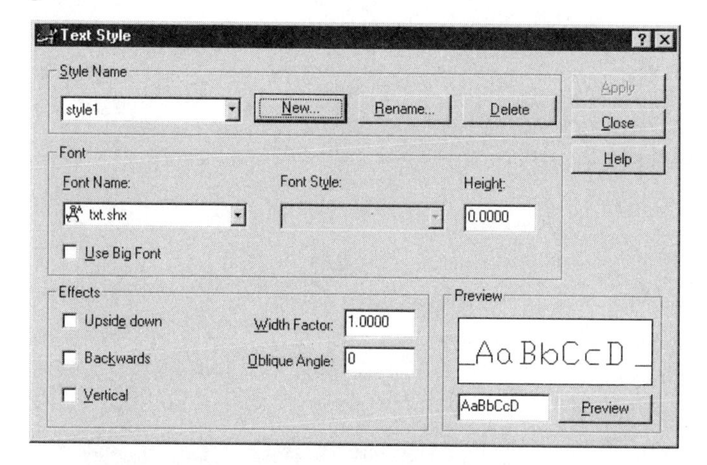

TEXT STYLE DIALOG BOX

Style Name options

Style Name selects an existing text style.

New creates new text styles; displays dialog box:

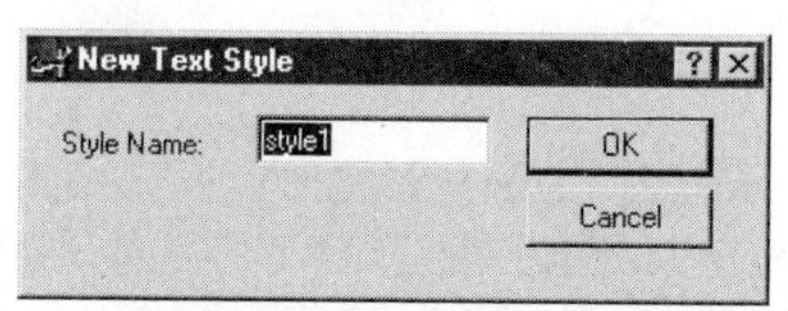

Rename renames existing text styles; displays dialog box:

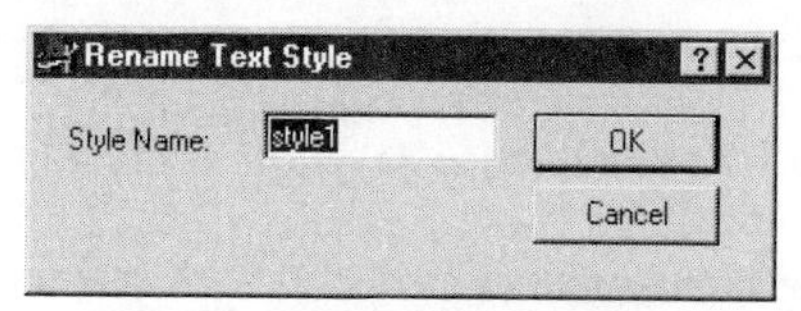

Delete deletes text styles.

Font

Font Name specifies the names of AutoCAD SHX and TrueType TTF fonts.

Font Style selects from available font styles, such as **Bold**, *Italic*, and ***Bold Italic***.

Height specifies the text height.

☐ **Use Big Font** specifies the use of a big font file, typically for Asian alphabets.

Effects *(not available for all fonts)*

☐ **Upside Down** draws text upside down:

☐ **Backwards** draws text backwards:

☐ **Vertical** draws text vertically.

Width Factor changes the width of characters.

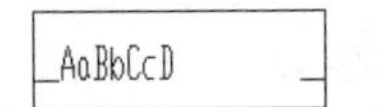

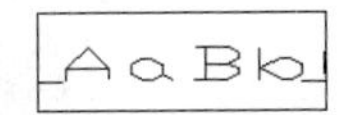

Left: *Width factor = 0.5.* ***Right***: *Width factor = 2.0.*

Oblique Angle slants characters forward or backward:

Left: *Oblique angle = 30.* ***Right***: *Oblique angle = -30.*

Preview previews the effects on the style.

Apply applies the changes to the style.

Close closes the dialog box; in some cases, you can click the **Close** button before clicking the **Apply** button; then AutoCAD displays the following warning dialog box:

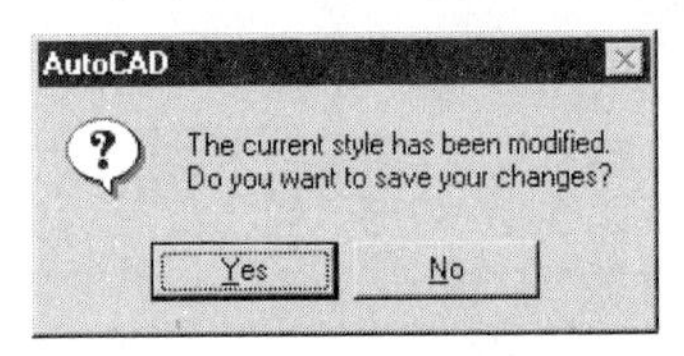

-STYLE Command

Command: -style

Enter name of text style or [?] <STANDARD>: *(Enter a name, or type* **?***.)*

Specify full font name or font filename (TTF or SHX) <txt>: *(Enter a name.)*

Specify height of text <0.0000>: *(Enter a value.)*

Specify width factor <1.0000>: *(Enter a value.)*

Specify obliquing angle <0>: *(Enter a value.)*

Display text backwards? [Yes/No] <N>: *(Type* **Y** *or* **N***.)*

Display text upside-down? [Yes/No] <N>: *(Type* **Y** *or* **N***.)*

Vertical? <N> *(Type* **Y** *or* **N***.)*

"STANDARD" is now the current text style.

COMMAND LINE OPTIONS

Enter name of text style names the text style; maximum = 31 characters (default = "STANDARD").

? lists the names of styles already defined in the drawing.

Specify full font name or font filename names the font file (SHX or TTF) from which the style is defined (default = *txt.shx*).

Specify height of text specifies the height of the text (default = 0 units).

Specify width factor specifies the width factor of the text (default = 1.00).

Specify obliquing angle specifies the obliquing angle or slant of the text (default = 0 degrees).

Display text backwards

Yes prints text backwards — mirror writing.

No prints text forwards.

Display text upside-down

Yes prints text upside-down.

No prints text rightside-up (default).

Vertical

Yes prints text vertically; not available for all fonts.

No prints text horizontally (default).

RELATED COMMANDS

Change changes the style assigned to selected text.

Purge removes any unused text style definitions.

Rename renames a text style.

MText places paragraph text.

Text places a single line of text.

RELATED SYSTEM VARIABLES

TextStyle specifies the current text style.

TextSize specifies the current text height.

RELATED FILES

.shp is Autodesk's format for vector source fonts.

.shx is Autodesk's format for compiled vector fonts; stored in *autocad 2007**fonts* folder.

.ttf are TrueType font files; stored in *windows**fonts* folder.

TIPS

- The Style command affects the font used with the Text and MText commands, as well as with dimension and table styles.
- A width factor of 0.85 fits in 15% more text without sacrificing legibility.
- The obliquing angle can be positive (forward slanting) or negative (backward slanting).
- To use PostScript fonts in drawings, use Compile to convert PostScript *.pfb* files into *.shx* format.
- A text height of 0 lets you specify the text height while you are adding text with the Text, DText, and MText commands. If you specify a text height other than 0 in the Style command, that height is always used with that particular style name.
- Text styles used in dimensions styles should have a height of 0 to allow DimScale to work.

StylesManager

2000 Displays the Plot Styles window.

Command	Alias	Ctrl+	F-key	Alt+	Menu Bar	Tablet
stylesmanager	...	...	...	**FY**	**File** ↳**Plot Style Manager**	**Y24**

You can also access this command through the Windows Control Panel: ***Autodesk Plot Style Manager.***

Command: stylesmanager

Displays window:

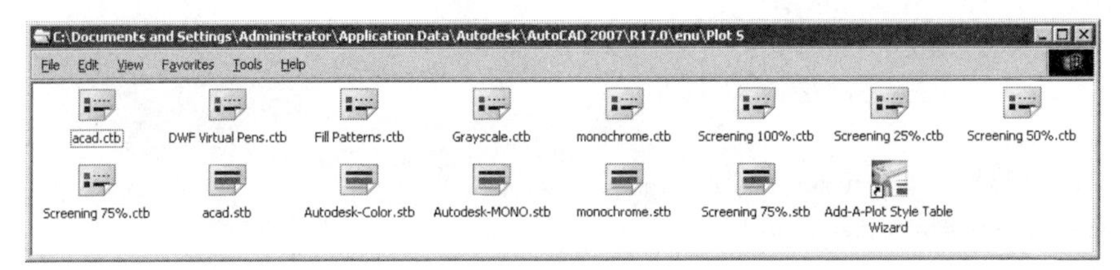

PLOT STYLES WINDOW

***.ctb** (color-table based) opens the Plot Style Table Editor dialog box.

*** .stb** (style-table based) also opens the Plot Style Table Editor dialog box.

Add-A-Plot Style Table Wizard opens Add Plot Style Table wizard; see the R14PenWizard command.

DEFINITIONS

StylesManager — the program that modifies plot styles stored in plot style tables.

Plot styles — when drawings have a plot style table attached to their model and layout tabs, any changes to the plot style changes the object using the plot style. Plot styles can be assigned by object or by layer. See the PlotStyle and Layer commands.

Color-table based (.cbt) — assigns colors to objects and layers, as used by older releases of AutoCAD. For example, the color of the object specifies the width of pen. This style of controlling the plot is now called "color dependent."

Style-table based (.stb) — newer releases of AutoCAD can control every aspect of the plot through "plot styles." By changing the *.stb* file attached to a layout tab, you immediately change the plot style for all objects and layers in the layout. This, for example, allows a quick change from monochrome to color plotting. A single drawing file can contain multiple plot styles.

PLOT STYLE TABLE EDITOR DIALOG BOX

General Tab

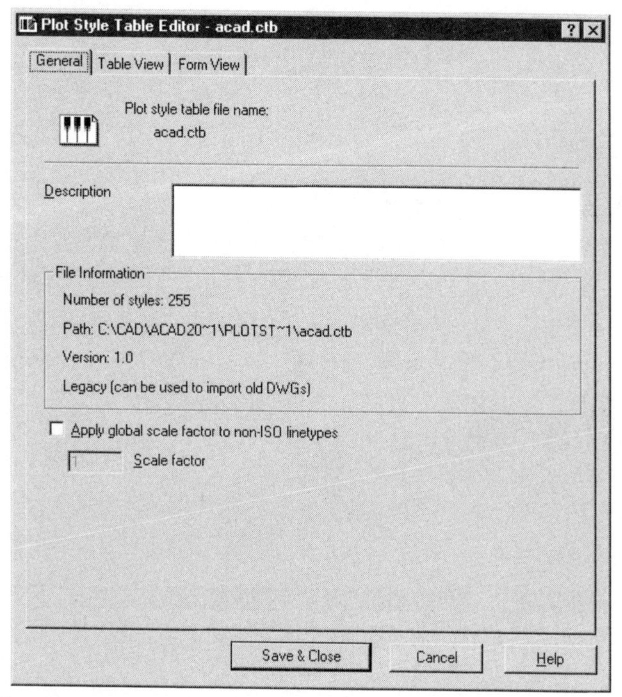

Description describes the plot style.

☐ **Apply global scale factor to non-ISO linetypes and fill patterns** applies the scale factor to all non-ISO linetypes and hatch patterns in the plot.

Scale factor specifies the scale factor.

Save and Close saves the changes, and closes the dialog box.

Cancel cancels the changes, and closes the dialog box.

CBT File Table View Tab

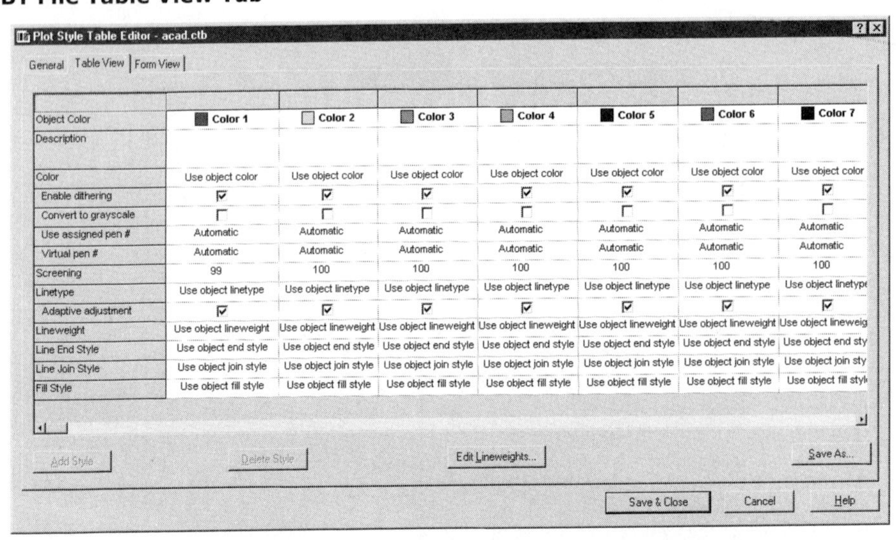

See the following pages for options.

CBT File Form View Tab

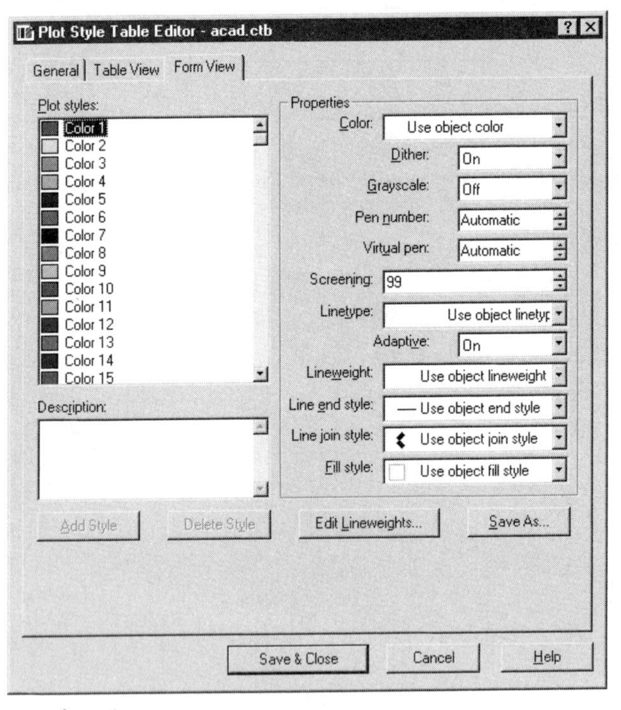

See the following pages for options.

STB File Table View Tab

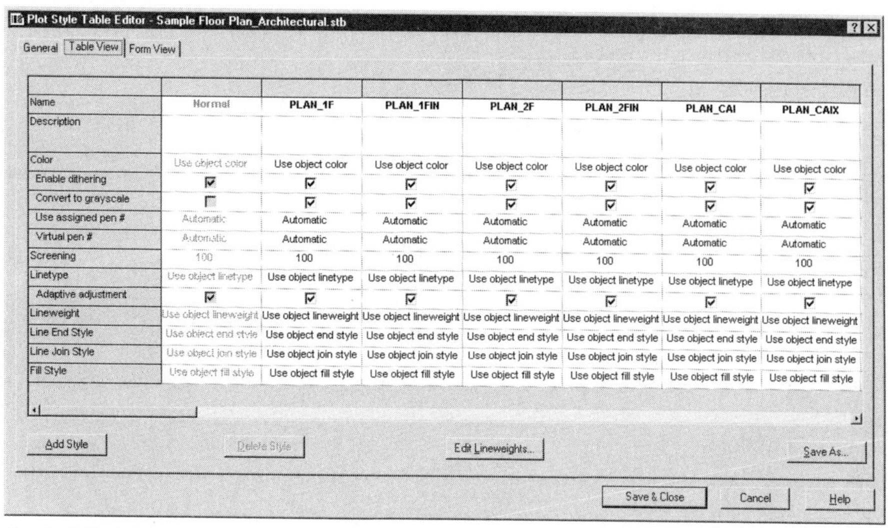

See the following pages for options.

STB File Form View Tab

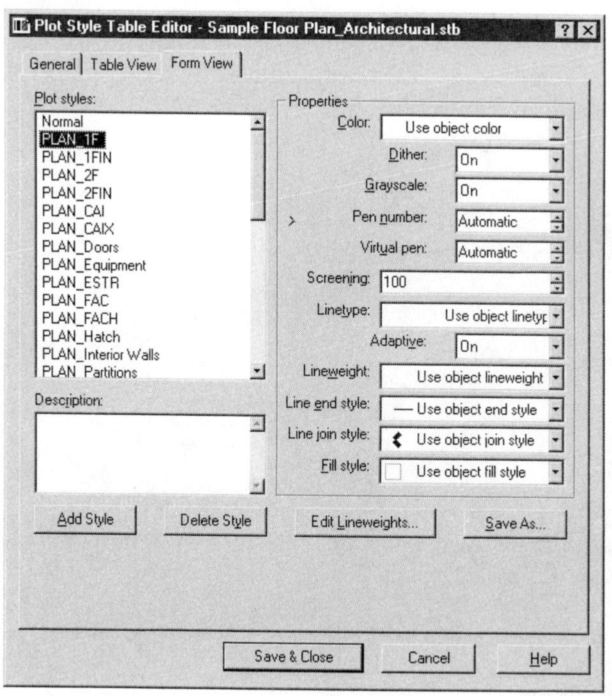

See the following pages for options.

Table View and Form View options are identical

Object color specifies the color of the object.

Description describes the plot style.

Color specifies the plotted color for the objects.

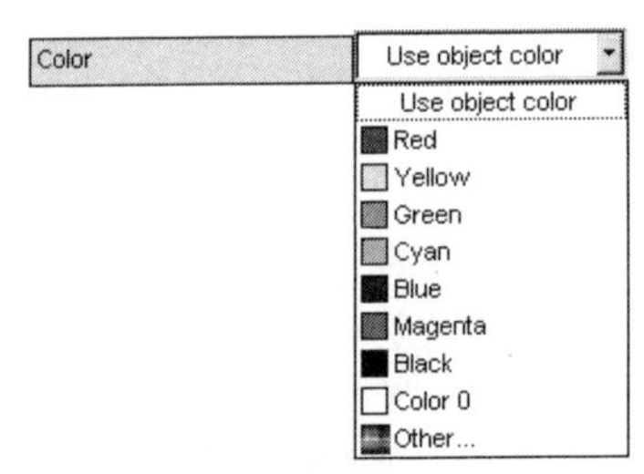

Enable dithering toggles dithering, if the plotter supports dithering, to generate more colors than the plotter is capable of; this setting should be turned off for plotting vectors, and turned on for plotting renderings.

Convert to grayscale converts colors to shades of gray, if the plotter supports gray scale.

Use assigned pen # specifies the pen number of pen plotters; range is 1 to 32.

Virtual pen number specifies the virtual pen number (default = Automatic); range is 1 to 255. A value of 0 (or Automatic) tells AutoCAD to assign virtual pens from ACI (AutoCAD Color Index); this setting is meant for non-pen plotters that can make use of virtual pens.

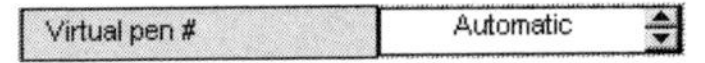

Screening specifies the intensity of plotted objects; range is 0 (plotted "white") to 100 (full density):

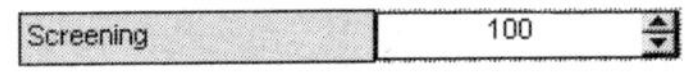

Linetype specifies the linetype with which to plot the objects:

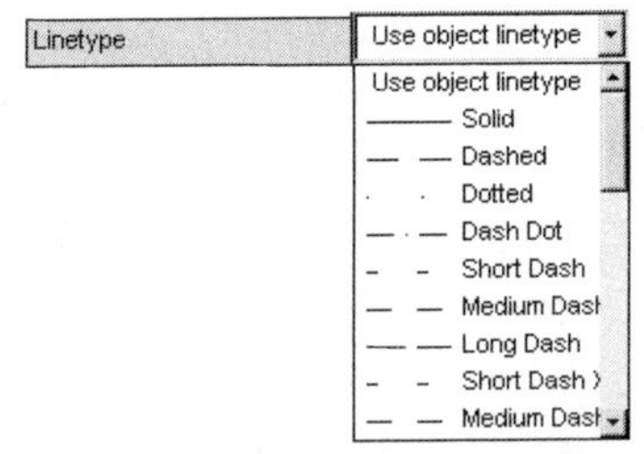

Adaptive adjustment adjusts linetype scale to prevent the linetype from ending in the middle of its pattern; keep off when the plotted linetype scale is crucial.

Lineweight specifies how wide lines are plotted, in millimeters:

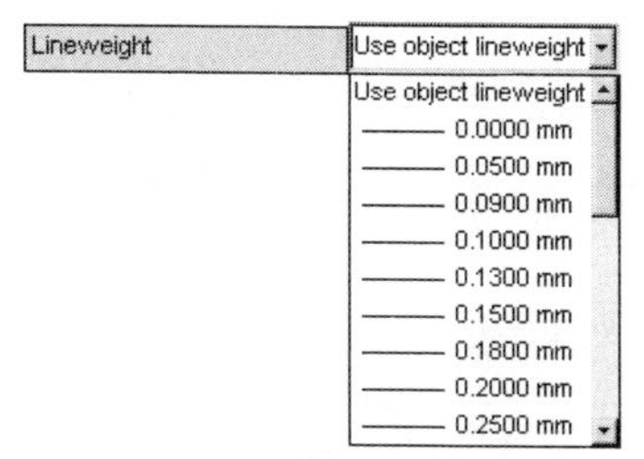

Line end style specifies how the ends of lines are plotted:

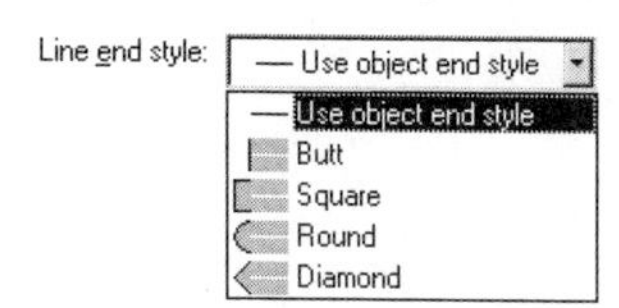

Line join style specifies how the intersections of lines are plotted:

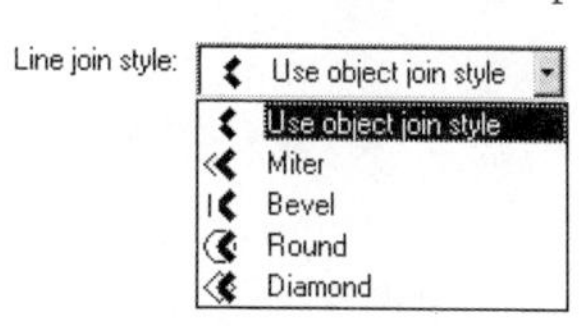

Fill style specifies how objects are filled:

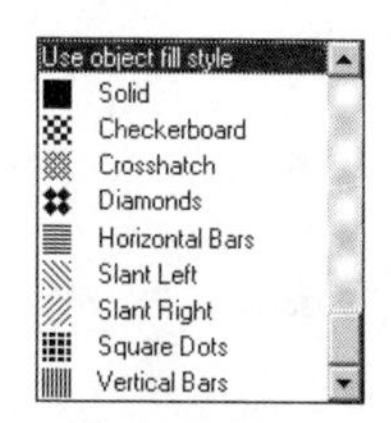

Add Style adds a plot style.

Delete Style removes a plot style.

Edit Lineweights displays the Edit Lineweights dialog box.

Save As displays the Save As dialog box.

TIPS

- To configure a printer or plotter for *virtual pens*:
    1. In the **Device and Document Settings** tab, open the PC3 Editor dialog box.
    2. In the **Vector Graphics** section, select **255 Virtual Pens** from **Color Depth**.
- **CTB** is short for "color-dependent based" style table, which is compatible with earlier versions of AutoCAD.
- **STB** is short for "style-table based."
- You can attach a different *.stb* file to each layout and to the model tab in a drawing.

Subtract

Rel.11 Removes the volume of one set of 3D models or 2D regions from another.

Command	Alias	Ctrl+	F-key	Alt+	Menu Bar	Tablet
subtract	su	...	...	MNS	Modify ↳Solids Editing ↳Subtract	X16

Command: subtract
Select objects: *(Select one or more solid objects.)*
Select objects: *(Press Enter to end object selection.)*
1 solid selected.
Objects to subtract from them...
Select objects: *(Select one or more solid objects.)*
Select objects: *(Press Enter to end object selection.)*
1 solid selected.

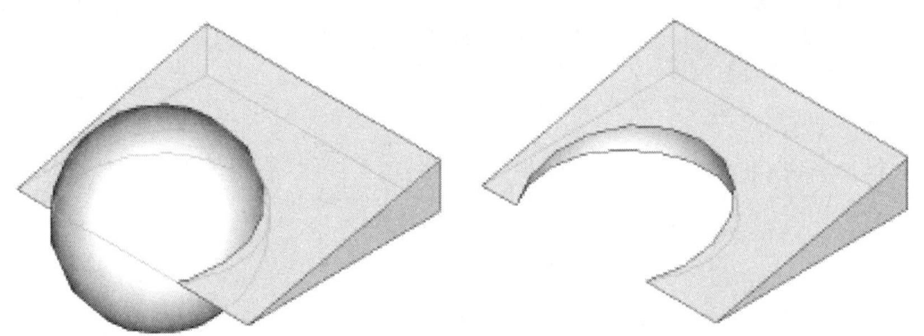

Left: *Original objects.* ***Right:*** *Sphere subtracted from wedge.*

COMMAND LINE OPTION

Select objects selects the objects to be subtracted.

RELATED COMMANDS

Interfere find the volume common to two or more solids.

Intersect removes all but the intersection of two solid volumes.

Union joins two solids.

RELATED SYSTEM VARIABLE

ShowHist toggles the display of solids' history

TIPS

- AutoCAD subtracts the objects you select *second* from the objects you select *first.*
- You can use this commands on 3D solids and 2D regions.
- To subtract one region from another, both must lie in the same plane.
- When one solid is fully inside another, AutoCAD performs the subtraction, but reports, "Null solid created — deleted."

 SunProperties, SunPropertiesClose

2007 Opens and closes the Sun Properties palette for controlling the sun light.

Command	Alias	Ctrl+	F-key	Alt+	Menu Bar	Tablet
sunproperties	...	...	...	**VELU**	**View** ↳**Render** ↳**Light** ↳**Sun Properties**	...
sunpropertiesclose						

Command: sunproperties

Displays palette:

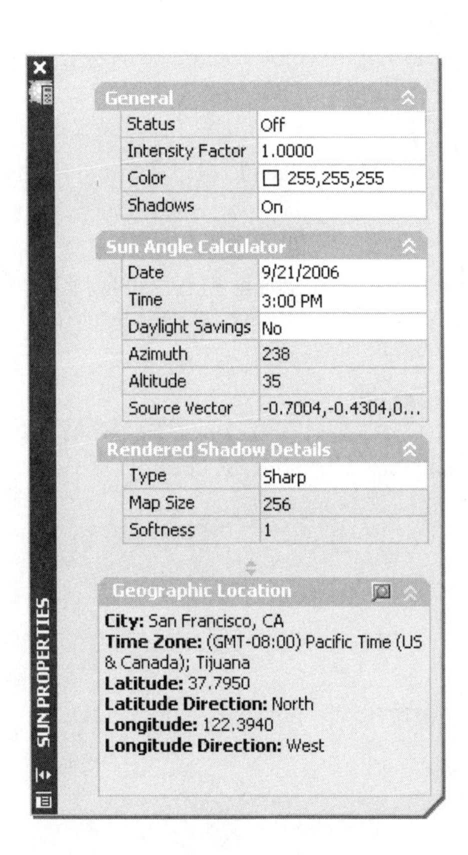

PALETTE OPTIONS

General options

Status toggles the sun light; inoperative when lighting is not enabled in the drawing.

Intensity Factor specifies the sun brightness; ranges from 0 (dark) to a very large number, like 9.999E+99 (bright).

Color specifies the color of the light.

Shadows toggles shadows generated by the sun.

Sun Angle Calculator options

Date specifies the day and month.

Time sets the time of day.

Daylight Saving toggles daylight savings time.

Azimuth reports the angle of the sun clockwise from due north; read-only.

Altitude reports the angle of the sun measured vertically from the horizon (90 = overhead); read-only.

Source Vector reports the direction of the sun; read-only.

Rendered Shadow Details options

Type selects the type of shadow, sharp or soft.

Map Size specifies the rendering area of soft shadows; range is 64 to 4096.

Softness specifies the appearance of shadow edges.

Geographic Location option

Edit Geographic Location displays the Geographic Location dialog box; see the GeographicLocation command.

RELATED COMMANDS

GeographicLocation displays the Geographic Location dialog box.

Render renders drawings, showing the effect of the sun light.

Light places lights in drawings.

RELATED SYSTEM VARIABLES

SunPropertiesState reports whether the Sun Properties palette is open.

SunStatus reports whether the sun is turned on or off.

Sweep

2007 Creates swept 3D solids and surfaces.

Command	Alias	Ctrl+	F-key	Alt+	Menu Bar	Tablet
sweep	...	...	...	**DMP**	**Draw** ↳**Modeling** ↳**Sweep**	...

Command: sweep
Current wire frame density: ISOLINES=4
Select objects to sweep: *(Select one or more objects.)*
Select objects to sweep: *(Press Enter.)*
Select sweep path or [Alignment/Base point/Scale/Twist]: *(Pick another object, or enter an option.)*

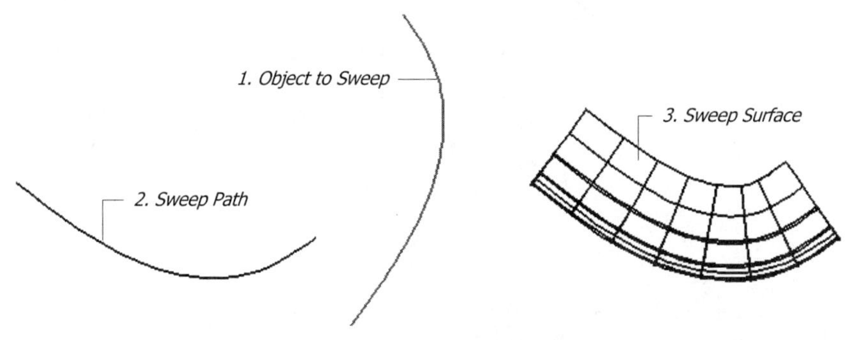

If the sweep object is closed (such as a circle), then this command produces a 3D solid; if open, such as an arc, then ut generates a 3D surface.

COMMAND OPTIONS

Select objects to sweep selects the objects to be swept.

Select sweep path defines the path along which the objects are swept.

Alignment aligns the path normal (at 90 degrees) to the tangent direction of the path.

Base point relocates the sweep's base point.

Scale resizes the sweep object.

Twist specifies the twist angle.

Twist options
Enter twist angle or allow banking for a non-planar sweep path [Bank]<0.0>: *(Enter an angle, or type* B.*)*

Twist angle specifies the angle through which the sweep twists, from one end to the other.

Bank causes the sweep to rotate along 3D curved paths: 3D polyline, 3D spline, or helix.

TIPS

- You can use the following objects as the sweep path: lines, arcs, circles, ellipses, elliptical arcs, 2D and 3D polylines, splines, helixes, and edges of surfaces and solids.
- The following objects can be swept: lines, traces, arcs, circles, ellipses, elliptical arcs, 2D polylines, 2D solids, splines, regions, flat 3D faces, flat solid faces, and planar surfaces.
- The figures below illustrate normal and twisted sweeps:

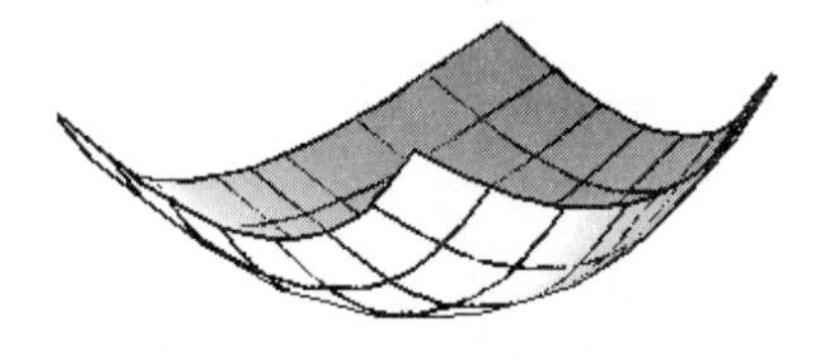

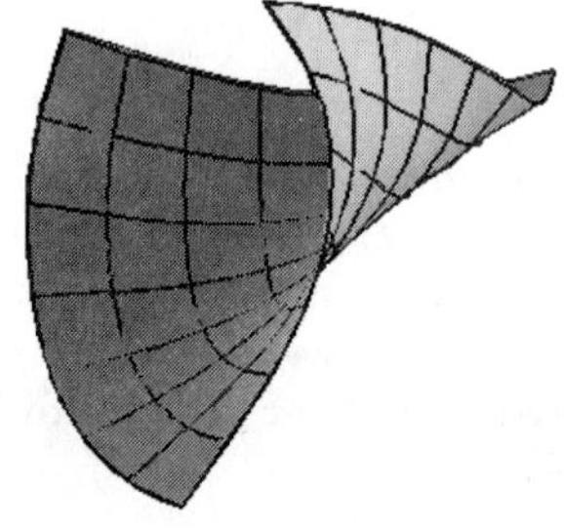

Left: *Normal sweep.*
Right: *Twisted sweep.*

- Apply this command to a helix to produce a spring:

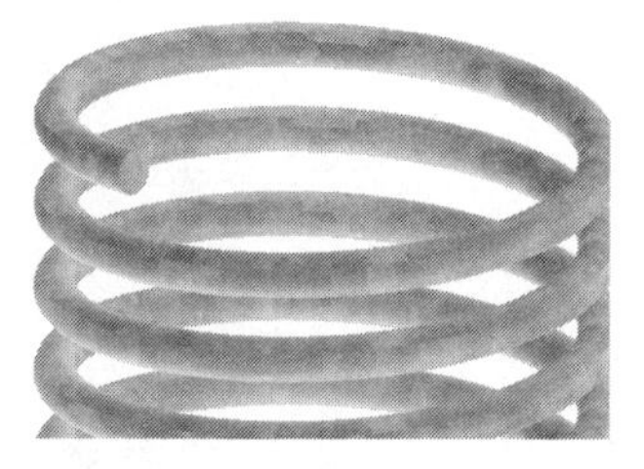

- The **Twist** option does not work when the path is a closed loop.

SysWindows

Rel.13 Controls multiple windows (short for SYStem WINDOWS).

Command	Alias	Ctrl+	F-key	Alt+	Menu Bar	Tablet
syswindows	...	**F6**	...	**W**	**Windows** ⇨*varies*	...

Command: syswindows

Enter an option [Cascade/tile Horizontal/tile Vertical/Arrange icons]: *(Enter an option.)*

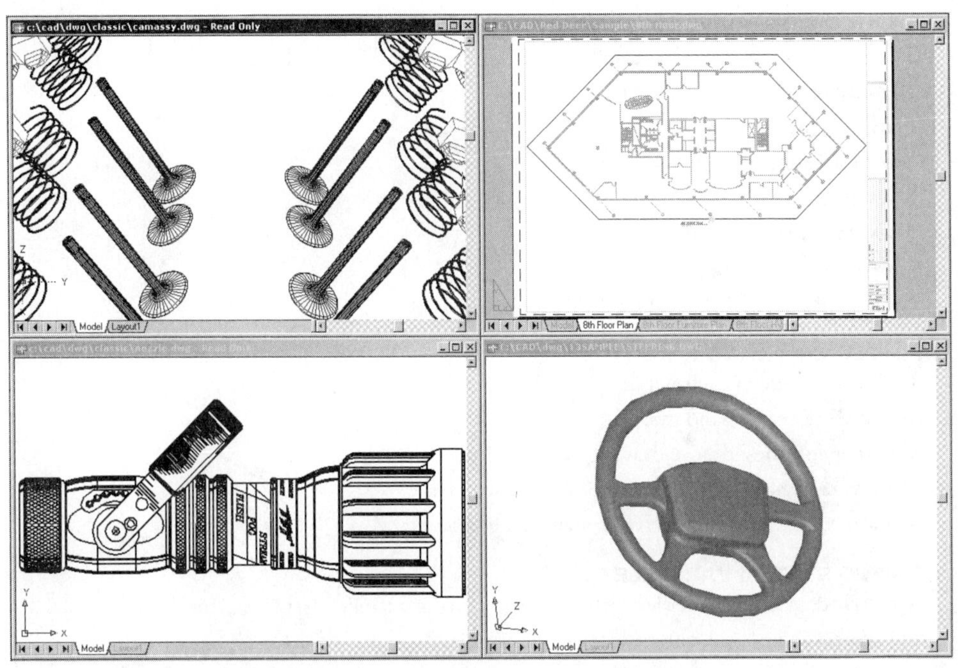

AutoCAD displays drawings in four tiled windows.

COMMAND LINE OPTIONS

Arrange icons arranges icons in an orderly fashion.

Cascade cascades the window.

tileHorizontal tiles the window horizontally.

tileVertical tiles the window vertically.

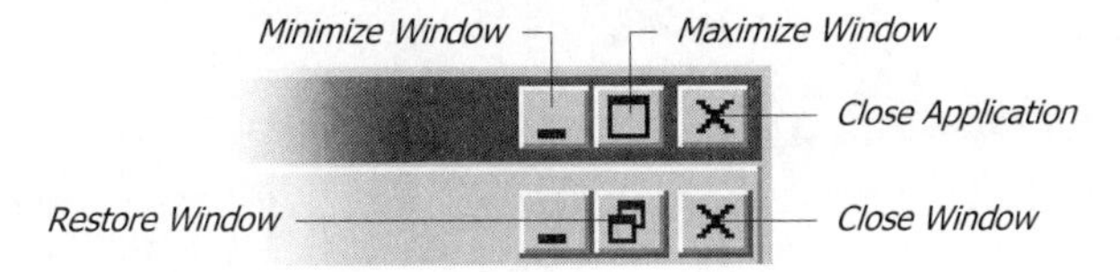

TITLE BAR MENU

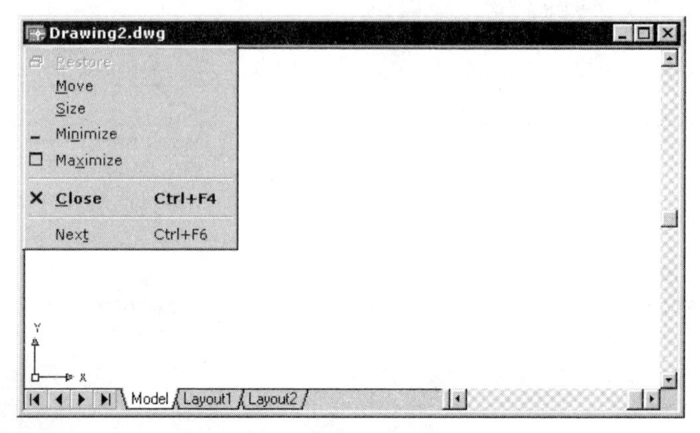

Restore restores the window to its "windowized" size.

Move moves the window.

Size resizes the window.

Minimize minimizes the window.

Maximize maximizes the window.

Close closes the window.

Next switches the focus to the next window.

RELATED COMMANDS

Close closes the current window.

CloseAll closes all windows.

Open opens one or more drawings, each in its own window.

MView creates paper space viewports in a window.

Vports creates model space viewports in a window.

RELATED SYSTEM VARIABLES

Taskbar determines whether each drawing opens in a single session of AutoCAD.

SDI forces AutoCAD to load just one drawing at a time.

TIPS

- The SysWindows command had no practical effect until AutoCAD 2000, because AutoCAD Release 13 and 14 supported a single window only.
- Press CTRL+F6 to switch quickly between currently-loaded drawings.
- Press CTRL+F4 to close the current drawing

Table

2005 Inserts formatted tables in drawings.

Command	Alias	Ctrl+	F-key	Alt+	Menu Bar	Tablet
table	**tb**	...	...	...	**Draw ↳Table**	...
-table						

Command: table

Displays dialog box:

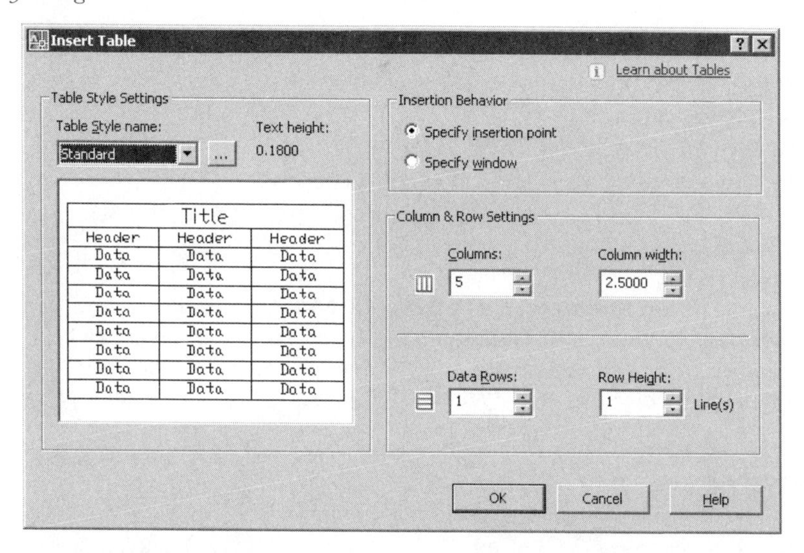

INSERT TABLE DIALOG BOX

Table Style Settings

Table Style Name selects from a list of pre-formatted styles; new drawings contain just one style, named Standard.

... displays the Table Style dialog box to edit or create new styles; see **TableStyle** command.

Insertion Behavior

⊙**Specify insertion point** locates the table by its upper-left corner; after you click **OK**, AutoCAD prompts:

Specify insertion point: *(Pick a point.)*

AutoCAD inserts the table in the drawing (default = 2 rows x 5 columns):

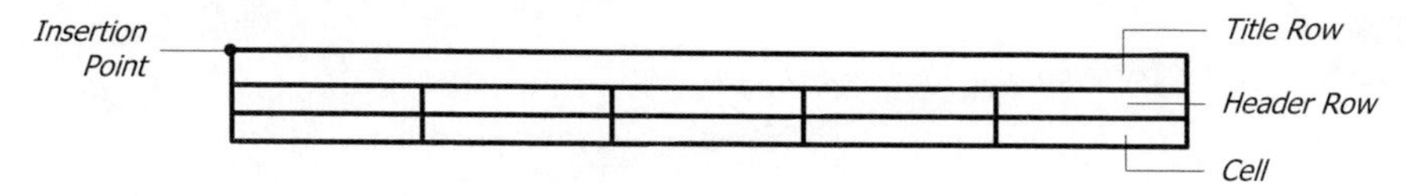

- **Specify window** fits the table to the window; after you click **OK**, AutoCAD prompts:

  Specify first corner: *(Pick point 1.)*

  Specify second corner: *(Pick point 2.)*

  AutoCAD places the table at the first pick point, and fits it to the second pick point.

1. First Corner

2. Second Corner

Column & Row Settings

- **Columns** specifies the initial number of columns.
- **Column width** specifies the initial width of columns.
- **Data Rows** specifies the initial number of "data" rows; the style determines whether the header and title rows are also placed.
- **Row height** specifies the initial height of rows.

-TABLE Command

Command: -table

Current table style: Standard Cell width: 2.5000 Cell height: 1 line(s)

Enter number of columns or [Auto] <5>: *(Type* **A***, or enter a number.)*

Enter number of rows or [Auto] <1>: *(Type* **A***, or enter a number.)*

Specify insertion point or [Style/Width/Height]: *(Pick a point, or enter an option.)*

COMMAND OPTIONS

Auto creates columns and rows to fit the table.

Style selects the table style name; see **TableStyle** command.

Width specifies the initial width of the table.

Height specifies the initial height of the table.

RELATED COMMANDS

MatchCell matches the style of one cell to other cells.

TableEdit edits text in table cells.

TableExport exports tables in CSV format.

TableStyle defines table styles.

RELATED SYSTEM VARIABLES

CTableStyle names the current table style.

TableIndicator toggles the display of row numbers and column letters.

TIPS

- To copy tables between drawings, use DesignCenter.
- Whether the table extends up or down from the insertion point depends on its style.

TablEdit

2005 Edits the text in cells; other methods edit the table itself.

Command	Alias	Ctrl+	F-key	Alt+	Menu Bar	Tablet
tabledit	...	...	...	...	...	...

As an alternative to the ***TablEdit*** *command, double-click cells to edit their content.*

Command: tabledit

Pick a table cell: *(Pick inside a cell.)*

AutoCAD highlights the cell, and then displays the mtext editor (see the ***MText*** *command):*

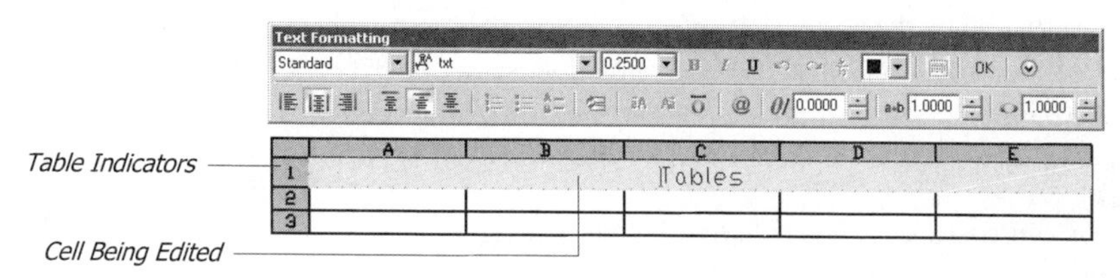

To edit other cells, press the arrow keys; to move to the "next" cell, press ***Tab.***

SHORTCUT MENUS

Click the center of a cell, right-click, and then select options from the shortcut menu:

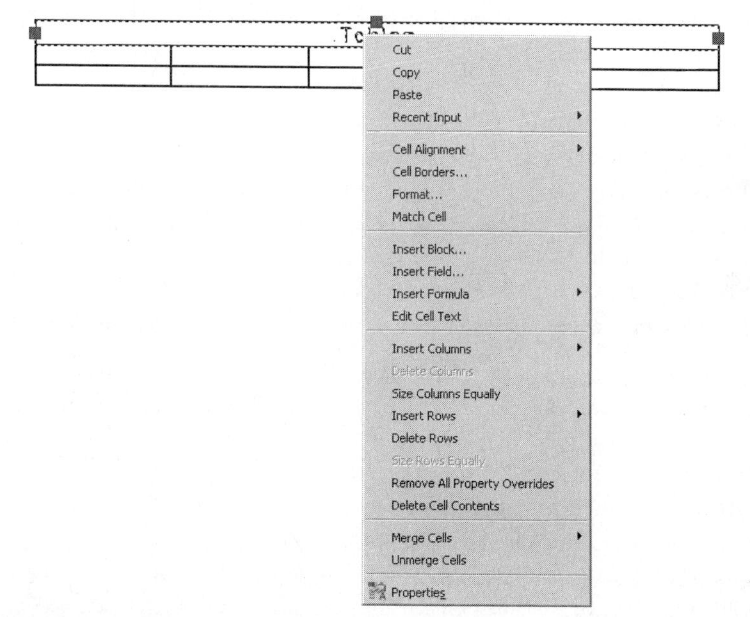

Cell Alignment displays a submenu of alignment options, such as Top Left.

Cell Borders displays the Cell Border Properties dialog box; change the thickness and color of cell border lines, and then click **OK**.

Format formats angles, dates, and coordinates; displays the Table Cell Format dialog box (*new to AutoCAD 2007*).

Match Cell copies the formatting of the current cell to other cells; see the MatchCell command.

Insert Block displays the Insert a Block in a Table Cell dialog box; select a block, set its options, and then click **OK**. Only one block can be inserted per cell; see the TInsert command.

Insert Field displays the Field dialog box; see the Field command.

Insert Formula inserts an arithmetic function; select a function from the submenu:

- **Sum** adds all values in selected cells.
- **Average** calculates the average of values in selected cells.
- **Count** counts the number of cells in the selected range.
- **Cell** references cells in spreadsheet notation, such as =B2.
- **Equation** allows arbitrary spreadsheet equations to be entered.

Edit Cell Text displays the mtext editor; see the MText command.

Insert Columns inserts one column to the left or right of the current cell; does not add columns to the title row.

- **Right** inserts the columns to the right of the selected cells.
- **Left** inserts the columns to the left of the selected cells.

Delete Columns deletes the current column without warning; use U command to undo.

Size Columns Equally makes selected columns the same width; available only when two or more cells are selected.

Insert Rows inserts one row above or below the current cell:

- **Above** inserts the rows above the selected cells.
- **Below** inserts the rows below the selected cells.

Delete Rows deletes the current row without warning; use U command to undo.

Size Rows Equally makes selected rows the same height; available only when two or more cells are selected.

Remove All Property Overrides does just what it says (*new to AutoCAD 2007*).

Delete Cell Contents erases text and blocks from the selected cell without warning.

Merge Cells joins the selected cells into a single cell; available only when two or more cells are selected. *Warning!* Only the text or block of the first cell is kept; the content of the other cells is erased.

- **All** merges the selected cells into a single cell.
- **By Row** merges the selected cells into a row.
- **By Column** merges the selected cells into a column.

Unmerge Cells splits merged cells back into the original number of cells; erased cell contents are not returned; available only when a previously merged cell is selected.

Properties displays the table properties.

Table Editing

Select a table, right-click, and then select an option from the shortcut menu:

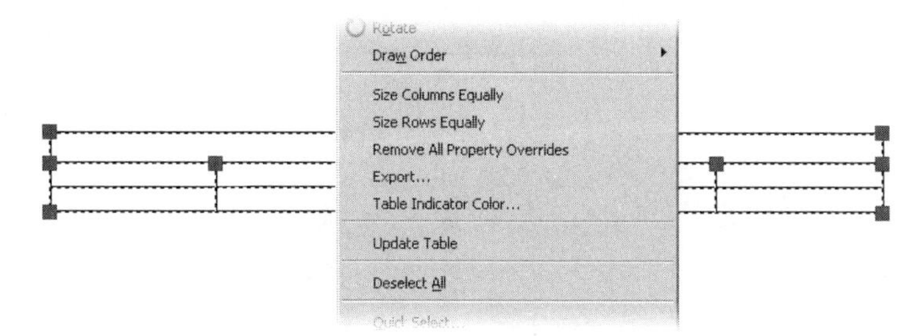

Command specific to table editing:

Size Columns Equally returns all columns to their original width.

Size Rows Equally adjusts the height of all rows to the height of the tallest row.

Remove All Property Overrides returns the table to its original style settings.

Export exports the table in CSV (comma separated values) format; see **TableExport** command.

Table Indicator Color changes the color of the row and column indicators; displays Select Color dialog box. See the **Color** command.

Update Table updates attribute-linked values. In addition, the tray shows a table icon:

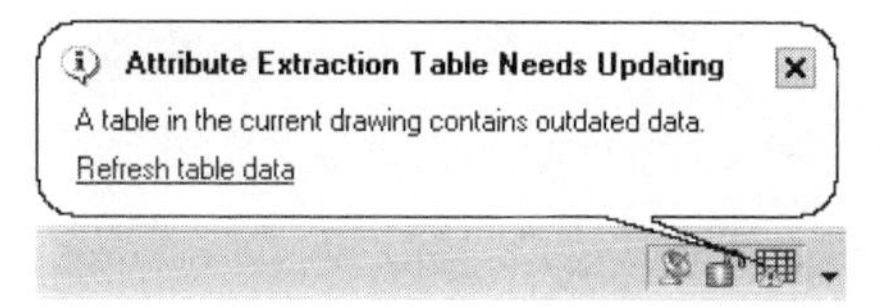

Grips Editing

Select a table, a cell, or several cells, and then use grips editing:

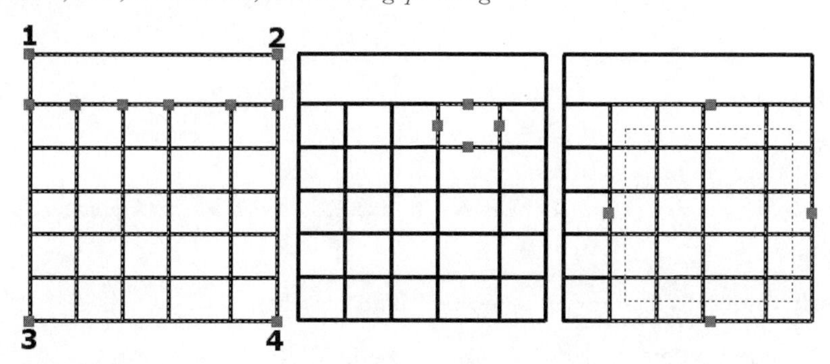

Table Grips Editing

Grip 1 enlarges and reduces the entire table.

Grip 2 widens and narrows all columns.

Grip 3 increases and decreases the height of all rows.

Grip 4 enlarges and/or reduces the entire table, depending on how you move the cursor.

Other grips widen and narrow individual columns.

Cell Grips Editing

Grips on vertical borders change the width of the cell(s).

Grips on horizontal borders change the height of the cell(s).

Properties Editing

Select one or more cells or tables, right-click, and then select Properties from the shortcut menu:

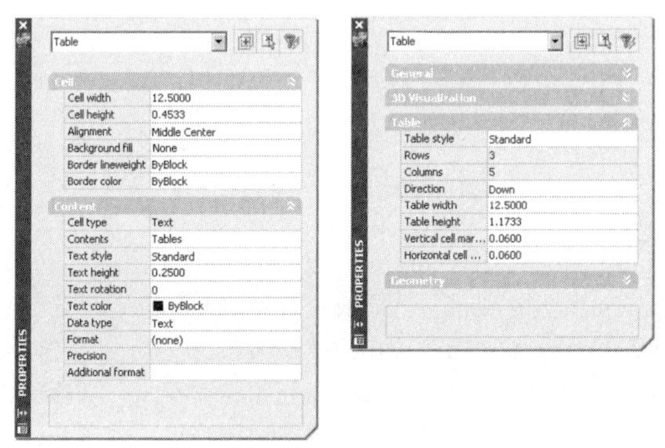

***Left**: Properties of cells.*
***Right**: Properties of tables.*

CELL BORDER PROPERTIES DIALOG BOX

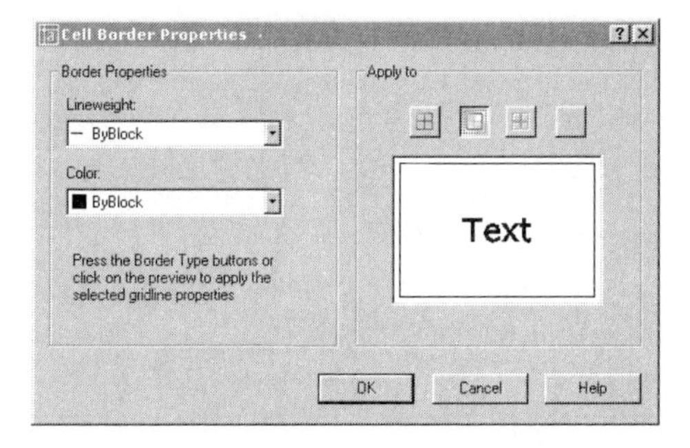

Border Properties

Lineweights sets the width of cell borders; see the Lineweight command.

Color sets the color of cell borders; see the Color command.

*To change the background color of cells, use the Properties command, and then change **Background Fill**.*

Apply To

All Borders applies properties to all cell borders.

Outside Borders applies properties to outer borders.

Inside Borders applies properties to inner borders.

No Borders applies properties to no borders.

INSERT A BLOCK IN A TABLE CELL DIALOG BOX

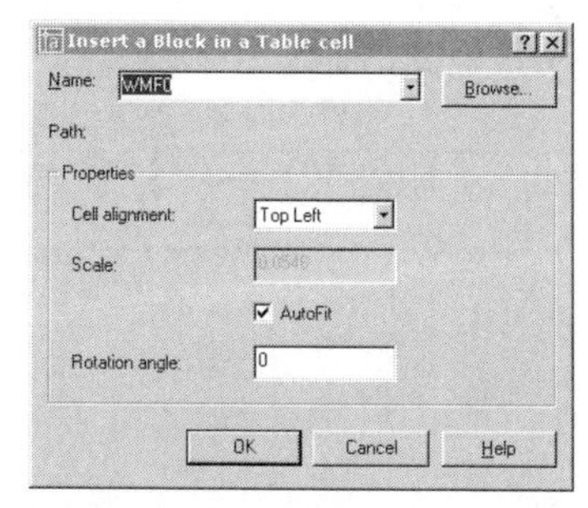

Name names the block; the droplist shows the names of blocks defined in the drawing.

Browse displays the Select Drawing File dialog box; select a *.dwg* or *.dxf* file, and then click **Open**.

Properties

Cell Alignment positions the block in the cell: top, middle, bottom, left, center, or right.

Scale specifies the scale factor; not available when **AutoFit** is turned on.

☑**AutoFit** scales the block to fit the cell.

Rotation Angle rotates the block.

TABLE CELL FORMAT DIALOG BOX

New to AutoCAD 2007:

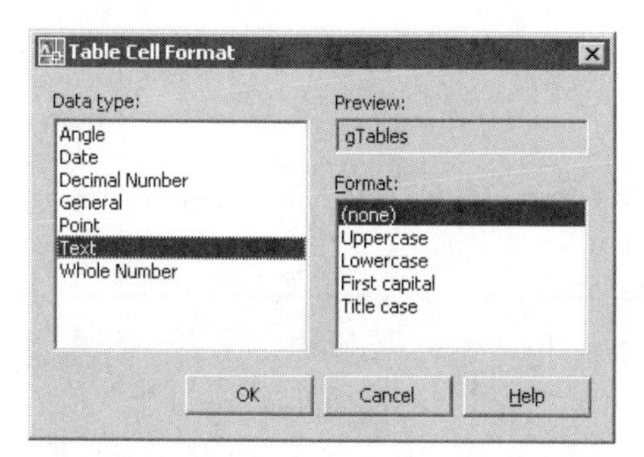

Data type selects the type of data to be formatted.

Format specifies the type of formatting.

RELATED COMMANDS

MatchCell matches the style of one cell to other cells.

Table creates tables.

TableStyle defines table styles.

Properties changes table style names, and overrides style settings.

RELATED SYSTEM VARIABLES

CTableStyle names the current table style.

TableIndicator toggles the display of row numbers and column letters.

TableExport

2005 Exports tables as CSV data files.

Command	Alias	Ctrl+	F-key	Alt+	Menu Bar	Tablet
tableexport	...	...	...	...	...	...

Command: tableexport
Select a table: *(Select one table.)*

Displays the Export Data dialog box.

Enter a file name, select a folder, and then click ***Save.***

TIPS

- *CSV* is short for "comma-separated values." Each table row is one line of data, and each cell is separated by a comma. This format can be read by spreadsheet and database programs.
- To import tables from Microsoft's Office XP spreadsheet, copy and paste it into the drawing. AutoCAD converts the spreadsheet into a table object.

  To import tables from other spreadsheet software:

  1. In the spreadsheet, copy the data to the Clipboard with CTRL+C.
  2. In AutoCAD, use the **Edit | Paste Special** command to paste the data:

Format	Meaning
Picture (Metafile)	Pasted as an OLE object.
Device Independent Bitmap	Pasted as an OLE object.
AutoCAD Entities	Pasted as a table object.
Image Entity	Pasted as a raster image.
Text	Pasted as mtext.
Unicode Text	Pasted as mtext.

- AutoCAD cannot import tables from CSV files.
- To copy tables between drawings, use DesignCenter.
- You can also access this command by selecting the table, right-clicking, and then choosing Export from the shortcut menu.

TableStyle

2005 Creates and edits table styles.

Command	Alias	Ctrl+	F-key	Alt+	Menu Bar	Tablet
tablestyle	**ts**	**...**	**...**	**FB**	**Format** ↳**Table Style**	**...**

Command: tablestyle

Displays dialog box:

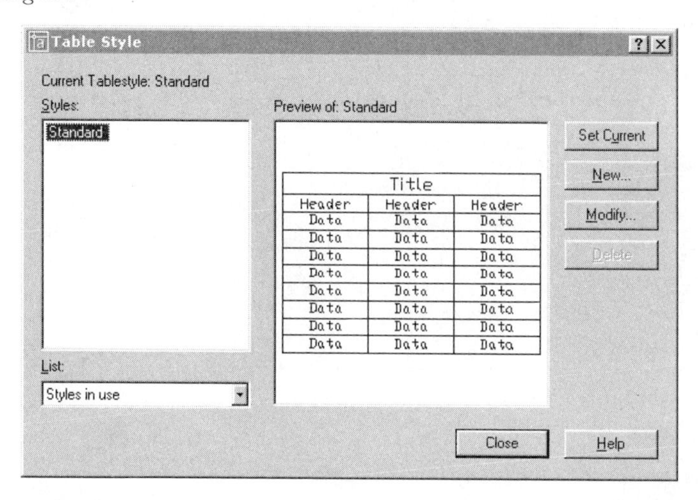

TABLE STYLE DIALOG BOX

Styles lists the styles defined in the drawing; every new drawing has the "Standard" style.

List filters the style names:

- **All** lists all styles.
- **Styles in use** lists the styles used by tables in the drawing.

Set Current sets the selected style as current.

New creates new table styles based on the settings of the current style; displays Create New Table Style dialog box.

Modify changes the settings of the table style; displays the Modify Table Style dialog box.

Delete erases the style from the drawing; unavailable if the "Standard" table style is selected.

CREATE NEW TABLE STYLE DIALOG BOX

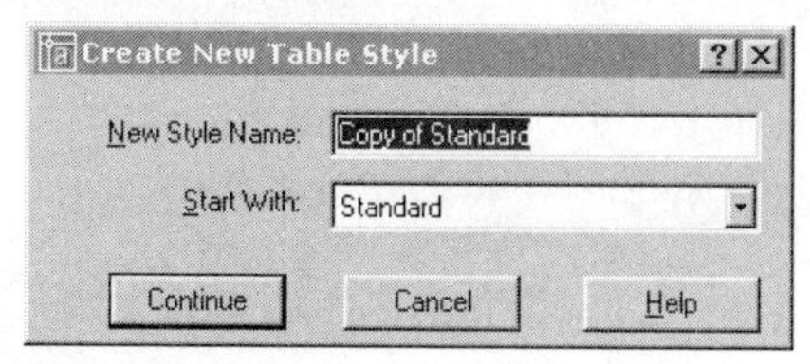

New Style Name names the new table style.

Start With selects the existing style upon which to base the new style.

Continue displays the New Table Style dialog box (identical to the Modify Table Style dialog box).

MODIFY TABLE STYLE DIALOG BOX

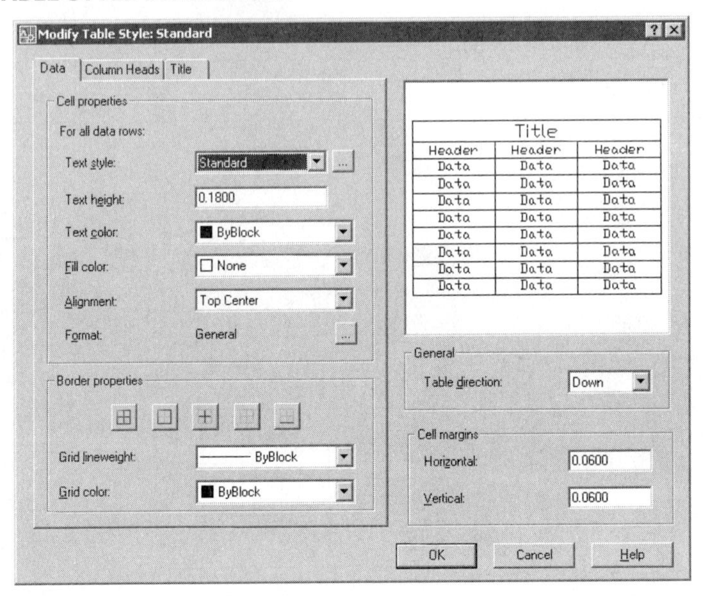

General

Table Direction determines how the table grows relative to the insertion point:

- **Down** (insertion point is at upper-left corner).
- **Up** (insertion point is at lower-left corner).

Cell Margins options

Horizontal specifies the horizontal margin between cell borders and text (default = 0.06).

Vertical specifies the vertical margin (default = 0.06).

Data Tab

Data are also known as "cells."

Cell Properties

Text Style selects a predefined text style.

... displays the Text Style dialog box; see the Style command.

Text Height specifies the height of the text; unavailable if the height is defined by text style.

Text Color specifies the color of the text; select **Select Color** for additional colors; see the Color command.

Fill Color specifies the background color of the cell.

Alignment determines the justification of the text within the cell: top, middle, bottom, left, center, or right.

Format formats angles, dates, and coordinates; displays the Table Cell Format dialog box (*new to AutoCAD 2007*). See the TablEdit command.

Border Properties

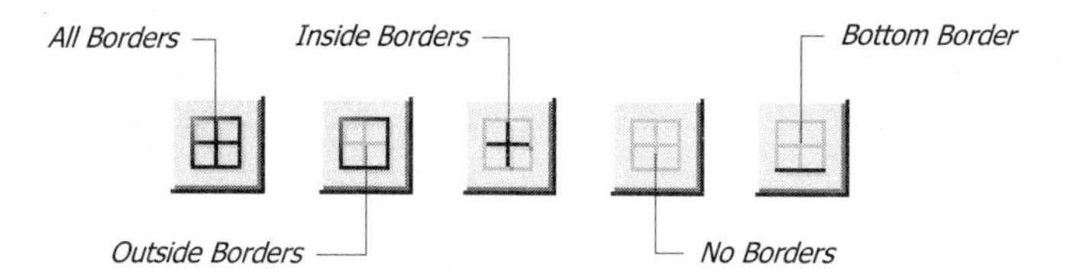

Grid Lineweight selects the width of cell borders; see the Lineweight command.

Grid Color selects the color of cell borders; see the Color command.

Column Heads Tab

Identical to the Data tab, except for this option:

☑**Include Header Row** determines whether tables have a column head row; use this row to provide titles for columns.

Title

Identical to the Data tab, except for this option:

☑**Include Title Row** determines whether tables have a title row; use this row to provide a title for the table.

SHORTCUT MENU

Right-click a style name to display shortcut menu:

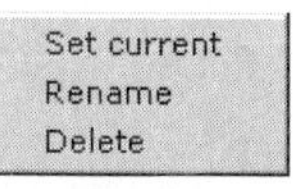

Set Current sets the selected style as current.

Rename renames the table style.

Delete erases the style from the drawing.

RELATED COMMANDS

Table places tables in drawings.

MatchCell matches the style of one cell to other cells.

TableEdit edits text in table cells.

Rename renames table styles.

Purge purges unused table styles from drawings.

DesignCenter shares tables styles between drawings.

Properties changes table style names, and overrides style settings.

RELATED SYSTEM VARIABLE

CTableStyle names the current table style.

TIPS

- To copy table styles from other drawings, copy a table with CTRL+C, and then paste it into the drawing with CTRL+V. Erase the table; the style remains in the drawing.
- To format individual cells, select the cell(s), right-click, and then choose Properties.
- To revert tables to original styles, select table, right-click, and select **Remove All Property Overrides.**

Tablet

V. 1.0 Configures and calibrates digitizing tablets, and toggles tablet mode.

Command	Alias	Ctrl+	F-key	Alt+	Menu Bar	Tablet
tablet	**ta**	**T**	**F4**	**TB**	**Tools** ↳**Tablet**	**X7**

Command: tablet

When no tablet is configured, AutoCAD complains, 'Your pointing device cannot be used as a tablet.'

When a digitizing tablet is configured, AutoCAD continues:

Enter an option [ON/OFF/CAL/CFG]: *(Enter an option.)*

COMMAND LINE OPTIONS

CAL calibrates the coordinates for the tablet.

CFG configures the menu areas on the tablet.

OFF turns off the tablet's digitizing mode.

ON turns on the tablet's digitizing mode.

RELATED SYSTEM VARIABLE

TabMode toggles use of the tablet:

0 — Tablet mode disabled.

1 — Tablet mode enabled.

RELATED FILES

acad.cui is the customization source code that defines the functions of tablet menu areas.

tablet.dwg is an AutoCAD drawing of the printed template overlay.

TIPS

- To change the tablet overlay, edit the *tablet.dwg* file, and then plot it to fit your digitizer.
- Tablet does not work if a digitizer has not been configured with the Options command.
- AutoCAD supports up to four independent menu areas; macros are specified by the Tablet Menu section of the *acad.cui* file.
- Menu areas may be skewed, but corners must form a right angle.
- Projective transformation is a limited form of "rubber sheeting": straight lines remain straight, but not necessarily parallel.

DEFINITIONS

Affine Transformation — requires three pick points; sets an arbitrary linear 2D transformation with independent x,y scaling and skewing.

Orthogonal Transformation — requires two pick points; sets the translation; the scaling and rotation angles remain uniform.

Residual Error — proves largest where mapping is least accurate.

Outcome of Fit — reports on the results of transformation types:

Outcome	Meaning
Exact	Enough points to transform data.
Success	More than enough points to transform data.
Impossible	Not enough points to transform data.
Failure	Too many colinear and coincident points.
Canceled	Fitting canceled during projective transformation.

Projective Transformation — maps a perspective projection from one plane to another plane.
RMS Error — specifies root mean square error; smaller is better; measures closeness of fit.
Standard Deviation —when near zero, residual error at each point is roughly the same.

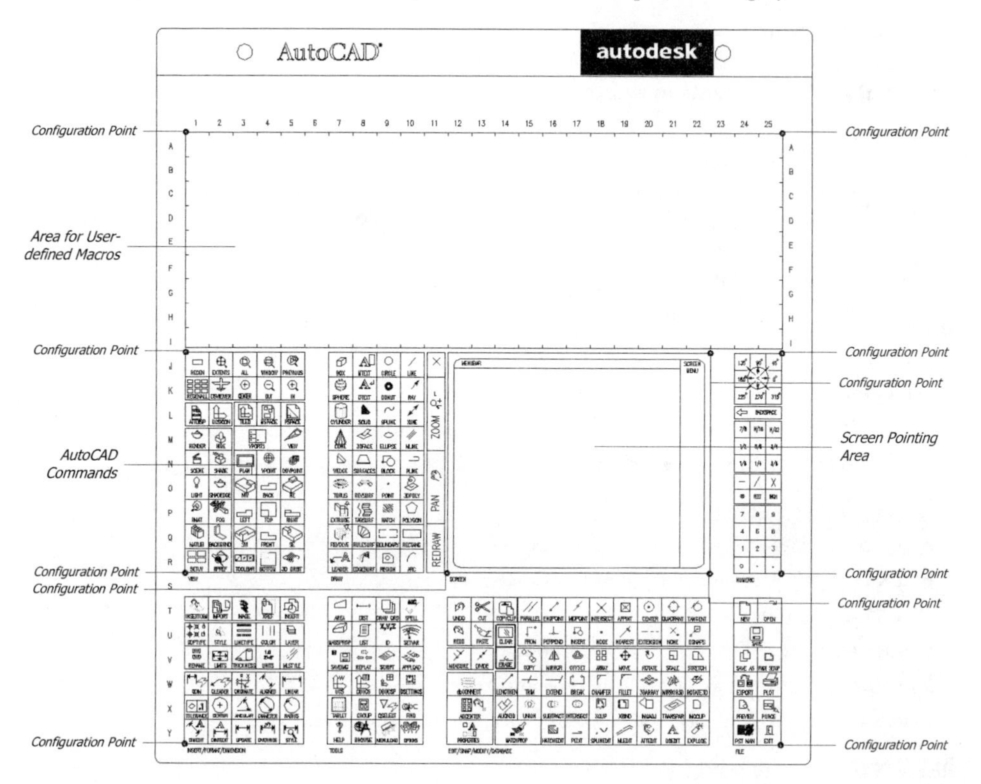

TabSurf

Rel.10 Draws tabulated meshes (not surfaces) defined by path curves and direction vectors (short for TABulated SURFace).

Command	Alias	Ctrl+	F-key	Alt+	Menu Bar	Tablet
tabsurf	...	...	...	**DMMT**	**Draw**	**P8**
					↳Modeling	
					↳Meshes	
					↳Tabulated Surface	

Command: tabsurf
Select object for path curve: *(Select an object.)*
Select object for direction vector: *(Select another object.)*

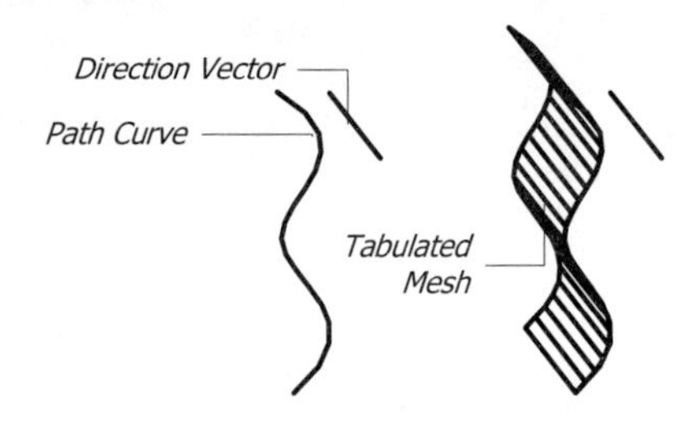

COMMAND LINE OPTIONS

Select object for path curve selects the object that defines the tabulation path.

Select object for direction vector selects the vector that defines the tabulation direction.

RELATED COMMANDS

Edge changes the visibility of 3D face edges.

Explode reduces tabulated meshes to 3D faces.

PEdit edits 3D meshes, such as tabulated ones.

EdgeSurf draws 3D meshes between boundaries.

RevSurf draws revolved 3D meshes around an axis.

RuleSurf draws 3D meshes between open or closed boundaries.

RELATED SYSTEM VARIABLE

SurfTab1 defines the number of tesselations drawn by TabSurf in the *n*-direction.

TIPS

- The path curve can be an open or closed object: line, 2D polyline, 3D polyline, arc, circle, or ellipse.
- The direction in which you draw the *direction vector* determines the direction of the extrusion; the length of the vector determines the thickness of the extrusion.
- The number of *m*-direction tabulations is always 1, and lies along the direction vector.
- The number of *n*-direction tabulations is determined by system variable SurfTab1 (default = 6) along curves only.

Text

V. 1.0 Places text, one line at a time, in drawings.

Command	Alias	Ctrl+	F-key	Alt+	Menu Bar	Tablet
text	dtext	...	...	...	Draw	K8
	dt				↳Text	
					↳Single Line Text	

Command: text

Current text style: "Standard" Text height: 0.2000

Specify start point of text or [Justify/Style]: *(Pick a point, or enter an option.)*

Specify height <0.2000>: *(Enter a value.)*

Specify rotation angle of text <0>: *(Enter a value.)*

Displays in-drawing text box:

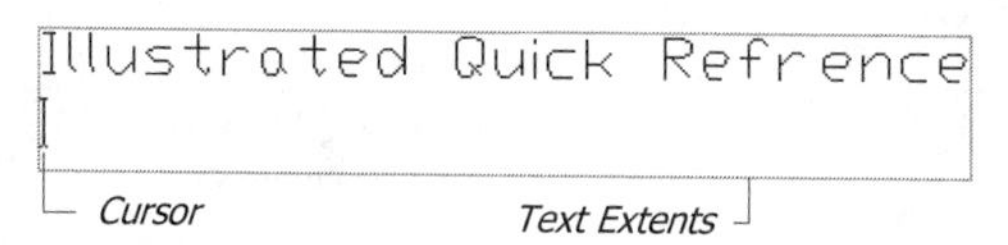

(Press Enter twice to exit command.)

COMMAND LINE OPTIONS

Specify start point of text indicates the starting point of the text.

Justify selects a text justification.

Style specifies the text style; see the Style command.

Specify height indicates the height of the text; this prompt does not appear when the style sets the height to a value other than 0.

Specify rotation angle of text indicates the rotation angle of the text.

Enter continues text one line below the previously-placed text; press Enter twice to exit.

Justify options

Enter an option [Align/Fit/Center/Middle/Right/TL/TC/TR/ML/MC/MR/BL/BC/BR]: *(Enter an option.)*

Align aligns the text between two points with adjusted text height.

Fit fits the text between two points with fixed text height.

Center centers the text along the baseline.

Middle centers the text horizontally and vertically.

Right right-justifies the text.

TL top-left justifies the text.

TC top-center justifies the text.

TR top-right justifies the text.

ML middle-left justifies the text.

MC middle-center justifies the text.

MR middle-right justifies the text.

BL bottom-left justifies the text.

BC bottom-center justifies the text.

BR bottom-right justifies the text.

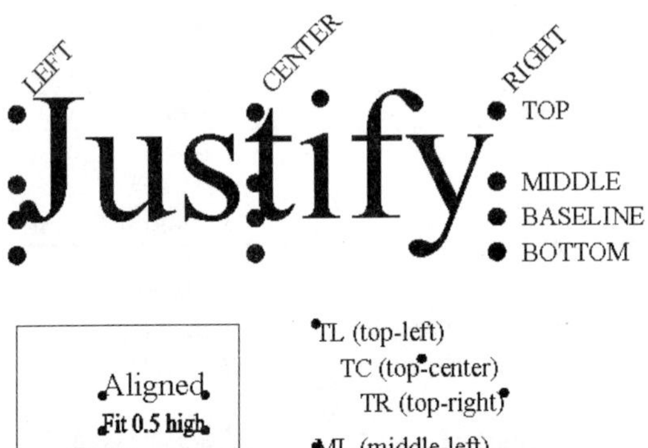

The dot indicates the text insertion point.

Style options

Enter style name or [?] <Standard>: *(Enter a name, or type* **?***.)*

Style name indicates a different style name.

? lists the currently-loaded styles.

SHORTCUT MENU

Double-click text, and then right-click to display the shortcut menu:

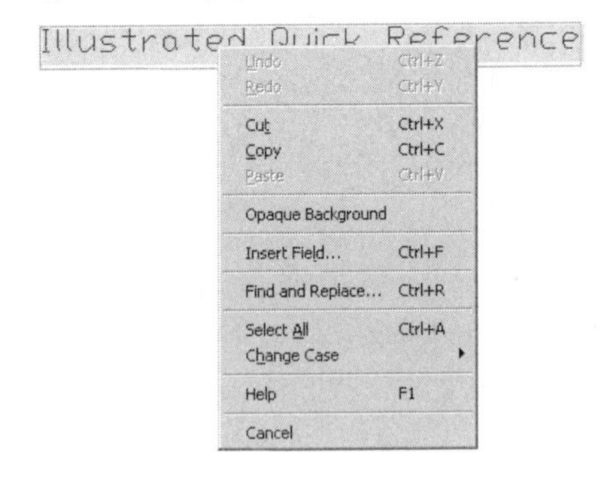

Options specific to text:

Opaque Background adds a gray background to the text editing area.

Insert Field inserts field text; displays the Field dialog box. See the Field command.

Find and Replace find and optionally replaces text. See th eFind command.

Select All selects all text in the editing area.

Change Case changes the case of selected text to all UPPERCASE or all lowercase.

SPECIAL SYMBOLS

%%c draws diameter symbol: Ø.

%%d draws degree symbol: °.

%%o starts and stops overlining.

%%p draws the plus-minus symbol: ±.

%%u starts and stops underlining.

RELATED COMMANDS

DdEdit edits text.

Change changes the text height, rotation, style, and content.

Properties changes all aspects of text.

Style creates new text styles.

MText places paragraph text in drawings.

RELATED SYSTEM VARIABLES

DTextEd toggles between the new in-place text editor and the old DText editor:

- **0** — displays the text box editor and cursor.
- **1** — displays "Enter text:" prompt; allows you to click elsewhere in the drawing to start new text blocks.
- **2** — displays the text box editor and cursor; allows you to click elsewhere in the drawing to start new text blocks (*new to AutoCAD 2007*).

TextSize is the current height of text.

TextStyle is the current style of text.

ShpName is the default shape name

TIPS

- You can enter any justification mode at the 'Start point' prompt.
- When DTextEd is set to 1, you can use the Text command to place text easily in many locations in the drawing. It displays text on screen as you type. Press **TAB** to move to the next text block, or **SHIFT+TAB** to the previous block.
- Press **Enter** at the end of a line to continue entering text; there is no need to repeat the command. To exit the command, press **Enter** twice.
- You can erase text by pressing **BACKSPACE**, or by selecting text and then pressing the **DEL** key.
- *Warning!* The spacing between lines of text does not necessarily match the snap spacing.
- Transparent commands and shortcut keystrokes do not work during the Text command.

TextToFront

2005 Places all text and dimensions visually on top of overlapping objects.

Command	Alias	Ctrl+	F-key	Alt+	Menu Bar	Tablet
texttofront	...	...	...	...	**Tools**	...
					↳Draw Order	
					↳Bring Text and Dimensions to Front	

Command: texttofront

Bring to front [Text/Dimensions/Both] <Both>: *(Enter an option.)*

nnn **object(s) brought to front.**

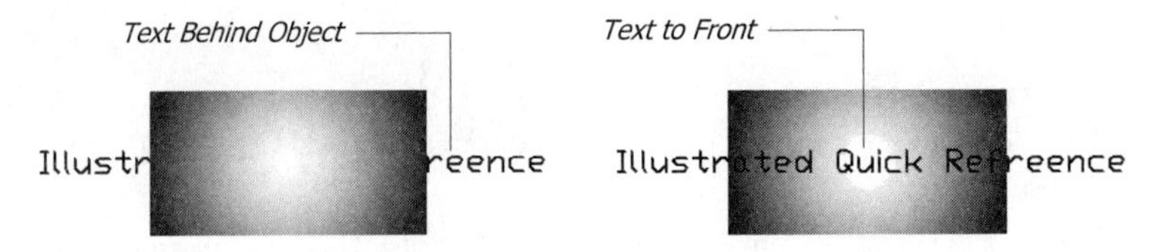

COMMAND LINE OPTIONS

Text brings all text to the front.

Dimensions brings all dimensions to the front.

Both brings both text and dimensions to the front.

RELATED COMMANDS

DrawOrder controls the display order of selected objects.

BHatch controls the display of hatch and fill patterns relative to their boundaries and other objects.

TIPS

- You must apply this command separately to model space and each layout.
- Newer objects are drawn on top of older objects in drawings; to change this behavior, change the DrawOrderCtrl system variable.

'TextScr

V. 2.1 Switches from the AutoCAD window to the Text window (short for TEXT SCReen).

Command	Alias	Ctrl+	F-key	Alt+	Menu Bar	Tablet
textscr	...	...	F2	VLT	View ↳Display ↳Text Window	...

Command: textscr

*Displays the **Text** window:*

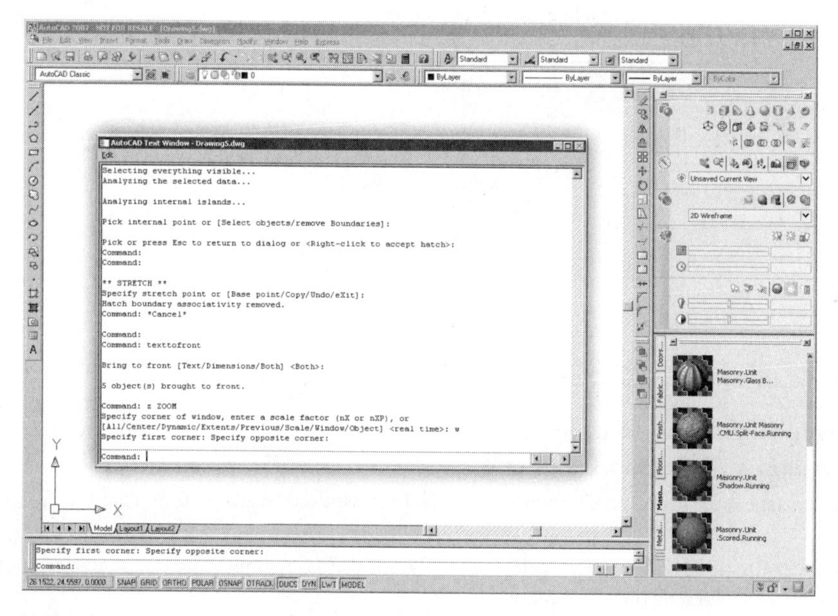

EDIT MENU

Paste to CmdLine pastes text from the Clipboard to the command line; available only when the Clipboard contains text.

Copy copies selected text to the Clipboard.

Copy History copies all text to the Clipboard.

Paste pastes text from the Clipboard into the Text window; available only when the Clipboard contains text.

Options displays the Options dialog box; see the Options command.

COMMAND LINE OPTIONS

Command window navigation:

Key	Meaning
←	Moves the cursor left by one character.
→	Moves the cursor right by one character.
↑	Displays the previous line in the command history.
↓	Displays the next line in the command history.
Page Up	Displays the previous screen of text.
Page Down	Moves to the next screen of text.
Home	Moves the cursor to the start of the line.
End	Moves the cursor to the end of the line.
Insert	Toggles insert mode.
Delete	Deletes the character to the right of the cursor.
BACKSPACE	Deletes the character to the left of the cursor.

SHORTCUT MENU

*Right-click the **Text** window:*

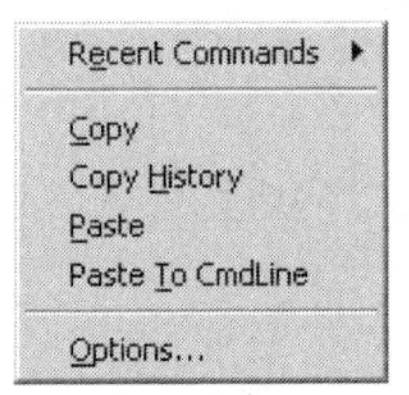

Recent Command displays a list of ten recently-used commands.

Copy copies selected text to the Clipboard.

Copy History copies all text to the Clipboard.

Paste pastes text from the Clipboard into the Text window; available only when the Clipboard contains text.

Paste to CmdLine pastes text from the Clipboard to the command line; available only when the Clipboard contains text.

Options displays the **Options** dialog box; see the **Options** command.

RELATED COMMAND

GraphScr switches from the Text window to the AutoCAD drawing window.

RELATED SYSTEM VARIABLE

ScreenMode reports whether the screen is in text or graphics mode:

0 — Text screen.

1 — Graphics screen.

2 — Dual screen displaying both text and graphics.

Removed Command

TiffIn was removed from AutoCAD R14. Use the **ImageAttach** command instead.

Thicken

2007 Creates 3D solids by thickening surfaces.

Command	Alias	Ctrl+	F-key	Alt+	Menu Bar	Tablet
thicken	...	...	...	M3T	Modify ↳3D Operations ↳Thicken	...

Command: thicken
Select surfaces to thicken: *(Select one or more surfaces.)*
Select surfaces to thicken: *(Press Enter.)*
Specify thickness <0.0>: *(Type a number.)*

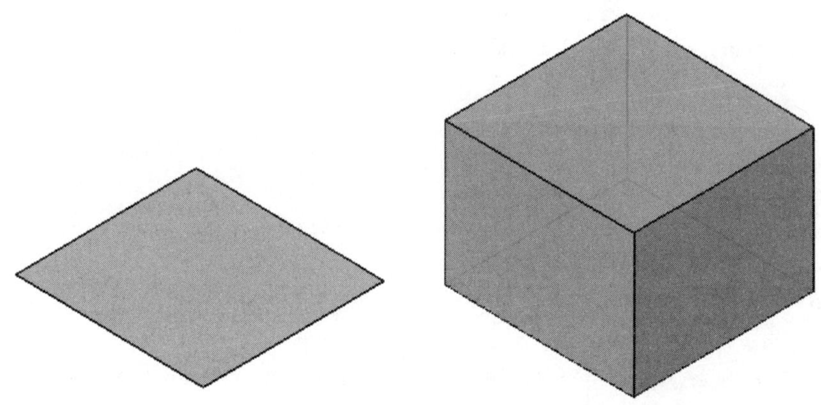

Left: *Surface before thickening.*
Right: *Surface turned into 3D solid.*

COMMAND LINE OPTIONS

Select surfaces to thicken selects one or more surfaces; this command does not work with meshes.

Specify thickness specifies the height of the thickened surface, such as 10. Enter a negative number, such as -10, to thicken the surface downwards.

RELATED COMMANDS

ConvertToSolid converts circles and polylines with thickness to 3D solids.

PolySolid converts existing lines, 2D polylines, arcs, and circles to 3D solids.

RELATED SYSTEM VARIABLE

DelObj determines whether or not the surface is erased following the thickening.

TifOut

2004 Exports the current viewports in TIFF raster format.

Command	Alias	Ctrl+	F-key	Alt+	Menu Bar	Tablet
tifout	...	...	...	...	...	...

Command: tifout

Displays Create Raster File dialog box. Specify a filename, and then click ***Save.***

Select objects or <all objects and viewports>: *(Select objects, or press Enter to select all objects and viewports.)*

COMMAND LINE OPTIONS

Select objects selects specific objects.

All objects and viewports selects all objects and all viewports, whether in model space or in layout mode.

RELATED COMMANDS

BmpOut exports drawings in BMP (bitmap) format.

Image places raster images in the drawing.

JpgOut exports drawings in JPEG (joint photographic expert group) format.

Plot exports drawings in TIFF and other raster formats.

PngOut exports drawings in PNG (portable network graphics) format.

TIPS

- The rendering effects of the VsCurrent command are preserved, but not of the Render command.
- TIFF files are commonly used in desktop publishing, but the format produced by AutoCAD cannot be read by some graphics programs.
- TIFF is short for "tagged image file format," and was co-developed by Microsoft and Aldus, the forerunner of Adobe.

'Time

V. 2.5 Displays time-related information about the current drawing.

Command	Alias	Ctrl+	F-key	Alt+	Menu Bar	Tablet
time	...	...	...	**TQT**	**Tools** ↳**Inquiry** ↳**Time**	...

Command: time

Display/ON/OFF/Reset: *(Enter an option.)*

Example output:

```
Current time:               Wednesday, April 19, 2016  4:36:28:562 PM
Times for this drawing:
  Created:                  Wednesday, April 19, 2016  3:19:27:359 PM
  Last updated:             Wednesday, April 19, 2016  3:19:27:359 PM
  Total editing time:       0 days 01:17:01:234
  Elapsed timer (on):       0 days 01:17:01:234
  Next automatic save in: <no modifications yet>
```

COMMAND LINE OPTIONS

Display displays the current time and date.

OFF turns off the user timer.

ON turns on the user timer.

Reset resets the user timer.

RELATED COMMAND

Status displays information about the current drawing and environment.

RELATED SYSTEM VARIABLES

CDate is the current date and time.

Date is the current date and time in Julian format.

SaveTime is the interval for automatic drawing saves.

TDCreate is the date and time the drawing was created.

TDInDwg is the time the drawing spent open in AutoCAD.

TDUpdate is the last date and time the drawing was changed.

TDUsrTimer is the current user timer setting.

TIPS

- The time displayed by the Time command is only as accurate as your computer's clock.
- The SaveAs command does not reset timings. The TdCreate system variable retains the original value, while the other timers continue accumulating seconds.

Removed Command

Today was removed from AutoCAD 2004; it was replaced by the Communication Center.

TInsert

2005 Inserts a block or a drawing in a table cell (short for Table INSERT).

Command	Alias	Ctrl+	F-key	Alt+	Menu Bar	Tablet
tinsert	...	...	...	...	...	...

Command: tinsert

Pick a table cell: *(Select a single cell in a table.)*

Displays Insert a Block in a Table Cell dialog box.

If the block contains attributes, the Enter Attributes dialog box is displayed.

COMMAND LINE OPTION

Pick a table cell specifies the cell in which to place the block.

INSERT A BLOCK IN A TABLE CELL DIALOG BOX

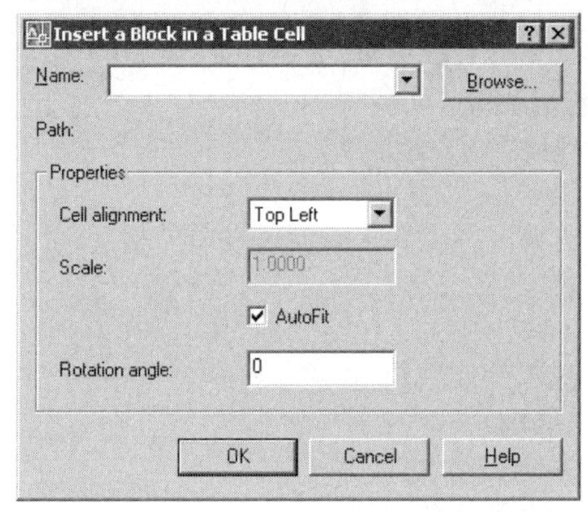

Name specifies the name of the block; the droplist names all blocks found in the current drawing.

Browse displays the Select Drawing File dialog box; select a *.dwg* or *.dxf* file, and click **Open**.

Properties

Cell Alignment specifies the position of the block within the cell, from top-left to bottom-right.

Scale sizes the block, and adjusts the cell to fit the block.

☑ **AutoFit** fits the block to the cell.

Rotation Angle specifies the angle at which to place the block.

RELATED COMMANDS

Table creates new tables.

Block creates blocks.

TIPS

- Alternatively, select a cell, right-click, and then choose **Insert Block** from the shortcut menu.
- A cell can contain at most one block.

Tolerance

Rel.13 Places geometric tolerancing symbols and text.

Command	Alias	Ctrl+	F-key	Alt+	Menu Bar	Tablet
tolerance	tol	...	...	NT	Dimension ↳Tolerance	X1

Command: tolerance

Displays dialog box.

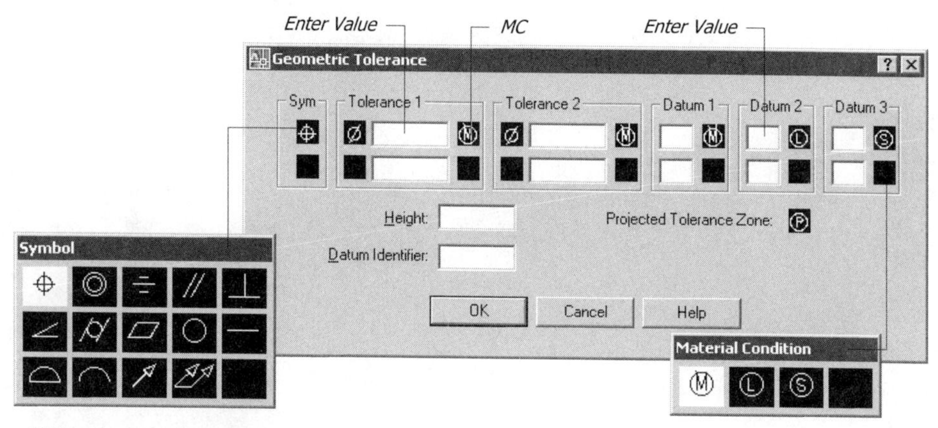

Fill in the dialog box, and then click ***OK****. AutoCAD prompts:*

Enter tolerance location: *(Pick a point.)*

GEOMETRIC TOLERANCE DIALOG BOX

Sym displays the Symbol dialog box.

Tolerance 1 and **Tolerance 2** specify the first and second tolerance values:

- **Dia** toggles the optional Ø (diameter) symbol.
- *Value* specifies the tolerance value.
- **MC** displays the Material Condition dialog box.

Datum 1, **Datum 2**, and **Datum 3** specify the datum references:

- **Value** specifies the datum value.
- **MC** displays the Material Condition dialog box.

Height specifies the projected tolerance zone value.

Projected Tolerance Zone toggles the projected tolerance zone symbol.

Datum Identifier creates the datum identifier symbol, such as -A-.

SYMBOL DIALOG BOX

Orientation Symbols

Position.

Concentricity and coaxiality.

Symmetry.

Parallelism.

Perpendicularity.

Angularity.

Form Symbols

Cylindricity.

Flatness.

Circularity and roundness.

Straightness.

Profile Symbols

Profile of the surface.

Profile of the line.

Circular runout.

Total runout.

MATERIAL CONDITION DIALOG BOX

(M) specifies maximum material condition.

(L) specifies least material condition.

(S) specifies regardless of feature size.

RELATED FILES

gdt.shp is the tolerance symbol definition source file.

gdt.shx is the compiled tolerance symbol file.

TIP

- You can use the **DdEdit** command to edit tolerance symbols and feature control frames.

DEFINITIONS

Datum — a theoretically-exact geometric reference that establishes the tolerance zone for the feature. These objects can be used as a datum: point, line, plane, cylinder, and other geometry.

Material Condition — symbols that modify the geometric characteristics and tolerance values (modifiers for features that vary in size).

Projected Tolerance Zone — the height of the fixed perpendicular part's extended portion; changes the tolerance to positional tolerance.

Tolerance — the amount of variance from perfect form.

-Toolbar

Rel.13 Displays and hides toolbars via the command line.

Command	Alias	Ctrl+	F-key	Alt+	Menu Bar	Tablet
-toolbar	...	...	...	...	...	...

Toolbar and TbConfig are aliases for the CUI command.

Command: -toolbar

Enter toolbar name or [ALL]: *(Enter a name, or type* **ALL**.*)*

Enter an option [Show/Hide]: *(Type* **S** *or* **H**.*)*

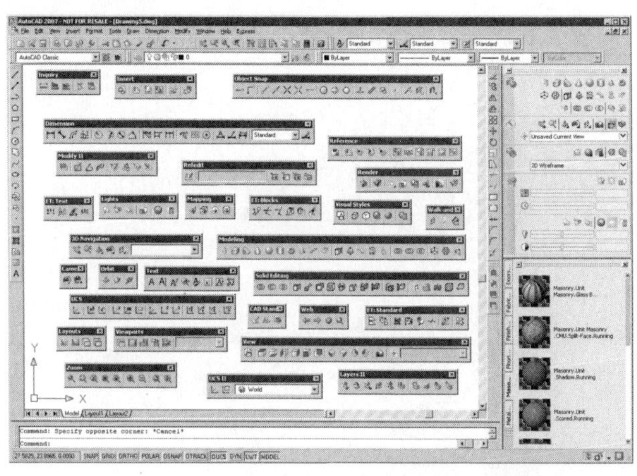

Opening AutoCAD's toolbars with the ***-Toolbar All Show*** *command.*

COMMAND LINE OPTIONS

Toolbar name specifies the name of the toolbar.

ALL applies the command to all toolbars; must be entered in all capital letters.

Show displays the toolbar.

Hide dismisses the toolbar.

RELATED COMMANDS

CUI customizes toolbars via a dialog box.

CuiLoad loads partial menu files, including toolbar definitions.

Tablet configures the tablet.

RELATED SYSTEM VARIABLE

ToolTips toggles the display of tooltips.

RELATED FILES

* ***.cui*** are user interface customization files.

* ***.bmp*** are bitmap files that define custom icon buttons.

'ToolPalettes, 'ToolPalettesClose

2004 Displays and closes the Tool palette.

Command	Alias	Ctrl+	F-key	Alt+	Menu Bar	Tablet
toolpalettes	**tp**	**3**	...	...	**Tools** **↳Palettes** **↳Tool Palettes**	...
toolpalettesclose		**3**				

Command: toolpalettes

The range of the Tool palette has increased significantly in AutoCAD 2007. Opens palette:

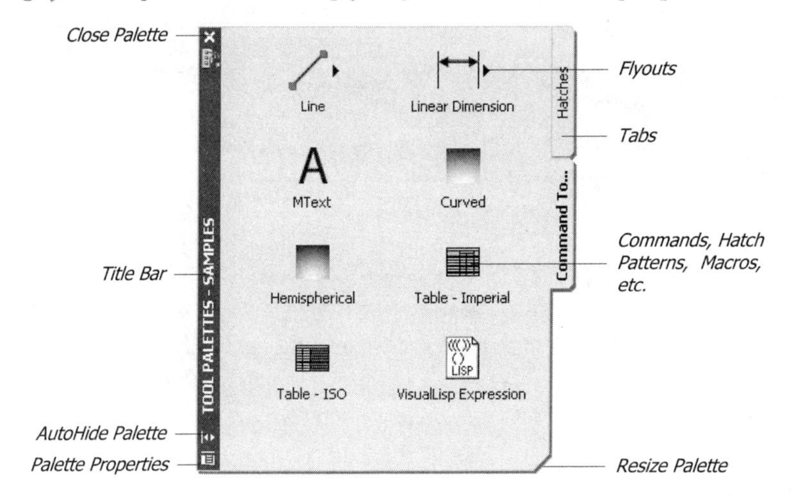

Command: toolpalettesclose

Closes the palette.

SHORTCUT MENUS

Tabs shortcut menu

Right-click an active tab:

Move up moves the tab up one position.

Move down moves the tab down one position.

View Options displays the View Options dialog box.

Paste pastes a tool previously copied from a palette.

New Tool Palette creates a new tab, and then prompts you for a name.

Delete Tool Palette removes the tab, after prompting you for confirmation.

Rename Tool Palette renames the tab.

Properties shortcut menu

Right-click the palette, or click the ***Properties*** *button:*

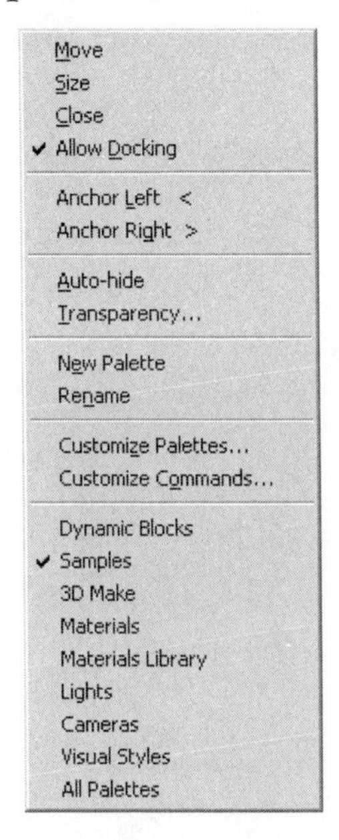

Move moves the palette.

Size changes the size of the palette.

Close closes the palette.

Allow Docking toggles whether the palette can be docked at the sides of the AutoCAD window.

Anchor Left < anchors (docks and minimizes) the palette at the left edge of AutoCAD (*new to AutoCAD 2007*).

Anchor Right > anchors the palette at the right edge of AutoCAD (*new to AutoCAD 2007*).

Auto-hide reduces the palette to the size of its title bar when the cursor is elsewhere.

Transparency displays the Transparency dialog box.

New Tool Palette creates a new tab, and then prompts you for a name.

Rename renames the selected tab.

Customize Palettes displays the Customize dialog box; see the Customize command.

Customize Commands displays the Customize User Interface dialog box; see the Cui command.

Palette shortcut menu

Right-click in the palette, away from any icon:

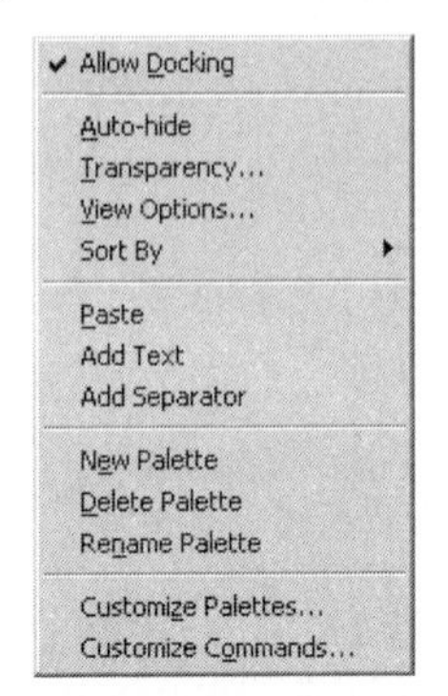

Allow Docking toggles whether the palette can be docked at the sides of the AutoCAD window.

Auto-hide reduces the palette to the size of its title bar when the cursor is elsewhere.

Transparency displays the Transparency dialog box.

View Options displays the View Options dialog box.

Sort by sorts the tools (doesn't seem to work correctly):

- **Name** of tool.
- **Type** of tool: Camera, Command, Face Settings, General, Insert, Materials Editor, Pattern, or Table.

Paste pastes a tool previously copied from a palette.

Add Text adds text to the palette.

Add Separator adds a horizontal line to the palette.

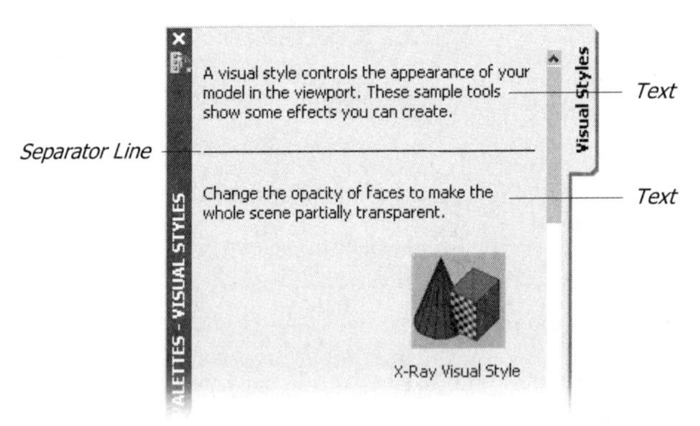

New Palette adds a new palette, names it "New Palette," and then encourages you to change the name.

Delete Palette removes the palette.

Rename Palette changes the palette's name, which appears on its associated tab.

Customize Palettes displays the Customize dialog box; see the Customize command.

Customize Commands displays the Customize User Interface dialog box; see the Cui command.

Icon shortcut menu

Right-click an icon:

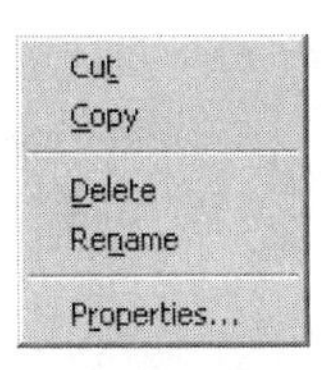

Cut copies the icon and its properties to the Clipboard, and then erases the icon.

Copy copies the icon and its properties, which can then be pasted to another palette.

Delete removes the icon.

Rename changes the icon's name.

Properties displays the Properties dialog box.

DIALOG BOXES

Transparency dialog box

Right-click the palette title bar, and then select Transparency:

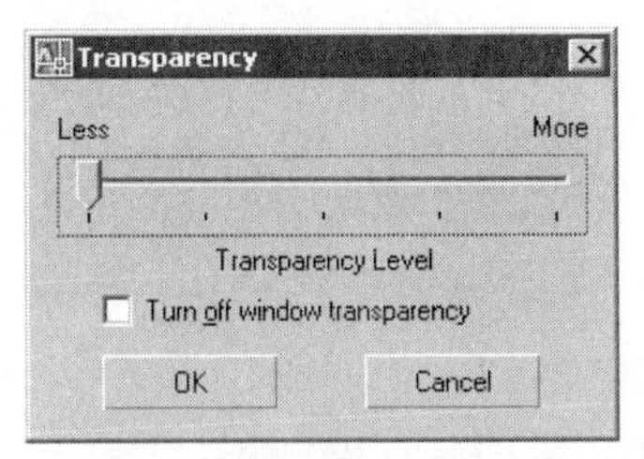

Transparency Level changes the transparency of the palette from Less (none) to More.

☐ **Turn off window transparency** makes the palette opaque.

View Options dialog box

Right-click a tab, and then select View Options:

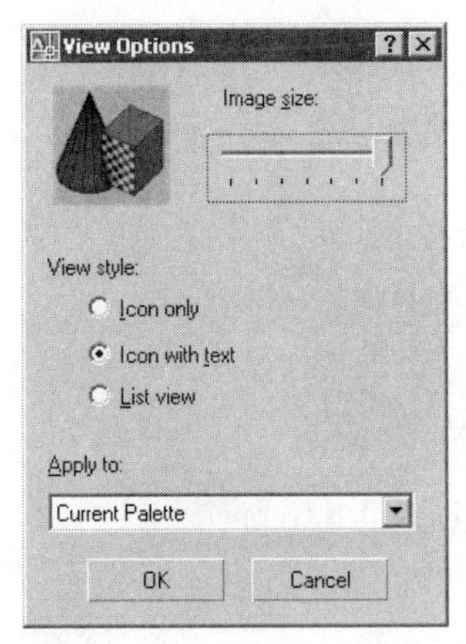

Image Size changes the size of icons from small (14 pixels square) to large (54 pixels).

View style toggles the display of icons and text:

- ○ **Icon only** displays icons only.
- ⊙ **Icon with text** displays icons with text below.
- ○ **List view** displays small icons with text beside.

Apply to determines if changes apply to the current palette or all palettes.

Properties dialog box

Right-click an icon, and then select Properties:

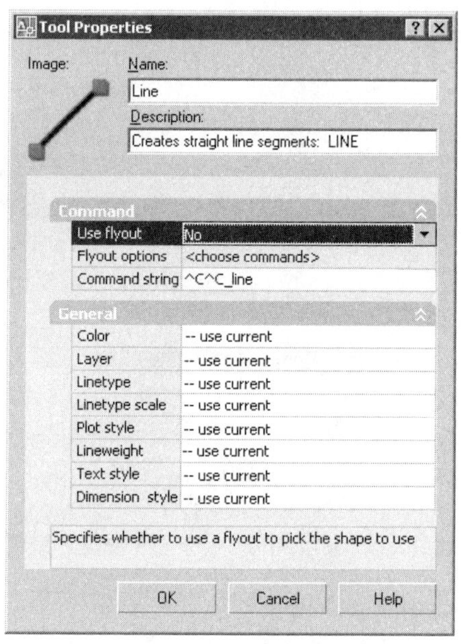

Image displays the icon associated with the tool. To change the icon, right-click it and then select **Specify Image**.

Name specifies the name that appears on the palette; it can be different from the Command String.

Description is the text that accompanies the icon on the palette; filling it in is optional.

The other items in this dialog box vary depending on the type of tool.

RELATED COMMANDS

Customize exports and imports Tool Palettes, and creates groups of palettes.

DesignCenter displays the content of drawings.

ToolPalettesClose closes the Tool palette.

RELATED SYSTEM VARIABLES

PaletteOpaque determines whether the Tools palette can be transparent.

TpState notes whether the Tools palette is open.

RELATED FILES

*.**xtp** stores the content of Tools palette in an XML-like format.

TIPS

- You can drag objects (while the right mouse button is held down) or toolbar buttons (while the Customize dialog box us open) onto palettes. Program code (macros, scripts, and so on) can also be stored on palettes by editing existing tools.
- To bring content from the DesignCenter into the Tool palette:

  1. In DesignCenter, right-click an item.
  2. From the shortcut menu, select **Create Tool Palette**.
- The *.xtp* files store content in XML (extended markup language) format.
- The Customize command allows you to create groups of palettes.

Torus

Rel.11 Draws 3D tori as solid models.

Command	Alias	Ctrl+	F-key	Alt+	Menu Bar	Tablet
torus	**tor**	...	...	**DMT**	**Draw** ↳**Modeling** ↳**Torus**	**O7**

Command: torus

This command has changed option prompts in AutoCAD 2007:

Specify center point or [3P/2P/Ttr]: *(Pick point 1.)*

Specify radius or [Diameter] <1.0>: *(Specify the radius, 2, or type* **D***.)*

Specify tube radius or [2Point/Diameter]: *(Specify the radius, 3, or type* **D***.)*

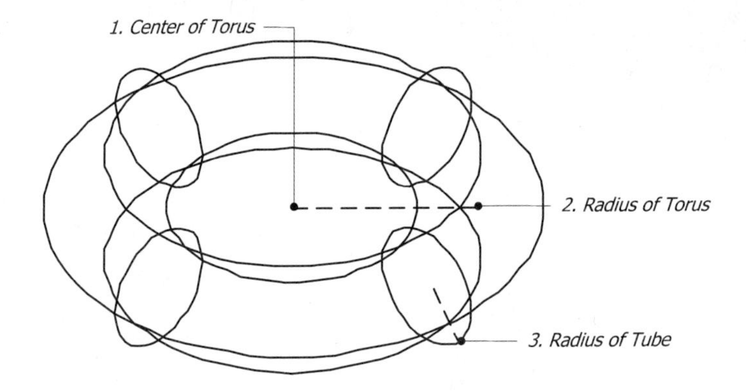

COMMAND LINE OPTIONS

Center point indicates the center of the torus.

3P specifies three points on the outer circumference of the torus (*new to AutoCAD 2007*).

2P specifies two points on the outer diameter of the torus (*new to AutoCAD 2007*).

Ttr specifies two tangent points plus the radius of the outer circumference of the torus (*new to AutoCAD 2007*).

Specify radius indicates the radius of the torus.

Diameter indicates the diameter of the torus.

Tube radius indicates the radius of the tube.

2Point asks for two points to determine the diameter of the tube.

RELATED COMMANDS

Ai_Torus creates a torus from 3D polyfaces.

Cone draws solid cones.

Cylinder draws solid cylinders.

Sphere draws solid spheres.

RELATED SYSTEM VARIABLES

Isolines specifies the number of isolines to draw on the torus.

DragVs specifies the visual style during tori construction.

TIPS

- When the torus radius is negative, the tube radius must be a larger positive number.
- A negative torus radius creates a football shape, as illustrated below:

Specify a tube diameter larger than the torus diameter to create a hole-less torus.

- After it is constructed, the torus can be edited by grips (*new to AutoCAD 2007*):

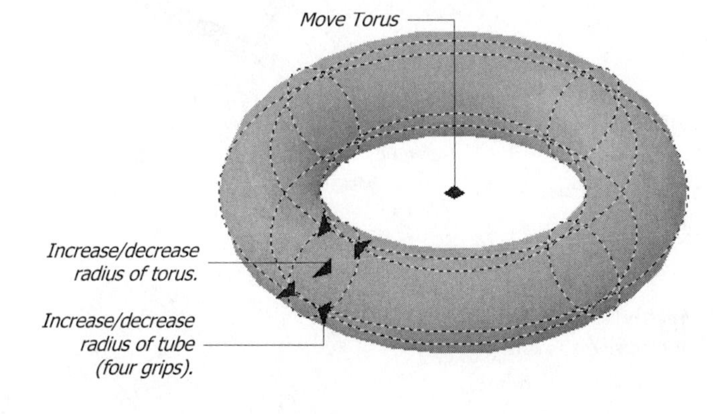

Trace

V. 1.0 Draws line segments with width.

Command	Alias	Ctrl+	F-key	Alt+	Menu Bar	Tablet
trace	...	...	...	...	...	...

Command: trace
Specify trace width <0.050>: *(Enter a value.)*
Specify start point: *(Pick a point.)*
Specify next point: *(Pick another point.)*
Specify next point: *(Press Enter to end the command.)*

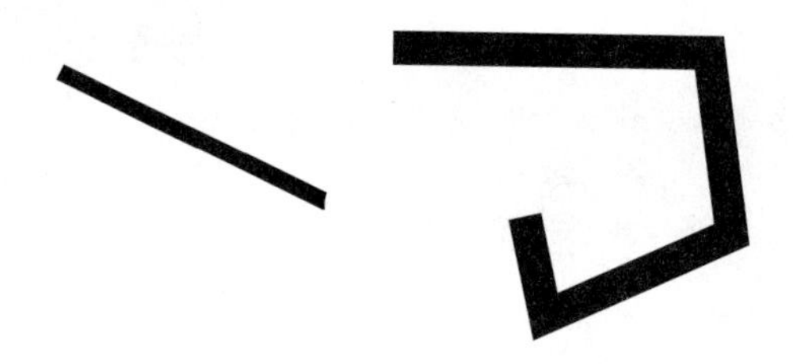

COMMAND LINE OPTIONS

Trace width specifies the width of the trace.

Start point picks the starting point.

Next point picks the next vertex.

ENTER exits the **Trace** command.

RELATED COMMANDS

Line draws lines with zero width.

MLine draws up to 16 parallel lines.

PLine draws polylines and polyline arcs with varying widths.

LWeight gives every object a width.

RELATED SYSTEM VARIABLES

FillMode toggles display of fill or outline traces (default = 1, on).

TraceWid specifies the current width of the trace (default = 0.05).

TIPS

- This command has largely been replaced by the PLine command; Trace was meant for drawing traces in printed circuit board designs.
- Traces are drawn along the centerline of the pick points. During drawing, the display of a trace segment is delayed by one pick point.
- During the drawing of traces, you cannot back up, because an **Undo** option is missing; if you require this feature, draw wide lines with the PLine command, setting the **Width** option.
- There is no option for controlling joints (always bevelled) or endcapping (always square); if you wish to modify these features, draw wide lines with the MLine command, after setting the solid fill, endcap, and joint options with the MlStyle command.

Tracking

Rel.14 Locates x and y points visually, relative to other points in the command sequence; *not* a command, but a option modifier.

Modifer	Alias	Ctrl+	F-key	Alt+	Menu Bar	Tablet
tracking	tk	...	...	...	...	T15
	track					

Example usage:

Command: line

Specify first point: *(Pick a point.)*

Specify next point or [Undo]: tk

Enters tracking mode:

First tracking point: *(Pick a point.)*

Next point (Press ENTER to end tracking): *(Pick a point.)*

Next point (Press ENTER to end tracking): *(Press Enter to end tracking.)*

Exits tracking mode:

Specify next point or [Undo]: *(Pick a point.)*

COMMAND LINE OPTIONS

First tracking point picks the first tracking point.

Next point picks the next tracking point.

ENTER exits tracking mode.

RELATED KEYBOARD MODIFIERS

Direct distance entry specifies an angle and relative distance to the next point.

from locates an offset point.

m2p finds a point midway between two picked points.

TIPS

- Tracking is not a command, but a option modifier. It is entered during commands without the ' transparent-command prefix.
- Tracking can be used in conjunction with direct distance entry.
- In tracking mode, AutoCAD automatically turns on ortho mode to constrain the cursor vertically and horizontally.
- If you start tracking in the x direction, the next tracking direction is y, and vice versa.
- You can use tracking as many times as you need to at the 'Specify first point' and 'Specify next point' prompts.

 Transparency

Rel.14 Toggles the transparency of pixels in raster images.

Command	Alias	Ctrl+	F-key	Alt+	Menu Bar	Tablet
transparency	...	...	...	**MOIT**	**Modify** ↳**Object** ↳**Image** ↳**Transparency**	**X21**

Command: transparency
Select image(s): *(Select one or more images inserted with the* **Image** *command.)*
Select image(s): *(Press Enter to end object selection.)*
Enter transparency mode [ON/OFF] <OFF>: *(Type* **ON** *or* **OFF***.)*

COMMAND LINE OPTIONS

Select image(s) selects the objects whose transparency to change.

ON makes the background pixels transparent.

OFF makes the background pixels opaque.

RELATED COMMANDS

ImageAttach attaches a raster image as an externally-referenced file.

ImageAdjust changes the brightness, contrast, and fading of a raster image.

TIPS

- This command works only with raster images placed in drawings with the **Image** command.
- This command is meant for use with raster file formats that allow transparent pixels.
- When on, transparent pixels allow graphics under the image to show through.
- For this command to work correctly, you must first specify the color of pixel to be transparent. Use a raster editor, such as PaintShop Pro: when saving the image, designate the color as transparent.

TraySettings

2004 Specifies settings for the Communications Center, located at the right end of the status bar (the "tray").

Command	Alias	Ctrl+	F-key	Alt+	Menu Bar	Tablet
traysettings	...	...	...	...	...	...

Command: traysettings

Displays dialog box:

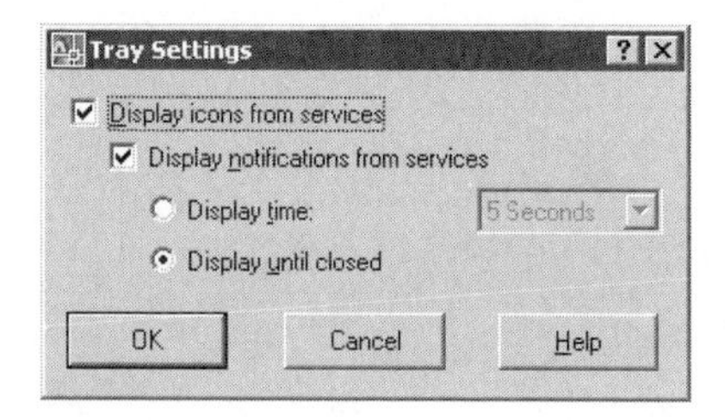

TRAY SETTINGS DIALOG BOX

Display icons from services:

☑ Displays "tray" at the right end of the status line; see below.

☐ Turns off the tray.

Display notification from services:

☑ Displays balloon notifications from a variety of services; see below.

☐ Turns off notifications:

○ **Display time** specifies the duration that a notification balloon is displayed.

⊙ **Display until closed** specifies that the notification balloon is displayed until closed by user.

RELATED SYSTEM VARIABLES

TrayIcons toggles the display of the tray on the status bar.

TrayNotify toggles whether notification balloons are displayed.

TrayTimeOut determines the length of time that notification balloons are displayed.

TIP

- Examples of notification balloons are illustrated below. Click underlined links for more information, or click **x** to dismiss.

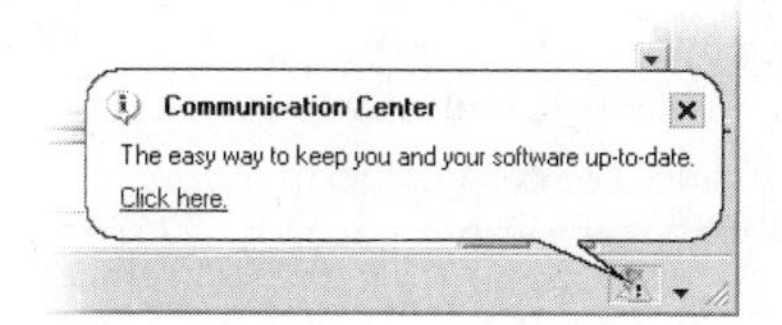

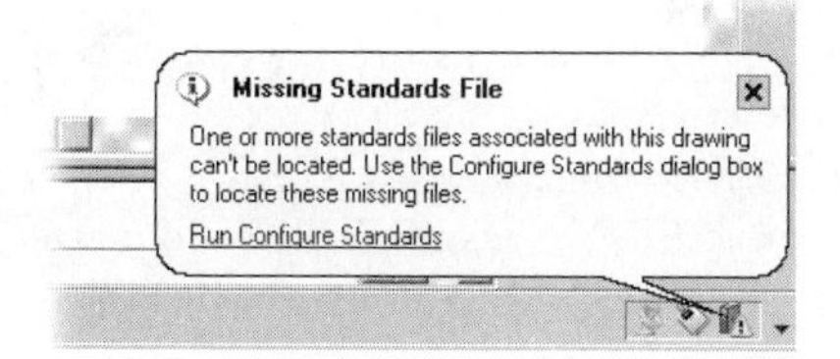

'TreeStat

Rel.12 Displays the status of the drawing's spatial index, including the number and depth of nodes (short for TREE STATistics).

Command	Alias	Ctrl+	F-key	Alt+	Menu Bar	Tablet
treestat	...	...	...	...	...	...

Command: treestat

Sample output:

```
Deleted objects: 12

Model-space branch
------------------
Oct-tree, depth limit = 30
Subtree containing objects with defined extents:
    Nodes: 21   Objects: 8   Maximum depth: 23
    Average objects per node: 0.38
    Average node depth: 14.24   Average object depth: 17.00
    Objects at depth 14: 2   17: 5   23: 1
    Nodes with population 0: 18   1: 1   2: 1   5: 1
Total nodes: 24   Total objects: 8

Paper-space branch
------------------
Quad-tree, depth limit = 20
Subtree containing objects with defined extents:
    Nodes: 1   Objects: 0
    Average objects per node: 0.00
    Average node depth: 5.00
    Nodes with population 0: 1
Total nodes: 4   Total objects: 0
```

RELATED SYSTEM VARIABLES

TreeDepth specifies the size of the tree-structured spatial index in *xxyy* format:

Depth	Meaning
xx	Number of model space nodes (default = 30).
yy	Number of paper space nodes (default = 20).
-xx	2D drawing.
+xx	3D drawing (default).

TreeMax is the maximum number of nodes (default = 10,000,000).

TIPS

- Better performance occurs with fewer objects per oct-tree node. When redraws and object selection seem slow, increase the value of TreeDepth.
- Each node consumes 80 bytes of memory.

DEFINITIONS

Oct Tree — the model space branch of the spatial index, where all objects are either 2D or 3D. *Oct* comes from the eight volumes in the x,y,z coordinate system of 3D space.

Quad Tree — the paper space branch of the spatial index, where all objects are two-dimensional. *Quad* comes from the four areas in the x,y coordinate system of 2D space.

Spatial Index — objects indexed by oct-region to record their position in 3D space; it has a tree structure with two primary branches: oct tree and quad tree. Objects are attached to *nodes*; each node is a branch of the *tree*. It's not clear that anyone understands this.

Trim

V. 2.5 Trims lines, arcs, circles, 2D polylines, and hatches to existing and projected cutting lines and views.

Command	Alias	Ctrl+	F-key	Alt+	Menu Bar	Tablet
trim	tr	...	...	MT	Modify ⇘Trim	W15

Command: trim
Current settings: Projection=UCS Edge=None
Select cutting edges ...
Select objects or <select all>: *(Select one or more objects.)*
Select objects: *(Press Enter to end object selection.)*
Select object to trim or shift-select to extend or [Fence/Crossing/Project/Edge/eRase/Undo] *(Select object, or enter an option.)*
Select object to trim or shift-select to extend or [Fence/Crossing/Project/Edge/eRase/Undo] *(Press Enter to end command.)*

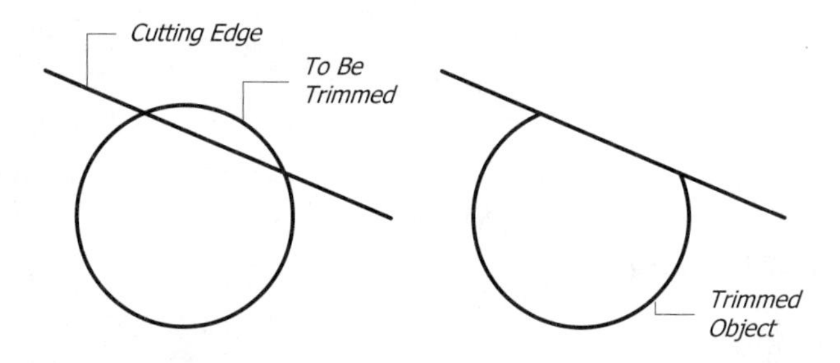

COMMAND LINE OPTIONS

Select objects selects the cutting edges.

Select all selects all objects in the drawing .

Select object to trim trims the object; picked portion is trimmed.

Undo untrims the last trim action.

Fence options

Specify first fence point locates the first point for fence selection mode.

Specify next fence point locates subsequent points; press Enter to exit selection mode.

Crossing options

Specify first corner locates the first corner for crossing selection mode.

Specify opposite corner locates the second corner.

Project options

Enter a projection option [None/Ucs/View] <Ucs>: *(Enter an option.)*

None uses only objects as cutting edges.

Ucs trims at the x,y plane of the current UCS.

View trims at the current view plane.

Edge options

Enter an implied edge extension mode [Extend/No extend] <No extend>:
(Type **E** *or* **N***.)*

Extend extends the cutting edge to trim object.

No extend trims only at an actual cutting edge.

Erase option

Select objects to erase or <exit> selects objects to erase; press ENTER to exit object selection mode.

RELATED COMMANDS

Change changes the size of lines, arcs and circles.

Extend lengthens lines, arcs and polylines.

Lengthen lengthens open objects.

RELATED SYSTEM VARIABLES

EdgeMode determines whether this command projects cutting edges.

ProjMode determines how cutting edges are projected in 3D space.

TIPS

- This command also trims hatches.
- You can select all objects in the drawing by pressing ENTER at the 'Select objects' prompt.
- To trim intersecting lines, it can be easier to use the Fillet command with **Radius** set to **0**.
- To have this command extend objects (rather than trim them), hold down the SHIFT key while selecting objects to extend; the opposite occurs in the Extend command.
- Cutting edges can be trimmed, too.

U

V. 2.5 Undoes the most recent AutoCAD command (short for Undo).

Command	Alias	Ctrl+	F-key	Alt+	Menu Bar	Tablet
u	...	Z	...	EU	Edit ↳Undo	T12

Command: u

COMMAND LINE OPTIONS

None.

RELATED COMMANDS

Oops unerases the most-recently erased object.

Quit exits the drawing, undoing all changes.

Redo reverses the most-recent undo, if the prior command was U or Undo.

Undo allows more sophisticated control over undo than U.

RELATED SYSTEM VARIABLE

UndoCtl toggles the undo mechanism.

TIPS

- The U command is convenient for stepping back through the design process, undoing one command at a time.
- As of AutoCAD 2006, multiple zooms and pans are grouped into a single undo. This feature can be toggled through the **Combine Zooms and Pans** option found on the User Preferences tab of the Options dialog box.
- The U command is the same as the Undo 1 command; for greater control over the undo process, use the Undo command.
- The Redo command redoes the most-recent undo only; use MRedo otherwise.
- The Quit command, followed by the Open command, restores the drawing to its original state, if not already saved.
- Because the undo mechanism creates a mirror drawing file on disk, disable the Undo command with system variable UndoCtl (set to 0) when your computer is low on disk space.
- Commands that involve writing to file, plotting, and some display functions (such as Render) are not undone.

Ucs

Rel.10 Defines new coordinate planes, and restores existing UCSs (short for User-defined Coordinate System).

Command	Alias	Ctrl+	F-key	Alt+	Menu Bar	Tablet
ucs	**dducs**	...	...	**TW**	**Tools** ↳**New UCS**	**W7**
				TH	**Tools** ↳**Orthographic UCS**	...

Command: ucs

The prompts are significantly changed in AutoCAD 2007. For compatibility with earlier versions of AutoCAD, you may enter any of the older options at the first prompt.

Current ucs name: *WORLD*

Specify origin of UCS or [Face/NAmed/OBject/Previous/View/World/X/Y/Z/ ZAxis] <World>: *(Enter an option.)*

COMMAND LINE OPTIONS

Specify origin of new UCS moves the UCS to a new origin.

Face aligns the UCS with the face of a 3D solid object.

NAmed saves, restores, and deletes named UCSs.

OBject aligns the UCS with a selected object; see tips.

Previous restores the previous UCS orientation.

World aligns the UCS with the WCS.

View aligns the UCS with the current view.

X rotates the UCS about the x axis.

Y rotates the UCS about the y axis.

Z rotates the UCS about the z axis.

ZAxis aligns the UCS with a new origin and z axis.

Face options

Select face of solid object: *(Pick a face.)*

Enter an option [Next/Xflip/Yflip] <accept>: *(Press Enter, or enter an option.)*

Select face selects a face to which the UCS will be applied.

Next selects the next face.

Xflip flips the x axis.

Yflip flips the y axis.

accept accepts the selected face (press Enter to accept.)

NAme options

Enter an option [Restore/Save/Delete/?]: *(Enter an option.)*

Restore restores a named UCS.

Save saves the current UCS by name.

Del deletes a saved UCS.

? lists the names of saved UCS orientations.

Hidden options

The following options also work with this command:

New defines a new user-defined coordinate system.

Move moves the UCS along the z axis, or specifies a new x,y-origin.

orthoGraphic selects a standard orthographic UCS: top, bottom, front, back, left, and right.

Apply applies the UCS setting to a selected viewport, or all active viewports.

3point aligns the UCS with a point on the positive x-axis and positive x,y plane.

STATUS BAR OPTION

New to AutoCAD 2007:

Click DUCS to turn on dynamic UCS mode, where the UCS matches the orientation of the selected face:

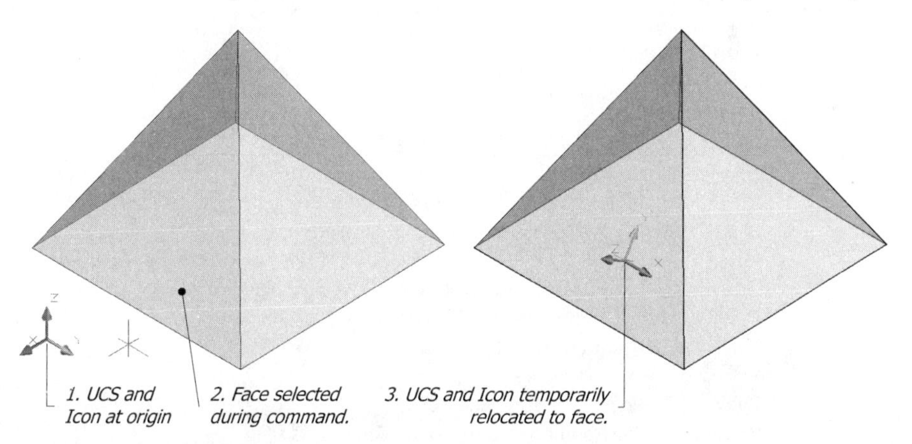

RELATED COMMANDS

UcsMan modifies the UCS via a dialog box.

UcsIcon controls the visibility of the UCS icon.

Plan changes to the plan view of the current UCS.

RELATED SYSTEM VARIABLES

GridDisplay forces the grid to follow the orientation of the dynamic UCS, when set to 8 (*new to AutoCAD 2007*).

UcsAxisAng specifies the default rotation angle when the UCS is rotated around an axis using the **X**, **Y**, or **Z** options of this command.

UcsBase specifies the UCS that defines the origin and orientation of orthographic UCS settings.

UcsDetect toggles dynamic UCS (*new to AutoCAD 2007*).

UcsFollow aligns the UCS with the view of a newly-activated viewport automatically.

UcsIcon determines the visibility and location of the UCS icon.

UcsOrg specifies the WCS coordinates of UCS icon (default = 0,0,0).

UcsOrtho specifies whether the related UCS is automatically displayed when an orthographic view is restored.

UcsView specifies whether the current UCS is saved when a view is created with the View command.

UcsVp specifies that the UCS reflects the UCS of the currently-active viewport.

UcsXdir specifies the X direction of the current UCS (default = 1,0,0).

UcsYdir specifies the Y direction of the current UCS (default = 0,1,0).

WorldUcs correlates the WCS to the UCS:

0 — Current UCS is WCS.

1 — UCS is the same as WCS (default).

TIPS

- Use the UCS command to draw objects at odd angles in 3D space.
- Although you can create UCSs in paper space, you cannot use 3D viewing commands in paper space.
- Each viewport can have its own UCS.
- UCSs can be aligned with these objects:

Object	UCS Origin	X Axis Passes Through...
Arc	Center	Endpoint closest to the pick point.
Attributes	Insertion point	Extrusion direction.
Blocks	Insertion point	Extrusion direction.
Circle	Center	Pick point.
Dimension	Midpoint of dimtext	Parallel to the X axis of the dimension.
Line	Endpoint nearest to pick	Parallel to the X axis of the line.
Point	Point	Anywhere.
Polyline	Start point	Start point to next vertex.
2D solid	First point	First two points.
Trace	First point	Centerline
3D Face	First point	First two points.
Shape	Insertion point	Extrusion direction.
Text	Insertion point	Extrusion direction.

- UCSs do *not* align with these objects: mlines, rays, xlines, 3D polylines, splines, ellipses, leaders, viewports, 3D solids, 3D meshes, and regions.
- When the GridDisplay system variable is set to 8, the grid matches the orientation of the dynamic UCS.

DEFINITIONS

UCS — user-defined 2D coordinate system oriented in 3D space; sets a working plane, orients 2D objects, defines the extrusion direction, and the axis of rotation.

WCS — world coordinate system; the default 3D x,y,z coordinate system.

UcsIcon

Rel.10 Controls the location and display of the UCS icon.

Command	Alias	Ctrl+	F-key	Alt+	Menu Bar	Tablet
ucsicon	...	...	...	**VLU**	**View** ↳**Display** ↳**UCS Icon**	**L2**

Command: ucsicon

Enter an option [ON/OFF/All/Noorigin/ORigin/Properties] <ON>: *(Enter an option.)*

COMMAND LINE OPTIONS

All applies changes of this command to all viewports.

Noorigin displays the UCS icon in the lower-left corner at all times.

OFF turns off the display of the UCS icon.

ON turns on the display of the UCS icon.

ORigin displays the UCS icon at the current UCS origin, or lower left corner of the viewport if the origin is off-screen.

Properties displays the UCS Icon dialog box.

UCS ICON DIALOG BOX

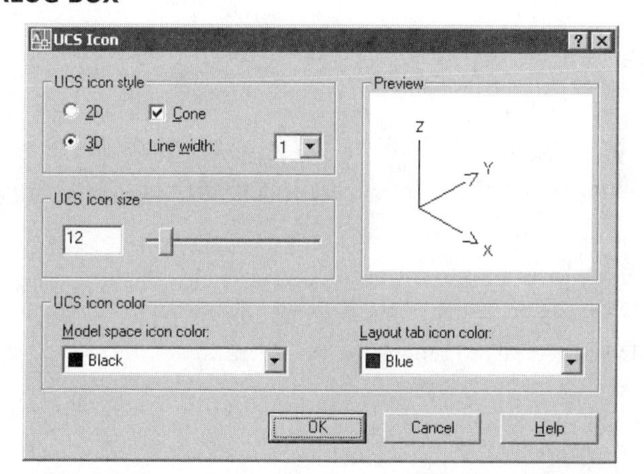

UCS Icon Style

○ **2D** displays flat UCS icon.

⊙ **3D** displays tripod UCS icon.

Cone option (available only with 3D style):

- ☑ Displays arrowheads as 3D cones.
- ☐ Displays plain arrowheads.

Line width changes the line width from 1 to 2 or 3 pixels.

UCS Icon Size

Slider bar changes the icon size from 5 to 95 pixels.

UCS Icon Color

Model space icon color selects the icon's color in model space.

Layout tab icon color selects the icon's color in layout (paper space).

RELATED SYSTEM VARIABLE

UcsIcon determines the visibility and location of UCS icon:

0 — UCS icon not displayed.

1 — UCS icon displayed in lower-left corner.

2 — UCS icon displayed at the UCS origin, when possible.

3 — UCS icon displayed at UCS always (default).

RELATED COMMAND

UCS creates and controls user-defined coordinate systems.

TIPS

- The UCS icon varies, depending on the current viewpoint relative to the active UCS..
- When AutoCAD switches from 2D wireframe mode to one of the VsCurrent command's 3D options, the UCS icon changes to a rendered 3D icon:

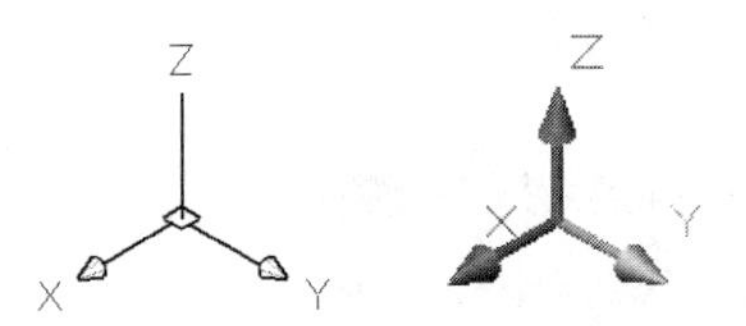

- There is generally no need for the UCS icon in 2D drafting, and it can be safely turned off.
- The Options command's 3D Modeling tab has options for toggling the display of the UCS icon:
  - Display in 2D model space.
  - Display in 3D parallel projection.
  - Display in 3D perspective projection.

UcsMan

2000 Displays the UCS dialog box.

Commands	Aliases	Ctrl+	F-key	Alt+	Menu Bar	Tablet
ucsman	dducs	...	...	TU	Tools	W8
	uc				↳Named UCS	
+ucsman						

Command: ucsman

Displays dialog box.

UCS DIALOG BOX

Named UCSs tab

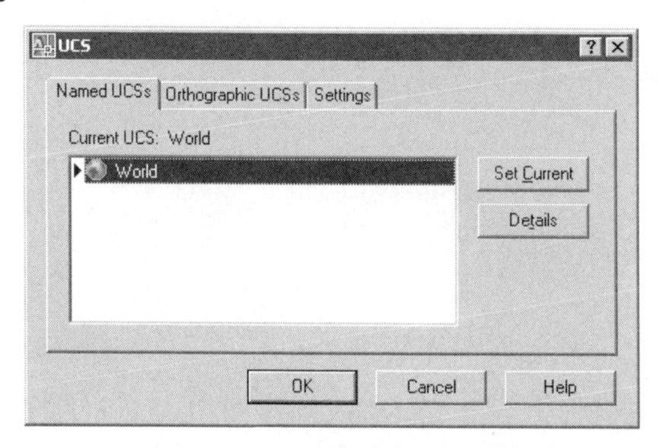

Named UCSs lists the names of the AutoCAD-generated and user-defined coordinate systems of the current viewport in the active drawing; the arrowhead points to the current UCS.

Set Current restores the selected UCS.

Details displays the UCS Details dialog box:

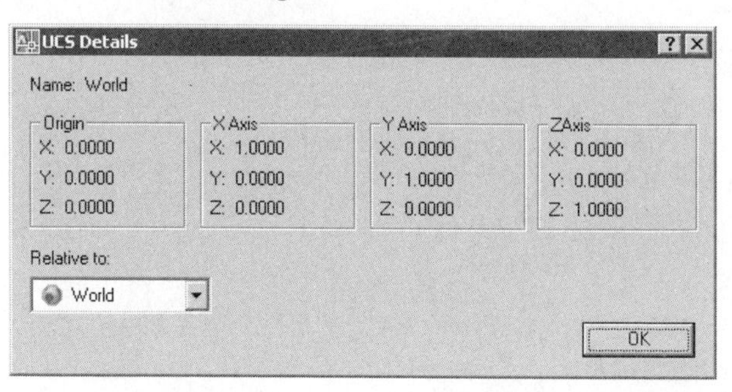

Orthographic UCSs tab

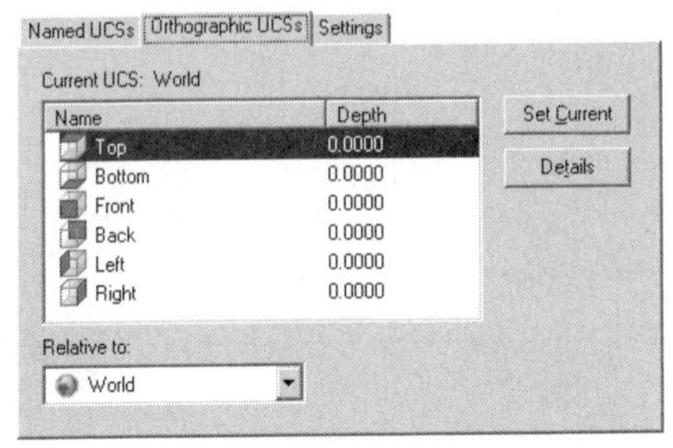

Name lists the six standard orthographic UCS views: top, bottom, front, back, left, and right.

Depth specifies the height of the UCS above the x,y plane.

Relative to specifies the orientation of the selected UCS relative to the WCS or to a customized UCS.

Set Current activates the selected UCS.

Details displays the UCS Details dialog box.

Settings tab

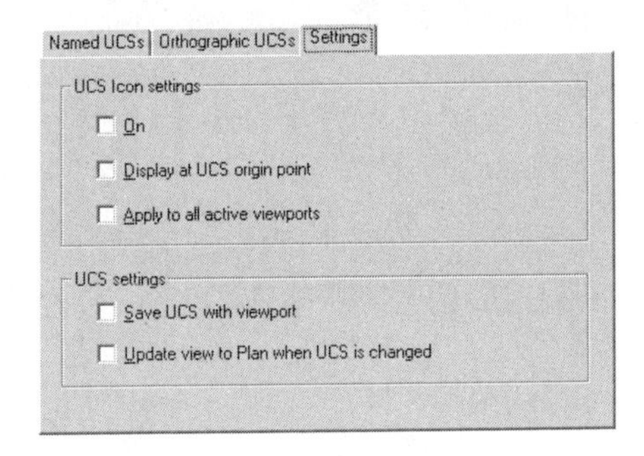

UCS icon settings options

☑ **On** displays the UCS icon in the current viewport; each viewport can display the UCS icon independently.

Display at UCS origin point:

☑ Displays the UCS icon at the origin of the current UCS.

☐ Displays the UCS icon at the lower-left corner of the viewport.

Apply to all active viewports:

☑ Applies these UCS icon settings to all active viewports in the current drawing.

☐ Applies to the current viewport only.

UCS settings options

Save UCS with viewport:

☑ Saves the UCS setting with the viewport.

☐ Determines UCS settings from current viewport.

☑ **Update view to Plan when UCS is changed** restores plan view when the UCS changes.

SHORTCUT MENUS

*Right-click the list in the **Named UCSs** tab:*

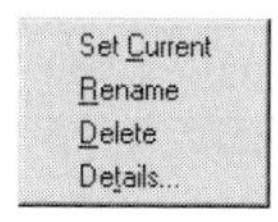

Set Current sets the selected UCS as active.

Rename renames the selected UCS; you cannot rename the World UCS.

Delete erases the selected UCS; you cannot delete the World UCS.

Details displays the UCS Details dialog box.

*Right-click the list in the **Orthographic UCS** tab:*

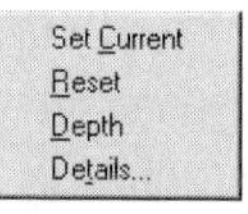

Set Current sets the selected UCS as active.

Reset restores the origin of the selected UCS.

Depth moves the UCS in the z direction; displays dialog box:

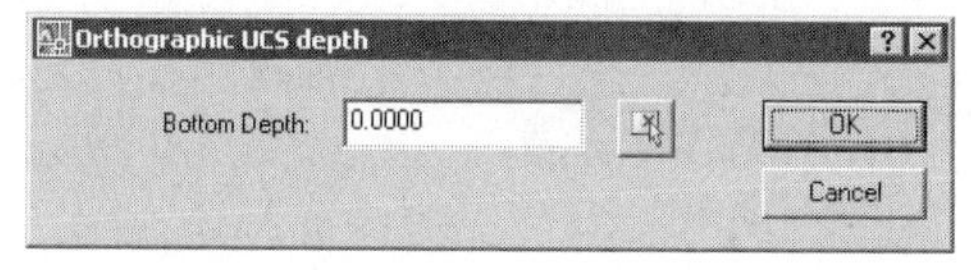

Details displays the UCS Details dialog box.

+UCSMAN Command

Command: +ucsman

Tab index <0>: *(Type* **1**, **2**, *or* **3**.*)*

COMMAND LINE OPTION

Tab index displays the tab related to the tab number:

- **0** — Named UCS tab.
- **1** — Orthographic UCS tab.
- **2** — Settings tab.

RELATED SYSTEM VARIABLES

*See the **UCS** command.*

RELATED COMMANDS

UCS displays the UCS options at the command line.

UcsIcon changes the display of the UCS icon.

Plan displays the plan view of the WCS or a UCS.

TIPS

- Functions of this command were formerly carried out by the DdUcs and DdUcsP commands.
- **Unnamed** is the first entry, when the current UCS is unnamed. **World** is the default for new drawings; it cannot be renamed or deleted. **Previous** is the previous UCS; you can move back through several previous UCSs.

Undefine

Rel. 9 Makes AutoCAD commands unavailable.

Command	Alias	Ctrl+	F-key	Alt+	Menu Bar	Tablet
undefine	...	...	...	...		

Command: undefine
Enter command name: *(Enter name.)*

Example usage:
Command: undefine
Enter command name: line
Command: line
Unknown command. Type ? for list of commands.
Command: .line
From point:

COMMAND OPTIONS

Enter command name specifies the name of the command to make unavailable.

. *(period)* is the prefix for undefined commands to redefine them temporarily.

RELATED COMMAND

Redefine redefines an AutoCAD command.

TIPS

- Prefixing undefined command with a period bypasses the effect of this command. For example:

  Command: .line

- This command allows AutoLISP and ObjectARX to override native AutoCAD commands.
- Commands created by programs cannot be undefined, including the following programming interfaces:

  AutoLISP and Visual LISP.
  ObjectARx.
  Visual Basic for Applications.
  External commands.
  Aliases.

- In menu macros written with international language versions of AutoCAD, precede command names with an underscore character (_) to translate the command name into English automatically.

 Undo

V. 2.5 Undoes the effect of the previous command(s).

Command	Alias	Ctrl+	F-key	Alt+	Menu Bar	Tablet
undo	...	...	...	...	...	...

Command: undo

Enter the number of operations to undo or [Auto/Control/BEgin/End/Mark/Back] <1>: *(Enter a number, or an option.)*

COMMAND LINE OPTIONS

Auto treats a menu macro as a single command.

Control limits the options of the **Undo** command.

BEgin groups a sequence of operations (formerly the Group option).

End ends the group option.

Mark sets a marker.

Back undoes back to the marker.

number indicates the number of commands to undo.

Control Options

Enter an UNDO control option [All/None/One] <All>:

All turns on full undo.

None turns off the undo feature.

One limits the **Undo** command to a single undo.

RELATED COMMANDS

Oops unerases the most-recently erased object.

Quit leaves the drawing without saving changes.

Redo undoes the most recent undo.

MRedo undoes multiple undoes.

U undoes a single step.

RELATED SYSTEM VARIABLES

UndoCtl determines the state of undo control:

0 — Undo disabled.

1 — Undo enabled.

2 — Undo limited to one command.

4 — Auto-group mode.

8 — Group currently active.

UndoMarks specifies the number of undo marks placed in the **Undo** control stream.

TIPS

- Since the undo mechanism creates a mirror drawing file on disk, disable the Undo command with system variable UndoCtl (set it to 0) when your computer is low on disk space.
- There are some commands that cannot be undone, such as Save, Plot, and UndoCtl.

Union

2000 Joins two or more solids and regions into a single body.

Command	Alias	Ctrl+	F-key	Alt+	Menu Bar	Tablet
union	uni	...	...	MNU	Modify ↳Solids Editing ↳Union	X15

Command: union
Select objects: *(Select one or more solid objects.)*
Select objects: *(Select one or more solid objects.)*
Select objects: *(Press Enter to end command.)*

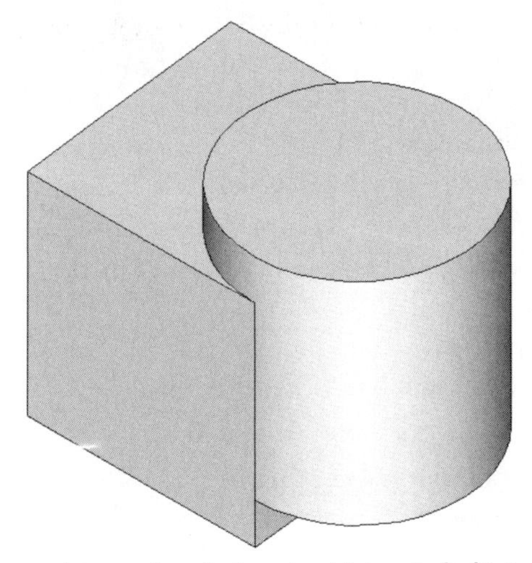

A box and a cylinder unioned into a single object.

COMMAND LINE OPTION

Select objects selects the objects to join into a single object; you must select at least two solid or region objects.

RELATED COMMANDS

Intersect creates a solid model from the intersection of two objects.

Subtract creates a solid model by subtracting one object from another.

RELATED SYSTEM VARIABLES

ShowHist toggles display of history in solids (*new to AutoCAD 2007*).

SolidHist toggles the retention of history in solids (*new to AutoCAD 2007*).

TIPS

- You must select at least two solid or coplanar region objects. The two objects need not overlap for this command to operate.
- History, if enabled, allows you to work with the original objects.

'Units

V. 1.4 Controls the display and format of coordinates and angles, as well as the orientation of angles.

Commands	Aliases	Ctrl+	F-key	Alt+	Menu Bar	Tablet
units	un	...	...	OU	Format	V4
	ddunits				↳Units	
-units	-un					

command: units

Displays dialog box:

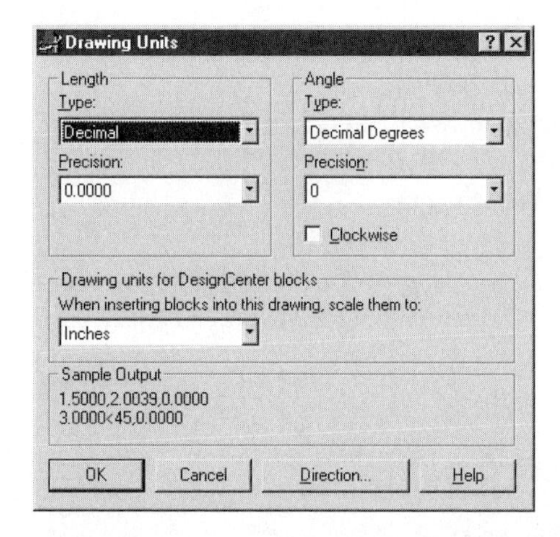

DRAWING UNITS DIALOG BOX

Length

Type sets the format for units of linear measure displayed by AutoCAD: Architectural, Decimal, Engineering, Fractional, or Scientific.

Precision specifies the number of decimal places or fractional accuracy.

Angle

Type sets the current angle format.

Precision sets the precision for the current angle format.

☐ **Clockwise** calculates positive angles in the clockwise direction.

Drawing units for DesignCenter blocks specifies the units when blocks are inserted from the DesignCenter.

Direction displays the Direction Control dialog box.

DIRECTION CONTROL DIALOG BOX

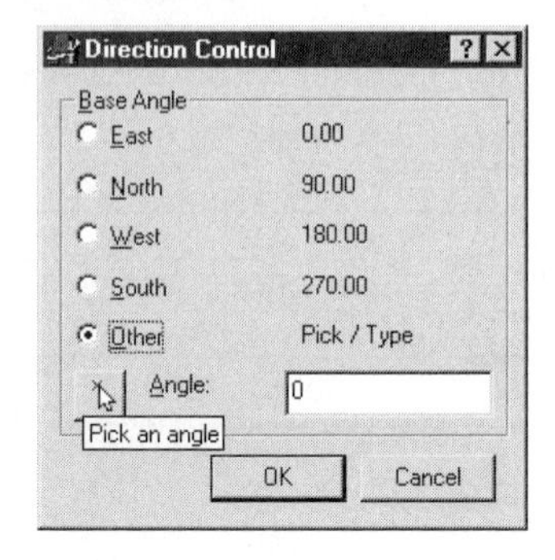

Base Angle

○ **East** sets the base angle to 0 degrees (default).

○ **North** sets the base angle to 90 degrees.

○ **West** sets the base angle to 180 degrees.

○ **South** sets the base angle to 270 degrees.

⊙ **Other** turns on the **Angle** option.

Angle sets the base angle to any direction.

Pick an angle dismisses the dialog box temporarily, and allows you to define the base angle by picking two points in the drawing; AutoCAD prompts you 'Pick angle' and 'Specify second point:'.

-UNITS Command

Command: -units

Report formats:	**(Examples)**
1. Scientific	**1.55E+01**
2. Decimal	**15.50**
3. Engineering	**1'-3.50"**
4. Architectural	**1'-3 1/2"**
5. Fractional	**15 1/2**

With the exception of Engineering and Architectural formats, these formats can be used with any basic unit of measurement. For example, Decimal mode is perfect for metric units as well as decimal English units.

Enter choice, 1 to 5 <2>: *(Enter a value.)*

Enter number of digits to right of decimal point (0 to 8) <4>: *(Enter a value.)*

Systems of angle measure:	**(Examples)**
1. Decimal degrees	**45.0000**
2. Degrees/minutes/seconds	**45d0'0"**
3. Grads	**50.0000g**
4. Radians	**0.7854r**
5. Surveyor's units	**N 45d0'0" E**

Enter choice, 1 to 5 <1>: *(Enter a value.)*

Enter number of fractional places for display of angles (0 to 8) <0>: *(Enter a value.)*

Direction for angle 0:
East 3 o'clock = 0
North 12 o'clock = 90
West 9 o'clock = 180
South 6 o'clock = 270
Enter direction for angle 0 <0>: *(Enter a value.)*

Measure angles clockwise? [Yes/No] <N> *(Enter **Y** or **N**.)*

COMMAND LINE OPTIONS

Report formats selects scientific, decimal, engineering, architectural, or fractional format for length display.

Number of digits to right of decimal point specifies the number of decimal places between 0 and 8.

Systems of angle measure selects decimal degrees, degrees/minutes/seconds, grads, radians, or surveyor's units for angle display.

Denominator of smallest fraction to display specifies the denominator of fraction displays, such as 1/2 or 1/256.

Number of fractional places for display of angles specifies the number of decimal places between 0 and 8.

Direction for angle 0 selects the direction for 0 degrees from east, north, west, or south.

Do you want angles measured clockwise?

- **Yes** measures angles clockwise.
- **No** measures angles counterclockwise.

RELATED SYSTEM VARIABLES

AngBase specifies the direction of zero degrees.

AngDir specifies the direction of angle measurement.

AUnits specifies the units of angles.

AuPrec specifies the displayed precision of angles.

InsUnits specifies the drawing units for blocks dragged from the DesignCenter:

- **0** — Unitless
- **1** — Inches
- **2** — Feet
- **3** — Miles
- **4** — Millimeters
- **5** — Centimeters
- **6** — Meters
- **7** — Kilometers
- **8** — Microinches
- **9** — Mils
- **10** — Yards; 3 feet
- **11** — Angstroms; 0.1 nanometers
- **12** — Nanometers; 10E-9 meters
- **13** — Microns; 10E-6 meters
- **14** — Decimeters; 0.1 meter
- **15** — Decameters; 10 meters
- **16** — Hectometers; 100 meters
- **17** — Gigameters; 10E9 meters
- **18** — Astronomical Units; 149.597E8 kilometers
- **19** — Light Years; 9.4605E9 kilometers
- **20** — Parsecs; 3.26 light years

InsUnitsDefSource specifies source units to be used.

InsUnitsDefTarget specifies target units to be used.

LUnits specifies the units of measurement.

LuPrec specifies the displayed precision of coordinates.

UnitMode toggles the type of display units.

RELATED COMMAND

New sets up drawings with Imperial or metric units.

TIPS

- Because 'Units is a transparent command, you can change units during another command.
- The 'Direction Angle' prompt lets AutoCAD start the angle measurement from any direction.
- AutoCAD accepts the following notations for angle input:

Notation	Meaning
<	Specify an angle based on current units setting.
<<	Bypass angle translation set by Units command to use 0-angle-is-east direction and decimal degrees.
<<<	Bypass angle translation; use angle units set by the Units command and 0-angle-is-east direction.

- The system variable UnitMode forces AutoCAD to display units in the same manner that you enter them.
- At one time, the smallest fraction was 1/2048, but now the smallest is 1/256.
- Do not use a suffix — such as 'r' or 'g' — for angles entered as radians or grads; instead, use the Units command to set angle measurement to radians and grads.
- The Drawing units for DesignCenter blocks option is for inserting blocks from the DesignCenter, and especially when the block was created in other units.
- To prevent blocks from being scaled when dragged from the DesignCenter palette, select Unitless.

UpdateField

2005 Forces the update of selected fields.

Command	Alias	Ctrl+	F-key	Alt+	Menu Bar	Tablet
updatefield	...	...	...	TT	Tools ↳Update Fields	...

Command: updatefield
Select objects: *(Select one or more fields.)*
Select objects: *(Press Enter to end field selection.)*
***n* field(s) found**
***n* field(s) updated**

COMMAND LINE OPTION

Select objects selects one or more fields; press CTRL+A to select all objects in the drawing, including field text.

RELATED COMMANDS

Field places field text, which is automatically updated as its value changes.

Find finds fields.

DdEdit edits the properties of field text.

RELATED SYSTEM VARIABLE

FieldEval determines when fields are updated (default = 31, all on):

- **0** — Fields are not updated (static).
- **1** — Fields are updated when drawing is opened.
- **2** — Fields are updated when drawing is saved.
- **4** — Fields are updated when plotted.
- **8** — Fields are updated with the eTransmit command.
- **16** — Fields are updated when the drawing is regenerated.

TIP

- This command forces individual fields to update. Depending on the setting of the FieldEval system variable, all fields are updated automatically.

UpdateThumbsNow

2005 Forces the update of preview images in the Sheet Set Manager palette.

Command	Alias	Ctrl+	F-key	Alt+	Menu Bar	Tablet
updatethumbsnow	...	...	...	...	...	...

Command: updatethumbsnow

COMMAND LINE OPTIONS

None.

RELATED COMMAND

SheetSet controls sheet sets.

RELATED SYSTEM VARIABLE

UpdateThumbnail determines which thumbnails are updated (default = 15):

0 — Thumbnail previews not updated.

1 — Model view thumbnails updated.

2 — Sheet view thumbnails updated.

4 — Sheet thumbnails updated.

8 — Thumbnails updated when sheets and views created, modified, and restored.

16 — Thumbnails updated when drawing saved.

TIPS

- This command controls when preview images are updated; required when the drawing changes.
- Thumbnails are displayed in the Sheet Set Manager, as illustrated below:

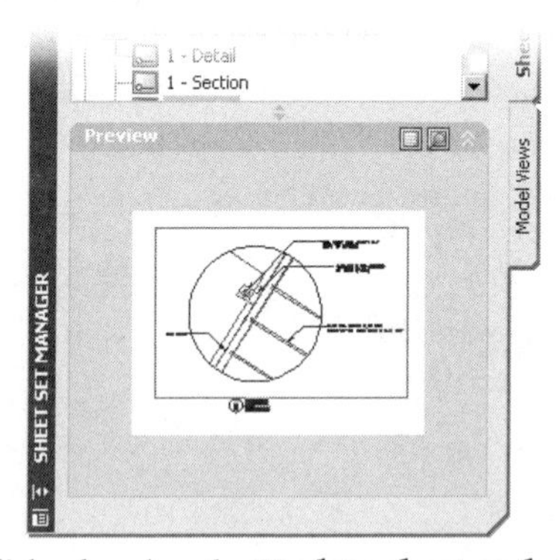

- The SaveAs command's dialog box has the **Update sheet and view thumbnails now** option.

VbaIde

2000 Displays the Visual Basic window (short for Visual Basic for Applications Integrated Development Environment).

Command	Alias	Alt+	F-key	Alt+	Menu Bar	Tablet
vbaide	...	**F11**	...	**TAV**	**Tools** ↳**Macro** ↳**Visual Basic Editor**	...

Command: vbaide

Displays window:

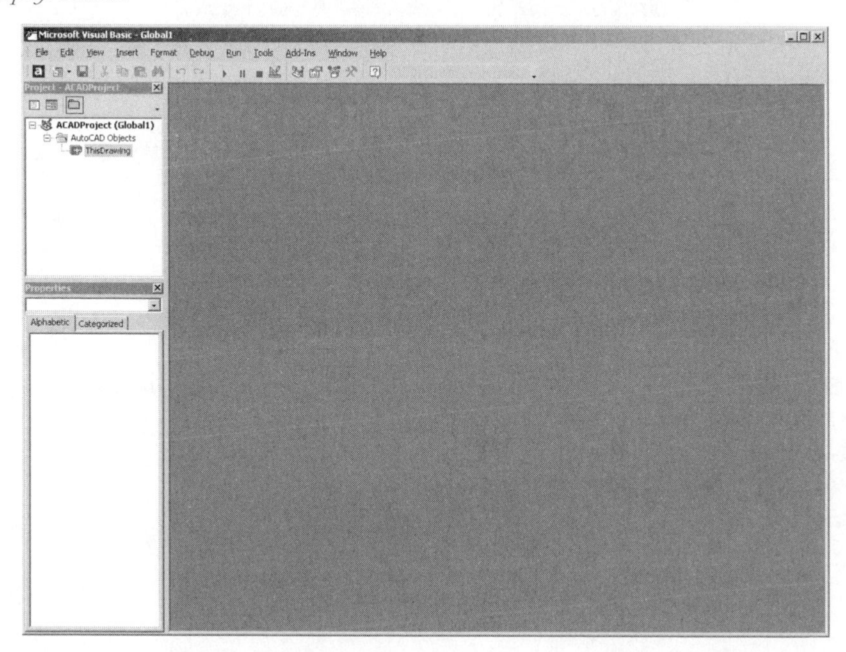

MENU BAR

Select ***Help | Microsoft Visual Basic Help*** *for assistance in using this VBA IDE window.*

RELATED COMMANDS

VbaLoad loads a VBA project; displays the Open VBA Project dialog box.

VbaMan displays the VBA Manager dialog box.

VbaRun displays the Macros dialog box.

VbaStmt executes a VBA expression at the command line.

VbaUnload unloads a VBA project.

RELATED SYSTEM VARIABLES

None.

TIPS

- *VBA* is short for "Visual Basic for Applications," a macro programming language common to a number of Windows applications; it is based on the Visual Basic programming language.
- AutoCAD contains sample VBA projects in the *\autocad 2007\sample\vba* folder.
- For more information about Visual Basic for Applications, read the online *ActiveX and VBA Developer's Guide* help files included with AutoCAD.
- Because many viruses can be spread via VBA macros, I strongly recommend that your computer have real-time virus protection to prevent infection.
- Because loading a macro from other sources into AutoCAD can also expose your computer to malicious viruses, AutoCAD displays the following warning:

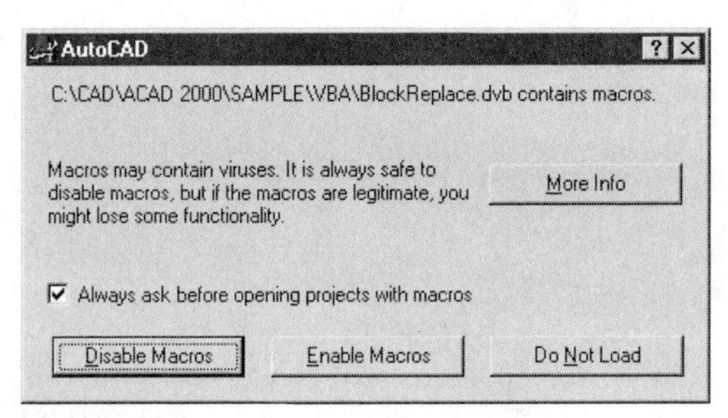

More Info displays AutoCAD's on-line help window.

Always ask before opening projects with macros:

☑ Displays this dialog box.

☐ Prevents this dialog box from being displayed; to turn back on, use the VbaRun command's Options dialog box; see VbaRun command.

Disable Macros loads a VBA project file, but disables macros; you can view, edit, and save the macros. To enable macros, close the project, and then open it again with Enable Macros.

Enable Macros loads the project file with macros enabled.

Do Not Load prevents the project file from being loaded.

VbaLoad, VbaUnload

2000 Loads a VBA project into AutoCAD (short for Visual Basic for Applications LOAD).

Commands	Alias	Ctrl+	F-key	Alt+	Menu Bar	Tablet
vbaload	...	...	...	TAL	Tools ↳Macro ↳Load Project	...
vbaunload						
-vbaload						

Command: vbaload

Displays the Open VBA Project dialog box; select a .dvd file, and then click ***Open.***

When the project contains macros, the AutoCAD dialog box is displayed; see the VbaIde command.

Command: vbaunload

Unload VBA Project: *(Enter project name.)*

COMMAND LINE OPTIONS

Unload VBA Project specifies the name of the VBA project to unload.

Enter unloads the active global project.

-VBALOAD Command

Command: -vbaload

Open VBA project <*projectname*>: *(Enter project name.)*

COMMAND LINE OPTION

Open VBA project specifies the project path and filename.

RELATED COMMANDS

Vbaide displays the Visual Basic for Applications development environment window.

VbaMan displays the VBA Manager dialog box.

VbaRun displays the Macros dialog box.

VbaStmt executes a VBA expression at the command line.

VbaUnload unloads a VBA project.

TIPS

- You may load one or more VBA projects; there is no practical limit to the number.
- This command does not load embedded VBA projects; these projects are automatically loaded with the drawing.
- When the project contains macros, AutoCAD displays the AutoCAD dialog box to warn you about protection against macro viruses; see the Vbaide command.
- For more information about Visual Basic for Applications, read the *ActiveX and VBA Developer's Guide* included with AutoCAD.

VbaMan

2000 Displays the VBA Manager dialog box (short for Visual Basic for Applications MANager).

Command	Alias	Ctrl+	F-key	Alt+	Menu Bar	Tablet
vbaman	...	...	...	**TAV**	**Tools** **↳Macro** **↳VBA Manager**	...

Command: vbaman

Displays dialog box:

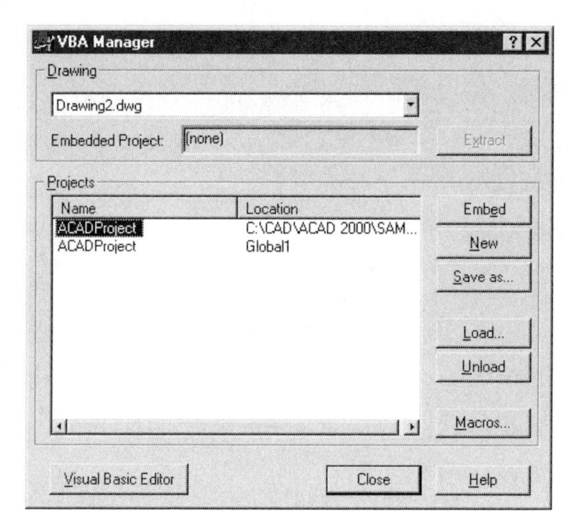

VBA MANAGER DIALOG BOX

Drawing

Drawing lists the names of drawings currently loaded in AutoCAD.

Embedded Project specifies the name of the embedded project.

Extract moves the embedded project from the drawing to a global project file.

Projects

Embed embeds the project in the drawing.

New creates a new project; default name = Global *n*.

Save as saves a global project.

Load displays the Open VBA Project dialog box; see the VbaLoad command.

Unload unloads the global project.

Macros displays the Macros dialog box; see the VbaRun command.

Visual Basic Editor displays the Visual Basic Editor; see the VbaIde command.

 VbaRun

2000 Displays the Macros dialog box (short for Visual Basic for Applications RUN).

Command	Alias	Alt+	F-key	Alt+	Menu Bar	Tablet
vbarun	...	F8	...	TAM	Tools ↳Macro ↳Macros	...
-vbarun						

Command: vbarun

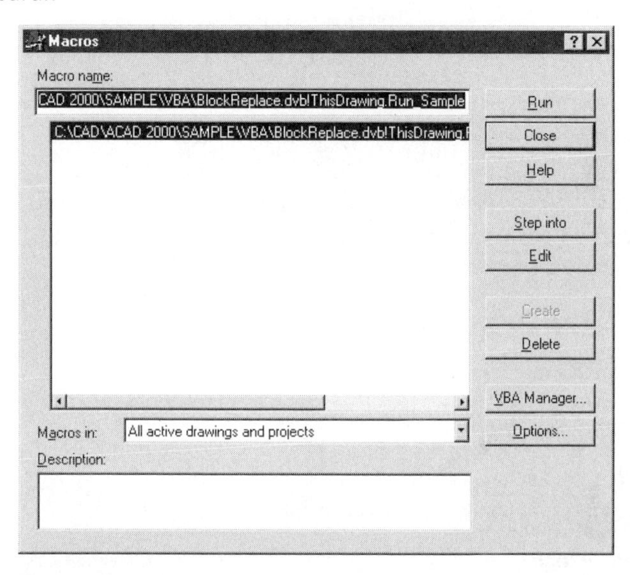

MACROS DIALOG BOX

Macro name specifies the name of the macro; enter a name or select one from the list.

Macros in specifies the projects and drawings containing macros in all active drawings and projects, all active drawings, all active projects, and any single drawing or project currently open in AutoCAD.

Description describes the macro; you may modify the description.

Run runs the macro.

Close closes the dialog box.

Help displays context-sensitive on-line help.

Step into displays the Visual Basic Editor, and executes the macro, pausing at the first executable line of code.

Edit displays the Visual Basic Editor with the macro; see the VbaIde command.

Create displays the Visual Basic Editor with an empty procedure.

Delete erases the selected macro.

VBA Manager displays the VBA Manager dialog box; see the VbaMan command.

Options displays the VBA Options dialog box.

VBA OPTIONS DIALOG BOX

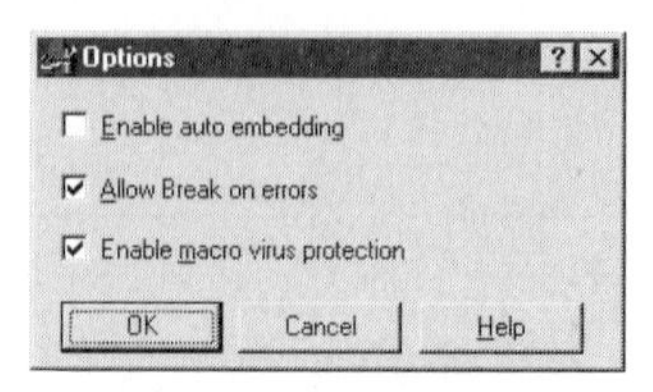

☐ **Enable auto embedding** creates an embedded VBA project for all drawings when you open the drawing:

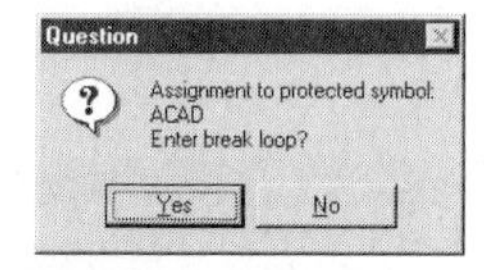

Allow Break on errors:

☑ Stops the macro, and displays the **Visual Basic Editor** with the code, showing the error in the macro.

☐ Displays an error message, and stops the macro.

☑ **Enable macro virus protection** enables virus protection, which displays a dialog box when VBA macros are loaded; see the **VbaIde** command.

-VBARUN Command

Command: -vbarun

Macro name: *(Enter macro name.)*

COMMAND LINE OPTION

Macro name treats a menu macro as a single command.

RELATED COMMANDS

Vbaide displays the Visual Basic for Applications integrated development environment window.

VbaLoad loads a VBA project; displays the Open VBA Project dialog box.

VbaMan displays the VBA Manager dialog box.

VbaStmt executes a VBA expression at the command line.

VbaUnload unloads a VBA project.

TIPS

- A *macro* is an executable subroutine; each project can contain one or more macros.
- When the macro's name is not unique among loaded projects, include the module's project and names in this format: **Project.Module.Macro**
- When the macro is not yet loaded, include the *.dvb* file name using this format:

  Filenamedvb!Project.Module.Macro

VbaStmt

2000 Executes a single line VBA expression at the command prompt (short for Visual Basic for Applications StaTeMenT).

Command	Alias	Ctrl+	F-key	Alt+	Menu Bar	Tablet
vbastmt	...	...	...	...	...	...

Command: vbastmt
Statement: *(Enter VBA statement.)*

COMMAND LINE OPTION

Statement specifies the VBA statement for AutoCAD to execute.

RELATED COMMANDS

Vbaide displays the Visual Basic for Applications integrated development environment window.

VbaLoad loads a VBA project; displays the Open VBA Project dialog box.

VbaMan displays the VBA Manager dialog box.

VbaRun displays the Macros dialog box.

VbaUnload unloads a VBA project.

RELATED SYSTEM VARIABLES

None.

TIPS

- A VBA *statement* is a complete instruction containing keywords, operators, variables, constants, and expressions.
- A VBA *macro* is an executable subroutine.
- At the 'Statement' prompt, enter a single line of code; use the colon (:) to separate multiple statements on the single line.

View

V. 2.0 Saves and displays views by name in the current viewport; creates view categories for sheet sets, live sections, and backgrounds for visual styles; and controls parameters for cameras.

Commands	Aliases	Ctrl+	F-key	Alt+	Menu Bar	Tablet
view	**v** **ddview**	**...**	**...**	**VN**	**View** **↳Named Views**	**M5**
-view	**-v**			**V3**	**View** **↳3D Views**	**O3-Q5**

Command: view

This dialog box has substantial changes in AutoCAD 2007:

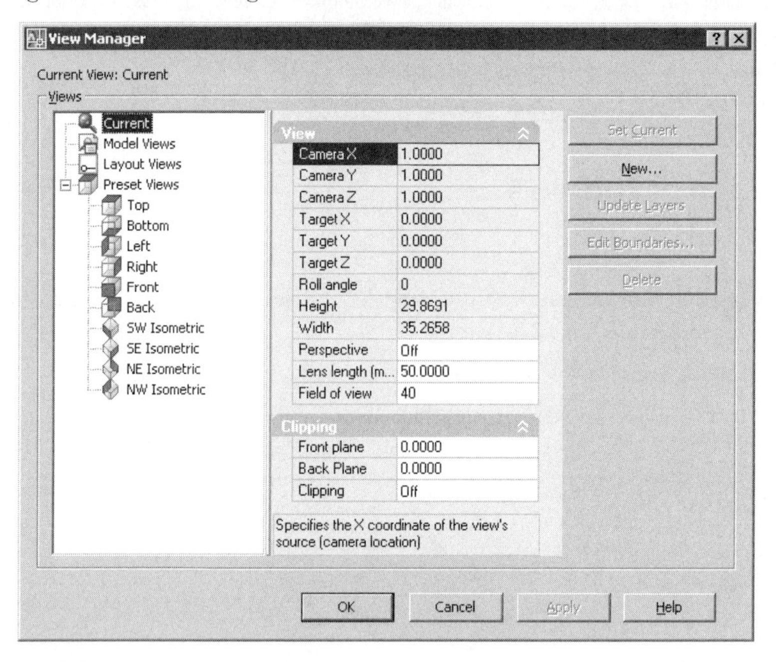

VIEW DIALOG BOX

Set Current restores the named view; alternatively, double-click the view name.

New displays the New View dialog box.

Update Layers updates view to match the layer visibility settings of the current viewport.

Edit Boundaries highlights the view, by graying out the area outside of the view.

Delete removes the named view from the drawing.

NEW VIEW DIALOG BOX

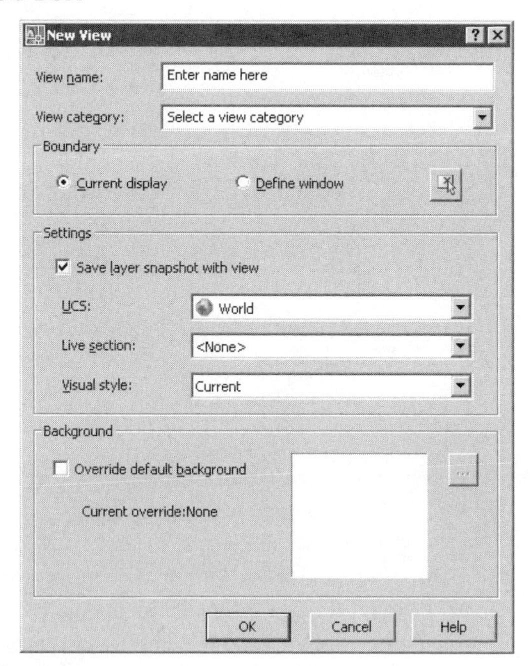

View name specifies the view name; up to 255 characters long.

View Category specifies the default prefix for named views. In the Sheet Set Manager, select the View List tab to see the names of views, under categories, as illustrated below:

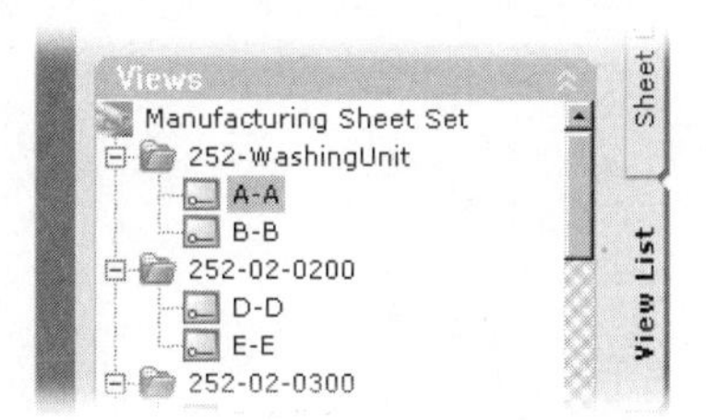

View categories, such as "A-A," listed in the View List tab of the Sheet Set Manager palette.

Boundary options

⊙ **Current display** stores the current viewport as the named view.

○ **Define window** stores a windowed area as the named view.

Define View Window dismisses the dialog box temporarily so that you can pick the two corners that define the view.

AutoCAD clears the dialog box, and prompts:

Specify first corner: *(Pick a point.)*

Specify opposite corner: *(Pick another point.)*

Specify first corner (or press ENTER to accept): *(Pres Enter to return to the dialog box.)*

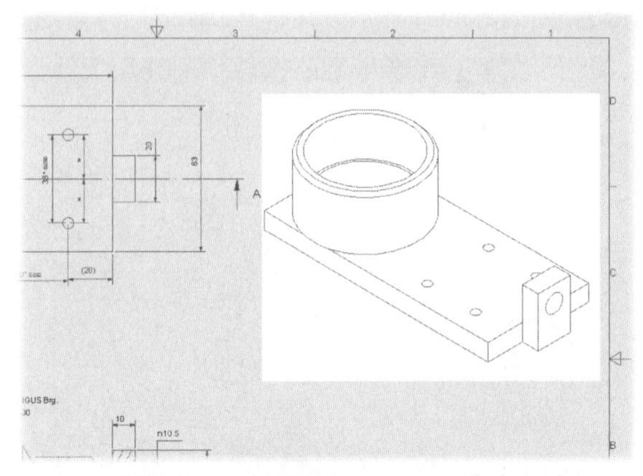

The white rectangle defines the area of the view window.

Settings

☑ **Save Current Layer Settings with View** toggles whether layer properties (such as freeze/thaw, lock/unlock, plot/no plot, and so on) are stored with the view. You must first use the Layer command to set layer properties.

UCS name specifies the name of a UCS to store with the named view. You must first create named UCSs with the Ucs command.

Live Section specifies the named live section to be displayed when the named view is restored. You must first use the SectionPlane and LiveSection commands to create the named section (*new to AutoCAD 2007*).

Visual Style specifies the named visual style to be displayed when the view is restored. You must first use the VisualStyles command to create the named visual style (*new to AutoCAD 2007*).

Background options

New to AutoCAD 2007:

☐ **Override default background** allows you to create a custom background for the view. Click to display the Background dialog box (*formerly the Background command*).

... displays the Background dialog box.

BACKGROUND DIALOG BOX

New to AutoCAD 2007.

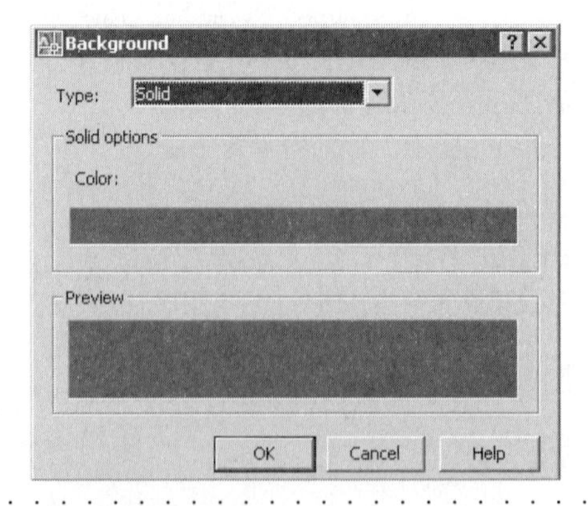

Type selects the type of background:

- **Solid** shows a single color; click the **Color** bar to display the Select Color dialog box.
- **Gradient** shows two or three colors.
- **Image** shows a picture.

Gradient options

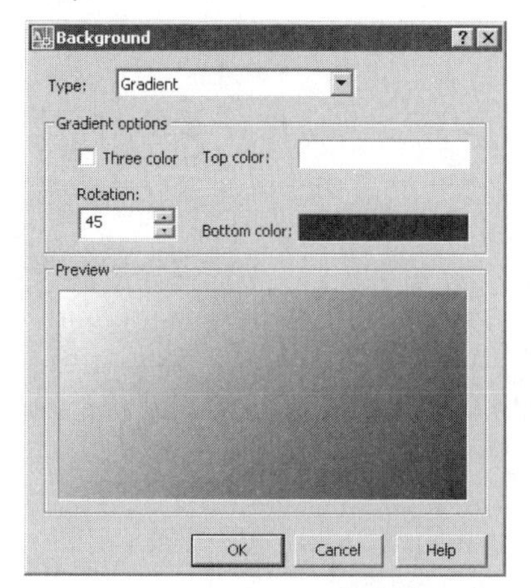

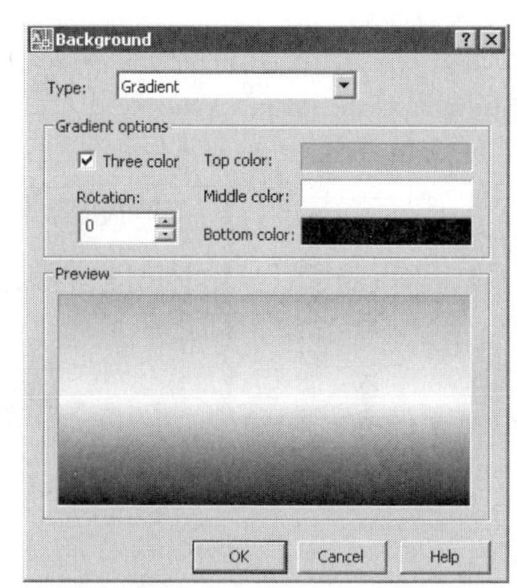

***Left**: Two-color gradient at 45 degrees.*
***Right**: Three-color gradient at 0 degrees.*

Three color toggles gradients between two and three colors.

Rotation displays the gradient at an angle; range is from 90 (vertical) to 0 (horizontal) to -90 degrees.

Top color, **Middle color**, and **Bottom color** display the Select Color dialog box.

Image options

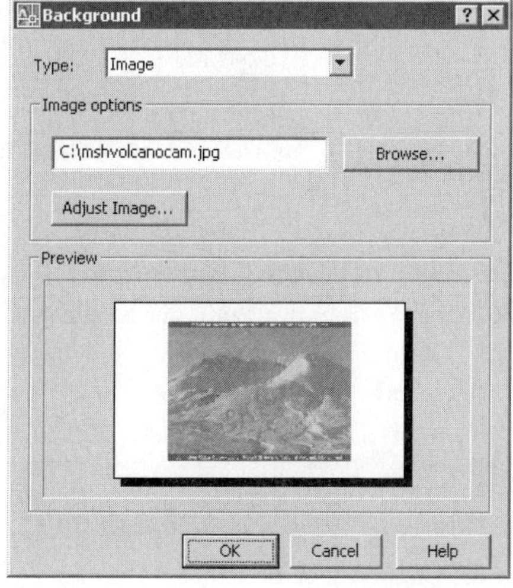

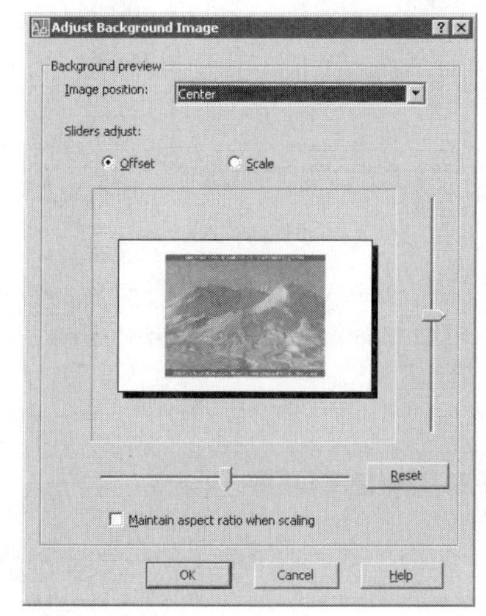

***Left**: Selecting the image background.*
***Right**: Adjusting the image.*

Browse selects the image from the Select File dialog box. You can chose from Targe, bitmap, TIFF, JPEG, GIF, and PCX formats.

Adjust image displays the Adjust Image dialog box.

ADJUST IMAGE DIALOG BOX

Formerly the SetUV command in previous releases of AutoCAD. New to AutoCAD 2007.

Image position:

- **Center** centers the image in the viewport.
- **Stretch** fits the image to the viewport.
- **Tile** repeats the image to fill the viewport.

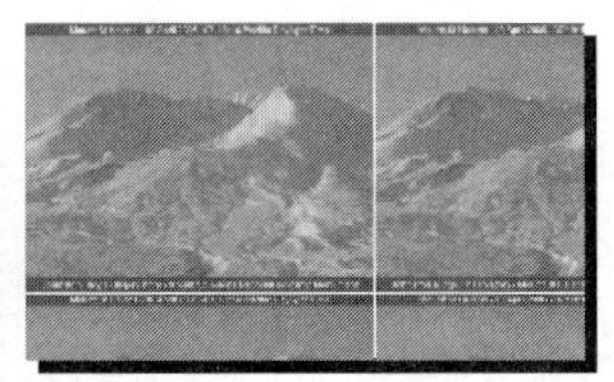

Left: *Image centered in viewport.*
Center: *Image stretched to fit viewport.*
Right: *Image tiled (repeated) to fit viewport.*
(Image is from Mount St Helens volcano-cam at www.fs.fed.us/gpnf/volcanocams/msh)

Sliders adjust:

Neither of the following options works in ***Stretch*** *mode:*

⊙ **Offset** moves the image relative to the viewport. When image goes outside the viewport boundary, its preview appears dimmed.

○ **Scale** changes the size of the image, making it larger or smaller. Negative x scale factors mirror the image; negative y inverts the image.

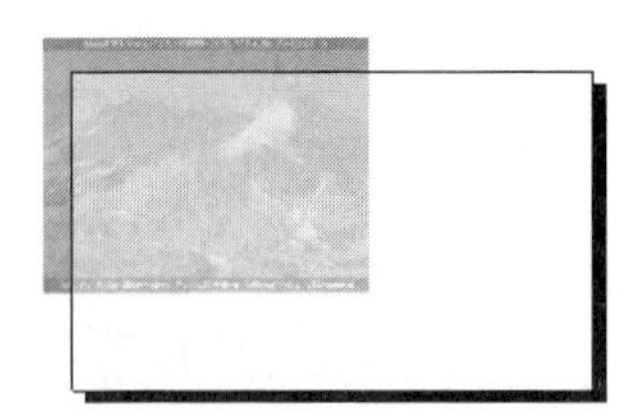

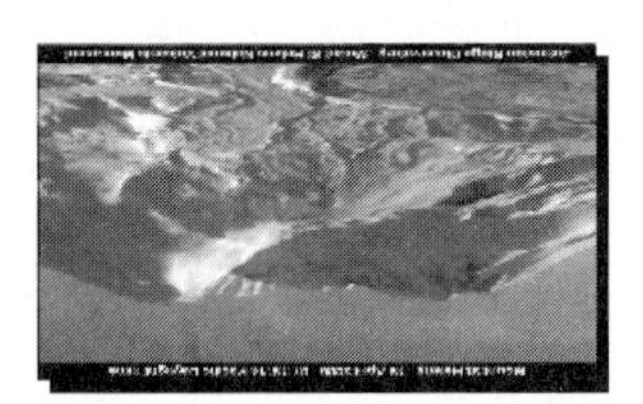

Left: *Image moved in viewport.*
Center: *Image scaled larger than viewport.*
Right: *Image upside down with negative y-scale factor.*

Maintain aspect ratio when scaling prevents the image from distortion due to unequal scaling.

Reset returns the image to its original size and aspect ratio.

VIEW PROPERTIES

The properties vary, depending on the type of view selected. New to AutoCAD 2007.

Properties must be set in the View dialog box, because the Properties palette cannot select views.

Properties in gray cannot be edited (read-only).

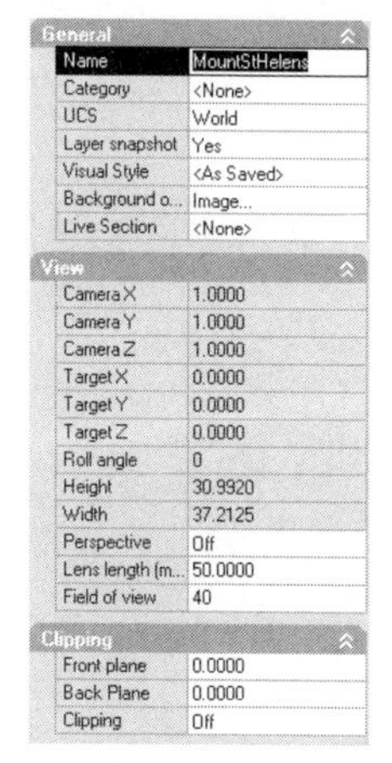

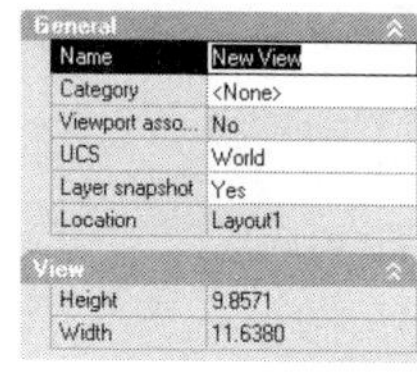

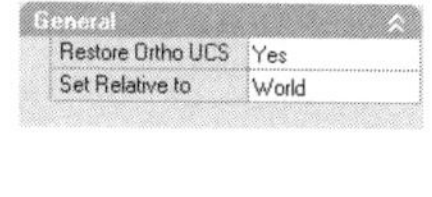

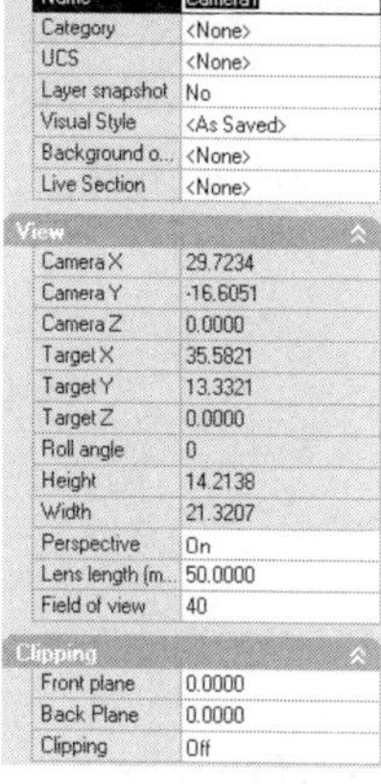

***From left to right**: Properties of views in model space, paper space (layout tabs), preset views, and cameras.*

Properties listed in alphabetical order:

Background specifies the type of background color or image: none, Solid, Gradient, Image, or edit.

Camera X, **Y**, and **Z** specifies the x, y, and z coordinates of the view's viewpoint (camera); read-only.

Category lists the names of view categories: none or user-defined.

Clipping selects clipping options: Off, Front On, Back On, or Front and Back On.

Field of View indicates the horizontal field of view angle; coupled with Lens Length.

Front Plane and **Back Plane** indicate the front (and back) clipping plane's offset distance.

Height indicates the view height above or below the x,y-plane; read-only.

Layer Snapshot toggles whether layer parameters are stored with the view: yes or no.

Lens Length (mm) indicates the lens length in millimeters; coupled with Field of View.

Live Section determines whether a live section is displayed when the view is restored: none or user-defined.

Location reports the layout associated with the view; read-only.

Name names the view.

Perspective toggles the projection between perspective and parallel: on or off.

Restore Ortho UCS restores the associated UCS when the orthographic view is current: yes or no.

Roll angle indicates the view's roll angle; read-only.

Set Relative To specifies the base coordinate system for the orthographic view: World or user-defined.

Target X, **Y**, and **Z** specifies the x, y, and z coordinates of the view's look-at point (target); read-only.

UCS names the user-defined coordinate system saved with the view: none, World, or user defined.

Viewport Association toggles the view's association with a sheet set's viewport.

Visual Style specifies the visual style to restore with the view: 2D Wireframe, 3D Wireframe, 3D Hidden, Realistic, Conceptual, or user-defined.

Width indicates the view width (field of view); read-only.

SHORTCUT MENU

Right-click the Views list:

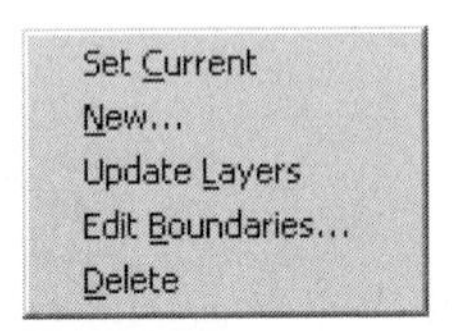

Set Current sets the selected view as active.

New creates a new view; opens the New View dialog box.

Update Layers updates the view to match the layer visibility settings of the current viewport.

Edit Boundaries highlights the view, by graying out the area outside of the view.

Delete erases the selected view; you cannot delete the Current view.

-VIEW Command

Command: -view

AutoCAD 2007 changes the option prompts:

Enter an option [?/Delete/Orthographic/Restore/Save/sEttings/Window]: *(Enter an option.)*

COMMAND LINE OPTIONS

? lists the names of views saved in the current drawing.

Delete deletes a named view.

Orthographic restores predefined orthographic views.

Restore restores a named view.

Save saves the current view with a name.

sEttings specifies view settings (*new to AutoCAD 2007*).

Window saves a windowed view with a name.

Orthographic options

Enter an option [Top/Bottom/Front/BAck/Left/Right] <Top>: *(Enter an option.)*

Select Viewport for view: *(Pick a viewport.)*

Enter an option selects a standard orthographic view for the current viewport: Top, Bottom, Front, BAck, Left, or Right.

Select Viewport for view selects the viewport — in either Model or Layout tab — in which to apply the orthographic view.

sEttings options

Enter an option [Background/Categorize/Layer snapshot/live Section/Ucs/ Visual style]]: *(Enter an option.)*

Background specifies the type of background for the view.

Categorize selects a sheet set category for the view.

Layer snapshot applies the current layer properties to the view.

live Section assigns a section to the view.

Ucs assigns a user-defined coordinate system to the view.

Visual style selects a visual style for the named view.

STARTUP SWITCH

/v specifies the view name to show when AutoCAD starts up.

RELATED COMMANDS

Rename changes the names of views via a dialog box.

UCS creates and displays user-defined coordinate systems.

PartialLoad loads portions of drawings based on view names.

Open opens drawings and optionally starts with a named view.

Plot plots named views.

SheetSet uses named views.

RELATED SYSTEM VARIABLES

DefaultViewCategory specifies the default name for categories.

ViewCtr specifies the coordinates of the center of the view.

ViewSize specifies the height of the view.

TIPS

- Name views in your drawing to move quickly from one detail to another.
- The Plot command can plot the named views of drawings; the Open command lets you select a view with which to display the drawing.
- Objects outside of the window created by the **Window** option may be displayed, but are not plotted.
- As of AutoCAD 2005, this command creates views for sheets.
- As of AutoCAD 2007, it controls the background for visual styles. Backgrounds do not appear in 2D wireframe mode.

Removed Command

The **+View** command was removed from AutoCAD 2007, because the redesigned dialog box no longer has tabs.

ViewPlotDetails

2005 Reports plotting errors and successes.

Command	Alias	Ctrl+	F-key	Alt+	Menu Bar	Tablet
viewplotdetails	...	...	...	**FB**	**File** ⇘**View Plot and Publish Details**	...

Command: viewplotdetails

Displays dialog box:

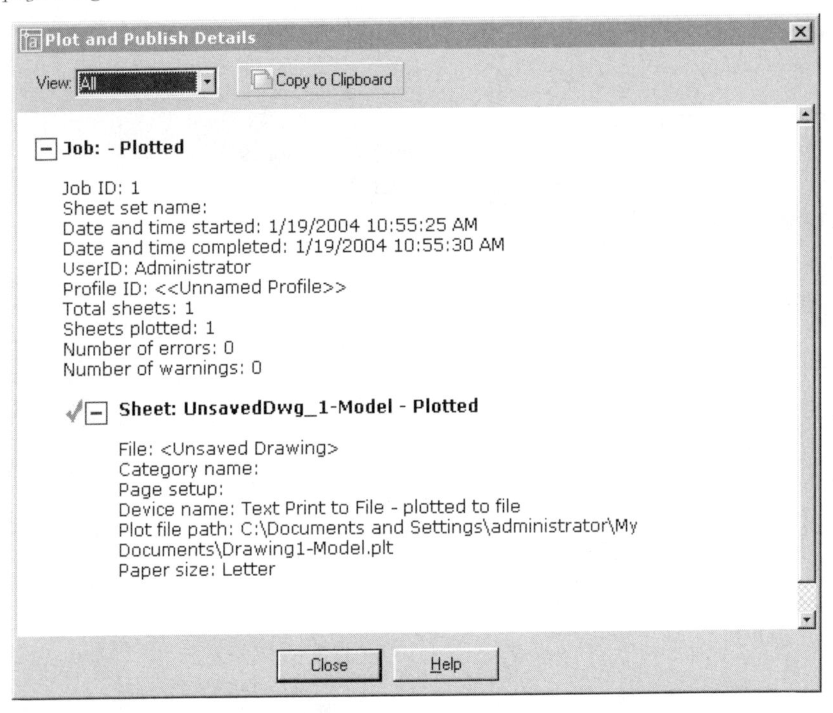

PLOT AND PUBLISH DETAILS DIALOG BOX

Green checkmark indicates successful plot.

Red "X" warns of plotting error.

View determines whether all messages are displayed, or just errors:

- All messages.
- Errors only.

Copy to Clipboard copies selected text to the Clipboard.

[–] Collapses text under heading.

[+] Expands text under heading.

SHORTCUT MENU

Select text in the dialog box, and then right-click:

Copy copies selected text to the Clipboard.

Select All selects all text in the dialog box; use the cursor to select portions of text.

Print displays Print dialog box; select a printer and its options, and then click **Print**. *All* text in the dialog box is printed, not just selected text.

The Cut and Paste options are unavailable.

RELATED COMMANDS

Plot plots drawings.

TraySettings toggles the display of icons in the tray.

TIPS

- When a plot is completed, AutoCAD displays a balloon. Clicking the blue underlined text displays the dialog box.

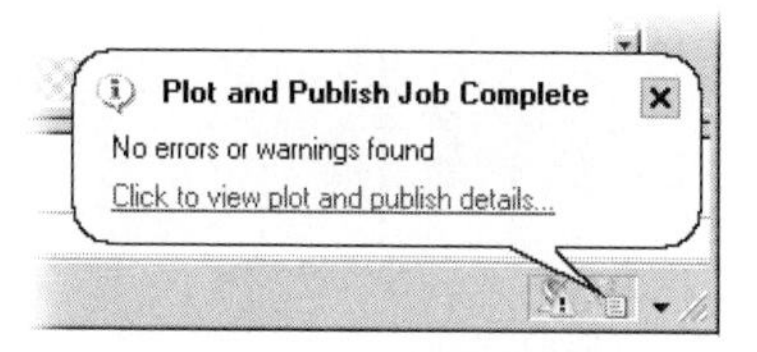

- You can also access this dialog box by right-clicking the Plotting icon in the tray, and then selecting **View Plot Details**.

ViewRes

V. 2.5 Controls the roundness of curved objects; determines whether zooms and pans are performed as redraws or regens (short for VIEW RESolution).

Command	Alias	Ctrl+	F-key	Alt+	Menu Bar	Tablet
viewres	...	...	...	...	...	...

Command: viewres
Do you want fast zooms? [Yes/No] <Y>: *(Type* **Y** *or* **N***.)*
Enter circle zoom percent (1-20000) <1000>: *(Enter a value.)*

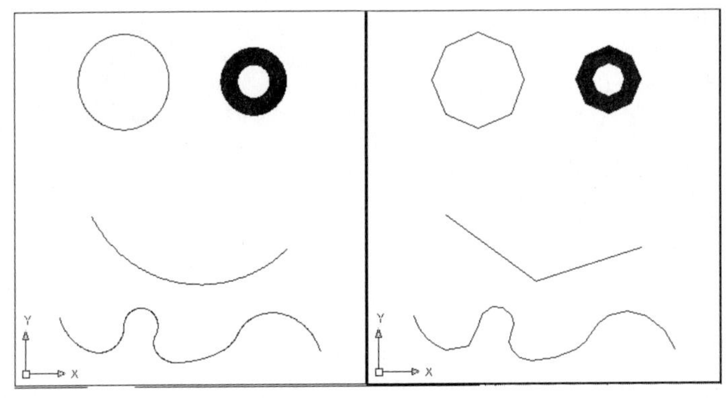

Left*: Circle zoom percent = 1000.*
Right*: Circle zoom percent = 1.*

COMMAND LINE OPTIONS

Do you want fast zooms? does not function; retained only for compatibility with macros and scripts.

Enter circle zoom percent specifies that smaller values display faster, but makes circles look less round (see figure); default = 1000.

RELATED SYSTEM VARIABLE

WhipArc toggles the display of circles and arcs as vectors or as true, rounded objects.

RELATED COMMAND

RegenAuto determines whether AutoCAD uses redraws or regens.

TIPS

- Setting WhipArc to 1 is recommended over increasing the value of ViewRes.
- This command was useful for speeding up the display when computers were much slower than today.
- Back in the days when "fast zooms" could be toggled, every zoom and pan caused a regeneration when the fast zooms were disabled. Today, AutoCAD tries to make every zoom and pan a redraw.

Removed Command

VlConv was removed from AutoCAD Release 14; use **3dsIn** instead.

VisualStyles, VisualStylesClose

2007 Creates and edits visual styles.

Command	Alias	Ctrl+	F-key	Alt+	Menu Bar	Tablet
visualstyles	**vsm**	...	...	**TPV**	**Tools ↳Palettes ↳Visual Styles**	...
visualstylesclose						
-visualstyles	**-vsm**					

Command: visualstyles

Displays palette:

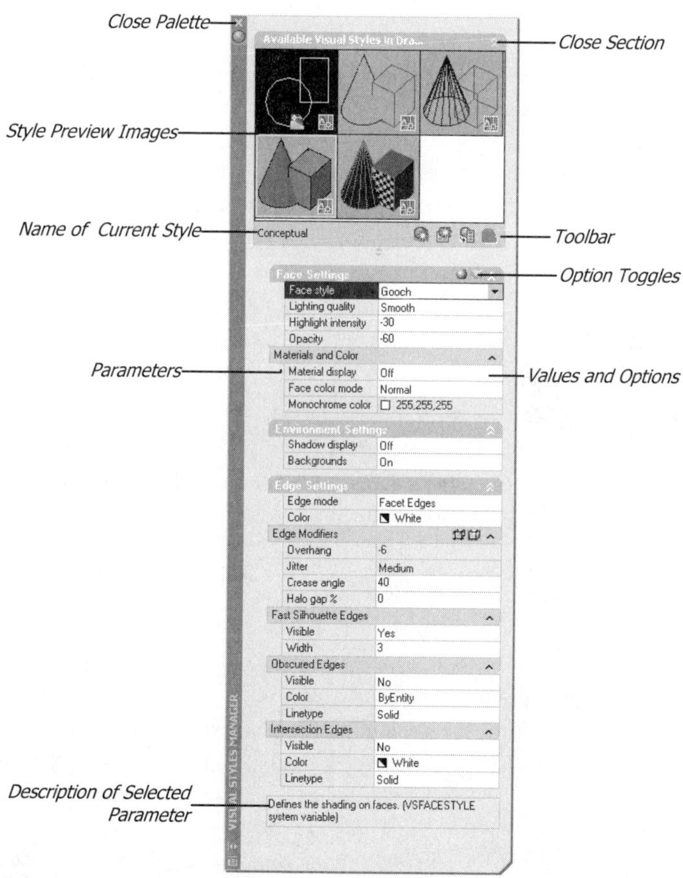

Command: visualstylesclose

Closes palette.

PALETTE OPTIONS

Negative values turn off the option.

Face Settings options

Face Style selects the style of face shading:

- **None** — applies no style to faces, like wireframe mode.
- **Real** — attempts to make the face appear real, like rendered mode; displays materials, if applied.
- **Gooch** — substitutes warm and cool colors for light and dark; allows faces to be seen more easily than with Real.

Lighting Quality toggles faceted and smooth; for curved surfaces only.

Highlight Intensity specifies the size of highlights on faces; for faces without materials. Range 0 (off) to 100. Larger numbers generate larger areas of intensity.

Opacity specifies the transparency of faces; range 0 to 100, where 0 is opaque, 1 is transparent, 2 - 99 are increasing translucent, and 100 (or any negative value) is again opaque.

Materials and Color options

Materials toggles the display of materials and textures, if previously applied with the MaterialAssign command. When other than **Off**, the Highlight Intensity and Opacity settings do not apply.

- **Off** — materials and textures are not displayed; visual style settings are applied to the model.
- **Materials** — materials are displayed, if assigned.
- **Materials and Textures** — materials and textures are displayed, if applied.

Face Color Mode specifies how colors are displayed:

- **Normal** — face colors displayed normally.
- **Monochrome** — face colors displayed in a monochrome color specified by the Monochrome Color option.
- **Tint** — face colors change their hue and saturation.
- **Desaturate** — face colors reduce their saturation by 30%.

Monochrome Color and **Tint Color** specify the color for monochrome and tint face colors. The color, selected from the Select Color dialog box, covers the model with a single color.

Environment Settings options

No lights need be placed in drawings for visual styles to cast shadows.

Shadow Display specifies the type of shadows to display.

- **Off** — objects cast no shadows. Shadow-casting slows down AutoCAD, so it's best left off unless needed for high-quality images or sun studies.
- **Ground shadows** — objects cast shadows on the shadow plane, but not on each other.
- **Full shadows** — objects cast shadows on the ground and on each other. The graphics board must be capable of full shadow-casting and hardware acceleration must be turned on; see the 3dConfig command.

Backgrounds toggles the display of a background in the viewport; backgrounds are created by the View command.

Edge Settings options

Edge Mode specifies which style of edge to display:

- **None** — facets, isolines, and edges are not displayed; this setting cannot be used when the Face Style set to None.
- **Isolines** — isolines are displayed on curved surfaces, and edges on all objects.
- **Facet Edges** — edges are displayed in a color.

Color specifies the color for edges; one color applies to all edges.

The following parameters appear only when Edge Mode = Isolines:

Number of Lines specifies the number of isolines drawn on curved surfaces; range is 0 to 2047, where the default value of 4 is too low. A good number to use is either 0 (no isolines) or 12; too many isolines completely blacken the curved objects. This parameter does not go into effect until after the Regen command is run.

Always on Top toggles hidden-line removal of isolines:

- **On** -- all isolines are displayed.
- **Off** — only foreground isolines are displayed; isolines located "around the back" are hidden.

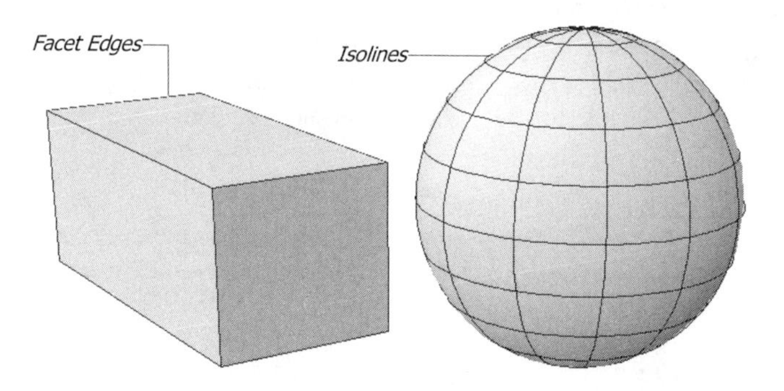

Edge Modifiers options

The Edge Modifiers settings apply to isolines and facet edges:

Overhang extends edges beyond their boundaries; range is 0 to 100 pixels.

Jitter skews edge lines to mimic a hand drawn effect by skewing the lines that define edges:

- **Off** — (0) no jitter lines.
- **Low** — (1) few jitters.
- **Medium** — (2) medium jitters.
- **High** — (3) high amount of jitter.

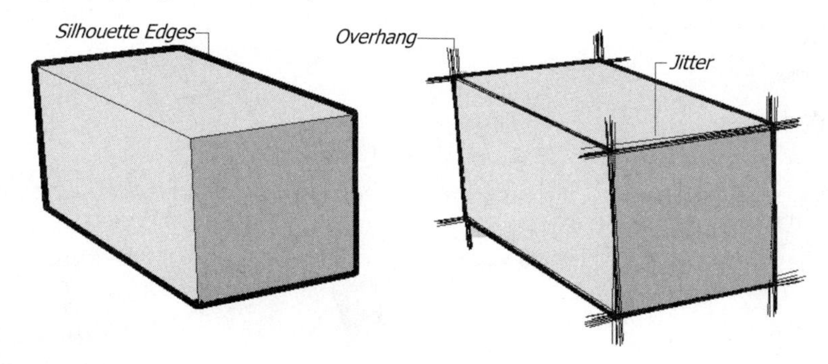

The following parameters appear only when Edge Mode = Facet Edges:

Crease Angle specifies the angle beyond which facet edges are not shown on curved surfaces, removing the edge lines; range is from 0 to 180 degrees. A high value, such as 180, turns off the display of facet edges.

Halo Gap % specifies the gap generated between visually overlapping objects; not available in visual styles based on wireframe modes. Range is 0 to 100 pixels; when greater than 0, silhouette edges are not displayed.

Fast Silhouette Edges options

Both of the following settings apply to all objects in the viewport equally:

Visible toggles the display of silhouette edges. AutoCAD does not display edges when the halo gap > 0, opacity > 0 (objects are translucent), or visual styles are wireframe.

Width determines the width of the silhouette edges; range is 1 to 25 pixels.

Obscured Edges options

The following parameters appear only when Edge Mode = Facet Edges:

Visible toggles the visibility of obscured edges and facets:

- **Yes** — hidden lines are shown through the model, although obscured edges are invisible on curved faces when a large crease angle prevents facet lines from showing.
- **No** — hidden lines are not shown.

Color specifies the color for visible obscured edges and facets.

Linetype specifies the line pattern for obscured (hidden) edges and facets. No scale factor can be applied to these linetypes, and you cannot use the linetypes accessed by the Linetype command or stored in the *acad.lin* file. The available linetypes are:

- Off
- Solid
- Dashed
- Dotted
- Short Dash
- Medium Dash
- Long Dash
- Double Short Dash
- Double Medium Dash
- Double Long Dash
- Medium Long Dash
- Sparse Dot

Intersection Edges options

The following parameters appear only when Edge Mode = Facet Edges:

Visible toggles the display of intersection edges. (Autodesk recommends leaving this setting turned off to increase performance.)

Color and Linetype specify the color and linetypes of intersection lines, as described above.

PALETTE TOOLBAR

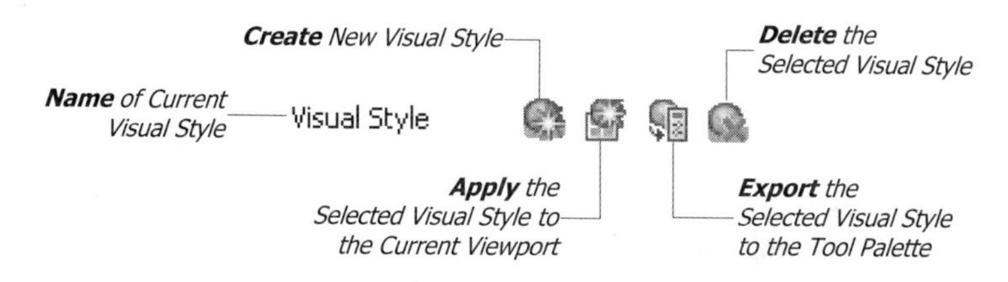

-VisualStyle Command

Command: -visualstyle

Enter an option [set Current/Saveas/Rename/Delete/?]:

set Current prompts for the type of visual style to apply: 2dwireframe, 3dwireframe, 3dHidden, Realistic, Conceptual, Other, or cUrrent. If Other, prompts for its name.

Saveas saves the current visual style; prompts for a name.

Rename renames a visual style; system styles cannot be renamed.

Delete deletes a visual style; system styles cannot be deleted.

? lists the names of visual styles in the current drawing.

RELATED COMMANDS

VsCurrent sets the visual style.

VsSave saves the current visual style by name.

ToolPalettes stores visual styles.

Dashboard accesses the Visual Styles control panel.

Rename renames visual styles.

Purge removes unused visual styles.

View applies visual styles to named views.

RELATED SYSTEM VARIABLES

CMaterial specifies the name of the material.

CShadow specifies the type of shadow cast by objects.

DragVs specifies the default visual style while creating 3D objects.

InterfereObjVs specifies the visual style for interference objects created by Interference.

InterfereVpVs specifies the visual style during interference checking.

Isolines specifies the number of isolines to display on curved solid models.

ShadowPlaneLocation locates the height of the invisible ground plane upon which shadows are cast; can be set to any distance along the z axis, including negative distances.

VsBackgrounds determines whether backgrounds are displayed in visual styles.

VsEdgeColor specifies the edge color; can be any ACI color.

VsEdgeJitter specifies the level of jitter effect.

VsEdgeOverhang extends edge lines beyond intersections.

VsEdges specifies the type of edge to display.

VsEdgeSmooth specifies the crease angle.

VsFaceHighlight specifies the color of highlights; ignored when VsMaterialMode is on.

VsFaceOpacity controls the transparency/opacity of faces.

VsFaceStyle determines how faces are displayed.

VsHaloGap specifies the "halo" gap (gap between intersecting lines).

VsHidePrecision specifies the accuracy of hides and shades.

VsIntersectionColor specifies the color of intersecting polylines.

VsIntersectionEdges toggles the display of intersecting edges.

VsIntersectionLtype specifies the linetype for intersecting polylines.

VsIsoOntop toggles whether isolines are displayed.

VsLightingQuality toggles the quality of lighting.

VsMaterialMode controls the display of material finishes:

VsMonoColor specifies the monochrome tint.

VObscuredColor specifies the color of obscured lines.

VsObscuredEdges toggles the display of obscured edges.

VsObscuredLtype specifies the linetype of obscured lines.

VsShadows determines the quality of shadows.

VsSilhEdges toggles the display of silhouette edges.

VsSilhWidth specifies the width of silhouette edge lines.

VsState reports whether the Visual Styles window is open.

...whew!

TIPS

- The ShadowPlaneLocation system variable allows you to change the elevation of the "ground" (shadow plane) upon which shadows are cast. The default is 0 units, and is measured in the z direction. This plane is independent of anything else.
- You can use the Dashboard's Light control panel to adjust the location and angle of shadow-casting.

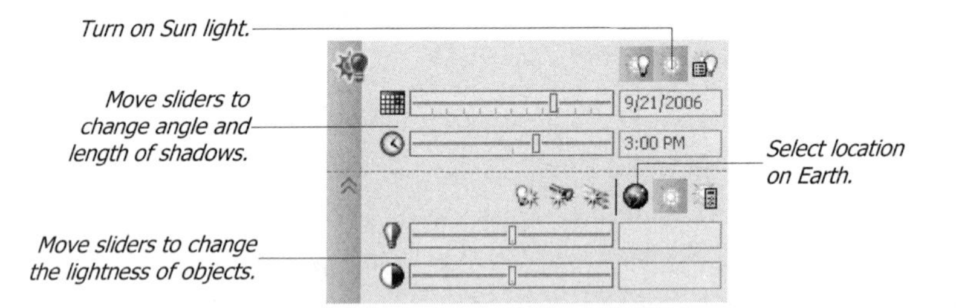
Turn on Sun light.
Move sliders to change angle and length of shadows.
Move sliders to change the lightness of objects.
9/21/2006
3:00 PM
Select location on Earth.

Click the Sun Status button, and then move the Date and Time sliders to rotate and lengthen the shadows. The Brightness and Contrast sliders make the objects brighter or more shadowy, They have no effect on the shadows, and work only when Face Style is set to Real.

'VLisp

2000 Opens the VLisp integrated development environment (short for Visual LISP).

Commands	Alias	Ctrl+	F-key	Alt+	Menu Bar	Tablet
vlisp	vlide	...	...	TSV	Tools ↳AutoLISP ↳Visual LISP Editor	...

Command: vlisp

Displays window:

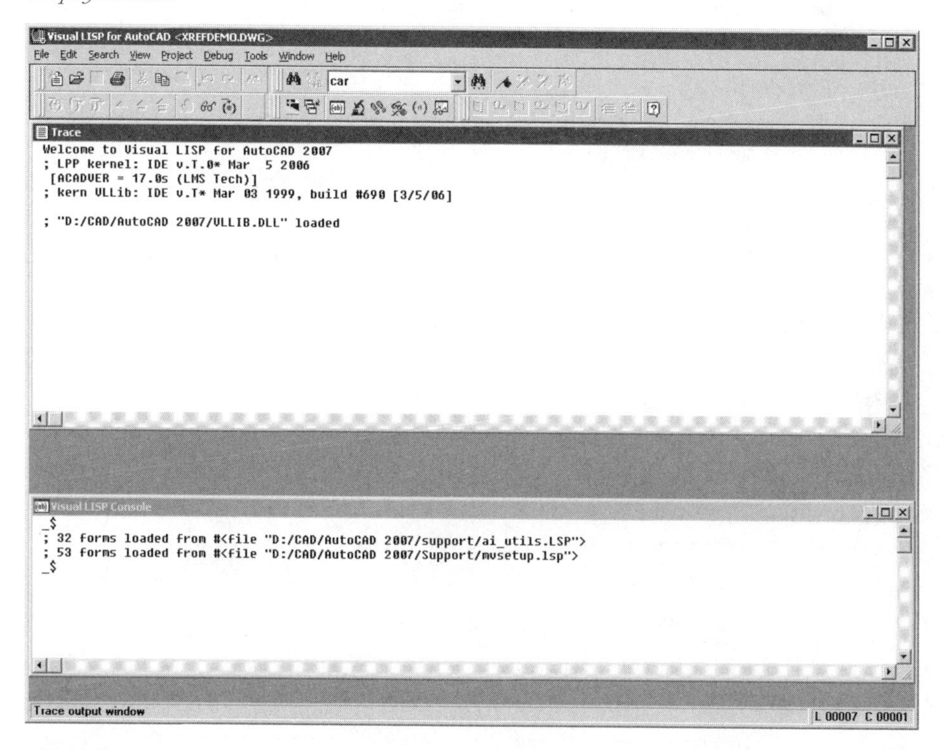

MENU BAR

Select **Help | Visual LISP Help Topics** *for assistance in using this VLISP window.*

RELATED COMMAND

AppLoad loads Visual LISP applications, as well as programs written in AutoLISP and other APIs.

TIP

- Sample VLisp code can be found in the *autocad 2007\sample\vlisp* folder.

VpClip

2000

Clips a layout viewport (short for ViewPort CLIPping).

Command	Alias	Ctrl+	F-key	Alt+	Menu Bar	Tablet
vpclip	...	...	...	**MCV**	**Modify** ⇨**Clip** ⇨**Viewport**	...

Command: vpclip
Select viewport to clip: *(Pick a viewport.)*
Select clipping object or [Polygonal] <Polygonal>: *(Select an object, or type* **P***.)*

The selected viewport disappears, and is replaced by the new clipped viewport.

COMMAND LINE OPTIONS

Select viewport to clip selects the viewport that will be clipped.

Select clipping object selects the object that defines the clipping boundary: closed polyline, circle, ellipse, closed spline, or region.

Polygonal Options

Specify start point: *(Pick a point.)*
Specify next point or [Arc/Close/Length/Undo]: *(Pick a point, or enter an option.)*
Specify next point or [Arc/Close/Length/Undo]: *(Type* **C** *to close.)*

Specify start point specifies the starting point for the polygon.

Arc draws an arc segment; see the **Arc** command.

Close closes the polygon.

Length draws a straight segment of specified length.

Undo undoes the previous polygon segment.

RELATED COMMAND

Mview creates rectangular and polygonal viewports in paper space.

TIPS

- This command does not operate in Model tab.
- The clipping polyline is placed on the current layer. When the layer is frozen, the viewport is not clipped until the layer is thawed. When the layer is off, the polyline is invisible, but still clips.

VpLayer

Rel. 11 Controls the visibility of layers in viewports, when a layout tab other than Model is selected (short for ViewPort LAYER).

Command	Alias	Ctrl+	F-key	Alt+	Menu Bar	Tablet
vplayer	...	...	...	...	...	...

Command: vplayer
Enter an option [?/Freeze/Thaw/Reset/Newfrz/Vpvisdflt]: *(Enter an option.)*
Select a viewport: *(Pick a viewport.)*

COMMAND LINE OPTIONS

Freeze indicates the names of layers to freeze in this viewport.

Newfrz creates new layers that are frozen in all newly-created viewports (short for NEW FReeZe).

Reset resets the state of layers based on the **Vpvisdflt** settings.

Thaw indicates the names of layers to thaw in this viewport.

Vpvisdflt determines which layers will be frozen in a newly-created viewport and default visibility in existing viewports (short for ViewPort VISibility DeFauLT).

? lists the layers frozen in the current viewport.

RELATED COMMANDS

Layer creates and controls layers in all viewports.

MView creates and joins viewports when tilemode is off.

RELATED SYSTEM VARIABLE

TileMode controls whether viewports are tiled (model) or overlapping (layouts).

VpMax, VpMin

2005 Maximizes or minimizes the selected viewport in the AutoCAD window (short for ViewPortMAXimize).

Command	Alias	Ctrl+	F-key	Alt+	Menu Bar	Tablet
vpmax	...	...	...	...	...	...
vpmin						

Command: vpmax

When the layout contains more than one viewport, AutoCAD prompts:

Select a viewport to maximize: *(Select a viewport.)*

AutoCAD maximizes the viewport to fill the entire AutoCAD window, and switches to model space for editing.

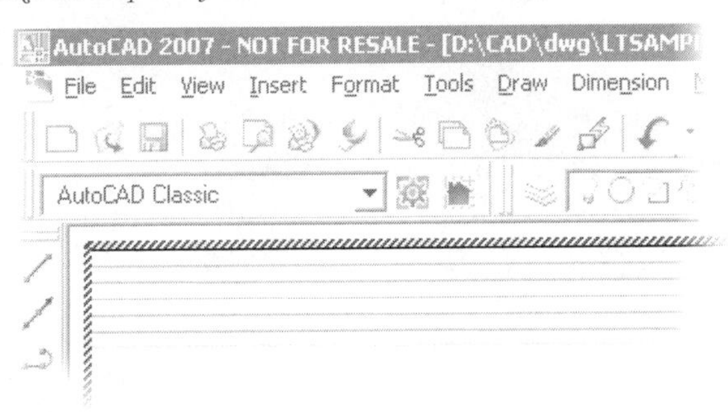

The red dashed border indicates AutoCAD is in VpMax mode.

COMMAND LINE OPTION

Select a viewport to maximize selects the viewport.

RELATED COMMAND

VPorts creates viewports.

RELATED SYSTEM VARIABLE

VpMaximixedState determines whether the viewport is maximized.

TIPS

- Use the VpMin command to return to the layout tab, and restore the viewport. Alternatively, you can double-click the red border to restore the viewport.
- When the viewport is maximized, AutoCAD displays the Minimize Viewport icon on the status bar. Click the arrows to move from one viewport to the next.

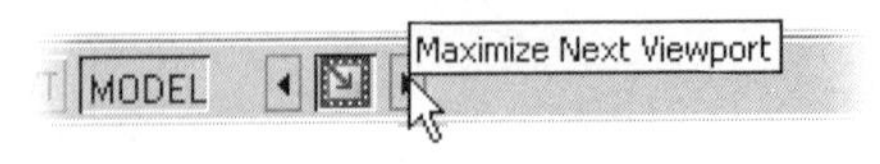

VPoint

V. 2.1 Changes the viewpoint of 3D drawings (short for ViewPOINT).

Command	Alias	Ctrl+	F-key	Alt+	Menu Bar	Tablet
vpoint	**-vp**	...	...	**V3V**	**View** **↳3D Views** **↳Viewpoint**	**N4**

Command: vpoint
Current view direction: VIEWDIR=0.0000,0.0000,1.0000
Specify a view point or [Rotate] <display compass and tripod>: *(Enter an option, or press Enter for the compass-tripod.)*

COMMAND LINE OPTIONS

Specify a view point indicates the new 3D viewpoint by coordinates.

Rotate indicates the new 3D viewpoint by angle.

Enter brings up visual guides (see figure below).

RELATED COMMANDS

DdVpoint adjusts the viewpoint via a dialog box.

3dOrbit rotates the viewpoint in real-time.

RELATED SYSTEM VARIABLES

VpointX is the x-coordinate of the current 3D view.

VpointY is the y-coordinate of the current 3D view.

VpointZ is the z-coordinate of the current 3D view.

WorldView determines whether VPoint coordinates are in WCS or UCS.

TIPS

- This command works only in model space, and is largely replaced by the 3dOrbit command
- The *compass* represents the globe, flattened to two dimensions: north pole (0, 0, z) is in the center; equator (x, y, 0) the inner circle; and south pole (0, 0, -z) the outer circle.
- As the cursor is moved on the compass, the *axis tripod* rotates showing the 3D view direction. To select the view direction, pick a location on the globe and press the pick button.

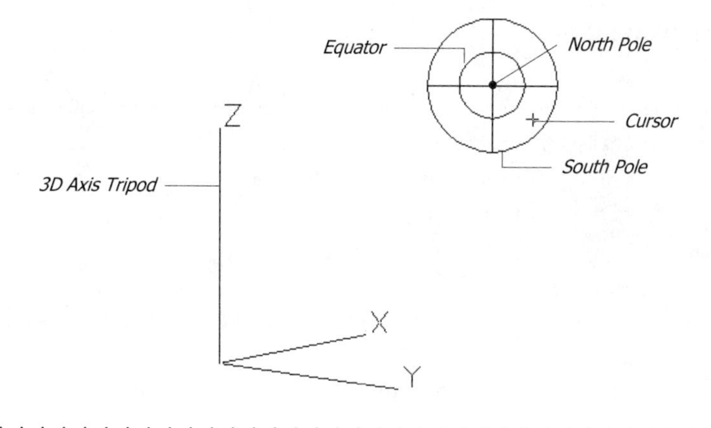

VPorts

Rel. 10 Creates viewports (short for ViewPORTS).

Commands	Alias	Ctrl+	F-key	Alt+	Menu Bar	Tablet
vports	**viewports**	...	...	**VV**	**View** ↳**Viewports**	**M3-4**
+vports						
-vports				**VV1**	**View** ↳**Viewports** ↳**1 Viewport**	

Command: vports

Displays tabbed dialog box.

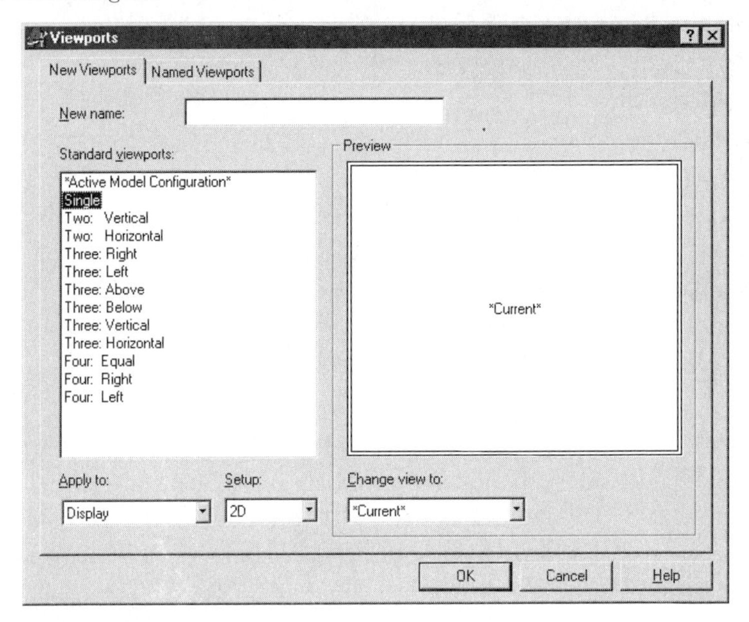

VIEWPORTS DIALOG BOX

In model space

New name specifies the name for the viewport configuration; can be up to 255 characters long.

Standard viewports lists the available viewport configurations.

Preview displays a preview of the viewport configuration.

Apply to applies the viewport configuration to:

- Display.
- Current Viewport.

Setup selects 2D or 3D configuration; the **3D** option applies orthogonal views, such as top, left, and front.

Change view to selects the type of view; in 3D mode, selects a standard orthoganal view.

In paper space

Viewport spacing specifies the spacing between the floating viewports; default = 0 units.

Named Viewports tab

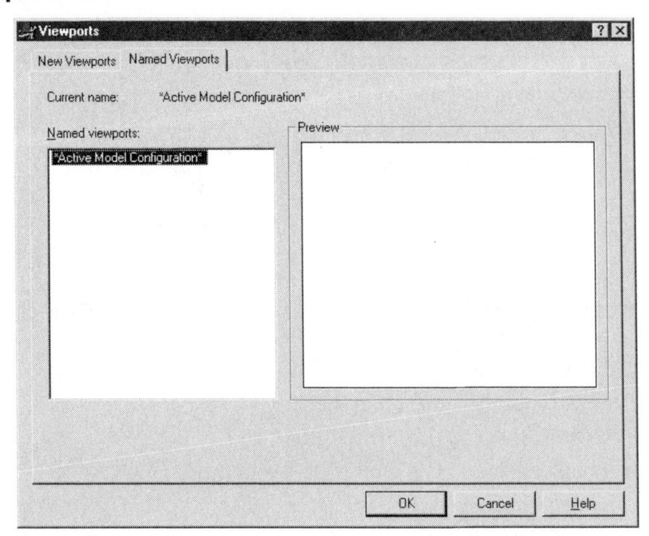

Named viewports lists the names of saved viewport configurations.

+VPORTS Command

Command: +vports

Tab index <0>: *(Type* **0** *or* **1**.*)*

Tab index specifies the tab to display:

0 — New Viewports tab (default).

1 — Named Viewports tab.

-VPORTS Command

Command: -vports

In layout mode, displays MView command prompts. In model space, prompts:

Enter an option [Save/Restore/Delete/Join/SIngle/?/2/3/4] <3>: *(Enter an option.)*

COMMAND LINE OPTIONS

In model space

Save saves the settings of a viewport by name.

Restore restores a viewport definition.

Delete deletes a viewport definition.

Join joins two viewports together as one when they form a rectangle.

SIngle joins all viewports into a single viewport.

? lists the names of saved viewport configurations.

4 divides the current viewport into four.

2 (Two Viewports) options

Horizontal creates one viewport over another.

Vertical creates one viewport beside another (default).

3 (Three Viewports) options

Horizontal creates three viewports over each other.

Vertical creates three viewports beside each other.

Above creates one viewport over two viewports.

Below creates one viewport below two viewports.

Left creates one viewport left of two viewports.

Right creates one viewport right of two viewports (default).

In paper space

Specify corner of viewport or [ON/OFF/Fit/Hideplot/Lock/Object/Polygonal/ Restore/2/3/4]<Fit>: *(Enter an option.)*

ON turns on the viewport; the objects in the viewport become visible.

OFF turns off the viewport; the objects in the viewport become invisible.

Fit creates one viewport that fills the display area.

Hideplot removes hidden lines when plotting in layout mode.

Lock locks the viewport to prevent editing.

Object converts a closed polyline, ellipse, spline, region, or circle into a viewport.

Polygonal creates an non-rectangular viewport .

Other options are identical to those displayed in model tab.

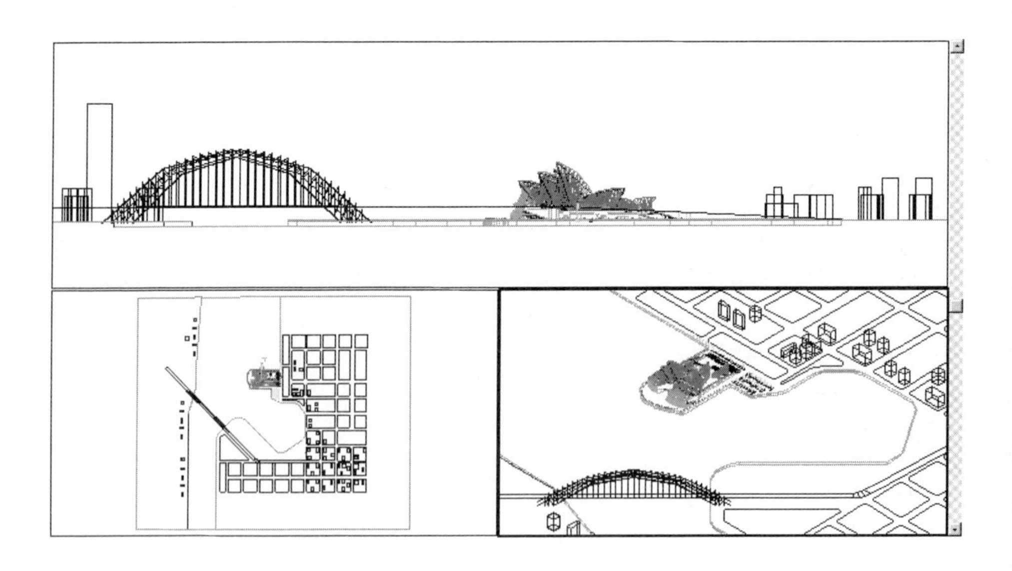

RELATED COMMANDS

MView creates viewports in paper space.

RedrawAll redraws all viewports.

RegenAll regenerates all viewports.

VpClip clips a viewport.

RELATED SYSTEM VARIABLES

CvPort identifies the current viewport number.

MaxActVp limits the maximum number of active viewports.

TileMode controls whether viewports can be overlapped or tiled.

TIPS

- The **Join** option joins two viewports only when they form a rectangle in model space.
- You can restore saved viewport arrangements in paper space using the MView command.
- Many display-related commands (such as Redraw and Grid) affect the current viewport only.
- Press CTRL+R to switch between viewports.

VsCurrent

2007 Sets the current visual style (short for Visual Style CURRENT).

Command	Aliases	F-key	Alt+	Menu Bar	Tablet
vscurrent	vs	...	vs	View	...
	shademode			↳Visual Styles	

Command: vscurrent
Enter an option [2dwireframe/3dwireframe/3dHidden/Realistic/Conceptual/ Other]: *(Enter an option, or type* **O**.*)*

COMMAND LINE OPTIONS

2dwireframe displays raster and OLE objects; linetypes, and lineweights are visible; this is AutoCAD's default display mode, and is not a visual style.

3dwireframe looks similar to 2D wireframe, but shows the shaded 3D UCS icon and optionally the compass.

3dHidden removes lines hidden by overlapping objects.

Realistic shades objects, smooths edges, and displays materials, if attached to objects.

Conceptual looks like Realistic, but ranges the colors from "cool to warm" to make the model easier to view.

Other prompts for the name of a user-defined visual style: "Enter a visual style name or [?]."

? lists the names of existing visual styles.

RELATED COMMANDS

VisualStyles creates and edits visual styles.

VsSave saves the current visual style by name.

ToolPalettes stores visual styles.

Dashboard accesses the Visual Styles control panel.

TIPS

- Visual styles can be applied through:
  - Using the VsCurrent command's options.
  - Dragging them from the Visual Styles palette into the drawing (see VisualStyles command).
  - Selecting them the Visual Styles toolbar.
  - Dragging them from the Tools palette into the drawing (see ToolPalettes command).
  - Selecting them from the Visual Styles control panel of the Dashboard (see Dashboard command).
- Visual styles work only in model space.
- Use the View command's background option to apply a background image or color to visual styles.

VSlide

V. 2.0 Displays slide files in the current viewport (short for View SLIDE).

Command	Alias	Ctrl+	F-key	Alt+	Menu Bar	Tablet
vslide	...	...	...	...	...	...

Command: vslide

Displays Select Slide File dialog box. Select an .sld file, and then click ***Open.***

COMMAND LINE OPTIONS

None.

RELATED COMMANDS

MSlide creates slide files of the current viewport.

Redraw erases slides from the screen.

RELATED AUTODESK PROGRAM

slidelib.exe creates an SLB-format library file of a group of slide files.

RELATED AUTOCAD FILES

****.sld*** stores individual slide files.

****.slb*** stores a library of slide files.

TIP

- The following applies when FileDia is set to 0, or when this command is used in a script:
  - For faster viewing of a series of slides, placing an asterisk before the VSlide command preloads the *.sld* slide file, as in:

    Command: *vslide filename

  - Use the following format to display a specific slide stored in an SLB slide library file:

    Command: vslide

    Slide file: acad.slb(slidefilename)

VsSave

2007 Saves the current visual style by name (short for Visual Style SAVE).

Command	Alias	Ctrl+	F-key	Alt+	Menu Bar	Tablet
vssave	...	...	...	...	...	...

Command: vssave

This command operates only in model space.

Save current visual style as or [?]: *(Enter a name, or type* **?***.)*

COMMAND LINE OPTIONS

Save current visual style as names the current visual style. If the name already exists, AutoCAD asks, "'*name*' already exists. Do you wish to replace the existing visual style? [Yes/No/Try again]."

? lists the names of existing visual styles, such as:

```
2D Wireframe
3D Hidden
3D Wireframe
Conceptual
Realistic
```

RELATED COMMANDS

VsCurrent sets the current visual style.

VisualStyles creates and edits visual styles.

Rename renames visual styles.

Purge removes unused visual styles from the current drawing.

ToolPalettes stores visual styles and shares them among drawings.

TIPS

- Every drawing contains the following visual styles predefined by Autodesk: 2D Wireframe, 3D Hidden, 3D Wireframe, Conceptual, and Realistic.
- To use a saved visual style in another drawing, drag it from the Visual Styles Manager palette onto the Tools palette. Open the other drawing, and then drag the visual style from the Tool palette into the drawing.

VtOptions

2006 Controls view transitions during pans and zooms.

Command	Alias	Ctrl+	F-key	Alt+	Menu Bar	Tablet
vtoptions	...	...	...	...	...	...

Command: vtoptions

Displays dialog box:

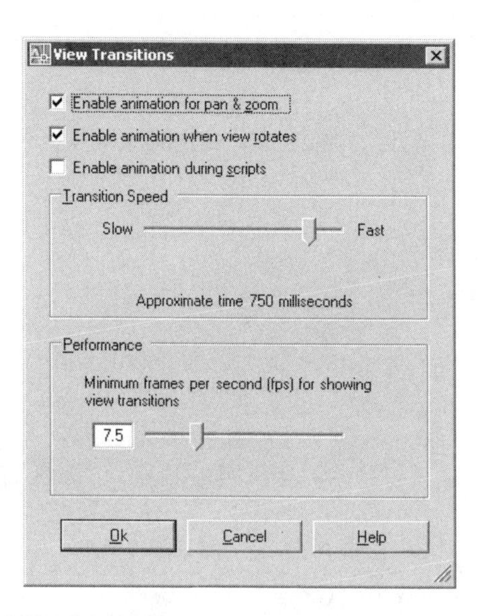

VIEW TRANSITIONS DIALOG BOX

☑**Enable Animation for Pan and Zoom** smoothes view transitions during pans and zooms.

☑**Enable Animation When View Rotates** smoothes view transitions during view angle changes.

☐**Enable Animation During Scripts** smoothes view transitions during scripts.

Transition Speed specifies the speed of view transitions; milliseconds.

Performance specifies the minimum speed for smooth view transitions; frames per second.

RELATED COMMAND

Zoom utilizes view transitions.

RELATED SYSTEM VARIABLES

VtDuration specifies the duration of view transitions; milliseconds.

VtEnable determines which commands use view transitions.

VtFps specifies the minimum view transition speed; frames per second.

TIPS

- View transitions are particularly useful when viewing in 3D.
- When AutoCAD cannot maintain the transition speed, it switches to an instant transition.
- View transitions make it easier to see where AutoCAD is zooming, but lengthen the time it takes to complete the zoom.

WalkFlySettings

2007 Presets walk and fly parameters.

Commands	Aliases	Ctrl+	F-key	Alt+	Menu Bar	Tablet
walkflysettings	...	...	...	VKS	View ↳Walk and Fly ↳Walk and Fly Settings	...

Command: walkflysettings

Displays dialog box:

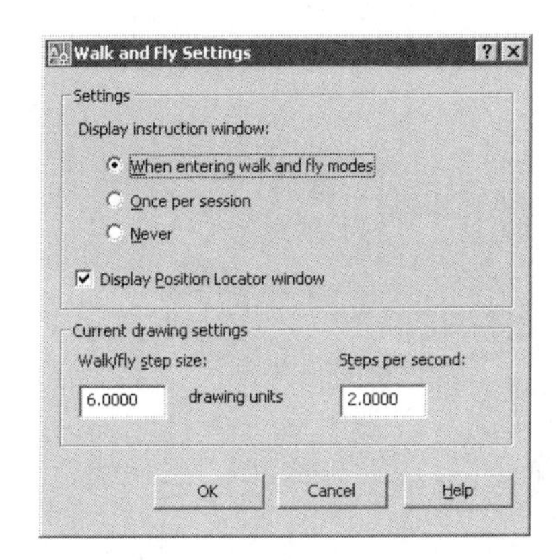

DIALOG BOX OPTIONS

⊙ **When Entering Walk and Fly Modes** displays the mappings window (see below) each time the 3dWalk and 3dFly commands are entered:

○ **Once Per Session** displays the mappings window just the first time the 3dWalk and 3dFly commands are entered in an AutoCAD session.

○ **Never** never displays the mappings window.

☑ **Display Position Locator Window** toggles the display of the Position Locator window.

Walk/Fly Step Size specifies the increment of position (step size) in drawing units.

Steps Per Second specifies the number of increments per second.

RELATED COMMANDS

3dFly and **3dWalk** move the viewpoint through 3D scenes.

RELATED SYSTEM VARIABLES

StepsPerSec specifies the number of steps per second.

StepSize specifies the distance per step.

WBlock

V. 1.4 Writes blocks or entire drawings to disk (short for Write BLOCK).

Commands	Aliases	Alt+	Menu Bar	Tablet
wblock	w	FE	File	...
	acadwblockdialog	↳DWG	↳Export	
-wblock	-w			

Command: wblock

Displays dialog box:

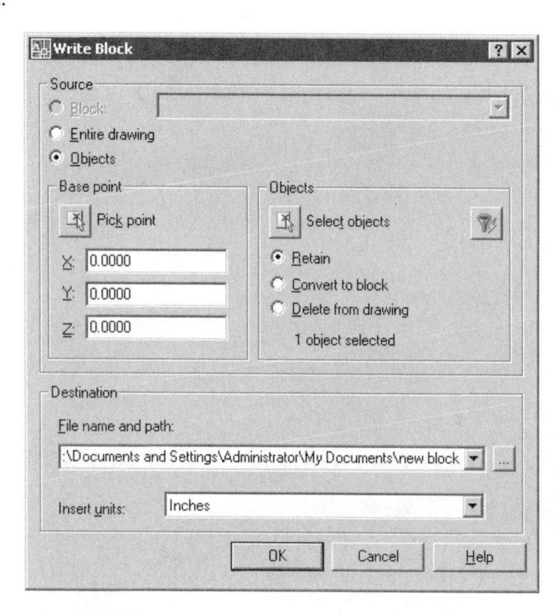

WRITE BLOCK DIALOG BOX

Source options

○**Block** specifies the name of the block to save as a *.dwg* file.

○**Entire drawing** selects the current drawing to save as a *.dwg* file.

⊙**Objects** specifies the objects from the drawing to save as a *.dwg* file.

Base point options

Pick Point dismisses the dialog box temporarily to select the insertion base point.

X specifies the x coordinate of the insertion point.

Y specifies the y coordinate of the insertion point.

Z specifies the z coordinate of the insertion point.

Objects options

Select Objects dismisses the dialog box temporarily to select one or more objects.

Quick Select displays the Quick Select dialog box; see the QSelect command.

⊙ **Retain** retains the selected objects in the current drawing after saving them as a drawing file.

○ **Convert to block** converts the selected objects to a block in the drawing, after saving them as a *.dwg* file; names the block under **File name** in the Destination section.

○ **Delete from drawing** deletes selected objects from the drawing, after saving them as a *.dwg* file.

Destination options

File name and path specifies the file name for the block or objects.

... displays the Browse for Folder dialog box.

Insert units specifies the units when the *.dwg* file is inserted as a block.

-WBLOCK Command

Command: -wblock

*Displays the Create Drawing File dialog box. Name the file, and then click **Save**.*

Enter name of existing block or

[= (block=output file)/* (whole drawing)] <define new drawing>: *(Enter a name, or use = and * options, or press Enter.)*

COMMAND LINE OPTIONS

Enter name of existing block specifies the name of a current block in the drawing.

= (*equals*) writes block to a *.dwg* file, using the block's name as the file name.

* (*asterisk*) writes the entire drawing to a *.dwg* file.

ENTER creates a *.dwg* drawing file of the selected objects on disk. (The selected objects are erased from the drawing; use the Oops command to bring them back.)

RELATED COMMANDS

Block creates a block of a group of objects.

Insert inserts a block or another drawing into the drawing.

RELATED SYSTEM VARIABLES

None.

TIPS

- Use the WBlock command to extract blocks from the drawing and store them on a disk drive. This allows the creation of a block library.
- Support for the DesignXML (extended markup language) format was withdrawn from AutoCAD 2004.

Wedge

Rel.11 Draws 3D wedges as solid models.

Command	Alias	Ctrl+	F-key	Alt+	Menu Bar	Tablet
wedge	**we**	...	...	**DMW**	**Draw** **↳Modeling** **↳Wedge**	**N7**

Command: wedge

The prompts change slightly in AutoCAD 2007:

Specify first corner or [Center]: *(Pick a point, or type* **C***.)*

Specify other corner or [Cube/Length]: *(Pick a point, or enter an option.)*

Specify height or [2Point]: *(Pick another point, or enter* **2P***.)*

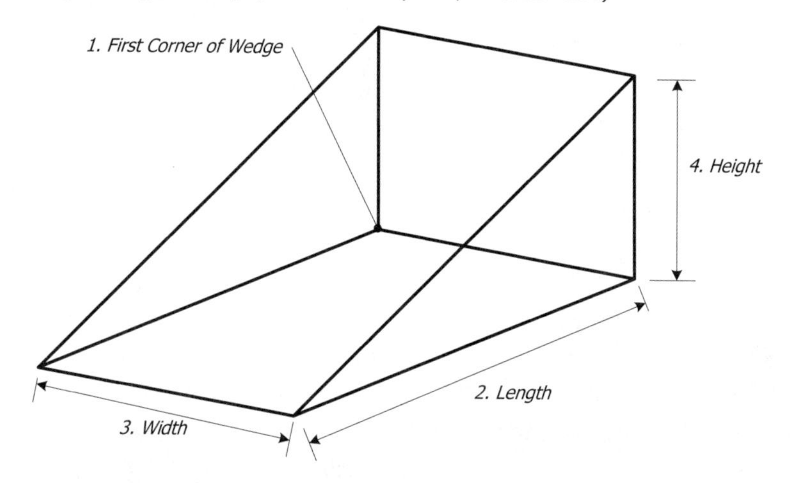

COMMAND LINE OPTIONS

First corner specifies the lower-left corner of the wedge.

Center draws the wedge's base about the center of the sloped face.

Cube draws cubic wedges.

Length specifies the length, width, and height of the wedge.

2Point picks two points to indicate the height (*new to AutoCAD 2007*).

Center options

Specify center: *(Pick a point.)*

Specify corner or [Cube/Length]: *(Pick a point, or enter an option.)*

Specify height or [2Point]: *(Pick a point.)*

Specify center of wedge indicates the center of the wedge's inclined face.

Specify opposite corner indicates the distance from the midpoint to one corner.

Cube options

Specify length: *(Specify the length.)*

Specify length indicates the length of all three sides.

Length options

Specify length: *(Specify the length.)*
Specify width: *(Specify the width.)*
Specify height: *(Specify the height.)*

Specify length indicates the length parallel to the x-axis.

Specify width indicates the width parallel to the y-axis.

Specify height indicates the height parallel to the z-axis.

RELATED COMMANDS

Ai_Wedge draws wedges as 3D surface models.

Box draws solid boxes.

Cone draws solid cones.

Cylinder draws solid cylinders.

Sphere draws solid spheres.

Torus draws solid tori.

TIPS

- *Length* means size in the x-direction.
- *Width* means size in the y-direction.
- *Height* means size in the z-direction.
- Use negative values for length, width, and height to draw the wedge in the negative x, y, and z directions.
- The IsoLines system variable has no effect on wedges.
- The wedge can be edited using grips:

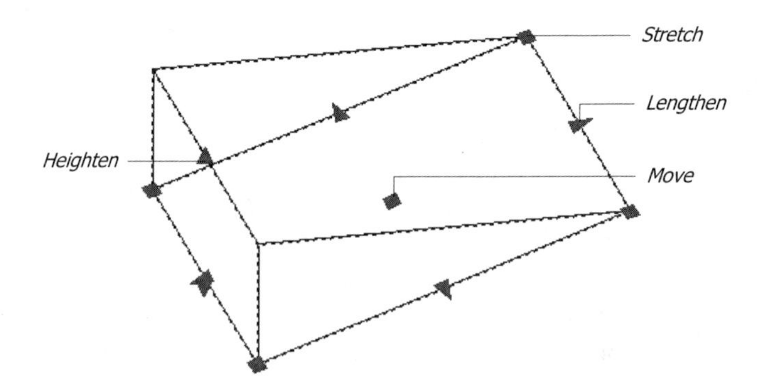

WhoHas

2000 Determines which computer has drawings open.

Command	Alias	Ctrl+	F-key	Alt+	Menu Bar	Tablet
whohas	...	...	...	...	...	...

Command: whohas

Displays the Select Drawing to Query dialog box. Select a drawing file, and then click **Open**.

When the drawing is open, reports:

Owner: ralphg
Computer's Name : HEATHER
Time Accessed : Monday, March 3, 2007 11:27:28 AM

When the drawing is not open, reports:

User: unknown.

COMMAND LINE OPTIONS

None.

RELATED COMMANDS

Open opens drawings.

XAttach attaches drawings that can be opened by other users.

TIP

- This command is meant for use over networks as a convenient way to find out which users are editing specific drawings.

WipeOut

2004 Fills areas with the background color to "wipe out" portions of drawings.

Command	Alias	Ctrl+	F-key	Alt+	Menu Bar	Tablet
wipeout	...	...	...	...	...	...

Command: wipeout
Specify first point or [Frames/Polyline] <Polyline>: *(Pick a point, or enter option.)*
Specify next point: *(Pick a point.)*
Specify next point or [Undo]: *(Pick a point, or type* **U**.*)*
Specify next point or [Close/Undo]: *(Pick a point, or type* **C** *or* **U**.*)*
Specify next point or [Close/Undo]: *(Type* **C** *to end the command.)*

COMMAND LINE OPTIONS

Specify first point specifies the starting point of the polygon.

Undo undoes the last segment.

Close closes the polygon.

Frames options
Enter mode [ON/OFF] <ON>: *(Type* **ON** *or* **OFF**.*)*

ON turns on the wipeout boundary polygon.

OFF turns off the boundary polygon.

Polyline options
Select a closed polyline: *(Pick a closed polyline.)*
Erase polyline? [Yes/No] <No>: *(Type* **Y** *or* **N**.*)*

Select a closed polyline picks a polyline that forms the wipeout boundary.

Erase polyline?

Yes erases the polyline.

No leaves the polyline in place.

RELATED COMMAND

DrawOrder displays overlapping objects in a different order.

TIPS

- The wipeout consists of an image object drawn with the background color. To create rectangular wipeout frames, use the Rectangle command.
- To create a wipeout under text, you may find it easier to use the **Background Mask** option of the MText command.
- Wipeout boundaries can be edited with grips; wipeouts cannot be edited when the **Frames** option is turned off. The **Frames** option applies to all wipeouts in the drawing; you cannot turn frames on and off for individual wipeouts.
- To make text appear above the wipeout, use the DrawOrder command to move the text to the **Front** — after creating the wipeout.
- Freezing or turning off the layer of the wipeout boundary stops the wipeout action.

WmfIn

Rel.12 Imports *.wmf* and *.clp* files (short for Windows MetaFile IN).

Command	Alias	Ctrl+	F-key	Alt+	Menu Bar	Tablet
wmfin	...	...	...	IW	Insert ↳Windows Metafile	...

Command: wmfin

Displays the Import WMF dialog box. Select a file, and then click **Open**.

Specify insertion point or [Scale/X/Y/Z/Rotate/PScale/PX/PY/PZ/PRotate]: *(Pick a point, or enter an option.)*

Enter X scale factor, specify opposite corner, or [Corner/XYZ] <1>: *(Specify a value, pick a point, or enter an option.)*

Enter Y scale factor <use X scale factor>: *(Specify a value, or press Enter.)*

Specify rotation angle <0>: *(Specify a value, or press Enter.)*

COMMAND LINE OPTIONS

Insertion point picks the insertion point of the lower-left corner of the WMF image.

X scale factor scales the WMF image in the x direction (default = 1).

Corner scales the WMF image in the x and y directions.

XYZ scales the image in the x, y, and z directions.

Y scale factor scales the image in the y direction (default = x scale).

Rotation angle rotates the image (default = 0).

RELATED COMMANDS

WmfOpts controls the importation of *.wmf* files.

WmfOut exports selected objects in WMF format.

RELATED FILES

**.clp* are Windows Clipboard files.

**.wmf* are Windows Metafiles.

TIPS

- The WMF image is placed as a block with the name **WMF0**; subsequent placements of *.wmf* files increase the number by one: **WMF1**, **WMF2**, and so on.
- The *.clp* files are created by the Windows Clipboard. After using CTRL+C to copy objects to the Clipboard, you can open the Clipboard Viewer, and then save the image as a *.clp* file.
- Exploding the WMF*n* block results in polylines; even circles, arcs, and text are converted to polylines; solid-filled areas are exploded into solid triangles. The *.clp* file is pasted as a block; when exploded, constituent parts are 2D polylines.
- Use WmfOut and WmfIn to explode splines, text, circles, and arc to short line segments. (This is useful for translation to CNC machining files.) Do not zoom or pan between WmfOut and WmfIn.
- WmfIn needs a scale factor of 2 to match the size of objects created by WmfOut.

WmfOpts

Rel.12 Controls the importation of *.wmf* files (short for Windows Meta File OPTionS).

Command	Alias	Ctrl+	F-key	Alt+	Menu Bar	Tablet
wmfopts	...	...	...	...	...	...

Command: wmfopts

Displays dialog box:

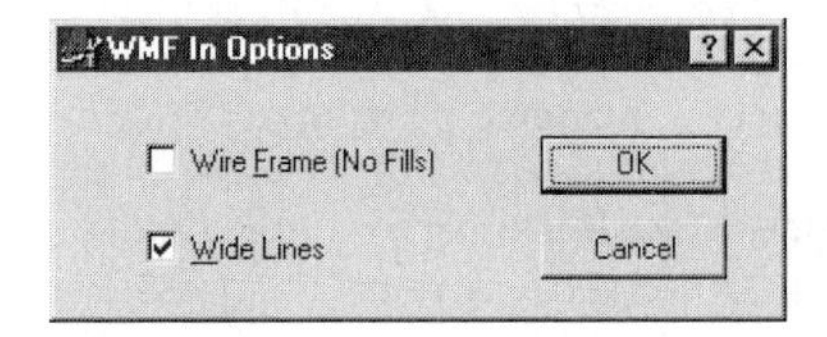

WMF IN OPTIONS DIALOG BOX

Wire Frame

☑ Displays the WMF images with lines only, no filled areas (default).

☐ Displays area fills.

Wide Lines

☑ Displays lines with width (default).

☐ Displays lines with a width of zero.

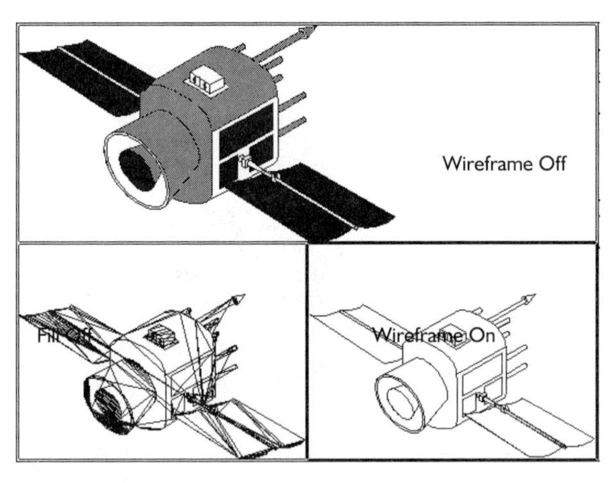

RELATED COMMANDS

WmfIn imports *.wmf* files.

WmfOut exports selected objects in WMF format.

WmfOut

Rel.12 Exports selected objects in WMF format (short for Windows MetaFile OUTput).

Command	Alias	Ctrl+	F-key	Alt+	Menu Bar	Tablet
wmfout	...	...	...	**FE**	**File**	...
				↳WMF	**↳Export**	
					↳Metafile	

Command: wmfout

Displays the Create WMF File dialog box. Enter a file name, and then click ***Save****.*

Select objects: *(Select one or more objects.)*

Select objects: *(Press Enter to end object selection.)*

COMMAND LINE OPTION

Select objects selects the objects to export. Press **Ctrl+A** to select all objects visible in the current viewport.

RELATED SYSTEM VARIABLE

WmfBkgnd toggles the background color of exported *.wmf* files:

0 — Transparent background.

1 — AutoCAD background color.

WmfForegnd switches the foreground and background colors of exported *.wmf* files as required:

0 — Foreground is darker than background color.

1 — Background is darker than foreground color.

RELATED COMMANDS

WmfOpts controls the importation of *.wmf* files.

WmfIn imports files in WMF format.

CopyClip copies selected objects to the Clipboard in several formats, including *.wmf*; also called "picture" format.

TIPS

- The *.wmf* files created by AutoCAD are resolution-dependent; small circles and arcs lose their roundness.
- The **All** selection does not select all objects in the drawing; instead, the WmfOut command selects all objects *visible* in the current viewport.
- Autodesk needs to update its WMF-related commands to handle EMF (enhanced meta format) files.

WorkSpace

2006 Controls workspaces and settings at the command-line.

Command	Alias	Ctrl+	F-key	Alt+	Menu Bar	Tablet
workspace	...	...	...	...	**Tools** ↳**Workspaces** ↳*workspace name*	...

Command: workspace
Enter workspace option [setCurrent/SAveas/Edit/Rename/Delete/SEttings/?] <setCurrent>: *(Enter an option.)*

COMMAND LINE OPTIONS

setCurrent makes a workspace current.

SAveas saves the current interface configuration as a named workspace.

Edit allows modifications to workspaces; displays the Customize User Interface dialog box; see the **CUI** command.

Rename renames workspaces.

Delete erases workspaces.

SEttings displays the Workspace Settings dialog box; see WsSettings command.

RELATED COMMANDS

Cui creates and modifies workspaces through a dialog box.

WsSave saves the current user interface configuration as a new workspace.

WsSettings specifies options for workspaces.

RELATED SYSTEM VARIABLES

WsCurrent names the current workspace.

STARTUP SWITCH

/w specifies the workspace to show when AutoCAD starts up.

TIPS

- This command is meant for use by scripts and programs.
- The WsCurrent system variable can be used to change workspaces quickly at the command line.
- You can also change workspaces through:
  - The Workspaces item on the Windows menu.
  - The droplist on the Workspaces toolbar.
- The default workspace is named "AutoCAD Classic." If it has not been edited, you can select this workspace to return AutoCAD to the way it looked when it was first installed on your computer.

WsSave

2006 Saves the current user interface as a named workspace (short for WorkSpace Save).

Command	Alias	Ctrl+	F-key	Alt+	Menu Bar	Tablet
wssave	...	...	...	...	**Tools** ↳**Workspaces** ↳**Save Current As**	...
-wssave						

Command: wssave

Displays dialog box:

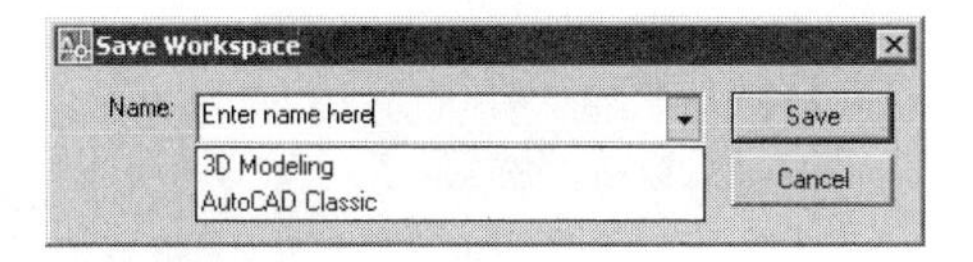

SAVE WORKSPACE DIALOG BOX

Name names the workspace. If the name already exists, AutoCAD warns, "Workspace already exists. Do you wish to replace it?" Click **Yes** or **No**.

-WSSAVE Command

Command: -wssave

Save Workspace as <AutoCAD Default>: *(Enter a workspace name.)*

OPTIONS

Save Workspace as names the workspace.

RELATED COMMANDS

Cui creates and modifies workspaces through a dialog box.

Workspace creates, saves, and controls workspaces at the command line.

WsSettings specifies options for workspaces.

RELATED SYSTEM VARIABLES

WsCurrent specifies the name of the current workspace.

TIP

- Workspaces are saved in *.cui* files, and can be edited with the Cui command.

WsSettings

2006 Specifies which workspaces appear in the Workspace menu.

Command	Alias	Ctrl+	F-key	Alt+	Menu Bar	Tablet
wssettings	...	...	...	...	**Tools** ↳**Workspaces** ↳**Workspace Settings**	...

Command: wssettings

Displays dialog box:

WORKSPACE SETTINGS DIALOG BOX

My Workspace selects the default workspace.

Move Up moves the selected item up the list.

Move Down moves the selected item down the list.

Add Separator adds a gray line above the selected item, which shows up in the Workspaces menu.

⊙ **Do Not Save Changes to Workspace** does not save user interface changes to the current workspace before switching to another workspace.

○ **Automatically Save Workspace Changes** saves changes to the user interface before switching to another workspace.

RELATED COMMANDS

Cui creates and modifies workspaces through a dialog box.

Workspace creates, saves, and controls workspaces at the command line.

WsSave saves the current user interface configuration as a new workspace.

XAttach

Rel.14 Attaches externally-referenced drawings to the current drawing (short for eXternal reference ATTACH).

Command	Alias	Ctrl+	F-key	Alt+	Menu Bar	Tablet
xattach	xa	...	...	ID	Insert ↳DWG Reference	...

Command: xattach

Displays the Select File to Attach dialog box. Select a file, and click ***Open.***

Displays dialog box:

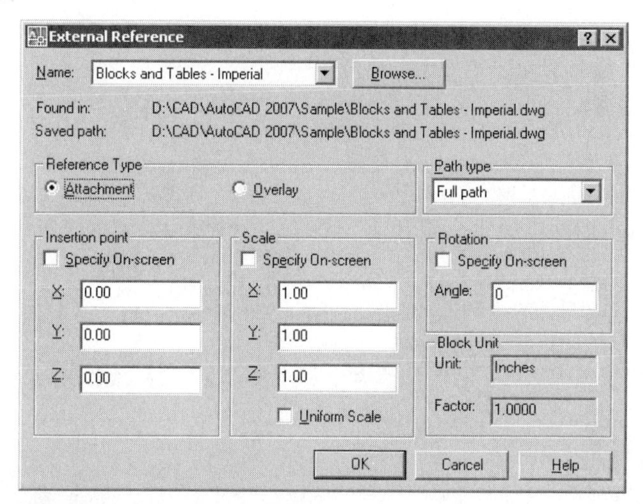

After you click ***OK****, AutoCAD confirms at the command line:*

Attach Xref FILENAME: C:\filename.dwg
FILENAME loaded.

EXTERNAL REFERENCE DIALOG BOX

Name specifies the file name of the external *.dwg* file to be attached; the drop list shows the names of currently-attached xrefs (externally-referenced files).

Browse displays the Select File To Attach dialog box.

Retain Path

☑ Saves the xref's filename and full path in the *.dwg* file.

☐ Saves only the filename of the xref.

Reference Type options

⊙ **Attachment** attaches the xref.

○ **Overlay** overlays the xref.

Insertion Point options

☑ **Specify On-screen** specifies the insertion point of the xref in the drawing.

*After you click **OK** to dismiss the dialog box, AutoCAD prompts:*

Specify insertion point or [Scale/X/Y/Z/Rotate/PScale/PX/PY/PZ/PRotate]: *(Pick a point, or enter an option.)*

Scale sets the scale factor for the x, y, and z axes.

X sets the x-scale factor.

Y sets the y-scale factor.

Z sets the z-scale factor.

Rotate specifies the rotation angle.

PScale presets the scale factor for the x, y, and z axes.

PX presets the x-scale factor.

PY presets the y-scale factor.

PZ presets the z-scale factor.

PRotate presets the rotation angle.

Scale options

☐ **Specify On-screen** specifies the scale of the xref in the drawing.

X sets the x-scale factor.

Y sets the y-scale factor.

Z sets the z-scale factor.

☐ **Uniform scale** sets the Y and Z factors equal to X.

*After you click **OK** to dismiss the dialog box, AutoCAD prompts:*

Enter X scale factor, specify opposite corner, or [Corner/XYZ] <1>: *(Enter a value, or enter an option.)*

Enter Y scale factor <use X scale factor>: *(Enter a value, or press* **ENTER***.)*

X scale factor scales the xref in the x direction.

Corner indicates the x,y scale factor by picking two points of a rectangle.

XYZ specifies the scale factor in the x, y, and z directions.

Y scale factor scales the xref in the y direction.

Rotation Angle options

☐ **Specify On-screen** specifies the rotation of the xref in the drawing.

*After you click **OK** to dismiss the dialog box, AutoCAD prompts:*

Specify rotation angle <0>: *(Enter a value, or press* **ENTER***.)*

Rotation angle specifies the rotation angle of the xref.

RELATED SYSTEM VARIABLES

XRefType determines whether xrefs are attached or overlaid, by default.

XEdit determines whether the drawing may be edited in-place, when being referenced by another drawing.

XLoadPath stores the path of temporary copies of demand-loaded xref drawings.

XRefCtl controls whether *.xlg* external reference log files are written.

ProjectName names the prject for the current drawing (default = "").

DemandLoad specifies if and when AutoCAD demand-loads a third-party application when a drawing contains custom objects created by the application:

0 — Turns off demand loading.

1 — Loads application when drawings contain proxy objects.

2 — Loads application when the application's commands are invoked.

3 — Loads application when drawings contain proxy objects, or when the application's commands are invoked.

IdxCtl controls the creation of layer and spatial indices:

0 — Creates no indices (default).

1 — Creates layer index.

2 — Creates spatial index.

3 — Creates both layer and spatial indices.

VisRetain specifies how the layer settings — on-off, freeze-thaw, color, and linetype — in xref drawings are defined by the current drawing:

0 — Xref layer definition in the current drawing takes precedence.

1 — Settings for xref-dependent layers take precedence over xref layer definition in the current drawing.

XLoadCtl controls the loading of xref drawings:

0 — Loads the entire xref drawing.

1 — Demand loading; xref is opened.

2 — Demand loading; copy of the xref is opened.

RELATED COMMANDS

XOpen edits xref drawings.

XBind binds portions of xref drawings to the current drawing.

XClip clips the display of xrefs.

ExternalReferences controls xrefs.

TIP

- When AutoCAD cannot find an xref, it searches in the following order:
    - The folder of the current drawing.
    - The project search paths defined in the Options dialog box's Files tab and the ProjectName system variable.
    - The support search paths defined in the Options dialog box's Files tab.
    - The Start In folder specified in the shortcut that launched AutoCAD.

XBind

Rel.11 Binds portions of externally-referenced drawings to the current drawing (short for eXternal BINDing).

Commands	Aliases	Ctrl+	F-key	Alt+	Menu Bar	Tablet
xbind	**xb**	...	...	**MOEB**	**Modify** ↳**Object** ↳**External Reference** ↳**Bind**	**X19**
-xbind	**-xb**					

Command: xbind

Displays dialog box:

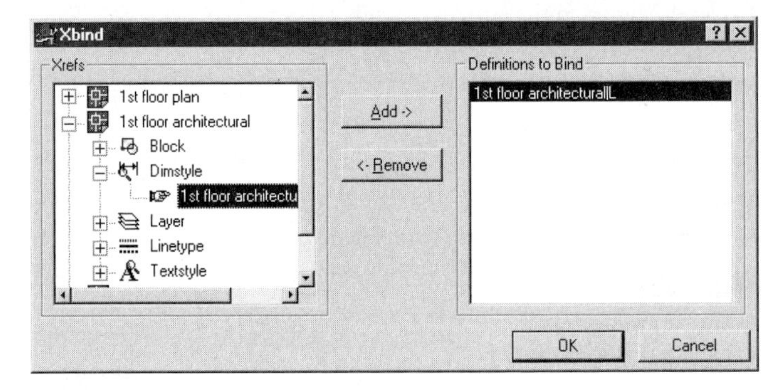

XBIND DIALOG BOX

Xrefs lists externally-referenced drawings, along with their bindable objects: blocks, dimension styles, layer names, linetypes, and text styles.

Definitions to Bind lists definitions that will be bound.

Add adds a definition to the binding list.

Remove removes a definition from the binding list.

-XBIND Command

Command: -xbind

Enter symbol type to bind [Block/Dimstyle/LAyer/LType/Style]: *(Enter an option.)*

Enter dependent name(s): *(Enter one or more names, separated by commas.)*

COMMAND LINE OPTIONS

Block binds blocks to the current drawing.

Dimstyle binds dimension styles to the current drawing.

LAyer binds layer names to the current drawing.

LType binds linetype definitions to the current drawing.

Style binds text styles to the current drawing.

Enter dependent names specifies the named objects to bind.

RELATED COMMANDS

RefEdit edits xref drawings.

ExternalReferences controls xrefs.

TIPS

- The **XBind** command lets you copy named objects from another drawing to the current drawing.
- Before you can use the XBind command, you must first use the XAttach command to attach an xref to the current drawing.
- Blocks, dimension styles, layer names, linetypes, and text styles are known as "dependent symbols."
- When a dependent symbol is part of an xrefed drawing, AutoCAD uses a vertical bar (|) to separate the xref name from the symbol name, as in *filename|layername*.
- After you use the XBind command, AutoCAD replaces the vertical bar with **0**, as in *filename***0***layername*. The second time you bind that layer from that drawing, XBind increases the digit, as in *filename***1***layername*.
- When the XBind command binds a layer with a linetype (other than Continuous), it automatically binds the linetype.
- When the XBind command binds a block — with a nested block, dimension style, layer, linetype, text style, and/or reference to another xref — it automatically binds those objects as well.

XClip

Rel.12 Clips portions of blocks and externally-referenced drawings (short for eXternal CLIP; formerly the XRefClip command).

Command	Alias	Ctrl+	F-key	Alt+	Menu Bar	Tablet
xclip	xc	...	...	...	Modify ↳Clip ↳Xref	X18

Command: xclip
Select objects: *(Select one or more blocks or xrefs.)*
Select objects: *(Press Enter to end object selection.)*
Enter clipping option
[ON/OFF/Clipdepth/Delete/generate Polyline/New boundary] <New>: *(Enter an option.)*

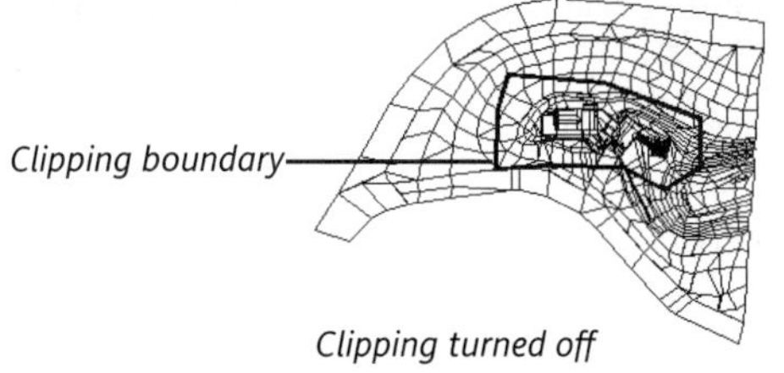

Clipping turned off

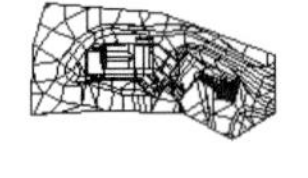
Clipping turned on

COMMAND LINE OPTIONS

Select objects selects the xref or block, *not* the clipping polyline.

ON turns on clipped display.

OFF turns off clipped display; displays all of the xref or block.

Clipdepth sets front and back clipping planes for 3D xrefs and blocks.

Delete erases the clipping boundary.

generate Polyline extracts the existing boundary as a polyline.

New boundary places a new rectangular or irregular polygon clipping boundary, or creates an irregular clipping boundary from an existing polyline.

RELATED COMMANDS

XBind binds parts of the xref to the current drawing.

ExternalReferences controls xrefs.

RELATED SYSTEM VARIABLE

XClipFrame toggles the display of the clipping boundary.

TIPS

- This command works for both blocks and xrefs.
- A spline-fit polyline results in a curved clip boundary, but a curve-fit polyline does not.

XEdges

2007 Extracts wireframes from the edges of 3D solids and surfaces, and 2D regions (short for eXtract EDGES).

Command	Alias	Ctrl+	F-key	Alt+	Menu Bar	Tablet
xedges	...	...	...	**M3E**	**Modify** ↳**3D Operations** ↳**Extract Edges**	...

Command: xedges
Select objects: *(Select one or more solids, surfaces, or regions. Hold down the **Ctrl** key to select individual edges.)*
Select objects: *(Press Enter.)*

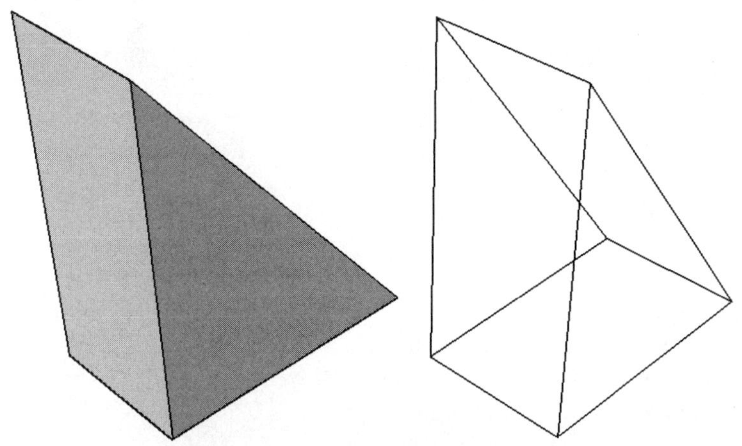

***Left**: Original 3D solid.*
***Right**: Extracted edges forming 3D wireframe object.*

TIPS

- This command does not work with spheres and tori, because they have no edges. Similarly, only the base of cones can be extracted.
- The command leaves edges "on top" of the original solid. Use grips editing to move the solid out of the way.
- Straight edges become lines; curves become arcs, circles, or splines.
- Xedges are created on the current layer.

XLine

Rel.13 Places infinitely long construction lines in drawings.

Command	Alias	Ctrl+	F-key	Alt+	Menu Bar	Tablet
xline	**xl**	...	...	**DT**	**Draw** ↳**Construction Line**	**L10**

Command: xline

Specify a point or [Hor/Ver/Ang/Bisect/Offset]: *(Pick a point, or enter an option.)*

Through point: *(Pick a point.)*

Through point: *(Press Enter to end the command.)*

Draws xlines, as illustrated below

COMMAND LINE OPTIONS

Specify a point picks the midpoint for the xline.

Through point picks another point through which the xline passes.

Ang places the construction line at an angle.

Bisect bisects an angle with the construction line.

From point places the construction line through a point.

Hor places a horizontal construction line.

Offset places the construction line parallel to another object.

Ver places a vertical construction line.

ENTER exits the command.

Angle options

Enter angle of xline <0> or [Reference]: *(Enter an angle, or type* **R***.)*

Enter angle of xline specifies the angle of the xline relative to the x-axis.

Reference specifies the angle relative to two points.

Bisect options

Specify angle vertex point: *(Pick a point.)*

Specify angle start point: *(Pick a point.)*

Specify angle end point: *(Pick a point.)*

Specify angle vertex point specifies the vertex of the angle.

Specify angle start point specifies the angle start point.

Specify angle end point specifies the angle endpoint.

Offset options

Specify offset distance or [Through] <1.0000>: *(Enter a distance, or type* **T***.)*

Select a line object: *(Select a line, xline, ray, or polyline line segment.)*

Specify side to offset: *(Pick a point.)*

Specify offset distance specifies the distance between xlines.

Through picks a point through which the xline should pass.

Select a line object selects the line, xline, ray, or polyline to offset.

Specify side to offset specifies the offset side.

RELATED COMMANDS

Properties modifies characteristics of xline and ray objects.

Ray places semi-infinite construction lines.

RELATED SYSTEM VARIABLE

OffsetDist specifies the current offset distance.

TIPS

- Use xlines to find the bisectors of triangles (using **MIDpoint** object snap), or to create intersection snap points (using **INTersection** object snap).

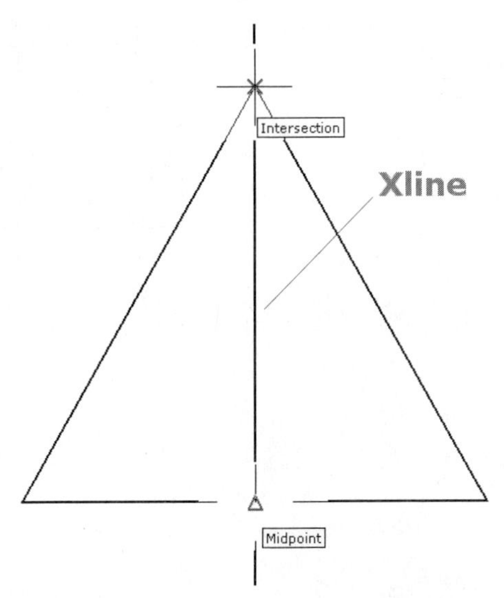

- Ray and xline construction lines are plotted; they do not affect the extents.

XOpen

2004 Opens externally-referenced drawings in new windows (short for eXternal OPEN).

Command	Alias	Ctrl+	F-key	Alt+	Menu Bar	Tablet
xopen	...	...	...	...	...	...

Command: xopen
Select xref: *(Select an externally-referenced drawing.)*

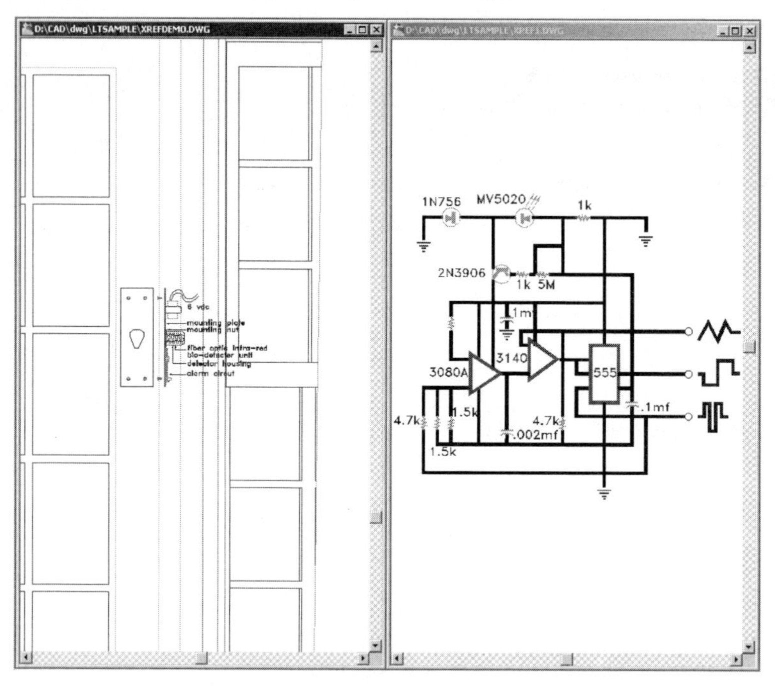

***Left**: original drawing.*
***Right**: Opened xref in separate window.*

COMMAND LINE OPTION

Select xrefs selects the xref to open; the xref must be inserted in the current drawing.

TIPS

- This command does not work with blocks.
- When you select a non-xref object, AutoCAD complains, 'Object is not an Xref.'

'Xplode

Rel.12 Explodes complex objects into simpler objects, with user control (short for eXPLODE).

Command	Alias	Ctrl+	F-key	Alt+	Menu Bar	Tablet
xplode	xp	...	...	...	...	...

Command: xplode
Select objects to XPlode.
Select objects: *(Select one or more objects.)*
Select objects: *(Press Enter to end object selection.)*
Enter an option [Individually/Globally] <Globally>: *(Type* **I** *or* **G***.)*
Enter an option [All/Color/LAyer/LType/Inherit from parent block/Explode] <Explode>: *(Enter an option.)*

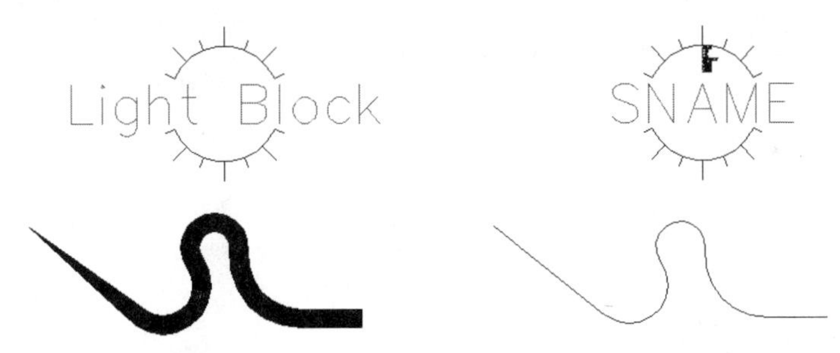

***Left**: Block and polyline.*
***Right**: Exploded objects lose attribute data and width .*

COMMAND LINE OPTIONS

Select objects selects objects to be exploded.

Individually allows you to specify options for each selected object.

Globally applies options to all selected objects.

All sets the color, layer, linetype and lineweight of exploded objects.

Color sets the color of objects after they are exploded: red, yellow, green, cyan, blue, magenta, white, bylayer, byblock, or any color number.

LWeight specifies a lineweight.

LAyer sets the layer for the exploded objects.

LType specifies any loaded linetype name for the exploded objects.

Inherit from parent block assigns the color, linetype, lineweight, and layer of the exploded objects, based on the original object.

Explode reduces complex objects into their components.

RELATED COMMANDS

Explode explodes the object without options.

U reverses the explosion.

TIPS

- Examples of complex objects include blocks and polylines; examples of simple objects include lines, circles, and arcs.
- Mirrored blocks can be exploded.
- The **LWeight** option is not displayed when the lineweight is off.
- The 'Enter an option [Individually/Globally]:' prompt appears only when more than one valid object is selected for explosion.
- Specifying **BYLayer** for the color or linetype means that the exploded objects take on the color or linetype of the object's original layer.
- Specifying **BYBlock** for the color or linetype means that the exploded objects take on the color or linetype of the original object.
- The default layer is the current layer, not the exploded object's original layer.
- The XPlode command breaks down complex objects as follows:

Object	Exploded into
Attribute	Attribute values are deleted; displays attribute definitions.
Block	Component objects.
Helix	Spline.
Leader	Line segments, splines, mtext, and tolerance objects; arrowheads become solids or blocks.
Mtext	Text.
Multiline	Line and arc segments.
PlaneSurf	Region.
Polyface mesh	Point, line, or 3D faces.
Polysolid	Region.
Region	Lines, arcs, and splines.
Table	Lines.
2D polyline	Line and arc segments; width and tangency are lost.
3D polyline	Line segments.
3D solid	Planar surfaces become regions; nonplanar surfaces become bodies.
3D body	Single-surface body, regions, or curves.

- The **Inherit** option works only when the parts were originally drawn with color, linetype, and lineweight set to BYBLOCK, and drawn on layer 0.

-XRef

Rel.11 Controls externally-referenced drawings in the current drawing at the command line (short for eXternal REFerence).

Commands	Alias	Ctrl+	F-key	Alt+	Menu Bar	Tablet
-xref	-xr	...	...	...	...	...

Command: -xref
Enter an option [?/Bind/Detach/Path/Unload/Reload/Overlay/Attach] <Attach>: *(Enter an option.)*

COMMAND LINE OPTIONS

? lists the names of xref files.

Bind makes the xref drawing part of the current drawing.

Detach removes xref files.

Path respecifies paths to xref files.

Unload unloads xref files.

Reload updates the xref files.

Overlay overlays the xref files.

Attach attaches another drawing to the current drawing.

RELATED COMMANDS

ExternalReferences displays a palette of all xrefs, image files, and DWF underlays.

Insert adds another drawing to the current drawing.

RefEdit edits xrefs.

XBind binds parts of xrefs to the current drawing.

XClip clips portions of xrefs.

RELATED SYSTEM VARIABLES

See the XAttach command.

TIPS

- *Warning!* Nested xrefs cannot be unloaded.
- The ExternalReferences command replaces the dialog-based XRef command, as of AutoCAD 2007. Use the undocumented ClassicXref command to access the old dialog box.

'Zoom

V. 1.0 Makes drawings larger or smaller in the current viewport.

Command	Aliases	Ctrl+	F-key	Alt+	Menu Bar	Tablet
zoom	z rtzoom	...	...	VZ	View ↳Zoom	K11

Command: zoom
Specify corner of window, enter a scale factor (nX or nXP), or [All/Center/Dynamic/Extents/Previous/Scale/Window/Object] <real time>:
(Pick a point, enter an option, or press Enter.)

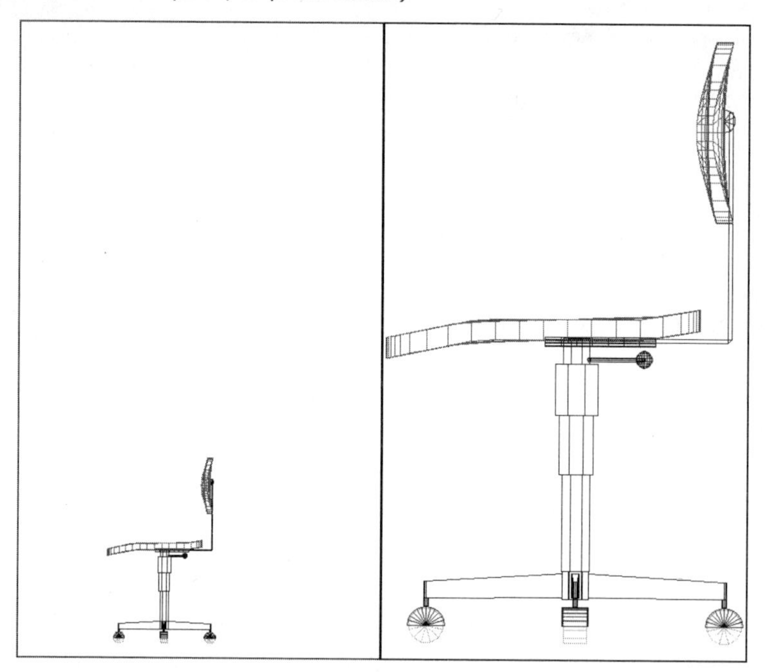

COMMAND LINE OPTIONS

(pick a point) begins the **Window** option.

realtime starts real-time zoom.

Enter *or* ESC ends real-time zoom.

All displays the drawing limits or extents, whichever is greater.

Dynamic brings up the dynamic zoom view.

Extents displays the current drawing extents.

Previous displays the previous view generated by the Pan, View, or Zoom commands.

Vmax displays the current virtual screen limits (short for Virtual MAXimum; undocumented).

Window indicates the two corners of the new view.

Object zooms to the extents of selected objects.

Center options

Specify center point: *(Pick a point.)*

Enter magnification or height <>: *(Enter a value.)*

Center point indicates the center point of the new view.

Enter magnification or height indicates a magnification value or height of view.

Left options (undocumented)

Lower left corner point: *(Pick a point.)*

Enter magnification or height <>: *(Enter a value.)*

Lower left corner point indicates the lower-left corner of the new view.

Enter magnification or height indicates a magnification value or height of view.

Scale(X/XP) options

*n***X** displays a new view as a factor of the current view.

*n***XP** displays a paper space view as a factor of model space.

RELATED COMMANDS

DsViewer displays the Aerial View palette, which zooms and pans.

Pan moves the view.

View saves zoomed views by name.

VtOptions controls smooth zoom transitions.

3dZoom performs real-time zooms in perspective viewing mode.

RELATED SYSTEM VARIABLES

ViewCtr reports the coordinates of the current view's center point.

ViewSize reports the height of the current view.

VtDuration specifies the duration of view transitions.

VtEnable turns on smooth view transitions.

VtFps specifies the speed of view transitions.

TIPS

- A magnification of 1x leaves the drawing unchanged; a magnification of 2x enlarges objects (zooms in), while 0.5x makes objects smaller (zooms out).
- Transparent zoom is *not* possible during the VPoint, Pan, DView, View and 3dOrbit commands.
- During real-time zoom, right-click in the drawing to see the shortcut menu displayed by the 3dOrbit command.

3D

Rel.11 Draws 3D mesh primitives with polygon meshes (short for three Dimensions).

Command	Alias	Ctrl+	F-key	Alt+	Menu Bar	Tablet
3d	...	...	...	...	...	**N8**

Command: 3d
Enter an option
[Box/Cone/DIsh/DOme/Mesh/Pyramid/Sphere/Torus/Wedge]: *(Enter an option.)*

See the Ai_ commands for details, such as Ai_Box and Ai_Wedge.

RELATED COMMANDS

Ai_Box draws 3D mesh boxes and cubes.

Ai_Cone draws 3D mesh cones.

Ai_Dish draws 3D mesh dishes.

Ai_Dome draws 3D mesh domes.

Ai_Mesh draws 3D meshes.

Ai_Pyramid draws 3D mesh pyramids.

Ai_Sphere draws 3D mesh spheres.

Ai_Torus draws 3D mesh tori.

Ai_Wedge draws 3D mesh wedges..

RELATED SYSTEM VARIABLES

SurfTab1 controls the mesh density in the M direction of mesh objects.

SurfTab2 controls the mesh density in the N direction of mesh objects.

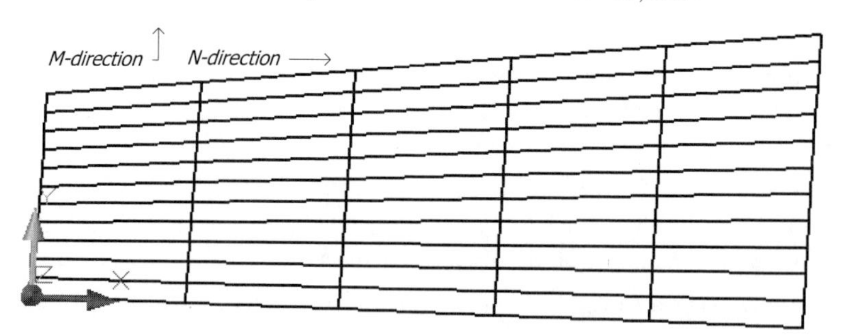

TIPS

- This command creates 3D objects made of 3D polygon meshes, and *not* of 3D solids or surfaces.
- You cannot perform Boolean operations on 3D mesh models.
- To draw cylinders with endcaps, apply thickness to circles.
- Use the Ucs or Align command to place 3D mesh models in space; use the VPoint and 3dOrbit commands to view mesh models from different 3D viewpoints.
- You can apply the Hide and Render commands to 3D mesh models.

3dAlign

Moves and rotates objects in 2D and 3D space.

Command	Alias	Ctrl+	F-key	Alt+	Menu Bar	Tablet
3dalign	**3al**	...	...	**M3A**	**Modify** **↳3D Operation** **↳3D Align**	**W20**

Command: 3dalign
Select objects: *(Select one or more objects.)*
Select objects: *(Press Enter to end object selection.)*

Specify source plane and orientation ...
Specify base point or [Copy]: *(Pick a point, or type* **C***.)*
Specify second point or [Continue] <C>: *(Press Enter, or pick another point.)*
Specify third point or [Continue] <C>: *(Press Enter, or pick a third point.)*

Specify destination plane and orientation ...
Specify first destination point: *(Pck a point.)*
Specify second destination point or [eXit] <X>: *(Press Enter, or pick another point.)*
Specify third destination point or [eXit] <X>: *(Press Enter, or pick a third point.)*

COMMAND LINE OPTIONS

Select objects selects one or more objects to be aligned.

Specify base, **Second**, and **Third point** specify the base, x axis, and y axis, respectively.

Copy copies the objects, instead of moving them.

Continue jumps ahead to the "Specify first destination point" prompt.

Specify First, **Second**, and **Third destination point** specify the basepoint and orientation of the moved/copied/rotated objects.

eXit exits the command.

RELATED COMMANDS

Align does the same, but in 2D space.

3dMove move objects in 3D space with the assistance of grip tools.

3dRotate rotates objects in 3D space with the assistance of grip tools.

TIP

- Use dynamic UCS to help position the moved/rotated objects more easily.

3dArray

Rel.11 Creates 3D rectangular and polar arrays.

Command	Alias	Ctrl+	F-key	Alt+	Menu Bar	Tablet
3darray	3a	...	...	M33	Modify ↳3D Operation ↳3D Array	W20

Command: 3darray
Select objects: *(Select one or more objects.)*
Select objects: *(Press Enter to end object selection.)*
Enter the type of array [Rectangular/Polar] <R>: *(Type **R** or **P**.)*

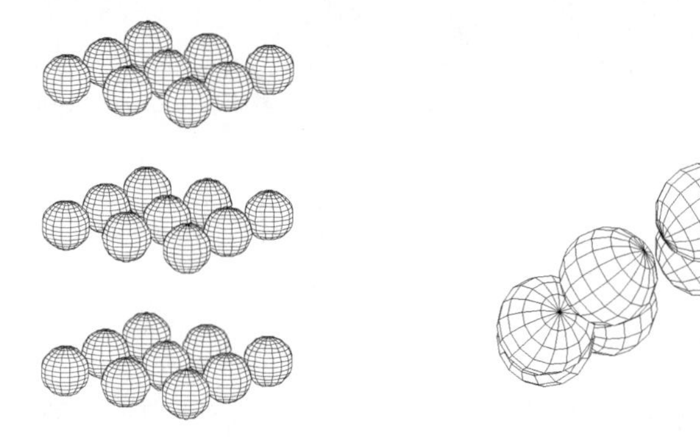

3D rectangular array *3D Polar array.*

COMMAND LINE OPTIONS

Select objects selects the objects to be arrayed.

R creates rectangular 3D arrays.

P creates a polar array in 3D space.

Rectangular Array options
Enter the number of rows (---) <1>: *(Enter a value.)*
Enter the number of columns (|||) <1>: *(Enter a value.)*
Enter the number of levels (...) <1>: *(Enter a value.)*
Specify the distance between rows (---) <1>: *(Enter a value.)*
Specify the distance between columns (|||) <1>: *(Enter a value.)*
Specify the distance between levels (...) <1>: *(Enter a value.)*

Enter the number of rows specifies the number of rows in the x direction.

Enter the number of columns specifies the number of columns in the y direction.

Enter the number of levels specifies the number of levels in the z direction.

Specify the distance between rows specifies the distance between objects in the x direction.

Specify the distance between columns specifies the distance between objects in the y direction.

Specify the distance between levels specifies the distance between objects in the z direction.

Polar Array options

Enter the number of items in the array: *(Enter a number.)*

Specify the angle to fill (+=ccw, -=cw) <360>: *(Enter an angle.)*

Rotate arrayed objects? [Yes/No] <Y>: *(Type **Y** or **N**.)*

Specify center point of array: *(Pick a point.)*

Specify second point on axis of rotation: *(Pick a point.)*

Enter the number of items specifies the number of objects to array.

Specify the angle to fill specifies the distance along the circumference that objects are arrayed (default = 360 degrees).

Rotate arrayed objects?

Yes objects rotate so that they face the central axis (default).

No objects do not rotate.

Specify center point of array specifies the center point of the array and one end of the axis.

Specify second point on axis of rotation specifies the other end of the array axis.

ESC interrupts the drawing of arrays.

RELATED COMMANDS

Array creates a rectangular or polar array in 2D space.

Copy creates one or more copies of the selected object.

MInsert creates a rectangular block-array of blocks.

TIP

- This command does not operate in paper space.

'3dClip

2000 Performs real-time front and back clipping (short for three Dimensional CLIPping).

Command	Alias	Ctrl+	F-key	Alt+	Menu Bar	Tablet
3dclip	...	...	...	...	...	...

Command: 3dclip

Displays window:

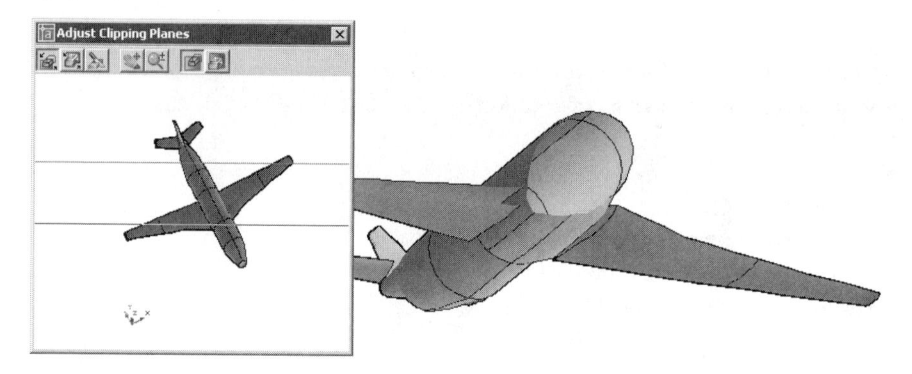

SHORTCUT MENU

*Right-click **Adjust Clipping Planes window**:*

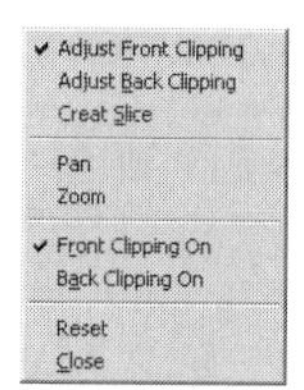

Adjust Front Clipping switches to front clipping mode.
Adjust Back Clipping switches to back clipping mode.
Create Slice switches to slicing — ganged front and back clipping — mode.

Pan moves the model within the window.
Zoom enlarges and reduces the view of the model within the window.

Front Clipping On toggles on and off front clipping.
Back Clipping On toggles on and off back clipping.

Reset removes clipping planes, and returns view to original state.
Close closes the Adjust Clipping Planes window.

TOOLBAR

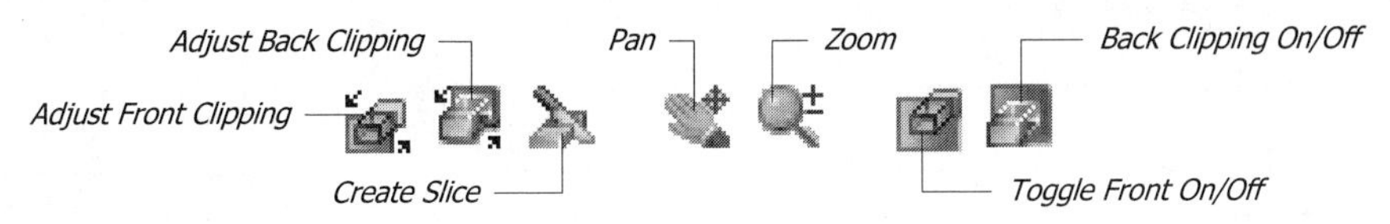

RELATED COMMANDS

3dOrbit provides real-time 3D viewing of the drawing.

3dPan performs real-time 3D sideways panning.

3dZoom performs real-time 3D zooming.

TIPS

- Use this command to hide objects in the front of 3D scenes, or to expose the interior of 3D models.
- Follow these steps to use this command effectively:
    1. Click **Toggle Front On/Off** to turn on front clipping.
    2. Drag the front clipping line over the model to cut off its front.
    3. Repeat, if necessary, with the back clipping plane.
- When you exit the Adjust Clipping Planes window, AutoCAD remains in 3dOrbit mode, and retains the clipped view. Press ESC to exit to the 'Command:' prompt.
- Clipping planes are relative to the viewport. Clipping planes do not follow when using 3dOrbit or other commands to rotate the view.
- The DWF Viewer is also capable of real-time clipping of 3D models exported in 3D DWF format by the **3dDwf** command.

'3dConfig

2004 Configures display characteristics of your computer's graphics board.

Command	Alias	Ctrl+	F-key	Alt+	Menu Bar	Tablet
3dconfig	...	...	...	...	...	...
-3dconfig						

Command: 3dconfig

Displays dialog box (new to AutoCAD 2007).

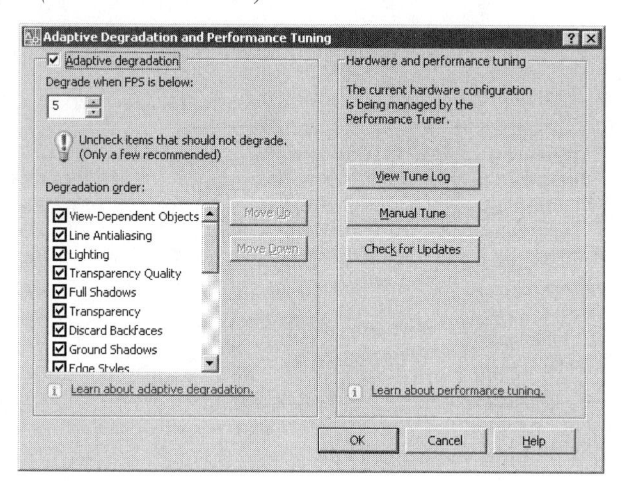

DIALOG BOX OPTIONS

☑ **Degrade When FPS Is Below** determines how AutoCAD degrades the screen image when motion falls below the specified FPS (frames per second), such as during the 3dOrbit command. Default is 5; range is 1 to 60 FPS.

Degradation Order specifies which visual effects are removed from display to prevent further degradation. The first item is degraded or removed first.

Move Up/Down changes the degradation order.

View Tune Log shows the results of AutoCAD's automatic evaluation of your computer's graphics board; displays the Performance Tuner Log dialog box.

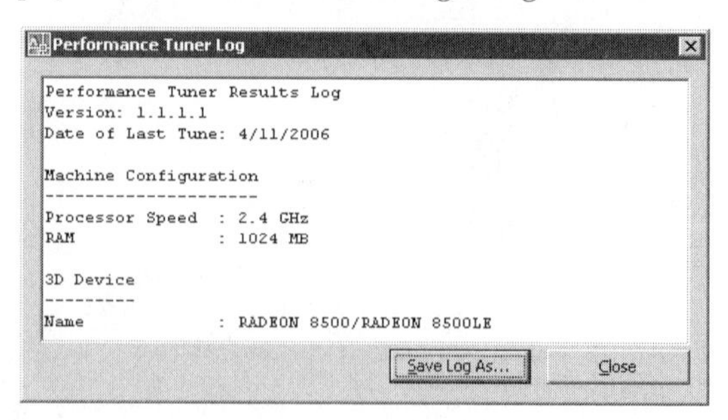

Manual Tune allows you to set the graphics board's capabilities; displays the Manual Performance Tuning dialog box.

Check for Updates launches Internet Explorer to access Autodesk's certification site.

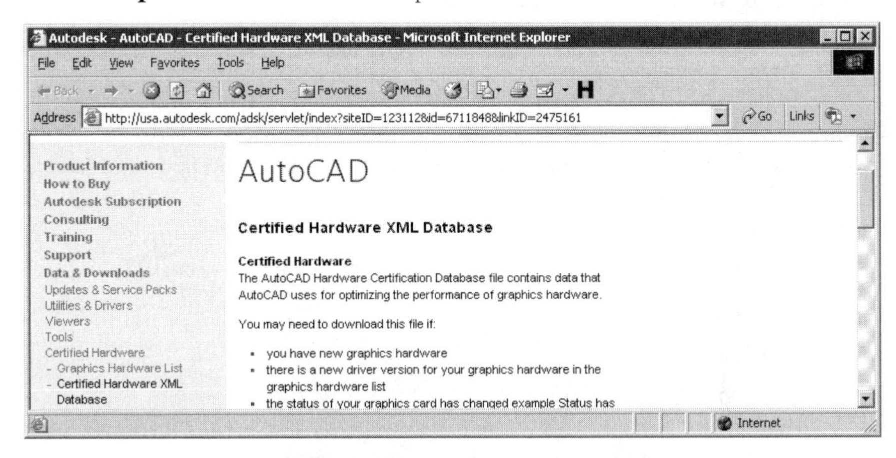

Manual Performance Turning dialog box

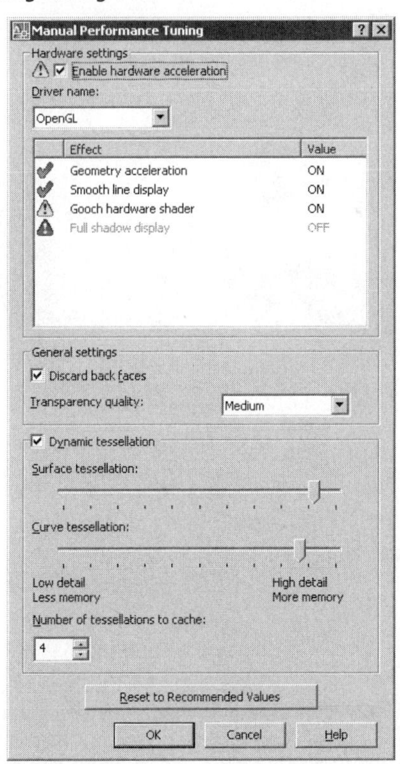

☑ **Enable Hardware Acceleration** toggles graphics generated by AutoCAD (software) or the graphics board (hardware). Usually hardware is faster, but if the graphics board lacks the capabilities listed under Effect, then it is better to turn off this option.

Warning! When Geometry Acceleration is turned off, full shadows cannot be displayed.

✔ (green) — Graphics board is capable of displaying all forms of graphics required by AutoCAD; hardware acceleration should be turned on.

⚠ (yellow) — Graphics card will run most but not all graphics; turn on hardware acceleration at the risk of AutoCAD locking up.

⚠ (red) — Graphics card is not capable; hardware acceleration is unavailable.

Driver Name selects the device driver for the graphics board. AutoCAD provides two drivers, and graphics board manufacturers may include their own:

- **OpenGL** — uses *wopengl9.hdi* driver file.
- **Direct 3D** — uses *direct3d9.hdi* driver file; required for Windows Vista.

☑ **Discard Back Faces** does not draw the backs of faces, which are not seen anyhow.

Transparency Quality:

- **Low (Faster)** — dithers the image to simulate transparency.
- **Medium** — blends the image for improved quality.
- **High (Slower)** — highest quality but slower speed (default for hardware acceleration).

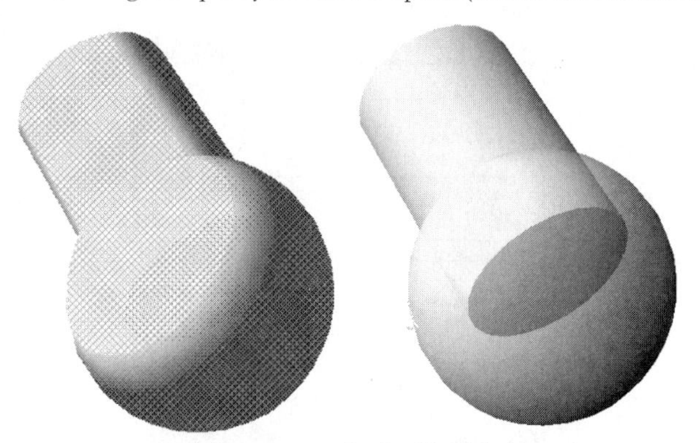

Left: *Transparency simulated by dithering.*
Right: *Hardware-accelerated transparency.*

☑ **Dynamic Tessellation** determines the smoothness of curved 3D objects.

Surface Tessellation specifies the level of surface details; less tessellation uses less memory.

Curve Tessellation specifies the level of curve details.

Number of Tessellations to Cache:

- **1** — tessellation is the same for all viewports; some objects may be regenerated during zooming.
- **2** or more — required when drawings have two or more viewports with different views.

Reset to Recommended Values reverts all options to their defaults

Command: -3dconfig
Configure: 3DCONFIG
Enter option [Adaptive degradation/Dynamic tessellation/General options/ acceLeration/eXit] <Adaptive degradation>: *(Enter an option, or type* **V** *for undocumented adVanced options.)*

COMMAND LINE OPTIONS

Adaptive degradation allows AutoCAD to switch to lower-quality rendering to maintain display speed.

Dynamic tessellation specifies the smoothness of faceted objects in 3D drawings.

General options handles options not dependent on the graphics board.

Geometry specifies the display of isolines and back faces in 3D drawings.

acceLeration specifies hardware or software acceleration.

adVanced lists undocumented options.

eXit exits the command.

Adaptive degradation options
Enter mode [ON/OFF] <ON>: *(Enter* **ON** *or* **OFF***.)*

ON or **OFF** turns on or off adaptive degradation.

When on, the following options become available:

Enter speed to maintain (0-60fps) <5>: *(Enter a value.)*

Maintain speed fps specifies the speed at which to display, in frames per second; range is 5fps to 60fps.

Dynamic tessellation options
Enter option [Surface tessellation/Curve tessellation/Tessellations to cache/ eXit] <Surface tessellations>: *(Enter an option.)*

Surface tessellation specifies the number of tessellation lines to display on surfaces; ranges from 0 to 100 (default = 92).

Curve tessellation specifies the number of tessellation lines to display on curved surfaces; ranges from 0 to 100 (default = 87).

Tessellations to cache specifies the number of tessellations to cache; ranges from 1 to 4 (default = 4).

General options

Formerly named Geometry options.

Enter option [Discard backfaces/Transparency quality/eXit] <Discard backfaces>: *(Enter an option.)*

Discard backfaces:

On discards back faces to enhance performance.

Off displays back faces.

Configure: Transparency - Enter mode sets the quality of transparency to low, medium, or high.

acceLeration options

Enter option [Hardware/Software/eXit] <Hardware>: *(Enter an option.)*

Hardware configures graphics card to perform 3D display rendering, the faster option.

Software configures 3D display through software, if the graphics board is unable to.

Enter option [Driver name/Geometry acceleration/Antialias lines/Shadows enabled/eXit] <Driver name>:

Driver name allows you to select a specific driver for the graphics board.

Geometry acceleration allows a more precise display, if turned on and supported by the graphics board.

Antialias lines uses anti-aliasing to drawn smoother lines.

Shadows enabled enables shadow-casting, if the graphics board is capable. (This option is not displayed when **Software** is selected.)

adVanced

Redraw on window expose toggles redrawing windows when they become visible.

RELATED COMMAND

3dOrbit provides real-time 3D viewing of the drawing.

'3dDistance/Zoom/Swivel/Pan

2000 This collection of commands is a subset of 3dOrbit.

Command	Alias	Ctrl+	F-key	Alt+	Menu Bar
3ddistance	...	...	...	VCA	View ↳Camera ↳Adjust Distance
3dzoom	...	...	...	...	...
3dswivel	...	...	...	VCS	View ↳Camera ↳Swivel
3dpan	...	...	...	...	...

Command: 3ddistance

Interactively changes the 3D viewing distance, closer to and further from the target.

Command: 3dzoom

Interactively zooms the 3D view, just like the 3dDistance command.

Command: 3dswivel

Interactively twists the 3D view about the target; the effect is like a rotating pan.

Command: 3dpan

Interactively pans the 3D view side to side; the effect is flat movement.

COMMAND LINE OPTIONS

ESC and **Enter** exit the command.

RELATED COMMANDS

3dClip performs real-time 3D front and back clipping.

3dOrbit performs real-time 3D viewing of the drawing.

3dOrbitCtr specifies the target point for 3D orbiting views.

TIPS

- You can use the 3dPan and 3dZoom commands when the current drawing is in perspective mode; when you try to use the regular Pan and Zoom command, AutoCAD complains, '** That command may not be invoked in a perspective view **.'
- Autodesk notes that the following commands exist but are not documented:
  - 3dZoomTransparent
  - 3dSwivelTransparent
  - 3dPanTransparent

  These appear to be used by the 3dOrbit command's shortcut menu.

3dDwf

2006 Saves drawings in 3D DWF format.

Command	Alias	F-key	Alt+	Menu Bar	Tablet
3ddwf	3ddwfpublish	...	...	...	...

Command: 3ddwf

This command operates only in model tab.

Displays Export 3D DWF dialog box. Enter a file name, and then click ***Save****.*

To specify options, select ***Tools*** | ***Options*** *from the file dialog box.*

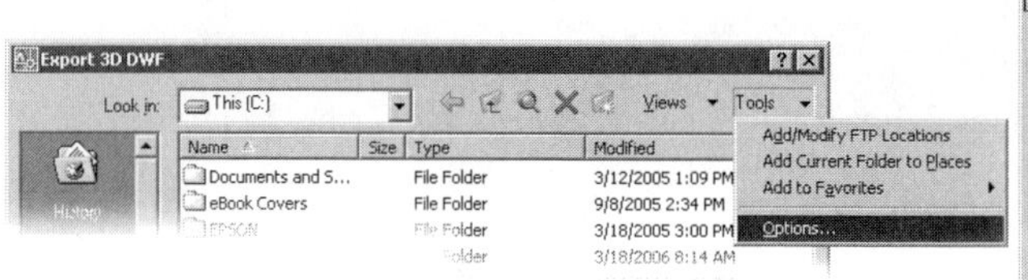

3D DWF PUBLISH DIALOG BOX

Objects to Publish

⊙ **All Model Space Objects** saves the entire drawing.

○ **Select Model Space Objects** saves selected objects.

removes dialog box for selecting objects. AutoCAD prompts, "Select objects."

3D DWF Organization

□ **Group by Xref Hierarchy** sorts objects by external-reference hierarchy in the DWF Viewer. This option is grayed out when the drawing has links to externally-referenced drawings.

Options

☑ **Publish with materials** includes material definitions for objects (*new in AutoCAD 2007*).

RELATED COMMANDS

Plot exports drawings in 2D DWF format.

Publish exports drawing sets in 2D DWF format.

RELATED SYSTEM VARIABLE

3dDwfPrec specifies the precision of 3D *.dwf* files:

1 — 1.000 units.
2 — 0.500
3 — 0.200
4 — 0.100
5 — 0.010
6 — 0.001

TIPS

- This command works only in model space; it cannot create *.dwf* files of layouts (paper space).
- This command was renamed from 3dDwfPublish in AutoCAD 2007.
- The level of precision defined by the 3dDwfPrec system variable affects file size and visual accuracy. For example, as precision is increased from level 1 to 6, the *steering.dwf* file size increases 11x, from 45KB to 513KB, while the visual accuracy of curves improves significantly:

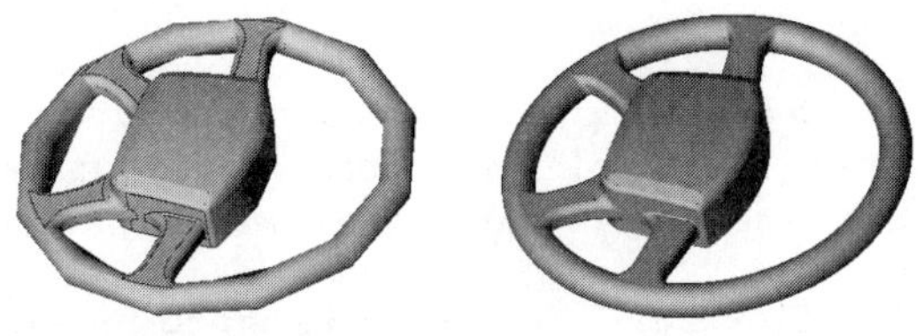

Left: *3D DWF file created with precision set to 1.*
Right: *Same drawing exported with precision set to 6.*

- When done saving the drawing in 3D DWF format, AutoCAD asks, "Filename.dwf was published successfully. Would you like to see it now?" Click **Yes** to view in DWF Viewer, which is included free with AutoCAD:

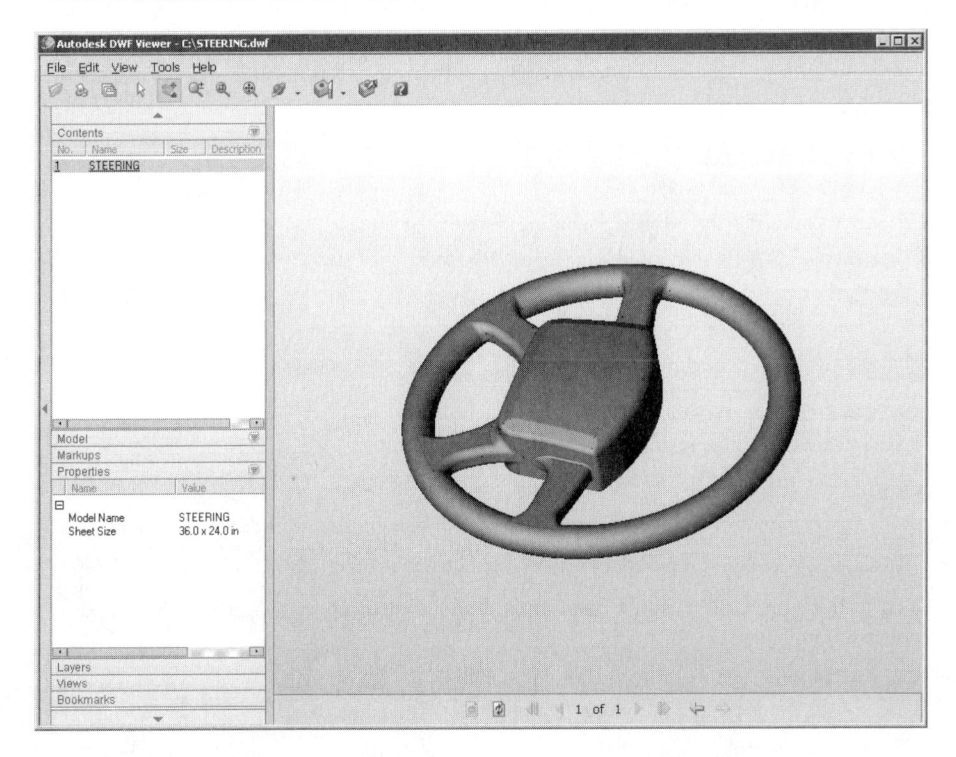

Renamed Command

3dDwfPublish was renamed **3dDwf** in AutoCAD 2007.

3dFace

V. 2.6 Draws 3D faces with three or four corners.

Command	Alias	Ctrl+	F-key	Alt+	Menu Bar	Tablet
3dface	3f	...	...	DMF	Draw ↳Surfaces ↳3D Face	M8

Command: 3dface
Specify first point or [Invisible]: *(Pick a point, or type* **I** *followed by the point.)*
Specify second point or [Invisible]: *(Pick a point, or type* **I** *followed by the point.)*
Specify third point or [Invisible] <exit>: *(Pick a point, or type* **I** *followed by the point.)*
Specify fourth point or [Invisible] <create three-sided face>: *(Pick a point, or type* **I** *followed by the point.)*
Specify third point or [Invisible] <exit>: *(Press Enter to exit the command.)*

COMMAND LINE OPTIONS

First point picks the first corner of the face.

Second point picks the second corner of the face.

Third point picks the third corner of the face.

Fourth point picks the fourth corner of the face; or press ENTER to create a triangular face.

Invisible makes the edge invisible.

RELATED COMMANDS

3D draws a 3D object: box, cone, dome, dish, pyramid, sphere, torus, or wedge.

Properties modifies the 3d face, including visibility of edges.

Edge changes the visibility of the edges of 3D faces.

EdgeSurf draws 3D surfaces made of 3D meshes.

PEdit edits 3D meshes.

PFace draws generalized 3D meshes.

RELATED SYSTEM VARIABLE

SplFrame controls the visibility of edges.

TIPS

- A 3D face is the same as a 2D solid, except that each corner can have a different z coordinate.
- Unlike the procedure for **Solid**, corner coordinates are entered in natural order.
- The **i** (short for invisible) suffix must be entered before osnap modes, point filters, and corner coordinates.
- *Invisible* 3D faces, where all four edges are invisible, do not appear in wireframe views; they hide objects behind them in hidden-line mode, and are rendered in shaded views.
- You can use the Properties and Edge commands to change the visibility of 3D face edges.
- 3D faces cannot be extruded.

3dMesh

Rel.10 Draws polygon 3D meshes.

Command	Alias	Ctrl+	F-key	Alt+	Menu Bar	Tablet
3dmesh	...	...	...	**DMM**	**Draw** ↳**Modeling** ↳**3D Mesh**	...

Command: 3dmesh
Enter size of mesh in M direction: *(Enter a value.)*
Enter size of mesh in N direction: *(Enter a value.)*
Specify location for vertex (0, 0): *(Pick a point.)*
Specify location for vertex (0, 1): *(Pick a point.)*
Specify location for vertex (1, 0): *(Pick a point.)*
Specify location for vertex (1, 1): *(Pick a point.)*

COMMAND LINE OPTIONS

Enter size of mesh in M direction specifies the m-direction mesh size (between 2 and 256).

Enter size of mesh in N direction specifies the n-direction mesh size (between 2 and 256).

Specify location for vertex (*m*,*n*) specifies a 2D or 3D coordinate for each vertex.

RELATED COMMANDS

3D draws a variety of 3D objects.

3dFace draws a 3D face with three or four corners.

Explode explodes a 3D mesh into individual 3D faces.

PEdit edits a 3D mesh.

PFace draws a generalized 3D face.

Xplode explodes a group of 3D meshes.

TIPS

- It is more convenient to use the EdgeSurf, RevSurf, RuleSurf, and TabSurf commands than the 3dMesh command. The 3dMesh command is meant for use by AutoLISP and other programs.
- The range of values for the m- and n-mesh size is 2 to 256.
- The number of vertices = **m** x **n**.
- The first vertex is (0,0). The vertices can be at any point in space.
- The coordinates for each vertex in row **m** must be entered before starting on vertices in row **m+1**.
- Use the PEdit command to close the mesh, since it is always created open.
- The SurfU and SurfV system variables do not affect the 3D mesh object.

3dMove

2007 Moves objects along the x, y, or z axis.

Command	Alias	Ctrl+	F-key	Alt+	Menu Bar	Tablet
3dmove	3m	...	...	M3M	Modify ↳3D Operations ↳3D Move	...

Command: 3dmove
Current positive angle in UCS: ANGDIR=counterclockwise ANGBASE=0
Select objects: *(Pick one or more objects.)*
Select objects: *(Press Enter.)*
Specify base point or [Displacement] <Displacement>: *(Pick a point, or type **D**.)*
Specify second point or <use first point as displacement>: *(Pick another point.)*

COMMAND LINE OPTIONS

Select objects selects the objects to be rotated.

Specify base point indicates the starting point for the move.

Displacement specifies relative x,y,z displacement when you press Enter at the next prompt.

Specify second point of displacement indicates the distance to move.

3D Move grip tool

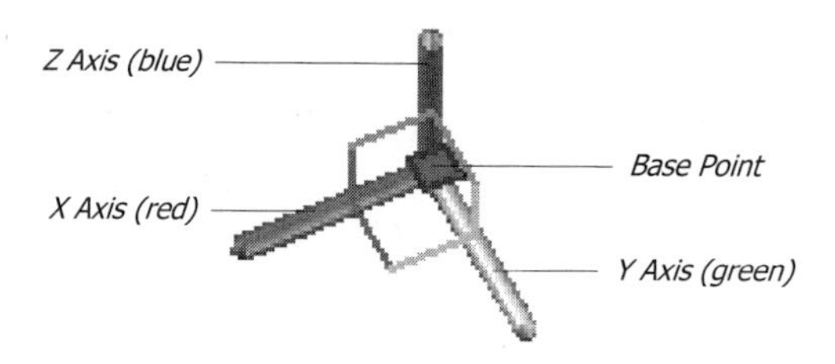

RELATED COMMANDS

Move moves objects in 2D (current UCS) and 3D, but without the grip tools.

3dRotate rotates objects in 3D space.

RELATED SYSTEM VARIABLES

GtAuto toggles the display of grips tools.

GtDefault determines whether the 3dMove or Move command are activated automatically in 3D space when you enter the Move command.

GtLocation locates the grip tool at the UCS icon or on the last selected object.

TIPS

- You can drag the base point (in the middle of the grip tool) to relocate the base point.
- When dynamic UCS is turned on, the grip tool aligns itself with the selected face. (Press CTRL+D or click **DUCS** to toggle dynamic UCS).

 '3dOrbit, '3dCOrbit, '3dFOrbit

2000 Provides interactive 3D viewing of drawings.

Command	Alias	Ctrl+	F-key	Alt+	Menu Bar	Tablet
3dorbit	**orbit** **3do**	**...**	**...**	**VB**	**View** **↳Orbit** **↳Constrained**	**R5**
3dcorbit	**...**	**...**	**...**	**VB**	**View** **↳Orbit** **↳Continuous**	**...**
3dforbit	**...**	**...**	**...**	**VBF**	**View** **↳Orbit** **↳Free Orbit**	**...**

Command: 3dorbit *or* 3dcorbit *or* 3dforbit

Press ESC or ENTER to exit, or right-click to display shortcut-menu. *(Move the cursor to change the viewpoint, and then press Enter to exit the command.)*

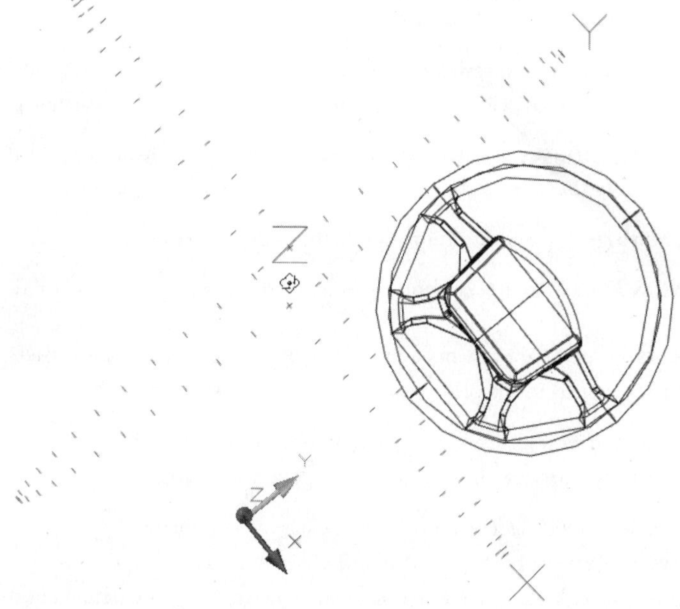

These commands are substantially changed in AutoCAD 2007:

- ***3dOrbit** changes from free to constrained, where the rotating view is locked to the x,y-plane or the z axis.*
- ***3dFOrbit** is new to AutoCAD 2007, replacing the free orbiting function of the 3dOrbit command.*
- *The shortcut menu has changed significantly.*

SHORTCUT MENU

Right-click the drawing:

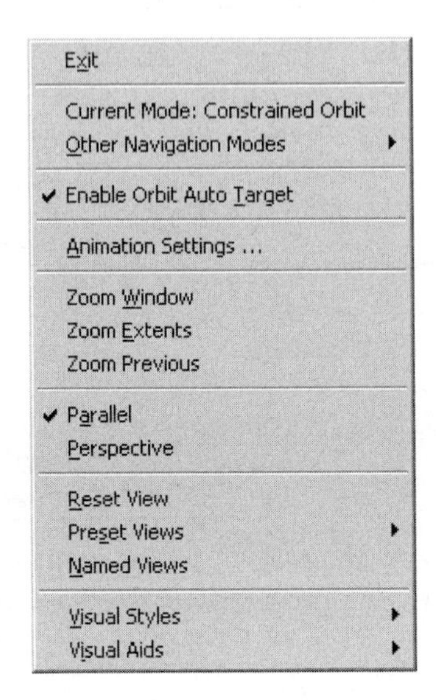

Exit exits the command.

Current Mode: Constrained Orbit reports the current mode; not a command option.

Other Navigation Modes displays a submenu that accesses other 3D viewing commands.

Enable Orbit Auto Target maintains the view on a target point instead of on the center of the viewport (*new to AutoCAD 2007*).

Animation Settings displays the Animation Settings dialog box.

Zoom Window zooms into a rectangular area defined by two pick points (see the **Zoom** command).

Zoom Extents zooms to the extents of the drawing to display all visible objects.

Zoom Previous returns to the prior view.

Parallel displays the view in parallel projection; more accurate.

Perspective displays the view in perspective projection; more realistic.

Reset View resets to the view when you first began the command.

Preset Views displays a submenu of standard views, such as top and side.

Named Views displays a submenu of user-named views, if any are defined in the drawing (see the **View** command).

Visual Styles displays a submenu for selecting the type of visual style (formerly the **Shading Modes** option; see the VisualStyles command).

Visual Aids displays a submenu of visual aids for navigating in 3D space.

Other Navigation Modes submenu

To switch between navigation modes, you can press the number keys 1 - 9 at any time during this command, without having to access the following shortcut submenu.

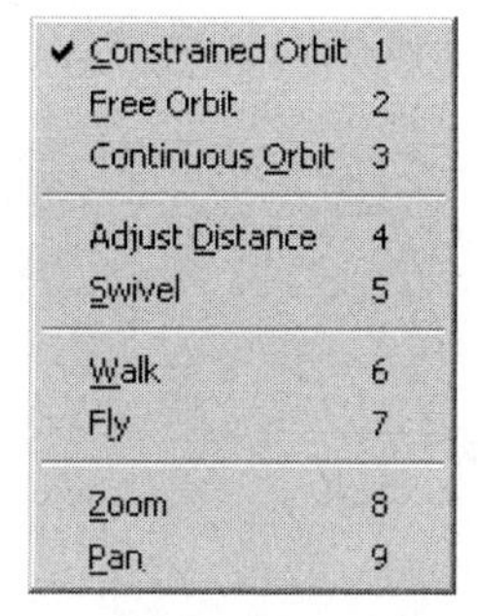

Constrained Orbit constrains movement to the x,y plane or the z axis (see the 3dOrbit command).

Free Orbit freely rotates the 3D view (3dFOrbit).

Continuous Orbit keeps the view spinning in the direction determined by dragging the mouse across the model (3dCOrbit).

Adjust Distance zooms the viewpoint nearer and further away (3dDistance).

Swivel pans the viewpoint from side to side (3dSwivelTransparent).

Walk moves through the 3D model at a fixed height (3dWalk).

Fly moves through the 3D model at any height (3dFly).

Zoom zooms in and out (3dZoomTransparent).

Pan pans around (3dPanTransparent).

Animation Settings dialog box

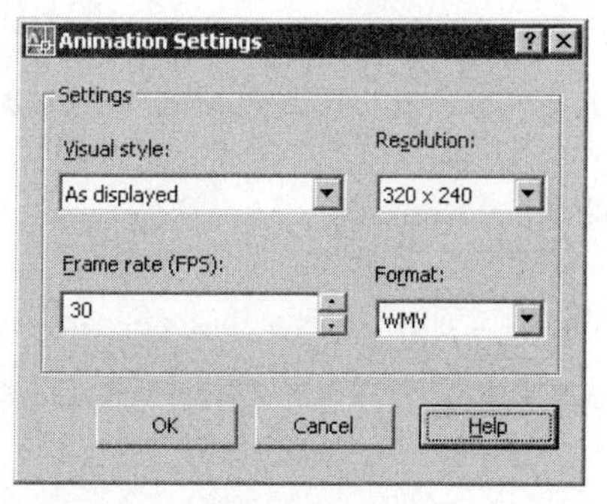

Visual Style selects the visual style by which to save the animation.

Resolution specifies the image quality; ranges from 160x120 (low quality, small file) to 1024x768 (high quality, large size).

Frame rate (FPS) specifies the frames per second, ranging from 1 to 60fps.

Format selects the movie format in which to save the file: Microsoft AVI, QuickTime MOV, generic MPG, or Windows WMV.

*To record an animation to file during the 3dOrbit command, click the **Record** button in the Dashboard's Navigation control panel:*

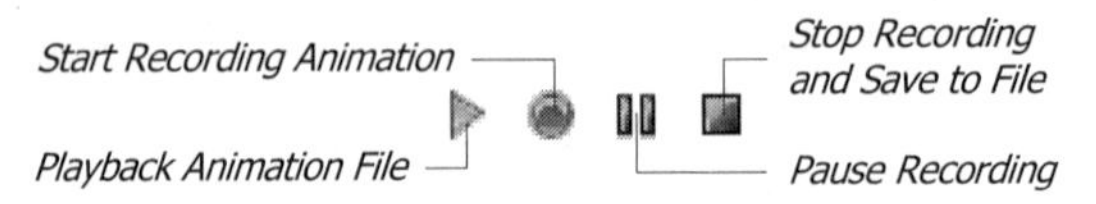

1. *Click **Start Recording Animation**.*
2. *Move cursor to animate the view.*
3. *Click **Stop Recording**, and then name the file.*

Visual Aids options

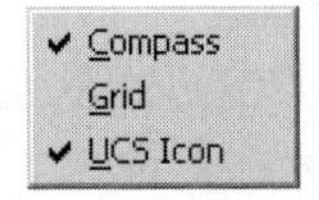

Compass toggles the display of the compass to help you navigate in 3D space.

Grid toggles the display of the grid as lines, to help you see the x,y-plane; see the Grid command.

UCS Icon toggles the display of the 3D UCS icon; see the UcsIcon command.

RELATED COMMANDS

Vpoint creates static 3D viewpoints.

3dClip performs real-time 3D front and back clipping.

3dDistance performs real-time 3D forward and backward panning.

3dPan and **3dZoom** perform real-time 3D panning and zooming.

3dWalk and **3dFly** move through 3D views.

AniPath saves flythroughs to movie files.

RELATED SYSTEM VARIABLE

Perspective reports whether the viewport is displaying a parallel or perspective projection.

TIPS

- If select one or more objects before invoking this command, they will appear while orbiting.
- To activate this command transparently during other commands, hold down the SHIFT key while pressing the mouse wheel.
- No transparent commands can be entered while this command is active.
- This command is meant to replace the DView command.

'3dOrbitCtr

2004 Specifies the target point for 3D orbiting views.

Command	Alias	Ctrl+	F-key	Alt+	Menu Bar	Tablet
3dorbitctr	...	...	...	...	...	...

Command: 3dorbitctr
Specify orbit center point: *(Pick a point.)*

The command then enters 3dOrbit mode.

COMMAND LINE OPTION

Specify orbit center point specifies the center of view rotation.

RELATED COMMANDS

3dCOrbit places the drawing in real-time, 3D, continuous orbit mode.

3dFOrbit performs real-time 3D viewing.

3dOrbit performs real-time 3D viewing fixed on the x,y-plane.

3dPoly

Rel.10 Draws 3D polylines (short for 3D POLYline).

Command	Alias	Ctrl+	F-key	Alt+	Menu Bar	Tablet
3dpoly	**3p**	...	...	**D3**	**Draw** **↳3D Polyline**	**O10**

Command: 3dpoly
Specify start point of polyline: *(Pick a point.)*
Specify endpoint of line or [Undo]: *(Pick a point, or type* **U***.)*
Specify endpoint of line or [Undo]: *(Pick a point, or type* **U***.)*
Specify endpoint of line or [Close/Undo]: *(Pick a point, type* **U** *or* **C***, or press Enter to end the command.)*

COMMAND LINE OPTIONS

Specify start point indicates the starting point of the 3D polyline.

Close joins the last endpoint with the start point.

Undo erases the last-drawn segment.

Specify endpoint of line indicates the endpoint of the current segment.

ENTER ends the command.

RELATED COMMANDS

Explode reduces a 3D polyline into lines and arcs.

PEdit edits 3D polylines.

PLine draws 2D polylines.

TIPS

- Because 3D polylines are made of straight lines, you can use the PEdit command to spline-fit the polyline.
- 3D polylines do not support linetypes and widths.
- You may use lineweights to fatten up a 3D polyline.
- Use the .xy point filter to place pick points with object snaps, and then specify the z coordinate.

3dRotate

2007 Rotates objects about the x, y, or z axis.

Command	Alias	Ctrl+	F-key	Alt+	Menu Bar	Tablet
3drotate	**3r**	...	...	**M3R**	**Modify**	...
					↳3D Operations	
					↳3D Rotate	

Command: 3rotate
Current positive angle in UCS: ANGDIR=counterclockwise ANGBASE=0
Select objects: *(Pick one or more objects.)*
Select objects: *(Press Enter.)*
Specify base point: *(Pick a point; locates the grip tool.)*
Pick a rotation axis: *(Pick one of the three axes.)*
Specify angle start point: *(Pick a point.)*
Specify angle end point: *(Pick another point.)*

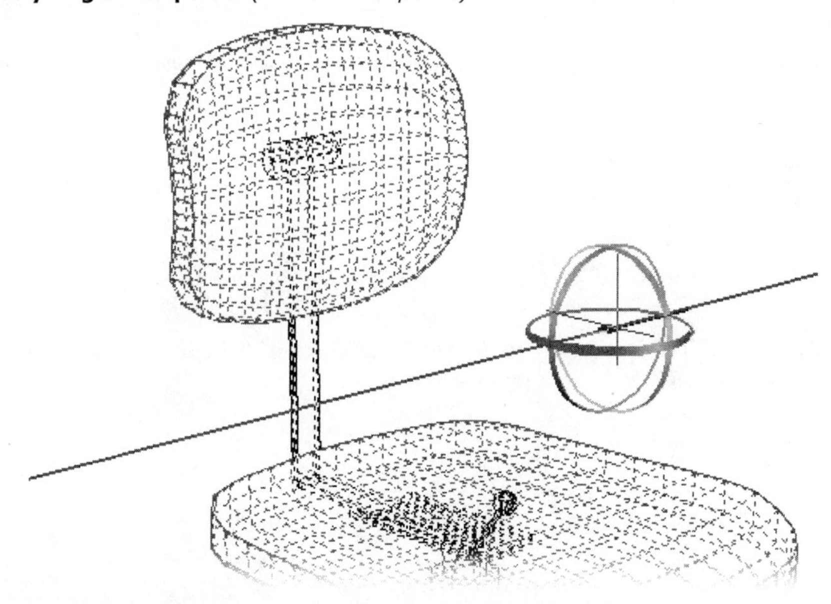

COMMAND LINE OPTIONS

Select objects selects the objects to be rotated.

Specify base point picks the point about which the objects will be rotated.

Pick a rotation axis selects the x (red), y (green), or z (blue) axis about which to rotate; the selected axis turns yellow.

Specify angle start point picks the starting point for rotation.

Specify angle end point picks the ending point for rotation.

3D Rotate grip tool

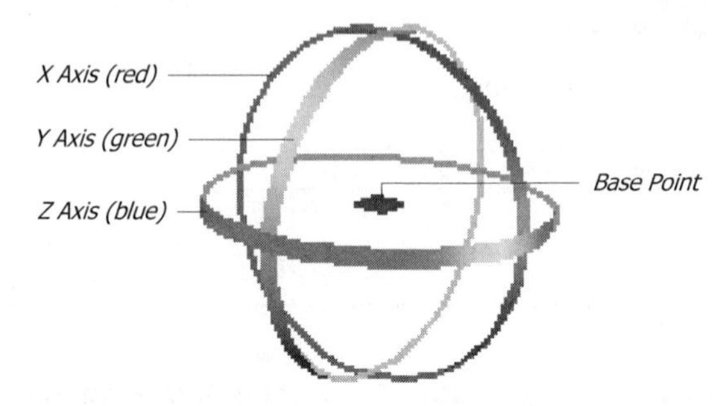

RELATED COMMANDS

Rotate rotates objects in 2D space (current UCS).

Rotate3d rotates objects in 3D space without the grip tool.

3dMove moves objects in 3D space.

RELATED SYSTEM VARIABLES

GtAuto toggles the display of grips tools.

GtDefault determines whether the 3dRotate or Rotate command is activated automatically in 3D space when you enter the Rotate command.

GtLocation locates the grip tool at the UCS icon or on the last selected object.

AngDir specifies the direction of positive angles (default =counterclockwise).

AngBase specifies the 0-angle relative to the current UCS (default = 0).

TIPS

- After starting the 3dRotate command, press the Spacebar to switch to the 3dMove command.
- When dynamic UCS is turned on, the grip tool aligns itself with the selected face. (Press CTRL+D or click **DUCS** to toggle dynamic UCS).

3dsIn

Rel.13 Imports *.3ds* files created by 3D Studio and other applications (short for 3D Studio IN).

Command	Alias	Ctrl+	F-key	Alt+	Menu Bar	Tablet
3dsin	...	...	...	**I3**	**Insert** ↳**3D Studio**	...

Command: 3dsin

Displays the 3D Studio File Import dialog box. Select a .3ds file, and then click ***Open.***

AutoCAD displays dialog box:

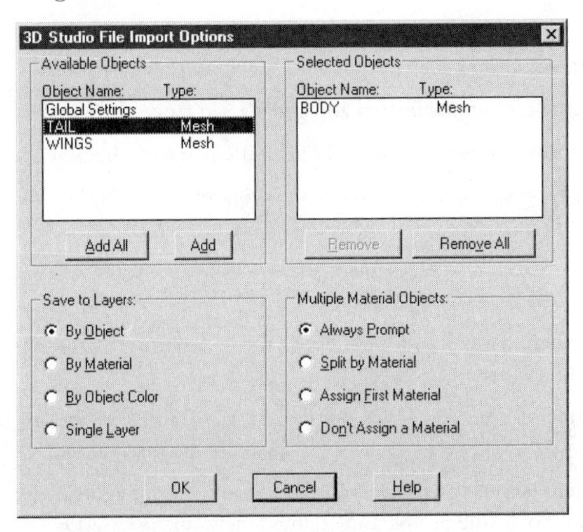

3D STUDIO FILE IMPORT OPTIONS DIALOG BOX

Available and Selected Objects options

Object Name names the object.

Type specifies the type of object.

Add adds the object to the Selected Objects list.

Add All adds all objects to the Selected Objects list.

Remove removes the object from the Selected Objects list.

Remove All removes all objects from the Selected Objects list.

Save to Layers options

⊙**By Object** places each object on its own layer.

○**By Material** places the objects on layers named after materials.

○**By Object Color** places the objects on layers named "Color*nn*."

○**Single Layer** places all objects on layer "AvLayer."

Multiple Material Objects options

⊙ **Always Prompt** prompts for each material.

○ **Split by Material** splits objects with more than one material into multiple objects, each with one material.

○ **Assign First Material** assigns the first material to the entire object.

○ **Don't Assign to a Material** removes all 3D Studio material definitions.

RELATED COMMANDS

None.

RELATED FILES

* ***.3ds*** are 3D Studio files.

* ***.tga*** are converted bitmap and animation files.

TIPS

- You are limited to selecting a maximum of 70 3D Studio objects.
- Conflicting object names are truncated and given a sequence number.
- The **By Object** option gives the AutoCAD layer the name of the object.
- The **By Object Color** option places all objects on layer "ColorNone" when no colors are defined in the 3DS file.
- 3D Studio assigns materials to faces, elements, and objects; AutoCAD assigns materials only to objects, colors, and layers.
- 3D Studio bitmaps are converted to *.tga* (Targa format) bitmaps. Converted *.tga* files are saved to the *.3ds* file folder.
- Only the first frame of animation files (CEL, CLI, FLC, and IFL) is converted to Targa bitmap files.
- 3D Studio "ambient lights" lose their color; 3D Studio "omni lights" become point lights in AutoCAD; 3D Studio "cameras" become named views in AutoCAD.

Removed Command

3dsOut was removed from AutoCAD 2007.

'3dWalk, '3dFly

2007 Moves through 3D models interactively.

Command	Alias	Ctrl+	F-key	Alt+	Menu Bar	Tablet
3dwalk	**3dw**	...	...	**VKK**	**View**	...
	3dnavigate				**↳Walk and Fly**	
					↳Walk	
3dfly	...	...	...	**VKF**	**View**	...
					↳Walk and Fly	
					↳Fly	

Command: 3dwalk *or* 3dfly

Press ESC or ENTER to exit, or right-click to display shortcut-menu. *(Drag mouse to move around, or enter these keystrokes:)*

Movement	Left Side	Right Side of Keyboard
Move forward	W	Up-arrow
Move back	S	Down-arrow
Move left	A	Left-arrow
Move right	D	Right-arrow
Toggle Walk-Fly mode	F	...
Display mappings	Tab	...
Exit walk/fly mode	Esc	Enter

Both commands display the same palette:

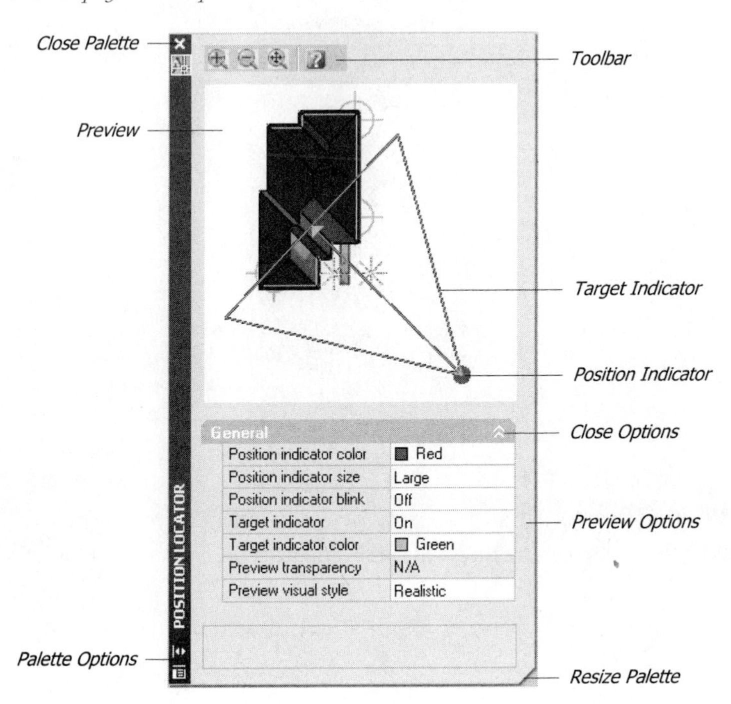

PALETTE OPTIONS

Position Indicator Color specifies the dot color; the dot that shows the current position in the 3D model.

Position Indicator Size sets the size of the indicator between Small (the default), Medium, and Large.

Position Indicator Blink toggles blinking of the position indicator.

Target Indicator toggles the display of the target.

Target Indicator Color specifies the color of the target indicator.

Preview Transparency changes the translucency of this palette, if supported by the graphics board; range is 0 (opaque) to 95 (nearly transparent).

Preview Visual Style specifies the visual style of the preview image.

PALETTE TOOLBAR

The zoom and pan buttons apply to the preview image in the palette:

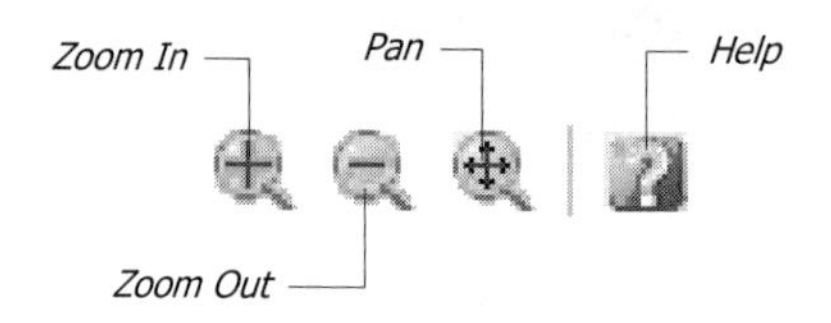

SHORTCUT MENU

The shortcut menu is nearly-identical to that of the 3dOrbit command:

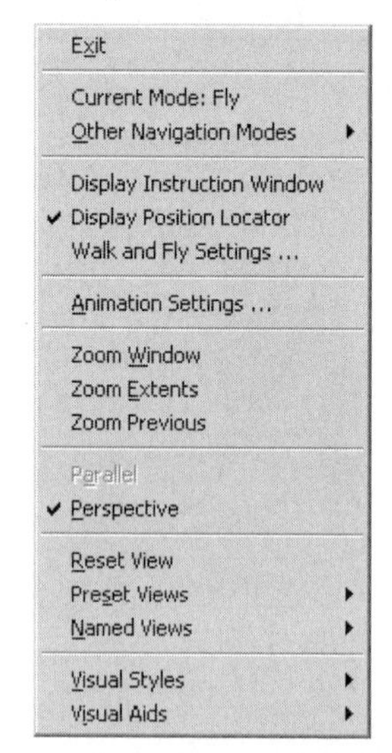

Unique to the 3dWalk and 3dFly commands are these options:

Display Position Locator toggles the display of the Position Locator palette.

Walk and Fly Settings displays the Walk and Fly Settings dialog box (WalkFlySettings command).

RELATED COMMANDS

WalkFlySettings presets the step size and rate.

AniPath saves flythroughs to movie files.

3dOrbit circles 3D models.

RELATED SYSTEM VARIABLES

StepsPerSec specifies the number of steps per second.

StepSize specifies the distance per step.

TIPS

- The 3dWalk command keeps the elevation fixed, while 3dFly is independent of elevation.
- If AutoCAD is not in perspective mode, this command displays the following dialog box:

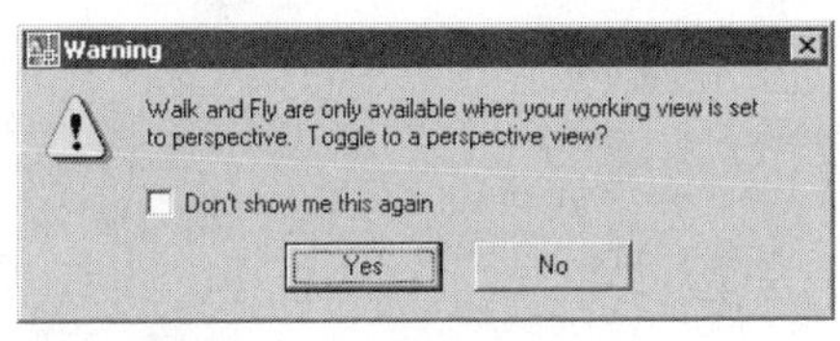

- In addition to moving the cursor in the drawing, you can also click-and-drag the view in the Preview area of the Position Locator palette. Notice the hand cursor in the figure below:

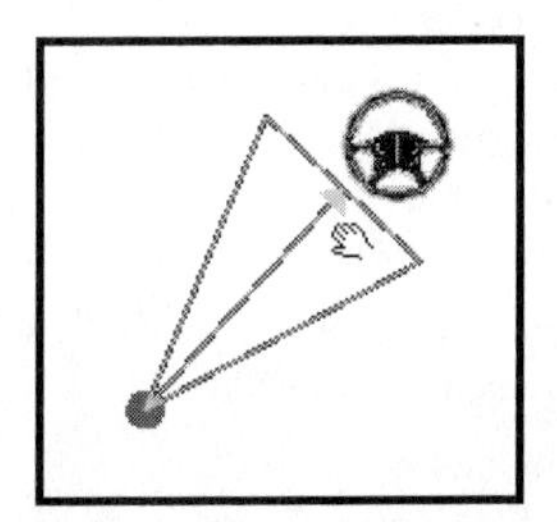

- Press **TAB** to show the window displaying walk'n fly keystrokes; they cannot be customized.
- You can switch between viewing modes using these keystrokes:

  1 — constrained orbit (same as the 3dOrbit command).
  2 — free orbiting (3dFOrbit).
  3 — continuous orbiting (3dCOrbit).
  4 — adjust distance (3dDistance).
  5 — swivel (3dSwivel).
  6 — walk (3dWalk).
  7 — fly (3dFly).
  8 — zoom (3dZoom).
  9 — pan (3dPan).

APPENDIX A
Express Tools

The following commands are included with the Express Tools add-on.

Command	Description
AcadInfo	Reports on the status of AutoCAD, and stores the data in *acadinfo.txt*.
AliasEdit	Edits the aliases stored in *acad.pgp*.
AlignSpace	Aligns model space objects, whether in different viewports or with objects in paper space.
ArcText	Places text along an arc.
AttIn / AttOut	Imports and extracts attributes in tab-delimited format.
BCount	Counts inserted blocks, and then generates a report.
BExtend	Extends to blocks.
BlockReplace	Replaces all inserts of one block with another.
BlockToXref	Convert blocks to xrefs.
Block?	Lists objects stored in blocks.
BreakLine	Creates the break-line symbol.
BScale	Scales blocks from their insertion points.
BTrim	Trims to blocks.
Burst	Explodes blocks, and converts attributes to text.
CdOrder	Display order by color.
ChUrls	DdEdit-like editor for editing hyperlinks (URL addresses).
ClipIt	Adds arcs, circles, and polylines to the XClip command.
CloseAll	Closes all open drawings.
CopyM	Copy command with repeat, divide, measure, and array options..
DimEx / DimIm	Exports and import dimension styles to/from ASCII files.
DimReassoc	Restores measurement to overridden text.
EditTime	Pauses the timer when not editing.
ExpressMenu	Loads the menu for Express Tools.
ExpressTools	Loads the express tools collection and its menu.
ExOffset	Adds options to the Offset command.
ExTrim	Trims all objects at the specified cutting line.
ExPlan	Adds options to the Plan command.
Flatten	Reduces 3D drawings to 2D.
FS	Selects objects that touch the selected object (renamed from FastSelect).
GAttE	Changes attributes globally.
GetSel	Selects objects based on layer and type.
ImageApp	Specifies the external image editor.
ImageEdit	Launches the image editor to edit selected images.
LayoutMerge	Combines layouts into a single layout.
LMan	Saves and restores layer settings.
Lsp	Provides AutoLISP function searching utilities.
LspSurf	Provides a LISP file viewer.
MkLtype	Creates linetypes from selected objects.
MkShape	Creates shapes from selected objects.
MoCoRo	Moves, copies, rotates, and scales objects.

Command	Description
MoveBak	Moves *.bak* files to specified folders.
MPEdit	Acts exactly like the PEdit command.
MStretch	Stretches with multiple selection windows.
NCopy	Copies objects nested inside blocks and xrefs.
OverKill	Removes overlapping duplicate objects.
Plt2Dwg	Imports HPGL files into the drawings.
Propulate	Updates, lists, and clears drawing properties.
PsBScale	Sets and updates the scale of blocks relative to paper space.
PsTScale	Sets text height relative to paper space.
QlAttach	Associates leaders with annotation objects.
QlAttachSet	Associates leaders with annotations.
QlDetachSet	Disassociates leaders with annotations.
QQuit	Closes all drawings, and then exits AutoCAD.
RDirMode	Sets options for the ReDir command.
ReDir	Changes paths for xrefs, images, shapes, and fonts.
RepUrls	Replaces hyperlinks.
Revert	Closes the drawing, and re-opens the original.
RtEdit	Edits remote text objects.
RText	Inserts and edits remote text objects.
RtUcs	Changes UCSs in real time.
SaveAll	Saves all drawings.
ShowUrls	Lists URLs in a dialog box.
Shp2blk	Converts a shape definition to a block definition.
Ssx	Creates selection sets.
SuperHatch	Uses images, blocks, external references, or wipeouts as hatch patterns.
SysvDlg	Launches an editor for system variables.
TCase	Changes text between Sentence, lower, UPPER, Title, and tOGGLE cASE.
TCircle	Surrounds text and multiline text with circles, slots, and rectangles.
TCount	Prefixes text with sequential numbers.
TextFit	Fits text between points.
TextMask	Places masks behind selected text.
TextUnmask	Removes masks behind selected text.
TFrames	Toggles the frames surrounding images and wipeouts.
TJust	Justifies text created with the MText and AttDef commands.
TOrient	Re-orients text, multiline text, and block attributes.
TrEx	Works like the Trim and Extend commands.
TScale	Scales text, multiline text, attributes, and attribute definitions.
TSpaceInvaders	Finds and selects text with overlapping objects.
Txt2Mtxt	Converts single-line to multiline text.
TxtExp	Explodes selected text into polylines.
VpScale	Lists the scale of the selected viewports.
VpSynch	Synchronizes viewports with a master viewport.
XData	Attaches xdata to objects.
XdList	Lists xdata attached to objects.
XList	Displays properties of objects nested in blocks and xref.

APPENDIX B

Obsolete & Removed Commands

The following commands have been removed or replaced since AutoCAD v 2.5. As of AutoCAD 2007, some operate only in 2D wireframe mode, such as Hide.

Command	Introduced	Removed	Replacement	Reaction
AcadBlockDialog	2000	2004	Block	Displays the Block dialog box
AcadWBlockDialog	2000	2004	WBlock	Displays the Write Block dialog box
AmeLite	R11	R12	Region	"Unknown command"
AscText	R11	R13	MText	"Unknown command"
Ase...	R12	R13	ASE...	"Unknown command"
	(Most R12 ASE commands were combined into fewer ASE commands with R13)			
Ase...	R13	2000	dbConnect	"Unknown command"
AseUnload	R12	R14	Arx Unload	"Unknown command"
Axis	v1.4	R12	*none*	"Discontinued command"
Background	R14	2007	View	"Unknown command"
BHatch	R12	2006	Hatch	Displays Hatch&Gradient dialog box
BMake	R12	2000	Block	Displays Block Definition dialog
BPoly	R12	2004	Boundary	Displays the Boundary dialog box
CConfig	R13	2000	PlotStyle	"Discontinued command"
Config	R12	R14	Options	Displays Options dialog
Content	2000	2004	AdCenter	Display DesignCenter window
DdAttDef	R12	2000	AttDef	Displays Attribute Def dialog
DdAttE	R9	2000	AttEdit	Displays Edit Attributes dialog
DdAttExt	R12	2000	AttExt	Displays Attribute Ext dialog
DdChProp	R12	2000	Properties	Displays Properties window
DdColor	R13	2000	Color	Displays Select Color dialog
DdEModes	R9	R14	Object Properties	Dialog box explains change
DdGrips	R12	2000	Options	Displays Options dialog
DDim	R12	2000	DimStyle	Displays Dim Style Mgr dialog
DdInsert	R12	2000	Insert	Displays Insert dialog
DdLModes	R9	R14	Layer	Displays Layer Manager dialog
DdLType	R9	R14	Linetype	Displays Linetype Mgr dialog
DdModify	R12	2000	Properties	Displays Properties window
DdOSnap	R12	2000	DSettings	Displays Drafting Settings dialog
DdPlotStamp	2000i	2004	PlotStamp	Displays Plot Stamp dialog box.
DdRename	R12	2000	Rename	"Unknown command"
DdRModes	R9	2000	DSettings	Displays Drafting Settings dialog
DdSelect	R12	2000	Options	Displays Options dialog
DdUcs	R10	2000	UcsMan	Displays UCS dialog

Command	Introduced	Removed	Replacement	Reaction
DdUcsP	R12	2000	UcsMan	Displays UCS dialog
DdUnits	R12	2000	Units	Displays Units dialog
DdView	R12	2000	View	Displays View dialog
DText	v2.5	2000	Text	Executes Text command
DLine	R11	R13	MLine	"Unknown command"
DwfOut	R14	2004	Publish	Executes Plot command.
DwfOutD	R14	2000	DwfOut	"Unknown command"
End	R11	R13	Quit	"Discontinued command"
EndRep	v1.0	v2.5	Minsert	"Discontinued command"
EndSv	v2.0	v2.5	End	"Discontinued command"
EndToday	2000i	2004	*none*	"Unknown command"
ExpressTools	2000	2002	ExpressTools	Restored in AutoCAD 2004.
Files	v1.4	R14	*Explorer*	"Discontinued command"
FilmRoll	v2.6	R13	*none*	"Unknown command"
FlatLand	R10	R11	*none*	Accepts only 0 as a value
Fog	R14	2007	RenderEnvironment	
GifIn	R12	R14	ImageAttach	"No longer supported"
HpConfig	R12	2000	PlotStyle	"Discontinued command"
IgesIn	v2.5	R13	*none*	"Discontinued command"
IgesOut	v2.5	R13	*none*	"Discontinued command"
Image	R14	2007	ClassicImage	Displays External References palette.
InetCfg	R14	2000	*none*	"Unknown command"
InetHelp	R14	2000	Help	"Unknown command"
InsertUrl	R14	2000	Insert	Displays Insert dialog.
ListUrl	R14	2000	QSelect	"Unknown command"
MakePreview	R13	R14	RasterPreview	"Discontinued command"
MenuLoad	R13	2006	CuiLoad	Load Customizations dialog box
MenuUnload	R13	2006	CuiUnload	Unoad Customizations dialog box
MeetNow	2000i	2004	*none*	"Unknown command"
OceConfig	R13	2000	PlotStyle	"Discontinued command"
OpenUrl	R14	2000	Open	Displays Select File dialog
OSnap	v2.0	2000	DSettings	Displays Drafting Settings dialog

Command	Introduced	Removed	Replacement	Reaction
Painter	R14	2004	MatchProp	Runs MatchProp command
PcxIn	R12	R14	ImageAttach	"No longer supported"
PsDrag	R12	2000i	*none*	"Unknown command"
PsIn	R12	2000i	*none*	"Unknown command"
Preferences	R11	2000	Options	Displays Options dialog box
PrPlot	v2.1	R12	Plot	"Discontinued command."
QPlot	v1.1	v2.0	SaveImg	"Unknown command"
RConfig	R12	R14	*none*	"Unknown command"
RendScr	R12	2007	RenderWin	Displays Render window.
RenderUnload	R12	R14	Arx Unload	"Unknown command"
Repeat	v1.0	v2.5	Minsert	"Discontinued command"
Replay	R12	2007	*none*	"Unknown command"
RMat	R13	2007	Materials	Displays Materials palette
RmlIn	2000i	2006	Markup	"Unknown command"
SaveAsR12	R13	R14	SaveAs	"Unknown command"
SaveUrl	R14	2000	SaveAs	Displays Save Drawing As dialog
Scene	R12	2007	*none*	"Unknown command"
SetUV	R14	2007	MaterialMap	Runs MaterialMap command.
ShadeMode	2000	2007	-ShadeMode	Runs VsCurrent command.
ShowMat	R13	2007	List	Runs List Command
Snapshot	v2.0	v2.1	Saveimg	"Unknown command"
Sol...	R11	R13	*(AME commands lost their SOL-prefix.)*	
Stats	R12	2007	*none*	"Unknown command"
TbConfig	R12	2000i	CUI	Displays CUI dialog box
TiffIn	R12	R14	ImageAttach	"No longer supported"
Today	2000i	2004	*none*	"Unknown command"
Toolbar	R13	2000i	CUI	Displays CUI dialog box
+View	...	2007	View	"Unknown command"
VlConv	R13	R14	3dsIn & 3dsOut	"Unknown command"
XRef	R14	2007	ClassicXref	Displays External References palette
3dDwfPublish	2006	2007	3dDwf	Displays Export 3D DWF dialog box
3dLine	R9	R11	Line	"Unknown command"
3dsOut	R13	2007	*none*	"Unknown command"

APPENDIX C

System Variables

AutoCAD stores information about its current state, the drawing, and the operating system in over 400 *system variables*. These variables help users and programmers — who often work with macros and AutoLISP — determine the state of the AutoCAD system.

CONVENTIONS

The following pages list all documented system variables, plus several more not documented by Autodesk. The list uses the following conventions:

Bold	System variable is documented in AutoCAD 2007.
Italicized	System variable is not listed by the SetVar command or Autodesk's documentation.
~~*Strikethru Italic*~~	System variable was removed from AutoCAD.
[keyboard icon]	System variable must be accessed through the SetVar command.
[new icon]	System variable is new to AutoCAD 2007.
Default	Indicates the default value, as set in the *acad.dwt* template drawing.

TIPS

- The SetVar command lets you change the value of all variables, except those marked read-only (R/O).
- You can get a list of most system variables at the 'Command:' prompt with the **?** option of the SetVar command, as follows:

  Command: setvar
  Variable name or ?: ?
  Variable(s) to list <*>: *(Press* ENTER.*)*

- *Toggle* means that the system variable has one of two values: 0 (off, closed, disabled) or 1 (on, open, enabled).
- *ACI color* means the system variable takes on one of AutoCAD's 255 basic colors; ACI is short for "AutoCAD color index."

Variable	Default	Meaning
~~_LInfo~~		*Removed from AutoCAD 2004.*
_AuthorPalettePath	*varies*	*Path to the customized Tool palette folder.*
_PkSer (R/O)	*varies*	*Software serial number, such as "117-69999999".*
_Server (R/O)	0	Network authorization code.
_ToolPalettePath	*varies*	*Path to the Tool palette folder.*
_VerNum (R/O)	*varies*	*Internal program build number, such as "Z.54.01".*
A		
AcadLspAsDoc	0	Controls whether *acad.lsp* is loaded into: **0** The first drawing only. **1** Every drawing.
AcadPrefix (R/O)	*varies*	Specifies paths used by AutoCAD search in **Options | Files** dialog box.
AcadVer (R/O)	"17.0"	Specifies the AutoCAD version number.
AcisOutVer	70	Controls the ACIS version number; values are 15, 16, 17, 18, 20, 21, 30, 40, or 70.
AcGiDumpMode	*0*	*Value of 0 or 1.*
AdcState (R/O)	0	Toggle: reports if DesignCenter is active.
AFlags	0	Controls the default attribute display mode: **0** No mode specified. **1** Invisible. **2** Constant. **4** Verify. **8** Preset. **16** Lock position in block.
AngBase	0	Controls the direction of zero degrees relative to the UCS.
AngDir	0	Controls the rotation of positive angles: **0** Clockwise. **1** Counterclockwise.
ApBox	0	Toggles the display of the AutoSnap aperture box cursor.
Aperture	10	Controls the object snap aperture in pixels: **1** Minimum size. **50** Maximum size.
ApState	*0*	*Toggle: reports the state of the* ***ApBox*** *variable.*
AssistState (R/O)	0	Toggle: reports if Info Palette is active.
Area (R/O)	0.0	Reports the area measured by the last **Area** command.
AttDia	0	Controls the user interface for entering attributes: **0** Command-line prompts. **1** Dialog box.

Variable	Default	Meaning
AttMode	1	Controls the display of attributes: **0** Off. **1** Normal. **2** On; displays invisible attributes.
AttReq	1	Toggles attribute values during insertion: **0** Uses default values. **1** Prompts user for values.
AuditCtl	0	Toggles creation of *.adt* audit log files: **0** File not created. **1** File created.
AUnits	0	Controls the type of angular units: **0** Decimal degrees. **1** Degrees-minutes-seconds. **2** Grads. **3** Radians. **4** Surveyor's units.
AUPrec	0	Controls the number of decimal places displayed by angles; range is 0 to 8.
AutoSnap	63	Controls the AutoSnap display: **0** Turns off all AutoSnap features. **1** Turns on marker. **2** Turns on SnapTip. **4** Turns on magnetic cursor. **8** Turns on polar tracking. **16** Turns on object snap tracking. **32** Turns on tooltips for polar tracking and object snap tracking.
AuxStat	*0*	*-32768 Minimum value.* *32767 Maximum value.*
~~AxisMode~~	*0*	*Removed from AutoCAD 2002.*
AxisUnit	*0.0*	*Obsolete system variable.*

B

Variable	Default	Meaning
BackgroundPlot	2	Controls background plotting and publishing (ignored during scripts): **0** Plot foreground; publish foreground. **1** Plot background; publish foreground. **2** Plot foreground; publish background. **3** Plot background; publish background.
BackZ	0.0	Controls the location of the back clipping plane offset from the target plane.
BActionColor	"7"	ACI text color for actions in Block Editor.
BDependencyHighlight	1	Toggles highlighting of dependent objects when parameters, actions, or grips selected in Block Editor: **0** Not highlighted. **1** Highlighted.

Variable	Default	Meaning
BGripObjColor	"141"	ACI color of grips in Block Editor.
BGripObjSize	8	Controls the size of grips in Block Editor; range is 1 to 255.
BgrdPlotTimeout	*20*	*Controls the timeout for failed background plots; ranges from 0 to 300 seconds.*
BindType	0	Controls how xref names are converted when being bound or edited: **0** From **xref|name** to **xref0name**. **1** From **xref|name** to **name**.
BlipMode	0	Toggles the display of blip marks.
BlockEditLock	0	Toggles the locking of dynamic blocks being edited: **0** Unlocked. **1** Locked.
BlockEditor (R/O)	0	Toggle: reports whether Block Editor is open.
BParameterColor	"7"	ACI color of parameters in Block Editor.
BParameterFont	"Simplex.shx"	Controls the font used for parameter and action text in the Block Editor.
BParameterSize	12	Controls the size of parameter text and features in Block Editor: **1** Minimum. **255** Maximum.
BTMarkDisplay	1	Controls the display of value set markers: **0** Unlocked. **1** Locked.
BVMode	0	Controls the display of invisible objects in the Block Editor: **0** Invisible. **1** Visible and dimmed.

C

Variable	Default	Meaning
CalcInput	1	Controls how formulas and global constants are evaluated in dialog boxes: **0** Not evaluated. **1** Evaluated after pressing **ALT+Enter**.
CameraDisplay	0	Toggles the display of camera glyphs.
CameraHeight	0	Specifies the default height of cameras.
CDate (R/O)	*varies*	Specifies the current date and time in the format YyyyMmDd.HhMmSsDd, such as 20010503.18082328
CeColor	"BYLAYER"	Controls the current color.
CeLtScale	1.0	Controls the current linetype scaling factor.
CeLType	"BYLAYER"	Controls the current linetype.

Variable	Default	Meaning
CeLWeight	-1	Controls the current lineweight in millimeters; valid values are 0, 5, 9, 13, 15, 18, 20, 25, 30, 35, 40, 50, 53, 60, 70, 80, 90, 100, 106, 120, 140, 158, 200, and 211, plus the following: **-1** BYLAYER. **-2** BYBLOCK. **-3** Default, as defined by **LwDdefault**.
CenterMT	0	Controls how corner grips stretch uncentered multiline text: **0** Center grip moves in same direction; opposite grip stays in place. **1** Center grip stays in place; both side grips move in direction of stretch.
ChamferA	0.5	Specifies the current value of the first chamfer distance.
ChamferB	0.5	Specifies the current value of the second chamfer distance.
ChamferC	1.0	Specifies the current value of the chamfer length.
ChamferD	0	Specifies the current value of the chamfer angle.
ChamMode	0	Toggles the chamfer input mode: **0** Chamfer by two lengths. **1** Chamfer by length and angle.
CircleRad	0.0	Specifies the most-recent circle radius.
CLayer	"0"	Current layer name.
CleanScreenState (R/O)	0	Toggle: reports whether cleanscreen mode is active.
CliState (R/O)	1	Reports the command line palette.
CMaterial	ByLayer	Sets the name of the current material.
CmdActive (R/O)	1	Reports the type of command currently active (used by programs): **1** Regular command. **2** Transparent command. **4** Script file. **8** Dialog box. **16** Dynamic data exchange. **32** AutoLISP command. **64** ARX command.
~~*CmdDia*~~	*1*	*Replaced by* ***PlQuiet*** *in AutoCAD 2000.*
CmdEcho	1	Toggles AutoLISP command display: **0** No command echoing. **1** Command echoing.
CmdInputHistoryMax	20	Controls the maximum command input items stored; works with **InputHistoryMode**.
CmdNames (R/O)	*varies*	Reports the name of the command currently active, such as "SETVAR".

Variable	*Default*	*Meaning*
CMLJust	0	Controls the multiline justification mode: **0** Top. **1** Middle. **2** Bottom.
CMLScale	1.0	Controls the scale of overall multiline width: *-n* Flips offsets of multiline. **0** Collapses to single line. *n* Scales by a factor of *n*.
CMLStyle	"STANDARD"	Specifies the current multiline style name.
Compass	0	Toggles the display of the 3D compass.
Coords	1	Controls the coordinate display style: **0** Updated by screen picks. **1** Continuous display. **2** Polar display upon request.
CPlotStyle	"ByColor"	Specifies the current plot style; options for named plot styles are: ByLayer, ByBlock, Normal, and User Defined.
CProfile (R/O)	"<<Unnamed Profile>>"	Specifies the name of the current profile.
CpuTicks *(R/O)*	*592020023071334.1*	*Reports the number of CPU ticks.*
CrossingColor	3	ACI color of crossing rectangle.
CShadow	0	Shadows cast by 3D objects, if graphics board is capable (see **3dConfig** command): **0** Casts and receives shadows. **1** Casts shadows. **2** Receives shadows. **3** Ignores shadows.
CTab (R/O)	"Model"	Specifies the name of the current tab.
CTableStyle	"Standard"	Specifies the name of the current table style name.
~~CurrentProfile~~	*"<<Unnamed Profile>>"*	*Removed from AutoCAD 2000; replaced by* ***CProfile***
CursorSize	5	Controls the cursor size as a percent of the viewport size. **1** Minimum size. **100** Full viewport.
CVPort	2	Specifies the current viewport number.

D

Variable	*Default*	*Meaning*
DashboardState	0	Toggle: reports whether Dashboard is open.
Date (R/O)	*varies*	Reports the current date in Julian format, such as 2448860.54043252
DbcState (R/O)	0	Toggles: specifies whether dbConnect Manager is active.
~~DBGListAll~~	*0*	*Removed from AutoCAD 2002.*
DblClkEdit	1	Toggles editing by double-clicking objects.
DBMod (R/O)	4	Reports how the drawing has been modified: **0** No modification since last save. **1** Object database modified.

Variable	Default	Meaning
		2 Symbol table modified. **4** Database variable modified. **8** Window modified. **16** View modified. **32** Field modified.
DctCust	"sample.cus"	Specifies the name of custom spelling dictionary.
DctMain	"enu"	Controls the code for spelling dictionary: **ca** Catalan. **cs** Czech. **da** Danish. **de** German; sharp 's'. **ded** German; double 's'. **ena** English; Australian. **ens** English; British 'ise'. **enu** English; American. **enz** English; British 'ize'. **es** Spanish; unaccented capitals. **esa** Spanish; accented capitals. **fi** Finish. **fr** French; unaccented capitals. **fra** French; accented capitals. **it** Italian. **nl** Dutch; primary. **nls** Dutch; secondary. **no** Norwegian; Bokmal. **non** Norwegian; Nynorsk. **pt** Portuguese; Iberian. **ptb** Portuguese; Brazilian. **ru** Russian; infrequent 'io'. **rui** Russian; frequent 'io'. **sv** Swedish.
DefaultLighting	1	Toggles distant lighting.
DefaultLightingType	0	Toggles between new (1) and old (0) type of lights.
DefaultViewCategory	""	*Specifies the default name for View Category in the* ***View*** *command's New View dialog box*
DefLPlStyle (R/O)	"ByColor"	Reports the default plot style for layer 0.
DefPlStyle (R/O)	"ByColor"	Reports the default plot style for new objects.
DelObj	1	Toggles the deletion of source objects: **-2** Users are prompted whether to erase all defining objects. **-1** Users are prompted whether to erase profiles and cross sections. **0** Objects retained. **1** Profiles and cross sections erased. **2** All defining objects erased.

Variable	Default	Meaning
DemandLoad	3	Controls application loading when drawing contains proxy objects: **0** Apps not demand-loaded. **1** Apps loaded when drawing opened. **2** Apps loaded at first command. **3** Apps loaded when drawing opened or at first command.
DgnFrame	*1*	*Toggles display of frame around .dgn underlays.*
DiaStat (R/O)	1	Reports whether user exited dialog box by clicking: **0** **Cancel** button. **1** **OK** button.

Dimension Variables

Variable	Default	Meaning
DimADec	0	Controls angular dimension precision: **-1** Use **DimDec** setting (default). **0** Zero decimal places (minimum). **8** Eight decimal places (maximum).
DimAlt	Off	Toggles alternate units: **On** Enabled. **Off** Disabled.
DimAltD	2	Controls alternate unit decimal places.
DimAltF	25.4	Controls alternate unit scale factor.
DimAltRnd	0.0	Controls rounding factor of alternate units.
DimAltTD	2	Controls decimal places of tolerance alternate units; range is 0 to 8.
DimAltTZ	0	Controls display of zeros in alternate tolerance units: **0** Zeros not suppressed. **1** All zeros suppressed. **2** Include 0 feet, but suppress 0 inches . **3** Includes 0 inches, but suppress 0 feet. **4** Suppresses leading zeros. **8** Suppresses trailing zeros.
DimAltU	2	Controls display of alternate units: **1** Scientific. **2** Decimal. **3** Engineering. **4** Architectural; stacked. **5** Fractional; stacked. **6** Architectural. **7** Fractional. **8** Windows desktop units setting.
DimAltZ	0	Controls the display of zeros in alternate units: **0** Suppresses 0 ft and 0 in. **1** Includes 0 ft and 0 in. **2** Includes 0 ft; suppress 0 in. **3** Suppresses 0 ft; include 0 in.

Variable	Default	Meaning
		4 Suppresses leading 0 in dec dims. **8** Suppresses trailing 0 in dec dims. **12** Suppresses leading and trailing zeroes.
DimAPost	""	Specifies the prefix and suffix for alternate text.
DimArcSym	0	Specifies the location of the arc symbol: **0** Before dimension text. **1** Above the dimension text. **2** Not displayed.
DimAso	On	Toggles associative dimensions: **On** Dimensions are created associative. **Off** Dimensions are not associative.
DimAssoc	2	Controls how dimensions are created: **0** Dimension elements are exploded. **1** Single dimension object, attached to defpoints. **2** Single dimension object, attached to geometric objects.
DimASz	0.18	Controls the default arrowhead length.
DimAtFit	3	Controls how text and arrows are fitted when there is insufficient space between extension lines (leader is added when **DimTMove** = 1): **0** Text and arrows outside extension lines. **1** Arrows first outside, then text. **2** Text first outside, then arrows. **3** Either text or arrows, whichever fits better.
DimAUnit	0	Controls the format of angular dimensions: **0** Decimal degrees. **1** Degrees.Minutes.Seconds. **2** Grads. **3** Radians. **4** Surveyor units.
DimAZin	0	Controls the display of zeros in angular dimensions: **0** Displays all leading and trailing zeros. **1** Suppresses 0 in front of decimal. **2** Suppresses trailing zeros behind decimal. **3** Suppresses zeros in front and behind the decimal.
DimBlk	""	Specifies the name of the arrowhead block: Architectural tick: "Archtick" Box filled: "Boxfilled" Box: "Boxblank" Closed blank: "Closedblank" Closed filled: "" (default) Closed: "Closed" Datum triangle filled: "Datumfilled"

Variable	Default	Meaning
		Datum triangle: "Datumblank" Dot blanked: "Dotblank" Dot small: "Dotsmall" Dot: "Dot" Integral: "Integral" None: "None" Oblique: "Oblique" Open 30: "Open30" Open: "Open" Origin indication: "Origin" Right-angle: "Open90"
DimBlk1	""	Specifies the name of first arrowhead's block; uses same list of names as under **DimBlk**. . No arrowhead.
DimBlk2	""	Specifies the name of second arrowhead's block.
DimCen	0.09	Controls how center marks are drawn: *-n* Draws center lines. **0** No center mark or lines drawn. *+n* Draws center marks of length *n*.
DimClrD	0	ACI color of dimension lines: **0** BYBLOCK (default). **256** BYLAYER.
DimClrE	0	ACI color of extension lines and leaders.
DimClrT	0	ACI color of dimension text.
DimDec	4	Controls the number of decimal places for the primary tolerance; range is 0 to 8.
DimDLE	0.0	Controls the length of the dimension line extension.
DimDLI	0.38	Controls the increment of the continued dimension lines.
DimDSep	"."	Specifies the decimal separator (must be a single character).
DimExe	0.18	Controls the extension above the dimension line.
DimExO	0.0625	Specifies the extension line origin offset.
~~*DimFit*~~	*3*	*Obsolete: Autodesk recommends use of* ***DimATfit*** *and* ***DimTMove*** *instead.*
DimFrac	0	Controls the fraction format when **DimLUnit** set to 4 or 5: **0** Horizontal. **1** Diagonal. **2** Not stacked.
DimFXL	1	Default length of fixed extension lines.
DimFxlOn	0	Toggles fixed extension lines.

Variable	Default	Meaning
DimGap	0.09	Controls the gap between text and the dimension line.
DimJogAngle	45	Default angle for jogged dimension lines.
DimJust	0	Controls the positioning of horizontal text: **0** Center justify. **1** Next to first extension line. **2** Next to second extension line. **3** Above first extension line. **4** Above second extension line.
DimLdrBlk	""	Specifies the name of the block used for leader arrowheads; same as **DimBlock** **.** Suppresses display of arrowhead.
DimLFac	1.0	Controls the linear unit scale factor.
DimLim	Off	Toggles the display of dimension limits.
DimLtEx1	""	Linetype name for the first extension line.
DimLtEx2	""	Linetype name for the second extension line.
DimLtype	""	Linetype name for the dimension line.
DimLUnit	2	Controls dimension units (except angular); replaces **DimUnit**: **1** Scientific. **2** Decimal. **3** Engineering. **4** Architectural. **5** Fractional. **6** Windows desktop.
DimLwD	-2	Controls the dimension line lineweight; valid values are BYLAYER, BYBLOCK, or integer multiples of 0.01mm.
DimLwE	-2	Controls the extension lineweight.
DimPost	""	Specifies the default prefix or suffix for dimension text (maximum 13 characters): " " No suffix.
DimRnd	0.0	Controls the rounding value for dimension distances.
DimSAh	Off	Toggles separate arrowhead blocks: **Off** Use arrowheads defined by **DimBlk**. **On** Use arrowheads defined by **DimBlk1** and **DimBlk2**.
DimScale	1.0	Controls the overall dimension scale factor: **0** Value is computed from the scale between current model space viewport and paper space. **>0** Scales text and arrowheads.
DimSD1	Off	Toggles display of the first dimension line: **On** First dimension line is suppressed. **Off** Not suppressed.

Variable	Default	Meaning
DimSD2	Off	Toggles display of the second dimension line: **On** Second dimension line is suppressed. **Off** Not suppressed.
DimSE1	Off	Toggles display of the first extension line: **On** First extension line is suppressed. **Off** Not suppressed.
DimSE2	Off	Toggles display of the second extension line: **On** Second extension line is suppressed. **Off** Not suppressed.
DimSho	On	Toggles dimension updates while dragging: **On** Dimensions are updated during drag. **Off** Dimensions are updated after drag.
DimSOXD	Off	Toggles display of dimension lines outside of extension lines: **On** Dimension lines not drawn outside extension lines. **Off** Are drawn outside extension lines.
DimStyle (R/O)	"STANDARD"	Reports the current dimension style.
DimTAD	0	Controls the vertical position of text: **0** Centered between extension lines. **1** Above dimension line, except when dimension line not horizontal and **DimTIH** = 1. **2** On side of dimension line farthest from the defining points. **3** Conforms to JIS.
DimTDec	4	Controls the number of decimal places for primary tolerances; range is 0 to 8.
DimTFac	1.0	Controls the scale factor for tolerance text height.
DimTFill	0	Toggles background fill color for dimension text.
DimTFillClr	0	Background color for dimension text.
DimTIH	On	Toggles alignment of text placed inside extension lines: **Off** Text aligned with dimension line. **On** Text is horizontal.
DimTIX	Off	Toggles placement of text inside extension lines: **Off** Text placed inside extension lines, if room. **On** Text forced between the extension lines.
DimTM	0.0	Controls the value of the minus tolerance.
DimTMove	0	Controls how dimension text is moved: **0** Dimension line moves with text. **1** Adds a leader when text is moved. **2** Text moves anywhere; no leader.

Variable	Default	Meaning
DimTOFL	Off	Toggles placement of dimension lines: **Off** Dimension lines not drawn when arrowheads are outside. **On** Dimension lines drawn, even when arrowheads are outside.
DimTOH	On	Toggles text alignment when outside of extension lines: **Off** Text aligned with dimension line. **On** Text is horizontal.
DimTol	Off	Toggles generation of dimension tolerances: **Off** Tolerances not drawn. **On** Tolerances are drawn.
DimTolJ	1	Controls vertical justification of tolerances: **0** Bottom. **1** Middle. **2** Top.
DimTP	0.0	Specifies the value of the plus tolerance.
DimTSz	0.0	Controls the size of oblique tick strokes: **0** Arrowheads. >**0** Oblique strokes.
DimTVP	0.0	Controls the vertical position of text when **DimTAD** = 0: **1** Turns on **DimTAD** (=1). >**-0.7** *or* <**0.7** Dimension line is split for text.
DimTxSty	"STANDARD"	Specifies the dimension text style.
DimTxt	0.18	Controls the text height.
DimTZin	0	Controls the display of zeros in tolerances: **0** Suppresses 0 ft and 0 in. **1** Includes 0 ft and 0 in. **2** Includes 0 ft; suppress 0 in. **3** Suppresses 0 ft; include 0 in. **4** Suppresses leading 0 in decimal dim. **8** Suppresses trailing 0 in decimal dim. **12** Suppresses leading and trailing zeroes.
~~*DimUnit*~~	*2*	*Obsolete; replaced by* ***DimLUnit*** *and* ***DimFrac***
DimUPT	Off	Controls user-positioned text: **Off** Cursor positions dimension line. **On** Cursor also positions text.
DimZIN	0	Controls the display of zero in feet-inches units: **0** Suppresses 0 ft and 0 in. **1** Includes 0 ft and 0 in. **2** Includes 0 ft; suppress 0 in. **3** Suppresses 0 ft; include 0 in. **4** Suppress leading 0 in decimal dim. **8** Suppresses trailing 0 in decimal dim. **12** Suppresses leading and trailing zeroes.

Variable	Default	Meaning
DispSilh	0	Toggles the silhouette display of 3D solids.
Distance (R/O)	0.0	Reports the distance last measured by the **Dist** command.
~~*Dither*~~		*Removed from Release 14.*
DonutId	0.5	Controls the inside diameter of donuts.
DonutOd	1.0	Controls the outside diameter of donuts.
DragMode	2	Controls the drag mode: **0** No drag. **1** On if requested. **2** Automatic.
DragP1	10	Controls the regen drag display; range is 0 to 32767.
DragP2	25	Controls the fast drag display; range is 0 to 32767.
DragVs	""	Default visual style when 3D solids are created by dragging the cursor; disabled when visual style is 2D wireframe.
DrawOrderCtrl	3	Controls the behavior of draw order: **0** Draw order not restored until next regen or drawing reopened. **1** Normal draw order behavior. **2** Draw order inheritance. **3** Combines options 1 and 2.
DrState	0	Toggles the Drawing Recovery palette.
DTextEd	0	Controls the user interface of **DText/Text** command: **0** In-place text editor. **1** Edit text dialog box. **2** Can click elsewhere in drawing to start new text string.
DwfFrame	2	Display of the frame around DWF overlays: **0** Frame is turned off. **1** Frame is displayed and plotted. **2** Frame is displayed but not plotted
DwfOsnap	1	Toggles osnapping of the DWF frame.
DwgCheck	0	Toggles checking of whether drawing was edited by software other than AutoCAD: **0** Suppresses dialog box. **1** Displays warning dialog box. **2** Warning appears on command line.
DwgCodePage (R/O)	*varies*	Same value as **SysCodePage**.
DwgName (R/O)	*varies*	Reports the current drawing file name, such as "drawing1.dwg".
DwgPrefix (R/O)	*varies*	Reports the drawing's drive and folder, such as "d:\acad 2007\".

Variable	Default	Meaning
DwgTitled (R/O)	0	Reports whether the drawing file name is: **0** "drawing1.dwg". **1** User-assigned name.
~~DwgWrite~~		*Removed from AutoCAD Release 14.*
DynDiGrip	31	Controls the dynamic dimensions displayed during grip stretch editing (**DynDiVis** =2): **0** None. **1** Resulting dimension. **2** Length change dimension. **4** Absolute angle dimension. **8** Angle change dimension. **16** Arc radius dimension.
DynDiVis	1	Controls the dynamic dimensions displayed during grip stretch editing: **0** First (in the cycle order). **1** First two (in the cycle order). **2** All (as specified by **DynDiGrip**).
DynMode	-2	Controls dynamic input features (click **DYN** on status bar to turn on hidden modes): **-3** Both on hidden. **-2** Dimensional input on hidden. **-1** Pointer input on hidden. **0** Off. **1** Pointer input on. **2** Dimensional input on. **3** Both on.
DynPiCoords	0	Toggles pointer input coordinates: **0** Relative. **1** Absolute.
DynPiFormat	0	Toggles pointer input coordinates: **0** Polar. **1** Cartesian.
DynPiVis	1	Controls when pointer input is displayed: **0** When user types at prompts for points. **1** When prompted for points. **2** Always.
DynPrompt	0	Toggles display of prompts in Dynamic Input tooltips.
DynToolTips	1	Toggles which tooltips are affected by tooltip appearance settings: **0** Only Dynamic Input value fields. **1** All drafting tooltips.

E

Variable	Default	Meaning
EdgeMode	0	Toggles edge mode for the **Trim** and **Extend** commands: **0** No extension. **1** Extends cutting edge.

Variable	Default	Meaning
Elevation	0.0	Specifies the current elevation, relative to current UCS.
EnterpriseMenu (R/O)	""	Reports the path and *.cui* file name.
EntExts	*1*	*Controls how drawing extents are calculated:* ***0*** *Extents calculated every time; slows down AutoCAD, but uses less memory.* ***1*** *Extents of every object are cached as a two-byte value (default).* ***2*** *Extents of every object are cached as a four-byte value (fastest but uses more memory).*
EntMods *(R/O)*	*0*	*Increments by one each time an object is modified since the drawing was opened; ranges from 0 to 4.29497E9.*
ErrNo (R/O)	0	Reports error numbers from AutoLISP, ADS, & Arx.
ErState (R/O)	-	Reports display of the External References palette.
~~*ExeDir*~~		*Removed from Release 14.*
Expert	0	Controls the displays of prompts: **0** Normal prompts. **1** 'About to regen, proceed?' and 'Really want to turn the current layer off?' **2** 'Block already defined. Redefine it?' and 'A drawing with this name already exists. Overwrite it?' **3** **Linetype** command messages. **4** **UCS Save** and **VPorts Save**. **5** **DimStyle Save** and **DimOverride.**
ExplMode	1	Toggles whether the **Explode** and **Xplode** commands explode non-uniformly scaled blocks: **0** Does not explode. **1** Explodes.
ExtMax (R/O)	-1.0E+20, -1.0E+20, -1.0E+20	Upper-right coordinate of drawing extents.
ExtMin (R/O)	1.0E+20, 1.0E+20, 1.0E+20	Lower-left coordinate of drawing extents.
ExtNames	1	Controls the format of named objects: **0** Names are limited to 31 characters, and can include A - Z, 0 - 9, dollar ($), underscore (_), and hyphen (-). **1** Names are limited to 255 characters, and can include A - Z, 0 - 9, spaces, and any characters not used by Windows or AutoCAD for special purposes.

Variable	Default	Meaning
F		
FaceTRatio	0	Controls the aspect ratio of facets on rounded 3D bodies: **0** Creates an *n* by 1 mesh. **1** Creates an *n* by *m* mesh.
FaceTRres	0.5000	Controls the smoothness of shaded and hidden-line objects; range is 0.01 to 10.
~~*FfLimit*~~	...	*Removed from AutoCAD Release 14.*
FieldDisplay	1	Toggles background to field text: **0** No background. **1** Gray background.
FieldEval	31	Controls how fields are updated: **0** Not updated **1** Updated with **Open**. **2** Updated with **Save**. **4** Updated with **Plot**. **8** Updated with **eTransmit**. **16** Updated with regeneration.
FileDia	1	Toggles the user interface for file-access commands, such as **Open** and **Save**: **0** Displays command-line prompts. **1** Displays file dialog boxes.
FilletRad	0.5	Specifies the current fillet radius.
FillMode	1	Toggles the fill of solid objects, wide polylines, fills, and hatches.
Flatland (R/O)	*0*	*Obsolete system variable.*
FontAlt	"simplex.shx"	Specifies the font used for missing fonts.
FontMap	"acad.fmp"	Specifies the name of the font mapping file.
Force_Paging	*0*	*Ranges from 0 to 4.29497E9.*
FrontZ	0.0	Reports the front clipping plane offset.
FullOpen (R/O)	1	Reports whether the drawing is: **0** Partially loaded. **1** Fully open.
FullPlotPath	1	Specifies the format of file name sent to plot spooler: **0** Drawing file name only. **1** Full path and drawing file.
G		
GfAng	*0*	*Controls the angle of gradient fill; 0 to 360 degrees.*
GfClr1	*"RGB 000,000,255"*	*Specifies the first gradient color in RGB format.*
GfClr2	*"RGB 255,255,153"*	*Specifies the second gradient color in RGB format.*
GfClrLum	*1.0*	*Controls the level of gray in one-color gradients:* *0 Black.* *1 White.*

Variable	Default	Meaning
GfClrState	*1*	*Specifies the type of gradient fill:* *0 Two-color.* *1 One-color.*
GfName	*1*	*Specifies the style of gradient fill:* *1 Linear.* *2 Cylindrical.* *3 Inverted cylindrical.* *4 Spherical.* *5 Inverted spherical.* *6 Hemispherical.* *7 Inverted hemispherical.* *8 Curved.* *9 Inverted curved.*
GfAShift	*0*	*Controls the origin of the gradient fill:* *0 Centered.* *1 Shifted up and left.*
GlobCheck	*0*	*Controls reporting on dialog boxes:* ***-1*** *Turn off local language.* ***0*** *Turn off.* ***1*** *Warns if larger than 640x400.* ***2*** *Also reports size in pixels.* ***3*** *Additional information.*
GridDisplay	3	Determines grid display (sum of bitcodes): **0** Grid restricted to area specified by the **Limits** command. **1** Grid is infinite. **2** Adaptive grid, with fewer grid lines when zoomed out. **4** Generates more grid lines when zoomed in. **8** Grid follows the x,y plane of the dynamic UCS.
GridMajor	5	Number of minor grid lines per major line.
GridMode	0	Toggles the display of the grid.
GridUnit	0.5,0.5	Controls the x, y spacing of the grid.
GripBlock	0	Toggles the display of grips in blocks: **0** At block insertion point. **1** Of all objects within block.
GripColor	160	ACI color of unselected grips.
GripDynColor	140	ACI color of custom grips in dynamic blocks.
GripHot	1	ACI color of selected grips.
GripHover	3	ACI grip fill color when cursor hovers.
GripLegacy	*0*	*Toggle of unknown meaning.*
GripObjLimit	100	Controls the maximum number of grips displayed; range is 1 to 32767; 0 = grips never suppressed.
Grips	1	Toggles the display of grips.

Variable	Default	Meaning
GripSize	3	Controls the size of grip; range is 1 to 255 pixels.
GripTips	1	Toggles the display of grip tips when cursor hovers over custom objects.
GtAuto	1	Toggles display of grip tools.
GtDefault	0	Toggles which commands are the default commands in 3D views: **0** **Move** and **Rotate**. **1** **3dMove** and **3dRotate**
GtLocation	0	Controls location of grip tools: **0** Aligns grip tool with UCS icon. **1** Aligns with the last selected object.

H

Variable	Default	Meaning
HaloGap	0	Controls the distance by which haloed lines are shortened in 2D wireframe visual style; a percentage of 1 unit.
~~*Handles*~~ *(R/O)*	*1*	*Obsolete system variable.*
HidePrecision	0	Controls the precision of hide calculations in 2D wireframe visual style: **0** Single precision, less accurate, faster. **1** Double precision, more accurate, but slower (recommended).
HideText	0	Controls the display of text during **Hide**: **0** Text is neither hidden nor hides other objects, unless text object has thickness. **1** Text is hidden and hides other objects.
Highlight	1	Toggles object selection highlighting.
HPAng	0	Current hatch pattern angle.
HpAssoc	1	Toggles associativity of hatches: **0** Not associative. **1** Associative.
HpBound	1	Controls the object created by the **Hatch** and **Boundary** commands: **0** Region. **1** Polyline.
HpDouble	0	Toggles double hatching.
HpDrawOrder	3	Controls draw order of hatches and fills: **0** None. **1** Behind all other objects. **2** In front of all other objects. **3** Behind the hatch boundary. **4** In front of the hatch boundary.
HpGapTol	0	Controls largest gap allowed in hatch boundaries; ranges from 0 to 5000 units.

Variable	Default	Meaning
HpInherit	0	Toggles how **MatchProp** copies the hatch origin from source object to destination objects: **0** As specified by **HpOrigin**. **1** As specified by the source hatch object.
HpName	"ANSI31"	Specifies default hatch name.
HpObjWarning	10000	Specifies the maximum number of hatch boundaries that can be selected before AutoCAD flashes warning message; range is 1 to 1073741823.
HpOrigin	0,0	Specifies the default origin for hatch objects.
HpOrigMode	0	Controls the default hatch origin point:. **0** Specified by **HpOrigin**. **1** Bottom-left corner of hatch's rectangular extents. **2** Bottom-right corner of hatch's rectangular extents. **3** Top-right corner of hatch's rectangular extents. **4** Top-left corner of hatch's rectangular extents. **5** Center of hatch's rectangular extents.
HpScale	1.0	Specifies the current hatch scale factor; cannot be zero.
HpSeparate	0	Controls the number of hatch objects made from multiple boundaries: **0** Single hatch object created. **1** Separate hatch object created.
HpSpace	1.0	Controls the default spacing of user-defined hatches; cannot be zero.
HyperlinkBase	""	Specifies the path for relative hyperlinks.
I		
ImageHlt	0	Toggles image frame highlighting when raster images are selected.
ImpliedFace	1	Toggles detection of implied faces.
IndexCtl	0	Controls the creation of layer and spatial indices: **0** No indices created. **1** Layer index created. **2** Spatial index created. **3** Both indices created.
InetLocation	"www.autodesk.com"	Specifies the default URL for **Browser**.
InputHistoryMode	15	Controls the content and location of user input history (bitcode sum of): **0** No history displayed.

Variable	Default	Meaning
		1 Displayed at the command line, and in dynamic prompt tooltips accessed with Up and Down arrow keys. **2** Current command displayed in the shortcut menu. **4** All commands in the shortcut menu. **8** Blipmark for recent input displayed in the drawing.
InsBase	0.0,0.0,0.0	Controls the default insertion base point relative to the current UCS for **Insert** and **XRef** commands.
InsName	""	Specifies the default block name: **.** Set to no default.
InsUnits	1	Controls the drawing units when blocks are dragged into drawings from DesignCenter: **0** Unitless. **1** Inches. **2** Feet. **3** Miles. **4** Millimeters. **5** Centimeters. **6** Meters. **7** Kilometers. **8** Microinches. **9** Mils. **10** Yards. **11** Angstroms. **12** Nanometers. **13** Microns. **14** Decimeters. **15** Decameters. **16** Hectometers. **17** Gigameters. **18** Astronomical Units. **19** Light Years. **20** Parsecs.
InsUnitsDefSource	1	Controls source drawing units value; ranges from 0 to 20; see above.
InsUnitsDefTarget	1	Controls target drawing units; see list above.
IntelligentUpdate	20	Controls graphics refresh rate in frames per second; range is 0 (off) to 100 fps.
InterfereColor	1	Color of interference objects.
InterfereObjVs	"Realistic"	Visual style of interference objects.
InterfereVpVs	"Wireframe"	Visual style during interference checking.
IntersectionColor	257	ACI color of intersection polylines in 2D wireframe visual style: **0** ByBlock. **256** ByLayer. **257** ByEntity.

Variable	Default	Meaning
IntersectionDisplay	0	Toggles display of 3D surface intersections during **Hide** command in 2D wireframe: **0** Does not draw intersections. **1** Draws polylines at intersections.
ISaveBak	1	Toggles creation of *.bak* backup files.
ISavePercent	50	Controls the percentage of waste in saved *.dwg* file before cleanup occurs: **0** Slower full saves. **>0** Faster partial saves. **100** Maximum.
IsoLines	4	Controls the number of contour lines on 3D solids; range is 0 to 2047.
L		
LastAngle (R/O)	0	Reports the end angle of last-drawn arc.
LastPoint	*varies*	Reports the x,y,z coordinates of the last-entered point.
LastPrompt (R/O)	""	Reports the last string on the command line.
Latitude	""	Last-used angle of latitude.
LayerFilterAlert	2	Controls the deletion of layer filters in excess of 99 filters *and* the number of layers: **0** Off. **1** Deletes all filters without warning, when layer dialog box opened. **2** Recommends deleting all filters when layer dialog box opened. **3** Displays dialog box for selecting filters to erase, upon opening the drawing.
LayoutRegenCtl	2	Controls display list for layouts: **0** Display-list regen'ed with each tab change. **1** Display-list is saved for model tab and last layout tab. **2** Display list is saved for all tabs.
LazyLoad	*0*	*Toggle: unknown purpose.*
LegacyCtrlPick	1	Toggles function of **Ctrl**+pick: **0** Selects faces, edges, and vertices of 3D solids. **1** Cycles through overlapping objects.
LensLength (R/O)	50.0	Reports perspective view lens length, in mm.
LightGlyphDisplay	1	Toggles display of light glyph.
ListListState	0	Toggles display of Light List palette.
LimCheck	0	Toggles drawing limits checking.
LimMax	12.0,9.0	Controls the upper right drawing limits.
LimMin	0.0,0.0	Controls the lower left drawing limits.
LispInit	1	Toggles AutoLISP functions and variables: **0** Preserved from drawing to drawing. **1** Valid in current drawing only.

Variable	Default	Meaning
Locale (R/O)	"enu"	Reports ISO language code; see **DctMain**.
LocalRootPrefix (R/O)	"d:\docume..."	Reports the path to folder holding local customizable files.
LockUi	0	Controls the position and size of toolbars and palettes; hold down **Ctrl** key to unlock temporarily (bitcode sum): **0** Toolbars and palettes unlocked. **1** Docked toolbars locked. **2** Docked palettes locked. **4** Floating toolbars locked. **8** Floating palettes locked.
LoftAng1	90	Angle of loft to first cross-section; range is 0 to 359.9 degrees.
LoftAng2	90	Angle of loft to second cross-section.
LoftMag1	1	Magnitude of loft at first cross-section; range is 1 to 10.
LoftMag2	1	Magnitude of loft at last cross-section.
LoftNormals	1	Specifies location of loft normals: **0** Ruled **1** Smooth **2** First normal **3** Last normal **4** Ends normal **5** All normal **6** Use draft angle and magnitude
LoftParam	7	Loft shape: **1** Minimizes twists between cross-sections. **2** Aligns start-to-end direction of each cross-section. **4** Generates simple solids and surfaces, instead of spline solids and surfaces. **8** Closes the surface or solid between the first and last cross-sections.
LogFileMode	0	Toggles writing command prompts to *.log* file.
LogFileName (R/O)	"...\Drawing1.log"	Reports file name and path for *.log* file.
LogFilePath	"d:\acad 2007\"	Specifies path to the *.log* file.
LogInName (R/O)	"*username*"	Reports user's login name; truncated after 30 characters.
~~*LongFName*~~		*Removed from AutoCAD Release 14.*
Longitude	-122.3940	Current angle of longitude.
LTScale	1.0	Controls linetype scale factor; cannot be 0.
LUnits	2	Controls linear units display: **1** Scientific. **2** Decimal. **3** Engineering. **4** Architectural. **5** Fractional.

Variable	Default	Meaning
LUPrec	4	Controls decimal places (or inverse of smallest fraction) of linear units; range is 0 to 8.
LwDefault	25	Controls the default lineweight, in millimeters; must be one of the following values: 0, 5, 9, 13, 15, 18, 20, 25, 30, 35, 40, 50, 53, 60, 70, 80, 90, 100, 106, 120, 140, 158, 200, or 211.
LwDisplay	0	Toggles whether lineweights are displayed; setting saved separately for Model space and each layout tab.
LwUnits	1	Toggles units used for lineweights: **0** Inches. **1** Millimeters.

M

Variable	Default	Meaning
MacroTrace	*0*	*Toggles diesel debug mode.*
MatState	0	Toggles display of Materials palette.
MaxActVP	64	Controls the maximum number of viewports to display; range is 2 to 64.
MaxObjMem	*0*	*Controls the maximum number of objects in memory; object pager is turned off when value = 0, <0, or 2,147,483,647.*
MaxSort	1000	Controls the maximum names sorted alphabetically; range is 0 to 32767.
MButtonPan	1	Toggles the behavior of the wheel mouse: **0** As defined by AutoCAD *.cui* file. **1** Pans when dragging with wheel.
MeasureInit	0	Toggles drawing units for default drawings: **0** English. **1** Metric.
Measurement	0	Toggles current drawing units: **0** English. **1** Metric.
MenuCtl	1	Toggles the display of submenus in side menu: **0** Only with menu picks. **1** Also with keyboard entry.
MenuEcho	0 ...	Controls menu and prompt echoing (sum): **0** Displays all prompts. **1** Suppresses menu echoing. **2** Suppresses system prompts. **4** Disables **^P** toggle. **8** Displays all input-output strings.
MenuName (R/O)	"acad"	Reports path and file name of *.cui* file.
Millisecs (R/O)	*248206921*	*Reports number of milliseconds since timing began.*
MirrText	0	Toggles text handling by **Mirror** command: **0** Retains text orientation. **1** Mirrors text.

Variable	Default	Meaning
ModeMacro	""	Invokes Diesel macros.
MsmState (R/O)	0	Specifies if Markup Set Manager is active: **0** No. **1** Yes.
MsOleScale	1.0	Controls the size of text-containing OLE objects when pasted in model space: **-1** Scales by value of **PlotScale**. **0** Scales by value of **DimScale**. **>0** Scale factor.
MTextEd	"Internal"	Controls the name of the **MText** editor: **.** Uses default editor. **0** Cancels the editing operation. **-1** Uses the secondary editor. **"blank"** MTEXT internal editor. **"Internal"** MTEXT internal editor. **"oldeditor"** Previous internal editor. **"Notepad"** Windows Notepad editor. **":lisped"** Built-in AutoLISP function. *string* Name of editor fewer than 256 characters long using this syntax: *:AutoLISPtextEditorFunction#TextEditor.*
MTextFixed	0	Controls the mtext editor appearance: **0** Mtext editor is used. **1** Mtext editor remembers its location. **2** Difficult-to-read text is displayed horizontally at a larger size.
MTJigString	"abc"	Specifies the sample text displayed by mtext editor; maximum 10 letters; enter . for no text.
MyDocumentsPrefix (R/O)	"C:\Documents and Settings*username*\My Documents"	Reports path to the *my documents* folder of the currently logged-in user.

N

Variable	Default	Meaning
NodeName (R/O)	*"AC$"*	*Reports the name of the network node; range is one to three characters.*
NoMutt	0	Toggles display of messages (a.k.a. muttering) during scripts, LISP, macros: **0** Displays prompt, as normal. **1** Suppresses muttering.
NorthDirection	0	Angle of the sun relative to positive y axis.
NwfState	*1*	*Reports whether New Features Workshop displays when AutoCAD starts.*

O

Variable	Default	Meaning
ObscuredColor	257	ACI color of objects obscured by **Hide** in 2D wireframe visual style.

Variable	*Default*	*Meaning*
ObscuredLtype	0	Linetype of objects obscured by **Hide** in 2D wireframe visual mode. **0** Invisible. **1** Solid. **2** Dashed. **3** Dotted. **4** Short dash. **5** Medium dash. **6** Long dash. **7** Double short dash. **8** Double medium dash. **9** Double long dash. **10** Medium long dash. **11** Sparse dot.
OffsetDist	1.0	Controls current offset distance: <**0** Offsets through a specified point. >**0** Default offset distance.
OffsetGapType	0	Controls how polylines reconnect when segments are offset: **0** Extends segments to fill gap. **1** Fills gap with fillet (arc segment). **2** Fills gap with chamfer (line segment).
OleFrame	2	Controls the visibility of the frame around OLE objects: **0** Frame is not displayed and not plotted. **1** Frame is displayed and is plotted. **2** Frame is displayed but is not plotted.
OleHide	0	Controls display and plotting of OLE objects: **0** All OLE objects visible. **1** Visible in paper space only. **2** Visible in model space only. **3** Not visible.
OleQuality	1	Controls the quality of display and plotting of embedded OLE objects: **0** Monochrome. **1** Low quality graphics. **2** High quality graphics. **3** Automatically selected mode.
OleStartup	0	Toggles loading of OLE source applications to improve plot quality: **0** Does not load OLE source application. **1** Loads OLE source app when plotting.
OpmState	0	Toggles whether **Properties** palette is active.
OrthoMode	0	Toggles orthographic mode.
OsMode	4133	Controls current object snap mode (sum): **0** NONe. **1** ENDpoint. **2** MIDpoint. **4** CENter.

Variable	*Default*	*Meaning*
		8 NODe. **16** QUAdrant. **32** INTersection. **64** INSertion. **128** PERpendicular. **256** TANgent. **512** NEARest. **1024** QUIck. **2048** APPint. **4096** EXTension. **8192** PARallel. **16383** All modes on. **16384** Object snap turned off via **OSNAP** on the status bar.
OSnapCoord	2	Controls keyboard overrides object snap: **0** Object snap overrides keyboard. **1** Keyboard overrides object snap. **2** Keyboard overrides object snap, except in scripts.
OSnapHatch	0	Toggles whether hatches are snapped: **0** Osnaps ignore hatches. **1** Hatches are snapped.
OSnapNodeLegacy	*1*	*Toggles whether osnap snaps to mtext insertion points.*
OsnapZ	0	Toggles osnap behavior in z direction: **0** Uses the z-coordinate. **1** Uses the current elevation setting.
OsOptions	3	Determines when objects with negative z values are osnaped: **0** Uses the actual z coordinate. **1** Substitutes z coordinate with the elevation of the current UCS.

P

Variable	*Default*	*Meaning*
PaletteOpaque	0	Controls transparency of palettes: **0** Turned off by user. **1** Turned on by user. **2** Unavailable, but turned on by user. **3** Unavailable, and turned off by user.
PaperUpdate	0	Toggles how AutoCAD plots layouts with paper size different from plotter's default: **0** Displays a warning dialog box. **1** Changes paper size to that of the plotter configuration file.
0 1 2 3 4 32 33 34 35 36 64 65 66 67 68 96 97 98 99 100	0	Controls point display style (sum): **0** Dot. **1** No display. **2** +-symbol. **3** x-symbol. **4** Short line.

Variable	Default	Meaning
		32 Circle. **64** Square.
PDSize	0.0	Controls point display size: **>0** Absolute size, in pixels. **0** 5% of drawing area height. **<0** Percentage of viewport size.
PEditAccept	0	Toggles display of the **PEdit** command's 'Object selected is not a polyline. Do you want to turn it into one? <Y>:' prompt.
PEllipse	0	Toggles object used to create ellipses: **0** True ellipse. **1** Polyline arcs.
Perimeter (R/O)	0.0	Reports perimeter calculated by the last **Area**, **DbList**, and **List** commands.
Perspective	0	Toggles perspective mode; not available in 2D wireframe visual mode.
PFaceVMax (R/O)	4	Reports the maximum vertices per 3D face.
PHandle	*0*	*Ranges from 0 to 4.29497E9; unknown usage.*
PickAdd	1	Toggles meaning of **SHIFT** key on selection sets: **0** Adds to selection set. **1** Removes from selection set.
PickAuto	1	Toggles selection set mode: **0** Single pick mode. **1** Automatic windowing and crossing.
PickBox	3	Controls selection pickbox size; range is 0 to 50 pixels..
PickDrag	0	Toggles selection window mode: **0** Pick two corners. **1** Pick a corner; drag to second corner.
PickFirst	1	Toggles command-selection mode: **0** Enter command first. **1** Select objects first.
PickStyle	1	Controls how groups and associative hatches are selected: **0** Includes neither. **1** Includes groups. **2** Includes associative hatches. **3** Includes both.
Platform (R/O)	*"varies"*	Reports the name of the operating system.
PLineGen	0	Toggles polyline linetype generation: **0** From vertex to vertex. **1** From end to end.
PLineType	2	Controls the automatic conversion and creation of 2D polylines by **PLine**: **0** Not converted; creates old-format polylines.

Variable	*Default*	*Meaning*
		1 Not converted; creates optimized lwpolylines. **2** Polylines in older drawings are converted on open; **PLine** creates optimized lwpolyline objects.
PLineWid	0.0	Controls current polyline width.
~~*PlotId*~~	*""*	*Obsolete; has no effect in AutoCAD.*
PlotOffset	0	Toggles the plot offset measurement: **0** Relative to edge of margins. **1** Relative to edge of paper.
PlotRotMode	1	Controls the orientation of plots: **0** Lower left = 0,0. **1** Lower left plotter area = lower left of media. **2** X, y-origin offsets calculated relative to the rotated origin position.
~~*Plotter*~~	*0*	*Obsolete; has no effect in AutoCAD.*
PlQuiet	0	Toggles display during batch plotting and scripts (replaces **CmdDia**): **0** Plot dialog boxes and nonfatal errors are displayed. **1** Logs nonfatal errors; plot dialog boxes are not displayed.
PolarAddAng	""	Holds a list of up to 10 user-defined polar angles; each angle can be up to 25 characters long, each separated with a semicolon (;). For example: 0;15;22.5;45.
PolarAng	90	Controls the increment of polar angle; contrary to Autodesk documentation, you may specify any angle.
PolarDist	0.0	Controls the polar snap increment when **SnapStyl** is set to 1 (isometric).
PolarMode	0	Controls polar and object snap tracking: **0** Measure polar angles based on current UCS (absolute), track orthogonally; don't use additional polar tracking angles; and acquire object tracking points automatically. **1** Measure polar angles from selected objects (relative). **2** Use polar tracking settings in object snap tracking. **4** Use additional polar tracking angles (via **PolarAng**). **8** Press **SHIFT** to acquire object snap tracking points.
PolySides	4	Controls the default number of polygon sides; range is 3 to 1024.

Variable	Default	Meaning
Popups (R/O)	1	Reports display driver support of AUI: **0** Not available. **1** Available.
PreviewEffect	2	Controls the visual effect for previewing selected objects: **0** Dashed lines. **1** Thick lines. **2** Thick dashed lines.
PreviewFilter	1	Controls the exclusion of objects from selection previewing (bitcode sum): **0** No objects excluded. **1** Objects on locked layers. **2** Objects in xrefs. **4** Tables. **8** Multiline text. **16** Hatch patterns. **32** Groups.
Product (R/O)	"AutoCAD"	Reports the name of the software.
Program (R/O)	"acad"	Reports the name of the software's executable file.
ProjectName	""	Controls the project name of the current drawing; searches for xref and image files.
ProjMode	1	Controls the projection mode for **Trim** and **Extend** commands: **0** Does not project. **1** Projects to x,y-plane of current UCS. **2** Projects to view plane.
ProxyGraphics	1	Toggles saving of proxy images in drawings: **0** Not saved; displays bounding box. **1** Image saved with drawing.
ProxyNotice	1	Toggles warning message displayed when drawing contains proxy objects.
ProxyShow	1	Controls the display of proxy objects: **0** Not displayed. **1** All displayed. **2** Bounding box displayed.
ProxyWebSearch	0	Toggles checking for object enablers: **0** Does not check for object enablers. **1** Checks for object enablers if an Internet connection is present.
PsLtScale	1	Toggles paper space linetype scaling: **0** Uses model space scale factor. **1** Uses viewport scale factor.
PSolHeight	4	Default height of polysolid objects.
PSolWidth	0.25	Default width of polysolid objects.
PsProlog	""	*Specifies the PostScript prologue file name.*

Variable	Default	Meaning
PsQuality	*75*	*Controls resolution of PostScript display, in pixels:* *<0 Display as outlines; no fill.* *0 Displays no fills.* *>0 Displays filled.*
PStyleMode	1	Toggles the plot color matching mode of the drawing: **0** Uses named plot style tables. **1** Uses color-dependent plot style tables.
PStylePolicy (R/O)	1	Reports whether the object color is associated with its plot style: **0** Not associated. **1** Associated.
PsVpScale	0	Controls the view scale factor (ratio of units in paper space to units in newly-created model space viewports) 0 = scaled to fit.
PublishAllSheets	1	Determines which sheets (model space and layouts) are loaded automatically into the **Publish** command's list: **0** Current drawing only. **1** All open drawings.
PUcsBase (R/O)	""	Reports name of UCS defining the origin and orientation of orthographic UCS settings; in paper space only.

Q

Variable	Default	Meaning
QAFlags	*0*	*Controls the quality assurance flags:* *0 Turned off.* *1 The ^C metacharacters in a menu macro cancels grips, just as if user pressed* ***ESC****.* *2 Long text screen listings do not pause.* *4 Error and warning messages are displayed at the command line, instead of in dialog boxes.* *128 Screen picks are accepted via the AutoLISP (command) function.*
QaUcsLock	*0*	*Toggles; purpose unknown.*
QcState	0	Toggles whether QuickCalc palette is open.
QTextMode	0	Toggles quick text mode.
QueuedRegenMax	*2147483647*	*Ranges between very large and very small numbers.*

R

Variable	Default	Meaning
R14RasterPlot	*0*	*Toggle; purpose unknown.*
RasterDpi	300	Controls the conversion of millimeters or inches to pixels, and vice versa; range is 100 to 32767.
RasterPreview (R/O)	1	Toggles creation of BMP preview image.

Variable	Default	Meaning
RecoveryMode	2	Controls recording of drawing recovery information after software failure: **0** Note recorded. **1** Recorded; Drawing Recovery palette not displayed automatically. **2** Recorded, and Drawing Recovery palette displays automatically.
RefEditName	""	Specifies the reference file name when in reference-editing mode.
RegenMode	1	Toggles regeneration mode: **0** Regens with each view change. **1** Regens only when required.
Re-Init	0	Controls the reinitialization of I/O devices: **1** Digitizer port. *2 Plotter port; obsolete.* **4** Digitizer. *8 Plotter; obsolete* **16** Reloads PGP file.
RememberFolders	1	Toggles the path search method: **0** Path specified in desktop AutoCAD icon is default for file dialog boxes. **1** Last path specified by each file dialog box is remembered.
RenderPrefsState	0	Toggles display of the Render Preferences palette.
ReportError	1	Determines if AutoCAD sends an error report to Autodesk: **0** No error report created. **1** Error report is generated and sent to Autodesk.
RoamableRootPrefix (R/O)	"d:\documents and settings*username*\application data\aut..."	Reports the path to the root folder where roamable customized files are located.
~~*RIAspect*~~		*Removed from AutoCAD Release 14.*
~~*RIBackG*~~		*Removed from AutoCAD Release 14.*
~~*RIEdge*~~		*Removed from AutoCAD Release 14.*
~~*RIGamit*~~		*Removed from AutoCAD Release 14.*
~~*RIGrey*~~		*Removed from AutoCAD Release 14.*
~~*RIThresh*~~		*Removed from AutoCAD Release 14.*
RTDisplay	1	Toggles raster display during real-time zoom and pan: **0** Displays the entire raster image. **1** Displays raster outline only.

S

Variable	Default	Meaning
SaveFile (R/O)	""	Reports the automatic save file name.
SaveFilePath	"...\temp\"	Specifies the path for automatic save files.

Variable	Default	Meaning
SaveName (R/O)	""	Reports the drawing's save-as file name.
SaveTime	10	Controls the automatic save interval, in minutes; 0 = disable auto save.
ScreenBoxes (R/O)	0	Reports the maximum number of menu items supported by display; 0 = screen menu turned off.
ScreenMode (R/O)	3	Reports the state of AutoCAD display: **0** Text screen. **1** Graphics screen. **2** Dual-screen display.
ScreenSize (R/O)	*varies*	Reports the current viewport size, in pixels, such as 719.0000,381.0000.
SDI	0	Controls the multiple-document interface (SDI is "single document interface"): **0** Turns on MDI. **1** Turns off MDI (only one drawing may be loaded into AutoCAD). **2** Disables MDI for apps that cannot support MDI; read-only. **3** (R/O) Disables MDI for apps that cannot support MDI, even when **SDI**= 1.
SelectionArea	1	Toggles use of colored selection areas.
SelectionAreaOpacity	25	Controls the opacity of color selection areas; range is 0 (transparent) to 100 (opaque).
SelectionPreview	2	Controls selection preview: **0** Off. **1** On when commands are inactive. **2** On when commands prompt for object selection.
ShadEdge	3	Controls shading by **Shade** command: **0** Faces only shaded. **1** Faces shaded, edges in background color. **2** Only edges, in object color. **3** Faces in object color, edges in background color.
ShadeDif	70	Controls percentage of diffuse to ambient light; range is 0 to 100 percent.
ShadowPlaneLocation	0	Default height of the shadow plane.
ShortcutMenu	11	Controls display of shortcut menus (sum): **0** Disables all. **1** Default shortcut menus. **2** Edit shortcut menus. **4** Command shortcut menus when commands are active. **8** Command shortcut menus only when options available at command line. **16** Shortcut menus when the right button held down longer.

Variable	Default	Meaning
ShowHist	1	Toggles display of history in solids: **0** Original solids are not displayed. **1** Display of original solids depends on Show History property settings. **2** Displays all original solids.
ShowLayerUsage	1	Toggles layer-usage icons in Layers dialog box.
ShpName	""	Specifies the default shape name: **.** Set to no default.
SigWarn	1	Toggles display of dialog box when drawings with digital signatures are opened: **0** Only when signature is invalid. **1** Always.
SketchInc	0.1	Controls the **Sketch** command's recording increment.
SkPoly	0	Toggles sketch line mode: **0** Record as lines. **1** Record as a polyline.
SnapAng	0	Controls rotation angle for snap and grid; when not 0, grid lines are not displayed.
SnapBase	0.0,0.0	Controls current origin for snap and grid.
SnapIsoPair	0	Controls current isometric drawing plane: **0** Left isoplane. **1** Top isoplane. **2** Right isoplane.
SnapMode	0	Toggles snap mode.
SnapStyl	0	Toggles snap style: **0** Normal. **1** Isometric.
SnapType	0	Toggles snap for the current viewport: **0** Standard snap. **1** Polar snap.
SnapUnit	0.5,0.5	Controls x,y spacing for snap distances.
SolidCheck	1	Toggles solid validation.
SolidHist	4	Toggles retention of history in solids.
SortEnts	*96*	*Controls object display sort order:* *0 Off.* *1 Object selection.* *2 Object snap.* *4 Redraw.* *8 Slide generation.* *16 Regeneration.* *32 Plot.* *64 PostScript output.*
SpaceSwitch	*1*	*Either 1 or 9; purpose unknown.*

Variable	*Default*	*Meaning*
SplFrame	0	Toggles polyline and mesh display: **0** Polyline control frame not displayed; display polygon fit mesh; 3D faces invisible edges not displayed. **1** Polyline control frame displayed; display polygon defining mesh; 3D faces invisible edges displayed.
SplineSegs	8	Controls number of line segments that define splined polylines; range is -32768 to 32767. **<0** Drawn with fit-curve arcs. **>0** Drawn with line segments.
SplineType	6	Controls type of spline curve: **5** Quadratic Bezier spline. **6** Cubic Bezier spline.
SsFound	""	Specifies path and file name of sheet sets.
SsLocate	1	Toggles whether sheet set files are opened with drawing: **0** Not opened. **1** Opened automatically.
SsmAutoOpen	1	Toggles whether the Sheet Set Manager is opened with drawing (**SsLocate** must be 1): **0** Not opened. **1** Opened automatically.
SsmPollTime	60	Controls time interval between automatic refreshes of status data in sheet sets; range is 20 to 600 seconds (**SsmSheetStatus** = 2).
SsmSheetStatus	2	Controls refresh of status data in sheet sets: **0** Not automatically refreshed. **1** Refresh when sheet set is loaded or updated. **2** Also refresh as specified by **SsmPollTime**.
SsmState (R/O)	0	Toggle: whether Sheet Set Manager is open.
StandardsViolation	2	Controls whether alerts are displayed when CAD standards are violated: **0** No alerts. **1** Alert displayed when CAD standard violated. **2** Displays icon on status bar when file is opened with CAD standards, and when non-standard objects are created.
Startup	0	Controls which dialog box is displayed by the **New** and **QNew** commands: **0** Displays Select Template dialog box. **1** Displays Startup and Create New Drawing dialog box.
~~*StartupToday*~~		~~*Removed from AutoCAD 2004.*~~

Variable	*Default*	*Meaning*
StepSize	6	Length of steps in walk mode; range is 1E-6 to 1E+6.
StepsPerSec	2	Speed of steps in walk mode; range is 1-30.
SunPropertiesState	0	Toggles display of Sun Properties palette.
SunStatus	1	Toggles display of light by the sun.
SurfTab1	6	Controls density of m-direction surfaces and meshes; range is 5 to 32766.
SurfTab2	6	Density of n-direction surfaces and meshes; range is 2 to 32766.
SurfType	6	Controls smoothing of surface by **PEdit**: **5** Quadratic Bezier spline. **6** Cubic Bezier spline. **8** Bezier surface.
SurfU	6	Controls surface density in m-direction; range is 2 to 200.
SurfV	6	Surface density in n-direction; range is 2 to 200.
SysCodePage (R/O)	"ANSI_1252"	Reports the system code page; set by operating system.

T

TableIndicator	1	Toggles display of column letters and row numbers during table editing.
TabMode	0	Toggles tablet mode.
Target (R/O)	0.0,0.0,0.0	Reports target coordinates in the current viewport.
Tbaskbar	*1*	*Toggles whether each drawing appears as a button on the Windows taskbar.*
TbCustomize	1	Toggles whether toolbars can be customized.
TDCreate (R/O)	*varies*	Reports the date and time that the drawing was created, such as 2448860.54014699.
TDInDwg (R/O)	*varies*	Reports the duration since the drawing was loaded, such as 0.00040625.
TDuCreate (R/O)	*varies*	Reports the universal date and time when the drawing was created, such as 2451318. 67772165.
TDUpdate (R/O)	*varies*	Reports the date and time of last update, such as 2448860.54014699.
TDUsrTimer (R/O)	*varies*	Reports the decimal time elapsed by user-timer, such as 0.00040694.
TDuUpdate (R/O)	*varies*	Reports the universal date and time of the last save, such as 2451318.67772165.
TempOverrides	1	Toggles temporary overrides.
TempPrefix (R/O)	"d:\temp"	Reports the path for temporary files set by **Temp** variable.

Variable	*Default*	*Meaning*
TextEval	0	Toggles the interpretation of text input during the **-Text** command: **0** Literal text. **1** Read **(** and **!** as AutoLISP code.
TextFill	1	Toggles the fill of TrueType fonts when plotted: **0** Outline text. **1** Filled text.
TextQlty	50	Controls the resolution of TrueType fonts when plotted; range is 0 to 100.
TextSize	0.2000	Controls the default height of text (2.5 in metric units).
TextStyle	"Standard"	Specifies the default name of text style.
Thickness	0.0000	Controls the default object thickness.
TileMode	1	Toggles the view mode: **0** Displays layout tab. **1** Displays model tab.
TimeZone	-80000	Current time zone.
ToolTipMerge	0	Toggles the merging of tooltips during dynamic display.
ToolTips	1	Toggles the display of tooltips.
TpState (R/O)	0	Reports if Tool Palettes palette is open.
TraceWid	0.0500	Specifies current width of traces.
TrackPath	0	Controls the display of polar and object snap tracking alignment paths: **0** Displays object snap tracking path across the entire viewport. **1** Displays object snap tracking path between the alignment point and "From point" to cursor location. **2** Turns off polar tracking path. **3** Turns off polar and object snap tracking paths.
TrayIcons	1	Toggles the display of the tray on status bar.
TrayNotify	1	Toggles service notifications displayed by the tray.
TrayTimeout	5	Controls length of time that tray notifications are displayed; range is 0 to 10 seconds.
TreeDepth	3020	Controls the maximum branch depth (in *xxyy* format): *xx* Model-space nodes. *yy* Paper-space nodes. *>0* 3D drawing. *<0* 2D drawing.
TreeMax	10000000	Controls the memory consumption during drawing regeneration.

Variable	Default	Meaning
TrimMode	1	Toggles trims during **Chamfer** and **Fillet**: **0** Leaves selected edges in place. **1** Trims selected edges.
TSpaceFac	1.0	Controls the mtext line spacing distance measured as a factor of "normal" text spacing; ranges from 0.25 to 4.0.
TSpaceType	1	Controls the type of mtext line spacing: **1** At Least: adjusts line spacing based on the height of the tallest character in a line of mtext. **2** Exactly: uses the specified line spacing; ignores character height.
TStackAlign	1	Controls vertical alignment of stacked text: **0** Bottom aligned. **1** Center aligned. **2** Top aligned.
TStackSize	70	Controls size of stacked text as a percentage of the current text height; range is 1 to 127%.

U

Variable	Default	Meaning
UcsAxisAng	90	Controls the default angle for rotating the UCS around an axes (via the **UCS** command using the **X**, **Y**, or **Z** options; valid values limited to: 5, 10, 15, 18, 22.5, 30, 45, 90, or 180.
UcsBase	""	Specifies name of UCS that defines the origin and orientation of orthographic UCS settings.
UcsDetect	1	Toggles dynamic UCS mode.
UcsFollow	0	Toggles view displayed with new UCSs: **0** No change. **1** Automatically aligns UCS with new view.
UcsIcon	3	Controls display of the UCS icon: **0** Off. **1** On. **2** Displays at UCS origin, if possible. **3** On, and displayed at origin.
UcsName (R/O)	"World"	Reports the name of current UCS view: "" Current UCS is unnamed.
UcsOrg (R/O)	0.0,0.0,0.0	Reports the origin of current UCS relative to WCS.
UcsOrtho	1	Controls whether the related orthographic UCS settings are restored automatically: **0** UCS setting remains unchanged when orthographic view is restored. **1** Related ortho UCS is restored automatically when an ortho view is restored.
UcsView	1	Toggles whether the current UCS is saved with a named view.

Variable	Default	Meaning
UcsVp	1	Toggles whether the UCS in active viewports remains fixed (locked) or changes (unlocked) to match the UCS of the current viewport.
UcsXDir (R/O)	1.0,0.0,0.0	Reports the x-direction of current UCS relative to WCS.
UcsYDir (R/O)	0.0,1.0,0.0	Reports the y-direction of current UCS relative to WCS.
UndoCtl (R/O)	21	Reports the status of undo: **0** Undo disabled. **1** Undo enabled. **2** Undo limited to one command. **4** Auto-group mode. **8** Group currently active. **16** Combines zooms and pans.
UndoMarks (R/O)	0	Reports the number of undo marks.
UnitMode	0	Toggles the type of units display: **0** As set by **Units** command. **1** As entered by user.
UpdateThumbnail	15	Controls how thumbnails are updated (sum): **0** Thumbnail previews not updated. **1** Sheet views updated. **2** Model views updated. **4** Sheets updated. **8** Updated when sheets or views are created, modified, or restored. **16** Updated when the drawing is saved.
UseAcis	*0*	*Toggle involving ACIS.*
UserI1 *thru* **UserI5**	0	Five user-definable integer variables.
UserR1 *thru* **UserR5**	0.0	Five user-definable real variables.
UserS1 *thru* **UserS5**	""	Five user-definable string variables; values are not saved.

V

Variable	Default	Meaning
ViewCtr (R/O)	*varies*	Reports x,y,z-coordinate of center of current view, such as 15,9,56.
ViewDir (R/O)	*varies*	Reports current view direction relative to UCS (0.0,0.0,1.0 = plan view).
ViewMode (R/O)	0	Reports the current view mode: **0** Normal view. **1** Perspective mode on. **2** Front clipping on. **4** Back clipping on. **8** UCS-follow on. **16** Front clip not at eye.
ViewSize (R/O)	*varies*	Reports the height of current view in drawing units.

Variable	Default	Meaning
ViewTwist (R/O)	0	Reports the twist angle of current view.
VisRetain	1	Controls xref drawing's layer settings: **0** Xref-dependent layer settings are not saved in the current drawing. **1** Xref-dependent layer settings are saved in the current drawing, and take precedence over settings in the xref'ed drawing the next time the current drawing is loaded.
VpMaximizedState (R/O)	0	Toggle: reports whether viewport has been maximized by **VpMax** command.
VsBackgrounds	1	Toggles display of backgrounds in visual styles.
VsEdgeColor	7	Specifies the edge color.
VsEdgeJitter	-2	Specifies the level of pencil effect: **0** *or -n* None. **1** Low. **2** Medium. **3** High.
VsEdgeOverhang	-6	Extension of pencil lines beyond edges; range is 1 to 100 pixels; *-n* = none.
VsEdges	1	Specifies types of edges to display: **0** No edges displayed. **1** Isolines displayed. **2** Facets and edges displayed.
VsEdgeSmooth	1	Specifies crease angle; range is 0 to 180 degrees.
VsFaceColorMode	1	Determines color of faces.
VsFaceHighlight	-30	Specifies the size of highlights; range is -100 to 100. Ignored when **VsMaterialMode** is 1 or 2 and objects have materials attached.
VsFaceOpacity	-60	Controls the transparency/opacity of faces; range is -100 to 100 (fully opaque).
VsFaceStyle	1	Determines how faces are displayed: **0** None. **1** Real. **2** Gooch.
VsHaloGap	0	Specifies halo gap; range is 0 to 100 pixels.
VsHidePrecision	0	Toggles accuracy of hides and shades.
VsIntersectionColor	7	Specifies the color of intersecting polylines.
VsIntersectionEdges	0	Toggles the display of intersecting edges.
VsIntersectionLtype	1	Specifies the linetype for intersecting polylines: **0** Off. **1** Solid. **2** Dashed. **3** Dotted. **4** Short dash. **5** Medium dash.

Variable	Default	Meaning
		6 Long dash. **7** Double-short dash. **8** Double-medium dash. **9** Double-long dash. **10** Medium-long dash. **11** Sparse dot.
VsIsoOntop	0	Toggles whether isolines are displayed.
VsLightingQuality	1	Toggles the quality of lighting: **0** Facets displayed. **1** Facets smoothed.
VsMaterialMode	0	Controls the display of material finishes: **0** No materials displayed. **1** Materials displayed. **2** Materials and textures displayed.
VSMax (R/O)	*varies*	Reports the upper-right corner of virtual screen, such as 37.46,27.00,0.00.
VSMin (R/O)	*varies*	Reports the lower-left corner of virtual screen, such as -24.97,-18.00,0.0.
VsMonoColor	255,255,255	Specifies the monochrome tint.
VsObscuredColor	byentity	Specifies the color of obscured lines.
VsObscuredEdges	1	Toggles the display of obscured edges.
VsObscuredLtype	1	Specifies the linetype of obscured lines.
VsShadows	0	Determines the quality of shadows: **0** No shadows displayed. **1** Ground shadows displayed. **2** Full shadows displayed.
VsSilhEdges	0	Toggles the display of silhouette edges.
VsSilhWidth	5	Specifies the width of silhouette edge lines; range is 1 to 25 pixels.
VsState	0	Toggles the Visual Styles palette.
VtDuration	750	Controls the duration of smooth view transition; range is 0 to 5000 seconds.
VtEnable	3	Controls smooth view transitions for pans, zooms, view rotations, and scripts: **0** Turned off. **1** Pans and zooms. **2** View rotations. **3** Pans, zooms, and view rotations. **4** During scripts only. **5** Pans and zooms during scripts. **6** View rotations during scripts. **7** Pans, zooms, and view rotations during scripts.
VtFps	7.0	Controls minimum speed for smooth view transitions; range is 1 to 30 frames per second.

Variable	Default	Meaning
W		
WhipArc	0	Toggles display of circular objects: **0** Displays as connected vectors. **1** Displays as true circles and arcs.
WhipThread	1	Controls multithreaded processing on two CPUs (if present) during redraws and regens: **0** Single-threaded calculations. **1** Regenerations multi-threaded. **2** Redraws multi-threaded. **3** Regens and redraws multi-threaded.
WindowAreaColor	5	ACI color of windowed selection area.
WmfBkgnd	Off	Toggles background of *.wmf* files: **Off** Background is transparent. **On** Background is same as AutoCAD's background color.
WmfForegnd	On	Toggles foreground colors of exported WMF images: **Off** Foreground is darker than background. **On** Foreground is lighter than background.
WorldUcs (R/O)	1	Toggles matching of WCS with UCS: **0** Current UCS does not match WCS. **1** UCS matches WCS.
WorldView	1	Toggles view during **3dOrbit**, **DView**, and **VPoint** commands: **0** Current UCS. **1** WCS.
WriteStat (R/O)	1	Toggle: reports whether *.dwg* file is read-only: **0** Drawing file cannot be written to. **1** Drawing file can be written to.
WsCurrent	"AutoCAD Default"	Controls name of current workspace.
X		
XClipFrame	0	Toggles visibility of xref clipping boundary.
XEdit	1	Toggles editing of xrefs: **0** Cannot in-place refedit. **1** Can in-place refedit.
XFadeCtl	50	Controls faded display of objects not being edited in-place: **0** No fading; minimum value. **90** 90% fading; maximum value.
XLoadCtl	2	Controls demand loading: **0** Demand loading turned off; entire drawing is loaded. **1** Demand loading turned on; xref file opened. **2** Demand loading turned on; a *copy* of the xref file is opened.

Variable	Default	Meaning
XLoadPath	"...\temp"	Specifies path for storing temporary copies of demand-loaded xref files.
XRefCtl	0	Toggles creation of *.xlg* xref log files.
XrefNotify	2	Controls notification of updated and missing xrefs: **0** No alert displayed. **1** Icon indicates xrefs are attached; a yellow alert indicates missing xrefs. **2** Also displays balloon messages when an xref is modified.
XrefType	0	Toggles xrefs: **0** Attached. **1** Overlaid.

Z

Variable	Default	Meaning
ZoomFactor	60	Controls the zoom level via mouse wheel; range from 3 to 100.
ZoomWheel	0	Switches the zoom direction when mouse wheel is rotated forward: **0** Zooms in. **1** Zooms out.

3

Variable	Default	Meaning
3dDwfPrc	2	Level of precision in drawings exported as *.dwf* files: **1** 1 **2** 0.5 **3** 0.2 **4** 0.1 **5** 0.01 **6** 0.001

AutoCAD 2007 Keystrokes & Shortcuts

Object Snap

APP	Apparent intersection
CEN	Center
END	Endpoint
EXT	Extension
FROM	From
INS	Insertion point
INT	Intersection
MID	Midpoint
NEA	Nearest
NOD	Node (point)
PAR	Parallel
PER	Perpendicular
QUA	Quadrant
TAN	Tangent
QUI	Quick mode

Selection Sets

Ctrl+pick	Selects faces
(Pick)	Selects one object
ALL	Selects all objects
AU	AUtomatic: *(pick)* or BOX
BOX	Left to right = Crossing; Right to left = Window
C	Crossing
CP	Crossing polygon
F	Fence
G	Group
L	Last
M	Multiple (no highlighting)
P	Previous
SI	Single selection
W	Window
WP	Window polygon

Selection modes:

A	Add to selection set (default)
R	Remove from selection set
SHIFT	Remove from selection set
U	Undo change to selection set

Command Prefixes

' Specifies transparent command:
`From point: 'zoom`

'? Provides context-sensitive help:
`Command: line '?`

~ Forces display of file dialog box:
`Command: -insert ~`

- Forces display on command line:
`Command: -mtext`

+ Prompts for tab number:
`Command: +options`

. Forces use of undefined command:
`Command: .line`

_ Forces English cmd in int'l version:
`Command: _line`

multiple Automatically repeats command:
`Command: multiple circle`

(Begins AutoLISP function

) Ends AutoLISP function:
`Radius: (/ 3.2 2.0)`

$(Begins Diesel macro

Option Modifiers

tt	Tracking
m2p	Midpoint between two points
from	Offsets from temp ref point
	Direct distance entry: move mouse, and enter a distance

Color Numbers & Names

0	...	Background color
1	R	Red
2	Y	Yellow
3	G	Green
4	C	Cyan
5	B	Blue
6	M	Magenta
7	W	White/black
8-249	...	Other colors
250-255	...	Shades of grey
BYLAYER	...	Color from layer
BYBLOCK	...	Color from block

DATE DUE

Schooling the boys

EDUCATING BOYS, LEARNING GENDER

Series editors: Debbie Epstein and Máirtín Mac an Ghaill

This timely series provides a well articulated response to the current concerns about boys in schools. Drawing upon a wide range of contemporary theorising, the series authors debate questions of masculinities and highlight the changing nature of gender and sexual interactions in educational institutions. The aim throughout is to offer teachers and other practitioners grounded support and new insights into the changing demands of teaching boys and girls.

Current and forthcoming titles:

Madeleine Arnot: *Boy's Work: Teacher Initiatives on Masculinity and Gender Equality*
Christine Skelton: *Schooling the Boys: Masculinities and Primary Education*
Martin Mills: *Challenging Violence in Schools: An Issue of Masculinities*
Leonie Rowan *et al.*: *Boys, Literacies and Schooling: The Dangerous Territories of Gender-based Literacy Reform*

Schooling the boys

Masculinities and primary education

Christine Skelton

Open University Press
Buckingham • Philadelphia

Open University Press
Celtic Court
22 Ballmoor
Buckingham
MK18 1XW

email: enquiries@openup.co.uk
world wide web: www.openup.co.uk

and
325 Chestnut Street
Philadelphia, PA 19106, USA

First Published 2001

A catalogue record of this book is available from the British Library

ISBN 0 335 20695 6 (pb) 0 335 20696 4 (hb)

Library of Congress Cataloging-in-Publication Data
Skelton, Christine.
Schooling the boys: masculinities and primary education / Christine Skelton.
p. cm. – (Educating boys, learning gender)
Includes bibliographical references (p.) and index.
ISBN 0-335-20696-4 – ISBN 0-335-20695-6 (pbk.)
1. Boys–Education (Elementary)–Great Britain. 2. Underachievers–Great Britain. 3. Sex differences in education–Great Britain. I. Title. II. Series.
LC1390 .S54 2001
372.1821′0941–dc21 00-050199

Typeset by Graphicraft Limited, Hong Kong
Printed in Great Britain by Biddles Limited, Guildford and Kings Lynn

For Richard James Aston (1977–98)

Contents

Series editors' introduction

Educating boys is currently seen – both globally and locally – to be in crisis. In fact, there is a long history to the question: what about the boys? However, it was not until the 1990s that the question of boys' education became a matter of public and political concern in a large number of countries around the world, most notably the UK, the USA and Australia.

There are a number of different approaches to troubling questions about boys in schools to be found in the literature. The questions concern the behaviours and identities of boys in schools, covering areas such as school violence and bullying, homophobia, sexism and racism, through to those about boys' perceived underachievement. In *Failing Boys? Issues in Gender and Achievement*, Epstein and her colleagues (1998) identify three specific discourses that are called upon in popular and political discussions of the schooling of boys: 'poor boys'; 'failing schools, failing boys'; and 'boys will be boys'. They suggest that it might be more useful to draw, instead, on feminist and profeminist insights in order to understand what is going on in terms of gender relations between boys and girls and amongst boys. Important questions, they suggest, are: what kind of masculinities are being produced in schools, in what ways, and how do they impact upon the education of boys? In other words, there is an urgent need to place boys' educational experiences within the wider gender relations within the institution and beyond.

Despite the plethora of rather simplistic and often counter-productive 'solutions' (such as making classrooms more 'boy-friendly' in macho ways) that are coming from governments in different part of the English-speaking world and from some of the more populist writers in the area (e.g. Steve Biddulph), there is a real necessity for a more thoughtful approach to the issues raised by what are quite long-standing problems in the schooling of boys. Approaches for advice to researchers in the field of 'boys' underachievement' by policy makers and by teachers and principals responsible for staff development in their schools are an almost daily event, and many have already tried the more simplistic approaches and found them wanting. There is, therefore, an urgent demand for more along the lines suggested here.

This is not a series of 'how to do it' handbooks for working with boys. Rather, the series draws upon a wide range of contemporary theorizing that is rethinking gender relations. While, as editors, we would argue strongly that the issues under discussion here require theorizing, it is equally important that books in the area address the real needs of practitioners as they struggle with day-to-day life in schools and other places where professionals meet and must deal with the varied, often troubling, masculinities of boys. Teachers, youth workers and policy makers (not to mention parents of boys – and girls!) are challenged by questions of masculinity. While many, perhaps most, boys are not particularly happy inhabiting the space of the boy who is rough, tough and dangerous to know, the bullying of boys who present themselves as more thoughtful and gentle can be problematic in the extreme. We see a need, then, for a series of books located within institutions, such as education, the family and training/workplace and grounded in practitioners' everyday experiences. These will be explored from new perspectives that encourage a more reflexive approach to teaching and learning with reference to boys and girls.

We aim, in this series, to bring together the best work in the area of masculinity and education from a range of countries. There are obvious differences in education systems and forms of available masculinity, even between English-speaking countries, as well as significant commonalties. We can learn from both of these, not in the sense of saying 'oh, they do that in Australia, so let's do it in the UK' (or vice versa), but rather by comparing and contrasting in order to develop deeper understandings both of the masculinities of boys and of the ways adults, especially professionals, can work with boys and girls in order to reduce those ways of 'doing boy' which seem problematic, and to encourage those that are more sustainable (by the boys themselves now and in later life). Thus books in the series address a number of key questions: How can we make sense of the identities and behaviours of those boys who achieve popularity and dominance by behaving in violent ways in school, and who are likely to find themselves in trouble when they are young men out on the streets? How can we address key practitioner concerns about how to teach these boys? What do we need to understand about the experiences of girls as well as boys in order to intervene effectively and in ways which do not put boys down or lead them to reject our approaches to their education? What do we need to understand about gender relations in order to teach both boys and girls more effectively? How can we make sense of masculinities in schools through multidimensional explanations, which take into account the overlapping social and cultural differences (of, for example, class, ethnicity, dis/ability and sexuality), as well as those of gender? What are the impacts of larger changes to patterns of employment and globalization on the lives of teachers and students in particular schools and locations? The series, as a whole, aims to provide practitioners with new insights into the changing demands of teaching boys and girls in response to these questions.

As editors, we have been fortunate to be able to attract authors from a number of different countries to contribute to our series. Much of the work currently being carried out on boys and schooling has taken place in the secondary sector. In this book, Chris Skelton provides an overview and critique of the schooling of boys in the primary sector in England. Traditionally, there has been a tendency to focus on the feminization of primary schooling. In contrast, Chris explores the masculinization of this sector. Writing from a feminist perspective, she critically examines the ways in which school organization and classroom management-control strategies are shaped by dominant notions of masculinities. The text also considers teacher perspectives on contemporary gender and sexual arrangements within primary schooling. Importantly, Chris explores how girls, as well as boys, engage with the educational images of masculinity that are made available to them. From her theoretical investigation emerges an evaluation of past policies and an analysis of implications for future professional practice.

Debbie Epstein
Máirtín Mac an Ghaill

Reference

Epstein, D., Elwood, J., Hey, V. and Maw, J. (1998) Schoolboy frictions: feminism and 'failing' boys, in D. Epstein, J. Elwood, V. Hey and J. Maw (eds) *Failing Boys? Issues in Gender and Achievement*. Buckingham: Open University Press.

Acknowledgements

There are many people I would like to thank who have contributed towards this book.

The data reported in the case study chapters were collected during the course of my PhD research and would not have been possible without the generous cooperation of the teachers and pupils at the two schools.

The idea and initial encouragement to write the book came from Paul Connolly. Máirtín Mac an Ghaill was a great support and thanks also go to Debbie Epstein and Shona Mullen.

Some of the data discussed in Chapters 4 to 7 have been presented in journal articles. I am grateful to Taylor and Francis and the editors of *British Journal of Sociology of Education*, *Gender and Education*, *International Journal of Inclusive Education* and *Sport, Education and Society* for allowing me to reproduce material here.

I would like to thank my family (mum, dad, Trisha and Simon) and my female friends for always being there. Particular love and thanks to: Becky Francis for supporting the writing up of the book and for being a great colleague, collaborator and mate; Diane Pearce for long distance counselling and friendship on everything from using PhD material for a book to ongoing intellectual and emotional sustenance – and the gossip; the women I've known for more years than any of us care to remember and who I love to bits, Jane Aston, Gil Bennet, Barbara Thompson, Bea Reed and Carol Rigby, and those I've got to know more recently, Kathryn Ecclestone, Pat Sikes and Sally Troyna.

There are several male friends and colleagues who have helped, in various ways, to shape my ideas about men and masculinities: Peter Aston, Bruce Carrington, Mick Gowar, Frank Hardman, Chris Haywood, Steve Higgins, David Hustler, Máirtín Mac an Ghaill (again!), Steve Munby (the best male friend you could ask for), James Tooley and John Williamson.

And there are two men who are sadly no longer around to hear how much I appreciated them: Barry Troyna, my dear friend and mentor, who supervised my PhD research until his untimely early death in 1996; and my wonderful godson Richard Aston, who used to talk to me about being a young man in the 1990s. This book is dedicated to him.

Introduction

The fact that this book appears in a series focusing on boys and education reflects the growing interest in the area. Many of the writers contributing to this series have been researching and writing about masculinities and schooling for many years before the mid-1990s. I mention this as 1995 appears to mark the time when discussions on boys and education, which had largely been undertaken in academic journals, attracted the attention of the British media and politicians. The manner in which this interest was expressed was in terms of a moral panic as can be seen in such headlines as 'Where did we go wrong?' (*Times Educational Supplement* (*TES*) 14 February 1997). Such has been the concern that schools are 'failing boys' that there has been widespread response from the government, local education authorities (LEAs), schools and publishers offering strategies and materials for tackling boys' underachievement (Noble and Bradford 2000; Sukhnandan *et al.* 2000). One interesting quirk in the discussions of boys' education (and there are many as this chapter will show) is the apparent concern with boys in secondary schooling. Yet, those aspects of schooling which dominate the discussions, such as boys' performance in Key Stage tests, their disinclination towards academic work, negative behaviours in the classroom, and truancy, are as relevant to boys in primary school as they are to boys in secondary school. Lessons could be learned here from the gender equality work of the 1980s when Judith Whyte (1983: 8) pointed out that concentrating initiatives on girls in secondary schooling was misplaced as 'it is unlikely that crucial differences between the sexes suddenly make their appearance at the age of 13. Their roots are to be uncovered . . . in the primary years'.

To say that this book redresses this oversight by focusing on boys in primary schools is accurate but also too generalized. Such a claim masks the complexity of asking questions about what 'being a boy' in school means. For example, *who* is asking the question is an important factor; is a female teacher interviewing 6-year-old boys in her class about their attitudes to schoolwork likely to be given the same answers as a male researcher who has a less formal relationship with the boys? (See, for example, Connolly 1998; Swain 2000.) Also, what about primary schools themselves – what particular images of masculinity do they convey to pupils and are these images the same across all primary schools? And, what impact have educational policies had, both historically and currently, for the education of boys? This book aims to consider such issues in order to place the recent concerns about boys' underachievement in context and to offer relevant, practical solutions for schools to adopt.

I should state from the outset that this exploration of boys, masculinity and primary schooling is written by a feminist. I say this because my motivation for 'looking at boys' was prompted by my continuing concern for girls' experiences in schooling. Although statistics published by the Department for Education and Employment (DfEE) inform us that girls of all ages are outperforming boys in terms of their academic achievements (and it is largely for this reason that the subject of boys' underachievement has received such widespread attention) there is nothing new in this. After all, girls have *always* outperformed boys in primary schools as became evident in the 11-plus examinations when more girls scored higher marks than boys (Gipps and Murphy 1994; see also Chapter 1). Setting their superior performance in public examinations to one side, were there any other signs that girls, and indeed primary schooling generally, were trouncing boys? When I was visiting schools in the 1990s as a teaching practice supervisor it was still the case that the boys were the ones dominating teachers' time through misbehaviour or simply by demanding more attention. They were the ones who pushed girls to the margins of the playground with their football games. So little change there then from primary schooling written about by feminists in the 1970s and 1980s showing how girls tended to be marginalized in the classroom (Clarricoates 1978; Holly 1985).

At the same time, watching children on the playground or going about their lessons in the classroom showed that it was not just girls who were being pushed to the sidelines. There were some boys who were never included in the lunchtime football games and would hang around the edges trying to show that they were involved albeit as spectators. Equally, in the classroom, there were some boys whose attitudes and demeanours were such that they did not attract the attention of the teacher and would make no attempt to do so. The question was then who were these 'boys'? Obviously they were not all the same, acting in similar ways and adopting similar behaviours. But how did *some* boys come to act in particular ways to ensure they did manage to secure disproportionate amounts of the teacher's time and to harass girls, and what part did the school play in this? In order to answer these questions I carried out case studies of two primary schools. The information collected in these schools form the second part of the book.

The book is divided into two parts. The first part is the 'textbook' element which concentrates on educational policy, literature and theoretical issues relating to boys, masculinity and schooling. This will be of specific interest to those who are themselves writing and researching into boys and schooling as well as to those who have a general interest in understanding the background to the current concern with boys' underachievement. Chapter 1 provides a broad sweep of the literature into boys and schooling. It begins with a consideration of the way in which gender has informed educational policy and provision since the introduction of mass state schooling in 1870. It then goes on to trace how the concept of 'boys' has featured in studies of schooling and draws attention to how they have been perceived differently at different historical points. Chapter 2 concentrates on how masculinities have, and are, theorized. Here there will be a comparison of the explanations of masculinity which underpin the literature on 'boys' underachievement' to that informing 'masculinities and schooling' (a brief discussion on these is to follow). Chapter 3 sets the scene for the case study chapters by providing a description of the schools and the theoretical position adopted in collecting and analysing the data. I have attempted to avoid alienating readers by *not* presenting an impenetrable, jargonized and lengthy discussion. At the same time, a clear account of how the research was shaped and undertaken is needed so that readers can see how conclusions were reached. This chapter will also be useful to other females who are carrying out research on or with males.

The second part consists of four chapters which focus on different aspects of the case studies. Chapter 4 looks at the impact of the location of a school on the social relations in the school and with the wider community. All schools have been affected by the changes imposed upon them by the radical reforms in educational policy enshrined in the Education Reform Act (ERA) 1988. For some schools there have been additional pressures brought about by shifts in the economy and local labour markets where whole communities have been hit by mass male unemployment. This chapter sets out to show how these changes have had knock-on implications for constructions of masculinity in the local culture, which in turn impact upon the organization and management structures of a school.

Chapter 5 explores how boys deal with the differing expectations of what it means to be 'a boy' compared to 'a school boy'. Here lie many of the problems that schools experience in attempting to encourage boys to be more explicitly positive towards, and interested in, education and schooling. While being a boy revolves around identifying themselves as 'tough', 'independent', 'resourceful', 'competitive', schools expect their pupils to conform to rules and authority – to demonstrate the very opposite characteristics in fact.

Chapter 6 looks at a different aspect of masculinity in primary schools – that of the male primary teacher. The 'feminization of teaching' (Moir and Moir 1999) is one argument which has been put forward to explain the underachievement of boys and thus one suggested solution is to increase the number of men teachers. Although men teachers might be underrepresented in terms of numbers they are overrepresented in terms of occupying high status positions in primary schools. One of the reasons for this may be because, as men in a predominantly female profession, they seek out ways of demonstrating their masculinity. Conventional ideas of masculinity see men as managers, disciplinarians and authority figures so it is perhaps not surprising to find them occupying headteacher positions and teaching the oldest children in the school. This chapter looks at how the men teachers in one primary school developed a 'laddish' relationship with the boys through humour and a passion for football in order to show they were 'properly masculine'. This had implications for the dominant form of masculinity in the school as well as a knock-on effect for how the girls were managed.

Chapter 7 expands upon the issue of how girls interact with men teachers who adopt laddish approaches by looking at heterosexuality in the primary school. A part of acting out laddish behaviours is to have some interaction with the opposite sex whether it be flirting, teasing or harassing. The first section of the chapter considers how the girls in one class dealt with the flirtatious attitudes of their class teacher by reversing the 'male gaze'. It then goes on to look at romance relationships between boys and girls and explores how, with very young boys, their actions often become more exaggerated and could readily be interpreted as sexual harassment. A question which is raised is how and when do young boys start to learn about sexual harassment. The final section briefly takes up the issue of the sexual abuse of children by teachers. The fear of being seen as a paedophile or being accused of abusing a pupil is a major concern of men teachers which is out of all proportion to the number of incidents that actually occur. However, there is a resounding silence on the subject which will do nothing to help counter the anxieties of men at a time when the government is seeking to increase the number of male primary teachers. The concluding chapter will summarize the main findings of the research and provide ways of addressing similar issues in other schools.

A note on boys' underachievement

At the beginning of this Introduction reference was made to the relatively recent high profile given in the British media to *boys' underachievement* although published research on *masculinities and schooling* had made its appearance several years earlier (Walker 1988; Abraham 1989; Mac an Ghaill 1991). The emergence of research into masculinities and schooling stemmed from a growing interest in 'men's studies' and the development of more sophisticated theories of gender which challenged restrictive ideas of the 'male sex-role' (Carrigan *et al.* 1985). Hence, the use of 'masculini*ties*' to denote multiple ways of being and becoming 'male'. These more sophisticated theories also allowed schools themselves to be investigated as places which produced masculinity (see Chapter 2 for discussion of sex-role theory and recent theorizing of masculinities). The main function of these research studies was to look at how male pupils 'learned' to be boys/young men in school settings. While these studies have proved extremely illuminating about how boys *are* in school they have rarely offered *practical* advice to schools and teachers (and I include my own work in this criticism). Even before 'boys' underachievement' attracted the attention it has, those involved in research into masculinities and schooling were saying that it is important for programmes to be developed so that 'work on issues of masculinity should be felt to be going somewhere' (Connell 1989: 301; Kenway 1995). Although publications are now beginning to appear which pull together carefully conducted research studies with practical advice to teachers (see Francis 2000) these are recent initiatives. So, while the literature on 'masculinities and schooling' has the hallmark of rigorously conducted research which has provided rich insights into the school lives of boys, it has let down teachers and schools by failing to offer any practical advice for school policies or classroom strategies.

In contrast the 'boys' underachievement' literature is awash with practical ideas and recommendations for schools (Bradford 1997; Bleach 1998). The reason for this is that the boys' underachievement literature emerged in response to a particular 'problem'. The problem was that boys were seen to be underachieving in relation to girls. This 'failure of boys' was located in the environment schools found themselves in following the Education Reform Act 1988. The ERA placed schools into a marketplace where they had to compete against others to ensure their survival. They had to demonstrate to 'customers' (parents) that they could provide a better service than the school next door. In order to facilitate this competition the government devised the idea of league tables for England and Wales where the examination results of all schools were published and ranked. By focusing on examination results schools were quick to realize that there was a 'gender gap' between the achievements of boys and girls (although as will be shown in Chapter 1 this was largely due to a superficial reading of the statistics). Thus, prevailing wisdom says that if schools could improve the results of their male pupils then the overall performance rating of the school would increase thereby enhancing their position in the league tables. So schools were on the lookout for ways of tackling 'failing boys'.

The difficulty here was that 'solutions' were suggested which were piecemeal and unsubstantiated. Recommendations were made based on the particular circumstances of one school and which, in themselves, had not been devised in response to systematic research carried out in that school. The consequences of such arbitrary approaches can be seen in the apparent contradictory solutions suggested. On the one hand we have the suggestion that there should be single-sex subject teaching, 'as boys will be more prepared to be wrong in the absence of girls' (Kingston 1996: 4), and on the other that there should be 'mixed gender groups [to] encourage boys to develop better language skills from girls' (Terry and Terry 1998: 116). A further problem in devising solutions in the absence of systematic research is that they tend to be based on people's personal and professional instincts. These instincts are informed by all manner of factors such as political perspectives, social and cultural understandings and so on. Given that people hold different perspectives and numerous people offered numerous solutions to the issue of 'failing boys' then it is not surprising to discover that the discussions on boys' underachievement encompassed several different viewpoints.

Debbie Epstein and her colleagues (1998) have identified three different perspectives within the discussions on boys' underachievement, all of which have different implications for how it is addressed and what it means for girls. What they refer to as the *Poor Boys* discourse has also been labelled as the *Lads' Movement* (Kenway 1995). Here boys are seen as 'victims', specifically of single (fatherless) families, female dominated primary schooling, and feminism which has enabled girls' successes. Thus, one strategy which has been recommended is the recruitment of more men into teaching, particularly primary teaching, to provide boys with male role models. A second perspective is that of *Failing Schools, Failing Boys* where any school deemed to be failing in that it does not produce pupils with high levels of literacy and numeracy and above average passes in public examinations is seen as failing the boys (and presumably girls) who attend them. Unlike the *Poor Boys* perspective, it is rare for proponents of the *Failing Schools* discourse to make overt and direct attacks on feminism. Finally, there is *Boys Will Be Boys*, which conceives of boys in conventional, stereotypical ways and attributes these traditional characteristics to 'natural differences' as a result of biology and psychology. This discourse has much in common with that of *Poor Boys* in that boys are seen to have been made 'victims' because of feminist women's successes at promoting the female over men and maleness, thus challenging traditional ways of being a man. Epstein *et al.* (1998) have pointed to the contradictory nature of the 'boys will be boys' debate saying that:

> What is particularly interesting . . . is the way it manages, at one and the same time, to posit an unchanging and unchangeable 'boyness', which involves aggression, fighting and delayed . . . maturity and yet situates poor achievement at school as extrinsic to boys themselves.
>
> (Epstein *et al.* 1998: 9)

These different understandings of 'boys' and the causes of boys' underachievement are taken up and discussed further in Chapter 2, particularly in relation to the kinds of programmes developed for use in schools.

Here attention has been drawn to a distinction returned to at several points in this book between the literature on *boys' underachievement* which offers a plethora of catch-all solutions to teachers but is not based on systematic rigorous research, and that of *masculinities and schooling*, which until recently provided rich insights but no practical advice. The next chapter will set out the background to the recent concerns about boys and schooling by looking at the impact of educational policy, practice and provision in gendering education.

part / **one**

Context and theoretical perspectives

chapter / **one**

A history of boys' schooling

Introduction

The main concern of this book is to look at how primary schools construct particular dominant images of *school boys* and to consider the ways in which boys set about dealing with these prevailing forms. However, a chapter devoted to the literature on 'boys and *primary* schooling' would be a very short one indeed! Although the trend in the sociology of education in the 1970s shifted the spotlight away from 'political arithmetic' approaches (Halsey and Ridge 1980) and towards investigations of what was going on in classrooms, the number of studies of middle and secondary schools far outweighed those of primary schooling (Hammersley and Woods 1984; Hargreaves and Woods 1984; Walker and Barton 1989). Of those studies of primary classrooms which were carried out some make reference to 'sex differences' in terms of teacher attitudes towards pupils, the differing behaviours and approaches to the education of boys and girls, and how management and organizational practices were structured around gender (Sharp and Green 1975; Clarricoates 1978; King 1978). At the same time, 'boys' and 'girls' were spoken of as if they were homogenous groups and it was not until the latter part of the twentieth century that analyses of gender relations in primary schools considered gender *as* difference rather than gender differences (Thorne 1993; Francis 1998). That is, there was a move away from talking about 'boys' in relation to 'girls' and recognizing that there are as many, if not more, differences *between* girls as a group, and similarly between boys (Griffin and Lees 1997).

To begin to place the current debates about and studies of boys, masculinities and schooling into perspective demands that we go back further than the 1970s. Thus the first section of the chapter outlines the ways in which schooling has, historically, been informed by gender. At the same time, the nature of that gendered schooling changed, and continues to change, across time. The second section of the chapter goes on to look at studies which have focused specifically on boys and schooling. It covers those aspects of schooling which have been identified as contributing towards constructing 'masculinities' and are of particular relevance to the primary classroom. These aspects are brought together under the heading of 'authority patterns' and include discussions on the curriculum, monitoring and assessment, and control and management of pupils. Finally the chapter will look at the way in which the 'boys' underachievement' debate has been constructed and consider the various criticisms made of it.

Gender discourses in education and boys' schooling

Whatever is considered to be of concern in schools at any one historical point is a reflection of the existent relationship between the economy, culture and politics (Brown *et al.* 1997). A useful means of examining how concerns about gender have, not only, constantly shifted across time but how gender has been perceived in relation to other recognized forms of social inequality such as social class and 'race' has been provided by Weiner *et al.* (1997; see Table 1.1).

Table 1.1 starts in 1870 but concerns for the underachievement of boys predate this time. In 1693 the English philosopher, John Locke, wrote in his treatise *Some Thoughts Concerning Education* of how schools were failing to develop 'writing and speaking' skills in young gentlemen (M. Cohen 1998). Also, the idea that schools can, and do, influence dominant images of masculinity within individual school sites stretches across history. For instance, the issue for some writers and politicians today is that the 'anti-swot' culture of the majority of boys has been neglected and schools should challenge 'laddish' or 'cool' images of masculinity by providing positive images of learners and learning (Byers 1998; Noble and Bradford 2000). Similarly, public schools of early nineteenth century Victorian England were expected to promote acceptable 'codes of manliness' among a select group of boys (at that time dominant images of masculinity in the public schools revolved around the idea of the 'tough, decisive, courageous leader': Mangan 1987; Heward 1988). Thus, ideas that schools are 'failing boys' and that they construct and/or can challenge particular dominant images of masculinity obviously have a long history and are not phenomena of recent times.

Table 1.1 begins to trace discourses in gender and education from 1870 as this was the year when mass state schooling was introduced in Britain. Each of the noted time frames will be expanded upon in order to tease out predominant ideas about boys and schooling.

Table 1.1 Parallel educational discourses

Historical period	*Prevalent discourses of education*	*Prevalent discourses of gender and education*
1870 to early 1900s	Inequality of opportunity: provision informed by gender and social class	Informed by social class (boys' public roles; girls' domestic roles)
1920s, 1930s	Different but equal	Differentiation on basis of social class and 'natural' skills, abilities, etc.
1940s, 1950s	Equality of opportunity: IQ testing (focus on access)	Weak (emphasis on equality according to intelligence)
1960s, early 1970s	Equality of opportunity: progressivism/mixed ability (focus on process)	Weak (emphasis on working-class, male disadvantage)
Late 1970s, early 1980s	Equality of opportunity: gender, race, disability, sexuality, etc. (focus on outcome)	Equal opportunities/ anti-sexism (emphasis on female disadvantage)
Late 1980s, early 1990s	Choice, vocationalism and marketization (focus on competition)	Identity politics and feminisms (emphasis on femininities and masculinities)
Mid-1990s to date	School effectiveness and improvement (focus on standards)	Performance and achievement (emphasis on male disadvantage)

Source: adapted from Weiner *et al.* 1997: 622

From the 1870s to the 1940s

Concerns about the kind of schooling boys receive have a long history. At the same time, schooling in late Victorian England was stratified along social class lines and while upper- and middle-class boys would receive some kind of formal schooling in public boarding or day schools (as well as the attentions of private tutors and governesses), the education of working-class boys was more haphazard (Whitbread 1972). Then came the Forster Act 1870, which introduced a system of mass state schooling for all children between the ages of 5 and 10, thereby legitimizing the rights of the working class to a formal education. However, state schooling was not made compulsory until 1880 and not free until 1891 (Purvis 1995) and, as many parents could not afford the ninepence fee charged by some school boards, the education that working-class boys received tended to be somewhat piecemeal. (For girls it was even more ad hoc as parents were more likely to give any money they had towards the education of their sons: see Purvis 1991.) Further distinctions between boys in different social classes were apparent in the different types of curriculum they were provided with. Public schools offered the kinds of educational experiences seen as befitting future leaders and entrepreneurs (such as an emphasis on sports, the prefect system and a knowledge of Latin, literature and science) while working-class boys were provided (in elementary schools) with basic numeracy, literacy and technical skills required by factory workers and soldiers (Hunt 1987; Heward 1988).

Similarly, girls received an education which distinguished them by both gender and social class. For both middle- and working-class girls this meant being educated for their future domestic roles (Davin 1978). While boys were trained for their futures as leaders or soldiers, different forms of domestic curriculum were given to middle- and working-class girls in accordance with the dominant versions of 'respectable' ladies and women. With that end in mind, middle-class girls undertook a curriculum which included such subjects as deportment, drawing and callisthenics (Cobbe 1894) – these subjects being thought to make them attractive as potential wives who were able to look pleasing to their husbands and to organize their domestic world. Working-class girls received a heavily domesticated curriculum. The Education Department influenced the elementary curriculum through the provision of grants and in the early 1870s girls would spend as much as one-fifth of their time in school doing needlework (Sharpe 1994). This was followed in 1878 by domestic economy being made a compulsory subject for girls, then cookery in 1882 and laundry work in 1890.

Very little changed in the gendered (and social class) nature of schooling from late Victorian England into the first few decades of the twentieth century. Although during this time women received the vote, educational policy paid no attention to the concept of 'equality': this was to come in the 1940s.

The 1940s and 1950s

Although this period did not see any recognition in official circles for *gender* inequalities, concerns were raised about inequality in terms of *social class*. The Butler Education Act 1944 introduced compulsory free secondary schooling. It was intended to redress inequalities experienced by working-class children by not only providing them with a secondary education, but also directing them towards different types of schools which could cater for their ascribed abilities; these were grammar, technical and secondary modern schools. The decision as to which state school a child attended was based on their measured intelligence. Although the tripartite system of grammar, technical and secondary modern schools was intended to eliminate social class divisions (as all three types of schools were supposed to be seen as equal but specializing in different areas) it actually reinforced social divisions as the academic grammar schools were seen as having higher status. Gaining entry to the more prestigious grammar schools rested upon a child's performance in the 11-plus examination. At this time boys were seen as 'late developers' and to accommodate their later (and greater) potential, the majority of local education authorities provided more grammar school places for boys. The consequence of this was that girls had to score more highly in the 11-plus examination than boys in order to secure a grammar school place (Gipps and Murphy 1994).

Another explanation as to why boys had greater access to grammar schools than girls has been related to the need to accommodate men who had served in the military services during the Second World War (Dean 1995). At that time women who had worked in jobs created by war industries found themselves redundant; other working women had to return to the home due to the reduction in the number of day nurseries (from over 1500 in 1945 to 903 in 1949), while others responded to the government promotion of certain expert advice on child-rearing such as Bowlby's theory of maternal deprivation (Oakley 1981). Locating women (exclusively) back in the home was reflected in the school curriculum and strongly influenced the educational opportunities of girls and boys. Girls in secondary modern schools continued with a curriculum which was influenced by domesticity such as sewing and cookery. Boys' futures lay in the labour market and, as such, were seen as more in need of those subjects provided by a grammar school education which would lead on to higher education and a professional career.

The 1960s and early 1970s

A concern for social class, rather than gender or 'race' inequalities, continued into the 1960s and 1970s. In terms of educational policy, there was a shift during the 1960s towards comprehensive secondary schooling as a means of redressing the educational disadvantages experienced by the working class through the tripartite system. Although comprehensive education broadened the curriculum offered to pupils, quite different subjects were offered to boys and girls. For girls, this expansion meant they could take such subjects as childcare, typing and shorthand while boys were offered technology and metalwork (Byrne 1978).

The late 1970s and 1980s

Following the passing of the Sex Discrimination Act in 1975, and continuing into the 1980s, the predominant discourse in gender and education was on girls and the ways in which they experienced an unequal and discriminatory education in relation to that received by boys. Importantly, this was the first time that the gendered nature of schooling was explicitly investigated and discussed although boys were the 'shadowy other'.

It was during this time that the early underachievement of boys first became evident as large-scale studies of children's achievements at primary school pointed to the fact that, with the possible exception of mathematics, girls consistently outperformed boys (Gorman *et al.* 1988). This was a feature of primary schooling which had long been known (see Douglas 1964) but did not become the issue it was to become in the late 1990s because, in the same way that more boys' grammar school places were made available, it was assumed that boys would 'catch up' at secondary school.

While the research of this period generally spoke of 'girls' and 'boys' as homogenous, collective groups, there were indications in some studies of primary schooling that gender was more complex than this. Pollard's (1985) case study of an 8–12 middle school identified twelve friendship sets which he was able to place into three types of child groups: Goodies, Jokers and Gangs. Both boy and girl friendship sets appeared in each of the three groups, which suggested that stereotypes of boys at school as 'lads' and girls as 'good' were too simplistic. Similarly, feminist investigations of primary schooling which sought to 'clarify the nature of social relations *between* girls and boys' (Clarricoates 1987: 188) often had difficulties in accounting for what were differences *between* boys. For example, Clarricoates talked of boys who did not 'fit' conventional notions of masculinity as somehow unacceptable to the 'main boys group' despite the fact that there seemed to be many who were not members of that group:

> Ian did not conform to the dominant values of the boys . . . he liked his lessons, did not fight and, though adequate if called upon to play football, he was by no means gifted at the sport. He transcended his *peer group deviation* by helping them to beat the girls . . . Danny, however, was a shy boy, afraid of anyone who approached him aggressively . . . He therefore can be categorized as an *outcast* . . . Michael's overt lack of masculine qualities makes him an *outcast* despite his high academic achievement . . . Although Simon may at first appear to be an outcast he was, however, disruptive and aggressive, traits seen as 'boyish' and hence 'masculine' behaviour. His status was commuted to *outsider.*
>
> (Clarricoates 1987: 190–1, original emphasis)

The problem of what to do with the children that did not 'fit' into neat gender categories was something which was to be taken up by researchers at a later date (see next section). Research evidence from this period, where the focus was on identifying gender differences *between* boys and girls, showed that, generally, some boys received more of the teacher's attention, occupied more physical space in the classroom and playground, and were more likely to take 'risks', for example, putting themselves forward to respond to a teacher's question even when they did not know the answer (Spender 1982). Studies into teacher attitudes indicated that many teachers were influenced in their dealings with pupils by gender specific preconceptions whereby 'proper boys' were seen as dominant, demanding and difficult but rewarding to teach (Stanworth 1981; Clarricoates 1987). However, why there were boys who were 'exceptions to the rule' (Hough 1985: 20) was a question which had yet to be asked.

Late 1980s and early 1990s

The 1980s witnessed a series of moves within educational policy making aimed at changing the ways in which schools were managed culminating in the Education Reform Act 1988. The characteristics of this period were choice, vocationalism and marketization with the overriding discourse being that of competition (see Introduction). There was a place for notions of equality within New Right and government discourses but these were presented in the form of 'entitlements'. As such, the post-war function of education, in terms of its importance to the development of economic growth, equality of opportunity and social justice (Haywood and Mac an Ghaill 1996), was replaced by the concept of education as a means of enabling individual aspirations through the rough justice of market forces. In effect this meant that the implicit assumption in educational policy, as enshrined in the ERA and subsequent educational policies, was that equality of opportunity was important for girls and boys but that it was appropriate only at the level of the individual (Weiner 1994).

At the same time, the notion of the individual and individual differences began to become apparent in educational research. In terms of research into gender this meant that there was a shift away from considering 'boys' and 'girls' as homogenous groups and looking at the differences between males and females. Instead, the notion of *difference* provided a means of exploring male and female individual subjectivities (B. Davies 1989; A. Jones 1993). By the mid-1990s a number of studies of secondary age boys had appeared which looked at how schools produce a range of masculinities and young men's engagement with them (see, for example, Walker 1988; Mac an Ghaill 1994). This is the literature referred to in the Introduction as 'masculinities and schooling'. The ways in which schools are sites upon which a multiplicity of masculinities are played out will form the main discussion in the second section of this chapter.

Mid-1990s to date

In the Introduction it was shown how the concern about boys' underachievement had emerged from the marketization of schooling. In particular, school inspections, and the consequent emergence of the school effectiveness and school improvement movements, have forced schools to identify aspects of their practice and performance which can be improved upon. At the same time, equal opportunities policies have remained in situ and the complexity of how these 'work' alongside policies on educational reform has been commented on by Salisbury and Riddell (2000):

> Sometimes, the two sets of policies appeared to be working in harmony, for instance, the emphasis on examination attainment may well have encouraged middle-class girls whose values were already attuned to academic achievement. On the other hand, the focus on measurable outcomes and the neglect of classroom processes meant that many aspects of patriarchal culture continue unchallenged.
>
> (Salisbury and Riddell 2000: 14)

Such issues have generally been unacknowledged by the government and media and the educational achievement 'problem' of the 1990s was not some groups of *pupils* who were underachieving because of factors relating to social class, ethnicity and so on but *all boys*. Hence the 'gender problem' of the late 1990s and early part of the twenty-first century is boys per se. The various criticisms of how this 'problem' has been constructed and conveyed will be discussed at the end of the next section on 'Studies of boys and schooling'.

This purpose here has been to outline gender discourses in education at particular historical points. It has been shown that concerns about boys' schooling have arisen at various points but in relation to social class rather than gender. While the current literature on boys' underachievement has clearly struck a chord with the media and teachers, most likely because of the range of 'solutions' it offers to schools, there is another group of research studies which offers insights into where masculinity making takes place in schools. The next section provides an overview of the literature which has specifically focused on masculinities and schooling, a characteristic of which is its recognition of female pupils/teachers and *differences* between males. As Gilbert and Gilbert (1998: 112) have remarked, the most insightful research on boys is where gender has been looked at from 'a relational perspective, where the practice of various forms of masculinity is seen to be constantly constructed along with but in distinction from femininities'. It is useful, first, to explore the taken-for-granted notion evident in many of the studies of schooling carried out before the second wave of feminism and the passing of the Sex Discrimination Act 1975, that boys' educational experiences were the 'norm'. The section will then go on to consider the research which has explored the ways in which schools and schooling have contributed towards constructions of masculinities.

Studies of boys and schooling

Studies before the Sex Discrimination Act 1975 (to 1979)

Although it was the late 1980s and 1990s that witnessed a burgeoning literature on boys, masculinities and schooling, many earlier studies also had this as a focus. However, the difference between the studies that appeared in the late 1960s and 1970s and more recent accounts is that the former tended to ignore the significance of the fact that it was the educational experiences of *boys* which were being explored. A useful illustration is that of the Manchester studies (D.H. Hargreaves 1967; Lacey 1970; Lambart 1976). Two of the schools involved were boys' schools (Hargreaves' research was in a secondary school and Lacey's looked at a grammar school). The third study was of a girls' grammar but, for various reasons, the findings of this study were not so widely published (Lambart 1997).

The Hargreaves and Lacey studies became widely known, particularly for their insights into pro- and anti-school cultures and for Lacey's development of the concepts of 'differentiation' and 'polarization'. Basically, 'differentiation' was a term used to describe the way in which boys were positioned, mainly by teachers, in relation to the academic curriculum in terms of behaviour, attitudes and abilities. Largely as a result of this differentiation subcultures developed which stood in relation to the curriculum. Academically successful boys were most likely to hold positive attitudes towards the school while academically unsuccessful boys were likely to criticize, reject or sabotage the system (polarization). However, that these were findings of boys in boys' schools was ignored and the assumption that these could be generalized to include girls was evident in the way other educationalist researchers wrote of the theory in relation to *pupils* (Meighan 1981; Sharp 1981; Hammersley 1986). Later studies on the differentiation-polarization theory did include both boys and girls and took into account differences in social class (Quine 1974; Woods 1979; Ball 1981). But, as Abraham (1995: 13) has pointed out, 'studying mixed schools and samples of girls as well as boys in order to explore social-class differences is not the same as exploring sex differences and gender differences'.

The problem of such 'gender blindness' was illustrated by Audrey Lambart's (1997) reflections on how her findings in the girls' grammar school were different from those of Hargreaves and Lacey and which she argues were 'probably grounded in gender'. For example, Lacey argued that those boys ranked as school successes exhibited positive attitudes to the school and to academic achievement whereas those deemed as 'bottom ability' were the ones most likely to exhibit negative attitudes and behaviours towards schooling. However, in Lambart's study it was some of the *highest* achieving girls who posed the most behavioural problems. Also, Lacey posited that it was middle-class boys who were most likely to be academically successful and to hold the most positive attitudes towards the school while working-class boys were disproportionately labelled as 'failures' and, therefore, the ones most likely to hold negative attitudes. This clear distinction between middle-class/academically successful and working-class/academically unsuccessful was not substantiated in Lambart's research. The teachers in her study found that girls from *all* social classes might not live up to ideal expectations. Lambart (1997: 452) goes on to say that 'Besides not finding polarization, I found no general connection between deviance, achievement and social class'.

Alongside the investigations into schooling influenced by the 'new sociology of education' (such as the Manchester projects) were those undertaken by researchers working in radical youth cultural studies who were committed to challenging the discursive positioning of young men as destructive and a threat to society (*Adolescent Boys in the East End of London*, Wilmott 1969; *View from the Boys*, H.J. Parker 1974; *Knuckle Sandwich: Growing Up in the Working-Class City*, Robins and Cohen 1978). Rather, these researchers attempted to present adolescent male (white, heterosexual, working-class) youth in more sympathetic terms by viewing their actions as challenges to dominant sources of social power and, as such, social class was the focus of their analyses (Griffin 1998). Similarly, investigations into the educational experiences of working-class boys were carried out by those committed to radical schooling who were influenced by neo-Marxist ideology. Hence, Willis's (1977) study of a group of 'lads' in a secondary school linked their counter-school culture to that of the factory floor but, rather than perceiving it as being a derivative of shopfloor culture, saw it as a creative response of the boys to their experiences of school. Boys were also the focus of Paul Corrigan's (1979) study *Schooling the Smash Street Kids*, which provided an account of the culture of a group of white, working-class, heterosexual boys in the north east of England. Similarly to Willis, Paul Corrigan documented the boys' resistance to the perceived repressive authority of the school.

While all these studies have since been criticized by feminists for their focus on white, heterosexual, male adolescent youths, their significance lies in the articulation of social class as a variable in developing masculine identities in school settings. The influence of neo-Marxist ideas was also evident in a study by Sharp and Green (1975) of teachers working in a progressive primary school. Sharp and Green discuss individual children in detail in the chapter on the 'social structuring of pupils' identities' but no heed is given to these as *gendered* identities. The lack of attention to gender in this and other investigations of schooling is not surprising given that it was not recognized as an 'issue', especially not for schools, until the second wave of feminism (Rendel 1985). Sharp and Green's study of a primary school was published the same year as the Sex Discrimination Act 1975 was passed and, largely through the efforts of feminists, gender was poised to come to the fore in studies of schooling.

Studies of schooling and boys from late 1970s

As was said earlier, not all sociological studies of primary schools have considered gender relations and the feminist literature which has explored them has leant towards dichotomous categories of 'girls' and 'boys'. Radical feminists provided understandings of how gender-power dynamics operated in the classroom, and were among the first to raise questions about how masculinity is constructed but, at the same time, their emphasis on power *between* males and females in educational settings failed to grasp the multiple and complex ways in which power is manifested (see for example Clarricoates 1978; C. Jones 1985). Similarly, the emphasis in liberal feminism on sex-role theory helped to generate an image of a 'typical boy' who was dominant, aggressive and defined in comparison to girls:

> Boys are praised for toughness, for strength, for leadership, for organization, for adult behaviour, for initiative and originality. They are criticized for weakness, for 'cissy' behaviour . . . and [learn] to hide their emotions and stifle even the healthier of their fears. They learn that Brownies are for girls and Cubs for boys.
>
> (Byrne 1978: 84–5)

Such approaches to analysing the schooling experiences of pupils by assuming it is possible to talk about 'all girls' or 'all boys' was inevitably open to criticism. The influence of post-structuralism,[1] on ways of theorizing masculinity and femininity began to appear in the educational literature of the late 1980s and early 1990s (B. Davies 1989; A. Jones 1993). Post-structuralism allows for a more complex understanding of how identities are constructed by taking the notion of *difference* as its starting point. One of the concepts associated with post-structuralism is that of 'discourse' and how people are positioned and position others 'discursively'. The idea here is that patterns of language (discourse) allow a variety of ways of placing (positioning) a person. For example, discourses surrounding men teachers of young children have fluctuated quite significantly both across and within particular periods of time, culture, region and so on. The city I live in was rocked in the early 1990s by a series of child sexual abuse cases where the perpetrators were teachers or nursery workers. At this time men teachers of young children were discursively positioned as 'suspect' which was apparent through the implementation of a variety of surveillance tactics ranging from female teachers given the job of 'marking' their male colleagues to actively refusing male student teachers/nursery nurses into primary/nursery schools (Skelton 1998a). Today, in the same region (but also elsewhere in the UK), the government is providing grants to higher education institutions to encourage men to enter primary teaching as a means of tackling boys' underachievement. Where men teachers were discursively situated as 'suspect' in the early 1990s they are currently positioned as 'positive role models'.

To summarize the ideas: post-structuralist arguments demonstrate that gender identity is not 'fixed' but changes across sites, time, culture and so forth. Where post-structuralism is useful then, is that the notion of discursive positioning allows for an explanation of the differences *between* groups/individuals in specific situations. It also permits consideration of variables such as ethnicity and social class on identity. However, it is not without its problems. With its emphasis on *difference* and belief that an individual's identity is dependent upon how they are positioned in discourses and how they position themself in discourse precludes any idea that it is possible to talk about people collectively, say in terms of certain groups discriminated against because of structural inequalities: basically, it has an inability to politicize power struggles (see Chapter 3). For example, it does not have the facility to explain why more black boys *generally* are excluded from school than white boys (Ouseley 1998).

Where does this leave us in terms of current research into masculinities/gender and schooling? Is it that all accounts informed by structural theories such as radical or liberal feminism, or Marxism should be disregarded and recognition given only to those adhering to post-structuralism on the basis that they offer a more accurate representation of 'being and becoming' a primary teacher/pupil? But we have just seen the problems inherent in adopting a wholly post-structuralist approach. Rather than opting for one theoretical perspective over another many writers draw on a range of explanations in order to illustrate the place of schooling in relation to gender (see Chapter 2 for discussion of these). In particular many of the studies which will be referred to later are not solely on boys but are on gender relations (Epstein 1993; Thorne 1993; Connolly 1998; Francis 1998). So what have these more diverse ways of approaching masculinities (and femininities/gender) told us about boys and schooling?

At the moment far more research has been carried out on masculinities and secondary schooling than primary schooling. Here I want to touch upon the findings of the studies of secondary schooling in order to ascertain whether they are equally as applicable to the primary sector. Also, it will become evident that more attention has been given in the studies undertaken in primary schools to pupils' construction of their gender identities; that is, where it is reasoned that to understand boys and masculinities in any one school site we need to look at this in relation to girls and femininities. Although all these studies recognize the place of schooling in the construction of gender, less consideration generally has been given in the research in both primary and secondary schools to the processes through which schools themselves construct dominant images of masculinity (some exceptions are Connell *et al.* 1982; Mac an Ghaill 1994; Sewell 1997; Connolly 1998; see also Gilbert and Gilbert 1998, and Lingard and Douglas 1999 for overviews). We begin by looking at what the studies of secondary schooling have said about where 'masculinity making' occurs then lead on to considering these issues in relation to primary schools.

Schooling masculinities: authority patterns

Many of those writing on adolescent boys in the secondary sector have argued that schooling has only a minor part to play in the formation of masculine identities for most men (Walker 1988; Connell 1989; Mac an Ghaill 1994). More important, according to Connell (1989: 301), are the 'childhood family, the adult workplace or sexual relationships (including marriage)'. This observation may be accurate but the adult workplace and sexual relationships are not as immediately relevant to the lives of primary age boys as they are to adolescent boys so it might be assumed that schools play a different, and possibly more significant, role in the development of masculinities of young boys.

What both sociological and feminist accounts of educational processes and practices have shown us is that schools themselves are sites where male dominance is regulated, normalized and legitimized (Beynon 1989; Skeggs 1991a). However, in order to avoid the male per se notion, schools are regarded as sites where multiple forms of masculinity and femininity are present. Kessler *et al.* (1985: 43) use the term *gender regime* to describe 'the pattern of practices that constructs various kinds of masculinity among staff and students, orders them in terms of prestige and power, and constructs a sexual division of labor within the institution'. So the interrelationships between members of the school, organizational processes, influences of the local community, wider culture, educational policy and provision all configure to produce various forms of masculinities in any one school site. What the research into boys and secondary schooling has indicated is that 'masculinity making' appears most evident through authority patterns (Connell 1989). For the purposes of the discussion here 'authority patterns' have been subdivided into

- authority/knowledge: that which is embraced in the official subject curriculum
- authority/surveillance: those aspects of schooling which have been generated or reintroduced through the increased amount of assessment and testing of pupils (such as setting and streaming)
- authority/control: the forms of management and control employed to regulate pupils.

Authority/knowledge

In terms of authority/knowledge, some commentators have referred to the 'male grammar school' nature of the National Curriculum saying that the emphasis on maths and science, as two of the three core subjects, was simply legislating in favour of the gender-biased programmes already in operation (Kant 1987; Burton and Weiner 1990). There was a tendency in earlier feminist literature to assume that all boys had equal access to all the benefits of this 'malestream' education (Byrne 1978; Whyld 1983; C. Jones 1985). However, when factors such as social class and ethnicity are taken into consideration it becomes apparent this is not the case. Christine Heward (1988), in her study of a private boys' school, argues that the curriculum of elitist schools was framed around a specific, social class version of masculinity, but adds that this was not simply a case of reproducing elitist classes, as boys from a range of backgrounds attended the school and not all of them were academically successful. Similarly, writers on black masculinities (among others) have observed that the Eurocentric, middle-class nature of the National Curriculum places limitations on which groups can readily access and benefit from education (Mac an Ghaill 1988; Parry 1996; Sewell 1997).

In addition, new subjects have been introduced into the secondary curriculum as a result of central government concern with upskilling the labour force. Hence, new forms of authority/knowledge can be seen in the introduction of educational technologies into the curriculum such as information communication technology (ICT). As a consequence, the range of masculinities in schools has increased. In his study of male identities in secondary schools Mac an Ghaill (1994) observed that vocationalization of the curriculum placed a different spin on Lacey's polarization-differentiation theory. At the time of Lacey's study the emphasis was on boys' academic failure or success but the introduction of vocational subjects has impacted upon what constitutes a school's understanding of 'failure' and 'success'. For example, some vocational subjects are accorded a higher status than some academic subjects and boys negotiate their masculine identities in relation to this broad curriculum.

While such information provides interesting insights into the influence of curriculum innovation on masculine identities in the secondary school, vocational subjects are generally not featured in the timetables of primary schools so it is clear that the findings are not directly transferable to primary schooling. However, changes in terms of monitoring and assessment have impacted upon the ways in which primary schools are structured and managed (Lee and Croll 1995; Galton *et al.* 1998). Academic success (and failure) is achieved through examinations and pupils are frequently aligned in relation to their potential in these public examinations. This is carried out through the practice of setting or streaming hence the school forces divisions between boys (and girls) thereby enabling the construction of different modes of masculinity. Thus, the fact that not all boys can successfully access the benefits of a 'male grammar school' curriculum illustrates a point made in many studies of schooling that, as important as the *content* of the curriculum is, the way in which pupils are *organized* in relation to it is equally, if not more, important. This leads to the second of the areas of schooling identified as 'masculinity making'; that is, authority/surveillance.

Authority/surveillance

Secondary education at the time of the research by Hargreaves (1967), Lacey (1970) and Willis (1977) was concerned with pupils securing academic achievement and access to higher education and careers. Lacey's (1970) findings were discussed earlier but are briefly reiterated here so that it can be shown how masculinities are aligned in relation to streaming and setting practices. A defining feature of pro- and anti-school cultures in Lacey's research was the way in which boys were positioned, mainly by teachers, in relation to the academic curriculum in terms of behaviour, attitudes and abilities (differentiation). Largely as a result of this differentiation, subcultures developed which stood in relation to the academic curriculum, with those boys who were doing badly academically likely to criticize, reject or sabotage the system (polarization). So, for those boys who were at the extreme of polarization, status could be achieved among their peers by adopting such behaviours as being 'cheeky' to teachers, playing truant, smoking, drinking and not doing homework. Lacey's theory of differentiation-polarization has been applied in subsequent research (Quine 1974; Ball 1981; Abraham 1995) as has investigations into the links between masculinities, peer group cultures and schooling (particularly secondary schooling) (H.J. Parker 1974; Willis 1977; Mac an Ghaill 1994). What these studies make clear is that those boys who cannot access social power through academic success pursue alternative sources through claims to sporting abilities, physical aggression and sexual prowess.

At the same time, these various competing masculinities do not occupy equal status but are organized hierarchically in relation to the form of dominant (hegemonic) masculinity prevailing in the school (see Chapter 2 for a definition of hegemonic masculinity). As Haywood and Mac an Ghaill (1996) explain:

> As schools create the conditions for a hegemonic masculinity, differing meanings of masculinity will compete for ascendancy. The curriculum offers male students a resource to develop their masculinity, through a range of responses to it. At the same time, relations of domination and subordination become apparent, as some groups are able to define their meaning of masculinity over others. These definitions create boundaries which serve to delineate what appropriate maleness should be within this social arena. Transgression of these boundaries activates techniques of normalization, ranging from labelling through to physical violence, that ultimately act to maintain differences embedded in the ascendant definitions of masculinity.
>
> (Haywood and Mac an Ghaill 1996: 55)

Primary education in England, prior to the ERA 1988, was framed around child-centred ideas of learning where competitive testing and assessment of pupils was not a principal goal. Rather the development of the *individual* was the main concern of teachers (Alexander 1984). However, the introduction of Standard Assessment Tasks (SATs) and the publication of primary league tables (with or without 'value added') has reinvoked the institutionalization of academic failure via competitive grading and streaming not seen since the days of the 11-plus examination. As Arnot *et al.* (1999: 143) have said: 'the reintroduction of streaming and the promotion of setting by school subjects for many [working-class] boys would confirm their failure to succeed'. Indeed, several educationalists working in English schools who have been involved in strategies to tackle boys' underachievement have pointed to the fact that the top sets in both primary and secondary schools are nearly all female and the lower sets all male, who become progressively demotivated (Frater 1998; Noble 1998; Penny 1998).

Authority/control

The third aspect of schooling which provides opportunities for masculinity making is in the area of authority/control. In the recent literature on gender, two main themes are apparent. There is the issue of gender cultures and its interrelationship with child-centred pedagogy, and the question of where and in whom control/discipline is located. To take each in turn, first what implications do gender cultures have for the day-to-day management of primary pupils?

Bronwyn Davies (1989) used a post-structuralist approach in her work with pre-school children to find out their interpretations of anti-sexist fairy tales. She concluded that because gender discourses clearly divide males and females into two oppositional categories (what she calls male–female dualism), children are eager to identify themselves with their 'correct' gender. So on starting school they already have the knowledge of where they 'fit' in terms of their gendered self and actively engage in the process of confirming this identity. Davies refers to this as 'gender category maintenance work' whereby if a male or female is seen to adopt ways of acting out of type then children will endeavour to let 'the "deviants" know they've got it wrong' (Davies 1989: 29). Davies's work involved reading a number of feminist fairy tales to 4–5-year-old children. I replicated her approach with a class of 6–7-year-old children at Benwood Primary School (see Chapters 3 to 7) with the same results. In the story of *The Paper Bag Princess*, Elizabeth the princess is the heroine who takes on the dragon in order to rescue her handsome fiancé Prince Ronald. During the course of her rescue attempt her clothes are burned off by the dragon and all she can find to wear is a paper bag. By the time she reaches Ronald she is dirty and dishevelled. Rather than welcoming Elizabeth and thanking her for saving him, Ronald tells her off for the way she looks. Elizabeth indignantly chastises him for his mean mindedness and dances off into the sunset with the accompanying textual explanation that they didn't get married after all. The children, far from seeing Elizabeth as the heroine, viewed her as someone who was in the wrong and that it was Ronald, rather than she, who decided not to get married. The reason being that she was simply not a 'proper princess' who stood around looking beautiful and waiting for her prince to rescue her. Similarly with *Princess Smartypants*, who is portrayed as a young woman with an independent lifestyle and a fast motorbike. The boys in the class were all keen to draw her motorbike but none of them put the princess sitting on it. They told me that riding motorbikes was not what princesses did. When they were presented with an image of a princess (or prince) who did not 'fit' with what they had learned about how these characters behaved in stories then the children would make judgements based on this. So princesses who were not 'proper princesses' in the conventional mould were censured and marginalized.

On arrival at primary school, children are confronted with various organizational features in which their gender is a major indicator of where they can go and what they have access to. Indeed recent studies show that one of the most highly gendered aspects of primary schools is the control of pupils' use of physical space. There are toilets and changing rooms for boys and girls (Thorne 1993), areas of the school grounds and times of the school day allocated to boys' football (Skelton 2000; Swain 2000), and decisions made as to where children sit in the class (Warren 1997). This gendering of the physical space is set to become even more controlled if schools adopt the recommended strategies to tackle boys' underachievement such as seating boys next to girls and single-sex subject teaching (Noble and Bradford 2000). Then there is the ongoing identification of boys' control of girls' use of physical space in the classroom and, particularly, the playground (Renold 1997; Connolly 1998; Epstein 1998; Francis 1998; MacNaughton 2000). It is not unreasonable to suggest that the 'symbolic gender cultures' (Francis 1998: 47) generated through the organizational practices of the school mesh with children's desire to demonstrate publicly their gender identity (B. Davies 1989), which then serves actively to reinforce notions of gender dualism.

This issue of gendered practices in primary schooling is hardly 'new' and attempts have been made in the past to redress such forms of discrimination. If we look back to the 1980s and early 1990s at how equal opportunities were incorporated into the daily work of primary schools we find a list of initiatives aimed at eliminating gender divisions by androgynizing children. The idea was that equality would be achieved if pupils were given equal access to materials and places in the classroom such as ensuring both sexes had their fair share of time to play in the home corner and sand tray. Also books and equipment were checked to remove sexist and racist images and schools alerted to their more obvious forms of discrimination such as lining up boys and girls separately (Skelton 1989; Tutchell 1990). These ideas were underpinned by a belief in the importance of sex-role theory (see Chapter 2). What we are seeing now in the recent studies of primary schooling is how feminist post-structuralism has allowed for more sophisticated understandings of gender *relations* (that is, where masculinity and femininity are understood in relation to each other) rather than looking at gender differences.

When equal opportunities initiatives based on sex-role theory were being promoted, one or two dissenting voices were heard in the literature which challenged the efficacy of such approaches. These voices drew on post-structuralist ideas. While Davies (1989) focused on children's constructions of gendered identities, Walkerdine (1983) argued that the child-centred approach adopted by primary schools was highly gendered. Her position was that the emphasis in child-centred philosophy on 'developing the needs of the individual' might seem to be non-gendered but the characteristics used to describe children such as inventive, creative and enquiring were actually those associated with masculinity (see discussion on hegemonic masculinity in Chapter 2). Equally the teacher within child-centred philosophy does not actively teach but provides a 'facilitating environment'; that is one where the teacher is passive and nurturant (feminine characteristics) in order to produce active learners (who, as we have just said, are male). The problem that this created for female teachers was identified by Reay (1990) when she carried out equal opportunities work in a primary school with groups of boys aged 9–11:

> I was anxious that the groups be run democratically with some devolution of power, but saw taking on the role of just a facilitator in relation to the boys as an ideological minefield. It could easily reinforce widely-held commonsense views that it is only men who have knowledge and expertise.
>
> (Reay 1990: 270)

In addition, there are implications for girls who, within child-centred discourses, are seen as inadequate as learners because 'instead of thinking properly, girls *simply* work hard – if femininity is defined by passivity, good behaviour, rule-following – then the outcome cannot be "real learning"' (Walkerdine 1983: 84).

Becky Francis (1998) takes up Walkerdine's premise in her study of primary school pupils. She uses the concept of a dichotomous construction to represent the predominant forms of masculinity and femininity among primary school children. The masculine construction is that of 'silly-selfish' with its associative qualities of immaturity, messiness and naughtiness, while the feminine construction is 'sensible-selflessness'. Her work focused on the children's articulation of their gendered identities and the words 'silly' and 'sensible' were those most frequently used by the children themselves. Although Francis developed a model based on oppositional gendered constructions, she pointed out that these cultures were not fixed and children would frequently cross the boundaries between the two (see also Thorne 1993). What such studies undertaken by sociologists/feminists, and which to some extent use post-structuralism, draw attention to is why equal opportunities strategies have not proved to be fully effective in challenging traditional gender stereotyping.

Equal opportunities programmes were based on the idea that to disrupt traditional gender attitudes, behaviours and choices all teachers had to do was introduce children to alternative, non-sexist images of masculinity and femininity. By presenting boys who spent their time avoiding literacy tasks with various messages that told them 'normal boys like reading stories' would be enough for them to soak up the 'new' (correct) missive. This assumed that 'when the [child] sees the veil of distortion lifted from [their] eyes [they] too will want to engage in those activities from which [they have] been excluded by virtue of [their] gender' (Walkerdine 1990: 89). In contrast, recent strategies drawing on ideas of gender as relational argue that to tackle conventional and stereotypical constructions of gender by schools and pupils involves a questioning and challenging of gendered discourses. For example, teachers might challenge primary boys' and girls' perceptions of gender through storylines (Yeoman 1999); confront stereotypical attitudes and behaviours towards schoolwork (Whitelaw *et al.* 2000); invite girls to take on more active roles in play (Marsh 2000); and encourage nursery boys to recognize their violent and aggressive actions (MacNaughton 2000). That *equal* opportunities strategies did not work and different forms of tackling gender constructions in primary schooling are needed is quite evident (Davies and Banks 1995; Francis 1998). After all, if they had worked we would not today be talking about boys' underachievement because one of the perspectives in that debate is how boys do not relate to traditionally gendered areas of the curriculum such as reading and writing (Alloway and Gilbert 1997; Sukhnandan *et al.* 2000).

A second theme raised in the area of authority/control is that of where and in whom control/discipline is located.

The authority patterns referred to by Connell (1989) are those centring around discipline, such as school uniform, class registers, school assemblies and different forms of punishment, all of which involve assessing pupils and shaping the pupil population into 'what Foucault terms a coherent "normative order"' (Wolpe 1988: 23; see also Foucault 1977). These authority patterns carry with them particular implications for modes of masculinity (and femininity). For example, in primary schools the concept of authority (drawing on all three interpretations) has been associated with male teachers (Byrne 1978; Askew and Ross 1988; Evetts 1990). Men teachers are more likely to

- have responsibility for the high status areas of the curriculum such as maths and science
- occupy central roles in the school requiring decision making (headteacher, deputy headteacher)

- 'control' older pupils and generally maintain discipline and punishment throughout the school.

The idea that discipline and punishment are part of the role of being a male primary school teacher is one that has been raised in many studies of teachers (Sheppard 1989; Skelton 1991; Smedley 1998). At the same time there is also evidence to suggest that aggressive forms of discipline are related to the development of particularly 'tough' forms of masculinity. When teachers (male or female) adopt more authoritarian types of discipline with male pupils who are not academically successful, they are helping to create the 'macho' modes of masculinity identified in practically all studies of masculinities and schooling (Willis 1977; Corrigan 1979; Walker 1988; Connell 1989; J. Stanley 1989; Mac an Ghaill 1994). These writers argue that a violent discipline system, particularly one locked into an educational system of academic success or failure, invites competition in 'machismo' among the boys and, sometimes, between the boys and male teachers. These findings emerged from studies of masculinities in secondary schooling but recent research points either implicitly or explicitly to the authority structures of primary schools as integral to the construction of masculine and feminine identities (Jordan 1995; Warren 1997; Connolly 1998; Francis 1998).

These studies of boys in primary schools identify similar 'macho' forms of masculinity to those in the secondary sector such as Jordan's (1995) 'Fighting Boys', Warren's (1997) 'Working-class Kings' and Connolly's (1998) 'Bad Boys', where dominant and violent behaviours are regarded as acceptable, even desirable. Jordan (1995) suggests that the development of school resistant masculinities is partly a result of the contrasting demands of being a 'boy' and being a conformist pupil. The latter is expected to speak quietly, not seek attention or use domineering behaviour, not to express anger or impatience and to avoid body contact while the opposite is expected of 'real' boys (Alloway and Gilbert 1997). To tackle these behaviours teachers frequently resort to using the same methods as the perpetrators thereby unwittingly sanctioning aggression (Gilbert and Gilbert 1998; see also Chapter 4).

Not all boys, however, adopt these more extreme modes of masculinity. But even those boys who demonstrate a seemingly greater accommodation to schooling on the basis that it provides a means to an end (such as securing personal privileges or qualifications that afford access to higher education) show an ambivalence towards teachers and authority structures (Connell 1989; see also Chapter 6). There is evidence in the above mentioned studies of primary schooling which offers some insights into how certain teacher discourses can place boys, collectively, at the centre of classroom management strategies while simultaneously acknowledging a 'hierarchy' of masculinities which require more or less teacher control.

The oppositional categories identified by Francis (1998) of 'silly-selfish' (males) and 'sensible-selfless' (girls) also emerge in Connolly's (1998) ethnography of a multiethnic, inner-city primary school. This time they relate to teachers' positionings of different ethnic groups. There are several examples of teachers describing boys as 'silly' but clearly there were degrees of 'silliness'. The 'silly' behaviours of a group of 5–6-year-old African Caribbean boys were seen by teachers as serious threats to their control and the social order of the classroom. In contrast, the 'silly' behaviours of South Asian boys were downplayed and closely associated with their immaturity. Interestingly, the teachers also incorporated into their perceptions of this ethnic group notions of South Asian boys as 'hardworking and helpful' (Connolly 1998: 119). Such descriptions can be seen to draw on Francis's (1998) feminine construction and, indeed, Connolly (1998: 121) says that the discussion of South Asian boys as being 'quiet and "little" and therefore needing to be befriended and looked after' served actively to feminize them.

The significance of the teachers' positioning of African Caribbean boys as serious threats to classroom order and the feminization of the 'silly' behaviours of South Asian boys had implications for their disciplinary strategies. Teachers adopt a range of approaches intended to convey a sense of 'the omniscient eye' to pupils, one of which brings them into direct challenges with individual children, namely public chastisement (Wolpe 1988). The issue of public chastisement revealed some significant insights. First, that it tended to be the 'harder' group Connolly calls the Bad Boys that were most likely to receive this form of teacher attention thus raising questions of racial (and gender) stereotyping. Second, the attitudes and behaviours of the teachers in these incidents suggests that they were drawing on control strategies which 'resonate[d] quite closely with their general perception of the street-wise hardened male living on the estate' (Connolly 1998: 78). Thus teachers utilized a similar form of language and posturing in their disciplining of the boys as the ones they saw the boys themselves using thereby maintaining the high ranking accorded to dominant, aggressive actions. Also:

> It created the self-fulfilling prophecy where the Bad Boys were forced into more fights and were then identified and publicly vilified by the teacher for being more aggressive . . . The boys were then set up with an even stronger 'masculine' identity which other boys within the school felt it necessary to challenge.
>
> (Connolly 1995: 177)

Third, research into dominant, more 'macho' modes of masculinity shows that, far from acting to inhibit boys' rule-breaking, anti-authority behaviours, drawing attention to them in public arenas provides individuals with kudos within their peer group (Measor and Woods 1984; Campbell 1993). This research into constructions of masculinity and femininity by primary pupils themselves and teachers' gendered engagements with their pupils illustrates how 'boys' as a collective group come to occupy a particular place within discourses on classroom management; it is one which places 'masculinities' as, at once, of far more significance than 'femininities' and crucial to the design and instigation of control strategies (see also Chapter 4).

All of the above studies can be located within that category of literature labelled 'masculinities and schooling'. The second category is that which comes under the 'boys' underachievement' banner. The final section in this chapter considers the responses which have been made to the debates on boys' underachievement at school.

Boys' underachievement: alternative perspectives

Concerns about 'failing boys' have captured the attention of the government and media in a way that girls, ethnic minority and social class equity issues never have. From the start of this media panic there have been attempts to look carefully at *what* is actually meant by boys' underachievement and *who* these failing boys are (Kenway 1995; Weiner *et al.* 1997). The number of voices challenging the simplistic nature of the boys' underachievement debate have continued to grow and there is now a substantial body of literature which places the arguments into context (Epstein *et al.* 1998; Delamont 1999; Salisbury *et al.* 1999; Francis 2000). At the same time, these more balanced appraisals of boys' underachievement and schooling do not seem to have attracted the same publicity. The reason why 'boys' underachievement' became *the* gender problem for schools was outlined in the Introduction. What follows is a brief description of the questions that have been raised about the assumptions underpinning the debate. However, the reader should bear in mind that it is not possible to do justice to these factors here and much greater insights will be gained by reading the authors cited in this section.

Are all boys underachieving?

By couching the debate in terms of 'boys' underachievement' implies that it is *all* boys who are failing at school and obviously this is not the case (Murphy and Elwood 1998). At the same time there are particular groups of boys who are doing badly but not just because they are boys. Underachievement generally is classed and racialized and as Epstein *et al.* (1998: 11) point out, 'class and the associated level of education of parents (for both boys and girls) continue to be the most reliable predictors of a child's success in school examinations'. Chris Woodhead, the Chief Inspector of Schools, identified one such group of boys when he said 'the failure of white working class boys is one of the most disturbing problems we face,

Table 1.2 GCSE 1999 performance by ethnic group

Boys: ethnicity	*Candidates*	*5+ A* to C (GCSE)*
Bangladeshi	21	18%
Black African	32	18%
Black Caribbean	56	4%
Black Other	26	29%
Chinese	8	75%
Greek/Greek Cypriot	3	–
Indian	85	42%
Pakistani	124	14%
Vietnamese	2	–
White UK	3252	37%
White Other	43	53%
Unclassified	43	37%

Source: Pulis 2000

within the whole education system' (*TES* 15 March 1996). But what of ethnic minority boys? Pulis (2000) reproduced the 1999 GCSE results of boys in Leeds LEA (Table 1.2) where he demonstrated that white working-class boys are doing better than some ethnic groups but worse than others.

I feel here that I should apologize to readers for looking at tables of GCSE results rather than Key Stage 1 (KS1) and Key Stage 2 (KS2) statistics. The reason for this is that, first, it is only recently that government agencies have collected the kinds of data necessary to make these kinds of comparisons (Salisbury *et al.* 1999), and second, even with more detailed evidence, the data that are made public are not specific enough for adequate comparisons to be made (Delamont 1999). The absence of sophisticated datasets has not prevented conclusions being reached on the basis of statistical information. This will be considered next but, before moving on, it should be reiterated that government and media hyperbole has led to a distorted impression that boys' underachievement refers to *all* boys and is widespread across schools. As is shown by all the authors discussed in this section, the 'boys' underachievement' debate as reported in the majority of the media is over-exaggerated and misinformed.

Statistics and boys' underachievement

It is fair to say that the main focus of this whole debate has been boys' apparent underperformance in public examinations in relation to girls. The discussions have tended to centre on GCSE and A levels and where possible we will look more closely at the KS1 and KS2 tests. To begin with, queries have been raised about how the statistics have been used to demonstrate a 'gender gap'. Lynne Raphael Reed (1999) argues that we need to consider a number of issues relating to assessment and achievement and how these appear as statistical 'evidence':

- the use of data selected to make a point (see Gorard *et al.* 1999 discussed below)
- the inability of the statistical data to demonstrate how gender intersects with social class and ethnicity on levels of achievement
- the socially situated nature of assessment practices (where differences in achievement are related to the knowledge, experiences and preferred learning styles pupils bring to bear on the task and the way in which the task itself is presented).

She then goes on to make two points regarding how the concentration on statistics as a means of determining that boys are underachieving in relation to girls draws attention away from two important factors:

- the focus on achievement across KS 1–4 removes attention from the reversed gender differential apparent in post-16 (where certain subjects still recruit a significantly higher number of boys)
- 'disadvantaged' boys do not appear to carry this through to progress through employment hierarchies.

To develop the point on how data are selected to prove particular arguments, Gorard *et al.* (1999) have shown that the methods used have been misleading because there is a tendency to confuse percentages and percentage points. For example, say we want to find out if there is a gender gap in performance at KS2 English in one particular school. We can see in Table 1.3 that, in terms of percentage points, in 1994 there was a 20 point gap in favour of girls. By 1998 it appears the situation had become more extreme as girls were then 24 percentage points ahead of the boys. In fact the reverse was the case. The confusion arises as both boys and girls were improving their achievement. So in 1994 boys needed to improve their performance by 50 per cent to reach the same level as girls but by 1998 needed to do only 40 per cent better. (The mathematics necessary to identify how much the boys needed to climb up to reduce the gender gap are: [point gap ÷ boys' score] × 100.)

Table 1.4 One school's KS2 English SAT results

	Girls	*Boys*	*Point difference*	*Achievement gap*
1994	60	40	20	50%
1998	84	60	24	40%

Furthermore, Salisbury *et al.* (1999: 405) have shown in their study of the assessments in Wales from KS1 to A levels that 'there are no significant gender differences at the *lowest* level of any assessment . . . the problem is not chiefly one of low-achieving boys but mainly concerns mid to high-attainers'. Their study reflects the findings of research in Australia which shows that where there is a gender gap it is at higher levels of achievement and at the higher end of the socio-economic scale (Teese *et al.* 1995). Also, Gorard *et al.* (1999) observed that not only is the gender gap apparent at the higher levels and in some subjects (for KS1 and KS2 this just applies to English) but also the gender gap is shrinking. This is supported by the statistics for England where the gap at KS2 English fell from 16.2 per cent in 1998 to 10.7 per cent in 1999 (National Literacy Trust 1999). However, we need to bear in mind the above proviso regarding the problems with focusing on percentage points.

The feminized primary school

The observation that primary schools are feminized institutions is one which has been frequently discussed in the literature since the second wave of feminism in the 1970s. The issues which framed the discussions were the low status of early years teaching and its association with mothering (Steedman 1988; Aspinwall and Drummond 1989), and the disparate careers of male and female primary teachers (Burgess 1989; Evetts 1990). Today the feminization of primary schools is seen to be a prime factor in boys' underachievement. The reasons put forward are that the day-to-day routines and practices of primary classrooms favour females and female behaviours; low expectations are held of boys' abilities; there is an absence of male role models; and the way in which the curriculum is delivered and assessed favours girls' learning styles (Delamont 1999).

We have already seen in the previous section how current research into gender and primary schools has found a number of ways in which boys and girls continue to be differentiated but none talks of these discriminative practices as disabling boys in particular. Drawing on this research enables us to deal fairly rapidly with the factors associated with the 'feminized primary school'. The daily routines and forms of organization in primary schools have been found to be gendered but these reinforce traditional forms of gender identity rather than advantaging girls and disadvantaging boys (Connolly 1998; MacNaughton 2000). At the same time, in their examination of the attitudes and behaviours of secondary teachers Younger *et al.* (1999) conclude that the emphasis given to the poorer performance of boys has generated a shift in views and in classroom practices. Boys were often constructed as 'more able', as in having more potential, in teacher discourses of the 1980s (Clarricoates 1980; Walden and Walkerdine 1985), although they also received more disciplinary comments (Serbin 1983).

Younger *et al.* (1999) have found that while boys continue to draw teachers' attention when it comes to behaviour, girls are the ones who receive the most positive forms of learning support.

The issue of boys needing male teachers as role models in order to develop academic potential raises more questions than provides answers. What *kinds* of images of adult males do we want boys to have? For example, one of the case study primary schools discussed in later chapters was set in a community where there was a high proportion of single mothers. Yet the two most alienated, disruptive and violent boys were from notorious local families whose involvement in crime was managed by a hierarchy of fathers, uncles and cousins. These two boys had a surfeit of male role models! Also it has been shown how men draw on traditional, conventional models of masculinity in presenting themselves as 'a male teacher' (Francis and Skelton 2001). Another question is whether there is any evidence to show that role models are effective? If they are, would we not have seen a significant increase in the number of female scientists and mathematicians who were exposed to an increase in female teachers of these subjects in the 1970s and 1980s? Recent research into encouraging ethnic minority men (and women) teachers suggests that role models are not a major motivator (Carrington *et al.* 2000).

The extent to which learning styles affect boys' and girls' achievement has attracted wide consideration. There does seem to be some evidence that girls prefer open-ended tasks which are related to real situations while boys favour memorization of rules and abstract facts (Arnot *et al.* 1998a). However, not all boys and girls have the same preferred learning styles. Also there is not a 'problem' in that one learning style is *better* than another, although some difficulties can emerge in particular subjects (Gipps and Murphy 1994). That is, different subjects make different learning demands to which boys and girls respond differently. This can lead to differences in achievement as Downes (1999) suggested in comparing the 1998 and 1999 KS2 English tests. He observed that in 1998 pupils had to read an 850 word extract from a story and then respond to questions many of which required reflection and empathy (girls' preferred styles). In 1999 the reading test was split into three different passages about spiders; the text was printed in larger type and accompanied by illustrations and diagrams. Although some questions required an elaborative form of response most of the marks were given for factual comprehension (boys' preferred learning styles). Downes argues that this shift towards more 'boy-friendly' assessment explained the 14 per cent point increase in reading scores. His concern was not that the two tests accommodated different learning styles (although ideally both tests should have encompassed a range of learning approaches) but that 'if in 2000 the tests revert to the 1998 approach the boys will again be seen to be "under-performing" and it will be the teachers who get the blame' (Downes 1999: 24).

Two points need to be made here. The idea that forms of assessment are the sole reason for gender differences in achievement is not supported by any hard evidence. It was thought that the emphasis in GCSE on course work explained why girls were achieving well but when changes were made (and this element was reduced) girls continued to improve their performance (Arnot *et al.* 1999). Also, the main subject area that boys do less well in than girls is literacy and rather than assessment practices being seen as the explanation for their lower success, many commentators have identified the gendered identity of the subject as the problem. English is associated with feminized literacy practices out of school and the emphasis on emotions and personal experiences (Arnot *et al.* 1998a; Qualifications and Curriculum Authority (QCA) 1998).

Summary

A number of points have been raised in this chapter. In the first section it was shown how education has been both gendered and classed since the introduction of mass state schooling in 1870. The second section observed the main shifts there have been in sociological studies of pupils' schooling. First, there is an expectation today that researchers acknowledge the composition of the 'pupils' they are studying so there is far less possibility of findings generated from research into 'boys' to be extrapolated to encompass girls (and the same applies to ethnic minority and social class groups). Second, the influence of post-structuralist theories have encouraged a move away from consideration of 'boys' and 'girls' as homogenous groupings towards one where differences within genders is recognized as well as a greater emphasis on gender as relational.

A further observation has been the identification of different voices within the literature on boys and schooling. There is that which focuses on boys' underachievement which appears to be largely atheoretical; research accounts of the multiple forms of masculinities found in specific school sites; and critiques by both feminists and pro-feminists of the current constructions of and approaches to 'failing boys' (Epstein *et al.* 1998; Lingard and Douglas 1999).

Readers may be aware that in the discussion of the literature on where 'masculinity making' opportunities are provided in and by the school itself, little or no attention was given to the perspectives of the boys themselves. There are two reasons for this. The first one being that the focus in the research which forms the basis of the case study chapters was specifically on how schools and teachers construct dominant images of masculinity. The data collected on how the boys interacted with those images and the lens through which those data were analysed was from the perspective of someone working in the school (see Chapter 3). There is also currently far more available evidence on primary children's perceptions of gender than on the school processes which contribute to those constructions, which leads on to the second explanation. There are obviously limits as to what can be packed into a book and readers can find rich accounts of primary children's constructions of gendered identities in schools by referring to the books and journals cited throughout this book (see in particular Epstein 1993; Thorne 1993; Connolly 1998; Francis 1998). In addition, detailed discussions on boys' attitudes to the 'uncool' nature of schoolwork and perceptions of schoolwork as something which is only extrinsically valuable can be found in more general books on masculinity/gender and schooling (Gilbert and Gilbert 1998; Kenway and Willis 1998). Some of the issues regarding primary aged boys' perceptions of 'cool' behaviours in school will also be discussed in the final chapter of this book.

Note

1 The term post-structuralist as opposed to postmodernist is used throughout this book. As Griffiths (1998) has pointed out, there have been endless debates on these concepts, many of which are unnecessarily complex and writers frequently opt out by using the terms interchangeably (see also Skeggs 1991b; Lather 1994, for further discussion). Francis (1998: 6) has provided a very helpful distinction:

> 'Post-modernism' refers to a body of work which revels in the recent fragmentation of modernist narratives and claims to scientific truth. Its foremost exponent is Lyotard (1984). Post-structuralism, on the other hand, refers specifically to the structuralist movement in literary criticism. Structuralists argued that all human narratives are based on similar known story-lines and that because words (or 'signifiers') structure our thoughts we are trapped [in] a 'prison-house of language' (see Saussure 1916). A universal social order is constructed through language – notions of ourselves as autonomous individuals are simply discourses within which human lives are positioned.

chapter / two

Theorizing masculinities

Introduction

The lists of strategies appearing in books and articles advising teachers on how to counter boys' underachievement bring back memories for me of the early 1980s. At the time I was a member of a group of teachers who met to discuss ways of tackling gender inequality in our schools. At that time the focus was on girls and we found various ideas in the literature aimed at redressing their gendered attitudes and our sexist bias. For instance, we could change our reading schemes to ones which reflected more diverse roles for women than the conventional 'mother at home' and alter our classroom management strategies to avoid stereotyping (Chetwynd and Hartnett 1978; Marland 1983). Deciding which of these approaches we would use in our schools was a matter of talking to colleagues and deciding which ones we, as a staff, 'felt' the most comfortable with. It was not until I began a degree course where issues about gender and schooling were raised that I realized the strategies we had discussed were based on feminist theories. However naive this now seems it remains that, as teachers, we simply did not know that the strategies we were adopting in schools were informed, not by familiarity with and reflection on the research into gender discrimination, but by our personal inclinations! This is not to suggest that the approaches we adopted in our schools were 'wrong' but they had been chosen by chance rather than through a knowledge and understanding of the research into the educational experiences of girls.

It also came as a surprise that feminists had *different* perceptions of what were the causes of female subordination and so the solutions they offered to address gender inequalities in the classroom also varied. At the same time, much of the early literature on girls' schooling did not make it easy to 'spot the feminist perspective' as writers rarely stated openly what perspective they were writing from.[1] Looking at much of the literature on boys' underachievement it would seem to be a case of history repeating itself as few authors say what theories of masculinity underpin the approaches they recommend. One possible explanation for this is that the writers of these publications are not aware of the differing perspectives on masculinity and are offering 'common sense' solutions.

The intention in this chapter is to outline briefly the main theories on masculinity and then to explore how these perspectives inform the strategies for tackling the 'problems of boys and schooling'. As discussed earlier there are divisions between what I refer to as the boys' underachievement literature and that on masculinities and schooling. The aim here is to provide readers with the information that will help them to identify the ideological frameworks upon which the various strategies have been devised and thus to consider the implications of adopting particular programmes. The final section of the chapter looks at a recent evaluation of the approaches schools are adopting to tackle boys' underachievement.

Current theories of masculinity

There are a range of ways of conceptualizing masculinity. Different theorists use different terms but their understandings indicate broadly similar categories. Messner (1997) identifies eight approaches (men's liberation, men's rights, radical feminist men, socialist feminist men, men of colour, gay activists, the mythopoetic men's movement, and the Promise Keepers); Lingard and Douglas (1999) write of four groupings they refer to as men's rights, pro-feminism, masculinity therapy, and conservative, acknowledging that their categorizations omit the perspectives of gay men (which are encapsulated in the other categories). For the purposes of the discussion here Clatterbaugh's (1990) six perspectives on masculinity will be used for the reasons that they cover all the dimensions identified by other masculinity theorists and that, as one of the earliest writers on the subject, his categorization is one of the most widely referred to. These perspectives are summarized below:

- '*conservative*': the position of men is seen as maintaining their traditional roles as protector and provider of the family.
- '*men's rights*': masculinity is seen as restrictive in that it disables men. Men are presented as victims of: violence, shorter life-spans, health problems, divorce and custody laws.
- '*spiritual*': encompasses the 'mythopoetic movement' where it is argued that men should be in touch with their inner selves to recognize and maximize their male energy (Bly 1990); and the 'Promise Keepers', a religious group drawing on notions of 'Muscular Christianity', a term

used to describe those organizations which emerged around the turn of the century whose aim was to 'revirilize the image of Jesus and thus remasculinize the Church' (Kimmel 1996: 177).

- '*pro-feminist*': where masculinity is seen as socially and culturally constructed. Conventional notions of masculinity as aggressive, tough, competitive and so on need to be challenged and overcome.
- '*socialist*': is a blending of some of the ideas of radical and Marxist feminist. Here the locus of male alienation rests in the relations of production and class division.
- '*group-specific*': this covers the writing of minority masculine groupings such as gay and black activists who have their own particular agendas such as homophobia and racism.

While Clatterbaugh's description of six main approaches to theorizing masculinity is useful, it has to be said that not all writings can be neatly slotted into one or other of the categories. Also, Clatterbaugh, nor indeed any of the main theorists on masculinity, engage with *how* the concept of masculinity or masculinities is understood by the various perspectives. Neither is it clear whether or not there are any shared understandings within or across the different positions. Before looking at the main categories in any depth, a more considered exploration of *how* the concept of masculinity or masculinities is used in the literature is necessary in order to understand the kinds of 'solutions' to the 'problem of maleness' offered by each of the perspectives.

The term 'masculinity' or 'masculinities' (to convey the multiplicity of male 'ways of being') is currently used by a number of disciplines which have a relevance to education, such as psychology, social psychology, sociology, anthropology and history. That many disciplines find 'masculinity' (or 'masculinities') a useful concept to describe constructions and manifestations of 'maleness' is evident but there are a number of problems with the way in which it is used. Hearn (1996: 203) points out that these problems include:

- the wide variety of the uses of the concept
- the imprecision of its use in many cases
- its use as a shorthand for a very wide range of social phenomena, and in particular those that are connected with men and males but which appear to be located in the individual
- the use of the concept as a primary and underlying cause of other social effects.

As Wetherell and Griffin (1995) have shown in their research into how men are exploring masculinity, the key areas of dispute between psychologists, social psychologists, sociologists and others lies around their theorization of male power. Their interviews with four male psychologists revealed two differing conceptions of masculinity; one definition located masculinity as an 'essence' linked to experiences of living in a male body, and the other definition centred around notions of gender socialization and humanistic psychology. Both definitions draw heavily on sex-role theory. In contrast, male sociologists were convinced that sex-role socialization theories were inadequate and they were more persuaded by theories which expressed complex relationships between the psychic and the social, such as those located within post-structuralism. Male psychologists' use of the term 'masculinity' placed the emphasis on the individual, and hence the personal, while male sociologists were more likely to politicize their discussions of masculinity and made greater use of the term 'masculinities'. This research by Wetherell and Griffin indicates that although the conceptions of 'masculinity' or 'masculinities' convey different meanings and understandings within and across traditional disciplines, they are rarely explicitly defined.

The important point here is that education, as itself a pluralist discipline, makes use of the understandings of various concepts as defined by the 'traditional' disciplines, such as psychology and sociology, to provide a basis for its own understandings, policies and practices. Thus, the differing perspectives on 'masculinity' can be found in studies of gender relations in schooling and have subsequently generated different approaches to ways of addressing such issues as sexual harassment and boys' underachievement. These different approaches can be seen by considering sex role socialization theories and the more complex theories which deconstruct gender identity formation.

Sex-role socialization and identity

The idea that girls and boys are socialized into their gender roles occupied a central place in educational literature during the late 1970s and into the 1980s (Lobban 1978; Delamont 1980; Jacklin 1983). Indeed, one of the first effects of feminism was to increase the volume of sex-role and sex-difference research (Connell 1987). It was notions of sex-roles and socialization theories that came to underpin liberal feminist programmes whereby emphasis was placed on the rights of the individual. In a nutshell, 'equal rights' for liberal feminists meant placing reliance on legal reforms to ensure equal numbers of male and female pilots, plumbers and politicians. Out of the three main feminist agendas on educational equality (liberal, radical and socialist) it has been liberal feminism which has proved to be the most influential in terms of influencing educational policy (Weiner 1994). The importance accorded to concepts of sex roles in liberal feminist writing was mirrored in the literature associated with the 'men's movement' of the same period (Farrell 1974; Pleck 1976).

Role theories argue that children learn 'appropriate' ways of relating to the world around them through observation and/or experiencing a system of rewards and sanctions which reinforce such behaviours. According to the developmental psychologist Erik Erikson (1965), identity emerges as a result of an individual's capacity to trust the world, and achieves fruition during adolescence, but is more than the sum of childhood identifications. Within developmental psychology, identity is characterized as the result of the solution of conflicts in life.

With regard to sex roles, this means females learning and internalizing such traits as caring, nurturing and selflessness, while males acquire and demonstrate characteristics such as aggression, independence and competitiveness (Oakley 1972; Seidler 1989). Social-learning theorists expounded the view that gender identity was learned by children modelling their behaviour on same-sex images in family, peer group and the media (Sharpe 1976). Alternatively, cognitive-development theorists maintained that a child's conceptualization of gender was dependent upon his or her stage of cognitive development. This view was one forcefully argued by Lawrence Kohlberg.

Kohlberg (1966) drew on Piaget's work on cognitive-development to develop a theory of children's sexual cognition. He accounted for young children's avoidance of opposite sex behaviours not in terms of reinforcement strategies but in relation to Piaget's use of object constancy. So, in the same way that children at a particular cognitive stage believed that a piece of plasticine changed weight when it changed shape, they would also believe that if a child dressed or played in a sex-inappropriate way its sex also changed (Emmerich *et al.* 1977). According to researchers subscribing to Kohlberg's explanation of sexual cognition, children's need to maintain a secure gender identity ensured they would strenuously resist cross-sex behaviours in themselves and in other children. For example, in Hough's (1985) study of children in a reception class (5–6 years of age) a girl's growing frustration of one boy's love of 'sparkly dresses' was recorded. The girl began by showing 'disapproval and anguish . . . on her face every time she observed him' which moved on to verbalized disquiet in her comments '"Take off that dress" . . . "Take it off!" "Will you please take off that dress, boys don't wear dresses" . . . "Take that dress off, you look silly".' Eventually the girl 'jumped out of her seat, charged over to the little boy . . . and quickly unzipped his dress, helped him to take it off and returned to her seat with a satisfied smile' (Hough 1985: 20). The accompanying commentary suggested that Hough was drawing on sex-role theory which propounds the idea that sexism is at its peak in children aged 5–6 years.

However, a more considered analysis of this research suggests that it is at this age when children acquire the notion that gender is 'fixed' rather than fluid (Short and Carrington 1989; Lloyd and Duveen 1992). Children are aware from an early age of their biological sex but the process of bodily inscription, the 'trying out' of the language, attitudes and behaviours associative of the two genders is a longer process. In Chapter 1 the work of Brownwyn Davies (1989) was referred to where she suggested that 4–5-year-old children employ 'category-maintenance work' to ensure they, and others, act out the 'correct' gender. From reading various feminist tales to groups of young children she noted how the alternative characters such as the assertive princess or the boy who attended a dance school were seen as 'deviant' and so deserving of punishment by others.

The benefit of adopting sex-role theories as a means of explaining gender behaviours and relationships was that these theories offered the potential for change. Rather than differences between males and females being based solely on biological assumptions about masculinity and femininity, sex-role theories suggested that both sexes were oppressed through 'agencies of socialization': families, schools, media, peer groups (Pleck 1981). The solution offered was to change the expectations of traditional gender roles.

When sex-role theories were at their peak in the late 1970s and 1980s, the focus in schools was on changing *girls*' perceptions and expectations of themselves. Although different feminisms held differing views as to how to tackle gender inequalities in schooling, it was conceptions of traditional sex-role socialization (as in liberal feminism) which held sway. The limitations of this perspective for exploring gender relations have been widely rehearsed (Carrigan *et al.* 1985; Segal 1990; Connell 1995). A summary of the main limitations of this work in British schools has been provided by Arnot (1991) who argued that:

> Sex-role socialization, which held together a multitude of projects as diverse as changing school texts, and establishing gender fair teaching styles, non-traditional role models, unbiased careers' advice and girl-friendly schools was seen to have a lot to answer for . . . The simplicity of its portrayal of the processes of learning and of gender identity formation, its assumptions about the nature of stereotyping, its somewhat negative view of girls as victims had all contributed to the creation of particular school based strategies. These strategies although designed to widen girls' and boys' horizons, and give them more opportunities in life were somewhat idealistic in intention and naive in approach.
>
> (Arnot 1991: 453)

While feminists argued that remediation strategies based on sex-role socialization theories were inadequate it was not an argument heard by all masculinity theorists. Through the late 1970s and 1980s feminism branched into differing perspectives but held together collectively in terms of the political position regarding gender inequalities. However, as Kerfoot and Whitehead (1998: 3) state, 'As more men writers came to study the sociology of masculinity, increasing numbers distanced themselves from a feminist position to one which emphasized "role strain", the emasculation of men by feminism, and a "crisis of masculinity"'. This failure to recognize or interrogate the issue of gendered power continues to occupy an influential position in some of the recommendations for addressing boys' underachievement (see section on 'Strategies for tackling boys' underachievement').

Boys and schooling I: men's rights

In seeking to identify the ways in which 'boys' have been theorized in the recent literature on boys and schooling it is useful to group Clatterbaugh's (1990) six classifications (the conservative, the pro-feminist, the men's rights, the spiritual, the socialist and the group-specific) into two strands. These two strands can loosely be categorized as 'personal' and 'political' constructions of masculinity (see Table 2.1).

The 'conservative', 'men's rights' and 'spiritual' perspectives emphasize the 'personal' aspects of masculinity. What these share is a belief that there is an *essential* nature of 'man' and that schooling restricts or oppresses this 'maleness'. I want to pause here briefly to consider gender evolutionary theory as this has clearly influenced writers working from a 'personal' perspective who have incorporated ideas of biological differences into their explanations of boys' difficulties in and disaffection from schooling (Biddulph 1998; Pollack 1998; Moir and Moir 1999).

Gender evolutionary theory: biological explanations of gender difference

The 'nature versus nurture' debate last received widespread attention in the 1970s when girls' stereotypical subject choices and lower A level and degree achievements led to restricted options in the labour market (Sharpe 1976; Byrne 1978). At this point feminists argued that these differences between the aspirations and achievements of girls compared to that of boys were as a result of societal expectations and gender socialization and not a result of innate characteristics (Delamont 1980; Oakley 1981). However, contrary to what some evolutionary proponents are now arguing, feminists did not dismiss biological considerations out of hand. Leaving aside the

Table 2.1 A categorization of masculinity theories

Personal	*Political*
Conservative	Pro-feminist
Men's rights	Socialist
Spiritual	Group-specific

somewhat emotive, anti-feminist stance taken by Moir and Moir (1999) they do reflect the opinion evident in some of the literature supportive of evolutionary perspectives (Tooley 1997; Fukuyama 1999) when they say:

> The hardline feminists simply refuse to entertain the idea that boys just might be plain better at higher mathematics than girls . . . the accumulated evidence from years of research demonstrates that the sex difference . . . is firmly in the biological domain.
>
> (Moir and Moir 1999: 119)

A skim through earlier texts on gender inequalities suggests this was not the case. Oakley (1981: 53, 61) noted the 'relative importance of biology' but also 'nurture affects nature'; Licht and Dweck (1983: 84) observed in their discussion of sex differences in mathematics achievement that they did not 'wish to discount the possible contribution of biological factors'. Measor and Sikes (1992: 7) stated in their book on *Gender and Schools* that 'What is clear is that there is a continuous process of interaction between social and biological factors'. While gender essentialist writers are reinvoking biological explanations of difference based on more recent studies of the brain they are also revisiting factors which have been subjected to much debate and found, at best, to be inconclusive. This revisiting has been made to explanations of sex differences as a result of male/female chromosomes and hormones (Biddulph 1998; Pollack 1998).

The intention here is not to rehearse the research into chromosomes and hormones, as Head (1999) provides an excellent critique of the 'biological effects' literature. Head indicates that some proponents of evolutionary theory are selective as to which studies they pick and how they choose to read the findings in order to support their premise that boys are 'naturally' different. The conclusion which can be drawn from his review of the literature is that there is nothing new to be said. There are no new discoveries which might lead to a reflection and re-evaluation of previous notions that hormones and chromosomes do not explain why, for example, more men obtain first and third class degrees or why more boys than girls experience problems with literacy. It appears that Halpern's (1992: 244) findings of her comparative analysis of research continue to be as relevant today when she says 'neither biological nor psychosocial hypotheses have emerged as a clear winner in their ability to account for all of the cognitive sex differences'.

Of greater interest and more potential than discussions about chromosomes and hormones is the research into the ways in which the brain functions. One argument stretching back to Victorian times, when differences in brain function was one justification given by the medical, clerical and scientific professions for not educating females (Spender 1987), is that girls and boys are born with in-built sex differences. Where this innate argument falls down is that while motor functions can be linked to certain areas of the brain, those which have to be acquired through *learning,* such as the ability to do a mathematical equation, do not (Caplan 1997). As Head (1999) says:

> Memories are not encoded on a digital basis on individual neurones, but are stored on a connected set of several thousand neurones. Long filaments, or dendrites make up these connections and they allow the storage of a particular memory to be spread out over the whole brain.
> (Head 1999: 14)

Recent research into brain functioning has shown that there is a tendency for males and females to use different parts of their brain for different higher order mental functions (with the acknowledgement that there is an overlap between the sexes: see Kimura 1996; Govier 1998). For example, males' brain function tends to be localized while females use more of their brains whether engaged in mathematical or verbal tasks (Haier and Benbow 1995). Paechter (1998: 46) has noted that 'current thinking in neuroscience suggests that neuronal connections are selectively strengthened as a result of experience (Sacks 1993)'. Thus, it can be argued that sex differences in brain function in such areas as mathematics and literacy are not innate but are a result of being exposed to different experiences. This offers clear potential for development through ensuring young babies and children are introduced to a range of experiences which will facilitate a broad range of learning styles (Moseley *et al.* 2000). Although these ideas are interesting and provide one avenue to pursue in addressing boys' literacy and girls' visio-spatial skills, biological features are but one element of the much broader picture regarding gender. At this juncture it is appropriate to return to the main discussion by looking at the educational programmes arising from 'personal' perspectives on masculinity.

Educating boys: programmes based on 'personal' perspectives

As was said at the beginning of the section, gender evolutionary theory has influenced those theories of masculinity which emphasize the 'personal' aspects of masculinity. The 'conservative', 'men's rights' and 'spiritual' positions share a belief in the *essential* nature of 'man' and see schooling as restricting or oppressing this 'maleness'. However, there are differences between what they see schools should do to bring an end to their repressive practices. Those writing from a conservative position will usually begin by saying that schools have done a good job in tackling girls' underperformance in those areas of the curriculum where they were failing but the changes introduced to bring this about has proved detrimental to boys. They advocate that schools should now pursue programmes which focus exclusively on boys (Redwood 1994). For example, 'conservative' supporters argue that schools need to return to traditional forms of educational organization and teaching such as streaming, stricter discipline and reintroduce 'more competitive (virile) tests' (Moir and Moir 1999: 152).

Those writing from a 'men's rights' perspective agree with the view that schools have done a good job for girls and that boys have got left behind. However, their position is that the problems that boys experience in school should not be addressed in isolation but within a gender equity framework (Lingard and Douglas 1999). Here it is seen that not only should programmes be developed which build upon the knowledge gained over the years from working with girls but also they should promote equal relationships between boys and girls. This suggests an awareness and recognition of feminist insights into schooling but an interesting contrast emerges in how schooling is perceived.

While feminists start from the premise that schools are masculinizing agencies, the men's rights perspective is that schools are failing boys because they are both feminized and feminizing; that is, schools are staffed predominantly by women and employ teaching styles, classroom management practices and promote approaches to learning which are favoured by girls (Noble 1998; Mason 1999). The beliefs which underpin the strategies to tackle the 'problems of boys' are that boys need to become comfortable and secure with themselves as males but should be introduced to a broader range of acceptable masculinities. Basically, they argue that restrictive versions of masculinity push boys towards aggressive, competitive behaviours in interpersonal relationships, while simultaneously promoting *laissez-faire* approaches to school and academic work. It is the men's rights perspective which is most evident in the policies towards tackling boys' underachievement adopted by schools and include such strategies as mixed-sex seating arrangements, active learning, mentoring and male role models, target setting, praise and reward systems (see Table 2.2 on p. 56).

While positive gains may well be made by adopting these approaches it remains that scant attention is given to broader social inequalities *between* boys due to social class, ethnicity and/or sexuality. Those gender reform programmes that do take account of the concept of *difference* emerge from those theories categorized in Table 2.1 as 'political'. These theories are influenced by post-structuralism and focus on deconstructing the differences between males.

Boys and schooling 2: 'different' masculinities

The 'men's rights' approach to theorizing and addressing masculinities in educational settings contrasts with those positions to be found in the 'political' category (those perspectives Clatterbaugh (1990) refers to as 'pro-feminist', 'socialist' and 'group specific'). These 'political' positions can be found in the educational literature on masculinities which recognizes the imbalances in power between males and females, and males and males (Mac an Ghaill 1994; Connolly 1998). Here the focus is on broader social issues through the influence of social class, age, ethnicity and sexuality as well as the connections between them. One of the most influential discussions on the construction of masculinities has been offered by Connell (1987, 1995, 1997). It is my intention here to go into Connell's theorizing of masculinities in some depth as it is integral to the case studies of the two primary schools discussed in Chapters 4 to 7.

Hegemonic masculinity

> When Carrigan, Connell and Lee introduced the concept of hegemonic masculinity in 1985, they were anxious to distance themselves, and ultimately the sociology of masculinity, from the concept of 'male sex role', men's studies approach, and neutrality in respect of a critique of men's power.
>
> (Kerfoot and Whitehead 1998: 3)

Connell's (1995) starting point is that gender is a way in which social practice is ordered. His argument that 'Gender is social practice that constantly refers to bodies and what bodies do, it is not social practice reduced to the body' (1995: 71) seems initially to agree with sex-role theory. However, in sex-role theory socialization is transmitted from a culture to its inhabitants, and tends to be something of a one-way process. But, for Connell, social practice interacts with, and is responsive to, particular situations as well as being generated within definite structures of social relations. He argues that gender relations are one of the major organizing structures of all societies. The social practice which relates to this structure occurs as a result of people grappling with their historical situations as a group and is not the result of individual actions. Connell refers to this as *gender projects* which he defines as:

> processes of configuring practice through time, which transform their starting-points in gender structures . . . We find the gender configuring of practice however we slice the social world, whatever unit of analysis we choose. The most familiar is the individual life course, the basis of the common-sense notions of masculinity and femininity.
>
> (Connell 1995: 72)

The 'configuration of practice' here is based on psychoanalytic concerns with 'personality' and 'character' and, as Connell goes on to say, any theoretical approach to gender which focuses on one area exaggerates the coherence of practice that can be achieved at any one site. Rather, as Butler (1990), among others, has suggested, gender is an internally complex structure, where a number of different logics are superimposed. Thus Connell argues for a three-fold model of the structure of gender relations, which distinguish relations of *power*, *production* and *cathexis* (emotional attachment).

In drawing on all three areas Kenway (1997: 59) argues that 'masculine identities are not static but historically and spatially situated and evolving. They arise through an individual's interaction with both the dynamisms and contradictions within and between immediate situations and broader social structures'. This model provides a means of considering power relations that exist between men and men as well as between males and females. While Connell (1995) acknowledges that there are many modes of masculinity, it is possible to identify certain configurations of masculinity on the basis of general social, cultural and institutional patterns of power and meaning, and to discern how they are constructed in relation to each other. These masculinities are defined as *hegemonic*, *complicitous*, *subordinate* and *marginal*.

Hegemonic masculinity is a concept which draws on Gramsci's notion of hegemony. It is used to describe the mode of masculinity which at any one point is 'culturally exalted' (Connell 1995); that is, it refers to those dominant and dominating modes of masculinity which claim the highest status and exercise the greatest influence and authority. Hegemonic masculinity is a position which is achieved as a result of collective cultural and institutional practices, and asserts its authority through these practices particularly through the media and the state (Kenway 1997). At the same time, it is not 'fixed', is in a constant state of flux and constantly needs to be achieved by dominating, not obliterating, alternative patterns and groups. Of particular significance then is that hegemonic masculinity is constructed in relation to women and subordinated masculinities and is heterosexual. As such, hegemonic masculinity structures dominant and subordinate relations across and between the sexes, as well as legitimizing patriarchy. Kenway and Fitzclarence (1997) have suggested that certain characteristics can be associated with hegemonic masculinity:

> At this stage of Western history, hegemonic masculinity mobilizes around physical strength, adventurousness, emotional neutrality, certainty, control, assertiveness, self-reliance, individuality, competitiveness, instrumental skills, public knowledge, discipline, reason, objectivity and rationality.
>
> (Kenway and Fitzclarence 1997: 121)

What is important here is the phrase 'mobilizes around' as this indicates there is no *one* form of hegemonic masculinity although all forms may draw upon, exaggerate, modify and distort these aspects.

Hegemonic masculinity defines what it means to be a 'real' man or boy, and other forms of masculinity are seen in relation to this form. It is important to note here that hegemonic masculinity is not something embodied within individual male personalities. Following Connell's (1987) argument, the fantasy figures suggested by the action film characters of Sylvester Stallone and Arnold Schwarzenegger bear no relation to the personalities of the actors. Rather, hegemonic masculinity is the public face of male power. As such, the concept also provides insights into the relationship between masculinities and institutional life: 'It is not too strong to say that *masculinity is an aspect of institutions*, and is produced in institutional life, *as much as it is an aspect of personality* or produced in interpersonal transactions' (Connell 1997: 608, emphases in the original). Of course, not all men or boys attempt to engage with, or even wish to aspire to, the rigorous standards demanded by hegemonic masculinity. Nevertheless, all men benefit from the *patriarchal dividend* (Connell 1995: 79) which is the advantage men gain from the overall subordination of women without actually being at the forefront of the struggles involved with hegemonic masculinity. Connell refers to that cluster of masculinities whereby men reap the benefits of hegemonic masculinity without actively seeking or supporting it as *complicitous* masculinities.

Standing in direct contrast to hegemonic masculinity is *subordinate* masculinity. In this category are those masculinities which are oppressed and repressed by hegemonic masculinity. Such masculinities stand outside of the circle of the legitimate forms of maleness represented in hegemonic masculinity. For example, gay masculinity is a form of subordinate masculinity. Also, these forms of masculinity are likely to attract violence from men attached to other, more aggressively dominant forms of masculinity.

Hegemonic, complicitous and subordinate masculinities, as defined here, are concerned with, and related specifically to, the internal mechanisms by which gender is ordered. However, the interrelationship of gender with other major social structures such as social class and 'race' creates further complex associations between masculinities. To explain masculinities at the intersection of gender, 'race' and social class, Connell (1995) uses the concept of *marginalized* masculinities. He refers here to the relations between dominant and subordinated classes or ethnic groups. Marginalized masculinities are contingent upon the sanctioning of the hegemonic masculinity of the dominant group. Connell (1995) offers the following as examples of this dynamic process in operation:

> in the United States, particular black athletes may be exemplars for hegemonic masculinity. But the fame and wealth of individual stars has no trickle-down effect; it does not yield social authority to black men generally . . . The relation of marginalization and authorization may also exist between subordinated masculinities. A striking example is the arrest and conviction of Oscar Wilde . . . Wilde was trapped because of his connections with homosexual working-class youths, a practice unchallenged until his legal battle with a wealthy aristocrat, the Marquess of Queensberry, made him vulnerable.
>
> (Connell 1995: 81)

Connell's theorizing has then identified two forms of relationship through which specific modes of masculinity can be analysed. There is the relationship between hegemony, domination/subordination and complicity and also the relationship between marginalization/authorization.

The more complex analysis of masculinities offered by Connell and others (see Brittan 1989; Hearn and Collinson 1990; Morgan 1992) began to appear in male researchers' studies of masculinities and schooling in the United Kingdom around the mid-1990s (Mac an Ghaill 1994; Connolly 1995; A. Parker 1996). Feminist analyses of masculinities and schooling has an earlier history but the concern for them was with the impact of male power on girls and women teachers rather than theorizing masculinities (Arnot 1984; Askew and Ross 1988; Heward 1988). A comment is needed here to acknowledge that for feminists, there was and, for some, continues to be a concern about exploring men, boys and masculinities (Hanmer 1990; Skelton 1998b; Delamont 1999; see also Chapter 3). Having said that, it remains that those feminists researching masculinities and primary schooling draw to greater and lesser extents on the concept of hegemonic masculinity (Thorne 1993; Jordan 1995; Renold 1997). However, the general acceptance of the usefulness of hegemonic masculinity has begun to be criticized in that it is argued to restrict further conceptual explorations of masculinities (Francis 1998, 2000).

While Connell did point out that women and gay men can also perform hegemonic masculinity this idea has generally been overlooked (Cheng 1996; Heywood 1997). Also, post-structuralists (see Chapter 1) find difficulty with concepts that regard 'power' as a commodity which can be held by any person or group or that assumes it is possible to talk about 'men' and 'women' as categories. It is not the intention here to engage with the problems associated with the use of hegemonic masculinity as these are discussed in detail elsewhere (Kerfoot and Whitehead 1998; MacInnes 1998). However, I would suggest that although there are problems with hegemonic masculinity, not least that it does not engage with notions of individuals occupying multiple, contradictory, subject positions, much of the cause for criticism lies with the way in which it is often loosely used. It is not uncommon to find categories of masculinities being 'read off' the data and accompanied by references to hegemonic masculinity in such a way that suggests the concept has been embraced as an explanation in itself without attempting to engage with it as *part* of the *writer's* overarching theorizing of gender relations (see Chapter 3 for discussion on the subjectivity of the researcher). Setting to one side concerns about the theoretical efficacy of hegemonic masculinity, it has enabled interesting insights into boys, masculinities and schooling (Mac an Ghaill 1994; Connolly 1998). It was also acknowledged in Chapter 1 that this research has not offered strategies which schools and teachers might use to underpin their work with boys. However, the idea that boys adopt multiple positions and schools too need to recognize the various forms of masculinities it constructs has informed some programmes (Salisbury and Jackson 1996; Pickering 1997). The next section will consider the different approaches to boys' inequalities and gender relational inequalities offered by the 'men's rights' and boys' underachievement perspectives and those suggested by 'pro-feminist' programmes.

Comparing programmes for boys/masculinities and schooling

Although there are strengths about boys' underachievement and men's rights approaches to tackling boys' attitudes and behaviours towards schooling in that both teachers and boys are encouraged to explore relationships between masculinity, subject and career choice, achievement and discipline (Kenway 1997), there are also some problems. First, there is an assumption that *all* boys act, think and behave in the same way and, consequently, such strategies as adopting mixed seating arrangements and increasing the number of male primary teachers are based largely on sex-role theory. The problems associated with sex-role theory have already been discussed and what the development of these strategies does suggest is that little heed has been given to what we have learned about gender equity from feminist work with girls and schooling. Second, these approaches tend to ignore broader social structures and issues related to power, particularly structural inequalities between males and females. They are based on a 'competing victims' discourse whereby boys are seen to be as, if not more, oppressed than girls have been. Some of the proffered solutions to addressing the problems that boys may be experiencing with schooling can not only marginalize girls but also, in some cases, rehearse the gendered pedagogies and practices found to be operating in schools in the 1970s. For example, Clarricoates (1978) wrote an article entitled 'Dinosaurs in the classroom' in which she observed that primary teachers based curriculum activities around topics they believed would motivate boys and, as it was boys who were seen as requiring more discipline, would also facilitate classroom control. That girls were being marginalized and possibly deterred from learning was not apparently something the teachers were aware of. A similar lack of attention to the potential detrimental effects on girls in adopting certain classroom practices occurs in the following recommendations:

> ***Boy-friendly subject matter*** . . . We all know that there is not just one kind of boy and thus there cannot be a simple answer to what topics and what materials will stimulate every boy. But a school can be creative in developing eclectic classroom materials and covering a broad range of topics that will spark the interest of many boys . . . In a subject like history, making the class stimulating for boys might mean

> reading and telling stories not only about men but also from the perspective of men.
>
> (Pollack 1998: 266)

Similarly, an observation made by the working party involved with the National Numeracy Project (*TES* 12 December 1997) (which did not appear in the final report) suggested that girls in the primary school were underperforming during the whole class teaching of the numeracy hour. Again this raises the question of the extent to which different pedagogical approaches to particular curriculum subjects are influenced by gender and can impede children's progress (Gipps and Murphy 1994). In a research report by Bleach *et al.* (1996), recommendations were made that subvert the findings of feminists which show that girls' frequently experience difficulties in mixed sex groups where boys are monopolizing equipment, the teacher's time and demonstrating intimidatory or harassing behaviours (Clarricoates 1980; C. Jones 1985; Frith and Mahony 1994). For example:

> Various approaches are being explored . . . for encouraging . . . boys to maintain a positive attitude . . . These include . . . giving boys a high profile in showing visitors around or performing in public, pairing boys with girls in group work to expose them to the 'feminine' skills of language and reflection.
>
> (Bleach *et al.* 1996: 25)

In summary: those strategies for addressing boys' underachievement which are concerned with developing curriculum materials aimed at boys' interests, focusing teaching approaches on their perceived learning needs, increasing male role models in school and so on are informed mainly by the men's rights position but also by other theoretical perspectives concerned with the 'personal' aspects of masculinity. We have seen that the language associated with these approaches is of *boys* (as a homogenous group) and of *masculinity* in the singular and, as such, the basis of the actions suggested for use in schools are based on sex-role theories. The starting point for gender reform programmes of this nature is what Lingard and Douglas (1999: 133) call a 'male repair agenda'. This approach can be found in programmes offered by Geoff Hannan (1999) in the UK and *Boys in Schools* (Browne and Fletcher 1995) in Australia. Underpinning these programmes is the idea that boys are the victims of feminism; that is, the gender gap in academic performance is seen as a consequence of girls' success and the feminization of schooling (Epstein *et al.* 1998). Therefore such programmes as recommended in *Boys in Schools* are aimed at enabling boys to affirm themselves as males and to define themselves positively in relation to education and schooling.

Research studies, findings and suggested strategies which locate themselves within the politicized approach to masculinities seek to move away from universalist and essentialist forms of theorizing gender relations yet retain ways of analysing power dynamics and processes. This work also aims to maintain the importance of psychic subjectivities in any analysis of masculinities while avoiding the 'identity therapy' redolent of the 'men's rights' approach. Such a project has required recourse to both materialist and deconstructionist (post-structuralist) approaches. Materialist approaches are those such as liberal and radical feminism, and Marxism (see Chapter 1).

The reason for holding on to different ways of looking at the problems of gender relations in schooling is that they both have problems but used together many of these difficulties are reduced. On the positive side a materialist analysis allows for explorations of the interrelationships between agency (people) and structure (institutions) and enables an understanding of dominant and subordinate power relations in terms of social groups. It logically proceeds to argue that the nature of social beings can be 'read off' from institutional infrastructures. The implication here is that identity is simply a reflection of the dynamics of an institution; so boys' and girls' identities at school are shaped solely as a result of the gendered practices of the school. This interpretation of how social identities are constructed leads to the conclusion that, in schools, all that is required is for pupils and teachers to recognize the 'false consciousness' shaping their behaviours and that will be sufficient to bring about change in gendered identities. It was this kind of approach which informed much of the anti-sexist and equal opportunities work of the 1980s. And, as was also observed earlier, such approaches cannot account for the multiple, complex and constantly shifting nature of power relations. Conversely, deconstructionist (post-structuralist) approaches are useful in that they enable a researcher to explore the simultaneous relationships between such variables as age, sexuality, ethnicity, culture, social class and so on. Also, psychoanalysis, which cannot strictly be located within deconstructionism, lends itself to postmodernist/post-structuralist approaches in that it 'illustrates the limits of over-rationalist accounts of sexual politics that fail to acknowledge that what we *feel* is as important as what we *know* in relation to the maintenance of dominant gender . . . practices' (Haywood and Mac an Ghaill 1995: 233, emphases in the original). But, as we have seen, these approaches too have their drawbacks notably in their inability to politicize power struggles (see also Chapter 3).

So how have the insights provided by researchers who have found it useful to retain the tensions between materialist, deconstructivist and psychoanalytic theories to analyse masculinities in educational settings informed gender relational programmes (Connell 1989; Mac an Ghaill 1994; Redman 1996; Hallden 1997)? Many of the current books and packs concerned with boys' attitudes and behaviours suggest activities which directly address their emotional and psychological experiences. At first sight this material implies that it should be located in 'personal' (men's rights/conservative/

Table 2.2 Key aspects of two perspectives on boys and schooling

The issue	*Boys' underachievement*	*Masculinities*
Theoretical perspectives:	Men's rights Conservative	Pro-feminist
Schools seen as:	Feminized and feminizing	Sites where a range of masculinities operate
Strategies based on:	'Male repair' agenda	Gender as relational
Strategies:	• male role models • mixed-sex seating • mentoring • target setting • praise and reward	• presenting alternative images of masculinity • redefining male characteristics i.e. strong, admired • challenging gender oppositional categories

spiritual) perspectives but when the debates and arguments around these emotional and psychological experiences are related to broader issues connected with social power then other theoretical positions are being drawn on (Salisbury and Jackson 1996; Mclean 1997; see also work by feminists on masculinity, Arizpe and Arnot 1997; Kenway and Fitzclarence 1997). But yet again the majority of these interventionist strategies are aimed at working with boys in the secondary sector. Those that are aimed at primary age pupils and younger are only just appearing (MacNaughton 2000; Marsh 2000). Table 2.2 clarifies what has been said so far.

Using the strategies in schools and classrooms

The National Foundation for Educational Research (NFER) has published the findings of a two phase study of gender differences in achievement (Sukhnandan 1999; Sukhnandan *et al.* 2000). The findings were similar to those in research undertaken by Arnot *et al.* (1998b). The first phase involved a review of LEAs to ascertain the extent to which schools within their jurisdiction had adopted strategies for addressing gender provision. Just over half (55 per cent) of LEAs responded to the questionnaire seeking information on their provision. Of these 73 per cent said they were aware that schools had policies already in operation and 64 per cent confirmed that strategies were in place at LEA level (furthermore 74 per cent responded that they were aware of future plans to implement initiatives at LEA and/or school level). The strategies adopted by the primary schools in the survey are reproduced in Table 2.3 (note: new teaching methods refer to the National Literacy Strategy and National Numeracy Strategy).

Table 2.3 Strategies targeting primary school pupils

Types of strategy	*Number*	*Percentage*
New teaching methods	17	28
Parental involvement	10	17
Role modelling/mentoring	5	8
Staff training	4	7
Target setting	4	7
Single-sex classes/grouping	4	7
New forms of class organization	4	7
Other strategies	4	7
Learning support	4	7
Policy development	2	3
Mixed gender pairing	2	3

In the second phase nineteen case study schools (of which only three were primary) were involved in research to ascertain how effective teachers and pupils found the strategies. The research did not investigate all those listed in Table 2.3 but focused on single-sex classes/groups, mentoring and additional literacy support. Perhaps not surprisingly teachers and pupils found advantages and disadvantages with all three strategies. A few of their observations follow.

In terms of single-sex groups, the staff found that this form of organization gave them the opportunity to use more 'boy friendly' teaching strategies (such as competition and specific target setting). They also felt that the absence of girls meant the boys were better behaved and increased the boys' confidence to contribute to lessons. However, they also said that they preferred to teach mixed-sex groups and pupils missed out on the opportunity to hear the perspective of the opposite sex.

With regard to mentoring, the advantages included improving pupils' skills by providing them with additional support and tuition and it helped to challenge some boys' stereotypical anti-school, anti-learning attitudes. Conversely, it was an additional burden for teachers who had to find time to train as mentors and to conduct the mentoring sessions. The pupils themselves felt that the sessions were not long enough or frequent enough. Finally, the additional literacy support was welcomed as it took the pressure off class teachers to ensure that boys received the kind of support and attention needed to improve their levels of literacy. Also the boys received one-to-one support for concentrated periods of time which is not something teachers have the time to give. The improvements in their levels of literacy also had a knock-on effect for their achievement in other areas and in their behaviour. The disadvantages included teachers having to make sure the boys did not miss out on anything when they were withdrawn from class for sessions and some of the boys felt stigmatized by being taken out for separate teaching.

In evaluating the effectiveness of these strategies from the perspective of teachers and pupils a number of further issues were raised.

Will teachers start exploiting the differing strengths of males and females rather than concentrating on their weaknesses in order to raise achievement?

The authors express concern that single-sex grouping might reinforce traditional gender stereotypes. They give the example of schools using male-orientated texts to improve boys' attitudes to reading but which may reinforce their preference for particular reading and writing styles. Also, the use of male role models is intended to challenge boys' notions of 'feminized' subjects (such as English) but this might well have a backlash effect by undermining the authority of female teachers.

Will teachers concentrate on boys to the exclusion of girls?

The focus on boys by LEAs and schools was particularly evident in the survey evidence. Interventionist strategies which targeted boys and girls decreased from 71 per cent prior to 1995 to 52 per cent in the academic year 1997–98. In comparison, 14 per cent of strategies were aimed at boys only before 1995 but this increased to 41 per cent during 1997–98. The concern here is that teachers will put all their efforts into making the learning environment 'boy-friendly' while girls are not provided with the requisite opportunities to fulfil their potential. Indeed this was a point made earlier when looking at how certain recommendations based on the 'men's rights' perspective reinvoked the kinds of marginalizing practices recognized in the feminist literature of the 1970s, such as choosing topics which would appeal to boys' interests.

What if 'boys'' and 'girls'' learning styles are used to inform teaching approaches and to modify the learning environment?

Differences in learning styles were discussed in Chapter 1 where it was said that, although there is some evidence of some differences in learning styles, not *all* boys and *all* girls favour the same approaches. Learning styles are influenced not simply by gender but by a range of factors including personal experience and out-of-school activities as well as the pupils' own individual characteristics (Gipps and Murphy 1994).

This evaluation and the research by Arnot *et al.* (1998b) has helped to pinpoint an important issue: that we are still at the early stages of working out exactly what the questions are that LEAs, schools and teachers need to be asking in terms of gender (see Chapter 8 for discussion). It also helps to reinforce a point which was raised earlier. Although reference is made to recent research into gender, achievement and boys' underachievement throughout the two publications emerging from the NFER study, at no time is any consideration given to the competing theories of masculinity which underpin the particular strategies the research team are evaluating. Recommending strategies to a school, say mixed seating, without saying what the outcome might be (boys dominating girls) is akin to giving someone a kit of flatpack furniture with the instructions but no picture of what the assembled item will look like!

Summary

It would probably be satisfying for the reader if this chapter were to conclude by answering the question 'What are schools supposed to do given these different approaches to boys and schooling?' But I am reluctant to pre-empt this prior to setting out the findings of my own research into masculinities and primary schooling (see concluding chapter for discussion of the question). What has been shown here is that the various ways of conceptualizing masculinity or masculinities have, to greater and lesser extents, informed current strategies for dealing with the 'problem of boys and schooling'. Inevitably these different approaches have generated tensions and not only between opposing camps but within them. For example, 'men's rights' (boys' underachievement) programmes advocate greater emphasis on competitive sports while pro-feminist programmes suggest more emphasis on collaborative teamwork and getting more girls involved in traditionally masculine sports such as football and rugby. More confusing messages come out of the boys' underachievement literature where there is a call on the one hand for more emphasis on 'boy only' groups (Bleach *et al.* 1996) with a simultaneous request for teachers to introduce seating plans which often means putting boys next to girls (Noble and Bradford 2000). Of course, as was said earlier, some of this confusion may simply be because those recommending the strategies are not themselves aware of the conceptual fields they are drawing on. However, implementing boys' underachievement, or indeed pro-feminist, programmes without a firm grasp of the theoretical underpinnings upon which they are based can be limiting, if not dangerous, for schools.

Note

1 The 'clues' as to the stance of the writer came from the language used and their analytical frameworks. The three main feminist perspectives commenting on

education in the 1970s and 1980s were liberal, radical and Marxist feminism. So, a focus on male power in terms of gender relations in the classroom suggested that a radical feminist perspective was being used (Mahony 1985); discussions on how schools reproduced social class and gender inequalities pointed to socialist feminism; and, a concern for equal rights through legislative reform indicated a liberal feminist stance.

chapter / **three**

Boys and primary schooling: a feminist perspective

Introduction

It is not unusual today to find a book or article on boys, masculinities and primary schooling where the research has been undertaken by a feminist (Jordan 1995; Renold 1997; Marsh 2000). At the same time, the majority of feminists exploring the primary sector maintain the practice of looking at gender and its relational implications for classroom practices (Thorne 1993; Epstein 1998; Francis 1998; MacNaughton 2000). There are several possible reasons for this. First, there is the question of what feminist research is supposed to be about. In the early days of second wave feminism it was argued that feminist research agendas should be 'on, by and for' women (Roberts 1981; Duelli Klein 1983). It quickly became apparent that this interpretation was too restrictive and hindered the work of certain groups of feminists. For example, those feminists working in law, criminology and social work found themselves in a position where it was necessary to explore masculinity in order to analyse male power and so would often begin their accounts by justifying their reasons for researching men (Cain 1986; Gelsthorpe 1990). The difficulties and complexities which arose in defining feminist research have been widely debated and what 'counts' as feminist research today is related more to epistemological issues than methods or technical concerns (Maynard and Purvis 1994; Ribbens and Edwards 1998). In addition, the growth of interest in 'men's studies' and the warnings by some feminists to maintain a watchful eye on what men are writing about masculinity (Canaan and Griffin 1990; Hanmer 1990) means that a feminist focus on masculinities and male practices is seen as necessary rather than questionable. This is not to say the situation goes unremarked as recent studies by feminists of masculinities show that authors of the research are at pains to locate their work clearly within feminism (Willott and Griffin 1996; Alloway and Gilbert 1997; Sumsion 1999).

A second possible explanation as to why many feminists have focused on gender relations rather than masculinity in primary schooling is simply that this topic provides a much more coherent and relative context. Masculinity and femininity are not mirror images of each other and so boys and girls need to be understood in relation to each other. As Gilbert and Gilbert (1998) argue:

> The most useful work on boys and gender studies the construction of gender from a relational perspective, where the practice of various forms of masculinity is seen to be constantly constructed along with but in distinction from femininities, and where the focus is on both boys and girls as participants in this process.
>
> (Gilbert and Gilbert 1998: 112)

A third explanation for why there are fewer feminist studies of masculinities and primary schooling of the kind found in the literature on adolescent boys is one I shall be exploring in more detail later on. It is enough to say here that the gender of the researcher inevitably influences the types of data one has access to and the theoretical and ideological perspective adopted also informs the kinds of questions that are being asked. For example, the notion of empowering the research group is something central to feminist research. Yet the extent to which a feminist might wish to 'give a voice' to young boys when the girls in the class are not given the same opportunities is a moot point!

The aim of this chapter is to provide a background to the study of the two primary schools. This includes some reflection on methodological issues such as the one just given of the kinds of questions a feminist might ask in researching boys, masculinities and schooling. The next section will provide some details of the two primary schools and, alongside this, offer insights into what was feasibly possible to find out about 'masculinities' in each of the settings. The chapter will then move on to outline theoretical concerns which influenced how the data were analysed.

The schools

Benwood Primary and Deneway Primary schools were situated on the perimeter of the north east city of Oldchester.[1] The schools were only eight miles apart but were completely different in terms of the catchment areas they served which, in these days of market forces, had immense implications for how they operated and what benefits they could access for their pupils. Benwood Primary School was located in Wickon, a long established but economically disadvantaged area of the city. Deneway Primary School was situated in a suburban and relatively newly developed area of Oldchester. While the gender balance was even across both schools, there were no ethnic minority pupils at Benwood Primary and they made up less than 1 per cent of the school population at Deneway Primary.

Benwood Primary School

The school was built in the 1930s and was typical of the schools of that period. The infant and junior sections were divided by a hall which also served as a dining room. There were two playgrounds, one for infants and one for juniors, and a small grassed area. An impression of the locality could be gained by standing in the playground. The majority of houses surrounding the perimeter of the school grounds were either burned out or boarded up. As will became apparent in the chapters focusing on events in Benwood Primary School the description of the local culture in many ways reflects the image of the north east of England as it is portrayed in the national media.

Following the inner-city disturbances of the early 1990s, the media have repeatedly returned to this part of Oldchester as representative of those areas which house the 'underclass' (Arnot *et al.* 1999). Earlier sociological studies of the inhabitants of the north east have also tended to focus on the economic deprivation of the area. Gofton and Gofton (1984: 280) refer to 'The Giro Cities of the North East . . . where the 1930s have never really gone away' and while Frank Coffield *et al.*'s (1986) study of young people growing up in the area acknowledges the 'vibrant culture' and 'rich networks of mutual support' which exist, these positive features can easily be forgotten when faced with the far more frequent references to economic deprivation. For example, Coffield *et al.* (1986: 216) say that 'The results of a long period of persistent decline were to be seen all around' and note the local economy's 'entrenched deprivation', the 'dirty and dangerous' working conditions resulting in 'a marked concentration of health problems' and that the number of pupils staying on at school is the 'lowest of all the English regions'. The local culture of Benwood Primary was of particular significance in the construction of masculinities in the school and, as such, will be returned to in later chapters.

Benwood Primary school had approximately 370 children on roll, and was made up of 13 classes and a nursery. There were 17 teaching staff in total and included three male teachers. One of them, Terry Blake, taught infant children, while the acting male deputy headteacher and the Year 6 (Y6) teacher, who held a B Allowance for information technology, worked with the oldest children in the school. Terry Blake also received an A Allowance for assessment. (A and B allowances are additional payments relating to specific responsibilities in the school, for example, coordinating the teaching of ICT.) Thus, in keeping with national statistics (DfEE

1999a), the few men teachers in the school occupied proportionally more senior management positions than their female colleagues. The numbers of children at Benwood Primary constantly fluctuated due to families moving around and between the vicinity and its two neighbouring areas.

The main data were collected in a Year 2 (Y2) class (aged 6–7 years). There were 24 children in the class, 10 girls and 14 boys, all of whom were white and, with one exception, had been born in the north east. The class teacher was Terry Blake, who was in his third year of teaching.

Deneway Primary School

It is fair to say that the prominence given by the national media and sociological studies to the working-class, and more recently 'underclass', population of the north east has presented a distorted image of the social class composition of the area. Oldchester city has always had a large middle-class population. While in the 1820s the shipyard workers lived to the west of Oldchester in such areas as Wickon (where Benwood Primary is situated), the shipbuilders, shipowners and other magnates of industry lived to the north of the city in the affluent areas of Greenvale and Parkside. During the economic boom period of the 1970s and 1980s Oldchester expanded and new housing estates were built. To the north west of the city a large estate of moderately priced detached and semi-detached housing was developed called Deneway. Given its prime position on a direct metro-line route to the local airport seven miles away, as well as to the city centre and the central railway station, the estate attracted many young, professional families. As a result, Deneway Primary School was opened in 1989.

The school catered for children from Deneway estate but also served two neighbouring areas, only one of which was predominantly working class. Deneway Primary School is Oldchester LEA's 'showpiece' school. According to the school handbook:

> The school is a new concept in its design . . . the teaching areas centre around a book garden which also serves as a library . . . there are four playgrounds plus a playing field surrounding the building . . . in the near future the already established wild garden area/stream will be adapted and developed.

There were approximately 350 children on the school roll when the research took place, but these numbers were due to rise when the nursery opened. There were 14 teaching staff, three of whom were male; these were the headteacher and the Y5 and Y6 teachers. The market forces introduced into schools by the ERA 1988 had been used to spectacular effect by the young, entrepreneurial headteacher, Tom Kenning. The school was extremely well resourced, with much of the money being raised by parents. It was, and continues to be, a central attraction for visitors to the area including members of the Royal Family, overseas foreign dignitaries, Members of Parliament, LEA officials, and staff from other schools inside and outside the authority. Unlike the area in which Benwood Primary was situated, this part of Oldchester has not attracted the attention of the media or been the subject of any sociological studies. Setting aside the fact that Deneway Primary can provide better resources than many schools, and has effectively risen to the challenge of market forces in education, it is the type of school that can be found in the middle-income suburbs of any city in England.

The majority of the data were collected in a Y5 (9–10 year olds) class which consisted of 27 children, 12 girls and 15 boys. The majority of the class were white, but three Japanese pupils and one Asian pupil attended the school for some time over the course of the research. The intensive observation period took place when the children were in Y5 and further observations and interviews were held in the following autumn term when the class was in Y6. In Y5, the class teacher, Philip Norris, had been a mature entrant who was in his first year of teaching (aged 28). In Y6, the teacher was Bill Naismith, a younger man in his second year of teaching, who took up his appointment towards the latter part of the observation period.

Doing the research: looking at masculinities

Schools are busy places, and it rapidly became apparent when negotiating access to classrooms that my teaching experience was part of the 'bargaining' process. Although the headteachers of both schools were prepared to allow me to undertake research without taking up a teaching role, both class teachers made it evident they expected me to undertake some teaching responsibilities. Participant observation, where the researcher takes on the role of part-time teacher, has been the favoured approach in several studies of primary schooling (Epstein 1993; Pollard 1985). However, I had not intended to take on the role of part-time teacher; rather the aim had been to be an 'adult helper'.

Initially I was disturbed about having to undertake the research wearing the label of 'teacher'. I had read the accounts of other researchers who had adopted the part-time teacher/researcher role and noted how this had had significant implications for their relationships with pupils and teachers. For example, my aim in electing to adopt a non-teaching participant observer role was so that I could locate myself outside teacher power/authority positionings in order to explore the ways in which boys constructed masculine subjective identities. Although I had some doubts as to whether I, as an adult female talking to young boys, would be able to develop the kind of relationship necessary to accessing this kind of information, being positioned as a teacher made such a possibility even more unlikely. As D.H. Hargreaves (1967) found in his study of a boys' secondary school, while participant observation helped in terms of his relationships with the teachers and his awareness of their concerns, it affected his rapport with some of the pupils. As a result he stopped teaching and 'from that point my relations with the boys improved to a remarkable extent' (Hargreaves 1967: 203). Hargreaves's experiences suggest that adopting the role of teacher in a research setting militates against establishing trusting relationships with pupils on the grounds that the expression of certain views, or the reporting of certain events, may get the child into trouble (Corrigan 1979).

On reflection the idea that I, as an adult woman, would be able to gain information about the ways in which young boys constructed and negotiated their masculine identities seems rather naive. After all relationships in the field are established on the basis of not who the researcher *pretends* to be but rather on the constant construction and negotiation of personal identity. Some researchers have adopted rather novel approaches to undertaking studies of primary school children such as King (1978) attempting to disguise himself (a 6 foot man) by sitting in the 'Wendy House' and ignoring the children if they spoke to him, and Holmes's (1995) interactions with kindergarten children where she attempted to transform their perceptions of her as an adult woman to 'one of them'. She began by asking the teacher to introduce her to the 4–5 year olds by her first name and to tell them she was a new pupil:

> I took my seat, where a box of crayons with my name on it was placed. . . . the children addressed me freely by my first name from the moment I was introduced to them. I found this strategy effective because it seemed to . . . strengthen my position as one of their classmates. My authority was further weakened by the teachers who treated me like another student as much as possible . . . One day I was playing with a toy that, unbeknownst to me, was not available for the children at playtime. Eric ran over from his playgroup to warn me, 'Robyn, you're not supposed to play with this. Hurry and put it back before the teacher sees you'.
>
> (Holmes 1995: 107)

The extent to which children can be duped into accepting an adult as a child is highly questionable and equally, it occurred to me that my age and sex would not allow me to be a 'buddy' to the boys in the hope of gaining insights into masculine subjectivities. Also, as Connolly (1997) has pointed out in relation to Holmes's attempt to be 'one of the kids', adopting the role did not allow her to overcome the way in which her subjectivity constructed, informed and theorized her relationships with the children. In the same way, had I been allowed to take on the role as adult helper it would not have prevented me drawing on the understandings I have as a trained teacher about teaching and learning or managing and organizing children in the classroom.

Here lay a fundamental issue which was to dictate how I, as an adult female, acting as a part-time teacher, could research masculinities in primary schools. The original intention had been to explore how boys constructed their masculine subjectivities. However, subjectivity is constructed across a range of sites and even if I could have accessed the kinds of information about how masculinity is constructed in relation to other masculinities (as men researchers have been able to: see Walker 1988; Connell 1989; Mac an Ghaill 1994; Connolly 1998) this would still not have provided information about boys' positionings within other discourses such as son, friend or child. Added to this was the realization that, as researchers, we are located and positioned in many different ways but, at the same time, we also locate and position ourselves although this is always defined by one's history, nature, age, gender, 'race', sexuality, social class and so on (Skeggs 1994; Stanley 1997). These issues were, to a large extent, resolved by acknowledging that whatever data I could collect about how masculinities were constructed by boys in primary schools could only ever be seen as one part of a much bigger project; that is only a *partial* picture of the lived realities of the boys could be gained. Thus, as a part-time teacher it was possible to collect information which would allow insights into the construction of hegemonic masculinities at the level of the *school* and, from there, to explore how boys and men teachers engaged with these forms, as well as to gain an appreciation of how girls negotiated and challenged dominant modes of masculinity.

Theorizing masculinities in Benwood and Deneway Primary Schools

The different ways of theorizing masculinities were discussed in detail in Chapter 2 so what follows is an indication of where the research into masculinities at Benwood and Deneway can be located. The first part of this section considers the selection of an appropriate theoretical framework and the second part discusses the model generated from that theoretical structure.

Theoretical framework

As the research was undertaken from a feminist perspective then the central issue was that of male power. Thus political concerns regarding *how* boys engage with male power in the forms it takes in primary schools in order to add to feminist understandings was a crucial factor. As Kelly *et al.* (1994) argue in their work on child abuse:

> While studying the construction of masculinity is of key importance, what needs to be explored is not so much how men 'experience' this,

> or explicating different 'masculinities', but . . . the connections between the construction and practice of masculinity and women's (and children's) oppression.
>
> (Kelly *et al.* 1994: 34)

The interconnectedness of power with constructions of maleness is a central point made in the literature on masculinities (Brittan 1989; Seidler 1997). Similarly, where the characteristic of *authority* is associated with hegemonic masculinity (Connell 1987; Kenway and Fitzclarence 1997) it is also associated with the role of teachers. As Pollard (1987: 177) has pointed out, 'the power of the teacher to threaten the children [is] because of their role and authority'. However, power is a notoriously difficult concept to define as there is little agreement about what it is or how it is constituted and some have argued that to search for a unifying definition is a fruitless endeavour (Lukes 1986; Deem 1994). Lukes has suggested that one way different theories of power can be explored is by categorizing the various kinds of questions that researchers have asked about power, such as:

- Who can adversely affect the interests of whom?
- Who can control whom?
- Who can get what – where not all can get what they want or need.
- Who can secure the achievement of collective goods?
- Where is power located or whom to hold responsible for the effects of power?
- Who gains by bringing about, or helping to bring about, the outcomes of power?

Here Lukes (1986) conceives of power as a scarce resource with the implications that some people (and institutions) will be able to access power, and that it is positively enabling for those who can access it but disabling for those over whom it is used.

This approach to defining power has been criticized by post-structuralists, notably Foucault, on the basis that:

- power is not a property of people
- power is not inextricably linked to the relations of production
- power can be facilitating as well as repressive.

Deem (1994) has pointed out that these apparently differing arguments are more compatible than at first appears, with the only real point of departure being the post-structuralist position that power is not a property of people. This can be illustrated by considering an incident that occurred at Benwood Primary School which points to the constantly shifting nature and fluidity of power as any one event unfolds.

On this occasion the teacher, Terry Blake, had the previous day punished one of the boys in his class for 'talking back' by making him stand outside the staffroom door at breaktime. That evening the boy's father was in the pub where he was told by another parent that his son had been 'dragged round the playground by the teacher by his neck'. The following morning the father came into the school and, ignoring the procedure to first approach the headteacher, went straight to the classroom where Terry Blake was teaching and accosted him in front of the class. An 'objective' assessment of the power relations between Terry Blake, an educated, employed, middle-class, white, young, healthy male, and the father, a poorly educated, unemployed, white, middle-aged, rather unhealthy man, would suggest that the power relations lay with the former. However, such 'objective' criteria do not reveal the nuances of power relations within any given situation at a particular time. Terry Blake was threatened and attacked in an 'unsafe' environment. He was physically threatened and verbally attacked and made vulnerable in a situation where the support and sympathy of the community lay with the aggressive father. Also, the class of young children were witnesses but were not questioned later about the events. In addition, at that particular time, education policy makers were making a great play in the media on 'parent power'. In order to identify where power was located and to acknowledge the shifting nature of power suggested that the use of discourses would be helpful in analysing the data collected in the two schools. However, many feminists see post-structuralist approaches as problematic in that the concept of language/power residing within discourses denies the idea of women as a class and thereby undermines feminism as a political movement (Skeggs 1991b; S. Jackson 1992; Ramazanoglu 1993; Maynard 1995; Griffiths 1998). The extent to which post-structuralism could appropriately enable the data analysis of dominant masculinities constructed at Benwood and Deneway Primary Schools will be returned to later.

In searching through existing feminist research into girls' experiences of schooling where some consideration had been given to masculinity and male power, it became apparent that symbolic interactionism was seen as offering a useful framework. Similarly, many sociologists of education have also employed it in order to explain classroom relationships (King 1978; Delamont 1980; L. Davies 1984; Pollard 1985). Symbolic interactionism as a general theory of society has three basic foci: meaning, process and interaction (Blumer 1969; Plummer 1975). Symbolic interactionists argue that people 'perform' on the basis of meanings and understandings which they develop through interaction with others. Through their interactions with others, individuals are believed to develop a concept of 'self'; this is generated through their interpretations of the responses of others to their own actions. Consequently, the 'self' is not seen as static but as constantly being refined.

Studies of primary schooling using symbolic interactionism have provided explanations of how the perspectives of the participants in a particular situation are related to their immediate context; for example, Pollard's (1985) description of primary teachers 'coping strategies'. In a similar way, feminist studies of gender relations in primary schooling have used 'grounded theory' to generate descriptions of the differences in the educational experiences of girls and boys such as teacher approaches to classroom management and control (Clarricoates 1978; L. Davies 1984). More recently, studies of boys' underachievement which have focused on boys' motivations and perceptions of schooling have couched their explanations loosely within 'grounded theory' (Bleach *et al.* 1996; Qualifications and Curriculum Authority 1998). This means of theorizing has certain strengths in that it focuses on the individual in the context of the social group; it shows a concern for the meanings that people bring to social interaction; and it has a conception of the individual as actively constructing social meaning (Hammersley 1990; Stanley and Wise 1993). However, there are limitations to what can be argued for on the basis of using symbolic interactionism.

A focus on interactions within specific social groups does not offer any insights into what *produces* the context in which the action takes place (Shilling 1992; Layder 1993). Symbolic interactionism has no means by which it can deal with society at the macro level; that is, it cannot account for social structures and historical change. Classroom studies by radical, socialist and black feminists have been united in their political will to expose girls' and women's subordinated positions in broader social structures but the limitations of symbolic interactionism have been revealed in attempts to link classroom processes with grand theory. On the one hand, there are feminist small-scale studies of classrooms focused on eliciting girls' experiences and perspectives which have generated explanations based on observations and interviews and these have, in turn, been 'hooked' on to male power bases in wider social structures (Sharpe 1976; Lees 1986; Clarricoates 1987; Reay 1991). On the other hand, there are feminist studies which have not attempted to make connections between classroom interaction and broader social structures thus attracting the criticism that they are undertheorized (Arnot and Weiner 1987; Middleton 1987). As Walby (1986) has argued:

> the concept of patriarchy is really outside the range of concepts admissible in the symbolic interactionist's vocabulary, since it involves notions of social structures that interactionists expressly reject. Thus their analyses are stuck on a micro level and cannot deal adequately with important forms of social structuring and hence with analyses of general changes in sexual meaning.
>
> (Walby 1986: 67)

The difficulties in linking the findings of small-scale studies to grand-scale theories (the micro–macro debate) have occupied a central place in methodological discussions (Scarth 1987; Hammersley 1990; Shilling 1992). Poststructuralist approaches have proved useful in providing an alternative means of exploring situations by dismissing macro–micro dimensions and locating issues of power within discourses (Foucault 1977; Weedon 1987; Stronach and MacLure 1997). However, as was said earlier, while many feminists have found post-structuralist insights into conceptions of knowledge and power useful in their understandings of gender relations, the majority of those researching education have retained the principle that men as a class are constantly in the process of maintaining and sustaining power over women as a class (Middleton 1993; Deem 1994; A. Jones 1997; Weiner *et al.* 1997; Yates 1997). Resolving such tensions is at the heart of current debates into feminist (and others) epistemological concerns (Raphael Reed 1999; Francis 2000).

Several writers have argued for a theoretical framework which makes use of a range of positions (Troyna and Hatcher 1992; Haywood and Mac an Ghaill 1997; Skeggs 1997). In their work on masculine subjectivities Haywood and Mac an Ghaill (1997: 263) state that they have 'found it productive to hold together what we identify as materialist and deconstructionist identity epistemologies, in order to access the structures and the categories of identities within educational arenas'. In a similar way, Weiner (1994) describes a theoretical position which she calls *materialist feminism*. This is a fusion of radical, Marxist and post-structuralist feminist ideas. It is a

> category of feminist scholarship which emphasizes the shifting notions of womanhood and also its dialectical relationship to other social formations such as class, family, religion . . . that is, one which contends that all human action, including that of women, is the consequence of specific cultural, economic and social conditions and influences.
>
> (Weiner 1994: 21)

Thus the decision was made to draw on the theoretical position of materialist feminism. What was then required was a theoretical model for exploring masculinities in the two schools which retained the usefulness of post-structuralist deconstructions of power *and* the notion of 'maleness' as a set of collective practices which marginalized and subordinated 'other' forms of masculinity and women. In other words a model which would recognize that discourses order a domain of (hegemonic male) reality whereby the effect of them is to 'silence' certain voices through their ability to authorize only certain persons to speak in particular ways (May 1997). A model produced by Troyna and Hatcher (1992) for their work with primary school children and racism provided an overarching framework which allowed the relationships to be shown between social structures, ideologies and context in understanding masculinities and male practices in specific settings.

A theoretical model for analysing masculinities

The model produced by Troyna and Hatcher (1992) was informed by one developed by Waddington *et al.* (1989) called *Flashpoints*. The framework

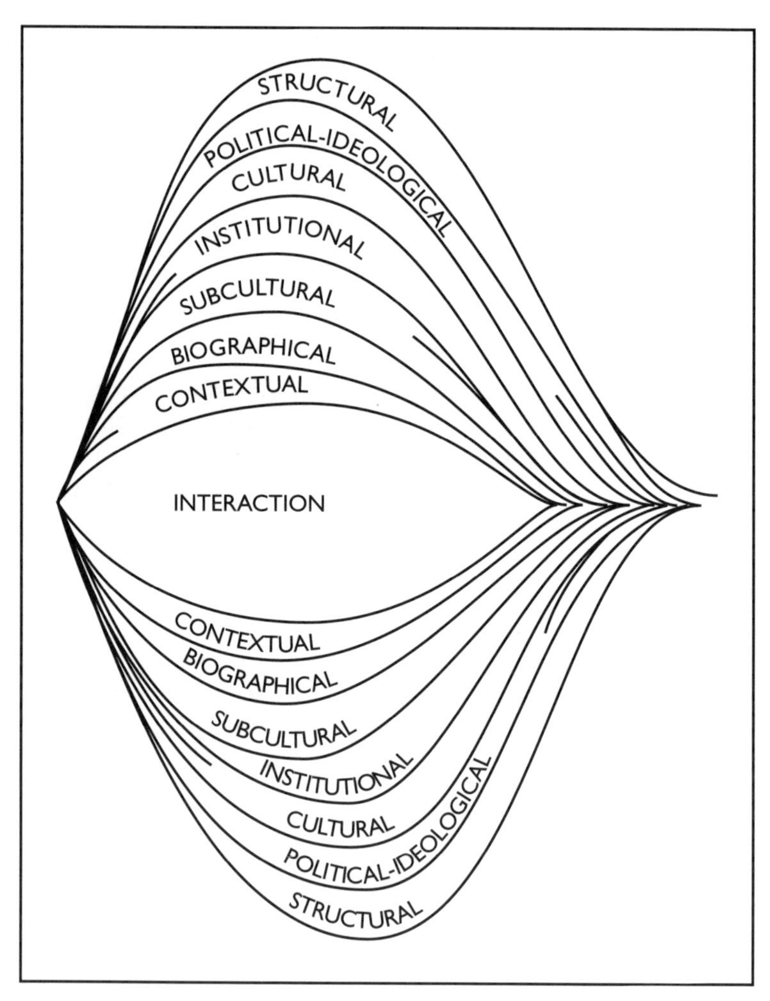

Figure 3.1 A model for analysing hegemonic masculinities in schools
Source: adapted from Troyna and Hatcher 1992: 40

devised by Waddington and his colleagues was intended to enable the exploration of public disorder associated with large-scale events such as the miners' strike. Troyna and Hatcher (1992: 39) pointed out that their focus was on 'incidents which tend to be far less spectacular; they are relatively ordinary and routine – "trivial incidents" to use the Home Affairs Committee term'.

Their modification of the *Flashpoints* model is one much more relevant to studies involving small-scale events in which only a few, sometimes just two, people are involved. The model developed by Troyna and Hatcher (1992) has eight levels: structural, political-ideological, cultural, institutional, subcultural, biographical, contextual and interaction (Figure 3.1).

They used the model to explore routine racist incidents in schools but retained the notion of a *flashpoint* in order to 'identify the social processes which come together in specific combinations in each racist incident whether they are an exception . . . or unexceptional and everyday' (Troyna and Hatcher 1992: 46). On this basis the concept of a *flashpoint* did not readily lend itself to my research on masculinities and primary schooling. The intention was not to focus on 'sexist' or 'genderist' incidents, but to consider how social processes came together in specific combinations for the construction of, and engagement with, hegemonic and other modes of masculinities. For this reason the notion of *flashpoint*s was replaced with the concept of *critical incidents*.

Tripp (1993) uses the term to describe how an event is *perceived*. He argues that *critical incidents* are produced by the way we look at a situation: that it is an interpretation of the significance of an event:

> The vast majority of critical incidents are not at all dramatic or obvious: they are mostly straightforward accounts of very commonplace events that occur . . . which are critical in the rather different sense that they are indicative of underlying trends, motives and structures. These incidents appear to be 'typical' rather than 'critical' at first sight, but are rendered critical through analysis.
>
> (Tripp 1993: 25)

This definition of *critical incidents* was used in this analysis of masculinities and primary schooling in order to revisit a (feminist) claim that schooling 'typically' upholds 'commonplace' or 'normative' conceptions of masculinity through structure, pedagogy and curriculum (Byrne 1978; Mahony 1985; Askew and Ross 1988; Thorne 1993; Murphy and Gipps 1996). The substitution of *critical incidents* for *flashpoints* does not compromise the model developed by Troyna and Hatcher (1992) as, in keeping with the understandings of Waddington *et al.* (1989), they state that what converts an incident into a flashpoint is not so much its inherent characteristics as the way the incident is interpreted at the time.

In terms of the model, the eight levels of analysis link with each other but there is no implication of chronological development or linear flow. To paraphrase Waddington and his colleagues:

> [A critical incident] does not 'begin' at the structural level and proceed through the others to the interactional level. Nor do we intend to imply that [a critical incident] is necessarily predetermined by the 'higher order' levels . . . Conversely, [a critical incident] can occur at the interactional level in the absence of pre-disposing factors at the other levels.
>
> (Waddington *et al.* 1989: 157)

These levels of analysis are described more fully below.

- *Structural*: Differences between the access that males and females have to power and resources. These differences occur both *within* and *between* groups of men and women and are further exacerbated by social class, sexuality and 'race' variables. Conflict can arise if subordinated groups cannot improve their position and, by dint of this position, have little stake in the existing institution of political and social order.
- *Political-ideological*: This level refers to the relationship of groups to political and ideological institutions. Egalitarianism is the espoused aim, but the real obstacles imposed because of sex and gender are deeply embedded.
- *Cultural*: The ways in which groups of people understand and experience their lives.
- *Institutional*: The ideologies, procedures and practices which a school sanctions, promotes and transmits.
- *Subcultural*: The children's and adults' worlds in which they construct, negotiate and reconstruct individual subjectivities.
- *Biographical*: Those factors and characteristics which are specific to the individual pupil or teacher involved in an incident.
- *Contextual*: The immediate history of an incident involving male power struggles.
- *Interaction*: The actual event regarding what was done and what was said.

The ways in which this model generated particular findings of masculinities and male practices in the two case study schools will be illuminated in the following chapters.

Summary

The intention in this chapter has been to set out the theoretical framework of the case studies of two primary schools. In so doing I have attempted to interrogate the research processes involved in reaching the findings which form the basis of the case study chapters. The main points that have arisen from this exercise are that, first, there are limitations on what 'knowledge'/ findings are produced given the positionings of, and positionings by, the researcher themselves. The whole research process was informed by the subjective knowledges and understandings of all the participants. As Harding (1991: 283) has observed: 'a necessary moment in understanding other people and my relations to them [is] understanding how I am situated in those relationships from the perspective of their lives.'

First, this means only a partial picture can ever be provided, and this is particularly so in this case where I, as an adult, female, part-time teacher and researcher was attempting to illuminate masculinities and male practices in school settings. Second, to gain insights into even a partial picture of masculinities and male practices in primary schooling requires a framework which enables these to be viewed across a number of different levels from broader structural variables to specific incidents in the classroom.

These questions enabled the identification of central themes which form the basis of the chapters which appear in the second part of the book.

Note

1 Pseudonyms are used throughout to preserve anonymity.

part / **two**

Inside the primary classroom

chapter / **four**

Primary schools and local communities

Introduction

One of the strengths of the 'boys' underachievement' literature is that it offers many practical strategies for teachers to try out with their pupils. The downside is the assumption that it is possible to find catch-all 'solutions' to boys' disaffected attitudes towards school. As any teacher knows, what works effectively with one group of pupils, whether it be a particular approach to classroom control or a way of teaching fractions, is no guarantee that it will work with another group of students. Therefore, just because a book or article says that English teachers in school X found mixed-sex seating in one year group improved boys' literacy skills it is no surprise to find that the same approach in school Y had a negative effect on the behaviour of boys (see chapters in Bleach 1998 for this contradiction). At the risk of stating the obvious, not all children (or all teachers) are the same and they bring with them into school attitudes and expectations about what is expected of them as *pupils* and also what they expect of the school. These expectations are themselves informed by pupils' knowledge of themselves as gendered beings and from a certain culture, religion, social class, region and so on, all of which contributes to, and helps shape, their schooling experiences. For example, several writers on masculinities have highlighted the significance of the *local* in the construction of masculinities (Mac an Ghaill 1994; Connolly 1998). As Connolly (1994) says when writing of the influence of ethnicity on pupils' perceptions and understandings:

> little if any attention has been given to looking beyond the school gates to exploring how the particular locale has provided a site through which broader, national political discourses on 'race' have been appropriated

> and re-worked, and how, consequently, this has impacted upon the nature of social relations in the school.
>
> (Connolly 1994: 1)

This chapter explores how teachers in one primary school drew on their experience and understandings of the local culture to inform their day-to-day management routines and practices. It will look at the ways in which approaches to classroom control at Benwood Primary School were partly informed by the main kinds of authority and control evident in the community and the forms of relationships that existed between the school and parents. Control within the local community was exercised by young men who constructed themselves around particularly violent forms of masculinity. The argument being presented here is one that underpins all the case study chapters; that in order to tackle 'the problems with boys' it is necessary for a school to identify the specific modes of masculinity operating within its own site and, importantly, to recognize the dominant pattern of masculinity the school itself constructs and performs.

The local community

Benwood Primary School is situated in Wickon, an economically deprived area of the north east city of Oldchester. It is a traditionally working-class area, but the decline of the coal and shipbuilding industries contributed to steep increases in male unemployment. Also, changes in social security legislation in the late 1980s meant that children leaving school at 16 had no rights to any form of benefit. At the time of the study, there were 432 teenage males in the Wickon area claiming dole; 180 of these were on government training schemes which left 252 with nothing to do (Campbell 1993). Visitors quickly become aware of Wickon's distinguishing characteristics: streets in which one-third of the houses are boarded up or burned out; shops which have heavy metal grids on the windows and across the serving counters; and the presence of children of all ages wandering the streets on schooldays. Given this situation, alternative, unconventional forms of 'work' have emerged. The 'cultural economy' became one in which petty thieving to more organized crime held a central place in the 'work lives' of many of the people living in the locale, and the ones committing these crimes were mainly groups of young men (Phillips 1993).

It was noted in Chapter 3 that the area has attracted a fair amount of national media interest, initially because of its involvement in the inner-city disturbances of the early 1990s. It has also been the subject of several studies which have considered the forms of masculinity evident among the men who live in the area. The dominant form of masculinity is that of the 'hard man' (Wallace 1992; Campbell 1993; Phillips 1993). According to

Campbell (1993: 201) this involved 'proving themselves by having bottle, being good drivers, getting into places, looking for fights all the time, being a bit crazier than everybody else, being able to get control of other people'. This 'hard man' image had particularly negative implications for women living in this part of the city. The potential and actual threat of sexual and physical attack by the most visible 'hard men', 14–19-year-old adolescent youths, restricted the freedom of women to leave their homes and move around the local vicinity:

> My mam was full of hell about it all, saying 'the bastards, rotten little bastards', she hates them that hang around the shops because she knows I can't go out there at night because it's too dangerous and I get frightened.
>
> (Female, aged 14, quoted in Wallace 1992: 48)

> It started where you couldn't go out the house in case you got run over by a hoisty [stolen] car, but now you're frightened even in the house in case you're burnt alive.
>
> (Female, aged 21, quoted in Wallace 1992: 46)

There was a high proportion of single mothers in the area (twice the national average) and many of those who had their own accommodation had to defend it from the 'lads' who tried to use it as a place to 'doss' or saw it as an easy target for burglary. Although the people living in the area were angry with the police's apparent inability to stem the tide of vandalism and burglaries, there was an awareness that it was the actions of the 'lads' which secured media attention and, consequently, the possibility of positive changes being brought about (Campbell 1995; Crowley 1995; Hadaway 1995).

Benwood Primary School and hegemonic masculinity

The significance of the social aspects of teaching on primary teachers' perceptions of their role has been widely discussed (King 1978; Nias 1985; Woods 1990). This may well be linked to the 'family-oriented' organization and ethos of primary schooling, where the physical environment reflects that of the home (play area, carpet, children's work displayed on the walls), re-creating parental relationships between teachers and pupils (Hartley 1985; Pollard 1985). However, similar to the situation found by Connolly (1998) in his study of an inner-city primary school, the teachers at Benwood Primary placed particular emphasis on the socializing/parental aspect of their role. Their perception appeared to be that the children needed to reach a certain level of social competence before they could be taught. For example, a Y2 class teacher said:

Terry Blake: I had certain academic aims for them when I first took the class over . . . I was surprised how uncommunicative they were and how little interest they showed in tasks . . . I felt very strongly that their overall language development and use of language is so poor that I had to get them talking more . . . to some extent I was trying to encourage them to be livelier . . . If you sit and read stories and sing songs for twenty minutes the children do get rather bored whereas they have learned very well to take turns and to listen to each other by and large . . . the expectations I had about their academic performance I've had to revise.

The importance placed on socializing the children before (academic) learning could take place was raised by another Y2 teacher, Mrs Smith, in relation to KS1 testing. One of the science tasks involved the children devising a time line setting out their own development. However, the task appears to have been conceived of with a traditional middle-class family in mind and the girls in Y2 at Benwood Primary apparently 'failed':

Mrs Smith: When it came to the time line in the science SATs most of the children got it wrong. The girls drew themselves at secondary school, then having a baby, then having another baby, then being a gran and then getting married! Well, of course, that's probably the situation for most of them. Tracey even drew herself with a baby sitting at a desk at secondary school! The boys didn't get on any better what with Wayne drawing himself leaving school at primary, then drawing himself in prison, then showing himself doing his first job . . . when I asked him he said they'd teach him to 'do up' motorbikes and cars in prison – his brother's been in a young offenders' institute so I guess that's what they taught him.

What the teachers at Benwood Primary aimed for in terms of their socializing/parental role was in marked contrast to their understandings of how parenting was undertaken in the local community. While attempting to develop a moral and ethical framework for the pupils to operate within when in the school building, the actual disciplinary strategies used for maintaining that framework drew on similar discourses of power and control found outside the school gates.

The gangs of adolescent youths who patrolled and controlled the streets of Wickon were referred to by locals as the 'lads'. The difficulties experienced by the local community through the behaviours of the lads were shared by the school. Significantly, the competitive, intimidatory, physically aggressive characteristics evident in the type of hegemonic masculinity manifest by the collective practices of the lads was reflected in the discipline and control strategies of the school. These strategies, in turn, served to inform the *gender regime* of the school. It should be said that approaches to pupil management evident in Benwood Primary which perpetuated the (symbolic) forms of violence associated with the masculinity of the 'lads' did not appear to have been wittingly chosen. Indeed those teachers who utilized aggressive, intimidatory classroom management strategies did not seem aware that they were emulating some of behaviours of the lads.

There were two aspects that could be identified in the general daily business of the school which helped to construct the pattern of hegemonic masculinity so similar to that operating outside the school gates. These were relationships with parents and the community, and changes in educational legislation during the 1980s which had eroded the powers of the school. To put this into context requires a consideration of the rule framework the school placed around itself to distinguish its moral values from that of the local community. The rules imposed were more or less fluid according to the position from which the school was operating. Each position adopted was rationalized in terms of enabling the teachers to 'do their job' in an increasingly difficult climate in which the school was becoming ever more vulnerable. Three positions became apparent in the ways in which the teachers 'managed' school rules:

- when there was a clear demarcation between school and 'outside behaviour'
- a blurring of the rules regarding parental interaction
- a merging of the boundaries when the school imitated life 'outside', in the local community.

Before discussing these three positions some exploration is required of the varying relationships the school had with parents and to establish how the school felt it was regarded in the eyes of the local community.

Parents and the school

The school brochure stated that 'Parents are welcome in school at any time'. Similar to most school brochures, it also advised parents of how they could get involved in their child's school life by attending assemblies, open days and having regular, informal chats with the school staff. The setting up of a Parents' Area with kitchen and toilet facilities seemed to be a sign that the school wished actively to encourage parents to come into the building. This might have been what the school would have liked had relationships between it and the local community been one of mutual support, trust and respect. However, this was not the case and, in reality, the school became a virtually impenetrable fortress during school hours. While the official voice of Benwood Primary offered a welcome to parents and expressed a desire to cultivate closer links between home and school, it also gave out clear messages to discourage them from entering the building. Five minutes after the bell had sounded at 8.55 a.m. all but two of the six entrances to the school were locked. Of the two remaining entrances, one provided access to the Parents' Area and the other gave admittance to the entrance lobby where the offices of the secretary, caretaker and headteacher were situated. The reason why the majority of entrances were locked was a concern for the personal safety of the teachers and securing school property.

The tragic events which have occurred in schools in recent years such as the murder of the headteacher, Philip Lawrence, the stabbing of a schoolgirl at a Middlesbrough comprehensive school and the shooting of children and staff at Dunblane Primary School, have prompted much tighter security. However, the research at Benwood Primary preceded these events and, at the time, the only schools in Oldchester with any form of security were those situated in this area of the city. The school had experienced problems with parents and other adults coming into the school, stealing school property and assaulting teachers in their classrooms. At least once a week the school was broken into or an attempt made to break into it. On one occasion during the time the study was being undertaken an attempted burglary was carried out while teachers were still in the school. The would-be burglar attempted to smash through a skylight window with a brick, although he could clearly see that the male acting deputy headteacher was observing him. He continued breaking and clearing glass away to give himself a safe entry and ran off only when the police car siren was clearly audible. It was because of incidents such as this that security cameras had been mounted around the building and heavy steel doors installed to cover the entrances. The high metal gate to the school car park was closed during school hours, but this did not prevent the staff's cars frequently being stolen or vandalized. The teachers themselves were vulnerable to both verbal and physical attack, mainly by men. In the course of a few weeks, one male teacher was confronted in the classroom on three separate occasions by angry fathers.

So, while parents were ostensibly welcome in the school 'at any time', there was understandable caution. As earlier research has shown, a part of teachers' technique is to use the environment, and this includes developing symbolic boundaries demarcating privacy, personal space and territoriality (Steele 1973; Woods 1983). Benwood Primary School partially achieved this by prohibiting access to many areas of the building. Also, by clearly identifying to parents the areas in the school to which they were granted access, the power relations that the school attempted to promote were explicitly confirmed (Eggleston 1977). Inevitably, the actual and potential threat of violence had implications for relationships between the school and community. The staff at Benwood Primary were obviously nervous because the building was a focus of attack and, as individuals, they were vulnerable. However, this was not the sole form of relationship between the school and local community. The school was seen as representative of authority by some of the parents who would call upon teachers to act as adjudicators in local disputes.

The acting headteacher, Mrs Masterson, estimated that for every 200 visits she received from parents only one would be connected with the academic development of their child or children. The majority of visitors were the mothers who came to school expecting her to sort out arguments between neighbours over their children. Frequently, the incidents which sparked the argument had not taken place in school:

> *Mrs Masterson:* Take for example Mrs X coming in on Monday . . . On Sunday Mrs Y had apparently accused Paul [Mrs X's son] of saying he was going to break Sharon's [Mrs Y's daughter] legs. She wanted her son's name clearing . . . I suppose she'd either got nowhere arguing with Mrs Y or she was frightened of getting a black eye so she thought she'd get the school to sort it out.

One of the Y2 class teachers also recalled several instances when parents had asked him to intervene:

> *Terry Blake:* Parents are always expecting the teacher to sort out problems they have with each other . . . Bethany's mother came in yesterday complaining that Wayne [in the other Y2 class] had slapped her across the face when they'd been out in the street the night before. She wanted me to punish him by slapping him back!

As the event described in Chapter 3 shows, when fathers came to the school it was more often to physically 'sort out' the teacher for attacking their child. Cortazzi (1991: 101) has noted that 'Awkward Parent narratives . . . focus on complaints, misunderstandings and exaggerations'. He goes on to point to the asymmetrical relationship between parents and teachers in saying that:

> Teachers have classes of children, whereas parents have, at most, a few individuals. While teachers do strive to teach children as individuals, their role imposes a more objective, achievement-oriented approach which is quite different to parents' subjective acceptance of their children, irrespective of standards. Where teachers emphasize social justice between many children, parents look after the welfare of their own family members.
>
> (Cortazzi 1991: 103)

At Benwood Primary the majority of parents were perceived of as 'awkward', and the particular form this took could be seen in teachers' discursive construction of parents in the local community as 'adult-children'. This was presented in a number of ways, namely, that some parents could not resolve problems concerning their own children, that some would steal from the school, and that others would adopt physical or symbolically violent attitudes rather than engage in verbal discussion with teachers:

> *Mrs Masterson:* The parents don't know how to communicate . . . they're still immature themselves. They don't see their responsibility as parents, for example, they'll swear at home in front of the kids. Our parents just won't take responsibility . . . Some land was given to them to garden and, at first, we [the teachers] worked with them on it and it won an award but as soon as we backed off to let them get on on their own they didn't bother with it. As long as there's somebody there to take responsibility they're all right but they won't take over responsibility.

Similarly, one of the Y2 teachers said when discussing why she found it more difficult to encourage the children at Benwood to recognize the importance of being accountable for their actions and behaviours than children she had taught elsewhere that:

> *Mrs Smith:* These children are only used to other children . . . as that's all their parents are . . . and they're not used to consistency.

The 'childlike' behaviours of the parents usually appeared to irritate the teachers, but it was sometimes used as a source of humour in the staffroom as when the LEA sent out a questionnaire requesting parental views on their children's experiences of the SATs. On the questionnaire parents were asked to 'Read the following statements and then circle one number which reflects how you feel on a 1 to 5 rating score'. On seeing this, the acting deputy male teacher commented, 'Knowing our parents they'll look at this and say "What's this! I'm not going to rat on my own kids!"'

The relationship with parents, then, was informed by fear/vulnerability on the one hand and a sense of maternalism/paternalism on the other. This perception fundamentally influenced the discipline and control strategies used in the school, in the sense that the 'school rules' were largely dependent upon the school's interaction with parents. At its most extreme, the teachers would identify different types of behaviour; those appropriate within the classroom and those they termed 'outside behaviour'.

Demarcating school space

A substantial body of ethnographic literature exists which considers classroom control, particularly that focusing on teacher–pupil power relations as in discussions of 'coping' or 'survival' strategies (A. Hargreaves 1978; Pollard 1985). An important point emerging from this work is that the nature of classroom control issues, and the nature of the strategies developed to remedy these issues, are shaped by the circumstances in which they occur (Denscombe 1985). While teachers have ostensibly more power than pupils, given their legal and institutional authority, teacher control strategies are not enacted upon a passive group of children.

It was said that the teachers at Benwood Primary seemed to feel simultaneously threatened yet despairing of the parenting by adults in the local community. So, when it came to matters of classroom control, teachers would utilize 'indulgence' (Woods 1979) as a way of coping. Denscombe (1985: 14) describes 'indulgence' as 'a teacher strategy in which pupils are allowed to go beyond normally accepted bounds of behaviour and where teachers decline to enforce general classroom rules'.

Teachers at Benwood would make reference to what they called 'outside behaviour' in order to modify children's actions. The notion of 'outside behaviour' seemed to focus on physical contact. Whether or not the physical contact would lead to actual aggression seemed to have been incidental. For example, when Terry Blake was asked what constituted 'outside behaviour' he listed fun fights, hand games and rolling around together on the floor. Mrs Smith, the other Y2 class teacher, more explicitly linked 'outside behaviour' to violence:

> *Mrs Smith:* I extend classroom to playground but as soon as they're out of school, well that's up to parents and because of the way our parents are, and that is, if somebody hits you then you hit them back, we say, in school, you're in a nice school, you follow my rules and I won't have fighting. I don't like fighting so you don't do it in school. What you do outside school is up to your parents but when you're in school you're in my charge and you do what I say.

However, it was not only actions involving physical contact which invoked the label 'outside behaviour'. Some of the children engaged in behaviours which would have been totally unacceptable in some schools, because such conduct was breaking the law let alone school rules:

> *Mrs Masterson:* I would not tolerate a child swearing at or kicking me. When it comes to thieving . . . I can't condone it but neither can I condemn it. It's a part of their values . . . their home background. We can't compensate for the values they get from home.

These 'school rules' were unofficial, yet closely adhered to by the staff. This contrasted with approaches taken towards the official school rules listed in the school brochure where two of these rules were continually broken by teachers, children and parents.

Blurring the boundaries

When the study of Benwood Primary was undertaken the National Curriculum had been in operation for three years. Although there were obvious pressures on teachers regarding the management of the curriculum in order to reach attainment targets with children, there was little evidence in the school brochure of a change in the philosophy of how children's learning would be organized. The school brochure indicated that the philosophy of Benwood Primary School was firmly grounded in child-centred ideology. National Curriculum subjects were briefly outlined, but parents were told that 'The experiences a child brings to the classroom are used to develop skills in all areas of life. A project centred approach is used as the basis for much of the work we do'.

With this in mind, parents were informed that the school rules were few, but necessary to the well-being of all the children on the school premises. Despite the claim that the rules were 'necessary to the well-being' of the pupils, two of these rules were broken on a daily basis, as specified in the 1992 school brochure:

> Children should arrive in school by 8.55 a.m. in time to enter the building with their teacher.
>
> Children are not allowed to bring sweets or biscuits to school although they are encouraged to bring fruit. We do have a Healthy Eating Tuck Shop which is open daily.

With regard to punctuality, the general trend was reflected in the attitudes to timekeeping of the Y2 class where only half to two-thirds were at school by 8.55 a.m. The majority would have arrived at school by 9.45 a.m. Terry Blake built this staggered beginning into his daily planning. Only on rare occasions did Terry ask why they were late. When asked about this he said he simply did not know how to deal with it. At his previous school he would have spoken to the parents, but he felt that Benwood parents were not particularly supportive of the school. Mrs Masterson's response to the same question was to say: 'Occasionally we send out a letter but punctuality isn't something we can do anything about. It's the parents' responsibility, not ours, and they hate the discipline of timing'.

The same reason was given for allowing children to bring sweets and crisps into school; that is, it was the parents' responsibility as to what they gave their children to bring to school to eat. It seemed that at Benwood School, when teachers talked about 'parental responsibility' it was a code for 'parental irresponsibility'.

The school saw itself as operating on a very different basis from that of the 'outside' community, so it attempted to impose various rule boundaries for the children between the school and local community. However, the school was a part of the community and the points of contact required a relaxing of official 'rules'. Furthermore, the perception of the school regarding the ways in which the local culture operated was a strong influence on the school's policies towards discipline and control. One aspect of the school's interaction with parents was located firmly in how gender divisions were played out in the local community. For example, the burglaries, arson attacks and assaults on teachers had been carried out by men, whereas the contacts involving requests for the school to mediate or adjudicate in disputes had, in the main, been sought by women. It was the men in the area who dominated, and Benwood School incorporated the competitive, intimidating and aggressive aspects of the hegemonic masculinity enshrined in the behaviours, attitudes and actions of the 'lads' into its control strategies.

The school and 'machismo'

As representatives of a power structure, teachers have power, but as individuals working in a community in which they and their workplace are frequently subject to physical attack, the form that power can take is significantly limited. For schools such as Benwood Primary this situation was made explicit in the educational legislation of the 1980s. The changes made throughout the 1980s which culminated in the Education Reform Act saw a devolution of power away from the existing power structure towards a centralized system. 'Parent power', as manifested in the Parent's Charter (Department for Education (DfE) 1991, 1994), was intended to act as an overseer to the work of teachers (Tomlinson 1988), although 'partnership' was, and is, the official term. However, there is now substantial evidence to show that parents have fewer 'choices' regarding their child's education than the Department for Education publicity suggested at the time (Gewirtz *et al.* 1995; Whitty *et al.* 1998). For some schools, a 'partnership' with parents was already operating, but this was not a universal situation (Pollard 1992). Affording parents a greater say in the running of schools assumed that some, if not all, parents in any one school's catchment area would want a closer involvement and that parents and teachers could work effectively together. Equally importantly, shifting power from LEAs to parents assumed a simple transfer of responsibilities but this could never actually be the case. Parents who already had access to the mechanisms of power would be, and are, in a much stronger position than those parents who had little or no chance of getting close to sources of power (Deem 1990; David 1993; Miles and Middleton 1995). Benwood Primary School's relationship with parents operated in different ways, but none of those ways came anywhere near fulfilling the type needed to enter into the kind of 'partnership' demanded by the legislation.

It has been argued that schools and middle-class homes have access to the same 'cultural capital' (Bourdieu 1986). A long-standing claim in the literature is that, for middle-class children, school is the place which encapsulates a mutually supporting set of family, school and peer practices (Becker 1952; Sharp and Green 1975; Furlong 1985). This is not the case for teachers working in areas with disenfranchised pupils. Authority is not something which is given to teachers by virtue of their position but something which has to be won. This puts individual teachers in potentially difficult positions and at Benwood Primary School this was certainly the reality. As Connell *et al.* (1982) have said:

> teaching is an emotionally-dangerous occupation. Authority is something they [teachers] construct in isolation, out of their own resources so it is a part and extension of themselves. To the extent that students resist, challenge or subvert their authority, so do they threaten them personally.
>
> (Connell *et al.* 1982: 103)

The teachers' ways of controlling children at Benwood was based upon the strategies they had experienced of parents' dealings with the school. As with teachers in all schools, the staff made it clear in interviews that they were very conscious of the need for the children to be aware of the hierarchy which had to exist in classrooms and the school generally. The way in which a school conveys how the hierarchy works and is maintained differs however and at Benwood this was achieved through an emphasis on intimidation.

The teachers felt that children at Benwood did not respect adults generally and, therefore, would not respect teachers. This had clear implications for control:

> *Mrs Smith:* In other schools the children are much more respectful of teachers. The children here don't respect their parents, seeing the way the children talk back and the way they answer parents back in a way I wouldn't have dared to and that is brought into the classroom and it's reflected in their behaviour. The fact that certain things they play up on, you wouldn't expect and it's because they don't treat you as an adult.

The way in which this 'respect' was gained was partly through 'fear', but the actual strategies varied among teachers. In an interview with Mrs Masterson she indicated that placing children in a position whereby they felt not only cared for but also anxious was necessary:

> *Mrs Masterson:* A relationship with these children has to be based on love, respect and fear. If you've got all three then you've got it made.

She gave equal weighting to all three elements but actually drew more heavily on the third. At the time of my interview with her a boy was waiting to see her as he had been stealing sweets from other children. Following the comment just quoted she asked if I would like to observe her dealing with the boy, to give me an idea of the policy she adopted. The boy came into the room and was told to sit on a low seat near the headteacher's desk. Mrs Masterson sat on the edge of the desk and bent over towards the boy so that there was a minimum amount of physical space between them. She looked into his eyes and asked him to look into hers. When he did she asked him what had happened, and every time he attempted to look away while offering an obviously implausible explanation, she reminded him to 'Look me in the eye'. When the boy had finished, Mrs Masterson, who never once moved her position, said slowly in a low tone which came across as quite menacing, 'I suggest you have given me a muddled story and I want you to sit outside and think about what you have said.' The boy left the room, to be dealt with after our interview had been completed. She went on to list other strategies which varied but always appeared to rely on intimidation, if not some overt physical action:

> *Mrs Masterson:* One child, a boy, was totally out of control of himself . . . kicking, hitting out . . . So I threw some cold water over him to cool him down. It was a physical shock and he calmed down. But there was no negotiating until he did what I asked. From that day on he realized who was pack leader.

Terry Blake provided a similar description of himself in terms of how he wanted the class to see him, but instead of the phrase 'pack leader' he referred to himself as 'the boss':

> *Terry Blake:* I mean, some children like Shane, it's clearly just a game with him. Whenever he's in a situation he likes to try and come out on top. He accepts when I'm in the class that I'm the boss but in any other situation he doesn't see that so he wants to try and play his game in whatever way he can.

One of the points made by several of the teachers was that parents' 'child-like' behaviour was often a barrier to effective communication. Terry Blake said of his experiences with the angry father, that he had tried to get him to explain fully the story behind the accusations, but was met with a continuous tirade of abuse. His experience of fathers' lack of negotiation skills found its way into his control strategies. One means of managing the children's behaviour was by *not* entering into negotiation with them:

> *Terry Blake:* I found it more difficult here to get the children to listen and to do things altogether, things like lining up and going somewhere have been more difficult and I've had to try different strategies for that, one of them being just to stand there and keep repeating what you want until the children actually do what you want. Shouting doesn't really seem to make much difference, . . . you have to find something that works with the children here.

What cannot be shown here is how the confrontational body language, similar to that used by Mrs Masterson, was essential to the success of this strategy; that is, the teacher standing or sitting close to the child and looking down into the child's eyes. The connection of intimidating or aggressive behaviours with authority could be found in other aspects of the school's practices.

Assembly time has been shown to be a pivotal venue where 'what the school is about' is laid down (Woods 1990). Many of the practices identified in the literature on gender and schooling were evident during assemblies at Benwood Primary. For example, the men teachers would be called upon by the headteacher to move equipment or lead the singing; teachers who could not recall a child's name would refer, in the case of boys to 'you' or 'that boy', but if it was a girl, to 'sweetheart' or 'darling' (Delamont 1990; Gaine and George 1999). When the images of 'maleness' that emerged in assembly time and at other points across the school, such as in classroom displays and stories, were examined closely, a particular type or types of masculinities were being highlighted. These masculine images were occasionally those of academic or sporting achievers, but more frequently portrayed maleness as physically violent, competitive and generally aggressive. Such images were at their most pointed in the stories read to the children during assembly.

These stories always had a moral, and tended to focus on boys and the problems they encountered or engendered. For example, a story about boys' friendships had implicit messages about 'tough' masculinity (*We are Best Friends*: Brandenburg 1984). The story revolved around two best friends who were always 'scrapping', and when one of them has to move away he is asked by his friend, 'Who will you fight with? Nobody fights like best friends'. So violent behaviour was given a legitimate face because it was done within the context of friendship. Similarly, the idea that boys and girls defer to different rules was suggested in a tale about a 'good' girl who was tempted to go to the shop by a boy before going to school. She does as he asks and continues the pattern until one morning she is late and misses the coach for the school trip. She is so upset she confesses what she has been doing to the headteacher, who tells her that through missing a treat he doubts she will ever tell a lie again. The boy's behaviour was not brought into question.

During the Friday morning assembly, 'Achievement Awards' would be given out for good behaviour and good work. The recipients of these awards were mainly girls and, to a lesser extent, the younger boys in the school. Occasionally, whichever teacher was taking the assembly would make a reference to one of the boys who were more likely to be found standing outside the staffroom at breaktimes than at the front of the assembly hall being praised. These comments were never made in a condemnatory way, but in a manner which suggested that boys' negative behaviours were condoned. For example, when a Y3 girl received an award for good behaviour the male acting deputy headteacher commented:

> *Mr Bolan:* Where's Stephen [girl's brother]? Oh there you are . . . Tania obviously doesn't take after you! I can't see you ever getting an award for good behaviour [*laughing*] but we live in hope, we live in hope! [*Spoken in a humorous way and all teachers, pupils, including Stephen, laughing*]

Stephen was a member of the school football team, the progress of which was a regular feature of assembly time. The status accorded to the football team in the school, and the amount of time and effort given to it by the (male) acting deputy headteacher and the (male) Y6 teacher, were useful in controlling and disciplining some of the more troublesome older boys, who were passionate about football. However, while football promoted camaraderie among the boys and men teachers, and status was given to the team players by all teachers at assembly time, any interest the girls may have had in taking part in some way was overlooked and they were thus excluded (see Chapter 6).

It was not only through stories that an image of masculinity celebrating violence, and 'toughness' was singled out. The videos which were shown to children during wet playtimes tended to be violent cartoons such as *Ninja Turtles* and *Tom and Jerry*. Also, in one of the Y2 classrooms Mrs Jones, a part-time teacher, put up a large display based on Anthony Browne's *Champ* books and labelled, 'Don't be a Wimp, be a Champ'. It was an interactive frieze which meant the children's attention was constantly being drawn to it as they resequenced the events and sentences.

When the implied messages in the stories are considered in relation to the concept of 'outside behaviour' then the children were, in effect, being told that male physical violence is understandable, acceptable and, by definition, a 'skill' which boys should learn to develop.

Summary

One of the main aims in looking at how schools construct specific modes of masculinity was to reflect on the proposition, put forward in much of the feminist literature into education written in the 1980s, that schools uphold normative conceptions of masculinity through curriculum, pedagogy and structure (Spender 1982; Mahony 1985; Skelton 1989). By employing deconstructionist theories and drawing on the insights provided by recent explorations of masculinities and schooling (Mac an Ghaill 1994; Francis 1998) to analyse the information collected at Benwood Primary, the complexity of how schools develop particular *gender regimes* became apparent and highlighted the simplicity of earlier notions of schools operating *a* masculine 'norm'. Although the National Curriculum brought the subjects taught to 5–16 year olds into line, *how* they were to be delivered was not (in 1991) proscribed. In keeping with existing research on primary teachers (Acker 1999), the teachers at Benwood Primary placed great emphasis on the socializing aspect of their role as, in their case, this was seen as instrumental to enabling their teaching/instruction role.

The stress placed on the socializing function of teaching was a response to the perceived inadequacies of the parents in the local area. The relationship between teachers and parents was characterized on the part of the teachers by, on the one hand, maternalism and paternalism and on the other hand anxiety and fear. The parent–child conceptualization of the relationship between teachers and *working-class* parents has been noted by Cortazzi (1991):

> In the working-class areas social, rather than academic, problems were emphasized.
> - It's very much a social job that we do as well as teaching, helping them out, following their court cases and divorces.
> - Some of them are involved in all sorts of (social) services, there's someone telling them about this and someone advising them about that, and they come to us when they've just had enough of these people.
>
> (Cortazzi 1991: 107)

The social class differences, which appeared to inform relationships between teachers and parents, were further intersected by gender. It was said earlier that mothers would visit the school to discuss perceived arguments or injustices, while the less frequent visits by fathers were characterized by threats of physical violence against teachers. Cortazzi (1991: 108) also alludes to this gender difference when he cites the words of a working-class mother saying to a headteacher: 'I'll send my husband round if this doesn't stop'. The potential for violent attacks on the school and the teachers placed Benwood Primary in a defensive position across a number of fronts.

In order to ward off the threat of physical aggression, the school and the teachers adopted the kinds of defensive strategies identified by men in their accounts of developing masculinity (Kaufman 1987; D. Cohen 1990; D. Jackson 1990). Hence, the school was armed with defensive weapons (security alarms, high fences, surveillance cameras) and the teachers' bodily stances and verbal control methods implied that they were constantly on their guard (Seidler 1991). In addition, the particular control and management strategies used by teachers reflected the intimidatory, aggressive aspects of the hegemonic masculinity evident in the local community (Phillips 1993; Campbell 1993, 1995). The concept of 'outside behaviour' was used in an attempt to differentiate its form of control from that same control exercised in the local culture, particularly by the 'lads'. In effect, teachers used the same masculine forms of authority to control all pupils, not just the boys, because both male and female teachers felt that this was the only kind available to them. An important issue here then is that women can be bearers of masculinity too (Connell 1995). Also, although the intention was to distinguish between what the school expected and what was expected of children 'outside', the perpetuation of stereotypical images of 'good, quiet girls' and 'tough, naughty boys' could be seen in assemblies, wall displays, stories and the attitudes of some teachers.

As Connell (1995) reminds us, at any given time, one form of masculinity is culturally exalted over others, thus bringing into question the idea of schools perpetuation of *a* male 'norm'. The issues raised in this chapter suggests that although schools are sites where a multiplicity of masculinities are constructed, negotiated and reconstructed, the modes of masculinity are shaped, informed by, and dependent upon, access to power. Benwood Primary School is in an area of Oldchester in which certain forms of masculinity are inaccessible and/or rejected, and the existing images available influence the type of hegemonic masculinity projected by the school. In schools such as Benwood, which has a problematic, if not uneasy, relationship with parents and the local community, the school staff are in a more vulnerable position than teachers in a school where there are shared values with the home. This vulnerability makes it more likely that tactics which appear to be effective in maintaining some form of control in the local community will be drawn on to reinforce the control and discipline strategies used inside the school.

Changes in educational legislation of the 1980s compounded problems for vulnerable schools like Benwood Primary. Parents in economically deprived areas do not have access to mechanisms of social power and so cannot fight the school's corner effectively (Deem 1990). Teachers in these schools are likely to find themselves associated with institutions at the bottom of league tables, which could inevitably affect their career prospects and job security.

It has been shown here how the pattern of practices which together reflected aggressive types of masculinity at Benwood Primary School was partly informed by the modes of masculinity found in the wider community and the forms of relationships that existed between the school and parents. The next chapter will consider how the infant boys in the school constructed, negotiated and reconstructed masculine identities in relation to, and intertwined with, the dominant version of masculinity of the school.

chapter / **five**

Being a (school) boy

Introduction

In common with several other research studies, this book argues that schools are places where a multiplicity of masculinities are played out (Mac an Ghaill 1994; Connolly 1998; Martino 1999). Importantly, these studies have indicated the way in which education influences constructions of masculinities; that is, masculinity is organized on a macro scale around social power, but the education system in this society is such that access to social power, in terms of entry to higher education and professional careers, is available only to those who possess the appropriate 'cultural capital' (Bourdieu 1986). Some commentators have argued that those boys who are unable to obtain entry to the forms of social power schooling has to offer, then seek alternative means of publicly demonstrating their masculinity, such as through the use of violence or demonstrating sporting prowess (Segal 1990; Back 1994).

It was said in Chapter 1 that there has been a tendency to focus on masculine identities in secondary education, particularly the impact of the ways in which boys are *organized* in relation to the curriculum (for example streaming; academic versus vocational subjects). Also, focusing on primary boys challenges the argument that schooling is in many ways incidental to the formation of masculine identities. Connell's (1989) interviews with a group of Australian males who had just left school may have indicated that sexual relationships and the adult workplace have a greater significance but these factors are in the future for primary age boys. In addition, infant boys are at the beginning of their school careers and have yet to see themselves as 'school successes' or 'school failures'. Boys starting school have already begun the process of constructing and negotiating their masculine identities in the home and among friends in the local culture. But schooling will also have an impact on the shaping of their masculine identities although in different ways at different points in their progression through school.

This chapter explores the ways in which a group of 6–7-year-old boys in one primary class at Benwood Primary negotiated their masculine identities in the school setting through various discursive positions such as being a *boy*, *white*, *child*, *school pupil*, and a member of the so-called '*underclass*' (Morris 1994; Collier 1995). To do this means providing a profile of the boys and their friendship networks.

The boys

There were four boys in the class at Benwood Primary who were immediately noticeable: John, Shane, Luke and Robert. John and Shane made their presence known to any newcomer, adult or child, to the class. They took the lead in encouraging behaviours among the other boys designed to 'suss out' (Beynon 1984) unfamiliar teachers; for example, responding to a query or instruction by ignoring or answering back, and engaging in general 'tomfoolery' rather than work. During the course of the year these two boys were, at various times, the best of friends and the worst of enemies. The tensions between them occurred largely as a result of their intense rivalry for what they referred to as the 'first boss' position. Both boys came from families who had a high profile in the local community in terms of their notoriety. During the time of the research both families had substantial contact with the police. At the beginning of the observation period one of Shane's brothers was in a remand home. He escaped from the remand centre on three separate occasions, and each time was rearrested at the house of one or other of his aunts. John's father was also in court, involved in a protracted trial (John's mother provided this information but she did not given any indication of the nature of the crime).

Luke also made his presence known. Although keen to associate himself with the activities of John and Shane this was not the reason why he secured attention. Rather, his *lack* of success at being seen as one of their accomplices resulted in jealous behaviours on his part which consequently attracted the attention of teachers. Also, the fact that he was the largest and most unkempt boy in the class gained him some unwelcome attention from the other children. Luke spent a substantial part of his time in the class trying to be accepted by the others, as he seemed universally disliked by boys, girls and teachers. In interviews the other children listed the reasons for disliking him as 'fat', 'smelly', 'snotty-nosed' and 'having skid marks on his pants'. One of the ways in which he attempted to compete and, at the same time, gain kudos with John and Shane was to compare the activities of his male relatives with theirs; for example, when Shane announced in 'news time' that his brother had been recaptured by the police, Luke announced that his cousin had been arrested along with his 'mates' for doing a 'hoistie' (stealing a car).

Robert stood out in a totally different way. He had recently moved into the Wickon area from another part of Oldchester, also economically deprived, but which did not have the same reputation for crime. In contrast to all the other boys, he seldom indulged in disruptive behaviours and certainly never instigated them. When the class was taken by a part-time or supply teacher, disruption on a mass level occurred with children 'fun-fighting', running round the classroom and generally making a lot of noise. On these occasions Robert would frequently offer reassurance to the harassed teacher by explaining why the others were acting in such a way (because she was not their classteacher). He was a boy who, if not keen to work, would comply and make some attempts to complete the tasks he was set. Robert was described by both Terry Blake and Mrs Smith in terms of being a 'really nice boy'.

As there did not appear to be any tight friendship groupings among the boys, it might be argued that this reflected the literature which says boys tend to form large, loosely connected groups (Lever 1978; Woods 1987). However, in common with Thorne (1993), I am aware of the risk that by simply comparing the overt characteristics of boys' friendship groups with those of girls overlooks the complexities of boys' friendships. For example, while Luke was an obvious outsider, Carl and Rick were also marginalized within the boys' group because of their physical appearance and personal habits; Rick because he was 'smelly' and Carl because he was 'slavery' (he dribbled). As a result of being avoided by many of the children in the class, they tended to seek each other out at 'choosing time'.

The remaining eight boys can be loosely grouped in terms of their relationship with John and Shane. As was said earlier, it tended to be these two boys who instigated the majority of competitive, challenging behaviours both to authority and among their peer group. Gary, Tommy and Matt were always the first to join in with John and Shane and, occasionally, attempted to initiate and take the lead in various challenging actions. Bobby, Adam, Dean, Sean and Martin always took part in any group actions but were unlikely to lead. Observing the way in which the boys tended to take up the same positions in the group whenever a challenge was mounted to a teacher's authority or for a confrontation with boys from another class, reminded me of the organization of the army during the Second World War.

Although a very crude analogy, it provided a useful image of the boys' metaphorical and often literal positions in the group:

- The generals (John and Shane), who organized the action and led the initial assault.
- The regular soldiers (Luke, Gary, Tommy and Matt), who were quick to see what was required and proficient in supporting the actions of the leaders.

- The conscripts (Bobby, Adam, Dean, Sean and Martin), who realized they had to join in but their involvement was minimal and they literally positioned themselves on the periphery of the action.
- The group who could have been seen as conscientious objectors (Robert, Rick and Carl), although for different reasons. Robert's preferred style of challenge was verbal while Rick and Carl, recognizing their marginal position among the boys, preferred to avoid contact with the rest of the group. At the same time Robert, Rick and Carl (and indeed the 'conscripts') always took some role in any action possibly because the alternative was more personally threatening.

As has been pointed out in feminist research and by several writers on masculinities, for many boys school days are characterized by avoidance; specifically, avoiding showing emotion or any sign of 'weakness' (Mahony 1985; D. Jackson 1990; Salisbury and Jackson 1996). Although writing of boys' public schools, Heward (1991) has encapsulated the essence of this literature when she says of schoolboy culture that:

> acceptance into friendship groups and the power structure of the [group] depended on conformity to its rigid set of norms. The aim of baiting [a] victim was to exploit his weaknesses to the point where he broke down and showed any sort of emotion, anger, fear, distress, pain. Such outbursts [are] then ridiculed.
>
> (Heward 1991: 21)

The important issue here is that the boys could be discussed as a 'group', on the basis of a shared relationship to the dominant masculinity of the local culture/school where they constructed and negotiated masculine identities at various levels. Although the boys in the class were too young to access the forms of power utilized by the 'lads' in the local community, they were aware of the power and status the 'lads' held and, as males themselves, knew they would eventually gain entry to this fraternity. Indeed, as will be shown, the actions of the boys, particularly those of Shane and John, can be interpreted as ways of 'working' themselves into older forms of masculinity in preparation for this time (Redman 1996).

Boys, masculinities and the local culture

The 'hard man' image of masculinity occupied a high status in the Wickon area, particularly among the 'lads'; gangs of 14–16 year olds whose actions and presence on the streets in the local community attracted constant police attention (Campbell 1993; Phillips 1993). The intention here is to consider the implications of being a 'hard man' for the interrelationships of the 'lads'. The purpose of this is to provide insights into the particular dominant masculinity of the 'lads' which the boys in the class would have become familiar with; a familiarity which would help shape their own masculine identity.

Being a 'lad'

The 'lads' hung around together, but information from studies of the area indicates that there was always a leader or leaders who managed the alternative, unconventional forms of 'work' which they engaged in. The 'cultural economy' was one in which petty thieving to more organized crime held a central place. A mother living in the area said of the 'lads':

> There is a hard core of 14–19 year olds and usually an older man, about 25 or 30, who encourages them. It's dead hard keeping your kids away from street culture. These boys belong with each other. They bond to each other. They brag about how they get away with what they do . . . They hear, so the story goes, that they can make up to £300 a week with petty thieving and far more through organized crime. These kids see that as a legitimate goal.
>
> (Phillips 1993: 33)

Many of the local criminal fraternities were well known. They were often made up of extended families and were 'fortified by their access not only to an arsenal – guns, crossbows, catapults – but also to a battalion of cousins and uncles, and orbiting around them, their courtiers, admirers and apprentices' (Campbell 1993: 176).

One of the ways for these young men to achieve high status within their group was through confrontations with authority, specifically the police. Particular kudos could be gained if the lads' activities attracted media coverage. So, local and national television footage of high speed car chases and, on one occasion, the theft of a well-publicized 'thief-proof' police car from the police station car park, brought credibility and status among peers and 'admiration and respect from younger ones' (Wallace 1992: 28). The intra-group struggle for a position in the hierarchy has to be seen in relation to inter-group, cultural/regional contestations. The inner-city disorders of September 1991 highlighted the lads' competitiveness. The street disturbances began in an area seven miles to the east of Oldchester. Residents of Wickon felt that a combination of the lads' competitiveness and the media attention succeeded in spreading the disorder from the east of the city to their own area:

> I knew for a fact that there would be bother here after [area] went up. The lads here man, they think they're the hardest and they were sick of [area] being in the paper and on the telly. They wanted [Wickon] to be in to show they were just as good, if not better.
>
> (Female, aged 17, quoted in Wallace 1992: 30)

Being a 'lad' also had racist overtones. For those males in Wickon who were anything other than white meant being positioned as a form of *marginalized* masculinity (see Chapter 2). There is a history of National Front and British Movement involvement in the Wickon area, and racism is evident in young men's views of the predicament of the Asian community during the disturbances of September 1991:

> Most of the buildings that were torched in Wickon were empty, and burning an Asian's house or shop doesn't count.
>
> (Male, aged 17)

> The darkies [Asians] are always whinging about something. It's us [whites] who should be though, we've got to live beside them.
>
> (Male, aged 21, quoted in Wallace 1992: 34)

Given the large family networks living in Wickon, it is likely that the boys in the class were familiar with what was entailed in being a 'lad'. The characteristics that allowed access to 'laddish' culture were being tough, competitive and white. Being a 'lad' also involved taking on a certain role as a member of a gang or group. The boys in the class had probably observed their male relatives engaged in struggles over hierarchical positionings in their various gangs or groups. Indeed, the fact that the males in the families of both John and Shane occupied a 'ringleader' status was reflected in their own struggles to be head of the boys' group.

Media, machismo and the 'underclass'

Among the sources which informed boys' knowledge of gender relations were the media. The images of masculinity transmitted through the television programmes and videos the boys enjoyed were particularly 'macho', for example action movies by Sylvester Stallone and horror films such as *Nightmare on Elm Street*. Ironically, while right-wing dogma of the early to mid-1990s argued that one of the reasons for juvenile crime was the increase in single parent (mother) families (Dennis 1997), the problem for Wickon youth was not that they were starved of male role models, it was that they were saturated with them!

Similar to Connolly's (1998) 'Bad Boys' whose masculine identities were constructed and negotiated in a culture concerned with day-to-day survival, so too the boys in the class were engaged in constructing masculinities informed by discourses of boyhood and childhood particular to the local context.

Discourses on being a boy and childhood

In discussing discourses on *boys* and *childhood* prevalent in the local culture, the intention is not to create an artificial divide, as clearly there is a dialectic. The amount children invest in establishing themselves, personally and publicly, as a member of their gender group was discussed in Chapter 2 (B. Davies 1989; Francis 1998). Being a male in Wickon was associated with forms of power (albeit subversive/anti-authority forms). So while young boys were not in a position to access those avenues of power, it was something that would be available to them in the future.

On the other hand, being a young child in the Wickon area appeared to be a necessary but undesirable phase; that is, where babies were proudly paraded round the streets in their prams and adolescence heralded new found status, childhood was framed in a similar way to that of working-class children in Victorian times and earlier, where the emphasis was on practising to be an adult (Cunningham 1991). For example, even the youngest pupils at Benwood Primary sported hairstyles, clothes and, occasionally with the girls, make-up, which reflected the fashions of adolescents. Children might also be encouraged to drink alcohol and smoke. Shane informed the rest of his Y2 class one 'news time' that he had been to the local pub where his dad and his mates had given him some beer. While there have to be some reservations regarding the accuracy of such comments given the boys' desires towards 'laddishness', some observations suggested these were not necessarily fabricated accounts. One morning Adam came into school looking pale and saying he felt sick. He said his uncle had put a 'tab' in his mouth and told him to suck it. Adam had done this but it 'tasted horrible so I chucked it on the floor and stood on it. Me uncle was cross 'cause it was a new 'un.'

These factors contributed to shaping the boys' knowledge, awareness and construction of their own masculine identities. This construction, negotiation and reconstruction of personal masculine identities was continued in the more public arena of the school. Here the boys were confronted with ambiguities between the school and local cultural discourses on childhood. Furthermore, they also encountered the tensions involved in discourses on *being a boy* and a *school pupil*.

Primary schooling and boys

The 'nature' of childhood

Beliefs and assertions about childhood and the nature of children have been, and continue to be, important elements in the professional ideology of primary school teachers (Pollard 1987). One line of argument suggests that these beliefs contain ambiguous conceptions of children; that is, between teachers' societal and individualistic aims (Ashton *et al.* 1975). From this perspective, teachers' societal aims position children as immature, irresponsible and dependent and, therefore, pupils need to be taught certain things for the benefit of society. On the other hand, teachers' individualistic aims stress the importance of personal growth and self-expression for their pupils. While this is clearly only one view of teachers' professional ideology, there was evidence of some ambiguity when the teachers discussed their aspirations for their pupils. Teachers linked the development of self-esteem (individualistic aims) with the ability of pupils to take responsibility for themselves and their actions (societal aims):

> *Terry Blake:* One of my goals for next year will be to try and raise their self-esteem more . . . At the beginning of the year when I gave them a task there were a good eight or nine children who would just sit at the table and do nothing until I actually spoke to them individually . . . that happens less and less . . . I encourage them to work and think for themselves . . . I suppose that's perhaps the aim, to get them to think more for themselves about what they're doing . . . to take responsibility for their learning.

> *Mrs Masterson:* My aims are for the children to come to some sort of ableness . . . to think for themselves, to engage in their own learning. That means raising their self-esteem, their self-confidence as the parents don't. I'm not bothered about the National Curriculum . . . they'll get that . . . it's *how* they learn, getting them to think, getting them to take responsibility for themselves, that's important.

The stress given here to developing a sense of 'responsibility' in the children seemed to have resonances with parental aims. Encouraging 'responsibility' could be interpreted as the school's desire for pupils to develop independence and show maturity. Similarly, those parents who dressed their children as adolescents and encouraged adult behaviours such as smoking and drinking might also be seen to be promoting maturity or independence. However, as implied so far, the reasons for, and means by which, independence and maturity were fostered, differed between teachers and parents. In fact, the strategy teachers employed to encourage boys to take responsibility had an unintentional consequence, as will be shown in the section on 'Self-esteem and masculinities'.

As can be seen from the quotations above, teachers linked the development of pupils' self-esteem with a greater sense of responsibility. Not only was self-esteem believed to facilitate pupils' abilities to take responsibility for their learning, but also much of the psychological literature on classroom behaviour links negative behaviour with low self-esteem (McIntire 1984; Fontana 1988). Therefore, it could be argued that if pupils acquired higher self-esteem then a consequence would be increased conformity to classroom rules. However, the boys in the class were positioned by multiple discourses and a particularly powerful discourse was of being a 'lad' or, in their case, an 'apprentice lad'. The tension created was that being a *school pupil* and being a *lad* demanded conflicting behaviours. While developing self-esteem might encourage conformity to classroom rules in the boys as *school pupils*, the opposite (lack of conformity) was expected of the boys as 'apprentice lads'.

School boys and 'being a boy'

For the teachers, taking responsibility meant pupils recognizing and conforming with the school authority structures; but for the boys *as* boys, 'responsibility' had implications for power dynamics:

> Although boys may maintain a privileged position in relation to girls in the school context, the schoolboy nevertheless is required to accept inferior status to the teacher, to experience powerlessness in the face of adult rule. . . . While boys are required to comply with the school's construction of the regulated student, the social construction of hegemonic masculinity promotes masculine subjectivity as less regulated, less conforming, and less compliant than schooling practices accommodate. As opposed to the constitution of the schoolboy as student, hegemonic masculinity ultimately refuses to be regulated or controlled.
>
> (Alloway and Gilbert 1997: 55–6)

The question here is how the tensions between discourses on *school pupil* and those in the local culture of *being a boy* were manifested. A useful starting point is to consider the boys' understandings of what was important to them, as boys, in the classroom. The following conversation suggested their main concern related to their position in the masculine hierarchy:

Luke: Who's the boss?
Shane: I'm the boss!
John: *I'm* the boss!
Luke: One can be first boss and one can be second boss.
Shane: Who's first boss and who's second boss?
Robert: You can take it in turns.
Shane and John [*in unison*]*:* I'll be first boss!
Robert: Shane is first boss and John can be second boss.
Luke: Yes.
Shane: I'll be first boss, John'll be second boss . . .
Luke: I'll be third boss and Robert can be four boss . . .
John: Shane'll be first boss, I'm second boss, Luke three boss and Robert four boss.
Robert: I don't want to be four boss.

Whether Robert's protests were because he did not want to be *fourth* boss, or because he did not want to be considered a *boss* at all, is not known, but either way it is not relevant to the issue raised here. It can be argued that two different agendas were in place. For the teachers, the aim was to foster a sense of responsibility both academically and behaviourally in the boys (and girls) as *school pupils*, while for the boys the main concern was to establish their masculine identity and place in the male hierarchy. The next section will explore this by focusing on how the ways in which teachers attempted to develop self-esteem in Shane inadvertently contributed towards his struggle for 'first boss' position.

Self-esteem and masculinities

Sociological studies of schools in working-class communities have identified the tensions between the culture of the school and that of the locality in which it is situated (Willis 1977; Connolly 1998). The arguments presented suggest that schools are seen by the local populace as ineffective and out of touch with the 'real' world which they inhabit, while a powerful strand within state education has been to see its role as transforming the culture of working-class children (Corrigan 1979; Dubberley 1993). Shane and John did not aspire to secure the approval of their teachers for conformity and 'good' behaviour, but to gain the recognition of the other boys and adults in the school of their potential as a 'lad'.

Mac an Ghaill (1994) has shown how the 'Macho Lads' in his study linked teacher and police authoritarianism and, as a consequence, developed their particular 'tough' version of masculinity around collective strategies of counter-interrogation, contestation and survival. Although it is fair to say that the boys in the class were probably aware of the similarities in the roles of the police and teachers in terms of control and discipline from an early age, the latter did occupy a different position. Terry Blake and Mrs Masterson referred in their interviews to the ways in which mothers would use the school as a threat by saying to pre-schoolers they would not be able to act in 'that' way when they went to school, and to older children, that they would 'tell the teacher' about the child's behaviour. It was shown in Chapter 4 how relationships between the teachers in the school and the local community were partly based on notions of maternalism and paternalism and partly by fear and vulnerability; a similarly conflictual position could be observed in relationships between the boys and teachers. As will be shown in the section on 'Inter-group conflict' the young age of the boys meant that they would look to adults (teachers) as a source of comfort and protection. At the same time, the way in which the 'lads' would stage events or engage in activities that would enable them to demonstrate superior skills, such as by driving faster than the police in stolen cars, the boys in the class would 'try out' the authority of the teacher.

Terry Blake remarked that a substantial part of his time was spent attempting to contain the behaviours of Shane and John. He said that their most prevalent misbehaviours were 'talking out of turn' and 'hindering other children' (Wheldall and Merrett 1988). This may well have been the case, but observations indicated that the greatest proportion of his time was spent in confronting behaviours which directly challenged his authority as teacher. For example:

> Terry Blake is reading a story to the children who are sitting on the carpet. Adam, Vicki and Katy have each asked over the last few minutes if they can go to the toilet (the rule is only three children are allowed in the toilets at any time). John stands up and walks past Terry Blake saying he is going to the toilet. Terry Blake calls him back and says he must wait. John sits back down. After about ten seconds he stands up and goes to the toilets even though none of the other three have returned. He stays out of the room for about ten minutes but continues to pop his head round the door to smile at the others (Terry Blake has his back to the door).
>
> (Field notes)

> It is a games session outside on the playground. The children are lined up against the wall to be sorted into teams. Terry Blake asks them to stand tall. Shane sits down on the ground.
>
> (Field notes)

Terry Blake recognized that occasionally the boys, but Shane in particular, would challenge his authority. (Terry was quoted in Chapter 4 saying how Shane's tactics were unsuccessful with him as Shane recognized him as 'the boss'.) Research on teacher–(male) pupil strategies have discussed how such behaviours as demonstrated by Shane are seen as 'deviant' (Furlong 1985) or representative of anti-school cultures (D.H. Hargreaves 1967; Lacey 1970). More recent work has suggested that anti-school behaviours, carried out by boys, should not be viewed solely in the light of teacher–(male) pupil power struggles but may also be related to processes involved in constructions of masculine identities (Mac an Ghaill 1994; Sewell 1997; Connolly 1998). Also, teacher–(male) pupil struggles do not take place in isolation from broader contexts of power; for example, the language used by Terry Blake to describe Shane's actions and his own responses drew on concepts associated with male hegemony, such as adventurousness, competitiveness, assertiveness and control (Kenway and Fitzclarence 1997). Feminist post-structuralists have argued that language and power are intertwined and, in western society, inscribed by patriarchy (Weedon 1987; Ramazanoglu 1993). So, while on one level Terry Blake's comments regarding Shane's contestation of his (teacher) authority can be read as a description of teacher–pupil relations, on another they can be interpreted as part of the process of normalization of masculinity through the regulative practice of accepting and rejecting certain forms of *gendered* pupil behaviour. This needs to be explained in more detail.

While Terry Blake was clearly aware of conventional male-teacher stereotypes and took steps to avoid perpetuating such images, there were incidents where a shared gender could influence the interpretation of relationships, for example when he accepted Shane's actions and challenges to his teacher authority. Similar attention-grabbing actions engaged in by one of the girls, Cheryl, resulted in her always being ignored. This was explained as follows:

> *Terry Blake:* With Cheryl, she breaks rules differently to the others [girls] . . . at breaktime she wandered into the junior end again and she told Mrs X that she just wanted to offer her a crisp . . . In the classroom she never sits down when she should and, I used to ask her, and she'd have some reason like she was 'tidying up'. Now I just ignore her.

The different approaches used to tackle similar forms of challenging behaviours by Shane and Cheryl lead to a consideration of how teachers achieved the apparent opposite of what they set out to achieve through encouraging pupil self-esteem. As the year progressed it became evident to teachers that it was Shane who would consistently attempt to bend, rather than always break, classroom rules, while John often seemed simply too tired to bother. To tackle Shane's behaviour teachers attempted to work on his self-esteem by involving him more centrally in classroom life with the result that he was constantly being singled out for attention:

> Terry Blake shows the book the children have made this week to them while they are sitting on the carpet. The book is called 'On Monday Afternoons' which is about their visits to the swimming baths. Two of the photographs on consecutive pages are of Shane accompanied by the text 'Shane is getting ready to swim' and 'Shane is putting his toe in the water'. The teacher adds 'I seem to remember you got into trouble for that!' Children laugh, including Shane. No other child appears either on their own or more than once in the book.
>
> (Field notes)

> Mrs Cooper [part-time teacher] reads 'The Enormous Crocodile' to the class. The children have enraptured expressions on their faces. She shows the illustrations to them and asks 'Which one is the crocodile? Shane can you show us?' Shane smiles, stands up, comes across the carpet, points to the crocodile then roars, pretends to be the crocodile and runs up to other children to 'eat' them. When he eventually calms down, he sits down. No other child is asked to contribute throughout the story.
>
> (Field notes)

These strategies may well have been building up Shane's self-esteem; however, the assumption that high self-esteem is inextricably linked with the conformist behaviours that schools value is ignoring wider contexts. As Dorothy Rowe has argued:

> It's absolutely true that to survive you have to have something you think you're good at. But you can see this in schools where kids who don't achieve find they *are* good at getting away with things and not getting caught. To say that's not self-esteem as you define it, is just imposing white, middle-class values.
>
> (Rowe, quoted in Grant 1994: 23)

The real achievement for Shane was in being able to demonstrate to his peers that not only could he take on and outwit the teachers, but also they actively appeared to sanction his behaviours by giving him more attention. Even being caught in attempts to outmanoeuvre teachers added to his self-esteem because it was important, as a 'lad', to be seen by his mates as having attracted attention to himself. As Campbell (1993) has argued in her study of the Wickon area, being caught and prosecuted, particularly if it involved a court appearance, was not seen as failure but added to the individual lad's status among the others.

The constant challenging behaviour of Shane towards teacher authority earned him a significant amount of teacher attention, and thereby the attention of the other boys. Attempts by teachers to encourage conformity through developing his self-esteem placed him even more centrally in the spotlight. Indeed, Shane's aim of being 'first boss' appeared to be enabled by these actions which pointed to the contradiction between the intended and actual aims of the teachers as a source of critical incidents. As will be shown in the next section, Shane was able to utilize teachers' attempts to promote his self-esteem in securing his 'first boss' place among the other boys.

Boys' relationships with each other

A boy among boys

Dominant modes of masculinity do not reside within individual persona but are 'ideal' constructions which few can actually achieve. The specific form of dominant masculinity in the local culture revolved around being a 'real' (hard) man, and this inevitably demanded rigorous, exacting standards of the boys. The boys in the class were engaged in running the risks involved with being in what Connell (1995: 79) has referred to as the 'frontline troops of patriarchy'. A note of caution needs to be sounded here to avoid conflating the terminology Connell uses to define hegemonic masculinity with the particular situation in which it was manifested in the local culture. That is, it is not appropriate to 'read off' the idea that dominant (violent) modes of masculinity are specific to working-class masculinities. Such a perception fails to take into account the fine-grained and complex ways in which masculinities are constructed. Indeed as was evident, the fact that some of the boys organized themselves in *relation* to the particularly dominant mode of masculinity practised by John and Shane suggested that alternative forms of masculinities were operating around and within it.

Taking into account the above proviso, it can be argued that as the 'hard man' mode of dominant masculinity centred around violence, aggression and competitiveness, then struggles to construct and negotiate one's individual male identity within this frame involved constant confrontations and challenges *between* men/boys. This positioning in a masculine hierarchy can be seen by looking at the interpersonal relationships of the boys in the class.

Shane's success in outmanoeuvring the authority of the teacher was made evident when others would join in with a situation which he initiated. The boys in the class did not constitute a gang as defined in the literature (Short and Strodtbeck 1965; W.B. Miller 1982; Goldstein 1994). Also, there is no one set of characteristics associated with being a gang leader (Patrick 1973), but one esteemed study has argued that 'The leader is usually . . . the best organizer and planner of delinquent activities' (Haskell and Yablonsky 1974: 174). On this basis, the following incident demonstrates Shane's successful ability in positioning himself as 'first boss':

> Terry Blake is carrying out a maths activity on the carpet with the whole class. They are sitting around in a semi-circle with a number of shapes on the carpet in front of them. Terry Blake describes a shape and the children have to decide which one it is. He asks Shane to start and say the name of the shape he has just described. He does this correctly. Terry Blake then describes a second shape and asks the children to put up their hands if they know which one it is. Several children put up their hands but Shane crawls to the middle of the carpet and points to the one Terry has described. The teacher says 'No Shane, you've had a turn'. A few minutes later he says 'No, Shane, Gary (who has joined in the attempt) . . . I've said move back'. He describes another shape and again Shane moves and points to it. This time Terry Blake ignores him and asks Bethany for the answer. Before she can say anything Gary shouts out 'It's a triangle!' The teacher says, 'Gary, have you changed your name to Bethany?' He then asks Charmain to choose a shape to describe. John and Matt both start to shout out a description of one of the shapes. Terry Blake says 'No, I've asked Charmain'.
>
> (Field notes)

What happened here was Shane was given the privilege of starting the game (to develop his self-esteem), but he then attempted to hold on to centre stage. When Gary started to join in, Shane moved away and simply acted as observer having 'proved' that the teacher was prepared to allow him to contravene his instructions. When Gary, and then John and Matt, attempted to mimic Shane's initial success they were immediately curtailed. At this point, Shane decided to challenge the teacher's control again but alters his strategy:

> The teacher has turned to deal with a message that has been brought in. Shane does press ups on the floor behind the teacher's back. Martin joins in, and Shane then stands up and starts demonstrating karate kicks. Martin again joins in but Terry Blake sees him and tells him (Martin) to sit down. Shane had been keeping his eye on the teacher all the time and had quickly sat as Terry Blake had turned back towards the children on the carpet.
>
> (Field notes)

> The carpet has been cleared of the shapes as it is coming up to lunch-time. The children are sent to wash their hands and are being sent out according to the month of their birth. Shane attempts to go out on the first month Terry Blake calls out. He walks past Terry smiling at him. Terry realizes what he is doing and tells him to sit back on the carpet. Shane repeats the action on the second month Terry calls out, which again isn't his birthday month. Terry lets him go. Matt and Luke attempt to join him. Terry Blake realizes immediately and tells them they will be the last to leave.
>
> (Field notes)

Some studies have found a hierarchical structure to the gangs in some communities with boys progressing from 'toddler' gangs through to 'heavy' teams (Kobrin 1962; Patrick 1973). Certainly the struggles between Shane and John in seeking 'first boss' position, and the willingness of the other boys to participate in subversive classroom activities, suggested they were taking on similar kinds of behaviours to those that were observable in the gangs of older 'lads'. And in the same way that the older lads used various means of locating themselves within the group, competing with each other to show who had the abilities to outmanoeuvre the authority of the teacher was not the sole means through which the boys negotiated their place in the masculine hierarchy.

The role of humour

While the 'lads' went to great lengths to demonstrate their individual skills at undermining police authority, many of their activities were generated as the result of collective practices so establishing a shared basis was equally important. One way of achieving this was for the 'lads' to 'hang out' and have a laugh together (Campbell 1993). A significant feature of the studies of working-class masculinities is the importance of humour to 'macho' forms of masculinity (Willis 1977; Corrigan 1979; Mac an Ghaill 1994; Sewell 1997). Indeed, as Kehily and Nayak (1997) have argued, (hetero-sexual) masculinities are regulated through humour. Their research was undertaken with working-class secondary school boys and, therefore, the actual practices differ but the principles underpinning the humour remain the same. Kehily and Nayak (1997) note that a common style of interaction was the elaborate use of game-play, incorporating ritualized verbal and physical assaults. This meant language and physicality were used in competitive ways 'where the "game" became the arena for competing masculinities' (Kehily and Nayak 1997: 71). The 'rough-and-tumble' forms of play that boys of all ages have been found to engage in were evident and were labelled by the boys themselves as 'fun fights' (Humphries and Smith 1984).

The favoured verbal game-play was 'telling the teacher'. This was where a boy would attempt to get another into trouble by 'telling':

> Dean and Bobby are sitting next to each other during a wet lunch time watching a *Ninja Turtles* video. They are fun fighting. Dean starts to grab at Bobby's legs and tickling his neck. Bobby pushes him and shouts out 'Miss, Miss, he's hitting me!' The teacher responds 'Oh Dean! Stop that now!' Both boys laugh and turn their attention back to the video.
>
> (Field notes)

> Adam, Matt and Shane are doing their maths work in the maths corner which is transformed into a bookshop complete with money till. They have been told to tidy up. They come out of the bookshop smiling and walk past me. Shane turns back and says 'He's [Adam] pinched some money . . .' All three laugh and I ask Adam to show me what's in his pockets. He attempts to take the money out by hiding it in his hand, then passing it behind his back and dropping it on the floor. All the time all three are laughing. Adam says 'Fair cop!' and the three walk away.
>
> (Field notes)

'Telling the teacher' as a form of humour and 'fun fighting' offered a means through which the boys could establish bonds with each other, and this enabled the kind of collective practices towards authority discussed earlier. At the same time, the boys did not always get on together and occasionally, unlike older boys in the school, the teacher's authority would be sought.

Inter-group conflict

As Phillips (1993) has said, there has been a tendency for feminists to ignore the fact that boys are different at ages 5 and 12 and, therefore, critical periods of masculine development have been ignored. For example, although Shane was undoubtedly the most enthusiastic 'apprentice lad' he was only 6 years old. His struggle to construct a masculine identity as 'first boss' sometimes appeared to overwhelm him and he would call on Terry Blake for support. Observations of what happened between boys after the class teacher had been asked to intervene suggested that the boy whose behaviour had been complained about always made some retaliatory gesture, as in the following episode:

> John takes one of Gary's words he is using to make a sentence. Gary crosses the room and tells Terry Blake. Terry turns and shouts to John that he must give Gary his word back, which he does. Gary sits back at the desk. John watches Terry Blake and, when he is involved with another child, snatches Gary's pencil and throws it on the floor.
>
> (Field notes)

These attempts to have the final say were common, and generally the boys appeared not to harbour any resentment. Terry Blake was aware that the boys did not value his way of dealing with their complaints, saying that 'When they come to me and say so-and-so has hit them they expect me to hit the child who hit them . . . they see it as being unfair if I don't. So if I don't sort it out the way they want they'll sort it out themselves.'

With Shane, retribution was far more protracted and vicious. As was said earlier, the research into gang leaders has identified different characteristics. Some research has found that gang leaders have to have the ability to get on with people (Haskell and Yablonsky 1974; Short 1990), while others have noted the physical and verbal aggression of the leader (Patrick 1973). The insights provided by recent research into adolescent male peer group cultures have enabled a greater understanding of these findings, particularly how significant *local* forms of masculinity are. The 'hard man' form of dominant masculinity in the Wickon community was inscribed with violence and aggression and, hence, it was more likely that leaders would draw on these to assert and retain their position. Shane, in his securing of 'first boss' position, not only used similar forms of violent behaviours but also, importantly, and unlike the majority of others in the class, would keep up a sustained attack which drew in other boys. The following incident took place over the course of an afternoon but its effects were more far reaching:

> Shane and Dean have been close friends for several days. This morning, at carpet time, Dean came in and Shane grabbed his legs. Dean fell down on top of Shane and they were both laughing. (After lunch) Shane comes into the classroom looking very upset and goes immediately to Terry Blake saying 'Dean's hit me with a stick'. Terry Blake tells him to sit down on the carpet and if he has something to say he

must put his hand up. Shane throws himself onto the carpet at Terry Blake's feet and puts his hand up. Terry listens to Shane's story and then asks Dean for his. After they had finished, Terry said, 'I don't think either of you are being very nice to each other . . . now sit down'. Shane turns and goes across the carpet to sit with John. After Terry has taken the register Shane shouts out 'Dean has got his sweatshirt on back to front!' Several of the children laugh. Dean looks embarrassed but says nothing. Terry says, 'Perhaps he wants to wear it like that'. As the children move to their activities Dean pulls off his sweatshirt and puts it back on the right way.

(Field notes)

(Later that afternoon). An incident occurs when Terry Blake observes Shane hitting Dean in the face. They are both in the same group and it is their turn to play in the home corner area. Shane is dressed in a long coat and tells me he is playing a 'man' 'wiv me tabs and me New-key Broon (drink) doen the Shambles' [street in Oldchester famous for its nightlife and as a rendezvous for young people]. Dean has been playing with the jigsaws and Shane approached him and hit him. Terry calls Shane over to him and asks why he is hitting Dean. Shane says his hand had 'slipped'. Shane is sent back to the home corner.

(Field notes)

(Later again). Dean is standing near the box of bricks. Shane comes up behind him and pushes him roughly to the floor. Terry Blake turns at the noise and, before he can say anything, Shane says, 'I'm only playing'. Terry Blake asks 'Does Dean know you're playing?' In response Dean stands up smiling which implies that what Shane has said is true.

(Field notes)

The children have been told to tidy away. Dean is kneeling on the carpet playing with the cars. Shane walks past him, puts his hand on the back of his head and forces his face down into the carpet. He quickly walks away. Dean looks up, rubs his chin and stares at Shane but says nothing.

(Field notes)

Having set up a situation in which Shane punished Dean persistently over the course of an afternoon, he built upon this over the next few days by isolating Dean from other boys who seemed to be offering any form of friendship to Dean. For example, Bobby was helping Dean with his maths and Shane said, 'Man, man come over here . . . this is real hard work, not easy-peasy stuff like that.' As suggested here, the main strategy used by Shane was that of ridicule. By ridiculing Dean, as in the comment about the sweatshirt, Shane implied that anyone who hung around with him was also ridiculous. Throughout all this Dean made no attempt to defend himself.

Within days other boys had joined in the attacks, justifying their actions on the basis of Shane's alleged mistreatment by Dean. For example, a few days after that afternoon Terry Blake told Dean off for copying his maths work into his weather diary. Adam told me that he, Gary and Martin had told Dean to do it:

CS: Why did you do that?
Adam: Because we wanted him to get wronged by the teacher.
CS: Why? What had he done?
Adam [*shrugs*]*:* We think he hit someone. Shane it was.

This section has attempted to outline the ways in which the boys negotiated their masculine identity within the classroom culture. Similar to the 'lads', it appeared they needed to demonstrate both individuality as well as group cohesion, and their effectiveness at this secured a place in the male hierarchy, albeit at various levels in relation to the 'boss'.

Summary

This chapter has been concerned with the ways in which masculine identities were constructed, negotiated and reconstructed by the infant boys in the school setting. The knowledge and awareness the boys brought into school with them of the dominant masculinity in the local culture appeared to inform their own behaviours and relationships. Where the police provided a focus for anti-authority activities for the 'lads', an awareness of the control and discipline aspect of the teaching role partly informed the boys' relationships with their teachers. In fact, shortly after the end of the observation period Shane and John had made their first attempt at 'getting one over' the police. Terry Blake said that to his knowledge what had happened was that John and Shane had been caught by the police after being seen removing a steering wheel from a car. According to an older boy, they had tried to drive it but neither of them could get their feet on the pedals. Further attempts with one steering and one operating the pedals had also failed, so, rather than go away empty-handed (as reported by Shane at 'news time'), they had taken the steering wheel off, even though they knew they had been seen by a police patrol.

It has also been shown that the process of normalizing masculinity takes place around and within a framework of discourses which the boys drew from and were located within. However, the 'grid of possibilities' (Skeggs 1991a) offered by this framework was itself constructed through power/knowledge positions (Heath 1982). The available discourses had differing relationships to power, so the discourse on *being a boy* drew on and incorporated greater access to power than discourses on *childhood*, or being a *school pupil*; therefore, discourses on the former were much more powerful (influential). The emphasis has been on those boys who were part of Connell's (1995) 'frontline troops', and it needs to be reiterated that there were boys in the class like Bobby, Dean, Sean, Adam, Martin and Robert who appeared to join in with the activities of the more forceful, challenging boys because it could have been potentially more personally damaging *not* to have colluded. Indeed, in much of the autobiographical literature written by men on masculinities, writers recall the fear they felt at school of being accused of being a 'poofter', 'wimp' or 'a girl' (D. Cohen 1990; D. Jackson 1990). At the same time it is important to emphasize two important points: that violent modes of dominant masculinities are not the 'preserve' of working-class male practices and indeed the responses of boys like Bobby, Sean and others indicate that alternative, if not resistant, patterns of masculinities are often operating within a more explicitly evident hegemonic framework. Also, the situation at the school in terms of the articulation of these differing modes of masculinities were particular to this school site and cannot be generalized to other primary schools located in similar economically situated areas.

A problem here is that in saying that the findings of research into one primary school cannot be generalized to all schools means that it can be criticized in a similar way to much of the literature on masculinities and schooling. That is, the problems have been identified but what about the solutions? Epstein *et al.* (1998) have suggested that the reason why there has been a reluctance on the part of ethnographic researchers to come up with strategies for teachers to use is because 'the issues are multi-faceted, the research complex, and it would be premature to suggest firm directions for others to follow' (Epstein *et al.* 1998: 14). While I support this thesis it does leave teachers somewhat stranded. Actual recommendations based on the findings of this research appear in the concluding chapter but an important point has been raised here and in Chapter 4. That is, individual schools need first of all to recognize the modes of masculinity operating within the school in order to develop responses and strategies which are relevant to their particular situation and needs.

chapter / six

Male teachers and primary schools

Introduction

The spotlight in Chapters 4 and 5 has been on how *schools* construct forms of masculinity through its authority patterns and how boys negotiate their own ways of 'being a boy' in relation to these. But what of men teachers in primary schools? As Kessler *et al.* (1985: 38) have argued, 'men teachers have a particular responsibility and opportunity . . . because what they say and do influences what kind of masculinity is hegemonic in the school'. The relatively low number of men teachers in primary schools has been identified by some commentators as instrumental in boys' disinclination towards academic work (Biddulph 1997; Bleach 1998). Government agencies are evidently taking this situation seriously and a number of projects are being funded by the Teacher Training Agency to encourage more men into primary teaching. Certainly the 'feminized' primary school is an idea which has attracted media attention in the boys' underachievement debate. Reports in the press use 'commonsense' (rather than research based) evidence to explain what has happened to boys in primary schools because of an absence of men teachers. For example, the predominance of female teaching staff has meant they have 'unknowingly [been] moulding education and assessment to suit their gender' (Budge 1994: 12). Similarly, a feminized profession explains why schools are excluding more boys so greater efforts should be made to encourage men teachers 'and to let them strut across the classroom as figures of authority, not as nannies' (Shakespeare 1998: 6).

These 'dangers' associated with a 'feminized' teaching profession are clearly debatable. After all, as was shown in Chapter 4, it is not only men who are able to manifest masculinity and women femininity. It was evident in Chapter 4 that women teachers drew on masculine control and management strategies as much as the men teachers. As Connell (1995: 230) has argued: 'Though most discussion of masculinity is silent about the issue, it follows from both psychoanalytic and social construction principles that women are bearers of masculinity as well as men'.

Also, increasing the numbers of men primary teachers per se sidesteps questions and issues which need to be considered. For instance, what kind of male role models do we want to provide boys with? The majority of those calling for more male primary teachers do not seem to have considered this and those that have present a pretty grim picture of the existing male primary teaching force. Biddulph (1998) argues that men teachers need 'special training' as:

> Readers will recall their own schooldays when more men worked in primary education and how many of these men were maladjusted (and even sadistic) individuals. Today many male teachers can be more interested in the career track than the genuine needs of children.
>
> (Biddulph 1998: 147)

Hardly an image of positive role models for boys, although Steve Biddulph's view is highly subjective and might not stand up to scrutiny!

Also, what are the implications for the hierarchy of primary schools if the numbers of men teachers are increased? Would we be exacerbating the current images children have of predominantly men in powerful positions such as headteacher while women do the 'housework' of the school (classroom assistants, lunchtime supervisors, cleaning staff)? Clearly there are issues which need thinking through but, more importantly, the discussions on men primary teachers do serve to raise the issue that teachers are gendered beings and that this influences the nature of relationships and interactions within the school as well as individuals' feelings of personal identity.

The intention here and in Chapter 7 is to identify current discourses on men teachers and to ascertain the ways in which men construct themselves as 'properly masculine' in school settings. For example, discourses on men teachers locate them as effective disciplinarians and emphasize the managerial aspect of the job hence an expectation on men to occupy positions of authority whether or not they wish to go for promotion (Skelton 1991). Men teachers have been shown to locate themselves as 'properly masculine' by drawing on gender relations such as aligning themselves with 'the lads' or questioning the masculinity of others (Mac an Ghaill 1994; Francis and Skelton 2001). This chapter begins by ascertaining where men actually are in teaching. The next section is concerned with placing current discourses on men teachers into context by looking at the historical background on men and the teaching profession. It has been said at various stages in this book that one perspective on gender is a consideration of how masculinities and femininities are constructed in relation to an 'other' (gender as relational). The third section explores this in terms of the discourses on men teachers in one primary school.

A key factor in exploring hegemonic masculinity is the centrality of heterosexuality. This will be considered in relation to men teachers in Chapter 7; here in this chapter the focus is on how the teachers located their masculine identities by associating themselves with the boys. Two aspects are explored: the place of humour and a shared commitment to football in generating exclusively male spaces within the school.

Where men are in primary schools

In some ways there is a reason to be concerned about the low numbers of men in the nursery/primary sector as government statistics note a decline from 1994 to 1998 while the number of women teachers increased slightly then stabilized. Taken together, the indications are that there is a problem with teacher supply generally and not just with men primary teachers (see Table 6.1).

The overall drop in men teachers involved in the nursery/primary sector is further reflected in the number of headteacher posts occupied by males. However, these need to be seen in context. While Howson (1998) reported that between 1992 and 1996 the number of male primary heads fell by 1000 and the number of women heads rose by about 500 the disproportionate number of men headteachers remains. The figures for headships in the nursery/primary sector for 1998 are given in Table 6.2.

To put these statistics more crudely, one male primary teacher in four is a headteacher. The chances of becoming a female headteacher is approximately one in thirteen. The result of more men in positions of authority is that the average salary of men teachers in the primary sector in 1998 was £21,030 and the corresponding income for women was £20,900 (DfEE 1999b).

Where there has been a slight increase is in the number of men in nursery schools, which has gone up from ten in 1985 to forty in 1998. At the same time, these male nursery workers are clearly in the minority in comparison to the 1400 female nursery staff (DfEE 1999b).

Table 6.1 Full-time teachers in maintained nursery/primary schools March 1994–8

Teachers	*1994*	*1995*	*1996*	*1997*	*1998*
Men	32,700	32,300	31,900	31,000	29,900
Women	147,700	149,400	150,100	150,400	150,300

Source: DfEE 1999a

Table 6.2 Headteachers in maintained nursery/primary schools 1998

Nursery/primary heads	*Number*	*Percentage*
Men	8,700	42.8
Women	11,700	57.2

Source: DfEE 1999a

Men and teaching: an historical perspective

It may well be the case that the teaching force is increasingly female but the idea that teaching always has been predominantly female masks a more complex picture. To gain a clearer insight into the development of the teaching profession, the significance of social class as well as ideologies of masculinity and femininity at any given historical point needs to be considered. For much of the Victorian period the most likely recipients of any formal education were upper-class and, to a lesser extent, middle-class boys. Although many of these upper-middle-class boys were privately tutored by governesses or masters when they were young, the majority of their education took place in one of England's public schools. The task of the school masters was to prepare boys for their future as leaders in politics, the military, the Church and the professions. It is interesting to note that 'education' was not classified as one of the professions appropriate for elitist males although some renowned and well-respected reformers emerged from this field such as Dr Thomas Arnold, headmaster of Rugby School, and Dr Edward Benson, headmaster of Wellington College. Rather, it was an occupation undertaken alongside a more reputable profession, for example, many, such as Arnold and Benson, had also been ordained into the Church (Trevor 1973). As Clarke (1991: 28) says of the historian and essayist Thomas Carlyle, who attempted to earn an income from, first, hack writing and then private tutoring, 'neither of these pursuits rated as "an honest calling"; each was rather a source of shame. They held in them no prospect of future status or security.' So, while teaching upper- and upper-middle-class boys was carried out exclusively by school masters who were themselves from the same social class background and who were products of the public school system, it was not an occupation held in high esteem. The situation was different for male working-class and lower-middle-class school teachers.

Although formal education for upper- and upper-middle-class boys occurred from about the age of 9–10 years, that of working-class boys began and finished earlier in elementary schools. The introduction of mass state schooling in 1870 (see Chapter 1) generated a need for more trained teachers. Prior to this point there were approximately equal numbers of men and women teachers (Partington 1976). However, this number included the school masters teaching in the public schools as well as teachers in elementary schools. The majority of elementary school teachers were drawn from the industrial or rural working classes although most commonly from the better off families (J. Miller 1996). The attraction of teaching for males and females was that it was 'the most accessible middle-class occupation for able children from working-class homes and thus served as an important avenue of social mobility' (Morrison and McIntyre 1969: 43).

School masters in the public schools had received a university education and were, therefore, not expected to undertake any professional training to become a teacher. In contrast, the introduction of mass compulsory schooling prompted the state's concern to secure a supply of properly qualified elementary teachers. From the mid- to late nineteenth century the majority of teachers entered through the pupil-teacher route which offered the possibility of some form of higher education through the Queen's Scholarships. Eligibility for these scholarships was obtained by satisfactorily completing a five year apprenticeship in schools after which time pupil-teachers received a certificate entitling him or her to sit the public examination held annually for the award. Prior to 1846, teacher trainees were expected to pay fees but after this time those students who passed the Queen's Scholarship entrance examination had four-fifths of their college place subsidized by the government. Those who qualified were given annual maintenance grants of £25 for men and £20 for women. Those who did not gain a scholarship could work as assistant teachers and after three years were able to obtain their certificates by taking the 'acting teachers' examination'. More men students than female students qualified through the scholarship rather than the acting teacher route, with the reverse being the case for women. Teaching provided working-class men with the opportunity to gain a form of higher education and reasonable pay in relation to other, manual, working-class occupations (although the salary was low in comparison to that of middle-class occupations). Given these advantages, when and why did teaching become a predominantly female occupation?

There are a number of interrelated factors which explain historical (and indeed current) discourses around men teachers. First, the expansion in the teaching profession after the 1870 Act was in the lower status elementary schools, not the prestigious public schools. Elementary teaching offered opportunities to women such as independence and a chance to be economically self-sufficient. These benefits were taken for granted by lower-middle-class and working-class men, who were then less likely to have a positive view of teaching as a profession (John 1983). Second, working with young children was then, as now, seen as female work. Ideologies of 'manliness' as a code of conduct for men in the nineteenth century centred around 'moral courage, sexual purity, athleticism and stoicism' (Roper and Tosh 1991: 2). While the majority of male occupations from the military, the Church and the law to working in industry allowed men to demonstrate one or more of these characteristics, irrespective of social class, teaching elementary children allowed for none.

A third point is that the position of women, specifically middle-class women, changed significantly during the latter part of the nineteenth century into the early part of the twentieth century. It was said earlier that in the mid-nineteenth century teaching attracted roughly equal numbers of boys and girls, mainly from the working class, who were recruited as pupil-teachers. Marriage was the assumed 'vocation' of women but there was a growing awareness that not all women did get married and that the tendency was towards later marriage for middle-class men. This, together with a number of other factors, served to make teaching an acceptable job for, initially, lower-middle-class women and, later, women from the professional classes (John 1983; Widdowson 1983; Purvis 1991). These factors were the introduction and relative cheapness of training colleges; the idea that the elementary teacher's role of socializing children could be seen as a practice for becoming a 'real' mother; and the pioneering work of such women as Emily Buss and Dorothea Beale, who campaigned for better education for girls and women. The increasing numbers of women entrants to elementary teaching coupled with ideas of 'manliness' and suitable careers for men made this an unattractive occupation for men.

The introduction of compulsory secondary state schooling in 1902 brought with it further implications for status within the profession. Figures on secondary teachers in post in 1913 show that out of 5000 men, 180 had been trained for secondary schooling, 1970 for elementary and the rest had not been trained at all (Statistics of Public Education 1912–13). The explanation for the large number of elementary trained men is that they were working in higher grade elementary schools before the 1902 Act which then became secondary after this time. Gosden (1972) speculates that many of these trained elementary men teachers had done so because facilities for secondary training had not been available. In terms of status:

> the big public schools were regarded as the aristocracy of secondary school teachers; they were the best educated, the best paid, teaching under the best conditions and in some ways embodying much of what was regarded as best in English education . . . public school masters had little need of 'training' in the formal sense, but their disregard of it and the low opinion they showed for it was certainly one of the main obstacles to the creation of a system of training for secondary school teachers.
>
> (Gosden 1972: 216)

He goes on to provide examples of the distinction made between those who were trained as teachers and those who were not, with the latter occupying a higher status; for example, a letter from the Revd Dr Abbott, appearing in the *Spectator* in 1877, asked 'Do you want to degrade the teachers of our public schools to the level of certificated masters?' His view was shared by the headmasters of the public schools, who were indifferent, if not hostile, to men training for teaching saying that those who did were 'men who had tried to teach, and had failed, or men who had not the confidence in their own powers' (Select Committee on the Teachers' Registration and Organisation Bill 1891). Such views are encapsulated in a letter to *The Times* in 1887 from the headmaster of Cheltenham College who spoke of

> a certificate which gives no real guarantee that its holder will not fail when he comes to meet his class . . . It is surely impossible to certify by examination that any man is a practically efficient teacher, and if he is not his theory is worthless. The subtle influence over boys which characterizes a 'good disciplinarian' is a quality which cannot be imparted by lectures. We cannot tell how it is acquired . . . Anyone who has been educated at a public school has had a previous training which, if he is a man of intelligence and sound common sense, must be of service to him as a teacher.
>
> (quoted in Gosden 1972: 221–2)

Furthermore, these ideas that men teachers were 'natural' products of their own (public school) education and therefore should be using their skills with older pupils (boys) was supported by official policy. A government report on the training of elementary teachers described it as 'a field of effort for the girl of average intellectual capacity and normal maternal instincts' but cautioned that 'for a man to spend his life teaching children of school age is to waste it in doing easy and not very valuable work, he would not do it if fit to do anything else' (Board of Education 1925: 34, 41). So, while training was seen as appropriate for women and women themselves responded to such opportunities, it was this, together with the fact that the first group of teachers to be formally instructed were elementary teachers, which helped locate training to teach, and teaching younger pupils, as inappropriate for men from particular social classes. Inevitably the 'stigma' would influence the decisions of all men considering entering the teaching profession.

This historical evidence illustrates how teaching as a career was clearly articulated as stratified along gender and class lines particularly from the introduction of compulsory schooling when the issue of teacher recruitment and supply became a government issue. For men teachers this involved positioning themselves as obviously different from women teachers: in terms of the age group they taught, the irrelevance of training for men, differential pay structures and the functions they served within the profession. Littlewood's (1995) analysis of the documents produced by the National Association of Schoolmasters (NAS) has discovered some amusing examples of the strategies adopted by the union to argue for differences in men and women teachers pay. In the early 1920s the NAS argued that as teaching was relatively low paid in relation to other male professions it would attract only 'the unambitious man of low mental power and low attainment' (*Equal Pay* 1920). This meant that, in relation to girls who were taught by better educated women teachers, boys would have 'not only an inferior type of teacher but [these] teachers [have] to work under greater mental stress and consequently lessened resiliency of mind' (*New Schoolmaster* 1922) (cited in Littlewood 1995: 48). That is, they were arguing the opposite of what is usually entailed in negotiating higher pay which relies on the worker being able to show superiority over others. Instead the NAS put forward the case that it was because men teachers were inferior to women teachers in terms of education and from a lower social class that the former should be paid more! As Littlewood (1995) points out, this is an alternative reading of the actual reasoning offered by the NAS, which premised its call for differential pay not on the incompetence of male teachers but on the roles men and women played within the family; that is, men needed more money to keep a family at home.

Feminist historians of the teaching profession and the teacher unions have shown the impact discourses on sexuality had in informing constructions of women teachers (Littlewood 1995; Oram 1999). To explain briefly: well into the first half of the twentieth century the marriage state was seen to be women's 'natural' vocation. Local education authorities introduced a marriage bar (which ran from the 1920s to the early 1930s whereby when a female teacher married she relinquished her contract) claiming that women could not fulfil their professional and personal responsibilities. In actuality the marriage bar was an economic device designed to prevent vast numbers of newly qualified teachers finding themselves unemployed. At around the same time there was an increased interest in sexuality and experts such as the sexologist Havelock Ellis and, later, Marie Stopes, argued that it was normal for women to experience sexual desire but that this needed to be exercised within the 'normal family' (Oram 1999). Women teachers were placed in an extremely difficult position: single women were portrayed as frigid or 'abnormal' (that is lesbians) while married women teachers were seen as more fulfilled but were also unemployed or undertaking the equivalent of supply teaching. What implications did this have for constructions of men teachers?

The *New Schoolmaster* (November 1922) observed that 'There are few men and few women so gifted as teachers that they would not be improved through marriage and parenthood'. In later years, the union, operating on the premise that masculinity was located in the body and required men teachers to foster and guide that essential manliness, cautioned that 'the biggest danger of having women as teachers of boys is that the mother-dependant type will become more common' (July 1935). Littlewood (1995: 48) has commented that such claims 'placed all men firmly in a particular relation of heterosexuality. There was no place for the heterosexual man who wished not to marry, and more importantly there was no room for the homosexual.'

The general tenor of the language used in the *New Schoolmaster* during the 1920s and 1930s provides a clear indication of how men teachers positioned themselves as heterosexual, family men. This not only justified the case for a family wage but also provided some insight into how masculinity was understood; that is, masculinity was 'achieved' by actively rejecting 'the feminine' in terms of rebelling against 'spinster authority' and instead boys would 'invite the leadership of a man' (*New Schoolmaster* December 1936). This clear identification of men teachers as *different* from female teachers and emphasizing their 'manliness' as central to their work as a teacher may also have been developed to address how they themselves were perceived by wider society.

Waller (1932) observed that men teachers were constantly aware of the gulf between them and other men in the community. To illustrate this, he recounts an incident in a barber's shop where the swearing and telling of ribald tales by the men stopped when the teacher walked in. When the teacher left, the barber said: 'Boys I enjoy a good story myself but this is a public place and we've got to treat such men as the one that left with respect. *Besides that I have many women customers*' (original italics). Waller goes on to say:

> the obviously artificial conversation . . . the assimilation of the teacher to the female character ideal . . . the suppression of normal activity when the teacher entered the room . . . It has been said that no woman and no negro is ever fully admitted to the white man's world. Possibly we should add men teachers to the list of the excluded.
>
> (Waller 1932: 50)

Here Waller is raising an issue which has continued to inform discourses on constructions of men teachers, specifically that of masculinity and sexuality. This issue will be explored more closely in Chapter 7.

Before moving on it is worth reiterating the key points made here. Our knowledge of teachers and teacher training prior to the mid-nineteenth century when the pupil-teacher scheme was introduced (by Kay-Shuttleworth in 1846) and before state intervention in teacher recruitment and supply is somewhat sketchy. However, there is evidence to suggest that the number of men and women teachers were roughly equal (Partington 1976; J. Miller 1996). From the introduction of compulsory schooling and the need to train elementary teachers, the teaching profession became clearly stratified along lines of social class and gender. School masters in the public schools were the products of the same system, hence drawn from the upper social classes, although such work was of lower status than those of other male professions. School teachers in elementary schools were increasingly likely to be female and lower-middle class, and later, middle class. Men elementary teachers originated from the working classes and, with the introduction of state secondary schooling, more likely to be found teaching this age of pupils.

Ideologies of masculinity and femininity informed the constructions of male and female teachers. The rapid shift in constructions of femininity at around the turn of the nineteenth century meant that while marriage and motherhood were seen as a woman's main role, there was a greater recognition and acceptance that some middle-class women might need to work (working-class women had a much longer history of paid employment). For working-class men teaching offered the opportunity for some form of higher education and a job which paid on a par with, but for shorter hours than, manual occupations. The negative aspect was that being a man teacher of young children provoked questions about an individual's masculinity. Thus within the teaching profession, school masters made efforts to distinguish between teachers of different age groups arguing the irrelevance of training for men teachers in the public schools, which spilled over into the grammar and secondary schools (Gosden 1972). In the same way that upper-middle-class male teachers distanced themselves from their working-class counterparts in the state system so the documentation produced by the NAS exemplified some of the ways in which men school teachers clearly articulated and emphasized *differences* between themselves and women teachers.

The next section considers current discourses on men teachers in primary school.

Men primary teachers: working in a woman's world

As Acker (1995) points out in her extensive review of the literature on teachers' work, the influence of gender has been minimized, even in studies of primary school teachers. She goes on to add: 'there is a small literature making problematic gender issues for those men who teach in elementary schools' (Acker 1995: 106). At the same time, despite the contribution of post-structuralism in terms of providing a means through which *difference* can be explored, the diversity *among* men (and women) teachers has yet to be taken into account. In a similar way, Abbott (1993: 197) has observed that although there are studies that compare the sexes, and many studies that examine differences among subgroups of women, 'varieties of women are [never] compared to varieties of men'. He continues: 'The antinomy between the simple male–female opposite and the diversity of women when considered alone is the basic conundrum of this literature, indeed, of the gender and work literature as a whole' (Abbott 1993: 197). What this means is that there is a substantial amount of information about men teachers yet very little about the *masculinities* of men teachers.

As yet, only a small amount of literature exists which explores the problematic nature of masculinity for men who work with primary and nursery children (Seifert 1988; Sheppard 1989; Skelton 1991; Allan 1993; Murray 1996; Cameron 1997; Thornton 1998; Smedley 1998, 1999; Sumsion 1999). However, some of the literature on masculinities and secondary school teachers offers opportunities for comparisons and differences between how men in the two sectors of education might construct and negotiate masculine identities. For example, in their consideration of masculinities and secondary teachers, Haywood and Mac an Ghaill (1996) note there are two interlinked areas that demonstrate the ways in which teachers' masculinities are produced. The first is concerned with teacher ideologies and their relationship to the labour process, and the second involves the use of discipline in teaching styles.

Teaching has changed substantially over recent years, with greater emphasis placed on specialization. At the same time, there has been a deskilling of the job represented by a move away from liberal-humanist constructions of teaching towards technicist approaches. This has had implications for the way in which masculinities are worked out. Mac an Ghaill's (1994) study identifies three groups of male teachers whose masculine identities were shaped by their different responses to educational reforms: 'Professionals', 'Collectivists' and 'New Entrepreneurs'. The 'Professionals' manifested a masculine style that revolved around discipline and control; the 'Collectivists' drew on masculine forms that supported equality in terms of antiracist and anti-sexist stances; and the 'New Entrepreneurs' supported conventional forms of masculinity centring on upward mobility achieved through processes of appraisal, accountability and effective management.

Given that, first, no comparable study exists of men primary teachers and, second but related, few primary schools have sufficient numbers of males for such patterns to be discerned then it is not possible to draw parallels with groupings within primary schools. A word of caution here as neither Mac an Ghaill's work nor the tenor of the discussion here is meant to suggest that such behaviours 'belong' to men teachers and their masculine identities nor rule out the notion that women may be bearers of masculinities too (Connell 1995; see also Chapters 4 and 7). Setting these issues to one side, links can be made to the particular form of masculinity demonstrated by the 'Professionals' which calls on the second area identified by Haywood and Mac an Ghaill where teachers' masculinities are reinforced, specifically through their legitimation of different teaching styles.

Masculinities have to operate, or be competent at operating, some degree of power and authority (Brittan 1989). Thus, male teachers' identities, ideologies and pedagogical styles are constructed around certain modes of masculinities which are intended to demonstrate what kind of men they are (Haywood and Mac an Ghaill 1996) and, in terms of teaching, that means making explicit forms of discipline and control (Beynon 1989). As was said earlier, the issue of control and management of pupils was, and continues to be, of significance for male teachers in the primary school. If teaching generally is seen as a 'soft option' in the list of male occupations (Connell 1985), and primary education in particular is related to femininity, then male teachers in this sector are aware of others' attention to their maleness (Thornton 1997). Studies have shown that for men working with young children this results in continual negotiation of their masculine identities (Skelton 1991; Penn 1996). For example, Allan (1993) argues:

> They must assert – and especially model – 'being a real man' in ways that are personally sustainable, that have integrity, and that are also acceptable to those who evaluate them on this important job criterion and control their careers. At the same time, they feel pressure to conform to stereotypically feminine qualities to establish the sensitive, caring relationships necessary to effectively teach children. For these men, gender is highly problematized, and they must negotiate the meaning of masculinity every day.
>
> (Allan 1993: 114)

One way of handling these contradictions is for men to emphasize those aspects of teaching that are more compatible with conventional masculinity (Connell 1985). This may partly explain why men primary teachers are concentrated in the upper years of primary school where they have responsibility for the management and control of the oldest pupils and occupy senior management positions. Also, the idea that men teachers generally may emphasize traditional male characteristics adds weight to those findings which show that male teachers identify with exaggerated forms of masculinity among boys in secondary schools such as the sexual prowess of the 'lads' (Willis 1977) while rejecting other forms, labelling them as effeminate, as with the anti-sports, anti-violence 'Goths' (Abraham 1989).

In primary schools, men teachers have been shown to place great emphasis on, and demonstrate deep commitment to, shared masculine activities, particularly football (Connolly 1998). The next section will explore the influence of football in one primary school in more detail. Before going on to look at the dominant form of masculinity evident in the discourse on men teachers in one of the case study schools, some indication is needed on how the discussion is to be shaped. It has been noted earlier that men teachers need to identify themselves as 'properly masculine'. This involves locating themselves in relation to discourses on men teachers as efficient managers, effective disciplinarians, knowledge specialists and so on. At the same time, identifying themselves as 'real men' demands recourse to heterosexuality. The significance of heterosexuality in discourses on men teachers is of particular importance given the added power inequalities of adult–child relationships. Because of this the intention here is to tease apart the specific dominant discourse on men teachers in one primary school so that

Chapter 7 will explore issues relating to heterosexuality and the discussion which follows looks at how the male teachers drew on the gender dichotomy to align themselves as 'one of the boys'.

Constructing hegemonic masculinity at Deneway Primary

There were three male teachers at Deneway Primary out of a total of 14 teaching staff at any one point in the research. The three men were the headteacher, Mr Kenning, a Y5 class teacher, Philip Norris, and a Y6 teacher, Stephen Coles. At the end of the year Philip Norris left and moved out of the area; his position was filled by Bill Naismith, who taught a Y6 class. Deneway Primary School is located in a middle-class area of Oldchester and the majority of pupils came from homes where both parents had professional careers (for details of the school see Chapter 3). Equality of opportunity was something referred to several times in the school brochure and this espoused commitment was supported by curriculum activities. For example, a Y6 class undertook a topic on 'gender', which involved the pupils in a range of activities such as carrying out an observational study at the local Tesco's to see who did the shopping, while one of the girls compiled a questionnaire for teachers to answer which attempted to find out their knowledge of 'women in society'. Added to this, curriculum opportunities were provided for cooperative games, knitting, sewing and cooking for all pupils. Furthermore, the men teachers would refer to gender equality with pupils and teachers. For example, on one occasion the school secretary came into a Y5 class and asked for two strong boys to help move some furniture to which Philip Norris responded, 'We have strong girls here too, perhaps they would like to volunteer'. However, an outward support for 'equality' did not tackle the subtle ways in which gender relational practices actively privileged a particular form of masculinity to which the men teachers aligned themselves.

The form in which the dominant masculinity at Deneway Primary appeared was reminiscent of an exclusive male sports club where the members bond through shared humour and shared commitments to, and interests in, 'their' sport. In many ways the form of masculinity adopted by the men teachers was a middle-class, adult male version of the 'lads'. This will be shown by focusing on two aspects: the role of humour in classroom management and the centrality of football in the school.

Men teachers and the use of humour

Some of the ways in which Deneway Primary demonstrated a commitment to gender equality were mentioned earlier. However, *how* curriculum subjects are taught is as, if not more, significant than the content material

(Murphy 1996). Also, in a study of secondary school boys Kehily and Nayak (1997) have shown how making supposedly 'humorous' comments consolidates heterosexual masculinities. This can been seen in looking at the events of one afternoon when Bill Naismith was teaching the 'Gender' topic to his Y6 class. It appeared that he was using humour as a means of subverting some of the serious issues raised by the subject.

At an early stage in the session the class were given a passage entitled 'Jobs for Boys' and were asked to look for, then change, words they believed were sexist, such as 'manning', 'chairman' and 'ladies'. When they had completed the task Bill Naismith went through the answers with them. After the first few words and phrases had been identified and amended he began to ridicule the task by interspersing 'humorous' comments:

> *Bill Naismith:* I have a friend who thought his name was sexist so he changed it. He used to be called Guy Chapman and now he's Person Personperson.

This was followed a few minutes later by an explanation of the non-sexist title 'Ms'. Here again humour was used to deride the inaccurate and misleading explanation he provided:

> *Bill Naismith:* Ms is used by some ladies who don't want people to know they are married. Hands up those of you who think it's important (several girls' hands go up) . . . now, those of you who don't think it's important (a few boys' hands and a couple of girls' hands go up) . . . and (in a mock exasperated tone) those of you who don't care!

At this, the majority of the class put their hands up and there was much laughter from several children and Bill Naismith. At this point the bell sounded to indicate the end of the school day. Bill Naismith had clearly not intended his statement to be used as a teaching point as he did not draw the class's attention to his own use of 'ladies' or the implicit sexism in the definition he had provided. Instead he began to dismiss the class by calling out the names of individual children. When four or five children had left the room he called back two of the boys:

> *Bill Naismith:* Gentlemen! Come back here and tidy this table for me. It has to be a gentleman . . . we can't ask the girls can we? (Laughter from himself and the rest of the class).

Kenway (1995) has argued that the challenges posed by feminism through gender reform practices, such as the topic on 'Gender', brings into question male teachers' personal and/or professional histories and specific masculine identities and frequently results in feelings of anger, resentment and/or anxiety. A way of dealing with these feelings is to subvert the serious messages that feminism has to offer by 'sending them up' thereby assuaging any anxiety but also, as was said earlier, making 'humorous' comments *consolidates* heterosexual masculinities. This was evident in the classroom management style of Philip Norris.

The term 'fraternization' has been used to describe the sorts of behaviours where teachers identify more closely with their pupils than their teaching colleagues (Woods 1990). In so doing these teachers are demonstrating signs of alienation from, although not posing direct challenges to, the official culture of the school. In Philip Norris's case he occasionally allied himself as a 'school pupil' in the way suggested in the following observation:

> The class teacher (Philip Norris) has the children sitting around him on the carpet . . . He tells the children they will have to get on with the plays they are producing for Class One on their own after breaktime, 'I'm on playground duty and Mr Kenning [headteacher] wants to see me . . . I'm probably in trouble again . . . don't know what I've done this time.' One pupil shouts, 'You'll be getting the sack!' and another, 'You'll have to stand outside the staffroom at playtimes for a week!' Teacher groans, 'Not again'.
>
> (Field notes)

More often, however, his attitudes and behaviours were heavily gendered and framed within a 'laddish' form of masculinity. The issue of Philip Norris's heterosexualized interactions with the girls will be taken up and discussed in detail in Chapter 7; here just one example is offered:

> (A Y5 class are coming back from the swimming baths on the coach.) Saskia gets out some photographs of herself and shows Maggie. Philip Norris is in the seat in front and he turns and takes them from her. He says 'This is Saskia smiling' and uses his fingers to pull his mouth out into a distorted grin. Saskia hits him on the shoulder and says 'Don't!' He looks at another photograph and attracts one of the boy's attention saying laughingly 'Mac, look at this one!' Saskia bends over the seat towards the teacher, hits him again and says quietly, 'Shut up'. The teacher hands her the photographs back laughingly.
>
> (Field notes)

On a personal level I experienced occasions when the area I was researching (masculinities in primary schools) generated 'humorous' behaviours which could be interpreted as men teachers consolidating their heterosexual identities. For example, one afternoon Philip Norris and a male teacher adviser on outdoor education were with a group of Y5 children in an area of the school grounds left to grow wild. Just prior to this I had been talking to the teacher adviser about my research into constructions of masculinity in the school. He had made it clear that he thought it was a waste of time, and that his PhD research (which he had abandoned several years previously) had been on 'real concerns'. However, this was all done in a 'jokey' way. My field notes record:

> The girls wandered off to collect some plant samples and I was temporarily standing alone in the clearing. Just then Mr N [class teacher] and Mr H. [teacher adviser] came out of the wooded area heading towards me. They were holding hands and skipping. When they arrived Mr H. said, 'We thought we'd give you some different sort of data!'

The only conclusion I can draw from this somewhat bizarre behaviour was that by 'humorously' presenting themselves as effeminate (skipping) and gay (holding hands), they were suggesting how 'ridiculous' effete men are in contrast to (their) 'normal', usual presentations of heterosexual masculinity. They seemed to be saying 'perhaps you'd prefer us like this?' (with the confidence that all 'normal-minded' people would laughingly shudder in the negative). Their reaction was fundamentally defensive. Simultaneously, their 'humorous spoof' of non-masculinity serves to 'other' non-masculinity, and hence to construct and confirm their own 'normal' version (Francis and Skelton 2001).

This discussion has given some indication of how 'humour' is used by men teachers to carry out similar kinds of 'gender boundary' work of the same kind as that identified in studies of school children (B. Davies 1989; Francis 1998). Clearly the use of humour was central to the dominant form of masculinity at Deneway Primary and was utilized by all the male staff. Of even greater significance was the place that football held in the maintenance of this particular hegemonic masculinity.

The place of football in constructing masculinities in the primary school

If we return to the school brochure produced by Deneway Primary School we can see one of the ways in which 'equal opportunities' can come to occupy a precarious position. Despite the emphasis given to equality, wittingly or otherwise, the school brochure implied that boys' sports were of higher status than those activities traditionally seen as appealing to girls:

> EXTRA ACTIVITIES
> On the extracurricular front, the children will be given the opportunity to participate in a number of activities including:
>
> Football
> Rugby
> Athletics/Cross Country
> Dance
> Nature Club
> Cooking
> Recorders, etc.

Although the school had been open only a short time, Deneway Primary had already established its strong commitment to football. This was not surprising given that football occupies a central place, literally and metaphorically, in the heart of Oldchester City. The local team is in the Premier division of the Football League and attracts support from males and females. This may be one of the reasons why parents of boys in the football team never commented on the consequences that leaving school early twice a week to travel to other schools for a match might have on their sons' academic development.

The game's high profile in the school could be seen from the regular football practices and inter-school matches and the setting up of a football club. Also, a concern with the performance of the school team was a regular feature of assemblies. However, despite the school's official policy on equal opportunities, football was ring fenced as a male pursuit. The official school team was totally male, and the team was coached and taken to matches by the Y5 and Y6 teachers Philip Norris and Stephen Coles. It was unusual *not* to see the men teachers, Philip Norris in particular, having a 'kick around' with the boys on the playground or school field at lunchtime. Opportunities to play football were grabbed whenever possible by the men teachers and the boys such as waiting for the coach to go to the swimming baths or when waiting for everyone to change before and after physical education (PE) lessons. Before going on to consider what it is about football that it occupied such a crucial place in the day-to-day routines of Deneway Primary and, according to the literature, in practically every other study of masculinity/gender and schooling, an exploration of the particular form it took and implications it had for gender relations will be made.

Football at Deneway Primary was central to the construction of a particular mode of middle-class masculinity which emphasized intelligence and proficiency. This was achieved in two ways: first, by differentiating between the boys at Deneway and those from other schools; and second, by giving status to certain groups of boys in relation to girls and 'other' boys within the school.

All the boys in the class (and indeed in the other KS2 classes) were supporters of Oldchester football team. The only football strip evident in the playground was that of the local team, and conversations about football always appeared to revolve around its current place in the premiership. The boys' and men teachers' exclusive support for the team was often used as a means of indicating their superiority over boys from other schools, who may have supported alternative local teams. In particular, reference would be made to Deneway boys' superior 'coolness' and 'intelligence'. On one such occasion, the class were travelling on a coach to the swimming baths. The driver's cab was covered in football stickers of another local team, also in the premiership. A few of the boys in a Y5 class start to discuss this:

> Once seated Malcolm said to Nigel, 'If [Oldchester] supporters had got on and seen those stickers there'd be nothing left of this bus by the end of the road!' Lee overhears and shouts to Philip Norris, 'Hey Mr N. – have you seen those [Bloxteth] stickers by the bus driver? Oldchester fans'd 'ave him and this bus!!' Philip Norris stands up and moves to where they are sitting. He says, 'Do you really think [Oldchester] supporters are that thick? Would you, for instance, do something so pathetic and stupid? Of course not, only [Bloxteth] supporters would be so dense'.
>
> (Field notes)

On another occasion, the school football team was scheduled to play a school team from the south of the city. Stephen Coles came into the classroom to ask for the boys who were playing that day:

> Philip Norris observes that they will be playing against a team who seemed to be having a run of success and had won every game in the last eight. Stephen Coles says loudly to the whole class '[School] team won't be a problem for us, will they lads? After all we're [Oldchester] fans and have picked up on their moves on the pitch . . . there are kids at [school] who support [Bloxteth] and even [Chesboro]!! [Two local teams, one in the Premiership and one in the First Division.] What chance will they have against us – we're the ones with the brains!'

The 'them' (dense) and 'us' (intelligent) was continued within the school site, but this time the (dense) 'other' was girls. One way this was achieved was through the use of coded language. This was used mostly by Philip Norris, in communicating with the boys in his Y5 class. For example, he would usually write a greetings message on the chalkboard every morning which would say things like, 'Only 25 school days to half-term'; but occasionally there would be more cryptic messages such as, 'Sad old them, clever old us'. On this occasion two of the girls, Beatrice and Holly, were trying to puzzle out what this meant when the class teacher came into the room. They asked what it was about, to which Philip Norris said, 'Ask Dougie or someone'. When it came to registration, Philip Norris asked Smittie to 'Put the girls out of their misery and tell them what this is all about'. Several boys spoke at once, with the gist being that the message meant: 'Sad old Bloxteth, clever Oldchester', as the former had been beaten 3-0 in a local match the night before. At another time, the teacher was starting the class activities for the afternoon when he saw the caretaker walking past the classroom:

> Philip Norris calls out 'Bob!' The caretaker does not hear him. He says 'Well we've got to know what he thinks haven't we? Who'll go and ask him?' The girls (and I) are puzzled. Smittie and several other boys shout out 'I'll go'. He sends Smittie and Mac to catch up with Bob.

> When they return they say 'He said he'll get over it but had to drown his sorrows at the pub'. It seems that a goal scored for Oldchester in the match the night before had been disallowed.

When girls did play football it was in PE and, even then, men class teachers would use these sessions as an opportunity of providing the boys with more practice. Usually the girls would be asked whether or not they wanted to take part. The majority refused and would sit on the sidelines and observe. Those girls who decided to join in the game were placed at a disadvantage, as indeed were some of the boys, simply by the accepted forms of communication. The majority of the boys in Philip Norris's class played football every breaktime together, and frequently with their teacher. In playing football on the playground, various nicknames or shortened versions of names were used to attract each other's attention. Not only were these boys not used to calling to people outside of their exclusive group to 'pass the ball', but also shouting out 'Hilary' or 'Beatrice' did not have the same intimacy and intensity as 'Mac' or 'Ossie'. When Philip Norris attempted to include the girls by giving them nicknames, this had the reverse effect and further established football as a 'male only' game. For example, when Hilary joined in she was given the name 'Thumper'. She appeared quite traumatized by this, as every time the nickname was used she would stand still for a moment. After several PE sessions when this had occurred she, and the other couple of girls who took part, stopped joining in. In the absence of research evidence on the effect of girls' responses to teachers' use of nicknames, it is difficult to make generalizations about how the girls in this Y5 class reacted. However, it is fair to surmise that, as pre-adolescent girls, they were sensitive to the physical changes they were beginning to go through and did not want attention being drawn to any aspect of their bodies. Hence, references to 'Thumper' (who could kick the ball hard but also was a particularly solidly built girl) and 'Lanky' (the tallest girl in the class) were not welcomed and indeed actively avoided.

The role of male teachers in seeking to preserve football as a male-only activity will be discussed later. What is of significance here are the means by which the men teachers in the school sought to exclude girls from football. This was achieved in two ways: by preventing them gaining access to the means of playing the game and by denying the possibility of girls possessing requisite skills.

The girls in Philip Norris's class had been made very aware that football was not something they were supposed to take part in, even though the official word was that they could. Several girls referred to the barriers that were placed in their way. In one interview two of the girls began by noting the way in which boys refused to let them join in a game but went on to note how teachers implicitly supported girls' exclusion simply by making no attempt to intervene:

Beth: Well on the play yard they never let girls play football!
CS: Wouldn't you start up your own game of football then?
Beth: We haven't got a ball . . .
Maggie: We're only allowed one ball per class and the boys always get it . . .
Beth: And it's on the top yard, and Year One's got a space and Year Two's got a space and the boys say it's their space.
CS: Have you ever been and spoken to anyone?
Beth: Well the teachers just say 'Let the girls have the play yard as well, it's not just for boys' but nobody does anything. Mr N's [Philip Norris] always playing football with the boys.

In another interview, Holly, Marie and Saskia spoke about how the boys' attempts to prevent the girls joining in with their football games at lunch times were supported by the male teachers:

Holly: The boys are sexist, because Mr Naismith [Y6 teacher] like, not trains 'em, but says boys would prefer playing football . . . and sometimes I feel like having a little kick around but we can't.
Marie: Mr Kenning says it's like for the girls too, but it isn't like that . . . He only lets us have one ball a class and the boys give it to Bob [the caretaker] to look after and they rush their dinner so they can get out first.

Also, the school offered a Sports Club and a Football Club outside school hours. At first, both clubs were held on the same night until some of the girls protested that the arrangement meant neither they nor the boys would be able to do both. Consequently, the Sports Club moved to another night but this did not prevent the domination of the Football Club by a masculine ethos the girls found off-putting:

Beatrice: They're more boys in the clubs. Is it 'Geordie Boys' who Ossie and Smittie play for? Well it wouldn't be very good for a girl walking in. Sometimes Mr Coles [who was running the Football Club] sticks up for the girls but not very often . . . you don't feel right.

A second means of deterring girls was by assuming their lack of skills for football. During one PE session, Deborah asked Philip Norris why it was that boys always got to be in the nets (goalkeeper), to which he replied, 'You want someone to keep the other team's goals out don't you, so let's go for the best!' Given that there were more children in the class than in two football teams, the girls were always given 'add-on' positions, usually as additional defenders. As Epstein (1998) commented in her study of a primary school, far more latitude for incompetence at football was given to boys by teachers, and by boys to other boys, than to girls. This observation is particularly pertinent given the significance of football to the boys' relationships at Deneway Primary. Even those boys who were not seen, or did not see themselves, as 'good at football' were still aligned with football (Skelton 2000).

Being proficient at football granted that select group of boys access to a range of benefits which were not available to the less competent footballers. These privileges included greater personal freedom such as being able to control their own time (as when the class teacher would say: 'I'll leave it to you to finish when you want. I know you want to be out practising for the game on Tuesday'). They also experienced privileges through being a member of a high status group within the school (as when their football successes were applauded in school assemblies when other sporting teams' achievements were not). However, the question remains as to what the men teachers gained from spending so much of their time occupied in various ways with football? In order to answer this we need to look at what it is about football that attracts such passionate male attention.

When Epstein (1998: 7) observed in her study of a primary school that 'football is a major signifier of successful masculinity', she was identifying the close association of football to hegemonic masculinity. The literature on football written by male sociologists demonstrates that the ways in which football has been conventionally 'played out' reflects those practices associated with attaining and sustaining hegemonic masculinity (Robins 1982; Dunning *et al.* 1988; Murphy *et al.* 1990). The majority of these accounts are couched in language which is evocative of the characteristics currently associated with hegemonic masculinity. For example, Kenway and Fitzclarence (1997) have suggested that:

> At this stage of Western history, hegemonic masculinity mobilizes around *physical strength*, adventurousness, emotional neutrality, certainty, *control*, *assertiveness*, self-reliance, *individuality*, competitiveness, *instrumental skills*, public knowledge, discipline, reason, objectivity and rationality.
>
> (Kenway and Fitzclarence 1997: 121, my emphasis)

These same characteristics are present in the following quotation from Dunning *et al.* (1988) in their argument that football requires

> a fine balance between a number of interdependent polarities . . . the polarity between *force* and *skill*, that between providing scope for *physical challenge* and *controlling* it, that between *individual* and team play and that between *attack* and *defence*.
>
> (Dunning *et al.* 1988: 4, my emphasis)

The significance of football in allowing a space in which aspects of masculinity can be elaborated has been discussed by a number of writers (Westwood 1990; Hornby 1992; Miedzian 1992). Connell (1983: 18) has argued that it instructs men in two aspects of power: the development of force ('the irresistible occupation of space') and skill ('the ability to operate on the objects within that space, including other humans'). The rules of football, where territorial control is important, almost literally conform to this definition.

Other commentators have considered football in relation to further aspects of hegemonic masculinity; specifically, that it is constructed in relation to women and subordinated masculinities and is heterosexual (see Chapter 7). As we saw earlier, it is not just boys who attempt to keep football as an all-male preserve as other studies also point to similar attitudes among male primary teachers (Renold 1997; Connolly 1998). One female student in Mac an Ghaill's (1994: 123) study said: 'It's like they're [men teachers and boys] in a club and girls can't join'. Another female student stated: 'One minute they're telling the boys off and the next they're talking together in a close way about telly programmes and football. They're always on about football'.

The discourses the men teachers at Deneway Primary School located themselves within drew on football and humour to identify themselves as 'properly masculine'. However, it was not simply a case of humour and a shared interest in football being used to generate camaraderie between the boys and the men teachers. Football was significant in terms of the gender regime of the school (see Chapter 1); that is, football was important in defining relationships between male teachers and boys, between boys and boys, and between girls and boys. Particular privileges were granted to some of the most proficient football players which marginalized the other boys and all of the girls.

Summary

This chapter has made three key points about discourses on men primary teachers. The first concerns the argument whereby the absence of men teachers in the primary sector is seen as part of the reason why boys are disinclined towards school work. Furthermore, their apparent lack of achievement in relation to girls is seen as a result of the femininization of primary schooling; this encompasses all aspects of school life from approaches to teaching and learning styles to organizational and management procedures. Yet women do take on 'masculinity' as we saw in the control strategies adopted by female as well as male teachers at Benwood Primary School (Chapter 4). We have seen that the predominance of female teachers is not a phenomenon of recent times but has been the case since the introduction of state schooling in the nineteenth century. It is more likely that the recent concern for 'underfathered' boys (Biddulph 1998) has found an echo in what the popular press present as the 'masculine-free zone' of the primary school.

A second key point is the importance particularly to men teachers in primary school of locating themselves as 'properly masculine'. When primary schools are perceived to be female environments, men teachers need to demonstrate their masculine credentials both to themselves and to others (colleagues, parents, pupils, friends and family). Doing this may involve exaggerating various aspects of masculinity and thus presenting themselves as 'laddish' through using humour and demonstrating a passion for football (see also Chapter 7).

A final point is the significance of the way in which men primary teachers present themselves as 'properly masculine' to the construction of a dominant masculinity in the school. This is something which has been raised in many studies of masculinities and secondary schools (Kessler *et al.* 1985; Connell 1989; Mac an Ghaill 1994). However, in secondary schools various forms of hegemonic masculinity may be on show as we saw with Mac an Ghaill's (1994) typology of three types of men teachers (the Professionals, the Collectivists and the New Entrepreneurs). The fact that primary school staffs are much smaller and there are fewer men teachers in this sector severely limits the forms of masculinity available to young boys to engage with. This brings the discussion full circle and back to the question raised in the introduction to this chapter: 'What kinds of male role models do we want men primary teachers to present?' This will be returned to in the concluding chapter.

chapter / **seven**

Heterosexuality in the primary classroom

Introduction

Ask any primary teacher to list the kinds of things boys and girls tease each other mercilessly about and near the top of the list will be romantic relationships. Francis (1998) has observed that just a girl and a boy playing together is enough to attract accusations of 'fancying' each other. Teachers will also undoubtedly be able to recount instances of more intrusive behaviours of boys towards girls, such as lifting their skirts up or making lewd comments about their bodies. There are many examples of name-calling in the primary playground where boys call each other 'sissy', 'faggot', 'poofter', 'gayboy', 'queer' (Thorne 1993; Redman 1996; Epstein 1998). What we are talking about here are the ways heterosexuality makes its appearance in various ways in day-to-day relationships in primary schools. Yet being aware of heterosexual practices is one thing, actually naming them as *sexual* is another. One reason for this is that linking sexuality with primary schooling is not 'comfortable', possibly because the occasions when it is discussed openly is usually the result of teachers sexually abusing children.[1]

Of course primary schooling is associated more publicly with sexuality through government policies on education, where schools are charged with responsibility for providing children's sex education, developing procedures to tackle sexual harassment and, more contentiously, as 'moral guardians' (for example the controversies generated by Section 28). There appears then to be a division between what is appropriate to discuss openly (policy on sexual issues) and what is not (romance/seduction). Epstein (1994) has suggested that one reason why there is a discomfort around, and segregation of some, aspects of sex and schooling is due to the public/private divide where schooling falls within the public domain but sexuality is perceived to be of a private nature.

This chapter will focus on heterosexuality in the primary classroom. As was said above, heterosexuality covers a wide spread of behaviours from romantic relationships between pupils to sexual harassment and homophobic bullying. It also includes a much less researched area which is the perpetuation of heterosexist norms in the classroom by male teachers. The first section continues the discussion in Chapter 6 by looking at the various ways in which men teachers use discourses of gender and sexuality to construct their masculinity. Here it will be shown how the girls negotiated the heterosexualized 'male gaze' (Skeggs 1991a).

Men teachers and heterosexual discourse

The three men teachers at Deneway Primary identified themselves as 'properly masculine' through positioning themselves as 'one of the boys'. In many ways these laddish behaviours drew on those patterns of masculinity found in studies of working-class adolescent male pupils at school which have been referred to variously as the 'lads' (Willis 1977), the 'footballers' (Walker 1988) and the 'Macho Lads' (Mac an Ghaill 1994). In the case of the teachers, the characteristics of these similar forms of masculinity were reworked into a middle-class adult male (teacher) version and revolved around athletic prowess, having a laugh, (not) looking smart, and having a good time with mates (pupils). Out of the three teachers, Philip Norris in particular exhibited these characteristics and brief examples of each of these behaviours are:

> *Athletic prowess:* A Y5 class are using large apparatus in the hall. Three of the boys are having a race to see who can reach the top of the climbing frame first. Philip Norris runs across the hall and clambers to the top of the frame, beating the boys.
>
> (Field notes)
>
> *Having a laugh:* It is the end of the afternoon. The Y5 children have their coats on ready to go. Philip Norris says, 'We've had a really good afternoon . . . even Dougie smiled even though I'd just fallen backwards through the door . . . or perhaps it was because he'd just farted!' Children laugh.
>
> (Field notes)
>
> *(Not) looking smart:* This morning a children's author is visiting the school and will work with Class Y after break. At milk time, the children collect their milk and sit on the carpet. Philip Norris gets out a tie and puts it on saying, 'I asked in the shop for something that would impress . . . trouble is it's so long since I've worn one!' He forms the tie into a bow around his neck.
>
> (Field notes)

> *Having a good time with mates:* The children are in the hall for gymnastics. The boys are ready first. Philip Norris tells them to get a ball and have a 'kick about'. Mark comes out of the changing room and asks Philip Norris what they are doing. He replies, 'Having a good time . . . improvising . . . hanging out while we wait for the girls'.
>
> (Field notes)

While all these strategies serve to demonstrate masculinity, a key feature is that it is heterosexual. Now it was shown in Chapter 6 how teaching generally is seen as a career which is a 'soft option' for men (Connell 1985). Male primary teachers are aware of others' attention to their masculinity (or lack of it) (Thornton 1997) and studies have shown that for men working with young children this results in continual negotiation of their masculine identities (Skelton 1991; Allan 1993; Penn 1996). As one of the elementary teachers in Allan's (1993) study argued:

> He [the elementary teacher] had better not be the least bit feminine. I mean they expect a male teacher to be a man . . . If a man were perceived as feminine, I'm sure it would be a problem . . . You need to be a male role model. Be the opposite of being feminine.
>
> (quoted in Allan 1993: 123)

Occasionally male teachers would draw on male–female difference as a way of policing the boys' (hetero)sexual masculine identities, for example:

> The children are filing past the headteacher to go into the hall for assembly. Michael has an evident fragrant scent about him which smells like aftershave. As he passes, Mr Kenning asks loudly, 'Are you wearing perfume lad?'
>
> (Field notes)

> Stephen Coles [Y6 teacher] has gone into the boys' changing rooms to hurry the boys along. It is the dress rehearsal for the school production. When they come out he says to them, 'There's a terrible smell in the boys' changing rooms. It's not the usual unpleasant body smells of sweaty boys . . . instead I could smell hair gel, hairspray, deodorant, aftershave. I thought for a minute I'd gone into the girls' changing rooms! When I asked him about this comment he said, 'I don't want to encourage them . . . I don't use anything like that myself . . . a good wash in the morning is enough'.
>
> (Field notes)

Such strategies are not unusual. There is a substantial amount of evidence of boys and men controlling others' masculine behaviours by questioning their (heterosexual) 'maleness' (Hough 1985; Askew and Ross 1988; Lee 1993). Indeed one of the worst insults for a boy is for his attitudes and behaviours to be likened to those of girls (Seidler 1991; Miedzian 1992).

However, hegemonic masculinity is not simply a matter of policing other males' sexualized behaviours; it also requires individual males to demonstrate their own heterosexual identity (Connell 1995). The use of humour offered one avenue through which this could be achieved.

The significance of humour in establishing relationships with the boys was discussed in Chapter 6 and it has been shown elsewhere how the humour of male bonding relationships centres around sexual and aggressive banter (Hearn 1985; Lyman 1987; Kehily and Nayak 1997). Often the boys in the Y5 class at Deneway Primary would initiate opportunities for heterosexual banter with the teacher. Inevitably such events were played out in front of or involved the girls, who often appeared to be embarrassed or uncomfortable by these displays of 'male humour':

> The class is in the hall playing a game in which one person chooses another person and gives them a task to act out. Michael is 'on' first. He says 'Mr N. Putting a condom on!' Philip Norris laughs but makes no move. Several girls are staring at the floor, others look uncomfortable. Nobody speaks. After about a minute, Michael says 'Be a Viking'. Philip Norris runs across the circle to Saskia, picks her up and runs back across to his place.
>
> (Field notes)

The intention here is not to present Philip Norris or any of the other men teachers at Deneway Primary as consciously adopting sexual and sexist practices. Rather what is being suggested is that men teachers, particularly those in primary schools, draw on wider discourses on hegemonic masculinity in order to define themselves as 'properly masculine' to themselves and in the eyes of others. At the same time, the girls who were on the receiving end of these heterosexist discourses were far from passive in their responses. The next part of the discussion will focus specifically on the Y5 girls in Philip Norris's class in order to explore how they negotiated the 'male gaze'.

The girls' group

The girls were taught in Y5 by Philip Norris and in Y6 by Bill Naismith. Both teachers distinguished between what they perceived to be two 'types' of girls in the class: those they labelled as 'Quiet' and those they saw as 'Confident'. For example:

> Following a class discussion about the parts the pupils will play in the school production I asked Philip Norris how he decided who would do what. He replied: 'There's always the ones who push themselves forward . . . both boys and girls that is . . . but there's the quiet girls like . . . like Beth, Ruth and Hilary who prefer to keep out of the spotlight'.
>
> (Field notes)

> Speaking to Bill Naismith about Maggie's enthusiasm for the project she was doing for the Gender topic he said, 'Yes, that's Maggie for you . . . "Equal Rights for Women" . . . she'd have made a great Suffragette! She's definitely very forthright . . . in everything really not just this topic. But, you know, there are plenty of confident girls in this class who speak their mind . . . look at how Marie, Sarah, Beatrice . . . um . . . Saskia, Deborah . . . I could go on . . . they're all ready to speak up'.
>
> (Field notes)

Those girls who were described at various times by Philip Norris and Bill Naismith as 'Quiet' were a pair of identical twins, Ruth and Rachel, together with Hilary, Kyoko and Beth. Kyoko was Japanese and had only recently joined the class. She had some functional competence in English but still did not find conversation easy, which may be one reason why she was seen as 'Quiet'. Whether cultural stereotyping was brought into play in the teachers' interpretation of her behaviour is not known. The girls labelled as 'Confident' by the men teachers were another pair of identical twins, Deborah and Marie, and Beatrice, Saskia, Sarah and Holly. Maggie too was considered 'Confident' but in a rather different way; that is, she was perceived by the teachers as 'difficult'.

While not wanting to adopt the teachers' crude and inappropriate categorization of the girls in the class, it is useful to note that the application of these labels suggest that there were gender-appropriate discourses in operation in the school. That is, *how* girls *could* be in the school was framed within specific expectations of 'girls' behaviours'.

Being a 'normal pupil' or 'being feminine'?

Although a general description of schools might be that they are sites which deploy male structures and practices, as was suggested in Chapter 4 the ways in which girls and boys are *positioned* will vary across school sites. As Alison Jones (1993) explains, in a school which emphasizes feminine decorum, boisterous girls may be seen as 'naughty' or 'difficult' but in another school where physicality is encouraged, these same girls may be perceived as admirably 'stroppy' or, at least, competent. The question here is what did Deneway Primary School consider to be a 'normal pupil' (Brown 1987) and to what extent was this complementary with what is seen as valuable and desirable in 'being feminine' (Walkerdine 1990; Hey 1997)?

In Deneway Primary School's 1992 school brochure one of the aims in relation to equal opportunities was to develop a system which did not discriminate on any basis and which valued 'all pupils as individuals in their own right'. The concept of 'pupils as individuals' was outlined in more detail in the school's official 'Statement of Intent' – a short document displayed in the school and distributed to parents. The 'Statement of Intent' identified what it sought to achieve in its pupils. That is, for pupils to 'accept responsibility for their own learning', 'develop independence' and to 'assess their own achievements' which the school would develop by 'valuing individual aspirations' and 'promoting self-interest and esteem'. Now, it might be suggested that this image of an independent, self-motivated, critically reflective pupil sits uneasily with the demands made on early adolescent girls to 'become feminine'. Rather, those attributes the school hopes to foster in its pupils are precisely those associated with dominant masculinities; that is, *self-reliance*, *reason* and *individuality* (Kenway and Fitzclarence 1997). While 'being feminine' does not mean an outright rejection of those characteristics, they do not fit with images of femininity where value is placed on cooperation, empathy and nurturance (B. Davies 1989; Walkerdine 1990). The girls in the class were faced with the dilemma described by Hey (1997) where they were in the position of having to work out

> how they are to become simultaneously a 'normal' schoolgirl and a 'proper young woman' within the respective cultural institutions of (compulsory) schooling and (compulsory heterosexuality). One institution denies difference while the other is fundamentally invested in producing it.
>
> (Hey 1997: 132)

The expression a 'normal pupil' is used here to describe those who are middle class and white and whose culture is positioned centrally in the education they receive (Kenway *et al.* 1996). Such girls are positioned positively in a school but, at the same time, *how* they can be and *what* they can do are all worked out within a 'male gaze'. That Philip Norris's approach to classroom management drew explicitly on a heterosexual masculine style was acknowledged when, after a PE session in which he told the girls to get changed quickly because he had his 'sexy shorts' with him and accompanied the comment by wiggling his hips, he said, 'I know I flirt . . . it's not sexual but gently flirtatious. It isn't harmful to them.'

At the beginning of the Y5 academic year some of the girls responded to their teacher's 'laddish' approach by flirting in return and would often use physical contact. For example, on one occasion when the class was playing rounders, Philip Norris said, 'Right everybody we're going to get Deborah out!' The boys got her out on the first post, whereupon she walked up to the teacher and smacked him on his bottom. His reaction to this was to laugh. That Philip Norris did indeed 'flirt' with all the girls and the reactions of some of them to this behaviour, for example, when Deborah smacked him and similarly when Saskia hit him for passing her photograph round on the coach coming back from the swimming baths, could be seen as challenging. At the same time, it can also be argued that such reactions are also couched within heterosexual relations.

It has been argued that when girls tackle the kinds of masculine heterosexual strategies employed by boys/men, such as flirting, by flirting they are becoming implicated in the normalization of masculinity and the policing of their own behaviours (Mahony 1985; Halson 1989; Kelly 1989). However, as Skeggs' (1991a) study of females in a further education college showed, the students refused to be rendered powerless. A similar process was happening here in that over the course of the year, those girls who had responded in kind to Philip Norris's behaviour began to withdraw their consent to his flirtatious actions and started to use masculine strategies to negotiate some degree of power. In effect, strategies were adopted which subverted the 'male gaze' (Skeggs 1991a). Where females have been the object of gaze, research literature demonstrates instances of girls reversing this technique. In his interpretation of this literature, Woods (1990) has suggested that teachers' lack of respect for pupils might be reflected in their appearance:

> Davies' (1984) girls were 'incensed by "dirty" teachers – any who wore scruffy suits . . . whose hair stuck up on end' or who showed a lack of propriety in appearance or behaviour (p. 29). Dubberley's (1988: 191) girls, similarly, criticized a teacher for being 'dead scruffy . . . filthy . . . Greasy hair – nobbut Oxfam clothes'.
>
> (Woods 1990: 18)

When the original texts from which these extracts are examined it appears that, in both cases, these are comments made by *girls* of their *male* teachers. So an alternative interpretation of these comments is that the girls showed disapproval of the attitudes or behaviours of their male teachers by reversing the heterosexual, hegemonic gaze. The central means through which the girls in the class demonstrated disapproval of Philip Norris was by focusing on his physical appearance. Given that a feature of his 'laddish' masculinity was not dressing along conventional, professional teacher lines, he responded to the girls' frequent comments about his clothes either by laughing or by ignoring what was said. At the same time, these criticisms of his appearance appeared to be interrelated with his authority as a 'proper teacher'. One afternoon Marie and Sarah were lying on the grass, having opted out of playing cricket in PE. Marie was watching Philip Norris:

Marie: Mr N. never tucks his shirt in.
Sarah: He never does any ironing either.
CS: How do you know?
Sarah: 'Cos he's lazy.
Marie: He always wears jogging pants, never trousers.
CS: Does that matter?
Marie: Yes . . . we had to fill in a form once and it asked what you would change about your teacher. I said he should wear smarter clothes. Mr Coles [the Y6 male teacher] wears trousers and ties.

Sarah: He does wear trousers sometimes.
Marie: Only cords . . . he only wears proper clothes when he's going for an interview.
CS: Why do you want him to wear proper clothes?
Marie: Because it's embarrassing . . . if someone comes into your classroom and sees him and everyone's carrying on and he's letting them get away with it . . .
Sarah: Yeah . . . not like a proper teacher would.

The discussion here has implied that *all* the girls engaged in heterosexual discourses in their relationships with Philip Norris. This was not the case, however, as one girl, Maggie, refused to be drawn into any sexualized 'bantering', choosing to either ignore or directly challenge Philip Norris. For example, one morning two Y1 girls came into the classroom as the children were sitting on the carpet and Philip Norris was taking the register. The girls handed him a note which he read and then started to laugh:

Maggie: What's so funny?
Teacher: Just a little love note from Mrs Morris. It says 'Darling Mr N, I can't wait to have a cup of coffee with you at breaktime . . .'

At this point several of the boys and a few girls are laughing. The two girls from Mrs Morris's class went red faced and looked at their shoes. Maggie interrupted Philip Norris's reading of the 'love letter' by saying:

Maggie: You've got a girlfriend.
Teacher: It's just a joke.
Maggie: Everything's just a joke to you.

This challenging by Maggie placed her in the position of being actively disliked by both the teacher and the other girls. The reasons offered by Philip Norris and the girls in the class for their dislike centred on her 'otherness' (Epstein 1993):

Philip Norris: the biggest problem in the class is Maggie, who doesn't tend to get chosen [for team activities], and she must be quite sensitive of that, but she brings it on herself . . . she's a little bit bossy and can be aggressive.

Several girls also implied her 'aggressiveness' rendered her as 'unfemale' and, therefore, unacceptable. One illustration of this came up in an interview which took place with some of the girls when they were in Y6. They began to talk about their impressions of Philip Norris who, by this time, had left the school:

Holly: He wouldn't tell you off even if we stomped out of the room and saying all these swear words . . .

Saskia [*shocked*]: I never did anything like that . . . nor you!
Holly [*laughing*]: No, we wouldn't, none of us girls would, but someone did!
Marie: Maggie did . . . she just uses the teacher, she goes 'No' and things like that.

All the girls in the class argued that aspects of the boys' behaviours they disliked were 'showing off' and bullying or being bossy. The dislike of Maggie was articulated in similar terms; however, the fact that Maggie also couched her dislike of many of the boys and some of the girls in the same way suggests that it was not the actual actions themselves, rather the strategies adopted in dealing with those actions that generated feelings of 'inappropriate or unfeminine' behaviours. For example, all of the girls commented that one of the ways the boys 'showed off' was always having to have the last word whereas they would ignore what was said or walk away. They adopted the same response to Maggie's actions:

CS: Why do you say you don't like Maggie . . . that she 'uses' you?
Holly: She's horrible . . . dead bossy.
Beatrice: She acts like you're a kid and tells you what to do.
CS: What do you do when that happens?
Beatrice: Avoid her.

In contrast, Maggie would stand her ground. Interestingly, the fact that some of the girls did not like her did not result in mutual dislike:

CS: Who do you like in the class?
Maggie: Deborah, Holly, Beth.
Beth: Yes, Holly's fun.
CS: What about the others?
Maggie: Well, no names, but I don't like some because they boss everybody around and everything.
CS: How do you deal with it if they boss you around?
Maggie: I just tell them to 'Shut up'.
Beth: Yes, you do, I just ignore them.
CS: What about you Hilary, how do you deal with it?
Hilary: I just try to ignore 'em.

It would seem then that Maggie did not conform either to the class teacher or to the other girls' notions of what was demanded in 'being feminine' although she was living up to the ideals of the school in terms of demonstrating those characteristics described in the school's 'Statement of Intent'. However, those characteristics were, as was argued earlier, relative to the dominant mode of masculinity of the school. It could be that the reason why she was disliked by the majority of girls in the class was because, as

Hey (1997) argued, girls' practices have as their major aim the making of feminine identity or reputation and insist on making each other into acceptable selves. This takes place within the superordinate gaze of hegemonic masculine culture so 'in very many respects they did the work of that culture among and between themselves in positioning each other into particular places' (Hey 1997: 131).

This section has looked at how men teachers draw upon heterosexuality in their relationships with pupils as a means of positioning themselves as masculine. This is not to say that *all* men teachers act in the same way as Philip Norris but the significance of a heterosexual identity is of particular importance for men primary teachers. At the same time, the section has looked at how girls are not passive in these interactions. Rather, the girls were engaged in negotiating their position in the classroom and their relationships with each other from their position as 'female'. Deneway Primary's espoused ideals regarding what they hoped the *pupils* would achieve (outlined in the 'Statement of Intent') drew on characteristics which have been associated with dominant forms of masculinity and therefore did not sit easily with those attributes linked with 'being feminine'. If recognition and acceptance is given only to those females who outwardly align themselves in accordance with, and what is expected by, the 'hegemonic gaze' (Connell 1987) then girls such as Maggie run the risk of being seen as an 'outsider' or 'difficult' by both teachers and peers. To summarize this idea: being a 'normal schoolgirl' is inscribed with heterosexual meaning. Being and becoming a 'normal schoolgirl' at Deneway Primary was worked out within a 'male gaze' and although the girls developed strategies which provided them with some degree of power it was ultimately constrained and they were unable to challenge the prerogatives of power.

This argument is in keeping with those made in other studies of schooling which have identified how men teachers and boys are able to retain a privileged position with regard to women teachers and girls through 'policing femininity' (Lees 1993; Kenway 1995). This leads to a consideration of another form which heterosexuality takes in primary schooling, that of sexual harassment. Sexual harassment conjures up an image of violent, aggressive actions and behaviours although the actual definition encompasses a range of behaviours including 'looks' and 'jokes' (Mott and Condor 1997). I raise this as adults know that sexual harassment is understood not so much by the actual action itself as by how it is received. However, younger primary age children do not have this level of sophisticated understanding and in the two case study schools boys often treated, and spoke of, girls in confused and contradictory ways, for example, conflating girls as friends with the idea of girlfriends. Because of the age of the children, romance relations will be considered alongside sexually harassing behaviours as it seemed that (and particularly with the Y2 boys) the one can easily tip over into and/or be confused with the other.

Primary boys: romance and sexual harassment

The idea that there is a fine line between romance relations and sexual harassment has been implied in several studies of primary schools. Children's teasing of 'who-fancies-who' can easily turn into ridicule (Francis 1998) and games of 'kiss-chase' can easily encapsulate both pleasurable and violent experiences (Thorne 1993). In his research into children's peer cultures, Hatcher (1994) noted that romance relationships occurred among high status groups at primary school, and these alliances reinforced individual and group status. A similar conclusion was reached by Redman (1996) in his exploration of heterosexual masculinities in boys in Y5 and Y6. He argued that boys at the 'top end' of primary school occupy a high status in the hierarchical organization, and this has implications for the reconstruction of masculine identities:

> Practising for heterosexuality is arguably one way in which boys collectively explore the newly available forms of authority and autonomy conferred by their position at the 'top of the school', and construct for themselves new forms of older child identity that negotiate and made sense of their new position within school structures.
>
> (Redman 1996: 178)

Generally, however, it is girls rather than boys who tend to take the lead in romance relationships. It is girls who initiate them and girls who spend more time talking about them, and pursuing and maintaining them (Hatcher 1995). Much of this evidence has been drawn from older age pupils in primary schools but as Connolly's (1998) research shows, even younger children engage in heterosexual relationships (for a rare study of heterosexual relations in primary schooling see Renold 2000). In his study of a multicultural primary school, Connolly describes how a group of 5- and 6-year-old Black boys placed a great deal of emphasis on girlfriends in forging their identities. This created conflict with white boys in terms of who 'owned' the girls. He also notes how this group of Black boys were unique among the groups of 5- and 6-year-old boys he observed in terms of their understandings of sexual relationships. Indeed, the discussions on boyfriends/girlfriends by the Y2 pupils at Benwood Primary indicated that romantic relationships among very young children usually amounted to nothing more than saying you were 'going out' with someone.

Y2 boys at Benwood Primary School

Many of the girls and boys in the Y2 classes at Benwood Primary had known each other from a very young age either because they lived in the same street or their parents were friends or they were related. This meant that several cross-gender friendships had been established and, although rarely evidenced in school time, these children continued to play together outside school. Occasionally the boys would talk about the girls as 'girlfriends' but, with the exceptions of the two boys competing to be 'first boss', Shane and John, most of the boys seemed to have different notions of what a 'girlfriend' was:

CS: So what about girlfriends?
Sean: Donna!
John: Worr I fancy 'er!
CS: Robert, have you got a girlfriend?
Robert: I used to have – Donna.
Martin: She's chucked him in for . . .
Robert: She's chucked everyone . . .
Martin: She hasn't! She hasn't chucked me or Gary 'cause she told us this morning.
CS: So what's nice about Donna? Why do you like her?
Robert: She keeps up with her work and her weather diary.
Martin: She doesn't tell on yer all the time.

While a majority of the boys experienced confusion with the notion of 'girlfriends' and 'girls as friends' in terms of girls of their own age, as the year progressed there was a shift towards the use of sexualized behaviours as a way of disturbing or shocking the girls. At the beginning of the year a feature of the boys' arguments or attempts to assert themselves over the girls was to make reference to bodily functions. For example, the most common form of insult was 'smelly arse'. Later, incidents began to occur when overt sexualized behaviours were adopted in their interactions with the girls:

The children are sitting on the carpet looking at books. There is a class book on 'Clothes' which Charmain is looking at. One of the pictures is of Sharon, Kylie and Lindsay wearing swimming costumes. Tommy notices this and says to Charmain 'They're being sexy', grins at her and makes kissing noises.

(Field notes)

Bethany is talking about her birthday cards to Charmain and Kylie. She looks across to the next desk and says to them 'Look, Matt's being rude'. Matt is making a masturbatory gesture.

(Field notes)

Kate, Lyndsay and Charmain are pretending to wash their dolls in the sink in the home corner. Tommy and Gary are playing nearby with the zoo animals. Tommy goes into the home corner but is chased out a few minutes later by Katy who says he was touching her doll's 'bottom' (pointing to the vagina).

(Field notes)

There is little published research on the topic of very young children's sexuality at school (Best 1983; Lloyd and Duveen 1992) so it is difficult to try to draw any conclusions about why such incidents began to appear when they did. It may be that as these writers have argued, children are aware of the unease their sexuality generates and, as a consequence, keep it secret. One aspect of the incidents which seemed to be fairly typical was the consistent way in which the boys appeared to target Charmain. Charmain had recently returned to the school after a long period of absence. She had been sexually abused by her father and the family was being monitored by social services. As with many sexually abused children, Charmain exhibited sexual behaviours towards both adults and peers and this may partly explain why she was often involved in the harassing behaviours of the boys.

When the incidents such as those described above began, the girls would report them to teachers. However, the approach adopted by the school and teachers to do with issues of control helped to establish the idea that sexual harassment was somehow acceptable because it was not potentially disruptive. As was shown in Chapter 4, teachers at Benwood Primary adopted a policy of 'outside behaviour' for behaviours they could not accept but also felt they could not challenge. Any action on the part of the children which was not seen as significantly disruptive would be allowed to pass by, or in some cases, dismissed. Unfortunately, all too frequently these were associated with a girl complaining about the actions of a boy:

> John and Shane are in the toilets. Lyndsay tells the teacher that they are together in the same cubicle and they'd been trying to look over at her when she'd been to the toilet. The teacher ignores what she has said and tells her to sit down.
>
> (Field notes)

> Luke pushes Kylie roughly onto her back on the floor and straddles her, sitting on her stomach. Kylie struggles free and reports what has happened to the teacher. Luke hovers behind her. The teacher ignores Kylie despite the fact that she repeats several times that 'Luke pushed me over and hurt me'. Luke smiles and walks away.
>
> (Field notes)

By the third term many such incidents were going unreported by the girls. There is now a substantial body of evidence which shows that schools, by failing to address boys' verbal and/or physical violence towards girls, make girls an easy target for boys who are flexing their male power muscles (Mahony 1985; Kelly 1989; Lees 1993).

What appeared to be taking place over the course of the year was that the boys were developing socially, physically and conceptually; their behaviours were constantly being reconstructed. Children become aware at an early age of the gender *category* to which they belong; however, the *meaning* of their gender *identity* evolves more slowly through the negotiation of gender discourses and practices (see Chapters 1 and 2). Interaction in the peer group is one way in which gender discourses and practices are negotiated and particularly significant is the playing out of confusions and ambiguities from the adult world in ways that make them familiar (Corsaro 1988). This leads back to the discussion in Chapter 5 on how children were positioned in different discourses and brought with them into school notions of masculinity and femininity from the local culture. As was shown, the hegemonic masculinity of the local culture was characterized by violence which was repeated in a modified version in the management and control strategies of the school.

In observing the boys' relationships with the girls shift over the course of the year from 'friend' to incorporate sexualized or violent behaviours, it occurred to me my own culpability and that of the other adult women in the school in fostering the boys' knowledge and awareness of the power of such strategies. Part of the bargaining process in gaining access to the school was that I would assist occasionally with some teaching. My experience of working with the children and observations of Mrs Brown (supply teacher), Mrs Jones (a 0.2 teacher supporting the other Y2 teacher) and Ms Lewis (second year BEd student) initially suggested that the boys in the class were all too familiar with sexually subordinating behaviours. It has to be said that all the children in the class adopted rebellious behaviours whenever an unfamiliar female teacher took them but the boys adopted different approaches which would have been less effective if used by the girls. The boys' actions drew upon violence and verbal aggression:

> At the end of a session Mrs Jones [0.2 teacher] says to Rick, 'You've worked really well today'. Rick responds with 'I'll smash your head in'.
>
> (Field notes)

> Shane is in the home corner with Bethany, Matt and Michelle. He picks up a pencil, puts it in his mouth and pretends to smoke. He realizes Bethany and I are watching him. He 'stubs out' his 'cigarette' on the back of his hand. He bares his teeth at us as he is doing this.
>
> (Field notes)

> Ms Lewis [student teacher] attempts to read Prince Cinders to a group of children. Gary has tantrums, rolls on the floor and mimics her. He then hits his fist on the floor while continually glancing up at her.
>
> (Field notes)

In addition the boys would use confrontational and sexualized bodily postures such as standing with legs apart, hands on hips and pelvis thrust forward. Conversations would be held in front of the teacher where they would refer to us as 'she' or 'her'. Some of the behaviours adopted by the boys appeared to be an attempt to show adult women that they, as males, 'knew' what role women had to play in their lives; that is to look after them. Time and again incidents were recorded when boys who had earlier been uncooperative, if not hostile in the ways given above, would approach an adult female and demand to have their shoes or trousers fastened. On the occasions when supply teachers or lunchtime assistants pointed out they were capable of doing it for themselves, they would laugh, turn away and fasten the shoelace or buckle themselves.

Now, in the same way that the girls' requests for help from teachers after experiencing some boys' aggressive or intimidatory actions had gone unheeded, I and other adult women would ignore the boys' violent behaviours when personally confronted with them. Sexism remains deeply embedded at an ideological level and adult women know what sexual harassment is and how best to protect ourselves. In keeping with research findings into sexual harassment and schooling we tried to ignore it or pretend that it was not happening (De Lyon 1989; Herbert 1992). When I spoke to Mrs Brown and Mrs Jones about the incidents recorded above, each responded with comments which rationalized the behaviour of the boys and, in so doing, avoided addressing the personal unease it causes us as women. So they argued that 'It's their age . . . they're trying out what they see their dads' doing' and 'It's what they see outside. What can you do?'

The important issue here is that the incidents which could easily have been 'read off' as boys putting into operation their knowledge and awareness of patriarchal rules were *my* understandings of such actions as a female researcher/part-time teacher. A more appropriate perspective would be to see them as the actions of young boys who were in the *process* of constructing their masculinities within a specific site. So, as the boys went about negotiating their masculine identities in the school setting, the inability of the school (or me!) to intervene and articulate what was happening in any of these relationships managed, by default, to sanction the boys' active use of violent/sexualized practices. However, their actions could readily have been challenged (see concluding chapter for suggestions).

The information collected at Benwood Primary with a group of young boys indicated some confusion over what was involved in having a 'girlfriend' rather than a friend who is a girl. Some actions of the boys which might have been intended as 'flirting' would often turn into behaviours which could (and were) seen as tormenting. While the study of Benwood Primary enabled an unpacking of issues relating to heterosexual masculinities and young boys, the fact remains that the incidents recorded above reflected the situation found in several studies of secondary schools where sexual harassment has been shown to impact upon girls' schooling experiences (Herbert 1989; Lees 1993; Larkin 1994). The study of Deneway Primary School was notable for its apparent absence of sexually harassing behaviours by the boys. What was significant in the study of Deneway Primary is that the findings suggest that sexual harassment is not an inevitable fact of school life for girls but is dependent on the interplay between the masculine self and wider social relations of domination and subordination.

Y5/6 boys at Deneway Primary School

The pupils at Deneway Primary School were from a different social class background from those at Benwood Primary School. Where the trend at Benwood Primary was for pupils to come from families where unemployment was in its second if not third generation, children at Deneway Primary had both parents working in professional occupations. Where children at Benwood Primary lived in houses rented by private landlords (and therefore in worse repair than council properties) their counterparts at Deneway resided in newly built detached or semi-detached dwellings.

A striking feature of the boys in the Y5 class was that they appeared to be an homogenous group but not in the way that the infant boys at Benwood did (see Chapter 5). These boys exhibited those modes of masculinity identified in studies of secondary schools as the 'ear 'oles' (Willis 1977), the 'Cyrils' (Connell *et al.* 1982) and the 'Swots' (Connell 1989; J. Stanley 1989). While inevitably there were some tensions between some of the boys, as a group, they could be considered 'ordinary' in that their progression through school involved doing some work, and not trying to dominate or sexually harass females in the school (Brown 1987; Wolpe 1988; Lees 1993). Over the course of the research there were no instances of sexual harassment by the boys nor did the girls make any reference in interviews to sexist or sexual behaviours by boys. How can this be explained?

The importance of football to the male camaraderie in the school was discussed at length in Chapter 6. The fact that because of their proficiency at football some boys were more 'favoured' than others did not appear to cause conflict. Studies of male friendship groups have pointed to the importance of friendships to pre-adolescent males (Fine 1980; Leahy 1983), a key feature of which is the 'cooperation, integration and sharing' of joint activities (Hall and Jose 1983: 269). Football provided opportunities for developing and reinforcing friendship bonds among all the boys in the class regardless of their actual sporting skill. However, simply because those boys who were 'Stars' in terms of football were able to access privileges more readily than other groups, such as girls and the less skilled footballers, does not mean to say they immersed themselves totally in the hegemonic masculinity of the school. Rather, the boys passively resisted a total collusion with the culturally exalted mode of masculinity in the school. Also, this apparently total commitment to football was not as all-consuming as it first appeared.

Boys in other studies have been noted as, in some way, refusing to 'buy into' the obsession that is football. For example, Martino (1999: 249) records the comments of one boy who relates the situation of two boys who joined the school, one of whom had never played football before while the other was very proficient, 'so instantly Joel was popular and David wasn't . . . so a person who was good at football, he became popular. It doesn't make sense but that's what happens'. Similarly, the boys in Pattman's (1998) research, after saying they liked football, then attempted to reposition themselves by distancing themselves from it following the girls' critique of boys who took the game 'so seriously'. Most interestingly, the four boys who were the 'Star' footballers at Deneway Primary all professed their prime interests were in individual rather than team sports; their favourite sports were judo, running, squash and swimming. This was supported when the class were allowed 'choice' in PE activities, Ossie and Dougie would jog around the school field, while Smittie would play with a tennis bat and ball, and Mac would choose a skipping rope rather than pick up one of the many footballs which were always available. These behaviours suggest that the boys were negotiating the dominant masculinity of the school which was modelled by the men teachers. This was informed by a 'laddish' form of masculinity based around shared humour, a love of football and flirting. The boys in Y5 were evidently prepared to engage with certain aspects of the dominant form of masculinity in Deneway Primary so why did the boys *not* attempt to access those heterosexualized attitudes and behaviours associated with it?

A possible reason to explain why the boys resisted this aspect of the dominant masculine discourse can be found by considering the interaction between the locale of the school, the age of the boys and the studies which have considered how boys see schoolwork as 'uncool'.

Locale of Deneway Primary

We have already seen how Benwood Primary School was situated within a local culture in which images of masculinity and femininity were clearly inscribed by gender power relations. The contrast in the home backgrounds of the two groups of pupils meant that the Y5 boys encountered a range of male and female subject positions; not only did they see their mothers holding down high powered careers but also women were active in the life of the school, not only as parent-helpers but as members of the school governing body. They were also frequently exposed to the challenges launched by many of the girls in the class to the flirtatious and sexist comments and actions of the male teachers. Examples of this are when several girls called a halt in a rehearsal for a school production to request that a word in one of the songs be changed from 'men' to 'people', and pointing out to the headteacher the implications for girls of having the Sports Club and Football Club on the same evening.

Age of the boys

Fine (1980) argues in his research into pre-adolescent male friendship groups that three types of friendship content seem salient during this period: work related activity, talk about sex and sexuality, and aggression. With regard to the second category, Fine points out that when it comes to sex and sexuality, the emphasis is on talk rather than on action. While discussing sex may be a high priority for pre-adolescent boys, there is a substantial body of research which shows how interacting with girls is positively avoided (Hallinan and Tuma 1978; Best 1983; Lloyd and Duveen 1992). Certainly observations showed that the Y5 boys rarely initiated any interactions with the girls although on those occasions when they were expected to collaborate they would do so apparently affably.

Although during interviews a few of the boys purported not to know the names of all the girls in the class by reverting to descriptions such as 'the one with the short hair who wears glasses', it also emerged that some cross-gender friendships did exist. These friendships were with girls who demonstrated the same three sets of principles they had listed as the ones they valued in male friends, that is, camaraderie (cooperation, helpfulness), (not) bullying, and who were 'Stars' (good at football). In terms of the latter, boys did not necessarily have to be any good at football, they just had to be classified as such by their mates. So while the boys listed 'moodiness', 'sarcasm' and 'snootiness' as reasons why they disliked many of the girls, those they liked, Holly and Emily, were described in terms of 'having a laugh', 'sticking up for you' and 'being good in a (fun) fight'. Although the boys did not engage in sexual or sexist discourses with the girls, and in some cases seemed puzzled when a girl was mentioned as if they had not come across this particular species, many justified their likes and dislikes in terms of the girls' physical appearance:

CS: You were saying you didn't like Saskia . . .
Lee: Yes . . . and Maggie, Kyoko . . .
Malcolm: Ruth and Rachel . . .
CS: What is it about those girls you don't like?
Lee: They're horrible, they don't look nice.

An interview with the 'Star' football players who were the most popular group of boys in the class with both the girls and the boys revealed a similar appraisal of the girls:

CS: You've talked about the girls you don't like . . . are there any girls you get on with?
Ossie: I li . . . [*pause*] I think Holly, Emily.
Mac: Yes, Emily's one of mine as well . . . nice legs.

The boys were at an age then when male friendships are particularly important and relationships with girls are generally avoided.

The third key factor which helps to explain why boys either as a group or individually did not take part in sexually harassing behaviours can be found in the recent research on boys' perceptions of what is 'cool' and what is not.

Schooling as 'uncool'

Several commentators have shown that one of the reasons why boys are not as successful in terms of academic achievement as girls is that it is not 'cool' to be seen to be working at, or achieving, good results (Alloway and Gilbert 1997; Francis 2000). This, it is argued, is a feature shared by all male pupils including those considered to be conformist. It is not something which has appeared in the attitudes of boys only in recent years, as in his study of working-class boys and schooling, Willis (1977) observed:

> It is not so much that they [school conformists] support teachers, rather they support the *idea* of teachers. [They have] invested something of their own identities in the formal aims of education and support of the school institution.
>
> (Willis 1977: 13)

As Willis suggests, the Y5 boys at Deneway Primary may have shared in many of the aims and values of the school culture (and some were able to access personal privileges from doing so) but this is not the same as imitating, aspiring to and identifying with the hegemonic masculinity of the school. Although the boys were able to draw upon the dominant version of masculinity in the school in order to negotiate and reconstruct their own masculine individuality, relations between the masculine identities of the boys in the class and their male teachers had necessarily to be 'different' in order for the boys to distance themselves from the school authority structures and personnel. It was not 'cool to collude'.

The combination of these three factors, the boys exposure to the gender–power relations in the local community, the nature of peer group friendships at the ages of 9 and 10, and the distancing from the authority of school and teachers suggests a reason why the boys did not engage with sexualized or sexist behaviours. If being a 'swot' meant being seen to imitate and conform to teacher authority, then some measures had to be adopted whereby differences could be discerned between the attitudes and actions of teachers and pupils (Alloway and Gilbert 1997). For the boys in the class this meant sharing their interest in football as not only was it central to the maintenance of peer group friendships and also brought some benefits but also it involved maintaining a distance from the girls in the class and the actions and behaviours of their male teachers.

This section has concentrated on the heterosexual engagements of primary age boys. It seems that as the boys got older they were more likely to verbalize a distinct lack of interest in girls often seeming not even to know their names. This may well have partially had something to do with the fact they were talking to me and a male researcher might have been given a different response. In Redman's (1996) research he found that it was during Y5 and Y6 when 'girlfriends' became a salient issue for boys. However, he argues that boys inhabit heterosexual discourses differently at primary and secondary level and whereas Y5 and Y6 boys are 'practising' for heterosexuality, Y8–Y10 are occupying it.

A third aspect of heterosexuality in the primary classroom is one for which no actual research evidence exists; that is, child sexual abuse.

Child sexual abuse in the primary school

> There is something about the combination of children and men and a caring environment which is seen, by some, and particularly in Britain and North America . . . as outlandish to the point of being 'a risk' or even dangerous to children's health and wellbeing.
>
> (Cameron *et al.* 1999: 132)

It is only a tiny minority of men primary teachers who are convicted of sexually abusing pupils but it is an issue which is high on the list of concerns of men who are thinking about or involved in teaching primary and nursery pupils. Research into the perceptions that sixth-formers have of teaching shows that males were anxious that if they displayed an inclination to work with young children, they may be seen as potential child abusers (Johnston *et al.* 1998). Similarly, the BEd students in Thornton's (1998) study indicated that one of the reasons why the majority of men avoided primary teaching was their fear of being seen by others as 'perverts'. It is fair to say that these fears have some foundation in that there is evidence to suggest that men teachers of young children are viewed with suspicion by those inside and outside the teaching profession (Aspinwall and Drummond 1989; Skelton 1991; Coulter and McNay 1993).

A number of questions may be considered in relation to men teachers working with primary and nursery age children.

How do we make sense of this reaction to men teachers as potentially dangerous to primary children when the actual accusations and conviction rates are so low?

One reason may be that men's involvement with children attracts considerable media attention; an analysis of stories in the *Guardian* and *Observer* newspapers in 1996 discovered that they carried 78 stories of 'men and children'. Of these 39 were about men taking responsibility for children and 26 of these involved violence towards children (Cameron and Moss 1998). Another possible reason is that masculinity is closely linked with sexuality (Metcalf and Humphries 1985; Connell 1995). Hearn (1988) has argued that child abuse is often discussed as an entity within itself without acknowledging that its roots lie in the construction of masculinity:

> child abuses, violences and sexualities to young people are . . . a close development of 'normal' masculinity and 'normal' male sexuality, itself characterized by power, aggression, penis orientation, the separation of sex from loving emotion, objectification, fetishism, and uncontrollability.
> (Hearn 1988: 541)

The argument put forward by Hearn (1988) is that the potential for violence (sexual and otherwise) is encoded in the way men are defined as men (see also Brittan 1989; Seidler 1991; Kenway and Fitzclarence 1997). On that basis *all* men have the potential for sexual violence but only *some* men sexually abuse young children.

What procedures are in place to protect primary children from the perceived risks of men?

There is one formal approach to ensuring the protection of children and a second, less formalized, set of practices adopted by schools and individual teachers. The formal approach involves police checks. The aim of these police checks is to identify those with the propensity to harm children and included here are those who have had contact with the police for 'serious sexual, violent, drug or drink offences' (DfE 1993: para. 36b). However, research has shown that police checks are an inefficient and ineffective way of trying to identify child abusers. In their research into child sex abuse in day care centres, Finkelhor *et al.* (1988) discovered that of all those convicted, 20 per cent had some prior police contact of which 12 per cent had actual convictions. These convictions were not, in the main, for sexual offences but for such offences as drink-driving. Of the 8 per cent who did have a police record for sexual transgressions these were not directly related to children; for example, one female perpetrator had been arrested for prostitution.

It is important to recognize that such a small return on such a large investment is a clear indication that police checks are not 'the' answer to preventing child sexual abuse in schools. Child sexual abusers frequently do not have police records. They are often adept at winning the confidence of both adults and children by demonstrating those 'person skills' which are needed when working with young people. Indeed a male teacher involved in a study I carried out in the early 1990s on the careers of men teachers of young children proved particularly skilful in duping others (see Sikes 2000 for a discussion of this).

An informal set of approaches has been adopted by some schools and individual teachers. These take the form of men teachers not putting themselves or children in a vulnerable position, hence, they would not change a child who had wet their knickers, cuddle a child who was hurt or sick, or tend to cuts and grazes if it involved removing trousers, skirts or pants. Following the conviction of a student for abusing children on a nursery placement, I interviewed early years men nursery and primary teachers to ascertain what impact the case had, if any, on their working lives. Lawrence, an experienced nursery teacher, said:

> We had staff review sessions on policy . . . about accompanying children to the toilet; should I do it, should I be in a position to get children changed? We had to adopt safeguards so that as a staff we weren't vulnerable. There was a sense of disbelief among us that one member of staff had to mark another and nobody really wanted to mark me. There was this period of awkwardness.

While such actions protect the child and the male teacher they do not, as Lawrence suggests, make for comfortable working conditions, relationships with colleagues or personal identity. Also what are we saying to children about what are appropriate behaviours for men teachers to demonstrate as opposed to those of women teachers?

What strategies are recommended for addressing child sexual abuse in primary schooling?

There is no suggestion that men should not work with young children (Hunt Report 1994). Pringle (1995) has argued all men are potential abusers of children so rather than looking at institutional change men should be doing 'preventative work on themselves'. However, this is a somewhat controversial approach and we know that any programme which begins from the premise that men and boys should feel guilty is not effective (see concluding chapter). Rather than seeing the prevention of child sexual abuse in school as something which involves simply identifying the 'high risk' case or dangerous individual (Parton 1990) a more appropriate response is where 'the protection of workers and the protection of children . . . [go] hand in hand' (Bateman 1998: 187). What this actually means in terms of developing an approach to child sexual abuse in a nursery or primary school can be seen in the following frequently cited comments of two male workers:

> The power of abuse lies in secrecy. If the culture of an institution is based upon empowerment and openness where adults and children are encouraged to speak up for themselves then the potential for abuse is minimized. When the structure, philosophy and practice of an institution proactively addresses issues of free communication where people

> of all ages and status are encouraged to listen to each other with respect, the abuse of power in relationships between adults, children, men, women, management and workers . . . will be avoided.
>
> (Chandler and Dennison 1995: 44)

Schools would begin this by developing whole school policies which encourage open communication on a number of 'difficult' issues (including all aspects of child abuse – verbal, physical, sexual). One approach to bullying has been through schools establishing 'pupil councils' where younger pupils can talk to older children. Such a forum would allow children the opportunity to discuss a wide range of issues which they may find difficulty in talking about with adults. 'Circle time' is another means through which some schools provide opportunities for children and teachers to communicate more openly with each other.

Higher education institutes might also engage more broadly with child sexual abuse. Anecdotal evidence suggests that the majority of one year primary postgraduate certificate in education courses have, at most, one or two lectures on the subject of child abuse. The main obstacle to covering the issue in more depth is both the limited time and the emphasis now placed by the National Curriculum for ITT on 'standards' demanded of newly qualified teachers (NQTs). As has been argued elsewhere, this emphasis on objective standards by which to measure a NQT's professional competencies lean heavily towards the perception of a teacher as a technicist and minimizes the personal and social role of teaching (Siraj-Blatchford 1993; Cole *et al.* 1997).

As was said at the beginning of this section, one of the most significant factors which influences men's avoidance of working with younger children is anxiety about being seen by others as 'perverts' (Thornton 1998). The centrality of (hetero)sexuality to dominant constructions of masculinity is thoroughly documented (Connell 1987; Mac an Ghaill 1994; Messner 1997), thus such fears need to be engaged with rather than avoided. As Penn (1998: 246) reminds us, 'men do not necessarily modify their views on coming into a woman's profession; they bring their masculinity . . . with them'. In the interviews carried out with early years men teachers following the conviction of the student nursery worker for child sexual abuse, Lawrence commented:

> Women are allowed to enjoy the sensuality of children in ways which men are not. Also most men confuse sex with sensuality and then feel guilty about any feelings they have towards children and this really needs to be discussed on teacher training courses . . . to unpack the differences between sexuality and sensuality but these sorts of conversations just aren't being had.
>
> (Skelton 1998a: 25)

This section has considered the largely unexplored issue of child sexual abuse in the primary school. There is a silence about this issue where it is rarely discussed openly in initial teacher training programmes, continuing professional development courses, studies on primary schooling or government policy. In official guidance to schools there is a leap from advising how to screen out inappropriate recruits (DfEE 1995a) directly into instructions as to what should be done when a teacher has been accused of sexually abusing a pupil (DfEE 1995b). However, it is clearly something which requires further, and open, discussion. Although the focus has been on issues to do with *men* teachers this is not to assume that it is only men who sexually abuse, but it is men teachers who are worried about the potential for accusations to be levelled against them. As Cameron *et al.* (1999) conclude, 'in the British context, gender is just below the surface in any discussion of risk in childcare', but as Lawrence, the nursery teacher, said more explicitly, it is constructions of masculinity that are at the heart of the problem.

Summary

This chapter has focused on heterosexuality in the primary school. Various aspects have been examined: heterosexuality in male teacher–female pupil interactions; romance relationships and (hetero)sexual harassment among pupils; and child sexual abuse.

The first section looked at how the dominant form of 'laddish' masculinity in one school, and the particular 'flirtatious' approach of one male teacher, produced a tension for the girls between being a 'normal pupil' and 'being feminine'. The girls were not passive in this process and some of them challenged the actions of the teacher by reversing the male gaze and positioning him as 'not a proper teacher'.

The second section considered romance relationships which, in the case of younger boys, often tipped over or were conflated with sexual harassing behaviours. It was argued that these young boys did not simply 'know' how to go about sexual harassment but rather drew on their knowledge of masculinity operating in the local culture. Rather, at the younger age boys are 'learning' various social processes and a key point here was the way in which adult females in the school reinforced any behaviours which could be constituted as sexual harassment (myself included). In contrast, the older boys in a different primary school were not observed at any time in the course of the year or reported by any of the girls in interviews as carrying out any form of sexually harassing behaviours. The reasons for this were argued to be a combination of the boys' knowledge of dominant feminine and masculine constructions which they brought with them into the school setting, their age – which dictated that girls were persona non grata – and their attitudes towards schooling. With regard to this, the boys enjoyed 'matey' relationships with the men teachers but, at the same time, did not wish to appear as if they were 'selling out' and a distancing strategy was not to imitate or join in with the flirtatious strategies of their teachers.

The final section touched upon issues relating to child sexual abuse *in* primary schooling. It is only a tiny minority of men teachers who are accused and convicted of sexually abusing their primary pupils yet it occupies a central place in the concerns of male teachers. This brings us back to a consideration of the impetus coming out of the boys' underachievement debate for an increase in the number of men teachers. If a major disincentive for potential male primary teachers is the fear of being seen as a 'pervert' then perhaps now is the time we should be having more open discussions of, and policies on, the subject. And not least for the sake of the children and women teachers in primary schools.

Note

1 Although such incidents are rare they make headline news. In the north east of England between 1993 and 1997 three male primary teachers, two male and one female nursery workers were imprisoned for involvement in five separate and unrelated incidents of abusing children in their classes.

chapter / **eight**

Conclusion: gender in the primary classroom

At the beginning of this book I pointed out that the reason for doing a study of boys and men teachers in primary schools was to get a clearer picture of how 'masculinity' affected girls' experiences of schooling. The boys' underachievement debate did not begin to gain momentum until the mid-1990s by which time my research was well underway. It was immediately apparent that the arguments put forward in the media, the way in which the government couched their concerns about boys' schooling and the catch-all solutions to boys' underachievement produced by some authors revealed a fundamental problem which had potentially hazardous consequences. The problem, quite simply, was that the idea of 'failing boys' provided politicians with a snappy soundbite and newspapers with banner headlines and in so doing instantly confined any appropriate discussion. Of course, the issue of the narrow parameters of the debate was raised by several feminists but little or no attention was given by politicians or the media to their reasoned observations (Kenway 1995; Weiner *et al.* 1997; Yates 1997; Epstein *et al.* 1998). Teachers are too busy to go through academic books and journals where this information can be found and have to depend instead on professional newspapers and public broadcasting for their information. As a result of not having the full picture, many schools and LEAs have policies in place specifically targeting boys which can have potentially damaging consequences both for girls and many boys.

This chapter, first of all, sets out some key issues which will, together, present a context for looking at boys in school in terms of gender relations. It will then suggest some strategies for schools to incorporate in order to address those questions which teachers are currently being told to find answers to, such as 'what do I do about boys in my class who don't like reading or aren't good at writing?' 'What can I do with those boys who persistently misbehave in the classroom or are constantly interfering with and disturbing the other children?'

Boys' underachievement: the issues

The arguments against all boys being seen as underachievers

The ways in which 'boys' are not underachieving in the way claimed in the popular press were set out in Chapter 1 but they are worth briefly reiterating here:

- The debate has centred around statistical differences which mask a complex picture. It is not *all* boys who underachieve in relation to girls but some boys *and* girls are underperforming in examinations. The focus on the 'gender gap' is actually concentrated at the high achieving end – so although there *is* a gender gap it is not one of boys being trounced by girls. Much has been made of girls now gaining more A grades at A level. However, boys have always gained more in the past and the difference now is insignificant (17.8 per cent of boys compared to 18.2 per cent of girls). Boys obtain more first class (and unclassified) degrees; girls gain more of the higher attainment levels in KS1–4 in English but there are no achievement gaps in maths and science; the achievement gaps that do exist (as in KS1 and KS2 English) are declining year by year (Gorard *et al.* 1999).
- The form of the assessment and the knowledge and understandings pupils bring with them to the task influence the grade they will achieve. An example of this was given in Chapter 1, when Downes (1999) alerted educationalists to the possible implications and inferences of the 1999 KS2 English test where the subject and framing of the assignment was more 'boy friendly'.
- Equity issues are not simply about boys versus girls but are significant in terms of particular groups of pupils. For example, boys from some ethnic minority groups are underachieving such as Black Caribbean and Pakistani boys but those from Chinese and, to a lesser extent, Asian Indian backgrounds are doing well (Pulis 2000). Also, there are ethnic minority groups where both boys and girls are underachieving, as in Bangladeshi pupils (Gillborn and Gipps 1996).
- Gender and ethnicity are the two factors that are focal to any discussions on achievement and it appears less 'fashionable' to talk about the impact of social class (for exceptions see Teese *et al.* 1995; Arnot *et al.* 1999). Yet social class would seem to be the most significant indicator of how successfully a pupil performs at school. Teese and his colleagues have observed how gender differences are most acute among pupils from socially disadvantaged backgrounds and at their least in children from

affluent homes. Obviously when social class is intertwined with ethnic and gendered identities then further stratifications become evident.
- The question has been raised that if schools disadvantage boys then why is this not reflected in higher education and the workplace (Raphael Reed 1999)? Equal numbers of men and women obtain university places (although these are predominantly people from middle-class backgrounds) and the average wage or salary of a man is higher than that of a woman.

Focusing on boys: the implications for schools

A key issue emerging from the case studies of Benwood Primary and Deneway Primary was the need for schools to identify the dominant image of masculinity it constructs and the kinds of masculinities operating in the school site. At both institutions the dominant forms of masculinity were incorporated into the control and management strategies and so obviously affected the ways in which boys and girls were seen by, and negotiated their places within, the school. The reason for schools to recognize what kinds of masculinities are operating and what form the dominant mode takes is so that appropriate strategies can be developed which challenge the particular 'problems' the boys in that establishment are having, whether these are laid back approaches to work or violent behaviours in the classroom. However, a major word of caution is *that no school policy should be developed which focuses exclusively on boys*. The importance of this is frequently given lip-service in books which are evidently based on the 'men's rights' perspective (see Chapter 2). Such publications offer recommendations whereby the classroom environment places a focus on boys while at the same time making placatory noises about the initiatives being designed with a view to improving girls' opportunities too (Bleach 1998; Noble and Bradford 2000). This is not to deny that there are real concerns about boys' abilities in English and that some action needs to be taken to tackle their lack of enthusiasm for this subject. However, the context in which initiatives aimed at promoting literacy among boys in school should be a more holistic one than studies indicate are currently in operation (see section on gender equity programmes).

The surveys of current practice on gender equality in LEAs and schools found that emphasis is being given to boys (Arnot *et al.* 1998b; Sukhnandan 1999). In the Arnot *et al.* (1998b) investigation 40 initiatives were targeted on boys only and of the 35 projects focusing on both sexes, boys' underachievement was mentioned as a particular issue. The authors of both surveys stressed that by far the main driver of these initiatives was gender differentiated performance in SATs. A second impetus was the concern by some schools for the obstacles boys encountered in their local communities such as those observed at Benwood Primary where there were high levels of unemployment and lack of positive male role models. Now this might be argued from some commentators writing from 'personal' masculine perspectives that the focus on boys is addressing the inequalities generated by having made schooling too girl-friendly. As we saw in Chapter 2, such a view fails to recognize the differences between boys because of social class, ethnicity, religion, sexuality and so on, and that many girls continue to experience difficulties in schooling. A simple concentration on 'boys' underachievement' according to an unsophisticated reading of performance data diverts attention away from looking at gender equality – and concepts of underachievement – more broadly.

It is all very well to make general comments about why programmes aimed specifically at boys are not helpful but a much clearer picture can be gained by looking at what has been said about specific programmes and how they have been received by boys and teachers. In the UK the talks and workshop sessions offered by Geoff Hannan seem to be most widely known about in schools (Pulis 2000). Hannan's (1999) position draws on both the conservative and men's rights perspective and so emphasis is given to the biological and psychological differences between boys and girls. Armed with this information, teachers are supposed to be equipped to develop teaching styles and approaches to classroom management which cater for these differences. The feelings of most teachers towards these particular approaches and their impact on boys and girls in the UK have yet to be evaluated but we can look to Australia where a range of programmes have been appraised.

Four main themes running through the programmes have been identified (Connell 1996; Gilbert and Gilbert 1998):

- boys' academic failure
- an apparent alienation and disinclination on the part of boys towards education and schooling
- the attributes of boys where they are seen as fragile and anxious simply because they are boys
- relationships with other people which are increasingly becoming intolerant and aggressive. This, when they are older, will lead to lifestyles which are destructive to themselves and others.

How these issues are explored tend to fall into the two types of programmes discussed in Chapter 2, that is *recuperative masculinity* and *pro-feminist* programmes. Browne and Fletcher's (1995) *Boys in Schools* incorporate a range of strategies including organizing boys into single-sex groups to talk about their relationships with each other – who are friends, who are enemies – and in order to do so they must first draw up a list of rules to prevent put-downs, conflict or aggressive behaviours. Other recommendations are boys taking on role-play activities which deal with bullying and various forms of harassment. Further strategies include exercises aimed at boosting boys' self-esteem. One of the attractions of such approaches for teachers is that they tap into the kinds of ideas we have become familiar with through equal opportunities work with girls as it uses the same language: where girls were oppressed, boys are victims; where schooling was seen as perpetuating the 'male as norm', primary schools are talked about as feminine and feminizing. But a major problem of these approaches is that they do not encourage either schools or boys to acknowledge that generally boys and men (and 'maleness') are in a more privileged and powerful position in the schooling system (Lingard and Douglas 1999). The results of this privileging was evident both in Benwood Primary School, where control and management strategies reflected those to be found in the local dominant form of masculinity, and at Deneway Primary in the celebration and centralization of male camaraderie.

In contrast are *pro-feminist* programmes such as *Boys-Talk* (Friedman 1996). This consists of a series of ten sessions which looks at the construction of gender, relationships, sexuality and alternatives to violence and is obviously aimed at an older age group. However, as we shall see in the next section, similar issues can be raised with primary pupils. The compilers of this programme recommend that it is taught by men teachers to demonstrate to boys that they are interested and keen to change certain aspects associated with masculinity such as violence. Other pro-feminist programmes would also draw on role play, as well as drama and dance/movement as alternatives to conversation techniques. An advantage of these programmes is they recognize male power and differences between men. A disadvantage is that they tend to make boys and men teachers feel bad about themselves! As Vogel (1997: 2) cynically commented in his review of *Boys-Talk*: 'Perhaps the expectation is that boys will feel good about getting a pat on the back from women?' The tendency of pro-feminist programmes to engender attitudes in boys and men teachers ranging from anger to guilt to disquiet and uncertainty has been observed by a number of commentators (Kenway 1995; Mclean 1997; Mills 2000). And it is not just males: Kenway (1995: 68) found that the girls in her research schools thought that in some ways the schools were 'sexist against boys'.

It seems then that there are problems in programmes based on both men's rights and pro-feminist perspectives. There are, in addition, two further difficulties associated with these programmes as a whole. First, as Lingard and Douglas (1999) have argued, existing programmes based on either approach concentrates on interpersonal relationships. Therefore they are not going to help in encouraging children to try out and become comfortable with a range of learning styles or take up activities they associate with the opposite sex such as girls playing football or boys choosing reading. A second difficulty is that these programmes locate boys as the problem and they are seen as the subjects of change. However, as the case studies of Benwood Primary and Deneway Primary clearly showed, the school itself constructs and maintains dominant images of masculinity which boys and girls have to negotiate with. There is no space in these programmes for schools to identify their role in the development of and engagement with various modes of masculinity. Furthermore, because these programmes are based on fundamentally opposed positions it seems a fruitless endeavour to try and identify some common ground so that schools are not put in the position of having to choose between two problematic approaches! At the same time, both perspectives have something to offer and strategies can be reworked into gender equity programmes (see section on this later).

Focusing on boys: the implications for pupils and teachers

Several of the reasons for not adopting strategies which focus exclusively on boys have already been raised. For example, there is an inbuilt assumption that all boys are the same and require similar treatment; that the problems many girls continue to experience in schooling will be ignored; the initiatives may reinvoke those practices identified in primary schools in the 1970s and 1980s which push girls to the margins of the classroom. Most importantly, projects which concentrate on boys reinforce gender differences and encourage teachers to think in terms of the strengths of children apropos their sex and accommodate, rather than address their weaknesses (Sukhnandan *et al.* 2000). Another consequence is that they invite boys and girls to redouble their efforts to establish their separate gendered identities. There are also a number of questions which require some consideration for their potential implications. Take the emphasis given to football as a means of increasing boys' motivations for schoolwork: in what ways are girls accounted for?

The government sponsored scheme *Playing for Success* aimed at primary and secondary pupils links school standards to football. The scheme is part of the government's £200 million 'Study Support' strategy and involves clubs in the Premier and first divisions. Each club provides premises and perks for the children who use the opportunity to do one or two sessions a week enhancing their literacy, numeracy and information technology skills. The incentive is, of course, that the pupils gain some association with 'their' high profile football club. There are also those strategies aimed at encouraging boys and literacy such as the idea that they 'read a team' where they read a selection of books chosen by the team of a high profile football club (Millard 1997). A similar government backed strategy is the introduction of a fantasy football league 'to make maths fun and end the innumeracy of many of Britain's soccer-obsessed *children*' (Wintour and Bright 2000, my emphasis).

The NFER is evaluating the *Playing for Success* scheme and early reports are that it is proving popular and effective with both boys and girls (Haigh 1999) and at the launch of the fantasy football league, David Blunkett pre-empted any criticism by pointing to a school in the pilot scheme where the best manager was a girl. While nobody is denying that girls are interested in football, it is not 'just a game' but is intersected by social class, ethnicity and heterosexuality (Epstein 1998; Skelton 2000). We saw in Chapter 6 how Deneway Primary marginalized girls from the game and how this was a common scenario in primary schools (Renold 1997; Connolly 1998; Swain 2000). Simply making moves to allow girls greater access is not tackling more fundamental questions about how the game is played out in primary schools:

> For the boys, to accommodate the girls would mean a reduction of the virility of the game as they understand it. For the girls, participation would mean accepting that their bodies, as well as the ball, would be the target of play.
>
> (Alloway 1995: 40)

Of course, it is not just girls who need to be considered as some boys are not proficient at football and this can lead to their lack of popularity among other boys and feelings of being marginalized from the main group (Martino 1999; Skelton 2000).

Now within a recuperative masculinity programme male-only football would be encouraged as it allows boys the opportunity for vigorous, robust exercise within a competitive framework. That is, it provides boys with a means through which they can enjoy a 'healthy expression of their masculinity' (Lingard and Douglas 1999: 139). Pro-feminists would discourage football as it is currently played on the grounds that it is presented as a war-like battle where toughness and aggression are celebrated. Obviously neither of these approaches is very helpful in addressing gender issues and alternatives are discussed in the gender equity section.

A further question for consideration is whether the emphasis on boys' underachievement has influenced the way in which teachers assess the abilities of pupils according to gender? Clarricoates' (1983) research into four primary schools was carried out at a time when boys were seen to be slow starters initially but caught up in the secondary years of schooling. One of the reasons the primary teachers in her study put forward for spending more time with the boys was that they found them more creative, and therefore more intelligent, and this needed careful nurturing to enable them to 'take off' academically as they matured. The same arguments arose in other studies where boys in primary schools were seen as more talented and girls as hardworking (Walden and Walkerdine 1985). Indeed, Francis (2000: 120) characterizes the 1980s as the period 'where the male pupil . . . personif[ied] all that was seen as right and proper in a learner'.

More recent studies have noted a shift in teacher attitudes towards their expectations of pupils. One study has found that although teachers claimed they did not treat boys and girls differently this was far from the case

(Younger *et al.* 1999). From their observations the authors found that boys and girls occupied very distinctive roles in the classroom with some boys dominating in terms of asking and responding to direct questions. However, this was tied up with teacher–male pupil power struggles with teachers directing more questions at boys in order to maintain class control. Their findings were that, increasingly, teachers were defining their 'ideal student' as female:

> Moreover . . . certain characteristics which girls have always exhibited but which have sometimes been subtly denigrated, such as diligence and application, are being re-evaluated as they come to be associated with a range of perceivably newer and more positive characteristics – for example, that girls are more confident, are self-learners, are more articulate and are more inclined to take the initiative in promoting their own learning.
>
> (Younger *et al.* 1999: 327)

The teachers in Connolly's (1998) multiethnic primary school expected girls to achieve more than boys although it has to be said these were racialized. Black girls were seen as particularly accomplished in sport and music, and South Asian and white girls as academically sound. In contrast, boys, as learners, were viewed by teachers as 'stubborn' (Black boys) and 'lazy' (South Asian boys) (Connolly 1998: 119). Yet similar to the situation in the 1980s, the teachers tended to make more of boys' successes than those of the girls. Connolly argues that boys' work was singled out and highlighted around the school partly so that teachers could demonstrate their success at getting good work from difficult pupils. With today's emphasis on teachers being able to prove their abilities publicly and the introduction of performance related pay it is hardly surprising to find such tactics being used!

It may well be that these changes in teacher attitudes to girls as the talented learners are partly due to the public emphasis on the gender gap and this shift in expectations could well feed back into classroom interactions. Indeed this is precisely what Younger *et al.* (1999: 327) found in their research and they warn against seeing girls as 'natural' learners and taking for granted increasing gender differentiation in girls' favour as this 'may all too easily generate a self-fulfilling prophecy'.

In her book *Boys, Girls and Achievement*, Becky Francis (2000: 129) asks an important question, the answer to which should inform any gender equity programme: 'Do we want boys to change?' There are many aspects of some boys' attitudes and behaviours which can bring their pleasures and amusements as well as sometimes proving frustrating and irritating. For example, the behaviour of Shane at Benwood Primary meant that teachers constantly had to maintain a vigilant eye wherever he was in the classroom – some of his less contentious actions involved constantly removing the carbon paper from my notebook but his other ventures included one (alleged) incident when he removed the key to the fire escape door from the teacher's desk and sold it on to one of the older boys in the school. Such events were provocative but also raised wry smiles from the teachers; Shane was undoubtedly a likeable 'character' and popular among the class. As Francis (2000) points out, it is hardly surprising to find that 'laddish' boys are popular as it is precisely this kind of 'boys behaving badly' form of masculinity which currently receives favour in society (as illustrated in TV sitcoms such as *Men Behaving Badly*, the teenage male characters in *EastEnders* and comic duos such as Baddiel and Skinner). If teachers and other pupils can see some positive aspects of those masculinities which are targeted for change in boys' programmes, then what of the boys themselves? To what extent do they want, or might be willing to engage with, activities that change the ways in which they have constructed themselves as 'proper boys'?

Martino's (1999) study of middle-class boys shows that being one of the lads is not a prerogative of the working class as appeared to be the situation at an earlier time (Pollard 1985; Clarricoates 1987). This being the case, and we know that being a lad involves having a laugh, being 'cool' at school and generally doing the opposite of what girls do, then it is likely that those boys whose attitudes schools want to change, are going to resist. Also, there is evidence to show that young boys are aware of women's changed role in society (that they are likely to work rather than stay at home to look after the family) but remain remarkably fixed in terms of their own role and tend to see themselves in conventional terms of being the 'breadwinner' and not taking equal shares in childcare (Lees 1993; Francis 1998). (Interestingly, girls' constructions of their gender identity have changed much more significantly over recent years: see Sharpe 1994; Arnot *et al.* 1999.) We have seen from the work of Davies (1989), Thorne (1993) and Francis (1998) that children invest time and energy in ensuring that their 'correct' gender is on public display. The discussion so far has provided a catalogue of potential limitations, disappointments and damaging results in pursuing existing programmes aimed at boys. Having depressed readers (and myself) with this scenario of doom and gloom for the future of boys and masculinities, we can now turn to a more promising view of gender equity and primary schooling.

Gender equity programmes: a way forward

From this point on there will be a move away from referring to projects for boys and talk instead of gender equity programmes. Now this does not mean that we should be looking for ways of treating primary pupils the same and/or trying to get boys to act like girls or vice versa. The strategy of down-playing if not eradicating gender differences seems to have been one of the aims of equal opportunities policies of the 1980s.[1] So while teachers ensured boys and girls were given equal shares of time in the home corner or with the Lego, they used the equipment in gendered ways. The boys would use the home corner as a castle to be attacked or acted as robbers while the girls played out domestic events (Whyte 1983); conversely the boys made vehicles with the Lego and the girls built houses (Burn 1989). This is not to say that equal opportunities policies were unhelpful as they did a great deal in raising awareness of the ways in which pupils were routinely organized and managed along gender lines; the traditional stereotypes to be found in reading schemes and other school resources; and the predominance of men as headteachers and working exclusively with older pupils. What children cannot be fooled into thinking is that gender does not matter nor can they be persuaded that it is acceptable, even desirable, to traverse gender boundaries – at least in the school setting, but don't try this in the street, boys and girls! An alternative and more appropriate agenda is to put into place initiatives that encourage children to think about their own position – to get them to question some of the more taken-for-granted aspects of what they see, hear, read, think, say and act out.

Starting with the school

To make any inroads into pupils' gendered ideas of identity, a whole school policy is needed. This policy would not be a checklist of 'things to do in the school or classroom' but would centralize four main questions which should be revisited on a regular basis to accommodate any changes. These questions are listed below and then briefly discussed:

- What images of masculinity and femininity are children bringing with them into school and what types are they acting out in the classroom and playground?
- What are the dominant images of masculinity and femininity that the school itself reflects to the children?
- What kinds of role models does the school want and expect of its teachers?
- What kinds of initiatives/strategies/projects should teachers be undertaking with children to question gender categories?

What images of masculinity and femininity are children bringing with them into school and what types are they acting out in the classroom and playground?

Teachers need to identify what ideas about men and women the children hold and where they are accessing this information. At Benwood Primary the children learned about aggressive forms of the dominant masculinity of the local culture from the youths and young men in the area, from videos of adult films, from violent cartoon films and television coverage of 'their' deprived area. This information then helps teachers to build up a picture of the various modes of masculinity and femininity, including those they meet in school, which children are negotiating and reconstructing.

What are the dominant images of masculinity and femininity that the school itself reflects to the children?

The school is itself a gendered institution which will convey specific dominant images of masculinity and femininity to children. It is important for the school to be aware of what those images are. The teachers at Benwood Primary wanted to provide pupils with alternatives to the 'outside behaviour' of the 'lads' but subconsciously incorporated the aggressive forms of dominant masculinity evident in the local community into their control and management strategies as a survival strategy. Pickering (1997) provides a number of activities for teachers to undertake as a staff to help them identify the various forms of masculinity in the school site. A similar set of activities can be devised to explore images of femininity.

What kinds of role models does the school want and expect of its teachers?

Although the question raises issues about role models for both sexes, a major concern at the moment is men teachers and what images they might present. It was argued in Chapter 6 that very little attention has been given to this question. Those commentators who have considered it, such as Biddulph (1998), give an impression of existing men primary teachers as deficient or cruel or so career minded they do not have time for their pupils! Another factor is how schools can ensure men teachers and children are able to work together in a safe environment where issues about child sexual abuse have been placed on the agenda and accommodated appropriately in the day-to-day running of the school.

The male teachers at Deneway Primary established friendly or matey relationships with the boys in particular based on shared masculine pleasures but, while effective for some, this did marginalize several of the less competent boys and the girls who were firmly positioned as female/'other' to their maleness. The claim currently being made within the boys' underachievement debate is that we 'need' more men teachers to act as a counterbalance to the predominantly female teaching force, which as we saw in Chapter 6 has resulted in the assertion that girls and girls' styles of learning are favoured. Now clearly we do not want to get into a debate about the reasons why we might or might not 'need' more men teachers – for one thing the idea that we might be arguing for more male nurses to counterbalance a female dominated profession on the basis that men can offer 'tough love' and 'sternness' (Biddulph 1997, 1998) to male patients sounds ridiculous! It is enough to say that it is clearly preferable for children to see adults working across a range of occupations.

Mills (2000) has developed one specific role that men teachers might positively adopt in working with boys. He makes the case that men teachers could work most effectively with boys on their aggression and similarly negative behaviours to draw out the links between violence and dominant forms of masculinity. Mills says that the aim of working with boys exclusively is so that their more privileged position (as males) can be challenged. Most importantly he raises the issue of the age of the boys as an important factor as 'while . . . boys have to be held accountable for their actions, they should not be held responsible for the ways in which masculinity has become associated with violence' (Mills 2000: 236). On this basis he advocates that men teachers adopt a respectful approach towards the boys rather than an arrogant, severe demeanour which they are more commonly associated with. Mills raises important issues here as boys' aggression and how to tackle it, and the attitudes of teachers, are highlighted in other recent research into primary schooling.

What kinds of initiatives/strategies/projects should teachers be undertaking with children to question gender categories?

Following on from Mills (2000), an issue which is raised over and again is the crucial importance of the teacher in enabling children to challenge gender categories and the need for them to intervene actively in teaching/learning/play situations (Yeoman 1999; MacNaughton 2000; Marsh 2000). The message in this literature is that we cannot rely on ideas within child-centred practices which suggest adult intervention in early play or learning should be minimal. To illustrate this, two current concerns will be looked at in more detail: boys' behaviour and literacy. As I write this I am aware of the irony of choosing two main targets of the boys' underachievement debate having said a few paragraphs ago that no school policy should focus exclusively on boys. The aim in identifying boys' behaviour and literacy practices is because, first, these are real concerns for teachers and they are not simply going to go away by pretending they do not need to be dealt with. Second, what follows will illustrate how many of the current problems identified with boys can be tackled, and more appropriately, within a gender equity framework.

Boys behaving badly

As we saw earlier, Mills (2000) has argued that boys' aggression and violence are areas which do need to single out boys for particular work and are aspects of schooling where male teachers can make a positive contribution as 'role models'. He stresses that it is not a case of men teachers working with boys on behavioural matters because men can do this work better than women, but because it is the responsibility of men to challenge the existing gender order where maleness occupies a privileged position. The success that women teachers have had in getting even very young boys to reflect and change their aggressive behaviours is shown in MacNaughton's (2000) study. She first explains how some well-intentioned strategies can go wrong. Here she cites recommendations suggesting that boys should be offered alternative versions of non-violent masculinity such as images of brave/admirable men (responding to community emergencies like lost children, major transport accidents); strong/admirable men (working in conservation or local campaigns to improve life for others); admirable/likable men (caring for family members, being childcare workers). MacNaughton (2000: 153) argues that such an approach is problematic on three counts: 'it takes time and energy away from a focus on the girls; boys resist; and there may be a backlash from parents'.

Boys' resistance is largely based on the fact that dominant ways of being masculine include a range of 'bad behaviours' from being aggressive to just plain naughty or mischievous and are, therefore, very attractive to boys. So instead of offering boys alternatives which they do not want to relate to MacNaughton (2000: 159) suggests that teachers find ways of:

- talking with boys about which masculinities they find desirable and why
- exploring with them the difficulties involved in different ways of being masculine
- exposing the occasions when boys make choices about how to be boys and the knock-on effects of those choices
- curtailing those boys who are violent and aggressive and offering strong support to those who are non-violent
- providing different images of masculinity and help them to develop definitions of masculinity that redefine what it means to be brave, strong, admirable and so on.

An example of a nursery teacher intervening with one boy who was causing particular concern because of his aggressive actions illustrates how this advice works in practice. The teacher observed that the boy, John, was persistently violent towards the girls in ways ranging from pushing past them to swearing to lifting up their dresses. He also refused to let them take on male associated roles such as firefighters in play sessions. The teacher challenged his behaviour by actively involving him in caring, domestic storylines. On one occasion when John was standing next to her in the home area she handed him one of the 'babies' (dolls) with the request 'Could you look after my baby for me?' John took the doll although he appeared embarrassed, which the teacher ignored and proceeded to tell him how to nurse the baby, burp it and so on. He became involved in the play and the next day voluntarily took up the 'baby' – and pretended he was giving birth to it! Over the course of seven months, with the teacher initiating similar storylines, his behaviour changed.

Another area where 'tough' forms of masculinity are particularly desirable and offer high status for boys is football. It was said earlier that neither recuperative masculinity or pro-feminist programmes offer a feasible way forward; it is seen, respectively, as a game which provides boys' 'natural' aggression with a structured outlet where 'healthy' competition can thrive, or it is a game which should be banned because it encourages those behaviours which are seen as negative. Football within a gender equity framework would be played by both boys and girls in mixed teams thus providing boys with a different means of conceiving the game (Salisbury and Jackson 1996). Also the way in which sports were constructed in any school site would appear differently from the more conventional sporting hierarchy with football at the top and athletics at the bottom (see Chapter 6). This would mean the valorization of football and its players would be 'replaced with a more balanced recognition of students' sporting and other achievements' (Gilbert and Gilbert 1998: 249). The popularity of football as a means of tackling primary aged boys disaffected attitudes to schooling is unlikely to dissipate; however, schools might consider how the game is constructed and played out in their own locations and make appropriate interventions to promote gender equity.

Gender and literacy practices

Generally girls are more interested in reading than boys (Millard 1997) but both genders are keen on role play and we know that boys in particular are more drawn to action based literacy tasks (QCA 1998). Marsh (2000) has identified fantasy play as a means of encouraging literacy practices and disrupting conventional gender categorization. Superheroes provide a rich source of play for young boys (Paley 1984; Jordan 1995). As Marsh (2000) indicates, these superheroes (Superman, Batman, Power Rangers) offer a narrow and extreme version of masculinity which is aggressive, strong and antisocial. There are female versions running alongside and they are not only brave but also attractive! Also interesting is that the male superheroes are 'men' as in *Batman* and *Superman*, while the female versions are placed in a more childlike position as in *Batgirl* and *Supergirl*. Marsh, among others, says that although it has been argued that girls are not interested in superhero stories this is inaccurate and it is the context in which the stories take place that deter girls. Superhero stories tend to be placed in very masculine environments and storylines where females are very much at the edges. Placing girls more centrally in the devising of superhero storylines encourages their participation. Marsh (2000) explains how a 'Bat cave' was set up in a classroom:

> The cave contained two desks, a computer, writing materials (notepads, pens, pencils, lined and unlined paper, two blank books labelled 'Batman's Diary' and 'Batwoman's Diary') and reading materials (maps, comics, messages, instructions). There was a dressing-up rack which contained home-made tabards, commercially produced Batman outfits, a cloak and a hat.
>
> (Marsh 2000: 212)

Much of the early discussions between teachers and pupils centred around the possibilities that the cave had to offer. This was especially important given that, with adult guidance on how to use literacy-related materials in role-play areas, young children will increase the number of literacy-related events. The children looked at books and watched videos on Batman and discussed them as a class. The sexist nature of several of the storylines was picked up on even by 6–7-year-old children. A great deal of work was put in by the teacher before and during children's use of the cave in order to help them disrupt conventional gender categories (see Marsh 2000 for details). The outcomes suggest that this endeavour was successful in encouraging boys (and girls) to get involved with literacy practices and in challenging traditional gender dichotomies. During a ten-day period girls engaged in 371 literacy activities and boys in 357 – a relatively low level of significance. Inevitably there was evidence of boys and girls taking up literacy differently. Girls made greater use of their Bat diary than the boys did of theirs; boys were more likely to write messages, and girls to write letters. However, the important missive here is that, by intervening, the teacher was able to disrupt conventional ways of boys and girls play and their involvement in literacy practices.

The use of a variety of texts to rewrite gender scripts has been shown to be both revealing and productive as we saw in the work on feminist fairy tales by Bronwyn Davies (1989). All those advocating text (film, plays, oral stories, pictures, books) as a means of encouraging children's questioning emphasize that teachers cannot rely on non-stereotypical characters and non-sexist curricula as children just reject them out of hand. These do not make 'sense' to a child and they do not tap into the child's imaginative world (Wing 1997; Yeoman 1999). Girls might be encouraged to identify with an independent heroine but the extent to which she does depends on her age as both Bronwyn Davies (1989) and I found that younger girls were as anti the heroine in the *Paper Bag Princess* as the boys. However, under no circumstances are boys prepared to align themselves with weak and silly princes or other wimpy characters (Westland 1993). Thus teachers should focus on providing real and fictional characters of the kind children can identify with (for specific ideas on critically reading texts with primary age children see Westland 1993; Wing 1997; Yeoman 1999).

I want to end this book with an anecdote which for me encapsulates the need to intervene in children's conceptual understandings of gender categories. It also points to the importance of not taking for granted that, in today's world of high tech communication and where the television provides images of events across the globe, children have access to a wide range of masculine and feminine ways of being. The incident occurred when I was interviewing a small group of 6-year-old girls at Benwood Primary. I was asking questions about what they liked doing at school and what they wanted to do when they left school. As is often the case for females interviewing girls, there was a conversational flavour to the interview when they were asking me questions about my own school days. One girl, Katy, said very little but sat watching me with a look of curiosity on her face. At one stage she interrupted the discussion and asked me, 'Have you got any bairns?' I replied no and continued talking to the other girls. A few moments later Katy interrupted again this time saying, 'Have you got any babies?' Again I replied in the negative. A few more minutes passed by in conversation with the other girls when Katy leapt to her feet and stood in the middle of the group, stared at me and said, 'Well, have you got any grown-up kids then?' When I replied 'No' her look of curiosity was replaced by one of complete incomprehension and she asked in a baffled tone, 'How?' I was the first adult woman she had come across who did not have children. Disrupting children's own conventional images of masculinity and femininity does not necessarily always mean introducing radical alternatives – sometimes it means simply telling them how it is.

Note

1 An alternative to equal opportunities policies were those based on anti-sexist initiatives (Weiner 1985). Anti-sexist policies placed girls at the centre of schooling processes; consideration was given to where women's contribution to society appeared in the school curriculum, 'girl only' spaces in the school and gender-power dynamics, for example, issues to do with sexual harassment. The overwhelming majority of primary schools opted for equal opportunities policies (or no policy at all!). Only a tiny minority of primary schools and nurseries – based in the Inner London Education Authority – adopted anti-sexist programmes (ILEA 1986; see also Myers 2000).

References

Abbott, A. (1993) The sociology of work and occupations, in J. Blake and J. Hagen (eds) *Annual Review of Sociology*, 19: 187–209.

Abraham, J. (1989) Gender differences and anti-school boys, *Sociological Review*, 37(1): 65–88.

Abraham, J. (1995) *Divide and School.* London: Falmer.

Acker, S. (1995) Gender and teachers' work, in M. Apple (ed.) *Review of Research in Education 1995–1996.* Washington, DC: American Educational Research Association.

Acker, S. (1999) *The Realities of Teachers' Work.* London: Cassell.

Alexander, R. (1984) *Primary Teaching.* Eastbourne: Holt, Rinehart and Winston.

Allan, J. (1993) Male elementary teachers: experiences and perspectives, in C. Williams (ed.) *Doing Women's Work.* London: Sage.

Alloway, N. (1995) Playing at gender? Young children struggling to get it right, in Ministerial Advisory Committee on Gender Equity in Education, *Proceedings of the Challenging Perspectives, Building Partnerships Conference.* Brisbane: Queensland Department of Education.

Alloway, N. and Gilbert, P. (1997) Boys and literacy: lessons from Australia, *Gender and Education*, 9(1): 49–59.

Arizpe, E. and Arnot, M. (1997) The new boys of the 90s: a study of the reconstruction of masculinities in relation to economic change. Paper presented to Gender and Education Conference, University of Warwick, 16–18 April.

Arnot, M. (1984) How shall we educate our sons?, in R. Deem (ed.) *Co-education Reconsidered.* Milton Keynes: Open University Press.

Arnot, M. (1991) Equality and democracy: a decade of struggle over education, *British Journal of Sociology of Education*, 12(4): 447–66.

Arnot, M. and Weiner, G. (eds) (1987) *Gender and the Politics of Schooling.* London: Unwin Hyman.

Arnot, M., David, M. and Weiner, G. (1999) *Closing the Gender Gap?* Cambridge: Polity.

Arnot, M., Gray, J., James, M. and Rudduck, J. (1998a) *A Review of Recent Research on Gender and Educational Performance.* OFSTED Research Series. London: The Stationery Office.

Arnot, M., Millen, D. and Maton, K. (1998b) *Current Innovative Practice in Schools in the United Kingdom.* Final Report. Cambridge: University of Cambridge for the Council of Europe.

Ashton, P.M., Kneen, P., Davies, F. and Holley, B.J. (1975) *The Aims of Primary Education.* London: Macmillan.

Askew, S. and Ross, C. (1988) *Boys Don't Cry: Boys and Sexism in Education.* Milton Keynes: Open University Press.

Aspinwall, K. and Drummond, M.J. (1989) Socialized into primary teaching, in H. De Lyon and F. Migniuolo (eds) *Women Teachers.* Milton Keynes: Open University Press.

Back, L. (1994) The 'White Negro' revisited: race and masculinities in south London, in A. Cornwall and N. Lindisfarne (eds) *Dislocating Masculinity.* London: Routledge.

Ball, S. (1981) *Beachside Comprehensive.* Cambridge: Cambridge University Press.

Bateman, A. (1998) Child protection, risk and allegations, in C. Owen, C. Cameron and P. Moss (eds) *Men as Workers in Services for Young Children: Issues of a Mixed Gender Workforce.* London: Institute of Education, University of London.

Becker, H. (1952) The career of the Chicago public school teacher, in M. Hammersley and P. Woods (eds) *The Process of Schooling.* London: Routledge and Kegan Paul.

Best, R. (1983) *We've All Got Scars: What Boys and Girls Learn in Elementary School.* Bloomington, IN: Indiana University Press.

Beynon, J. (1984) 'Sussing out' teachers: pupils as data gatherers, in M. Hammersley and P. Woods (eds) *Life in School.* Milton Keynes: Open University Press.

Beynon, J. (1989) 'A school for men': an ethnographic case study of routine violence in schooling, in S. Walker and L. Barton (eds) *Politics and the Processes of Schooling.* Milton Keynes: Open University Press.

Biddulph, S. (1997) *Raising Boys.* London: Thorsons.

Biddulph, S. (1998) *Manhood.* Stroud: Hawthorn.

Bleach, K. (ed.) (1998) *Raising Boys' Achievement in Schools.* Stoke-on-Trent: Trentham.

Bleach, K., with Blagden, T., Ebbutt, D. *et al.* (1996) *What Difference Does it Make? An Investigation of Factors Influencing the Motivation and Performance of Year 8 Boys in a West Midlands Comprehensive School.* Wolverhampton: Educational Research Unit, University of Wolverhampton.

Blumer, H. (1969) *Symbolic Interactionism: Perspective and Method.* Englewood Cliffs, NJ: Prentice-Hall.

Bly, R. (1990) *Iron John: A Book about Men.* New York: Addison-Wesley.

Board of Education (1925) *Report of the Departmental Committee on the Training of Teachers for Public Elementary Schools.* London: HMSO.

Bourdieu, P. (1986) *Distinction: A Social Critique of the Judgement of Taste.* London: Routledge.

Bradford, W. (1997) *Raising Boys' Achievement.* Kirklees: Kirklees Education Advisory Service.

Brandenburg, A. (1984) *We are Best Friends*. London: Picolo.
Brittan, A. (1989) *Masculinity and Power*. Oxford: Blackwell.
Brown, P. (1987) *Schooling Ordinary Kids: Inequality, Unemployment and the New Vocationalism*. London: Tavistock.
Brown, P., Halsey, A.H., Lauder, H. and Stuart Wells, A. (1997) The transformation of education and society: an introduction, in A.H. Halsey, H. Lauder, P. Brown and A. Stuart Wells (eds) *Education: Culture, Economy and Society*. Oxford: Oxford University Press.
Browne, R. and Fletcher, R. (eds) (1995) *Boys in Schools*. Sydney: Finch.
Budge, D. (1994) A world made for women? *Times Educational Supplement*, 24 June, 12.
Burgess, H. (1989) 'A sort of career': Women in primary schools, in C. Skelton (ed.) *Whatever Happens to Little Women?* Milton Keynes: Open University Press.
Burn, E. (1989) Inside the Lego house, in C. Skelton (ed.) *Whatever Happens to Little Women?* Milton Keynes: Open University Press.
Burton, L. and Weiner, G. (1990) Social justice and the National Curriculum, *Research Papers in Education*, 5(3): 203–28.
Butler, J. (1990) *Gender Trouble: Feminism and the Subversion of Identity*. London: Routledge.
Byers, S. (1998) *Co-ordinated Action to Tackle Boys' Underachievement*. Speech presented to the Eleventh International Congress for School Effectiveness and Improvement, University of Manchester Institute of Science and Technology, 5 January.
Byrne, E. (1978) *Women and Education*. London: Tavistock.
Cain, M. (1986) Realism, feminism, methodology and law, *International Journal of the Sociology of Law*, 14: 255–67.
Cameron, C. (1997) Promise or problem? A review of the literature on men working in childcare services. Draft working paper provided by Thomas Coram Research Unit.
Cameron, C. and Moss, P. (1998) Men as carers for children: an introduction, in C. Owen, C. Cameron and P. Moss (eds) *Men as Workers in Services for Young Children: Issues of a Mixed Gender Workforce*. London: Institute of Education, University of London.
Cameron, C., Moss, P. and Owen, C. (1999) *Men in the Nursery*. London: Paul Chapman.
Campbell, B. (1993) *Goliath*. London: Methuen.
Campbell, B. (1995) Little Beirut, *Guardian Weekend*, 1 July: 14–21.
Canaan, J. and Griffin, C. (1990) The new men's studies: part of the problem or part of the solution?, in J. Hearn and D. Morgan (eds) *Men, Masculinities and Social Theory*. London: Unwin Hyman.
Caplan, P. (1997) *Gender Differences in Human Cognition*. New York: Oxford University Press.
Carrigan, T., Connell, B. and Lee, J. (1985) Toward a new sociology of masculinity, *Theory and Society*, 14: 551–604.
Carrington, B., Bonnett, A., Demaine, J. *et al.* (2000) *Being and Becoming a Teacher: The Perceptions and Experiences of Ethnic Minority PGCE Students*. Interim Report no. 3 to the Teacher Training Agency. London: Teacher Training Agency.

Chandler, T. and Dennison, M. (1995) Should men work with young children? *The Abuse of Children in Day Care Settings* (report of a conference held at the NSPCC National Training Centre, Leicester). Leicester: NSPCC.

Cheng, C. (ed.) (1996) *Masculinities in Organizations*. London: Sage.

Chetwynd, J. and Hartnett, O. (eds) (1978) *The Sex Role System*. London: Routledge and Kegan Paul.

Clarke, N. (1991) Strenuous idleness, in M. Roper and J. Tosh (eds) *Manful Assertions*. London: Routledge.

Clarricoates, K. (1978) Dinosaurs in the classroom: a re-examination of some aspects of the 'hidden curriculum' in primary schools, *Women's Studies International Quarterly*, 1: 353–64.

Clarricoates, K. (1980) The importance of being Ernest . . . Emma . . . Tom . . . Jane. The perception and categorization of gender conformity and gender deviation in primary schools, in R. Deem (ed.) *Schooling for Women's Work*. London: Routledge and Kegan Paul.

Clarricoates, K. (1983) Classroom interaction, in J. Whyld (ed.) *Sexism in the Secondary Curriculum*. London: Harper and Row.

Clarricoates, K. (1987) Child culture at school: a clash between gendered worlds, in A. Pollard (ed.) *Children and their Primary Schools*. Lewes: Falmer.

Clatterbaugh, K. (1990) *Contemporary Perspectives on Masculinity: Men, Women and Politics in Modern Society*. Washington, DC: Westview.

Cobbe, F.P. (1894) *Life of Frances Power Cobbe*, 3rd edn. London: Richard Bentley and Son.

Coffield, F., Borrill, C. and Marshall, S. (1986) *Growing Up at the Margins*. Milton Keynes: Open University Press.

Cohen, D. (1990) *Being a Man*. London: Routledge.

Cohen, M. (1998) 'A habit of healthy idleness': boys' underachievement in historical perspective, in D. Epstein, J. Elwood, V. Hey and J. Maw (eds) *Failing Boys? Issues in Gender and Achievement*. Buckingham: Open University Press.

Cole, M., Hill, D. and Shan, S. (eds) (1997) *Promoting Equality in Primary Schools*. London: Cassell.

Collier, R. (1995) *Masculinity, Law and the Family*. London: Routledge.

Connell, R.W. (1983) *Which Way is Up?* Sydney: Allen & Unwin.

Connell, R.W. (1985) *Teachers' Work*. Sydney: Allen & Unwin.

Connell, R.W. (1987) *Gender and Power*. Cambridge: Polity.

Connell, R.W. (1989) Cool guys, swots and wimps: the interplay of masculinity and education, *Oxford Review of Education*, 15(3): 291–303.

Connell, R.W. (1995) *Masculinities*. Cambridge: Polity.

Connell, R.W. (1996) Teaching the boys: new research on masculinity, and gender strategies for schools, *Teachers College Record*, 98(2): 206–35.

Connell, R.W. (1997) The big picture: masculinities in recent world history, in A.H. Halsey, H. Lauder, P. Brown and A. Stuart Wells (eds) *Education: Culture, Economy and Society*. Oxford: Oxford University Press.

Connell, R.W., Ashenden, D.J., Kessler, S. and Dowsett, G.W. (1982) *Making the Difference*. Sydney: Allen and Unwin.

Connolly, P. (1994) Theorising racism in educational settings: reintroducing the work of Pierre Bourdieu. Paper presented at British Educational Research Association Conference, University of Oxford, 8–11 September.

Connolly, P. (1997) In search of authenticity: researching young children's perspectives, in A. Pollard, D. Thiessen and A. Filer (eds) *Children and their Curriculum.* London: Falmer.
Connolly, P. (1998) *Racism, Gender Identities and Young Children.* London: Routledge.
Corrigan, P. (1979) *Schooling the Smash Street Kids.* London: Macmillan.
Corsaro, W. (1988) Routines in the peer culture of American and Italian nursery school children, *Sociology of Education*, 61: 1–14.
Cortazzi, M. (1991) *Primary Teaching: How It Is.* London: David Fulton.
Coulter, R. and McNay, M. (1993) Exploring men's experiences as elementary school teachers, *Canadian Journal of Education*, 18: 398–413.
Crowley, N. (1995) Letter to *Guardian Weekend*, 8 July: 5.
Cunningham, H. (1991) *The Children of the Poor.* Blackwell: Oxford.
David, M. (1993) *Parents, Gender and Education Reform.* Cambridge: Polity.
Davies, B. (1989) *Frogs and Snails and Feminist Tales.* London: Allen & Unwin.
Davies, B. and Banks, C. (1995) The gender trap: a feminist poststructuralist analysis of primary school children's talk about gender, in J. Holland and M. Blair with S. Sheldon (eds) *Debates and Issues in Feminist Research and Pedagogy.* Clevedon: Multilingual Matters.
Davies, L. (1984) *Pupil Power: Deviance and Gender in School.* Lewes: Falmer.
Davin, A. (1978) Imperialism and motherhood, *History Workshop Journal*, 5: 9–65.
Dean, D.W. (1995) Education for moral improvement, domesticity and social cohesion: the Labour Government, 1945–1951, in L. Dawtrey, J. Holland and M. Hammer (eds) *Equality and Inequality in Education Policy.* Clevedon: Multilingual Matters.
Deem, R. (1990) The reform of school-governing bodies: the power of the consumer over the producer?, in M. Flude and M. Hammer (eds) *The Education Reform Act 1988.* Lewes: Falmer.
Deem, R. (1994) Researching the locally powerful: a study of school governance, in G. Walford (ed.) *Researching the Powerful in Education.* London: UCL Press.
Delamont, S. (1980) *Sex Roles and the School.* London: Methuen.
Delamont, S. (1990) *Sex Roles and the School*, 2nd edn. London: Routledge.
Delamont, S. (1999) Gender and the discourse of derision, *Research Papers in Education*, 14(1): 3–21.
De Lyon, H. (1989) Sexual harassment, in H. De Lyon and F. Migniuolo (eds) *Women Teachers.* Milton Keynes: Open University Press.
De Lyon, H. and Migniuolo, F. (eds) (1989) *Women Teachers.* Milton Keynes: Open University Press.
Dennis, N. (1997) *The Invention of Permanent Poverty.* London: Institute of Economic Affairs.
Denscombe, M. (1985) *Classroom Control.* London: Allen & Unwin.
Department for Education (1991) *The Parent's Charter.* London: HMSO.
Department for Education (1993) *Protection of Children: Disclosure of Criminal Background of Those with Access to Children*, Circular 9/93. London: HMSO.
Department for Education (1994) *Our Children's Education: The Updated Parent's Charter.* London: HMSO.
Department for Education and Employment (1995a) *Protecting Children from Abuse: The Role of the Education Service*, Circular 10/95. London: DfEE.

Department for Education and Employment (1995b) *Misconduct of Teachers and Workers with Children and Young Persons*, Circular 11/95. London: DfEE.

Department for Education and Employment (1999a) *Full-time Teachers in Maintained Nursery/Primary, Secondary and Special Schools*, Table 26. www.dfee.gov.uk/tpr/126a.htm

Department for Education and Employment (1999b) *Statistics of Education: Teachers 1999*. London: The Stationery Office.

Douglas, J.W.B. (1964) *The Home and the School*. London: MacGibbon and Kee.

Downes, P. (1999) 1999 reading test more boy-friendly, *Times Educational Supplement*, 15 October: 24.

Dubberley, W. (1988) Social class and the process of schooling – a case study of a comprehensive school in a mining community, in A. Green and S. Ball (eds) *Progress and Inequality in Comprehensive Education*. London: Routledge.

Dubberley, W.S. (1993) Humour as resistance, in M. Hammersley and P. Woods (eds) *Gender and Ethnicity in Schools*. London: Routledge.

Duelli Klein, R. (1983) How to do what we want to do: thoughts about feminist methodology, in G. Bowles and R. Duelli Klein (eds) *Theories of Women's Studies*. London: Routledge and Kegan Paul.

Dunning, E., Murphy, P. and Williams, J. (1988) *The Roots of Football Hooliganism*. London: Routledge and Kegan Paul.

Eggleston, J. (1977) *The Ecology of the School*. London: Methuen.

Emmerich, W., Goldman, S., Kirsh, B. and Sharabany, R. (1977) Evidence for a transitional phase in the development of 'gender constancy', *Child Development*, 48: 930–6.

Epstein, D. (1993) *Changing Classroom Cultures: Antiracism, Politics and Schools*. Stoke on Trent: Trentham Books.

Epstein, D. (ed.) (1994) *Challenging Lesbian and Gay Inequalities in Education*. Buckingham: Open University Press.

Epstein, D. (1998) Stranger in the mirror: gender, ethnicity, sexuality and nation in schooling. Paper presented at Multiple Marginalities: Gender Citizenship and Nationality in Education Conference, Nordic-Baltic Research Symposium (NORFA), Helsinki, August.

Epstein, D., Elwood, J., Hey, V. and Maw, J. (eds) (1998) *Failing Boys? Issues in Gender and Achievement*. Buckingham: Open University Press.

Erikson, E. (1965) *Childhood and Society*. Harmondsworth: Penguin.

Evetts, J. (1990) *Women in Primary Teaching: Career Contexts and Strategies*. London: Unwin Hyman.

Farrell, W. (1974) *The Liberated Man*. New York: Random House.

Fine, G. (1980) The natural history of pre-adolescent male friendship groups, in H. Foot, A. Chapman and J. Smith (eds) *Friendship and Social Relations in Children*. Chichester: John Wiley.

Finkelhor, D., Williams, L. and Burns, N. (1988) *Nursery Crimes: Sexual Abuse in Day Care*. Newbury Park, CA: Sage.

Fontana, D. (1988) *Psychology for Teachers*. London: Macmillan.

Foucault, M. (1977) *Discipline and Punish: The Birth of the Prison*. London: Allen Lane/Penguin.

Francis, B. (1998) *Power Plays*. Stoke-on-Trent: Trentham.

Francis, B. (2000) *Boys, Girls and Achievement*. London: Routledge/Falmer.

Francis, B. and Skelton, C. (2001) Men teachers and the construction of heterosexual masculinity in the classroom, *Sex Education*, 1(1): 9–21.
Frater, G. (1998) Boys and literacy, in K. Bleach (ed.) *Raising Boys' Achievement in Schools*. Stoke-on-Trent: Trentham.
Friedman, B. (1996) *Boys-Talk: A Program for Young Men about Masculinity, Non-violence and Relationships*. Adelaide: Men Against Sexual Assault, Kookaburra.
Frith, R. and Mahony, P. (1994) *Promoting Quality and Equality in Schools*. London: David Fulton.
Fukuyama, F. (1999) *The Great Disruption*. London: Profile.
Furlong, V.J. (1985) *The Deviant Pupil*. Milton Keynes: Open University Press.
Gaine, C. and George, R. (1997) *Gender, 'Race' and Class in Schooling*. London: Falmer.
Galton, M., Hargreaves, L. and Comber, C. (1998) Classroom Practice and the National Curriculum in Small Rural Primary Schools, *British Educational Research Journal*, 24(1): 43–61.
Gelsthorpe, L. (1990) Feminist methodologies in criminology: a new approach or old wine in new bottles?, in L. Gelsthorpe and A. Morris (eds) *Feminist Perspectives in Criminology*. Buckingham: Open University Press.
Gewirtz, S., Ball, S. and Bowe, R. (1995) *Markets, Choice and Equity in Education*. Buckingham: Open University Press.
Gilbert, R. and Gilbert, P. (1998) *Masculinity Goes to School*. London: Routledge.
Gillborn, D. and Gipps, C. (1996) *Recent Research on the Achievements of Ethnic Minority Pupils*. London: HMSO.
Gipps, C. and Murphy, P. (1994) *A Fair Test? Assessment, Achievement and Equity*. Buckingham: Open University Press.
Gofton, L. and Gofton, C. (1984) Making out in Giro City, *New Society*, 70(1144): 280–2.
Goldstein, A. (1994) Delinquent gangs, in J. Archer (ed.) *Male Violence*. London: Routledge.
Gorard, S., Rees, G. and Salisbury, J. (1999) Reappraising the apparent underachievement of boys at school, *Gender and Education*, 11(4): 441–54.
Gorman, T.P., White, J., Brooks, G., Maclure, M. and Kispal, A. (1988) *Language Performance in Schools: Review of APU Language Monitoring 1979–1983*. London: HMSO.
Gosden, P.H.J.H. (1972) *The Evolution of a Profession*. Oxford: Basil Blackwell.
Govier, E. (1998) Brainsex and occupation, in J. Radford (ed.) *Gender and Choice in Education and Occupation*. London: Routledge.
Grant, L. (1994) Positive thinking doesn't work, *Independent on Sunday*, 8 May: 23.
Griffin, C. (1998) Representations of youth and the 'boys' underachievement debate: just the same old stories? Paper presented at Gendering the Millennium Conference, University of Dundee, 11–13 September.
Griffin, C. and Lees, S. (1997) Editorial, *Gender and Education*, 9(1): 5–8.
Griffiths, M. (1998) *Educational Research for Social Justice*. Buckingham: Open University Press.
Hadaway, P. (1995) Letter to *Guardian Weekend*, 8 July: 5.
Haier, R.J. and Benbow, C.P. (1995) Sex differences and lateralization in temporal lobe glucose metabolism during mathematical reasoning, *Developmental Neuropsychology*, 11(4): 405.

Haigh, G. (1999) Football crazy, football mad . . . , *Times Educational Supplement*, 5 March: 7–8.
Hall, W. and Jose, P. (1983) Cultural effects on the development of equality and inequality, in R. Leahy (ed.) *The Child's Construction of Social Inequality*. New York: Academic Press.
Hallden, G. (1997) Competence and connection: gender and generation in boys' narratives, *Gender and Education*, 9(3): 307–16.
Hallinan, M. and Tuma, N. (1978) Classroom effects on change in children's friendships, *Sociology of Education*, 51: 270–82.
Halpern, D. (1992) *Sex Differences in Cognitive Abilities*, 2nd edn. Hillsdale, NJ: Lawrence Erlbaum.
Halsey, A. and Ridge, J. (1980) *Origins and Destinations: Family, Class and Education in Modern Britain*. Oxford: Clarendon.
Halson, J. (1989) The sexual harassment of young women, in L. Holly (ed.) *Girls and Sexuality*. Milton Keynes: Open University Press.
Hammersley, M. (ed.) (1986) *Controversies in Classroom Research*. Milton Keynes: Open University Press.
Hammersley, M. (1990) *Classroom Ethnography*. Buckingham: Open University Press.
Hammersley, M. and Woods, P. (eds) (1984) *Life in School: The Sociology of Pupil Culture*. Milton Keynes: Open University Press.
Hanmer, J. (1990) Men, power and the exploitation of women, in J. Hearn and D. Morgan (eds) *Men, Masculinities and Social Theory*. London: Unwin Hyman.
Hannan, G. (1999) *Improving Boys' Performance*. London: Folens.
Harding, S. (1991) *Whose Science? Whose Knowledge?* Buckingham: Open University Press.
Hargreaves, A. (1978) The significance of classroom coping strategies, in L. Barton and R. Meighan (eds) *Sociological Interpretations of Schooling and Classrooms: A Reappraisal*. London: Nafferton.
Hargreaves, A. and Woods, P. (eds) (1984) *Classrooms and Staffrooms*. Milton Keynes: Open University Press.
Hargreaves, D.H. (1967) *Social Relations in a Secondary School*. London: Routledge and Kegan Paul.
Hartley, D. (1985) *Understanding the Primary School*. London: Croom Helm.
Haskell, M. and Yablonsky, L. (1974) *Juvenile Delinquency*. Chicago: Rand McNally.
Hatcher, R. (1994) Children's lives: a study of children's peer cultures with special reference to 'race'. Unpublished PhD thesis, Department of Education, University of Warwick.
Hatcher, R. (1995) Boyfriends, girlfriends: gender and 'race' in children's cultures, *International Play Journal*, 3: 187–97.
Haywood, C. and Mac an Ghaill, M. (1995) The sexual politics of the curriculum: contesting values, *International Studies in Sociology of Education*, 5(2): 221–36.
Haywood, C. and Mac an Ghaill, M. (1996) Schooling masculinities, in M. Mac an Ghaill (ed.) *Understanding Masculinities*. Buckingham: Open University Press.
Haywood, C. and Mac an Ghaill, M. (1997) Materialism and deconstructivism: education and the epistemology of identity, *Cambridge Journal of Education*, 27(2): 261–72.

Head, J. (1999) *Understanding the Boys.* London: Falmer.
Hearn, J. (1985) Men's sexuality at work, in A. Metcalf and M. Humphries (eds) *The Sexuality of Men.* London: Pluto.
Hearn, J. (1988) Child abuse: violences and sexualities towards young people, *Sociology*, 22(4): 531–44.
Hearn, J. (1996) Is masculinity dead? A critique of the concept of masculinity/masculinities, in M. Mac an Ghaill (ed.) *Understanding Masculinities.* Buckingham: Open University Press.
Hearn, J. and Collinson, D. (1990) Unities and differences between men and masculinities: the categories of men and the case of sociology. Paper presented to British Sociological Association Conference, Social Divisions and Social Change, University of Surrey, 2–5 April.
Heath, S. (1982) *The Sexual Fix.* London: Macmillan.
Herbert, C. (1989) *Talking of Silence: The Sexual Harassment of Girls.* Lewes: Falmer.
Herbert, C. (1992) *Sexual Harassment in Schools.* London: David Fulton.
Heward, C. (1988) *Making a Man of Him.* London: Routledge.
Heward, C. (1991) Public school masculinities: an essay in gender and power. Private correspondence.
Hey, V. (1997) *The Company She Keeps: An Ethnography of Girls' Friendship.* Buckingham: Open University Press.
Heywood, J. (1997) The object of desire is the object of contempt: representations of masculinity in straight to hell magazine, in S. Johnson and U.H. Meinhof (eds) *Language and Masculinity.* Oxford: Blackwell.
Holly, L. (1985) Mary, Jane and Virginia Woolf: ten-year-old girls talking, in G. Weiner (ed.) *Just a Bunch of Girls.* Milton Keynes: Open University Press.
Holmes, R. (1995) *How Young Children Perceive Race.* London: Sage.
Hornby, N. (1992) *Fever Pitch.* London: Victor Gollancz.
Hough, J. (1985) Developing individuals rather than boys and girls, *School Organization*, 5(1): 17–25.
Howson, J. (1998) Is the glass ceiling beginning to crack? *The Times Higher Education Supplement*, 30 October.
Humphries, A. and Smith, P. (1984) Rough-and-tumble in preschool and playground, in P. Smith (ed.) *Play in Animals and Humans.* Oxford: Blackwell.
Hunt, F. (ed.) (1987) *Lessons for Life: The Schooling of Girls and Women, 1850–1950.* Oxford: Blackwell.
Hunt, P. (1994) *Report of the Independent Inquiry into Multiple Abuse in Nursery Classes.* City Council of Newcastle upon Tyne.
Inner London Education Authority (ILEA) (1986) *Primary Matters.* London: ILEA.
Jacklin, C.N. (1983) Boys and girls entering school, in M. Marland (ed.) *Sex Differentiation and Schooling.* London: Heinemann.
Jackson, D. (1990) *Unmasking Masculinity.* London: Unwin Hyman.
Jackson, S. (1992) The amazing deconstructing woman, *Trouble and Strife*, 25, winter: 25–31.
John, V. (1983) Foreword, in F. Widdowson, *Going Up into the Next Class.* London: Hutchinson.
Johnston, J., McKeown, E. and McEwan, A. (1998) A study of factors influencing choice of teaching in primary schools as a career. Paper presented at the British

Educational Research Association Conference, Queen's University, Belfast, 27–30 August.

Jones, A. (1993) Becoming a 'girl': post-structuralist suggestions for educational research, *Gender and Education*, 5(2): 157–66.

Jones, A. (1997) Teaching post-structuralist feminist theory in education: student resistances, *Gender and Education*, 9(3): 261–9.

Jones, C. (1985) Sexual tyranny: male violence in a mixed secondary school, in G. Weiner (ed.) *Just a Bunch of Girls*. Milton Keynes: Open University Press.

Jordan, E. (1995) Fighting boys and fantasy play: the construction of masculinity in the early years of school, *Gender and Education*, 7(1): 69–86.

Kant, L. (1987) National curriculum: notionally equal? *NUT Education Review*, 1(2): 41–4.

Kaufman, M. (ed.) (1987) *Beyond Patriarchy*. Oxford: Oxford University Press.

Kehily, M. and Nayak, A. (1997) 'Lads and laughter': humour and the production of heterosexual hierarchies, *Gender and Education*, 9(1): 69–87.

Kelly, L. (1989) Our issues, our analysis: two decades of work on sexual violence, in C. Jones and P. Mahony (eds) *Learning Our Lines: Sexuality and Social Control in Education*. London: The Women's Press.

Kelly, L., Burton, S. and Regan, L. (1994) Researching women's lives or studying women's oppression? Reflections on what constitutes feminist research, in M. Maynard and J. Purvis (eds) *Researching Women's Lives from a Feminist Perspective*. London: Taylor and Francis.

Kenway, J. (1995) Masculinities in schools: under siege, on the defensive and under reconstruction, *Discourse*, 16(1): 59–79.

Kenway, J. (1997) Point and counterpoint: boys' education in the context of gender reform, *Curriculum Perspectives*, 17(1): 57–61.

Kenway, J. and Fitzclarence, L. (1997) Masculinity, violence and schooling: challenging 'poisonous pedagogies', *Gender and Education*, 9(1): 117–33.

Kenway, J. and Willis, S. (1998) *Answering Back*. London: Routledge.

Kenway, J., Blackmore, J., Willis, S. and Rennie, L. (1996) The emotional dimensions of feminist pedagogy in schools, in P. Murphy and C. Gipps (eds) *Equity in the Classroom*. London: Falmer/UNESCO.

Kerfoot, S. and Whitehead, S. (1998) Whither hegemonic masculinity? Paper presented at Gendering the Millennium Conference, University of Dundee, 11–13 September.

Kessler, S., Ashenden, D.J., Connell, R.W. and Dowsett, G.W. (1985) Gender relations in secondary schooling, *Sociology of Education*, 58: 34–48.

Kimmel, M. (1996) *Manhood in America: A Cultural History*. New York: Free Press.

Kimura, D. (1996) Sex, sexual orientation and sex hormones influence human cognitive function, *Current Opinion in Neurobiology*, 259.

King, R. (1978) *All Things Bright and Beautiful*. Chichester: John Wiley.

Kingston, P. (1996) The failing sex, *Guardian Education*, 12 March: 4.

Kobrin, S. (1962) The impact of cultural factors on selected problems of adolescent development in the middle and lower class, *American Journal of Orthopsychology*, 32(3): 387–90.

Kohlberg, L. (1966) A cognitive-developmental analysis of children's sex-role concepts and attitudes, in E. Maccoby (ed.) *The Development of Sex Differences*. Stanford, CA: Stanford University Press.

Lacey, C. (1970) *Hightown Grammar*. Manchester: Manchester University Press.
Lambart, A. (1976) The sisterhood, in M. Hammersley and P. Woods (eds) *The Process of Schooling*. London: Routledge and Kegan Paul.
Lambart, A. (1997) Mereside: a grammar school for girls in the 1960s, *Gender and Education*, 9(4): 441–56.
Larkin, J. (1994) Walking through walls: the sexual harassment of high school girls, *Gender and Education*, 6(3): 263–80.
Lather, P. (1994) Fertile obsession: validity after poststructuralism, in A. Gitlin (ed.) *Power and Method: Political Activism and Educational Research*. London: Routledge.
Layder, D. (1993) *New Strategies in Social Research*. Cambridge: Polity.
Leahy, R. (ed.) (1983) *The Child's Construction of Social Inequality*. New York: Academic Press.
Lee, C. (1993) *Talking Tough: The Fight for Masculinity*. London: Arrow.
Lee, J. and Croll, P. (1995) Streaming and subject specialism at Key Stage 2: a survey in two local authorities, *Educational Studies*, 21(2): 155–65.
Lees, S. (1986) *Losing Out*. London: Hutchinson.
Lees, S. (1993) *Sugar and Spice: Sexuality and Adolescent Girls*. Harmondsworth: Penguin.
Lever, J. (1978) Sex differences in the complexity of children's games, *American Sociological Review*, 43: 471–83.
Licht, B. and Dweck, C. (1983) Sex differences in achievement orientations: consequences for academic choices and attainments, in M. Marland (ed.) *Sex Differentiation and Schooling*. London: Heinemann.
Lingard, B. and Douglas, P. (1999) *Men Engaging Feminisms*. Buckingham: Open University Press.
Littlewood, M. (1995) Makers of men, in L. Dawtrey, J. Holland and M. Hammer with S. Sheldon (eds) *Equality and Inequality in Education Policy*. Clevedon: Multilingual Matters.
Lloyd, B. and Duveen, G. (1992) *Gender Identities and Education*. Hemel Hempstead: Harvester Wheatsheaf.
Lobban, G. (1978) The influence of the school on sex-role stereotyping, in J. Chetwynd and O. Hartnett (eds) *The Sex Role System*. London: Routledge and Kegan Paul.
Lukes, S. (ed.) (1986) *Power*. Oxford: Blackwell.
Lyman, P. (1987) The fraternal bond as a joking relationship: a case study of the role of sexist jokes in male group bonding, in M. Kimmel (ed.) *Changing Men*. London: Sage.
Lyotard, J.F. (1984) *The Postmodern Condition: A Report on Knowledge*. Manchester: Manchester University Press.
Mac an Ghaill, M. (1988) *Young, Gifted and Black: Student–Teacher Relations in the Schooling of Black Youth*. Milton Keynes: Open University Press.
Mac an Ghaill, M. (1991) Schooling, sexuality and male power: towards an emancipatory curriculum, *Gender and Education*, 3(3): 291–309.
Mac an Ghaill, M. (1994) *The Making of Men: Masculinities, Sexualities and Schooling*. Buckingham: Open University Press.
MacInnes, J. (1998) *The End of Masculinity?* Buckingham: Open University Press.

McIntire, R.W. (1984) How children learn, in D. Fontana (ed.) *The Education of the Young Child.* Oxford: Blackwell.

Mclean, C. (1997) Engaging with boys' experiences of masculinity: implications for gender reform in schools, in J. Kenway (ed.) Point and Counterpoint: Boys' Education in the Context of Gender Reform, *Curriculum Perspectives*, 17(1): 57–78.

MacNaughton, G. (2000) *Rethinking Gender in Early Childhood Education.* London: Paul Chapman.

Mahony, P. (1985) *Schools for the Boys?* London: Hutchinson.

Mangan, J.A. (1987) Social Darwinism and upper class education in late Victorian and Edwardian England, in J. Mangan and J. Walvin (eds) *Manliness and Morality: Middle-Class Masculinity in Britain and America, 1800–1940.* Manchester: Manchester University Press.

Marland, M. (ed.) (1983) *Sex Differentiation and Schooling.* London: Heinemann.

Marsh, J. (2000) 'But I want to fly too!': girls and superhero play in the infant classroom, *Gender and Education*, 12(2): 209–20.

Martino, W. (1999) 'Cool boys', 'Party animals', 'Squids' and 'Poofters': interrogating the dynamics and politics of adolescent masculinities in school, *British Journal of Sociology of Education*, 20(2): 239–63.

Mason, R. (1999) Giant steps for mankind, *Times Educational Supplement*, 14 May: 19.

May, T. (1997) *Social Research: Issues, Methods and Process*, 2nd edn. Buckingham: Open University Press.

Maynard, M. (1995) Beyond the 'Big Three': the development of feminist theory into the 1990s, *Women's History Review*, 4(3): 259–81.

Maynard, M. and Purvis, J. (eds) (1994) *Researching Women's Lives from a Feminist Perspective.* London: Taylor and Francis.

Measor, L. and Sikes, P. (1992) *Gender and Schools.* London: Cassell.

Measor, L. and Woods, P. (1984) *Changing Schools.* Milton Keynes: Open University Press.

Meighan, R. (1981) *A Sociology of Educating.* Eastbourne: Holt, Rinehart and Winston.

Messner, M.A. (1997) *Politics of Masculinities: Men in Movements.* London: Sage.

Metcalf, A. and Humphries, M. (1985) *The Sexuality of Men.* London: Pluto.

Middleton, S. (1987) The sociology of women's education as a field of academic study, in M. Arnot and G. Weiner (eds) *Gender and the Politics of Schooling.* London: Unwin Hyman.

Middleton, S. (1993) A post-modern pedagogy for the sociology of women's education, in M. Arnot and K. Weiler (eds) *Feminism and Social Justice in Education.* London: Falmer.

Miedzian, M. (1992) *Boys Will Be Boys.* London: Virago.

Miles, S. and Middleton, C. (1995) Girls' education in the balance: the ERA and inequality, in L. Dawtrey, J. Holland and M. Hammer with S. Sheldon (eds) *Equality and Inequality in Education Policy.* Clevedon: Multilingual Matters.

Millard, E. (1997) *Differently Literate.* London: Falmer.

Miller, J. (1996) *School for Women.* London: Virago.

Miller, W.B. (1982) *Crimes by Youth Gangs and Groups in the United States.* Washington, DC: National Institute of Juvenile Justice and Delinquency Prevention.
Mills, M. (2000) Issues in implementing boys' programmes in schools: male teachers and empowerment, *Gender and Education*, 12(2): 221–38.
Moir, A. and Moir, B. (1999) *Why Men Don't Iron.* London: Harper Collins.
Morgan, D. (1992) *Discovering Men.* London: Routledge.
Morris, L. (1994) *Dangerous Classes.* London: Routledge.
Morrison, A. and McIntyre, D. (1969) *Teachers and Teaching.* Harmondsworth: Penguin.
Moseley, D., Merrell, C. and Cranston, G. (2000) *A Study of KS1 Pupil Progress in Five Newcastle Schools 1996–99.* University of Newcastle upon Tyne.
Mott, H. and Condor, S. (1997) Sexual harassment and the working lives of secretaries, in A. Thomas and C. Kitzinger (eds) *Sexual Harassment.* Buckingham: Open University Press.
Murphy, P. (1996) Defining pedagogy, in P. Murphy and C. Gipps (eds) *Equity in the Classroom.* London: Falmer/UNESCO.
Murphy, P. and Elwood, J. (1998) Gendered learning outside and inside school: influences on achievement, in D. Epstein, J. Elwood, V. Hey and J. Maw (eds) *Failing Boys?* Buckingham: Open University Press.
Murphy, P. and Gipps, C. (eds) (1996) *Equity in the Classroom.* London: Falmer/UNESCO.
Murphy, P., Williams, J. and Dunning, E. (1990) *Football on Trial.* London: Routledge.
Murray, S. (1996) 'We all love Charles': men in child care and the social construction of gender, *Gender and Society*, 10: 368–85.
Myers, K. (ed.) (2000) *Whatever Happened to Equal Opportunities in Schools?* Buckingham: Open University Press.
Nash, R. (1973) *Classrooms Observed.* London: Routledge and Kegan Paul.
National Literacy Trust (1999) *The Literacy Hour won't Solve the Gender Gap.* www.literacytrust.org.uk/Database/lhboys.html
Nias, J. (1985) Reference groups in primary teaching: talking, listening and identity, in S. Ball and I. Goodson (eds) *Teachers' Lives and Careers.* Lewes: Falmer.
Noble, C. (1998) Helping boys do better in their primary schools, in K. Bleach (ed.) *Raising Boys' Achievement in Schools.* Stoke-on-Trent: Trentham.
Noble, C. and Bradford, W. (2000) *Getting it Right for Boys . . . and Girls.* London: Routledge.
Oakley, A. (1972) *Sex, Gender and Society.* London: Temple Smith.
Oakley, A. (1981) *Subject Women.* London: Fontana.
Oram, A. (1999) 'To cook dinners with love in them'? Sexuality, marital status and women teachers in England and Wales, 1920–39, in K. Weiler and S. Middleton (eds) *Telling Women's Lives.* Buckingham: Open University Press.
Ouseley, H. (1998) Presidential address, North of England Education Conference, 5–7 January.
Paechter, C. (1998) *Educating the Other: Gender, Power and Schooling.* London: Falmer.
Paley, V.G. (1984) *Boys and Girls.* Chicago: University of Chicago Press.
Parker, A. (1996) The construction of masculinity within boys' physical education, *Gender and Education*, 8(2): 141–57.

Parker, H.J. (1974) *View from the Boys.* Newton Abbot: David and Charles.

Parry, O. (1996) 'Schooling is fooling': why do Jamaican boys underachieve in school? *Gender and Education*, 9(2): 223–31.

Partington, G. (1976) *Women Teachers in the 20th Century.* Windsor: NFER.

Parton, N. (1990) Taking child abuse seriously, in The Violence Against Children Study Group (eds) *Taking Child Abuse Seriously.* London: Unwin Hyman.

Patrick, J. (1973) *A Glasgow Gang Observed.* London: Methuen.

Pattman, R. (1998) 'Boys have a heart but they don't think that much': maturity, immaturity and how boys and girls position themselves and each other in interviews. Paper presented at Gender and Education Conference, University of Warwick, 28–30 April.

Penn, H. (1996) Three men and a baby, *Nursery World*, 19 September.

Penn, H. (1998) Summary: men as workers in services for young children, in C. Owen, C. Cameron and P. Moss (eds) *Men as Workers in Services for Young Children: Issues of a Mixed Gender Workforce.* London: Institute of Education, Bedford Way Paper.

Penny, V. (1998) Raising boys' achievement in English, in K. Bleach (ed.) *Raising Boys' Achievement in Schools.* Stoke-on-Trent: Trentham.

Phillips, A. (1993) *The Trouble with Boys.* London: Pandora.

Pickering, J. (1997) *Raising Boys' Achievement.* London: Network Educational Press.

Pleck, J.H. (1976) The male sex role: definitions, problems, and sources of change, *Journal of Social Issues*, 32(3): 155–64.

Pleck, J.H. (1981) *The Myth of Masculinity.* Cambridge, MA: MIT Press.

Plummer, K. (1975) *Sexual Stigma: An Interactionist Account.* London: Routledge and Kegan Paul.

Pollack, W. (1998) *Real Boys.* New York: Owl Books.

Pollard, A. (1985) *The Social World of the Primary School.* Eastbourne: Holt, Rinehart and Winston.

Pollard, A. (1987) *Children and their Primary Schools.* Lewes: Falmer.

Pollard, A. (1992) Teachers' responses to the reshaping of primary education, in M. Arnot and L. Barton (eds) *Voicing Concerns.* London: Triangle.

Pringle, K. (1995) *Men, Masculinities and Social Welfare.* London: UCL Press.

Pulis, M.V. (2000) Boys' under-performance in an inner-city multi-ethnic environment. Paper presented to the Third European Conference, Network on Intercultural and Multicultural Education in Europe, Haugesund, Norway, 19–22 May.

Purvis, J. (1991) *A History of Women's Education in England.* Buckingham: Open University Press.

Purvis, J. (1995) Women and education: a historical account 1800–1914, in L. Dawtrey, J. Holland and M. Hammer with S. Sheldon (eds) *Equality and Inequality in Education Policy.* Clevedon: Multilingual Matters.

Qualifications and Curriculum Authority (QCA) (1998) *Can Do Better: Raising Boys' Achievement in English.* London: QCA Publications.

Quine, W.G. (1974) Polarised cultures in comprehensive schools, *Research in Education*, 12: 9–25.

Raphael Reed, L. (1999) Troubling boys and disturbing discourses on masculinity and schooling: a feminist exploration of current debates and interventions concerning boys in school, *Gender and Education*, 11(1): 93–110.

Ramazanoglu, C. (ed.) (1993) *Up Against Foucault*. London: Routledge.
Reay, D. (1990) Working with boys, *Gender and Education*, 2(3): 269–81.
Roper, M. and Tosh, J. (eds) (1991) *Manful Assertions: Masculinities in Britain since 1800*. London: Routledge.
Reay, D. (1991) Intersectors of gender, race and class in the primary school, *British Journal of Sociology of Education*, 12(2): 163–82.
Redman, P. (1996) Curtis loves Ranjit: heterosexual masculinities, schooling and pupils' sexual cultures, *Educational Review*, 48(2): 175–82.
Redwood, F. (1994) Now let's give boys a boost, *Daily Telegraph*, 7 December: 23.
Rendel, M. (1985) The winning of the Sex Discrimination Act, in M. Arnot (ed.) *Race and Gender: Equal Opportunities Policies in Education*. Oxford: Pergamon.
Renold, E. (1997) 'All they've got on their brains is football': sport, masculinity and the gendered practices of playground relations, *Sport, Education and Society*, 2(1): 5–23.
Renold, E. (2000) 'Coming out': gender (hetero) sexuality and the primary school, *Gender and Education*, 12(3): 309–26.
Ribbens, J. and Edwards, R. (eds) (1998) *Feminist Dilemmas in Qualitative Research*. London: Sage.
Robins, D. (1982) Sport and youth culture, in J. Hargreaves (ed.) *Sport, Culture and Ideology*. London: Routledge and Kegan Paul.
Robins, D. and Cohen, P. (1978) *Knuckle Sandwich: Growing Up in the Working-Class City*. Harmondsworth: Penguin.
Roper, M. and Tosh, J. (eds) (1991) *Manful Assertions: Masculinities in Britain since 1880*. London: Routledge.
Sacks, O. (1993) Making up the mind, *New York Review of Books*, 8 April: 42–7.
Salisbury, J. and Jackson, D. (1996) *Challenging Macho Values*. London: Falmer.
Salisbury, J. and Riddell, S. (eds) (2000) *Gender, Policy and Educational Change*. London: Routledge.
Salisbury, J., Rees, G. and Gorard, S. (1999) Accounting for the differential attainment of boys and girls at school, *School Leadership and Management*, 19(4): 403–26.
Saussure, F. (1916) *Course in General Linguistics*, translated by B. Wade. London: Fontana.
Scarth, J. (1987) Teaching to the exam? The case of the Schools Council History Project, in T. Horton (ed.) *The GCSE: Examining the New System*. London: Harper and Row.
Segal, L. (1990) *Slow Motion: Changing Masculinities, Changing Men*. London: Virago.
Seidler, V. (1989) *Rediscovering Masculinity*. London: Routledge.
Seidler, V. (1991) *Recreating Sexual Politics*. London: Routledge.
Seidler, V. (1997) *Man Enough*. London: Sage.
Seifert, K. (1988) The culture of early education and the preparation of male teachers, *Early Child Development and Care*, 38: 69–80.
Select Committee on the Teachers' Registration and Organisation Bill (1891) *Report and Evidence*. Evidence of Dr R. Wormell, headmaster of the City Middle Class Schools, p. 28.

Serbin, L. (1983) The hidden curriculum: academic consequences of teacher expectations, in M. Marland (ed.) *Sex Differentiation and Schooling*. London: Heinemann.
Sewell, T. (1997) *Black Masculinities and Schooling*. Stoke-on-Trent: Trentham.
Shakespeare, S. (1998) Consign the cissy culture to history, *Daily Mail*, 5 January: 6.
Sharp, R. (1981) Review of Stephen Ball's 'Beachside Comprehensive', *British Journal of Sociology of Education*, 2(1): 278–85.
Sharp, R. and Green, A. (1975) *Education and Social Control*. London: Routledge and Kegan Paul.
Sharpe, S. (1976) *Just Like a Girl*. Harmondsworth: Penguin.
Sharpe, S. (1994) *Just Like a Girl*, 2nd edn. Harmondsworth: Penguin.
Sheppard, N. (1989) *Equal Opportunity in the Nursery: An Investigative Study*. North Tyneside: North Tyneside Education Authority.
Shilling, C. (1992) Reconceptualising structure and agency in the sociology of education: structuration theory and schooling, *British Journal of Sociology of Education*, 13(1): 69–87.
Short, G. and Carrington, B. (1989) Discourse on gender: the perceptions of children aged between six and eleven, in C. Skelton (ed.) *Whatever Happens to Little Women? Gender and Primary Schooling*. Milton Keynes: Open University Press.
Short, J. and Strodtbeck, F.L. (1965) *Group Process and Gang Delinquency*. Chicago: University of Chicago Press.
Short, J.F. (1990) New wine in old bottles? Change and continuity in American gangs, in C.R. Huff (ed.) *Gangs in America*. Newbury Park, CA: Sage.
Sikes, P. (2000) 'Truth' and 'lies' revisited, *British Educational Research Journal*, 26(2): 257–70.
Siraj-Blatchford, I. (ed.) (1993) *'Race', Gender and the Education of Teachers*. Buckingham: Open University Press.
Skeggs, B. (1991a) Challenging masculinity and using sexuality, *British Journal of Sociology of Education*, 12(2): 127–39.
Skeggs, B. (1991b) Postmodernism: what is all the fuss about? *British Journal of Sociology of Education*, 12(2): 255–67.
Skeggs, B. (1994) Situating the production of feminist ethnography, in M. Maynard and J. Purvis (eds) *Researching Women's Lives from a Feminist Perspective*. London: Taylor and Francis.
Skeggs, B. (1997) *Formations of Class and Gender*. London: Sage.
Skelton, C. (ed.) (1989) *Whatever Happens to Little Women?: Gender and Schooling*. Milton Keynes: Open University Press.
Skelton, C. (1991) A study of the career perspectives of male teachers of young children, *Gender and Education*, 3(3): 279–89.
Skelton, C. (1998a) Men teachers, young children and sexual violence. Paper presented at the British Educational Research Conference, Queen's University, Belfast, 27–30 August.
Skelton, C. (1998b) Feminism and research into masculinities and schooling, *Gender and Education*, 10(3): 217–27.
Skelton, C. (2000) 'A passion for football': dominant masculinities and primary schooling, *Sport, Education and Society*, 5(1): 5–18.

Sluckin, A. (1981) *Growing Up in the Playground.* London: Routledge and Kegan Paul.
Smedley, S. (1998) Perspectives on male student primary teachers, *Changing English*, 5(2): 147–59.
Smedley, S. (1999) 'Don't rock the boat': men student teachers' understanding of gender and equality. Paper presented to the British Educational Research Association Conference, University of Sussex, 2–5 September.
Spender, D. (1982) *Invisible Women: The Schooling Scandal.* London: Writers and Readers.
Spender, D. (ed.) (1987) *The Education Papers.* London: Routledge and Kegan Paul.
Stanley, J. (1989) *Marks on the Memory.* Milton Keynes: Open University Press.
Stanley, L. (1997) Methodology matters!, in V. Robinson and D. Richardson (eds) *Introducing Women's Studies.* London: Macmillan.
Stanley, L. and Wise, S. (1993) *Breaking Out: Feminist Consciousness and Feminist Research*, 2nd edn. London: Routledge.
Stanworth, M. (1981) *Gender and Schooling.* London: Hutchinson.
Statistics of Public Education, 1912–13, Part 1.
Steedman, C. (1988) 'The mother made conscious': the historical development of a primary school pedagogy, in M. Woodhead and A. McGrath (eds) *Family, School and Society.* Milton Keynes: Open University Press.
Steele, F. (1973) *Physical Settings and Organization Development.* Reading, MA: Addison-Wesley.
Stronach, I. and MacLure, M. (1997) *Educational Research Undone.* Buckingham: Open University Press.
Sukhnandan, L. (1999) *An Investigation into Gender Differences in Achievement: Phase I., A Review of Recent Research and LEA Information on Provision.* Slough: NFER.
Sukhnandan, L., Lee, B. and Kelleher, S. (2000) *An Investigation into Gender Differences in Achievement. Phase 2: School and Classroom Strategies.* Slough: National Foundation for Educational Research.
Sumsion, J. (1999) Critical reflections on the experiences of a male early childhood worker, *Gender and Education*, 11(4): 455–68.
Swain, J. (2000) 'The money's good, the fame's good, the girls are good': the role of playground football in the construction of young boys' masculinity in a junior school, *British Journal of Sociology of Education*, 21(1): 95–109.
Teese, R., Davies, M., Charlton, M. and Polesel, J. (1995) *Who Wins at School? Boys and Girls in Secondary Education.* Melbourne: Department of Education Policy and Management, University of Melbourne.
Terry, B. and Terry, L. (1998) A multi-layered approach to raising boys' (and girls') achievement, in K. Bleach (ed.) *Raising Boys' Achievement in Schools.* Stoke-on-Trent: Trentham.
Thorne, B. (1993) *Gender Play: Girls and Boys in School.* Buckingham: Open University Press.
Thornton, M. (1997) Gender issues in the recruitment, training and career prospects of early years and primary school teachers. Paper presented to the British Educational Research Association Conference. University of York, 11–14 September.

Thornton, M. (1998) Save the whale, sorry – male: wasted role models in primary ITE. Paper presented at the British Educational Research Association Conference, Queen's University, Belfast, 27–30 August.

Tomlinson, J. (1988) Curriculum and market: are they compatible?, in J. Haviland (ed.) *Take Care, Mr Baker*. London: Fourth Estate.

Tooley, J. (1997) Discourse of difference: some implications of feminist research. Paper presented at the British Educational Research Association Conference, University of York, 11–14 September.

Trevor, M. (1973) *The Arnolds*. London: Bodley Head.

Tripp, D. (1993) *Critical Incidents in Teaching*. London: Routledge.

Troyna, B. and Hatcher, R. (1992) *Racism in Children's Lives*. London: Routledge.

Tutchell, E. (1990) *Dolls and Dungarees: Gender Issues in the Primary School Curriculum*. Buckingham: Open University Press.

Vogel, P. (1997) *Review of Boys-Talk*. www.boysed@halibut.pnc.com.au, 14 February.

Waddington, D., Jones, K. and Critcher, C. (1989) *Flashpoints: Studies in Public Disorder*. London: Routledge.

Walby, S. (1986) *Patriarchy at Work*. Cambridge: Polity.

Walden, R. and Walkerdine, V. (1985) *Girls and Mathematics*. London: Bedford Way Papers, Institute of Education, University of London.

Walker, J. (1988) *Louts and Legends*. Sydney: Allen & Unwin.

Walker, S. and Barton, L. (eds) (1989) *Politics and the Processes of Schooling*. Milton Keynes: Open University Press.

Walkerdine, V. (1983) It's only natural: rethinking child-centred pedagogy, in A.M. Wolpe and J. Donald (eds) *Is There Anyone Here From Education?* London: Pluto.

Walkerdine, V. (1990) *Schoolgirl Fictions*. London: Verso.

Wallace, M.C. (1992) West End riots 1991: young people's perceptions. Unpublished MA thesis, University of Northumbria.

Waller, W. (1932) *The Sociology of Teaching*. London: John Wiley.

Warren, S. (1997) Who do these boys think they are? An investigation into the construction of masculinities in a primary classroom, *International Journal of Inclusive Education*, 1(2): 207–22.

Weedon, C. (1987) *Feminist Practice and Poststructuralist Theory*. Oxford: Blackwell.

Weiner, G. (1985) *Just a Bunch of Girls*. Milton Keynes: Open University Press.

Weiner, G. (1994) *Feminisms in Education*. Buckingham: Open University Press.

Weiner, G., Arnot, M. and David, M. (1997) Is the future female? Female success, male disadvantage, and changing gender patterns in education, in A. Halsey, H. Lauder, P. Brown and A. Stuart Wells (eds) *Education: Culture, Economy, Society*. Oxford: Oxford University Press.

Westland, E. (1993) Cinderella in the classroom: children's responses to gender roles in fairy tales, *Gender and Education*, 5(2): 237–49.

Westwood, S. (1990) Racism, black masculinity and the politics of space, in J. Hearn and D. Morgan (eds) *Men, Masculinities and Social Theory*. London: Unwin Hyman.

Wetherell, M. and Griffin, C. (1995) Feminist psychology and the study of men and masculinity: assumptions and perspectives, in M. Blair and J. Holland with

S. Sheldon (eds) *Identity and Diversity: Gender and the Experience of Education.* Clevedon: Multilingual Matters.

Wheldall, K. and Merrett, F. (1988) Which classroom behaviour problems do primary school teachers say they find most troublesome? *Education Review*, 40: 13–27.

Whitbread, N. (1972) *The Evolution of the Nursery-Infant School.* London: Routledge and Kegan Paul.

Whitelaw, S., Milosevic, L. and Daniels, S. (2000) Gender, behaviour and achievement: a preliminary study of pupil perceptions and attitudes, *Gender and Education*, 12(1): 87–113.

Whitty, G., Power, S. and Halpin, D. (1998) *Devolution and Choice in Education.* Buckingham: Open University Press.

Whyld, J. (ed.) (1983) *Sexism in the Secondary Curriculum.* London: Harper and Row.

Whyte, J. (1983) *Beyond the Wendy House: Sex Role Stereotyping in Primary Schools.* York: Longman.

Widdowson, F. (1983) *Going Up into the Next Class.* London: Hutchinson.

Willis, P. (1977) *Learning to Labour.* Aldershot: Saxon House.

Willott, S. and Griffin, C. (1996) Men, masculinity and the challenge of long-term unemployment, in M. Mac an Ghaill (ed.) *Understanding Masculinities.* Buckingham: Open University Press.

Wilmott, P. (1969) *Adolescent Boys in the East End of London.* London: Harmondsworth.

Wing, A. (1997) How can children be taught to read differently? *Bill's New Frock* and the 'hidden curriculum', *Gender and Education*, 9(4): 491–504.

Wintour, P. and Bright, M. (2000) Fantasy football helps schoolboys score at maths, *Observer*, 23 January: 7.

Wolpe, A.M. (1988) *Within School Walls: Discipline, Sexuality and Curriculum.* London: Routledge and Kegan Paul.

Woods, P. (1979) *The Divided School.* London: Routledge and Kegan Paul.

Woods, P. (1983) *Sociology and the School: An Interactionist Viewpoint.* London: Routledge and Kegan Paul.

Woods, P. (1987) Becoming a junior: pupil development following transfer from infants, in A. Pollard (ed.) *Children and their Primary Schools.* Lewes: Falmer.

Woods, P. (1990) *Teacher Skills and Strategies.* London: Falmer.

Woods, P. (1996) *Researching the Art of Teaching: Ethnography for Educational Use.* London: Routledge.

Yates, L. (1997) Gender equity and the boys debate: what sort of challenge is it?, *British Journal of Sociology of Education*, 18(3): 337–47.

Yeoman, E. (1999) How does it get into my imagination? Elementary school children's intertextual knowledge and gendered storylines, *Gender and Education*, 11(4): 427–40.

Younger, M., Warrington, M. and Williams, J. (1999) The gender gap and classroom interactions: reality and rhetoric?, *British Journal of Sociology of Education*, 20(3): 325–42.

Index

FAILING BOYS?
ISSUES IN GENDER AND ACHIEVEMENT

Debbie Epstein, Jannette Elwood, Valerie Hey and Janet Maw

Failing Boys? Issues in Gender and Achievement challenges the widespread perception that all boys are underachieving at school. It raises the more important and critical questions of which boys? At what stage of education? And according to what criteria?

The issues surrounding boys' 'underachievement' have been at the centre of public debate about education and the raising of standards in recent years. Media and political responses to the 'problem of boys' have tended to be simplistic, partial, and owe more to 'quick fixes' than investigation and research. *Failing Boys?* provides a detailed and nuanced 'case study' of the issues in the UK, which will be of international relevance as the moral panic is a globalised one, taking place in diverse countries. The contributors to this book take seriously the issues of boys' 'underachievement' inside and outside school from a critical perspective which draws on the insights of previous feminist studies of education to illuminate the problems associated with the education of boys.

This will be a key text for educators, policy makers, students and teachers of education, sociology, gender studies and cultural studies and others interested in gender and achievement.

Contents
Part I: Boys' underachievement in context – Part II: Different constructions of the debate and its undercurrents – Part II: Boys, which boys? – Part IV: Curriculum, assessment and the debate

208pp 0 335 20238 1 (Paperback) 0 335 20239 X (Hardback)

WHATEVER HAPPENED TO EQUAL OPPORTUNITIES IN SCHOOLS?

Kate Myers (ed.)

Whatever Happened to Equal Opportunities in Schools? is an edited book which makes an important contribution to the current debate about equal opportunities. Today the dominant concern is about boys' achievement but it was not always thus. Contributors trace events relating to schools since the introduction of the Sex Discrimination Act and the establishment of the Equal Opportunities Commission in the mid-1970s. Prior to the advent of the National Curriculum it was common practice for boys and girls to take different subjects and be offered very different opportunities and experience through the overt and covert school curriculum. Initiatives emerging from central government, quangos, trade unions, local education authorities, and individual schools are described. The book discusses: how much has really changed; the extent to which credit should be given to earlier initiatives concerned with the raising of girls' aspirations and achievement; what we can learn from these initiatives; and what we should really be concerned about now. The book also addresses the question of boys' achievement both past and present and will be important reading for all educators with an interest in promoting gender equality in schools.

Contents

256pp   0 335 20303 5 (Paperback)   0 335 20304 3 (Hardback)